VERNACULAR AND EARTHEN ARCHITECTURE TOWARDS LOCAL DEVELOPMENT

Proceedings of 2019 ICOMOS CIAV & ISCEAH International Conference

乡土未来:面向地方发展的乡土和土质建筑保护

2019 年国际古迹遗址理事会乡土建筑科学委员会与土质建筑遗产科学委员会
联合年会暨国际学术研讨会论文集

Editor-in-Chief:
主编:

邵　甬(SHAO Yong)
(挪)吉斯勒·亚克林(Gisle Jakhelln)
(葡)玛丽安娜·科雷亚(Mariana Correia)

中国·上海

图书在版编目(CIP)数据

乡土未来 ：面向地方发展的乡土和土质建筑保护 ＝ VERNACULAR AND EARTHEN ARCHITECTURE TOWARDS LOCAL DEVELOPMENT：英文 / 邵甬，（挪）吉斯勒・亚克林（Gisle Jakhelln），（葡）玛丽安娜・科雷亚（Mariana Correia）主编. —— 上海 ：同济大学出版社，2019.9

ISBN 978-7-5608-8656-5

Ⅰ. ①乡… Ⅱ. ①邵…②吉…③玛… Ⅲ. ①乡村－古建筑－保护－国际学术会议－文集－英文 Ⅳ. ①TU-87

中国版本图书馆 CIP 数据核字(2019)第 154049 号

乡土未来：面向地方发展的乡土和土质建筑保护

邵　甬　（挪）吉斯勒・亚克林　（葡）玛丽安娜・科雷亚　主编

责任编辑　由爱华　　责任校对　徐春莲　　封面设计　钱如潺　　版式设计　朱丹天

出版发行　同济大学出版社　　www.tongjipress.com.cn
（地址：上海市四平路 1239 号　邮编：200092　电话：021－65985622）
经　　销　全国各地新华书店
印　　刷　江苏凤凰数码印务有限公司
开　　本　787mm×1092mm　1/16
印　　张　49.75
印　　数　1—1100
字　　数　1242000
版　　次　2019 年 9 月第 1 版　　2019 年 9 月第 1 次印刷
书　　号　ISBN 978-7-5608-8656-5

定　　价　200.00 元

2019 ICOMOS CIAV&ISCEAH INTERNATIONAL CONFERENCE on "Vernacular and Earthen Architecture towards Local Development"

Organized by

- International Council on Monuments and Sites-International Committee on Vernacular Architecture (ICOMOS-CIAV)
- International Council on Monuments and Sites-International Scientific Committee on Earthen Architectural Heritage (ICOMOS-ISCEAH)
- Chinese National Committee for the International Council on Monuments and Sites (ICOMOS-CHINA)
- World Heritage Institute of Training and Research for the Asia and the Pacific Region under the Auspices of UNESCO (WHITRAP Shanghai)
- The Academic Committee of Historical and Cultural City Planning, the Urban Planning Society of China (ACHCCP-UPSC)
- Committee of Urban and Rural Built Heritage-ASC (CURBH-ASC)
- Tongji University, China

Co-Organized by

- Shanghai Tongji Urban Planning & Design Institute Co., Ltd. (TJUPDI)
- Tongji Architectural Design (Group) Co., Ltd. (TJAD)

Academic Supported by

- Urban Planning Society of China
- Architectural Society of China
- *Heritage Architecture*
- *Built Heritage*

Hosted by

- Pingyao County People's Government

2019 ICOMOS CIAV & ISCEAH International Conference Scientific Committee

Chairs(A to Z)

Gisle Jakhelln **Norway** | Mariana Correia **Portugal**

Members(A to Z)

Aase Tveitnes **Norway** | Amanda Rivera Vidal **Chile** | Camilla Mileto **Spain**
CHANG Qing **China** | Claudia Cancino **USA** | DAI Shibing **China**
Deirdre McDermott **Ireland** | DONG Wei **China** | Elo Lutsepp **Estonia**
Erin Guerra **USA** | Fernando Vegas **Spain** | Gouhar Shemdin **Canada**
HE Yi **China** | HOU Weidong **China** | Ione Stiegler **USA** | Ivan Enev **Bulgaria**
Jorge Tomasi **Argentina** | Julieta Barada **Argentina** | Lassana Cissé **Mali**
LYU Zhou **China** | Maddalena Achenza **Italy** | Maribel Beas **USA**
Marwa Dabaieh **Egypt** | Miles Lewis **Australia** | Naima Benkari **Algeria**
Natalia Turekulova **Kazakhstan** | Natsuko Akagawa **Japan**
Oussouby Sacko **Mali/Japan** | Pamela Jerome **USA** | Pilwon Han **R. O. Korea**
Prikko Marika **Finland** | Rasool Vatandoust **Iran** | Rawiwan Oranratmanee **Thailand**
Saleh Lamei **Egypt** | SHAO Yong **China** | Valeria Prieto **Mexico**
Venkatarama Reddy **India** | WANG Lijun **China** | WANG Xudong **China**
ZHANG Guanghan **China** | ZHANG Jianwei **China** | ZHANG Jie **China**
ZHANG Peng **China** | ZHANG Song **China** | ZHOU Jian **China**

Preface 1

CIAV is an international platform for the dialogue and cooperation between professionals, experts, academics and students of vernacular heritage. We work primarily through our annual meetings and scientific conferences. CIAV fosters discussions and activities on national and regional levels.

CIAV consists of members with established expertise in the field of Vernacular Architecture. Our annual conferences are held to exchange experiences and new knowledge. This is of utmost importance for us professionals in order to further develop our competence in our fields of work. The term *Vernacular Architecture* includes a great variety of building types and sites, making it necessary for us to have knowledge in nearly all kinds of building construction and the land use methods.

We aim to ensure a multidisciplinary approach to the vernacular heritage by encouraging interaction between different disciplines within the framework of the annual CIAV scientific meetings and through CIAV membership. Establishing strategic alliances with other ISCs whose field of work are represented in vernacular architecture is an important aspect. This year's conference organised together with ISCEAH, International committee on Earthen Architectural Heritage, is an illustration of this work.

CIAV offers support for the cause of conservation for vernacular architecture around the world on different levels: international, regional, national and local. This may be moral support or exercising pressure to save endangered buildings and built heritage. It may also be technical support on international level to UNESCO and NGO's such as WMF and national and local levels to relevant authorities and organisations upon request.

CIAV was founded in 1976 following a resolution from the international conference for the conservation of vernacular architecture held in Plovdiv, Bulgaria, in 1975. There were 12 founding members from the 12 national committees (countries) and an additional 10 associate members. Today, there are 120 members (including 8 honorary members) within CIAV, from 52 countries.

At the moment CIAV is headed by architect Gisle Jakhelln (Norway) as President, architect Valeria Prieto (Mexico) as Vice President, architect Maria Inés Subercaseaux (Canada) as Vice President and architect Ivan Enev (Bulgaria) as Secretary General.

CIAV's field of work is wide. Over the years the committee has learned to enlarge its understanding of what 'vernacular' is — from single farmsteads and traditional village units to urban vernacular areas and settlements, to cultural landscape areas, and the links between the vernacular heritage and the geomorphologic conditions of the landscape.

This conference underlines the wide field of CIAV's work. I am thankful for the keen interest in organising this conference by ICOMOS China and in particular Professor SHAO Yong.

Gisle Jakhelln
President of ICOMOS-CIAV
June 26th, 2019

序 1

国际古迹遗址理事会乡土建筑科学委员会(下简称 CIAV)是一个供专业人士、专家、学者和学生之间就乡土遗产进行对话与合作的国际平台,主要通过年度会议和学术研讨会开展工作,在国家与地区层面促成讨论与活动。

CIAV 由在乡土建筑领域拥有丰富专业知识的成员组成。CIAV 每年会举办年度会议,旨在交流经验和最新技术。这对于专业人员进一步提升在工作领域的竞争力至关重要。“乡土建筑”一词涵盖了广泛的建筑类型和场地,这就要求委员会有必要了解几乎所有类型的建筑建造和土地使用方式。

委员会的目标是在 CIAV 成员之间、年度学术研讨会的框架之下,通过鼓励跨学科的交流互动,建立一个研究乡土遗产的多学科方法。与诸多研究特定类型乡土建筑的 ICOMOS 科学委员会建立战略联盟,也是 CIAV 的重要工作之一。今年的会议与土质建筑遗产科学委员会(ISCEAH)一起举办,即是一次较好的合作范例。

CIAV 为全世界范围的乡土建筑保护事业提供支持,涵盖国际、区域、国家和地方各个层面。这些支持或是道义上的支持,或是为了拯救濒危建筑与建成遗产施加压力,或是根据有关机构与组织的要求提供技术支援。服务对象既包括国际层面的联合国教科文组织、非政府组织(如 WMF),也有国家和地方层面的有关当局和机构。

CIAV 根据 1975 年保加利亚普罗夫迪夫举办的乡土建筑保护国际会议的相关决议,于 1976 年正式成立。初成立时,只有来自 12 个国家委员会的 12 名创始成员和 10 名准成员。今天,CIAV 已有来自全球 52 个国家的 120 名成员(包括 8 名荣誉委员)。

目前,CIAV 由挪威建筑师吉斯勒・亚克林担任主席,墨西哥建筑师瓦莱里亚・普里托和加拿大建筑师玛丽亚・伊内斯・叙贝卡索担任副主席,保加利亚建筑师伊万・埃涅夫担任秘书长。

CIAV 的工作领域很广。数年来,委员会不断扩展对“乡土”一词的内涵释义,从单一的农庄、传统的村落单元到城市中自发形成的区域与居民点,再到文化景观区以及乡土遗产与地理环境之间形成的强烈联系。

本次会议正凸显了 CIAV 工作的广泛领域。在此,我要十分感谢中国古迹遗址保护协会,特别是邵甬教授对这次会议组织的热情与持续投入。

吉斯勒・亚克林
国际古迹遗址理事会乡土建筑科学委员会主席
2019 年 6 月 26 日

Preface 2

The International Scientific Committee on Earthen Architectural Heritage is a Committee from ICOMOS with 140 experts and associate members, from 41 countries. ISCEAH contributes to the development of better practice and methods for the protection and conservation of the world's earthen architectural, archaeological and cultural landscape heritage.

The objectives of the scientific program of ISCEAH are focused on: (1) Conserving and studying the standing, and in use, architectural heritage; (2) Conserving and studying the earthen archaeological environment; (3) Understanding the historic/ traditional techniques of earthen structures, including its impact on new earthen construction; (4) Researching the contribution of earthen architectural heritage to cultural landscapes and its relation to the intangible heritage and living traditions; and (5) Researching ancient/ historic a-seismic techniques, in addition to current research, to inform retrofitting of existing structures and appropriate new construction.

At present, ISCEAH is headed by Mariana Correia (Portugal) as President; Maddalena Achenza (Italy) as Vice-President; Pamela Jerome (USA) as Secretary-General; Shao Yong (China) as Treasurer; Ione Stiegler (USA), as In Use Chair; Jorge Aching (Peru), as Archaeology Chair; Bakonirina Rakotomamonjy (Madagascar), as Technology Chair; Ishanlosen Odiaua (Nigeria), as Landscape Chair; and Claudia Cancino (Peru/USA), as Seismic Chair. The International Committee has been a platform of dialogue and knowledge exchange between professionals and academics from the 5 continents.

For almost 50 years, earthen architecture experts have gathered on different continents. The first symposium in 1972 in Iran, grew from a group of less than two-dozen people to an International Conference with 750 people attending TERRA 2016, in Lyon, France. In the near future, TERRA 2021 will take place in Santa Fe, USA, and TERRA 2024 in Cuenca, Ecuador, and more than a thousand experts will be expected to attend the events that became now a World Congress, under the aegis of ISCEAH. The Committee now contributes for World Heritage assessments regarding desk-reviews, technical evaluation missions and reactive monitoring missions to World Heritage earthen properties, and addresses assessments for the World Monument Fund regarding cultural nominations.

The organization of the ICOMOS CIAV-ISCEAH 2019, Joint Annual Meeting and International Conference "Vernacular & Earthen Architecture towards Local Development" reveals the commitment of ICOMOS-ISCEAH and ICOMOS-CIAV, with ICOMOS-China, to work together with WHITRAP and Tongji University, to protect cultural heritage and enhance best practices in the conservation of vernacular and earthen sites, and in particular, the World Heritage city of Pingyao. The successful collaboration of international and national institutions organizing CIAV-ISCEAH 2019 was just possible due to the commitment of professionals from ICOMOS-China and in particular of Prof. Shao Yong and her team. Thank you to all.

The International Scientific Committee of Earthen Architecture Heritage (ICOMOS-ISCEAH) joins efforts with other institutions to continue supporting capacity-building initiatives for and with communities: in Pingyao and around the world.

Mariana Correia
President of ICOMOS-ISCEAH

序　2

国际古迹遗址理事会土质建筑遗产科学委员会(以下简称 ISCEAH)由来自 41 个国家的 140 位专家委员与委员组成。ISCEAH 致力于为世界范围内的土质建筑、考古和文化景观遗产寻求更好的保护实践和方法。

ISCEAH 科学计划的目标着重于五大主题:(一)保护和研究既存且处于使用中的建筑遗产;(二)保护和研究土质考古环境;(三)通过研究物质特性及其对新的土质构造的影响,理解土质结构相关的历史及传统技术,并在这一过程中进行合作;(四)研究土质建筑遗产对文化景观,以及其与非物质遗产和生活传统之间关系的促进作用;(五)研究古代及历史上的抗震技术,并在当前研究中应用这些技术为既有结构的改造和适应性新建造提供借鉴。

目前,ISCEAH 由玛丽安娜·科雷亚(Mariana Correia,葡萄牙)担任主席,玛达莱娜·阿西娜(Maddalena Achenza,意大利)担任副主席,帕梅拉·杰罗姆(Pamela Jerome,美国)担任秘书长,邵甬(中国)担任司库。艾奥尼·斯蒂格勒(Ione Stiegler,美国)负责遗产利用主题,豪尔赫·阿辛(Jorge Aching,秘鲁)负责考古主题,巴科尼丽娜·拉库图马蒙吉(Bakonirina Rakotomamonjy,马达加斯加)负责技术主题,伊斯汉洛森·奥迪欧(Ishanlosen Odiaua,尼日利亚)负责景观主题,克劳迪娅·坎西诺(Claudia Cancino,秘鲁/美国)负责抗震主题。ISCEAH 已成为五大洲专业实践和学术界人士进行对话和知识交流的平台。

近 50 年以来,土质建筑领域的专家在世界各地齐聚一堂。1972 年在伊朗举办的首届研讨会还只是 20 人左右的团体,到 2016 年法国里昂举办的世界生土建筑大会(以下简称 TERRA 大会),已有 750 余名参与者。不久之后即将在美国圣达菲和厄瓜多尔昆卡举办的 2021 和 2024 年 TERRA 大会将会吸引一千余名参与者,成为在 ISCEAH 指导下的一个真正意义上的世界大会。如今,ISCEAH 还致力于世界遗产评估,包括针对土质世界遗产的书面材料的审查、技术评估考察和反应性监测考察,同时还为世界文物基金会的文化遗产保护申请进行评估。举办 2019 年国际古迹遗址理事会乡土建筑和土质建筑遗产科学委员会联合年会暨“面向地方发展的乡土和土质建筑保护”国际学术研讨会,体现了国际古迹遗址理事会乡土建筑和土质建筑遗产科学委员会,以及中国古迹遗址保护协会与高水准机构如亚太遗产中心和同济大学等共同合作保护文化遗产并对乡土和土质遗产地保护最佳案例(特别是世界遗产城市平遥)进行优化的承诺。2019 年国际古迹遗址理事会乡土建筑和土质建筑遗产科学委员会联合年会能够成功举办离不开各国际机构和国家机构通力合作,离不开中国古迹遗址保护协会的专家,特别是邵甬教授及其团队的辛勤付出。在此向所有人表示感谢。

国际古迹遗址理事会土质建筑遗产科学委员会将同其他机构一起共同努力,继续支持平遥乃至全世界针对社区的能力建设倡议和举措。

玛丽安娜·科雷亚

国际古迹遗址理事会土质建筑遗产科学委员会主席

Preface 3

The ancient city of Pingyao is an outstanding example of Han cities in the Ming and Qing dynasties (from the 14th to 20th century). It retains all the Han city features, provides a complete picture of the cultural, social, economic and religious development in Chinese history, and it is of great value for us to study the social form, economic structure, military defense, religious belief, traditional thinking, traditional ethics and dwelling form.

Since 1980s, China has started the establishment of the conservation theory of historic cities and the practices. The effort to conserve Pingyao ancient city, in which I was engaged then as programme director, is a typical and important example of integrated conservation in China. The inscription of the ancient city of Pingyao on the World Heritage List in 1997 has brought the city into a new phase of development. Pingyao County People's Government has worked with Tongji University, Shanxi Urban Planning & Design Institute and other universities and institutions since 2000 to carry out continuous research and practice to deal with the dual objectives of human-habitat heritage conservation and urban development. In addition to important and officially protected cultural relics, a large number of vernacular heritage have been rescued and protected through the effort, which also led to a multidimensional development in social, economic, environmental and cultural terms of urban and rural areas.

I am very pleased to know that the ICOMOS CIAV & ISCEAH 2019 Annual Meeting & International Conference on "Vernacular & Earthen Architecture towards Local Development" will be held in Pingyao. I hope that this conference will provide Pingyao with new development opportunities in the future, as well as the advanced theories, methods and techniques from international experience to conserve its abundant vernacular and earthen heritage resources. And importantly, this is an occasion where Pingyao would make its own contribution to the international community of heritage conservation, by building an important platform for ICOMOS and international experts, scholars, managers and technicians to exchange experiences and discuss challenges on the conservation of vernacular & earthen heritage.

Ms. SHAO Yong, a professor at Tongji University and my student, is in charge of the organization of conference and the publication of the conference proceedings, etc., which makes me feel so proud. I am glad to write a preface for this proceeding to express my sincere wishes for the inheritance of the spirit of heritage conservation and the undertaking from generation to generation, and I also expect more and more successful practices in heritage conservation that will emerge in the future.

RUAN Yisan
Professor of College of Architecture and Urban Planning, Tongji University
Director of National Historic Cities Research Center
Advisory Board Member of the Academic Committee of Historical and Cultural City Planning, the Urban Planning Society of China (ACHCCP-UPSC)
June 24th, 2019 in Tongji University

序　3

平遥古城是汉族城市在明清时期的杰出范例(14—20 世纪),它保存了汉民族城市的所有特征,为人们展示了一幅中国历史非同寻常的文化、社会、经济及宗教发展的完整画卷,并且对研究这一时期的社会形态、经济结构、军事防御、宗教信仰、传统思想、伦理道德和人类居住形式等都具有重要价值。

1980 年代以来,中国开始了保护历史城市的理论探讨和实践,我当年主持的平遥古城保护即为典型的完整保护的重要案例。平遥古城在 1997 年成为世界遗产,从而进入新的历史发展阶段。从 2000 年以后,针对人居遗产保护与城市发展的双重目标,平遥县政府和同济大学、山西省城市规划设计院等高校、研究机构进行了持续不断的保护研究和实践的工作,除了重要的文物保护单位,也抢救和保护了大量的乡土遗产,并且也获得城乡社会、经济、环境和文化等方面的发展。

我非常高兴地看到 2019 年 ICOMOS 乡土建筑科学委员会与土质建筑遗产科学委员会的联合年会暨面向地方发展的乡土和土质建筑保护学术研讨会在平遥召开,希望平遥古城不仅通过此次会议获得新的发展契机,从国际经验中获得针对平遥丰富的乡土遗产和土质遗产保护的理论、方法和技术参考,同时也为国际的遗产保护贡献自己的力量,成为 ICOMOS 和国际专家、学者、管理者、技术人员等就乡土遗产和土质遗产的保护经验和挑战进行交流的一个重要平台。

同济大学邵甬教授主持了本次国际研讨会的组织、论文集的编辑出版等,作为她的老师我甚为宽慰。值此学术研讨会论文集出版之际,欣然为之序,祝愿遗产保护精神和事业能够代代传承,涌现更多的遗产保护成功案例。

阮仪三
同济大学建筑与城市规划学院 教授
国家历史文化名城研究中心主任
中国城市规划学会历史文化名城保护规划学术委员会 顾问委员
2019 年 6 月 24 日于同济大学

Contents

Subtheme 2: Challenges and Possible Solutions

Subtheme 3: Contemporary conservation and technical innovation

Subtheme 4: Adaptive reuse and revitalisation towards local development

Special Theme: Heritage Conservation Going Public: Case Studies of Pingyao International Workshop

Subtheme 1:

The architectural features, values and conservation

Earth-Timber Hybrid Constructions

Miles Lewis

University of Melbourne, Melbourne Australia

Abstract Hybrids of earth and timber, such as wattle and daub, are found in almost every part of the world but there are a source of great confusion to archaeologists and vernacular scholars. Archaeologists are misled by the remains, in which the earth may have survived but the timber usually does not. Techniques such as lehmwickel are never recognised in archaeological work. There is no common system of naming or classification, and English speakers regularly describe any combination of earth and timber in walling as 'wattle and daub'. Many of these techniques have been changed by the impact of modern technology. Pole and pug construction is a case in point. It is often confused with wattle and daub, though it is quite different in principle. There are related forms, but pole and pug is essentially a nineteenth-century development because it relies upon plentiful and cheap nails. It is found in Ireland, in Australia, in Mexico and elsewhere in North America, where it has been associated with the Papago Indians near the Mexican border of the United States, with 'German-Russian-Americans' in Nebraska, and with Ukrainian settlers in both the United States and Canada. In Australia, it seems to evolve in quite a different way from mining works.

Keywords earth-timber hybrids, wattle and daub, lath and pug, pole and pug, nails

1 Palisade and Pug

For obvious reasons most primitive structures of earth and timber have vanished. But earthfast posts (those with their ends in the ground) may leave traces, because if they are burnt the ash remains. The same is true of palisade construction (rows of slabs or poles set in the ground). Although posts and palisades are sometimes pointed to help set them in place, they are not driven into the ground like piles but set into an excavated hole or trench, respectively. Therefore the disturbed soil used in backfilling may also be observable by the archaeologist.

2 Wattling

Wattling is a basketwork of woven twigs and laths used in many cultures for walls of structures such a grain stores, which require ventilation at the same time as protection from vermin. It can be set within a timber frame by boring or mortising holes in the horizontal members to receive the vertical laths or larger members, while the twigs or horizontals are sprung into grooves in the sides of the timber verticals of the frame. In more modern times with nails more readily available, the horizontals may be nailed to either side of an attached cleat, rather than sprung into a groove.

Fig. 1 Gavin house site, Penrith, Australia, evidence of posts and palisade walling. Archaeology and Heritage P/L

In a Danubian long house at Köln-Lindenthal one end is palisaded, and would probably have been chinked with moss and mud to make it weatherproof, while the other end was wattled and probably accommodated livestock and produce.

Wattling was used all over the world

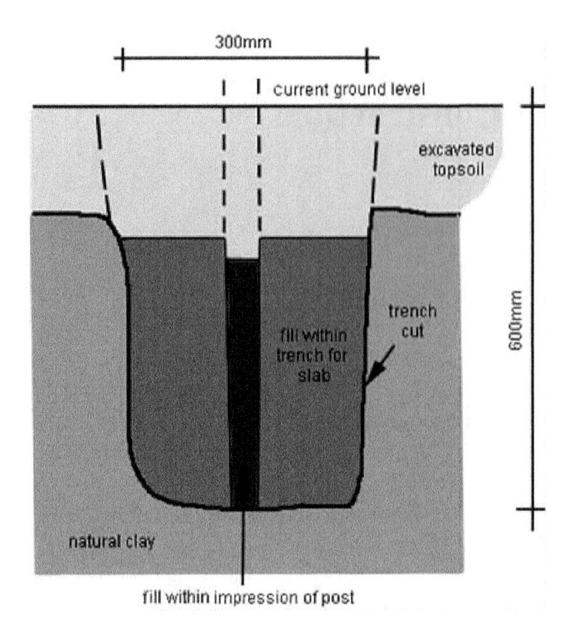

Fig. 2 Gavin house site, Penrith, Australia, palisade wall base, section drawing by Franz Reidel AAH. Archaeology and Heritage P/L

Fig. 3 Model of a Danish slave house in the West Indies, Nationalmuseet, Copenhagen © Miles Lewis

where permeable walls were required, but even more common were walls of wattling plastered over to seal it, which are described as wattle and daub.

3 Wattle and Daub

Wattle and daub is the best-known hybrid technique, used in most cultures across the world, but the term is commonly misused. It applies only to wattling—which is a basketwork of woven twigs — which has been daubed or plastered with some form of mud or lime composition.

Wattle and daub is said to have been used as a building material at Middle Stone Age sites as far apart as Belgium and Palestine. But its origins are obscured by the lack of clarity in archaeological reports. A structure at Tell es-Sultan, Jordan, of about 9000 BCE, was said to be built on a platform and consisted of posts and mud, indicating wattle-and-daub construction (Van Beek 2007). But that inference is completely unwarranted, for there are many alternative interpretations.

Fig. 4 Wattle and daub at Cumalikizak, Turkey © Miles Lewis

We are on safer ground in Europe, though even there it is likely that the remains of other forms of construction, such as *lehmwickel* (totally unknown to archaeologists) have been misinterpreted as wattle and daub. The remains of some of the earliest British wattle and daub buildings have been found in Somerset, at Glastonbury lake village (Bradford 1954). and nearby at Meare, from of about 200 BC (West no date). Wattle and daub was also a form of construction known to the Romans (Pliny 79, book xxxv, ch xlviii) and is commonly assumed to have been the infill in *opus craticum*, Roman half timbered construction. Since Roman times it has been indigenous to many parts of Britain: as the stud and mud of Leicestershire; the clam, staff and daub of Lancashire; the freeth or vreath of the West; the rad and dab or raddle and daub of Cheshire; and the rice and stower of the North (Innocent, 1916; Cook, 1954; Papworth, 1853—1892). The typical English method consisted of vertical rods of hazel sprung into prepared grooves in the framing, between which thinner rods were woven in and out horizontally to form a basketwork, and both sides of the basketwork daubed with a mixture of clay, water and straw, sometimes with cow dung. In East Anglia the vertical hazel sticks, known as 'rizzes' or 'razors', were left with their bark on and no horizontals were used (Cook, 1954). Variants include 'stake and rice' in which reportedly slabs or poles are interwoven with ropes or brush bush.

In Ireland 'sally rods', osiers, twigs, switches, hay or straw ropes, or briars were interwoven between stronger timbers, either vertically or horizontally. (Danachair, 1957).

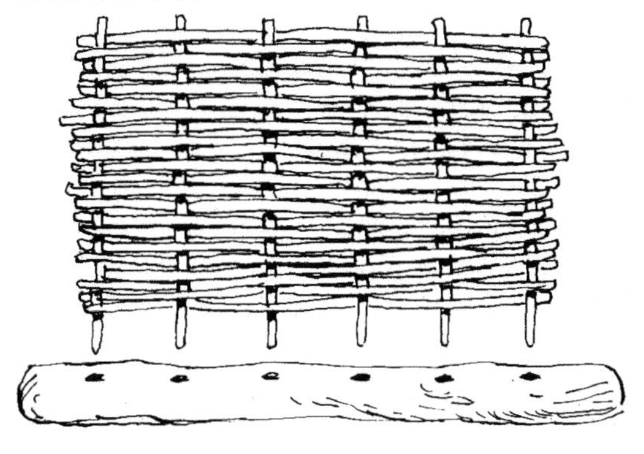

Fig. 5 Wattling for wattle and daub in China: Edwards 1974,73

In France an 18th century description of how to build a beehive reads:

Pour faire les murs des côté s et du devant, en enfonce quelques morceaux de bois dans la terre àla distance d'un pied et demi environ, et les tenant aussi elévés que les poteaux qui soutiennent l'édifice: pour les mieux fixer et les rendre plus solides, on en met deux ou trois en travers, Qu'on cloue aux poteaux: en entrelace ces bois avec des branches de saules ou de tout autre bois, et on applique extérieurement de la terre grasse battue avec de l'eau, pour en faire une espèce de mortier; au défaut du terre glaise on emploie la la terre commune, qu'on m le avec un peu de chaux, pour qu'elle lie mieux. (*Rozier*, 1793—1800). It is not surprising that the technique migrated to French colonies, but what is surprising is that the term used in Québec, *gasparde* (Arthur & Witney, 1998), does not seem to be used in France itself. Wattle and daub is also known in China, Mexico, and elsewhere.

4 Lehmwickel

Lehmwickel is a more specialised type, in which mud and straw are wrapped around timber stakes, which are then packed into the panels of a *fachwerkbau* or half-timbered wall. It is German in origin, and although it became widespread, from Brazil to Australia, and apart from early examples in France, it was always confined to German settled areas. I have considered it elsewhere, and it is not necessary to do so here.

5 Cane and Daub

A number of methods using canes, bamboos or timber rods resemble wattle and daub, but are not in fact wattled. And this distinction is itself often unclear. The *bajareque* of the Dominican Republic as reported is a form of wattle and daub (Vicioso, no date) whereas the *bahareque* of Cuba is simply a grid of rods, daubed over (Melero, no date).

The simplest of these methods is a medieval one in which the studs are so close together that small horizontal rods can span directly between them.

If the spacing is larger there needs to be some sort or grid or mesh, combining

Fig. 6 Eynsham Abbey, Oxfordshire, UK, reconstruction of part of an Anglo-Saxon wall, probably before the 10th century, based upon the imprints left in the plaster, Woodstock Museum, Oxfordshire: Miles Lewis

Fig. 7 Cane and daub wall, Uchiko, Japan, interpretative display showing the layers of mud plastering: Miles Lewis

vertical and horizontal members. As these are not woven they need to be fixed together in some way to develop any strength. In Mexico, Japan, and elsewhere, round rods or canes are tied at some or all of the intersections. In Japan the daub is commonly reinforced with rice straw.

6 Lath and Pug

It is convenient to use the term 'lath and daub' for construction in which split laths (rather than poles) are nailed right across the face—or both faces—of the wall, and the cavity is filled with pug. I distinguish pug from daub in that the material, though similar, is a filling rather than a coating.

This seems to be largely a British technique, and would not merit consideration here were it not for the fact that it is the antecedent of pole and pug construction. In England in 1775 Nathaniel Kent advised that 'Where pollards are plenty, and bricks scarce, it will sometimes be proper to prefer the wooden-lath and rough-cast cottages (Kent, 1775).' In 1805 William Atkinson advocated the use of timber posts of about four inches [100 mm] diameter at fifteen inch [780 mm] spacing, clad on both faces with plastering laths. The outside was to be plastered in clayey mud mixed with chopped straw, and the inside with mud or with lime and hemp (Atkinson, 1805). Loudon refers to this as 'rab and dab', a means of constructing internal partitions, implicitly in a building with external walls of cob: 'instead of brick-nogging for partitions cob is used for filling in the framework which is previously lathed with stout slit oak or hazel branches in place of laths (Loudon, 1846)'.

According to McCann, 'mud and stud' buildings, which he equates with timber and laths, were built in many parts of England, but the majority of survivors are in Lincolnshire (McCann, 2004). He seems to be conflating two different things, the widespread tradition of horizontal lathing, which is particularly common in Essex, and the inappositely named 'mud and stud' using vertical rods or split pieces, known only in Lincolnshire (as discussed above). In the early twentieth century, according to Ketteridge and Mays, practice in Essex was to use split ash poles as laths, and a coating of clay reinforced with straw and chaff (Ketteridge & Mays, 1972).

A number of examples of this technique are known in Australia, mainly of the early twentieth century

7 Pole and Pug

There is a major change in principle when instead of filling within the panels of a timber frame, poles or saplings are nailed horizontal to both faces of the frame, and then filled with mud. It is more a question of packing the mud within the timbers, than of plastering or daubing it on the outside faces. This is a widespread form of construction, but seems to appear only in the nineteenth century.

Fig. 8 Jamiltepec, Oaxaca, Mexico: Walter Reuter. Prieto 1994,32

A method of this sort is reported in Ireland (in addition to true wattle and daub): light timbers are nailed to the face of the uprights at intervals of about 300 mm, said to resemble the rungs of a ladder (Danachair, 1957). A similar construction is found in Mexico, Canada, Australia, and elsewhere. There seems no doubt that this was brought about by the availability for the first time of cheap nails, and probably the wire nail in particular.

Although split laths are also dependent upon nails, they would be familiar from internal plastering, and therefore may be older than round poles in exterior walling, despite the cost.

The mass production of nails began with the mechanisation of various parts of the nail making process from the 1790s onwards especially in the United States Precisely when wrought nails began to be produced entirely by machinery is unknown, but in 1834 Thomas John Fuller obtained a British patent for apparatus which produced both 'square-pointed' and 'flat pointed' [probably chisel pointed] nails, using hammers acting alternately to form the points, so as to retain the fibrous texture, and then finishing them with rollers (Mechanics' Magazine, 1839).

By the 1850s Usonian factories were making what was described as wrought iron nails, on an improved system in which the fibre of the iron lay in the direction of the nail rather than transverse to it. For the smaller nails the machinery was 'self-acting' or automatic, but for the larger ones a boy had to supply the iron to the machine (Builder, 1856). The critical development was probably the wire nail, which had been long known in France but was not widely manufactured until the mid-nineteenth century. By the 1870s it was probably available in most parts of the world and at considerably less cost than traditional wrought nail. Hence it is from about this time that we should expect the appearance of pole and pug construction. And, indeed, that seems to be the case.

Such construction is found in the United States in two quite different contexts. One is that of Ukrainian settlers, who undoubtedly brought the technique from their homeland, and from Galicia in particular, but another seems to be a spontaneous local development. A Ukrainian settler of the early twentieth century recalled his family house in Boryskivtsi, 'built by digging poles into the ground. Between the poles we constructed a grille-like framework and filled it with earth.' Another said:

'Our homes were made with poles set up in the ground and fastened smaller poles to the upright posts and the cracks were filled or plastered with mud mixed with hay or straw. They were finished with mud alone so it would be smoother.'

Construction of this sort characterises a 930 square kilometre area of western North Dakota in which nearly forty such structures have been found (Martin, 1989).

Christopher Martin describes the horizontal members as laths, but it is clear from an illustration that they are saplings in the round (though there are later variants using sawn lathing). The posts were of cedar, about 150 mm in diameter and 0. 9 to 1. 2 m apart, the horizontals were of willow, and the filling was a mixture of earth, dung and straw, locally known as 'gumbo'. Sometimes small rocks were placed in the bottom of the frame, to resist erosion.

In Boyd County, Nebraska, buildings by what are described as 'German-Russian-Americans' used much the same system, with willow uprights, willow poles, and a packing of mud and straw (Murphy, 1989). Other examples of this construction have also been identified amongst Ukrainian buildings in Saskatchewan, Canada, though log construction is more common (Martin, 1989). The same form of building is found amongst the Papago Indians, near the Mexican border. The buildings are flat-roofed and the posts, which are relatively lightweight and closely spaced, have been selected with natural forks to hold the roof beams. In some cases the poles run horizontally, exactly in accordance with the method already discussed, but in others about four poles run horizontally, and the close-set poles are vertical (Nabokov & Easton, 1989).

It can by no means be inferred that this is an indigenous tradition, for the Indians in the region had been in contact with Europeans since 1540, and it is worth bearing in mind how much more difficult such a building technique would have been before nails were available. A very similar horizontal pole and pug technique has been found in the Great Basin area of Nevada, where the buildings are known as 'mud', 'pole' or 'Indian' houses, and certainly there is no Ukrainian component. However, Blanton Owen sees them as having been a spontaneous local development rather than something derived from the Indians (Owen, 1989). Whether or not this is the case, this does seem to be a form of construction which any settler might invent for himself, as a matter of common sense, and it must in any case be a fairly modern one, given that it is more or less dependent upon nails.

In Australia, too, this construction may have evolved spontaneously, or have been influenced by Aboriginal practices, given that it is prominent in South Australia, where the Aborigines used saplings and mud for building (Elkin, 1979). However the indications are that this was not the case, because split laths rather than round poles were used in the earliest versions.

In contrast, the use of round poles and pug in Australia seems to have been an autochthonous development arising from a similar form construction in mine workings where the packing was of stone, not of mud.

Fig. 9 Duer's barn, Maldon, Australia, 1864: view & detail. Hatton 1964; Terry Williamson 1972

Pole and stone construction appears in a number of mining areas. In a house built in South Australia in 1851 by a German carpenter (who was generally following local rather than German practice), '1 inch wide slats would be nailed onto the beams [sic] and the space between these and the outer wall would be filled with earth, small stones and even wood shavings.' (Listemann, 1851) An example at Onkaparinga is of studwork solidly filled with rubble and finished in lath and plaster, and is also distinctive for being built of earthfast studs, with no ground plate (Young, 1988). Similar construction was used prior to 1880 at Moonta and Kadina:

'The favourite building materials of the miners were pines and battens, cut from the adjoining scrub, formed into a framework, and filled in with mud and stones, all of which were to be found close at hand.' (Franklyn,1880)

Surviving examples can be seen at sites in Victoria such as Duer's barn, Maldon. of 1860.

Generally it seems that pole and pug appear first in similar mining areas, such as the Central Goldfields of Victoria, and Hill End and Gulgong in New South Wales, and after that it spreads more widely.

Fig. 10 Hut at Wattle Flat, New South Wales. Warwick Forge

8 Conclusion

Hybrid earth and timber construction has been used all over the world, in the forms of palisade and pug, lehmwickel, wattle and daub, and cane and daub. The choice has been governed by the materials available and by cultural traditions. But there has been no overall taxonomy or understanding of these methods, and they are often confused. In the nineteenth century a new form appeared, still clearly vernacular, but much less dependent upon local factors. Pole and pug is found in many parts of the world and has been generated mainly by the availability of cheap nails. This is the clearest case, although far from the only one, of a vernacular tradition resulting from the use of industrially produced materials.

References

ARTHUR E, Dudley W, 1988. The Barn: a vanishing landmark in North America. New York: Arrowood Press.

ATKINSON W, 1805. Views of picturesque cottages with plans. London: T Gardiner.

BRADFORD J, 1954. Building in Wattle, Wood and Turf' // Charles Singer, et al. A history of technology, volume I, from early times to fall of ancient empires. Oxford: OUP.

BURKE K, 1972. Gold and silver: an Album of Hill End and Gulgong photographs from the Holtermann collection. Melbourne: Heinemann.

COOK O, No date [1954]. English cottages and farmhouses. London: Thames & Hudson.

DANACHAIR CÓ, 1957. Materials and methods in irish traditional building. Journal of the royal society of antiquaries of Ireland, 87: 61-74.

EDWARDS R, LIN Weihao, 1984. Mud brick and earth building the Chinese way. 2nd ed. Kuranda [Queensland] authors.

ELKIN A P, 1979. The Australian aborigines: how to understand them. Sydney: Angus & Robertson.

FRANKLYN H M, 1881. A glance at Australia in 1880. Melbourne: Victorian Review Publishing.

HATTON W, 1964. Maldon. 5 vols, Arch thesis. Melbourne: University of Melbourne.

INNOCENT C F, 1916. The development of English building construction. Cambridge: CUP.

KENT N, 1775. Hints to gentlemen of landed property. London: J Dodsley.

KETTERIDGE C, Spike M, 1972. Five miles from Bunkum London: Eyre Methuen.

LISTEMANN G, 1851. Meine Auswanderung nach Sued-Australien und Rueckkehr zum Vaterlande; ein Wort zur Warnung und Belehrung fuer alle Auswanderunglustige. // Lothar Brasse. The German Contribution. Historic Environment, VI, 2 & 3 (1988):48.

LOUDON J C, 1846 [1833]. An encyclop dia of cottage, farm, and villa architecture and furniture, &c. 2nd ed. London: Longman, Brown.

LUCAS A T, 1960. Furze: a survey of its history and uses in Ireland, Dublin National Museum of Ireland.

MARTIN C, 1989. Skeleton of settlement: ukrainian folk building in Western North Dakota// Thomas C, Herman B L. Perspectives in vernacular architecture, III. Columbia [Missouri]: University of Missouri Press: 86-98.

MCCANN J, 2004 [1983] Clay and cob buildings. Princes Risborough [Buckinghamshire], Shire Publications.

Mechanics' magazine, museum, register, Journal and gazette (London), 1839.

MELERO N, No date. Performance and current state of latin american rural vernacular architecture: a challenge to be addressed//Dabaieh M,

Prieto V. Vernacular architecture reflections: challenges and future. Lund [Norway], Media-tryck: 218-230.

MURPHY D, 1989. Building in clay on the central plains //THOMAS C, HERMAN B L, Perspectives in Vernacular Architecture, III. Columbia [Missouri]: University of Missouri Press: 74-85.

Nabokov P, Robert E, 1989. Native american architecture. New York: OUP.

OWEN B, 1989. The great basinmud' house: preliminary findings (abstract)//Thomas C, HERMAN B L. Perspectives in vernacular architecture, III. Columbia [Missouri]: University of Missouri Press: 245-6.

PAPWORTH W, 1853—1892. The dictionary of architecture. 6 vols. London: Architectural Publication Society.

PLINY, the Elder [Caius Plinius Secundus], Uncompleted 79. Naturalis historia.

PRIETO V, 1994. Vivienda Campesina en México. 2nd ed. Mexico City: Beatrice Trueblood.

ROZIER F P, 1793-1800. Cours Complet d'Agriculture. 10 vols. Paris: Les Libraires Associés.

Singer, Charles, et al. , 1954. A history of technology, volume 1, From Early Times to Fall of Ancient Empires. Oxford: OUP.

VAN BEEK G, VAN BEEKO, 2007. Glorious mud. Washington: Smithsonian Institution.

VICIOSO E P, No date. Dominican Republic: origins and evolution//MARWA D, VALERIA P. Vernacular architecture reflections: challenges and future. Lund [Norway], Media-Tryck: 88-109.

WEST Trudy, No date. The Timber-frame house in England. Newton Abbot [Devonshire]: David & Charles.

YOUNG G, 1988. Onkaparinga heritage. Adelaide: South Australian Centre for Settlement Studies.

World Heritage: an Approach to Vernacular and Earthen Architecture

Mariana Correia, Gilberto Carlos, Teresa Bermudez

Escola Superior Gallaecia, Portugal

Abstract World Heritage is an international concept established by the World Heritage Convention. However, what it means, how is evaluated, and conserved is still unclear even for heritage professionals and conservation experts. This article aims to clarify what does Outstanding Universal Value means and how its foundation stands to establish three key-pillars: criteria, integrity and authenticity, protection and management. Their compliance justifies why a nominated site becomes a World Heritage property. These three key-pillars are also analysed in the case of vernacular and earthen architecture World Heritage properties.

The paper introduces the fact that vernacular architecture is underrepresented in the World Heritage List. As a result, research was addressed regarding the analysis of different types of vernacular categories. This paper introduces the different parameters that were identified considering the thematic approach introduced by the vernacular and earthen architecture framework: the landscape setting; the urban layout; the architecture features; and the building culture. A deeper comparative analysis regarding the diversity existing in vernacular and earthen architecture is being undertaken to contribute to the rising of nominations in this thematic area, therefore contributing to a more balanced, representative and credible World Heritage List.

Keywords World Heritage, vernacular, earthen architecture, outstanding universal value, criteria, authenticity, integrity

1 Introduction

UNESCO formally established the concept of World Heritage, in 1972, trough the "Convention for the Protection of World Cultural and Natural Heritage", which was already signed by 193 State Parties (UNESCO-WHC, 2017a). To support the practical application of this Convention, a set of regulations, known as the "Operational Guidelines for the Implementation of the World Heritage Convention" (UNESCO-WHC, 2017b), were established by the World Heritage Committee. The procedural document, which is provided in different languages is revised and updated regularly.

In 2019, at the 43rd session of the World Heritage Committee, in Baku, the Republic of Azerbaijan, 29 new World Heritage properties were approved, from which 24 were cultural, 4 were natural and 1 was a mixed site. As of 11 July 2019, the World Heritage List consists of 1121 properties from 167 countries, of which 869 are cultural, 213 are natural, and 39 are mixed properties. At present, the World Heritage Committee has 53 Properties on the List of World Heritage in Danger (UNESCO-WHC, 2019a).

2 Outstanding Universal Value (OUV)

For a property to be considered World Heritage, it must contain exceptional significance and sufficient attributes to support its Outstanding Universal Value (OUV). If adequate OUV evidence is demonstrated during the nomination process, the property is eligible to be classified as a World Heritage Site, as its cultural and natural significance are so outstanding that it transcends the national boundaries of the State Party.

Therefore, the decision to list a site as World Heritage means that a property is of such great importance to current and future generations and has such universal status that the legislative protection and the management system of the site must ensure that the attributes and values by which the sites are listed as World Heritage, are preserved as much as possible. Therefore, the OUV Declaration of each World Heritage property should guide the preservation of the authenticity and integrity of the site, as well as its conservation, protection and effective long-term management.

A property is listed as World Heritage if the World Heritage Committee, upon the technical evaluation proposed by the Advisory Bodies, considers that the property Outstanding Universal Value is established. To ensure that the OUV is recognised, ICOMOS International addresses the assessment of Cultural sites and IUCN addresses the assessment of natural sites.

3 The Three Pillars That Define the OUV

The technical evaluation completed by the Advisory Bodies regards the assessment of the three pillars of World Heritage, which will grant the Outstanding Universal Value to the nominated property. First, the State Party has to present evidence that the property has sufficient justification to match one of the ten CRITERIA of universal value; secondly, the nomination has to provide enough proof regarding the justification of the PRINCIPLES of integrity and authenticity (in the case of a natural property nomination, just integrity needs to be substantiated, as Nature already has authenticity); and thirdly, the nomination file has to present evidence regarding PROTECTION of the property and that a MANAGEMENT system is in place (or at least being implemented).

3.1 Criteria Establishing the OUV

The World Heritage Committee has set ten precise criteria for the classification of properties on the World Heritage List, following the establishment that the property has Outstanding Universal Value. The first six criteria classify cultural heritage. The following four criteria classify natural heritage. For a site to be approved as a World Heritage mixed property, it must have at least two approved criteria: one criterion must be between (i) and (vi), and one other criterion must be listed between (vii) and (x).

According to the Operational Guidelines (UNESCO-WHC, 2017b), the criteria established for sites of cultural value should:

Fig. 1 The World Heritage property of Chan Chan Archaeological Zone in Trujillo, Peru

(i) "*Represent a masterpiece of human creative genius*". Under this criterion is listed the Great Wall of China, a vast and ambitious undertaking, considered an absolute masterpiece, also for its construction. This is a criterion that is not frequently observed matching vernacular and earthen architecture, due to the dimensions and masterwork that are needed to comply with it. One of the few examples of a World Heritage earthen site that conforms to criterion (i) is Chan Chan Archeological Zone (Fig. 1). The largest earthen city of pre-Columbian America is a masterpiece of town planning. "*Rigorous zoning, differentiated use of inhabited space, and hierarchical construction illustrate a political and social ideal which has rarely been expressed with such clarity*" (UNESCO-WHC, 2019b).

(ii) "*Exhibit an important interchange of human values, over a span of time or within a cultural area of the world (...)*". This is the case of the vernacular earthen city of Timbuktu, in Mali, with its mosques and holy places. "*Timbuktu played an essential role in the spread of Islam in Africa at an early period*". The Historic Centre of the earthen vernacular city of Agadez in Niger, also complies with criterion (ii). "*From the 15th century, Agadez, 'the gateway to the desert', became an exceptional crossroads for the caravan trade. It bears witness to an early historic town, forming a major centre for trans-Saharan cultural interchanges. Its architecture embodies a synthesis of stylistic influences in an original urban ensemble, made entirely of mudbrick and which is specific to the Aïr region*" (UNESCO-WHC, 2019b).

(iii) "*Bear a unique or at least exceptional testimony to a cultural tradition or to a civilization which is living or which has disappeared*". This criterion is applied at the Ancient City of Pingyao, in China. The Ancient City "*was a financial center in China from the 19th century to the early 20th century. The business shops and traditional dwellings in the city are historical witnesses to the economic prosperity of the Ancient City of Ping Yao in this period*" (UNESCO-WHC, 2019b).

(iv) "*Be an outstanding example of a type of building, architectural or technological ensemble or landscape which illustrates (a) significant stage (s) in human history*". Under this criterion is the Ksar of Ait-Ben-Haddou, in Morocco, a traditional pre-Saharan habitat comprised by a group of earthen buildings still inhabited and surrounded by earthen walls. "*The Ksar of Ait-Ben-Haddou is an eminent example of a ksar in southern Morocco illustrating the main types of earthen constructions that may be observed dating from the 17th century in the valleys of Dra, Todgha, Dad s and Souss*". The old town of Djenn, in Mali, inhabited since 250 B. C. is comprised of earthen traditional houses, which represent "*an outstanding example of an architectural group of buildings illustrating a significant historic period*" (UNESCO-WHC, 2019b).

(v) "*Be an outstanding example (...), which is representative of a culture (or cultures), or human interaction with the environment, especially when it has become vulnerable under the impact of irreversible change.*" This is the case of Iran's earthen Historic City of Yazd, which witnesses the use of limited resources, particularly in architecture and agriculture, as a means of survival in the desert. Also worth mentioning is the Cultural Landscape of Honghe Hani Rice Terraces, in China. A good example of a landscape with adobe building scattered through the rice terraces. The landscape reflects "*in an exceptional way a specific interaction with the environment mediated by integrated farming and water management systems, and underpinned by socio-economic-religious systems that express the dual relationship between people and gods and between individuals and community, a system that has persisted for at least a millennium*" (UNESCO-WHC, 2019b).

(vi) "*Be directly or tangibly associated with events or living traditions, with ideas, or with beliefs, with artistic and literary works of outstanding universal significance.*" This is a criterion that is not very common to find specifically associated with vernacular and earthen architecture. Under this criterion, it can be found Koutammakou, the land of Batammariba, in Togo. The Koutammakou landscape is home to the Batammariba, whose remarkable Takienta earthen tower-houses become the symbol of Togo. Criterion (vi) is revealed through the Koutammakou, "*an eloquent testimony to the strength of spiritual association between people and the landscape, as manifested in the harmony between the Batammariba and their natural surroundings*" (UNESCO-WHC, 2019b).

Vernacular and earthen heritage can be found across the six criteria that established Outstanding Universal Value in cultural sites. Notwithstanding, the criteria that bear the most of vernacular and earthen heritage are criteria (ii), (iii), (iv), and (v).

3.2 Principles of Integrity and Authenticity

For the classification of a World Heritage site, it is essential that the property has imbedded the principles of Integrity and Authenticity. Their justification should be complementary to the assessment of the Outstanding Universal Value of the World Heritage property.

Regarding Integrity, this is the principle that refers to the state of the listed property as a whole, which is complete and indivisible. This principle is referred to in Articles 7 and 8 of the Venice Charter. In this sense, integrity should be considered in relation to the characteristics identified in the proposed property's OUV. Thus, a clear understanding of the definition and nature of the site attributes, of its characteristics, which are central to its OUV, and the boundaries of ownership are needed to consider the overall integrity of the property.

The principle of Authenticity, which is expressed in the Nara Document of Authenticity, will depend on the type of cultural heritage and its context. Classified sites satisfy the conditions of authenticity if they have a variety of tangible and intangible attributes, expressed by form, material, construction, use, function, traditions, management system, place spirit, etc. The notion of authenticity is expressed through distinct attributes in very different cultural contexts, which may have different traditions of heritage conservation, notably in the Far East, Africa or Western Europe (Correia, 2018).

3.3 Site Protection and Management

When considering the protection and management of a World Heritage site, it is paramount that the State publishes appropriate legislative, regulatory, institutional and traditional protection for the classified site. This will ensure its safeguard in the long term. It is equally essential that the World Heritage site includes boundaries of the listed property, and of its buffer zone. Therefore, regulatory legislation should be published, in order to protect as much as possible, the attributes that define the property's criteria and principles of integrity and authenticity.

For vernacular and earthen heritage this is often difficult to attend, as several of these sites are still inhabited and frequently, they do not hold legislative protection and specific regulations about what to do and not to do in regards to intervention. Furthermore, vernacular and earthen architecture are often not covered by existing cultural heritage protection laws, which are a wide threat for effective protection.

Many of the threats that currently affect listed World Heritage properties are due to the lack of protective legislation, the lack of its implementation, or the lack of mechanisms for monitoring the state of conservation of the property. Also of note is the importance of developing and implementing management systems that ensure the efficient and sustainable management of existing resources (Correia, 2018).

4 World Heritage Vernacular Architecture is an Underrepresented Category

In 2004, ICOMOS published the document "The World Heritage List: Filling the Gaps - an Action Plan for the Future" (ICOMOS, 2004), which addressed a revision of the properties that had been listed from 1978 to 2003, in order to contribute for a more balanced and credible World Heritage List. Vernacular architecture was one of the typologies that was mentioned has been underrepresented on the World Heritage List.

In this assessment, It was highlighted the need to include in the World Heritage List, properties of outstanding universal value from rural settlements and vernacular architecture. It was also emphasized that the 1972 World Heritage Convention included mostly physical evidence, and there was a need to acknowledge intangible aspects (ICOMOS, 2004). Intangible knowledge is part of the authenticity of vernacular and earthen architecture. It can be foreseen for instance, in Western Africa, where traditional construction is inter-linked with local know-how. An active traditional local knowledge is needed to continuously improve the inhabited dwellings maintenance and the rural landscape where they are located.

The Filling the Gaps document defines vernacular architecture as the "use of traditionally established building types, application of traditional construction systems and crafts"(ICOMOS, 2004). Considering this conceptual definition, the typological framework that was contemplated in the overall study was based in a multi-category approach. The most represented categories were by order of World Heritage List representation: Historic Buildings (architectural properties), Settlements (historic towns), Religious properties, and Archaeological heritage. These constitute 69% of the cultural properties on the List. Following were Landscapes (with almost 100 sites), Military heritage, agricultural properties, and finally vernacular architecture (with a little higher than 50 properties) (Fig. 2, ICOMOS, 2004).

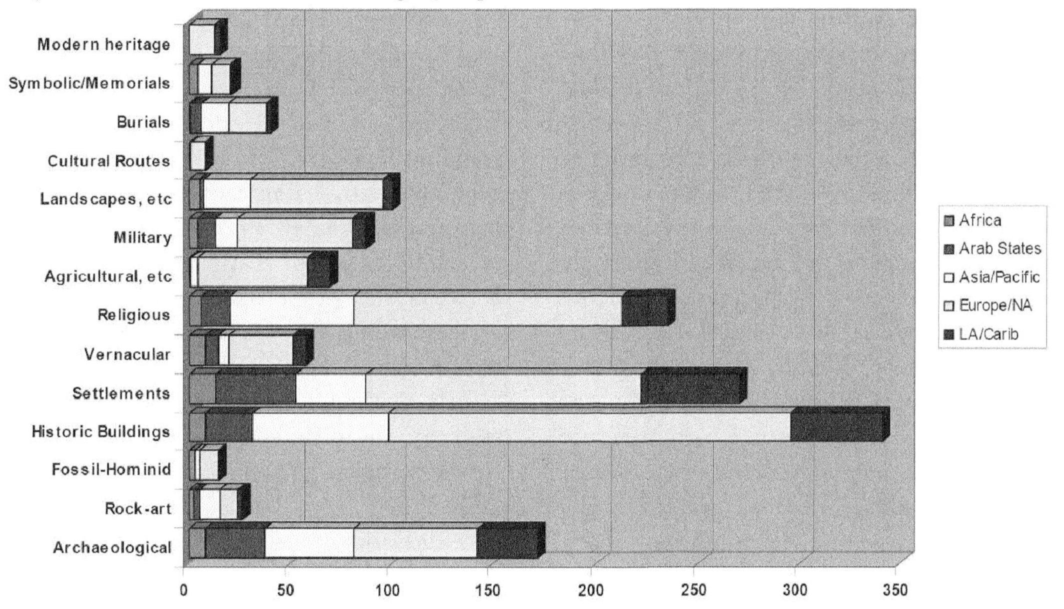

Fig. 2 World Heritage List classification, by category (ICOMOS, 2004)

In the case of vernacular architecture, the study established the following categories (ICOMOS, 2004):

- Monuments: traditionally established building types using traditional construction systems and crafts;
- Groups of buildings: groups of traditionally established building types; and
- Sites: cultural landscapes with vernacular settlements.

The study also revealed that Africa, Asia-Pacific, Latin America and the Caribbean had a low representation of vernacular properties, when it was expected for this category to be stronger on these international regions.

5 WHEAP Inventory of Earthen Architecture

The World Heritage Earthen Architecture Programme (WHEAP) was developed from 2007 to 2017, following its approval at the 31st Session of the World Heritage Committee. Under the framework of the programme, a World Heritage Inventory of Earthen Architecture (CRAterre-ENSAG, 2012) was developed and published in English and French. This inventory considered that 10% of the World Heritage properties integrated earthen structures. It also highlighted that in 2011, 24% of the sites inscribed on the World Heritage List in Danger were earthen sites (UNESCO-WHC, 2019b).

The 2012 inventory established that most of the earthen World Heritage properties were historic centres, followed by archaeological sites, historic towns or urban centres, and historical architectural buildings. These are very similar results also observed on the 'Filling the Gaps' study addressed by ICOMOS, in 2004.

Regarding earthen constructive typologies and ways of building, the 2012 inventory recognized that adobe was the most used technique worldwide, with 50% of adobe being identified in earthen World Heritage properties, followed by 24% of daubed earth, 20% of rammed earth, 6% of cob, and 39% of other regional techniques (CRAterre-ENSAG, 2012). This assessment considers that several of the earthen techniques could be applied simultaneously or in a combined way (for instance *entramado* in Spain, is both a structure technique, with wood filled-in with adobe).

This earthen World Heritage inventory was based in classifying earthen architecture through three main building typologies: monolithic walls (rammed earth, cob, etc.), masonry walls (adobe, CEBs, etc.), and as a secondary element (earth as infill of wattle and daub, earth filling in of floors and ceilings, flat roofs, mortars, paintings, etc.). As a result, the inventory included World Heritage sites that are mainly built in earth (as was the case of the citadel of Bam in Iran; or Bahla Fort, in Oman), but it also included properties with almost no observed earth material (this was the case of the Historic Centre of Porto, in Portugal; or the Historic Centre of Cuenca, in Spain; in both cases, mortars are mostly done in lime and sand).

6 World Heritage Vernacular and Earthen Architecture Study

Several publications have addressed the concept of vernacular heritage and earthen architecture, such as Correia, Carlos, & Rocha (2014); Correia (2015); Mileto, Vegas, & Garc a-Soriano (2017); among others. However, research specifically focusing in both vernacular and earthen architecture in World Heritage has not been yet addressed.

Considering this operational definition and in order to better understand the reason why vernacular architecture is underrepresented in the UNESCO's World Heritage List, a particular study regarding vernacular and earthen World Heritage was addressed and a detailed classification of the properties was established. The research was addressed based on UNESCO's criteria for site selection, as well as its integrity and authenticity. Data was provided by UNESCO platform, addressing the Outstanding Universal Value of the property.

The study being carried out on vernacular and earthen architecture will contribute to establishing correlations and differences between the properties, which will allow understanding the parameters that originated its outstanding universal value. This research methodology allows deepening the study of the different parameters that featured vernacular and earthen architecture (Fig. 3).

The study that was addressed classified vernacular heritage in two main categories focusing on the character of the heritage value and its nature. The main focus of the first category was accurately related to vernacular architecture has being the principal typology contributing to the OUV of the property; whereas the second category concentrated on vernacular architecture as a secondary value. This last category also differed from the previous one, as other attributes were valued, such as the traditional use of the land, the landscape, the routes, the community social system and the artistic works between other assessment parameters, among others(Fig. 4).

The study analysis of the vernacular sites considered the following parameters:

• Landscape setting: a traditional form of the landscape, land use patterns, outstanding landscape, etc.

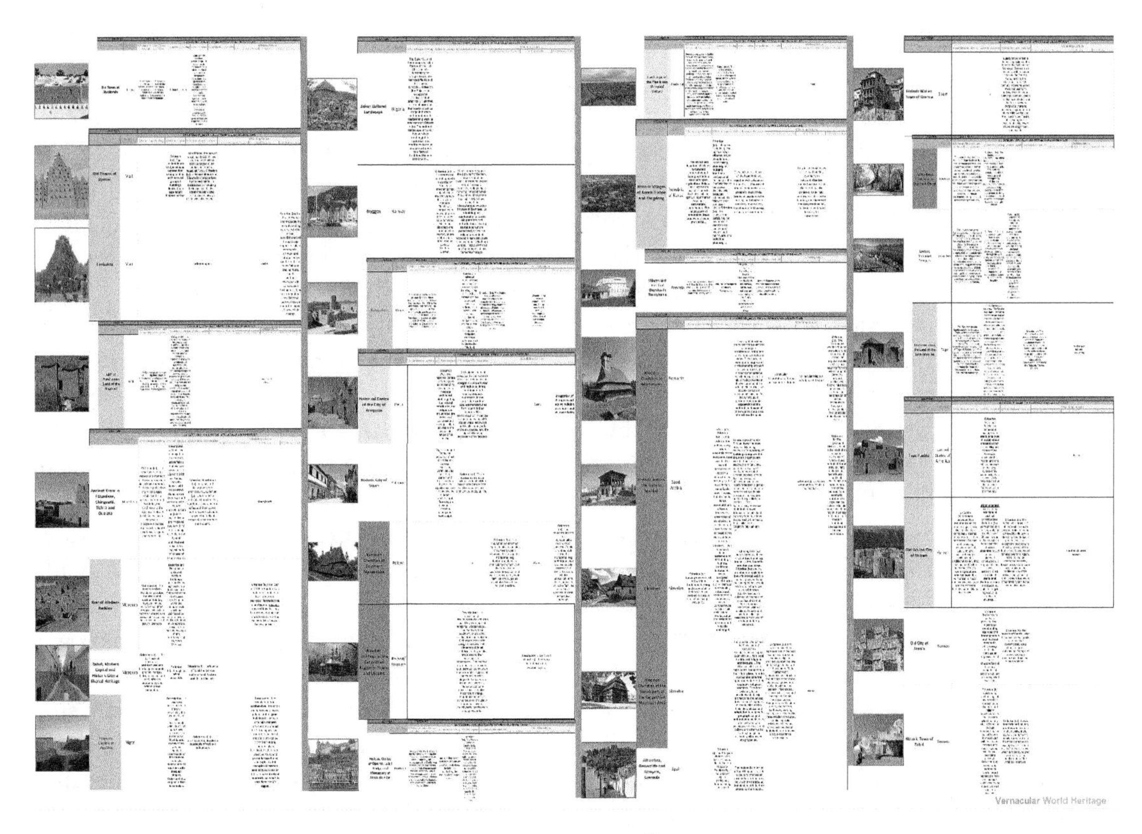

Fig. 3　Example of the Vernacular World Heritage analysis © Ci-ESG, Escola Superior Gallaecia, 2019

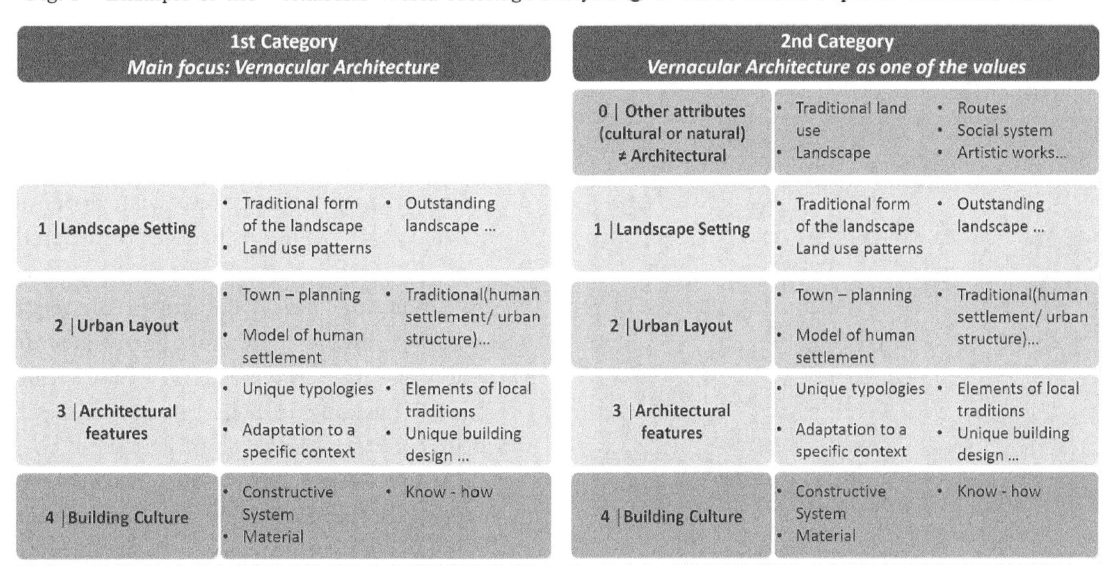

Fig. 4　World Heritage List classification by categories and parameters © Ci-ESG, Escola Superior Gallaecia, 2019

• Urban layout: town planning, model of human settlement, traditional human settlement, urban structure, etc.

• Architectural features: unique typologies, adaptation to a specific context, elements of local traditions, unique building design, etc.

• Building culture: constructive systems, material, know-how, local knowledge.

Despite the difference in the nature of the heritage value, vernacular sites classification as World Heritage properties were analysed for both categories, according to their value as landscape setting, urban layout, architectural features or building culture.

According with the established research methodology, despite most of the earthen World Heritage properties having their specific importance in their own vernacular architecture, there are sites which count with vernacular architecture as one of the values that define the relevance of the site, sharing its importance with other tangible or intangible elements of the ensembles like the landscape, urban layout, social system. Most of the vernacular and earthen architecture cases are specifically listed as cultural sites, while there are a few mixed sites that combine cultural and natural World Heritage categories. This latest is the case of Cliff of Bandiagara (Land of the Dogons), in Mali, which has an Outstanding Universal Value related to the landscape of cliffs and its earthen vernacular architecture, specifically regarding its geological, archaeological and ethnological interest (Fig. 5, criterion iii and vi).

Fig. 5 Cliff of Bandiagara Mixed property, in Mali © Mariana Correia, 2008.

7 Analysis of the Study Parameters

7.1 Landscape Setting

In sites where landscape setting possesses Outstanding Universal Value, cultural landscapes are intrinsically related to vernacular settlements, as the population that lives in the vernacular sites depends on the landscapes to survive. ICOMOS even suggests in regard to vernacular settlements that "*consideration should be given to the possibility of extending nominations to adjacent landscapes if these preserve qualities and features associated with the settlements*" (ICOMOS, 2004).

The importance of the landscape setting in vernacular and earthen architecture World Heritage is related to the interaction of man and nature in a specific environment, an outstanding natural landscape of unique and exceptional beauty that acts as a natural monument with architecture integrated harmoniously. In this case, it could be a settlement that illustrates a traditional earthen, stone, wooden or straw habitat; a settlement located to provide both physical and spiritual nourishment from their surrounding landscape, strength of spiritual association between people and landscape, which is manifested in the harmony between settlement and their natural surroundings. A cultural landscape is remarkable due to associations between architecture, people and landscape(Fig. 6).

The cultural landscape of Honghe Hani Terraces in China is an appropriate example. The property stands up because of its authenticity, in relation to the traditional form of the landscape and also the continuity of landscape function. Unfortunately, one of the threats that has been encompassed at the property is the fact that adobe traditional building is being rapidly substituted by concrete blocks painted in sand colour. As the main focus of the State Party monitoring is the traditional management of the rice terraces, in this case, the traditional way of building is being lost, except for the 3 protected vernacular settlements existing at Honghe Hani Terraces.

7.2 Urban Layout

There are several features and attributes that contribute fo the importance of the urban layout in the OUV of a World Heritage property. The following variables are

Fig. 6 Cultural landscape of Honghe Hani Rice Terraces in China © Mariana Correia, 2017

the most prominent: Ancient fabric as an outstanding example of an architectural group of buildings illustrating a significant historic period; group of earthen buildings which represent a traditional habitat being a striking example of the architecture and urban layout of a specific context; specific street pattern that is still in place today as an original urban example.

Vernacular buildings could demonstrate a distinctive settlement pattern related to a specific location, planning and building traditions that represent an exceptional testimony of a period or religion and also represent a particular ensemble. Traditional unique type of architectural ensemble from a specific period that retains most of its traditional forms up to the present day. This could be an accomplished example of traditional urban architecture both in the grid layout of streets and squares and also an example of urban planning based on multi-storeyed construction.

This is the case of the Old Walled City of Shibam, in Yemen. In this paradigmatic example, urban planning is based on the principle of vertical construction, but also considering the adaptation to the desert climate and environment. The earthen tower structures gave to Shibam the name of 'Manhattan of the desert'. It is an example of traditional Hadrami urban architecture, because of the grid layout of its streets and squares, and also the visual impact of its vertical form rising out of the flood plain due to the height of its adobe tower houses. Because of this and other relevant selection criteria, Shiban constitutes and Outstanding Universal Value that illustrates the key period of Hadrami history. Due to the unrest of war in Yemen, Shibam walled city was listed in the in Danger World Heritage List.

7.3 Architectural Features

The relevance of vernacular architecture is strongly related to the particular features that characterise earthen architecture. Traditional buildings that are a testimony of a unique architectural style of a specific context, location or period, or of architecture that embodies a synthesis of stylistic influences, traditional decoration motifs with symbolic meaning for a specific and unique community. Traditional buildings are featured by a special particularity, by outstanding examples of an architectural typology or by an ensemble of a specific period that could reflect a singular culture. These buildings can be an outstanding example of an extraordinary masterpiece and their ensemble can establish a traditional urban settlement with exceptional OUV(Fig. 7).

One of the examples regarding outstanding architectural features is the Old town of Djenné and in particular its mosque. This exceptional building is located in Mali, and represents the importance of the identity, strengthen by local architecture of a specific community.

7.4 Building Culture

Vernacular heritage building culture is

Fig. 7 Great Mosque of Djenné, in the old town of Djenné, in Mali © Mariana Correia, 2008

reflected in many tangible and intangible aspects like the constructive systems that were employed to build the structure, the applied materials, as well as the know-how. In this last case, it means the practical and technical intangible knowledge held by a community, and their ability to develop a traditional construction process, embedded in local knowledge and traditions that are still active until nowadays.

The richness and diversity of the unique building culture, the perpetuation in time of a building culture related to the continuity of the use regarding local traditional building systems, the traditional techniques and materials, among several other reasons are crucial for the establishment of OUV. The conservation of the ancestrally inherited know-how that belongs to a specific community and contributes to its identity is also part of this rich heritage.

The know-how is reflected in the original construction process, as well as in the conservation, restoration and maintenance techniques that inhabitants have inherited from their ancestors, and still apply nowadays. This know-how process evolves constantly. It results in very efficient constructive systems and techniques that are perfectly adapted to the climatic conditions and the local resources. Due to the vulnerability of earthen architecture, maintenance works are needed regularly. In general, local communities organise maintenance works for special events, respecting traditional know-how, strengthening the authenticity of the site, using vernacular techniques through traditional constructive systems. This engagement of the inhabitants in the continuity of the traditional culture results in the perpetuation of intangible values associated with the building culture and the traditional events(Fig. 8).

Fig. 8 Ksar of Ait-Ben-Haddou, in Morocco © Mariana Correia, 2006

Ksar of Ait-Ben-Haddou, in Morocco, illustrates a building culture that is adapted to the climatic conditions and the cultural context, but at high risk of degradation. The lack of conservation, as a result of lack of use and of maintenance from some of its inhabitants, contributes to the increase of the vulnerability of earthen buildings.

8 Reflections

Vernacular and earthen World Heritage is defined by its Outstanding Universal Value as a testimony of the past and present of a community that has retained its identity, authenticity and integrity. The exceptional attributes and the efficiency of its features constitute the accumulation of an evolved knowledge that works perfectly adapted to climatic conditions, local resources and cultural expressions. Vernacular architecture means a specific interaction with the environment with numerous associations between people and landscape, where traditional society has adapted its life to a particular context. All of these reasons reinforce, the importance of vernacular and earthen architecture properties, as part of the World Heritage List, acting as a testimony of exceptional cultures or civilizations.

Following the study analyses, it is evident that the vulnerability of vernacular and earthen World Heritage properties are due to urban pressure, globalization, climate change and extreme weather threats. In order to avoid a possible degradation and increase loss of this heritage, it is extremely important to maintain the link between local people and their traditional architecture, allowing and encouraging them to carry out the conservation, maintenance and restoration works with traditional techniques that were developed by communities since antiquity. This maintenance works of the vernacular built heritage allow the preservation of the authenticity of the site, across its spectrum, including tangible and intangible aspects that define the character of the heritage.

Due to the lack of use of some of these World Heritage properties and to their abandonment by some of its inhabitants, the conservation of vernacular properties is in danger and at risk. The use of the heritage by the community should be encouraged, in order to achieve the full preservation of the property. In addition, it is crucial to inspire and support activities regarding heritage and its educational value, in order to preserve its features, harmony and identity, and at the same time, fomenting its protection of external influences, which could cause the loss of its unique value. This is just possible, if communities and their vernacular and earthen heritage settings are placed at the centre of sustainable development respecting their traditional way of life.

Taking into account the little number of vernacular and earthen architecture in the World Heritage List, in comparison with other categories, there is a need to really reflect on their importance. A deeper comparative analysis regarding the diversity existing in vernacular and earthen architecture continues to be addressed. It aims to contribute for the rising of nominations in this thematic area, therefore contributing for a more balanced, representative and credible World Heritage List. To increase vernacular and earthen architecture research, legislative protection and site preservation will contribute to human cultural diversity and the conservation of exceptional ways of living that are deemed to disappear if no proactive approach is undertaken to protect these exceptional sites.

Acknowledgments

This paper was developed in the framework of "3dPast: Leaving and virtual visiting European World Heritage", a European research project approved and funded by the programme Creative Europe and sub-programme Culture, with the reference 570729-CREA-1-2016-1-PT-CULT-COOP1 (2016—2020). The European project is coordinated by Escola Superior Gallaecia, in Portugal (Project-leader: Mariana Correia); in partnership with Polytechnic University of Valencia [Universitat Politecnica de Valencia], in Spain; and University of Florence [Universita Degli Studi di Firenze], in Italy.

References

CORREIA M, 2015. Criteria and methodology for intervention in vernacular architecture and earthen heritage//CORREIA M, LOURENÇO P, VARUM H. Seismic Retrofitting: Learning from Vernacular Architecture. London: CRC Press, Taylor & Francis Group.

CORREIA M, 2018. Avaliação de sitios culturais património mundial//Fórum do porto | património, cidade, arquitectura. Porto: CEAU-FAUP.

CORREIA M, CARLOS G, ROCHA S, 2014. Vernacular heritage and earthen architecture: contributions for sustainable development. London: CRC Press, Taylor & Francis Group.

CRAterre-ENSAG, 2012. World Heritage: Inventory of World Heritage. https://www. scribd. com/document/126199699/Wheap-Inventory.

ICOMOS, 1994. The nara document on authenticity. https://www. icomos. org/charters/nara-e. pdf.

ICOMOS, 2004. The World Heritage List: filling the gaps — an action plan for the future an analysis by ICOMOS. https://whc.unesco.org/document/102409.

MILETO C, VEGAS F, GARCÍA-SORIANO L, 2017. Vernacular and earthen architecture: conservation and sustainability. London: CRC Press, Taylor & Francis Group.

UNESCO-WHC, 2017a. Convention concerning the protection of the World Cultural and Natural Heritage. https://whc.unesco.org/en/convention/.

UNESCO-WHC, 2017b. Operational guidelines for the implementation of the World Heritage Convention. https://whc.unesco.org/en/convention/.

UNESCO-WHC, 2019a. New inscribed properties. https://whc.unesco.org/en/newproperties/.

UNESCO-WHC, 2019b. World heritage platform. https://whc.unesco.org.

The Dynamics of Dai Cultural Landscape and Vernacular Architecture in Asia

Rawiwan Oranramanee

Faculty of Architecture Chiang Mai University, Chiang Mai, Thailand

Abstract Cultural landscape and vernacular architecture represent the relationship between human society and environment in place and time. The dynamic changes of both landscapes and architecture exhibit human adaptation to constraints in particular locales and their continuing evolution generates the cultural meanings of different places. This paper explores the dynamics of Dai cultural landscapes and architecture in Asia. Dai are indigenous groups living across the continuing cultural boundaries from upper Southeast Asia, southern China and northeastern India. Based from field studies in three sites namely Shan State in Myanmar, Dehong Prefecture in China and Assam Region in northeastern India, this paper identifies and compares the similarities and differences of the localized patterns of Dai settlements and houses, and discusses the issues of cross-cultural adaptation and evolution over time in relation to migration factors and cultural assimilation. The knowledge from this paper contributes to an understanding of the production of landscape and architecture in the cross-cultural contexts of Asia thereby provides the comparative worldviews for the studies of rural landscapes in other parts of the world with similar constraints and conditions.

Keywords cultural landscape, vernacular architecture, Dai, ethnic group, Asia

1 Introduction

1.1 Problems and Significance

Cultural landscapes and vernacular architecture share some commonalities as the tacit knowledge rooted in the daily lives of ordinary peoples around the world. The interrelationship between cultural landscape and vernacular architecture from diverse societies can provide lessons to be learned for both the conservation of traditional characteristics and applied design knowledge for contemporary societies that are evolving along with the global and local dynamics of change. This paper explores the characteristics of Dai cultural landscape and vernacular architecture in Asia so as to provide a comparative worldview of the peasantry landscape in the cross-cultural context. The reviews of the existing research work identify the predominant characteristics of the Dai cultural landscape based on the wet-rice culture that is deep-rooted in Asian societies (Waterson, 1990). As Dai live in a vast area in Southeast Asia, India Dai and China, the production of Dai landscape and architecture is strongly influenced by these local cultures. The study of the landscape and architecture of Dai can contribute significantly to the knowledge about Dai ethnic identity and their dynamic response due to social contact and cultural assimilation.

2 Reviewed Concepts of Dai Landscape and Architecture

2.1 The Dai Ethnic Group

Dai are indigenous ethnic groups who settled in the lowland river basin in northeastern India, Myanmar, Thailand, Laos, Vietnam and southern China. They consist of an approximate number 120 million population, most of which speak their indigenous languages in the Tai-Kadai language family.

As lowland dwellers, Dai sustain their lives on rice farming using the traditional method of wet-rice cultivation that are long rooted in Asia. In terms of a social system, Dai bears some assimilated beliefs of Animism and Buddhism. In terms of kinship,

Dai lives in an extended family structure, with patrilineal lineage and patrilocal residence (Dodd, 1997; Sai Aung Tun, 2009).

2.2 Some Rules and Concepts of Dai Landscape and Architecture

The reviews about cultural landscape and architecture of the Dai (Milne, 1970; Panin, 1996) pinpoints some predominant rules and concepts including: (Fig. 1)

1) Lowland, water-based peasantry landscape of rice farmers,

2) Settlement system consisting of urban town center called *meng* or *muang* (literally means town) and a collective rural village settlement called *baan* (literally means village),

3) A village comprises of a defined inhabited area for house clusters and some social buildings, including a Buddhist temple and a spirit house, agricultural fields and forest,

4) Stilt house type made of wood called italic font (literally means house) built for an extended family living together under one roof.

Fig. 1 Dai cultural landscape taken from Shan State in Myanmar © the author

3 Field Study

3.1 The Sites and Methods

The Dai population distribute across the large geography as shown in Fig. 2. This paper draws together the findings from three consecutive research works conducted by the author from the year 2014 to 2019. The research sites are in three areas including:

1) Assam Region in northeastern India (Site A),

2) Shan State in Myanmar (Site B) and,

3) Dehong Prefecture in southern China (Site C).

These three sites are considered as the homeland of western Dai geography, known as the Greater Dai, Tai Yai or Shan.

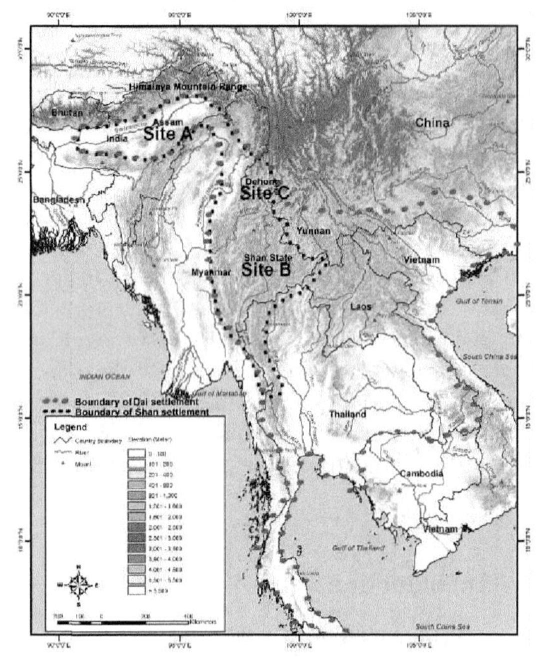

Fig. 2 Location of research sites © The author

The researches were based on architectural surveys and qualitative enquiries. In each site, the researches conducted the surveys in three scopes including a preliminary survey of township settlement, a detailed survey of village settlement and an in-depth survey of houses. Mapping techniques were used to collect data on townships and villages while architectural surveys and measurements were made with house samples. Analysis of data comprises of three levels including:

1) An overview of lowland peasantry settlement,

2) Spatial characteristics of villages,

3) Architectural characteristics of houses.

The detailed findings of three research works have been published in a conference proceeding and two journals including —

Site A in India (Oranratmanee and Saicharoent, 2017), Site B in Shan State (Oranratmanee, 2018) and Site C in Dehong (Oranratmanee, forthcoming 2019—2020). The results in this paper bring the key results from these publications and synthesize them through cross-case comparison.

4 Identical Characteristics of Dai Cultural Landscape and Architecture

4.1 Background of the Case Studies

According to the Dai Chronicle, the Dai who lives in Assam Region of India, Shan State of Myanmar and Dehong Prefecture of China originated from the same homeland in southern China. During the long past, they settled as a small ethnic group this dispersed along the river valleys. During the 12th to 13th century, the Dai principality rose in power and established their feudal state in the continuing area from Dehong to Shan State, known then as Meng Mao, or in full name, Meng Mok Kao Mao Long, literally means the city of white mist and large river named Mao. As Dai State got larger, a small group of Dai began migrating further to India and settled another feudal state there, known then as Ahom Kingdom. In the 15th century, the Meng Mao was defeated by the Chinese Dynasty and due to political dynamics over several centuries Meng Mao has been divided, one became Dehong Profecture of China and the other became Shan State of Myanmar. Similarly, the Ahom Kingdom became a part of Assam Region of India. From then until now Dai have assimilated to local cultures and thereby their characteristics have become the localized patterns.

4.2 Lowland Peasantry Settlements

4.2.1 Pattern of Settlement

Dai in Assam, Shan and Dehong live in a similar lowland, river valley geography. In India, they settle in the upper Brambutra River Valley in today's area of upper Assam and Arunachal Pradesh. Their ancient Ahom Kingdom was in today's Sibsagar City. Dai in Shan State settle in the river valleys in the highland Shan Plateau. On the other hand, Dai in Dehong live in the river valleys in between mountain ranges (Fig. 3).

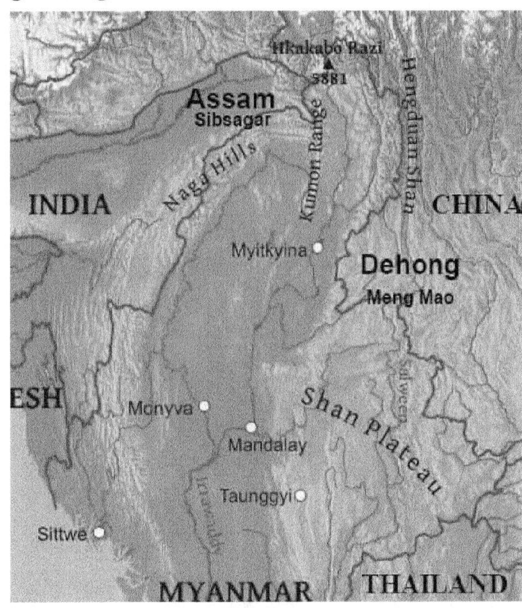

Fig. 3 Settlement area of Dai in Assam, Shan and Dehong

As seen in the Figure, the lowland areas are in between mountain ranges. The Dai settlements in lowland are river valleys vary, depending on the sizes and geographic boundaries of the valleys. According to the surveys, large valleys, including those in Assam and lower Shan State, can accommodate more than one million population who live densely in a city and more disperse in rural village areas. Comparatively, small river valleys in longitudinal shape located between the mountain ranges are found in Dai settlements in upper Shan State and Dehong. The limit of the spatial configuration of lowland valleys allows the smaller sizes of linear settlements, accommodating between 10000 to 100000 populations.

4.2.2 Settlement Pattern

Most Dai settlement patterns found from three sites are usually of linear settlement pattern (see an example of settlement in Mengla or Yingjiang in Dehong in (Fig. 4). To settle in the longitudinal spatial configuration of river valley, the Dai live along the sides of the river valley, name their town as same as the river, and set their rules in relation to the river flow direction. Due to animistic belief, the settlement, or meng, is considered as animated body which consists of three parts including a head, a body or midpoint, and a foot or endpoint. A head is positioned in the upstream while an endpoint is in the downstream direction. We can usually find some spiritual, social and economic nodes, including ancestral a worship place, a main Buddhist temple, a market and a plaza.

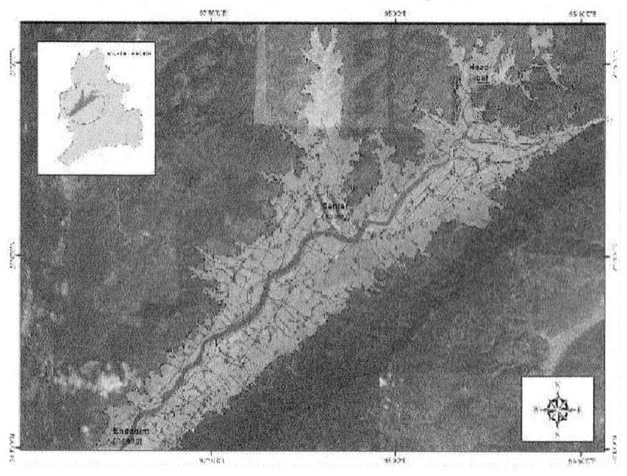

Fig. 4 Example of Dai's linear settlement pattern from Dehong, China © the author

The Dai system of land use in lowland peasantry settlement divided largely into two areas including urban and rural. The urban area is found in the center, usually at the midpoint of the river valley while a rural area is found outside. While urban area functions as the center of township governance, the rural area is the core of dwelling units and economic production of peasant society. Most rural areas comprise three land uses including the rice fields, village hamlets, and natural forests, which are found along the foothill of the mountainous areas where the highland hill tribe peoples live.

4.3 Characteristics of Villages

As is common to peasantry society, Dai live a sedentary settlement with a strong tie to their village. A village, or baan, can mean a central place where daily life activities and social life of villagers take place, and also the whole range of cultural landscape of peasants including the village, rice fields, gardens, rivers and mountains. Like the settlement, the village is considered as a living body with three signified parts, head body and foot/end (Fig. 5).

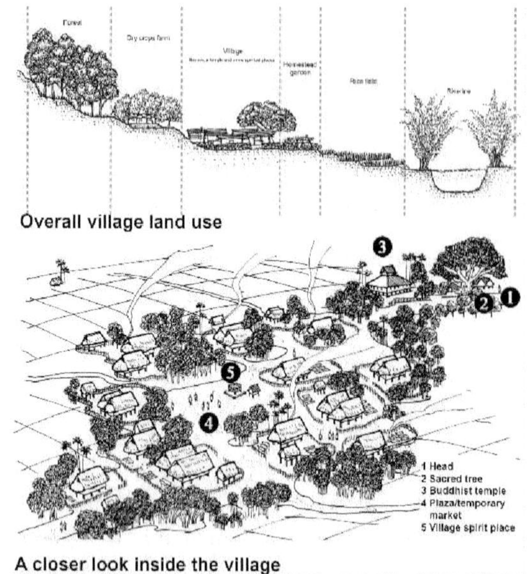

Fig. 5 Aerial view of village © the author

From the surveys in the villages from three sites (including 15 villages in Assam, 25 villages in Shan and 13 villages in Dehong), it was found that the villages of Dai appear in similar number of houses and population with three ranges in sizes: small size (less than 100 houses/500 people), medium size (100～200 houses/500～800 people) and large size (200～400 houses/800～1500 people. The underlying factor of these sizes is the availability of land. To ensure adequate land and water resource sharing, Dai tend to limit the size of houses and their population in accordance to the land available. Another factor is related to state policies for rural land use. In Assam and Shan State where rural lands are not strictly controlled by the States, each household tends to have two to four acres, while in Dehong where rural land is allocated by the state on per head basis, each household tends to have lesser land, approximately one to three acres.

4.4 Architectural Characteristics

Although the characteristics of settlement and village tend to share some commonalities of the peasantry landscape, those of architectures appear to have some localized patterns due to influences from local cultures including Indian in Assam, Burmese in Shan and Chinese in Dehong, especially on the religious and public buildings. Except for the Dai Ahom People in Assam who have respected Hinduism, the rest of Dai people in Assam (those who migrated to Assam after the Ahom), Shan and Dehong are Buddhists. The Buddhist temples which are in every village tend to have similar architectural characteristics in planning and raised-floor structure but different in building forms, structure, construction, materials and decoration, as shown in Fig. 6.

Fig. 6 Localized styles of architecture of Dai in three sites © the author

Houses of Dai are generally built on piles, with a steep roof. The orientation of a house responds to the river flow and spatial configuration of the river valley. The front veranda always faces south while the bedroom faces north. Inside the spaces are divided into three parts — outside, middle and inside. Buddha shrine, if any, and ancestral spirit post is located in the east side.

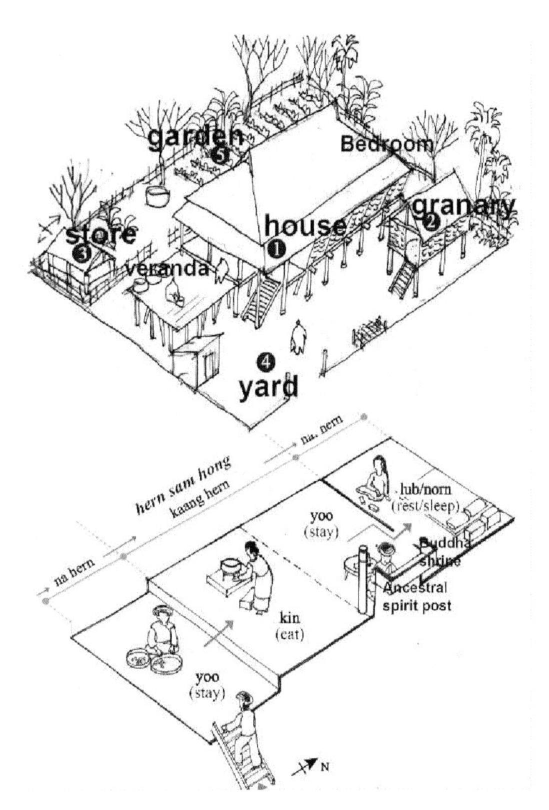

Fig. 7 Traditional Dai house © the author

The rules as exhibited on Fig. 7.

From surveys of Dai houses in three sites (45 houses in Assam, 111 houses in Shan and 40 houses in Dehong), there are some commonalities and differences among them. The common characteristics are the directional references to geography, spiritual beliefs and spatial relation inside the houses. In most houses being surveyed, the rules of orientation, three-room composition and movement in spaces are quite similar. Variations appear in building forms, fabrics and material technology. In Assam, houses can be largely divided into two types including ① the earthen, ground house of Ahom who was the first who migrated to Assam and over time assimilated into Indian cultures and ② the wooden, stilt house of the latter migration group (Fig. 8, first row). In Shan, most houses tend to remain Dai traditional characteristics to the larger extent than those in Assam and Dehong but with a spatial change from a single roof form to double and triple roof forms (Fig. 8, second row). In Dehong, houses appear in two patterns, similar to Assam. Traditional house form, similar to those found in Shan, are found in the southern part of Dehong which borders Shan State while the emerging forms due to Chinese courtyard house influences are found in the northern Dehong (Fig. 8, third row).

Fig. 8 The variation of house forms in three sites © yhe author

5 Conclusion

The findings have exhibited the traditional characteristics and the dynamics of changes in cultural landscape and houses of Dai in three geographies including India, Myanmar and China. Dai not only represent lowland, peasantry dwellers that are found throughout most south, south east and east Asia but also exhibit how indigenous ethnic minority chose to maintain their own tradition meanwhile embrace some local cultures as part of their cultural heritage.

Throughout the paper, the relationship to the geographic and ecological system is clearly evident, being the common norms of Dai ethnic group. The traditional rules and everyday life practices exhibit the strong relationships between human habitat, social systems, and natural environment. These relationship are interpreted into built forms of settlements, villages and houses with embedded social values and meanings. However, the history of Dai exhibits that Dai tend to migrate vastly across the upper mainland Southeast Asia, they undeniably must confront the dominant Indian and Chinese cultures. As a smaller group, Dai assimilated into both cultures both by forces through state politics and voluntary through intermarriages and personal choices.

The production of Dai cultural landscape and architecture in Asia exhibit several key concepts noted in architectural and social studies including dualistic identity, hybrid and eclectic architecture due to cultural assimilation. It also demonstrates the merging of different built forms of ground house and stilt house, shared use of building technologies between earthen and wooden materials. What makes these exchanges possible is perhaps the similar lowland geographies and the long-term cultural exchange between Dai, Indian and Chinese themselves. This paper thereby contributes not only to the study of the Dai ethnic group but also to other studies seeking to understand the dynamics of cross-cultural study about landscape and architecture.

Acknowledgements

This paper is part of researches funded by the National Research Council of Thailand and Thailand Research Fund Contract No. 5980063.

References

DODD W C, 1997. Tai race: elder brother of Chinese. Bangkok: White Lotus.

MILNE L, 1970. Shans at home. New York: Paragon Books.

ORANRATMANEE R, SAICHAROENT S, 2017. The dynamics of Tai vernacular architecture in Northeast India. Proceedings from the 13th international conference on Thai Studies Chiang Mai Thailand, July: 1181-1192.

ORANRATMANEE R, 2018. Vernacular houses of the Shan in myanmar in the Southeast Asian context. 49(1): 99-120.

ORANRATMANEE R, 2019. Cultural geography of vernacular architecture in a cross-cultural context: houses of the Dai ethnic minority in South China. Journal of cultural geography.

PANIN O, 1996. Tai architecture. Bangkok: Department of Fine Arts.

SAI Aungtun,2009. History of Shan State from its origin to 1962. Chiang Mai: Silkworm Books.
WATERSON R, 1990. Living house: anthropology of architecture in Southeast Asia. Oxford: University Press.

Arab Values: Towards Regional Guidelines for ICOMOS Doctrinal Documents in Arab Countries

Hossam Mahdy

Freelance Consultant on the Conservation of Built Heritage, British & Egyptian, Oxford, UK

Abstract Conservation of the built heritage is practiced and regulated in the Arab Region by national laws, institutional structures and formal mechanisms that strive to be aligned with international best practices as guided by ICOMOS doctrinal documents, with different degrees of success.

Informal attitudes are, however, a totally different story. Local communities often fail to appreciate values that formal practices propagate. In some extreme cases, the local community is considered the major threat facing formal efforts to conserve and manage heritage resources.

It is ironic that conservation of cultural heritage could be rejected by mostly traditional conservative societies. The premise of this paper is that there is a need for a deep understanding of values related to the built heritage and its conservation, particularly in traditional societies, such as rural or historic quarters in urban settings.

The paper will examine different levels and aspects of values in the Arab Region, such as geographic and environmental factors, Islamic worldview and value system, Arabic language, local traditional norms, premodern systems of management, conservation and development as well as modernization processes, including the impact of Orientalism, colonialism and the establishment of modern nation states.

The indicators resulting from the proposed in-depth investigations should inform the compilation of regional guidelines for the Arab Region of ICOMOS doctrinal documents such as the CIAV Charter on the Built Vernacular Heritage and the IFLA-ICOMOS Principles Concerning Rural Landscapes as Heritage.

Keywords Arabic, Islamic, ICOMOS, values, regional guidelines

1 Introduction

The inhabitants of the historic quarters of Arab cities are known for being conservative communities taking pride in their traditional Arab values. However, their present attitudes towards Arab-Islamic built heritage are passive, careless or hostile.

Some Islamic historic buildings have been effectively conserved and managed for more than ten centuries and up to modern times, only to be vandalized or misused in the last few decades. Did historic buildings lose their meaning and values in the eyes of local communities? The problem is much deeper than mere technical or financial issues.

The premise of the present paper is that modern conservation and management of the built heritage in the region are formally carried out according to a philosophy that clashes with traditional Arab-Islamic worldview and values.

The paper sheds light on two major issues in order to develop an in-depth understanding of the present challenges facing the effective implementation of internationally accepted best practices according to ICOMOS doctrinal documents. The first issue is the traditional Arab-Islamic views on conservation of the built heritage. The second is the impact of modernization on the way historic buildings are conserved, managed and perceived in Arab countries.

The aim is to identify indicators for the changes that are needed in order to address the causes of the current challenges, which should inform regional guidelines of ICOMOS doctrinal documents for the Arab Region.

2 Traditional Arab Values

2.1 Arabic and Islam

The significance of the word within traditional Arab culture cannot be overemphasized. Arabic poetry was the main cultural achievement and pride of the Bedouins who lived in the Arabian Peninsula before Islam. The significance of the word was even more emphasized later by Islam, with the utmost importance given by Muslims to the Qur'an, the exact words of God. The spread of Islam carried Arabic language far beyond the Arabian Peninsula. The connection between Arabic and Islam is unique. On the one hand Islam guaranteed the spread and survival of Arabic as a living language throughout time and place. On the other hand, Arabic contributed to the preservation of Islam as a religion, worldview and value system by enabling Arabic-speaking Muslims to have direct access to the Qur'an as well as other Islamic texts and rituals. Arab values have been formed and impacted by both Arabic language and the religion of Islam for more than fourteen centuries up to the present, without suppressing diverse local values and traditions across the region(Musallam, 1983).

2.2 Diversity

The Arabs of today are not a unified race. What is known today as the Arab Region includes a great diversity of races, languages, religions, geophysical areas and geopolitical entities. Arabic is not the only spoken language in the region. Nubian, Kurdish, Imazighen, Swahili and other languages are spoken by different ethnic groups throughout the Arab Region. Nevertheless, the impact of Arabic across the region is overwhelming. Not only because it is the language of education and mainstream cultural activities, but more importantly because it is the language of Islam and the Qur'an. The text of the Qur'an has not been changed for more than fourteen centuries. Muslims have read and memorized Qur'anic texts for centuries and up to the present. Consequently, the understanding and use of Arabic were preserved without major changes. Although Islam is not the only religion practiced in the Arab Region, the influence of its worldview, value system and cultural framework is manifest in all communities across the region, including the followers of other religions and those whose mother tongues are not Arabic (Mahdy, 2014).

2.3 Unity and Diversity

Despite being agents for unity, Islam and Arabic endorsed and encouraged diversity across the Islamic world. Islamic instructions respected different local traditions and customs as far as they did not conflict with Islamic values. Therefore, the spread of Islam in different regions did not wipe away the diverse cultures and traditions of different regions and communities. This could be seen, for example, in mosque architecture across the world. All mosques are directed towards Makkah, have mihrabs, minibars, minarets and ablution places. Nevertheless, the styles of mosque architecture in different countries and regions belong to their cultural and environmental localities. Thus, the architectural style of an Ottoman mosque in Istanbul is closer to the style of Anatolian basilica churches than to Egyptian mosques. And the architectural style of a West African mosque is closer to local pagan vernacular architecture than to Indian mosques. The Islamic views on Arabic further encourages cultural diversity. From the Islamic point of view, being an Arab does not indicate a race, an ethnicity or a nation, but simply indicates that a person speaks Arabic. Therefore, many influential Muslim scholars who taught and wrote in Arabic were not ethnically or racially Arabs, such as Al-Bukhari and Al-Tirmithy as well as many others.

2.4 The Impact of Arabic and Islam

Arab traditional views and concepts regarding historic buildings and their conservation were formed by the characteristics of Arabic language and rooted in the Islamic worldview and values system. For example, the Arabic word *haram* expresses a concept of protection that is used for buildings and places. Generally, *haraam* means forbidden by Islamic law. In the case of buildings and places, it means a protected entity that should be treated according to a specific set of rules. Hence Al-Aqsa Mosque in Jerusalem is called Al-Haram Al-Sharif, meaning the noble protected place. The same concept is used in many other ways. It is used regarding places, times, persons and actions. *Haraam* is used pertaining to specific months of the year, during which, fighting wars is forbidden. It is also used pertaining to specific persons, who should be treated according to certain rules. Therefore, places assigned to women were called *hareem*, meaning that these places are protected and should not be treated like common places. Linguistically, the root '*h r m*' means forbidden. Accordingly, many words and concepts generated from this root are used to express the meaning of 'forbidden' or 'protected', for example:

Al-Bayt al-haram: The protected/ forbidden house, referring to al-Ka'bah;

Al-balad al-haraam: The protected/ forbidden city, referring to Makkah;

Al-haram: The protected mosque, referring to al-Ka'ba, the Prophet's Mosque, or Al-Aqsa Mosque;

Al-haramain: The two protected mosques, referring to al-Ka'ba and the Prophet's Mosque;

Ihraam: Wearing the assigned clothes for performing hajj (pilgrimage to Makkah);

Ahram: Entered an area or a period that is controlled by certain rules;

Hareem al-masjid: The buffer zone around a mosque;

Manteqah haram: An area that is managed by certain rules, and cannot be accessed without permission;

Haram Aamen: A protected place;

Hareem al-bi'r: The buffer zone around a well;

Haram al-rajul: A man's wife;

Al-Shahr al-haraam: The forbidden month (one of four months in the Islamic calendar);

Hurmah: What is forbidden.

The above-mentioned examples highlight the characteristics of Arabic and the unique way the language carries Islamic concepts with regard to protection of places. An Arabic word is often charged with religious, historic and cultural meanings beyond the face value meaning. Although conservation is practiced in modern times as a secular activity, Islamic dimensions of Arabic words remain relevant when commonly used outside the circles of heritage professionals.

2.5 Conservation Principles in Arab Context

The adoption and implementation in the Arab Region of conservation principles that are rooted in European languages, history and culture, should not be mere literal translations into Arabic words. For such concepts to make sense in the minds of traditional Arab communities, in-depth investigations are required to reconcile them with Islamic values and to find the right Arabic words that express them correctly and meaningfully without conflicting with deep-rooted Arab-Islamic values and concept.

The Arab Region has a long history of built heritage conservation. The adoption and implementation of international best practices should take into consideration pre-modern approaches and practices. For example, the waqf system was a formidable mechanism for effective conservation and sustainable management of historic buildings throughout the Arab Region for many centuries. Waqf is an Islamic system of an endowment. Accordingly, a founder of a charitable or a religious building integrated the building into a waqf arrangement by allocating the funds to establish a revenue-generating project. The revenue was spent in the maintenance and running costs of the building and in financing all its functions. A waqf arrangement was documented by a waqf deed (waqfeyyah) and legally enforced by Islamic courts. No changes were permitted to be made to a waqf arrangement after its establishment unless approved by the judge and only if the change would enhance the state of conservation of the relevant building and/or improve the efficiently of its functions. According to Islamic law, no one had the power to cancel a waqf or to make changes to its arrangements. Arab-Islamic historic buildings are standing today thanks to the waqf system.

3 Modern Eurocentric Values

3.1 The Modernization Process

The modernization process in the Arab Region was difficult and diverse. Some Arab countries like Algeria were modernized by 19th century European colonial powers. Other countries, like the United Arab Emirates were modernized by national elites as late as the 1970s. Whereas Egypt and the Levant were modernized by a mix of both European colonizers and national elites throughout the 19th and 20th centuries. Modernization was also a process of Europeanization or Westernization, which was imposed on the majority by the controlling powerful minority. Hence, modernization was not willingly accepted by the majority. Not only because Western culture and values are alien to the Arab mind, but more importantly because of the image and place of the Arab and the Muslim within the European mindset of the 19th and early 20th centuries, which identified the Arab and the Muslim as "the other", "what the European is not", or "a man of inferior culture and abilities".

3.2 The Impact of Orientalism

A huge body of Western literature, visual arts and historical essays developed on the subject of the "Orient" and formed what is commonly known as Orientalism. Orientalism produced a portrait of the "Oriental" people, their culture, manners, arts and architecture, which was often not factual, objective, or sympathetic. The "Oriental" Muslim Arab was often portrayed as a mysterious, aggressive, sensual and superstitious, among other imagined and exaggerated characteristics, which were often inspired by the fantastical stories of *The Arabian Nights*.

The modernization of the Arab Region produced national elites who acquired modern education according to European worldview, structure and content. They saw their own peoples, lands, cultures, traditions and values through the Orientalist lens. This elite strove to modernize their communities and peoples according to the European model. They often saw their communities' premodern traditions and values with embarrassment, dismissing them as "backward", "not modern", or "not scientific". The independence of the Arab countries did not bring a better understanding or endorsement of Arab-Islamic values and identity. The European-educated Arab elite replaced European colonizers without essential changes to the educational and legal systems or the state's institutional structure. Arab independent nation states strove to follow the European model of modern nation states, including their formal cultures, educational systems, institutional structures, and legal systems. Globalization and international mass tourism during the last decades of the 20th century further complicated the situation by creating a big market for shallow views on Arab cultural heritage and its values. The economic benefits from mass tourism encouraged Arab communities and authorities to present to foreign tourists what they expected and were ready to pay for: the "Oriental" image of their cultures.

3.3 Orientalist Approach to Conservation

The process of modernization in the Arab Region instated and formalized Orientalists' clichés, while producing bad copies of Western modern practices. Meanwhile, there are many indications that traditional Arab values remained prevalent for the majority of populations throughout the Arab Region, even if informal. The result is two-tier societies with Westernized formal attitudes adopted by elites, and traditional Arab-Islamic attitudes adopted by most Arabs. This duality is strongly manifested in the field of conservation for cultural heritage. For example, historic sabils (drinking water fountains) in historic Cairo have been conserved by restoring the buildings' fabric and discontinuing the function of offering water to the passersby. This attitude was initiated by foreign conservators in the early 20th century and continued after independence by Egyptian conservators up to the present. Historic sabils were parts of waqf arrangements. According to Islamic Law, it is forbidden to discontinue a function that was established by waqf. As far as local communities in historic Cairo are concerned, conservation officials have transgressed on others' rights (i. e. the rights of the founders of restored sabils who established waqf to secure their sustainability).

3.4 Formal-Informal Duality

For the locals, formally conserved sabils are meaningless structures. While, the function of offering water to passersby is still needed. Offering water to the thirsty is considered by Islam a great act of charity. Therefore, local communities produced, and continue to produce, informal sabils in the form of pots of water or water dispensers in many streets and public spaces in Cairo.

Another example of the rift between formal and informal attitudes is the demolition of centuries-old vernacular settlements in the vicinity of ancient archaeological sites and grand monuments with the justification of clearing the context for the benefit of international tourists, with no consideration for the interests or the wellbeing of affected local communities. The World Heritage Site of Thebes and its Necropolis in Egypt witnessed a number of such practices: The old vernacular village of Gourna on the West Bank was partially demolished in 2008 to "protect" the ancient Egyptian tombs throughout the site. Furthermore, in 2019 the vernacular settlement close to Karnak Temple on the East Bank was demolished to clear the path for reconstructing the avenue of sphinxes between Karnak Temple and Luxor Temple. Furthermore, the village of Nazlet Al-Semman was bulldozed in 2018 for being in the vicinity of the Giza Pyramids within the World Heritage Site of Memphis and its necropolis. These practices clash with the Islamic attitude towards peoples' livelihood as opposed to the presentation of historic buildings and sites. As far as local communities in Luxor and Giza are concerned, the authorities have been committing grave injustices in the name of conservation of cultural heritage.

3.5 Deceiving Appearances

The modernization process in the Arab Region created a misleading appearance of modern conservation practices that do not seem to be much different than those currently practiced in the West, while the philosophy of conservation behind formal Arab practices remains 19th century European in essence, which is not appreciated by local communities, whose majority remain traditional at heart. The adoption and implementation of international best practices should strive to guide Arab conservation professionals and decision makers to rectify this situation.

4 Present Challenges

4.1 Causes

The adoption of international best practices in the field of heritage conservation, as outlined by ICOMOS doctrinal documents, in the Arab Region faces many challenges. Most of which, are caused by the modernization process following European examples while ignoring or dismissing centuries of Arab-Islamic traditional philosophy, approaches and practices. It is ironic that European philosophy and practices of conservation have been progressing in the West for decades and continue to progress in line with changes in cultural contexts, while 19th century European philosophies and practices continue to be followed religiously in the Arab Region since colonial times. A simple explanation is that European, or Eurocentric, conservation philosophies and practices have always been unrelated to Arab cultural contexts. Therefore, the changes in cultural contexts did not prompt major changes to conservation philosophy or practices. Arab professionals and decision makers have been, and continue to be, consumers rather than producers of conservation philosophy, theory, practices and techniques. The laws, the institutional structures and conservation education and training form a vicious circle that makes the status quo difficult to change. The situation is further complicated by the lack of democracy, aggressive development pressures and the threat of armed conflicts.

4.2 Laws

In most Arab countries, laws on protection and conservation of cultural heritage often go back to colonial times. Today, laws protect "antiquities" or archaeology and single monumental buildings, leaving out vernacular heritage, industrial heritage, historic urban fabric, and cultural landscapes, to name a few categories of the built heritage that have been included within the category of built heritage by ICOMOS over the last few decades. Even when laws are updated or rewritten, they remain out of touch with both international developments in conservation philosophy and traditional Arab-Islamic values.

4.3 Institutional Structures

Institutional structures are another colonial legacy that lives up to the present day in many Arab countries. The roles and mandates of different institutions often defy the effective implementation of recently developed concepts such as multidisciplinary approach, participatory conservation, stakeholders' engagement, authenticity as redefined by Nara and Nara +20, to name a few. Another challenge caused by institutional structures is the dismissal or undermining of traditional Arab-Islamic practices. For example, historic mosques are managed by at least two institutions: Religious aspects are managed by the Ministry of Awqaf and Islamic Affairs, while their maintenance and conservation are managed by the Ministry of Culture or the Ministry of Antiquities. Furthermore, the Ministry of Tourism and the Ministry of Interior may be also involved to manage tourists' visits to a historic mosque. These different ministries often do not fully coordinate their plans or activities.

4.4 Education and Training

Education and training in the field of heritage and conservation in the Arab Region prepare specialists with convictions, ideas and skills that are fit for old school approaches, which focus on "antiquities", single monumental buildings, and archaeological sites. Most conservation specialists in the Arab Region are not trained to deal with other categories of the built heritage. They are not trained to engage local communities or stakeholders. They often dismiss traditional Arab-Islamic values and attitudes as backward or unscientific. As a result, local communities and the general public are often absent from the decision-making process. Hence their lack of interest and low awareness of their heritage and its conservation, which remains an elitist interest. Foreign mass tourism does not help as it further alienates local communities from their own heritage, reducing it to a mere source of financial gain.

5 Towards Regional Guidelines

5.1 Philosophy and Theory

There is a need to revise approaches and attitudes that were established during the modernization process. Colonial legacies should be deconstructed and examined in-depth for suitability of Arab contexts. Definitions, principles, concepts and practices of conservation for cultural heritage should be reconciled with traditional Arab-Islamic worldview, values and concepts whenever appropriate.

5.2 Arabic Language and Terminology

Glossaries of Arabic terminology should be developed, discussed and used with the careful study of historic, cultural and religious aspects of Arabic concepts and meanings. Local communities and stakeholders will be engaged and empowered only if Arabic is used in all aspects and phases of the conservation and management of their heritage.

5.3 Legislation

Laws protecting the built heritage should be revised with a view to endorsing up-to-date concepts, principles, and processes of conservation for cultural heritage. Furthermore, mechanisms for continuously revising and adapting laws and regulations should be put in place in order to accommodate for the continuous development of conservation thought.

5.4 Institutions

The existing formal institutions should be critically analyzed with the view of adapting their aims, mandates and structures to effectively protect, manage and present the built heritage according to the above-mentioned revised philosophy, theory and legislation.

5.5 Education

The right balance should be sought between international development of conservation theory and practice on the one hand, and on the other hand local and national cultural contexts, social needs and traditional Arab-Islamic views.

5.6 Waqf

The waqf system merits to be studied, revised, updated and considered for reconciliation with internationally accepted methods and tools for the conservation and management of the built heritage.

5.7 Geographic, Environmental and Cultural Diversity

Regional guidelines for the Arab Region should be considered high-level guidelines that do not override the diverse geographic, environmental and cultural diversity within the region.

5.8 Armed Conflicts

The built heritage in the Arab Region is hugely impacted by armed conflicts due to wars or civil unrest. Preparedness, mitigation and risk management plans should be prepared and updated.

6 Conclusion

Effective implementation of ICOMOS doctrinal documents in the Arab Region requires an in-depth understanding of traditional Arab-Islamic values and views on conservation for the built heritage. In addition, the impact of modernization following European examples should be carefully analyzed.

The rift between formal and informal attitudes as well as the absence of clear and convincing philosophy for conservation practices in the Arab Region should be addressed and guided by Regional guidelines for ICOMOS doctrinal documents on national and regional levels.

Effective and sustainable changes in the right direction require major changes to the philosophy and theory of conservation, the legislations, Institutional structures, education and training curricula. Careful attention should be paid to Arabic language and terminology, the waqf system, the geographical, environmental and cultural diversity throughout the Arab Region and careful planning and response to risks caused by armed conflicts.

References

AMIN M, 1980. The Waqfs and social life in Egypt 648-923AH/ 1250-1517 AD. Cairo: Dar Al Nahda Al Arabeyyah.

ASSAD M, 2000. The road to Mecca. Louisville: The Book Company.

GHAZALEH P, 2011. Held in trust. Waqf in the Islamic World. Cairo: AUC Press.

HAKIM B S, 1986. Arabic-Islamic cities. building and planning principles. London: KPI.

Ibn Manzur, 1883. Lissan Al Arab. Cairo: Bulaq.

MAHDY H, 1998. Travellers, colonisers and conservators//STARKEY P, STARKEY J. Travellers in Egypt, London: I. B. Tauris Publishers.

MAHDY H, 2008a. Glossary of Arabic Terms for the conservation of cultural heritage. Rome: ICCROM. https://www.iccrom.org/sites/default/files/2017-12/iccrom_glossary_en-ar.pdf.

MAHDY H, 2008b. Raising awareness of the value of earthen architecture for living and working in the Nile Valley, Egypt. The 10th international conference on the study and conservation of earthen architectural heritage, Mali, TERRA.

MAHDY H, 2014. Proposed Arabic-Islamic contributions to the theory of cosnrvation for cultural heritage. Heritage and landscape as human values

conference proceedings, ICOMOS, Florence.
MAHDY H, 2017. Approaches to the conservation of Islamic Cities. Sharjah: ICCROM-ATHAR. https://www. iccrom. org/sites/default/files/2017-12/web-email_hussam_book_17112017. pdf.
MICHELL G, 1978. Architecture of the Islamic World. Its history and social meaning. London: Thames and Hudson.
MUSALLAM B, 1983. The Arabs: a living history. London: Collins / Havrill.

Local Conceptions for Conservation Practices. Earthen Architectural Heritage in the Andean area from Participatory Experiences

Jorge Tomasi, Julieta Barada

Laboratorio de Arquitecturas Andinas y Construcción con Tierra, Universidad Nacional de Jujuy, Argentina

Abstract Conservation practices are associated with certain conceptions of temporality, that is, a way of defining and perceiving time. The tension that exists between hegemonic notions of temporality and those of many local communities is a relevant issue to problematize the approaches to heritage and its conservation. Contemporary reflections lead to the establishment of new methodologies to achieve effective participation of communities and their ways of thinking about architecture over time. The ignorance of local knowledge and practices has often led to interventions that have altered the values of these heritages and have distanced communities from their architectures.

This paper will present three experiences of participatory conservation that have been developed with different communities in the highlands of the province of Jujuy (Argentina), in particular Puna and Quebrada de Humahuaca areas, the latter inscribed in the World Heritage List. Architectures that have been treated in these projects have a high significance, historical and religious, for these communities, which have sustained the continuous use of buildings, and also have different types of heritage declarations. In all cases, these are earthen architectures, with different techniques such as adobe for the walls and mud roofs. Knowledge associated with these techniques is currently part of the daily practices of the population.

The objective of the presentation is to contribute to the theoretical and methodological discussions from the analysis of these case studies, considering forms of community participation in the framework of conservation. In turn, technical problems existing in the three buildings and the intervention strategies generated from the proposed theoretical approach will be described.

Keywords indigenous communities, participatory methodologies, andes, ethnography, adobe

1 Introduction

In recent years, heritage conservation works from an approach that involves greater participation of local populations has been gaining consensus in disciplinary debates (Correia, 2007; Johnston & Mayers, 2009; Joffroy, 2005; Pendlebury, 2009). This is associated with a comprehension of heritage that seeks to transcend a focus on the object, to consider the contexts of production and use. Beyond this emphasis, in practice there is not so much consensus about the implications and scope should have a participatory approach. In some projects, this refers to prior consultation with local populations, such that they have sufficient information about the work to be carried out, and, eventually, provide their consent. In other cases, participation implies recognition of the existence of a certain set of local knowledge that is necessary for an adequate intervention. Consequently, various working instances are generated to recover a certain amount of data, generally through interviews, with the most experienced people in a certain community. Finally, in some other cases, participation refers to a deeper immersion in the field, which necessarily implies a revision of the criteria that will be used in conservation works, understanding that these must be permeated by the diverse senses given to their heritage by of the local communities. These approaches, necessarily, imply a revision of the conceptual presuppositions associated with conservation, historically consolidated from global hegemonic institutions, to allow the entrance of other ontologies, as different forms of being in the world (Jones, 2010; Muñoz Vi as, 2011; Yarrow, 2018). If this is valid for architectural heritage in general, it is particularly true for vernacular productions, considering the particular relationship that these architectures tend to have with the societies that produced them and, in many cases, continue producing.

In the last decade, a series of research projects have been developed on the architectural production in northern Argentina, particularly in what is known as Puna (Altiplano) and Quebrada de Humahuaca. These projects, developed from an ethnographic approach based on systematic fieldwork, have been aimed at recognizing the way in which these architectures model and are modeled within the framework of social relations and certain ways of understanding the world (Barada, 2015; Tomasi, 2013). Recently, these studies have begun to be inserted in concrete actions for heritage conservation in different communities of the region, specifically in the cases of the colonial residence called "Casa del Marques", in Yavi, the Church of the Santa Cruz and San Francisco de Paula, in Uquía, and the Church of Tabladitas. It is about three buildings with a high significance for their communities, both in historical and current terms. I fact, the first two also have declarations as a National Historic Monument. These are architectures built between the seventeenth and nineteenth centuries, with different subsequent interventions, and were built in adobe, with stone and mud foundations, and roofed with mud over wooden structures. In the three cases, in recent decades, different projects and interventions have been developed for their conservation, which involved significant transformations that, in turn, have caused various pathologies that must be remedied. These new proposals for its restoration have been prepared under the support of the Secretary of Cultura of the province of Jujuy, Argentina.

This paper will present the characteristics of these projects, particularly considering the theoretical and methodological approaches used for the recognition of local conceptions and ways of doing. After a presentation of the places and case studies, the main conservation issues that have arisen in these projects will be presented and how, from a participative approach, it has been possible to generate proposals with the existing techniques and materials, will be analyzed.

2 Places and Cases in the Andean Area

What is known as "Andean area" refers to a heterogeneous region in South America that currently corresponds to part of the territories of Peru, Bolivia, Ecuador, and northern Chile and Argentina. Historically, from pre-hispanic moments-before the Spanish colonization in the sixteenth century-, this area presented multiple and continued occupation of different social groups, which were sustained with very important changes during the colonial period, and even in republican times from of the nineteenth century. Although the regional variations are very important, it is possible to observe certain shared features, which include, among other aspects, architectures.

Throughout the whole area, earthen construction has a fundamental role in architectural traditions, which includes a variety of techniques such as adobe, rammed earth and wattle and daub for walls, and a variety of procedures and techniques for roofs (Viñuales, 1991). Many of these practices have been recorded for pre-hispanic moments in different archaeological sites, being that these traditions intersected in different ways with the characteristic techniques of the Iberian Peninsula. An important aspect in this regard is that most of these techniques continue to have a very prominent place today, with many local communities, particularly in the rural world, which sustain these traditions, within the framework of constant transformations (Tomasi & Rivet, 2011).

In northern Argentina, the working area of these projects, earthen building techniques have a significant persistence in their use, beyond the constant stigmatization processes that have been developed by different state agencies. In fact, there is still an important institutional rejection to earthen construction, looking for its replacement by other technologies, or the partial incorporation of certain materials, generally incompatibles. As it will be seen, this not only refers to new constructions but has also involved significant transformations for heritage, even in those cases that do have formal declarations as National Historical Monuments. These stigmatizations associated with the constructive are framed in discourses and civilizational practices, which have denied the ontologies of many communities, including their architectures.

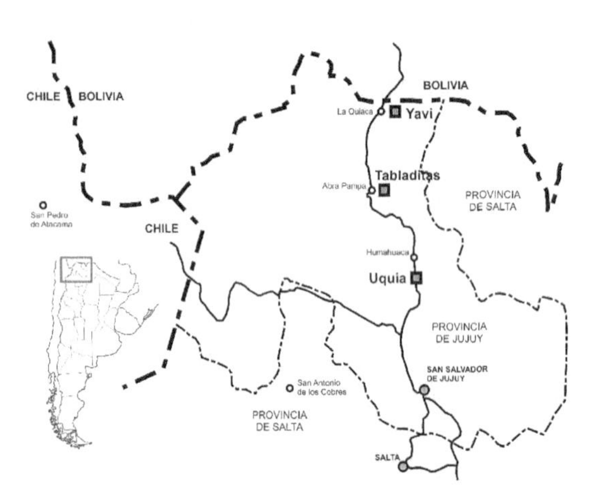

Fig. 1 Location of case studies in the province of Jujuy © authors.

Case studies that have been worked on within the framework of this project are located two (Tabladitas and Yavi) in what is known as Puna, and the remaining (Uqu a) in the Quebrada de Humahuaca. The Puna corresponds with a height plateau, with an average of 4000 mosl with a population that historically has had in pastoralism their main productive activity. Quebrada de Humahuaca is a north-south corridor that links the lowlands with the Puna, where agriculture has played a more important role. In 2003, it was incorporated into the World Heritage List, which has involved very important transformations. Regarding specific cases, the Church of Uqu a was built in the 17th century and is a Catholic temple of great importance in the area, being a National Historic Monument. Both its single nave and the exempt tower were built with adobe and have a remarkable wooden structure with reinforcements on the roof. The Church of Tabladitas is also a Catholic temple, very relevant for its community, which was built in the 19th century, using adobe and covered with straw and mud (waylla). Casa del Marques, in Yavi, it is a large-scale domestic architecture that was the residence of an owner and colonial authority in the 18th century, but then had had different uses to the present as a museum.

Fig. 2 Yard of Casa del Marques, in Yavi © authors

Fig. 3 Exterior of the Church of Uqu a with its exempt tower © authors.

Fig. 4 Exterior of the Church of Tabladitas © authors.

3 Theoretical and Methodological Approaches

As said in the introduction, in recent years, participatory approaches to the conservation of architectural heritage have been established as a methodology to bring different social actors closer to interventions on historic buildings (Johnston & Myers, 2009). These approaches encompass a diversity of practices ranging from the registration of local knowledge to the development of collective instances of work for the communication of projects or even the joint design of proposals (Cox Aranibar, 1996). Although these actions have implied a contribution to conservation actions, in general they tend to develop from the conceptual and methodological assumptions established as universal values in international conventions and chapters, and therefore, arise from a certain cultural framework and certain conception of time and space.

Indeed, conservation is linked to ideas of duration and continuity, which implies a certain temporality, as a conception of time. In this way, the understanding of what conservation implies is subject to the particularities of the conceptions of the different social groups. In other words, architectural conservation practices are collectively defined within the framework of certain relationships between subjects and objects, historically situated. The notion of conservation, and its practices, constitutes itself a category to be problematized and resignified, in order to put into play a multivocality that contemplates the existence of a variety of ontologies. The constructions on the concept have historically been traversed by hegemonic logics that have tended to make other narratives and possible forms of action invisible (Alonso Gonz lez, 2015). The established criteria for conservation not only offer general frameworks but also conditions specific interventions on buildings, often modifying their senses (Jones & Yarrow, 2013).

Ethnographic studies, particularly those associated with the study of architecture, have brought to light local practices that are linked to the duration of buildings within the framework of certain temporalities, which in themselves put into question the meanings of authenticity, integrality and minimum intervention as guiding criteria. Indeed, many architectures that can be defined as patrimonial are subject to constant processes of change, with the addition or withdrawal of parts, being this transformation an inherent condition of its existence (Jones & Yarrow, 2013; Yarrow, 2018).

On the other hand, the contributions generated from the anthropology of technology (Lemonnier, 1992) allow us to recognize how constructive techniques cannot be treated as an objective field, but must be observed as part of the social networks. As proposed by Dietler & Herbich (1998: 235), "techniques are embedded in and conditioned by social relations and cultural practice". In this sense and also in the actions of conservation, it is necessary to recognize how historical techniques are inserted in a more complex web, and particularly, which are the social implications of the potential changes that are incorporated.

The methodology adopted in the ongoing projects is based, in the first place, on the experiences generated in more than a decade of ethnographic fieldwork in the region, which allowed to recognize the density and complexity of the local construction systems, beyond the strictly technological senses, to consider all their social implications and signification (Barada, 2015; Tomasi, 2013; Tomasi & Rivet, 2011). From this, the different stages of each project: survey, diagnosis and generation of proposals, were carried out from different workshops with representatives of the communities, in which the different local views and interests could be expressed for the development of a common proposal. The objective of these workshops is to generate a symmetrical space of knowledge construction, where the academic one can dialogue from equality with local experiences. These are not merely informative instances regarding the decisions to be made in cabinets, but rather the same places where the intervention strategies are defined.

Fig. 5 One of the activities during the workshops in Yavi © authors

4 Conservation Challenges, Between the Social and the Technical

The diagnosis constructed jointly with the communities was organized around the recognition of structural damages, pathologies of the materials and the existence of diverse interventions that could generate both alterations in the characteristics of the property and different pathologies that could have multiplied the damages (Rainer, 2008). Although in analytical terms these problems can be considered independently, it is necessary to consider that they are phenomena that are intimately linked.

In structural terms, studies were undertaken, including analysis with finite element method, have allowed us to observe a certain recurrence of damages that are, in some cases, intrinsic to the constructive system. First, problems associated with the link between the roofs and the structure of the walls must be considered. In the case of the Church of Tabladitas, for example, deformation in trusses caused horizontal stresses to be transferred to the walls, generating a horizontal and longitudinal crack, causing an important vertical loss that puts the stability of the wall at risk. A similar situation could be recorded in the Casa del Marques, in Yavi, where the result was a series of vertical cracks in the transverse wall encounters, in addition to the cracking struts in the supports of the roof beams. The Church of Uquía, on the other hand, provides tools to think about the potential solutions, as it presents a structure of the roof that rests on a wooden upper beam that helps the linking of the trusses with the walls, spreading the efforts.

Fig. 6 Structural damage in one of the corners in Casa del Marques © authors.

Main pathologies associated with materials are related to the affectation by humidity, which causes not only the degradation and loss of cohesion in the earth materials, particularly in the adobes, but also phenomena of putrefaction in organic elements, such as wood from the structure of the roofs. In this regard, both pathologies have been diagnosed due to rising humidity and leaks due to roof failures. In the first case, a recurrent problem is represented by the lack of protection elements in the lower sections of the walls, particularly due to the lack of elevation of the stone foundations that should prevent the contact of the adobes with the soil moisture. This is a situation that has been registered in both the Casa del Marques and the Church of Tabladitas. Regarding the leaks in the roof, in the cases analyzed, they are linked to the lack of the necessary maintenance in the earthen roofs or with recent alterations that have significantly changed the materiality of the roofs, as will be seen below.

Fig. 7 Changes in the materiality of the roof in Uquía © authors

Both buildings that have national protection declarations as Historical Monuments, Church of Uqu a and Casa del Marques, have been subject to different conservation actions during the second half of the 20th century generated by official organisms. As a result of prejudices about earthen techniques and the lack of joint work with local communities, these interventions were characterized by the introduction of substantial modifications in the characteristics of the techniques that affected their behavior. One of the main transformations has been the incorporation of reinforced concrete in different sectors in an isolated form. This has included braces in corners or beams in the upper sections of the walls that form a solution to the supposed weakness of the material. The incompatibility of the materials, far from solving pathologies, has caused new structural problems. The second field with great affectations is that of the roofs. In both casas as a consequence of doubts from the specialists on the mud covers (torteado), different systems were designed including foreign materials, such as cement mortars, liquid membrane, among others. This has caused overloads on the walls and has favored serious leaks in both cases.

5 Intervention Proposals

The joint intervention proposals were developed based on the following objectives:

a. Favor active participation of different local actors in the survey, diagnosis, generation of executive project and intervention, within the framework of their own conceptions.

b. Value local practices and knowledge, considering the diversity of existing procedures.

c. Achieving structural consolidation through the use of solutions that are coherent with materials and techniques present in the buildings.

d. Restitute the constructive characteristics, based on the existing documentation, survey and history of the community.

e. Generate sustainable solutions capable of being conserved by local communities in the long term, with minimal intervention, based on actions that are reversible in the future.

Building technologies used for the elevation of these three buildings do not differ substantially from those that have historically been used in the domestic architecture of the region, and in many cases are still maintained. This has a series of implications that have guided the intervention proposals. In the first place, knowledge, practices and experiences of the members of the communities are inescapable for the establishment of the intervention criteria as they are coherent with the constructive techniques historically used. This also implies framing actions within the conceptions of these communities. Secondly, it is necessary to take into account that actions carried out for the consolidation and restoration may have implications beyond the protection of these specific buildings, favoring good practices for the maintenance of other heritages in the region and for the improvement of the structural and habitability conditions of current houses. Finally, the implementation of solutions based on local knowledge allows the establishment of preventive maintenance dynamics, feasible to be sustained over time with locally available resources. To this end, workshops were established as spaces for mutual learning and review of best solutions for each of the detected problems.

The general criterion in these proposals has been to assume as a starting point the technological knowledge of local communities, understanding that it is part of the broader body of knowledge that is linked to dimensions that exceed, although they include, architecture. In more concrete terms, in all cases it has been sought to resolve damages and pathologies following the proposals and recommendations of local builders. Particularly in the case of the roofs, where the main alterations have been generated, it has been sought to recompose the original technology associated with the recognition of the different elements that a good *torteado* must include, particularly as regards the preparation of the mud. This implies assuming a temporality of the materials, which leads to a periodic replacement to replenish their properties. In this sense, authenticity as a criterion does not go through the conservation of existing material, which certainly has been replaced many times throughout the history of the building, but rather by the maintenance of certain constructive practices, within the dynamics of the social. The central argument does not go through the object itself, but through the set of practices associated with its production.

This positioning does not go against the possibility of introducing reinforcement systems in order to consolidate buildings, but these must arise from a detailed study of each case and other similar testimonies. Incorporation of upper wooden beams to link roofs with walls, improving their behavior, as it will be done in Casa del Marques and Church of Tabladitas, arises from the academic experiences in this respect, but also from the observation of its use, for example in the Church of Uqu a. The same can be referred to the use of wooden keys for improvement of encounters in sectors where damage was caused by deficiencies in masonry. These solutions are also common in many houses in the region, and as such are applied by the builders.

6 Conclusions

Criteria established in international documents for architectural conservation provide a point of departure arising from consensus among specialists and have allowed to developing shared forms of intervention on heritage, encouraging good practices. Theoretical discussion on this issue generated by many professionals has called attention to the need to consider in discussions that conceptions about conserving are not necessarily universal, and this has led to the establishment of methodologies that seek to incorporate in different ways local actors in intervention processes.

Through this paper we have sought to briefly present the characteristics of three conservation projects that are ongoing, in different community in the north of Argentina. The buildings studied have a high incidence of earthen building techniques, and in turn have a high significance for each communities. Methodologies associated with proposed theoretical frameworks have been established, which seek to favor the development of spaces for discussion and exchange from a symmetrical perspective, with ethnography as an approach for the recognition of local points of view. Rather than propose a solution, the objective of this text has been to describe a possible strategy that, by definition, will need modifications based on the same practices.

References

ALONSO G P, 2014. Patrimonio y ontologías m ltiples: hacia la coproducción del patrimonio cultural. In: Patrimonio y Multivocalidad. Teoría. práctica y experiencias en torno a la construcción del conocimiento en Patrimonio. Montevideo: CSIC.

ARNOLD D, 2009. Using ethnography to unravel different kinds of knowledge in the Andes. Journal of Latin American cultural studies: Travesia, 6-1.

BARADA J, 2015. Entre casas, departamentos y viviendas. Un análisis etnográfico sobre la producción del espacio doméstico en la Puna de Jujuy. Master Thesis. IDES / IDAES / Universidad Nacional de San Martín.

CORREIA M, 2007. Teoría de la conservación y su aplicación al patrimonio en tierra. Apuntes, Vol. 20 N 2.

COX A R, 1996. El saber local. Metodologías y técnicas participativas. La Paz: Nogug-Cosude.

DIETLER M, HERBICH I, 1998. Habitus, techniques, style: An integrated approach to the social understanding of material culture and boundaries. In: The archaeology of social boundaries. Washington DC: Smithsonian Institution Press.

JOFFROY T, 2005. Les pratiques de conservation traditionelles en Afrique. Roma: ICCROM.

JOHNSTON C, MAYERS D, 2009. Resolving conflict and building consensus in heritage place management: Issues and challenges. In: Consensus building, negotiation and conflict resolution for heritage place management. Los Angeles: Getty Conservation Institute.

JONES S, 2010. Negotiating authentic objects and authentic selves. Journal of material culture 15: 181 203.

JONES S, YARROW T, 2013. Crafting authenticity: an ethnography of conservation practice. Journal of material culture, 18(1), 3-26.

LEMONNIER P, 1992. Elements for an anthropology of technology. Anthropological papers, museum of anthropology, university of Michigan, No. 88. Michigan: Ann Arbor.

MU OZ V S, 2011. Minimal intervention revisited //Conservation: principles, dilemmas and uncomfortable truths. London: Routledge: 47-59.

PENDLEBURY J, 2009. Conservation in the age of consensus. Abingdon: Routledge.

RAINER L, 2008. Deterioration and pathology of earthen architecture//Terra Literature review. An overview of research in earthen architecture conservation. The Getty Conservation Institute.

TOMASI J, 2011. Geografías del pastoreo. Territorios, movilidades y espacio doméstico en Susques (provincia de Jujuy). PhD Thesis. Universidad de Buenos Aires.

TOMASI J, RIVET C, 2011. Puna y arquitectura. Las formas locales de la construcción. Buenos Aires: CEDODAL.

VI UALES G, 1991. La arquitectura en tierra en la región andina. Anales del Instituto de Arte Americano, 27-28.

YARROW T, 2018. How conservation matters: Ethnographic explorations of historic building renovation. Journal of material culture.

The Involvement of Local Communities into Conservation Process of Earthen Architecture in the Sahel-Sahara Region— The case of Djenne, Mali

Oussouby Sacko

Graduate School of Architecture, Kyoto Seika University, Kyoto, Japan

Abstract Recently, many cultural aspects are in danger of being lost due to the cultural disruption, as well as due to non-adaptive construction technics and lack of adequate conservation systems and strategies, in some African countries. This includes the well-known manuscripts of Timbuktu (Tombouctou) but also architecture, languages, beads, textiles, costumes and other cultural objects as well as immaterial heritages such as oral history, traditional music, instruments and dance.

In the north-western part of Africa, the so-called Sahel-Sahara region, earthen architecture has played an important role as cultural identity, as well as a key item for community establishment. This architecture heritage includes a wide variety of creations which goes from simple houses, granaries, and palaces to religious buildings, urban centers, cultural landscapes and archaeological sites. By studying North-western African cultural heritage and cultural exchanges during the Trans-Saharan Trade, we are able to benefit from the traditional knowledge and use the teachings to build a sustainable knowledge system on earthen architecture conservation. This work is urgent since these heritages are in danger of being lost, destroyed and bad conservation, and not documented.

The aim of this research is to point out, through anthropological approach, the importance of local communities involvement into the conservation process. This research is based on a comparative study about earthen architecture conservation situation, between different spaces within the above-mentioned region that have been important and was influenced by cultural exchanges during the Trans-Saharan Trade. In those historical towns, some of the architectural heritages are abandoned and in ruins, while some are being preserved solely as tourist attractions. In this presentation, we will introduce the case of Djenné to share the approach of our research. This discussion paper can be an opportunity to learn about earthen architecture conservation issues, and how local communities have preserved cultural heritages with tangible and intangible methods.

Keywords earthen architecture, tangible, intangible, Djenné, conservation, Mali

1 Purpose

In this discussion paper the complex relation ships between cultural conservation and cultural heritage preservation will be discussed based on observation of restoration projects in Djenné. In most of the preservation process, some of the traditional techniques and know-how are revalorized, rehabilitated while some of them are denied and lost in the same process. On the other hand, preservation experts usually lead the projects in places where local builders are much more experienced. In this paper I will try to point out the social issues restoration projects raised. The relation and the meaning of preservation will be analyzed as well through many aspects. The field surveys were conducted between 2010 and 2017. Some of the information will be based on the surveyed conducted in Djenné, about its town and architecture between 2004 and 2010.

2 Overview of Mali

2.1 Mali: History and Geography

Mali is a landlocked country situated in the heart of West Africa. Mali was a French colony for about hundred years and use to be called French Sudan. It became independent in September 22nd, 1960. Mali is bordered on the north by Algeria, on the east by Niger, and Burkina Faso, on the

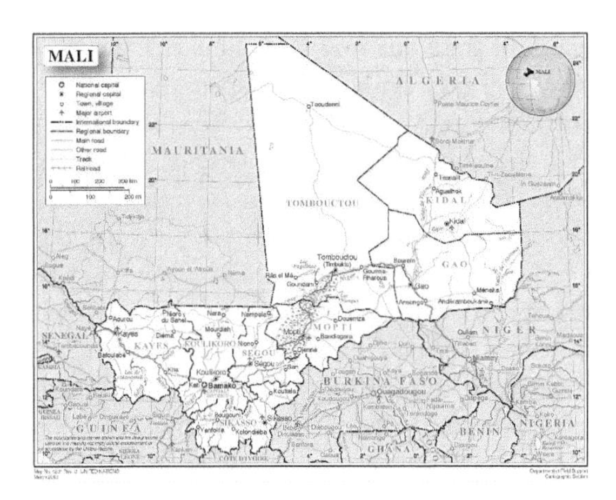

Fig. 1 Map of the Republic of Mali
http://www. un. org/Depts/Cartographic/map/proffile/mali. pdf(2019. 3)

south by Côte d'Ivoire and Guinea, and on the west by Senegal, and Mauritania. It is a relatively large country with a surface area of 1, 240, 192km^2, and the Sahara desert covers 65% of its territory.

Mali has 4 World Cultural Heritages sites, Djenné (since 1988), Timbuktu (since 1988) and Tomb of Askia (since 2004) as cultural heritages and Cliff of Bandiagara (Land of the Dogons) (since 1989) as mixed cultural heritage.

2.2 Introduction of Djenné

Djenné (also Jenne), a historically and commercially important small city in the Niger Inland Delta of central Mali, became a market center and an important link in the trans-Saharan gold trade. Inhabited since 250 B. C. , Djenné is a historically and commercially important small city located on the internal delta of the Niger in Mali, at the crossroads of the major trade routes of West Africa. Djenné became a market center and an important link in the trans-Saharan gold trade. From the 13th century, Djenné developed as the distribution point for everyday commodities such as rice and corn, and also as a center of arts, learning and religion. In the middle of the old city stands a great Sudanese-style mosque, built in 1220 and rebuilt in 1907. Djenné covers some 50 hectares on the banks of the river Bani. Djenné has an ethnically diverse population of about 12,000 (in 1987) and 20,000 (in 2007). It became famous for its mud brick (adobe) architecture. The inhabitants of Djenné mostly speak Djenné Chiini, a variety of Songhay, but the languages spoken also reflect the diversity of the area. In the surrounding villages, Bozo, Fulfulde, or Bambara are also spoken.

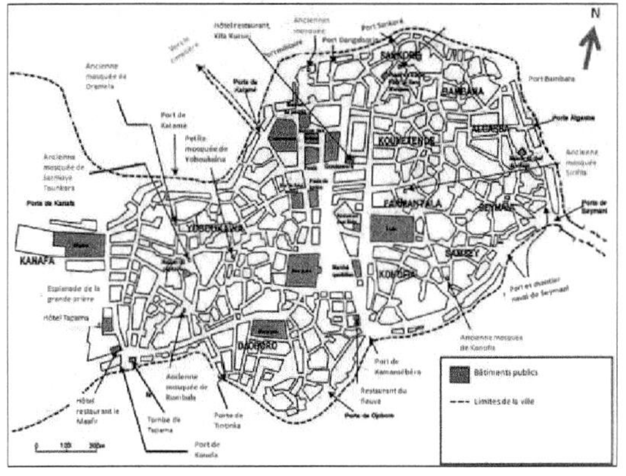

Fig. 2 Town of Djenné
Source: Plan de Conservation et de Gestion des "Villes anciennes de Djenné" (DNPC 2008)

Djenné with its mosque became a world heritage site in 1988. Since, many restoration projects, conducted by cultural agencies, non-profit organizations and foreign government aids are held. The restoration of the great mosque, which starts in 2008 by a foreign trust, raised a lot of discussion about cultural conservation and heritage preservation.

Fig. 3 The Great Mosque and market of Djenné
Source: Oussouby SACKO (2004. 3)

2.3 The Town Organization

Djenné is organized around ethnic/professional quarters along a major axial sys-

Fig. 4 Sudan architecture style of Djenné
Source: Oussouby SACKO (2011. 8)

tem, at the center of which the mosque and the market square form an imposing urban space. Different ethnic groups coexist in the region and have different occupations and specializations. Hence, the Bozos are fishermen, the Peulh (Fulani) raise cattle, the Bambara (Bamanan) are farmers, and the Sarakolé (Soninke or Maraka) are merchants. It is mostly the Bozo, the fishermen, who also provide the masons. This ethnic diversity is very clear on market days, when each guild has its own place. Djenné offers spectacular scenery from the Bani River because of the distance between houses and the uniqueness of its buildings results from the plastic quality given by mud. Indeed, the major construction material of the whole region is banco, the local name for the mud used in blocks, mortar and plaster.

2.4 Houses Typologies

Three major typologies can be identified in the city houses: "Moroccan", "Tukulor" and "Plain". These typologies relate to a certain extent to different historical periods and socio-economic levels, but they refer specifically to the way the main facade of a building is treated and how this corresponds to the spatial organization of the house. Most houses are two-storey, with roof terraces. In all three-house typologies, the spatial organization reflects a strict separation of the sexes and the relative social positions of the inhabitants. One of the aspects in the Sudanese style house is the vestibule. The vestibule is a space used as an entrance hall, located between the street and the courtyard or the veranda. Usually there is a door which can be opened and closed, and connects to the courtyard depending on the way the inhabitant wishes to use it. In some cases the vestibule is divided into two spaces, as it is possible with the courtyard. In Djenné, the vestibule is used as a Koranic school space in most of the Marabout's houses, as an atelier in artisan's houses, as a meeting place in the Chief's house. It is therefore clear that the role of the vestibule depends more to the household head's social status and occupation. In the case of Djenné, it is still difficult to say that the vestibule's role is relating to a particular ethnic group or race. Nevertheless, it plays a very important role in socializing between different families in the community; therefore, it can be seen as a key in the cohabitation between different ethnic groups.

Fig. 5 Djenné's Townsscape
Source: Oussouby SACKO (2011. 8)

2.5 Construction Materials

There are two types of mud blocks. The older type, now no longer used except for specific restoration work, is called Djenné ferey (bricks of blocks of Djenné) and consists of roughly cylindrical pieces. From the 1930s, it became common to shape the mud in rectangular block forms (toubabou ferey which means foreign blocks). During the dry season builders transform the riverbanks into pits for the

Fig. 6 Bird view of Djenné (by Helicpter)
Source: Oussouby SACKO (2004. 5)

itself is a kind of family and community affair shaped by the special ties that bind the masons and the families that own the houses. A family has "their mason", as much as the mason has "his family". The relationship goes from father to son on both sides (the son of the house-owner's mason is the mason of the house-owner's son) and lasts for the whole of their life: the mason of the house also builds the house-owner's grave. For a mason to work for a different client requires the agreement of both the family and his fellow masons. Masons (barey) are organized by a professional body, the barey ton, which guarantees their pro-

Fig. 7 The Different types of House Façade in Djenné
Source: Oussouby SACKO (2010. 02)

preparation of banco, the mud that forms construction blocks and rendering mortar. The mortar for mixing mud with rice husks makes rendering and the mixture is then covered with water and stirred occasionally, the same quantity again of rice husks being added gradually. The mixture then rests for two to three weeks to be fermented. Wood is used for the construction of floors, ceiling and roofs. It is also used for toron, this natural architectonic elements and details so specific to the region.

2. 6 Mason Organization (Barey-ton)

In this region the construction process

fessional training and establishes codes of conduct and support with other professions. Apprenticeship begins at the age of around seven. The apprentice goes through a clearly codified structure of training during the course of which he becomes familiar with tools and materials, building techniques, building conception and the supervision of construction, until finally, in his mid-twenties; he is officially accepted as a barey. Magic plays an important role both as a means of protection against professional risks and as part of the code of relations between all the participants in the creation of

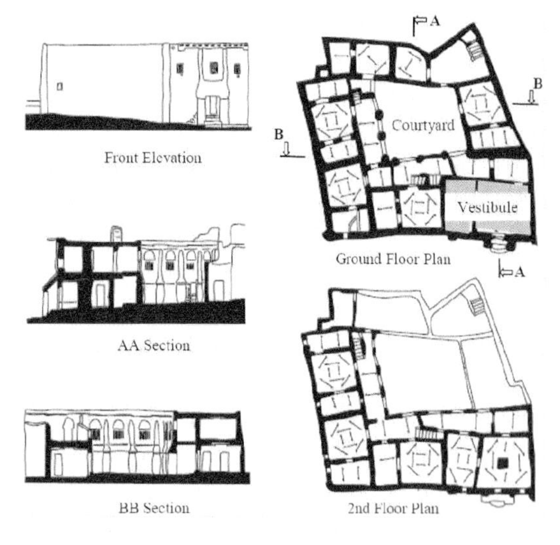

Fig. 8 House plan of Djenné (Source: Plan de Conservation et de Gestion des "Villes anciennes de Djenné" — Pp. 48 Mali2008—2012 (DNPC 2008) rewritten by Oussouby SACKO

Fig. 9 Mortar preparation in for house and buildings covering
Source: Oussouby SACKO (2009. 3)

Fig. 10 Old style mud bricks in Djenné (Djenné Ferey) (left) and new type of bricks (toubabou ferey) (right)
Source: Oussouby SACKO (2009. 8)

a house. It is quite interesting to note that most masons in the region start by first "drawing" the facade. The type of facade is the first issue for the mason and the client to agree on, since it seems to determine the whole spatial organization of the house.

3 Conservation Projects in Djenné

3.1 Djenné Conservation Projets

3.1.1 Houses Restoration Projects by the Netherlands Found

Fig. 11 The chief manson of Djenné
Source: Oussouby SACKO (2004. 5)

Fig. 12 Mosque recoring site by Djenné's inhabitants
Source: Oussouby SACKO (2004. 5)

In 1995 and 1996, joint missions were constituted. The mission, acknowledging the special quality of the built space of Djenné, also recognized that if the city had not suffered serious aggression from "modernity", it was in part due to its isolation and to the stagnation of economic activity, which were at the same time causes of the collapse of an alarming number of older structures. It was thus envisaged to undertake a short-term project with the goal of "conserving this unique monument for the present and future generations", which focused on the rehabilitation of 168 of the monumental houses considered to be the most representative of the "national cultural identity". The intervention ranges from minor repairs and wall rendering to total reconstruction, based on existing documents or relying on the descriptions of those who remember.

Tab. 1 Conservation projects in Mali

Dates	Description	Partnership
Some National Initiative Conservation Projects		
Dates	Description	Partnership
1996	Workshop and Training of West African's Cultural Sites Mangements	Financed by UNESCO
1995	Survey about Tourism Impact on Cultural Sites by Cultural Mission Agencies of Djenné and Banddiagara	Financed by UNESCO
1993	Creation of Djenné Culural Mission Agency by Ministerial Decision (90-203 P-RM of 1993 June 11th, review by Ord. No. 01-032/P-RM of 2001 August 3rd).	Malian Government Body
1988	Subscription of Djenné on the UNESCO World Cultural Heritage list	Malian Government Committee World Heritage
Architectural Education Projects		
1996—2006	Djenné's Architecture Restoration Project. Restoration of 130 Houses. Construction of New Public Projects and Valoriser of Djenné's Masons Know-How	Netherlands
2008—2012	Great Mosque renovation and Training	Aga Khan Trust for Culture

Fig. 13 Mosque conservation by Aga Khan Trust, before and after images
Source: Oussouby SACKO

3.2 Mosque Restoration Projects by the Aga Khan Trust for Culture

Beginning in 2004, under a public-private partnership, the Aga Khan Trust for Culture (AKTC) began working to revitalize the centers of three cities in Mali. AKTC started with the restoration of the Great Mosques of Djenné and Mopti and the Djingereyber Mosque in Timbuktu, as well as the public spaces around them. The mosque restorations became the most visible part of a multidisciplinary program aimed at improving the quality of life in the cities. These efforts included the installation of new water and sanitation systems, street paving, early childhood education, training, health care and economic development. The Trust's work relies on close co-operation with local institutions and stakeholders and the participation of experienced local masons and specialists in restoration.

3.3 Issues Rose From Projects

In the case of Djenné, the foreign agencies and experts lead preservation project and local population is much more like an observer of what is supposed to be theirs. In some cases, new techniques were introduced for making mortars true that some old techniques were revalorized. With its long construction tradition, the preservation and restoration project in Djenné seems to create a gap between cultural conservation and preservation. This discussion paper will be aim to be an opportunity to rethink about cultural conservation, which is lost in the process of preservation process.

4 Remarks From Djenné's Fieldworks

4.1 Issues Raised from House Restoration Projects

In the previous chapters, I mentioned the views of the government body about house conservation projects. Those reports mentioned that the conservation of Djenné has been a success as a living cultural heritage and also that it helps to conserve and promote some specific "know-how" on earth architecture. From the economic point of view, Djenné was promoting as a center of cultural tourism raised the prestige of "earth architecture". Some training and workshops of young people in local construction techniques were held. Efforts of collaboration with actors such as the corporation of masons《Barey Ton》, the Djenné Heritage Association, the Association of Guides, the customs authorities and opinion leaders was seen. But on the other hand, the property continued to suffer changes as regards its buildings. The reason for these changes is essentially due to the modification of the original plans of the houses, the escalation in cost of the materials used for traditional livelihood (rice and millet, baobab fruit, Shea butter). The introduction of inappropriate solutions using cement and other materials is quite popular; the abandonment of numerous buildings that have fallen into ruin is increasing. The appearance of new constructions in the inscribed periphery is quite frequent.

4.2 Issues Rose from Mosque Restoration Project

In the conservation site report and review mentioned in several chapters, it was said that the Mosque restoration saves Djenné's architecture, saves the national identity, which is the architecture and construction, and gives the population a sense of conservation. The continuation of restoration and conservation projects will contribute to Djenné's socio-economic and tourism development

On the other hand the restoration projects raised some issues, such as the gap between traditional and modern techniques or approaches during of construction. The population mentioned that AKTC didn't keep the promises of providing material for the following years; they are some confusion between the conservation technics and system and traditional way of preservation. But, in the case of Djenné mosque, the populations start to be much unified and took responsibility for their own cultural heritage conservation.

4.3 Issues About Town Shape and Construction Material Problems

Recently new constructions in concrete and cement blocks at the entry of Djenné start to be popular. One of the explanations was that part is not included within the conserved area. Those buildings mostly belong to successful Djenné's born who live in the capital city Bamako and are building second houses. The use of new material and technics can be discussed, but the design of those buildings doesn't reflect at all the essence of Djenné's architecture or building shapes. This means a lack of control in preserving the construction system by local and central government bodies. In the conserved area, almost 700 houses are recovered by fired clay including 4 municipal buildings. The lack of using the essence of Djenné's housing plan or construction spirit needs to be pointed out. The local group of Djenné Patrimoine did researches on new conservation and building technics. Djenné Patrimoine conservation group was saying that the conservation in a cultural evaluative context is needed in this town.

4.4 Post Malian North Region Crisis in Djenné and the Deterioration of Tourism

Djenné was one of the most visited tourist sites in Mali. But, the principal hotel infrastructure in Djenné (Campement — Maafir — Chez Baba — Rèsidence Tapama, Kita Kuru, Djenné — Djeno, Hotel Welingara) shows the number of tourists decreasing between 2003 and 2007, and since 2012 due to the rebellion in the north of Mali and the series of foreign kidnapping by several groups of bandits in the same area.

Fig. 14 Town shape with new construction
Source: Oussouby SACKO (2015. 3)

During my recent visit to Segou, Djenné and Mopti, I saw a lot of empty hotels, guest-houses, restaurants and bars. It seems that tourism was the main business target in those regions for local and young population. Some government-supported hotels have few foreign guest and public workers for seminars, but most of them have been ruined. Guides and young people I have known for many years, lost their jobs, a lot of them are probably in the capital of Bamako, and some in neighborhood countries to continue the same jobs. Some return to their villages. Tourist facilities also are suffering the same problem because people don't feel the necessity of maintain them and a lot of buildings collapse. This is to say how big was the tourist business, but people are losing the courage of conserving their own heritages designated as world heritage. We should come again with the question of for whom the world heritages should be preserved.

Fig. 15 Tourist facilities in ruins
Source: Oussouby SACKO (2015. 3)

5 Conclusion

The question of why Djenné is being preserved, for whom and for how long into the future needs to be addressed on a number of levels. If the town is being preserved exclusively for its residents, in order to ensure the protection of their "identity" - a nebulous term which would make reference to knowledge transmission, cultural pride and continuity - then an adaptation of the local architecture may help residents to meet the demands of the upkeep of their homes and fulfill a desire for better housing conditions and modernity.

For the people who were lucky enough to have their houses restored, the benefits were primarily functional. Further difficulties came from the fact that people saw the restoration project as an opportunity to improve their homes. Furthermore, the use of fired clay tiles on houses to protect them from the rain, a practice condemned by UNESCO but seen as a potential solution by Djenné residents, is described in a functional, not aesthetic, way. In the case of Djenné, therefore, housing is at once a basic human need, and right - the right to shelter- but also considered "world heritage" and falls in to new thinking about cultural rights and identity. What is contradictory in this process is by imposing rules for cultural heritages protection and at the same time denying the basic human right - the right to shelter, and freedom from constant anxiety about it- should be given more weight than the protection of world heritage.

References

BERBARD G, PIERRE M, 1995. Geert mommersteeg, Bintou sanankoua [Djenné, il ya cent ans], KIT Publications.

SNEIDER R, The Great Mosque at Djenné-Its impact today as a model.

NA MA Chabbi-Chemrouk, 2007. On site report revive-conservation of djenné, djenné, mali.

World Heritage Earthen Architecture Programme (WHEAP). http://whc. unesco. org/en/earthen-architecture/ (2012. 04).

Study on the Protection and Utilization of Architectural Cultural Heritage of Rammed Earth Folk Manor Tower Houses in Southwest China

SHU Ying

Sichuan Fine Arts Institute, Chongqing, China

Abstract Chinese tower houses are unique in the cultural heritage of local architecture in the world. A large number of earthen manor tower houses exist in southwest China, which can be called "living fossils of traditional tower buildings in China." It is an indispensable part of the Chinese residential tower system and has obvious protection value. Due to the general inadequate attention to the historical and cultural value of rammed earth tower houses, this characteristic residential building is at risk of vanishing in the process of urbanization. Therefore, the in-depth study of rammed tower houses in southwest China will be a great supplement and perfection to the integrity of Chinese residential towers as an integrated world cultural heritage. At the same time, pay attention to the inheritance of architectural skills and the ecological and environmental protection of materials research, development, promotion and utilization, will maximize the effective activation and full use of architectural cultural heritage.

Keywords southwest region, rammed earth, manor towers, protection and utilization

1 Historical Formation and Cultural Protection of Tower Residential Buildings in China

Tower house is a closed multi-story building built by different nationalities to resist external threats during the historical period, which exists all over the world. With the development of the society, the external factors of tower defensing against natural disasters, beasts, tribal vendetta, war and bandits are gradually eliminated, and the special functions of the cold weapon era are gradually declining. Most of the tower houses have become architectural relics, only a very few of which have followed and developed. Most of the type of architecture in the west is the defense ancillary facilities of castles and military fortresses, while the towers in the form of residential buildings are splendid in China, some of which has entered the field of vision of the protection of the world cultural heritage, which has aroused widespread attention and concern. However, the historical and cultural value and the ways of protection and utilization of the special type of architectural heritage still need to be further studied.

Tower houses are constructed in different ethnic areas around China, among which five types are the most typical: Han traditional manor towers, southwest Tibet & Qiang towers, Jiangxi, Fujian and Guangdong Hakka traditional towers, Tubao towers and Wuyi overseas Chinese hometown towers. According to the historical background of their emergence and development, we can clearly sort out the complete historical change process of "settlement defense — family defense — house defense". For thousands of years, they have continued to blend with each other, learn from each other, and jointly improve the cultural connotation of this special type of architecture in China.

In the past ten years, Fujian Tulou and Kaiping Diaolou in the architectural sequence of Chinese tower houses, have been included in the "World Cultural Heritage Protection list", West Sichuan Tibet & Qiang Towers and villages have also entered the "preliminary list of Chinese World Cultural Heritage". Domestic scholars also have actively studied the tower houses as Jiangxi, Sichuan, Chongqing, Guizhou and other places, which have not yet included in the world cultural heritage property protection list but have the value of precious protection.

So far, under the guidance of the government, various localities have issued corresponding protective plans and regulations for the towers that have been listed in the protection of the world cultural heritage, and implemented building maintenance and protection and environmental renovation, so that under the modern background of the gradual loss of their basic functions, As a symbol of local historical tradition and regional cultural image, it continues to play a role in cultural protection, tourism and other aspects as far as possible, and serves the social and economic development, but there are still many problems in the actual operation. In addition to the inevitable natural decline caused by the loss of basic functions and the difficulty of space activation and utilization, the impact of the destruction of the surrounding landscape environment on the overall architectural style, the great impact of man-made commercial activities, and so on. All aspects restrict the protection and utilization of the tower building heritage, which makes it difficult for the government to lead and the whole people to protect. For example, after the successful application of Kaiping building, it fell into a predicament, and the gap of protection money reached 230 million, which triggered the cold thinking of the academic circles after the heat of heritage protection and development. Its essence is still the lack of activation and utilization measures.

2 Historical Cultural Background and Present Situation of Tower Houses in Southwest China

Southwest China is the area where the tower houses are relatively concentrated. The Han traditional manor tower, the Tibet and Qiang tower and the Tunbao tower coexist in this geographical unit. On the surface, the natural and social background conditions of the three appear to be different. But in fact, they are all human settlements with the combination of the north migration and the environment. The latter two have also received more attention in recent years, and the rammed earth manor tower of the Han nationality has not only failed to attract social attention in the past few decades. On the contrary, it is still disappearing rapidly, which can be said some kind of regret.

The general trend of geographical distribution of the rammed earth towers of Han nationality in southwest China is that Bayu is more than western Sichuan, and southern Sichuan is more than northern Sichuan. Specifically, it is mainly distributed in southern Sichuan and Shangba-Yu areas, such as Fuling, Dianjiang, Ba County, Nanchuan, Qijiang, Jiang Jin and Qianjiang areas in Chongqing, as well as Gao County, Gong County, Yibin, Xuyong, Gulin, Hejiang and Naxi areas in southern Sichuan. In addition, it is also scattered in Yilong, Bazhong, Emei, Hongya, Mabian, Muchuan, Linshui, Guangan, Dazhu, Renshou, Jingyan, Weiyuan, Zhongjiang and other counties. Throughout the current situation of its distribution, there are many relics and rich types in Chongqing, so this paper will mainly take Chongqing as an example to carry out in-depth analysis.

According to statistics, in the 1990s, there were still more than 1000 towers of various types in various districts in Chongqing. According to the statistics of the relevant data of the third national cultural relics and the field survey in recent years, there are now less than 300. Most of them are naturally damaged and artificially demolished. In the corresponding subject study undertaken by the author, after conducting a field sampling survey in the districts where the number of towers is concentrated and the use of the towers is better, the statistical results show that the main body is in good condition, and 55% of the users continue to use it, and the main body still exists. 29% were abandoned, 7% were demolished, 6% were rebuilt or planned to be rebuilt, and 3% were newly built. Among them, 57.4% have been listed as cultural relics, which is one of the main measures for the listing and protection of cultural relics, but the level of cultural protection is low and the protection is weak. In addition, there are no more means of protection and utilization. This data is only a sampling of areas with better protection status, and the abandoned state of more district and county tower buildings should be much higher than this proportion.

3 Analysis on the Current Use Types of Rammed Earth Manor Tower

According to massive data from the literature analysis and the author's generous field survey, field mapping, we found that there are much important information. Compare with another tower buildings, we found there have been many immigrants in the history of Chongqing, so many types of towers have been gathered, and many "miniature versions" of towers have been constructed, which can more completely reflect the complete evolution of Chinese towers from ancient times to the present. Among them, the most rare is to retain the structure, which is deeply influenced by the spread of northern culture and immigration, the rammed earth manor tower, which can be called the living fossil of Chinese traditional residential tower architecture.

On the basis of our field survey in Chongqing area during recent years, there is a delighted phenomenon that with the progress of the construction in the new countryside, the tower building has attracted some attention in the planning, construction and cultural circles, and some districts and counties have carried out related work around the protection of the tower house. The creation of traditional village tourism attractions, the promotion of local culture, has achieved certain economic and cultural effects, but on the whole, the scope of protection and utilization is still very narrow, means and measures are also extremely limited. Here, we analyze the types around the current situation of use by some typical examples of our field investigation:

3.1 Sorts of Well Preserved and Sustainable Use

Jiangjin Huilong manor and Bishan Hanlin manor are typical examples of good protection and utilization of rammed earth manor towers in Chongqing at present. The two buildings have been built for more than 200 years, the manor covers a wide area, there are not less than two towers to implement defense, the main towers are typical of the traditional watchtower (Fig. 1, 2). Because the two manors are located in deep mountains, have good rammed earth craftsmanship, are less influenced by nature and society, and the architectural space is well preserved, Huilong manor has been built as the "first village in the southwest", which is used to display the life culture of the manor in the southwest of the late Qing Dynasty. Hanlin Villa is currently a private mansion, with stable maintenance and protection, some of which are used for cultural management.

Fig. 1 Huilong Manor tower in Jiangjin © WANG Qian

3.2 Sorts of Listings by Governmental Protection, Yet Leaved Unused in Practice

Banan Yang's Garden and Nanchuan Wang's ancestral temple are typical examples. The former has the tallest existing tower in Chongqing (Fig. 3), while the latter is the only well-preserved rammed earth ancestral hall (Fig. 4). The two manor tow-

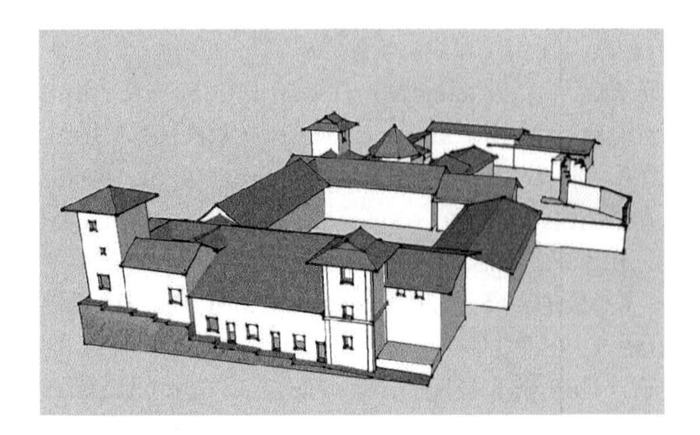

Fig. 2 Hanlin Manor in Bishan aerial view model © SHU Ying

ers have been built for more than 100 years, and the courtyard and towers have the functions of both defense and long-term residence, and the transformation and utilization are operable. They belong to the protection objects of outstanding historical buildings listed at the municipal level, and there are actually no maintenance and protection measures. Basically idle, the whole tends to be dilapidated; Great hidden dangers in safety.

Fig. 3 The tallest tower in Banan Shilong Yang's Manor © HE Zhiya

Fig. 4 Wang's Ancestral temple towers of Shi'xi, Nanchuan © SHU Ying

3.3 Sorts of Decaying Gradually Without any Treatment

Jiangjin Yanghui Villa is a typical examples. The villa was built during the period of the Republic of China. The traditional watchtower is a rare combination of traditional with Chinese and Western architectural decoration. It was originally a closed courtyard with three towers responding to each other (Fig. 5). At present, the courtyard has been destroyed. There are only four floors of rammed earth main towers, which are generally well preserved and beautifully detailed (Fig. 6). Without the protection of any government departments, the construction is idle and dilapidated, there are serious safety risks.

3.4 Sorts of Regional Natural Formation Without Special Protection

In villages known as building towers, there are a large number of low-grade towers built before and after the Republic of China for residential use, forming a certain scale and regional features of the rural landscape. Such as Fuling Dashun area around the farm houses because of the retention of

Fig. 5 Restoration image of Jiangjin Yanghui villa © SHU Ying

Fig. 6 The main tower of Yanghui villa with its top diagonal brace and decoration © SHU Ying

more towers and get the name "Bayu Tower of the hometown." These tower villages and gentle slope terraces form a beautiful natural rural scenery, showing the unique pastoral style of Bayu mountain houses. This is compared with the current situation of the dilapidated and idle buildings of the larger manors in the mountains. Under the background of the construction of beautiful villages, it provides an idea for the protection of rammed earth tower manors in Chongqing.

4 Conclusions

The best protection of architectural cultural heritage is to make use of it. The historical and cultural roots and construction techniques of rammed earth tower manor houses are the aggregations of previous wisdom. It is the mission of the towers to guard the security of their homes during the war years, and nirvana is reborn in the form of ecological construction in peacetime. It is necessary to follow the way of historical and traditional cultural rejuvenation and rural construction of green environmental protection. For this reason, in view of the current situation of rammed earth tower manor houses in southwest China, it is suggested that the following protection and utilization plans should be carried out for the remaining more than 300 blockhouses and the area on which they rely:

4.1 Fixed Preservation

Formulate preservation standards, classify and list the existing rammed earth manor tower houses, and carry out rescue protection for endangered tower buildings with high historical and cultural value, so as to avoid collapse again and cause the disappearance of architectural cultural heritage. Then consider the follow-up activation and utilization, such as Banan Shilong Yang's Garden tower, Jiang Jin Simian Mountain Yanghui Villa tower and so on.

4.2 Preservation by the Regional Division

For the distribution of a large number of rammed earth tower houses, the single value is not high or the completion time is relatively late, but also formed a certain style of the area to focus on the protection, retain the original local traditional culture genes, Such as Jiang Jin Siping Town, Tongnan Guxi Town, Beibei Tianfu Town and other areas can become rammed earth tower manor residential demonstration area.

4.3 Features Construction

The transition of tower buildings from defense to residence encourages the construction of traditional tower buildings in areas with historical and cultural origins, such as Dashun, Longtan, Qingyang, and other places in Fuling, so as to create a historical and humanistic space in rural areas, which is conducive to the inheritance of traditional craftsmanship. It is also conducive to the construction of rural characteristics, but also provides a harmonious space background for the protection of key towers.

4.4 Sightseeing and Tourism

The tower house building has its own tourism product attribute, especially the rammed earth tower manor residence has more unique advantages in this respect, which can be relied on to carry out rural tourism planning, so that the tower house building in southwest China can go out of the mountain area of valley with the unique folk culture to attract attention, the introduction of tourist flow, to promote the development of rural economy.

4.5 Make Livings Applicable and Flexible

The building of rammed earth tower manor is flexible and plastic, has strong adaptability, has the advantages of seeking area from space, natural environmental protection with materials, warm winter and cool summer in living micro-environment. The way of activation and utilization can start with improving the function of use. Or consider the rammed earth materials in the maintenance and protection of modern technical means to support functional transformation, give full play to its advantages of ecological environmental protection, improve the quality of rural livable, with a scientific way to retain the tradition, retain nostalgia memory.

References

DU Fanding, ZHANG Fuhe, 2005. History study of Kaiping Towers in Gongdong. Beijing: Tsinghua University.

JI Fuzheng. 2007. Three gorges classical town. Chengdu: Southwest Jiaotong University Press.

SHU Ying, 2015. Chongqing existing tower buildings general survey research. Chongqing: Chongqing Urban Planning Bureau Key Policy Decision Consultation Project.

LI Xiankui, 2009. Series of Chineseresidential architecture. Sichuan residential architecture. Beijing: China Architecture& Building Press: 169.

SHU Ying, 2017. Research on tracing the source of construction technology about the rammed earth folk manor tower houses and protection and utilization in Bayu. Chongqing: Chongqing Urban Planning Bureau Key Policy Decision Consultation Project.

SHU Ying, LIU Zhiwei, 2018. Study on historical evolution, conservation and utilization of bunker architecture in Bayu area. Chongqing social sciences, (10): 131.

Classified Strategies for Conservation of Vernacular Villages in Coastal Areas of Shandong Peninsula

MA Jinjian[1], *XU Dongming*[2], *GAO Yisheng*[3]

1 Shandong Provincial Conservation Engineering Institute of Vernacular Heritage, Jinan, China
2 National Key Laboratory of Vernacular Heritage Conservation of Chinese State Administration of Cultural Heritage (SACH), Shandong Jianzhu University, Jinan, China
3 School of Architecture and Urban Planning, Shandong Jianzhu University, Jinan, China

Abstract Seagrass thatched house is a unique vernacular housing form in the coastal area of Shandong Peninsula, China. It features distinct oceanic culture and is considered as precious and non-renewable resource. During the high-speed urbanization process in recent decades in China, a large group of young and middle-aged people migrate to the city for work, leaving behind only aged people in the empty villages. As a result, many seagrass thatched houses have long been vacant and further abandoned. Since the "Regulations for the Conservation of Seagrass Thatched Houses in Rongcheng City (on trial)" was issued in 2006, the conservation practice has achieved remarkable accomplishments, and the decay and disappearance of the houses in vernacular villages have been contained to a certain extent in the past decade. However, many misunderstandings still exist in the conservation and utilization practice of the vernacular villages, which leads to many natural and man-made damages. Therefore, it requires an approach to in-depth research and theoretical guidance on both the value assessment and the methods for conservation and use. Based on the fieldwork and studies of over one hundred vernacular villages of traditional seagrass thatched houses in the coastal areas of Shandong Peninsula, this paper summarizes the features and values of the seagrass thatched houses and the vernacular villages. Further, it proposes the strategies for classified conservation of vernacular villages along the coastal areas of Shandong Peninsula in the hope of making contributions to understanding the value and carrying out the conservation work of these villages.
Keywords seagrass thatched house, vernacular village, feature, conservation and use

1 Introduction

Seagrass thatched houses are unique residential structure in the coastal areas of Shandong Peninsula. These typical villages reflect the life wisdom and cultural consciousness of the coastal villagers. However, the urbanization and construction of new countryside have put the vernacular-village-featured seagrass thatched houses under the threat of decline and disappearance. The conservation practice starting in 2006 has slowed down the extinction process of these villages. Nevertheless, given the large quantity of such villages, lacking in-depth research in this regard makes it very challenging to conserve the villages. Current research on seagrass thatched houses and vernacular villages focuses primarily on the features of the vernacular residence buildings, construction techniques, ecological characteristics, folk culture and so on. Scant research concentrates on the conservation of vernacular villages, especially the overall evaluation of the seagrass thatched houses in the coastal area of Shandong Peninsula. The paper takes the macro and regional perspective to explore the features and values of the vernacular villages of seagrass thatched houses, summarizes the existing problems, and proposes the strategies for classified conservation by taking the comprehensive value assessment into consideration.

2 Village Characteristics and Values

2.1 Formation and Evolution

The vernacular villages with seagrass

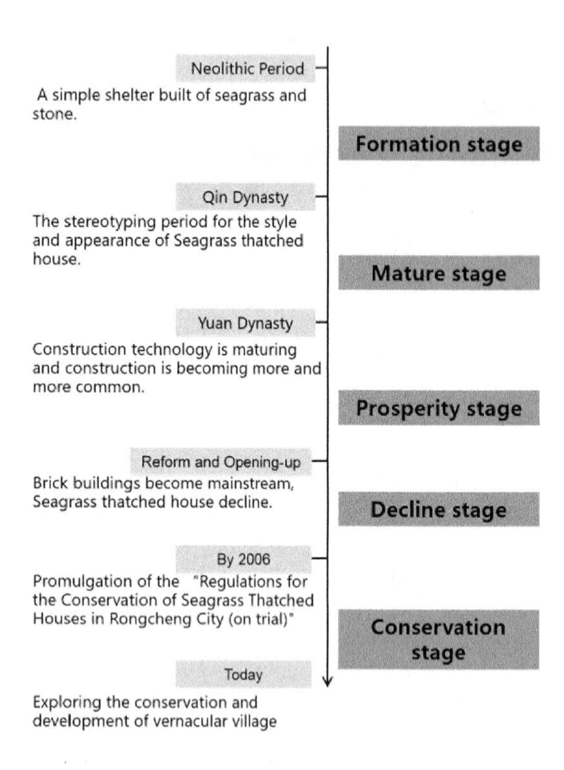

Fig. 1 The evolution of seagrass thatched houses along the coastal area of Shandong Peninsula © MA Jinjian

thatched houses have gone through five stages of evolution: the formation stage, the mature stage, the prosperity stage, the decline stage, and the conservation stage (Fig. 1). Its earliest form can be dated back to Neolithic period when a simple shelter built of seagrass came into shape, which can hardly be considered as the vernacular buildings in the real sense(Liu, 2008). The Warring States period (475—221 BC) witnessed many fisherfolk and salt producers settled down here and promoted the development of seagrass thatched houses (Liu and Chen, 2015). After China's first emperor Qin Shi Huang (259—210 BC) unified the country, the emperor made two east patrols, which hugely promoted the economic and cultural prosperity of the area and the construction of seagrass thatched houses. During that time, the style and appearance of the seagrass thatched houses came into the basic prototype. In the Yuan Dynasty (1271—1368)the construction technology of building seagrass houses was advancing and became mature, leading to the prevalence of seagrass thatched houses. The large number of migrants flushing into the coastal areas of east Shandong in the beginning of the Ming Dynasty (1367—1644) gave rise to the incredible increase of seagrass thatched houses, especially a great number of villages serving the military defense purposes during the reign of Emperor Hongwu (1368—1398)(Liu and Chen, 2015). The Ming Dynasty saw a huge increase of seagrass thatched houses and the peak of the construction techniques, procedures, and folk etiquettes. After the foundation of the People's Republic of China in 1949, a small number of brick houses began to appear in the Shandong coastal areas, but they did not affect the dominant position of seagrass thatched houses in the villages. However, following the reform and opening-up (1978), brick houses became the main stream and seagrass thatched ones began to decline (Zhang, 2012). During the high-speed urbanization process in recent decades in China, a large group of young and middle-aged people migrate to the city for jobs, leaving behind only aged people in the empty villages. As a result, many seagrass thatched houses have long been vacant and further abandoned. Since the "Regulations for the Conservation of Seagrass Thatched Houses in Rongcheng City (on trial)" was issued in 2006, the continuous conservation practice has achieved fruitful results, and the decay and disappearance of the houses in vernacular villages have been contained to a certain extent.

2.2 Site Selection and Layout

The vernacular villages of seagrass thatched houses are located along the coastal areas of Shandong Peninsula and primarily concentrated within 15 to 20 kilometers to the coastal line along the mountains of Weide Shan, Shi Shan, Huimu Ding Shan, and Jiazi Shan of Rongcheng, Weihai. Most vernacular villages of seagrass thatched houses are close to mountains and near the river. Statistics show that 21% of seagrass thatched houses are close to the mountains,

44% close to the sea, and 52% near the rivers. Constrained by the terrain and limited land for building houses, seagrass thatched houses take the advantage of the mountainous terrain to economize the land use and reduce the loss of heat. Given this layout consideration, the vernacular villages of seagrass thatched houses are of high construction density and characterized by neat and compact appearance.

2.3 Buildings and Courtyards

The seagrass thatched houses are arranged in the form of a courtyard (Fig. 2) with each building detached to others yet remaining compact and flexible. The wing gables can partly hide the principal suite and the back wall is aligned with the gable of the principal suite. The gable is only one meter away from the principal suite and reverse rooms. The right rooms are about 3 to 4 meters from the left rooms. Trees (mainly fig trees) are growing in the courtyard. The entrance and exit of the courtyard are in the direction of southeast with a few of them facing the north. Such layout can dehumidify and create coolness in summer while blocking the cold and keeping the warmth in winter.

Fig. 2 Bird view of seagrass thatched house courtyard © MA Jinjian

The principal suite consists of 3 to 4 rooms with a small depth of 3 to 5 meters, which is attributed to the limited adhesion of loess and the heavy seagrass roof. The room with a front door is the outer room of the principal suite and is about 3 meters in width with stoves on both the right and the left sides for heating the brick beds in the bedrooms (inner rooms) on the east and west sides. Usually the east room is longer than the west room. In addition to the opening door of the seagrass thatched houses, only the outer room opens the clerestory facing the north and the inner rooms open wooden windows facing the south for lighting and ventilation. The small windows, doors, and thick stone walls and grass tops greatly minimize the heat loss in the winter.

2.4 Value Evaluation

The vernacular villages of seagrass thatched houses are the result of the joint action of local natural, geographical, and climate environment and the production and lifestyle of local people(Yang, 2011). As a village with strong regional characteristics, the vernacular villages possess great ecological, research, cultural, artistic, and economic value. First, the construction materials coming from nature scarcely produce nondegradable construction trash; the sea stone walls and seagrass roof greatly reduce the heat exchange and generate thermal insulation effect; the courtyard layout and the village structure align with the local natural conditions of wind, light and water, reflecting the strong ecological and adaptable abilities. Second, the thatching technique is the wisdom accumulated and developed by the ancient craftsmen with the constant effort of generations, which plays a significant part in completing the history of China's vernacular residence and ancient architectural techniques. The soft seagrass roof, mottled stone walls, and the hillsides near the fish villages constitute a splendid picture, drawn by many painters, photographers, and writers to create works and giving full play to the cultural and artistic value of the seagrass thatched houses. The harmony between people and nature and the unique oceanic culture have attracted many urban people to experience the local life and culture and thus promoted the local economic development.

3 Current Issues

3.1 Strong Invasion of Modern Landscape

The improvement of living standards has transformed villagers' ideologies and concepts on residence. Housing has not only assumed the function of residence but also reflected people's economic conditions, wisdoms, values, and other ideologies. The low seagrass thatched houses can no longer satisfy people's desire for wealth, changes, and advancement. Gradually these seagrass thatched houses have been changed into bricks houses, and then storied buildings. Moreover, the urbanization takes up more land and swallows up the land close to towns and cities where seagrass thatched houses used to stand. The invasion of urban landscape has transformed the landscape of the traditional seagrass thatched houses. Most of these changes occurred at the edge of the villages where multi-storied buildings dominate the land with low vacancy rate and young people are as the majority residents. In contrast, within the villages are mostly seagrass thatched houses with high vacancy rate and seniors of over 60 years old as the majority residents. With no buffer landscape zone between the two types of buildings, the modern buildings on the edge of the villages contrast sharply with the vernacular seagrass thatched houses within the villages.

3.2 Construction Materials in Short Supply and Construction Techniques in Loss

Seagrass is the indispensable resource to build seagrass thatched houses. However, since the 1980s, the ecological environment for seagrass has lost the balance and the wild seagrass is on the verge of extinction, because the shallow beaches around villages of seagrass thatched houses have been taken up by seafood farms, coupled with excessive marine fishing and industrial pollution(Wu, 2008). The reduction in the amount of seagrass gives rise to its soaring price, going up from a few cents per kilo to more than 20 CNY per kilo. Consequently, villagers could not afford the high cost of construction and repairment. When the roof is broken, the light damage is repaired while the heavy damage is replaced by tiles. To protect the seagrass roof, people use red and blue color steel tiles to cover it. Therefore, the vernacular villages of seagrass thatched houses display a mixed landscape of red bricks tiles and red and blue color steel tiles coexisting with the gray seagrass roof(Fig. 3). The decrease in the demand for construction and repair makes it very hard for the craftspeople to sustain their life and they are forced to change careers. The heavy labor and limited income make the technique unprofitable to young people and the construction technique is subject to get lost.

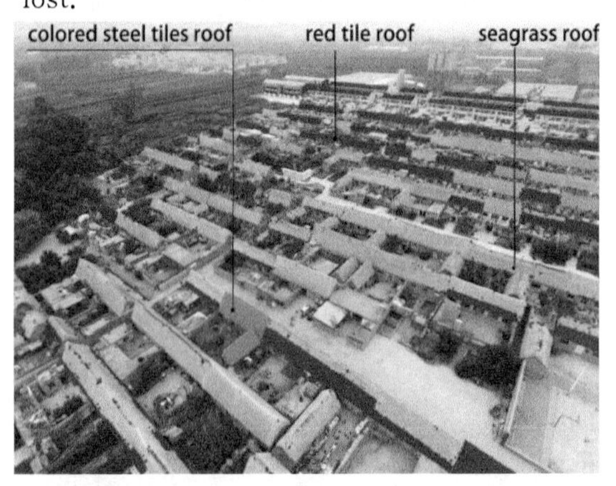

Fig. 3 A mixed townscape of red ceramic tile roofs, red and blue colored steel tile roofs, and seagrass thatched roofs in a traditional village in 2016 © MA Jinjian

3.3 Conservation Facing Severe Challenges

Statistics show that the coastal areas of Shandong Peninsula have 430 vernacular villages with nearly 100 000 seagrass thatched houses, accommodating 23 000 households. More than half of the seagrass thatched houses should be repaired. Varied degrees of damage may ask for varied repair time and cost. Repairing one seagrass thatched house needs 1 to 4 craftspeople working for 1 to 10 days at the cost of a few hundred CNY to over ten thousand CNY.

Such cost coupled with the investment in renovating the village environment, improving infrastructure and service facilities makes it very challenging to support the conservation of seagrass thatched houses under the current policy.

Problems also sustain for villages under the conservation. From 2008, China began to attach importance to the conservation of vernacular villages and provided financial support on this regard. Over 30 vernacular villages of seagrass thatched houses in the coastal areas of Shandong Peninsula have enjoyed the financial support from the government and begun to explore the approaches for conservation and development. Nonetheless, the conservation and development practice for the seagrass thatched houses is far from enough and mature. Most practitioners work in a scattered manner and have few collaborations between each other. Missing instructional quality standard and management system, the construction teams show inconsistent work quality and lack standardized completion acceptance and approval standard. As a result, many seagrass thatched houses suffered protective damage, affecting the village landscape and the value of the houses.

4 Classified Strategies for Conservation

4.1 Why Classified Conservation Strategy

Given the large number of vernacular villages of seagrass thatched houses, the existing differences among them, and the limited labor, materials and financial resources, the focus and mode of the conservation should be defined from the overall and macro perspective. Priority should be given to villages of outstanding value and such villages should be preserved as the central villages to promote the conservation and protection of the neighboring villages. For the village of high value, a further priority should be given to houses with higher value within the village in collaboration with similar villages nearby to obtain win-win results. As for villages of average value, conservation priority should be given to only one or two most featured streets as the axis for village development and the conservation should support the development of the central villages and the collaborative villages.

4.2 Classification Method and Conservation Strategies

Detailed surveys were carried out to make scientific and rational assessment of the comprehensive value of vernacular villages of seagrass thatched houses by adopting the Value Assessment Mode (Tab. 1) in which both AHP (Analytic Hierarchy Process) and Delphi methods are used. Villages are rated based on the survey results and the higher its score is, the greater its value is.

The assessment results show that the number of seagrass thatched houses in 63 villages along the coastal areas of Shandong Peninsula accounts for over 10% of the total villages (Fig. 4). Among the 63 villages, over 10 villages, including Weiwei village, Dazhuang Xujia village, Chen Fengzhuang village, and Zhima Tan village have the highest comprehensive value and are listed as village under integrated preservation. Villages such as Dongchu Island, Liucun, Xiaoxi, and Xiangjia Zhai have above average comprehensive value and are listed as the village under partial conservation, which means the conservation applies to sections where seagrass thatched houses are mostly concentrated. Villages such as Linxing Jia, Xilin, Xingnan Tai, and Datuan Linjia have average comprehensive value and are listed as villages with some allies under protection. The conservation applies to alleys featuring seagrass thatched houses and characteristic landscapes.

Different types of conservation require different conservation scopes. Villages under overall conservation are to be conserved as the entire unit, including the tangible and intangible elements and cultures. The conservation scope for such villages applies to all construction land of the villages and the village's administrative scope is also the construction control zone. The conservation scope for villages under partial conservation

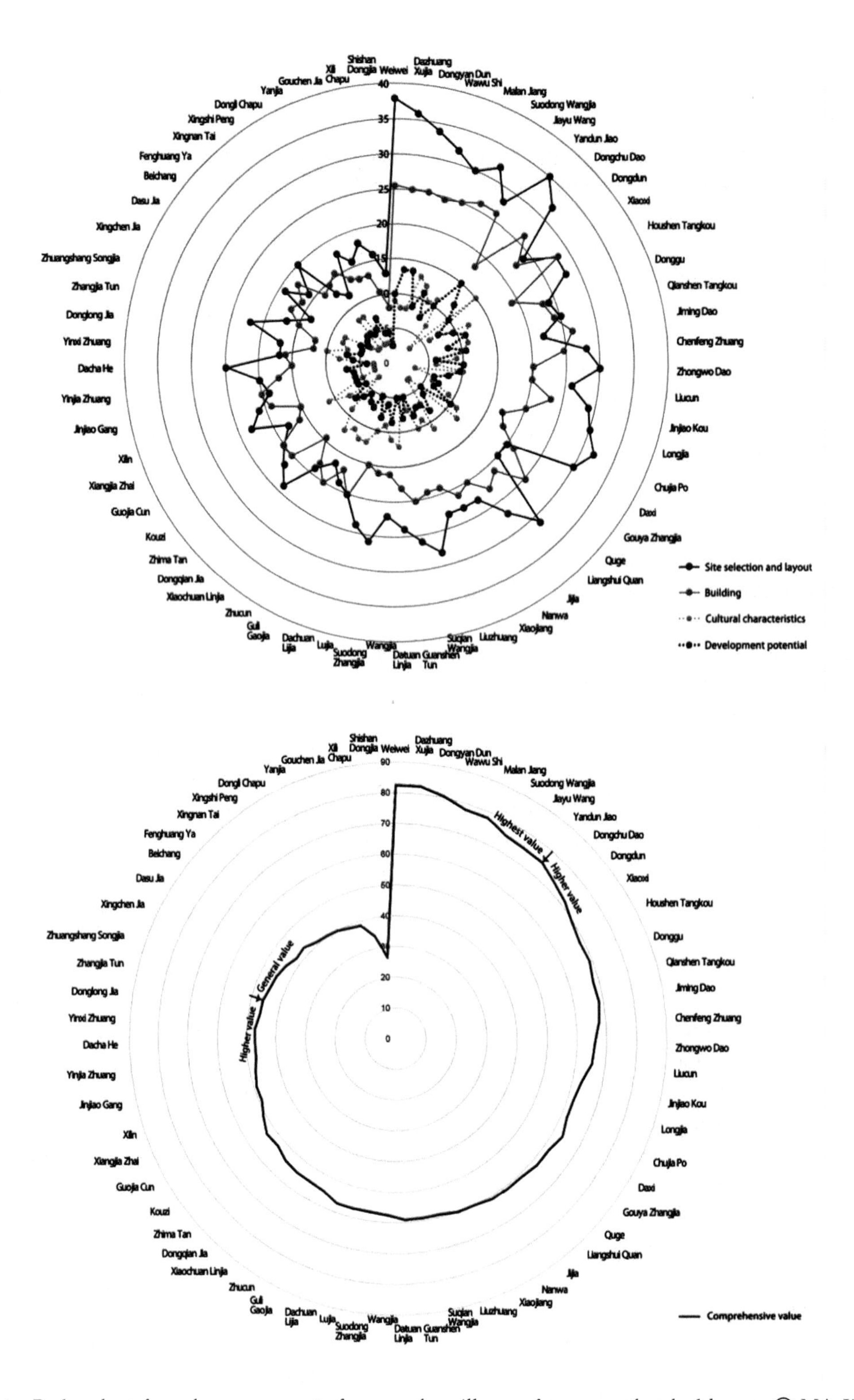

Fig. 4 Radar chart for value assessment of vernacular villages of seagrass thatched houses © MA Jinjian

applies to sections where seagrass thatched houses are mostly concentrated. The conservation scope for villages with some alleys under conservation should cover alleys, streets, and sections featuring the seagrass thatched houses. Construction control scope should apply to 10 to 20 meters outside of the conservation scope for villages under partial conservation and villages with some alleys under conservation to control the landscape of the area.

Tab. 1 Assessment form for conservation value of vernacular villages of seagrass thatched houses

Items	Assessment indicator	Scoring criteria and definitions	Full score
Site selection and layout(30 points)	Formation age	5 for the Yuan Dynasty; 4 for the Ming Dynasty; 3 for the Qing Dynasty; 2 for the Republic of China; 1 for after the founding of the P. R. C.	5
	Site selection	6—10 for having typical local features and coexisting with surrounding environment in harmony; 1—5 for having some local features; 0 for having no local features	10
	Conservation of traditional layout	6—10 for intact conservation; 3—5 for average conservation; 0—2 for poor conservation.	10
	Historical elements	0.5 given to each historical element	5
Building (40 points)	Number of courtyards	10 for more than 200; 5 for 100—200; 2 for 50—100	10
	Distribution	10 for concentrated distribution; 5 for comparatively concentrated distribution; 2 for comparatively dispersed distribution; 0 for dispersed distribution	10
	Construction age	10 for the Yuan Dynasty; 8 for the Ming Dynasty; 6 for the Qing Dynasty; 4 for the Republic of China; 2 for after the founding of the P. R. C.	10
	Aesthetic value	6—10 for high value; 3—5 for average value; 1—2 for low value; 0 for no value	10
Cultural characteristics (15 points)	Intangible cultural heritage	10 for national intangible cultural heritage; 8 for provincial intangible cultural heritage; 6 for Weihai intangible cultural heritage; 4 for Rongcheng intangible cultural heritage; 0 for no intangible cultural heritage	10
	Historical events and figures	1 for each historical event or figure	5
Development potential(15 points)	Location	2 for close to city or town; 1 for having important traffic line; 0 for inconvenient traffic	2
	Population	2 for more than 1 500 people; 1.5 for 1 200 to 1,500 people; 1 for 900 to 1 200 people; 0.5 for 600 to 900 people; 0 for less than 600 people	2

Continued

Items	Assessment indicator	Scoring criteria and definitions	Full score
Development potential(15 points)	Demographic composition	2 for balanced composition of seniors, middle-aged, and young people; 1 for visible sign of aging population; 0 for serious aging population and few young people	2
	Vacancy rate	2 for vacancy rate less than 10%; 1.5 for vacancy rate between 10% and 20%; 1 for vacancy rate between 20%~30%; 0.5 for vacancy rate between 30%~40%; 0 for vacancy rate more than 40%	2
	Village collective income	2 for more than 10 million CNY; 1.5 for 2 million to 10 million CNY; 1 for half million to 2 million CNY; 0.5 for 200 000 to 500 000 CNY; 0 for less than 200 000 CNY	2
	Villager personal income	2 for more than 9 000 CNY; 1.5 for 8 000 CNY to 9 000 CNY; 1 for 6 000 CNY to 8 000 CNY; 0.5 for less than 6 000 CNY	2
	Infrastructure	0.5 for each infrastructure and 2 in total	2
	Conservation list	1 for national historical cultural village or national vernacular village; 0.5 for provincial historical cultural village, provincial vernacular village, provincial cultural heritage conservation unit; 0 for none of these	1

5 Conclusion

The paper explores how to assess the value of the vernacular villages based on the site selection, layout, buildings, culture characteristics, and development potentials of seagrass thatched houses by taking into consideration the field investigations of over one hundred vernacular villages of such houses and the existing problems. It further proposes the classified conservation strategies of overall conservation, partial conservation, and alleys conservation for vernacular villages of different values in the hope of contributing to the conservation of seagrass thatched houses. Conserving vernacular villages of seagrass thatched houses is a long-term and challenging job, asking for collaborative wisdom, intelligence, and effort to make further explorations.

Acknowledgements

Our special thanks go to Ms CHEN Jianfen for taking on the great labor of text proofreading on the manuscript of this paper.

References

LIU Huanyang, CHEN Aiqiang, 2015. General introduction to Jiaodong culture. Jinan: QiLu Press.

LIU Zhigang, 2008. Visiting China's rare vernacular dwellings: seagrass thatched houses. Beijing: Ocean Press.

WU Tianyi, 2008. Study on seagrass thatched houses in Weihai. Jinan: Shandong University.

YANG Jun, 2011. Regional dwellings choice of materials and applications: ecological seaweed house in Jiaodong peninsula for example. Architectural journal, S2: 152-155.

ZHANG Yong, 2012. Harmonious dwelling: windows and doors of vernacular dwellings in Qilu Area. Jinan: Shandong Fine Arts Publishing House.

Style Analyses of Mendai Hmong Village at Laer Mountains in Western Hunan

LONG Lingege, *GAN Zhenkun*, *ZHANG Dayu*

Beijing University of Civil Engineering and Architecture, Beijing, China

Abstract Mendai Hmong Village at Laer Mountains in western Hunan is a typical representative of traditional Hmong villages in mountainous alpine regions, reflecting the diligence and wisdom of the Hmong people. Firstly, through the analyses of the overall layout of villages, it is pointed out that the settlement as a whole is adapted to local conditions, forming a characteristic spatial form of stepped-up rivers, paddy fields, villages, and mountains, representing the Hmong people's cosmic view of harmony between nature and human. Secondly, the study carries out an in-depth exploration of street space and building construction. It points out that streets and lanes change with the changing terrain. Alleys of different width and buildings of varying height compose abundant and changeable areas. The main structure of the building is column and tie construction and building envelope materials are mainly adobe bricks and stones to adapt to the humid and cold climate in alpine regions. At the same time, the use function of different spaces in the Hmong village combining with the customs and habits of the nationality is also analyzed. Finally, stresses that as an essential carrier of national culture and farming culture, Mendai Hmong Village has bred distinct cultural connotations, and shows the unique charm of the Hmong village style and living scenes to the world.

Keywords La'er Mountains, Hmong Villages, human and geological features, harmony between nature and human

1 Introduction

As an essential carrier of Wuchu culture, the traditional villages of Hmong nationality in western Hunan are witnesses and bearers of the blending of minority culture and the Han culture for thousands of years. As a kind of material form culture, the traditional village style of Hmong is reflected in the natural environment of rolling hills and crossing rivers, in the layout of villages in accordance with the situation of mountains and precise sequence, in the street space with rough mineral and appropriate scale, in the wisdom of building houses that follow nature and are skillful in choosing, and also in the area and place closely related to the people's lives of Hmong nationality.

La'er Mountain Platform has been the core area where Hmong people live together in Hunan, Guizhou and Sichuan (Chongqing) border areas since ancient times. The construction of the Great Wall of the South in ancient China was to resist the local "seedling" people. At that time, the ruling class adopted the policy of military conquest and compulsory assimilation of the "living world," which triggered a large-scale uprising of the Hmong people, in which the Qianjia Hmong people's revolution was ignited in the Laer Mountains area. On the one hand, the Hmong people in Laer Mountains have been oppressed by the rulers for a long time, forcing them to form strong national centripetal force and national anti-foreign psychology. The Hmong people adopt the model of "one family with difficulties and one hundred families to help" at home, while they maintain a resentful attitude toward the outside world. As a result, they are simple, hardworking and sturdy. On the other hand, because of their remote location and extremely backward social and economic development, the ancient Hmong people were less influenced by the ruling class in ideology and culture and largely retained their cultural characteristics.

The traditional villages of Hmong nationality in La'er Mountains are the material carrier of history and culture. They record the values of Hmong people's desire for stability and pursuit of peaceful living, and also form the traditional village residential culture with regional and local flavor and become the social epitome of the self-sufficiency production and lifestyle of Hmong nationality.

As one of the typical traditional Hmong villages in the La'er Mountain platform area of Western Hunan, the Hmong village of Mengdai is a suitable entry point for studying the typical Hmong villages in Alpine mountainous regions. By analyzing its traditional features, we can get the main characteristics of the material space of the traditional Hmong villages in this region, as much as explore the history, culture, and value of the ancient Hmong village of Mengdai. Therefore, this study systematically analyzed the traditional features of the ancient Hmong village in the La'er Mountains by elaborating the characteristics of its landscape habitat, the village layout structure, historical environment elements, and residential construction.

2 Village Generating Environment

2.1 General Situation

Mangdai Village is located in Yayou Town, Huayuan County, Xiangxi Tujia and Miao Autonomous Prefecture, Hunan Province. It borders Fenghuang County in the east, Wudou and Xiushui Village in the south, Yama Village in the northwest and Gaowu Village in the northeast. "Mengdai" is a Hmong-language paraphrase, "Meng" refers to massive, "Dai" symbolizes broad. This village names Mengdai because it is near the Mount of Mengdai. Now under the jurisdiction of two native communities, the ancient village is situated at the foot of Mt. Mengdai, and the new one is at the foot of Mt. Mengdai. This research takes the ancient village of Mengdai as the primary research object (Fig. 1).

The ancient village was formed around the Yuan Dynasty, and gradually showed the scale of communities in the Ming Dynasty. At present, the population is about 400; the villagers are all Hmong, mainly engaged in traditional agriculture. Mangdai lives by the mountain situation, the overall style is coordinated and unified, the pattern is preserved intact, the stone road is winding, the stone walls stand on both sides, and traditional dwellings are scattered in different places. Stones enclose most of the houses. The corners of stone houses, the lintels of stone carved doors and windows are countless in Mangdai. Also, there are still many historical relics such as ancient wells, land temples, old bridges, ancient trees and so on, and intangible cultural heritage such as traditional Hmong sacrificial activities such as "Cone Cattle" and festival celebrations such as "Catching up with the Autumn Festival" are particularly abundant.

Fig. 1 Aerial view of Mengdai Village © GAN Zhenkui

2.2 Natural Environment

Mangdai Village is situated in Laer Mountain Platform of Wuling Mountains. It extends from Yunnan-Guizhou Plateau. Its elevation is about 800 — 1200 meters. The annual average temperature is 14℃. It belongs to the region's high-cold area. There are mountain winds in summer and sunshine in winter, and that forms a microclimate of warm winter and cool summer. The village is on the mountainside, and the whole terrain rises from east to west step by step. In the north, the Zhuiba River separates the Mandai Village from Gaowu Village, and the Mandai Stream converges from west to east to the Zhuiba River.

2.3 Natural Environment

In the aspect of historical change, Hmong people were forced to choose remote mountainous areas as living places because of the imbalance of social development caused by cruel war and tortuous migration. After the fifth Hmong migration in the Tang and Song Dynasties, the Hmong compatriots gradually formed relatively stable settlements in the Laer Mountains Platform, and the village of Mengdai became the birthplace of the "Long" surname of the Hmong people in the cold areas of Western Hunan.

In terms of cultural practices, the long-persecuted Hmong people place their hopes on religious beliefs. They worship nature, believe that all things are spiritual, and believe that gods have irresistible power. Therefore, in the construction of the homeland, they always maintain the awe of nature, which naturally forms the cultural concept of harmony between nature and human. At the same time, land represents the Hmong people's folk belief in land worship. Under the natural conditions of the mountainous Laer Mountains with few fields, the land is directly related to the survival and development of the group. Hmong people cherish the earth very much. Therefore, every year, they have to choose auspicious days to hold land worship activities.

In terms of social structure, the village of Mangdai keeps the tradition of one family name and one community, with simple social relations and a patriarchal-centered small family system. Families are mostly two generations living together. Children live separately from their parents after they get married. The living function and space requirements are simple, and they only need to meet the daily living requirements.

3 Spatial Form Characteristics of Villages

3.1 The Overall Layout

To facilitate living together and agricultural irrigation, the flat area near the Zhuiba River, is reserved for cultivation. Settlements grow out of the narrow zone between rivers and mountains and extend along the contours parallel to the mountains, with the overall shape of a curved belt. The terraces are distributed in the canyons, and in the distance are the clear running river, which constitutes the specific spatial form of water, fields, villages and mountains rising step by step(Fig. 2).

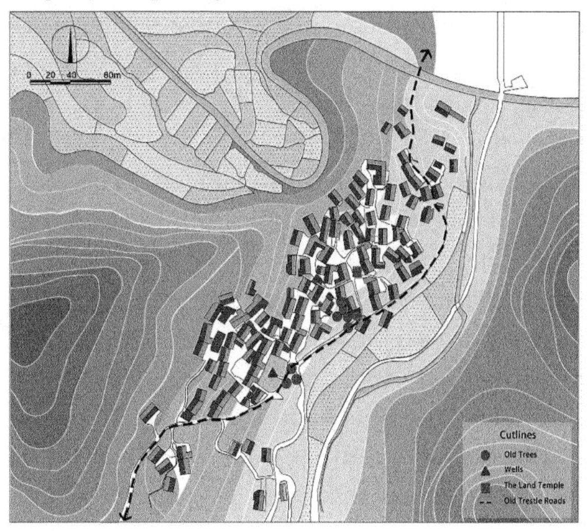

Fig. 2 The Overall layout of Mengdai Village © LONG Lingege

The first level is the Zhuiba River, the Mandai Stream, and the paddy fields. The paddy fields are along the direction of the river. Most of them are concentrated in the areas on both sides of the river, and a small part is terraced along the contour line. The second level is the ancient village of Mengdai, whose dwellings fall from west to east towards the bunny slope. Folk dwellings conform to the layout form of the terrain, not only reproduce the natural landform with the ups and downs of the buildings but also reduce the amount of earthwork in the process of building construction and avoid natural disasters such as landslides caused by constructive destruction. The third level is Mt. Mengdai, which hides the ancient village stockade in the dense forests and exposes little black tile roofs in the green ocean. Against the rolling hills and overlapping paddy fields, the village is integrated with the natural environment is integrated, rough, pure and simple, just like the character of the Hmong people, which embodies the inherent naivety of "primitive" and "wild", and keeps the internal relationship between the natural environment, people and villages.

3.2 The Spatial Organization Structure

The main road and three circular main roadways form the framework of the spatial structure of Mengdai. It creates the structural system of the settlement space together with the village public space such as a pond, land temple, wells, old tree forest, node space of traditional elements, and three dwelling groups.

The main streets and lanes are connected in series with private groups, which ensures the accessibility between groups. The long branches extending around the skeleton roads are related to the residential buildings in the groups. The pond is the catchment pool in the rainy season, and also a valuable living space in the village. The villagers enjoy the cool by the pond on weekdays. Every Hmong Festival celebration, it becomes a small theatre for lion and dragon dancing. The Land Temple, under a big Platycarya tree, is the place of worship for Hmong people to pray for happiness and work (Fig. 3).

Fig. 3 The spatial organization structure © GAN Zhenkui

Ancient wells are the source of water for raising the villagers from generation to generation. Old trees have witnessed the rise and fall of villages. Ancient land temples protect the peace and tranquility of communities. Old trestle roads and ancient city walls reflect the diligence and wealth of the ancestors of Mengdai. Although human beings make the space of the midpoint, line, and surface of the village, it is just like that created by nature. The artificial area is organically integrated into the landscape, presenting the construction philosophy of harmony between man and nature.

3.3 Street Space

The adaptation of the ancient village to the topography enriches the spatial level of communities, streets, and lanes with the changing topography. The street space is tortuous when the parallel contour line is used, and the street space is undulant while the vertical contour line is used. These changes are in harmony with the original landform so that people can continuously change the landscape in front of them during their journey, avoiding the monotony and dullness of a glance, and increasing the continuity of space. The buildings are arranged on both sides of the street, forming a plurality of street and lane side interfaces. They are enclosed by the gables or courtyard walls of the houses on both sides, surrounded by the houses and corn fields, and the street crossings with Hmong characteristics (Fig. 4).

The road paved with bluestone slabs, which is abundant in the area, is the main body of the space bottom interface of the ancient village streets in Mengdai. To better adapt to the terrain, village roads adopt a combination of ramps and steps. Most of them are lined with bluestone slabs, and both sides have horizontal pavement, which has a potential guiding effect on pedestrians.

Fig. 4 Street space © LONG Lingege

4 The Construction Mode of Dwellings

4.1 Spatial Form of Courtyard

The first step in the construction of residential buildings is to flatten the land.

The common practice of the village is to excavate the hill back, fill the excavated earth in front of the building, and build a retaining wall with stones in the periphery, eventually forming a flat platform to meet the needs of construction. The courtyards are arranged tightly, the layout does not pursue rectification, and the building orientation does not require sitting in the north to the south. The construction logic is flexible and changeable, showing a romantic, open, and flexible spatial form.

The courtyards of each family in Mangdai are composed of buildings and courtyard flats, with a small number of courtyard walls. The yards are mostly irregular platform spaces, commonly known as "Shai guping," paving stone slabs to facilitate the drying of grain, is a vital activity space for production and life. The courtyard organization mode is mainly "I" courtyard and "L" courtyard, and a small part is "U" courtyard. The "I" courtyard is a courtyard with only one main building, some of which have subsidiary buildings on both sides, side by side with the main house. The patio of "L" type has a right angle between the main house and the compartment. The functions of the main house and the compartment complement each other, and the internal space is interconnected. The "U" type courtyard is a kind of courtyard dam space which is formed based on the "L" type courtyard by adding compartments at the other end and is relatively close with the main roof.

4.2 Functional Layout

The dwelling takes a single house as a basic unit. The main building relies on Mengdai Mountain and opens doors and windows facing the front courtyard flat. The principal residences are three rooms with concave entrance, which has more design function. On the one hand, it has the geomantic meaning of "gathering treasures and making money", on the other hand, it also considers the use of functions, The relatively deep space under the eaves can be used as the space for daily rest and is conducive to the enclosure of wooden boards(Fig. 5).

Although the courtyard layout is free, the single building layout is orderly. The middle room is a hall house, which is a place for daily living, work, visitors, and sacrifices. In most cases, the hall house is not over-furnished, because many of the Hmong people's sacrificial activities are carried out indoors, and the hall room needs enough space for its use. The bays on both sides of the hall are side rooms, which are separated into front and back parts by thick cloth and black curtains. The front part of the right Bay is the kitchen, the front part of the left Bay is the fire pond, and the back part of the two side rooms is the bedroom. Apart from the need for ventilation and exhaust directly above the fire pond, the other two floors are usually used as storage space or as bedrooms for young people or guests.

The compartment is usually on the side of the main house. Single-story buildings are generally used as toilets or livestock railings. The first floor of the auxiliary rooms of double-story buildings is often enclosed with stones or boards, which are used for storing articles or raising livestock, and the second floor is used as dormitories or guest rooms of unmarried family members.

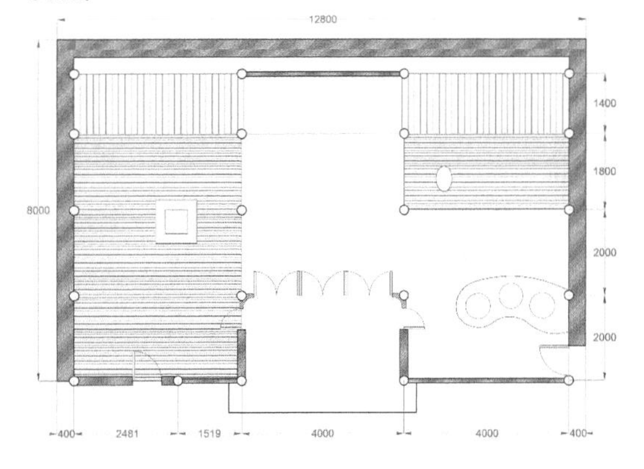

Fig. 5 Plan of Long Jun's house © LONG Lingege

4.3 Dwelling Structure

The residential buildings in Mengdai are built on the hills, and the buildings are made of local materials. The main structure of the dwelling is column and tie construction, which not only cost-effective but also lightweight and elegant. It has the characteristics of Chinese traditional wooden structure building, that is, The wall falls, and the house does not fall. The most common existing form of structure is five columns, eight short columns, and three rooms; there are also variants of " five columns, six short columns " and "three columns, six short columns. " There is no strict restriction on the number of columns and short columns and no uniform regulation on space and height dimensions of houses.

They can be adjusted according to the land use situation and the requirements of householders. Generally, they pursue favorable aspects. Also, to block underground moisture, residential dwellings will raise the indoor floor and make the bottom overhead about 30 cm above the ground (Fig. 6).

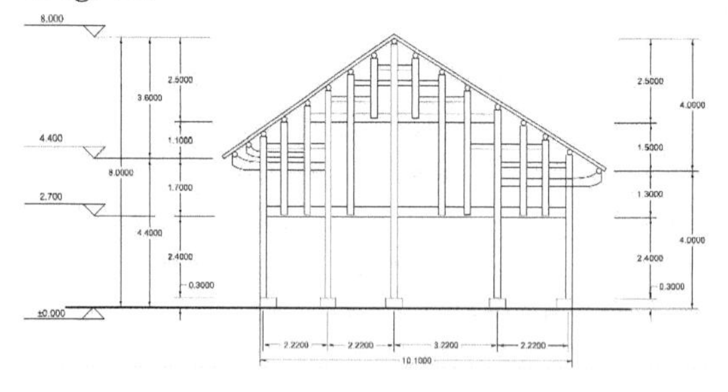

Fig. 6 Section of dwelling © LONG Lingege.

4.4 Building Enclosure Material and Decoration

In response to a wet and cold climate, the building envelopes are mainly using insulated and moisture-proof materials such as stone or adobe bricks which are abundant in the local area. The roof uses a "Cold Booth Tile Roof" method so that indoor moisture can be released from the roof as soon as possible. The overall color is mainly composed of the yellowish-brown color of stone and soil and the dark gray color of black tiles, which complements the surrounding landscape environment and exudes the charm of simplicity and agility.

There are some decorative elements in the window and roof, which are generous and straightforward. Geometric patterns separate the windows. The 2 ~ 3cm wide strips of wood are riveted in different ways to form a square, diamond, or polygon shapes, which are combined repeatedly according to specific rules and rhythms. The central part of the main roof ridge is pressed by "ridge flower. " Its primary purpose is to tighten the roof tiles to prevent the outside pieces from blowing away and loosening. It also achieves aesthetic attainments, these ridge flowers, which are made of black tiles, are assembled into triangles, copper coins, flowers, diamonds, and other patterns. They are the repository of the Hmong people's wishes for auspiciousness, well-being, prosperity, and longevity.

5 Conclusion

Influenced by nature and the human environment, the ancient village of Mengdai displays its unique characteristics from the village to dwellings. In terms of village construction, whether as a whole or as a single, they are subordinate to the natural scale and attached to the natural landscape, reflecting the worship of nature and the characteristics of mysterious Wuchu cultural. Terms of street space, the vast and narrow alleys and the high and low buildings constitute a productive multi-level space, emerging an intense Romanticism. In the aspect of building construction, the application of natural materials such as stone, wood, and earth is the original, rough, and primitive flavor of the dwellings. As a model of traditional villages in Laer Mountain area, the ancient village of Mandai is a concentrated reflection of the spatial form of traditional villages of Hmong nationality, showing the unique charm of the village style and living scenes.

References

GAN Zhenkui, LONG Lingege, 2017. Style analyses of Youma Hmong village at Huayuan county in western Hunan. Journal of uuman settlements in west China, 32(05):21-26.

GUAN Xueying, 2012. Research on the protection and inheritance of architectural art of Hmong dwellings. Beijing:Minzu University of China.

JIANG Baoyu, 2015. Research on the decoration of

traditional architecture at Fenghuang county in western Hunan. Xiangtan: Hunan University of Technology.
LI Sihong, 2009. Researches on characteristic of mountian villages and towns in west Hunan. Changsha: Hunan University.
LI Xiaofeng, 2009. Dwellings in Hunan and Hubei Provinces. Beijing: China Architecture & Building Press.
LI Zhe, 2011. The research on modern adaptation of traditional wooden houses of minority in western of Hunan. Changsha: Hunan University.
LIU Su, 2008. Dwellings in in western Hunan. Beijing: China Architecture & Building Press.
LONG Lingege, 2018. Spatial form analysis of Hmong traditional villages at Huayuan county in western Hunan Province. Beijing: Beijing University of Civil Engineering and Architecture.
QIN Zhongying, 2010. Family education and traditional culture inheritance of Hmong nationality from the perspective of anthropology. Beijing: Minzu University of China.
SHEN Hui, 2001. Research on the development of traditional settlements and traditional dwellings in Xiangxi area under the influence of social's progress. Changsha: Hunan University.
WEI Yili, 2010. Vernacular architecture in the west part of Hunan. Tianjin: Huazhong University of Science and Technology Press.
WU Jihai, 2015. Visit traditional villages in western Hunan. Changsha: Hunan Fine Arts Press.

Value of Cultural Inheritance of World Heritage Site Tajima Yahei Sericulture Farm

CHENG Sweet Yee, *Ono Satoshi*

Yokohama National University, Yokohama, Japan

Abstract Located in Shimamura, Isesaki city, Gunma prefecture, Tajima Yahei Sericulture Farm is a sericulture-space-cum-dwelling farmhouse built by Tajima Yahei in 1863 where high quality silkworm eggs were bred. The structure is a large-scale two-storey building with tiled gable roof built on a tall stone platform, the second floor is used entirely for sericulture activities, with openings on all four walls, and the roof is designed with a large ventilation system. The architecture style of this building is a concrete representation of the theory of *seiryō-iku* technique that was advocated by Tajima Yahei. This architecture style is a revolutionary result of Japanese sericulture farmhouses where sericulture activity making use of attic space of thatch roof dwellings was mainstream. However, the current second floor sericulture space does not retain the form when it was originally built. In other words, the sericulture space was built and renovated in response to the development of sericulture theories and technologies. Therefore, to investigate the original form of the building, it is vital to grasp as much as possible the process of modification and the reason behind it. This paper examines the above issue in detail. In addition, similar large-scale farmhouses exist densely around the Farm. The village and its surrounding landscape is substantial proof which reflects the prosperity of Shimamura area that had provided a large amount of high-quality silkworm eggs in the former major industry of sericulture in Japan. This paper will discuss the above points and the importance of village landscapes.

Keywords sericulture building, world heritage, preservation, value of inheritance, Tajima Yahei Sericulture Farm

1 Introduction

1.1 Sericulture in Japan

Silk production had started in Japan during the Nara Period from as early as 8th century. While Fukushima prefecture was an important sericulture region in early Edo period, sericulture in Nagano and Gunma prefecture had begun to develop around end of 17th century.

The innovation of breeding new silkworm species was considered at its peak around mid-18th century and several new hybrid Mulberry leaf species were introduced into sericulture. Sericulture techniques were then introduced early 19th century and small-scale silkworm rearing became popularized as it did not require a large plot of land and can be carried out in attic or mezzanine floors of a peasant's dwelling.

Raw silk was said to account for 65% of Japanese exports when the Yokohama port was opened in 1859 and the 19th century was a dynamic period where it marked the mechanization of the silk industry in Japan.

1.2 Regional Characteristics of Shimamura

Shimamura is located on the floodplains of midstream of Tone River. People have been engaged in water transportation and upland cropping. Croplands are damaged by flood every year and sandy land had limited the types of crops.

These disadvantages are in fact an advantage in Mulberry cultivation. A vital point in Mulberry cultivation is that the breeding of Uzi fly can be suppressed by a flood. These flies lay eggs on Mulberry leaves and its larvae are parasitic in silkworms, after ingesting the eggs, the larvae will eat through the cocoon. While Mulberry cultivation can be carried out even on sandy soil, the distribution of silkworm eggs and cocoon through water transport was also merit.

1.3 Silkworm Egg Production in Shimamura and Tajima Yahei

Taking advantage of regional characteristics, leading farmers in Shimamura began to work on silkworm egg production in the late 18th century. Incorporating production techniques learnt from advanced sericulture area of Date region in Fukushima prefecture in early 19th century, Shimamura had grown into the core of silkworm egg production in Japan by late 19th century. One of the leading figures of prospering the village of Shimamura is Tajima Yahei (second generation).

From an early age, Yahei travelled with his father to advanced sericulture areas and gained skills and experience in silkworm breeding, production and sales which lead to his establishment of a silkworm production theory.

2 History of Tajima Yahei Sericulture Farm Omoya

2.1 General Sericulture Farmhouses at the End of Edo Period

An overview of existing sericulture farmhouses across the nation in 18th and mid-19th centuries shows that these farmhouses were developed from attic or mezzanine floor of the omoya or main building used as a sericulture space (Miyazaki; Kurotsu; Toriumi; Takahashi; Ono; Nakamura, 2011). Depending on areas, mezzanine floors where modified into proper second floors.

While thatch was mainly used as roofing material of sericulture farmhouses to create a large attic space, plank and bark roofs with gentle slopes were also popularized.

2.2 Devising a Sericulture Space in Tile Roof Storage Room

Common knowledge regarding sericulture farmhouses is that sericulture can be carried out most effectively under a thatch roof, second by a plank roof, explained by Yahei in a sericulture guidance book.

While tile roof is fire resistant as compared to thatch and plank, heat accumulating under the tile roof space is considered unsuitable for sericulture. However, Yahei had desired to build the omoya with a tile roof for fire prevention purposes as his farmhouse had experienced fire twice. Also, to fulfil the social desire of using high-quality tile roof as wealthy farmers do.

As a result, Yahei built a two-story storage in 1856 and used the second floor as a sericulture space. Sericulture in the first year was not successful and Yahei had an epiphany that interior of the sericulture space should be kept as close as possible to natural climate. In the following year, two monitors were added to the roof and he had produced good-quality cocoon due to increased ventilation efficiency. In 1858, openings were expanded and windows installed on all four sides, further success was achieved. Based on this experience, Yahei in his *seiryō-iku* theory, believes that to nurture healthy silkworms and high-quality cocoons, it is vital to consider the ventilation of sericulture spaces through a roof ventilation system and openings on four sides.

2.3 Technological Innovation and New Sericulture Space in the Current Omoya

Based on the successful storage space, Yahei then built a two-storey farmhouse in 1863. The residence is the omoya that exists today, the second floor is a large room spanning 25.4m by 9.4m; front and back fa ade facing East and West respectively are installed with windows. A 1.9m wide monitor was built on the 25.4m long tile roof, with windows installed on both sides of the monitor for ventilation purposes. This roof ventilation system is described as a *mado* (window) or *nukimado* (window vent) in Yahei's book, but is now commonly referred to as *sō-yagura* (combined monitor) in Shimamura. While Yahei tried experimenting the storage with *futatsu-yagura* (two monitors), the omoya was built with *sō-yagura*, a long yagura spanning across the ridge of the roof (Miyazaki; Kurotsu; Toriumi; Takahashi; Ono; Nakamura, 2011).

Based on his *seiryō-iku* theory which prioritizes indoor ventilation, the 240m^2 large sericulture space on second floor of the omoya was installed with windows on four sides and a *sō-yagura* to control indoor temperatures and sunlight into the sericulture space through opening and closing of windows of the sericulture space and *yagura* (Isesaki City Board of Education, 2012).

In 1873, a large-scale two-storey building specially dedicated for sericulture known as *shin-sanshitsu* was built located East of the *omoya*. Then, the second floor of omoya and shin-sanshitsu was connected by a covered corridor. Although the *shin-sanshitsu* was demolished in 1953, a part of the covered corridor was remained connected to the *omoya*.

2.4 Remodeling of the Omoya Sericulture Space and Its Implications

When the connecting corridor was installed during the construction of the *shin-sanshitsu*, east windows of the *omoya* sericulture space were renovated into a side entrance. This is considered to be the first modification of the *omoya* sericulture space (Ono; Kurotsu; Takahashi; Ishii, 2016).

Many modifications have been done to the *omoya* sericulture space up until today as shown below.

① Windows on the front facade was modified into full height windows with transom windows or *ranma* installed above; balcony and handrails were installed (Fig. 1). A 0.5*ken* and 1*ken* opening located west of the front facade is combined into a 1.5*ken* wide opening*. This modification was for the convenience of transporting large tools directly in and out of the second floor from the front and used as a place to dry straw mats used in sericulture.

② Height of rear windows was increased and ranma was installed above the windows.

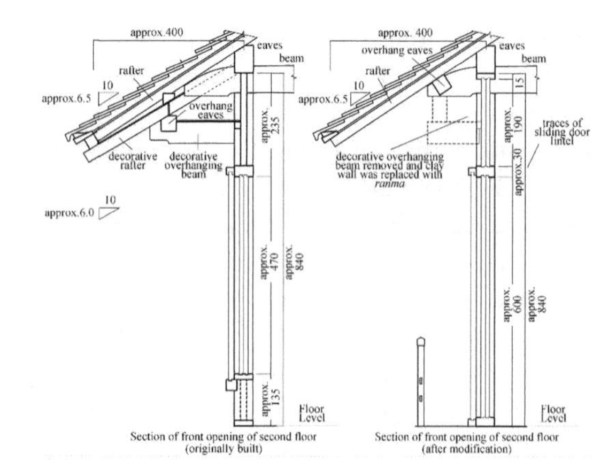

Fig. 1 Section of the front opening of the second floor of Tajima Yahei Sericulture Farm © second author

③ In order to modify the ranma above the front and rear windows, lavishly-patterned decorative rafters were substituted with simple-patterned decorative rafters. In other words, to enhance openings, the architectural style was sacrificed.

④ For east-facing windows, the threshold was lowered and sill height increased to expand area of the opening to ensure its opening is similar to west-facing windows. Ranma was not installed on the east and west sides.

⑤ West-facing windows are assumed left disused around 1925 when the shinzashiki was built attached to the east of omoya.

⑥Bracing was installed to certain walls and the roof truss. This modification may be due to the Kantō earthquake in 1923, however it is unconfirmed.

⑦ Addition of microscope room on the rear east side. Microscopes were purchased by Yahei in late 1870s during his voyage to Italy to sell silkworm eggs, it is believed that microscopic examinations have been performed since then, however it is uncertain when exactly was the microscope room added.

⑧ The second floor sericulture space was originally a single room but was later

* *Ken* is a traditional Japanese measurement system used for column spacing. In this case, 1ken is equivalent to 1.879m.

partitioned into smaller rooms with ceilings. A 1*ken* passageway was partitioned on the east, a 0. 5*ken* passageway at the rear, and the remaining 12. 5*ken* by 4*ken* was divided into 6 rooms. The frame and additional columns are remained and is assumed that the partition was made of Shōji sliding doors and ranma Shōji, a duckboard ceiling has also remained. There are also evidence of weatherstripping sealants found on the ceiling and ranma partition and assumed that these rooms were divided and were temperature controlled using thermal power. This indicates that at a certain period, other than *seiryō-iku*, the secchu-iku technique was also carried out in these rooms.

The above modifications listed are organized as follow.

First, devices which increase ventilation efficiency of the sericulture space (①, ②, ④). This modification was executed while architecture style was sacrificed, this proved the devotion in improving the function of the sericulture space in accordance to the seiryō-iku technique. At the same time, the modification of front windows (expanded openings and installation of balcony and handrails) had largely contributed to improving the efficiency of silkworm breeding (①). Furthermore, the addition of the microscope room is important to ensure the quality production of silkworm eggs through examinations (⑦).

On the other hand, west-facing openings disused around 1925 (⑤) meant that Yahei had corrected his theory of having openings on four sides of the sericulture space. Also, the division of the sericulture space using passages meant the recognition of secchu-iku technique which also emphasizes temperature control (⑧).

Installation of bracings on walls and roof truss showed positive effort to ensure safety of building (⑥).

3 Relationship Between Tajima Yahei Sericulture Farm and Surrounding Silkworm Egg Production Farmhouses

3. 1 Yahei's Former Residence and Its Vicinity

Currently the Tajima Yahei Sericulture Farm, mainly referring to the omoya (main residence), related historical buildings such as the Shoin (drawing room), Mulberry leaf barn, bessō (summer house), a roofed water well, silkworm egg storage, a front gate, an East Gate, a village shrine and sericulture equipment storehouse are remained and is altogether designated as a Historic Site in 2012. In 2014, it became a component of the World Heritage "Tomioka Silk Mill and Related Sites".

Located nearby the Tajima Yahei Sericulture Farm are several farmhouses built between late Edo period and early Meiji period (Fig. 2). These farmhouses share similar basic features of a large, two-story tile roof building with a roof ventilation system, but are different in terms of its shape, position and modification methods of openings, windows and roof ventilation systems.

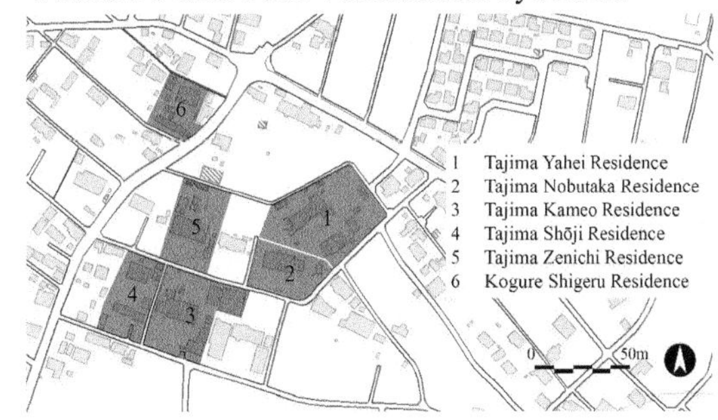

Fig. 2 Location of residences in Shimamura. Source: map from Geospatial Information Authority of Japan © first author

3. 2 Features of Yahei's Former Residence from Perspective of Similar Remains Nearby

Matters pointed out in the previous section will be explained through concrete examples as follows.

There are two examples of two-story sericulture space with openings on all four sides which are similar to Yahei's former residence, these are Tajima Nobutaka residence (built in 1863) and Tajima Tatsuyuki residence (built in 1866). These three farmhouses located next to each other are closely related and may have devotedly carried out Yahei's theory. Nobutaka residence and Tatsuyuki residence are larger than

Yahei's former residence as they have more windows on front and rear of the sericulture space. East-facing openings were disused when a shed was attached on that side of the Tatsuyuki residence, Nobutaka residence is the only example which persisted with having openings on four sides until its retirement from sericulture in 1980, the yagura was removed and omoya was scaled down. There are many examples which residences were originally built with openings on three sides, these include Tajima *Shōji* residence (built in 1861), Tajima Zenichi residence (built around end of Edo period), Tajima Kameo residence (built around 1868) and Kogure Shigeru residence (built between 1874 and 1877). The west wall of the sericulture space in Kameo residence is an *ōkabe* (finished wall with no exposed columns). It is believed that it was done for fire protection due to winter monsoon winds and temperature rise due to west sun. Generally, sericulture farmhouses with openings on front and back facades are considered a national standard but openings on three sides are considered rare, while in Shimamura, farmhouses with openings on three sides are standard while four sides are rare. Other than Yahei's theory, it is believed that farmhouses in Shimamura have also received influences from others (Ono; Mizuno; Arai; 2017).

Next, the roof ventilation system of the yagura. Tatsuyuki residence and Zenichi residence adopted the sō-yagura while Kameo residence, Shōji residence, and Shigeru residence adopted a *mitsu-yagura* (three monitors) (Fig. 3). In the case of Nobutaka residence, it is assumed that the roof was built with plank roofing without a yagura in 1863. However in 1870, it was replaced with *tile roofing* and a *sō-yagura*. Around 1897, the *sō-yagura* was replaced with six cylindrical vents. Later around 1907, the vents were replaced with mitsu-yagura as observed from archives and old photographs.

Regarding the expansion of windows, residences where full height windows, balcony and handrails were installed on the front facade similar to Yahei's former residence include *Shōji* residence and Zenichi residence, whereas Nobutaka residence and Tatsuyuki residence added a corridor on the ground floor which its roof acts as a balcony for the second floor. Contrarily, front windows of Kameo residence and Shigeru residence remained its original form.

With regards to partitions and ceilings in the sericulture space, it is certain that all of the above examples are retrofits and were originally single rooms. However, it is still uncertain when the installation or modification took place.

Residences where structural reinforcement such as bracings were added include Shōji residence, Nobutaka residence, Kameo residence and Shigeru residence. Amongst them, Kameo residence reinforced its floor, frame and roof trusses with iron hoops.

Regarding Tatsuyuki residence, an eaves was extended on the east side of sericulture space and a microscope room was added to the rear in 1928.

These farmhouses around Yahei's former residence were gradually built one after another between late Edo and early Meiji period, while basic features of these farmhouses are similar to Yahei's former residence, all farmhouses are not identical and are unique on its own.

3. 3 The Completed Form of Sericulture Space with Accordance to *Seiryō-iku* Technique

From examples in the previous section, there are many unclear points as to how long the *seiryō-iku* continued to be carried out in Shimamura, when was the peak of the *seiryō-iku*, and when did Shimamura started to adopt the *secchu-iku*.

Amongst them, the Kurihara Toshishige residence located in the west neighbouring village was proved to have been rebuilt in 1887. It is clear that the sericulture space was originally a single room but was partitioned into smaller rooms with ceilings, the sericulture space was originally built with

Fig. 3 Types of roof ventilation system on sericulture farmhouses in Shimamura

(Above: *sō-yagura* on Tajima Yahei Sericulture Farm omoya; Below: mitsu-yagura of Kameo Residence omoya and futatsu-yagura of Kameo Residence san-shitsu). © first author

openings on three sides, excluding the west, full height windows, *ranma*, balcony, handrail were installed on the front. This farmhouse is an ancient branch family of the Kurihara family who was the first to fully engage in sericulture in Shimamura. It is certain that they have been engaged in sericulture since the end of 18th century and has been a powerful silkworm egg production family.

From the above, the trend of silkworm egg production farmhouses as of 1889 could be understood as below.

① Sericulture space built according to *seiryō-iku* theory was mainstream.

② Installation of full height windows, balcony and handrails on the front of the sericulture space was popularized.

③ *Ranma* were installed above the windows on the front facade.

④ Openings were not installed on the west facade.

Toshishige residence is the complete form of a silkworm egg production farmhouse built in accordance to the *seiryō-iku* theory. However, there are still unanswered questions such as whether the combination of *seiryō-iku* and secchu-iku was carried out at the same time and the details of partitioned rooms. The adoption of *secchu-iku* technique and its actual condition will be examined in future researches.

3.4 Necessity of the Existence of Surrounding Remains as a Whole

As describe above, it is impossible to grasp the actual condition of farmhouses and its relationship to the beginning, development, and transition of seiryō-iku technique by looking only at Tajima Yahei Sericulture Farm, and therefore is essential to do comparative studies of the surrounding remains. And although the surrounding remains may seem similar at first glance, the original form and history of each remains has its individuality. Important buildings such as omoya are built on a tall foundation for flood prevention, other buildings such as gates, storehouse, Mulberry leaf barn and water well blends in well with the hedges, creating an attractive landscape. A couple of these attractive farmhouses form a captivating village and Shimamura is made up of a couple of these captivating villages. In this aspect, we can observe the characteristics of Shimamura which played a core role that supported the silk industry during the early modern period of Japan.

4 Preservation and Utilization of Sericulture Farmhouses in Shimamura

4.1 Current Plans and Proposals for Sericulture Farmhouses in Shimamura

The nomination of a buffer zone is to add an additional layer of protection to the actual site (Agency for Cultural Affairs, 2013). While Yahei's former residence is a World Heritage Site and Historic Site and is receiving financial assistance, remaining residences are located in the buffer zone and is not entitled to financial assistance for maintenance. These residences may have changed due to natural deterioration and disasters. Therefore, to receive financial assistance for maintenance, these residences are under the process of applying for designation of National Cultural Properties.

All properties are privately owned and are currently inhabited. Restricted areas and areas open to the public are discreetly segmented to protect the owner's privacy but also to allow visitors to experience the spaces. Regardless, the owners are cooperative to utilize the sites to promote cultural activities. In celebration of the annual chrysanthemum exhibition, Yahei Residence was used as a location to house arts and craft workshops and events such as handcrafting cocoon straps, pin badges and 1 : 125 scale paper model of the omoya. Mulberry leaf barn of Yahei Residence is currently a gallery to showcase information regarding the invention of *seiryō-iku* theory. The public is allowed to enter the internal space of *sō-yagura* in Zenichi Residence (see Fig. 4).

Fig. 4 Internal space of *sō-yagura* in Zenichi Residence. © first author

Discussion and meetings between locals, researchers and university students are arranged extensively to discuss preservation and utilization ideas for the residences. Proposals of utilization ideas for Kameo Residence was carried out in spring 2018 by students of Yokohama National University. Ideas proposed include reusing the san-shitsu as a community space for locals and tourists, local produce processing factory and to house club activities. The Shimamura Harvest Festival and a harvest schedule were proposed where visitors may experience harvesting local produce (Fig. 5) in different seasons. Walking tours and stamp rally events with the residences as pit stops for visitors to explore the village by foot.

Fig. 5 Students experiencing corn harvest during Summer in Shimamura. © the first author

4.2 Revitalization of the Village of Shimamura

Infrastructures in Shimamura are replanned to accommodate to education and tourism purposes include reusing the Shimamura Primary School as a visitor information centre. A Sakai Shimamura Hospitality Plaza and the Ferry Boat Fiesta was set up to sell local produce and local handcrafts. While people in the Taisho period used to travel up north by crossing the river, today, free boat-rides allow visitors to cruise through Tone River enjoying the scenic view of Shimamura.

5 Future Prospect and Conclusive Remarks

With Yahei Residence as a World Heritage Site, the awareness of Shimamura locals on the importance of preserving their local history and culture has increased ever since. With ongoing preparation for the designation of these residences as Important Cultural Properties, this successful designation will pave way for the designation of the entire village as a Preservation District for Groups of Historic Buildings. Maintenance works of buildings in the site of these residences will be carried out as soon as possible while utilization plans are currently under division. Next, local tourism resources should be looked into such as reuse of sericulture related buildings for tours, workshops, accommodation, etc.

In light of the above, it is necessary to

recognize the cultural value of not only the Tajima Yahei Sericulture Farm but also the surrounding silkworm egg production farmhouses, and to recognize the importance of sustainability of the omoya and its landscape. And all together, the preservation and inheritance of rural landscape were devised, and these became the characteristics of Shimamura that was highly valued as a World Heritage. In other words, Shimamura is the "region which had created a technological innovation for the production of high quality silkworm egg species" and should be conveyed to future generations.

For that to happen, it is necessary to continue the investigation of these silkworm egg production farmhouses and their relationships with one another.

References

Agency for Cultural Affairs, 2013. Justification of inscription. Tomioka silk mill and related sites: world heritage nomination,13-164.

Isesaki City Board of Education. 2012. Survey report of Tajima Yahei's former residence. Isesaki: Isesaki City Board of Education.

MIYAZAKI T,KUROTSU T,TORIUMI S,et al., 2011. Buildings of Shimamura: Survey report of Sakai Shimamura sericulture farmhouses. Isesaki: Isesaki City Board of Education.

ONO S, KUROTSU T, TAKAHASHI M, et al., 2016. Improved process of window ventilation of second floor of sericulture farmhouses built in the late 19th century in gunma prefecture shimamura. AIJ Journal of technology and design, 22: 777-782.

ONO S,MIZUNO T,ARAI Y, 2017. Silkworm egg production farmhouses in Shimamura (Isesaki city). Yokohama: Yokohama National University Ono Satoshi Laboratory.

The Value and Conservation of Rural Landscape in Mountain Area of Zhaoyuan

WANG Shengnan, *WANG Jianbo*

Shandong University, Jinan, China

Abstract According to the elaboration on the heritage value of rural landscape in the *ICOMOS-IFLA principles concerning rural landscapes as heritage*, at the western foot of mount Luo, the rural landscapes are divided into two parts, according to if they are located in the villages. The rural landscape which is outside the village is mainly composed of stone weir terraced or orchards collectively built in the 1970s with agricultural water conservancy facilities and ancient persimmon trees retained since the Ming and Qing dynasties. The rural landscape in the village is centered on the rural settlements and includes other ancillary buildings, for example, the Earth temples, forming the traditional architectural, street and public space landscape with local stones as materials and reflecting people's belief in the land. These two parts are interrelated and unified, reflecting the interaction between humans and in the development process, forming the unique rural landscape. However, the reduction of rural nature population not only leads to the loss of the labor force, but also results in the brain drain. Nowadays, the unscientific development mode, such as the excavation of mountains, the cutting of ancient trees, the demolition and reconstruction of traditional buildings, has brought threats and challenges to the local rural landscape. In view of these factors, the fundamental problem is to solve the problem of rural development. First of all, the villagers should be encouraged to actively establish villager organization and participate in the development work of the village. At the same time, the knowledge of rural landscape heritage and successful rural development cases should be widely popularized among rural areas. And then, the government should act as the leader, on the basis of providing policy guarantee, to integrate social resources to promote the development of rural economy and community. The ultimate goal is to protect the overall environment of the rural landscape with the exploration of the best combination of nature and human development.

Keywords rural landscape, heritage value, Zhaoyuan city, mountain area

1 Introduction of the Village in Mountain of Zhaoyuan

1.1 Introduction of Zhaoyuan City

Zhaoyuan is located in the west of Yantai city, in the east of Shandong province (Fig. 1). During the Han dynasty, it was part of Qucheng county. Zhaoyuan county was set up in the ninth year of Tianhui in the Jin dynasty (1131). Zhaoyuan city was established in 1991, continuing to use the original name. County Explanation records "Zhaoyuan means to attract people from afar." (Jia &Li, 2005) The terrain here is dominated by mountains and hills. Luo mountain in the northeast is the highest peak of Zhaoyuan city. And its scenery is beautiful, lofty magnificent. (Niu &Bai, 2006) It is located in the warm temperate monsoon zone, which is a continental semi-humid climate, four distinct seasons, significant heat and cold. (Shan, 1991) Here only summer precipitation more, but also with a drought. In addition, the surface rivers are mostly seasonal in Zhaoyuan. (Shan, 1991) Therefore, for the peasants, most of them basically live on the weather. Moreover, it is recorded in Zhaoyuan county annals written during the reign of emperor Shunzhi of the Qing dynasty that "Zhao is a remote place with many mountains. Its soil is barren and its people are poor. The customs are honest and simple." (Zhang &Zhang, 1846) The study area of this paper is the Kouhou area of the western foot of the Luo mountains. As Zaoyang top east extension of the vein of a mountain pass on a stack of three stacked stones, so this

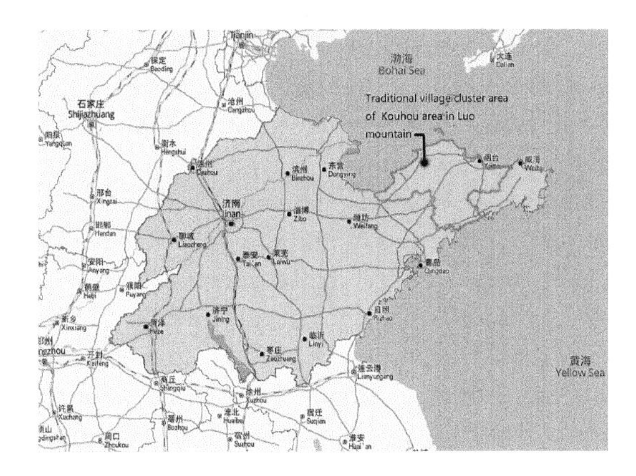

Fig. 1 Location of traditional village cluster area of Kouhou area in Mount Luo © http://www.gscloud.cn/search

mountain pass is called Duoshikou. The northern part of the Duoshikou is called the Kouhou area.

1.2 The Problems of Rural Landscape in Zhaoyuan

With the rapid development of cities in recent decades, the hollow phenomenon appears in the countryside. Thus, the countryside lacks vigor and vitality. No one is willing to inherit traditional knowledge and production skills. And the cultivated land is facing lying waste. In addition, the natural environment the local villages are dependent on is constantly being consumed and undergoing irreversible changes. (Ren &Han &Yang, 2018) For example, many quarries were built in Kouhou area. Moreover, the countryside is faced with homogenization of landscape, including spontaneous imitation and top-down unified "planning". These activities not only damaged the local rural landscape, but also affected the local development.

2 The Value Analysis of Rural Landscape in Mountainous Area of Zhaoyuan City

2.1 Analysis on the Value of Rural Landscape Heritage Outside the Village

According to the definition of rural landscape heritage in the *ICOMOS-IFLA principles concerning rural landscapes as heritage*, at the western foot of mount Luo, the rural landscapes are divided into two parts. The rural landscape which are outside the village is mainly composed of stone weir terraced or orchards and agricultural water conservancy facilities collectively built and ancient persimmon trees retained since the Ming and Qing dynasties. The value of its rural landscape is shown in the following aspects.

Firstly, it was the product of the period from the late Qing dynasty to the People's commune. As early as the end of the Qing dynasty, some villages began to renovate their land. In 1955, the pilot project began, and the whole county entered the climax of land renovation in 1965. To the end of the 1980s, the Kouhou area formed a stone weir terraced landscape. Besides, persimmon trees have been planted since ancient times. In the 1950s, villagers planted the trees on hillsides, beside ditches and dikes. (Shan, 1991) There are still thousands of persimmon trees over a hundred years old. In addition, irrigation and water diversion projects were carried out simultaneously. The Shangyuan reservoir was completed in September 1966 and the Chenjia reservoir in June 1977. The main canal starts from Chenjia reservoir, passes 12 hilltops, 37 river trenches and ends at the west wall of Hulongdou. The total length is 18.5 kilometers, forming the joint irrigation area of Shangyuan-Chenjia reservoir. (Shan, 1991) These structures are still standing today.

Fig. 2 Terraced orchard of Kouhou Hanjia village © WANG Jianbo

Secondly, it is a mountainous rural agricultural landscape with Jiaodong characteristics. First of all, this area is mainly stone weir terraced orchards. The terraces are made of locally abundant yellow stone and are mainly distributed around villages. Terraces are planted with apricots, peaches and apple trees. The entire terraced fields are stacked on top of each other, grand and spectacular (Fig. 2). In spring, flowers bloom like snow and sea. In addition, there are patches of persimmon forests in the Kouhou area, such as persimmon forests in Jiegou Jiangjia village and persimmon forests in the north terraced fields of Kouhou Hanjia village. And there are scattered in the mountain stream, slope, terraces in the ancient persimmon. In late autumn, leaves fall down, leaving only the branches and red persimmon dye all over the mountains. Also, the reservoir blends into the surrounding natural landscape. For example, Xijian reservoir of Kouhou Hanjia village is built with local mixed stones for slope protection. The reservoir is surrounded by high and steep mountains, fantastic rocks and waterfalls. (Shan, 2014)

Thirdly, it is a successful example of local residents' efforts to improve the environment. The reservoir meets the local agricultural needs, improves the living environment, and successfully integrates into the villagers' lives. Today, for example, Xihe reservoir in Zhaike village is not only a place for poultry to play, but also a place for women to wash clothes (Fig. 3).

Fig. 3 Xihe reservoir of Zhaike village and Mount Fenghui © ZHANG Piji

Importantly, these landscapes are dependent on the protection of the natural environment. The area is mainly composed of Hanyang mountain, Fenghui mountain, Shanqian mountain, Hucun mountain and so on. Here steep mountains, crisscross the peak, rock stand, dense forests, streams sweet, sometimes birds and animals. Historically, Duoshikou is a famous pass on the ancient post road from the southern plain to Huang county. (Shan, 2015) In addition, there are historical relics of different periods on the mountain, such as the polder wall built during the Nien invasion to the east, the ruins of Yaowang temple, and the water conservancy aqueduct and water tunnel in the period of the people's commune.

Furthermore, from time immemorial, Chinese landscape has been closely related to local culture. Accordingly, each hill or mountain has its name and relevant common saying. For example, there are Fenghui mountain, steamed bun peaks, tiger mouth named according to the shape. There are Huangmu peaks and Wangfu stone named after folk tales. And by the name of the surrounding landscape, such as Xishan depression and Xijian depression and so on. In addition, there are also three mountains, nine peaks and twelve hollows in Zhaike village, which express the site selection and Fengshui advantages of the village. Moreover, there are related legends, such as the legend of the ship-shape stone. Legend has it that in the second year of Hongwu in the Ming dynasty, it floods occurred in Kouhou area. As the waters receded, a wooden boat ran aground and then turned to stone. Now, every mountain god festival, people will come here to offer incense and sacrifice. (Shan, 2018)

In China, many folk stories and traditional customs are based on its landscape environment. The villagers name them according to their experience, making the natural landscape a part of life. These names are the unique characteristics of rural areas and are closely related to their life and spiritual pursuit.

2.2 Analysis on the Value of Rural Landscape Heritage in the Village

Due to inconvenient transportation and slow economic development, there are still a large number of traditional villages. Therefore, the rural landscape in the village is centered on the rural settlements, forming the traditional architectural landscape with local stone as the material, the traditional street and public space landscape, and the ancillary architectural landscape of villages such as the earth temple. The analysis of its heritage value is mainly carried out from the following aspects.

Firstly, in terms of architecture, its rural landscape is mainly reflected in the appearance and style of architecture. The area abounds in marble and yellow stones. Early dwellings were built of warm yellow stone, which was simply treated. From the end of the Qing dynasty, the villagers turned to use well-cut blue-white marble to build houses. The buildings are scattered along the topography, simultaneously spreading outward around the center of the clan, or along the river. In addition, the alley is paved with yellow gravel, pebbles or marble. Moreover, villagers often build fences or vegetable (flower and grape) stands outside their buildings, and build them into shapes. Green vines coil around the eaves and outer walls. In this way, *hutongs*, buildings and plants are integrated.

Fig. 4 The Guaiguai *hutong* of Xujia village. © WANG Shengnan

Secondly, there are many kinds of streets and alleys. For example, the Guaiguai hutong in Xujia village, built for the purpose of enhancing the defense function, is distributed in the north-south direction and is long and far-reaching (Fig. 4). At present, it is mainly concentrated in the north of Hou street of the old block of the village. Chuanli Linjia village retains several alleys with gate and north-south longitudinal distribution. The alley is not only sheltered from the wind and rain, but also sheltered from the sun. The overall terrain of Zhaike village is high in the north and low in the south. The old block is located on the platform in the northwest of the village. Because it is built along the east-west river and along the terrain, these long streets are mostly east-west. Due to the topography, walking in the alleys, you can fully appreciate the surrounding mountains and village landscape.

Finally, the public space is probably the most rural landscape. For example, the public space centered on the great pagoda tree (Fig. 5) and the Guandi temple in Chuanli Linjia village, the public space of the site of Guandi temple in Zhaike village, and the public space of the ancient persimmon tree in Xujia village. Most of them are the public space along the main street or river and the ancestral temple. For example, the folk dwellings are built along the Y-shaped river course in Zhaike village, and white marble is placed outside their courtyards for sitting and resting.

Besides, the entrance to the village is dotted with earth temples and other ancillary buildings. The local earth temple is built of marble stones and is surrounded by pine and cypress trees. The temple of Kouhou Hanjia village is located in the village east cliff slope. There used to be three cypress trees around the temple, but now they are dead, with only two branches left. The original earth temple was demolished in the 1970s. The existing earth temple is made of granite and the roof is made of stone and four corners pavilion roof (Fig. 6). The statue of the land grandpa and land grandma and the stone incense

Fig. 5 The public space of the great pagoda tree in Chuanli Linjia village © WANG Shengnan

burner are still in the temple. (Shan, 2018) Earth temples are a legacy of China's traditional agricultural society. It is the hunger for a good harvest in agriculture at a time of low productivity. It is not only a simple traditional folk belief, but also the god worship of the closest to the village.

Fig. 6 The earth temple of Kouhou Hanjia Villag © Shengnan Wang

In general, the rural landscapes within these villages are interconnected. The buildings are arranged and help to form alleys and public spaces. The vegetation decorates the street buildings and public Spaces. These materials are taken from nature and built in accordance with nature. Therefore, the value of the rural landscape in the village lies in: first, it is the harmonious creation of human to the environment and the reflection of the traditional view of the unity of nature and man. Second, it is the unique landscape features for this area about the building materials and street forms. Third, the traditional landscape is the reflection of the local historical development, and the reflection of one or several families seeking for a suitable living environment and sustainable development. In a word, they are closely related to the life of residents, the memory of generations. And they are the product of the historical period at that time and the reflection of the traditional clan-centered settlement rural landscape.

3 Suggestions on Rural Landscape Protection in Zhaoyuan

3.1 Introductions to Relevant Protection Concepts and Practices

The theory and practice of rural landscape conservation can be used for reference from the theory of rural conservation. For example, community planning in Japan, namely the idea of creating community atmosphere, preserves local characteristics, shapes settlement forms and improves the quality of living environment from the perspective of (grassroots) community participation. (Nishimura, 2007) Subsequently, Taiwan borrowed its ideas and combined them with traditional Chinese rural society to develop a rural regeneration plan. (Peng, 2018) Germany, on the other hand, is an integrated strategy of rural renewal guided by the concepts of urban-rural equivalence and sustainable development, historical and cultural protection. Integrate existing laws, regulations, financial support and policy tools from the government level to promote the overall development quality of rural areas. (Yi & Schneider, 2013) The British government tries to maintain a high-quality environment and quality of life in rural areas through piecewise and orderly policy guidance, attach importance to rural characteristics, protect the rural natural landscape and continue historical environmental features, and promote sustainable development of rural economy and society with local culture as the core.

3.2 Enlightenment on the Protection of Rural Landscape in China

In fact, the protection of rural landscape can ultimately be attributed to the biggest problem facing Chinese villages at present -- the problem of development. Nowadays, it is difficult to find a way out in rural areas, especially for traditional villages. At present, what we really need to do is to excavate the historical and cultural connotations and characteristics of villages, and then carry out protection. In this way, we won't lose our rural features. And the work requires multi-disciplinary and multi-cooperation. In addition, what is most important is the enthusiasm of the villagers for the recognition and protection of their place of residence. (Nishimura, 2007) The success of Japan and Taiwan lies in the villagers' love and concern for their hometown. A reanalysis of Germany and Britain shows the importance of multilateral guidance and safeguards, especially in the policy. Finally, it should be noted that protection and development cannot be accomplished in one day. Thus, there needs to be persistent multi-participation.

Therefore, based on the existing experience above and combined with the existing problems of the local rural landscape, the following Suggestions are put forward.

Firstly, the key to the rural landscape outside the village is the protection of the ecological environment. The survival of villagers and all their activities, including production and management, are based on the local natural environment. Terraces, reservoirs and ancient persimmon trees are all part of the natural landscape. Therefore, we should avoid destructive development and protect local landscape features. It is not conducive to the development of villages to excavate mountains, cut down trees and introduce urban green vegetation.

Secondly, the key to the rural landscape in the village is the protection of the traditional spatial pattern and traditional architecture, as well as the rural vegetation and people. If a rural landscape heritage lost the villagers, then even if well protected, it is only a shell, a dead museum. Therefore, the most important thing for a village is to know what the rural landscape is, and not to use urban greening means at will. Therefore, it is necessary to do a good survey, understand the local vegetation and building materials, and understand the local history and culture, as well as the belief and spirit.

Finally, the conservation of the rural landscape actually involves the whole area — the conservation across multiple villages. So, it is essential to collaborate among villages. Therefore, the first thing we should do is to arouse the enthusiasm of the villagers, set up villagers' organizations, and carry out the popularization education of heritage knowledge and related successful cases. Let the villagers themselves really know the place where they live and take part in the development planning of the village. Secondly, the state provides financial policy support. At the same time, relevant laws and regulations should be improved to confirm the rural landscape law before tourism development. For example, many villages in Japan have their landscape regulations to prevent tourism from damaging the rural landscape. (Nishimura, 2007) Simultaneously, we need to integrate social resources to promote rural development as well. In addition, experts provide long-term guidance to villagers and set up cooperative organizations. Ultimately, helping to improve residents' living conditions while preserving traditional rural landscapes.

4 Conclusion

This paper analyzes the value and protection of the rural landscape in the mountainous area of Zhaoyuan city by taking the Kouhou area as an example. Firstly, in terms of rural landscape outside the villages, it is the product of the period from the late Qing dynasty to the people's commune. It is man's transformation of nature, which conforms to environment and nature. This transformation not only successfully forms a good landscape in the local area, but also successfully integrates with the production and life of the residents. Secondly, for the rural landscape in the interior of the village, it reflects the communication between man and nature since the village was established in the Ming and Qing dynasties. The change of building stone is just its reflection. People use this kind of stone for different construction, such as paving roads, earth temples, stone benches and so on, to form different streets and Spaces, and dig flowers and plants from the mountains, with the help of persimmon trees, conifers to express their aesthetic, life, faith and culture. These constitute the unique landscape of the local countryside. Because the rural landscape in this area has its commonality, it is essential to establish a villager organization during the villages, mobilize the enthusiasm of villagers, make them truly realize the value and significance of their village, and help them find a suitable development way. Preserve its rural landscape and help to explore the harmonious atmosphere of its community between human and nature.

References

Institute of cultural heritage of Shandong university. Beijing architectural design co. ltd. 2014. Protection and development plan for traditional villages in Xujia village, Zhaoyuan city (2014-2025):7.

Institute of cultural heritage of Shandong university, 2015. Protection and development plan for traditional villages in Kouhou Wangjia village, Zhangxing town, Zhaoyuan city (2015—2025): 5-6.

Institute of cultural heritage, Shandong university, 2018. Protection and development plan for traditional villages in Kouhou Hanjia village, Zhangxing town, Zhaoyuan city (2018—2030):21.

JIA Wenyu, LI Yin, 2005. Chinese gazetteer etymology. Beijing: Huaxia publishing house.

LIU Zhaoqi, LIU Kewei, 2016. Research on the experience of rural regeneration in Taiwan based on the conceptual practice and enlightenment of rural planning. Modern city research, 06:54-59.

NISHIMURA Yukio, 2007. Regeneration of charming hometown — a story of rebirth of traditional blocks in Japan. Beijing: Tsinghua University press.

NIU Guodong, BAI Jinyuan, 2006. Qilu lansheng. Jinan: Shandong map publishing house.

PENG Haodong, 2018. Research on the application of Taiwan community construction idea in rural planning and construction. Development of small cities & town, 06: 65-66.

REN Wei, HAN Feng, YANG Chen, 2018. Sustainable development of rural landscape heritage in UK-Taking chastleton house and garden in England as an example. Chinese landscape architecture, 11(16):15-19.

Shandong Zhaoyuan county annals compilation committee, 1991. Zhaoyuan county annals. Beijing: Hualing press.

YI Xin, SCHNEIDER C, 2013. Integrated rural development strategy and cultural identity cultivation in Germany. Modern city research, 06: 51-59.

YUAN Lin, LI Min, 2015. Cultural value of the hakka rural landscape in Guangdong based on the world heritage value criteria. Guangdong landscape architecture, 02:8-12.

ZHANG Zuoli, ZHANG Fengyu, 1846. Zhaoyuan county annals. (Shunzhi of Qing dynasty)

The Potential for Conserving Moken Ethnic Houses in the Surin Islands of Phang Nga Province, Thailand

Monsinee Attavánich[1], *Ayako Fujieda*[2], *Hirohide Kobayashi*[3], *Puttapot Kuprasit*[4]

1 Faculty of Architecture, King Mongkut's Institute of Technology Ladkrabang, Thailand
2 Kyoto University ASEAN Center, Kyoto University, Japan
3 Graduate School of Global Environmental Studies, Kyoto University, Japan
4 Department of National Parks, Wildlife and Plant Conservation, Thailand

Abstract In Southern Thailand, ethnic minority groups of Sea Gypsies can be divided into three subgroups: Moken, Moklen, and Urak Lawoi. In the past, they lived in boats, building temporary houses on islands during the monsoon season. In recent decades, they have settled permanently on land. However, the traditional ethnic houses have rapidly disappeared due to new house reconstruction after the 2004 Tsunami and the influence of modernisation. A Moken village in the Surin Islands of Phang Nga Province has kept its traditional-style ethnic houses more than other Sea Gypsy villages. Based on field surveys conducted in 2018 and 2019, this study illustrates the current living conditions of Moken in the Surin Islands and examines the factors affecting ethnic house characteristics.

The Moken houses in the Surin Islands have kept their traditional characteristics with some modifications. These stilt houses are built with natural materials, based on nine posts with a floor step to separate sleeping from the living and kitchen area. It is also observed that many houses have been extended for storage or to provide a balcony. In some houses, space underneath is enclosed by a wall and used as a shop. It can be concluded that three factors encourage the Moken to maintain their traditional-style house and characteristics of the Moken ethnicity. Firstly, the limited access to electricity on the islands. This limitation pushes the Moken to adapt and maintain their traditional ways of living and housing knowledge with no basic infrastructure as before. Secondly, the Surin Islands National Park office strongly encourages the Moken to use natural materials for building their houses. Finally, tourism supports their economy for living and the ethnic houses attract tourists.
Keywords sea gypsies, moken, ethnic house, living conditions, Surin Islands

1 Background

1.1 Current Situation of Sea Gypsies Housing in Thailand

Sea Gypsies are an ethnic minority, generally known as *Chao Lay* (people of the sea). They lead a nomadic life and travel by sea. To avoid the strong wind and waves in monsoon season, they build temporary shelters on the shore of several islands in the Andaman Sea and on the coast of Thailand and Myanmar. Although there is no clear classification for *Chao Lay*, they are commonly divided into three subgroups: the Moken, Moklen, and Urak Lawoi.

In recent decades, they have settled permanently on the coasts in temporary houses. The influence of modernisation has transformed the gypsies' traditional houses into a local modern style. In particular, after the Indian Ocean Tsunami in 2004, the destroyed ethnic houses were replaced by new style dwellings. This has resulted in the rapid disappearance of Sea Gypsies' ethnic houses. A survey from 2013 — 2017 shows that the characteristics of the Sea Gypsies' ethnic house have changed considerably since industrial building materials such as concrete and steel are affordable and more durable. The use of modern materials can be found in many Sea Gypsy villages such as Tubpla and Tungwa in Phang Nga Province (Attavanich & Kobayashi, 2016). However, despite such changes, the Moken village on the Surin Islands in Phang Nga Province has managed to retain its traditional characteristics.

The objective of this study is to establish the factors affecting the Moken village on the Surin Islands where the traditional house characteristics have been successfully retained, how modernisation has affected them, and what the future holds for such housing.

1.2 Methodology

Three field surveys were conducted; the first of which was in February 2018 to collect information on overall living conditions in the villages. However, a fire incident on the night of February 3, 2019, resulted in 61 houses out of 81 being burned, leaving only 20 houses. In the middle of February 2019, a survey was conducted to collect information on these 20 houses by taking measurements and recording living conditions of the villagers in each household with a questionnaire about their current lifestyles, incomes, daily lives, and housing, including house construction and improvement. Interviews were also conducted with related persons such as National Park officers concerning the policy for the Moken village. This information will subsequently be analysed to examine the factors affecting the villagers' existence and how it relates to the Moken ethnic houses.

Fig. 1 Moken village on the Surin Islands

2 Moken on the Surin Islands

2.1 Moken

The Moken, or *Chao Lay*, live in traditional boats and travel around the Andaman Sea during the north-east monsoon season from November to April. The Moken also use boats for commuting, fishery, and other sea-related harvesting to barter and trade with middlemen for rice and other necessities.

In the south-west monsoon season (rainy season) from May to October, when the sea is rough, the Moken build houses on the beach on the east side of the islands to escape from strong winds and high waves. Twenty or more families tend to live together, forming a temporary village during the rainy season (Ivanoff, 1999). Moken communities can be found in the Mergui Archipelago in Myanmar and islands in Phang Nga Province including the Surin Islands and Phuket Province in Thailand. Moken from the Surin Islands have remained a relatively traditional tribe compared to the other groups (Narumon, 2008).

2.2 Settlement

The Surin Islands are in the Kuraburi district of Phang Nga Province. The distance between the mainland and these islands is approximately 60 kilometres or one hour and 10 minutes by ferry from Kuraburi Pier or four hours by small boat. The Surin Islands consist of two main islands; North Surin Island (Ko Surin Nua) and South Surin Island (Ko Surin Tai) where the Moken village is located. It takes 15 minutes to reach from the National Park office at Chong Kad Bay on Ko Surin Nua (Fig. 2).

A small group of Moken had already settled on the Surin Islands prior to the declaration of the Surin Island National Park in 1981 and gave away their settlement to accommodate the infrastructure of the National Park and moved to another part of the island. The elder Moken recall that the government staff treated them in a good manner, were well-behaved (Arunotai, et al., 2007), and never forced them to leave the Park as indigenous people are often required to do so (Henley, et al., 2013).

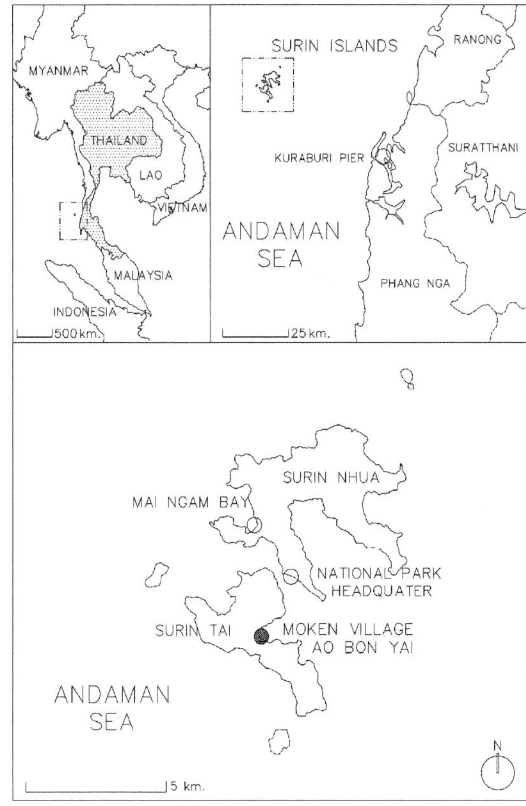

Fig. 2 Location of the Moken village

During this time, additional numbers of Moken began to settle on the Surin Islands as well. One reason for this transformation was that several areas of the Andaman Sea were declared as National Parks. This limited the Moken from cutting trees for building boats and houses, as well as hunting and gathering among themselves, requiring the Moken to change their way of living (Suzuki, 2016).

Prior to the 2004 Tsunami, the Moken lived in two communities, one of which consisted of 16 households, located at Ao Sai En on Ko Surin Nua. The second community, consisting of 30 households, was located at Ao Bon Lek on South Surin Island (Arunotai, et al., 2007).

After both villages were destroyed by the 2004 Tsunami, the Park officials decided that only one large village should be rebuilt. The new settlement of Ao Bon Yai was built further into the head of the bay. Nowadays, the village lifestyle and its traditional houses have become famous, attracting tourists who visit the Surin Islands. Nowadays, there are 80 households in the village (Fig. 3).

2.3 Lifestyle

The Moken specialise in fishing and catching marine life as a way to earn a living. Nowadays, the main income resources of the village come from tourism. In the high season (November to April), this income is used to buy rice and dry food in preparation for living during the monsoon season from May to October. In the monsoon period, the Moken mostly stays on the island and occasionally travels to the mainland to buy additional supplies. Their travel frequency is based on the weather and sea conditions. They breed some animals such as chickens to exchange for gasoline from the fishing boats that come close to the village to avoid the strong wind and waves of the sea.

2.3.1 Tourism

The number of tourists has continued to increase from the 1990s to 2000s. An average of 25 tourists per day visited the Surin Islands by the middle of the 1990s (Arunotai, et al., 2007), increasing by eight times in 2004. Although tourism was badly affected by the 2004 Indian Ocean Tsunami which destroyed the Park's facilities and Moken village houses, tourists began to return from 2006 onwards. In order to respond to the high demand for tourism activities, the Moken started working for the National Park and tour companies.

Moken who works for Mu Koh Surin National Park are hired as both permanent and part-time employees. There are currently 43 staff members including five Moken who hold Thai identity cards. These staff members are divided and assigned to

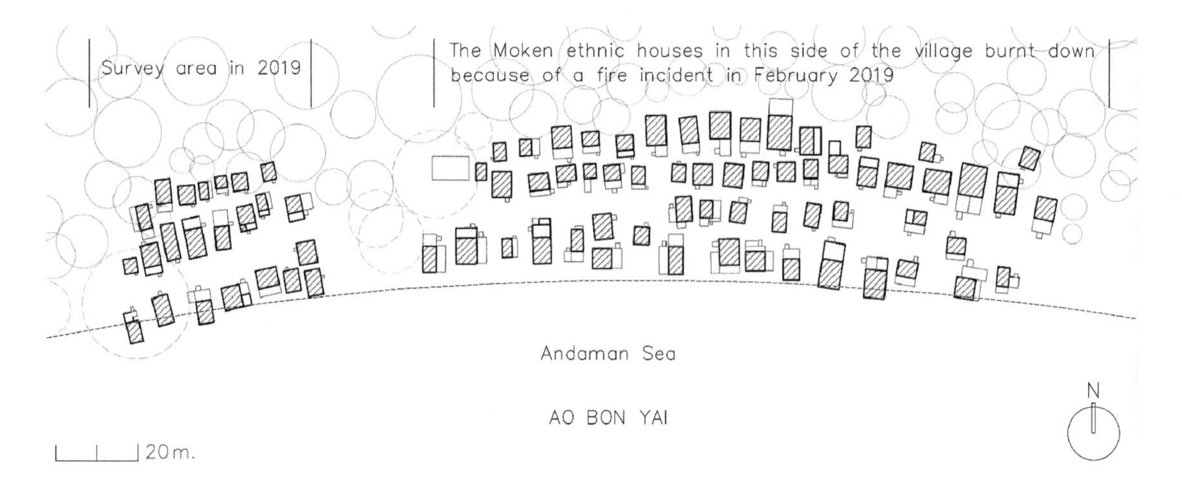

Fig. 3 The Moken village in 218

three locations: Chong Kad Station, Ao Mai Ngam Station, and the office at Kuraburi Pier to operate the Park's facilities. In addition, approximately 20 Moken work as housekeepers.

Apart from employment by the National Park, many Moken men work for tourism companies. The majority work as boat drivers, tourist guides, and tourism divers. Moken women and children stay in the village and earn income by selling handcrafts. Most of their customers are tourists visiting the village. Almost all Moken households now engage in tourism activities.

2.3.2 Income

According to the household survey, the Moken earns 300 baht (9.5 USD) per day working for the National Park. Earnings for Moken male working for tour companies are around 6,000—9,000 baht (95 to 125 USD) per month, topped-up with an additional 400 baht per trip. Moken female income is around 3,000—4,000 baht (95 to 125 USD) per month from selling handcrafts. The total household income is around 12,000—15,000 baht/month.

This level of income is quite high for living on the island in comparison to the Moken's normal lifestyle, so they prefer not to move to other places. It also attracts nearby Moken people to come to the village in the high season.

2.3.3 Daily Life

The Moken villagers start work at the same time as the National Park, which is from 7 am for housekeepers. From 9 to 11 am, the boat drivers take tourists to the National Park and work as guides, transporting them to diving points. The ferry back from the National Park leaves at 2:30 pm, in accordance with the time schedule of the sea level. At 3 pm, the sea water level is too low for the boats and ferry to sail, so most of them go back to the village at this time and stay there during the night. They can use electricity from 6 to 10 pm from shared central generators in the village. This concept is similar to the National Park electricity limit regulation which allows the use of electricity only from 6 pm to 6 am each day.

3 Moken Houses and Living Conditions

3.1 Moken Ethnic Houses

3.1.1 House Characteristics

The Moken ethnic house is a temporary dwelling built for the monsoon season. Since the village is located on an island in Southeast Asia, the house design is considered to be tropical architecture and the natural materials for building the house are taken from the nearby area.

The house plan is mostly based on nine posts. The inside area is divided mainly into two levels in order to separate its use. The highest step leads to the sleeping area while the kitchen is located from the lowest step since it is always wet because of cooking and washing activities.

The high stilt house allows for underfloor ventilation, and the highly sloping gable roof is good for quick rainwater drainage. The space under the roof acts as air gap insulation, causing the area under the roof to have a lower temperature, considered to be good conditions for living.

The main structure is built from a local tree called *chobalak*, while the secondary structure also comes from local trees such as *ping*, *kolen*, or *tutung*. The floor is mostly built from bamboo. The fan palm (*ta-lor*) or nipa palm (*cha-la*) is widely used for the roof and can be found in the local forest. These natural materials allow air to pass through the house. The wind from the sea helps to ventilate the dwellings in the village.

The above-mentioned factors help the house to provide good living conditions and complement the Moken lifestyle. In addition, the Moken can build and repair the houses themselves, helping to conserve housing knowledge.

3.1.2 House Modification

The Moken have modified the houses by adding an extension to include a kitchen and balcony. The house can be used for around four to five years but due to the different levels of the durability of natural materials, the roof may need to be changed every two years. The Moken consider changing some parts of the house or making repairs until it becomes impossible to do so, at which time a new house will be built.

3.1.3 National Park Policy in Relation to Moken Houses

According to National Park Act, policy to conserve natural resources in National Park area affects Moken housing. Large trees are not allowed to be cut since this affects the environment.

However, the villagers can collect natural materials for their housing, but this must be with the understanding that the villagers be concerned with the environmental conditions of the island. In particular, villagers are encouraged to use natural materials for their houses in accordance with the natural environment.

After the 2004 Indian Ocean Tsunami, the Park authority decided to build post-tsunami houses using materials from the mainland. Since the materials were donated and transported to the island, most were not traditional like those the Moken would have used before. For example, a wood board was used to build floors. Even though this material is more durable than bamboo, it must be brought from the mainland, which increases the building cost instead of the villagers sourcing the housing materials for free.

3.2 Living Conditions

The survey in 2018 and 2019 (the survey area in 2019 shown in Fig. 4) shows that living conditions have responded to the modern lifestyle. Three issues arise, namely house modification and area use, material, and living equipment.

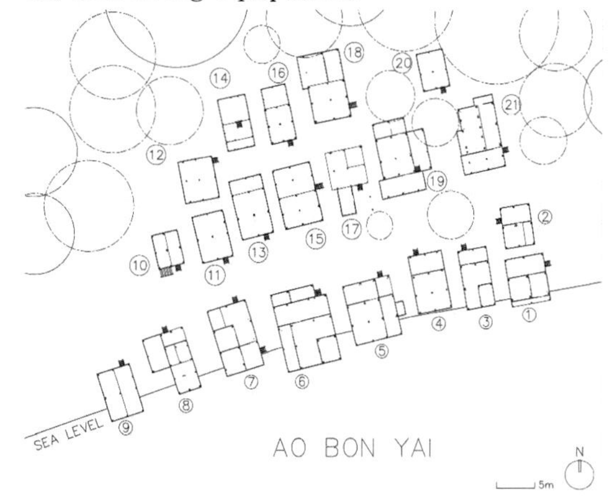

Fig. 4 The survey area in 2019

House modification and area use: the houses are modified by extending the kitchen and balcony to support long-term use such as with bigger storage. Some houses add an extension to the front area of the house on the ground floor for a shop for selling products in the village. Three houses have developed the frontal area of the house into a shop.

The area under the house is used for keeping things such as a boat, boat machinery, and breeding chickens. Most of the area is used for selling handcrafts to tourists.

Examples from the survey are shown in Fig. 5.

Material use: the houses are built from natural materials, but modern materials have also been introduced due to their affordability.

At present, the limitation of material collection to build a house is affected by the policy of the National Park and the use of traditional materials is decreasing. The records for 20 houses (Fig. 4) in Tab. 1 show that in the structure of the house (consisting of posts, beams, and rafters, including the floor, wall, and roof frames) mostly the same materials are used as in the traditional Moken houses found on the Surin Islands.

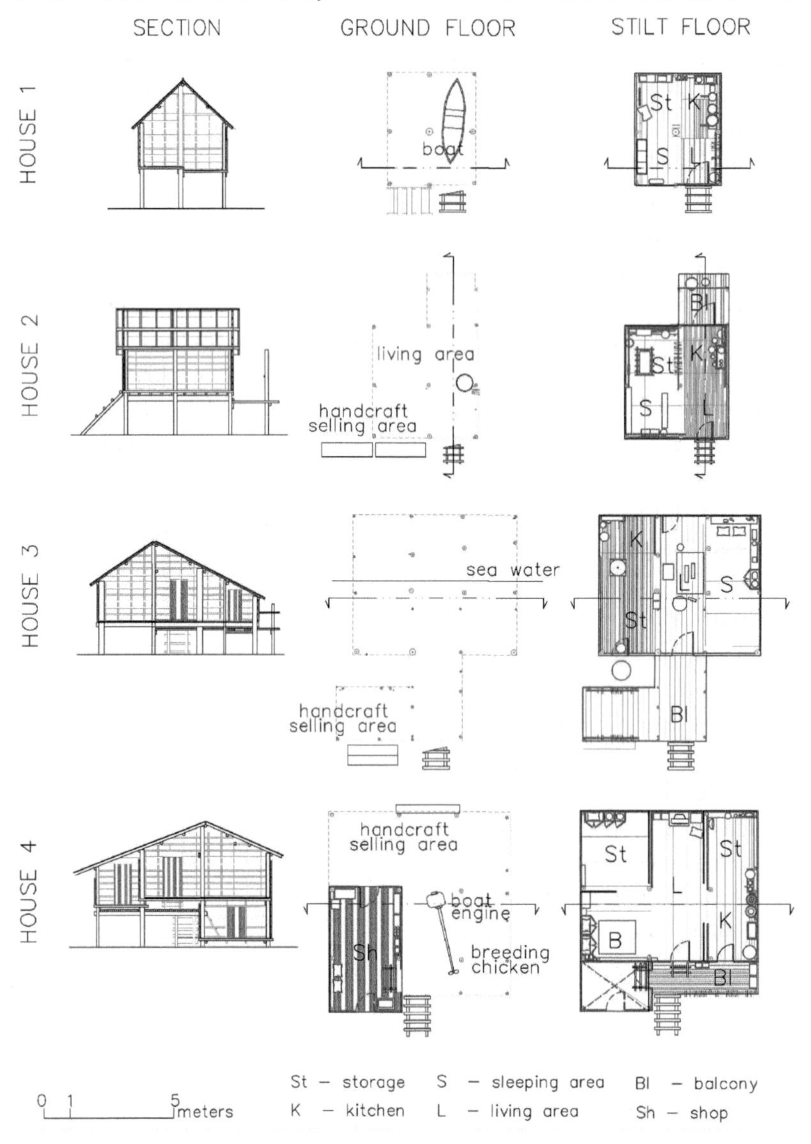

Fig. 5 Example of the living conditions in Moken houses

The survey found that timbers are used for the beams in some houses. For flooring, walls, and roofing, the materials vary from house to house. For example, timber boards brought from the mainland have been used to build the floor. Donated wood board or pieces of wood collected from the ruins of broken boats are other flooring options instead of the traditional bamboo. Bamboo is still used in the house in proportions or one, three, or four in the kitchen area. Wooden poles cut in half are also used as a flooring material.

For the walls, panels of ready-made nipa palm leaves can be bought in the mainland. Kor palm leaves and bamboo are also options for wall and exterior partitions and can be collected from the island. Some houses still adopt these local materials for their external walls. Apart from wall material application, the Kor palm leaves and ready-made nipa palm are also used as roofing materials in many houses because the roof is the first part of the house needing to be changed every two years. This shows that although materials can be collected from the islands to build part of the house, local resources are not in sufficient supply to build the whole house.

Tab. 1 Amount of material used in the houses

House No.		Post	Beam		Rafter	Wall /Roof frame		Floor			Wall			Roof	
Material		Wood pile	Wood pile	Timber	Wood pile	Wood pile		wood board	Bamboo	other	Nipa Palm leaves	Palm leave	Bamboo	Nipa Palm leaves	Palm leave
1	Structure			–			Material for wall and roof				–				
2		100%	75%	25%	100%	100%		75%	25%		100%			100%	
3		100%	100%		100%	100%		75%	25%		80%		20%	100%	
4		100%	100%		100%	100%		70%	30%				100%	100%	
5		100%	70%	30%	100%	100%		80%	20%		10%		90%	100%	
6		100%	70%	30%	100%	100%		70%	30%			80%	20%	50%	50%
7		100%	100%		100%	100%		60%	40%			100%			100%
8		100%	90%	10%	100%	100%		20%	80%			80%	20%		100%
9		100%	100%		100%	100%		50%	50%		90%		10%	90%	
10		100%	100%		100%	100%		100%			100%				100%
11		100%	100%		100%	100%			100%			100%			100%
12		100%	100%		100%	100%			100%		100%			100%	
13		100%	100%		100%	100%			100%			100%			100%
14		100%	100%		100%	100%			100%			40%	60%	100%	
15		100%	100%		100%	100%			20%	80%	100%			100%	
16		100%	60%	40%	100%	100%		30%	70%			100%		100%	
17		100%	100%		100%	100%		75%	25%		50%	40%	10%	100%	
18		100%	90%	10%	100%	100%		100%				70%	30%	100%	
19		100%	100%		100%	100%		70%	30%		85%	5%	10%	100%	
20		100%	100%		100%	100%			100%			100%		100%	
21		100%	75%	25%	100%	100%		80%	20%		60%	40%			100%

Amount of material use: 0-14% | 15-29% | 30-44% | 45-54% | 55-69% | 70-84% | 90-100%

Remark: House no.1 was canceled because it burnt by fire.

Therefore, modern materials from the mainland offer an alternative to help the villagers maintain and repair their houses. However, since the same original materials collected from the islands are used for the main posts, a future plan should be prepared to cope with the scarcity of resources.

Living equipment: the records for living conditions show that the Moken make shelves inside the house and most have a fire stove for cooking. There is some furniture inside as well such as a cabinet, bed, etc. Traditional Moken houses are also mixed and blended with modern electric appliances such as televisions, audio equipment, and fans powered by batteries and solar cells.

The majority of electrical appliances are used for the purpose of entertainment such as televisions and audio equipment. The survey found that 65% of all houses have them. Only two houses have fans, representing 10% of the village. It can, therefore, be assumed that most villagers are satisfied with their living conditions and do not require any equipment to make themselves comfortable in their houses.

One of the interesting findings from the survey is that 45% of the houses in the village have solar cells for power generation purposes. The village itself also has shared generators for supplying electricity to its members. The villagers share the cost of the gasoline to be able to use the shared electricity.

This means every house has electricity for lighting and electrical appliances. The Moken mostly uses electricity from 6 to 10 pm each day. The limitation of the electricity supply forces them to use electricity for very short periods for necessary use.

4 Factors of Moken House Conservation

The study found three factors for encouraging the Moken to maintain their traditional-style houses and ethnic characteristics: limited electricity accessibility, policy of the Surin Islands National Park, and tourism (Fig. 6).

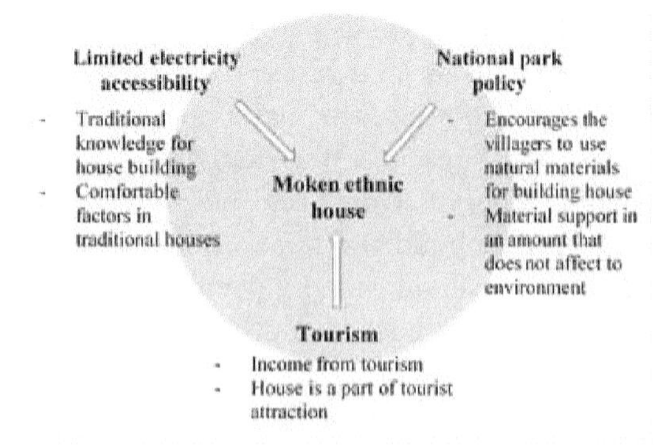

Fig. 6 Three factors affecting Moken ethnic houses

Limited electricity accessibility. From the study of lifestyles and living conditions in Moken houses, the Moken lifestyle was found to be transformed into a more modern style. Since the Surin Islands are far away from the mainland, access to electricity is limited. The villagers can use electricity from 6 to 10 pm each day. In addition, the Moken live in the village during the monsoon period and it is difficult to buy gasoline for generators. In addition, the electrical appliances in the house are mostly used for entertainment rather than comfort. This shows that the Moken ethnic house responds well to the climate and environment of the island, meaning that it can support the villagers to live well.

Surin Islands National Park Policy. According to the policy of Surin National Park, the Moken are encouraged to use natural materials for building houses in accordance with the natural environment. According to the study, in the National Park area of the Surin Islands, the building materials is small amoun t can be collected. The study of material used in each of the 20 houses shows the variety of materials adopted. It also shows that some materials can be brought from the mainland to replace the traditional resources. However, the amount of traditional material from the island such as posts and wood poles for making frames is also considered. The villagers are encouraged to use natural materials since it is important to retain their traditional house-building knowledge.

Tourism: tourism plays an important role in the Moken village and affects the villagers' lifestyles. A large number of tourists visiting the Surin Island brings income to the village. The Moken uses the money from tourism to live in the modern context and enables them to buy things to support their lifestyle. The ethnic houses in the group make the village unique. It is representative of traditional Sea Gypsy housing and has turned the village into an important tourist attraction.

5 Conclusion

Three factors: limited electricity accessibility, the policy of the Surin Islands National Park, and tourism illustrate how the Moken retain their traditional housing characteristics. These factors co-operated to preserve the traditional village in the modern context. This study also shows that the villagers have gradually changed their lifestyles towards modern living. However, if one of these factors is missing, ethnic traditional house conservation might not succeed. For example, if the villages can gain unlimited access to electricity, they can find many other ways to make life more comfortable. The traditional-style house will no longer be important to them and their characteristics will vanish as well. If there is no policy to control or support construction materials, the Moken may need to buy and use modern materials to rebuild their houses. In addition, if there is no touristic income for the Moken village, their lifestyles and living ideas may change. They may not realise how important the ethnic characteristics of the house are to tourism and may not maintain the traditional-style house.

Acknowledgements

Authors would like to thank JSPS Scholarship for the funds to support this research and the research team from Kyoto University and King Mongkut's Institute of Technology Ladkrabang, Mr Kridsada Pollasap, Mr Taksin Chergroo, Ms Sirayu Vijitrattanakit, Ms Jidapa Janjamsai, Ms Sutatta Janpan, Ms Sudarat Sahagaro, and Ms Araya Hongprayoon.

References

ARUNOTAI N, 2008. Saved by an old legend and a keen observation: the case of moken sea nomads in Thailand. Indigenous Knowledge for Disaster Risk Reduction, 72-78.

ARUNOTAI N, WONGBUSARAKUM S, ELIAS D, 2007. Bridging the gap between the rights and needs of indigenous communities and the management of protected areas case studies from Thailand// Coastal region and small island papers. UNESCO: Bangkok: 22.

ATTAVANICH M, KOBAYASHI H, 2016. Living conditions in post-tsunami houses: the case study of Moklen Ethnic People in Tungwa Village, Phang Nga Province, Southern Thailand, International journal of disaster risk reduction, 12-21.

HENRY T, KLATHALAY G, and KLATHALAY J, 2013. Courage of the Sea. Last Stand of the Moken. Bangkok: [s. n.].

IVANOFF J, 1999. The Moken Boat: Symbolic Technology. Bangkok: White Lotus

SUZUKIY, 2016. 現代の＜漂流民＞津波後を生きる海民モーケンの民族誌. Tokyo: Mekong－publishing.

The department of national park, wildlife and plant conservation, statistical data of national park, wildlife and plant (DNP), Thailand. [2018-02]. http://www.dnp.go.th/statistics/dnpstatmain.asp.

Analysis on the Architectural Heritage of Jin Mountain Compound — Take Guo Family Residence in Guanyao Village as an Example

LI Shiwei, *WANG Jinping*, *XU Chengying*

Taiyuan University of Technology, Taiyuan, China

Abstract In the west of Xiaoyi city, Shanxi Province, there are numerous "storey-style mountain courtyards" along Xiapu river basin. Influenced by the two typical folk dwellings, the cave dwelling in the west of Shanxi and the courtyard in the middle of Shanxi, the building types with unique regional characteristics are formed. This paper, taking Guo's house in Guanyao village, Xiaoyi city as the research object, through the way of field investigation and literature review, on the basis of clarify the village form, from the house of "clan culture, the compound shape, build skills" three levels of the Ming and Qing dynasties floor type mountain compound architectural remains analysis and interpretation, and The protection principles and activation measures of Guo's dwelling houses are proposed, hoping to provide a reasonable basis for the protection and activation of Guo's residence and other similar buildings in Guanyao village.

Keywords Jin system, mountain courtyard, Guanyao village, Guo house, protection, activation

1 Introduction

As a wonderful flower in the development of China's architectural history, residential architecture has its unique charm. Traditional residential buildings are the most primitive architectural forms that can best represent the traditional Chinese style, as well as the architectural types that can best reflect local characteristics and ethnic customs. Residential buildings record the development of society and reflect the culture of different regions and nationalities. The Guo family mansion in Guanyao village, Xiapao town, Xiaoyi is taken as a sample of the storey-style mountain compound, and its heritage research and protection are conducive to the continuation of the traditional architectural culture of Xiaoyi, and at the same time to create a living environment with regional characteristics in the contemporary architectural creation.

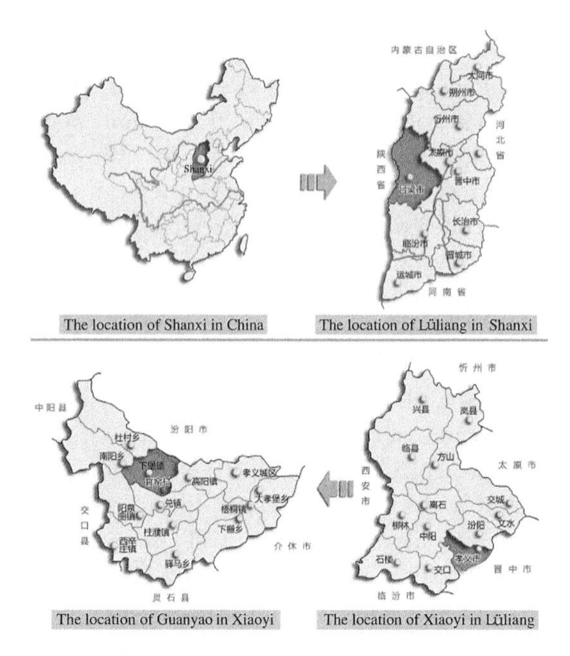

Fig. 1 Location map of Guanyao village

2 Overview of Residential Houses in West Jin and Middle Jin

2.1 Cave Dwellings in Western Jin Dynasty

Jinxi region is located in the west of Shanxi Province, in the east of the Jinshan grand canyon of the Lvliang mountains most areas. It belongs to the Loess plateau landform, which is high in the east and low in the west. Most of the earth's surface is covered by "Isolated Loess", with ravines and sparse vegetation. Here the climate is dry, precipitation concentrated, cold climate.

The geographical condition of plateau Loess and the climate characteristic of dry and little rain are the natural conditions for cave dwellings in western Jin, and cave dwellings dominated by cliff kiln, half pit kiln and occlusion kiln are produced.

2.2 Residential Houses in Jinzhong Compound

Jinzhong region is located in the central part of Shanxi Province, bordering Taihang in the east and Hebei in the west, bordering Fenhe river in the west and Lvliang mountain in the west of Jin. It has been the political, economic and cultural center area of Shanxi since ancient times. The terrain is mainly dominated by the basin and the hilly platform around the basin, which is relatively flat, and the Fenhe river system provides irrigation facilities. The soil in the hometown of Jin merchants in the plain area is fertile. Spring dry sand, hot summer rain, autumn days cool, cold winter snow.

During the northern and southern dynasties, old JinYang city flourished in the ancient Jin dynasty and reached its peak in the Ming and Qing dynasties. Jinzhong merchants with huge wealth built houses on a large scale in their hometown.

Under the influence of special geographical and climatic conditions as well as the political, economic and cultural conditions in the later period of the feudal society, the residential form of Jin merchants' courtyard with very distinctive characteristics was formed. In order to adapt to the geographical and climatic characteristics, cave dwelling and narrow courtyard have become the main characteristics of the residential houses in Jinzhong district. The house culture and exquisite carving are the results of the materialization of the wealth of merchants in Jinzhong district in Ming and Qing dynasties.

2.3 Dwelling Houses in Xiaoyi Mountain

Xiaoyi is located in the southwest margin of Shanxi platform anticline and Jinzhong basin, and the east wing of Lvliang mountain platform anticline. Economic development, Xiaoyi and Pingyao, Qi county and other contiguous, Ming and Qing dynasties Jin merchants also had a certain development, economic prosperity. Xiaoyi city is subordinate to Lvliang city.

Influenced by the compound culture in Jinzhong and combined with the similar landform and landform characteristics in the west of Jin, some unique storey-style compound buildings appeared in the gully area of Xiaoyi city. The Guo family mansion located in Guanyao village of Xiapu town is a typical representative among them.

3 Overview of Guanyao Village

3.1 Location and evolution

Guanyao village is affiliated to Xiapu town, Xiaoyi city, Shanxi Province. It is located at 111°52′ east longitude and 37°12′ north latitude. It is 18 kilometers from Xiaoyi new town and 3.1 kilometers from Xiapu town.

Guanyao village has a long history, dating back to the Ming dynasty and more than 600 years ago. Ancient Guanyao village rich in coal resources, coal variety is complete, by the government organization mining. In ancient times, the office of the coal mine was also called the official kiln. Coal miners lived along the mountain, and gradually formed a village, and the village name continued to be the official kiln. Guanyao village was prosperous in history, and reached its heyday in the collective period after liberation. Later, due to the mountainous area, the traffic is blocked, the social development is blocked and the economy is poor, but the historical and cultural resources of the village are preserved in the original ecology. In December 2018, Guanyao village was included in the fourth batch of Chinese traditional village protection list.

3.2 Natural Environment and Village Pattern

Guanyao village, located in the hilly area of the east wing of the Lvliang mountain anticline, is a Loess mountain area with rolling hills and ravines. It has unique mountain scenery and typical climate characteristics of Jinzhong region.

Fig. 2 Guanyao village pattern map

The layout of the village is based on the mountain, extending along the east and west slope layout, north and south long east and west narrow, back mountain water, hide wind and gather gas, like a phoenix spreading its wings. The building is built along with the trend, layer by layer, strewn at random, forming a three-dimensional intersecting three-dimensional layout form, integrated with the surrounding natural environment. At the bottom of the ditch in the center of the village, there is a north-south road that doubles as a waterway, dividing the village into east and west slopes. The southern section ends at the foot of Fenghuang mountain. Southwest side of the slope, from west to east distribution of Dadao ditch, ShenTaoditch, Zhaitou ditch, Daotouditch; In ancient times, there was a river temple at the entrance to the village, Sanguan temple and Yuhuang temple on the phoenix mountain, Guanyin temple in Dongpo, and a mountain temple in big head ditch. Thus forms "the round mountain report water, one mountain one bottom, one heart two, three slope four ditch, five temples five gates" the terrain pattern.

3.3 Street and Lane System

Street and lane system is an important part of village pattern. Guanyao village has distinct roads with different functions, which are connected and interact with each other to form a complete traffic system.

The distribution of the traditional roadway is built according to the terrain and connected up and down. The village is divided into the main road at the bottom of the ditch and secondary streets that follow the trend of the slope from north to south. East and west streets are auxiliary roads that connect the north and south streets of different heights.

Road function, terrain and other reasons, shape and size are different. The intersection of the main road at the bottom of the ditch and Xipo south road needs to relieve the crowded people, so it is relatively open and often used as a gathering and distributing place, which is now a public fitness place. The horizontal street of the tandem courtyard has a wide view and some pass through the cave roof of the lower floor.

Most of the courtyards and roads are staggered in an open space, which has both public and private properties. The upper and lower streets are steep and narrow.

4 Guo's House

Guanyao village has gone through the construction of the Ming dynasty, the prosperity of the Qing dynasty and the decline of modern times.

The residential construction of Guanyao village follows the principle of "local materials and measures according to local conditions". Through the adaptive reconstruction of Pingchuan compound in the mountainous terrain, the unique architectural complex of "storey-style compound in the mountainous area" is formed. The five groups of Guo compound are the most representative living fossils.

4.1 Clan Culture

Due to the geographical relationship, Guanyao village has been a family structure with the main family name since it was founded. At most, Zhang, Wang, Cheng, Guo, Tian and Gao coexist, among which Zhang clan and Guo clan are the largest. Their confrontation and integration, development and decline make Guanyao village show its unique vitality in the process of survival and development.

Early on, Zhang was a farmer for a living, and was the first family to rise. In Chongzhen years, there were ancestral graves, known as the eight family members of Zhang. By the Ming and Qing dynasties, Guo's business was booming and his family was prospering, and his status in the village was on the rise. However, Zhang's family was mainly engaged in physical labor such as coal mining, which gradually formed a social organization model in which Guo was the "management" and Zhang was the "labor". However, with the development of The Times, the coal business gradually withered, and before and after the liberation, the original organizational structure gradually collapsed, most of the Guo clan went out to make a living, and the Zhang clan gradually became the village manager, and once again became the main family name of Guanyao village. The current clan pattern of Guanyao village is formed in the process of the gradual rise and fall of the two clans, the Zhang clan and the Guo clan.

During the prosperity of Guo family, the family style of carrying forward simplicity, pragmatism and learning can also be seen through architectural plaques. In the process of doing business for a long time, Guo, as a businessman in Jin, knew that Confucianism could help his family, so when building houses, he integrated the clan concept into the building construction and decoration. In addition, he did not forget to donate money and help the villagers, which further improved Guo's reputation in Guanyao village and even Xiapu town.

4.2 House Shape

Under the influence of terrain, climate, use function and blood clan, Guo's house have formed three kinds of house layout, such as dispersed, master-subordinate and compound.

4.2.1 Distributed Layout

The main gate and the second gate are composed of a series of courtyards, arranged in line with the two sides of the ditch road, which are spectacular, convenient transportation and advantageous geographical location. According to functional requirements, the two courtyards are distributed with prominent emphasis.

The gate is composed of four courtyards — the upper court, the middle court, the lower court and the south court. Four courtyards are arranged side by side along the ditch bottom road, among which the middle courtyard is the second courtyard, mainly used for receiving and receiving visitors. The other three courtyards are all one into the courtyard, respectively for the mill, storage, servants living. The two gates are also juxtaposed with other courtyards, and the layout is basically the same as the gate.

Fig. 3 The first compound combination

4.2.2 Master-Slave Layout

Sanmen courtyard is located in the middle of Dongpo, near ditch road, south side of the road built three doors with the courtyard, the two houses a large one small, the master from the layout, each other cleverly foil.

As the main courtyard, sanmen is a typical storey-style building with four floors and three courtyards. There is no direct connection between the courtyards and they need to enter from the outside road. The layout is unique. There is an open space in front of the courtyard. Once entering the main room of the courtyard, there are five cave dwellings built on the platform, with three east and west wings. The second entrance courtyard is located on the main room of the first entrance courtyard, with wing rooms on both sides and five cave dwellings in the main room. The main room is on the second floor, with five caves on the first floor, five small windows on the second floor, and two side rooms on each side. Second, third to enter the hospital through the south side door. The courtyard on the south side is the servants' residence. Once entering the courtyard, the main room has five cave dwellings, and the two side rooms have three each. The function is perfect and the layout is simple.

Fig. 4 The third courtyard combination

4.2.3 Courtyard Layout

Four gates and five gates are located at the entrance of Guanyao village. The area of the two Chambers is similar, and the shape and structure of the two Chambers are exactly the same. Four doors, five doors are five into the courtyard layout, inside and outside clearly, the hierarchy. Once entering the courtyard, it is a transition space without wings, and the entrance is located on one side of the courtyard. Two into the courtyard east and west wing three, wing with a flat roof, a courtyard door connected to the first and third into the courtyard; Three into the courtyard, the main room of five caves, eaves house, Ming for ancestral hall, east and west wing for the flat roof, on both sides of the ear, the main room built a step can reach the upper four into the courtyard; Four into the courtyard in front of the 1.5 meters wide walkway, the main room five, east and west wing for single-pitched roof; Five into the courtyard only two side rooms.

The two courtyards move up step by step, with clever layout and clear streamline, reflecting the traditional clan culture, which is a typical representative of the mountainous compound of Jin merchants.

Fig. 5 The fourth and fifth courtyards courtyard combination

4.3 Construction Technique

In terrain conditions formed under the influence of floor type mountain courtyard building, to make full use of flat land, the lower the upper cave first built the upper brick and wood wing or occlusion kiln of masonry. Thus, the kiln building isomorphism technology of "kiln building on kiln" and "house building on kiln" is formed by mixing brick and stone issuing structures with wooden structures.

The isomorphic technology of building kiln on top of the kiln is mainly brick layer kiln in Guo's house, which is embodied by cliff masonry. The third courtyard of the four-door courtyard is a Three-section compound pattern, with 5 kiln surface width and 3 east and west kilns symmetrical layout. According to the south side Xiayao gable masonry steps, leading to the second floor. Because Zhengyao is much higher than Xiayao, the steps at the top of the wing space conversion, go west a few steps to the upper courtyard. The kiln house on the front of the upper courtyard has a large retreat distance, and the kiln legs are not superimposed on the wall of the lower cave house, so the lower cave house no longer bears the load of the upper cave house. Therefore, although the appearance gives people the feeling of Multi-storey building, there is only one floor, which is not the real sense of kiln house. In the upper cave house, the roof of the cave house on the first floor is used as the courtyard. Combined with the brick and wooden wing houses on the left and right sides in front of the kiln, the terrace courtyard is created just like the flat courtyard, which is convenient for dividing rooms and households to live in, and also for the production and life style of traditional farming.

In Guo's house, brick and wood mixed structure wing house is built above the cave, and the key technology lies in the treatment of kiln roof and wood structure foundation. Similarly, taking the third kiln in the fourth courtyard and the summer kiln in the fourth courtyard as examples, the roof of the kiln in the third courtyard was left with a thickness of about 1m when the roof was sealed, and the wooden structure foundation was set at the same time. First, determine the upper side of the wooden structure bearing column position, in the column position to choose a large volume, the shape of the rubble masonry layer, other parts are filled with gravel, finally grouting. The treatment of the upper cave courtyard ground is the same as that of panax notoginseng ash cushion and screed layer. The method for the column capstone, column and upper beam frame of the brick and wood mixed structure on the kiln is similar to that of other wooden structures.

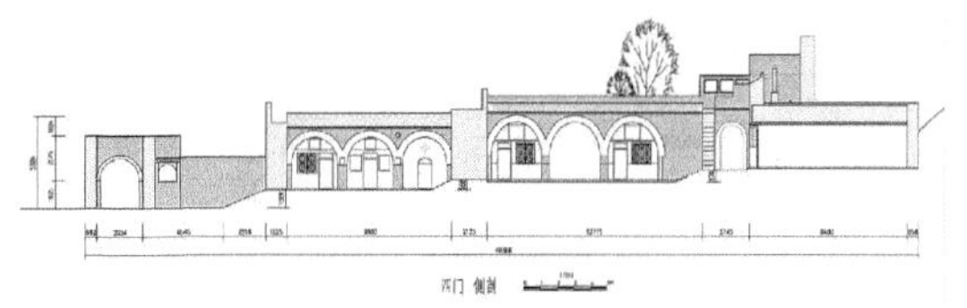

Fig. 6 Elevation plan of compound section of the fourth gate

5 Protection and Utilization

Traditional villages are the important carriers of traditional Chinese cultural landscape and agricultural civilization landscape, and these cultural landscapes are the important basis of national identifications and identity, as well as the essence of national individuality and differentiation. In the context of urbanization, numerous villages are disappearing and losing their own cultural attributes. Therefore, the protection of traditional villages such as Guanyao village represented by Guo's house is of great significance.

The protection of Guanyao village should follow the principles of authenticity, integrity and sustainability. In accordance with the principle of authenticity, the historical information carried by Guan kilns is protected by means of overall protection, hierarchical protection and classified protection, so that the traditional features of villages are truly protected. In accordance with the principle of integrity, attention should be paid to the integrity of historical features, the unity of material culture and non-material culture, the continuity of history, culture and natural landscape, and the integrity of the overall features and pattern of villages in the protection process. Follow the principle of sustainable, according to the change of society, economy, culture, the protection and development of the method of coordinate and promote each other, in the premise of guarantee and the overall historical and environmental coordination, to modern life must be some functional space, selectively darning into traditional villages in the original space structure, improve the living environment, maintain the continuation of traditional villages in function.

The use of Guanyao village should adhere to the combination of modern tourism and traditional culture. Guanyao village has been prosperous due to coal resources in history, but now the coal mine is declining. Traditional villages and traditional buildings left over from history have become unique and valuable cultural resources in Guanyao village, and become important tourism resources that can be used in the development process. In addition to direct tourism development, traditional cultural activities, such as bowls and bowls, shadow puppetry, and puppetry, should be gradually restored in the principle of overall protection to increase employment opportunities. Attract young people to return to rural areas to start businesses and solve the hollowing out and aging problems in villages.

6 Conclusion

Under the social background of economic transformation and development, how to rejuvenate the traditional cultural heritage is a topic concerned by all circles. Based on the field investigation and research analysis of Guo's storey-style mountain compound in Guanyao village, Xiaoyi, this paper summarizes that Guo's house is a comprehensive mountain building community with three aspects: cultural characteristics of "integration of Confucianism, commerce and agriculture", characteristics of the living environment of "integration of nature and man", and construction characteristics of "kiln house isomorphism". To study, protect and utilize it is an important premise to make it become the driving force of regional economic transformation.

References

LI Yu, 2016. Architectural research of Jinzhong compound. Chongqing: Chongqing university.

LIU Xingiyan, Li Yaguang, 2007. The China traditional local-style dwelling houses building form elements research. Shanxi building, (7) : 3-4.

Ren Fang, 2011. Comparative study of cave dwellings in western Shanxi and northern Shanxi. Taiyuan: Taiyuan university of science and technology.

WANG Jinping, XU qiang, HAN weicheng, 2010. Shanxi folk houses. Beijing: China building industry press.

WANG Jinping, 2016. Research on isomorphic building construction technology of kiln houses in Ming, Qing and Jin dynasties. Taiyuan: Shanxi university.

WU Bihu, 2016. Protection and activation of traditional villages based on rural tourism. Social scientist, (2) :7-9.

Space Analysis of Traditional Settlements of the Fishing and Hunting Nationality in East Asia —Take the Ewenki for Example

ZHU Ying[1], *WU Yating*[1], *LI Honglin*[2]

1 School of Architecture, Harbin University of Technology; Key Laboratory of Human Settlement Environment Science and Technology, Ministry of Industry and Information Technology, Harbin, China
2 Architecture Design and Research Institute of HIT, Harbin, China

Abstract Migrating from the shores of Lake Baykal in Russia to the Greater Khingan Range area of China, the Ewenki people retain the traditional style of safari and living together. Nowadays, they still migrate between China and Russia to fishing and hunting, but they have experienced a huge transformation from residence to settlement in modern times. The torrent of modern civilization has caused an unprecedented impact on the Ewenki lifestyle and living culture, which is not only a loss of cultural heritage but also a loss of architectural heritage for the unique fishing and hunting ethnic group in East Asia. From the perspective of anthropology, the thesis uses systematic theory to regard the cultural living space of the Ewenki traditional settlement as a whole, combined with the method of architecture. By taking individual-family-collapse as the starting point of different levels, the thesis regards the individual, group and social cultural space as cultural genes at corresponding levels in the system. Moreover, the thesis not only excavates and integrates the information flow of the cultural evolution of the Ewenki settlement in the hierarchical relationship of the nested layers, but purifies and screens the cultural space prototypes in the three levels, which finally presents a complete cultural genes of a living space that a fishing and hunting ethnic group settles in. The extraction and restoration of the cultural gene is an application of national heritage, which also provides a reference and solution for those endangered ethnic groups to the challenge of maintaining the cultural authenticity in contemporary civilization.
Keywords cultural gene, Ewenki, settlement space, system theory

1 Tracing Its Origin — The Historical and Cultural Extension of Traditional Ewenki Settlements

As a fishing and hunting nationality in the north, Ewenki has a long and long cultural history. The origin of national culture comes from Baikal area about two thousand BC. The position, style and material of ornaments worn by ancient people excavated by archaeology are very similar to those used by modern Ewenki people. In the self-claim of our nation, the original meaning of "Ewenki" is "people living in the mountains" or "people living in the southern slope of the mountains", which shows the close attachment between Ewenki people and nature. The Chinese Ewenki people moved to China from the shores of the outer Lake Baikal and the upstream of Lena River in Siberia in the 16th and 17th centuries. Based on different economic and cultural types, the Ewenki can be divided into three branches: distributed in the Ewenki autonomous banner and other places, mainly engaged in agricultural economic production "Solon" "Ewenki" "Solon" Ewenki who are distributed in the Ewenki autonomous banner and mainly engaged in agricultural economic production; distributed in the border between China and Russia Prairie Chenbarhu banner mainly engaged in animal husbandry production of the "the Tungus Ewenki" who are distributed in the border between China and Russia Prairie Chenbarhu banner and mainly engaged in animal husbandry production; As well as the research object of this paper: distributed in the Inner Mongolia Genhe City, take the hunting as the main production mode "Сахалар" Ewenki people, also known as "use deer Ewenki people" as well as "Сахалар" Ewenki people or "use deer Ewenki people", the research object of this paper, who are distributed in the Inner Mongolia Genhe City and mainly engaged in hunting production. Due to the particularity of the production mode of the hunting nationality, its traditional dwellings are different from other nationalities. Ewenki's living space is made of dozens of different thick and thin wood as the skeleton, and people change — different materials used for shelters according to different seasons. Ewenki people used to live in groups, but did not form a fixed location and form of villages, is a natural resource-oriented, dynamic residential mode, namely "settlement". The cultural content contained in this dynamic and flexible settlement mode is unique and abundant.

2 Cultural Boundary — Ethnic Cultural Genes Derived from Settlements

As a fishing and hunting ethnic group migrating across the border, Ewenki is mainly distributed in the cold areas of the eastern border of east Asia of Russia and China. The region is characterized by mountains, dense forests, cold weather and abundant resources. The challenges of relying on such primitive production methods as hunting and fishing in such a frigid region are obviously huge, such as how to resist the extreme cold, how to resist predators and how to ensure the continuity of resource access. Based on such natural environment background, the formation of Ewenki people's living, production and lifestyle were determined, and then formed the spatial characteristics of "large scattered, small group living" settlements. The traditional social structure was composed of gens, families and families. The living culture space formed by different hierarchical structures is the form of the interaction between different system factors to form the whole. This living cultural space can be regarded as a system, while the cultural space at different system levels constitutes the internal cultural gene of this nation.

2.1 Clan Cultural Space at the Settlement Level

The Ewenki people's living environment and production mode determine the characteristics of their group life. In the social relations of ues deer Ewenki people, the institutional culture of the clan organization "omouk" had existed for a long time. In the initial stage of the settlement hierarchy, the clan culture became the source of the downward fission of social culture. In the change of historical development, the clan system is dying out day by day, the turning point of which is the time node of the transition from the social and cultural space of matrilineal clan to the cultural space of paternal clan. However, the transformation of this social relationship system is a long time information transfer, is a time and space scale mutation. As early as in the era of hunting in the Lena river basin, the reindeer Ewenki people organized hunting and production in the clan commune, which lasted until the early days of the founding of new China. "Omouk" is the name of "gens" by the reindeer Ewenki hunters, and it is the basic social unit of this group which has lasted for thousands of years. In the whole range of Ewenki nationality distribution, the living space with clan as a unit presents a dispersive trend. The life between gens and gens is relatively scattered, even far apart, while the settlement of settlements also tries to avoid settling in the same area in case of conflicts on the struggle for resources. In the clan, the patriarch was the chief to deal with matters within the clan, solve conflicts and disputes within the clan, and divide hunting areas to avoid conflicts in resource allocation and life.

In such clan cultural relations, the social system of the nation is relatively stable and harmonious, and the integrity and unity of the system can be maintained for a relatively long time, which is a reflection of the stability of other hierarchical systems in the big system of ethnic relations. The culture of the clan is reflected in the way of conduct and moral standards passed down from generation to generation, thus forming a public consciousness, which is reflected in the common ancestor, group relationship, customs and habits, religious belief, national character and so on recognized by different clans. Under the background of such cultural space, the public consciousness formed shows a kind of compulsion and power, which is the embodiment of cultural inheritance created by collective wisdom. The collective intelligence of this group reflects the emergence of complex systems, that is, complex systems have many components and structures to form a whole, and the whole has the characteristics that its components do not have, which is fully reflected in this hierarchical cultural space.

2.2 Commune Cultural Space at the Family Level

Under the clan commune"ulilin", composed of several families, is the next level of systematic organization of the downward fission of Ewenki ethnic cultural space. As an integral part of the clan hierarchy, the ulilin commune is relatively small in number and scale, composed of several families, and still follows the collective use and possession in terms of production and lifestyle. Not only in the life level of interdependence and contact, in the production of hunting activities are mainly carried out by the commune. The Ewenki rarely went hunting alone, and in practice and experience they had a common understanding of collective action. Ulilin maintains the social relations of mutual help among nationalities in social functions. Living with several families in the same urinen and giving special care and attention to the elderly and children enables the system to achieve close internal connections. In production and life, Ewenki people formed a set of habits and experiences, which were passed down through oral transmission. These experiences and habits evolved into genes in the family group culture and continued in the cultural inheritance. The religious belief and inheritance of shamanism of Render Ewenki fully reflects the influence of culture on the settlement space. The Ewenki people's religious belief composition is more pluralistic, but makes the Render Ewenki people believe in shamanism mainly. The word shaman originated from the ancient Ewenki language and means "excited and reveling" people. Shaman is a highly respected group in the ethnic group. People believe that shaman knows everything and cures diseases and disasters. In the spiritual world shamanism becomes the other shore and sustenance of people's souls; In the settlement life, the shaman in the ethnic group becomes the guidance of people's behavior. It is precise because of the consistency of production needs and beliefs that different families, or system "units", are tightly integrated into a cluster system, resulting in a cluster effect. Under the condition of natural environment, the cluster system can fully reflect the advantages of the group, such as flexibility, adaptability, creativity and self-organization, which is reasonable and inevitable for the nation with the culture of group living.

2.3 Family Cultural Space at Individual Level

In the initial stage of ethnic formation, families were the living unit, and layers of social networks were established in the continuous production contacts between people. With the continuous development of time and natural environment, the production mode of wandering, hunting and mountain gathering was formed. Due to the nonlinear characteristics of complex systems, a large number of hierarchical systems in the system restrict and depend on each other, presenting a layer upon layer nested nonlinear relationship of mutual influence. Therefore, family cultural space is also an important part of national cultural space. In the evolution of the settlement of Ewenki, people became the key factor in the development of the settlement in time and space. The composition of the system is inseparable from the aggregation of elements, while the composition of clusters is inseparable from the organization of units. For all in hunting as a nation of resource access, individual due to the limited power and strength cannot resist the impact of the external environment, especially in the cold region of vast forest to survive, so people can together live together, use of the individual system of contact with the outside world to form the next level, so as to realize information, energy, technology and material and so on a series of exchange and interaction in order to keep the sustainable development of itself.

Different from the scattered regional space of the clan and the concentrated group living space of the ulilin commune, the family lives in the " organ-pipe cactus " of traditional Ewenki dwellings. A small organ-pipe cactus is not only space for each family to inhabit, but also a systematic organization composed of individuals. Starting from the individual, the organizational system of "individual — organ — pipe cactus — commune — clan" was formed. The cultural characteristics of each level of units constitute the cultural gene of a nation. In the cluster system, individuals have a certain degree of autonomy, just as people in the settlement must follow the common way of life, complete collective activities and goals together, but there is still space and field for individual activities, creating the culture that individuals can derive. At the same time, due to the interconnectedness between units, individuals can produce culture and influence others at the same time, so that the group will be more or less affected by "ripples".

3 Native-soil Crisis — The Impact of Modernization of Traditional Settlements

3.1 The Impact of Modern Civilization on Traditional Settlements

With the dramatic transformation of modern human production and life, some indigenous people who originally belonged to nature walked out of the forest and into modern society. In 1957, in order to provide the who had been wandering in the forest with a place to settle, the government set up Qiqian township. The Ewenki reindeer began to settle on the Erguna river. Later, due to the low-lying land in Qiqian area and the flood disaster, the settlement of Qiqian township from Qiqian township to manguya township, which is close to the hunting ground and adapted to the life of reindeer, realized the concentrated settlement life and had a village for the first time. The two did not change their hunting and herding reindeer production way of life, in August 2003, reindeer Ewenki nationality hunters in the government organized a settled a historic move, move to the outskirts of root river, known as the aoluguya ewenke hunters who saemaul undong, reindeer Ewenki nationality people for the first time the whole left the forest hunting, began the transformation of national economic development.

Ewenki's hunting culture has created rich cultural resources in the long history, and its culture carries the collective memory of the ethnic group from ancient to modern times, which is a kind of cultural identity and identity construction. However, as these intangible cultural heritages are attached to the first economic form and belong to the primitive culture, their survival ability is weak. However, under the strong impact of globalization, urbanization and new rural construction, the old traditional cultural system is rapidly disintegrated and traditional cultural values are out of order. If not rescued and protected in time, these intangible cultural heritages will soon disappear, which stimulates the deep significance and long-term value of this paper's research. Since the 1950s, the settlement policy has been implemented for the fishing and hunting ethnic groups in northeast China, and large-scale agricultural development has begun in the greater and lesser xingan mountains, resulting in a fragmented original ecological environment. In the process of transformation and continuous exploration, the traditional settlement mode of Ewenki is gradually replaced by the unified planning villages, and the cultural gene in the national blood line is also facing atrophy and disappearance.

3.2 Renewal and Restoration of Villages in Contemporary Context

In the study of village protection, we often regard modern civilization as a "flood beast" that erases the authenticity of the nation, and think that it is the rolling modern civilization with barbarism and brutality that impacts those ancient cultures and civilizations. However, it is undeniable that modern civilization not only brings shock and destruction to national culture, but also can bring vigor and vitality. The key lies in how we combine the two and continue the cultural gene of a nation, which is the right direction for a nation to survive and continue. Heritage protection of minority villages has always been a worldwide problem. Under the current situation that various countries pay more attention to the problem of ethnic minorities, how to continue and inherit these ethnic genes is necessary for us to find the answer of its particularity in the universality of the problem. Heritage protection is a common problem. We can extract similar dilemmas and contradictions from these problems, but we need to proceed from the situation of each ethnic group, formulate reasonable countermeasures for protection and recovery when it comes to specific protection measures and method so as to better solve the problems faced by villages of different ethnic groups.

In terms of a theoretical framework, we can establish each nation's own information base by means of modern big data integration, and preserve the history, culture and skills of the whole nation in a comprehensive and three-dimensional way through modern technology. In addition, we can find out the cultural gene of a nation by establishing a theoretical model, and dig out the information variables in the living space of a nation by constructing the prototype space of each nation.

At the level of action, a combination of policy guidance and relevant theories is more needed. For a long time, the government, experts, designers and villagers have been facing the lack of several key links in the protection of villages. The result is that the implementation and application of some protections have not been really combined with either the actual needs, or various advantages. The key to solving this problem lies in how to build an interactive platform at the level of government, experts, designers and villagers to maximize the effectiveness of information and optimize the activation of villages. In the process of activating and protecting villages, it is also necessary to effectively identify which can be recycled through some means and how to achieve regeneration. What is facing extinction and cannot regenerate, and how do we retain information are questions that we need to consider on a practical level.

4 Conclusion

Meme is a kind of cultural inheritance connected by social relations, which will not be excluded from the system or blocked by external natural forces. In the history of many nations, the formation and evolution of social culture often need a long period, and will also retain the original culture in a new phase of social development. For a nation faced with social space transformation, the constant, stable and key parameters of a pure gene is the culture of a nation. This kind of "change" and "unchanging" also indicates that the superstructure will not die out soon, and the overlap of social and cultural space will not be achieved overnight, but will undergo the process of gradual decomposition and transformation of internal elements. From material to spirit, from objective to subjective, from the program to the tradition, from the living to life, this kind of cultural practices have become invisible rules of and even restrictions on people's production and living, daily behaviors, and habits in a particular area. The invisible rules are also main influence factors of internal and external space form, and they influence the settlement of the shape, layout and space evolution, and become the external reflection of inner spiritual demands.

The cultural attribute of a nation is not lost due to constant migration, but has new meanings in the constantly changing paths. Many problems will eventually return to national culture. According to the theory of complex systems, the spatial form of settlements is the external characteristic of the long-term accumulation of natural environment and history and culture of settlements, and the form and significance of various spatial elements formed into a whole through certain organizational relations. Therefore, the external manifestation of traditional village settlement is the result of the interaction of social economy, geographical environment and cultural customs of the settlement, which respectively constitute the parametric variables affecting the evolution of the whole village system. How to reshape and restore the national vitality is actually a problem faced by every ethnic minority. For multi-ethnic countries, maintaining the vitality of ethnic minorities and inheriting the culture of these ethnic groups is an important part of the cultural composition and continuation of the whole country. Under the tide of modernization drive and convergence, how to maintain the cultural gene of the nation is a common problem faced by the national living space, especially for the fishing and hunting nation with no writing but language, the unique fishing and hunting culture of the dynamic settlement should be recognized and concerned by us. By analyzing and interpreting the phenomena, rules and causes of this special form of folk houses, we can dig out the answers to the general social problems such as the reservation of contemporary traditional folk houses, the interaction of communication and the inheritance of culture.

Acknowledgements

1. Key Research Topics on Economic and Social Development of Heilongjiang Province in 2018. (Project No. 18208)

2. Heilongjiang Philosophy and Social Sciences Research Program in 2018. (Project No. 18SHB074)

3. Independent Research Project of Heilongjiang Key Laboratory of Cold Architectural Science in 2016. (Project No. 2016HDJ2-1203)

References

ANDERSON D G, 2017. Review of reclaiming the forest: the ewenki reindeer herders of aoluguya by Åshild Kolås and Yuanyuan Xie. Pastoralism, 7 (1):17.

CAO Zijia, 2017. Research on hunan traditional clan village under the perspective of Meme. Huazhong Architecture (8).

NADEZHDA M , 2016. Review of Åshild Kolås and Yuanyuan Xie, Reclaiming the forest. The ewenki reindeer herders of aoluguya. Nomadic peoples, 20(1).

RUAN C, 2018. Reclaiming the forest: the Ewenki reindeer herders of Aoluguya.

WANG Haining, 2008. Cultural gene analysis of settlement morphology: a case of the Qingyan Town of Guizhou Province//WANG Haining. Planners, 24(5):61-65.

WANG Xiaobin, HAOMengquan. The design of organic renovation of traditional village houses: taking Inner Mongolia Tumdright banner Meidai Bridge Village for example.

YANG Dayu, 2011. Inheritance of traditional residence and its architectural Memes. South architecture, (6):7-11.

YANG Dayu, 2015. The inheritance of culter genes of regional architecture and contemporary innovation. New architecture (5).

YIN Jing, AN Yong, 2018. Study on protectingtraditionalfolk houses in Western Hunan with cultural genes under the background of new urbanization. Chinese and overseas architecture, (10).

The Reconstruction of the Sense of Place: Vernacular Architectural Heritage in the Rural Temples in the Context of Authenticity

WANG Warunee[1], *WANG Huiying*[2]

1 Faculty of Architecture, Khon Kaen Univerity, Khon Kaen, Thailand
2 Faculty of Humanities and Social Sciences, Khon Kaen Univerity, Khon Kaen, Thailand

Abstract In some rural temples of northeast Thailand, vernacular architectural heritages called Sim (Buddhism ordination hall) still exist. The study combined the theory of Genius Loci and the principle of authenticity, selected adobe Sims each remained in rural Chase and Phochai temples of the Sawathi village, Khon Kaen province, as a case study to conduct fieldwork. Based on authenticity, the study found a different existing pattern of them. In fact, the architectural authenticity did not only refer to the form and materials of a single building, but the complex of spaces related to it, including the emotional space, folkloristic space, aesthetic space, and the discourse space, which cultivate the sense of place of the villagers. According to the fieldwork, Sim in Chaisi temple had been renovated without integration with the surrounding landscape and developed towards tourism. The local authority provided an unduly package of folk activities nearby the Sim in order to cater to tourists. Another Sim in Phochai temple was in dire straits, and lack of the creation of emotion and folkloristic space because of ignorance from the authorized heritage discourse. When the villagers entered the place that did not abide by the principle of authenticity, the sense of place would be lacking or weakened, which was composed of the identity, belongings, historic sense, and aesthetic interest. In the context of authentic, rationally plan and layout of vernacular heritage in the rural temples must be considered when it is renovated or reused. Meanwhile, the restoration practice plays a positive role in improving the regional image, cultivating the villagers to spontaneously respect for their local culture, then scientifically and rationally reconstruct the sense of place.

Keywords vernacular architecture, authenticity, sense of place, renovation of Sim

1 Introduction

Sawathi village locates in the Northeast of Thailand with 2 temples, Chaisi and Phochai temple. It is one of a few villages in the region to have remained vernacular architectural heritage called Sim (An ordination hall in Thai-Lao language, main structure generally built from adobe wall and wooden roof structure). It also may be the only village in the region to have Sim appears in both temples, while almost all temples were demolished and replaced by concrete structures and change from local to Bangkok style. In fact, these heritages still play a part in the villagers' daily life. But the local authority does not attach enough importance to the Sim of Phochai temple because of their disrepair and simple decoration, which in other words, lack of potential for tourism. Thus, financial support for the renovation and maintenance are all given to the Chaisi temple due to its architectural value. However, during the process of renovation, due to the lack of scientific planning and rational guidance, its unique characteristics seldom comply with the principle of authenticity nor integrate with the surrounding landscape and built environment.

So, it is an important issue to make vernacular architectural heritage achieve its maximum level of social and aesthetic function for the villagers' daily life when conduct renovation, reuse and even develop them.

Therefore, based on the context of authenticity and the sense of place, the study selected these two Sims, which locate 750-

meters far away from each other, as a case study. The study compared their existence and discussed on the embody of authenticity in the belief space, folkloristic space, aesthetic space and the discourse space when the Sim is renovated in order to create an internal resonance between the perception of the villagers and environments, so that the sense of place can be more prominent on the aspect of natural ecology, historic culture and regional development in the contemporary society.

2 Literature Review

"Authenticity" means original, true, non-replicating. The explicit statement in the Operational Guidelines for the Implementation of the World Heritage Convention (short of OGIWHC) is " to meet the conditions of authenticity of their cultural values are truthful and credibly expressed through a variety of attributes including: form and design; materials and substance; use and function; traditions, techniques and management systems; location and setting; language, and other forms of intangible heritage; spirit and feeling; another internal and external factors" (The article 82, 2017). However, for the protection and conservation of architectural heritage, the new trend happened in the application of the concept. For example, in the region of East Asia, the researcher emphasized on recovering and reshaping the original cultural spirit to effectively inherit through the reconstruction of cultural sites. In fact, the authenticity of the architecture is a method to study its characteristics that the natural, social and human environments of the building are taken into account as comprehensive factors. Authenticity, not only includes the architectural form, building materials, but also involves the complexity of the natural environment, cultural landscape, lifestyle, traditional customs, and national cultural psychology that closely relates to the building in the original space. Many research emphasized the conservation for a single building of vernacular heritage, seldom unaware to deep in the authenticity which will arouse the sense of place.

Through exploring the building, Norberg Schulz (1980) viewed the sense of place as the core content of architectural phenomenology. He integrated the space and characteristics into the building, emphasized the orientation and identification of the building, while the memory, value, and experience (social attributes) and places (material entities) interacted each other in the living space which is composed of the building, and demonstrated the unique disposition that was the sense of place. He articulated in other research that "to dwell implies the establishment of a meaningful relationship between human and a given environment (Norberg-Schulz, 1985)." Holl (2013) explored new approaches to "integrate an organizing idea with the programmatic and functional essence of a building. Rather than imposing a style upon different sites or pursued irrespective of program, the unique character of a program and a site becomes the starting point of an architectural idea." In addition, many world-class architects such as Karsten Harries, Tadao Ando, and other architects have deeply studied the relationship between human and nature, human and architectural space with the help of phenomenology.

In general, memory, value, and experience of a person will interact with the place; during the process, people generate the emotional dependence which becomes an important link between people and the place. On the premise of authenticity, the architectural symbols of vernacular heritages could enhance the social, religious, cultural and aesthetic features. This is the basic standard when the sense of place is reconstructed in the heritage.

3 Research Methods

Due to the objective is to study the improvement of the sense of place in rural temples with the vernacular architectural heritage in northeast Thailand based on the perspective of authenticity. The follow-up fieldwork was conducted in Chaisi temple and Phochai temple, which have retained Sims with a history of more than 100 years①. Sim in Chaisi temple is renovated; while the old one in the Phochai temple has not been renovated. Participant observation was adopted in the study and participants of the Songkran festival were interviewed as key informants.

4 Result and Discussion

In these temples with the vernacular architectural heritage, the emotional space, folkloristic space and aesthetic space around the architectural entity constitute the place and cultivate the sense of place. According to fieldwork, Sim in Chaisi temple had been restored without integration with the surrounding landscape and developed towards the orientation of tourism. Especially during the Songkran festival (Thai New Year), the local authority provided folk activities next to the Sim in order to cater to tourists. While Sim in Phochai temple was in dire straits, and lack of the creation of emotion and folkloristic space because of the neglecting of the authorized heritage discourse. Therefore, no matter those Sims are renovated or not, the common phenomenon is that the sense of place based on authenticity is neglected to improve during the creation process of the emotional, religious, folkloristic and aesthetic space affected by the authorized heritage discourse.

4.1 The Carrier of the Emotional Space: Different Existence of Vernacular Architectural Heritage

Although the sense of place may be very personal, they are not entirely the result of an individual's feelings and meanings; rather, such feelings and meanings are shaped in larger part by the social, cultural and economic circumstances in which individuals find themselves (Ros, 1995). For historic sites, the identification and belongingness are important characteristics of the sense of place. In northeast Thailand, the old Sims with a long history, normally afford spiritual identification and belongingness for local people, so they might become a more everlasting being. If the old Sim could be renovated based on authenticity, the belongingness and identification of local people could be furthermore motivated. Of course, the reconstruction of its place should keep pace with the times and meet the demand of villagers in modern life.

Fig. 1 Sim of Chaisi temple during Sia kror ceremony in Songkarn festival. Sim of Phochai temple is not used for daily rituals

The Sim of Chaisi temple built in 1865 and still in-use until now. Therefore, it is relatively intact as a historic site. The natural color painting and fresco on the exterior wall show the unique character of local craftsmenship of that time. This Sim is renovated by the local in 1982 without advice from an expert by changing the roof from

① Generally, villagers will go to make merit at the temple near to their houses. Thus, villagers in village no. 8 of Sawathi village usually go to Chaisi temple while ones in village no. 6, 7, and 21 go to Phochai temple.

local to central Bangkok style. In the same year, the Fine Art Department added wing roofs and supporting columns. However, due to its architectural significance, the Sim becomes an icon of Khon Kaen city's vernacular architecture. More of visitors coming to participate in an annual religious activity regularly. In recent years, the Songkran festival has been held by a support of local authority and the Tourism Authority of Thailand. At this time, the identification and belongingness of the villagers have been enhanced and improved.

The old Sim in Phochai temple is facing a different situation. Its function is abandoned and replaced by the new Sim which built next to its front stair. Thus, it is in the cramped space between the new Sim and a pavilion. Its foundation is also in a bad condition, which causes cracking in the adobe wall while the wooden roof structure is damaged by rain and termite. Unlike the Chaisi temple, the wall has no fresco outside or painting inside and an old Buddha statue in the Sim is simpler in decoration. During the construction of a pavilion, the relevant parties did not consider leaving enough space for approaching the old Sim. As a result, its usage value is not embodied. At present, the villagers can only observe the old Sim from its front door due to a lack of confidence in safety. It also does not have an important role in peoples' daily life anymore, compared with the one in Chaisi temple.

4.2 Folkloristic Spaces of Vernacular Architectural Heritage: Orientation of Tourism Commercialization vs. the Lack of Folkloristic Space

In the historical development, a variety of architectural forms become the tangible carrier of intangible heritage. Visible concrete symbols are condensed in the building materials, architectural form, and style. They provide a place for intangible heritage, such as a festival, and trigger the recurrence of the sense of place. In fact, invisible space around the Sim, especially folkloristic space, as a kind of representation of culture, still needs to abide by the authenticity in modern society. The cultural elements of authenticity, mainly include three aspects: the sense of the times, the function of architecture, the local attribute of the building. These aspects are unified into the folkloristic space which is formed by the building entity and folk activity. Although it is not physical architectural space, folkloristic space is a necessary part of developing and implementing the architectural function indeed. Meanwhile, it carries the expression and publicity of cultural significance and folk life of the villagers.

In a process of modernization, the tourism economy is gradually developing in a rural area. Tourism, as Thailand leading economic industry, has unconsciously influenced, intervene and reconstruct rural economic and community life. Therefore, the local authority arranged a budget and available resources to promote tourism during the traditional festival held in the rural temple with vernacular heritage. Under the operation of the local authority, traditional festivals are "packed" in the folkloristic space to attract visitors.

Since 2015, the authority organized traditional activities that used to be done by the villagers during Songkran festival in the Chaisi temple for visitors. On April 13th, besides soaking Buddha statue and building sand towers in a temple as usual, the authority organized a large-scale blessing ceremony called Sia Kror (exorcise) around the Sim. Besides preparing for their families annually, the villagers were also responsible for preparing Gratong gao chong (sacrifice plate) for visitors. Around 2 pm. on that day, they sat inside the defensive wall of the Sim, and the abbot presided over the blessing ceremony. Visitors also took part in the ceremony with the Gratong Gao Chong under the shady shade of the pavilion built next to the Sim as multi-purpose buildings. The door of Sim will be open during this time of the year and allow men to get inside to pay respect to the Buddha statue, but all women are strictly prohibited from entering.

In Phochai temple, this blessing ceremony Sia Kror was arranged in the main hall around 9 am. on April 14th. The local villagers said that it was in accordance with the traditional time of the Songkran festival. Not many villagers made Gratong Gao Chong as parts of their families' sacrifice, but usually prepared incense, candles and auspicious yarns for the ceremony. During the Songkran festival and others, the religious activities mainly took place in the main hall, while the old Sim only opened for a short time during this period to welcome the villagers and visitors for a glance visit.

The two blessing ceremonies during Songkran festival showed the different status of Sims in the villagers' daily life. The ceremony in Chaisi temple was more complicated, and took place in a folkloristic space built around the Sim, while the colorful folk activities were held for tourism. The simpler activities and lacked the folkloristic in Phochai temple, which needed to depend on physical space formed by the old Sim.

4.3 Aesthetic Space of Vernacular Architectural Heritage: The Unharmonious with The Surrounding Landscape

Allen Carlson advocates viewing the human environment as the main field of everyday life and uses the perspective of natural ecology for reference, proposes an ecological method of architecture, places the building into everyday life to observe and study (Parsons and Carlson, 2008). Therefore, only those buildings are good ones that maintain a harmonious relationship with the surrounding environment or landscape. The idea subverts the criteria that architects pursue pure appearance and "functional adaptation" in the past.

The architectural form of vernacular heritage can be placed into the relationship of functional adaptation in the larger environment in order to obtain more aesthetic pleasure and value. Aesthetic pleasure, not only depends on the physical properties of the building itself, but also relies on the harmony and unity of the overall landscape and feasible reuse of the building space. In other words, the landscape is the basis and the carrier of aesthetic pleasure in the rural temple, the aesthetic pleasure is the attachment and extension of the architectural landscape, finally, the two parts unifies in the sense of place.

The Sim of Chaisi temple not only plays an important role in people's faith life, but also enables the villagers to gain an aesthetic pleasure. Especially, in the folkloristic space during the festival, people get the aesthetic pleasure from physical of built environment mixed with the dynamic performance and the solemnity of the ceremony. However, the renovated and reuse of the Sim, as a single building is not the right approach. The authority should stress more on the integration with the surrounding built environment and landscapes of the whole temple.

At the same time, without renovation, the architectural space of the old Sim in Phochai temple has been offended. The new Sim, imitating from the central Thai style, stands less than one meter of distance in front of it and a pavilion with Buddha statues in the back. The sense of belongingness and sublimity is difficult to embody in the old Sim due to its poor-arranged layout. Consequently, the villagers could not gain aesthetic pleasure through using the architectural space or appreciating the appearance of the old Sim.

4.4 The Discourse Space Related to Vernacular Architectural Heritage: Excessive Packaging Against Being Ignored

The authenticity of the vernacular architectural heritage in rural temples is reflected in the renovation of appearances and the function of buildings for various purposes. Its renovation and reuse is closely related to the authorized heritage discourse. Authorized heritage discourse plays a vital role in the direction of tourism. In the main activity space of Chaisi temple, traditional customs and folklore shows are deliberate "packaging", not only through the advocacy of the government and publicity by mass media but also the attention and intervention of scholars. This kind of "packaging" depends on the space created by the vernacular heritage which is composed of the physical form of the Sim and traditional folklore. Space also provides more possibility and operability for tourism.

Therefore, the transplantation, integration, and re-invention of folklore have been applied to folk tourism. The additional contents are absorbed into the original folk activities, so the content and structure of folklore are more complicated than before. As for Chaisi and Phochai temple, folkloristic space is the extension, and the revitalization of the use function of the Sim conducted by authorized heritage discourse. In Chaisi temple, the cultural department of local government begun to add some folk activities since 2015 and promoted through various websites before the Songkran Festival. Since then, the city governor of local government visited the temple to preside over the opening ceremony every year, which held on the lawn next to the Sim. Participants will be able to see the traditional costume contest, local game performances, and also Sia Kror ceremony. In 2015, as the first year to have tour groups came to join the activity, the government subsidized villagers to make Gratong Gao Chong for visitors. Therefore, the villagers positively made them and actively participate in the blessing ceremony. For the next couple of days, traditional games for children held at the same place. However, the authority had adapted some games from central Thailand into these folk activities, which presented an "excessive" performance and becomes traditional new-image to serve tourism.

The old Sim of Phochai temple has still not been renovated because lack of fund from the government; naturally, it will not attract tourists' attention. The favor of scholars and local government as well as the media to Chaisi temple had caused dissatisfaction with such attitude from some villagers in communities around the Phochai temple and expressed it out in public during Songkran festival at Phochai temple. They worried that the popularity of the Sim at Chaisi temple was constantly improving, but the old Sim of Phochai temple was still unknown and ignored. In addition, they considered their *Sia Kror* ceremony was the original one. Moreover, the traditional schedule for *Sia Kror*, which was important for a folk living has changed in order to serve tourism. Obviously, the influence of the government's power of discourse and intervenes caused the different existing status between the two Sim. It also leads to the different orientation of the folkloristic space in the two temples. One is towards commercialization of tourism. Another is trying to adhere to the tradition, but lack of support from the government to renovate and reuse the old Sim. The existence of two Sim has affected the identification and belongingness of the villagers, then obviously affects the integration of ecological and social resources, as well as the harmonious development of social and interpersonal relationships among villagers and vice versa.

5 The Advice on Vernacular Architectural Heritage in the Rural Temple

The conservation of vernacular architectural heritage should focus on dynamic protection, for example, the place of heritage is reconstructed and reused in daily life. This is also consistent with the core concept of architectural heritage conservation in Western which emphasizes the preservation of the original form of the building and endows the building practical function. So the study puts forward the advice, as follows:

5.1 To Highlight the Authenticity of Place Through the Practice of "Renovate to its Original State"

To renovate vernacular heritage, they should strengthen architectural materials, form, colors, symbols with the historic sense based on authenticity as much as possible. The renovation should adhere to the principle of "restoring it to its original state".

5.2 To Strengthen the Aesthetic Value of Authenticity by Integrating with the Surrounding Landscape

The renovation should not transform, deform or split the architectural symbols and materials at will according to the modern aesthetic value, but properly recover the historical beauty of vernacular heritage. More importantly, it should be integrated and harmony with the architectural landscape and surrounding built environment, in order to create greater aesthetic value.

5.3 To Enhance the Belongingness and Identification of the Villagers Based on the Normalization of Development

Under the guidance of the normalizing development, the current religious belief, emotional and folkloristic space of the heritage should be recovered and reused once again. The villagers would acquire more belongingness and identification via the reconstructed space in daily life.

5.4 To Popularize the Conscious Planning and Strengthen the Participation Behavior of Architects and the Villagers

The temple council and the relevant departments of the local authority should ask for advice from experts, collect the ideas from villagers then conduct the renovation of vernacular heritage so that the heritage could achieve more prominent social and religious function, aesthetic value and cultural characteristics.

6 Conclusion

As for vernacular architectural heritage, space, form, and style constitute the spatial characteristics of the place; the historical, religious belief and aesthetic sense of the building generates a spiritual place. When the heritages are renovated, authenticity is the important standard to cultivate and promote their sense of place. However, in northeast Thailand, the authenticity of the architecture is ignored at a large extent no matter the heritage is renovated or not. The overlook is not only embodied in the architectural entity and landscape, but also in the creation of emotional, folkloristic and aesthetic space. In addition, the power of discourse created by the local authority and scholars inevitably affects the orientation and development of authenticity of the heritage. When the villagers enter into the place which formed by the architectural entity, landscape and folk activity that lack of authenticity, their identification, belongingness, historical sense, and aesthetic interest will be weakened or discounted. That is to say, it is difficult to achieve a high unification of emotion, aesthetic, environment and culture. So, it is indispensable to reconstruct the sense of place during the renovation of the heritage.

On the premise of the renovation of architectural heritage, various powers, especially from the local authority, the scholars and local craftsmen should focus on all kinds of architectural vernacular heritage, rationally plan and layout space, conduct the integration of pluralistic space based on authenticity when the heritage is renovated or reused. This practice may improve the living quality of the villagers; reinforce their identity and strength the trend of aesthetic of everyday life, finally, rationally reconstruct the sense of place. Meanwhile, the practice plays a positive role in improving the regional image, cultivating the villagers to spontaneously respect for the local culture. It also provides a new way for the renovation, planning, and development of vernacular architectural heritage in other regions of Thailand.

Acknowledgments

This article is supported by Faculty of Architecture, Khon Kaen University and partially supported by the Center for Research on Plurality in the Mekong Region (CERP), Faculty of Humanities and Social Sciences, Khon Kaen University.

References

CARLSON A, 2008. Nature and landscape: an introduction to environmental aesthetics. Columbia: Columbia University Press.

HOLL S, 2013. Architects. https://www.archi-

tectmagazine. com/firms/steven-holl-architects.
NORBERG-SCHULZ C, 1984. Genius Loci: toward a phenomenology of architecture. Taiwan: Shanglin Press.
Operational Guidelines for the Implementation of the World Heritage Convention, WHC. 12/01, July 2012. http://whc. unesco. org/archive/opguide12-en. pdf.
PARSONS G, CARLSON A, 2008. Functional beauty. New York: Oxford University Press.
ROSE G, 1995. Place and identity: a sense of place//MASSEY D, JESS P, A place on the world. Milton Keynes: The Open University.
SEAMON D, MUGERAUER R, 1985. Dwelling, place and environment: towards a phenomenology of Person and world. Leiden, Boston: Martinus Nijhoff Publishers.
SMITH L ,2006. Use of heritage. London and New York: Routledge.

The Cities and Buildings in Hakka Area Under Influence of Yang Junsong

WU Qingzhou

School of Architecture & State Key Laboratory of Subtropical Building Science, South China University of Technology, Guangzhou, China

Abstract This paper states that the Master of Fengshui, Yang Junsong, came to Ganzhou to teach disciples about geomantic omen towards the end of Tang Dynasty, and that the cities and buildings of Hakkas area have been greatly influenced by Fengshui. It takes Ganzhou City, Meizhou City and Weilongwu as examples to explain this viewpoint. Ganzhou City was planned and constructed by Yang Junsong according to the shape of a tortoise and became an unshakable iron city. Meizhou is also a turtle city, and now they are the state historical and cultural cities. Weilongwu embodies the Chinese philosophical ideas and wisdom, it's Huatai being unique over China, which express the idea of reproductive worship and also created by Yang Junsong. There leave a great number of traditional houses in Meizhou. Conservation of these cultural heritage is difficult. The native governments adopt some measure for protection of these houses and get effective results.

Keywords the Master of Fengshui, Yang Junsong, Ganzhou City, Meizhou City, Weilongwu

1 Yang Junsong Selecting the Site and Planning the Tortoise City of Ganzhou

Yang Junsong (about 847—937), was the master of Fengshui. *The Records of Ganzhou Prefecture* says:

Yang Junsong is of Douzhou, an official of the post Jingzhi Guanglu Dafu in the reign of Xizong (875—888) of Tang Dynasty. He was in charge of astronomy, geography and geomancy. When Huang Chao captured the capital Changan, he cut off his hair and went to Kun Lun Mountain. Then, he came to Quanzhou (Ganzhou) and teaching practical geomancy to his students, and died in Yudu Yaokouba, Quanzhou.

The Collection of Ancient Books. Ode to Ganzhou records: "The city walls and moats of Ganzhou: Gao Yan, the commanding officer, constructed them at the confluence of Zhang River and Gong River in the 7th year of the Yonghe Reign of the Eastern Jin Dynasty. Lu Guangchu, the commanding officer, developed the city to the south, and excavated moats on its eastern, western and southern boundaries."

The initial city walls of Ganzhou were made of earth in the 5th year of the Yonghe Reign (349). Being the governor of the area, Lu Guangchu asked Yang Junsong to choose the site and build the city. Yang Junsong planned the city with the idea of a tortoise to replace the current city form, with the head of the tortoise being the South Gate, its tail at the confluence which has been called Guiweijiao (i. e., the corner of the tortoise's tail) until today, its legs being the gates on the eastern and western city walls facing the rivers. As a result, Ganzhou was constructed into an unshakable iron city with its boundaries being surrounded by water in three directions (Fig. 1). Lu Guangchu occupied Ganzhou with a powerful army and made himself the King of this region for more than 30 years.

Tortoise is one of the four sacred animals in ancient China. Tortoise symbols the Heaven, the Earth, and the Man. Thus, tortoise itself becomes a universal model. Out of ordinary, the clan of Emperor Huangdi had a totem of tortoise, resulting that the culture of tortoise worship was combined with ancestor worship. It is still more important that the divinatory symbols

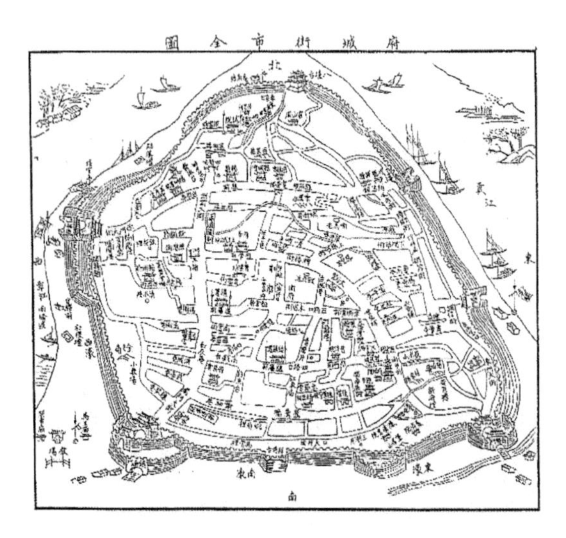

Fig. 1 Plan of Ganzhou City

of *The Book of Changes* originated from the veins of tortoise carapaces. Therefore, the culture of tortoise worship is the source of philosophy and wisdom of *The Book of Changes* of Chinese people. Tortoise has a long life, thus being the symbol of life worship for the purpose of perpetual rejuvenation.

There are several reasons for Yang Junsong to built Ganzhou City with the idea of a tortoise:

① The city will develop into a political, economic, and cultural center with the inspiration of tortoise.

② Inspired by the protective function of tortoise's solid shells, Ganzhou imitated the tortoise as the idea for construction, which became an iron city of solid defense by being surrounded by water in three directions;

③ As tortoises live in the water with a bodily form of round and curved shell, the city walls built in the shape of tortoise shell may reduce impulsive force of floods and is thus beneficial to the flood resistance.

④ As tortoise has a long life, Ganzhou was expected to be able to develop continuously for thousands of years.

The model of the tortoise Ganzhou is a turtle against the current, or a tortoise towards the Heaven.

The pattern of Ganzhou's Fengshui is ten snakes getting together around a tortoise.

Ganzhou City has been developing for more than one thousand years and becomes the second large city of Jiangxi and a state historical and cultural City. The facts prove that it is very successful for Yang Junsong to build the tortoise City Ganzhou.

Yang Junsong selected the site and planned Ganzhou City in 903. Being old and infirm, and overworked for planning the city, he died.

Yang Junsong trained and brought up a lot geomancer of Kejia, and giving deep influences over the cities and buildings in Kejia area.

2 The Construction of Meizhou City

Jiayinqzhou Citywalls were built in the Reign of Huang You (1049—1054) in the Song Dynasty.

Jiayingzhou City (now Meizhou City) was of a shape of tortoise(Fig. 2).

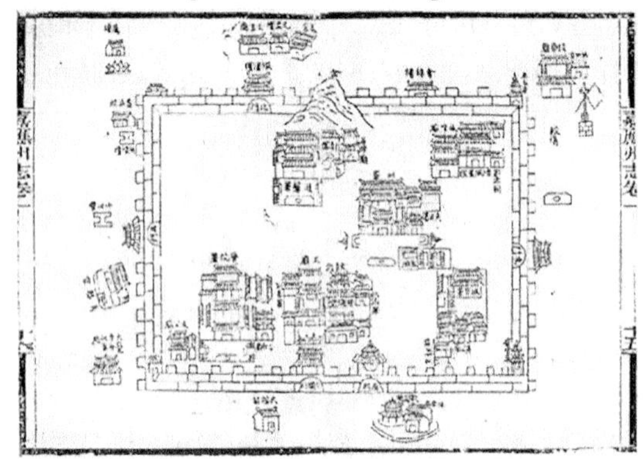

Fig. 2 Plan of Jiayingzhou City
Source: The Records of Jiayingzhou Prefecture During Guangxu Reign of (1875—1908) Qing Dynasty

The Records of Jiayingzhou Prefecture says that the shape of the city is a turtle running towards the river.

Jiayinzhou City situates on the north bank of Meijiang River, it's shaped being like a turtle running towards the river, it's head being the South Gate. To keep the tortoise safe, Wang Zhepu, the official in charge of Jiayingzhou during the Reign of Quanlong (1736—1795) in Qing Dynasty, built a magnificent tower Hang Guanglou outside the South Gate, and a pavilion Guanlanting beside the tower. He also used a big turtle to support one of the pillars of the pavilion. In 1987, the pavilion was repaired, people found that this turtle was still alive. Wang Zhepu also built a pagoda of Fengshui to repress the turtle of Meizhou City. The Tower, pavilion and pagoda are the landscape of Fengshui in Meizhou. After more than one thousand years of development, Meizhou becomes the capital of Kejia, and also as Ganzhou, a state historical and cultural city.

3 The *Huatai* of *Weilongwu* was Created by Yang Junsong

Weilongwu is another example. In the Northern Song Dynasty (960—1127) there was a master of philosophy Zhou Dunyi (1017—1073), who wrote a famous treatise *Tai Ji Tu Shuo*. It's main viewpoint as follows:

Wu Ji comes to Tai Ji(Fig. 3).

Fig. 3 Graph of Tai Ji in ancient time

Movement of Tai Ji produces yang. After moving it rests and produces yin. After rest it comes to move. Movement and rest are of mutual causality. Yin and yang are called Liang Yi.

Tai Ji is metaphysical *Dao* (the Way'); yin and yang being material objects.

Yang changes and combines with yin, as a result, water, fire, wood, metal and earth being produced. The five elements spreads everywhere and the four seasons appear.

Weilongwu, is a type of the Hakkas houses in Meizhou, Guangdong Province, it's form being the symbolic expression of the philosophic theory *Tai Ji Tu Shuo* (Fig. 4).

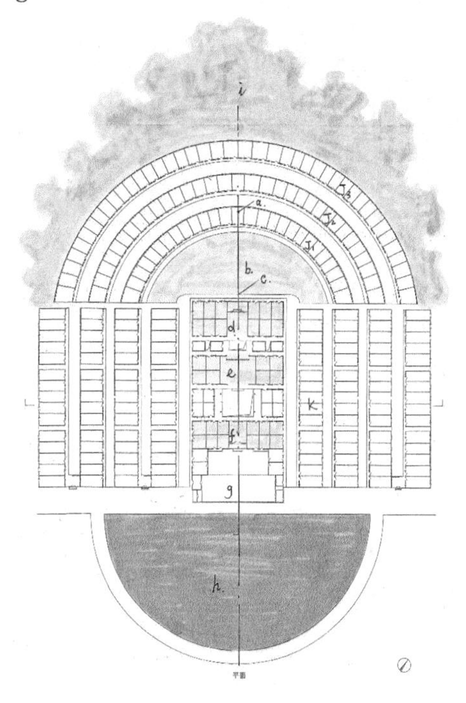

Fig. 4 Plan of a Weilongwu
a. rear hall b. Huatai c. five element stones
d. ancestral hall e. middlehall f. entrance hall
g. sunning ground h. semicircular pool
i. forest for Fengshui j. Weilong k. hengwu

The semicircular pool symbolizes yin; the semicircular mound (*huatai*) and the multi-semicircular houses mean yang. They combine into one circle symbolizing heaven, the rectangle between the two semicircles meaning "earth". This composition expresses the idea of heaven and earth, or yin and yang, being respectively combined into one harmony.

Another feature of Weilongwu is the *huatai* or *taitu* (the earth of embryo), which being a semicircular and shaped as a curvature of the spinal column behind the ancestral hall.

There are five element stones at the base of *huatai* just on the axis(Fig. 5).

These five element stones express the theory:

Yang changes and combines with yin, as a result, water, fire, wood, metal, and earth being produced.

There are thousands upon thousands of

Fig. 5 Five element stones

cabblestones on huatai.

These cabbles express the theory of Zhou Dunyi：

Fig. 6 Huatai

The attributes expressed by Qian constitute the male; those expressed by Kun constitute the female. Qian combines with Kun and producing all things on earth. All things move and change for ever, and becoming more and more.

Huatai symbolizes the womb of earth. The population of the family will grow more and more as the cabblestones on the huatai.

We can see that Weilongwu is a universal model to express the philosophical theory of Zhou Dunyi, *Tai Ji Tu Shuo*.

Huatai expresses the culture of reproductive worship.

Only in Kejia houses we can find Huatai. Huatai was created by Master Yang Junsong. In kejia area, such as Meizhou, Xingning of Guangdong, a book of Yang Junsong spreads there. My wife's father had a hand-written copy. There is a plan of a Kejia house in this copy called plan of Huatai in house created by Master Yang (Fig. 7).

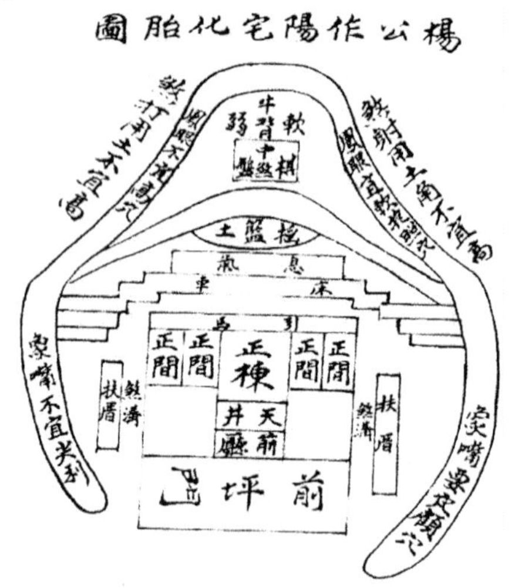

Fig. 7 Plan of *Huatai* in house created by Master Yang Juhsong

Master Yang worship is very popular in Kejia area. Master Yang's altar is near the corner behind the ancestral shrine. There are three tablets respectively for Master Yang, Master Zheng, and Master Liao.

The former residence of Cai Mengji (1245—1276), a Jin Shi and a hero of resisting Yuan invansion, is a Weilongwu with Huatai and the semicircular pool, being built in Southern Song Dynasty (1127—1279).

Luo's Dongshengwei is a Weilongwu built in Yuan Dynasty (1271—1368) with a more than 700 Years History .

Pan's Qiuguangdi at Nankou in Mei County was built during the reign of Jiajing (1522—1566) in Ming Dynasty with a history of more than 400 Years. The above examples are able to prove the fact that Weilongwu appeared in Mei County from Southern Song Dynasty, developed in Yuan and Ming Dynasties, and become very popular in Qing Dynasty(1636—1912).

The Huatai of Weilongwu was created by Master Yang Junsong, the whole composition of Weilongwu should be the work made by Master Yang and his disciples.

In the kejia area of Fujian, there are

many five phoenix towers with Huatai but without Weilong and five element stones. These Kejia residences might be built according Master Linmu's plan(Fig. 8).

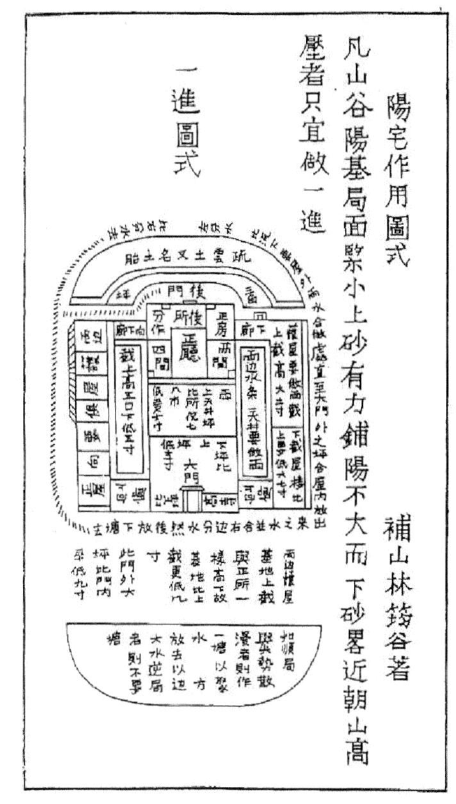

Fig. 8 Plan of residence, Linmu. Yang Zhai Hui Xin Ji. Jiaqing 16th years (1811) block printed edition

4 Distinguishing Features of *Weilongwu*

Weilongwu is a building combing ancestral hall and residence into one, it being exotic flowers embodying both the thought of ceremony system and maral principles of confucianism, symbolizing philosophy of Taiji, Ying and yang, and the five elements, standing for the ideas of Fengshui, and architectural arts. Weilongwu is cultural and artistic treasure of Kejia buildings. It's distinguishing features are as follows.

4.1 Plan Following the Regulation of Axial Symmestry

Primary elements of weilongwu are laid along the axis, and secondary elements being laid two sides of the axis.

Along the axis, we can see the rear hall, Huatai, the five element stones, ancestral hall, middle hall, entrance hall, sunning ground, and semicircular pool.

There is the forest for Fengshui outside the weilong. Plan of Weilongwu follows the regulation of axial symmestry as the imperial palace.

4.2 Organic Building Able to Grow up

A lot of Weilongwu have only a weilong at the beginning of construction.

However, there leave some space around the house for later construction.

After many years, the population increased, the second weilong appears.

Wenrenhou ancestral hall, is a Weilongwu with eight heng, three weilong, built during Jiajing Reign (1522—1566) of Ming Dynasty.

Up to now, Pipatang old house at Yetang, Xingning with seven Weilong is the upmost number for all the Weilongwu.

4.3 Ecological Building Adaptive to Mountain Area

There are forests for Fengshui behind Weilong. Weilongwu is adaptive to mountain area, being able to reduce soil erosion and to resist landslide and mud-rock flow (Fig. 9).

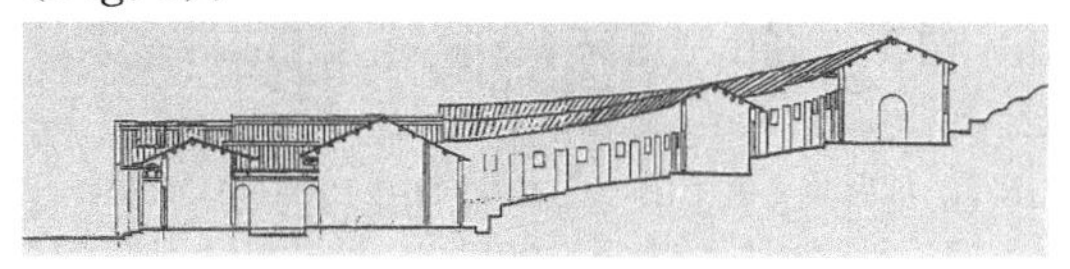

Fig. 9 Section of Pan's De Qingtang. It's huatai and weilong with the slope as the hill, are able to resist landslide and mud-rock flow, and the forest of Fengshui can reduce soil erosion.

There were a lot of Weilongwu in Hong Kong area before urbanization. Hong Kong's Vegetation underwent a great damage in the urbanization. Fortunately, some best forests have been preserved on account of Kejia's Fengshui faith. These forests for Fengshui spread in more than 600 old villages, their areas being of 600m^2 to 6 ha. There are more than 600 kinds of plants in these forests, those wood's ages from 50 years to more than 300 years, including camphor trees, evergreen chinquapin, banyan, huang tung tree, wild jujube, Chinese sweet gum, Chinese tailow tree, litchi, longan, mango, carambola, and so on. Weilongwu of Hong Kong leaves a huge ecological heritage to the city.

4.4 Of all Residences in China, Only Hakka Weilongwu and Five Phoenix Towers with Huatai, only Weilongwu with Five Element Stones

Hakkas live in mountain area, Master Yang Junsong created Huatai as an very important element of Kejia house to adapt mountain area environment. It is a great devotion to China's vernacular architecture. Huatai also expresses the idea of reproductive worship and life worship of Kejia.

Weilongwu was created by Master Yang and his disciples, There are five element stones in *Weilongwu*, which embody Master Zhou Dunyi's theory Tai Ji Tu Shuo. They are unique of all the houses in China.

4.5 Ecological Functions and Fengshui Idea of the Semicircular Pool

The semicircular pool of Weilongwu is of multi-functions, such as, storage of rain water, fire protection, using as a fishpond, breeding, and so on. It also expresses the idea of Fengshui that with the same shape as the pool of confucian temple, there will be more Kejia people passing imperial examinations and to be officials.

4.6 Adoption as an element in Yinzhai (Grave)

The graves in South China are of the shape as a round-backed armchair. They are of the some composition as Weilongwu.

For example, the grave of Xiong Fei and his wife (Fig. 10), built at the end of the Southern Song Dynasty, is of the same composition as Weilongwu. It's front is a semicircular Danci as semicircular pool of Weilongwu. It's Baishi is of a square shape, being similar to three tang and two heng of Weilongwu. Behind the square, there is a part, it's shape as a curvature of the spinal column, like huatai of Weilongwu.

Another example, graves of Ye Yongqing family are the same composition, they also being built at the end of the Southern Song Dynasty.

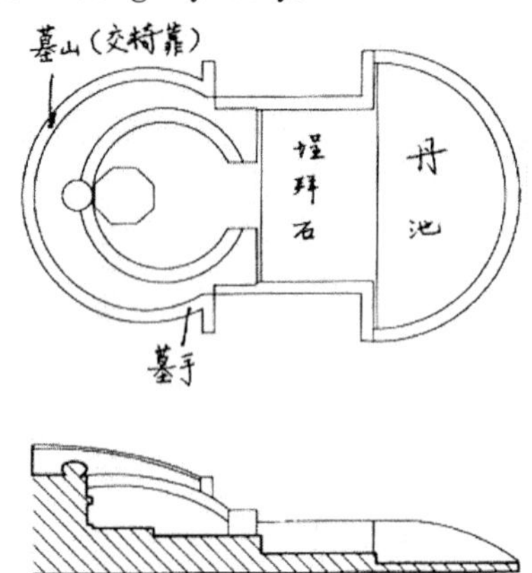

Fig. 10 Plan and section of grave of Xiong Fei and his wife (source: Relics of Dongguan, building Industry Press, 2005:31)

5 Cultural Heritage in Hakka Area after Master Yang

Master Yang Junsong selected the site and planned the turtle city Ganzhou, he teaching practical geomacy to his student's, and creating huatai for Kejia houses. Under Master Yang's influence, many cities in Kejia area become thriving and prosperous. Beside Ganzhou and Meizhou, Huizhou, Changting, Ruijin are the state historical and cultural cities.

Yang Junsong was also called Yang Jiupin, which meaning to save the poor people. Kejia people lives in the poor mountain area, if their residences could not adapt to the hill environment, they would suffer from disasters, such as soil erosion, mountain torrents, landslide and mud-rock flow, and so on. Master Yang created a residence for Kejia with huatai, he and his disciples designed Weilongwu to be adaptive to the hill surroundings. Weilongwu is both an organic building and an ecological building, it also embodying Chinese traditional philosophy and wisdom, standing for ideas of Fengshui and architectural arts. Under his influence, Kejia people of talent coming forth in large numbers. From Tang to Qing Dynasties(618—1911), there are many Kejia people of Meizhou passed imperial examinations as follows: the number of Jin Shi being 121, of them, the number of secleted to be Hanlinyuan Xueshi being 35, and the number of Ju Ren being 1645, Jie Yuan being 17.

There are many outstanding Kejia people, such as famous poet and calligrapher Song Xiang, famous educator and poet Qiu Fengjia, state man Ding Richang, diplomat and poet Huang Zunxian, state man Ye Jiangying, artist Lin Fengming and so on.

6 Conservation of Traditional Houses in Meizhou

The government of Meizhou pays great attention to protection of cultural heritage. The conservative measures are as follows:

a. Selecting outstanding Kejia traditional houses to apply for protective units of cultural relics.

As we know, there are 16,461 traditional houses in Meizhou.

According to Guandong Cultural Heritage: List of Immovable Relics, there are 5 traditional houses of the state units of preservation of cultural relics, and 50 of the provincial units, and 109 of the county units in Meizhou. The number of protective units is 164, being 1% of the total number 16,461.

b. Making full use of these protective units as the educative bases to unfold revolutionary traditional education, patriotism education, and so on.

For example, home of Ye Jiangying, the state unit of protective cultural relics, is used as a base of revolutionary traditional education.

The former residence of Qiu Fengjia, a famous educator and poet, and a national hero of resisting Japanese invasion, is of the state unit of protective cultural relics. The home of Xie Jinyuan General, who died for our country, is of the state unit of protection. Their homes are used as the bases of patriotism education.

c. Most protective units are used to show the traditional culture and architectural arts to the visitors, and used for tourism.

Conservative measures for non-protective units of traditional dwellings are the following:

a. Selecting traditional houses with features in Xingning

There are 4,751 traditional houses in Xingning County. The cultural office of Xingning launched an activity of selecting traditional houses with features in Xingning in 2010 to 2011. It fanned the flame of native people, and developed like a raging fire. Finally, 135 traditional houses with features were selected, and every house gains a license board carving "Traditional dwelling with features in Xingning". The number of traditional houses with features is about 2.84% of the total number 4,751.

b. Encouraging common traditional houses to run as hotels, museums, restaurants, offices, and so on.

c. In April, 17, 2018, the Forum of China's Country Reviving was held in Songkou Town, Meizhou, for the purpose to revive the economy of the country-side in Meizhou.

These measures of conservation above are of effectiveness.

7 Conclusion

This paper explores how the cities and buildings in Kejia area developing under influence of Yang Junsong. It finally discusses how to protect the cultural heritage, for example, the great number of traditional houses in Meizhou.

References

CAO Jingzheng, 2015. The dream of Suona horn, Beijng:China Nationality University Press.

CHEN Zhihua. Li Qiuxian, 2007. Three villages in Mei county. Beijng:Qinghua University Press.

Cultural Relics Bureau of Guangdong, 2013. Guangdong cultural heritage: list of immovable relics. Beijng: Science Press.

Dongguan Culture Bureau, Dongguan Relic Management Committee, 2005. Relics of Dongguan, Beijng:China's Architecture & Building Press.

HU Yuchun, 2004. Yang Jiupin and the origin and dissemination of the culture of Fengshui of the Hakkas in the Southern Ganzhou. Relics from south: 4, 67-70.

LIU Yujiang, 1992. The turtle divinative culture in ancient China. Guilin: Guangxi Normal University Press.

Record of Jiayingzhou prefecture during the Guangxu reign (1875. A. D. -1908. A. D.) of Qing dynasty. Volum 4.

Record of Jiayingzhou prefecture during the Guangxu reign (1875. A. D. -1908. A. D.) of Qing dynasty. Volum 9.

The collection of ancient books. Zhifang Standards.

The records of Ganzhou prefecture during Daoguang reign.

The relics management committee of Guangdong province. The hyistorical and cultural cities of Guangdong. Guangdong Map Press, 1992:102.

WU Qingzhou, 2008. Architectural culture of Hakkasin China, Wuhan: Hubei Educational Press.

WU Qingzhou, 2015. How the Chineseancient cities constructed with the ideas of bionics and pictographs. Beijng: China's Architecture & Building Press.

WU Qingzhou, 2011. Study on urban construction history and culture of the Tortoise City of Ganzhou, China city planning review, 20.

WU Qingzhou, 2012. Weilongwu and the form of Taiji Huasheng Tu//JIA Jun. Architectural history. Beijng: Qinghua Univesity Press.

YANG Sen, 1996. A dictionary of scenic spots and historical sites in Guangdong. Beijing: Beijing Yanshan Press.

YI Zhi, 2011. The homeland of hakkas. Beijing: Commercial Press.

ZHAO Guichang, MA Luhua, WEI Yuane, 2005. The significance of study on the Fengshui Ling's soil in Hong Kong//LIU Yizhang. HongKong hakkas. Guilin: Guangxi Normal University Press.

Rethinking the Nature and Values of Chinese Traditional Villages underpinned by Current Understandings of Heritage Authenticity

SHI Xiaofeng[1], *Beau B. Beza*[1], *David S. Jones*[1], *CUI Dongxu*[2]

1 School of Architecture & Built Environment, Deakin University, Geelong, Australia
2 School of Architecture & Urban Planning, Shandong Jianzhu University, Jinan, China

Abstract The nature of cultural heritage expresses the values of this heritage. Values determine the nature of the heritage elements to be conserved and the conservation approaches to be adopted. Under the tendency of contemporary cultural heritage understandings, which lean towards identifying and conserving more intangible orientation elements and expanding heritage values. This is despite the need to conserve Chinese traditional villages towards safeguarding China's spiritual vernacular homes and the living heritage of Chinese civilization recognising that "Asian heritage concerns more on the spiritual aspects than fabric" from debates about authenticity globally. However, it is hard to reach the aims above and even lead to "protective damage" in current conservation practice, due to the hysteretic and mechanistic steps in the selection of elements and the adoption of approaches. The root of these issues lies in understanding the nature of heritage. Based upon an analysis of the worldwide heritage authenticity debate, this paper examines Chinese heritage values and expectations assigned to traditional villages, and compares these heritage characteristics with China's current understanding of cultural heritage. In addition, the paper re-examines the nature of heritage towards improving the conservation element system recognizing discussions about authenticity, with a view to contributing to the improvement of Chinese traditional village conservation discussions and to providing experience and insights to the worldwide heritage conservation community.
Keywords traditional villages, China, heritage nature, authenticity, heritage value, conservation practice

1 Introduction: the Involving Ideology of Heritage and Authenticity

The Burra Charter (Australia-ICOMOS, 1979) is considered to be an important milestone in the ideological evolution of cultural heritage internationally. The *Burra Charter* put forward the notion of "the significance of culture" that has thereupon guide heritage value assessment and identification in Australia (Ahmad, 2006; Waterton, et al., 2006). The *Burra Charter* changed Australia's understanding of heritage values from specific historical, aesthetic and scientific values, which are naturally possessed by tangible entities as advocated by the *Venice Charter* (ICATHM①, 1964), to a wider range of cultural values including social, spiritual, emotional values, etc. (Ruggles & Silverman, 2009; Goetcheus & Mitchell, 2014).

More importantly, the *Burra Charter* advocated this ideology perspective whereby heritage is structured by subjective logic based upon its perceived and endowed values rather than objective logic that assumes that heritage is determined by the intrinsic representation of its fabric (Vecco, 2010). Such not only opens up a context for intangible elements in heritage sites to be appreciated, but also redefines the concept of heritage to some extent. In 1987, when the *Washington Charter* (ICOMOS, 1987) expressed the significance of protecting historical cities, this Charter was viewed as being

① Formed in the second International Congress of Architects and Technicians of Historic Monuments, 1964, Venice. Approved by ICOMOS, 1965 innovative because it recognized both tangible and intangible values as objects of conservation (Ruggles & Silverman, 2009).

With the introduction of concepts such as "cultural landscapes" and "living heritage", the connotation of current cultural heritage internationally has been greatly enriched, and the types and characteristics of heritage have multiplied. In particular, the use of value and living characteristics of contemporary heritage are now more widely accepted and emphasized drawing upon the experience and lessons from ICCROM's practice of heritage conservation in Asian areas (Poulios, 2014).

Debate about "authenticity" is another important factor that has inspired the ideological evolution of cultural heritage. There has been a long-standing debate over whether Western heritage philosophy and approaches are applicable in other parts of the world, especially in Asia (Winter, 2012). The initial cultural heritage philosophy, that documents and systematically categorises the modern heritage movement originated in Europe and was drafted for Western heritage definitions and Western heritage contexts (Taylor, 2017). Therefore, Western heritage and tangible celebratory places and sites are preferred from the perspective of ICOMOS' heritage categories, and material authenticity is pursued more in conservation practice and has been the mainstream cultural heritage agenda for a long time (Winter, 2012).

However, the importance of material heritage is limited in many cultures. For example, due to the decay of materials used in the construction and fabrication of structures in Asia, the nature of its conservation traditions has tenuously embraced a common renovation practice: update the legacy materials with minimal intervention and reversibility instead of maintaining the authenticity of the materials.

The Nara Document (ICOMOS, 1994) examined authenticity, a crucial conservation principle, in 1994. *The Nara Document* observed that the definition of authenticity should be based upon the relevant cultural context. Traditional Western philosophies and approaches on conservation are embodied in historical relics, while Eastern philosophies and approaches traditionally try to conserve spirits that are represented by those relics (Winter, 2012). In Asia, many cultures have a spiritual rather than a material view to place and or building, in which objects and places are important tools to convey deeper spiritual meanings (Winter, 2012).

In 1996, *the Declaration of San Antonio* (Americas-ICOMOS, 1996) continued the international discussion about all aspects of authenticity. It observed that authenticity includes more than material integrity and is the sum of a series of material, spiritual and relational elements. Different types of heritage have different compositions of authentic contents. The key is to conserve the most important authentic contents of the heritage and to balance them (Americas-ICOMOS, 1996).

Obviously, authenticity here points to the nature of cultural heritage, which is determined by the values endowed by people when defining a heritage legacy. To the contemporary heritage, the nature, value and authenticity in the conservation of a heritage legacy together structure possess a linkage of mutually supporting and deciding variables (Fig. 1). Among them are the contents that need to maintain authenticity, being the nature of the heritage that needs to be conserved mostly, that are the basic elements and characteristics of the heritage legacy that is generally acknowledged as bearing "heritage values".

2 Issues in the Current Conservation of Heritage

When the values of a heritage legacy are no longer inherent and self-evident in objective material entities, this change of values is dependent upon people's perceptions, understandings and endowments about a place (de la Torre, 2013; Ashworth, 2011; Ashworth, 1997). The key point of

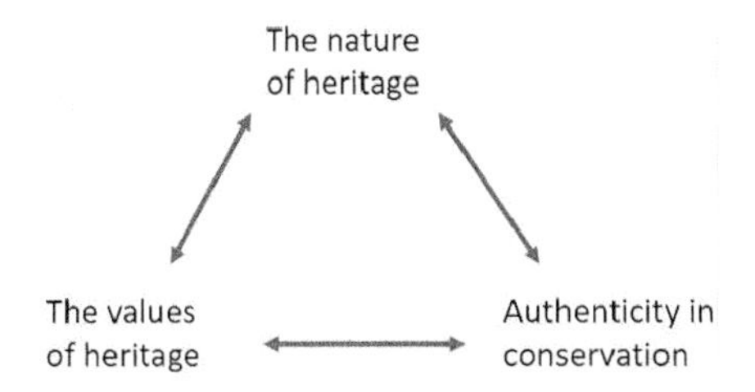

Fig. 1 The relationship diagram of the nature, value and authenticity of a heritage legacy. Source: authors

conservation in this context rests upon authenticity, being the need for it to be seriously discussed, identified and (re)defined. At present, the concept and many conservation approaches to cultural heritage are developing recognising the evolution of heritage values, types, connotations and changes in the understandings of authenticity. However, this phenomena cannot appropr iately deal with the nature of heritage that is present globally (Poulios, 2014; Fredheim & Khalaf, 2016).

The following two case studies evidence this conclusion.

The first is the Chaco Culture National Historical Park, a world heritage in the state of New Mexico in the United States of America (USA). In order to conserve the archaeological remains, the Park authority moved the remaining Navajo Indigenous communities and peoples, whose ancestors established this site, out of the designated Park area or lands. Though in the last decades the Park authority has shown a consistent interest in consultation and participation with Navajo communities in the conservation and management of this site, however they refused to permit and enable a religious group named "new age" to perform rituals on this site and without religion discrimination, all rituals were (and are) forbidden there including those practiced by the Navajo Indigenous communities. Therefore, the community who established and developed strong family, cultural and religious ties to the site (inside the Chaco Culture National Historical Park) has lost the right to perform rituals on the site (Poulios, 2014).

A similar phenomenon has happened at the Honourable Society of the Middle Temple in London in the United Kingdom (UK). The "Middle Temple" is one of the four English Inns of Court that hold the exclusive right to call their members to the Bar as barristers and has been synonymous with both a cultural organisation and a place for centuries (Fredheim & Khalaf, 2016). The "Middle Temple" possesses not only buildings and a built environment that needs to be conserved as heritage, but also exquisite interior decoration, furniture and instruments. As argued by Fredheim & Khalaf, "Conserving the Middle Temple appropriately, therefore, involves considering the significance of traditional practices at the Middle Temple, alongside its buildings, collections and urban context. However, existing value typologies in conservation practice have not been designed to accommodate such heritage diversity. Established approaches to objects conservation would recognise the value of the historic interiors and collections and regard their continued use as a threat, thereby margina-lizing and delegitimising the significance of the traditional use of the Middle Temple" (Fredheim & Khalaf,2016).

The authentic contents in current conservation practice struggle to be aligned with diversified heritage values and rich heritage essences, which leads to a series of problems caused by conservation, including even leading to "protective damage" to some extent (Hu, et al. , 2017) . The conservation of traditional villages in China encounters similar issues.

3 The Conservation and Issues of Chinese Traditional Villages

Traditional Villages' in China refer to villages that still retain a large number of traditional characteristics, customs and traditions of Chinese civilization in their contemporary daily life. "Traditional Villages" is a recognized term and concept in Chinese heritage practice. These villages are considered to be important carriers and representations of Chinese traditional culture, the spiritual homelands of the Chinese nation, and offer narratives about the foundations of the Chinese nation (MOHURD, et al., 2012; Tu, et al., 2016). More importantly, as the carriers and representations of contemporary villager's lives, they are living fossils (MOHURD, et al., 2012; Liu, 2014; Yang & Liu, 2017), possessing and representative of living heritage (Zhang et al., 2017).

However, due to China's rapid urbanization, modernization and cultural globalization, the number of Traditional Villages in China continues to decrease rapidly (Feng, 2013). In 2012, the Ministry of Housing and Urban-Rural Development (MHURD) and the State Administration of Cultural Heritage (SACH) jointly launched a selection process to support and enable the conservation of nationally recognised Traditional Villages. As of April 2019, according to their selection criteria[①], a total of 6,799 villages have been selected. These villages receive a payment allocation of ¥3 million each underpinned by national finances set aside for their conservation and restoration activities (MOHURD, et al., 2012). Thus, this "stop-loss" approach to conservation for Traditional Villages has achieved remark-able results in this conservation arena at present. The loss of Traditional Villages was initially prevented better enabling the telling of their contributions towards the maintenance of global cultural diversity and for traditional Chinese culture.

However, although the state has invested financial and other resources, issues still persist in conservation practices for Traditional Villages. A large amount of funds is still used for the renovation of tangible houses and streets, but hardly for the conservation of intangible heritage aspects. Moreover, in order to conserve these villages, especially the typical traditional buildings and spatial patterns, the conservation planning of those villages has been often designed to migrate the villagers out of their original houses, while the old villages are "renovated" or developed into tourism venues (Fig. 2). In other words, the actual resident villagers have had to move out of their homes because the houses they have lived in for generations were being "conserved" (Xu & Wan, 2015; Hu, et al., 2017).

Fig. 2 Plan of a national traditional Village in north China

In addition, some intangible heritage projects are often being used to enable

① Evaluation and Identification Index System for Traditional Villages (Trial) (Jiancun [2012]125)

tourism and profit rather than supporting the continuity of daily life practices and traditions (Hu, et al. ,2016).

Thus, in China's conservation practice towards Traditional Village's, which aims at safe-guarding the living heritage of Chinese culture, a de-activation and dislocation between tangible and intangible heritage is occurring precisely because of the intervention of "protection" policies, despite these villages being identified and being emphasized as living carriers of Chinese culture, or living cultural fossils. When probing into the causes of this issue, an examination and comparison of the authenticity contents possessed by Traditional Village conservation methods is warranted, and such needs to embrace the endowed heritage values and the nature of this heritage.

4 Examining the Heritage Values and Conservation Approaches Towards Chinese Traditional Villages

There are mainly three Chinese government documents related to the conservation of Chinese Traditional Villages. They are:

- *Guidance on Strengthening the Conservation and Development of Traditional Villages* (Jiancun [2012] 184) [GSCD];
- *Evaluation and Identification Index System for Traditional Villages* (Trial) (Jiancun [2012]125) [EIIS]; and,
- *The Basic Requirements for the Conservation and Development Plannig of Traditional Villages* (2013 Trial Version) [BRCDP].

The GSCD is the overarching guiding document for the identification and conservation of Traditional Villages as heritage. It presents the purpose, significance and expectations for the conservation of Traditional Villages. In this document, diversified values are embodied explicitly and implicitly, such as: historic values, cultural values, scientific values, aesthetic values, social values, economic values, spiritual values, emotional values, living values and use values (Tab. 1). Among them, are expressions like "living transmission" that are quoted many times by the other two documents and Chinese academia.

Tab. 1 Relevant values of traditional villages in the lens of government documents

Related statements	Source	Values present
Villages with high historic, cultural, scientific, artistic, social and economic values	Guidance on Strengthening the Conservation and Development of Traditional Villages (Jiancun [2012]184)	Historic value, cultural value, scientific value, aesthetic value, social value, economic value, spiritual value, emotional value, living value and use value
Carrying the essence of Chinese traditional culture, it is the non-renewable cultural heritage of agricultural civilization.		
Embodying the spirit of the Chinese nation and is the link to maintain the cultural identity of Chinese descendants.		
The foundation of prospering and developing national culture		
The principle of "living transmission"		
The authenticity, integrity and sustainability of cultural heritage should be maintained.		

Continued

Related statements	Source	Values present
It has high scientific, cultural, historical and archaeological values	Evaluation and Identification Index System for Traditional Villages (Trial) (Jiancun [2012]125)	Scientific value, the value of technological aesthetics, cultural value, historic value, archaeological value, living value
The architectural details and decorations are very exquisite, with high artistic value		
"Liveliness" in the part of intangible heritage		
The principle of "living transmission"	The Basic Requirements for the Conservation and Development Plannig of Traditional Villages (2013 Trial Version)	Historic value, aesthetic value, scientific value, social value, living value
Assessing its historical, artistic, scientific and social values		

The EIIS is mainly used to guide the selection of villages (determine who can be a Traditional Village and the nature of the heritage to be conserved). The EIIS also provides the basis for the selection of specific heritage elements and the implementation of conservation practice in the later period of conservation works. In other words, it is responsible for translating heritage claims to the specific contents of conservation.

The last one, the BRCDP, offers specific directions as to how to draft a conservation plan.

By comparing the three documents, it has been concluded that the relevant provisions of EIIS narrows down those in the GSCD in terms of the significance of Traditional Villages as possessing heritage (Tab. 1). In relevant government documents, such as GSCD and discussions in academia about these documents, it is not difficult to find that cultural values, living values, use values, spiritual values and emotional values are repeatedly emphasized in distinguishing Traditional Villages from other Chinese rural heritage. However, some of the values above have not been translated into heritage elements or requirements for conservation in the EIIS, such as emotional values, spiritual values, contemporary use values. This means that there is no specific conservation contents or means to echo these values in the conservation process. In addition, some values are confined. For example, "living transmission" is only used for intangible heritage in EIIS. Aesthetic values, historic values, scientific and values are emphasized in the EIIS.

However, drawing from a comparison of the selection criteria between Chinese Traditional Villages and Chinese Famous Historic and Cultural Villages, another type of rural heritage, the requirements for the aesthetic, historic and scientific values of Traditional Villages are lower than those of Chinese Famous Historic and Cultural Villages. In other words, these values are not the key points that determine a Traditional Village to be deemed as possessing a heritage legacy, but might become the key points of conservation practice because the provisions lack a clear description of the village's basic characteristics and the performance measures used in those heritage projects. Thus, it is hard to conserve some necessary intangible projects in practice.

Moreover, from actual conservation processes and referenced documents, the conservation of Traditional Villages finally presents a situation in which the conservation of physical built environments are given priority with their tangible historic, aesthetic and scientific values being emphasized. In addition, as the selection threshold of intangible heritage items is higher than that of material heritage from the policies in the EIIS, it is difficult to implement the conservation of these items in conservation practice.

First-hand investigating some of these designated Traditional Villages by the authors confirms this conclusion. It is clear that conservation of authentic contents pursued by mainstream Chinese conservation methods are not able to addresses the nature of this heritage and the values that are attributed to it. In substance, China's concern and emphasis upon heritage values of Traditional Villages demonstrate a comprehensive understanding of such heritage under current ideology. However, in conservation practice, similar to heritage conservation issues globally, such as evidenced in the Honourable Society of the Middle Temple in the UK, and the Chaco Culture National Historical Park in the USA, the authenticity presented by heritage conservation methods does not adequately enable a respectful conservation of the nature and values of that heritage.

5 Conclusion

When authenticity in conservation methods cannot be aligned with the nature of the heritage, the direct consequence is that some important heritage elements and features may be ignored in conservation practice. For Traditional Villages in China, as a typical continuing cultural landscape, the human elements and some non-material elements constituting that heritage, such as the interactive relationships between humans and their natural environment, as well as the characteristics of liveliness and integrity, have been neglected in current practice.

In addition, it is relevant to recognise that development and change are also charac-teristics of living heritage (Poulios, 2014). For Traditional Villages, as a kind of living heritage, it is neither practical nor realistic to completely and statically protect (preserve) buildings in these villages, but rather conservation policy and practice attention should be given to the spiritual, emotional and use values that a particular village bears and narrates. It is a great challenge in China to develop a creative and culturally respectful conservation approach towards Traditional Villages in the contemporary setting to align with the nature of living heritage evident in Traditional Villages. It will also be a momentous achievement in the conservation of cultural heritage in China and abroad.

In addition, there is a need for a clear logical path to be followed when drafting conservation methods and a conservation plan for this type of heritage. When a heritage legacy is defined by its heritage values, its basic elements and characteristics should be sorted out from the perspective of its important values. That is, we need to translate an understanding of the nature of that heritage. The authenticity of that heritage should be identified by this process. Finally, the possible heritage conservation method should be designed based upon the authenticity of this type of the heritage.

References

AHMAD Y, 2006. The scope and definitions of heritage: from tangible to intangible. International journal of heritage studies, 12(3): 292-300.

Americas-ICOMOS, 1996. The declaration of San Antonio. San Antonio: Americas-ICOMOS.

ASHWORTH G, 2011. Preservation, conservation and heritage: approaches to the past in the present through the built environment. Asian anthropology, 10(1): 1-18.

ASHWORTH G J, 1997. Conservation as preservation or as heritage: two paradigms and two answers. Built environment, 23(2): 92.

Australia-ICOMOS, 1979. The Burra charter: the Australia ICOMOS charter for places of cultural significance. Australia: Australia-ICOMOS.

DE LA TORREM, 2013. Values and heritage conservation. Heritage & Society, 6(2): 155-166.

FENG Jicai, 2013. The dilemma and outlet of traditional villages. Folk culture forum, 218(1): 7-12.

FREDHEIM L H, KHALAF M, 2016. The significance of values: heritage value typologies re-examined. International journal of heritage studies, 22(6): 466-481.

GOETCHEUS C, NORA M, 2014. The Venice charter and cultural landscapes: evolution of heritage concepts and conservation over time. Change Over Time, 4(2): 338-357.

HU Binbin, LI Xiangjun, WANG Xiaobo, 2017. Investigation report on the protection of Chinese traditional villages. Beijing: Social Sciences Academic Press (China).

ICOMOS, 1987. Charter for the conservation of his-

toric towns and urban areas (the Washington charter). Washington DC: ICOMOS.
ICOMOS, 1994. Nara document on authenticity. Nara: ICOMOS Symposia.
LIU Peilin, 2014. Traditional Settlement cultural landscape gene. [s. l.]: The Commercial Press.
MOHURD, MCT & MF, 2012. Guiding opinions on strengthening the protection and development of traditional villages Jiancun. [2012]. http://www.mohurd.gov.cn/wjfb/201212/t20121219_212337.html.
POULIOSI, 2014. Discussing strategy in heritage conservation: living heritage approach as an example of strategic innovation. Journal of cultural heritage management and sustainable development, 4(1): 16-34.
RUGGLES D F, HELAINE S, 2009. From tangible to intangible heritage. Intangible heritage embodied. [s. l.]: Springer.
TAYLOR K, 2017. Landscape, culure and heritage, Changing perspectives in an Asian Context. Victoria: Deakin University.
TU Li, ZHAO Pengjun, ZHANG Chaorong, 2016. Theory of conservation of traditional villages. Urban development studies, 23(10): 118-124.
Vecco M, 2010. A definition of cultural heritage: from the tangible to the intangible. Journal of cultural heritage, 11(3): 321-324.
WATERTON E, SMITH L, CAMPBELL G, 2006. The utility of discourse analysis to heritage studies: the Burra charter and social inclusion. International journal of heritage studies, 12(4): 339-355.
WINTER T, 2012. Beyond Eurocentrism? Heritage conservation and the politics of difference. International journal of heritage studies, 20 (2): 123-137.
XU Chuncheng, WAN Zhiqin, 2015. Argumentation on basic thoughts of traditonal village protection. Journal of Huazhong Agricultural University (social science edition), 120(6): 58-64.
YANG Liguo, LIU Peilin, 2017. The inheritance and its evaluation system of traditional village culture: a case study of traditional village in Hunan province. Economic geography, 37 (12): 203-210.
ZHANG Haolong, CHEN Jing, ZHOU Chunshan, 2017. Research review and prospects of traditional villages in China. City planning review, 41(4): 74-80.

Nomad Vernacular Architecture in the Heritage City: the Re-assemblage of Traditional Anhui Architecture in the Low Yangtze River Area

Plàcido Gonzàlez Martínez

Tongji University, Shanghai, China

Abstract The production of identity is consubstantial to urban development, where vernacular architecture has been reproduced and recreated with the aim to provide roots to the characterless urban periphery. Some studies have reflected in the production of identity in the peripheral areas of Shanghai in the shape of historically-inspired architecture, opening a longstanding discussion about authenticity and fake in contemporary architecture. This paper studies a completely different phenomenon; not pointing at the copy or interpretation of historic buildings, but to the dissasemblage of vernacular architecture in the Anhui province of China and its reconstruction in suburban enclaves of the Low Yangtze River megalopolis with the same purpose of identity production. Despite the parallelism with the typology of the open air museum, it is our contention that this constitutes a completely novel phenomenon that deserves particular attention in the wider framework of a global city like Shanghai with a particular thirst for heritage. This paper will offer an insight into two recent real estate and touristic developments in the lower course of the Yangtze River that share this procedure of transplantation of vernacular architecture from rural areas to suburban metropolitan locations. A practice that resulted from lack of regulations prior to 2005, the study of this transfer of building stock reveals important heritage questions around authenticity, integrity and identity in the framework of the contemporary heritage city. This paper wants to highlight what Chappell characterized as "mixed feeling" around this practice, introducing an analysis of the reconstruction projects, with an insight towards the methodologies of documentation and conservation, together with an outlook of the new urban layouts that transplanted architecture in two projects (The Ahn Luh Hotel in Zhujiajiao and the Ahn Luh Lanting in Shaoxing) are recreating, pointing to the possibility of enhancing their interpretation through minimal interventions.

Keywords heritage authenticity, heritage reconstruction, Huizhou architecture, suburban identity, heritage city

1 Introduction

This research departs from a trivial hotel inquiry for a family visit to Shanghai. A search on "the most exclusive hotels in Shanghai" in Tripadvisor in the summer of 2016 caught my attention: The top result was the Ahn Luh Hotel in Zhujiajiao: Strikingly for a heritage conservation specialist, the hotel featured an ancestral hall and a theatre stage, both dating from the 18th century and originally from the neighbouring Anhui Province.

Rather coincidentally, a few months later the author received an invitation from the Qinsen Group, owner of Ahn Luh, to visit the projects that the Company was undergoing. The visit included the hotel, plus the new development of Shaoxing and the general storage center of the Qinsen Group in an unknown location.

The interest on the case was later heightened by further conversations with specialists on heritage conservation in China, who referred to the extent of the phenomenon of architectural nomadism in the country. Authors such as Knapp (2010) have referred to the development of open-air village-like museums in China which have fundamentally educational and tourism purposes; but the conversations pointed towards more entrepreneur-oriented aims, such as hotels and residential compounds, posing multiple questions on conservation values and on the production of identity in suburban locations.

This paper aims to share thoughts on the initial stages of this research. Although similar examples are spread all over the country, it focuses on two sites developed by the Qinsen Group: Ahn Luh Zhujiajiao and Ahn Luh Shaoxing. The author has visited the two sites, plus developed a series of semi-structured interviews with managers, designers and heritage specialists of the Qinsen Group as well as with researchers on heritage conservation in China. The author has also performed an intensive literature review on the issue of open air museums internationally and in China, steering the reflection about issues of authenticity; integrity; the continuation of traditions; and urban development that stand at the base of this research.

This research poses the following questions:

How does relocation affect established notions of heritage authenticity? Corbin (2002) and Gregory (2008) have reflected on open-air museums; but the case studies represent mostly entrepreneurial rather than educative interests, with a crucial conceptual shift among fields: what could be interpreted as shortcomings in heritage authenticity could result in enhanced urban identity. What other non-heritage notions of authenticity are applying here, and how are they affecting the understanding of heritage authenticity itself?

How does this entrepreneurial practice constitute an alternative to intangible heritage conservation? ICOMOS (2003) refers to the importance of the continuity of skills: Are the opportunities that the relocation process gives to trained artisans a way to compensate loss of material authenticity? Does this also imply a homogenization of traditions, in case this constitutes the only reservoir for the practice of crafts in the future?

Is it possible to add further layers that build upon the understanding of heritage authenticity in the realm of heritage interpretation? As international conservation documents state, vernacular heritage authenticity is based on the relationship with its original surroundings. Once relocation breaks this relationship, we argue how it can be reestablished by interpretation means.

The paper will first introduce the open air museum as an immediate precedent; as it rises many relevant questions around the motivations, the techniques and the outcomes of the relocation of vernacular architecture. It will later reflect on the production of identity in the outskirts of the city. Then it will analyze the two cases, pointing at the issues of authenticity and narrative as core concepts for the discussion, and will introduce a proposal to complete the narrativization of the displaced structures, acknowledging contemporary prerequisites on heritage interpretation.

2 Open Air Museums

Corbin (2002) traces the origins of open air museums to the aristocratic hameaus built in palaces like Versailles in France during the late 18th Century. Such reconstructions followed ideals of a simple rural life as a stronghold against the threats of the urban environment (Gonzàlez Martínez, 2010). Moolman (1996), Chappell (1999) and Corbin (2002) refer to how later on, 19th-century World Exhibitions displayed on site vernacular villages with propagandistic purposes in the framework of colonialism. The same authors also point at nationalism in Europe as a key to understand the origins of Skansen, the first open air museum in the world, inaugurated in Stockholm in 1891.

As Moolman (1996) states, "an open air museum comes into existence when specific museum objects (buildings) are not preserved in situ anymore, but are moved to a new site where they can be better preserved and interpreted". From this perspective, displacement causes what Chappell (1999) refers to as "mixed feelings", as the aim of preservation is confronted with established notions in heritage conservation such as the 1957 ICOM Doctrine, mentioning that "the preservation of a specimen of popular architecture on the site where it was erected is the most satisfying solution from the historic and artistic point of view, and should always be given priority" (Moolman, 1996).

The Venice Charter adds on: "a monument is inseparable from the history to which it bears witness and from the setting in which occurs" (ICOMOS, 1964). The later contribution from Gregory (2008) is relevant to this discussion, as its extensive review of conservation documents validates disassemblage and reconstruction when they result the only way to preserve the heritage asset (ICOMOS, 1964, ICOMOS Canada, 1983, ICOMOS New Zealand, 1992, ICOMOS, 2003), or when the surrounding site is no longer of particular significance (ICOMOS New Zealand, 1992).

Accordingly, the decision to displace a structure is based on an arithmetical evaluation: to the "folkloristic Lord Elgins" (Chappell, 1999), the original place is perceived of lesser historical significance than the potential increase of value in the future. By these means, the loss of the original setting is compensated through the increase in educational value. Displaced structures serve then as "sacrificial elements" to enhance public awareness about vernacular architecture, promoting its conservation elsewhere according to established conventions.

Quantifying almost one quarter of all visits to museums in a global scale, open air museums are "major venues for public encounters with architectural history" (Chappell, 1999), allowing for the practice of traditional building construction methods, thus joining tangible and intangible heritage dimensions. From this perspective, the practices of disassemblage and reconstruction would offer the possibility of not only extending the life of buildings, but also of a specific knowledge on crafts that would otherwise be at risk.

The discussion about authenticity is subject to multiple interpretations under the lenses of open air museums. We would focus on the blend of (1) Edition; (2) Scale and (3) Narrative that Corbin (2002) has characterized as main obstacles for the authentication of open air museums. According to Corbin, the "edition" of vernacular architecture results in sanitized versions of life. Also, the reduced "scale" and number of the rebuilt elements, mainly representing residential typologies, lack the functional diversity that characterizes a living settlement. Finally, the "narrative" that is outlined from this selection is simplified, easy to understand by visitors, especially due to the affinities that residential spaces develop between house museums and the public.

3 New Typologies of Displaced Architecture in China

The recent boom of open air museums in China is the outcome of several factors. Firstly, an accelerated hollowing of rural settlements and the massive shift from rural to urban population (Pola, 2019). Secondly, a recovery in the appreciation of Chinese culture and traditions as part of rising nationalism in the framework of globalization (Dirlik, 2016). Thirdly, the social changes in China promoting leisure culture and the rapid growth of the tourism industries (Silverman and Blumenfield, 2013).

Depopulation of rural areas means also the dereliction of large tracts of vernacular houses and buildings, which appear as highly coveted economic and cultural assets. The condition of this building stock has been generally deemed as dilapidated, after years of neglect, lack of proper infrastructure and poor living conditions by the population. Not subject to any maintenance works, not to mention conservation works, the main heritage question rises as according to the international doctrine, we may be facing a situation on which the displacement of vernacular architecture can be, in fact, the sole means of survival. At the moment we cannot confirm if this was the sole option due to the impossible traceability of the buildings that have been reassembled in our case studies, but recent literature (Pola, 2019) highlights this possibility.

As in many cases in China, a blend of preservation and economic interests stand at the origins of this process. Even if the Law of the People's Republic of China on the Protection of Cultural Relics passed in 1982 adhered to international principles preventing from the displacement of cultural relics, the slow advancement of listing of cultural relics since the 1980s, as well as the lack of popular and institutional appreciation, enabled this practice. This offered the opportunity for the sale of buildings to connoisseurs and collectors at a yet unknown scale.

A piecemeal inquiry on this process reveals multiple developments in different locations in China: from the reconstructions in the rural areas that intensify the vernacular flavour of small towns with touristic purposes; to pure open air museum displays in world events like the 2019 International Horticultural Exhibition in Beijing. In our paper we just focus in one developer, the Qinsen Company, due to the intrinsically different character of its work.

Interviews with the Qinsen Group reveal how from the 1980s to 2005, the company acquired around 600 buildings, later stored for future reconstruction. From the legal perspective also; even if the existing regulation and increased resident's awareness now forbids the acquisition and disassemblage of buildings, owners of already disassembled structures are still entitled to store them and rebuild them in their preferred locations.

Our research focuses on suburban areas in the Yangtze River Delta megalopolis, which incorporates Shanghai, Suzhou and Hangzhou as main metropolitan areas. Historically, suburban locations are the preferred locations for this kind of processes as they offer in-between spaces where urbanization is on the making, offering the opportunity to experiment with cultural assemblages in architecture with the goal of producing a local identity. This comes as no news: studies on suburbanization (Fishman, 1987; Hayden, 2004; Gonzalez Martinez, 2010) have highlighted how in the process of urbanization, real estate developers; urban planners and architectural designers used the paradigms of culture, the small town and vernacular architecture to facilitate the popular appropriation of new urban developments and therefore, increase the economic performance in the real estate market.

The referred authors also mention how architectural historicisms would be the preferred tools to recreate ancient architecture, making the urban periphery a living catalogue of the continuity of styles through the process of modernization. What we witness in the case studies is how the importance of authenticity, not only for heritage conservation but also to other industries like tourism and the real estate market, has produced a counter effect.

The reproduction of architectural styles is identified as representative of low culture and depicted as "fake" (Den Hartog, 2010; Piazzoni, 2016), therefore depreciating their real estate value and their touristic reputation, particularly for the most affluent clientele and also for administration cadres. It is our contention that the emergence of authenticity as a key concept in the fields of tourism and urban studies has fuelled the search for a more intense, real connection with the users; therefore finding in "real" historic architecture the raw material for (entrepreneurial, educational) success, that legitimizes the practice of architectural dislocation, such as in the following case studies:

3.1 Ahn Luh Pinzhen Garden in Zhujiajiao, Shanghai

Located in the west of Shanghai, the compound incorporates a luxury hotel plus 35 residential villas. Satellite pictures show a collage of different real estate developments linked to new roads, scattered among the characteristic lakes and canals of the lower course of the Yangtze River and with which the compound establishes no relationship.

Fig. 1 Satellite image of the site © Baidu

The urban layout strategy produces an urban image towards the two main roads; Kezhiyuan Road to the south and Zhuhu Road to the east, through the abstract fa ades of two-floor buildings incorporating commercial facilities. The other two sides are defined by the 35 villas facing the canals. The historic structures an ancestral hall and a theatre stage- are located in the core of the block and constitute the highlights of the hotel. They are surrounded by the 40 hotel rooms, which have pedestrian access from the ancestral hall. The villas are accessible by car through an internal private road that separates them from the hotel rooms. The height and size of the villas creates effectively a buffer zone between the surrounding environment and the hotel facilities.

The hotel incorporates two main elements from the Qing Dinasty: an ancestral hall following the Five Phoenix Tower model, and a theatre stage located in front. The open space between both buildings constitutes the main entrance to the hotel, and is occasionally used for open air theatre plays. Ahn Luh uses now the main open courtyard of the 1,400 sq. m. ancestral hall as open-air lobby occasionally hosting art exhibitions; an issue that reinforces its perception as a high-end touristic facility. The second courtyard of the building houses now the reception area itself, with its main lobby in the ground floor and meeting rooms and a library in the second floor.

The reconstruction is a faithful one, and the Qinsen company proudly refers to the support of the Ruan Yisan Foundation, one of the most important private institutions in the field of heritage in China of

Fig. 2 The ancestral hall © The author

which the chairman of Qinsen is also director, as a legitimizing fact. In fact, the authentication of the structure is deemed possible thanks to the intervention of trained craftsmen. Their skills were necessary not only for the reassemblage of the structure, but also for the production of construction and decoration elements and the later maintenance of the building.

Fig. 3 The theatre stage © The author

The historic structures have an iconic role, acting as motivators for the historically-inspired modern design of the other buildings in the block. The hotel rooms are basically one-floor structures resembling the shape of traditional roofs that are interpreted in an abstract way and built with modern construction materials. The belt of high standing residential villas incorporates some displaced elements too, such as the stone carved gates, and their layout enables an interpretation of traditional residential architecture as they feature also the traditional courtyards, even though the main interior spaces of the villas are facing the waterways in the north and west sides of the plot in order to increase the sense of isolation.

3.2 Ahn Luh Pinzhen Garden in Shaoxing

Located in the east of Hangzhou, Ahn Luh Lanting spreads in a picturesque valley setting of the neighboring Kuaiji hills. Satellite pictures show the relatively isolated location, as the valley is only accessible from the neighbouring Yangming Road, which runs in a north-south direction following the Pingshuidongjiang watercourse. A cherry tree park acts as a buffer area between the road and the compound. The topography and the urban layout contribute to the production of this enclave, which is divided into two parts. The 99-room hotel is developed along a new artificial watercourse running in the west-east direction; whereas the private residential compound with 22 residential units is located to the north, closer to the main access to the site.

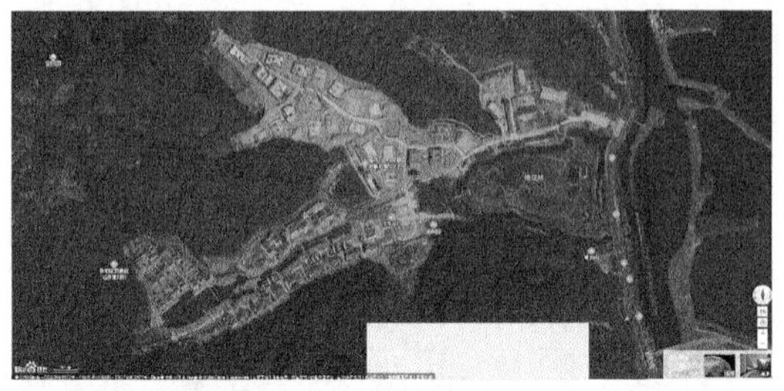

Fig. 4 Satellite image of the site and its surroundings © Baidu

The artificial creek originates in a reservoir, which is located in the uppermost part of the compound acting also as a scenic pond. The design layout aims to reinforce the picturesque character of the pond and the creek, using a two half fish spine road system in the north and the south, accessing to the hotel rooms. Classified as a high end hospitality facility by Tripadvisor, the hotel incorporates reassembled Huizhou-style residential structures of two types; a group of traditional courtyard houses in the north side, including the reception and administration area; plus a group of single-bay residential units in the south.

Fig. 5 The pond and the hotel rooms in the southern fringe © The author

The website of the Qinsen Group highlights eight major historic structures in Ahn Luh Shaoxing, dating from the Ming and Qing dynasties and ranging between 400 and 117 sq. m. in size. These eight buildings originally had both private and public functions, and their integration in the complex offers a recreation of the diverse size and complexity of a small town. As in Ahn Luh Zhujiajiao, the reconstruction was developed by the craftsmen of the company, answering to the quest for authenticity aimed by the Company. Furthermore, we also detect an added procedure for the authentication of the site, which is the conscious analogy to the UNESCO World Heritage sites of Hongcun and Xidi in the neighbouring Anhui Province, which appears as a clear reference in the landscape design of the site.

Fig. 6 The carved building © The author

4 Discussion

As Corbin (2002) has shown, the authentication of the case studies by their guests and residents and lies in three fundamental factors. The tangible ones refer to the display of original materials, which after being reassembled, still show the required patina and a verisimilar appearance. The associational ones appear in their relative position and layout; particularly when they reproduce famous picturesque settings of rural China, and also when they are tagged in an effort to show the scientific rigour of the reconstruction work. The supporting ones appear when the buildings express a notion of variety, which is achieved through the vernacular reassembled elements as well as by the modern incorporations.

Despite this successful contribution to the authentication of the reconstruction, the editing, scale and narrative challenges to the authenticity of the site also described by Corbin (2002) operate similarly in the two cases. Both are now sanitized versions of the past, which have erased the important lapse of time during which these vernacular structures became home for an important number of families. Furthermore, the lack of a representative variety of urban functions in such villages is still a challenge for its interpretation. Adding to this, the whole discourse is simplified for there is a single understanding of the nature and character of this houses which leads to their high class origins.

A reflection on this process must also consider the intrinsic character of this architecture consisting of easy to assemble elements. Acknowledgement to this fact, as it was traditionally and in an extreme way-depicted by ICOMOS through the Ise Temple as the conceptual origin of the Nara Document, is fundamental in order to understand the ordinary, simplified assumption of East Asian architecture as almost movable assets. In light of this, and even though dissassemblage may have a traditional justification, international principles of conservation that inspire regulations such as the current Law on Cultural Heritage in China, still acknowledge the importance of the attachment of built structures to their original sites.

Paradoxically, the particularities of the recent history of China, particularly the distribution of the housing stock among the population in the 1950s, are now making that conservation practices promote a return to their affluent origins. In this sense, and acknowledging the legitimate practice of the Qinsen Group, we would propose the addition of an additional layer of information in the interpretation of this process. This layer could be included as attached information that beyond the current descriptions used in the Company s website, could be completed with the original location of the houses, as well as graphic information about their originally surrounding landscape.

A last, and not minor point of this discussion refers to the important effort that the Qinsen Group is making for the continuity in the practice of traditional building crafts. It is undeniable that the artisans that the Company is currently training are key for the preservation of a great and most valuable intangible heritage: further research will aim to shed light on how the Company works with local artisans in the original communities and helps to their future economic development.

5 Conclusion

This research has shown how the notion of authenticity is a constantly evolving one, reflecting on its multiple layers as a key for the production of identity in suburban locations in the Lower Yangtze area, for a variety of different urban and architectural typologies.

Our findings show how the Company is pointing towards an important way of future development, where entrepreneurial initiative is a key for the development of the heritage field. This incorporates other values, mainly economic ones, which do not necessarily stay at the core of conventional practice outside of China.

Nevertheless, the definition of authenticity should remain as a guide for the future development of projects. As we have seen, the perception of users and residents is a key to the authentication of the historic structures as heritage. The incorporation of additional layers of information will not only enhance their scientific rigour, but most of all, highlight their important past as home to hundreds of working families who struggled and tried to prosper in the fields, surrounded by alien walls.

Acknowledgements

The author is grateful to the Qinsen Group for their hospitality and attention during the site visit and the interviews.

References

CHAPPELL E, 1999. Open-air museums: architectural history for the masses. Journal of the society of architectural historians (special issue), 58: 3, 334.

CORBIN C, 2002. Representations of an imagined past: fairground heritage villages. International journal of heritage studies, 8(3): 225-245.

DEN H H, 2010. Shanghai new towns - searching for community and identity in a Sprawling Metropolis. Rotterdam: 010 Publishers.

DIRLIK A, 2016. Modernity and revolution in Eastern Asia: Chinese socialism in regional perspective. Translocal Chinese: East Asian Perspectives, 10: 13-32.

FISHMAN R, 1987. Bourgeois Utopias. The rise and fall of Suburbia. New York: Basic Books.

MART NEZ G P, 2010. La Plaza de los Trofeos. Arquitectura y paisajes para el lugar com n de la periferia. Seville: University of Seville.

GREGORY J, 2008. Reconsidering relocated buildings: ICOMOS, Authenticity and Mass Relocation. International Journal of Heritage Studies, 14(2): 112-130.

HAYDEN D, 2004. Building Suburbia: green fields and urban growth, 1820-2000. New York: Vintage Books.

ICOMOS, 1964. International charter for the conservation and restoration of monuments and sites. [2019-06-30]. http://www.icomos.org/charters/venice_e.pdf.

ICOMOS, 2003. Principles for the analysis, conservation and structural restoration of architectural heritage. [2019-06-30]. https://www.icomos.org/charters/structures_e.pdf.

ICOMOS Canada, 1983. Appleton charter for the protection and enhancement of the built environment. [2019-06-30]. https://www.icomos.org/charters/appleton.pdf.

ICOMOS New Zealand, 1992. Charter for the conservation of places of cultural value. [2019-06-30]. http://www.gdrc.org/heritage/icomos-nz.html.

KNAPP R G, 2010. China: Open-air village and town museums, list for vernacular architecture atlas. [2019-06-30]. https://www2.newpaltz.edu/~knappr/ChinaOpenAir.html.

MOOLMAN J H, Site museums: their origins, definition and categorization. Museum management and curatorship, 15(4): 387-400.

PIAZZONI M F, 2018. The real fake, authenticity and the production of space. New York: Fordham University Press.

POLA A P, 2019. When heritage is rural: environmental conservation, cultural interpretation and rural renaissance in chinese listed villages. Built heritage, 3(2): 64-80.

SILVERMAN H, BLUMENFIELD, 2013. Cultural heritage politics in China: an introduction//BLUMENFIELD, SILVERMAN H. Cultural heritage politics in China. New York: Springer.

ZHOU M, CHU S, DU X, 2019. Safeguarding traditional villages in China: the role and challenges of rural heritage preservation. Built heritage. 3(2): 81-93.

Research on Ecological Experience and Applied Design Model of Local Dwelling Houses in Turpan Under the Strategy of Rural Revitalization

Ruziahong Paerhati, *ZHAO Xue*

College of Architecture and Urban Planning, Tongji University, Shanghai, China

Abstract Traditional Chinese folk dwellings are the labor products formed by the ancients in the long-term architectural practice of dealing with natural and social constraints. They contain a lot of valuable experience in ecological construction such as abstract design concepts and design thinking. In the era of accelerating globalization, Xinjiang, such as the protection and renewal of rural dwellings in western China, has also achieved remarkable results. At the same time, rural traditional building construction is facing enormous challenges and new development opportunities. This study takes the rural revitalization strategy as the background, takes the settlement of rural houses and surrounding environment in Turpan as the research object, selects the typical natural rural settlements for investigation, and analyzes the local experience of local folk houses from the macro, meso and micro perspectives. Based on the current situation of practice and existing problems, the ecological design thinking of traditional rural houses built on "whole", "situation" and "strain" is summarized, and corresponding construction and protection proposals are proposed.

Keywords traditional native dwellings, eco-experience, design thinking, architecture with eco-character, architecture creation

1 Introduction

1.1 Research Background

Traditional Chinese folk dwellings are the labor products formed by the ancients in the long-term architectural practice of dealing with natural and social constraints, and they contain a lot of experience in ecological construction. These valuable experiences are not only represented by specific design measures, but also abstract design concepts and design thinking.

In the era of accelerating globalization, China's urbanization level is developing rapidly. The construction and protection of rural dwellings in western China, such as Xinjiang, has also achieved remarkable results. Rural living conditions in rural areas have been greatly improved. Traditional local architecture construction is also facing enormous challenges and new development opportunities, especially as an important area for climate dryness and multi-ethnic cohabitation and multi-cultural convergence. The construction of rural local human settlements directly affects the ecological security of desert oasis in the region. National unity and social stability.

1.2 Research Questions and Methods

This study takes the rural revitalization strategy as the background, takes the settlement of rural houses and surrounding environment in Turpan as the research object, selects the typical natural rural settlements for investigation, and analyzes the local experience of local folk houses from the macro, meso and micro perspectives. Based on the status of practice and existing problems, the ecological design thinking of traditional rural residential buildings was proposed, and corresponding construction and protection proposals were formulated.

1.3 Overview of the Rural Landscape in Turpan

The Turpan Basin in Turpan is located in the eastern part of Xinjiang Uygur Autonomous Region, including 27 towns and towns in one city and two counties, including Turpan City, Shanshan County and

Toksun County. The Turpan area covers an area of 70,049 square kilometers and is surrounded by mountains. In the middle of the Turpan Basin is the Aiding Lake, which is known as the "navel" of the earth. It is the third lowest altitude in the world. The rural landscape of Turpan is north high and low south, narrow west wide and east wide, with an average altitude of 32. 8 meters. It is a unique warm temperate continental arid desert climate. It is located in the basin and surrounded by high mountains. It has rapid heat generation and slow heat dissipation. It has the characteristics of long sunshine, high temperature, large temperature difference between day and night, less precipitation, and strong wind power. It is known as "Fire State" and "Wind Pool". Because the Turpan area is located in the desert edge area, it is equivalent to a desert oasis and also affects residents. The key factors for survival are also the constraints that determine the development model of different rural forms. Because the flame mountain is lying in the middle of the basin, the basin is divided into two parts. The building materials are also thicker and lighter, and the architectural color is more saturated by low chroma and low brightness. This is reflected in the geographic and cultural location of the various regions of Turpan. It is also an important basis and theoretical starting point for the study and overall control of the rural landscape in Turpan.

2 Macro: Regional Distribution Level

2.1 Overview of Turpan Village Villages

The formation of rural villages in Turpan cannot be separated from the existence of Kaner Well. The series of ecological environment formed by Kaner Wells has become the homeland for the villagers to survive, and it contains certain ecological design experience. The main problems of the rural villages in Turpan are as follows:

1) Most villages have no clear construction planning regulations, which leads to the unreasonable layout of village land. The random construction in the village is mainly reflected in the high density distribution of residential buildings along the main roads in rural areas, and the layout of homesteads is chaotic and the land is irregular;

2) The current building is dominated by civil structures, and most of the houses have not completed the renovation of dilapidated houses;

3) Public service facilities are not sound enough, commercial outlets are scattered and small in scale, and there are no large-scale public activities venues within the village. Although there are small public events venues in combination with public buildings, the quality is not high enough to meet the living needs of the villagers;

4) There is no public green space in the village, and the quality of the living environment is poor;

5) The municipal infrastructure is not matched or perfect.

2.2 Village Type Analysis

The rural form that is integrated into nature is subject to certain constraints on the structure of its internal public space. For example, due to the influence of geographical terrain factors, the distribution of Kaner well has a great influence on the village form. Under normal circumstances, the public space is mostly formed along the direction of the Kaner well water flow; the plain valley-type village is less affected by the terrain, and the village mainly along the river channel is unfolded, and most of the public space is banded or grouped; the plain water network is mainly affected by the shape of the water network, and the public space is mostly core or multi-center. It can be summarized as a whole, and can be roughly divided into several types of villages, such as core type, multi-center type, strip type, and branch type.

1) Core type: The specific dominant central space is the main body of the public space, often close to the main public facilities and serving the entire village. (Fig. 1)

2) Multi-center type: It consists of several medium-sized public spaces, each of

Fig. 1 Core Village © Google Maps

which mainly serves residents within the service radius; the central space of the village level may exist or not exist. (Fig. 2)

Fig. 2 Multi-center type Village © Google Maps

3) Strip type: Adapted to the shape of the village, the public space is distributed along the sand dunes, Kaner wells or highways (Fig. 3).

Fig. 3 Strip type Village © Google Maps

4) Branch type: With a main trunk strip as the main body, several strips of space are extended to multiple sides, and the layout is flexible(Fig. 4).

2.3 Current Issues and Strategies

The different characteristics of rural

Fig. 4 Branch type Village © Google Maps

settlements in four types of public spaces determine their own problems.

1) Core type: poor accessibility; low utilization efficiency and lack of vitality.

Suggestions: A. Add trails to the connection between the space edge and the external interface; B. Optimize the site and increase the corresponding public service facilities.

2) Multi-center type: each public space lacks connection with each other, lacks integrity; the utility and accessibility are uneven

Suggestions: A. Add pedestrian roads to contact each space; B. Selectively add public space to areas with excessive service radius; C. Give play to the advantages of each space center and improve overall uniformity. (eg: old trees, ponds, open spaces, etc.)

3) Strip type: lack of rural characteristics on the river bank; destruction of rural style and natural environment by industrial plants and environment.

Suggestions: A. Set the joint space along the boundary of the belt space, such as waterfront belts and roads. B. Increase the landscape of green plants for line-of-sight blocking and air purification.

4) Branch type: public space is occupied by traffic function; effective active space is scattered.

Suggestions: A. Branch-type intersections as suitable scale space nodes; B. When conditions permit, the dendritic spaces can be connected to form a ring to improve space and utilization.

3 Meso: Settlement Space Construction

3.1 Overview of the Public Space in Turpan Settlement

The public space of settlements in Turpan is inseparable from the settlement form. Under normal circumstances, the shape of the village can directly reflect the shape of the public space. For example, the types of villages such as core type, multi-center type and strip type correspond to aggregate type and node. The type of public space and the type of public space, the corresponding public space functions are also roughly the same, but differ in scale. The main public activities are: festival activity gathering, production life gathering and leisure gathering. The settlement public space is generally located at the center of the settlement, and the houses are scattered around the public space. Under normal circumstances, the public space in the Turpan area is a space where a pool of water from the Kaner Well is formed to form a certain area of water, and trees are surrounded by water to form a comfortable public space.

3.2 Analysis of Settlement Public Space Types

The spatial types of rural settlements can be divided into different elements such as node space, linear space and aggregate (surface) space. In the rural rectification, it is necessary to rely on the existing elemental conditions to control the above spatial elements to make the public space form an open system as much as possible, avoiding closure and marginalization, and forming a systematic remediation strategy.

(1) Linear space Including roads and water systems (Kanerjing), such as the Kaner well water system, green belts, walking roads, small streets, etc., bear the functions of recreation, pedestrian transportation and space. The organization of these spaces is not necessarily horizontal and vertical, the form can be more free and irregular, and the combination needs to restore the natural boundary. To protect the public areas of the Kaner well water system, only local residential spaces can be extended to the boundary of the Kaner well water system and the desert area, and more than half of the natural boundaries should be preserved around the desert edge and the surrounding public belt around the Kaner well water system (Fig. 5).

Fig. 5 Linear space © The Author

(2) The node space is generally located in a partially enlarged part of the linear space, such as each door of the door, the entrance of the village, the road mouth, etc.; or the space for people to stay and interact, such as corridors, pavilions, and space under the trees; In addition, it also includes places with special significance, such as the space that usually works together (such as snoring), near the living production space, an ancient tree, an ancient well, the space where the ground water gathers, etc. Take appropriate protective measures and make them play a unique role in the public space system (Fig. 6).

Fig. 6 Node space © The Author

(3) Gathering space Generally speaking, it refers to a public activity place with a relatively large area in the village. It is often located near public service facilities and

can provide venues for gatherings, performances, sports and other activities for the whole village. The gathering space can bring about public changes through the change of scales, and the various scales are the embodiment of human nature. In addition, in the rectification of the gathering space, we should adhere to the principle of simple and simple, create a pleasant and intimate space experience, the scale is small and small(Fig. 7).

Fig. 7 Gathering space © the Autuors

3.3 Status and Strategy of Linear Space

1) Road: The increase of motor vehicles occupies the original living and communication function space of the street; the road hardening pavement has no difference, the form is single; the road side interface has a single level.

It is recommended to set a fixed parking space; reduce the speed of the vehicle through road form, material and texture design; the main road and some secondary roads can be used for hard paving, but the other secondary roads, household roads and walking roads should use local materials. Roads along the street can be used to enhance the diversity of the landscape through landscape greening;

2) Water system: The water network system based on the Kaner well has various degrees of damage, such as: landfill, pollution, cognac, etc. ; the interface of the water system composed of open channels, culverts and public waters lacks reasonable protection; domestic water and agricultural water Close contact, poor balance between practical and ornamental.

It is recommended to participate in multiple parties to jointly protect the existing Kaner well path and maintain its original function and style. The boundary of the water system is recommended to adopt a "breathable interface" that is composed of stone blocks and greening. The roads in the culvert water system should be narrow and should not be wide, and the structure should be simple and Into the nature, should not be abrupt; protect the public waters of the original Kaner well water system, control the domestic water use path and the amount of agricultural irrigation water through water management.

3.4 Node Space Status Issues and Strategies

1) Front space at the entrance of the household: Every household in front of the rural area is the communication space between the villagers and the place for information transmission.

The space in front of the house is also a place where everyone can do housework or farm work, such as: most residents have craters in front of them, moving beds for rest.

It is recommended to keep the space in front of the door, sum up the design pattern from the aspects of space proportion, scale, color, etc. , and optimize the space in front of the door accordingly.

2) Village entrance space: affected by transit traffic, the original village space for communication and rest gradually disappeared.

Combine the local climate of green plants and local materials to create the original interface of the village entrance.

3) Space under the tree: The comfortable space formed by the enclosed waters of the trees is mostly lacking in the waters of the green plants on the side of the stay. The function and form of the green plant interface are not clear.

Form reference: top can be closed or combined with vine plants to shield, but no walls, provide a rest seat stone table, using local materials; the highest usage rate of the villagers according to the use of functions for safety optimization, choose to stay Add other social features to the location.

4) Other spaces: There are also some corner spaces and transition spaces that play an important role in real life. They are flattened by bulldozers and exploited without regard to their special role in carrying people's lives and memories.

Combined with the memory of the majority of the districts, the interviews can be used to find out the places where they have the most stealing value. It can be an old tree, an ancient well, and a historically leaking building, which can reflect them and The fate of this land.

3.5 Aggregate Space Status Issues and Strategies

1) Scale: The public gathering space of most villages is basically consistent in scale; the aggregation space of different levels is low.

The aggregation space of different public properties can be distinguished according to its functionality to enrich the spatial experience; the hierarchy of the aggregation space can be divided according to the degree of its publicity.

2) Interface: lack of top interface; side interface form is single; ornamental is greater than practicality.

The use of the canopy of the green plant as the top interface is a common form in traditional villages, combined with the innovation of the current technology, enriching the top interface; the side space interface can be optimized through the walls of plants and buildings; the traditional rural construction materials and their construction patterns can be explored. Realize the contemporary interpretation of traditional materials through construction, splicing, and form conversion.

4 Micro: Building Space Construction

4.1 Turpan Local Architecture Overview

Due to historical and climatic reasons, the main material of the current local architecture in Turpan is still based on raw soil. Because of the raw material as the most primitive and convenient building material, its structure and performance are very suitable as building materials in the dry areas of the desert, due to its heat. It has good inertia, small heat transfer coefficient and good heat insulation effect. It uses its own physical properties to adjust the indoor temperature and humidity to increase the comfort of the occupants and ensure the house is warm in winter and cool in summer. Born in nature, it is natural, does not pollute and destroy the environment, and has the characteristics of closeness to nature, simple construction, energy saving, low cost, warm winter and cool summer, healthy and comfortable. Therefore, the earthen building is the main representative of the local architecture.

4.2 Turpan Local Building Type

The main types of local architecture in the Turpan area are mainly residential buildings and public-skinned buildings. In addition to this, there are some traditional soil relics, such as the Jiaohe Old City and the Gaochang Old City. The dwelling houses are divided into several different types of spaces, such as private spaces that satisfy personal activities such as sleeping and going to the toilet, semi-public spaces such as living rooms and kitchens that meet the public life of family members, and courtyard corridors. Auxiliary space such as transition space and storage livestock circle; the main types of public buildings of public nature include: production space such as drying room, space for living facilities such as Kaner Well, and religious space such as mosque.

Through the overall investigation and evaluation of the condition of the soil and soil in the Turpan area, it is concluded that the human settlements of the Turpan soil structure are affected by the ecological environment (geographical environment, climate characteristics, natural resources, etc.) and the human environment (historical background, economy, politics). Many influences such as influences, culture and religious beliefs have created the diversity of raw earth buildings.

4.3 Turpan's Local Architecture Problems and Strategies

4.3.1 Root Loss

Introducing a large number of new elements that have nothing to do with rural daily life and regional culture, making rural architecture "decorative", losing its connotation and losing the "real" side of rural wind. It is suggested that the study of a large number of traditional buildings in the Turpan area should be used to sort out the appropriate local architectural culture prototypes and use them as prototypes for change. (Fig. 8)

Fig. 8 Root loss © The Author

4.3.2 The Form is Cumbersome

In order to highlight each other's wealth, the building often used decorative elements to pile up, resulting in the loss of the rustic simplicity and diversity of the country buildings. The most important strategy is to suggest that the relevant departments can control the overall situation of the building from the policy level, and allow all households to freely play within the scope of the principle.

4.3.3 Scale Imbalance

In the large-scale construction in recent years, the scale and layout of urban buildings have been imitated, resulting in an imbalance in the scale of rural architecture. In view of such scale problems, it is suggested to sort out the centralized space model from the perspective of traditional building types, and then combine the needs of the current residents to comprehensively consider the scale of construction scale. (Fig. 9)

Fig. 9 Scale imbalance © The Author

5 The Ecological Design Law and Summary of the Local Folk Houses in Turpan

In the long-term historical development process, the Turpan area has formed ecological design rules such as "whole", "situation blending" and "strain" in the local design level.

1) The whole is reflected in the macro settlement level. Regardless of the size, each settlement has its own perfect settlement mechanism. For example, the size of the residential block is often positively correlated with the public (water area) in the village, and the population size also conforms to this law. The Kaner Well is the decisive factor in the size of the village and directly determines the composition of other public facilities or functions of the village.

2) Scenario blending is reflected in the settlement public space level of the Middle View. The water system of the Kaner Well is used as the water source of the entire village. There are ditches and green vegetation landscapes and water-type green vegetation landscapes. The green plants are generally distributed around the water system, and the ditches are mostly surrounded by. On the side of the residential area, the waters are generally located in the public space of the settlement. Therefore, the space environment of the green water system carries a lot of life memories of the villagers.

3) The strain is reflected in the microscopic architectural level. In response to the hot and dry ecological environment of Turpan in summer, the Turpan residents re-

sponded by opening small windows, increasing wall thickness and digging underground space.

In order to inherit the three ecological design rules of "whole", "situation" and "strain", the protection model can be started from the following points: maintaining the overall layout of the settlement, improving public facilities and functions, optimizing the function of building space and increasing the system of green water. Landscape, etc.

Throughout the long-term rural construction in Turpan, the process has been nonlinearly deepened, and the practice of rural construction has been sorted out through different professional discipline systems. At the same time of the accumulation of "quantity", there is an urgent need for a "quality". Summary of the phases. Therefore, the study of the guidelines for integrating the characteristics of rural settlements and the theory of space construction can provide a framework and basis for the construction of the medium and micro rural landscapes, and avoid the one-sidedness of the rural environment guidance caused by the single direction, and reduce the research from different angles. The difference in expression is contradictory and inefficient for rural construction investment. The study of the characteristics of rural settlements and the construction of space ideas and frameworks is based on the urban planning, the intersection of landscape and architecture disciplines and the long-term practice of rural landscape construction in Turpan, in order to play a certain role in the construction of the Turpan area and even the wider rural areas. Realistic guiding significance.

References

JIN Qiming, 1982. Geographical research on rural settlement— taking Jiangsu province as an example. Geography research, (3): 11-20.

LI Li, 2007. Rural settlement: form, type and evolution — taking Jiangnan area as an example. Nanjing: Southeast University Press.

LI Shengying, 2007. Study on the terrestrial buildings in Xinjiang — Tailing Turpan as an example. Xinjiang: Xinjiang University.

The Effect of Zoroastrianism on the Architecture of Chahar- Soffeh (four-sided) Houses in the Zoroastrian Village of Mazraeh Kalantar

Avisa Farzaneh

Architecture and Urban Planning Faculty, Shahid Beheshti University, Yazd, Iran

Abstract One of many forms of religion in the Zoroastrian architecture is the plan of the Chahar- Soffeh (four-sided) houses, which are called Chahar- Peskami in the Zoroastrian religion. The simulation of the quadruple pattern in Chahar-Soffe houses can be seen as a centered orientation and attention to the direction of the sky, which is the unchanging principle of the design of all spaces, both fire temples and residential spaces.

In this research, to find the main characteristics of the Chahar- Sofhe houses in the Zoroastrian village of Mazraeh Kalantar, a field survey was carried out on five existing houses in the village and then they were compared with each other.

In this village, Chahar- Sofeh houses are seen. Existence of features such as attention to the sky, centralism, philosophy of number four, embedding of a clean room and a large platform (The Largest Soffe called Peskam Mas), orientation of houses, etc. in these houses, which from a general law influence the whole and small parts Indicate the extent to whichReligion has been found in the culture and attitudes of the inhabitants. So that it can be said that the formation of spaces is originated from religious teachings and, finally, its desirable form is in order to conduct ritual ceremonies and religious beliefs.

Keywords Chahar-Soffeh (four-sided) houses, village of Mazraeh Kalantar, zoroastrian houses

1 Introduction

Yazd is one of the cities in Iran where various divine religions (Zoroastrians, Muslims, Jews, and Christians) coexist peacefully, and this is one of the reasons of Yazd being registered as a world heritage in UNESCO list. "The most important religion of Iranians before the arrival of Islam is Zoroastrian religion; Ashu Zarathushtra was the Prophet and he believed in the only God." (Memarian and Hashemi Taqar al-Jardadi, 2010). Their religious books is Avestan and Gathas. Their qiblah is light and fire temple is where they worship God. Yazd was one of the important centers for Zoroastrian to live both before and after Islam, and it has various Zoroastrian holy places, and Zoroastrian villages such as Mazraeh Kalantar, Cham, Sharifabad, etc.

"One of the religious expressions in Zoroastrian architecture is houses with chahar-soffeh plan called "chahar pesmaki". The history of the chahar-soffeh architecture returns at least to the architecture of the Achaemenid period. Also, during the Sassanid dynasty, this plan had a religious and residential function. This plan is the one appearing in the form of four iwan plan in the architecture of mosques and cross form in the architecture of churches. " (Jodaki Azizi and Mousavi Haji, 2014)

Mazraeh Kalantar is a village of Zoroastrian inhabitants in Iran and it is near the city of meybod in Azd Province. The year of the foundation of the village is around 250-300 years ago. "(Shahmardan, 1957). The houses in this village mostly have the chahar-soffeh plan consisting of 9 parts (chahar-soffeh, 4 rooms in the corners and the middle space central courtyard) and are introverted with central courtyard.

Despite the great literature on chahar-soffeh houses in Iran, there have been no significant studies on the role and effect of Zoroastrian religion in the formation of Zoroastrian chahar-sofeh houses in Mazraeh Kalantar. Since the author of this study is Zoroastrian and is originally from this village, she has managed to understand it deeply; therefore, she felt the necessity of this research to be carried out.

2 Research Methodology

To achieve the main goal of the research which is, finding the main constituent characteristics the spaces of the *Chahar-Soffeh* houses in the Zoroastrian village of Mazraeh Kalantar with an emphasis on the role of Zoroastrianism, The *Chahar-Soffeh* houses in the village were studied and reviewed by the author and five houses have been selected and studied as samples. The method of collecting information is field and documents, and libraries, and this research has been developed with a descriptive-analytical approach. It should be noted that Zoroastrian religious texts (such as *Gothic* and *Khordeh Avesta*) and dialogue with the priests (Zoroastrian religious leaders) and the villagers of *Mazraeh Kalantar* have been used.

To enter research objectives, initially, the locations of the five sample houses in the present (2019) in the village are shown. Then briefly the available spaces in the *Chahar-Soffeh* houses are discussed.

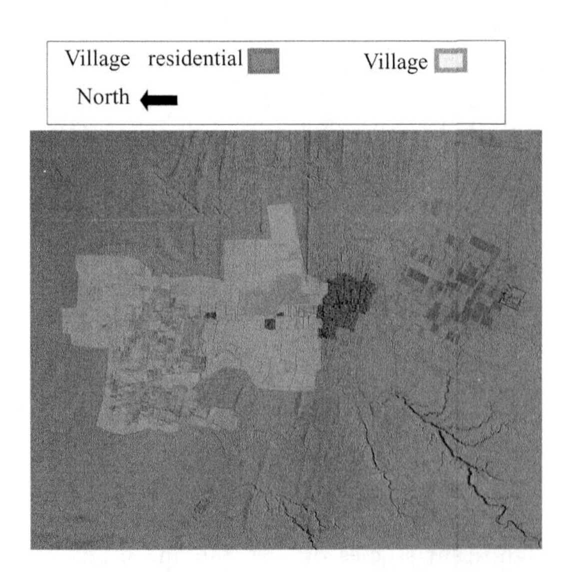

Fig. 1 Village aerial photography, year of 1967

Fig. 2 The position of the houses in the aerial photograph, year of 2019

The available spaces in the Chahar-Soffe houses in Mazraeh Kalantar can be divided into three areas:

1-The living space area, which is the *Chahar-Soffeh* of the house 2. Livestock area and service area including livestock keeping (livestock and barbers), storage area for forage and straw and ... and storage facilities. 3. A communication field that includes the entrance field and entrance corridor.

Residential spaces include:

The *Chahar-Soffeh*: Includes the southern side of the same large Soffeh (peskam mas), Northern Soffe and the western Soffeh and eastern *Soffeh*.

Central courtyard: In general, this space was without a roof and in the middle of it there is often a garden.

Room: bedroom and living room

Fig. 3 The pictures of vijoo room

Clean Room (*Otaq Sofreh*): Place for the praying and holding some religious ceremonies

Kitchen (Matbakh): The *Matbakh* has two ovens for cooking bread, a place to hold firewood and it also has an oven for cooking.

Vijoo Room: This room is a place to store food such as meat and bread and that stuffs inside a white cloth which is hanging from the ceiling by wooden logs which is called "*Vijoo*". This room also holds copper dishes.

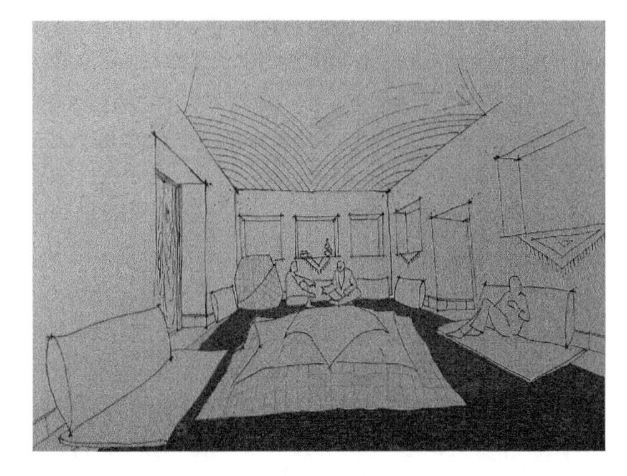

Fig. 4 The picture of the room

3 The Role and Effect of Zoroastrianism on the Architecture of Chahar-Soffeh Houses in the Mazraeh Kalantar Village

3.1 Holiness of Number Four in Zoroastrianism

Locate tables close to the first reference to The Zoroastrians believe that the universe consists of four *Akhshij* (four elements): Water, wind, fire, and dust which should always keep them away from contamination. And also believe that the universe consists in four directions (Main directions: East, West, North, South) and four sub-directions (northeast, southwest, northwest, southeast). That these four axes are sacred and honorable. Another reason for the sacredness of the number four is that they believe that "the soul (Faravahar) of man passes through Bridge of Chinavad (way bridge) on the dawn of the fourth day after death and enter another world" (Shahmardan, 2001). A religious ceremony for that dead man held on the fourth day after death. (Reminding you to be careful about your actions and behaviors in this world). *Ashuzartosht* says: "Your house and building should consist of four *soffehs* or porches". (Boluki and Okhovat, 2015)

3.1.1 The Manifestation of the Number Four in the *Chahar-Soffeh* house

The orientation and main arrangement of the *Chahar-Soffeh* houses is based on four main directions and four alternate directions with the two main axis are North-South and East-West and the two northeastern-Southwest axes and Northwest-South east that these axes intersect at a central point. The space which is located in the direction or end of the four axes must be sacred and honorable and the second space like a corridor or bathroom should not be placed there.

Therefore, they are aligned with the two main axes in the direction of the two axes, which are actually the same corners of space (Focusing on the center on the corners), the rooms are located which make *Chahar-Soffeh* house with a chalipaei plan and nine springs. This number four, is expressed in four *soffes* from the house.

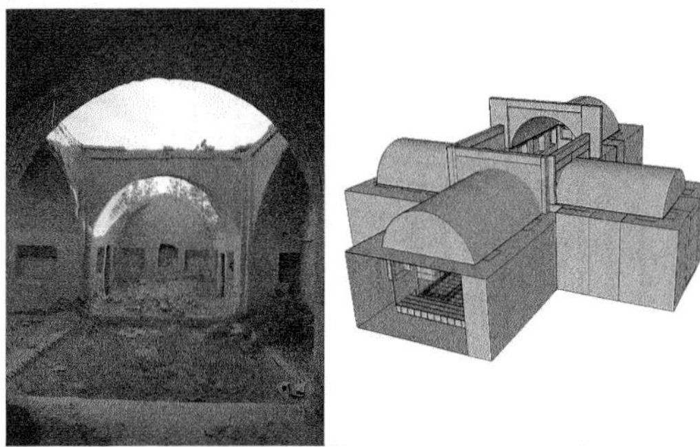

Fig. 5 Four numbers of soffes in the chahar-soffeh house

3.2 Attention to the Center

Place the In the Zoroastrian beliefs of the world and everything in it, is from a unique God and eventually returns to him and paying attention to the center is an indication of God's attention and connecting to him.

Tab. 1 The main and the secondary axes in the *Chahar-Sof feh* house

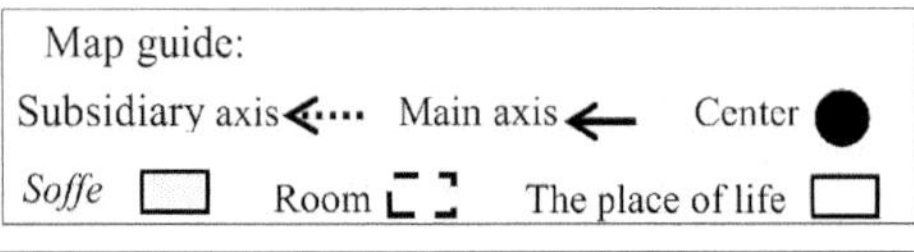

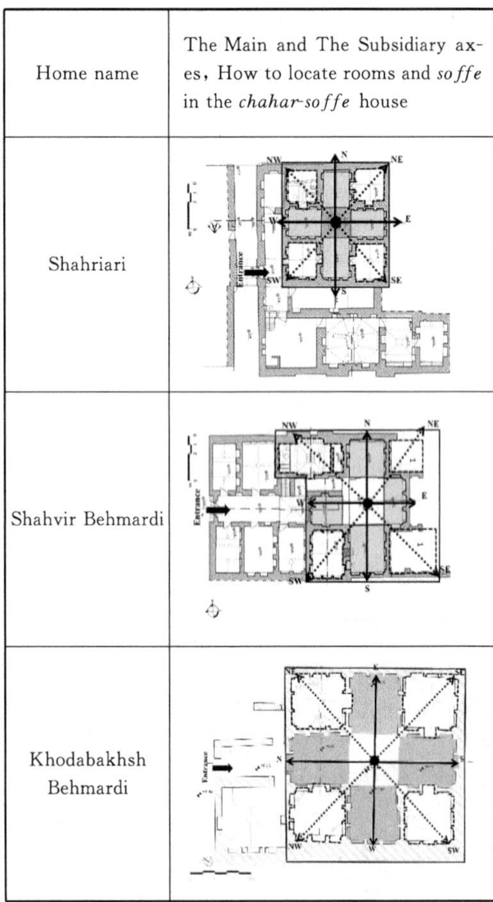

Home name	The Main and The Subsidiary axes, How to locate rooms and *soffe* in the *chahar-soffe* house
Shahriari	
Shahvir Behmardi	
Khodabakhsh Behmardi	

"The *Qibla* of Zoroastrian religion is the light, since the Zoroastrians know the light as manifestations of the existence of God on earth and they symbolize fire and symbol of it. As a result, they respect the fire. This source of lighting during the day, sunlight and at night the light of the moon." (Shahmardan, 2001) It is also mentioned in the Avesta Book that there are two parts in this book called "The Sun of Prayer" and "The Moon of Prayer". They put this fire (source of light) at the center of the building as the center of the universe that all the attentions are from each side towards this center and toward it, the religious ceremonies and the prayers of the God are done.

3.2.1 Manifestation of Attention to the Center in the Chahar-Soffeh house

According to religious beliefs, paying attention to the center and the need for a centralized space, is an immutable principle in the design of all spaces in fireworks as well as residential spaces.

Considering the comparison of the five sample houses we can conclude that these houses have four cross-axes which are two by two vertical and each one intersects with each other in the center and they create the central courtyard at the center of the space. In fact, these four axes go toward the center and also get out of the center that "this same pull in four directions, contrary to expectation, creates a kind of unimaginable space which emphasizes the importance of the center." (Navaei and Haj Qasemi, 2011). So according to what was discipline of the quadruple pattern in *Chahar-Soffeh* houses is a centrally oriented discipline.

Fig. 6 Four *soffes* in the house are not built facing back to sun (light)

According to religious beliefs in the design of *Chahar-Soffeh* houses there should be a space at the center of the building that meets the importance and attention of the center. So putting the courtyard that is the open field and the garden that is in the center of it, responds to this need and attention. This attention to the center is visible in the rooms with their roof covering, which is dome-shaped. The reflection of the sound in the dome roof is an indication of the reflection of man's actions on himself.

3.3 The Direction of the Sky

In the Zoroastrian religion they believe that God is eternal and deadly and ultimately, the human psyche (*Faravahar*) will return to God after death. And becomes eternal. This belief in immortality and restlessness has been revealed as one of the seven Amshaspandan (seven stages of mysticism to approach god and achieve salvation) called Amordad (immortality) Am?haspand. Which symbolizes the cypress tree, which is always green and also is the "tree of life (Zoroastrian sacred tree). This tree emphasizes the vertical axis (the direction of the sky). This axis shows the Zoroastrian belief in the presence of God and his unity. *Ashuzartosht* emphasizes respect for nature and plants. (Boluki and Okhovat, 2015)

3.3.1 Manifestation of the Sky Direction in Chahar-Soffeh House

In the center of the *Chahar-Soffeh* house there is a courtyard and four *Soffes* face to this center. The Cypress tree (symbolizing the direction of the sky) is planted in the middle garden of the courtyard. "The courtyard is the center of center that its vertical axis has been drawn to the sky and it can be said that the sky has found its way into this center." (Mohammadi, 2011). This attention to the center and sky, is in spaces like rooms with ceiling light in the middle of the dome-shaped roof and in fact the center of the room refers to this point. According to the above, Focusing on the center and the direction of the sky is created exactly in the center of the house (the central courtyard). It can be concluded that these two complement each other and have a mutual impact on each other.

Fig. 7 The dome of the room and ceiling lighter in the room

Fig. 8 A tree in the center of the gardenand central courtyard in the *chahar-soffeh* house

3.4 Special Place of Zoroastrian Religious Ceremonies in Chahar-Soffeh Houses

Given the high status of religious ceremonies among Zoroastrians, they committed themselves to doing it which will bring them an eternal reward. So they have a special place in their homes for doing so. This place should always absolutely clean and there is absolutely also enough concentration for people in this place. And distinct from other spaces in the house and it has privacy. Which is in the *Chahar- Soffeh* houses of *Mazraeh Kalantar* case studies this special place called the largest *Soffe* (*Peskam mas*) and the other is a clean room (otaq sofreh).

Tab. 2 Pay attention to the direction of the sky and centeredness

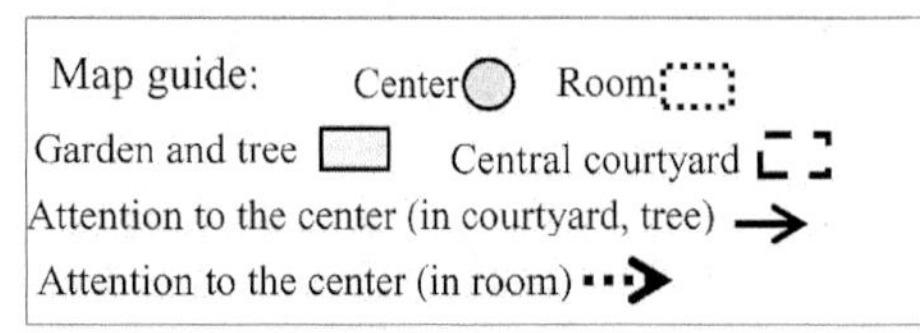

Home name	Centeredness (Central courtyard, garden, tree ceiling dome and ceiling light in the rooms)
Khodabakhsh Behmardi	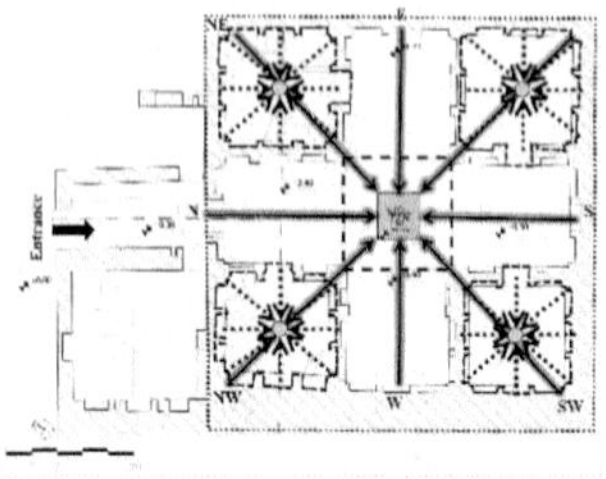
Shahriari	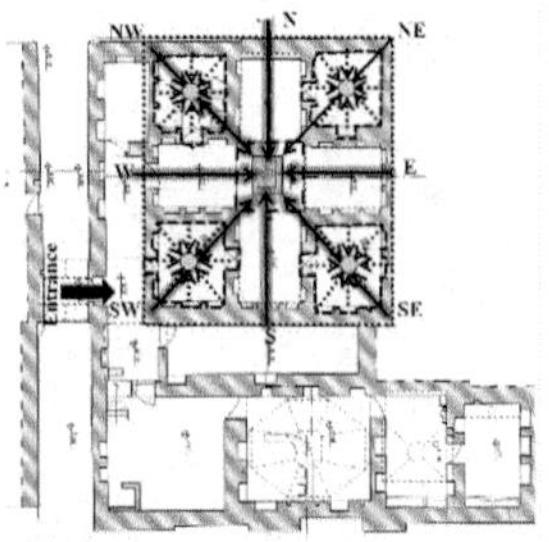

3.4.1 Peskam Mas (The largest soffeh)

One of the special places for the religious ceremony such as the *panjeh*, *gahanbar* and the ritual of the dead and some of the celebrations in these houses, is "*Peskam mas*" This *soffeh* is southern *Soffeh*. The span a bit toward the east and it's toward the dawn of the sun. That's a bit bigger than the rest of the *Soffeh* and sometimes higher, and in some way, give the stronger of the north-south axis feel to the viewer. This makes it possible to distinguish the largest *Soffeh* from other *Soffeh* of the house.

3.4.2 Clean Room

In these five sample houses, one of the

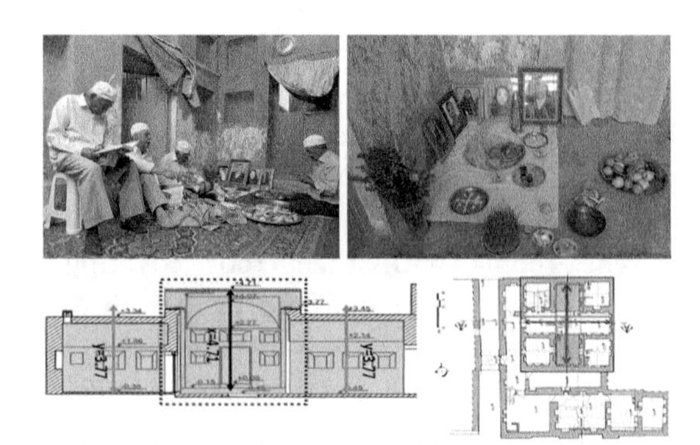

Fig. 9 The pictures of the peskam mas

rooms is a clean room or dining room. This is the special room for the Lord's praying and the table of some religious ceremonies in this room is widens. Because the direction of this place according to the Lord's worship is important and since the sunlight which is a symbol of the Zoroastrian Mysticism rising from the east then it is common for the room to be located on the east front of the house. The ceiling and walls of the room are Plaster plated and its covered it with white and blue ziloo. While the interior plated of the rest of the home space is from the lagoon this makes the clean room different from the rest of the house. the reason of use of white color in Zoroastrian religious spaces is the holy color and it shows the purity of inside and outside and It gives you peace and concentration while worshiping to the extent that Ashuzartosht emphasizes in the book of Avesta: "the necessity of thinking and knowing and praying God in peace and in the clean space" (Shahmardan, 2004). In this room, also spreading the religious table in the center of the room and put the fire in its center and the presence of the dome and its horny roof everything is showing the center and direction of the sky.

3.5 Privacy and Dignity in Zoroastrianism

Zoroastrians have always insisted on creating privacy, sanctity and hierarchy in their living spaces. In the other words; by passing in a pre-space string, the person enters the main space. Which exemplifies can be seen in clean room and Peskam mas. This creation of the privacy and sanctity of the whole spaces, can be seen in five sample houses.

Fig. 10 The picture of the clean room

Tab. 3 Special place for religious ceremony

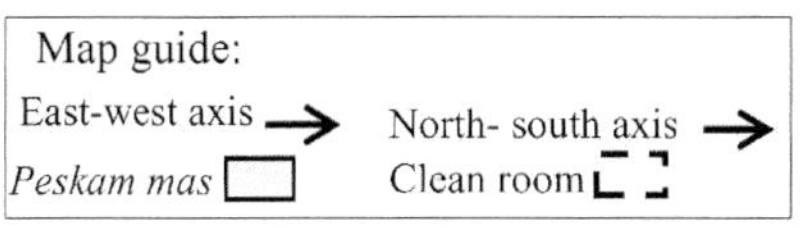

Home name	Placement of spaces on the plan
Javanmardi	
Shahriari	

3.5.1 Hierarchy of Entering the Residential Sector (Chahar-soffeh) from the Outside to the House

The parts available in each of these five sample houses can be divided into three arenas: 1- The arena of living part (part *chahar-soffeh* of the house) 2. The arena of Livestock and service part 3. The communication arena, which includes the input arena and entrance hall. This entrance hall is the space between the arena of living part and the livestock which divides these two arenas and creates privacy for the arena of living part. The hierarchy of entry from the outside into the building (Living quarters) is like this: entrance door, Entrance hall (closed space), Entrance soffeh (semi-open space), Central courtyard (outdoor).

Tab. 4 Entrance from the outside to the house

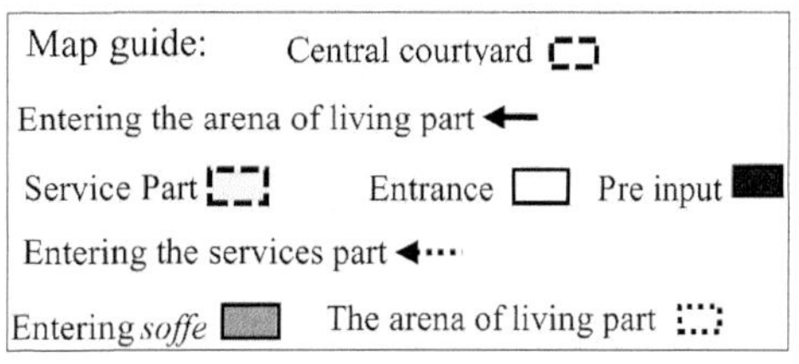

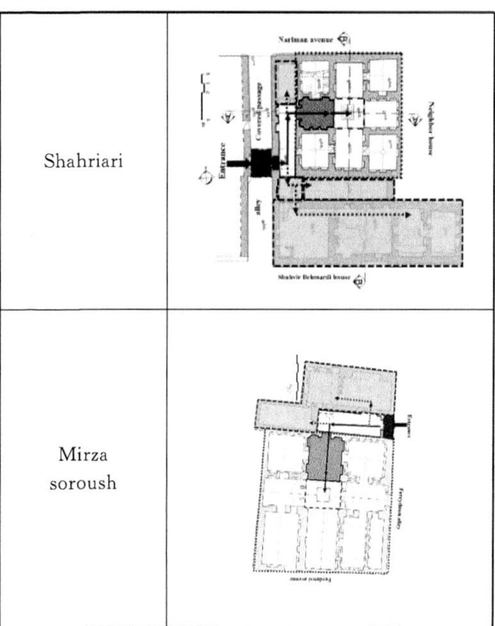

Shahriari	
Mirza soroush	

3.5.2 Hierarchy of Entry from Inside the Courtyard to the Interior Spaces of the House

Entrance to the interior spaces of these five sample houses also has a hierarchy that includes: Yard (outdoor), Interface and privacy (rows-semi-open space), Room (closed space).

Another aspect of this privacy and respect in the spaces can point out that there are two tenors inside the houses which is a tenor for baking bread for home use and another tenor is for baking bread for religious ceremonies.

Tab. 5 Internal circulation of the house

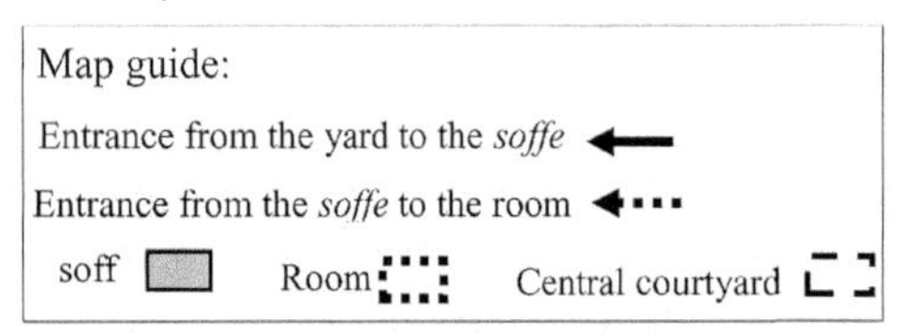

Home name	Plan house
Shahriari	
Javanmardi	

Fig. 11 Two tenors in the *Matbakh*

4 Conclusion

Based on the research, "according to the opinion of culture-oriented researchers, culture (the factor of religion) is the main factor in the formation of a house, and climate, economy, etc. are in the second place." (Memarian and Hashemi Taher al-Jardadi, 2010) It can be concluded that the factor of religion as well as religious beliefs and customs are directly related to the physical forms of a house, a clear example of which can be attributed to the effect of Zoroastrian religion on the formation of chahar-soffeh houses of the Mazraeh Kalantar village, including five sample houses. These religious effects influence the whole house, including the orientation and the placement of different spaces in the house, the roof covering and lighting, the floor, the way of accessing the spaces and the hierarchy of entering them, the color and the type of the materials applied. As a result, chahar-soffeh houses in the village of Mazraeh Kalantar can be generalized and explained as follows: these houses are formed with five axis (two main axis and two axes and a vertical axis [in the direction of the sky]), which cross each other in the center of the house (Center-oriented system) and form the central courtyard. Peskam mas and the cleaning room were places for holding religious ceremonies and praying.

Tab. 6　The effect of the Zoroastrian religion on the elements of the chahar-soffeh houses in Mazraeh Kalantar

<table>
<tr><td colspan="2">The elements of chahar-soffeh house</td><td>The cause and effect of Zoroastrian religion on the formation of home spaces</td></tr>
<tr><td colspan="2">Four soffehs in the house, four axes in house (placing soffes along the two main axes and the rooms at the end of the two subsidiary axes)</td><td>-Believing that the universe is consisted of four elements and it has four axes Believing that human's soul passes As-Chinavad bridge four days after the death</td></tr>
<tr><td rowspan="4">Paying attention to the center and the direction of sky</td><td>Central courtyard in the center of the house, the presence of a garden and planting a mort or cedar in it</td><td rowspan="4">- God is in the center of the universe and ultimately humans return to him.
-Placing the fire (as a symbol of light, which is the qiblah for Zoroastrians) in the center of the building as a manifestation of the presence of God and the praying toward it.
-Believing in oneness and immortality of God;
cedar and myrtle trees are the symbols of this belief
-The returning of human actions to his own (sounds being echoed under the dome)</td></tr>
<tr><td>Dome-shaped roof and ceiling light in the center of the ceiling of the room</td></tr>
<tr><td>VGDF4Center-oriented system of houses</td></tr>
<tr><td>Placing the fire in the center of the cleen room</td></tr>
<tr><td colspan="2">Cleaning room: the room for praying, the ceiling and wall being covered with plaster. (other spaces being covered with straw and clay)</td><td rowspan="2">-Creating a special place at home to pray and hold religious ceremonies that is clean, people can have a peace of mind and concentration and it is different from other spaces of the house.
-White color, the sacred color of the Zoroastrians, which is the symbol of clearity</td></tr>
<tr><td colspan="2">Peskam mas: the southern soffeh, which is larger than others, it is the place for holding religious ceremonies</td></tr>
<tr><td colspan="2">There are three kinds of spaces in these houses and the separation of the living space from the arena of Livestock through the entrance hall: Hierarchy of entry from the outside to the house and hierarchy of accessing the interior spaces:</td><td>The Zoroastrians' belief in the creation of privacy and hierarchy in their living spaces. Which can be seen in the cleaning room and the largest soffeh.</td></tr>
<tr><td rowspan="4">The orientation of the house and spaces</td><td>Unbiased houses and Center-oriented system</td><td>God is in the center of the universe and attention from each side to it</td></tr>
<tr><td>Stronger north-south axis</td><td>Larger and taller south soffeh</td></tr>
<tr><td>Emphasis on the eastern front of the house</td><td>Putting the clean room on the eastern front and the southern soffeh oriented toward the east, the emphasis is on the direction of the rising sun (respect for light)</td></tr>
<tr><td>Zoroastrian Qiblah is light (respect for light)</td><td>Four soffes in the house are not built facing back to the light</td></tr>
</table>

References

BOLUKI S, OKHOVAT H, 2015. Comparative study of similarities and distinctions between Muslim and Zoroastrian homes in Yazd. Quarterly journal of Iranian Islamic studies, 22:51-66.

FARZANEH A, 2017. Restoring and reviving a collection of Chahar-Soffeh houses in the Mazraeh Kalantar an approach to preserving the Zoroastrian existing customs. Iran: Shahid Beheshti University.

JODAKI A A, MOUSAVI H R, 2014. Chahar-soffeh pattern typology at Iranian architecture and its evolution. Scientific journal of Islamic architectural studies, 2: 64-88.

MEMARIAN G, HASHEMI T M, 2010. The impact of religious culture on the formation of the house: comparative comparison of the house in the neighborhood of muslims, Zoroastrians and jews in kerman. Quarterly journal of cultural research, 2:1-25.

MOHAMADI A, 2011. Investigating and explaining the pattern of the Iranian house based on the pattern of Chahar-Soffeh. Iran: Yazd University.

NAVAI K, HAJIGHASEMI K, 2011. Adobe and imagination. Tehran: Soroush Publishing.

SHAHMARDAN R, 1966. Zoroastrian worship. Bombay: Zoroastrian Youth Organization Bombay.

Shahmardan, Rashid. 2001. Khordeh Avesta. Tehran: Faravahar.

ZARGAR A, 2014. An introduction to the Iranian rural architecture. Tehran: Shahid Beheshti University.

Earthen Walled Villages in the Shanxi Province: a Heritage at Risk of Disappearing

Loredana Luvidi[1], *Fabio Fratini*[1], *Silvia Rescic*[1], *Laura Genovese*[1], *Roberta Varriale*[2], *ZHANG Jinfeng*[3]

1 National Research Council of Italy / Institute for the Conservation and Valorization of Cultural Heritage, Italy
2 National Research Council of Italy / Institute of Studies on Mediterranean Societies, Italy
3 Chinese Academy of Cultural Heritage, China

Abstract In the Shanxi province, there are many vernacular settlements enclosed by a square earthen wall that strongly characterize the rural landscape of the Loess Plateau. These farm villages are noteworthy examples of vernacular and earthen architecture but their high earthen walls are often lying in abandonment because they have lost their defensive role and therefore they have not been maintained, with the risk to definitely disappear. The urban tissue of these villages, made of earthen architecture, is the product of traditional building systems that should be recognized as a valuable witness of human capacity to adapt to the specific geographical and climatic conditions of the Plateau. An example of great attention to Chinese housing culture is present in the regeneration policies already carried out in Laoniuwan, a village sited between a bend of the Yellow River and a rugged terrain of the Loess Plateau, whose vernacular dwellings are partly dug into soft stone of the loess hillsides - yaodong - and partly completely above the ground. Although in antiquity this and other villages have known prosperity, many of the residents have moved to the big cities, with the consequence of a slow but inexorable deterioration of the buildings. Nonetheless, a rehabilitation project has recently allowed the dwellings reconstruction taking into account the traditional wisdom and modern materials. This is a good practice that needs to be codified and systematically extended elsewhere in order to avoid material and immaterial risks thus promoting sustainable conservation. The paper introduces our research aimed at evaluating solutions to ensure continuity of use and significance to these minor architectures of earthen walled villages and also analyzing those villages on the comparative base between China and Mediterranean countries.

Our belief is that the minor architecture, different in every single region, still gives us the desire to travel and to discover new constructive cultures and new landscapes.

Keywords earthen walled village, military settlements, conservation, enhancement, rural landscape

1 Introduction

In northern China, from the coast in the northeast of Hebei into the Yellow River meander areas among Inner Mongolia, Shanxi and Shaanxi, there is an area along the Great Wall known as Great Wall Rural Settlement Region. This area is characterized by rural villages, surrounded by earthen walls, that formed part of the fortification system of the Great Wall, reinforced until the Ming dynasty (Knapp, 1992; Wang Linfeng, 2018). This paper will focus, particularly, to those settlements sited in Shanxi' loessian area, in particular along the road from Laoniuwan village to Deshengbao Fortress (Fig. 1a-b). Those villages are characterized by a square earthen wall enclosing earthen architectures, in yaodong style, composing a very peculiar urban shape. In their current name is present the word "bao" or "pu" that means it was a defensive castle and "cun" means that it is now a common village. These villages were built during the Ming Dynasty and their high earthen wall had a defensive role. These military settlements are part of the minor architectures compared to the majesty of the Great Wall but no less important was their strategic role. Over time, farmers who did not need a defense wall have settled in the villages and therefore these high earthen walls have not received proper maintenance and are partly in ruins. Instead, both Laoniuwan and Deshengbao have been the subject of rehabilitation projects that have recently allowed the dwellings reconstruction taking into account the traditional wisdom and the modern materials. This good practice needs to be codified and systematically extended in order to avoid material and immaterial risks thus promoting a sustainable conservation. On a comparative basis between China and Mediterranean countries, our research is aiming, on one hand, at defining conservative protocols for the earthen structures, on the other, at evaluating solutions to ensure the continuity of use and significance to those sites in order to preserve local cultural identity (Genovese, 2019).

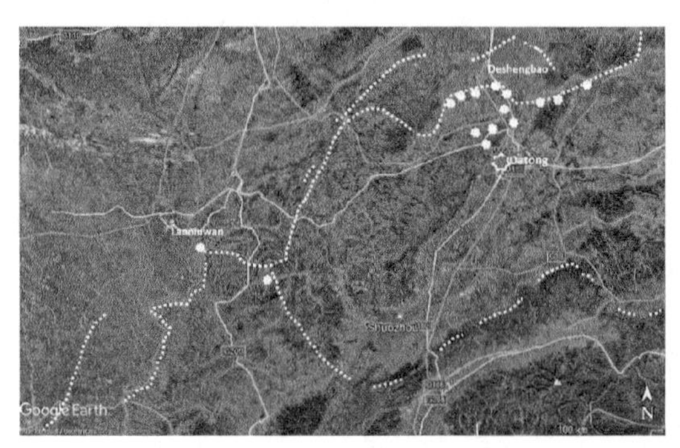

Fig. 1a Great Wall Rural Settlement Region: area from Laoniuwan village to Deshengbao Fortress. Walled villages are showed with an all-white star

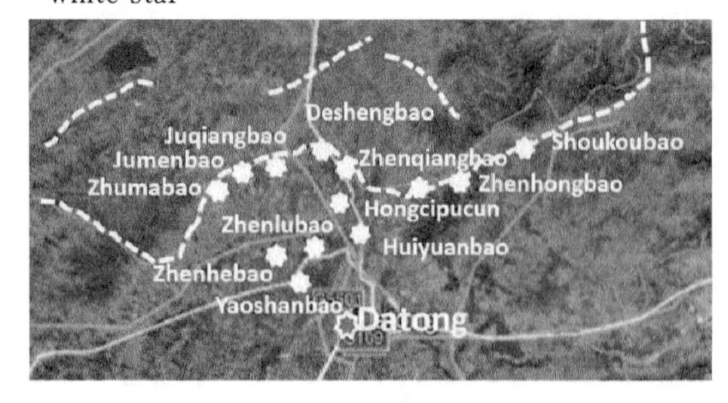

Fig. 1b Zoom in the area around Deshengbao where the earthen walled villages are located

2 Geographical and Historical Setting

The China's Loess Plateau is the largest accumulation of "dust" on Earth with a surface of 640 000 km^2. It covers most parts of the provinces of Shaanxi and Shanxi and it extends into parts of Gansu, Ningxia, and Inner Mongolia. It was formed by winds blowing from what is now the Mu Us Desert, depositing dust or removing dust over the last 2. 6 million years (Kapp, 2015).

With the exception of the stony mountainous areas, the average thickness of the loess deposit is 50 80 m in most parts of the Plateau it may reach 150 180 m at maximum in some areas (Loess Plateau Scientific Expedition Team, 1991).

The word "Loess" comes from the German "lose" (English loose) and it is referred to the easy erodibility of these lands constituted mainly of silt, a granular material with a grain size in-between that of sand and clay (0. 002 mm and 0. 063 mm) whose mineral origin is quartz, feldspar with only a little amount of clay minerals. Despite the easy erodibility, these terrains have the characteristic of being pseudo coherent, i. e. they can be excavated and the cavities are more or less self-sustaining like in the case of volcanic tuff. Thanks to this property, historically the loessian region has provided carved insulated shelters from the cold winters and hot summers, a kind of cave dwellings called yaodong, still inhabited today. Moreover, loess thanks to the pseudo coherent characteristics acquired by mixing with water and then dried, has been the material used to build locally the different portions of the Great Wall, fortresses and dwellings according to a particular technology that will be discussed later. The Plateau was highly fertile and easy to farm in ancient times, which contributed to the development of early Chinese civilization around it.

Centuries of deforestation and overgrazing, exacerbated by China's population increase, have resulted in degenerated ecosystems, desertification, and poor local economies.

In 1994 an effort known as the Loess Plateau Watershed Rehabilitation Project was launched to mitigate desertification; limited success has resulted for a portion of the Plateau area, where now trees and grass have turned green. A major focus of the Project was trying to guide the people living in the Plateau to use more sustainable ways of living such as keeping goats in pens not being allowed to roam free and erode the soft silty soil found in the plateau. Many trees were planted and nature is now reclaiming a portion of the Loessian area. Results have reduced the massive silt loads to the Yellow River by about one per cent (World bank, 2006).

In addition focusing on protecting the Plateau' landscape, great attention was paid to the remains of the Great Wall since this Wall - which is the most important historical monument and the national symbol - was declared a World Heritage Site by UNESCO in 1987.

In the late Warring States Period (476 BC-221 BC), many ancient states (Han, Zhao, Wei, Yan, Chu and Qi) built their own defensive wall along their borders to prevent attack from other states. When Qin state conquered those six states, creating the first empire in Chinese history, the different wall sections were put together. Since then, every dynasty repaired and reinforced the Great Wall until Ming Dynasty.

The Great Wall of China was not only a long defensive wall, but also a complete military defense system also consisting of beacon towers and military settlements where troops were placed.

The distance between two bao was about 20 km. At the beginning of the Ming dynasty, their preliminary role was to defend against the invasion of northern nomadic tribes. In the central period of the Ming dynasty, the commercial function increased. When the Qing dynasty took the place of the Ming dynasty, these fortresses totally lost their military function and became rural villages. Their earthen high walls were partly abandoned having lost their defensive role. Unfortunately, today most of those structures are partially destroyed or disappeared (Fig. 2).

Fig. 2 Ruins of earthen wall (Fratini, 2017)

3 Earthen Walled Villages and Conservation Issues

The Laoniuwan valley between a bend in the Yellow River and a rough yet green terrain of the Loess Plateau, is the place where partially underground settlements - yaodongs - are a diffused urban strategy which gave shape to a unique and long-lasting adaptation of people to local environmental elements. For centuries this place represented the outer limit of imperial rule and there remain portions of the Great Wall, which rises and falls according to the ancient erosion gullies of the earth. Along the mountain ridge, great circular embankments still stand where once beacons tower were lit to warn of invasion from Mongol horsemen.

The Laoniuwan village is a very ancient one. Since the XV cent. under the Ming Dynasty, it raised as fortification occupying a dominant position on an important landing along the Yellow River, playing a strategic role for the defense of borders and for trade with the Tartars, on the Mongolian side of the river. Under the Qing Dynasty (1644—1912), when emperors no longer stationed troops in Laoniuwan Castle, some of the soldiers remained here becoming farmers or retailers, continuing trades with Tartars, and building houses in or near the walls of the fortress. As an increasing number of people moved to Laoniuwan, the community grew into a village.

Today, Laoniuwan's historic urban

landscape still shows the traditional morphology of Chinese housing culture: is a typical example of Chinese vernacular settlement (Golany, 1992). Some of Laoniuwan' cave dwellings were carved out of the hillside, together with farming terraces, and here there are also stone walls remains subdividing each property (Fig. 3).

Fig. 3 Laoniuwan village, particular of yaodongs (Fratini, 2017)

This type of yaodong depends on of ground morphology: the presence of the bedrock exposed on top of the hill makes excavation somewhat difficult and motivates the construction of houses wholly above ground, while if these insist on the slope of the hill they are partially excavated. As a result, all the materials for the construction of a traditional yaodong can be found in the local area, making these dwellings among the most eco-friendly in China. On top of the hill, the rammed earth walls of the castle, are still there in a very poor state (Fig. 4). At the main entrance to the castle the masonry is realized mixed earthen and local stone.

In Laoniuwan village, the main activity of the dwellers was farming, and they were isolated from the rest of the world. In recent years, many of the residents have moved to big cities in search of work. Now, very few people still lives in Laoniuwan village.

However, in last years, local government has planned the reconstruction and restoration of vernacular structures using

Fig. 4 Laoniuwan village: remains of the fortress (Fratini, 2017)

local materials and traditional knowledge to reuse these dwellings with the goal of local development through a sustainable tourism. Laoniuwan has been faced the challenge of not losing its vitality, safeguarding its environment and culture, by attracting tourism.

Indeed, the role that these rural villages had in the past is still underestimated: these settlements are unique cultural landscapes full of character at the core of local identity, examples of living heritage.

Based on these local policies and the national project on the protection of the Great Wall zone, a conservation and enhancement plan was also addressed to Deshengbao village, a fortress built in 1539 (Ming Dynasty) near the Great Wall. The earthen defensive wall of the village was severely damaged throughout the course of many years by natural factors as rain, wind, storms, etc. and by lack of maintenance. Moreover, in the 1970s, when an agricultural development campaign began widespread, people living alongside the Wall dug out the bricks and earthen material and used them for building houses or building enclosure for their livestock, causing considerable damage.

Some actions to protecting and restoring it have been implemented and also has been promoting awareness among local villagers but much remains to be done. Along the road from Laoniuwan to the Deshengbao Fortress are placed many walled villages

(Tab. 1, Fig. 1a-b) built on flat slope and in the surrounding of the Great Wall.

Rammed earth is the main construction technique of the large rammed earth mound walls that encircled villages in the Shanxi Province. Rammed earth is a construction technique where the soil is taken from the ground and compacted to form structures. Removable formwork is installed, and the soil compacted within it. This technique was widely used in ancient constructions. The term "hangtu" is used by Chinese archaeologists to describe both rammed earth mounds and earth rammed between formwork (Jaquin, 2008). For earthen site exposed outdoor two main deterioration factors can be distinguished: natural and human. The human factor is due to technique of construction and engineering-related properties of the materials selected to build the site. The natural factor is due to rain, wind, solar radiation, mudslides, biological action etc. and they cause physical and chemical weathering (Li, 2009; Wang, 2004; Li, 2011; Du, 2017).

Tab. 1 Defensive castles of Ming Dynasty along the road from Laoniuwan to Deshengbao

	Earthen walled village	Year of build	Current state
1	Shoukoubao	1522—1525	Ruins
2	Zhenqiangbao	uncertain date	Village
3	Juqiangbao	1545	Village
4	Zhenhongbao	1546	Village
5	Zhenbianbao	1539	Ruins/ Touristic point
6	Huiyuanbao	1522—1567	Village
7	Zhenhebao	1524	Village
8	Zhenlubao	1539	Village
9	Jumenbao	uncertain date	Village
10	Zhumabao	1545	Village
11	Yaoshanbao	1524	Village
12	Laoyingcun	uncertain date	Village
13	Hongcipucun	1539	Village

Chemical deterioration is mainly due to enrichment of soluble salts that cyclically dissolve and crystallize leading to the destruction of cohesive forces and the erosion of earthen material. Physical weathering is erosion by wind and rain. In arid climate, precipitations are generally low but heavy, they contribute significantly to erosion, as they soften and disintegrate the earth. Especially in summer season, the high temperature, promote a high evaporation rate of the water leading to quickly dries earthen surface and turns the softened earthen material into scalelike crusts, which fall off under the combined action of wind and rain.

The following types of decay can be observed in earthen structure: erosion, exfoliation, honeycomb, scaling off (Cui, 2019), sapping, gulling, cracking, collapse and biodeterioration. Among these types of deterioration, exfoliation, cracking and erosion are the most frequent because wind and severe rainfall are the main impact factors of decay. All these types of decay mechanism can threaten the stability of the earthen site.

According to the "Principles for the Conservation of Heritage Sites in China", conservation intervention refers to all measures carried out to preserve the physical remains of sites and their authenticity. The conservation involves the identification and investigation of heritage sites for determining the values of a site, its state of preservation, and its management context trough analysis of historical document and on-site survey. Only if on the basis of the previous investigation the site is formally proclaimed as an officially protected entity the preparation of a preservation master plan is carried out. The first step of the conservation practice is the routine maintenance to slow deterioration and only if the sites is considered at risk of heavy damage the "minimal" conservation intervention is planned (Agnew, 2002).

The consolidation techniques for earthen sites include: surface consolidation (Li, 1995, 2009, 2011; Zhou, 2004; Wan,

2012; Zhao, 2016; Wang, 2016); grouting; mud bricklaying; anchor bamboos or wooden rods and building a new wall where the wall collapsed.

In the area of the walled villages investigated in this contribution, the remains of the rammed earth walls are affected by various climate-related conservation problems. In fact, in the area the rains are few but intense. Most of the rain falls in the summer period, in two months, when the temperatures are very high with consequent rapid evaporation which leads to the formation of cracks, exfoliation, honeycomb, sapping until the collapse of the foundation. The dry but very windy winter favors a strong superficial erosion. Also the presence of vegetation favors phenomena of disintegration.

These problems, together with a total lack of maintenance, are leading to the complete destruction of the rammed earth walls of the villages. Therefore an immediate action is necessary both to assure structures at risk of collapse and to preserve existing structures with surface and structural conservation methods with attention to the construction of drainage systems for rain.

4 Raw Earth Architecture in Italy

Since very long times, building houses, fortifications or entire settlements in raw earth has been the most typical habit for many populations in large sectors of the Mediterranean Basin and all over the World.

Earthen buildings can also be widely found in many Italian regions, particularly in rural context. Each region has its proper building technique: pisè and raw brick in Piemonte, ladiri in south Sardegna, massone in Abruzzo and Marche, and so on (Achenza, Cocco, 2015). Despite the millenary tradition and the importance that these buildings have had in the past, currently they suffer from maintenance and conservation problems, often not satisfying the needs of modern life. The progressive abandonment of building techniques and the loss of connections with modern civilization have led to the degradation and rapid extinction of a large part of the heritage in raw earth. Unfortunately, while in 1933 in Abruzzo 7012 buildings were registered, at the end of the last century only 806 were still standing and in other regions as Emilia-Romagna, Toscana, Basilicata and Calabria just a few structures remained (Arbib, 2012).

In the seventies of the last century, the spread of a renewed interest in raw earth buildings owed, among others, to the initiative of organizations such as ICOMOS which, in 1972, organized in Yazd, Iran, the "First International Symposium on Conservation of Raw Brick Monuments". This interest also involved the Italian academic world, which slowly began to orientate research towards the rediscovery of this architectural heritage, encouraging, among other things, the study of intervention methods to guarantee their conservation over time. The desire to define concrete actions aimed at safeguarding the heritage in raw earth has also manifested itself at the regulatory level with the enactment of the regional laws 2/2006 in Piedmont and 17/1997 in Abruzzo. These laws aimed to bring out the role of resource of the raw earth heritage, of which it was intended to promote, on the one hand, physical conservation and enhancement, on the other the actualization of knowledge and techniques in a perspective of environmental sustainability and cultural compatibility. Sardegna region, where the characteristics of its pre-industrial agro-pastoral culture are still alive, has preserved a large number of raw earth settlements (Angioni, Sanna, 1988). In particular in Sulcis, in the south-west of Sardegna region, settlements were historically built using mixed techniques — i. e. using both stone and adobe — which are known as *furriadroxius and medaus* (Sanna, Cuboni, 2008). Recently a "Territorial Museum of Scattered Settlements", in Santadi town, started a project aiming at preserving and promoting those particular kinds of architectures by involving local communities (Bianchi, Botto,

2019). More specifically, the "Territorial Museum" has helped local people to rediscover their past identity through the teaching of earthen building technologies in order to allow people to maintain the earthen structures. Thus, thank to social participation, some vernacular architecture were restored and converted into public spaces for artistic performances and/or training, or for residential uses. Thus, the visibility of those cultural objects and their new "uses" helped residents to be more aware of the advantages of safeguarding and enhancing them. In addition, the "Territorial Museum" has encouraged the creation of thematic itineraries throughout the Sulcis' territory by involving local museums. The final aim has been the creation a network of museums and cultural point of interest in order to stimulate, on one hand, the territorial synergies for the identity preservation, on the other, to stimulate a sustainable economic growth based on the development of cultural tourism. While the Italian context is still waiting for a comprehensive recovery plan, some projects — as the Sardinian one — show possible creative solutions that could became good practice to be codified and systematically extended elsewhere in order to preserve and enhance these peculiar settlements full of character at the core of local identity.

5 Conclusion

The conservation and enhancement of the earthen walled villages should be associated with the protection of the rural landscape as required by the Venice Charter (1964) which underlines the concept of conservation of historical monument but also the urban or rural setting in which is found the evidence of a particular civilization. In recent decades the Chinese government has paid much attention to the protection and restoration of the Great Wall and has implemented the "Great Wall Protection Ordinance" in 2006, which outlines actions concerning natural factors and human activities that seriously threaten the structure of the Great Wall (last updated on 22 September 2017).

Tangible results of this activity are highlighted in the village of Laoniuwan and in the Fortress of Deshengbao located near the Great Wall in Shanxi Province. Furthermore, the series of Ming Dynasty military settlements, whose inhabitants are currently farmers, have not received the same attention or only in some cases there are examples of projects for the conservation and maintenance of the characteristics of their architecture vernacular.

A reasonable solution should be a common strategic plan for conservation projects and reconstruction of earthen walled settlements that takes into account all the minor sites as well as the rural landscape. Only this more global vision will allow local development by promoting awareness among the inhabitants of local villages of the intrinsic values of their heritage. The paper aims at defining guidelines to be used in the promotion of this kind of heritage. The final objective is to support actions aimed at preserving the most important cultural heritage from the effects of extreme anthropic pressure and to monitor, preserve and promote the less visited minor rural villages.

Acknowledgements

We thank Dr. Hu Yihai for assistance and translation of information from Chinese reference source

References

ACHENZA M, COCCO C, 2015. A web map for italian earthen architecture//MILETO C, VEGAS F, et al., Earthen architecture. London: past present future CRC Press Taylor & Francis Group.

AGNEW N, DEMAS M, 2002. Principles for the conservation of heritage sites in China. Los Angeles: Getty Conservation Institute.

ANGIONI G, SANNA A, 1988. L'architettura popolare in Italia. Sardegna Laterza Rome-Bari.

ARBIB C, 2012. Case in terra: specie protetta in via di estinzione. In www. salviamoilpaesaggio. it Febbraio 2012.

BIANCHI M, BOTTO M, PASCI P, 2019. Raw

earth architecture in Sardinia: Sulcis as a case study//CNR. Past and Present of the Earth Architectures in China and Italy. Rome Italy.

CUI K, DU Y, ZHANG Y, et al., 2019. An evaluation system for the development of scaling off at earthen sites in arid areas in NW China. Herit Sci 7: 14. https://doi.org/10.1186/s40494-019-0256-z.

DU Y, CHEN W, CUI K, et al., 2017. A model characterizing deterioration at earthen sites of the Ming Great Wall in Qinghai province. China soil mech found eng, 53: 426. https://doi.org/10.1007/s11204-017-9423-y.

GENOVESE L, VARRIALE R, LUVIDI L, et al., 2019. Italy and China sharing best practices on the sustainable development of small underground settlements. Heritage, 253: 813-825. https://doi.org/10.3390/heritage2010053.

GOLANY G S, 1992. Chinese earth sheltered dwellings: indigenous lessons for modern urban design. Honolulu: University of Hawaii Press.

JAQUIN P A, AUGARDE C E, GERRARD C M, 2008. Chronological description of the spatial development of rammed earth techniques. International journal of architectural heritage 2:377-400. DOI: 10.1080/15583050801958826

KAPP P, PULLEN A, PELLETIER J D, 2015. From dust to dust: Quaternary wind erosion of the Mu Us Desert and Loess Plateau China. Geology, 43(9):835-838.

KNAPP R G, 1992. Chinese Landscape: The Village as place Honolulu: University of Hawaii Press:313.

LI Z, ZHANG H, WANG X, 1995. Reinforcement of ancient earthen structures. Dunhuang Studies3: 1-18.

LI Z, ZHAO L, SUN M, 2009. Deterioration of earthen sites and consolidation with PS material along Silk Road of China. Yanshilixue Yu Gongcheng Xuebao//Chinese journal of rock mechanics and engineering, 28:1047-1054.

LI Z, WANG X, SUN M, et al., 2011. Conservation of Jiaohe ancient earthen site. China journal of rock mechanics and geotechnical engineering, 3 (3): 270-281. https://doi.org/10.3724/SP.J.1235.2011.00270

SANNA A, CUBONI F, 2008. L'edilizia diffusa e i paesi il Sulcis e l'Iglesiente. I manuali del recupero dei centri storici della Sardegna DEI Tipografia del Genio Civile Cagliari.

WAN T, LIN J, 2014. A new inorganic-organic hybrid material as consolidation material for Jinsha archaeological site of Chengdu. Journal of central. South University, 21: 487.
https://doi.org/10.1007/s11771-014-1965-9

WANG X, LI Z, ZHANG L, 2004. Condition conservation and reinforcement of the Yumen Pass and Hecang Earthen Ruins near Dunhuang. // Neville Agnew. Conservation of Ancient Sites on the Silk Road Proceedings of the Second International Conference on the Conservation of Grotto Sites Mogao Grottoes Dunhuang People's Republic of China June 28, July 3, 2004:370-377.

WANG X, GUO Q, YANG S, et al., 2016. Nondestructive testing and assessment of consolidation effects of earthen sites. Journal of rock mechanics and geotechnical engineering 8(5): 726-733. https://doi.org/10.1016/j.jrmge.2016.06.001.

World Bank. 2006. Reengaging in agricultural water management Challenges and options:218.

ZHANG Y, LI X, SONG W, et al., 2016. Land abandonment under rural restructuring in China explained from a cost-benefit perspective in Journal of rural studies, 47:524-532.

ZHAO D, LU W, WANG Y, et al., 2016. Experimental studies on earthen architecture sites consolidated with BS materials in arid regions advances. Materials science and engineering:13. https://doi.org/10.1155/2016/6836315

Protection and Renewal of the Great Wall Settlement in Chicheng Area, Zhangjiakou

XIE Dan, *ZHANG Weiya*

Hebei University of Technology, Tianjin, China

Abstract As part of the Great Wall military defense system, the Great Wall settlement has a special military cultural connotation. However, due to social, economic, ecological and other factors, the development of the Great Wall settlement is facing the disappearance of cultural heritage. Through the excavation of the cultural heritage of the Great Wall settlement in Chicheng County, Zhangjiakou, this paper summarizes the value of the settlement of the Great Wall, and analyzes the characteristics of the site layout, defense system, castle spatial distribution and morphological characteristics of the Great Wall military settlement defense system in the region. The status quo and problems of historical humanities and natural landscape protection in Chicheng County were clarified. From the aspects of the overall protection of the Great Wall settlement, the establishment of community management mechanism and the sustainable development of rural renewal, suggestions for the protection and renewal of the Great Wall settlement were proposed.

Keywords Chicheng Country, Great Wall settlement, defense system, traditional village, protection method

1 Overview of the Great Wall Settlement in Chicheng County, Zhangjiakou

1.1 Regional Definition and Historical Evolution

Chicheng County is located in the eastern part of Zhangjiakou City, Hebei Province, upstream of the Baihe River. It is adjacent to Beijing in the east, Yanqing County in the south, and Chongli and Xuanhua in the west. Surrounded by mountains, the gullies are vertical and horizontal. There are many Great Walls in Chicheng. In addition to the historical materials, the Northern Wei, Northern Qi, Ming Great Wall, and the Great Wall have traces of the Tang and Song Dynasties. At present, there are 87 enemy buildings, of which only three are better, twelve are wrecked, and only 72 are left.

The Great Wall in Chicheng contains the Great Wall of the North Road, which is the defense system of Xuanfu Town. The Great Wall of the North Road is also called Du Shi Road, which is divided into Shangbei Road, Xiabei Road and Zhongbei Road. The zone is connected to Chaohe River in the east (now Chengde Fengning), west to Suoyangguan, south to Chang'anling, and north to felt hat mountain (now north of Du Shikou). The section of the Great Wall was built in the late Ming Dynasty. It was built on the basis of the original Great Wall of the Northern Wei Dynasty and the Great Wall of Tang Dynasty. It is 345 kilometers long and is ring-shaped around the whole city of Chicheng(Fig. 1).

In the long history of Chinese history, many feudal dynasties have built it many times in order to consolidate their rule. According to historical records, after the Northern Wei, Northern Qi or the Great Wall of the Tang Dynasty, they were reused in the Ming Dynasty. The system has been built for nearly 230 years.

Since ancient times, Hebei Chicheng has been a mountain corridor leading to the outside of the Central Plains. It is a strategic place for the nomadic tribes of the north and the Central Plains farming people to compete for thousands of years. During the Ming and Qing Dynasties, garrisons, such as the Guards, the Fort, and the Fort, were set up to defend the frontiers and the military struggles were fierce. According to historical records, due to the arduous task of building the Great Wall, in order to stabilize the military, the imperial court promoted the policy of "the real people of the migrants". During the construction of the Great Wall and for a long period of time, the families of the officers and men of the field were allowed to come and follow(Zhao,2009). Later, the descendants of the Great Wall defenders settled down the mountain and gradually evolved the village. As the years changed, due to historical, economic, political and other reasons, the military functions of the Great Wall were degraded and gradually developed into traditional villages with military cultural characteristics.

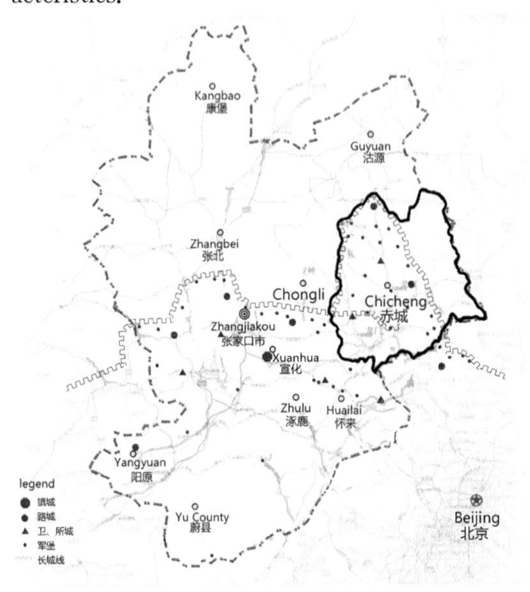

Fig. 1 Zhangjiakou Great Wall defense distribution location map

1.2 The Value of the Settlement of the Great Wall in Chicheng County

1.2.1 Villages along the Great Wall are an Important Component of the Great Wall Defense System

The Great Wall is one of the most magnificent cultural heritages in human society, carrying a heavy historical civilization. The Great Wall settlement, as part of the Great Wall military defense system, has a special military cultural connotation. The Great Wall is the dividing line between the Central Plains nation and the ethnic minorities, which has formed a national cultural resource with multiple ethnic groups. The cultural connotation and functional role of the Frontier Fortress is richer than the city wall itself. With the development of history, these military settlements carry a large amount of traditional information and have important cultural and historical values, which also have positive significance for the protection of the Great Wall in the context of today's cultural heritage.

As a special kind of traditional settlement, the Great Wall defensive settlements belong to the national-level defense system, which maintains the rise and fall of the country and the country and even between the dynasties. The ancient Great Wall's military settlement records the inheritance and information of history, contains the essence of profound cultural heritage, is an important component of the development of human settlements, and has a high academic research value. Nowadays, most of the traditional villages that have evolved have preserved the historical features or national characteristics of a certain period or periods, and they have a strong sense of historical atmosphere or national cultural characteristics in terms of material characteristics and spatial forms. The material form reflects the deep structure of political, cultural, economic, military and other aspects in a certain period and a certain region, showing strong historical characteristics, cultural inheritance and humanistic ideas. Therefore, military settlements along the Great Wall are inextricably linked to the Great Wall in history, landscape and culture. It is concluded that the protection of the Great Wall requires the overall protection of surrounding villages and the natural environment.

1.2.2 The Great Wall Historical Culture and Fortress Building in Chicheng County

The Great Wall is not only a defense system, but also a rich Great Wall culture. The traditional architecture, natural environment, religious beliefs and other cultural heritage along the Great Wall of Chicheng illustrate that Chicheng once occupied an important position in history. Chicheng has a superior geographical position and has become an important military position with the construction of the Great

Wall. Geographically, Chicheng is located at the intersection of the North China Plain and the Inner Mongolian Grassland, at the junction of nomadic culture and farming culture. Therefore, the Chicheng area has become a region where the two cultures collide and blend, and is rich in ethnic traditional culture.

The villages in Chicheng conform to the changes of the mountains, merge with the mountains and rivers, and are surrounded by the Great Wall. Because of its centuries-old history, the area has a rich historical connotation and regional culture, reflecting the landscape value of Chinese and nature. In addition to the Great Wall, Chicheng also includes defensive buildings such as the military fort, sampan and enemy stations, as well as religious buildings such as the Chicheng Drum Tower, the Relight Tower and the Thousand Buddha Caves, reflecting its unique cultural characteristics and architectural style. The villages in Chicheng County are rich in historical and cultural connotations. Natural landscapes and cultural heritages all have high economic value and important historical, scientific, artistic and social values.

In conclusion, according to the human landscape and traditional features of Chicheng, the Great Wall is connected with the village, and the overall study on the protection of the Great Wall settlement will become more important.

2 Distribution Characteristics of the Military Settlement Defense System of the Great Wall in Chicheng District

2.1 Location Layout

The Chicheng area is based on the northwest of Beijing, and the Great Wall has been built in the Northern Wei, Northern Qi, and Tang Dynasties, and the Ming Dynasty reached its peak. The Chicheng Great Wall settlement occupies an important position in the Xuanfu Town defense zone. For the location of the Great Wall, it is mainly built on the dangerous terrain of the mountains and rivers to control the enemy's vitality. Moreover, the location of the site is close to the water source to ensure daily life and farming.

For the location of the military fort, in the Chicheng area are mostly castles along the line, as well as castles arranged in depth. The location of the castles at all levels depends on the geographical location of the geographical defense area, the extent of the terrain, and the strategic and tactical value. Locations with tactical value, such as the Longmen Institute (Tan, 2010). An important regional setting with strategic value, such as Kaipingwei (now Du Shikou).

2.2 Defense System and Castle Space Layout Features

Establish a hierarchical defense system based on the management of the Ming Great Wall. The military defense zone of the Great Wall in the Jiubian Town of the Great Wall is divided into the town of town-road-Wei, Suo-fort according to the level of the high-low garrison castle. Therefore, not only the linear wall of the Great Wall includes the defense engineering systems of different levels of functions such as beacon towers (烽火台), watch towers (敌台), castle barracks (营城堡), passes (关隘), fortresses (堡寨), inns (驿站). The Great Wall settlements in Chicheng County are mainly distributed on Shangbei Road, Xiabei Road and Middle road. The castles of each road are based on the quartet, and the Three Great Walls form a triangular area. Du Shicheng on Shangbei Road has 12 castles under its jurisdiction; Xiabei Road is stationed at the Longmenwei, and it governs 8 castles. Jinjiazhuang Fort, Longmenguan Fort, Sanchakou Fort, Carved Fort, and Chang'anling are classified as the middle road (Tab. 1), which is led by Longmenwei. The whole consists of two cities: Du Shicheng and Longmen City, two cities and 24 military forts, with more than 1,300 beacon towers and watch towers, forming a complete military defense system. First of all, the military institutions in the area are complete, and the castles under each jurisdiction have corresponding defensive or defensive positions. The guards, military forts, watch towers are clearly defined, and the overall defense system is guarded at all levels. In addition, the distribution of the castle is dense and ranks. The distance between Fort City and the Great Wall is 30 to 40 miles. The distance from Fort City to the Great Wall is generally no more than 20 miles, so that when the enemy invades, the army can quickly enter the city. The distance between Wei and the city is about a hundred miles apart, and is arranged with the Fort City to effectively control the fortified city. The castle defense layout system based on the Great Wall also reflects the layout characteristics of the castle villages around the Great Wall. (Fig. 2)

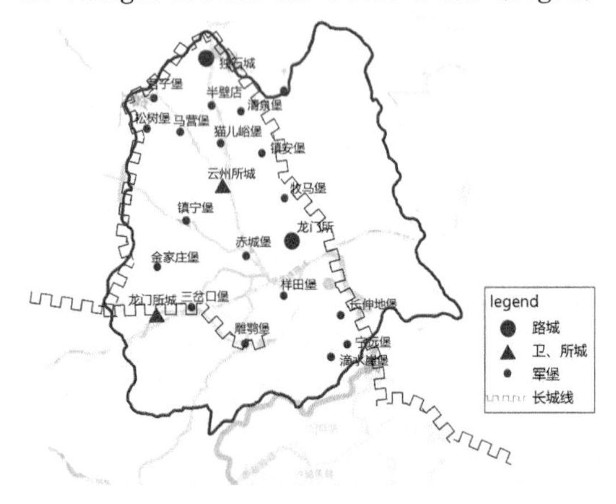

Fig. 2 Distribution map of the Chicheng Great Wall defense system

2.3 Morphological Characteristics and Functions of Castles Along the Line

The shape characteristics of the castle along the Great Wall can reflect the characteristics of military defense. The morphological characteristics of the military fort include the plane form and the structural relationship between the various components of the military fort. Its morphological characteristics are as follows: strict hierarchical division, the size of different castles is different; and the plane shape of the castle is mostly rectangular, and there are also free forms due to terrain or defensive factors.

The development and evolution of castles in different regions are different due to different natural conditions, geographical location and economic development. For example: Dashikou, resulting in relatively complete preservation, good traffic conditions, large development space, and large scale of settlement. Some settlement castles have disappeared, such as Ningyuan Fort.

Tab. 1 Important military fort, gate, morphological characteristics of the station, preservation of the status

Branch area	Name	Military institution	Morphological characteristics	Preservation status	Status picture
Shangbei Road	Dushikou fort 独石口	military fortresses		The city wall has been kept intermittently	
	Banbidian fort 半壁店	military fortresses		Brick coupon door arch remains, part of the soil wall is preserved	
	Maoeryu fort 猫儿峪	military fortresses		The wall is destroyed and the gate remains.	

Continued

Branch area	Name	Military institution	Morphological characteristics	Preservation status	Status picture
Shangbei Road	Junzi fort 君子堡	military fortresses		The bricks were removed and only the internal rammed walls remained	
	Songshu fort 松树堡	military fortresses	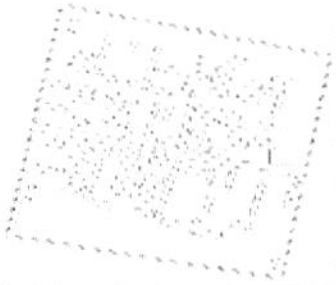	Destroyed into a pile of piers, there are 58 watch tower, of which three are better and nine are ruined	
	Mayingfort 马营堡	military fortresses		Visible castle base, residual wall	
	Zhenan fort 镇安堡	military fortresses		The wall is intermittent, the castle has a small amount of bricks, the pier and the bricks have been completely removed, and the gate remains	
	Yunzhou Wei, Suo 云州所	Wei, Suo		The outsourced bricks have been dismantled and the bauxite of the wall has been broken	
	Chicheng fort 赤城堡	military fortresses		Leave the Chicheng Drum Tower	
Middle road	Jinjiazhuang fort 金家庄堡	military fortresses		Leave the Chicheng Drum Tower	
	Longmenguan fort 龙门关堡	military fortresses	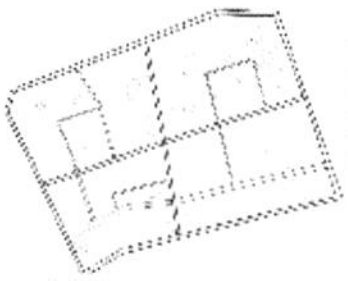	The remaining south wall, east wall and north wall, there is only a section of inner brick with bricks in the middle of the east wall	
	Sanchakou fort 三岔口堡	military fortresses		The entire wall has been demolished and the damage is most serious	
	Diaoe fort 雕鹗堡	military fortresses		The ruins of the South Gate are ruined, leaving a ticket still there, already bricked	

Continued

Branch area	Name	Military institution	Morphological characteristics	Preservation status	Status picture
Xiabei road	Longmen WeiSuo 龙门所	military for-tresses		The destruction of the city is very powerful, there are still traces to find	
	Muma fort 牧马堡	military for-tresses		Leave a residual wall, a city gate	
	Yang tian 样田堡	military for-tresses		The site is hard to find	
	Changshendi fort 长伸地堡	military for-tresses		The wall was damaged, leaving a door and the road was badly damaged. City wall and earthen bricks retain better	
	Ningyuan fort 宁远堡	military for-tresses		The department gate was de-molished, leaving a residual wall	
	Dishuiya fort 滴水崖堡	military for-tresses		The city is stone building, the collapse is quite serious	

3 The Current Situation and Problems of the Settlement of the Great Wall in Chicheng County

3.1 Protection Status

Through field research, it is found that in addition to Juyongguan, Badaling, Changchun City, and Shudao City, through the protection of tourism development, many numbers of Great Walls have not been effectively protected, but are in an extremely declining state. Some of the Great Walls in the Chicheng area of Zhangjiakou have not been effectively protected, and some villages are also in decline. The reason for its decline was mainly that after the Great Wall lost its military functions, the Great Wall settlement gradually lost its abiding position, and the ancient village gradually developed into a natural village. With economic development, social changes, and changes in people's production and lifestyle, especially in the 1980s, many villagers went out to work and seek employment, and the population plummeted, leading to economic recession. Coupled with limited land use and low income sources, the Chicheng area is relatively poor. The economic and cultural development of modern society has also caused changes in social concepts and living environment in traditional settlements. These factors have affected the redevelopment process of settlements, making the survival and development of traditional settlements face severe challenges (Ru & Yang, 2008). At the same time, the decline of the Great Wall settlements, carrying a large amount of historical and humanistic information will also disappear. In view of this, it is imperative to rescue and discover this non-renewable cultural heritage.

3.2 Existing Problems

3.2.1 Lack of Systematic Management System

Because the geographical position of the Great Wall is mostly in the difficult or remote location of the provincial and city junctions, the traffic is inconvenient, the administrative division has caused the isolation of protection management, and it is difficult to cooperate across regions; the management level of villages and towns is low, lack of effective measures, some ancient walls Ancient buildings and so on were demolished. Although the relevant departments have already conducted resource surveys on them, the specific protection methods are difficult to implement, and some areas have relatively commercial development, ignoring their cultural connotations.

3.2.2 Weak Awareness of Protection

Due to the transformation of the social economy, the original humanistic consciousness has subsided, and the traditional historical humanistic values are gradually changing. Nowadays, the tourism industry has gradually become a pillar industry in the Chicheng area, bringing a lot of benefits to the villagers. However, the villagers still have insufficient understanding of the value of the historical and cultural heritage itself. Some buildings with historical value have been destroyed and abandoned, and the buildings of historical style have been protected. The consciousness is weak.

3.2.3 Historical and Cultural Information is Seriously Damaged

The cultural relics in Chicheng are protected by the construction of the Great Wall, the military fort, and the shovel. Due to the influence of the natural environment, some ancient Great Walls have been severely damaged due to cracks, collapses and gaps in the walls. Repairing it requires a lot of manpower and resources, and some buildings are not better protected. For some traditional buildings, villagers adopt re-construction or alternative addresses, and the abandonment of traditional buildings has a certain degree of influence on traditional features.

4 Suggestions for the Overall Protection and Renewal of the Great Wall Settlement

4.1 The Overall Protection of the Great Wall and the Construction of the Heritage Corridor Protection Model

The Great Wall in Chicheng is a unified whole that includes the historical environment, spatial pattern and natural environment, and belongs to the cultural landscape cultural heritage. In the process of protection and utilization, it is necessary to protect and update the style, texture and form of the Great Wall and surrounding villages in a holistic manner. Therefore, the protection mode of the heritage corridor can be established, and the protection content is selected according to the distribution characteristics of the military settlement of the Great Wall of Chicheng analyzed above, so as to divide the three levels of regional-collapse-element (Cao, 2014). With the Great Wall Corridor as the link, the wall, the military fort, the health center and other related cultural contents together constitute the overall protection and utilization of a strip area. schematic diagram.

4.2 Establish a Community Management Mechanism

In the protection process of the Great Wall settlement in Chicheng District, the government, the economy and the villagers are the three main bodies of the Great Wall settlement protection (Chen, 2007). It is necessary to fully mobilize the power of these three, rationally divide the work, and supervise each other to form a sustainable development mechanism. First, under the guidance of the government, prepare protection plans, propose protection measures and management systems. In this process, we must fully mobilize the enthusiasm of local people and emphasize community participation. Secondly, let the villagers go deep into protection and development, which will help raise the awareness of local villagers. Encourage villagers to inherit and carry forward traditional culture, and adopt fund support to protect various cultural heritages, including intangible cultural heritage.

4.3 Renewal and Sustainable Development of Villages Around the Great Wall

The renovation and renovation of the villages around the Great Wall should develop the economy and improve the local living environment under the premise of protecting the traditional villages. In the protection plan for the Zhangjiakou Great Wall, due to the better preservation of individual castles, it is possible to coordinate tourism development and conduct linear tourism planning and integration. This kind of tourism planning of the Great Wall linear heritage by point and line can make people nostalgic, experience traditional culture, return to nature, explore novelty, etc. It is also a high-grade cultural tourism form that is different from the current tourism status, and can make "protection" and the concept of "sustainable development" runs through.

5 Conclusion

The Fortress-style settlement along the Great Wall is inextricably linked to the Great Wall in history, landscape and culture. It contains rich Great Wall culture and has important scientific and cultural heritage values. The main structure of the Great Wall in Chicheng is relatively intact and is an important part of the Ming Great Wall defense system. This paper analyzed the defense system of Ming Great Wall in Zhangjiakou Chicheng. The authors arrived at the following conclusions: the location of the Great Wall is mostly built according to the terrain; the defense system is complete; the spatial pattern and the morphological characteristics of the castle are characterized by clear classification and local conditions. However, many castles along the Great Wall have not been protected for a long time. The important cultural heritage of quite a few castles is seriously damaged, and the cultural information contained therein has disappeared. In view of the protection of the Great Wall settlements in Chicheng, the protection model of the Great Wall Heritage Corridor should be constructed with the goal of demonstrating complete military space characteristics and defense systems. Combined with the community management mechanism, the tourism planning of the Great Wall linear heritage will be established to realize the renewal and sustainable development of surrounding villages.

References

CAO Xiangming, 2014. The evolution of military fortress along the Ming Great Wall in Shanxi province and its protection and utilization model. Xi'an: Xi'an University of Architecture and Technology.

CHEN Ji, 2007. Preliminary study on morphological characteristics and protection strategies of Baozhai village along the Great Wall in Beijing Area. Xi'an: Xi'an University of Architecture and Technology.

TAN Lifeng, 2010. A probe into the military fortress system in Hebei in the Ming Dynasty. Journal of Tianjin University (Social Science Edition), 12 (06): 544-552.

ZHAO Lihong, 2009. Renovation plan of castle village based on Great Wall protection: Taking Baiyangtun Village in Qian'an as an example//China

Urban Planning Society. Urban Planning and Scientific Development — 2009 China Urban Planning Annual Conference Proceedings. China Urban Planning Association: China Urban Planning Association, 12.

Study on the Protection of Lingnan Vernacular Architecture Heritage in Guangdong-Hong Kong-Macao Greater Bay Area: From the Perspective of Ethnic Identity

LAU Gwokwai, *LU Qi*

School of Architecture, South China University of Technology,
State Key Laboratory of Subtropical Building Science, Guangzhou, China

Abstract First, unscramble the policy of *the Development Plan for Guangdong-Hong Kong-Macao Greater Bay Area* (The Development Plan). It is found that the government attaches great importance to the development of the Chinese cultural spirit and Lingnan culture in the construction process of the Greater Bay Area. The Guangdong, Hong Kong, and Macao regions have the roots of the clan patriarchal system of Chinese culture in terms of Ethnic Identity (Consanguinity, Geographical Overlap, Developmental History and Business Relationship). They also have the local characteristics of Lingnan culture and the vernacular architecture heritage with multiple values and unique spatial forms. The study found that the Clan Rule was a cultural concept that was reasonable in theory and could be implemented in the construction of the local architectural heritage area in the Greater Bay Area. Most rural areas in Guangdong still carry out production and construction and daily life in accordance with the Clan Rule. They are rich in traditional villages and vernacular architecture heritage. Some villages in Hong Kong still have a large number of clan organizations and even village alliances, and there are still many cultural heritage values worthy of excavation in the villages such as Sai Bin Wai and Lin Maa Haang. Some architectural heritage left by Macau. For example, Zhengjia Dawu and Lujia Dawu can also be further integrated into the cultural heritage research of the Greater Bay Area. Finally, by sorting out the different urban and rural cultural heritage management methods and policies in Guangdong, Hong Kong, and Macao, it is found that the criteria for cultural heritage assessment and management in three regions are not the same. A regional vernacular architecture heritage system should be established to further improve the research and practice of Local Architectural Heritage in Guangdong-Hong Kong-Macao Greater Bay Area.
Keywords vernacular architecture heritage, Guangdong-Hong Kong-Macao Greater Bay Area, ethnic identity, Lingnan culture, regional culture heritage

1 Cultural Construction in Guangdong-Hong Kong-Macao Greater Bay Area at the National Strategy

The "Humanistic Bay Area" is an important part of *the Development Plan* (Fig. 1) that is clearly proposed to serve the people, build livable, suitable, and enjoy a quality life circle. Hong Kong, Macao and the nine cities of the Pearl River Delta have the same cultural homogeneity, close kinship, similar folk customs and complementary advantages.

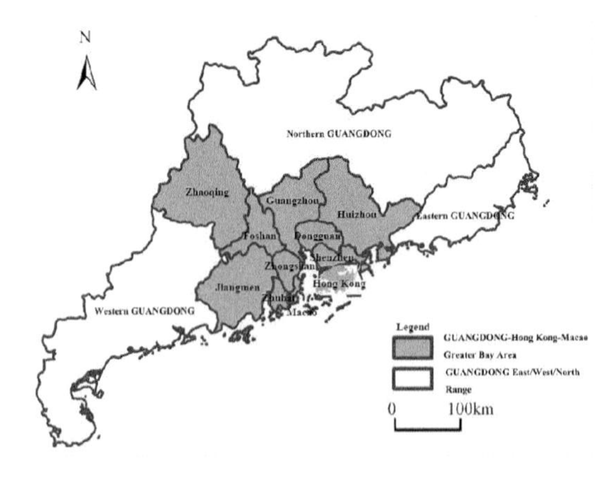

Fig. 1 The Location of the Guangdong-Hong Kong-Macao Greater Bay Area © CHEN Pinyu

The similarities between Guangdong, Hong Kong and Macao and the contextual relationship are the historical and cultural foundations for the construction of the Humanities Bay Area. The deeper sources involved are proposed by relevant scholars: based on the relationship between Ethnic

Identity (Consanguinity and Geographical Overlap, Developmental History and Business Relationship), Lingnan culture is the common cultural origin of the spirit of Guangdong and the spirit of Hong Kong. The historical development of Macao culture also has an open Lingnan cultural heritage. Lingnan culture plays a vital role in the regional cultural reconstruction of the Humanities Bay Area. In *the Development Plan*, "Lingnan culture" has been explicitly mentioned many times, and the Chinese cultural spirit on which it depends.

The spirit of Chinese culture has been explored for a long time. The breadth of its involvement involves the level of "cultural ecology". The social psychology and spiritual characteristics of the Chinese nation should be as far as the scholar Feng Tianyu said, "A patriarchal autocracy that relies on a semi-closed continental-coastal geographical environment to develop a natural economy and a nomadic economy". *The Development Plan* clearly proposes to encourage Hong Kong to promote China's excellent traditional culture and to support Macao to build a communication and cooperation base with Chinese culture as the mainstream. We should jointly promote the inheritance and development of the fine traditional culture of the Chinese mainland, Hong Kong and Macao.

As a part of Chinese culture, the Greater Bay Area should be a multicultural and harmonious culture with Lingnan culture as the main body. Since ancient times, Lingnan culture has been limited by political concepts and geographical categories. From the perspective of history, the period of the true formation of "Lingnan culture" is also the time of the Ming and Qing Dynasties, and it has been in the gestation period before. In fact, in *the Development Plan*, it is also clear that "support Lingnan culture", "to promote the integration of urban and rural areas in the nine cities of the Pearl River Delta", "to build a livable urban and rural area with Lingnan characteristics". The city orientation of Guangzhou, the capital of Guangdong Province, is clearly defined as the construction of the "Lingnan Cultural Center", "expanding the influence and expansion of Lingnan culture" [1]. Guangdong-Hong Kong-Macao Greater Bay Area at the national level should promote the construction of the Humanities Bay Area and its cultural heritage with the concept of "Lingnan culture".

2 Lingnan Cultural Foundation in Guangdong-Hong Kong-Macao Greater Bay Area from the Perspective of Ethnic Groups

By interpreting *the Development Plan for Guangdong-Hong Kong-Macao Greater Bay Area*, the state attaches great importance to the construction and development of the Chinese cultural spirit and Lingnan culture in the construction process of Guangdong-Hong Kong-Macao Greater Bay Area. In the development of specific regional cultural heritage protection, the selection and application of established cultural traditions have become a top priority. The study found that "Clan" and "Religion" are two concepts that are theoretically reasonable and can be implemented in urban and rural practice in Greater Bay Area. Here the clan is discussed as an example.

According to Nathan Glazer, an ethnic group refers to a group with its own cultural characteristics in a larger cultural and social system. The most prominent trait is the religious and linguistic characteristics of this group and its members or ancestors. They have material, ethnic and geographic origins.

Zhou Daming believes that ethnic groups often emphasize common succession and blood so that it is easy to form cohesive groups because of common ancestors, history and cultural origins. Traditional Chinese society is a clan-led home society, especially in rural areas. In a certain sense, ethnic groups are special communities, and traditional rural areas are mostly developed by the fusion of blood and ethnic groups, showing the intertwined integration of ethnic groups and communities. One of the inherent requirements of the return of rural nostalgia is the ethnic community.

Ancient China was a patriarchal dictatorship with the same structure as home. Hong Kong and Macao, like ancient Guangdong, were similar before being colonized. The regional cultural character they retained became an important component of the colonial urban character and has continued. The outstanding performance of the patriarchal autocracy is the patriarchal system and the continuation of the clan. The Guangdong region, especially the rural areas of Guangdong, is highly influenced by the Confucian culture and the clan system. There are clear clan activities and patriarchal relations. This patriarchal clan relationship has been successfully transformed into a substantive group in modern times for production and construction and daily life. Some villages in Hong Kong still have more clan organizations and even village alliances. John A. Brim also mentioned the study of Hong Kong township temples. "Village alliances (people in the New Territories, often referred to as 'townships') have long been a feature of social organizations in the region. ""According to the inscriptions on the temples, the village alliance originated in the mid-18th century to the 19th century... and today, these alliances still have a weak influence". The Macao region, especially in the 16th and 17th centuries, is still under the rule of the Qing Dynasty. The influence of Lngnan culture on Macao is also divided into two categories by scholar Liu Ranling. "There is mainly a strong regional Lingnan culture", and "Confucian culture related to the regime". The Macao culture at the regional level is subordinate to Lingnan culture, while the population structure of Macao is mostly Han, and the main body of the Macao Han is also Guangfu culture.

It can be seen from the above that the historical Guangdong, Hong Kong, and Macao regions have the roots of the clan patriarchal system with obvious Chinese culture in terms of Ethnic Identity (Consanguinity and Geographical Overlap, Developmental History and Business Relationship). They also have the local characteristics of Lingnan culture. They are cultural heritage areas with diverse values and unique ideas. It is unique in the country and the world.

3 Lingnan Vernacular Architecture Heritage in Guangdong-Hong Kong-Macao Greater Bay Area

The Vernacular architectural heritage has a special type of regulation in many categories of World Heritage. In 1999, the *Charter on the Build Vernacular Heritage* (1999) was called "the built vernacular heritage". Before the Vernacular architecture became a heritage, it was very important to protect its material form, the inheritance of intangible ideas and the construction of the value system. Because it determined whether he could meet the General Issues proposed by the Charter, and officially was assessed as a World Heritage.

The "Vernacular architecture heritage" stipulated in the Charter is a collective name for the whole settlement including the local architecture and the environment. The domestic use of "vernacular architecture" is also different. It contains traditional official buildings, residential buildings, academies and many factors such as temples, ponds, and even tombs. The existence of vernacular architecture is the formation of settlements. A variety of buildings of different types, functions, and properties are combined into a complete system in the settlement. Chinese Traditional Village and Folk Residential Architecture are the main research directions of this thesis. However, because the "Chinese traditional villages" selection site does not involve Hong Kong and Macao, the ancient buildings (groups) with traditional styles and regulations are discussed.

3.1 Lingnan Vernacular Architecture Heritage in Guangdong Province

As of today, the mainland has announced five batches of Chinese traditional villages, involving 6,819 national traditional villages. There are 263 traditional Chinese villages in Guangdong Province, and they all have the cultural and regional style of Lingnan. The clan ethnic groups are the main factors for the formation and development of traditional villages in Guangdong.

Since the Song Dynasty, people in the north had gradually moved southward, and the scale of the formation has formed between the Ming and Qing Dynasties. Coupled with the maintenance of overseas social organizations in Guangdong after the Ming and Qing Dynasties, the clan organizations in the region are very developed.

During the Ming and Qing Dynasties, the number of ancestral monks increased sharply in the Guangdong area. The culture is rich, the shape is complete, and the functions are diversified. It is a mature period in the development history of the Lingnan Temple. The development of the local architecture under the influence of the Guangdong clan concept continued until the end of the Republic of China. After the liberation, the agrarian revolution and economic reform organized by the government departments touched the foundation of the traditional clan organizations everywhere. Until modern times, the clan and the ancestral hall still played an important role.

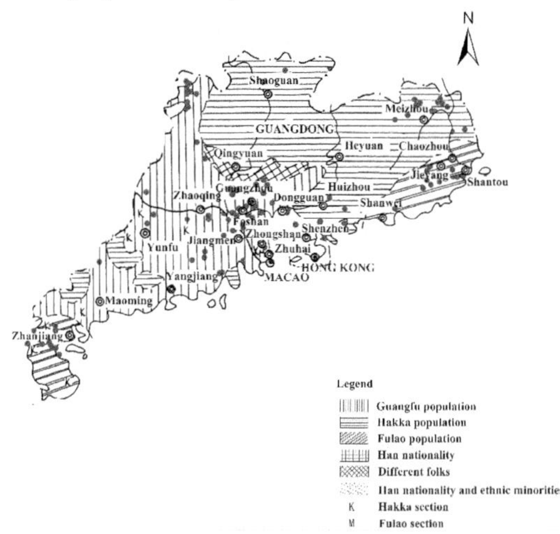

Fig. 2 Spatial distribution of Chinese traditional villages using ArcGIS in Guangdong

Taking the first four batches of national traditional villages in Guangdong Province as an example, We used ArcGIS software for analysis and found that the traditional villages in Guangdong Province are in three cluster areas, namely Guangzhou-Foshan Pearl River Delta, Meizhou-Chaoshan District in northeastern Guangdong and Minority areas in northwestern Guangdong (Fig. 2).

The Lingnan traditional village has strong regional characteristics. Taking the Xuri Village in Guangdong Province as an example. It is a scarce ancient architectural group that integrates the architectural culture of the Ming, Qing Dynasties and the Republic of China in a village. It is well preserved. Xuri Village has a history of more than 400 years. It has been praised by experts and scholars as "the model of Lingnan ancient residential architecture" and "the first ancient village of Luofu". Xuri Village is a typical Guangfu folk village. The village is built because of the traditional three-room two-lane plane form of Lingnan. It is combined into a quadrangle patio and a multi-passage patio. The specific village regulation and architectural form are shown(Fig. 3).

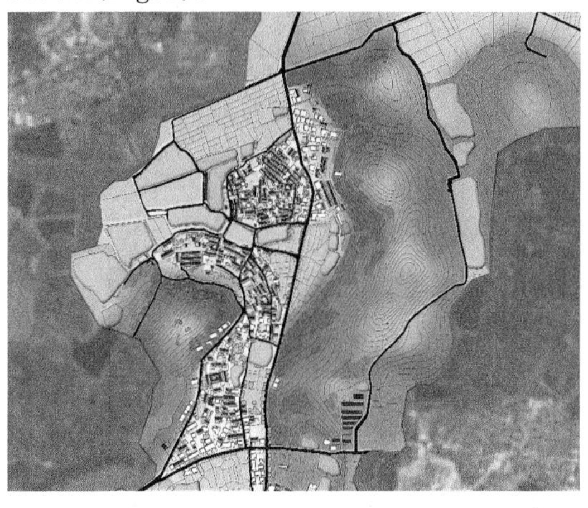

Fig. 3 Village plan of Xuri Village in Guangdong

3.2 Lingnan Vernacular Architecture Heritage in Hong Kong

Lingnan culture is one of the local cultural forms of China. It is the cultural foundation shared by Hong Kong, Guangdong, and other regions. It is also the conceptual basis of people's identity awareness in these areas. Hong Kong was originally a city of Xin'an County inthe Qing Dynasty. It is a place with a typical influence of Lingnan culture. Today, there are still many traditional buildings such as folk houses, ancestral halls, and temples. In the Qing Dynasty, Hong Kong residents experienced collective internal migration and relocation, and at the time of relocation, there were more Hakkas. They built many market cities and houses, such as Pingshan Heritage Trail, Daai Fu Dai, Gat Hing Wai and so on. Hong Kong's encirclement, for example, there are local villages such as Ngaa-cin Wai, Tyun-zi Wai, Cing-zyun Wai, and Gat Hing Wai. There are also Hakkas, such as Zeng Da Wu(Fig. 4), Bai Sha Ao, and Xin Wu Zi. In addition to the surrounding area, there are many ancestral buildings, such as Hong Kong's first-class historical buildings, such as Tao's Ancestral Hall, Mr. He's Ancestral Hall, Hakka's House, and Peng's Ancestral Hall.

Fig. 4 Vernacular architecture of Zeng Da Wu in Hong Kong

However, there are still many villages that have not yet entered the academic line of sight. For example, Hong Kong Lin Maa Haang(Fig. 5), near Shenzhen, is located within the restricted area of the New Territories North District. Most of the members of the village are Hakkas from Guangdong. There are about 200 buildings in Lin Maa Haang with good historical value. The village landscape garden is also built along with the ancient Chinese concept of Feng Shui. There are artificial natural structures such as Fengshui Forest, Fengshui River, and Fengshui Pool. There are also eight major ancestral halls that can represent the characteristics of the clan blood group. Among them, Mr. Ye's Ancestral Hall is the largest, and it also follows the imperial examination system formed by the influence of ancient Confucianism.

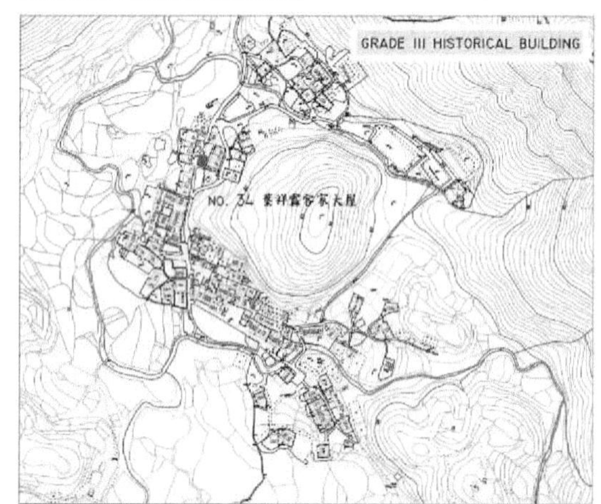

Fig. 5 Building research and case studies of Lin Maa Haang in Hong Kong © YEUNG Shan Yan Cindy

The village has a list of contemporary students "golden title". Hong Kong also has a typical "Guangfu residential" flat-shaped heritage building. For example, the Luo Wu Folk House, a legal monument in Caai Waan, is a residential structure of "three houses and two galleries". From the perspective of architecture, the local architecture with Lingnan style is built by Hakkas, but it has more Guangfu characteristics from the architectural style, and rarely forms the Tu Lou architectural style like the Hakka residential buildings in Fujian and Guangdong Province.

3.3 Lingnan Vernacular Architecture Heritage in Macao

The historic district of Macau was named a World Cultural Heritage in 2005, and its architectural entities with unique Lingnan cultural characteristics have been preserved. Based on the clan-led blood and geography groups mentioned above, it is the main reason for the formation of very few sites. Before Macao was colonized, it was under the jurisdiction of Xiangshan County. Its architectural form such as ancestral hall was still used as a feudal patriarchal system. The architectural expression of ethnic autonomy is the social basis of the ancient patriarchal system consciousness.

While having a traditional Chinese culture, Macao also has an open culture of Lingnan culture (Liu Ranling, 2008). The local architectural heritage of Macao, especially the traditional residential buildings, has obvious characteristics of Guangfu, which also implies that the traditional culture of Guangfu has a profound impact on Macao society, and the Cantonese style is rich(Liu Ranling,2008).

Fig. 6 Courtyard of Zhengjia Dawu in Macau

The Zhengjia Dawu and the Lujia Dawu are the only remaining residential buildings with the characteristics of Guangfu in the Macao area, and Zheng's House is the largest(Fig. 6).

Zhengjia Dawu is the house of Mr. Zheng Guan Ying in the late Qing Dynasty. The building complex has both architecture and gardens. It has many functions such as living, living, recreation, and meeting. There are three main living houses in Zheng's Family House, which are the different descendants of Zheng's family, namely, Ji Shantang and Yu Qingtang.

4 Vernacular Architecture Heritage Issues and Solutions in Guangdong-Hong Kong-Macao Greater Bay Area

4.1 Status and Issues

Different from the domestic Beijing-Tianjin-Hebei region and the Yangtze River Delta region, the construction of the Guangdong-Hong Kong-Macao Greater Bay Area is carried out under the political system of "one country, two systems", and the research cases are unique. Due to its unique geographical conditions and economic and cultural concepts, the number of Hong Kong and Macao cultural heritages during the period of the British colonial period and the Portugal colonial period was relatively large. The current methods and types of cultural heritage protection were also different from those of the Mainland.

In addition to the above types, the urban and rural cultural heritage types in the Mainland should also have the World Cultural Heritage Project and the type of "Chinese Traditional Villages" that began in 2012. The scenic spots, archaeological parks and modern parks in the Mainland have many Historical and cultural landscapes that are not included in the above system should also be included. The protection of cultural heritage in Hong Kong is mainly related to the protection of cultural relics. The nature conservation section of Hong Kong is similar to the scenic spots in the Mainland. The "special areas" in the "Country Parks and Special Areas" refer to government land with special and important values in terms of flora, fauna, geology, culture or archaeological features. (Hong Kong SAR Govenment, Planning Department,2017)

With some historical and cultural landscapes, it is also included in the scope of this study. Compared with the traditional villages in the Mainland, there are still some villages in Hong Kong that are worthy of further excavation. For example, the western boundary in Yuen Long and the Lin Maa Haang in the North District have strong cultural heritage value. Some cultural heritages in the Macao area are can also be further integrated like Zheng's house and Lu 's house. The criteria for the assessment and management of cultural heritage in Guangdong, Hong Kong, Macao are not the same. We need further research and practice to establish a regionally coordinated cultural heritage system.

4.2 System Establishment

"Clan has the function of consanguinity, geographical overlap, and interests"(Li Huayin,2019). The reason why we use ethnic groups, especially clan, to describe the article is here. With the advancement of current world cultural heritage projects in the Asian region, China is facing the construction and implementation of the "The Belt and Road" transnational and cross-regional heritage routes and the demand for the "Maritime Silk Road" offshore route project. The above questions urgently require in-depth research results at the academic level, with the aim of promoting practical development.

The academic construction of the Lingnan Vernacular Architecture heritage system is a fundamental subject for the practice of urban and rural planning in Guangdong, Hong Kong, and Macao, and its research directly determines the quality of construction. Drawing on the successful experience of the world's cultural heritage, the application of cultural heritage in the region and the construction of cross-regional heritage, due to the in-depth exploration of the historical appearance of cultural heritage and the revitalization of the city within and outside the region, have led to the pursuit of paradigms and methods on a global scale. The political system of "one country, two systems" in Hong Kong, and Macao has determined that the urban and rural cultural heritage of Greater Bay Area in Guangdong, Hong Kong and Macao is a regional cultural heritage network with "cross-regional characteristics". The cross-regional cultural heritage construction has been gradually paid attention to since the 1990s. This has reference significance for the current construction of the cultural heritage network in the Greater Bay Area.

5 Conclusion

The construction of the Lingnan Vernacular Architecture Heritage in Guangdong-Hong Kong-Macao Greater Bay Area needs to be put on the research agenda in a timely manner. Through the excavation of Chinese culture and Lingnan cultural resources in Guangdong, Hong Kong, and Macao, the Lingnan cultural genetic and cultural heritage resources in Hong Kong, and Macao are sought. Through the multidisciplinary research on cultural heritage, the Guangdong, Hong Kong and Macao cultural heritage think tanks are established. We should work hard to promote the construction of the World Maritime Heritage Culture and the construction of the cultural heritage of Guangdong, Hong Kong and Macao. It is suggested that scholars of various disciplines, especially architectural and cultural heritage scholars, should sort out the existing cultural heritage types of the three sides of the Taiwan Straits and establish a regionally coordinated cultural heritage system in Greater Bay Area, so as to carry out multi-regional and multi-dimensional regional comprehensive linkage research, and finally implement it in urban and rural space practice in Greater Bay Area.

Acknowledgements

This paper is one of the phased achievements of the National Natural Science Foundation of China (51278194) and the State Key Laboratory of Subtropical Building Science Project (2017KB06-x2jzC7170160).

References

CHEN Zhihua, ZHAO Wei, 2000. Reflections on charter on the built vernacular heritage. Time + Architecture, 3:20-24.

CPC Central Committee, 2019. The state council of the people's republic of China. The development plan for Guangdong-Hong Kong-Macao greater bay area. Beijing: People's Publishing House.

FENG Tianyu, 2010. Chinese culture history. Shanghai: Shanghai People's Publishing House.

GLAZER N, MOYNIHAN D P, 1975. Ethnicity-theory and experience. Boston: Harvard University Press.

Hong Kong SAR Government, Planning Department, 2017. Hongkong planning standards and guidelines. Hong Kong.

HUANG Yuexi, HUANG Chudan, 2018. A study of the relationship between Chinese spirit, Guangdong spirit and Hong Kong spirit. Journal of Shenzhen institute of information technology, 16: 1-5.

LI Huayin, 2019. Family-villa relationship: a new perspective on the understanding of social heterogeneity in Chinese villages. Journal of Guangxi University (philosophy and social science), 41:127-134.

LIU Ranling, 2008. Wen Ming De Bo Yi-A historical investigation of the long band of Macao culture in the 16th — 19th centuries. Guangzhou: Guangdong People's Publishing House.

WOLF P, 2014. Religion and ritual in Chinese society. Stanford: Stanford University Press.

XING Lijun, XU Haibo, 2015. On the construction of Lingnan cultural spirit and national consciousness of Hong Kong people. Journal of Guangxi Normal University (philosophy and social sciences edition), 51:20-27.

ZHOU Daming, 2014. Hakka ethnic group and ethnic awareness in turbulence. Journal of Guangxi University for nationalities (philosophy and social science edition), 27:13-20.

A Study on Architectural Characteristics and Structural Properties of Earth Structure Mazar Tombs in Southern Xinjiang Uygur Autonomous Region

XU Lei, *SUN Jingyuan*

Beijing University of Civil Engineering and Architecture, Beijing, China

Abstract Southern Xinjiang Uygur Autonomous Region (see later "southern Xinjiang region") is situated in the ancient Silk Road. The climate characteristics of arid and semi-arid bring up abundant vernacular architecture type. Mazar Tombs are extremely representative religious buildings, which are for Islam to mourn their ancestors. Mazar tombs embody a concentrated reflection of the specific historical period and area, religious worship, worship etiquette culture, and architectural art. Based on the studies of existing Mazar tombs in southern Xinjiang region, through trace back to the historical background and analyze its environmental characteristics, layout characteristics, choice of building materials, this paper summarizes the implied religious beliefs and worship soil consciousness, for the purpose of studying the reduction process that from packed to skillfully use soil bricks. This paper discusses the structural form and detailed characteristics of the single building of raw soil Mazar tombs, especially the detailed structural techniques such as arch roof and pointed door. This paper summarizes the structural performance of the regional adaptability of raw soil type Mazar tombs in southern Xinjiang region from the aspects of the development history and forms of technology.
Keywords Mazar, cultural connotation, construction, structural form

1 Introduction

The southern Xinjiang region refers to the area south of the Tianshan Mountains in Xinjiang. It has been a multi-ethnic area since ancient times. The area is located between the two main lines of the ancient Silk Road, where the multi-cultures of various ethnic groups are here to merge and converge, creating a unique regional culture and characteristics. The climate in southern Xinjiang is dry and rainless all the year-round. The arid and semi-arid climatic characteristics have caused the construction technology of raw soil to continue to be developed in the region for more than two thousand years. It is an important structural form of local dwellings, religious buildings and others. Mazar is a form of tomb construction that came into being after the introduction of Islam into Xinjiang. It originated from the tomb system of the Islamic royal family and gradually evolved into a form of tomb construction for religious belief in southern Xinjiang and even the whole Xinjiang region. Religious ritual architecture was built to mourn their ancestors. More mosques have been built near Mazar, gradually becoming a holy place for the religious beliefs and spiritual sustenance of the surrounding people.

2 General Situation of Natural Soil Mazar in Southern Xinjiang

2.1 Time Background

Mazar architecture has gradually evolved along with the development of Islam in Xinjiang. As a building carrier of religious institutions and sacrificial interests, Mazar carries multiple social functions. Since the ninth century, the Shaman dynasty that believed in Islam gradually occupied and ruled the southern Xinjiang region, bringing Islamic architectural culture into the region. This form of mausoleum architecture, which was built to commemorate the leaders of the Islamic dynasty or religious leaders, was gradually accepted by the people.

The construction technology of noumenon architecture in southern Xinjiang has experienced the evolution from native architecture to whole native architecture, and gradually to the civil mixed structure of civil engineering. Therefore, as early as the beginning of the construction of Mazar buildings in southern Xinjiang, the raw soil materials have become the best choice of building materials.

2.2 Environmental Characteristic

The southern Xinjiang region is located in the south of the Tianshan Mountains, in the hinterland of the Tarim Basin, surrounded by alpine plateaus. This closed inland basin is dominated by an extremely dry continental climate with annual precipitation below 100mm. The ecological characteristics of the raw soil are extremely adapted to the local climate, maximizing energy conservation and creating a comfortable natural environment. It cames from nature and returns to nature.

Due to its adaptability to the arid climate and energy-saving ecological characteristics and strong plasticity, raw soil has become the main building material in southern Xinjiang. It has been widely used in various architectural forms in southern Xinjiang and integrated into the multi-ethnic culture of the region. The architectural culture that meets the local conditions, thus creating a local architecture culture with regional characteristics according to local conditions and tailored conditions, is the most characteristic sacrificial public building in the region.

2.3 Distribution

Judging from the distribution and existing situation of Mazar architecture in southern Xinjiang, the Mazar buildings in this area are mainly Islamic architectural styles. According to the facade form, they can be divided into single dome type, multi-dome type, and flat roof type. The building materials include raw soil, adobe bricks, masonry, wood, etc(Tab. 1).

Tab. 1 Distribution of Existing Raw Soils in Southern Xinjiang

	Name	Region	Years	Grade	Main Style	Structure Type
1	Aleslan Khan Mazar	Kashgar Region, Kashgar City	Song	Provincial protection unit	Islamic architecture	Raw soil structure
2	Kajesak Ata Mazar	Kashgar Region, Kashgar City	Song, Liao, Jin	County-level protection unit	Islamic architecture	Raw soil structure
3	Yusuin Paizula Hoga Mazar	Kashgar Region, Kashgar City	Song, Liao, Jin	County-level protection unit	Islamic architecture	Civil structure
4	Ottoman Bugash Khan Mazar	Kashgar Region, Kashgar City	Yuan, Ming	County-level protection unit	Residential building	Civil structure
5	Ali Tiki Ken Mazar	Kashi Prefecture, Shaofu County	Wudai	County-level protection unit	Residential building	Civil structure
6	Aizire Tipachai Mazar	Kashi Prefecture, Shaofu County	Song, Liao, Jin	County-level protection unit	Islamic architecture	Civil structure
7	Suritan Ali Fu Mazar	Kashi Prefecture, Shaofu County	Song, Liao, Jin	County-level protection unit	Islamic architecture	Adobe structure
8	Hoga Ai Mu Hoga Mazar	Kashi Prefecture, Shaofu County	Qing	County-level protection unit	Bixiuke	Adobe structure
9	Takezi Riteng Mazar	Kashi Prefecture, Shaole County	Song, Liao, Jin	County-level protection unit	Residential building	Adobe structure

Continued

	Name	Region	Years	Grade	Main Style	Structure Type
10	Dos Buglak Mazar	Kashi Prefecture, Shaole County	Ming	County-level protection unit	Islamic architecture	Adobe brick structure
11	Habd Mohammed Mazar	Kashi Prefecture, Shache County	Qing	Provincial protection unit	Islamic architecture	Adobe structure
12	Garathani Mazar	Kashi Prefecture, Shache County	Qing	County-level protection unit	Bixiuke	Adobe structure
13	Sulitan Karasha Khal Atamu Mazar	Kashi Prefecture, Jiashi County	Song, Liao, Jin	County level protection unit	Islamic architecture	Adobe structure
14	Koktalek Ata Mazar	Kashi Prefecture, Maigai County	Wudai	County-level protection unit	Bixiuke	Adobe structure
15	Kizileqi Mazar	Kashi Prefecture, Maigai County	Song, Liao, Jin	County-level protection unit	Bixiuke	Adobe structure
16	Sietidun Mazar	Kashi Prefecture, Maigai County	Song, Liao, Jin	County-level protection unit	Bixiuke	Adobe structure
17	Karatag Mazar	Kashi Prefecture, Aksu County	Yuan	County-level protection unit	Courtyard architecture	Civil structure
18	Merana Eshtin Mazar	Kashi Prefecture, Kuche County	Yuan, Ming	Provincial protection unit	Islamic architecture	Civil structure
19	Eckham Ata Mazar	Aksu Prefecture, Shaya County	Undetermined	general survey	Enclosure building	Civil structure
20	Bozuruk Atata Mazar	Aksu Prefecture, Shaya County	Undetermined	general survey	Enclosure building	Civil structure
21	Kahan Bazi Ata Mazar	Aksu Prefecture, Shaya County	Undetermined	general survey	Islamic architecture	Adobe structure
22	Abutar Mazar	Aksu Prefecture, Wensu County	Yuan	County-level protection unit	Islamic architecture	
23	Notamsulitan Mazar	Aksu Prefecture, Baicheng County	Qing	County-level protection unit	Islamic architecture	Adobe structure
24	Kurban Ata Mazar	Aksu Prefecture, Baicheng County	Qing	County-level protection unit	Islamic architecture	Adobe structure
25	Ahetam Mazar	Aksu Prefecture, Awati County	Undetermined	County-level protection unit	Residential building	Civil structure
26	Caileda Wu Mazar	Aksu Prefecture, Awati County	Undetermined	County-level protection unit	Roof enclosure	Civil structure
27	Guginai Xietilik Mazar	Hotan area, Hotan County	Song, Liao, Jin	County-level protection unit	Enclosure building	Adobe structure

Continued

	Name	Region	Years	Grade	Main Style	Structure Type
28	Emma Zedin Mazar	Hotan area, Hotan County	Song, Liao, Jin	County-level protection unit	Relics	Raw soil structure
29	Sheitang Kanwimu Mazar	Hotan area, Yutan County	Song Liao Jin	County-level protection unit	Enclosure building	Adobe masonry
30	Karach-ache Mazar	Bayingolin, Mongolian Autonomous Prefecture, Luntai County	The Republic of China	County-level protection unit	Residential building	Adobe structure

3 Cultural Connotation of Native Soil Mazar in Southern Xinjiang

3.1 Worship Etiquette Under the Influence of Religion Gene

Islam, which is the origin of the sacrificial architecture, was promoted by the rulers of successive dynasties in its ruling area with its powerful appeal, in order to gradually realize the expansion and conquest of the territory, thereby achieving the purpose of expanding the territory and strengthening the centralized rule. In the early period, Mazar mainly offered sacrifices to the leaders and believers who died in the "jihad" in order to stimulate the enthusiasm of the people to participate in the "jihad". As sages, Masha's master is regarded as the messenger of communication between believers and Allah. Therefore, Masha has gradually evolved into a daily religious ritual place for daily worship and became a holy place for the spiritual support of the people.

Through the development and evolution in the folk, Mazar is also known as Gongbei. This kind of house with four-sided walls, doors on each side and dome roof is considered by Islamic believers as the best amphibious and ascending land for souls after death. As an important part and embodiment of regional social culture and religious culture, funeral culture is the most common form of tomb ritual in Mazar, which reflects the religious beliefs and rituals of Islam (Fig. 1).

Fig. 1 Aleslan Khan Mazar

3.2 The Earth Worship Under the Influence of Primitive Belief

As one of the most primitive building materials, raw soil is the summary of ancient ancestors' understanding of nature in the process of historical evolution. No matter what kind of religious genre belief or people's public cognition, they all have deep emotional sustenance to the soil. The concept of using earth as a raw material for human creation, which is shown in the Islamic classics, is the humanistic origin of raw soil being widely used in Islamic architecture (Fig. 2, Fig. 3).

Fig. 2 Mazar Group, Shenmuyuan, Tianshan

Fig. 3 Mausoleum of King Kadel

4 The Construction and Evolution of Native Soil Mazar in Southern Xinjiang

The construction of such special buildings in southern Xinjiang has experienced the development process from the construction of raw soils to the skillful use of earthen blocks. Raw soil Mazar refers to the construction which is completed by using raw-earth materials without the use of wood material. The roof is arched, and the walls and foundations are built by ramming soil layer or soil embryo(Fig. 4).

4.1 Ramming of Raw Earth

The raw materials come from nature, and the khaki has a sense of original color. It is coordinated and integrated into the natural environment. It has become the best choice of building materials for the earliest people to build Mazar. The raw soil itself has good plasticity, can be taken locally, and is very convenient to construct, which is convenient for manual operation. Because of the plasticity of raw soil, it has been widely used in architectural decoration, including walls, roofs, floors and so on. The pleasant, comfortable and unique living environment gives people a primitive sense of color and a local feeling of harmony with nature.

The rammed earth wall is a widely used wall situation in earth building. The root slab can be divided into two types according to the different molds of the building. One is plate construction, which consists of two wooden boards clamped together to form a template. The two boards are poured into uniform mixing soil, laminated and compacted, and the rammed earth wall is formed after the demolition of the moulds is completed. The other method of squatting is that the difference between the slab and the slab is that the stencils of the slabs are made of logs, and the walls of the plaques are left with obvious traces of logs.

4.2 Soil Blank Block

The earthen bricks were first used in the construction of buildings by the ancient Egyptian empires on both sides of the Nile River in 3000 BC. In the 6th century, the technique of the earthen wall building achieved unprecedented development. In the early archaeological excavations of the Xinjiang area, the Tomb No. 1 tomb and the city walls of the Jiaohe Old City and the temples in the city were constructed with earthen bricks, which shows the application of adobe technology in the prehistoric era in Xinjiang.

The technology of adobe wall brick is to use suitable raw soil mixed with an appropriate amount of straw, gravel, sand and other materials to mix evenly, and to be compacted and polished in a fixed-size wooden mould to make adobe, which is dried thoroughly before wall masonry. In places where the clay resources are abundant and the foundation is dry, mud is used to bond the masonry and surface.

As a masonry, adobe bricks are widely used in the walls of Mazar buildings or mixed with rammed earth. The production of the adobe is an innovation based on raw materials. It is also a development of building materials. It has improved the quality of the building and accelerated the construction process. It is one of the building materials widely used in the construction of Mazar in southern Xinjiang.

Fig. 4 Soil embryo

5 The Structural Form of Raw Soil Mazar in Southern Xinjiang

5.1 Arch Technology

Arch technology was first introduced into the Tarim Basin from Turkmenistan of Central Asia. Arch technology mainly refers to the load-bearing structure that spans the space built by the side pressure between materials. Adobe brick is the most common brick used for arching in the construction of Mazar in southern Xinjiang. When arching with earthen bricks, there is no strict and consistent requirement for the spacing of the arches in the equal span, so no formwork is required for the laying.

This method of "formless masonry" is a unique local characteristic of the soil arch masonry technology in southern Xinjiang. The specific method is to first determine the

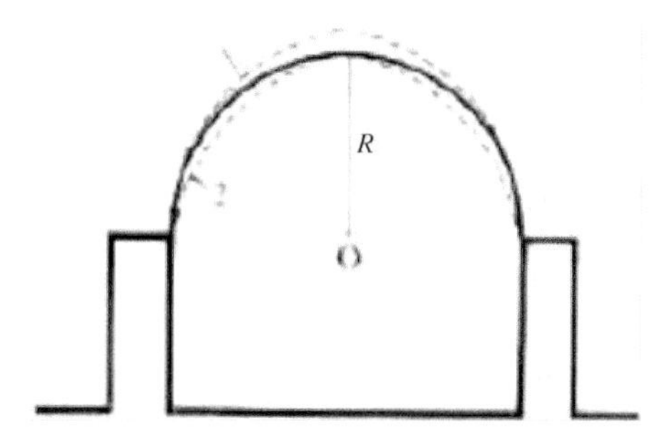

Fig. 5 Determine the arch line

wall base of the arched end, and take the half of the span as the radius, the center of the span as the dot, and draw the semicircular arch line on the back wall (Fig. 5). Aiming at the arch coupon line, the soil arch is built from the part of the arch, and the masonry is merged from the two sides of the arch to the middle. When the two sides are built to the middle position, it is found that the brick cannot be connected. The shape of the arch can be appropriately adjusted. raise or reduce the arch height to achieve the whole block masonry (Fig. 6). The grass mud is used as the bonding material between the soil embryos, and the overall flatness of the arch can also be adjusted.

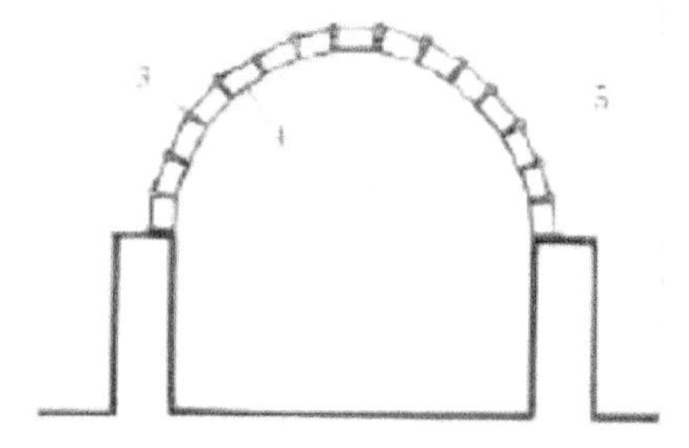

Fig. 6 Schematic diagram of moldless plastering

5.2 Dome Technology

The dome roof technology is generally used in the roof of the Mazar building in southern Xinjiang. The masonry dome is built from the dividing line of the quadrilateral base and is inclined inward from the bottom to the upper layer so that each layer forms a square arc. Each circle is gradually reduced until the center of the square plane forms a dome with a large vector height.

The earthen dome is not used for the construction of the dome, and the wooden beam is placed on the plane of the arch, and the center of the ball is found on the wooden beam, the nail is nailed, the rope is equal to the radius of the ball, and the layer is built around the circle. Stick it up. This method is common in Mazar buildings with simple early-stage properties such as octagonal, such as Nur Allanul Khan Mazar, Ai Di Di Weihan Mu Mazar(Fig. 7).

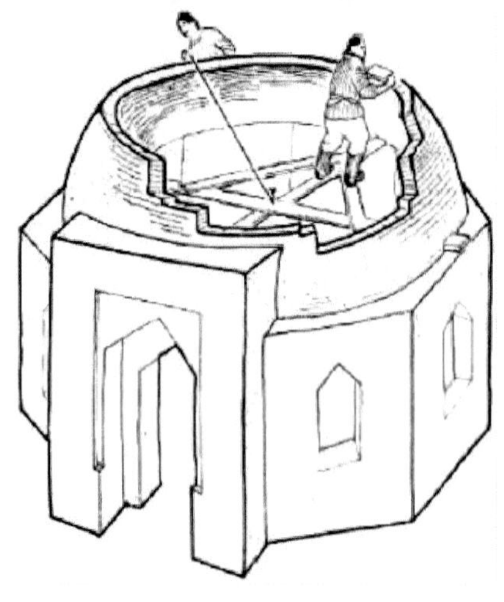

Fig. 7 Moldless topping (Image source: The History of Ancient Chinese Architecture Technology)

5.3 Overlapping Technology

Overlapping technology refers to the construction of buildings with adobe or masonry. The layers are stacked outwards or inwards. When each layer is lifted out, each layer is subjected to the gravity of the upper layer. It is a relatively traditional way of laying bricks. The overlapping technique is mostly used in the dome of the Mazar building, and the brickwork method of the brick-by-skin brick-level is used successively. The brick joints of the overlapping structure are horizontal and are topped in such a way that the brick layer is picked up. From the perspective of structural force, not only the pressure is affected by the shear pressure, but because of its low construction difficulty and saving construction materials, it is widely used in the Mazar buildings in southern Xinjiang(Fig. 8).

In the early stage of the development of Mazar architecture in southern Xinjiang, there were many forms of soil embryo domes with superimposed structure, and in the later stage, bricks and stones vouchers were often used(Fig. 9).

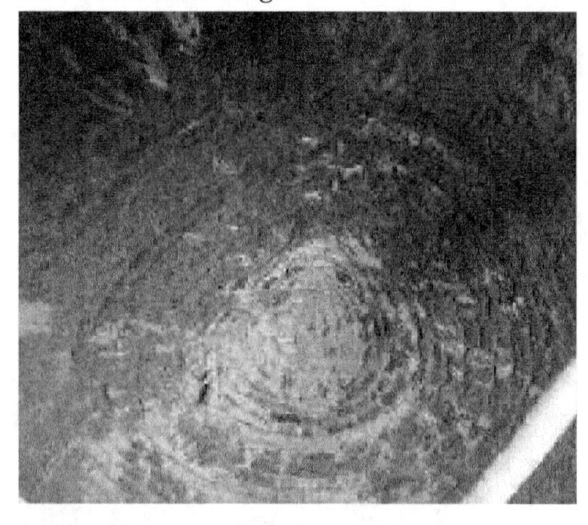

Fig. 8 Nur Aranur Khan Mazar roof stacking technology

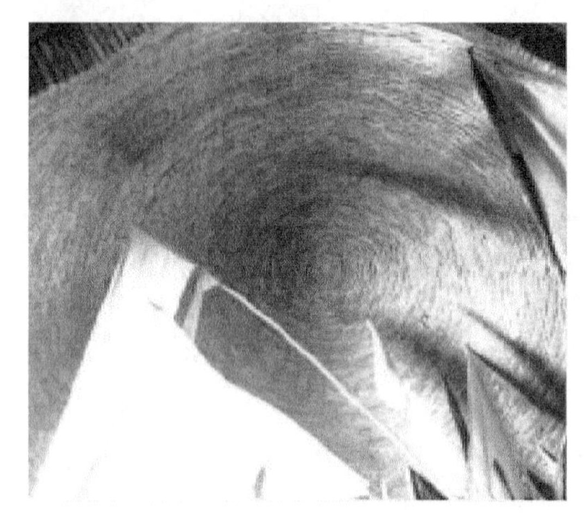

Fig. 9 Muhammad Sheriff Mazar Roof Overlay Technology

6 Conclusion

Due to its special geographical location in the middle and south branches of the ancient Silk Road and its arid continental natural climate, the southern Xinjiang region was born based on Islamic religious genes and primitive worship of raw soil. The special type of Mazar building, which is made of raw materials, provides important physical testimony to the study of the form of tombs in the southern Xinjiang region and the rituals of worship. When constructing a special building such as Mazar in southern Xinjiang, it experienced the development process from the construction of raw soil to the skillful use of earthen blocks. In the specific construction, it also used the techniques of arch vouchers, stacks, domes and so on. It is an important representative of public buildings in southern Xinjiang.

References

CHANG Qing, 1992. Historical view of architectural and cultural relations along the silk road. Journal of Tongji University (humanities and social sciences edition), (01):8-14+42.

Cultural Relics Bureau of Xinjiang Uygur Autonomous Region, 2011. Integration of the third national cultural relics survey in Xinjiang Uygur Autonomous Region. Beijing:Science Press.

GAO Xiang,2002. Exploration on the protection and renovation of raw earth architectural system blocks in Xinjiang: a case study of Kashgar. Chongqing:Chongqing University.

Institute of Natural Science History, Chinese Academy of Sciences, 2000. History of architectural technology in ancient China. Beijing: Science Press.

JOHN D H,YANG Changming,et al. , 1999. Islamic architecture. Beijing: China Construction Industry Publishing House.

LI Pengfei, 2004. Study on the value assessment system of Mazar Islamic architecture in Xinjiang. Urumchi: Xinjiang University.

LI Qun. Analysis on the coupling of settlement space in Kashgar ancient city. Journal of Nanjing academy of art (art in design edition),8:104-108.

LIU Zhiping, 2011. Islamic architecture in China. Beijing: China Construction Industry Publishing House.

QIU Yulan, YU Zhensheng, 1992. Islamic architecture in China. Beijing: China Construction Industry Publishing House.

SUN Jingyan,2018. Study on the architectural characteristics of Mazha tombs in Kashgar area of Xinjiang. Beijing:Beijing University of Architecture.

WANG Xiaodong, 1994. Orientation of Islamic architecture in Xinjiang. Journal of architecture, (03):49-53.

Analysis of Characteristics and Protection Strategies of "Rammed Earth House" Vernacular Architecture —Taking Hongqi Village in Daqing as an Example

SUN Zhimin, *SONG Tianqi*, *YE Yao*

School of Civil Engineering and Architecture, Northeast Petroleum University, Daqing, China

Abstract "Rammed earth house", as vernacular architecture, is a particular traditional residence in Northeast China. It adopts "one room bright and two rooms dark" plane layout and "soil and wood" combined construction, which is evolved from the traditional wall-building technology. It always uses the local material, such as wood, soil, and grass. And it also gradually forms a heating system combing heated bed, firewall, and chimney. So "rammed earth house" has some significant regional characteristics. "Rammed earth house" in Daqing continues some traditional characteristics of "rammed earth house" in Northeast China, it has developed from "first type of rammed earth house" to "third type of rammed earth house". Because it is a special product of the Daqing Oil Battle period, it has an important cultural value. In recent years, as the residents moved out, most of the rammed earth houses have been dismantled. And now there are only small numbers of rammed earth houses in Hongqi village, which are listed in the provincial key cultural relics protection units in Heilongjiang province. Unfortunately, because of being empty for many years and lacking of protection, most houses have been damaged in different levels. This paper takes rammed earth houses in Hongqi village as the research object and uses the methods of literature review and field research to analyze their plane layout forms and appearances, construction methods and heating methods. And then it interprets multiple values of rammed earth houses, such as the historical value, social value, and economic value. At last, this paper proposes the future sustainable protection methods.

Keywords rammed earth house, vernacular architecture, Hongqi Village, shape characteristics, protection

1 Introduction

"Rammed earth house" vernacular architecture is a unique form of traditional dwelling in Northeast China. It was built by local materials like soil, wood, and straw. It has the characteristics of warm winter and cool summer. Hongqi Village's "rammed earth house" inherits the original "rammed earth house" architectural characteristics of Northeast areas of China. It has undergone "three generations"rammed earth house. And combined with actual needs, it is different from the traditional "rammed earth house" architecture in terms of building layout, building structure according.

2 The Background of the Formation of the Local Architecture of "Rammed Earth House" in Hongqi Village

"Rammed earth house" in Hoingqi Village is a carrier of architectural culture in Northeast China, which is affected by the unique cold climate in the Northeast area of China. It is also the product of special historical factors in the development of Daqing Oilfield in the People's Republic of China. It is the result of the combination of regional natural factors and social factors.

2.1 The Natural Geographical Environment of Hongqi Village

Hongqi Village is located in the south of Longfeng District, Daqing, Heilongjiang Province. It has a continental monsoon climate and is in the middle temperate zone. It has the characteristics of continentality, large temperature difference, obvious monsoon, and less precipitation (Han, 2005). Winter lasts for more than half a year, and there is frozen soil. In summer, southeast wind prevails and the temperature is high and it is rainy. Daqing enjoys the "city with hundreds of lakes". The freezing period of rivers is about 160 days, of which the freezing closed period is 130 days(Zhou,2016).

The long and cold climate environment had a profound impact on the habits and customs of the residents of Hongqi Village: thick walls, slightly arched roofs, heated *kang*, and firewall were used to resist the cold; simultaneously, the indoor use of "one-character *kang*" and "L-shaped *kang*" made the *Kang* become the main living space of the people at that time, forming a kind of "life on the *kang*" of the residents of Hongqi Village. Residential customs have conspicuous regional characteristics.

2.2 Historical Background of Hongqi Village

At the beginning of 1960, on the deserted Nenjiang grassland, more than 60,000 workers from all over the country launched the famous oil battle in Daqing China in the history of the People's Republic of China. Because of the remote geographical location of Daqing Oilfield and its long distance from the town, the problem of accommodation for workers became particularly difficult to solve. Ouyang Qin, the first Secretary of Heilongjiang Province, proposed to Daqing Oil Exhibition Commander to build "rammed earth house". The Oilfield Capital Construction Command investigated and summarized the construction technology of the "rammed earth house" built by the local people. The Daqing Oilfield Design Institute proposed the design standard of the "rammed earth house"(Dong,2014).

At this stage, the building forms are basically "rammed earth house", brick-column adobe houses and brick bungalows. Today, Hongqi Village has the only natural ecological "rammed earth house" vernacular building complex in Daqing, and it is also the origin of the spirit of "rammed earth house", which is one of the "six treasures" of Daqing spirit(Yong,2018).

3 An Analysis of the Characteristic of "Rammed Earth House" Vernacular Architecture

The construction of the "rammed earth house" in Hongqi Village has experienced the development from "first-generation rammed earth house" and "second-generation rammed earth house" to the "third generation rammed earth". Many things from the layout of the building, building materials, and structure, or building heating methods have changed. Below is the analysis.

3.1 Layout and Appearance of Buildings

Traditional "rammed earth house" in the northeast area of China is a single courtyard building, which is the main form, and it is the most basic group unit form of traditional Chinese dwellings. Most buildings adopt the layout of "one room light and two rooms dark" (Fig. 1).

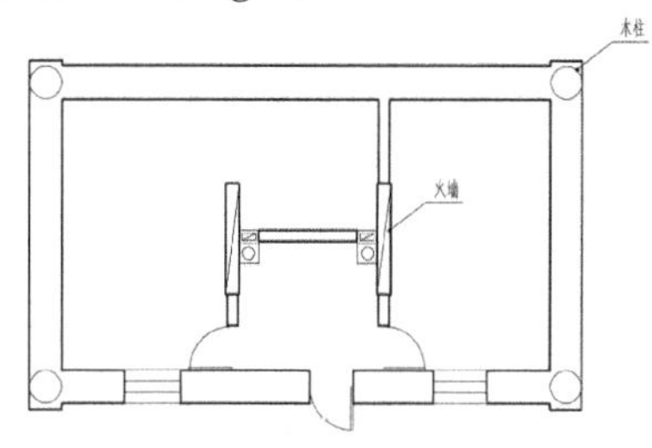

Fig. 1 The plane layout of traditional "rammed earth house" vernacular architecture

The indoor layout is symmetrical in the form of axes, with large bays and small entrance depth, so that more light can enter the house. The entrance is centered in the south, and windows are opened across the north and south. Entering the door, there is a kitchen with cooking bench, stove, water pot, and other daily necessities. The bedrooms on both sides are usually equipped with the Chinese character "一" shaped heated *Kang* to ensure the indoor temperature is comfortable. There is no bathroom in the room, and dry latrines are usually installed outdoors(Nanxi, 2016).

The plane layout of "first-generation rammed earth house" and "second-generation rammed earth house" in Hongqi Village is different from the traditional "rammed earth house" vernacular architecture (Fig. 2, Fig. 3). Most of them take four rooms and two houses as one unit. The two houses are symmetrically arranged and the doors are opened on both sides. The windows are at the north and south of the house, which are opposite to each other. The windows are also symmetrical on both sides of the gable. On the basis of the second generation of "rammed earth house", seven direction room was added in the third generation of "rammed earth house", with three households as one unit. The third room in the middle is one household, one on each side, the middle one opens to the south, and the east and west two doors open to the gable respectively. The south elevation has large windows for heating and the north elevation has small windows for ventilation (Fig. 4).

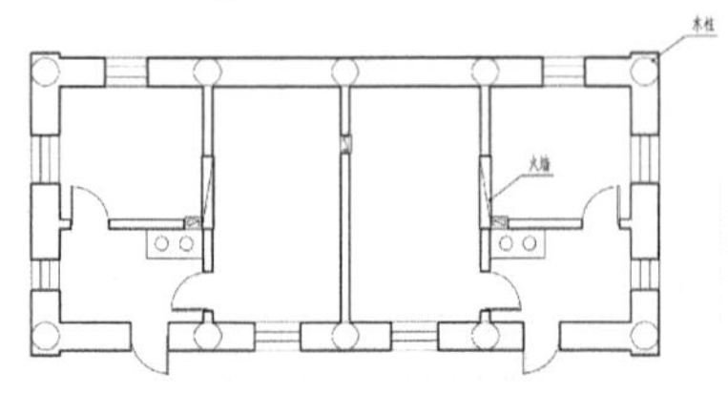

Fig. 2 The plane layout of "first-generation"

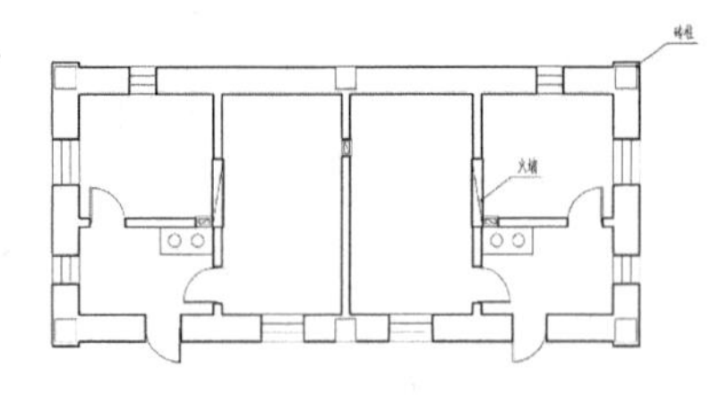

Fig. 3 The plane layout of "second-generation rammed earth house" vernacular architecture

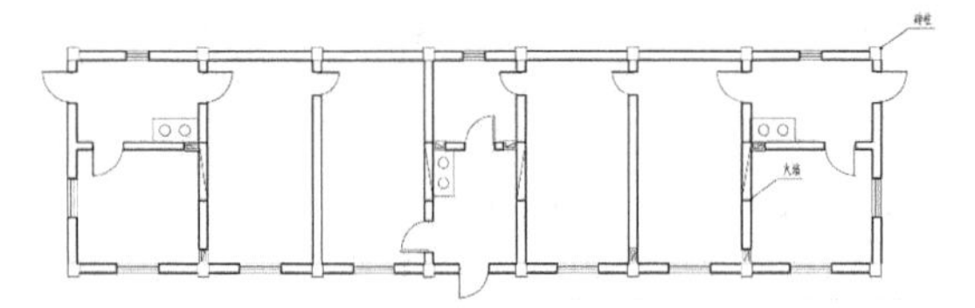

Fig. 4 The plane layout of "third-generation rammed earth" vernacular architecture

The roofs of the "first-generation rammed earth house" to the "third-generation rammed earth house" in Hongqi Village are all in the form of "hoarding roofs". The winter in Northeast China is long and there is always rainy or snowy weather. Therefore, the slopes of the roofs are designed to be gentle. On the one hand, the snow loads on the roofs can be reduced, on the other hand, it is convenient for the roofs to preserve snow in winter, which can improve the thermal insulation of the room, and also beneficial for resisting wind and sand(Zhou, 2016).

3.2 Materials Used in the Building and Method for Construction

"Rammed earth house" in Hongqi Village follows the traditional "rammed earth house" building civil structure practices in the northeast area of China, building materials are mainly soil, wood, grass. The material of the wall are mostly grass mud, the roofs are mostly grass roofs or mud roofs, and the materials of doors and windows are wood (Nanxi, 2016). Compared with the traditional "rammed earth house", the construction method of that in Hongqi village is relatively optimized.

The "First-generation rammed earth house" in Hongqi Village has the following improvement compared with traditional "rammed earth house". First, a 1:6 lime-soil foundation with a depth of 30 centimeters was made. The width of the lower wall is 60 centimeters, and the width of the upper side of the wall is 45 centimeters. Only the outer wall surface is processed, which increases the usable area of the room. Under the wall, a 1 : 5 residual soil moisture-proof layer is made (Fig. 5).

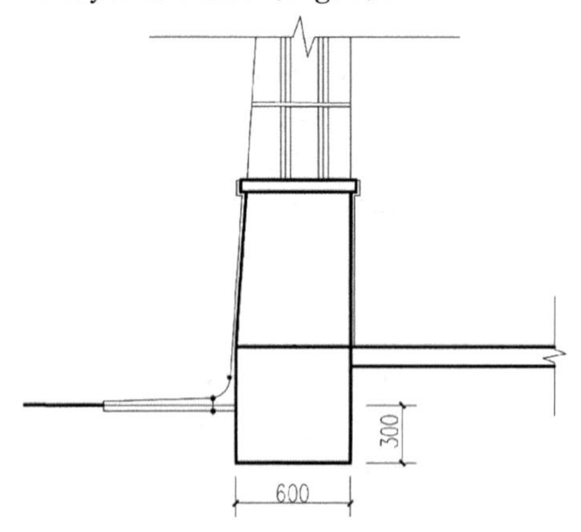

Fig. 5 Construction method of "first-generation rammed earth" vernacular architecture

The wall is usually built by artificial layered ramming, first building the gable, then building the front and back walls. Every layer of the rammed wall needs to be covered with a layer of leymus chinensis, which plays a reinforcing role and can facilitate ventilation and drying of the wall. The roof camber reaches 1/14-1/12 (the ratio of arch height to span), which is beneficial for drainage. Residual oil soil is used as waterproof layer material in the roof, which has strong impermeability and resistance to erosion. Wall painting can usually be started when the wall is dried to about 80% by air, and its material is usually grass mud. Simultaneously, the soil of grass mud is the same as the soil of the wall so that there can be better effect for the combination between the paint layer and the wall (Construction and Design of Daqing Oilfield Institute, 1966).

"The second-generation rammed earth house" is replaced by the wooden column structure system of "The first generation rammed earth house" with the brick column structure system. When "The third generation of rammed earth house" was developed, the building was still a brick-column structure system, and red brick was used as the main material for walls and roofs. The appearance of the building has not changed much.

3.3 Heating Characteristics of the Building

In order to adapt to the climate conditions in the northeast region of China, the "rammed earth house" in Hongqi Village adopts the heating mode of combining heated *Kang*, fire wall and fire stove. The air under the fire heated *Kang* is connected with the smoke outlet of the stove. The smoke generated when cooking is usually heated by the flue under the heated *Kang* to heat the heated *Kang* itself, so as to achieve the effect of increasing indoor temperature. The chimney is built on the outside of the building wall. This unique shape can extend the distance of smoke passing through the room so that the heat generated by "smoke" when cooking can be more uniformly left in the room while achieving heat preservation and saving energy, it also effectively reduces the incidence of fire (Nanxi, 2016).

A firewall is a hollow wall in the partition wall between the kitchen and dormitory. The firewall is connected with the stove, and the smoke from cooking can also go through the fire wall, making the indoor temperature more comfortable. Fire stoves are usually made of grass, mud, bricks or stones, which generate heat continuously to the outside and effectively improve the indoor temperature. When the weather is severely cold, we can also resist the cold using fire basins, stoves, and other heating methods (Nanxi, 2016).

4 Analysis of the Value of "Rammed Earth House" Architecture in Hongqi Village

As the diversity of the civilization of the world is gradually recognized, the evaluation of the value of "architectural heritage" has also expanded from the early "historical, artistic, scientific, social and cultural values". David Sorosby, the father of cultural economics, put forward the model of cultural capital accumulation in 1999, which inspired the idea of expanding heritage value evaluation from qualitative cultural value to quantitative economic value (Chen and Xu, 2018). Economic value evaluation should be treated differently from cultural value evaluation, which is also the opinion of the Bara Charter. Therefore, this paper divides the architectural value of "rammed earth house" into two categories: cultural value and economic value.

4.1 Analysis of Cultural Value

From the perspective of qualitative cultural science, this paper defines cultural value (where culture is broad) as historical value, artistic value, scientific value and social value (which is close to Mason and Zhang Yanhua's definition of cultural value evaluation). Faced with different research objects, the specific composition of cultural value and the proportion of its weight are different. The cultural value of the "dry battle base" complex in Hongqi Village is mainly manifested in two aspects: highlighting the historical value and social value

(Yong, 2018). Therefore, historical value and social value are analyzed in the following paper.

The "rammed earth house" of Hongqi Village is a special product of the Daqing Oilfield Battle. Its formation and development solved the accommodation of the vast number of workers in the oilfield at that time, and provided the necessary conditions for "people entering houses, machines entering houses, garages entering garages and vegetables entering cellars" in Daqing Petroleum Battle. It is of special political significance to ensure the smooth completion of the Daqing Oilfield Battle as scheduled. At that time, the circular water tower, circular shops and the auditorium of "Agricultural Science Big Village" built in the village of Hongqi also recorded the history of the early days of the Peoples' Republic of China. Therefore, the "rammed earth house" in Hongqi Village has special historical value.

The social cohesion generated by cultural and emotional identification and spiritual motivation has commemorative and educational significance (Li, Xu, Aoki Shinfu, 2019). Hongqi village is the birthplace of the spirit of "fighting hard" among the "six treasure spirits" of Daqing's "rammed earth house". It embodies the "struggling hard" spirit and the spirit of and workers in the early days of the Daqing Oilfield Battle (Yong, 2018). People have already closely linked the development of "rammed earth house" with oil. Therefore, in the beginning, it did not simply exist as an architecture but represented the "hard working" process of oil development. Its village culture is the inheritance and promotion of the spirit of "rammed earth house", which is the culture and charm of Hongqi Village and its social value.

4.2 Economic Value

From the quantitative economic point of view, the economic value of heritage lies in the time after the industrial heritage is marketized, it can reflect the market price and bring economic benefits (Chen and Xu, 2018). Two "rammed earth house" buildings cover a large area, involving dozens of single buildings. If people carry out reasonable protective usage, they can not only revitalize the village but also produce certain economic value.

In field survey, people found that the buildings in Hongqi second village are well preserved and can be reused as long as they are simply repaired and paved. However, the overall preservation of Hongqi Village is relatively poor, and the houses are damaged in varying degrees, some even collapsed. However, various types of public buildings such as auditorium, circular shops, water towers retain in the village. Therefore, according to the future cultural and tourism development plan of Daqing, new urban functions are injected into the "rammed earth house" building complex of Hongqi village, which leads to a historical dialogue between the old and new languages, and realized the "static heritage", which also reflects its economic value.

5 Protection Strategy of "Rammed Earth House" Architecture

Vernacular architecture can hardly be represented by single buildings. It is better for each region to protect vernacular by maintaining and preserving typical buildings and villages(Chen, Zhao, 2000). This paper discussed the protection of Hongqi Village on the basis of the principle of integrity protection and sustainability protection. This article guides the protection of Hongqi Village's vernacular architecture by identifying its value, so that its value can be further realized.

5.1 Problems of "Rammed Earth House" Vernacular Architecture in Hongqi Village

This paper draws the following conclusions through field surveying and mapping as well as visiting local residents:

(1) Most of the original residents of first Hongqi village and second Hongqi villages have moved out to urban areas, and the total number of residents in the village is less than 10, which makes them truly

empty villages.

(2) Because of the large number of villagers moving out in the past five years, the houses have not been repaired as scheduled, which leads to the great damage to houses, there are damages in walls, doors, windows, and roofs to varying degrees. Some houses even become ruins (Fig. 6).

Fig. 6 Current status of the house

(3) Because of the development of the oilfield at that time, the "rammed earth house" complex in Hongqi Village is far from the urban areas nowadays. The nearby road is adjacent to Daguang Expressway, but as an oil field operation road, the condition of the road is poor, there are little cars driving on the road, and there are little signs, which is difficult to attract people's attention (Yong, 2018).

(4) Hongqi Village was in the list of key cultural relics being protected in Heilongjiang Province in 2014. Only the stone tablet standing in front of the village introduces about it, while there is a lack of sustained attention of the government and related institutions and organizations.

5.2 Preliminary Survey on the Protection of "Rammed Earth House" Complex in Hongqi Village

As a kind of heritage, vernacular architecture is a special architectural form produced by different nationalities, regions, and cultures in different periods, and it is a manifestation of cultural diversity. The Charter on Rural Architectural Heritage, adopted in Mexico in 1999, points out the direction for the global protection of vernacular heritage and establishes the principles for the protection of vernacular architectural heritage (Zhang, 2017).

The protection of vernacular architecture is a full-cycle and sustainable process, which includes not only the integration of different types of literature, such as words, images and videos, the records of local residents' living habits, but also the restoration of damaged single architecture (Fig. 7, 8), site space, and the display of historical scenes in different periods. All these contribute to a better understanding of the Hongqi village. The value and significance of the "rammed earth house" vernacular architecture in villages is more helpful for people to put forward a scientific plan for protecting "rammed earth house".

Fig. 7 Restoration of the house

Fig. 8 Restoration of the house

5.2.1 Establishment of Open Basic Data Information

The first step to protect vernacular architecture is to establish a basic data information base. The basic data information base is to sort out the basic information (image, text, dictation) of the vernacular architecture, and to establish information tables according to the basic characteristics of the vernacular architecture, record the basic information and archive it. Finally, the basic information is displayed visually by using the digital method. Its purpose is to compare and analyze these data information so as to provide basic prerequisites for later protection.

5.2.2 Restoration of Historical Scenes in A Specific Period

At present, because the overall layout of the first Hongqi village and the second Hongqi village was blurred, some single buildings can not be fully presented, so that people can not fully recognize the historical value and social value of Hongqi village. Therefore, through the analysis of the data information base and the aid of graphics software, this paper simulates and repairs the existing houses in Hongqi Village. which is seriously damaged, and restoration of the road traffic and surrounding environment of Hongqi Village to the village style in the early 1960s is carried out (Fig. 9, Fig. 10). The restoration of historical scenes is helpful to control the trend of the overall protection of Hongqi Village from a macro perspective, so as to put forward more suitable strategies for protection in the future.

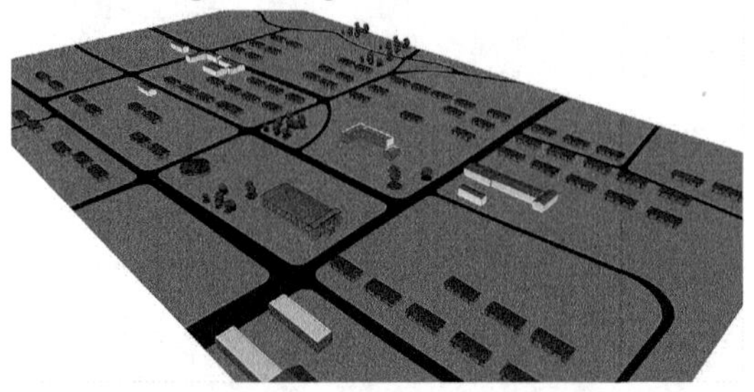

Fig. 9 Restoration of Hongqi Village 1

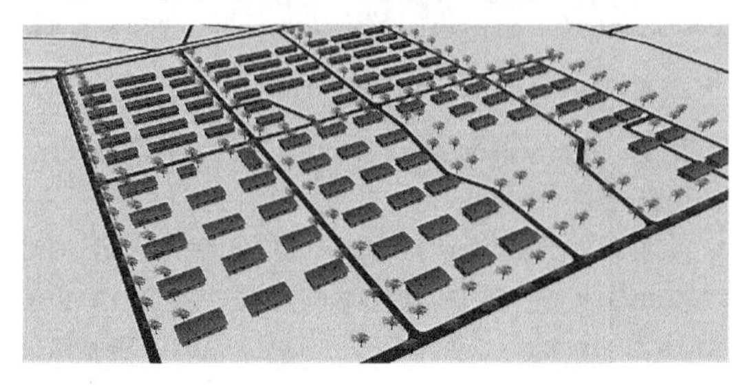

Fig. 10 Restoration of Hongqi Village 2

5.2.3 Protection Mode Combining Macro and Micro

"Rammed earth house" complex in Hongqi Village is an important part of Daqing oil's industrial heritage and a key project of Daqing's urban culture and tourism planning and construction in the future. Therefore, its protection involves not only the protection of buildings and the surrounding environment but also the protection strategy of Daqing petroleum industry heritage and urban future development planning strategy. It needs to adopt a combination of macro and micro protection methods.

First of all, from the mid-macro perspective, "rammed earth house" complex in Hongqi Village can be regarded as a landscape anchor of Daqing oil industry heritage system, which is related to other types of heritage of oil industry and forms Daqing oil industry heritage tourism route in a certain way, and it becomes an important cultural tourism project of Daqing in the future. This not only makes the spirit of "rammed earth house" and the culture of Hongqi Village better inherited and promoted but also sustainably return the economic value generated through the model of cultural tourism to local residents and society.

Secondly, from the micro perspective, the "rammed earth house" complex in Hongqi Village was a small worker village at that time. Therefore, the traditional grid layout pattern should be followed in the protection planning, preserving the spatial structure of the original village, while taking into account the natural and cultural environment of the region. Because individual buildings are damaged in different degrees, it is necessary to record and classify buildings according to the current situation, and then reinforce, repair, or demolish the buildings according to different circumstances. And through the investigation of local residents, architecture-workers and relevant government departments, this paper discussed the most reasonable protection plan. The best protection for the heritage itself is to make the "rammed earth house" complex of Hongqi Village a sustainable "living heritage".

Acknowledgements

Heilongjiang Youth Science Foundation (QC2017041), Research planning project of Heilongjiang Philosophy and Social Science (17YSB075), Cultivation Foundation of National Foundation of Northeast Petroleum University.

References

Construction and Design of Daqing Oilfield Institute, 1966. Design and construction of "rammed-earth house". Journal of architecture, (Z1): 30-32.

CHEN Jiamin, XU Subin, 2018. Preliminary study on the evaluation of economic value of cultural capital in industrial heritage after the transformation. Architecture and culture, (10): 41-43.

CHEN Zhihua, ZHAO Wei, 2000. Discussion arising from the charter on the heritage of rural architecture. Times architecture, (03): 20-24.

DONG Beizhi, 2014. Simple "rammed-earth house" because of bad conditions. Petro China, (01): 76-78.

HAN Ying, 2005. Climate change and meteorological services in Daqing area. Daqing social sciences, (03): 52.

LI Songsong, XU Subin, AOKI Shinfu, 2019. Interpretation of the value of industrial heritage under the background of cultural relics. Chinese cultural heritage, (01): 54-61.

NANXI, 2016. Brief discussion on the architectural characteristics of local dwellings in Northeast China and its ecological implications. Industrial design, (10): 147-148.

YONG Liang, 2018. Exploration on the protection of "rammed-earth house" complex in Hongqi village of Daqing from the perspective of heritage activation. Heritage and protection research, 3 (08): 34-37.

ZHOU Mengqi, 2016. On residential buildings in the northeast of China under the influence of climate. Changchun: Northeast Normal University.

ZHANG Song, 2017. Traditional villages as human settlements and their overall protection. Journal of urban planning, (02): 44-49.

The Water Ecological Wisdom of Ancient Cities in Semi-arid Area: the Case study of Three Ancient Cities in Jinzhong Area

WANG Haoyue[1], *WANG Sisi*[1], *LI Ang*[1], *SUN Zhe*[2]

1 Key Laboratory of Urban Stormwater System and Water Environment (Ministry of Education), Beijing University of Civil Engineering and Architecture, Beijing Advanced Innovation Center for Future Urban Design, Beijing, China

2 College of Architecture and Urban Planning, Beijing University of Civil Engineering and Architecture, Beijing, China

Abstract In the history of human development, the ancestors have formed rich experiences in dealing with water-related natural disasters such as drought and flood. These experiences are embodied in the exceptionally well-preserved ancient cities. Through exploring the wisdom of water management in ancient cities, it can provide useful experience for heritage protection and utilization, and modern urban rainwater and flood management. Cites of Pingyao, Qixian and Taigu, which located in semi-arid and semi-humid climate zones of China, are the representatives of traditional water management wisdom of ancient China. Taking these three cities as research objects, this paper studies the water management wisdom of ancient cities from the following aspects: urban location, vertical layout, storage and drainage system design, courtyard and architecture design. The vertical design of cities, the structure of rainwater collection and drainage system, and the features of traditional architecture and gardens are analyzed in detail by filed investigation and literature study. The research results not only explore the features and values of vernacular landscapes and architecture in coping with water and natural disasters but also give suggestions for constructing a more resilient city in the future.

Keywords ancient city, water management, ecological wisdom, drainage system, China

1 Introduction

Ecological wisdom refers to the survival and living wisdom that is understood and accumulated in the long process of co-evolution of human beings and nature. Water features are closely related to human production and life, and have the dual attributes of great potential and destructive, therefore water ecological wisdom is the focus of current research. Water ecological wisdom is the ability and wisdom of human beings to live in harmony with water and nature in the process of water use. College of Architecture and Urban Planning in Tongji University took the lead in setting up the "Ecological Wisdom and Practice Research Center", it is committed to researching the ecological wisdom of the ancients to solve the increasingly urban stormwater security problem; Che Wu and Li Zhenzi analyzed the design ideas and construction points of the ancient drainage system from the perspective of ancient city road and its drainage system; Wu Qingzhou studied the flood control and drainage of ancient towns from the perspective of urban infrastructure and layout; Zhao Hongyu and others conducted in-depth research on several ancient villages and explored the wisdom of water ecological practice in traditional villages in northern China.

The ecological wisdom and practical experience of water management in ancient cities is worth exploring and inheriting. Taking the ancient city of Jinzhong area in Pingyao, Qixian, and Taigu as examples, this paper analyzes the water ecological wisdom of ancient Chinese cities from the aspects of site selection, vertical planning, storage and drainage system, architecture and landscape design. It is aimed at exploring the "rainwater adaptability" of cities in the semi-arid area, letting the ecological experience and wisdom of ancient rainwater management continue to play a role in modern cities, effectively protecting the cultural heritage, and solving the frequent rain and flood problems in cities today.

2 Overview of Jinzhong Region

The Jinzhong area is located in the central part of Shanxi Province, at the eastern of the Loess Plateau of China (Fig. 1). The main soil type is cinnamon soil, the topography consists of three parts: mountain, hill and plain, and the surface water system belongs to the Weihe River and its tributaries. The whole region is a warm temperate continental monsoon climate with a dry climate and uneven annual rainfall distribution, which causes the region to face the double threat of drought and flood disasters in summer and autumn. As one of the cradles of Chinese civilization, Jinzhong still retains large-scale ancient cities of Ming and Qing Dynasties. The paper selects three representative ancient cities in the Jinzhong area: Pingyao, Qixian and Taigu (Fig. 2). The ancient city of Pingyao was built in the Western Zhou Dynasty that is the most well-preserved ancient county in China. It was listed by UNESCO in the World Cultural Heritage List in 1997 and retains the architectural style of the Ming and Qing Dynasties. Qixian was once known as the "Zhaoyu Ancient City", experienced historical periods from the Northern Wei Dynasty to the Eastern Wei to the Ming and Qing Dynasties, and the famous Courtyard of Family Qiao, which symbolizes the culture of Jin's merchants Courtyard, is located in Qixian. Taigu was built in the Northern Zhou Dynasty. During the heyday of the Shanxi merchants, Taigu was rich among the cities in Shanxi province. Its private gardens were large in scale and well-built. Now the ancient city walls were destroyed, and the historical districts still retain the appearance of the Ming and Qing Dynasties.

Fig. 1 Location of the Jinzhong area

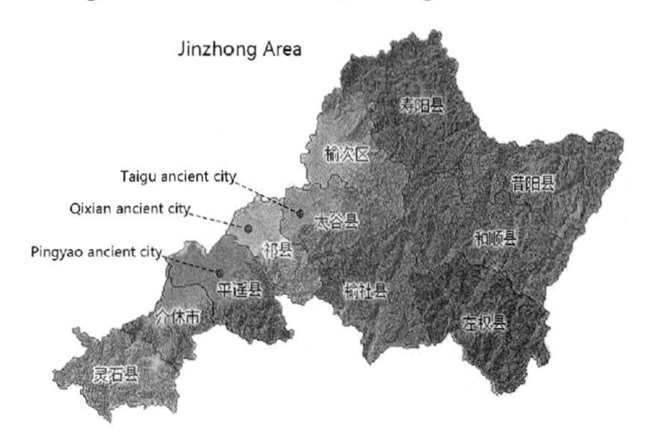

Fig. 2 Location of the three ancient cities

The three ancient cities are the core areas of Jin's merchants in the late Qing Dynasty and the early Republic of China, the representatives of the Shanxi merchant culture. They not only preserve the architecture and city pattern in the Ming and Qing Dynasties but also show the water ecological wisdom of local dwellings adapting to the arid climate and using rainwater resources, which can provide enlightenment for the layout of urban planning and water management in arid areas.

3 Water Management Wisdom of Ancient Cities in Jinzhong Region

3.1 Urban Site Selection

The Jinzhong area is surrounded by Taihang Mountain and Luliang Mountain, and the middle part is crossed by the Fenhe River to form the Fenhe Valley. The overall terrain is high in the east and west and low in the middle. The locations of three ancient cities adapt to the terrain. They are located in the plain of the Fenhe Valley on the west side of Taihang Mountain, close to the Fenhe River. The terrain is low and gradually decreases from the southeast to the northwest. In general, the ancient city is located in the sloping plain area of the piedmont, and the elevation of the site is about 10m above the Fenhe River in the west. The land with a higher elevation is selected as the city site to avoid the flood disaster caused by the rising water level in the summer and autumn.

3.2 Vertical Layout

The three ancient cities are located in the sloping plain in front of the mountain. The overall terrain slopes from southeast to northwest with a slope of 3.2%. The height difference makes the city form a fixed runoff direction, which is conducive to the drainage of rainwater to the low-lying river valley in the northwest to prevent water accumulation in the city. From the micro-topography, the ancient city of Pingyao and Qixian are both high in the center and low in the periphery, showing a tortoiseshell shape, and the height difference in Qixian can reach 3m. The micro-topography of Taigu Ancient City in addition to the slightly higher of the central drum tower, the whole is gradually decreasing from the southeast to the northwest. The runoff direction flows from the southeast to the northwest into the northwest corner of the ancient city.

3.3 Storage and Drainage System Design

The Jinzhong area is in a state of drought for most of the year, the average precipitation in the past 50 years is 479.6mm, and the annual average evaporation is 1,718.4mm, the average annual evaporation is 3.5 times of the average annual precipitation. Therefore, the storage and utilization of rainwater resources is crucial. The three ancient cities are all drained by surface drainage, and the roads in the city parallel to the direction of the city wall are used as the main drainage channels.

The ancient city of Pingyao retains the Ming City Wall, which is compacting of soil, and the bricks are outsourced. The wall is paved with water-resistant bricks on a slight slope, and its external is higher than internal. The inside of the city wall is provided with drainage ditch at an appropriate distance to collect rainwater into the city. The drainage of rainwater in the city is mainly adapted to the topography. Rainwater is drained from the center to the surrounding area and is discharged into the moat through the gates on four sides. The plane layout of the ancient city also considers drainage. For example, the north gate is located at the lowest point on the north by the west side, the Nanda street and the Da street are staggered, which are conducive to the smooth discharge of rainwater into the moat(Fig. 3).

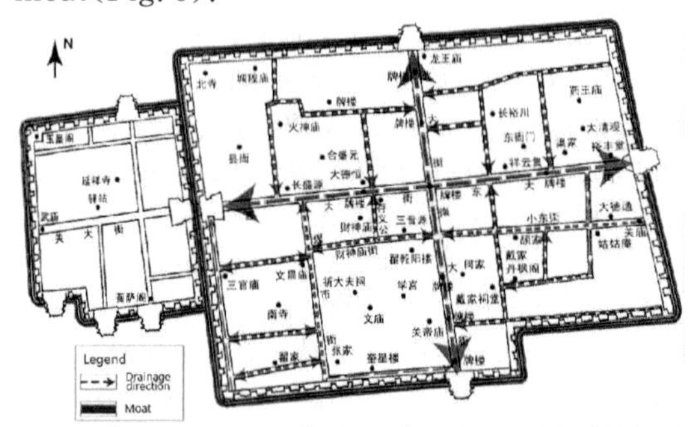

Fig. 3 The layout of drainage pattern in Qixian (Source: Author redrawn from *Shanxi Residential*)

The west side of the main city of Qixian is Xiguan, the outer periphery of the Xiguan and the ancient city excavate the city moat, the rainwater in the city is discharged into the moat from the center. It can not only resist the flood disaster caused by the flooding of the river, but also collect rainwater for the dry season and used for irrigation, life, and firefighting, to achieve multi-functional storage of rainwater resources.

Xiyuan Park in the northwest corner of Taigu Ancient City is the most important open space in the city, and Xiyuan Reservoir is the largest rainwater pond in this city. The overall vertical design of decreasing from the southeast to the northwest of Taigu Ancient City makes the surface runoff in the city gather in the Xiyuan reservoir. At the same time, a drainage channel is opened in the northwest corner of the ancient city to drain into the near valley. In summer and autumn, when the rain is relatively strong, the ponds accumulate rainwater to cope with the perennial drought while creating the waterscape of the ancient city (Fig. 4).

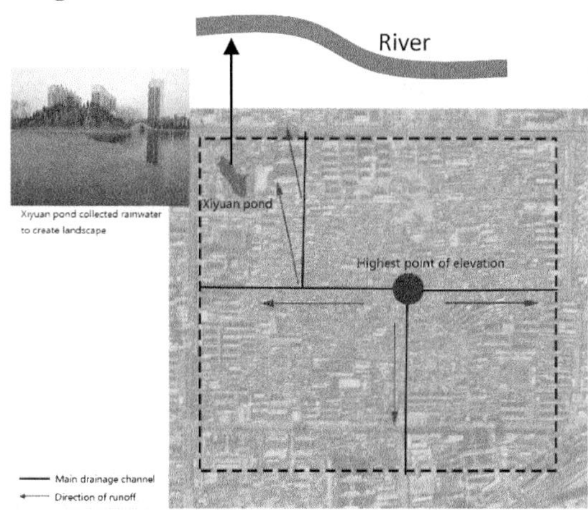

Fig. 4 Street layout and the direction of runoff

3.4 Courtyard and Architecture Design

The ancient city of Jinzhong area has displayed the courtyard culture of Shanxi merchants. The residences of three ancient cities are all quadrangle courtyards with brick walls and tile roofs. There are clear central axes, the walls are all brick wall without plastering, and the outer wall stone bricks are reinforced and waterproof. The process of collecting rainwater, infiltration, storage, and discharge is completed in the courtyard. Different from Beijing Siheyuan, the cabins on both sides are mostly single-slopes(Fig. 5), the roof is inclined to the courtyard, and the slope is more obvious, which helps to drain the rain into the courtyard, create a scene of "Si Shui Gui Tang". The surface drain of the courtyard adopts "Rao Men Shui". The specific method is to extend the rainwater convergence process by diagonally arranging the drainage outlets of adjacent yards and the surface runoff, which is beneficial to increase the air humidity and form a livable microclimate. In the center of the courtyard, the Taiping cylinder was used to store water and fireproof in ancient times, so that rainwater resources could be reused(Fig. 6).

Fig. 5 The courtyard uses a single slope to collect rainwater

Fig. 6 Taiping cylinder of the courtyard

4 Conclusion

Taking the three ancient cities in Jinzhong area as an example, from the aspects of urban site selection, vertical layout, storage and drainage system, courtyard and architecture layout, analyze the ancient people's diverse water ecological wisdom in dealing with rainwater and climate adaptation, formed an ecological rainwater management system adapted to drought-oriented and supplemented by drainage and flood control. Summarizing and analyzing the ecological wisdom of the ancient city is conducive to a better understanding of the experience and wisdom of the ancients in dealing with the relations of human and water. The water ecological wisdom is mostly reflected in the treatment of rainwater resources. In view of the shortage of water resources and the frequent floods in modern cities, it is necessary to integrate water ecological wisdom into the urban planning of city, and learn from the ecological practices of the ancients and nature to inherit and transform traditional water wisdom such as ancient city drainage, river network design, ditches and ponds, courtyard storage and drainage, and better promote the development of urban stormwater management practices.

In the protection of ancient cities and its heritage, in addition to protect the cultural heritage of the ancient city walls, buildings, street layouts, it is also necessary to protect the water cultural heritage such as water control, water storage and drainage wisdom. Although traditional ecological wisdom is not all applicable to the modern living environment, we should actively explore the ecological wisdom that can still be borrowed, utilized and evolved in modern civilization.

Acknowledgements

Funding: Supported by "The Fundamental Research Funds for Beijing Universities" (X18180).

References

CHE Wu, QIAO Mengxi, WANG Sisi,2012. Modern enlightenment of ancient Chinese rainwater management practice. Nanning: China civil engineering society national drainage committee 2012 annual meeting: 21-25.

LI Ang, WANG Sisi, YUAN Donghai,et al. ,2018. Performance characteristics and influence factors of water adaptive landscape in urban and rural under the threat of droughts and floods- the case study in Jingzhong area. Journal of arid land resources and environment, 32(4):183-188.

LI Zhenzi, CHE Wu, ZHAO Yang,2015. Analysis of major drainage system of urban road in ancient China and its modern enlightenment. China water&wastewater,2015(10): 1-7.

Qixian County Compilation Committee,1999. Qixian zhi. Beijing:Zhonghua Book Company: 52-168.

WANG Jinping, XU Qiang, HAN Weicheng,2009. Shanxi residential. Beijing: China Architecture & Building Press: 80-82.

WANG Shaozeng, XIANG Weining, LIU Zhixin, 2016. Exploring the answer to urban rainwater security and utilization from the perspective of ecological wisdom. Acta ecologica sinica,36(16) : 4921-4925.

WANG Shaozeng, XIANG Weining, PENG Zhenwei, 2016. Urban rainwater management practice guided by ecological wisdom. Acta ecologica sinica,36(16): 4919-4920.

WU Qingzhou, 2002. The historical experience and lessons of the flood control in ancient China. City planning review, 26(4): 84-92.

YAN Wentao, WANG Yuncai, XIANG Weining, 2016 . Urban rainwater management practice requires the guidance of ecological practice wisdom. Acta ecologica sinica,36(16): 4926-4928.

ZHAO Hongyu, XIE Wenlong, LU Ruifang, et al. , 2018. Ancient ecological practice wisdom in traditional villages in north China and its contemporary enlightenment. Modern urban research, (07):20-24.

Study on Stone Structural Villages in Jingxing Area

ZHANG Chao

Beijing University of Civil Engineering and Architecture, China

Abstract Traditional villages are an important historical and cultural heritage in China and the research on them is an important part of promoting the protection and development of historical towns in China. Jingxing is the fifth path of the Eight Paths of Taihang and the sixth of the Nine Fortresses of ancient China. Jingxing is located in a very important area. Since ancient times, Jingxing has been the gateway between Shanxi and Hebei, and has always been a battleground for military strategists. Many historical events show that Jingxing bears a heavy history and has a profound cultural accumulation. There are more than ten traditional villages in the middle area of Jingxing County. They were built from the Sui and Tang Dynasties to the Ming and Qing Dynasties. They almost spanned the entire feudal society of China. Most of the buildings in the village were built of stone. Even today villagers still farm and live here. By 2017, there were 7 villages in Jingxing County that were listed as national traditional villages, mainly distributed in the central area. In the central mountainous area, the traffic is remote and the culture is primitive. The villages, especially the traditional ancient villages, are in urgent need of research, protection, and development. Through the autoptical records in these traditional villages, this paper reinterprets the comprehensive value and the significance of traditional villages in today's society. Based on the study of the traditional stone village, this paper summarizes the importance to protect these precious patrimonies in the central of Jingxing area and puts forward some scientific protection measures.
Keywords Jingxing, traditional villages, the ponderance to protect the stone villages

Due to the rapid development of the commodity economy and the strong interference of the construction of infrastructural facilities, the ancient villages, which are the carriers of China's diverse material and intangible culture, are now faced with a situation in which the remaining area is narrowing, and the number of disappearances is increasing. There are more than 12,000 traditional villages in China, which account for only 1.9% of our country's administrative villages, 0.5% of natural villages, and less than 5,000 of them have high conservation value. Some scholars have studied and counted that 1.6% of traditional villages in China are dying out every day on average. Jingxing County, Hebei Province, located deep in the Taihang Mountains, there are still more than a dozen ancient villages with great research and conservation value. Most of them are built of stone. In the past this district was very hard to reach, for this reason, it is still well preserved. The main purpose of writing this article is to introduce these valuable villages to more people. At the same time, I hope that these villages will be more actively protected while receiving the attention of scholars.

1 Overview of Stone Village, Jingxing County, Hebei Province

1.1 The Historical Position of the Jingxing District

Jingxing County is located on the western border of Hebei Province, at the junction of Hebei and Shanxi, and the eastern foot of Taihang Mountain. It is known as "The Fifth Xing[①] of Eight Xings in Taihang

① Xing, is where the mountain breaks, the defile of a mountain. [Tang] Li Jifu: Records of Yuan He Jun Xian Zhi Volume 20, Hebei Dao I, records that: Taihang Mountain began in Hanoi, and there were eight Xings from Hanoi to Youzhou. The first Zhiguan Xingwas located in Jiyuan County, Henan Province. The second Taihang Xing and the third Bai Xing werelocated in Hanoi. Fourth, Fukou Xing. Fifth Jing Xing, Sixth Flying Fox Xing, the other name is Wangdu Guan, Seventh Puyin Xing, these three Xings were in Zhongshan. The eighth Jundu Xing, was in Youzhou.

and the Sixth Sai of Nine Sai in the world", and it is the vital thoroughfare of Ji district. Jingxing was the main battlefield for the well-known battle between Han Xin and other army groups before the Han Dynasty, known as the Hundred Regiments Battle during the Chinese Liberation War.

The important position of Jingxing in Chinese history can be seen from the evolution of China's history and geography in every dynasty. Tab. 1 is the arrangement of the prefectures and counties in the Jingxing area from the Qin Dynasty to the Ming and Qing Dynasties.

Tab. 1 The prefectures evolution of Jingxing area①

Dynasty	Administrative District
Qin	Hengshan Jun/ Changshan Jun, Jingxing County (The northwest region of today's Jingxing County.)
Han, Three Kingdoms	Ji Zhou Cishibu, Hengshan Jun (Country)/ Changshan Jun (Country), Jingxiang County
Wei, Jin, Southern and Northern Dynasty	Ji Zhou, Changshan Jun, Jingxiang County
Sui	Heng Zhou/Changshan Jun, Jingxing County
Tang	Hebei Dao, Jing Zhou, Jingxing County
Five Dynasties, Ten Kingdoms	Zhen Zhou/Heng Zhou, Jingxing County
Song	Hebei West Lu(960－997)/ Hebei Lu(997－1162), Zhen Zhou, Jingxiang County
Jin	Hebei West Lu, Wei Zhou, Jingxiang County
Yuan	Zhongshu Province, Guangping Lum, Wei Zhou, Jingxiang County
Ming	Jingshi, Zhending Fu, Jingxiang County
Qing	Zhili Area, Zhengding Fu, Jingxiang County(Fig. 1)

From Tab. 1, we can see that although the prefectures of Jingxing County in each dynasty had changed and many counties in North China had been abolished and annexed during its history, the name of Jinglong County never disappeared from the sovereign's territory. Jingxing has never even changed its name, which shows that the rulers of the past have never slackened their control over it.

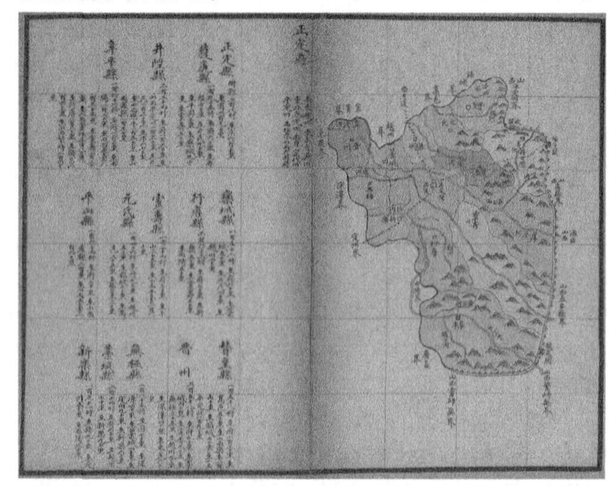

Fig. 1 Map of Zhengding Fu. This map is from an atlas painted by Xu Zhidao during the Qing Xianfeng period, 1859. The map is oriented with north at the bottom and uses the so-called Chinese "latticework" cartographic grid. A side of each grid represents 100 li written on the first general map. There is also a text legend for the symbols used in the first atlas. The administrative divisions, mountains, rivers, and the Great wall are sketched with an explanation for the symbols to explain the procedures used for compilation.

Source: Lin (2014)

The famous poets of the Tang Dynasty, Wang Wei and Du Mu, both wrote poems to describe the history and scenery of Jingxing. Gu Yanwu also wrote a poem *Jing Xing*② describing the dangerous terrain and its important military position in

① Squares indicate Fu(Prefecture); diamonds for Ting(Subprefecture); rectangles for Zhou(Department); double squares for Zhilizhou(Municipality); circles for Xian(Country).

Fu was an administrative division during the Yuan (1271—1368), Ming(1368—1644) and Qing(1644—1911) Dynasties of China. It was also called "Dao or Lu" during Tang (618—907) and Song(960—1279) Dynasties.

Jun was a traditional administrative division in China from the Warring States period. In the beginning it was smaller than a country, as time went on it became larger and larger.

② Jing Xing, a poem, was written by Gu Yanwu(1613—1683) during the end of the Ming Dynasty(1368—1644), and the beginning of the Qing Dynasty(1644—1911).

the history of China.

1.2 Stone Village in Jingxing District

Jingxing is the main thoroughfare to Shanxi province. Because it is located in the depths of the Taihang Mountain Valley. The geographical environment is secluded, so many ancient villages have been completely preserved. Most of these villages are located in the "Qin Huang Gu Yi Dao"①. They are situated according to the position of the mountains. Those villages' layout is deeply influenced by the concept of Chinese Feng Shui②. The number of traditional buildings in this district is numerous and concentrated. The construction of the buildings and environment are mostly based on local materials, such as earth, wood, stone and brick, which have obvious regional characteristics.

Liu Peilin, in his book *Ancient Villages — Harmonious Human Inhabitation Space*, divides ancient Chinese villages into five types: ① Primitive Settlement Type; ② Regional Development Type; ③ Ethnic Migration Type; ④ Evasive Migration Type; ⑤ Diachronic Immigrant Type③. The ancient villages in the western mountainous areas of Jingxing County include three types: Primitive Settlement Type; Evasive Migration type and Diachronic Immigrant Type. Tab. 2 shows the basic situation of several existing historical villages in the Jingxing area. It can be seen from the above that most of the villages that were built before the Tang Dynasty depended on the basis of "Qin Huang Gu Yi Dao", indicating that the construction of villages at that time relied on national infrastructure construction to some extent. However, the villages during Yuan and Ming in the relatively late periods were mostly which had been immigrated from other places, that is, the Diachronic Immigrant Type.

Tab. 2 Jingxing District Villages Survey

	Founding Age	Landform	Architectural form	Intangible Cultural Heritage	Type
Song Gucheng*	Han	Hilly	Brick and Wood Structure	Jingxing La Hua	Primitive Settlement Type
Bai Wang*	Han	Mountainous	Brick and Wood Structure, Seriously damaged	Legend of Qin Shihuang's Xie Ling Station	Primitive Settlement Type
He Taoyuan*	Han	Mountainous	Brick, Stone, Wood Structure	He Tao Yuan Noodles	Primitive Settlement Type
Xiao Longwo*	Sui	Hilly	Brick, Stone, Wood Structure, Grotto	Jin Opera	Primitive Settlement Type
Ban Qiao*	Tang	Hilly	Brick and Wood Structure	Jiuqu Yellow River Lantern Festival Qinhuang Ancient Music	Primitive Settlement Type
Chang Shengkou*	Tang	Mountainous	Brick, Stone, Wood Structure	Chang Shengkou Embroidery	Primitive Settlement Type
Liang Jia	Yuan	Mountainous	Brick and Wood Structure	Liang's Ying Nao Boxing	Primitive Settlement Type
Dong Nanzheng	Yuan	Hilly	Brick, Stone, Wood Structure	Jingxing La Hua	Diachronic Immigrant Type

① Qin Huang Gu Yi Dao, literally Qinhuang Ancient Post Road, east to Luquan City, West to Shanxi Province, through a number of villages and towns in the county. It is about a hundred miles long. Since the opening of this road in Qin Dynasty, it has been a battleground for military strategists.

② Feng Shui, also known as geomantic omen, and is a Chinese philosophy that seeks ways to harmonize humans with the surrounding environment.

③ Source: Liu (1998:56—81)

Continued

	Founding Age	Landform	Architectural form	Intangible Cultural Heritage	Type
Da Liangjiang	Yuan	Mountainous	Brick, Stone, Wood Structure	Da Liangjiang Unlocking Skills	Diachronic Immigrant Type
Tao Linping	Ming	Hilly	Brick, Stone, Wood Structure	Tao Linping Traditional Festivities with a Painted Face	Primitive Settlement Type
Lyu Jia	Ming	Mountainous	Brick, Stone, Wood Structure		Diachronic Immigrant Type
Nan Zhang jing	Ming	Hilly	Brick, Stone, Wood Structure	Nan Zhangjing Tiger Fire; Nan Zhangjing Rough Stone Wall Craftsmanship	Diachronic Immigrant Type
Di Du	Ming	Hilly	Brick, Stone, Wood Structure		Diachronic Immigrant Type
Yu Jia	Ming	Hilly	Brick, Stone, Wood Structure	Stone Building Skills	Evasive Migration Type

Note: The villages with * is "Qin Huang Gu Yi Dao" which passes through the village.

1.2.1 Primitive Settlement Type

The representatives of these villages are Song Gucheng Village, Baiwang Village, Xiao Longwo Village, and He Taoyuan Village. This type is the main type of village formation, but most of the villages originally formed in the Jingxing area are located along the "Qin Huang Gu Yi Dao", so they are different from the original human settlements which relied upon agricultural cultivation in the traditional sense. To a certain extent, this reflects the military and political influence of the country on the initial formation of traditional settlements. With the continuous development and expansion of the original settlements and the gradual saturation of population and resources, naturally some small branches of settlements would go out independently and establish new villages and towns, and the number of villages would gradually increase.

1.2.2 Diachronic Immigrant Type

This type of village is represented by Yujia village.

In Yujia Village, after Yu Qian, the national hero of the Ming Dynasty, was framed by some treacherous colleagues, his grandson, in order to avoid becoming a victim of court intrigue, led his family to escape to the vicinity of Baimiao Mountain (in the depths of the mountain in Jingxing County). At that time, Baimiao Mountain was still barren. He went through all kinds of hardships, reclaimed the wasteland and multiplied his descendants. So far, it's numbers 24 generations. The location of Yujiacun was influenced by its unique natural environment. It is surrounded by mountains and has a small hidden basin, which is in line with the thinking of its ancestors to avoid disasters.

1.2.3 Diachronic Immigrant Type

Nan Zhangjing Village, in the Ming Dynasty (1465—1787), the Zhang Family moved from Hongdong County, Shanxi Province to establish the village.

According to the genealogy of Lyu Jia village, Lyu Jia village was built at the end of the Yuan Dynasty, formerly known as Wang Er village. During the Ming Yongle period, Lyu Jia village, formerly located in Nanbaocheng County and Honghe Cao, moved here. After the prosperity of Lyu's people, the village changed its name to "Luy Jia village".

Dong Zhengnan Village, according to Yang's family tree: in the Yuan Dynasty, the Yang family moved from Hengjian County to live here.

The most typical of these types of villages is Daliangjiang Village. Daliangjiang Village was built in the late Yuan Dynasty and the early Ming Dynasty. It was originally called "Gantao Village". Several families the living in this village, they are Kang, Chen, and Wang. In the Ming Dynasty, it was under the jurisdiction of Pingding Prefecture in Taiyuan, Shanxi Province. During the Ming Dynasty, the Liang family moved to "Gantao Village", then the clan continued to grow stronger and stronger. In the middle of the Qing Dynasty, the Liang family has a Wu Juren① Liang Shen. Since then, the Liang Family has become a prominent clan in the village. The name of the village has also been changed from "Gantao Village" to "Daliangjia". The word "Jiang" comes from people's cherishment of water, on behalf of the villagers' expectations for favorable weather. "Daliangjia" then changed to "Daliangjiang". Daliangjiang Village is a village based on clan relations. The surname of Liang is gradually growing in the village. Since then, the Liang surname has accounted for 90% of the population of the village.

Because the topography of the Jingxing area is dominated by mountains and the environment there is relatively secluded, it avoids the likelihood of new residents and the original residents had to struggle to get the best pieces of land and resources for living. This is also the main reason why a large number of "Diachronic Immigrant Type" villages appeared in the Jingxing area at the end of the Yuan Dynasty and the beginning of the Ming Dynasty.

2 Exploration on the Method of Protecting Ancient Villages in the Jingxing Area

China's historical and cultural protection work has gone through the process of protecting the building's ontology to the protection of the building group as a whole, and then on to the overall protection of material, spiritual, cultural, ecological and other aspects. The protection work has gradually expanded from the study of "history" to the combination of "historical theory", now paying more attention to the study of "theory". The ancient village itself is a complex system, and the value of its protection is not only embodied in the material level but also includes the intangible, spiritual, cultural, atmospheric and other nonmaterial levels. Tab. 3 is the main content of the research on the protection of ancient villages as summarized by the author through this investigation. In this paper, we have only put forward some undeveloped views on the protection of the existing historical villages in the Jingxing area on the basis of the previous in-depth study.

Tab. 3 The Main Content of Ancient Village Protection

Village	Form: differences in size, shape, and layout of villages, differences in geographical location and institutional evolution The relationship between the village itself and other villages: the development of the village itself; the development of the branch settlements formed by the expansion of these villages Generative Mechanism: Primitive Settlement Type; Regional Development type; Ethnic Migration Type; Evasive Migration Type; Diachronic Immigrant Type
Environment	Natural Environment: Habitat Quantity; Habitat Quality Agricultural Environment: Main Crops; Production Methods; Farming Methods
Population	Age Structure, Sex Ratio, Education Level, etc.
Intangible Cultural	Drama, Handicraft, Legend, Tradition (including Family Instructions, Village Rules, etc.)

2.1 Pay Attention to the Protection of Intangible Cultural Heritage

The protection of ancient villages is not exactly the same as the protection of cultural relics, which constitutes the elements of

① A successfulmilitary candidate in the imperial provincial examination.

ancient villages. The existing buildings, streets and landscapes are inevitably important, but the more attractive and memorable ones are the elements of their tastes — culture and people, including non-material aspects such as folklore, production, emotions, festivals, etc. They are so diverse that each village has its own unique cultural genes.

There are many intangible cultural heritages in Jingxing County, which are rich and diverse. It covers folklore, folk literature, dance, music, traditional drama, athletics, etc. Among them, "Tao Linping Traditional Festivities in a Painted Face" is listed in the national intangible cultural heritage list. It originating in commemoration of a battle to defeat the enemy in Tao Linping Village which took place 800 years ago on Liang mountain, during the Three Kingdoms period. Tao Linping Traditional Festivities in a "Painted Face" is very peculiar in facial makeup. The performers are excellent in martial arts. During the Jiajing years, it was selected by the imperial court as a guardian. The stone buildings in Yujia Village are well known, and "Stone Building Skills" is listed as part of an intangible cultural heritage list in Hebei Province. Every household in this village has stonemasons and the skills are ancestral. Stone houses and stone courtyards, stone pavilions and stone tables and benches are filled with stones. Almost everything is made of stone. However, at present, there are dozens of stone craftsmen in Yujia Village, most of whom are over 45 years old. There are so few young people who are willing to integrate this skill that the inheritance of stone building skills is difficult to maintain. The precious patrimonies have been gradually abandoned and they are facing the crisis of inheritance and disconnection, which urgently needs to be saved.

For the protection of this intangible cultural heritage, we should not only pay attention to the support and encouragement of the inheritors of intangible cultural heritages and craftsmen with traditional skills but also be careful to ensure appropriate protection measures for the preparation, production, and collection of these related raw materials. In order to guarantee that traditional materials and crafts can continue to be inherited. What's more, the protection of intangible cultural heritage refers to the protection of the intangible cultural heritage itself and the material space on which it depends. Therefore, attention should also be paid to the protection of the inheritance sites of intangible cultural heritage, including the main routes, buildings and spaces of the activities. If our protection of the ancient villages is only subjective, one-sided, superficial, and only protects the material level, then even if some ancient villages are fortunate enough to survive, all that would remain are the shells of their bodies.

2.2 Encourage Civil Protection with Villagers as the Main Body

As the main body of the village, the local villagers themselves are part of the traditional villages and inherit the patrimonies of the traditional villages. Many inheritors of intangible cultural heritage are local villagers. Traditional cultural heritage is not only the physical space of traditional villages, such as residential buildings, public buildings and places but also the intangible cultural connotation of social structure, cultural heritage and local folklore.

Some villages in the Jingxing area have a long history. Many of them have the culture and rules inherited from their ancestors, which is not less demanding than the requirements of today's government. For example, the management standard system of Yu Jia Village has three levels. Family Instruction is the first management level, Family Law is the second management level, Clan Rules are the third management level, and the three levels go from bottom to top, and cooperate with each other to form a strict system. In fact, as early as the early days of the construction of the village, there were literacy rules. Parents would have to teach their children to learn common words. In the Qing Dynasty, the Yu family had more than 60 talented people in the literary and military fields, a family of seven generations which included 26 talented people, it created quite a sensation for a while amongst such villages with highly-skilled people. The villagers also have a relatively deep understanding of their own village culture. The government only needs to do a good job of guiding and supporting so that the villagers can self-govern and self-protect, and in that way achieve better results than they likely would through excessive intervention.

2.3 Application of Biodiversity Conservation Theory in Ancient Village Protection

The formation and development of biomes are very similar to the construction and evolution of villages. First of all, its initial formation requires a certain material basis. For organisms, it is sunlight, air, and water. For the initial human settlements, they are arable land, residences, and transportation. The initial villages built along the "Qin Huang Gu Yi Dao" in the pre-Qin period relied on the convenience of transportation. Secondly, the development process of the two is very similar. The development of biology has gone through a process of the sigmoidal curve, from the initial adaptation phase to the rapid development phase, and finally stabilization. The development of the village also follows this pattern. After a period of adaptation, rapid development and stabilization, an energy balance between production and consumption has been achieved. The original settlement will not continue to expand, but there will be small settlement branches that go out independently, find suitable places, and establish new settlements. Finally, in the face of invasion by alien species, the original biomes will be affected to some extent, and the original species may be destroyed. This situation happens to correspond with the formation characteristics of "Diachronic Immigrant" villages. The formation of "Lyu Jia Village" and "Da Liangjiang Village" is a genuine reflection of this situation.

The protection of biodiversity has been guided by the theories of "Island Biogeography" and "Metapopulation" as early as the last century. Many countries in the Western world have already carried out " integrative preservation " for their ancient villages. However, China's "overall protection" of ancient villages has been difficult to implement. The reason is that the importance of ancient village protection is not fully comprehended and the theoretical research has not yet penetrated into the essence. These are the roots of the problem. For creatures in nature, if a place is no longer suitable for its survival and reproduction, there are only two alternatives to leave or to migrate, which may lead to the heteromorphosis or extinction of the entire species. The same is true for the villagers in the protection of ancient villages. In today's China, where the economy and science and technology are developing at a rapid pace, we cannot expect the peasants to persist in poverty and the hard life, to just continue with the inheritance of a certain tradition. It is impossible to protect these ancient villages without the participation and support of local villagers. Only by improving the living environment of the villagers and allowing them to realize the value of protecting the villages they have lived in for generations, and by encouraging villagers to take the initiative to stay in the village, can they fundamentally protect these ancient villages.

3 Conclusion

For the protection of stone buildings, we should also focus on adopting a relatively positive attitude, to not be blinded by the romanticism of European protection of historical buildings. As for the houses which are gradually abandoned by the villagers after years of disrepair, we must not allow them to collapse. We should actively protect them and try to make use of them on the basis of maintaining the status quo.

For the protection of traditional villages, both material and non-material aspects should be taken into account, and that involves multi-disciplinary mutual learning and cooperation. As in the words of Mr. Sun Hua: "Although the main research objects and purposes of different subjects are different, their theories and methods can be used as a reference for the protection and utilization of traditional villages."

The well-known "Rivet Hypothesis" in biology believes that nature is like an airplane, and each species is equivalent to a rivet on the airplane. We don't know how many lost rivets would ultimately lead to the complete disintegration of the airplane. Likewise, on the plane of protecting human cultural diversity, we don't know which village will be the last "rivet", nor can we bear the disastrous consequences of the disintegration of the "plane". If maintaining the diversity of species is the prerequisite of maintaining the ecological stability of a region, then maintaining the diversity of a region or ethnic culture can also be regarded as the prerequisite of maintaining the stable development of its region and ethnic groups.

References

EHRLICH P R, WILSON E O, 1991. Biodiversity studies: science and policy. Science, 253: 758-762.

GU Yanwu, 2006. A collection of Gu Tinglin's poems variorum. Shanghai: Shanghai Classics Publishing House.

LIN Tiengjen, 2014. Reading imperial cartography-Ming-Qing historical maps in the library of congress. Taipei, China: Academic Sinica Digital Center.

LIU Peilin, 1998. Ancient villages-harmonious human inhabitation space. Shanghai: SDX Joint Publishing Company.

LUO Wencong, 2013. Consideration on the current situation and countermeasure of the protection of traditional villages in China. Urban construction theory research, 20.

MACARTHUR R H, 1972. Geographic ecology. New York: Harper&Row.

MACARTHUR R H, WILSON E O, 1967. The theory of island biogeography. Princeton: Princeton University Press.

SUN Chunjie, 2013. Research on traditional village investigation and protection of Jingxing county.

SUN Hua, 2015a. The natures and problems of traditional villages protection: a preliminary discussion on the protection and utilization of rural cultural landscape in China (I). China cultural heritage, 4: 50-57.

SUN Hua, 2015b. The disciplines and methods of traditional village protection: a preliminary discussion on the protection and utilization of rural cultural landscape in China (II). China cultural heritage, 5: 62-70.

SUN Hua, 2015c. The planning and action of traditional village protection: a preliminary discussion on the protection and utilization of rural cultural landscape in China (III). China cultural heritage, 6: 68-76.

WANG Fang, 2016. Geo-architecture and landscape in China's geographic and historic context. Singapore: Springer Science+ Business Media.

WILSON E O, 1984. Biophilia. Cambridge, MA: Harvard University Press.

WILSON E O, 1992. The diversity of life. Cambridge, MA: The Belknap Press of Harvard University Press.

ZHOU Zhenhe, 2017. General history of administrative regions in China. Shanghai: Fudan University Press.

A Study on Value Cognition and Heritage Composition of Fujian Tubao from the Perspective of World Heritage

ZOU Han[1,2], *HU Xiao*[1]

1 Hubei University of Technology, Wuhan, Hubei Province, China
2 Southeast University, Nanjing, Jiangsu Province, China

Abstract Tubao is a unique building type with clear defensive features in the central mountainous areas of Fujian Province. Geographically, Tubao mainly distributes in the central of Fujian Province, and intersects with eastern Fujian, southern Fujian, northern Fujian and Hakka-Gansu dialects area. Its regional cultural customs and habits are also integrated, presenting a pluralistic architectural cultural. Cultural heritage is an open concept that develops with the times. Cultural heritage types include monuments, sites and groups of buildings, and then identify special heritage including historical towns, heritage canals, cultural routes and cultural landscapes. UNESCO-led World Cultural Heritage promotes States parties to re-recognize their cultural achievements, encourages low-representative areas to promote social development by identifying heritage, and establishes links with the new Convention for the Protection of Intangible Cultural Heritage. Fujian Tubao has the feasibility of applying for Heritage in its artistic aesthetic value, scientific and technological value and historical and humanistic value. This paper integrates the resources of Fujian Tubao Architecture from the perspective of World Heritage. Moreover, according to the *Operational Guidelines for the Implementation of the World Heritage Convention*, the paper also identifies and targets the heritage composition and characteristics of Fujian Tubao.

Keywords Fujian Tubao, world heritage application, cultural heritage, heritage value, heritage composition and objects

1 Introduction

Fujian Tubao is mainly distributed in the central part of Fujian Province and belongs to the residential vernacular building. The predecessor of Bao can be traced back to the defensive walled residence of the primitive society. During the Wei, Jin, and Southern and Northern Dynasties, the architectural form of the Wubi was born in the turbulent social environment because of the demand for defense. At the end of the Western Jin Dynasty, as the Zhongyuan culture was introducd to Fujian Province on a large scale, the early form of the Wubi building was introduced, and different changes were made according to the surrounding of different regions and cultural environments. The existing Fujian Tubao is mainly built during the Ming and Qing Dynasties, and its defense features are obvious. As for the geographical location, Fujian Tubao is mainly located near the Daiyun Mountain range. Because the area is connected with eastern Fujian, southern Fujian, northern Fujian and Hakka dialect area, and different regional cultures and social customs are exchanged and integrated here. Thus forming a diverse architectural culture, representing the unique settlement structure, the characteristics, the spatial form and humanistic spiritand architecture of southwest China.

In 2008, Fujian Tulou vernacular dwellings were officially listed in the World Cultural Heritage List, and the influence of Fujian Tulou is also expanding. At the same time, however, the protection and development of Fujian Tubao remains at a low level and there is not enough concern. In the framework of the *Convention Concerning the Protection of the World Cultural and Natural Heritage*, and the 2017 edition of the *Operational Guidelines for the*

Implementation of the World Heritage Convention, this paper discusses the residential and intangible cultural values, and the composition of the heritage of Fujian Tubao. It is hoped that it will be helpful in applying for the World Heritage and the follow-up protection work.

2 Overview of Fujian Tubao

Tubao is mainly located in the middle of the mountain. The form of the Tubao is composed of a combination of the outer passenger-style dwellings and the combined houses or dwelling houses. Defensiveness is the main functional requirement of the Tubao, and the internal dwellings also have the function of living and offering. The fort wall is mainly built of earth-filled concrete with stones. The shape includes turrets, watchtowers, trenches, fences, racetracks, gates and gun holes. Some scholars have defined Fujian Tubao as: "*The main distribution is in the mountainous area of central Fujian, and the defensive strong earth-building and residential forms combined by the outer corridor-style castles and the internal courtyard-style houses or determinant houses.*"(Lin, 2017)

2.1 Background

Before the Han Dynasty, the Fujian area was mainly inhabited by the Baiyue tribe. After the Qin Dynasty unified China, and the Zhongyuan people migrating to the south, a part of the Baiyue tribe was merged by the Zhongyuan Han nationality who moved south. During the Northern and Southern Dynasties, due to the turbulent social environment, and at the beginning of the historical records "the families of scholar-officials migrated to the south, and eight big famililies moved to Min area(Fujian Province today)", the scholar-officials and the culture of Zhonyuan confessed to the south for four times, and entered Fujian on a large scale. Gradually, it was integrating with the regional culture of Fujian. The Han people who moved to the south formed a unique folk system in Hakka because of the complex geographical environment, and the cohesiveness between the clans was stronger. Therefore the early Wubi-shaped buildings were inherited and changed with the changes of the natural environment, geographical factors and regional culture, resulting in a new architectural culture. According to the administrative division, there are three counties (cities) included: Yong'an, Sanming and Shaxian. Most of the immigrants were people from the Zhongyuan area who avoided the turbulence. After migrating to Fujian, some of them chose to move further and reached the central part of Fujian Province. Due to the special social environment and the occlusion of the geographical environment, the residents of the central part of Fujian paid more attention to the inheritance and cohesion of their families, and gradually differentiated their own dialects. During the Southern Song Dynasty, the local area was rich in gold, silver, iron, lead, zinc, etc. the bandits were increasingly rampant while the people became richer. During the Ming and Qing Dynasties, according to the historical record, "*Though Bandits were everywhere, the government was too far to control it.*" which means the goverment was weak at that time compared to the bandits. In order to protect the families and village residents, "*resistance to bandits, preservation the lives, defence the locations*"—according to the records, which have given birth to the defensive architectural form called Tubao. After the Kangxi period of the Qing Dynasty, the social environment was relatively stable, and the construction activities of the Tubao were slowly weakened.

2.2 Geographic Location and Heritage Surrounding

Fujian Tubao is mainly located between Wuyi Mountain range and Daiyun Mountain range, with the north side crossing the Min River and expending to the southern part of the Jiufeng Mountain range, and with the south side to the southern Boping Ridge. The east reaches the coastal area, and the west reaches the Shaxi basin. Peaks, mountains, and rolling hills are rich in the area which are continuous, affected by the Xinhuaxia structure, forming the Minzhong Mountain Belt and riched in natural resources. Fujian Tubao mainly distributed in Datian, Youxi, Yong'an, Shaxian, Jiangle,

Ninghua, Qingliu, Mingxi, Jianning, Meilie and Sanyuan in the city of Sanming; Hunchun, Dehua, Nan'an and Anxi in Quanzhou (county, city); Fuqing, Fuqing and Yongtai (county, city); Gutian County of Ningde; and there are a small number of Tubao in Mindong and Weinan. The area where the Tobao is located is continuous, with basins and river valleys, and the natural landscape there is magnificent. The area is rich in rainfall, developed in water systems, developed green mountains, diverse in vegetation, and rarely dry, which is suitable for farming. Because the Minzhong area is located between the Daiyun Mountain range and the Wuyi Mountain area, the natural conditions are superior but the geographical environment is complex. The valley and the river are interlaced, and the water and land transportation are not convenient. It has always been difficult to interact with other places. Therefore, the residents of Minzhong have gradually formed a characteristic of contentment, a fondness for hometown and a mode of small-peasant economy as the custom "Take care of parents, and not too far from hometown." The residents lived in the central part of Min were more focus on family inheritance and cohesion, indifferent to fame, have a deep feeling to their hometown and really self-contained. Reflected to its style of residential houses, it has created the seclusion cultural style of traditional residential buildings, including its appearance is simple, abandoning luxury and paying attention to practicality(Da, 2011). The way to select the site of Tubao is divided into three categories: the Tubao built on hill, on sloping land, and in fields. It generally meets the requirements of *Fengshui* which including Shelter the building from wind and gas gathering, and is combined with nature. Apart from using the surrounding mountains to withstand the winter cold winds, natural water bodies are also used for drinking, washing and agricultural irrigation which is necessary for living and production.

2.3 Architectural Features

Fujian Tubao can be divided into two parts: outside Bao and inner residence. Outside Bao means corridor-style fort on the periphery of Tubao. Inner residence refers to courtyard house-style residence or linear-style residence inside the Tubao. The defensive function is concentrated in the outer ring-shape buildings, and the living functions are mostly placed in the internal dwellings, satisfying the requirements of the family or the village for the functions of defense, residence and sacrifice. According to the shape of the plane, the Tubao is generally divided into three types, the front part is square, the back part is circular, and the circle and other shapes. Usually, there are eighteen procedures required in building a Tubao, including site selection, logging, earthmoving, flat foundation, determined the position of Sang(It is a stone under the column), fixed beam, made doors and windows, closed foundation, tamped earth wall, erected the pillar, mounted beam, sawing, booking eaves, etc. (Liu, 2007) The fortified fortress wall of Tubao is the main body of architectural defense. The fort wall is constructed by the underlying stone made purlin and fixed eaves. The defensive main body is the tall strong wall of the Tubao. Generally, Tubao's wall is constructed by the underlying stone and the upper layer of rammed earth. The wall is thick, mostly thicker than 2 meters, and some can even be as thick as 4 meters to 6 meters. The wall of the Tubao is generally supported by the external wall and the wooden frame, and occasionally the wooden frame is used to bear the weight separately. Rammed earth walls and racetracks also have different combinations of load-bearing combinations. Tubao has a wide variety of wall materials, usually using local materials, focusing on economy and practicality. The main materials for the construction of courtyard houses are bricks and stones, and the main material for the construction of the outer walls is earth. Due to the lush forests and the natural environment, the wood is still the most widely used building material in its residential buildings. The various types of wood in the mountains are rich local building materials resources. Considering the location of geography, the outstanding defensiveness, the ecological nature of the local materials, and the unique hiatorical memory are the common architecture features of the Tubao.

2.4 Current Status of Tubao Building Site

Historically, there were nearly 2,000 Tubao distributed in the central part of Fujian. Currently, there are 43 Tubao, the status of which 10 are good, 10 are general, 16 are poor, and 7 are empty(Liu, 2007). In recent years, the protection and utilization of Fujian Tubao has made certain progress. Through the third cultural relics survey, Fujian

Tab. 1 Fujian Tubao protection status

Heritage protection level	Cultural relics name	Location
Key cultural relics site under the state protection	Anzhen Tubao	Yong'an City, Sanming City
	Anliang Tubao, Fanglian Tubao, Guangyu Tubao, Shaohu Tubao, Tai'an Tubao, Guangchong Tubao, Pipa Tubao (Datian Tubao Group)	Datian County, Sanming
Key cultural relics site under the province protection	Jukui Tubao, Shujing Tubao	Youxi County, Sanming
	Fuxing Tubao, Fulin Tubao	Yongan, Sanming
	Kanhou Tubao	Mingle County, Sanming
	Datian Tubao Group	Datian County, Sanming
	Shuimei Tubao Group	Shaxian County, Sanming
Other	All are county (city) level cultural relics protection or "third cultural relics census" cultural relics	

Province and the city of Sanming have organized special research, national academic seminars and landscape photography exhibitions(Tab. 1). The events increased the popularity of the Tubao. At the same time, Sanming has formulated the *Measures for the Management of Earthen Buildings*, *The Maintenance Plan for Tubao* and the *Regulations on Cultural Relics and Fire Management* based on the *Law of the People's Republic of China on the Protection of Cultural Relics* and the *Regulations on the Management of Cultural Relics in Fujian Province*. The Datian County Cultural and Sports Bureau is responsible for the maintenance of the Tubao, reflecting the government's strong sense of protection. With the rapid development of modern society, the ancient time and space relationship and the pattern of difference sequence of the central Min region have changed. The number of residents living in the Tubao has become less and less, and their cultural space has gradually disappeared. Only in Yongan, Datian and Youxi, a few inhabitants are still living in individual Tubao; only a few Tubao with historical value have been recognized the value of conversion, utilization and development of functions. More Tubao are in idle state, and the use of the Tubao is still in the status to be developed.

3 Cognition of Fujian Tubao Value in the View of World Heritage

The *Operational Guidelines for the Implementation of the World Heritage Convention* updated by the World Cultural and Natural Heritage stipulates that the nominated heritage must meet at least one of the following criteria, then the Commission considers it to be of outstanding universal value(Tab. 2). (i) represent a masterpiece of human creative genius;(ii) exhibit an important interchange of human values, over a span of time or within a cultural area of the world, on developments in architecture or technology, monumental arts, town-planning or landscape design;

Tab. 2 Fujian Tubao material cultural heritage composition

Composition Category	Heritage composition and benchmarking
Heritage surrounding	The mountainous area of central Fujian has dense forests and a suitable climate. It has a beautiful natural environment and landscape. (v)
Site selection	The site selection conforms to the trend of the mountain, and the terrain and wind direction are skillfully utilized. The shape is diverse. (v)
Construction	8 traditional procedures such as site selection, logging, earthmoving, flat foundation, determined the position of Sang(It is a stone under the column), fixed beam, made doors and windows, closed foundation, tamped earther wall, erected the pillar, mounted beam, sawing, booking eaves, etc. (ii), (iii), (iv)
Architect form	From the plane, often divided into 3 types: the front part is square, the back part is circular, and the circle and other shapes. Tubao usually has a cannon, a watchtower, a muzzle and a horse-racing road. Generally, there is only one gate, mostly a double door. There is also a fire-proof sink and water injection hole above the door. (iii)
Architectural decoration	The decoration includes the couplet, door and window carving, column foundation, and mural painting. (iii), (vi)

(iii) bear a unique or at least exceptional testimony to a cultural tradition or to a civilization which is living or which has disappeared; (iv) be an outstanding example of a type of building, architectural or technological ensemble or landscape which illustrates (a) significant stage(s) in human history; (v) be an outstanding example of a traditional human settlement, land-use, or sea-use which is representative of a culture (or cultures), or human interaction with the environment especially when it has become vulnerable under the impact of irreversible change; (vi) be directly or tangibly associated with events or living traditions, with ideas, or with beliefs, with artistic and literary works of outstanding universal significance. (The Committee considers that this criterion should preferably be used in conjunction with other criteria.) It can be learned that cultural heritage should have a certain cultural value of art, culture, history and science in the world. In the process of applying for the world cultural heritage, the work should not only summarize the characteristics of its material form but also recognize the unique historical values contained in cultural heritage and its role in the evolution of historical culture. This has become one of the key factors affecting the success of the application. In the framework of the *Implementation of the World Heritage Convention and the Operation Guides* of the 2017 edition, this paper takes Fujian Tubao residence as the research object, discussing the material and intangible cultural values and the composition of the heritage(Tab. 3). In the identification of heritage values, the site selection of Tubao follows the Fengshui architectural planning theory. It has a beautiful layout and strong defensiveness. It is located in the hills or fields. It is a masterpiece of man and nature, reflecting the integrity, originality and aesthetic value of the heritage. Traditional construction techniques take advantage of the terrain and use local materials such as earth, wood, and pebbles. The construction techniques are scientific and geographically distinctive.

Tab. 3 Fujian Tubao intangible cultural heritage composition

Heritage composition	Heritage composition and benchmarking
Historical origin	The construction form of Tubao originates from the construction of the *Wubi*; the continuous population migration since the Western Jin Dynasty has made the Baiyue tribe and the Central Plains culture blend. (iv)
Traditional folk activities	*Over-fire Bricks*, *Hundred Feasts*, *Wujia Punch etc.* (vi)
Cultural connotation	Including the Dwelling Culture, Hakka culture, Etiquette Culture and Seclusion Culture. (iii)

The plane of the building, the method of construction, the structure of the space, the decoration of the details, and the environmental relationship all indicate the strong family consciousness of the builder, which is unique. It has a clear relationship with the original culture in the Yellow River in the late Western Jin Dynasty. In terms of intangible culture, Fujian Tubao has a long history, inheriting modern folk activities and rich traditional cultural connotations. It is a living fossil that witnessed the traditional power structure in Fujian, population migration, cultural circulation, family inheritance, and changes in the pattern of difference.

4 Conclusion

The existing Fujian Tubao dwellings are unique in China and the world in terms of diversity, integrity and outstanding defensiveness. They have certain value in historical humanities, artistic aesthetics and science and technology, and have the feasibility of application. Fujian Tubao witnessed the literary allusion of the *families of scholar-officials migrated to the south* as a kind of defensive type of residential residence in special historical period, geographical environment and social form.

This kind of dwelling located between the Daiyun Mountain range and the Wuyi Mountain area in the central part of Fujian also witnessed the development of Fujian's seclusion culture, the integration of Baiyue tribe, the Zhongyuan culture, and the change of Fujian's social form. Its historical value and artistic value require people's attention, understanding and research. This paper systematically organizes and integrates the architecture features of Tubao in Fujian from the perspective of world heritage. At the same time, according to the *Operational Guidelines for the Implementation of the World Heritage Convention*, the composition and characteristics of its heritage have been sorted, identified, and it is considered that the application for Fujian Tubao is very valuable and feasible. In view of the current status of protection of the Tubao, it is recommended to speed up the work of the application and continue to increase protection. Applying for the World Heritage is a long-term job. While applying for the job, it not only improves the cultural identity and heritage protection awareness of the local residents but also improves the urban context. Continuously organize and improve the declared the world culture heritage also benefit in achieving sustainable protection, the development of cultural heritage, and finally achieve positive interaction with social development.

Acknowledgements

The authors would like to acknowledge the funding of China Postdoctoral Science Foundation (No. 2015M581700), Postdoctoral Science Foundation of Jiangsu Province (No. 1501015C).

References

DAI Zhijian, 2007. Minculture and its influence on Fujian vernacular architecture. South architecture, (6): 28.

LIN Xinhong, 2017. On the comparison of the architectural features of earthen fort in Fujian. Chinese & overseas architecture, (5):59-62.

LIU Xiaoying, 2007. Investigation of Sanming Hakka Tubao. Fujian provincial federation of social sciences, (5).

Strategies for Earthen Constructions in Armenia

Suzanne Monnot

National School of Architecture of Lyon, Lyon (ENSAL) researcher at the
Lyon Architecture Urbanism Research (LAURE) laboratory of ENSAL, UMR 5600 CNRS, France;
National University of Architecture and Construction of Armenia (NUACA) / ICOMOS, Armenia

Abstract As part of an architecture thesis in co-supervision with UNACA, the objective of this research is to raise awareness of the various actors of the society to the heritage of earthen architecture in Armenia, to the qualities of this primary building material and to its potential for sustainable territorial development. What strategies will transform the population's attitude; recognize this architectural "heritage" (in the meaning of legacy) as architectural "heritage" (with the meaning of patrimony) and this material as a new possibility in the development of eco-responsible architectures?

A first action research method involving Armenian and French university partners allowed for the beginning of an inventory of the many heritage of earthen architecture: archaeological remains and vernacular constructions. Regions rich in earthen habitats have been identified thanks to bibliographic sources and especially to interviews with inhabitants and heritage managers. But if the first intention was, through this inventory, to demonstrate to decision-makers the importance of this architectural heritage and its environmental issues, I realized that this goal would take a long time, without guarantee of efficiency!

This awareness led me to shift my strategy upstream so that the concerned actors get involved, appropriate and value this inventory to reconsider this material, which is beginning to be recognized in Europe. While questioning the epistemological basis of this approach, I sought the arguments for this strategy which combines research, societal transformation and alternative solutions -on the basis of the "principle of responsibility" (Jonas, 2013), applied also to scientists; but also to better understand how to develop and disseminate the strategies for their continuation. Based on different currents of thought — the pragmatic philosophies, the Chinese thought (F. Jullien) and the creative approaches of T. Ingold, I will propose a new definition of this research's modality that engages particular social skills.

Keywords earthen construction heritages, Armenia, sustainable architecture

1 Introduction

Architect, teacher-researcher in a militant posture and committed to a sustainable living in line with my values, I started in 2014 a thesis in international joint PhD on the earthen architectures in Armenia, their knowledge, and recognition. The goal is to change the way we look at these heritages and to allow their development — by reactivating local know-how and traditional knowledge — to promote innovative combinations of this material in contemporary constructions(Fig. 1).

Fig. 1 Example of contemporary constructions: Terra Nostra © ENSAG 2016

The article specifies the context of this research — deployed for a decade — in the framework of Franco-Armenian educational and scientific cooperation and discusses the actions initiated: university partnerships, institutional cooperation, political rapprochements ... The central part of the argument questions how this research has become "action research", how the adopted "pragmatic strategy" can be considered as a "*making*", in the sense of T. Ingold (Ingold, 2017). We will also seek to highlight the epistemological basis of this "adventurous" approach (Jullien, 2012) which can only be reported afterward and whose effectiveness has allowed the initiation of autonomous dynamics.

2 Franco-Armenian Context, Expected Results

2.1 Opening of a New Field of Investigation in Armenia

It should be noted that my dual Armenian-French culture, at the East-West interface, and an interest in China's practices, of its historic pragmatic culture — the Armenian culture being probably closer to the Chinese one that we can think *a priori* (Bedrosian, 1981), have fostered these synergies between Armenia and my environment Lyonnais / Rh ne-alpins. A dual training in architecture and philosophy will also help improve to take a step back from this action research's "making".

This research has opened a new field of investigation on Armenian territory, in the field of the architectural and technical cultures of this country's earthen constructions (Fig. 2). Indeed, "*Armenia has long been known and recognized for the very ancient tradition of mastering stereotomy and its uses in stone architecture. It has preserved until today among the oldest churches with stone dome and specific monasteries. Various recent historical and archaeological works show its significant contribution to this plan throughout Europe and the Middle East, both through the circulation of ideas and the fact of the wanderings of Armenian master builders on the continent.*" (Monnot, 2019)

But this excellence has probably been a brake on the taking into account of earthen constructions, for different reasons: first the stone is omnipresent, remarkably

Fig. 2 Armenia recognized for its tradition of stereotomy: Holy Hripsim Church VII-th century (stones and adobes) © Suzanne Monnot 2015

worked (tuffs rich in colors, finely carved basalt, travertines), dedicated to the queen of the arts that is architecture and more specifically to the churches, monasteries therefore to the religious sphere, that of the sacred, which devalues even more the other traditional materials, especially the earth. Indeed this one without protection disappears quickly enough and when it is protected by a coating, becomes invisible.

The maintenance of the religious heritage in stone — to restore — already a heavy load and is considered priority over earthen heritages. De facto even in European countries, the heritage recognition of this material is rather recent.

2.2 Objectives: Raise Awareness and Campaign in Favor of Earthen Architectures

In this context, the thesis problem raised this question: under which conditions and how earthen architecture could be reassessed in Armenia? The advanced research hypothesis is that by initially preserving and highlighting the *earthen archaeological heritage*, it would then become a lever for changing the way people look on *vernacular earthen heritages* — unjustly considered fragile and associated with the poverty (Fig. 3).

The thesis aims to raise awareness and consequently to mobilize the Armenian authorities on this topic, whether political, economic, cultural or academic — with dissemination strategies in progress (Monnot, 2019).

Fig. 3 Vernacular habitat in adobes © S. Monnot 2017

By thus drawing attention to the qualities of this ancestral material, it would be possible to open up new dynamics in the perspective of eco-responsible architectures — in reference to frugal innovation movements (Bornarel, 2018) that are developing in some international networks with the primary materials, also called bio-sourced or ecological (Anger, 2011).

We first explored different methods of training for the earthen architecture in Armenia and France in higher education (UNACA, ENSAL), but also more widely, with other public (elected, professional).

The objective being that eventually the knowledge generated spreads and induces the reactivation of know-how also in the local populations. At the same time, we have also begun surveys of these populations to collect data on constructive cultures.

2.3 First Step: Checking the Existence of These Heritages

A first state of the publications, interviews with architects and heritage managers were used to target communities rich in vernacular earthen habitats and a first step of the work was to go to the field, to check the existence of this heritage in Armenia.

Exchanges with the inhabitants have shown a real loss of constructive cultures and local know-how.

With techniques of location, surveys (plans/cuts) on the spot and photographs, we have described the diversity, richness, characteristics of these heritages, and alerted to the urgency to intervene, to keep at least some representative examples(Fig. 4).

Fig. 4 Norashen church in adobes (1879) in Ararat's plain © Suzanne Monnot 2016

2.4 What Effective Strategy to Transform the View on These Heritages

During these investigations, I also became aware that this time-consuming work, even if it is done as part of a co-supervised thesis, is likely to remain in a library fund and ultimately have little impact! This is often the case for French archaeological missions working in Armenia, whose excavation reports, written in French, are poorly translated and not taken into account (Deschamps, 2012).

My initial problem then shifted, questioning further upstream how could the action of a single person engaged in a thesis — beyond the production of knowledge, which remains one of the challenges of a thesis and relying on the legitimacy conferred the scientific approach — have an impact on a larger scale? In what kinds of ways to engage more efficient actions to make the actors concerned, to involve them in interactive processes? How to do this concretely so that these actions in Armenia (but also in France) then stimulate a set of transmissions' processes, researches, rediscovered, not only at the university levels but also more widely the managing institutions, the decision-making authorities and the public opinion...? What strategy should be adopted for synergies to take place in various spheres of society, so that others invest in these issues, otherwise the process may run out of steam, fall back(Fig. 5)?

Fig. 5 Ancient habitats in the center of Yerevan demolition © Suzanne Monnot 2014

2.5 Adoption of an Action Research Approach Developed by "Making"

These questions, the various opportunities following several months on the ground and many contacts, led me to a new awareness. It is certainly a transformation of the look towards this earthen building material that must be aimed at, with actions on a larger scale. By fostering original dynamics, synergies will be created; but it is also an objective to be considered in the long term, a diffusion on various fields (French and Armenian), with a sometimes indirect impact.

My actions have thus diversified: for example through the AURA region, the Metropolis of Lyon and the ENSAL, I was led to participate in major public events. Little by little, a web of relationships was formed (French Ministry of Foreign Affairs, French Embassy): we organized symposia and exhibitions with French and Armenian students, which in turn enabled the mobilization of architects, UNACA's teacher-researcher, management bodies and politicians...

Although the objectives and expected results have remained targeted, the approach has reoriented towards more listening: exploit the "situation potential" (Jullien, 2018), foster the use of occurrences "weaving" with them. It was no longer a planned strategy, applied with a goal defined in advance, but rather an approach that could be described as "adventurous" (Gaillard, 2015). It is a methodology that has transformed the mode of action of initial research into an "action research" of another order.

Therefore, the central part of this article will question the nature of this mode of action: how to think it? It will present arguments on the basis of knowledge formed in other disciplines: pragmatic philosophy, anthropology. Finally, it will seek to shed light on the epistemological basis of this new approach, its methodological coherence and its relevance in intercultural exchanges.

3 What Epistemological Posture?

3.1 Interest of an Epistemological Questioning

The academic interest of this research is to start from my specific intercultural situation and to carry out an epistemological reflection in the architectural field, epistemology being defined as "the way in which scientific knowledge is acquired and validated" (Hoang, 2018).

The objective will, therefore, be at two levels:

— at the individual level, this work would allow me to gain understanding and therefore clarification in relation to my research posture, because research is inseparable from an epistemological position (Hoang, 2018). In fact, it is also a way of presenting international cooperation in a French ENSA, but practiced by a teacher-researcher born in Armenia.

— at a more general level, this work seeks to enrich the scientific discussions on the epistemological support of action research in the ENSA (National Higher School of Architecture), the latter being still in a dynamics of disciplinary foundation because of its relative youth, in France at least: this applies in particular to the doctorate in Architecture, with major epistemological stakes.

3.2 Philosophical Rooting of This Research's "Making"

As mentioned above, methods mobilized in our research, have not been defined previously planned or been a theory upstream: they proved "in and through action". In fact, the necessary critical distance is difficult in an action-type search: how to rethink what one has often done "in the heat of the moment". How to identify and think the strategies "operative" in this "adventurous" approach?

We can therefore only think of it afterward (after the work done) as Hegel recalls for philosophy: the owl "takes flight at dusk" (Hegel, 1989). Allegory of philosophy, the owl represents the "delay" taken by the consciousness on the action. It is all the ambivalence of this reflection on the action research's approach (which should actually be called "action-recherche" and not "recherche-action" in French). So I will borrow arguments from the philosophy to clarify my posture and its evolution.

We can't discuss these approaches without citing pragmatic philosophies Pierce, James, and Dewey mainly to the experiences role: it is in Dewey's wake and from the 1970s that the "experiential knowledge" (Lochard, 2007) has become established in the French social discourse. Its success is indirectly linked to the support of associations that wanted to enhance the experience of the disadvantaged people, experience "lived", "action"...

Closer to home, the anthropologist Tim Ingold in his latest book defends the pedagogical, epistemological and creative value of situated practice. "Making" is seen as "*a process of growth that places from the outset one who makes as someone who is in a world of active materials*" (Ingold, 2017). The making process then consists of joining with these lives in the common task of fashioning a sustainable world. His image of the kite makes it possible to understand the interaction between the person running and the kite flying(Ingold, 2017): the activity of the person is listening and she acts in correspondence with the air, the environment, for *a dance with three*.

Our research has really progressed in this interaction "where activity and passivity are interlaced" (Ingold, 2017) objectives appeared, actions/strategies were deployed as and when meetings with actors, contextual opportunities, and their agentivity... But can we conceive of an empirical mode of action-research without questioning its effectiveness?

3.3 Efficacy: Cultural Notion and Issue for Action Research

It is in the philosophical approach of the sinologist Fran ois Jullien that this research finds arguments. It is a question of the nature of the effectiveness: "or how the human intervention manages to connect itself with the propensity of things and allows itself to be integrated?" (Jullien, 2014) — this posture is reminiscent of Montaigne's " -propos" strategy, thinker of the moment, of the "occurrence" (Jullien, 2018).

Traditional China has developed a relationship with efficiency, different from what the Greek tradition has transmitted (mainly with modeling and finality). "*The Chinese thought of efficiency, indirect and discreet, relies on the situation potential and induces "silent transformations"*. It always seeks to promote and associate with the natural course of things: the human action "seconds", it accompanies this transformation as a gardener accompanies the growth of plants and promotes the best conditions of its natural development (Jullien, 2014).

We refer here to a Mencius's metaphor: to accelerate the growth of his plants, a gardener pulls on them, obviously, they fade and die. Therefore, the well-informed gardener favors the rooting of his plants and their development by working their environment with patience and discretion, allowing time to work -that's how effective it is, suggests Mencius.

In *The practice of China*, André Chieng quotes Jullien: "*from the strategic point of view that is his, because it has not widened a cleavage between the world and consciousness, (...) all for it being case of process (...) the Chinese thought did not hesitate to think the manipulation in the upstream of the process*". And he adds: "*In other words, China does not just wait for luck, it provokes it, or at least it tries to work as far upstream as possible so that the process brings the desired result, not by a direct action, by a geometric design (the straight line is the shortest path between two points), but by a "hidden" action, which because it is hidden is all the more effective*" (Chieng, 2006) [Translation].

For our research, we have therefore tried to practice listening to the potential of the situation and to "connect with the propensity of things and allow ourselves to be integrated" by working upstream to initiate these processes: particularly in 2016, with more than six months on the Armenian field and contributions for the 2016 Terra Congress in Lyon, but also with participation in the 3rd meeting of decentralized cooperation in Yerevan with the involvement of UNACA, city officials and urban planning agencies of Lyon and Yerevan, etc(Fig. 6).

Fig. 6 Terra Lyon 2016 Congress: (Armenia guest of honor of Terra 2016) the Armenian delegation presented 5 scientific papers (Serge Monnot 2016)

3.4 Displacement: "l'écart" to Make Dialogue Cultural Relations/Tensions

With the story of Mencius gardener, one can conceive how Chinese thought understands the development of the plant, what a look it takes on its "living". The plant is apprehended as a transformation process in-between: the sky and the ground (two foreign terms). The interest of this way to see thinks lies in its ability to create tension and maintain the opposites in a vis-à-vis relationship. Thus this "relational tension" — disturbance, derangement — suggests "the thing" in its process of transformation (and internal regulation) with its environment.

André Chieng in the chapter *Virtue of Contradiction* (Chieng, 2006) discusses this aspect: "*China gives opposition-correlation signs an entirely different utility: that of representing the functioning of the world.*"

It is in this "écart" that a "between" appears — which the Chinese also call "center" — where "living" can naturally unfold (Jullien, 2012). Even if it is not obvious for the Western mentality to understand what the "between" is — emptiness acting — the Chinese attitude consists in looking for the situations of setting in vis--vis so that the "displacements/lags" work in an autonomous way (Feyertag, 2015).

By putting the question of two concepts or two cultures in these terms of "écart", F. Jullien uses this opposing/tension of the different ways of thought. Thus this activates a lively operation of thinking, the generation, and regulation of spontaneous processes which happen in this intercultural translation work.

4 What Kind of Action Research?

4.1 Redefinition of This "Action Research" with These Notions

In concrete terms, these different notions — accompanying the propensity of things, making work the "carts" and to remain available, — make it possible to better understand the operating methods mobilized in this research.

Very often the actions consisted in putting vis-à-vis the forces involved: the different actors, cultural, academic, and institutional. Different types of events, symposia, thematic exchanges, student workshops, exploited the double affiliations and confronted them — put in tension — from situated problems (the Kond district, teachings on earth material, the sustainable development of the territories, the ecotourism) to engage the different dynamics. For example, to bring together French students/teachers (Ensal) in Armenia with students/teachers from UNACA, the staff of the urban planning agency of Lyon with those of Yerevan Project or the rural mayors of France and Armenia, the city hall of Lyon with that of Yerevan ...

It is indeed all these different situations, their discrepancy, sometimes their opposition — so their interaction and tensions — that have sometimes made possible synergies, deployment in different institutions and environments. The result is the constitution of a network and/or its activation.

These meetings are always a formative experience, trigger processes on both sides, dynamics that then, often feed on themselves "naturally", but must also be maintained.

4.2 From Action Research to Transformation Research

This mode of understanding — valid for all domains in China — by the fact that it recognizes the phenomena observed, a natural ability to deploy autonomously, as soon as they result from a relation/ tension; gives human intervention another role — a "second" one — and thus helps to consider efficiency differently. We will recognize here what symbolizes Chinese thought, Wuwei -non-action or action through inaction.

But *if we do not act, what do we do? The key word of Chinese thought is "transformation" (hua). Not to act but to transform (...) Chinese thought thus leads us to conceive what transformation is in the face of action* (Jullien, 2014) [Translation].

The author systematically opposes them both: the action because it is momentary (even if this moment may last a long time), local and referring to a subject-I (which can be collective) it always stands out from the course of things. So she notices herself. But transformation, thought in an inverse mode, is global, extends over time, progressive and continuous: "*it refers less to a designated subject than it proceeds unobtrusively by influence on an ambient mode, pregnant and pervasive. So the transformation is not visible, we only see the results*". (Jullien, 2014) [Translation].

Finally, the key word of this research methodology is transformation: thanks to these epistemological precisions, we can characterize this approach not as an *action research but as a transformation research*.

4.3 What Interest to Think in Terms of Transformation Research?

Is it here a paradigm shift, a new way of seeing things? To think this type of approach with a theoretical perspective seems important to us for various reasons.

In this period that we call Anthropocene, and in a context often perceived as an ecological crisis, adding to the possibilities of the digital age (social networks, coworking, forums, participation, and so on...) access to knowledge and the research's making are deeply changed. Indeed, more and more scientists are not only seeking to produce knowledge but are also engaged in militant postures to guide research and support society towards "alternative" modes of thinking, of acting, using the resources of the planet. Among them, the territory planners and architects of the laboratories, such as CRAterre (Gandreau, 2017), Am co for example (Anger, 2011), as well as those mentioned at the beginning of our article (networks of frugal architectures ...) have taken steps which seem to us closer to *the transformation research than action research*.

For these recent modes of "scientific exploration", their performance being linked to the use of collective intelligence, to think these transformation/research processes is also to think of resources for creativity and to promote more eco-responsible solutions.

However, Western thought is not well equipped to think about transformations. Relying on other cultures — China, for example — allows for questioning and revealing our unthought; the articulation between the two cultures is also to be built... Finally, to account for transformations is difficult: how to represent silent transformations (Jullien, 2010). This issue of representation will be to think too.

5 Conclusion

Returning with Tim Ingold, we will highlight three differences between the two types of research:

— in the *transformation research*, after having learned what I had to learn, "*I trace my own path by moving forward while reflecting on this first experience*". Unlike in an "action" research framework "*I never stop to take a look back on the collected material to uncover trends and patterns*".

— in the transformation research, "*it is about studying with and to learn from, it opens a process of life that involves a transformation of the process itself* ". While "action" research: "*is a study of and learning about, whose results obtained in the long term are the result of a selective report responding to a documentary purpose*". (Ingold, 2017) [Translation].

"*The movement that guides the first project is primarily transformational, while the imperatives of the second project are essentially documentary*".

Acknowledgements

The research and actions were possible thanks to the financial support of Auvergne-Rh ne-Alpes Region, National University of Architecture and Construction of Armenia (NUACA), Ministry of Culture (France), Embassy of France in Armenia and Metropole de Lyon, and EVS-LAURe, ENSA of Lyon.

References

ANGER R, 2011. Thèse àl'INSA de LYON Approche granulaire et collo dale du matériau terre pour la construction Sous la direction d'Hugo Houben et Christian Olagnon.

BEDROSIAN R, 1981. China and the Chinese according to 5-13th century classical Armenian Sources. Armenian review, 34, n 1-133: 17-24.

BORNAREL A, GAUZIN-M LLER D, MADEC P. Pour une architecture frugale. https://reporterre. net/TRIB-Manifeste-pour-une-frugalite-heureuse-en-architecture-La-frugalite (consulté le 15-05-2018)

CHIENG A, 2006. La pratique de la Chine, en compagnie de Fran ois Jullien. Editions grasset & fasquelle: 127, 223.

DESCHAMPS S, et al., 2006-2012. Rapports de fouilles, Beniamin-2006/2007, Erébouni-2010/2012, Institut d'Archéologie et d'Ethnographie de l'Académie des Sciences de la République d'Arménie.

FEYERTAG K, 2015. In conversation with Fran ois Jullien making ambiguity fertile is the present mission of thought. Paris: In conversation with Fran ois Jullien.

GAILLARD F, RATTE P, 2015. Des possibles de la pensée. L'itinéraire philosophique de Fran ois Jullien, Hermann: 5.

GANDREAU D, 2017. Th se en Architecture, Patrimoines archéologiques en terre et développement local: enjeux interdisciplinaires et perspectives de formation.

HEGEL G W F, 1989. Principes de la philosophie du droit. Gallimard.

HOANG A-N, 2018. Clairage herméneutique de la posture épistémologique d'un chercheur en SIC face. l'entre-deux langagier et culturel vietnamo-fran ais, Revue fran aise des sciences de l'information et de la communication [Online], 12 | 2018, Online since 01 January 2018, connection on 12 June 2018. URL: http://journals.openedition.org/rfsic/3501; DOI: 10.4000/rfsic.3501.

INGOLD T F, 2017. Anthropologie, archéologie, art et architecture. Editions dehors: 24, 25, 26, 60, 208, 209, 211.

JONAS H, 2013. Le principe responsabilité, Flammarion.

JULLIEN F, 2010. Les Transformations silencieuses. Le Livre de Poche.

JULLIEN F, 2012. L'écart et l'entre. Le on inaugurale de la Chaire sur l'altérité. Galilée: 35, 51.

JULLIEN F, 2014. Conférence sur l'efficacité. PUF.: 47, 55, 56, 64.

JULLIEN F, 2018. Du Temps. Èléments d'une philosophie du vivre. Le Livre de Poche: 176.

LOCHARD Y, 2007. L'av nement des 《savoirs expérientiels》 La Revue de l'Ires 3 (n55). diteur I. R. E. S.

MONNOT S, 2019. Patrimoines en terre, matériau pour le développement durable des territoires. Presses Universitaires de Saint Etienne.

Subtheme 2:

Challenges and Possible Solutions

Restoration and Rehabilitation of Traditional Earthen Architecture in the Iberian Peninsula

Camilla Mileto, Fernando Vegas, Lidia García-Soriano, Valentina Cristini

Universitat Politécnica de Valéncia / IRP, Cam de Vera s/s, Valencia, Spain

Abstract The Iberian Peninsula has a huge amount of earthen buildings both monumental and non-monumental. The extension of this territory and its geographic and climatic heterogeneity as well as the variety of available materials and cultural diversity represent the main factors responsible for generating the many earthen constructive techniques employed throughout the centuries (rammed earth, adobe, half-timber, cob and their variants). This built heritage constitutes an important part of our culture both for its remote origin and for its technological diversity and adaptation to the natural and cultural milieu. Nevertheless, earthen architecture and its specific techniques have been disappearing, both neglected and substituted by new standardized techniques, above all, from the second half of the 20th century on.

Thus, the conservation of existing earthen buildings has been frequently undertaken with alien materials and techniques generating not only a cultural and constructive loss, but also phenomena of material, constructive and structural incompatibility. The ignorance and discredit began to be overcome in the last decades of the 20th century when pioneer research rescued the intrinsic values of this architecture related to heritage, cultural and bioconstructive factors.

In this context, the research project SOStierra, financed by the Spanish Ministery of Economy and Competitivity, aims to investigate on the possibilities of compatible, respectful and sustainable conservation and adaptive reuse of existing traditional non-monumental architecture built with earth in the Iberian Peninsula, avoiding alien and standardized solutions and privileging the options that may respect the technical and cultural diversity and their sustainability lessons for the future.

In this paper the main conclusions of the project will be presented, since the analysis and evaluation of the interventions carried out through a multidisciplinary methodology by experts in the different subjects, has allowed to draw up guidelines and tools to guarantee the real transfer of the project results to the technicians and society through the administrations and official organisms interested.

Keywords restoration, rehabilitation, earthen architecture, research project

1 Introduction

Knowledge of vernacular earthen architecture is the starting point for the promotion of compatible, respectful and sustainable restoration and rehabilitation processes, avoiding external standardized options and favouring those respecting technical and cultural diversity and providing lessons in sustainability for the future.

The research team for the SOStierra project has worked alongside the work group and scientific collaborators, specialists from different countries. This has made it possible to draw up a comprehensive inventory of traditional earthen architecture in the Iberian Peninsula.

In addition, other national and international institutes and research centres have collaborated in this project, alongside international organizations and public authorities.

2 Project Objectives

This project chiefly aims to contribute to the valorization of traditional earthen architecture in the Iberian Peninsula, as well as that of its materials and traditional techniques in order to encourage compatible and sustainable conservation and restoration of this heritage. To do so, specific objectives have been set:

Data collection on traditional non monumental earthen architecture in the Iberian Peninsula, based on different studies with various approaches: geographical study, study of constructive techniques and variants, and study of deterioration phenomena.

Comprehensive data collection and the creation of an intervention database for traditional non monumental earthen architecture in the Iberian Peninsula.

Fine-tuning and application in case studies of an analysis and assessment methodology for interventions carried out.

Drawing up of guidelines for the restoration and rehabilitation of traditional earthen architecture.

Transmission of knowledge acquired and training for professionals and students through the different actions planned (seminars and proceedings, book, exhibition, website ...).

3 Methodology

In order to meet these objectives a strict methodology was organized through the following actions:

Analysis of the current situation through data collection and the creation of a database for traditional earthen architecture and interventions in the Iberian Peninsula.

Analysis of the current situation through a multidisciplinary methodology based on select case studies from the most common interventions (geographical, social and cultural context; intervention analysis taking into consideration conservation, construction and technique used, as well as analysing material and structural compatibility, energy efficiency, bioclimate and accessibility).

Results comparison with other settings in Europe and worldwide.

Proposal and definition of guidelines for the current needs and objectives to be met by these interventions. Guidelines for the restoration and rehabilitation of traditional earthen architecture in the Iberian Peninsula.

Dissemination: organization of different events to promote the principal results of the project.

4 Geographical Study

The morphological, climatic and geological conditions of a given setting determine the different constructive earthen techniques, characteristics and variants. This study therefore aims to establish geographical correlations between constructive earthen techniques (rammed earth, adobe and half-timber) and the features of individual locations, comparing localization mapping for the different techniques with themed maps. The themed maps analysed, obtained from official sources (IGN, EAmeT, INE, etc.), make it possible to work with reliable contrasted information. A wide diversity of aspects of the territory has been studied to compare the characteristics of traditional earthen architecture with the different factors to which it is associated in varying degrees. These comparisons provide information from the conditions or factors benefiting or hindering the adoption of solutions with constructive earthen techniques in the traditional architecture of a specific locality.

This study examined a sample of 618 locations throughout the Iberian Peninsula identifying one or several constructive earthen techniques. Rammed earth was recorded in 323 of these 618 locations, while adobe was found in 293 and half-timber in 178. The geographical study for each individual location superimposed the geographical, climatic and geological maps over that showing all 618 locations, establishing a correlation between these locations and their individual position in the themed maps. This comparative study establishes a correlation between the techniques and factors benefiting and hindering their occurrence in a specific region (Mileto, et al., 2017)(Fig. 1, Fig. 2).

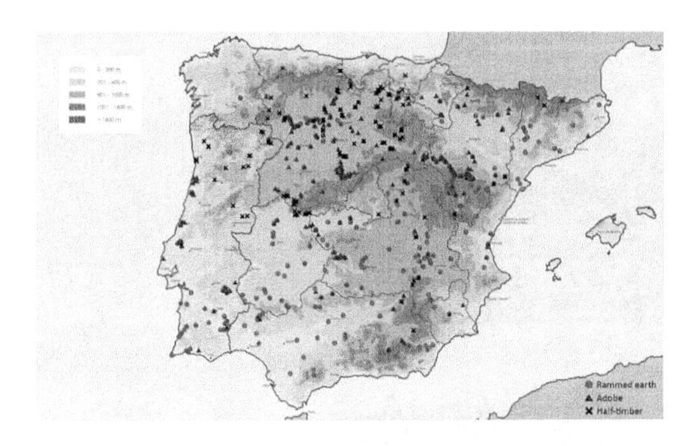

Fig. 1 Constructive techniques and relief. Source: Dirección General de Planificación Territorial, 1994 and Instituto Geográfico Português © SOStierra Project

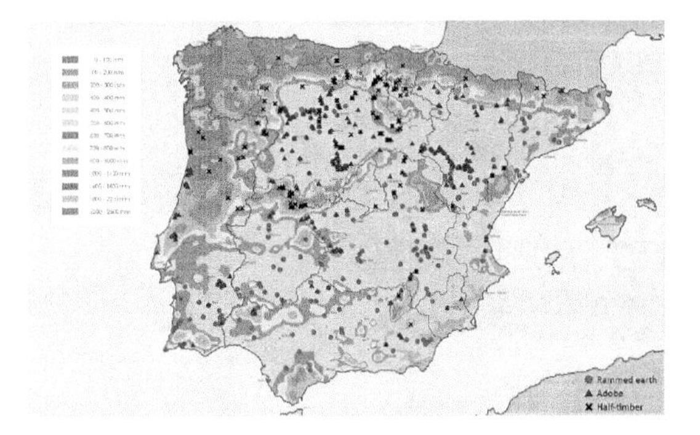

Fig. 2 Constructive techniques and rainfall. Source: Agencia Estatal de Metereolog a and Instituto Portugu s de Metereologia © SOStierra Project

5 Study of Constructive Techniques and Variants

This study presents an analysis of the different earthen construction techniques found in the Iberian Peninsula. Some of these — the different variants of adobe, rammed earth and half-timber — are still frequently found. In contrast, others like clay lumps or poured rammed earth, are much rarer or have disappeared almost completely, and are found almost exclusively in archaeological contexts. Classification reflects the ways in which earth is used as a constructive material and the techniques are defined as follows: monolithic earthen walls (rammed earth, poured earth, cob, piled earth), masonry walls with earth or pieces (moulded adobe, adobe moulded by hand, clay lumps, sod, marl), mixed earth and timber walls (half-timber, wattle and daub), as well as other uses in earthen vaults, earthen renderings and finishes (flooring, renderings, roofs), earth in auxiliary elements (mortars, reinforcement, filling and regularization elements), excavated architecture. 2,460 cases of traditional architecture throughout the Iberian Peninsula have been obtained from this fieldwork.

The classification of the dataset has allowed the most frequent constructive techniques and variants to be identified, in turn compiling a taxonomy as comprehensive as possible and a geographical account of their location in the Peninsula. Based on the above data it was established that the most common technique is that of rammed earth, which is widespread throughout almost all the Iberian Peninsula except for the Cantabrian corridor.

Adobe is also found in much of the territory, although its presence is greater in the northern half as is the case of half-timber with earth.

In addition to mapping the different constructive techniques through the analysis of the case studies catalogued, the work initiated in previous research projects has been completed with drawings and diagrams of the different variants (Fig. 3).

6 Study of Degradation Phenomena

Earthen constructions are highly resistant when their foundations and crowning are protected, as these are degradation hubs and the main entry points for water. Moreover, earthen walls must be protected from the rain with earthen, lime or gypsum mortars or any other type of rendering (MECD, 2017).

Therefore, degradation in earthen walls is mainly caused by direct and constant exposure to the elements (damp, water, wind, etc.), occasional structural deficiencies and biological and human agents such as growing vegetation; lack of maintenance and abandonment.

The use of incompatible materials such as cement, concrete or other non-breathable

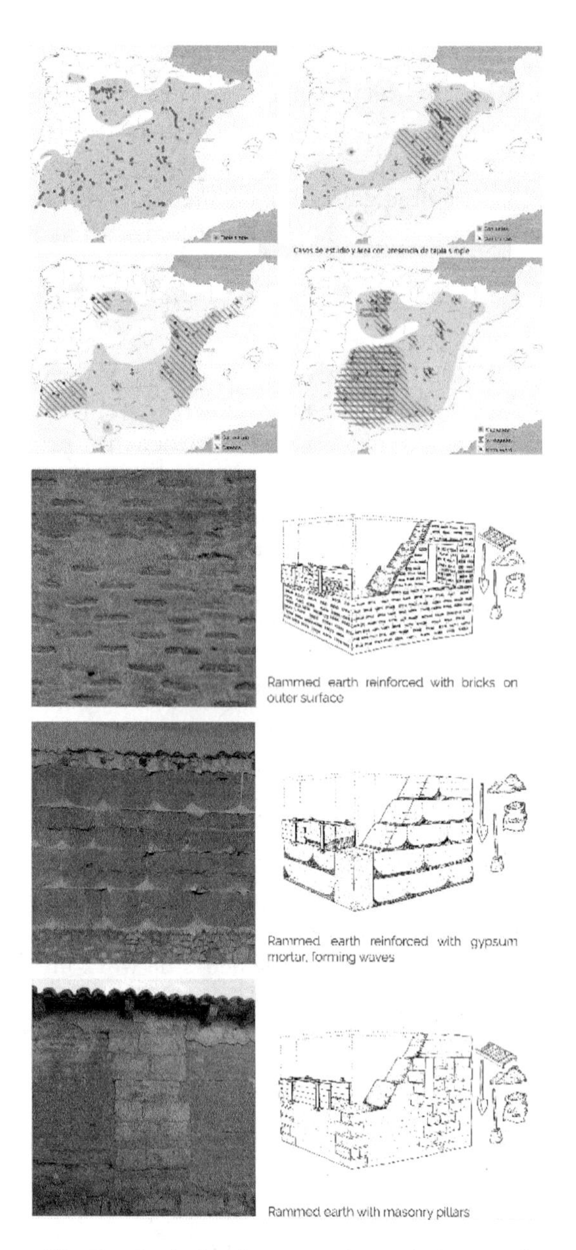

Fig. 3　Analysis of construction techniques and variants. Examples of maps and drawings of the rammed earth family © SOStierra Project

materials, which can cause medium-and long-term damage to the construction.

Every phenomenon observed in the building must be identified as an alteration requiring no intervention or a degradation to be resolved, paying special attention to the causes to prevent increasing damage from the mechanism (MECD, 2017).

Each of the effects observed, known as lesions, represent an evolving mechanism phase previously manifested as a specific

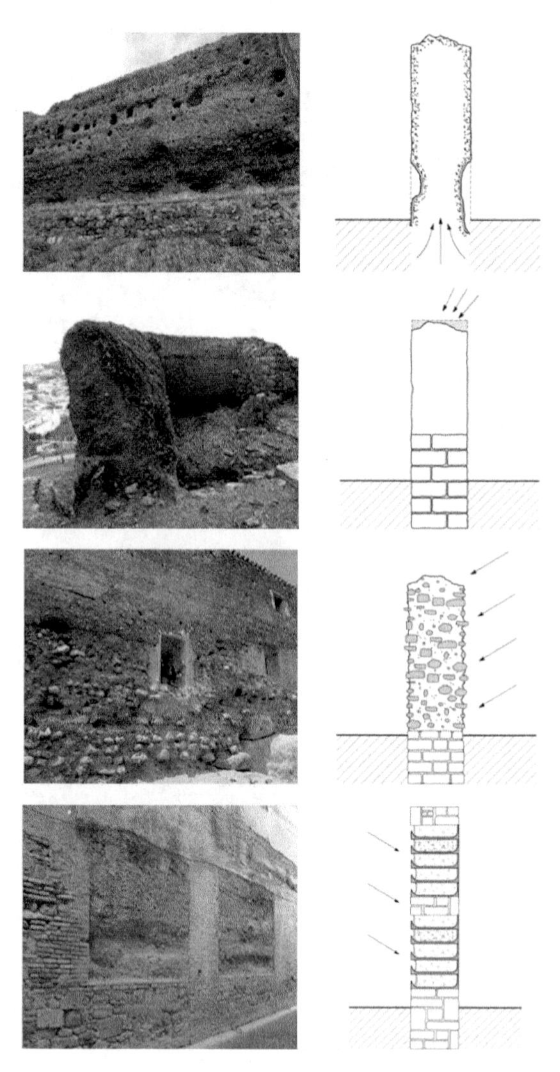

Fig. 4　Analysis of the degradation phenomena according to the construction technique, variant and area of the affected wall. Degradation examples of the rammed earth family © SOStierra Project

phenomenon. Different present and future phenomena create a chain in continuous evolution.

Degradation phenomena were analysed depending on the section of the wall in which they occur and the constructive variant of the wall analysed.

Due to the high number of abandoned constructions documented, it is common to find cases of constructions in ruins or severely deteriorated. Although in these buildings the effect of water is the main cause of serious lesions, it usually starts to act on the points of the wall which have lost protective elements. This therefore clearly shows that continuous maintenance is crucial to the survival of this architecture(Fig. 4).

7 Study of Interventions

The lack of specific legislation regulating interventions in traditional buildings, especially earthen buildings, in most population nuclei has led to the existence of a wide range of heterogeneous solutions depending primarily on the needs and will of the owners. Most interventions are generally the result of the new needs and changes in use, mostly carried out by permanent and temporary residents, and the need to repair existing damage. Therefore, the analysis of intervention cases in vernacular architecture must take into consideration the specific aspects of this type of architecture. Interventions in monumental architecture are based on projects which follow specific intervention criteria and related techniques. In most cases, interventions in vernacular architecturc do not follow a plan or design, and are constrained by economic necessities and limitations.

Accordingly, rather than examining the underlying intervention criteria, the analysis of the interventions in vernacular architecture is geared towards understanding which type of intervention is being carried out and the reasons behind it, as well as towards identifying the transformation dynamics of this architecture as a result of the interventions carried out. To do so, the information compiled in section 3 of the study fiche, whose data was extracted from numerical and percentage information on the type of general intervention and parts of the building, has been analysed. These data were cross-referenced with other parameters from the fiche, including the type and frequency of use in order to establish possible correlations of these factors with the number or type of interventions in the different constructions. There are 274 case studies in the sample. 85% (234 cases) of these are constructions which have undergone intervention and are therefore used as the general sample in this section of the study. These 234 cases have been sorted by constructive technique. Rammed earth is present in a higher number of cases (107) while adobe and half-timber were found in similar proportions (60 and 67 cases respectively).

As in the case of monumental architecture, the study of interventions and the specific intervention criteria is an even more arduous task in the case of traditional architecture. In most cases the interventions carried out are born from the need to prevent further deteriorations, change of use or merely aesthetic changes which usually have no established criteria, but rather attach importance to immediate execution and sourcing of the materials used. It should also be noted that in the case of interventions in vernacular architecture, the original condition of most of the buildings is not known. Therefore, prior lesions can only be located during occasional interventions or when the initial degradation mechanisms persist despite interventions. This has led to the need for simplification of intervention criteria, linked to intentionality and prior consideration, classed as spontaneous or planned interventions.

Interventions are not generally based on serious consideration of conservation or restoration. Most actions aim to eliminate problems or lesions immediately, without seeking compatible options which can be reversed or identified, etc. For this reason, the analysis criteria focused on the study of intervention types in relation to the technique and type of material used. The complexity of the sample made it possible to carry out partial studies of different case study groups with comparable characteristics such as the original constructive technique used in the building (rammed earth, adobe or half-timber).

These partial analyses have facilitated the identification and comparison of the unique aspects of each group. In addition, partial comparative analyses were linked to the general level of intervention of the different cases separating cases with only ma-

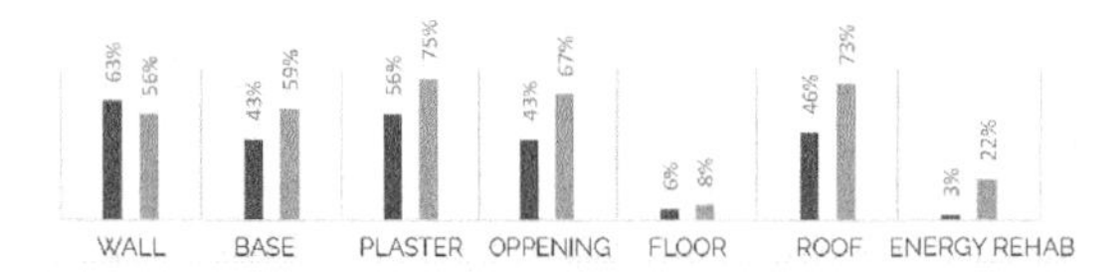

Fig. 5 Ratio of interventions per area depending on the type of general intervention: Group 1 (maintenance and repair) and Group 2 (further interventions) © SOStierra Project

intenance and repair interventions from other cases with a higher level of intervention (restoration, partial rehabilitation, expansion and demolition). As modernization interventions are mostly aesthetic they are linked to other types of intervention. Finally, it should be noted that in all cases the initial analysis was that of the type of general intervention in the building, followed by a study of the partial interventions in each area of the building recorded in the fiche: walls, plinth, rendering, openings, floor, ceilings and roof(Fig. 5).

8 Proposed Guidelines

The proposal for intervention must always be the result of a set of factors which unites the needs for conservation and consolidation arising from a process of rigorous study and research on the building, its setting and state of conservation. Furthermore, the potential of the values of local heritage and its unique aspects, and the convenient recommendations for use, management and valorization of heritage should be taken into account.

To ensure maximum reliability of results in an intervention these needs, opportunities and benefits must be united respecting general principles which developed at the heart of the field of conservation and restoration, and especially attempting to guarantee the material, social and cultural integrity of heritage assets.

The earthen architecture of the Iberian Peninsula presents highly heterogeneous characteristics classified into type and size of buildings; constructive techniques and variants; setting and location; and state of abandonment and transformation dynamics which affect the architecture itself.

The unique heritage of each case should be treated accordingly, following the principles and criteria guaranteeing the best conservation of the asset and ensuring compatibility with the needs for use: the respect and conservation of monumental or vernacular historic buildings; functional, material and structural compatibility; environmental, socio-economic and socio-cultural sustainability; and the expression of the intervention.

The project even proposed a series of typical interventions which are technical solutions representing possible specific solutions. These proposals focused on the earthen walls and the relationship or connections with other elements which could affect these directly.

Therefore, work only affecting other constructive elements (reinforcement of wooden elements, joinery, construction of roofs, etc.) was not examined as it was outside the scope of the objective set for this research (Fig. 6).

9 Dissemination of the Project

Different tasks were carried out to contribute to the dissemination of the project results:

a. The international congress SOS-TIERRA2017 (held at UPV, 14-17 September 2017).

b. Two seminars. The first, "The restoration of earthen architecture. First SOS-TIERRA seminar" (12-14 November 2015), organized at the Higher Technical School of Architecture in Valencia, and the second "International congress SOS-TIERRA. "The protection and restoration of built earthen heritage. Experiences and opportunities" (20-22 April 2016), organized at the Higher Technical School of Architecture of Madrid.

c. A specific website has been set up (http://sistuerra. blogs. upv. es)) compiling the main results of the project (cataloguing, case studies, mappings, etc.), publications, exhibitions, etc.

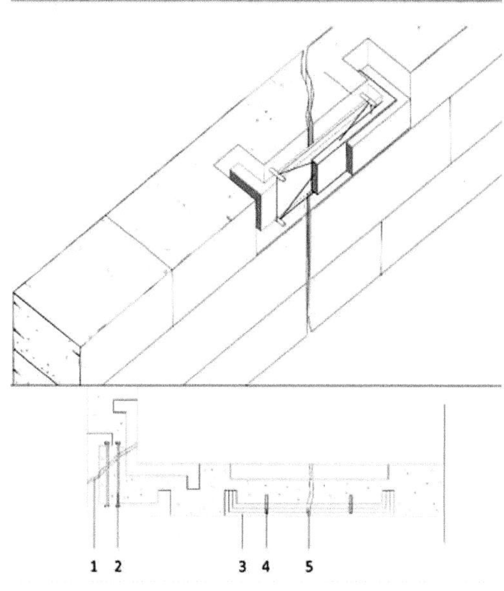

Fig. 6 Example of sewing cracks in an adobe wall (above) and in a mud wall (below) © SOStierra Project

d. An exhibition showcasing the final or intermediate results of the research.

e. In parallel, the results of the SOS-Tierra project have been published in different articles in indexed journals and will be published in a coordinated book, MILETO C, VEGAS F(eds.), Arquitectura de tierra. Restauraci n y rehabilitaci n en la pen nsula ib rica, Argumentum-TC (in preparation, 2019), with chapters from over 50 Spanish and international contributors.

10 Conclusions and Future Lines of Research

Built earthen architecture, global and rich in all its architectural, anthropological, landscape and constructive manifestations has values which have earned it progressive and growing recognition as cultural heritage. The historic and cultural values are linked to its extension in time and geography as an account of human cultural manifestations, economic, constructive and environmental values of earth as a material; intangible values linked to the transmission of knowledge and traditions; aesthetic values as witness to the creativity of the human being; socio-economic values linked to the development of the local economy. These values are the main reasons for the conservation of this heritage, potentially providing lessons to contemporary architecture.

However, earthen architecture continues to suffer, not only from the natural threats which translate into degradation factors, always solvable, but also from the social threats which are currently the greatest risk factor for this architecture. The biggest threats are abandonment, social discredit, loss of trades, use of industrial materials, and pressure from tourist development.

This study aims to research the built earthen architectural heritage by defining it, its scope, values and lessons which can be of use to the present, in order to propose guidelines for restoration and rehabilitation interventions.

Following extensive research, reflection and the overall taxonomy afforded by this work in the Iberian Peninsula and beyond, major advances and systematizations of knowledge have been achieved, opening up future paths of research. Currently, work in ongoing on the research project RISK-Terra -La arquitectura de tierra en la Península Ibérica: estudio de los riesgos naturales, sociales y antrópicos y estrategias de intervención e incremento de la resiliencia (Earthen architecture in the Iberian Peninsula: study of natural, social and anthropic risks and strategies to improve resilience), funded by the Ministry of Science and Universities, and is in some way seen as a continuation of the previous research projects.

Acknowledgements

The research project SOStierra "Restoration and rehabilitation of traditional earthen architecture in the Iberian Peninsula. Guidelines and tools for a sustainable intervention", was funded by the Spanish Ministry of Science and Innovation (Ref.: BIA2014-55924-R; main researchers: Camilla Mileto and Fernando Vegas L pez-Manzanares).

References

AA. VV., 2008. Terra Incognita. Lisboa: Argumentum.

AA. VV., 2011. Terra Europe: earthen architecture in the European Union. Pisa: ETS.

FONT A J, 2005. Earth construction in Spain and Portugal. //Fernandes M, Correia M. Earthen architecture in portugal. Lisboa: Argumentum: 119-123.

GÓMEZ-PATROCINIO F J, VEGAS F, MILETO C, et al., 2018. Techniques and characteristics of traditional earthen masonry. The case of Spain. International journal of architectural heritage. Londres: Taylor & Francis.

MECD, 2017. Proyecto coremans: criterios de intervención en la arquitectura de tierra. Madrid: Ministerio de Educación, Cultura y Deporte.

MILETO C, VEGAS F, VILLACAMPA L, et al., 2017. Vernacular earthen architecture in the Iberian peninsula. First phase of taxonomy and geographical distribution. Earth USA 2017. Proceedings. Adobe in Action: 102-107.

MILETO C, VEGAS L F, GARC A S L, et al., 2017. Vernacular and earthen architecture: conservation and sustainability (SosTierra 2017, Valencia, Spain, 14-16 September 2017). Londres: Taylor and Francis.

Dare to Build: Designing with Earth, Reeds and Straw for Contemporary Sustainable Welfare Architecture

Marwa Dabaieh

The Technical Faculty of IT and Design, Department of Architecture, Design and Media Technology, Aalborg University, Egypt

Abstract Earth, straw, reeds and wood are the main natural building materials in many parts of the world. These materials have several positive properties including thermal resilience, climatic adaptive performance, and a lower-impact on the environment, which have been tested and proven in vernacular architecture over the years. In contemporary practice there is still a very limited use of vernacular natural materials. Conventional industrial materials dominate, even when traditional materials offer the same quality with the same cost and performance, if not sometimes better. This study is part of a semester-long course in sustainable architecture for students completing masters. It will present students' hands-on experimental work for 8 different wall sections using wood, earth, reeds and straw in several combinations. The wall sections are built at a 1:1 scale and tested in a living laboratory environment consistent with the Danish climate. Energy performance and U-values were mathematically calculated to assure compliance with Danish energy-efficient building standards. Life cycle costs and a life cycle analysis were calculated as they were of prime concern. Thermal performance, time lag and heat coefficient values were modeled and simulated as well. Students also had to consider water and fire resistance and the formation of moisture in their design proposals. The study proved that using traditional materials can provide equivalent thermal performance outcomes as contemporary industrial materials while producing better indoor air quality and a lower impact on the environment through their minimal carbon footprint (based on cradle-to-cradle calculations). The paper concludes that there are diverse challenges that still hinder the use of vernacular thinking in contemporary practice.

Keywords dare to build, living labs, project based learning, natural materials, C2C, vernacular re-thinking

1 Introduction

A big percentage of the world's population lives or works in buildings built from natural materials — mainly earth and bamboo(Nunan, 2010). Vernacular buildings around the world exemplify the use of diverse types of low-impact natural materials and construction techniques that have managed to survive for centuries(Turan, 1990). Evidence-based examples of indigenous construction methods used in vernacular buildings have also demonstrated their resilience to extreme weather conditions, fulfilling locals needs through their relatively low cost to build and low impact on the environment (Piesik, et al., 2017). With today's climate challenges, there is a drive to look at climate-responsive building designs and construction. Natural materials like earth, reeds and clay have a big potential to reduce carbon emissions from the building industry, especially if they are abundant around the building site(Dabaieh, 2015).

This paper explains how, through the course of a design studio, students can gain practical experience in designing and building with natural materials for sustainable welfare buildings for the Danish climate. The study shows a prime example of how a pilot education program can apply project based learning for "dare-to-build" concepts. Challenges and lessons learned from this design studio are discussed together with an evaluation of the whole experience of building on campus.

1.1 Project Background

Sustainable welfare buildings were the

main theme for the second semester in the master's-level program, "Integrated Design for Sustainable Architecture and Tectonic Architecture". The main project over the course of the semester was meant to introduce students to the concept of welfare architecture with focus on healing architecture. The location of the project was chosen on Egholm Island in Aalborg City in Denmark which is expected to sink in less than 50 years due to sea-level rise (SLR). The Island is inhabited by less than 60 residents and the main activities include farming and tourism, especially in summer. Because of the farming that takes place on the Island, there is an abundance of agricultural waste, like straw, available for use as bio-based materials in building. There is also an abundance of reeds available along the Island's shore, in addition to seaweed, seashells, clay and timber.

The students' task was to design a 700m^2 healthcare center for trauma patients. The building needed to be designed with a low environmental impact to the island, and needed to be able to sustain imminent climate challenges due to SLR. The building was meant to be designed with disassembly in mind as in 50 years it is intended to be relocated before the Island sinks. If not suitable for disassembly, the building would need to withstand flooding.

Another design criteria was to look at the existing vernacular buildings on the island to better understand how to integrate vernacular methods into their building's construction. This would also serve as a source of inspiration for their contemporary designs overall. They were taught to calculate life cycle costs (LCC) and conduct a life cycle assessment (LCA) together with other calculations and modeling for building thermal performance and energy efficiency. Students were also taught about the other challenges of using natural materials that they would need to consider in their designs like protection from humidity and fire. Other design considerations, like using passive heating, passive cooling, natural ventilation and daylight strategies and other off grid energy production systems, were major foci of the course but will be not covered in this paper.

The outcome of the healthcare center project was presented at the end of the semester in the form of architectural drawings, working details for walls, roofs and floors. Together with rendered shots and other sketches which show the whole design process from its conceptual phases to the final design decisions. Additionally, students presented 1 : 200 models and the 1 : 1 wall section model for the design they proposed using natural local materials on the Island. Solutions presented were diverse and outcomes were evaluated with the help of jury members during the final semester examination.

2 Methodology

2.1 Project-based Learning and Integrated Design

Project-based learning (PBL) and integrated design were the main methodological approaches in this design studio. PBL is a teaching method in which students gain knowledge and acquire skills by working for an extended period of time (ideally a semester) to investigate and respond to an authentic, engaging and complex real-world problem, or answer a complex question or challenge (Dunkin, 1983; Barron, et al., 2007; Nakada, et al., 2018). PBL also prepares students to rise to the upcoming challenges in the world they will inherit (Bell, 2010). More specifically in the case of this project, students were charged with the task of responding to issues related to climate change (particularly, rapid SLR), the mounting increase in carbon emissions and scarcity of resources.

An integrated design process was a complementary component to the PBL. The integrated design process is an approach to building design that seeks to achieve high performance outcomes based on a wide variety of well-defined environmental and social goals (Hansen and Knudstrup, 2015). In this project, architectural design considerations were informed by modeling technical designs and verifying calculations. Students developed their designs several times showing different iterations based on the outcome of technical design support tools. Integrated design was meant to help students integrate knowledge from engineering and architecture to try to solve challenges associated with the design of sustainable buildings for a changing climate.

Through theory lectures in the first phase, students were introduced to different possible materials and techniques. Lectures also served as a source of inspiration through the discussion of case studies which used natural materials in similar climatic contexts. In the second phase, a hands-on workshop was conducted to teach students low-tech building techniques for using natural materials like straw, reed, clay, seaweed, etc. As part of the integrated design process, students were taught how to calculate LCA and LCC for the building materials they proposed in their design using LCCbyg and LCAbyg, which are Danish software. These tools helped them make informed decisions relatively early on-during the design-sketching phase. During the design process, students also had to calculate: the U-value of the wall sample they were to design and build, thermal conductivity and moisture formation. Their designs needed to consider fire-proofing and resistance to water. Moreover, aesthetics and acoustics were also design criteria considered in their final evaluations. All calculations needed to comply with Danish building regulations.

2.2 The Experimental Living Lab's 1 : 1 Model

Despite the often daunting new environmental, social and economic conflicts that are arising, there is a parallel growing trend in architecture education which teaches responsibility towards society to help develop innovative and holistic environmental solutions(Hagy, et al., 2017; Keyson, et al., 2016). Living labs and dare-to-build concepts are used as new tools for developing integrated and holistic education programs in sustainable architecture(Hagy, et al., 2017; Masseck, 2017)). Experimental living labs are an opportunity for students to gain practice-based knowledge through innovative green design processes(Dabaieh, et al., 2018).

Based on students' design solutions, 8 groups of students (4 — 5 students per group) were challenged over the course of 4 working days on campus to build 8 wall samples in full scale (1 : 1) models. The idea was to try different wall compositions using only natural materials and a focus on resources locally available on Egholm island. These resources included materials like straw, reed, earth, seaweed, seashells and wood. The students were instructed on how to thatch reeds and straw on walls with the help of an expert craftsmen. The involvement of multidisciplinary experts in a hands-on building experience, was an additional component of the design process.

3 Results and Discussion

Integrated design within the architecture living lab fostered synergies between teaching, research and innovation for sustainable welfare buildings. Students presented their resilient designs for disassembly building solutions for the healthcare center. Fig. 1 and Fig. 2 show the final outcomes with a focus on the 1 : 1 model and Fig. 3 shows part of the calculation and simulation outcome.

Fig. 1 Students building part of the 1 : 1 wall model during the living lab experiment

3.1 Design Decisions and Material Use

Students showed in their designs that using plant-based materials such as reed,

Fig. 2 8 wall sections outcomes of the 1 : 1 wall model

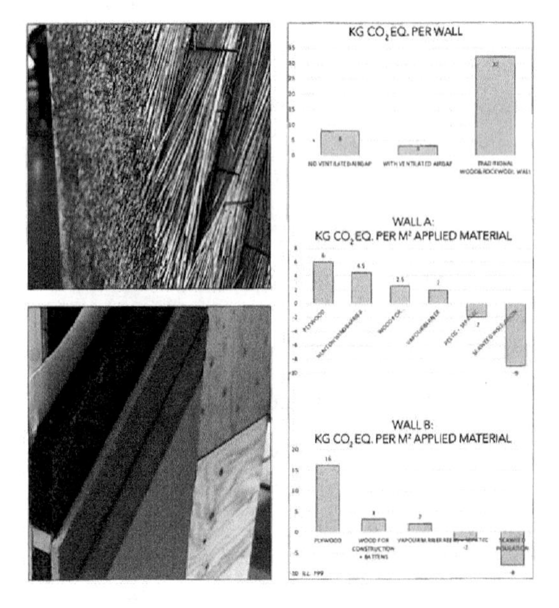

Fig. 3 Example of carbon emission calculations for different wall compositions

straw, seaweed and wood can help reduce a building's total carbon emissions. They proved through their calculations that materials like straw and reed capture carbon due to their sequestration properties in earlier life stages as plants. In many cases, their calculations also revealed negative carbon emissions in the life cycle of their buildings. The most successful proposals were designed as diffusion-open constructions, which not only improved the indoor air quality, but also resulted in a breathable construction without the need to use an air-tight vapor barrier. Some students used seaweed as an external cladding or insulation, while others used straw and reeds as cladding or straw as insulation, which helped in the overall thermal performance of the walls and roofs alike. Some designs used a mixed composition of rammed earth and Adobe together with straw as insulation which also enhanced the buildings' thermal performance both in summer and winter as earth naturally has heat-regulating properties. Other proposals used shells as a ground cover or capillary break layer since shells are hard, water resistant and light weight.

Students who integrated materials like straw and seaweed as insulation managed to reach a better thermal performance as the thermal conductivity could reach up to 0.07 W/mK for walls and 0.056 W/mK for roofs with adequate wall and roof thickness ranges from 40 to 55 cm. The LCA calculations were varied among students. Some were not able to find accurate and reliable values for straw, reeds and seashells as there are few natural materials databases for embodied carbon emissions. This resulted sometimes in illogical figures for the buildings' total carbon footprint, revealing one of the challenges of proving the benefits of using natural materials. The same held true for LCC calculations, where students had to do an inventory investigation to build their own material library in LCCbyg.

3.2 Design and Construction Challenges

This design studio is based on a "real world" challenge for a design-build project. As it varies significantly from typical studio-based architectural design, the challenges were diverse. Students faced several obstacles along the way, from the integrated design process to the final hands-on construction methods. As part of the second semester of their masters, they learned about the existence of natural building materials and the alternatives to reinforced concrete and industrial bricks. Students commented that they felt like they were learning architecture from the beginning and they were used to wood as the only natural material to use. Using diverse natural materials, they had to learn how to draw and represent their design through accurate construction detailing. They discovered that there are other precautions to consider since natural materials behave completely different than industrial ones. Moreover, they found it difficult to source reference drawings for construction details for contemporary architecture using natural materials. A lack of accurate updated materials databases was another big challenge to find values and mathematical calculations to use in their building modeling and simulations and for building energy or thermal performance, respectively.

Other challenges in planning this integrated and experimental living lab include the cost of materials. Natural materials in some contexts are more expensive than industrial materials. In Aalborg City in Denmark, where this design studio took place, Portland cement is more accessible and cheaper than reeds, straw and seaweed. Although this is counter-intuitive provided the abundance of reeds, straw and seaweed around the Island's shore, it was hard to find an adequate material supplier for treated reed bundles, seaweed-compressed boards or straw bales suitable for construction. Getting help from skilled craftsmen and laboratory assistants who have knowledge in building with natural materials using both traditional and modern techniques was yet another challenge as skilled labor was both hard to find and expensive. Finally, 1 : 1 model work needs space both to store the materials and for construction. Fire precautions also sometimes hinder conducting such experimental living labs on campus. The after-design studio work needs additional space for showcasing and storing students' work for future exhibitions.

The physical 1 : 1 experimental model work could be taken to another level by using a developed step of material lab testing, in order to get robust measurements of thermal conductivity and thermal transmission and accurate U-values for walls. This would help verify the simulation and mathematical calculations made by students throughout the design process.

4 Conclusion

This study demonstrates how an experimental, project-based learning studio experience, can use a hands-on 1: 1 living lab model as its main pedagogical tool. In this design studio, students gained a better understanding of vernacular materials and vernacular building techniques. They were tasked to design a building that could be disassembled due to the challenge posed by SLR. In this process, they learned about material performance and practical issues related to handling and treating natural materials. They were in direct contact with local craftsmen to learn the essence of traditional building methods. The outcomes of the design solutions and models proved that using natural local materials for contemporary designs is feasible. Moreover, the design solutions offered by the students comply with local standards and proved to be better for the environment in terms of emissions and environmental impact.

This experimental design studio equipped participating students for their future as careers as architects with the skills needed to design using natural materials. In particular, this design studio added the ingredient of tangible experience, which is more than just the direct physical contact with materials, but also exposes students to decision making processes, designing, and building in a real place with real challenges. They now have both the confidence and knowledge to start using low-impact materials as a norm in building professions. And despite the existing misunderstandings around the use of vernacular methods-especially using natural materials-in our contemporary architectural practices, this study hopefully contributes to the proof that it is possible.

Acknowledgements

The author would like to thank the whole teaching team, students and workshop assistants for their collaboration in the design studio.

References

BARRON B, SCHWARTZ D, VYE N, et al., 2007. Doing with understanding: lessons from research on problem and project — based learning. Journal of the learning sciences, 7:271.

BELL S, 2010. Project—based learning for the 21st century: skills for the future. The clearing house: a journal of educational strategies, issues and ideas, 83(2):39-43.

BROWN R, MAUDLIN D, 2012. Concepts of vernacular architecture. The SAGE Handbook of Architectural Theory.

DABAIEH M, 2015. More than vernacular: vernacular architecture between past tradition and future vision. Lund: Media-Tryck.

DABAIEH M, ELMAHDY D, MAGUID D, 2018. Living labs as a pedagogical teaching tool for green building design and construction in hot arid regions. Archnet-IJAR, 12.

DUNKIN MJ, 1983. A review of research on project-based learning. Higher Education Research & Development.

HAGY S, SELBERG P, TOUPS L, et al., 2017. Dare2build bt-living labs: design and assessment of sustainable living. KEYSONDV, GUERRA-SANTINO, LOCKTOND. Cham: Springer International Publishing.

HANSENHTR, KNUDSTRUPMA, 2005. The integrated design process (IDP): a more holistic approach to sustainable architecture. Action for sustainability, 05.

Keyson DV, GUERRA — SANTINO, LOCKTOND, 2016. Living labs: design and assessment of sustainable living. Living Labs: Design and assessment of sustainable living.

MASSECK T, 2017. Living labs in architecture as innovation arenas within higher education institutions. Energy procedia, 115:383-389.

NAKADA A, KOBAYASHIM, OKADA Y, 2018. Project-based learning. Journal of the Medical Society of Toho University.

NUNAN J, 2010. The complete guide to alternative home building materials & methods: including sod, compressed earth, plaster, straw, beer cans, bottles, cordwood, and many other low cost materials. Atlantic Pub. Group, Ocala, Fla.

PIESIK S, CHRUSZCZOW T, SOUCH C, et al., 2017. Habitat vernacular architecture for a changing planet. New York: Harry N. Abrams Inc.

TURAN M, 1990. Vernacular architecture: paradigms of environmental response, ethnoscapes; 4. Avebury, Aldershot.

WEBER W, YANNAS S, 2014. Lessons from vernacular architecture.

The Builders of Timbuktu Earthen Architecture

Ali Ould Sidi

Timbuktu Heritage, Malian

Abstract *The Builders of Timbuktu Earthen Architecture* deals with Timbuktu earthen architecture, challenges, threats and future. If focuses on Timbuktu's old architects called masons who've been protecting and preserving Timbuktu tangible heritage since the fourteenth century. They are deeply linked to Timbuktu historical monuments and are very proud about the how-know transmitted from their ancestors. Indeed, they have been struggling for centuries to preserve Timbuktu monuments, its OUV and got both national recognition as Mali Alive Human Beings Tresor in 2008; and an international recognition from the world heritage committee during its 40 session in Bonn, Germany in 2015.

Nowadays they are facing crisis and challenges due to both new construction materials like cement-iron, urban development, insecurity, lost of how-know. In fact tremendous architectural changes are being done inside the medina of Timbuktu like modern concrete buildings brooking down both the integrity and authenticity of the city, so far earth is being replaced by stone as alhore, cement and iron. The cultural property "Timbuktu" itself is threatened by the new buildings, covered by bright colors like red, yellow used on walls. Today, no dought about the commitment of Timbuktu masons to maintain and enhance the authentic architecture of the city, however, there are facing a series of challenges as follows:

— The jenne feray or koyra ferray: a round stone generally used for conical and pyramidal shapes such as mosque minarets and some mausoleums is disappearing,

— With new materials introduced like cement and iron, the knowledge and traditional know how of the mason are daily abandoned,

— With new styles and plans buildings, the requirements of urban development, the phenomena of insecurity and globalization, we notify a new generation of masons ignoring the use and the conservation of earthen architecture.

Keywords earthen architecture masons, how-know, threats, challenges

1 Introduction

The writing and the popularization of this work entitled *The Builders of Timbuktu Earthen Architecture* is for us a way of fulfilling a duty vis-à-vis the corporation of the masons of Timbuktu or Al-Banna-taray which for almost a quarter of a century, have been voluntarily committed, to our side, to preserve and enhance the cultural heritage of Timbuktu for the sustainable conservation of its architecture.

2 Methodology

It consisted of drawing inspiration from our book entitled *The Builders of Timbuktu Earthen Architecture*, itself composed of eight chapters as follows:

— The perception of the mason through the literature of Timbuktu,

— Literature review on Sudanese architecture,

— The types of architecture,

— History of masonry in Timbuktu,

— The role of masons in the maintenance process of mausoleums and mosques,

— The intangible heritage of the mason,

— Conservation problems , challenges and perspectives,

— Conclusion.

In relation with the topic chosen, namely the case study of Timbuktu, the methodological approach followed consists in emphasizing respectively the second and eighth chapters of the book. Thus, this article will deal with the reasons which contributed to the choice of the thematic local materials of architecture in the land of tombouctou.

As part of a preventive management of the traditional architecture of Timbuktu and to perpetuate the knowledge and traditional know-how of the mason, the study will formulate a serie of recommendations tending to the maintenance of the outstanding universal value of the cultural heritage Timbuktu through durability of materials and capacity building of masons.

3 Reasons for the Choice of the Topic

The masonry corporation known as Al-Banna-Taray in Timbuktu or Barry-ton in Djenne is made up of masons who are at the masonry corporation known as Al-Banna-taray in Timbuktu or Barry-ton in Djenn is made up by masons who are both feared and respected, because they are not only holders of knowledge, of secrecy, but, they are also upstream and downstream of the life of the communities: they are the builders of our houses in this world here, also the gravediggers of our graves for the afterlife.

Holders of traditional knowledge and skills that have consolidated and perpetuated the traditional practices of maintenance and conservation of earthen architecture, the masons of Timbuktu are characterized by a deep attachment to the cultural values inherited from their ancestors. For these reasons, their lives are also linked to the physical and functional health of the historic monuments of Timbuktu, world heritage properties to which they are fundamentally attached.

In Mali, seven natural and legal persons were proclaimed "Living Human Treasures", by the Ministry of Culture during the celebration of the National Cultural Heritage Week in April 2008. In the lot, there were the Hamane Hou families. (Sankoré district) and Koba Hou (Djingareyber) of Timbuktu, which are masonry corporations famous for their knowledge and know-how in the field of earthen architecture. The Masonry Corporation is an integral part of the Timbuktu community and features prominently in Timbuktu's three criteria for inclusion in the UNESCO Register. Thus Criterion (v) states that: "*The three mosques and mausoleums are exceptional testimonies of the urban establishment of Timbuktu, of its role as an important commercial, spiritual and cultural center on the southern borders of the Trans-Saharan trade route, and of his characteristic traditional construction techniques*". Today, the commitment of masons to maintain and enhance the architecture of Tombouctou land has earned them the special attention of the international community, such as the recognition of the World Heritage Committee for its award to the Timbuktu Mason Corporation at its 40th session in June 2016 in Bonn, Germany. One of the objectives of this book is to preserve the patrimonial dimension of all the traditions of maintenance of the mosques of Timbuktu through sustained actions in favor of the built heritage. The aim is to collect, to make known and to transmit the traditional practices of masons tending to preserve, to safeguard and to perpetuate the Timbuktu earthen architecture. Following the urban development of Timbuktu and its occupation in 2012, the masonry corporation of this city recorded many changes related to the introduction of new building materials, new styles and plans, the requirements of urban development, the phenomena of insecurity and globalization, as well as the advent of a new generation of masons who do not master traditional know-how.

The debate that the author invites his readers addresses the major challenges facing the bricklayer and architecture of Timbuktu, namely the loss of traditional knowledge and know-how, threats due to the introduction of new materials such as cement and iron.

4 Building Materials

4.1 Rocks

About the stones used in the construction, there are three types in Timbuktu:

• The jenne feray or koyra ferray: a round stone generally used for conical and pyramidal shapes such as mosque minarets and some mausoleums.

• The Toubabou ferray or French stone dates from the French period. For this category of stone, there are two molds: a rectangular wooden mold 50 cm long and 20 cm wide for walls 60 cm thick or first level of the two-storey house, then another 40 cm mold long and 20 cm wide for the wall of the floor. For the making of these bricks, a wooden mold is 50 or 40 cm is filled with banco kneaded and already prepared, as soon as it takes a tougher consistency, the mold is removed. The bricks are dried for several days in the sun and are then used as ordinary bricks for the construction of walls.

• Limestone or alhore: a whitish rock that outcrops in a radius of 20 to 30 km around Timbuktu. This stone according to its color is used for the outer facade of houses, to build pillars and reinforcements or to make latrines, it will change the architecture of Timbuktu and made a stone architecture. In the title still building materials feature prominently:

-Sand,

-The simple banco quarried in the quarries of Kabara and the banco-black of deep bottoms which comes from Koriomé about 12 km south of Timbuktu,

- Gum arabic which comes from acacias and used as adjut,

-Shea butter, a vegetable fat extracted from the fruits of shea, used to prepare the earth ocher Bourem, also used as adjuvant,

-Rice straw and baobab powder are used to enrich the banco,

-The agile extracted from the ponds used to make the bricks,

-Three types of wood are used in Timbuktu, namely:

* The gaulettes or stems of euphorbia that are arranged in the form of branches perpendicularly to the beams of the covers and participate in the ventilation of the roof and also serve as a kind of spring between the r niers and the roof, also they soften the weight of the roof and ensure the durability of the house.

* The Hasu wood or maerua crassifolia found on the pyramidal forms of mosques with a double importance; reinforcement of the built structure and serve as a natural scaffolding during mosque maintenance work.

* The okum wood used to make Timbuktu doors and windows in a special workshop called Diam Tend or blacksmith shop.

4.2 Windows and Doors

The Djenn -inspired Timbuktu earthen architecture has been enriched with new additions: the alhor stone discovered on the spot and the Moroccan craftsmanship that embellishes the local domain with doors and windows with very expressive patterns and varieties ranging from the most sober at most richly decorated(UNESCO,2015).

4.2.1 Doors

The entrance door and the facade, in the medina of Timbuktu, reflect the habitat and the social condition of the occupant. It is generally better held than that of the interior of houses. That's why front doors are usually well done. Previously it was still a single big, massive, studded fighter to discourage potential looters in an area prone to raids. Nowadays the doors are of different qualities with wooden locks and are presented as follows:

The decorations are variable depending on the type of place. In residential homes, the traditional door is established at the time of the marriage of the owner. In these types of doors, the decoration consists of four figures representing the dove and placed at the four ends of the door; the dove symbol of femininity and peace and the number four reflects the four legal wives that a Muslim can have. From the mouth of each dove comes out studded washers that are the flock of children to whom the woman can give birth. The doors of places of worship — mausoleums, mosques and oratories — are quite different; there is only one flying symbol of the oneness of God. The doors of the mausoleums have never been decorated and the knocker did not exist or consisted of a single metal round and very reduced. Moreover, in these places of devotion, there were never figures representing the dove but figures representing the crescent moon and a star, iconic figures of the Muslim religion.

5 Testimonials from Travelers and Visitors

Building materials are generally a function of the components of the physical environment, the building styles and the creative genius of the Master Mason. The building materials are varied and are described by travelers and historians who visited the bilad es-sudan at different times. Ould Sidi recalls that Leon the African who visited Tombouctou in 1512 gives the following description of the houses: "*the houses of Tombouctou are cabins made of piles plastered with clay with straw roofs. In the middle of the city is a temple built of stone masonry with lime mortar by an architect of the Betique*"(Ould Sidi, 2008).

Réné Caille (Caille, 1965) who visited the locality three centuries after Leon the Africans, describes the materials and techniques of construction of the city of Essadi. He even likens it to Jenné: "*The house they had given me for housing was not yet finished, so I had the opportunity to observe how to build masons in the country.*" One digs into the city itself a few feet deep, there is a gray sand mixed with clay with which one makes round-shaped bricks that are put to dry in the sun, these bricks are similar to those of Jenné. Masons work with as much intelligence as Jenné. Building materials have been enriched by the cultural contributions of the Sudanese metropolis with the Maghreb, Africa south of the Sahara and Europe. Thus in the 20th century, Paul MARTY(Marty,1920) former Administrator of the former French Sudan gives details on the adobe and mud now used in the architecture of Timbuktu: "*the building materials are sometimes this mixture of clay soil and mud straw said banco, the rammed earth of southern Algeria, sometimes of the simple clay of the country kneaded into small brick balls*".

Gaudio traces the variety of construction techniques according to styles imported from the Maghreb and Europe "the Tombouctienne house reflects its dual originality Sudanese and Maghreb". In the Middle Ages, thirteenth century, the immigrants of Birou and Djenné and Diaqui introduced the technique of Sudanese construction in banco. Then in the fourteenth century with the arrival in Mali of the Architect of Granada Ibrahim Es-Saheli, we adopted the art of building and decorating in Maghreb style. In the colonial era, the French introduced new materials which were not widespread because of the high price.

6 Problems of Conservation and Perspectives

6.1 Conservation

Timbuktu sites do face enormous constraints related to insecurity, socio-economic even and physical environment, among which we can mention:

• Deterioration of the building due to humidity by stagnation of the water at the base of the walls. This phenomenon is dangerous for the stability of the structure of earthen architecture of the city because of the absence of a drainage system.

• The silting of streets and sites due to the advance of the desert also to a social phenomenon that consists of replacing the fine sand inside the houses on the eve of the big holidays.

• Maintaining the authenticity of the site: use quality local materials to ensure sustainability (banco de koriom, stabilized with baobab powder, domiers or chawey, boundou as rôniers).

• Maintaining this authenticity also means reinforcing human capacities, enhancing the Mason's know-how , and creating a school for trades.

It is a question of ensuring that these

mosques and mausoleums that are cultural property of humanity can benefit from the contributions of modern technology while preserving their architectural originality. Consequently, in improving this traditional cultural work, one must always seek to maintain a balance between the Right to Development and the Duty to recognize the past and its transmission to future generations.

• The available funding has too often been allocated to "improvements" such as sound, lighting, ventilation, construction of new elements such as mihrab outside San-kor) to the detriment of a treatment of funds such as the replacement of degraded elements or the repair of plaster.

• Some conservation challenges also stem from the fact that available funding has too often been allocated to improvements that deform the integrity of historic buildings.

Today, new threats of various origins continue to weigh on this architecture:

• Anthropic threat related to religious considerations introduced by the fundamentalists during the occupation in 2012 who consider the existence of mausoleums and their function as symbols of idolatry;

• Slow but apparent changes in the Medina's physiognomy following structural dysfunctions perceptible through two new phenomena. The first of these phenomena is the mineralization of Timbuktu or transformation of the architecture into earth by the alhore or limestone; The second phenomenon is the ferralisation or construction of grotesque buildings with new, aggressive materials with bright colors both in the Medina of Timbuktu and in the buffer zone around the Medina.

• One has the impression of experiencing a concrete effect of concrete, a form of open competition is being made between the inhabitants themselves responsible for the conservation of classified property. In some cases, it's like being in Manhattan, California, as evidenced by the names on the stores in the Medina.

6.2 Perspectives

Timbuktu has always been a pole of attraction and even a favorite area for corporations, including masons with knowledge, skills, secrets and living cultural practices. This part of Mali has been marked many times by the seal of these masons through their colossal works represented by various architectural works that contributed to both the golden age of Timbuktu and its current cultural renaissance. Unfortunately, this "golden age" has faded for several reasons, including:

• The invasion of local markets by new construction materials such as cement, iron, which daily contribute to the mineralization and ferralisation of Timbuktu, thus distorting the urban landscape of Timbuktu and also leading to the loss of the traditional know-how of builder.

* In the media, the opening of Timbuktu to the modern world by TV and satellite dishes is often accompanied by foreign cultural aggression threatening the art, and even the authenticity of the corporation of masons who are more tempted by new buildings than the maintenance of earthen architecture. However, this corporation needs the support of cultural authorities and international institutions in charge of heritage preservation. In this regard, the future University of Timbuktu can envisage the creation of cultural centers of development where all the crafts and heritage craftsmen will be grouped: workshops or Tendé Hou masonry, carpentry, apprenticeship centers and conservation of earthen architecture can serve as a stimulus to economic activity and the creation of income and employment.

As for the prospects, the maintenance of the outstanding universal value of Timbuktu and its status as a world heritage city requires the updating of the Timbuktu Management Plan to integrate the management of natural, conflict and human-induced risks.

Under the current context characterized by a threat to the profession of the mason, its survival and that of its traditions, the author advocates their rehabilitation to the requirements of the modern world, their ability to maintain a balance between modernity and tradition, then urges a balance between the right to development and the duty to preserve heritage and pass it on to future generations.

References

CALLIE R, 1965. Diary of a trip to Timbuktu and Djenné, Volume II. Paris: Edition Atropos.

GAUDIO A, 1988. Mali. Paris: Edition Karthala.

MARTY P, 1920. Studies on Islam and the tribes of Sudan, Volume II. Paris: Edition Ernest-Leroux.

OULD S A, 2008. Le patrimoine culturel de Tombouctou: enjeux et perspectives. Bamako: La sahelienne.

UNESCO, 2015. Etude porte secr te de Sidi Yehia.

The Vernacular Architecture in Centre of Italy After the Big Earthquake

Salvatore Santuccio, *Enrica Pieragostini*

Università di Camerino, Scuola di Architettura e Design "E. Vittoria", Italy

Abstract In 2016, a serie of strong earthquakes hit central Italy, many towns were destroyed, some even completely. These towns mostly consisted of very old buildings made of local stone that crumbled under the shock. The quality of these towns was defined by the buildings in the urban space and the vernacular architecture of the individual houses. Before the earthquakes, VERNADOC Italia had studied precisely this quality with survey campaigns organized by the authors with students at the School of Architecture in Ascoli Piceno. After the terrible event and the destruction of many of these homes, the problem arose as to how to reconstruct these towns and in particular how to prevent them from becoming depopulated, since they are mostly located in the mountains in disadvantaged circumstances with respect to Italy's larger production and economic system. The residents immediately exhibited a desire to recreate the urban spaces and houses that were present before the earthquake, reconstructing the town "where and how it was" because the vernacular architecture of those places was a symbol of the places themselves. This reconstruction, however, is very difficult and economically very expensive. To support it, a real willingness to economically revive those places is needed. This is more a political and social problem than an architectural one. In Italy, however, there have been interesting examples of regenerating semi-abandoned mountain towns, for example, Riace in Calabria, where the town has been reborn by integrating migrants from Africa and Asia. This essay describes the problem of reconstruction using direct evidence of these events and involving the mayors and residents of the destroyed towns. With respect to images, those made in the affected towns during the VERNADOC camps alternate with images of the destruction caused by the earthquakes to recount the serious loss of architectural and environmental quality.

Keywords vernacular architecture, earthquake, reconstruction, drawing, symbols

1 The 2016 Earthquake and the Survey in the Historical Centers

In 2016, a strong earthquake hit central Italy, many towns were destroyed, some even completely. The tremors occurred for several days and were succeeded by more significant events for more than six months. The first earthquake occurred on 24 August 2016 at 3:36 am with a magnitude of 6.0 and an epicentre in the Tronto Valley between Accumuli and Arquata del Tronto; it strongly damaged historical cities such as Amatrice. On 26 October 2016, another two earthquakes were registered, the first at 19:11 with a magnitude of 5.4 and the second at 21:18 with a magnitude of 5.9, both centred on the Umbria-Marche border and affecting many towns in the Province of Macerata, some of which, such as Visso and Ussita, are very important.

The strongest tremor was recorded on 30 October 2016 with a magnitude of 6.5 and an epicentre between the towns of Norcia and Preci. Finally, on 18 January 2017, a new set of four earthquakes of magnitude greater than 5 occurred with an epicentre in Abruzzo. This entire series of events provoked 41,000 evacuees, along with 388 wounded people and 303 deaths. In addition, it damaged an exceedingly large area of small mountain towns stretching up to 150 kilometres.

Many of these towns boast a thousand-year history. Amatrice is recorded in the

Farfa register beginning in 1012[①]; Visso certainly existed in the Roman era[②]; Norcia is a small pre-Roman city conquered by the Romans in the third century BC[③]; Arquata del Tronto, albeit with uncertain origins, belonged to the Flavi family in the first century AD.

In sum, this entire area developed in the Medieval era with a compact, articulated urban morphology characterized by stone houses with wooden beams, along with important defensive bastions, sighting towers, and imposing cathedrals and civic buildings. Both residential and public buildings therefore combine to create great historical and cultural value, an undeniable vulnerability that was at the base of very serious damage in the area[④].

VERNADOC Italia documented these historical towns in central Italy well before the earthquake occurred, in collaboration with the School of Architecture at the University of Camerino in Ascoli Piceno. These extremely detailed surveys were shown to be essential in managing the knowledge of the buildings destroyed following the earthquake and imagining their reconstruction.

In 2015—2016 a survey campaign was made of the Medieval buildings in the high Macerata area[⑤]. Three types of buildings were studied: castles, fortresses, and sighting towers. The castles in the following towns were studied: Pitino, Torricchio, and Castelsantangelo sul Nera. The fortresses in Varano and Santa Lucia were studied. Finally, the guard tower in Visso was studied, along with towers in Carpignano, Castelraimondo, Castelfantellino, and Smeducci. As mentioned above, these are found near Macerata and were built by the duchies of Spoleto and Camerino.

In addition, the entire urban layout of Visso, a small historical town on the slopes of the Sibillini Mountains, was analysed. The survey of the city entailed the division of the whole area into urban sectors classified according to the degree of importance i.e., the presence of highly valuable architectural objects, leading to the following three sets: urban sectors in the "centre", "primary periphery", and "secondary periphery".

For each sector, surveys were made with direct and indirect methods and a laser scanner was used to survey the two squares of the centre, Piazza Pietro Capuzi and Piazza Martiri Vissani. Following the initial inspections, the research proceeded at the University of Camerino in Ascoli Piceno, where the data were processed through the use of graphical software and traditional

① I. Giorgi, U. Bazzani (Eds.), Il regesto di Farfa/compiled by Gregorio di Catino and published by R. Società romana di storia patria, Rome, 1879, 1914.

② The oldest and surest archaeological evidence recording the existence of a settlement in the Roman era is a funerary epigraph of a freedman dated between the first and second centuries AD. The freed slave evidently pertained to the Horatia tribe situated near Spoleto.

③ The city's foundation probably dates to the fifth century as the work of the Sabinis, who in the zone of Norcia identified today with the toponym Capo la terra, were located in the most northerly outpost of the territory they controlled. The name Norcia is likely related to the Etruscan name Northia after the Roman god Fortuna. The city was conquered by the Romans at the beginning of the third century BC, obtained Roman citizenship in 268 BC, combined with the Quirina tribe, and allied with Rome in the war with Carthage.

④ "According to an estimate by the National Association of Italian Municipalities (Associazione Nazionale Comuni Italiani, ANCI), around 200,000 buildings were damaged or made unusable in the areas of central Italy hit by the earthquake". See Agi (Agenzia Italia) Cronaca, 12 December 2016, at https://www.agi.it/cronaca/terremoto_centro_italia_i_numeri_del_sisma—1213113/news/2016-11-02/

⑤ The study of defensive architecture in the duchy of Camerino produced ten investigations differentiated by type: castles, fortresses, towers, and defensive walls. Two-and three-dimensional reproductions of the data were made. For the 2D data, the scale of representation ranged from 1∶500 to 1∶10. Descriptive sheets were created with bibliographic and archival references. This type of documentation led to a historical framework of each architectural object. The 3D data were processed to create 3D printouts, which was very laborious in terms of both modelling and printing.

representation techniques. These two survey campaigns were enriched with a very careful study of some architectural details such as doors and decorative elements made using the manual survey techniques characteristic of the VERNADOC culture.

After the 2016 earthquakes, a large part of the works that had been surveyed were destroyed or strongly damaged. The city of Visso experienced enormous damage and, in particular, many buildings in the two above-mentioned squares collapsed. With regard to the towers and defensive walls, many experienced serious damage and, in particular, the towers in Castelfantellino di Ussita and Castelsantangelo sul Nera practically disappeared from the urban landscape of their respective towns.

Fig. 1 Piazza Pietro Capuzi, in Visso represented before earthquake (January 2016) and after (February 2017).

These tragic events demonstrated the importance of surveying the buildings and cities and the survey campaigns were made available to those charged with reconstructing the damaged towns.

At the same time, new plans were made to survey the heavily damaged historical centres such as Visso with the idea of comparing the situations before and after on paper. In less-damaged towns, the idea was to acquire data on the current state in order to understand the historical heritage and preserve its memory in case of possible future developments(Fig. 2).

In Visso, a thorough survey campaign made with drones, a Topcon laser scanner, and traditional methods produced a vast amount of data that, when compared with pre-earthquake data, constituted an essential basis for reflecting on the reconstruction

Fig. 2 Damages in historical center of Visso: Prospective views of point cloud in rgb by laser scanner and drone, 2017.

of the historical centre.

2 Detect to Rebuilt

Many towns declared themselves open to the VERNADOC camps in their urban territories. These included Offida and Grottammare, which were identified as the next areas of study for VERNADOC camps in Italy, based on the experience made in Monteprandone, a town also affected by the earthquake, albeit marginally (Fig. 3,4).

Fig. 3 Historical gates in Monteprandone drawn by students of school of architecture of Ascoli during the Monteprandone Vernadoc camp 2019.

Indeed, in 2019 in Monteprandone the second VERNADOC camp in the Marche Region[①] was held; the experience was considered a very important opportunity for its future. At the end of it, the town's mayor declared: "the surveys were used by the administration, gathered in a publication, and will be used again for graphical operations to develop tourism. In addition, these works will be used as a memory archive for future generations."[②] But Monteprandone also made a substantial survey campaign of the historical centre composed of the VER-

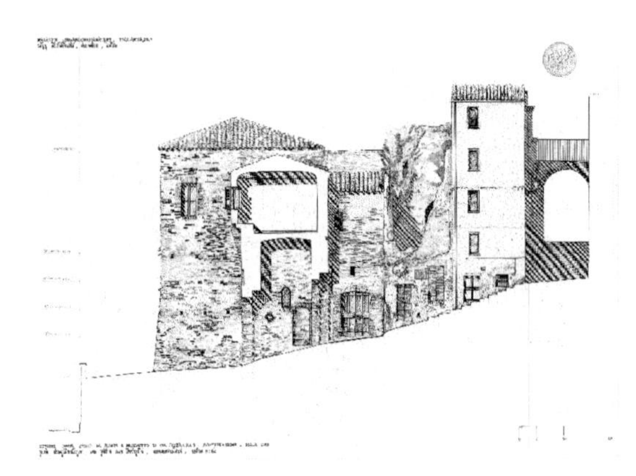

Fig. 4 Elevation of a building of Monteprandone drawn by Watanyoo Chompoo' Shivapakwajjanalert during the Monteprandone Vernadoc camp 2019.

NADOC experience and also with the use of terrestrial photogrammetry, direct surveys, and drones.

The same procedure was followed by the City of Offida, which also conducted a preventive knowledge campaign by surveying the building and architectural consistency of its historical centre(Fig. 5).

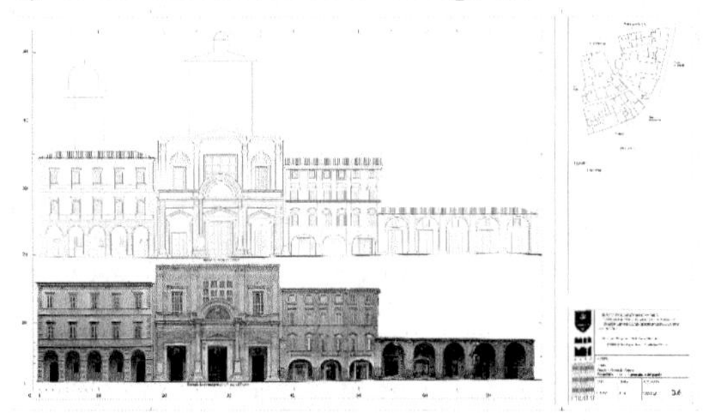

Fig. 5 Elevation of Piazza del Popolo in Offida, 2018.

In this case, a survey was made using a Topcon laser scanner and terrestrial photogrammetry. A VERNADOC campaign was also used for some important decorative details on buildings in the centre, which constitute a specific quality of this small city.

The most interesting case of post-earthquake surveys, however, regarded Arquata del Tronto[③]. Here, the earthquake caused an array of problems and very specific possible solutions, and two universities - Camerino and Roma Tre — contributed to studying both the documentation and possible reconstruction(Fig. 6).

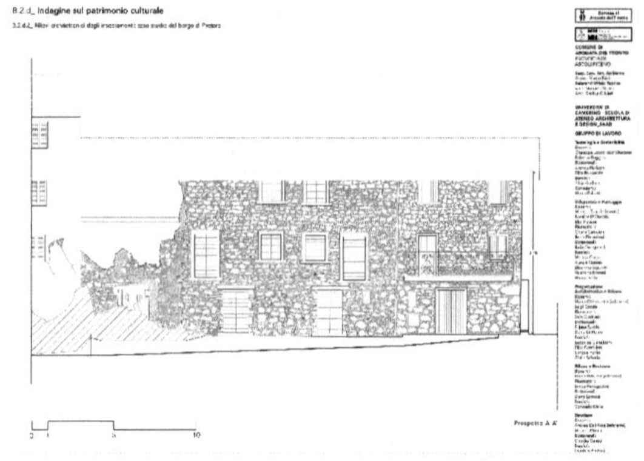

Fig. 6 Survey of damaged center of Pretare, an hamlet of Arquata del Tronto, 2018.

The Municipality of Arquata del Tronto encompasses many small hamlets[④], some of which experienced absolutely irreparable damage. The best known is Pescara del Tronto, for which the impossibility of rebuilding it where and as it was publicly declared to be impossible. In the newspaper Manifesto on 1 September 2017 we read this terrible sentence: "Pescara was the village most affected after Amatrice: the tuff and sandstone houses were literally flattened by the reinforced concrete roofs. Of nearly 200

① The first VERNADOC camp in the Marche Region took place in Amandola, 7-21 March 2015.

② Interview cited in A. Coccia, Italian Vernadoc, studenti da tutto il mondo a Monteprandone. Stracci: "Utile per le future generazioni", in Riviera Oggi, 17 January 2019, see https://www.rivieraoggi.it/2019/01/17/277332/italian-vernadoc-studenti-da-tutto-il-mondo-uniti-a-monteprandone-stracci-utile-per-le-future-generazioni/

③ An extensive bibliography exists regarding Arquata del Tronto. Some important recent texts regarding this city include: E. Giorgi, G. Paci, Storia di Ascoli dai Piceni all'etǎ Tardoantica, Ascoli Piceno, Lìbrati, 2014. S. Scacchia, Un gioiello marchigiano a pochi passi da noi. Un tesoro immerso tra due parchi., in L'Araldo Abruzzese, Anno CVIII, no. 2, 29 January 2012, p. 14.

④ The hamlets in the Municipality of Arquata del Tronto are: Borgo, Camartina, Capodacqua, Colle, Faete, Pescara del Tronto, Piedilama, Pretare, Spelonga, Trisungo, Tufo, and Vezzano. These are small, sparsely populated mountain villages, but in cases such as Pescara del Tronto, Spelonca, and Trisungo, are rich in ancient architectural buildings of great prestige, some of which were destroyed in the earthquakes.

residents, 49 did not survive the earthquake. Many remained under the rubble for hours before being rescued; others managed to flee in time only to see everything crumble. Now, passing near Pescara, one sees nothing, only rubble: a town demolished in the night by the first earthquake and then pulverized by the tremors in October and January. It is impossible to bring it back to life..."①. The mayor of the Municipality of Arquata, which oversees Pescara del Tronto, clearly declared: "We already knew that Pescara would never be rebuilt where it was before. Of course we hoped to the last minute that something would change, but the technicians have confirmed that it is not possible to recreate everything where it was"②. But for the other hamlets and the town of Arquata itself, the situation is slightly different although very compromised. Indeed, some hamlets such as Colle, Spelonca, and Trisungo, while experiencing enormous damage, are still considering how and where to rebuild, and the same can be said, while with greater caution, with regard to the administrative town of Arquata.

The reconstruction that should take place in these places is not what has been adopted as yet. Indeed, as of today, only small emergency homes have been built, complicating the relationship among the residents and only postponing the debate on how to rebuild these places. As a resident of Trisungo declared: "I miss my town so much. I get up in the morning and hear the sound of the bells, the chatter of passers-by under my house, the sound of the fountains flowing, the river, and the scent of burning wood in fireplaces ..."③. This need was, up to today, gathered only by addressing documentation on the current state, making a survey campaign with drones and laser scanners in order to trace the certain borders of the transfiguration of these historical centres and aiming to understand if and how to restore the urban quality present before the earthquake.

3 Conclusions: to Rebuild Buildings and to Rebuild People

The theme is obviously very complex and affects not only the building texture of the towns, but also the awareness of their inhabitants and the psychological repercussions they have undergone④.

There are two main issues involving these places. On the one hand, consideration of the opportunity to rebuild means, first, understanding the extent of the damage and if possible, comparing it to the prior documented state. This means understanding how much of the old town still exists, how much can be restored, and how much, once rebuilt, risks creating a historical, perhaps useful, imitation in which the drawings of building details acquired over the years play a very important role⑤. The historical imitation would be based on restoring that vernacular architecture that constituted the overall quality of these towns, but al-

① M. Di Vito, Ricostruzione impossibile, addio Pescara del Tronto, in Il Manifesto, 1. 9. 2017.

② Idem.

③ Idem.

④ "In a destroyed territory in which the face and style of daily life has changed, psychological support should be a focus of rebirth. [...] No one has assumed the responsibility of scrutinizing the psychological risks that can be triggered in the health of individuals affected by traumas such as the violent earthquake of 2016, ensuring an itinerant team that could periodically visit the people demonstrating a need. Various types of intervention are possible. The development of trauma could be based on the memory and narration of people through experiential or expressive laboratories, with valid individual and group psychological support". Ibid.

⑤ Massimiliano Tonelli already writes some days after the earthquake: "There are three possible choices for post-earthquake reconstruction". Building "where and how it was"(a hundred-year motto from Venice when the bell tower of San Marco crumbled), building "where but not how it was", or "building somewhere else and therefore creating a new city that only preserves the residents and name of the old town". M. Tonelli, Terremoto e ricostruzione. Vogliono trasformare Amatrice in un outlet? in Artribune, 28/8/2016.

so focusing on the vitally important seismic safety of the new buildings. This attitude boasts illustrious precedents. Perhaps the best known example is the total reconstruction where and as it was of the historical centre of Warsaw, which was razed to the ground by the Nazis during the Second World War①. Many people oppose this solution with justified and historically based reasons, instead trusting the historical tradition of the seventeenth-century reconstruction in southern Italy. Here, the towns destroyed by the earthquake of 1693 were rebuilt, often in different places, producing works of Sicilian baroque art such as Noto, Ragusa, Granmichele, Modica, Scicli, etc. ②, defining the reconstruction itself as an opportunity for contemporary architectural culture.

The answer is not simple and should vary according to the different situations. Regardless, any type of restoration that centres on rebuilding a system of urban and social relationships similar to those that have characterized these small villages necessarily requires knowledge that starts with careful surveys.

The second main issue, which is related to the first and no less important is the progressive depopulation of these places. The younger generations especially are discouraged by the reconstruction and tend to leave such places for more certain, more secure situations where the new generations can grow. This area thus sees a progressive increase in defections, it sees people closing their houses, and economic and production activities and take shelter in faraway places. These towns are left only to elderly people; this is another serious risk that neither architecture nor surveys can provide for.

Provisions for revitalization can be made on different levels to counter this abandonment and tediousness. One of the most interesting, which has also been discussed in the Marche Region, is the approach taken in Riace, Calabria, wherein the mayor, Mimmo Lucano, has experimented with integrating refugees and migrants in city life, offering them housing and work in exchange for reactivating some of the town's production activities and rebuilding a social life③

The resource of new residents and their desire for participation may be a flywheel in rebuilding towns undergoing abandonment, thus inverting this trend and favouring an urban reconstruction based on the fundamental graphical record of architectural quality.

---

① "From the start of the rebuild, the city's own rubble was utilised in the reconstruction process, and original fragments of Old Town buildings were recovered. Rubble from the former ghetto district was used to produce new bricks for the modern quarter, while architectural details from demolished buildings in the Old Town were put on to the reconstructed facades," explains Małgorzata Popiołek, an expert in heritage conservation at the Technical University of Berlin. While much of this work was done by construction workers and specialised builders, Małgorzata says local people were required to help clear the vast amounts of debris. 'The entire nation builds its capital "became the city's rallying cry. When the rubble that was to hand would not suffice, more material was imported from neighbouring ruined cities. And to ensure it was all put back in roughly the right place, Bellotto's cityscapes were used as references for key locations". See The Guardian, Story of cities #28: how postwar Warsaw was rebuilt using 18th century paintings, 20 May 2016 at https://www.theguardian.com/cities/2016/apr/22/story-cities-warsaw-rebuilt-18th-century-paintings

② See Giuffrè M. (2007). Barocco in Sicilia, Verona, Arsenale; Nobile M. R., & Piazza S. (2009). Architettura del Settecento in Sicilia, Storie e protagonisti del tardo Barocco, Palermo, Edizioni d'arte Kalós; Trigilia L. (2007). Un viaggio nella valle del barocco, Pantalica, Siracusa e le città del val di Noto "patrimonio dell'umanità", Catania, Sanfilippo.

③ "For decades emigration drained life from Riace, a village of 2,000 on the Calabrian coast. When a boatload of Kurdish refugees reached its shores in 1998, Lucano, then a schoolteacher, saw an opportunity. He offered them Riace's abandoned apartments along with job training. Eighteen years on, Mayor Lucano is hailed for saving the town, whose population now includes migrants from 20-some nations, and rejuvenating its economy. (Riace has hosted more than 6,000 asylum seekers in all.) Though his pro-refugee stance has pitted him against the mafia and the state, Lucano's model is being studied and adopted as Europe's refugee crisis crests". Fortune World's Greatest Leader, 2016. At http://fortune.com/worlds—greatest—leaders/2016/domenico—lucano—40/.

Comprehensive Protection of Living Heritage — Traditional Stilt-Style House in Jingmai Ancient Tea Garden of China

BI Yi[1], *ZOU Yiqing*[2]

1 Beijing Tsinghua Tongheng Urban Planning & Design Institute, Beijing, China
2 Beijing Guo Wen Yan Cultural Heritage Conservation Center, Beijing, China

Abstract Ancient Tea Garden of Jingmai Mountain is in Yunnan Province. It was listed in National Priority Protected Site and World Heritage tentative list in 2013. "Stilt style" (Two-storied pile) dwellings in the tea gardens built by native ethnic groups of Blang and Dai, as a significant composition in the heritage, reflect harmonious relationship between human and land. On the one hand, these residential houses are valuable site that represent regional culture which requires protection; on the other hand, obsolete building style cannot afford modern demands. The conservation working team chose to stay on site and discuss the project with the locals one by one, providing solutions depending on one's living demand, cultural conventions and religious belief. Besides, the team kept providing training service including preservation theory and repair techniques for local craftsmen and administrators to reactivate the traditions in the community. This article focuses on how the working team consider those buildings as living heritages, analyze their "value characteristic element", then seek the balance between protecting and retaining the locals to keep the cultural heritage in the whole conservation process.
Keywords conservation, cultural landscape, tea garden, traditional house

1 General Situation of the Traditional Stilt-Style House in Jingmai Ancient Tea Garden

Jingmai Ancient Tea Garden is located in Pu'er, Yunnan Province. It is one of the oldest, and largest cultivated ancient tea garden cultural landscape. Jingmai Ancient Tea Garden and its tea cultural system was nominated as the Pilot program of Globally Important Agricultural Heritage Systems by FAO in 2012, and was listed in the Preliminary List of China's World Cultural Heritage in the same year(Fig. 1).

Fig. 1 Aerial view of Jingmai ancient tea garden
Administration of Jingmai Ancient Tea Garden

In Jingmai Ancient Tea Garden, there are 10 traditional villages. Dai and Blang people live on this land for centuries, they are creators and inheritors of Jingmai tea culture. The lifestyle, customs, religion and land of these people, are the main influencing factors that shaped this cultural landscape. Therefore, these traditional villages are parts of the most important value carriers of this heritage(Fig. 2).

Fig. 2 Aerial view of the Nuogan village in Jingmai ancient tea garden

Dai and Blang ethnic dwellings, as the

main elements of traditional villages in Jingmai, are all made of wood, and build in the Stilt style, which means the second floor is elevated (built on stilts) and enclosed by wooden walls as main living space, the ground floor is just for storage. One side of the main building is usually equipped with a balcony as a space for drying tea leaves, the roof is sloping roof covered with black Burma tiles. Dai and blang ethnic dwellings are similar in structure, but differ in architectural details. For example, the ox horn shaped roof-decoration is used on Dai house, while the Blang use tea leaves shaped decoration instead. Due to the damp climate, these houses need to rebuild every 50 years or so.

Most of the existing traditional houses in Jingmai were built in the 1970s and 1980s. With the gradual change of people's lifestyle and living requirements, the form of traditional houses has undergone an evolving process. Most early residences are straw top with very deep eaves, basically don't set any window, and just use some furnitures or curtains to divide indoor space simply. As a result of the straw roof is easy to catch fire, it was gradually replaced by Burma tile. The plane form and facade shape of house also slightly changed by degrees with process of the modern way of life was introduced into the village, "two generation style", "three generation style" house gradually appeared — the height of wooden walls increased, windows, and partition walls also appeared. However, the big Xieshan style roof, elevated second floor, open fire-pit in the living room, balcony for tea leaf drying and the special roof decorations — these forms that adapted to the local climate, traditional living habits and beliefs of the residents — have been continued(Tab. 1).

2 Protection and Continuance of Traditional Stilt-Style House in Jingmai

These traditional houses carry value of this cultural landscape, meanwhile, they are closely related with villager's daily life, bound to be in a constant process of evolution as lifestyle changes. How should they be protected while they constantly change? The key is to understand their value based on respect of the evolution process, treat them as "living heritage".

Tab. 1 The "evolution" of the traditional house in Jingmai

Early traditional house: without any window on the wall, use straw roof.	
"First generation style" house: straw roof was substituted with Burma tile roof.	
"Second generation style" house: the second floor became a little bit higher.	
"Third generation style" house: the second floor became higher, and windows appeared.	

Evolution means that change is allowed, but at the same time, evolution is different from mutation, there is a need to maintain some "unchanged parts" to ensure qualitative change won't happen. For vernacular architecture of ethnic groups who live in Jingmai Mountain for generations, to achieve equilibrium between variants and invariants, it requires a synthetic analysis of different types of factors in terms of social, cultural, economic, technological and material, including architecture space, form, and relevant human activities; it also requires a comprehensive analysis of how these factors influence architecture, as well as the results they caused so that the attributes — so called "characteristic elements"— which should be protected and emphasized can be defined, and then other attributes can be regarded as elements which are allowed to be changed and improved along with the growth of life demands.

2.1 Protection of the Characteristic Elements

Technologies of "vertical utilization of land", "forest understory planting", and traditional settlement construction of Jingmai residents, reflect the wisdom of local people to make use of land efficient and harmonious, reveals the harmonious co-prosperity of ecological ethics, and accorded with the core essence of tea culture. These are where the heritage's values lives.

From the traditional house, characteristic elements in which the value reflected can be found. These elements can be summarized as having the following characteristics:

— Can reflect ethnical and regional characteristics, or closely related with the locals' traditional belief, custom: for example, roof of big Xieshan style, special roof decorations.

— Can reflect the harmonious relationship between man and nature: for example, the form of built on stilts, which means the ground floor is not paved, plants can grow there.

— Can reflect the relationship between man and the tea culture: for example, all the "tea space", such as the balcony where the tea leaves are dried.

Then all the characteristic elements of the traditional house can be listed (Tab. 2).

Tab. 2 The characteristic elements of the traditional house

Characteristic elements	Illustrations
The form of "built on stilts".	
Ground floor is used for tea making.	
The house is built on the floor directly, without change the landform much. The floor is usually unpaved (entirely or partially).	
Balcony beside the main house, is used to dry the tea leaves.	
Open fire-pit and "sacred pillar" in the living room.	
Staircase is directly connected with the exedra which can be used for tea drinking room.	
"Chuan Dou" form of structure, The appearance is simple and primitive	
All structure is made of wood, use timber walls and floors.	
Roof of Xieshan style, eave without rises at the tail.	
The roof is very big that can provides shade.	

Continued

Characteristic elements	Illustrations
Use of Burma tile, without eave tile and drip tile.	
Roof decoration: "tea leaf" style for Blang while "ox horn" style for Dai.	
Use of rough local stones for column base. The appearance is natural and crude.	

These characteristic elements are the essence that should be kept "unchanged", in this way, then make sure the traditional house won't "mutate".

2.2 Reasonable Improvement of the Traditional House

The main function of dwellings is to provide a safe and comfortable living place for people. With the continuous improvement of people's living standard, the understanding and requirements of "comfortable and livable" are also changing. The protection of traditional houses still in use should not only pursue the preservation of forms, but more importantly, protect the culture and life they carry. If one overemphasizes the preservation of forms at the expense of the normal and reasonable living conditions of the residents, long-term effective protection cannot be truly achieved. Especially for Jingmai ancient tea garden, only by retaining the aboriginals and making them live and work in peace and contentment, continuing to develop their traditional culture, can truly possess Jingmai's soul and vitality.

Due to its remoteness, backwardness and long history of cultural and economic underdevelopment, Jingmai residents have always had low requirements on living conditions. Even today, many houses still fail to meet basic safety and applicable standards. In addition to the damage caused by disrepair, there are also the following major problems on the traditional house:

Because of the lack of structure and the aseismatic knowledge, combined with rough materials and construction methods, many houses have some birth defects, such as poor structural stiffness, insufficient effective area of Fangs (a type of horizontal tension member, tiebeam), reduction of the columns in the wrong way, etc. directly affect the safety.

The living space is too low and undivided, basic comfort and privacy cannot be guaranteed. Many villagers have no beds and still sleep on the floor, poor indoor lighting and heat-insulating property, bad sanitary, made the condition even worse.

The ground floor has insufficient space height. In the past, the ground floor is used as hog pen or storeroom, so the ground floor doesn't have to be high(another reason is that low height can keep the house more stable), but now, residents need to make tea and park the tractor there, so the space height is not enough.

Lack of toilets and basic sewerage facilities, as well as livestock sheds on the main floor of some dwellings, have led to highly unsanitary(Fig. 3).

Fig. 3 Typical living situation in the traditional house of Jingmai

In recent years, with the development

of Jingmai tea economy, the economic conditions of villagers have been generally improved. Consequently, the demand for improving living conditions becomes more and more urgent, In particular, the younger generation of villagers can no longer accept the formerly humble and dirty living space. However, due to their lack of knowledge to improve the traditional residence with the modern construction method organically, they attempt to abandon traditions completely, replace traditional houses with concrete buildings, as a result, the survival of traditional house is under unprecedented threat. In this condition, if one still arbitrarily requires the villagers strictly maintain the original form in stead of allowing any improvement, residents will find ways to abandon or even destroy these traditional houses, finally, demise of them will be accelerated as a result.

We keep asking ourselves a question in the course of conservation work — would we want to live in a house like this? Therefore, the working team always takes improving people's livelihood as important as protecting traditional houses. We believe that on the premise of protecting the characteristic elements as far as possible, other forms should be allowed to be "variable", and appropriate adjustment and development should be made according to the problems that residents want to improve most, so as to meet the basic usage requirements, such as:

— Allow and even guide the residents to strengthen and improve the building in view of the structure defects. For example, adding additional columns and Fangs, use of diagonal bracing in concealed places to reinforce the connection between the roof truss and the lower main structure, use of "Bamboo seismic structural walls", to improve the building's stability and seismic performance(Fig. 4).

— On the premise of ensuring the stability, raise the height of ground floor and upper rooms, to meet the convenience of living activities and furniture arrangement.

Fig. 4 Use of diagonal bracing when necessary, to improve the seismic performance

— Improve the way of timber walls and floors, reduce the gap between planks, and enhance the effect of warm keeping and windproof.

— Add some transparent roof tiles and removable windows, to improve the lighting condition(Fig. 5).

Fig. 5 Adding of transparent roof tiles

— Put bathroom and kitchen in the ground floor, construct water supply and drainage system, to make the living condition more comfortable and sanitary.

— Allow residents to use some light partition walls in the main room, select proper filling materials in the walls to enhance the thermal insulation and sound insulation performance.

— Encourage villagers to send their livestock to a pen that can hold all the livestock together instead of breed them under the living room, to improve the sanitary condition.

Besides, the reasonable individual needs of each villager should be fully considered in the process of conservation. For example, in the Wengji village of Jingmai, there is one family with two members who suffer from Nanism, the stair of their house is too steep for them to climb, and they always want to change the slope of the staircase. After learning this situation, the working team redesigned the stair for them in the project, the householder is quite touched after knowing that. By doing this, the house became more responsive to the needs of each household so they are willing to live in the traditional house. Moreover, a relationship of mutual trust has been gradually built up between the conservation experts and the residents, they are more understanding and supportive of our conservation efforts.

3 Conclusion

Through the traditional house conservation project in Jingmai, the working team has a deeper understanding of the complexity of vernacular heritages. By working on site for several years, the more contact we have with the environment and people there, the more we understand that the residents' life is not easy, they eager to improve the quality of living. Besides, we have a better comprehension of the most precious and impressive essence of this cultural heritage. From the begining till now, our views on many issues are changing, some decisions broke the previous practice, sometimes even seemingly broke through the principle, but after serious thinking over and over again, we insisted that the change and breakthrough are correct and reasonable. The principle didn't change, it is just the understanding of the principle is developed with the process of knowing the objects better and better. The "unchanged parts"— namely the characteristic elements — can be protected more efficiently only when some other parts are allowed to be changed, so that the traditional house can keep evolving, keep alive. When villagers happily move into the renovated house to continue their traditional life and culture, the heritage will continue to glow with vitality and luster, then the purpose of protection was achieved(Fig. 6,7).

Fig. 6 Comparison of the NO. 4 house in Wengji village before and after renovation

Fig. 7 The villagers held a banquet to celebrate the completion of the house renovation

References

Editorial committee of County Annals of the Lancang County, 1996. County annals of the Lancang county. Yunnan: People's Publishing Press of Yunnan.

GAO Yun, 2003. Traditional house of Tai in Yunnan China. Peking: Peking University Press.

Analysis and Simulation of Original Status: a Study on Traditional Villages Planning Method Focusing on Settlement Ontology Monitoring

LU Shijia

Shenzhen University the Institute of Urban Planning Design & Research CO., LTD., Shenzhen, China

Abstract For holistic protection of traditional villages, it would be required to focus on both the inheritance of the historic morphology of settlements and the continuation of their historic information, and also entail the maintenance of rural landscape patterns from the perspective of ecological civilization. Considering the protection objective, more specific requirements are raised for coordinating protection study and development planning of traditional villages and aligning technologically with their protection strategies as a result of the expansion of protection value distribution domain, and improvement of the quality of settlements' sustainable development underpinned by the Rural Revitalization Strategy of China. This study, by creating a comprehensive collection of factors in four dimensions including the site selection logic of historic settlements, agreed form of rural-site spatial resource distribution, attributes of public spaces, and morphological-ontology pattern, attempts to fully scan measure, and analyze the "original status" of traditional villages' settlement ontology. It also intends to reveal protection value factor attribute of the historic settlement and its density distribution characteristics, and on this basis, produce a vector data that can be used to give a quantitative description of the correlated states among the morphological, historic and ecological information of settlements. When formulating a sustainable development strategy for villages, we need to seek for an optimized decision based on the construction requirements by comparing and screening the protection value attribute and its structure density and morphological data. With Shiqiao Village of Danzhai County in china as sample, the study results show that this method is expected to provide a technical monitoring scheme that simulates the original status growth for the protective and organic renewal of traditional villages when the adaptive adjustment is allowed.

Keywords traditional villages, settlement ontology, holistic protection, original-status analysis, morphology monitoring

1 Study Background and Key Issues

1.1 Traditional Village Protection and Development Study Under the Rural Revitalization Strategy of China

Rural settlements(Jin, 1998) are a type of morphology where human building system is most tightly bonded with the natural system. In an era when ecological civilization (Huang & Chen, 2002) is given high priority, how rural settlements can play a dominant role of revitalizing rural areas is an issue of common concern to the whole society.

In the town-country magnet (Ebenezer, 2000) interaction, the heterogeneous value of rural landscape is one of the crucial pillars for villages to develop as a mainstay, especially for those traditional villages that "*preserve the traditional pattern and historical style, and collectively showcase the cultural or ethnical features of local buildings*". And their morphological value goes well beyond their land, spatial and landscape values. Traditional villages, however, are too weak to bear the brunt of construction impact as they are large in number, small in scale and scattered in distribution. Therefore, it is imperative to give careful thought to the logic in the technologies used for traditional village protection and development under the Rural Revitalization Strategy.

1.2 Challenges Confronting Effective Protection and Development of Traditional Villages in China

The interesting thing about human settlements is that they constantly adjusting and trading up their environment in their own capacity. In this sense, even those traditional villages that fall under protection must also be allowed to "grow" moderately. To align results of their "growth" with the protection principles, top consideration must be given to the core value of the object of protection. Hence, we need to further discuss two basic problems associated with the development of traditional villages under the protection framework.

1.2.1 Settlement-based cultural landscape is "a masterpiece created by human and nature". and the technical method for delimiting the scope of historical settlement ontology that reflects the core value of settlement-based cultural landscape needs to be further established.

Reasonably determining the protection scope based on the value density of heritage ontology is a prerequisite for an effective protection of settlement-based cultural landscape. As the interactive morphologies of settlement and natural system are similar in concept but different in concrete form due to a cumulative effect of history, appropriate measurement dimensions and precise technical methods are needed to support a discerning and careful resolving ability to delimit the protection scope of settlement ontology.

1.2.2 For historic settlements with prominent cultural traits, It is necessary to identify their value as a "living heritage" system, and make clear the logics in technology update for their protection.

Unlike the preservation of cultural relics, "living heritage" is a core concept of the protection goal of settlement-based cultural landscape. How the heritage value presents its "phase" is, more often than not, subject to the vitality of settlements' compound system. In other words, the function for the compound system of a settlement to continuously and steadily provide functional and emotional resources is the cornerstone for the settlement to maintain its living status. Therefore, it is essential to reveal the operating mechanism of settlement's living status.

2 A Study Method to Analyze the "Original Status"

2.1 Study Methods and Data

The study uses the "original status" to describe a core value set of the historical settlement ontology, and uses the "original status" identification and measurement of historical settlement ontology of traditional villages as the basis for planning their protection and monitoring. The study analyzes the morphological structure by inputting measured data of sample settlement morphology and the location data of historical morphology into software including GIS and CAD, and describes it by using the correlated factor set of historical information and morphological elements of traditional villages. The measurement factor collection, which is relatively stable in structure, can be used for a long-term monitoring of sequence system of the same sample.

The "original status" refers to a compound system comprised of natural geomorphology, social network and building system that are carried by the settlement ontology, run normally with historical stratigraphic features and bless villages. The study divides this complicated compound system into four formation factor sets, namely①the site selection logic of historical settlements, ②the agreed form of rural-site spatial resource distribution, ③ the attributes of the public spaces ,④the morphological pattern of buildings.

2.2 Applicable Conditions of the Study Method

The "original status" analysis is a method that gives a quantitative description of the associated features of the study object's historical and morphological information, and that graphically presents them. On the study front, information of the study object's "original status" shall be authentic with sufficient historical and morphological features to yield fundamental analysis materials. On the application front, the necessity for the study object to "grow" towards the original status should be sufficient, that is, the study object must be sufficiently significant or demonstrative in terms of cultural landscape value protection and maintenance of rural landscape security pattern.

3 Original Status Analysis and Simulation with Shiqiao Village in China as a Sample

3.1 Introduction to Shiqiao Village

3.1.1 Shiqiao Village's Cultural Landscape Feature is Markedly Suitable for Original Status Protection

Shiqiao Village is located in Nangao Township, Danzhai County, Qiandongnan Miao and Dong Autonomous Prefecture, Guizhou Province, where Danzhai County is populated by the Miao people, a Chinese ethnic minority group. Shiqiao Village consists of Daboji Miao Village, Zhijie Miao Village, Huang Miao Village and Sanju Miao Village distributed on both sides of the Nangao River, which is called a Shiqiao Community in general. Shiqiao Village gets its name from a geological landscape of natural stone bridge, equivalent to "Shiqiao" in Chinese, which arches across the Nangao River and is surrounded by Miao Villages. The exquisite stone bridge weakens river turbulence and connects two sides of the river, endowing people with a great habitable place, like God's providence, and making it the first unique value of Shiqiao Village.

Nangao River, with abundant water resources and good water quality, and bordered by lush paper mulberries, is where an ancient road on which the foot of Leigong Mountain leads to Qingshui River passes through. In Dayanjiao site, a cliff of about a hundred meters in width and eighty meters in height forms a shelter from wind and rain with its body leaning forward, and a clear weakly alkaline spring flows through this village. Ancient paper-making, an intangible cultural heritage of craftsmanship, has been handed down from one generation to the next. And all these constitute the second unique value of Shiqiao Village.

Nangao River is surrounded by steep mountains on both sides, and the Shiqiao Community beside the river covers an area of about 30 hectares, with an established area of historical sites about 1 to 3 hectares. The settlement is small but close-knit and fully equipped where farmers, craftsmen and businessmen live in harmony. Compact and rich traditional production and lifestyle that retains its original status is the third unique value of Shiqiao Village.

Mountainous building featuring column and tie wooden construction in Daboji Miao Village and Zhijie Miao Village is the fourth unique value of Shiqiao Village.

In a word, the area where Shiqiao Community is located presents a rich cultural diversity for which a low-impact and protective development path should apply.

3.1.2 Appeals of Shiqiao Village

Danzhai is a national-level poverty-stricken county, and its upper-level administrative unit, the Qiandongnan Miao and Dong Autonomous Prefecture, is an area of extreme poverty. The opening of the highway in close proximity to Shiqiao Village poured renewed vigor into Shiqiao Community where construction activities have stagnated. On the one hand, Shiqiao Village seeks for quick alignment with the modern industrial system for a boost in its growth; on the other hand, land and space must be planned and allocated to accommodate facilities for the modern tourist industry system and community public service system bolstered by the Rural Revitalization Strategy. However, as Shiqiao Village is highly sensitive in regional ecosystem and fragile in historical morphology, a sufficient planning of technical schemes is required to avoid constructive destruction to it.

3.2 Technical Test of the Original Status Analysis

The study measures the "original status" of traditional villages in four dimensions including the site selection logic of historical settlements, agreed form of intra-site spatial resource distribution, attributes of the public space and morphological ontology pattern of buildings.

3.2.1 Site Selection Logic of Historical Settlements

The study focuses on a data-based description of the dependence relationship between traditional villages and natural environment and summarizes the correlated factors of upland settlement site selection as listed in Tab. 1:

Tab. 1 Analysis factors and standard of upland settlement site

Categories	Analysis Factors	Level
Emergency and disaster prevention	Security of the site (seasonal flood level + seasonal variation of hydrology), Stability of the site (geological condition + slope + soil and water conservation condition);	3 2 1
Subsistence support (Cultural memory)	Backup land that is easy to farm (flat + perennial irrigation/water storage), Primary natural morphology that maintains remarkable traits of the landscape (elevation + slope + rocky landscape or vegetation landscape), Natural runoff collection irrigation system;	3 2 1
Complexity of construction	Topographic slope, Ground elevation, Basic types of local buildings.	3 2 1

Source: collected by the author.

Data analysis of geographic information can reveal the interval grading results of Shiqiao Village's site selection logic during historical period regarding different assessment factors. By analyzing the cross description of factor grading results, we can obtain the graphical representation of the morphometric measurements. The combined use of the factor set of site selection technologies for historical sites and the description of experience in historical texts will be conducive to the consistent use of regional original logic to simulate a natural historical "growth" status, so as to provide a technical basis for planning a necessary "growth" space. (Fig. 1-3)

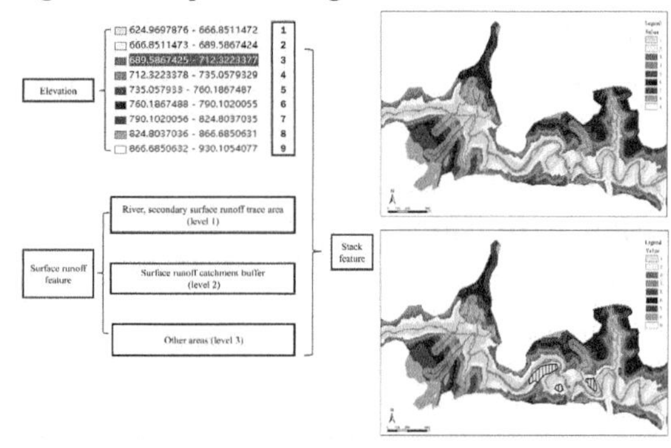

Fig. 1 Emergency and disaster prevention

Source: prepared by the project team

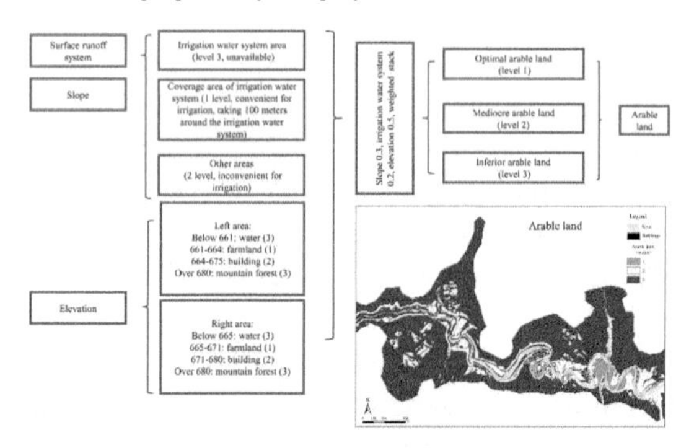

Fig. 2 Complexity of construction

Source: prepared by the project team

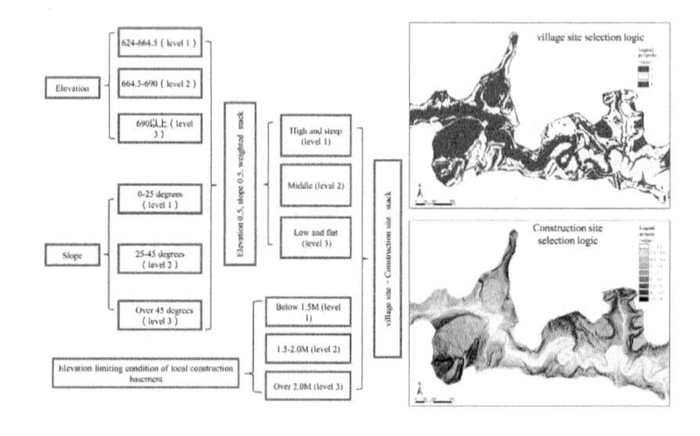

Fig. 3 Subsistence constraints

Source: prepared by the project team

3.2.2 Agreed Form of Rural-Site Spatial Resource Distribution

The agreed form of rural-site spatial resource distribution constitutes an integral part of the historical culture and social economy information in a region. The construction mode of spatial demarcation marker as a physical manifestation of econo-cultural forms, and its material evidence are at the core of the protection of historical and cultural heritage and also the holistic protection of cultural landscape in a region.

(1) Original Distribution of Spatial Resources

An empirical analysis on the chronological sequence of settlements' appearance and spatial geometry of settlement morphology shows that (Fig. 4), the Shiqiao settlement, with farming as an original means to earn a living, already had a clear spatial resource distribution agreement that imposes an absolute control over morphological transition in its historical formation and transition period.

Daboji Miao Village and Zhijie Village are two original growth units of the Shiqiao settlement. As shown in Fig. 4, two spatial power circles are simulated by using the two gates as the center and the distance between the two gates and the midpoint of a line connected by them as a radius. The legends in Fig. 4 clearly display the agreed form of rural-site spatial resource distribution between the two villages during the historical period. The Land Temple, Ancient Sweetgum Tree and Old Bridge, as spatial demarcation markers, are solid evidence of the agreement that regulates people's production and life in and between the villages and they have existed till today.

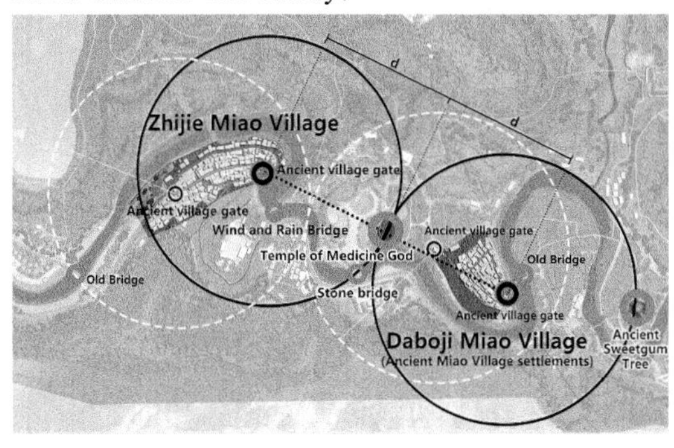

Fig. 4 Distribution of the existing markers as an agreed form of intra-site spatial resource distribution

Source: prepared by the project team

(2) Resource Distribution Form after Expansion and Growth.

As the population of the region began to wax and the original agricultural productivity failed to accommodate the additional population, some of the incremental demands that failed to be balanced and satisfied under the original spatial agreement logic would constantly seek and establish a new supplemental method in an interdependent reality. Taking the Shiqiao settlement as an example, a paper-making workshop (Dayanjiao workshop), was established outside the spatial power circle of Daboji Village in seek of supplementary support through handicraft, and a commercial street, Zhijie, was derived from it, both of which offer new resources for living and ensure a sustainable development of the settlements. From the settlement morphology shown in Fig. 5, we can see that the Scenery Village was propagated from Daboji and Huang Village from Zhijie. Afterwards, probably because of the intimate relationship between modern society and commercial civilization, or the inclusive feature of commercial settlements, public service facilities such as schools and village committees built lately in the settlement are in close proximity to Zhijie Miao Village, which is quite logical. The combined settlement morphology of Shiqiao Village that we see today has emerged from that moment on.

The settlement morphology and its intrinsic growth mechanism together constitute the "original status" of the cultural landscape of Shiqiao Village. The agreed form of rural-site spatial resource distribution is demarcated and inherited by the boundary-specific makers. Only when traditional villages with historical significance grow by simulating the original status based on their morphological information can a holistic protection of them be achieved.

3.2.3 Attributes and Morphology of Public Spaces

Periodic public behaviors and affairs in a settlement could, in historical period, solidify into a cultural landscape of the settlement that corresponds to a specific spatial bearing form and triggering rule. The cul-

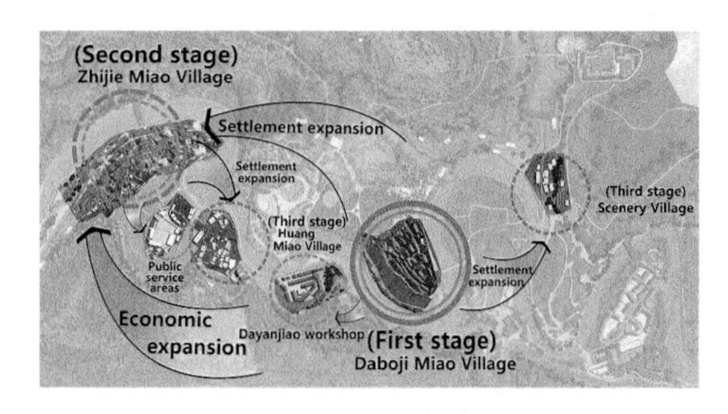

Fig. 5 Analysis of evolution of settlements and expansion of space
Source: prepared by the project team

tural symbolic meaning and level of publicity define the attributes of a public site. The way how the succession of the disappearing and updated parts of cultural connotation defined by the attributes of a public site presents itself is a mapping of changes in social values.

The study simplifies the mapping of the cultural attributes and highlights the inheritance requirements for their publicity level. This study maintains the highest publicity level of the public space defined (through documentary records or by word of mouth) in social lives of a settlement, for example, Shiqiao site, Dayanjiao workshop, and Zhijie Street . On the basis of the grading value (accessibility) of the public site obtained through spatial morphological analysis of the settlement, this study performs a publicity level-based identification, measurement and demarcation of site space, and develops a site publicity level of the original settlement through combination with the previous definition. The correspondence between site publicity level of historic morphology and publicity level of facilities in the public sites also becomes an essential basis to simulate the "growth" of the original status.

3.2.4 Morphology-Ontology Pattern of Buildings

Morphology-ontology patterns of traditional villages are a crucial achievement of traditional culture in academic inheritance. Subjective preference is inevitable when an adaptive pattern is chosen as criteria for design update based on the regional features and the settlement characteristics. However, the measurement levels of the historic value factors are based on the level of public spaces and inheritance requirements of their publicity level. This standard is relatively constant on the timeline. Hence, the protection factor about constraint conditions of architectural renewal design relies on the publicity and trait levels of morphological factor involved in the design object in the historic morphology factor set. The morphological factor set with a higher publicity level will have much more attribute factors in the original status measurement data than that with a lower publicity level. The same morphology factor will show constraint factors of different levels in different sections of the public sites with different publicity levels. Types of morphology factor sets that should be focused on are listed in Tab. 2 as follows.

Tab. 2 Analysis factors and standard of settlement morphology

Categories	Analysis Factors	Publicity value	Trait value
Public profiles of settlements	Interface outside the village building groups, landscape interface of waterfront coastline, enclosure surface of public vision (landscape enclosure surface of main vantage point), and sequence stack profile of settlement entrance's landscape	3 2 1	3 2 1
Streets system	Interface of settlement streets and important public site proportion	3 2 1	3 2 1
Combined form collection of buildings	Building (public) elevation form, roof form and its combined form, building plane form and its combined form, main structure type of regional building	3 2 1	3 2 1
Form collection of detail decoration	Regional materials and craft paradigms, colors, graphics, etc.	3 2 1	3 2 1

Source: prepared by the author

3.2.5 Study Conclusion of Shiqiao Village Sample Test

Through a sample study test of Shiqiao Village, this study analyzes the original status in four dimensions. We can completely produce technical data for the scanning and generation logic of the historical settlements in Shiqiao Village, thus laying a good foundation for building a factor set framework to simulate the historical original status growth in Shiqiao Village.

4 Planning Monitoring Strategy Focusing on Simulation of Original Status

The discussion on Shiqiao Village's sustainable development certainly pays attention to a renewal development path of long-term sequence. The factor collection that build on the original status analysis provide a basic data constraint condition closely related to the ontology protection value for the planning strategy, which helps further a discussion on the planning scheme that simulates original status growth in practice.

4.1 Delimitation of Historical Settlement Ontology

According to the analysis in 3.2.1 and 3.2.2, the land adaptability can be reviewed and corrected based on the contour line of overlaying area of the site selection level and agreed form of rural-site spatial resource distribution, and thus delimiting the historical settlements, which also represents the core landscape circle of the planning.

Landscape features including topographic relief, surface runoff trace and local vegetation texture should be strictly preserved within the scope of historical settlement ontology. It is forbidden to implement new construction and reconstruction projects on a large scale. Indispensable public service facilities can only be built after passing the test based on the technical solution of original status. It is required to designate tourism expansion circle and scenery related circle outside the settlement ontology, the function access conditions, landscape preservation strategy and facility construction or renovation conditions should be set up respectively (Fig. 6).

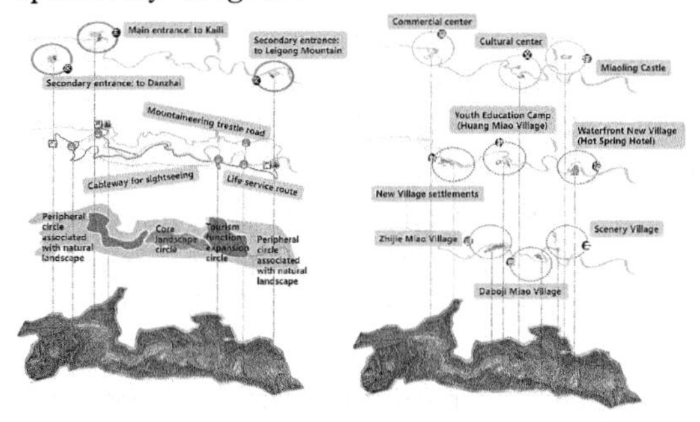

Fig. 6 Distribution of the three circles in the planning area of Shiqiao Village in china
Source: prepared by the project team

4.2 Production Space Synchronously Evolving with the Agreed Form of Spatial Resource Distribution

As traditional villages began to implant modern tourism, some parts of the land for agriculture, forestry and handicraft should be used for leisure tourism projects, which is equivalent to a redistribution of rural-site spatial resource during the historical period. The planning proposes that agricultural land within the scope of settlement ontology, namely the core landscape circle, should be retained in its original status, and the land used for handicraft should be retained at its original site. The land in the tourism expansion circle can be duly allotted for tourism. As for the land in the related circle of natural landscape (primarily the forestry land), priority is given to the maintenance of the original status of the side that highlights landscape characteristics. (See the analysis in 3.2.1)

4.3 Site Selection Related to the Inheritance of Its Attribute

A site selection of the Shiqiao Hotel, which covers an area of 1.5 hectares and an estimated construction area over 10,000 m^2, was accomplished under the planning, abiding by the basic principle of historical settlement site selection logic, the reservation principle of agreed form of rural-site spatial resource distribution and the inheritance principle of public site attribute (see the analysis in 3.2.1 and 3.2.2). The

Shiqiao Hotel project is a crucial part of the investment plan for tourism development programme. The developer selected a base near the stone bridge between Daboji and Zhijie Miao Village within the scope of settlement ontology, which went against the logic in the technologies for the original status "growth" and thus was not adopted according to the planning(Lu & Pan & He, 2018).

4.4 Building Form Protection Aligned with the Constant Form of Communal Cultural Memory for Public Sites

The planning proposed a grading requirement for how public sites are used and landscape is presented. The presentation of settlements' public site interface requires a grading control for related building system based on the publicity and trait grading results.

5 Conclusion

The study on the protection development strategy of traditional villages that is driven by the quality development of village ontology has grabbed wide attention under the Rural Revitalization Strategy of china. The constructive destruction that we are concerned about is usually caused by lack of accuracy class and data basis of the renewal technologies in different technical dimensions. The same project, if constructed with different strategies and constraint conditions, will have varying impact on the cultural landscape. The original status analysis of the settlements aims at providing an operable technical basis that covers a complexity description for the relatively constant characteristics of historic site morphology and public culture landscape memory form.

Traditional settlements are lively and diversified. The original status analysis provides a measurement method that helps locate the growth demands of villages on the path of "original status growth" based on facts, and this is probably a viable path for traditional villages to develop with inheritance in mind.

Acknowledgements

This study is funded by the National Natural Science Foundation of China: "Chinese Traditional Settlement Construction History and Presentational Protection of Original Status of Constructive Remains" (51678362).

References

EBENEZER H, 2000. Garden cities of tomorrow. Beijing: Commercial Press.

HUANG Guangyu, CHEN Yong, 2002. Ecological city theory and planning design method. Beijing: Science Press.

JIN Qiming, 1998. Study history of rural settlement geography in China and its recent trends. Acta geography sinica, 43(4): 311-316.

LU Shijia, PAN Heng, HE xiaotong, et al., 2018. Detailed construction planning of shiqiao village, Danzhai county, Guizhou Province. Shenzhen university urban planning and design institute Co. Ltd.

State Council of the People's Republic of China, 2018. Regulation on the protection of famous historical and cultural cities. Towns and villages, 7.

United Nations Educational, Scientific and Cultural Organization, Intergovernmental Committee for the Protection of the Cultural and Natural Heritage of Outstanding Universal Value, et al., 2015. Operational Guidelines for the Implementation of the World Heritage Convention.

Study on Strategies of Continuity of Regional Materials of Vernacular Architecture in Local Development — Taking the Cases of Jiaodong Peninsula as Examples

ZHANG Yun[1], *GAO Yisheng*[2], *XU Dongming*[3]

1 Shandong Provincial Conservation Engineering Institute of Vernacular Heritage, Jinan, China
2 Shandong Jianzhu University, Jinan, China
3 Key Scientific Research Base of Vernacular Culture Heritage Conservation
(Shandong Jianzhu University), State Administration for Cultural Heritage, Jinan, China

Abstract The regional materials have been widely adapted into the construction of vernacular architecture, shaped the local environment, and formed the identity of the place. They also play an important role to balance the conflicts between the preservation of vernacular architecture and local development. The paper focuses on the problems of the continuity of application of regional materials in Jiaodong Peninsula in China based on the field investigation of the existing cases of vernacular architecture in the area. It demonstrates the comprehensive assessment of regional construction materials including timber, stone, adobe, straw and so on. Thereafter, it puts forward the feasible strategies of continuity of regional materials of vernacular architecture in local development in the present transformation period in China.
Keywords regional materials, vernacular architecture, Jiaodong peninsula, strategies of continuity, local development

1 Introduction

With China's rapid urbanization since the twentieth century, many vernacular villages are facing the accelerated disappearance of vernacular architecture and the increasingly serious problem of "one thousand villages look the same". Protecting and sustaining vernacular architecture in the contemporary development to mitigate the adverse impact, including its damage, is a significant topic in the conservation of vernacular cultural heritage. Vernacular architecture follows the principle of taking advantage of local resources and contains the wisdom of human's adaptation to the geographical conditions of his residence. Regional materials are the vital carrier of the regionality of vernacular architecture, impacting the emergence, development, and evolution of vernacular architecture craftsmanship. Regional materials also represent the direct responsiveness of vernacular architecture to the environment. Over the past three years, the author has surveyed 96 vernacular villages in Jiaodong Peninsula and interviewed many local craftspeople. Based on this field study, this paper classifies and analyzes the regional materials of vernacular architecture and proposes the strategies on the continuity of regional materials of vernacular architecture amid development difficulties.

2 Overview of Regional Materials

Located in the east of Shandong Province, Jiaodong Peninsula is the largest peninsula in China. Facing the ocean on three sides, the peninsula is in the warm monsoon climate zone, boasting a long coastline. Though dominated by mountains and hills, it has coastal plains in the northwest and a vast area of shoal land along the coast. Such local natural environment and geographical location impact on the choice of regional vernacular architectural materials. The quantitative comparison, induction and summarization of survey data indicates that the regional materials for vernacular architecture can be classified into timber, stone, straw, adobe, as well as marine resources.

2.1 Timber

2.1.1 Sources

Jiaodong Peninsula is rich in forest resources, there is a large area of deciduous forest in the warm zone. The main species of trees include oaks, Chinese pine, catalpa wood, Chinese toon, poplar, etc. Residents usually plant a few trees in front of or at the backyard of the house to not only decorate the courtyards and create cool shade in summer, but also to possibly use them to make beams or furniture to reduce the cost in future house building. Wealthy families may purchase finer timber from south China such as spruce, sandalwood, and red pine. During the field studies, the author also found the practice of making beams with bamboo in the field research, but the examples were few and not systematic.

2.1.2 Uses

Timber is mainly used in beam frames, doors and windows in the vernacular buildings of Jiaodong Peninsula.

Beam frame is the main load-bearing structure in the vernacular architecture. To avoid easy fracture, timbers selected for beam frame are required to be long and straight with fine texture. The commonly used timber for beam frame includes pillar, Chinese toon, and pine. The trunks of such timber are processed and used for beams and pillars and their branches are used to make intermediate purlin, rafter, or other components. Bracket set, a typical flexible connecting component between column and beam also called "Dougong" in ancient Chinese architecture, is only used in the buildings of family temple and it is made of hardwood such as Chinese oak.

Hardwood with beautiful texture and high toughness such as Chinese locust is selected to make door and window frames and planks, and gratings. The exquisite woodcarving decorations of door thresholds and dangling flowers feature the regional vernacular architecture of Jiaodong Peninsula (Fig. 1).

Fig. 1 The exquisite woodcarving decorations of door thresholds © Zhang Yun

2.2 Stone

2.2.1 Sources

Sturdy and durable, stones are resistant to salt spray and weathering, and easy to obtain at a low price. They are therefore a commonly used building material in traditional construction. Jiaodong Peninsula has a rich variety of granite which can be divided into hill-stone and sea-stone according to their geographical distribution. The mining of hill-stone is mainly in the hilly areas such as Zhaoyuan of Yantai. Sea-stone is mainly produced in the eastern and southern coastal areas.

2.2.2 Uses

In Jiaodong Peninsula area, stone is generally used in foundations, walls, cornices and outdoor steps to bear and maintain the load. Large pieces can be processed to make ashlars and blocks, while small pieces can be used to build a stone wall or as a filling material. The combinations of stone, adobe, brick, and straw constitute the different practices of regional stone masonry.

The color change of the stone brings different color feelings to the vernacular architecture. The stone used in the coastal areas of Jiaodong Peninsula is mostly white, gray, blue, and of other cool colors and the stone used in the central hilly area is mostly in colors of warm red or yellow.

Stone is also used in vernacular rural architecture for decoration, i. e. stone carving, which is also a common type of carving in the vernacular construction of Jiaodong Peninsula. The survey data reveals stone carvings are primarily seen in socle, the drum stone, and the stone for tying horses with granite as the main material in light and monotonous pattern. In temples or more elaborate buildings, exquisite stone carvings are seen on the plinth, head porch, gable board, thresholds and outdoor steps.

2.3 Straw

2.3.1 Sources

Straw grass includes natural mountain grass and seaweed, as well as crop straws such as sorghum straw and wheat straw produced during farming activities.

Seaweed, also known as seaweed grass, is a special building material used in covering of roofs of the vernacular buildings of Jiaodong Peninsula. Seaweed grows in the offshore bays of tropical and temperate waters within the depth of 3-6 meters. The fallen seaweed in autumn and winter is pushed to the shore by the waves (Cong, 2009). The seaweed becomes purple-brown after dried. Containing a lot of salt and pectin, dried seaweed is moisture-proof and won't get burned easily, having good thermal insulation performance at low cost. It is thus an energy-saving and environmentally friendly roofing material (Chu, Xiong &Du, 2012).

Mountain grass is a straw resource that grows in mountainous or hilly areas, mainly including thatch and sorghum, etc.

2.3.2 Uses

In vernacular architecture of Jiaodong Peninsula, straw is commonly used as roofing materials. Weihai and Laizhou are the main areas featuring seaweed thatched houses in China with the towering roof. The roof boarding in the regional roofing practice employs the bundled mountain grass or sorghum straw on rafts. Straw is also used in wall materials. It is mixed with yellow mud to fill in the cracks of walls or make adobe bricks.

2.4 Adobe

2.4.1 Sources

Adobe is a common, easy-to-obtain, and basic building material in vernacular buildings. It is mainly used as bonding material and base material for producing other building materials. In addition to being used for building walls or roofing plaster, it can also be mixed with water and wheat straw before being pressed into adobe bricks.

2.4.2 Uses

Adobe plaster was the most common bonding material in vernacular buildings before cement and lime become widely applicable. Yellow mud is the main component of the adobe plaster because it is most viscous and relatively firm when used for bonding and jointing in making walls.

Fig. 2 The wall made of adobe bricks badly damaged by weathering © Zhang Yun

Adobe bricks are building materials that can be processed on site by using simple molds. Due to the distribution of regional resources, adobe bricks are made of pure loess or mixed with wheat straw, gravel, sand, and shell fragments. Because it has not been fired, adobe brick is vulnerable to weathering, and the wall made of such materials is less stable (Fig. 2). For this reason, adobe bricks are less used in the load-bearing wall of a main building. In coastal areas with heavier humidity, adobe bricks are hardly seen among building materials.

Jiaodong Peninsula boasts developed salt industry which consumes a large number of timbers. Because of this, workshops for firing clay to produce fired bricks are rarely seen and fired bricks are also poor in salt spray resistance. This explains why fired bricks are less used in vernacular buildings in Jiaodong Peninsula than in other places (Yang, 2011). Other adobe fired components include tiles and cresting.

2.5 Other Marine Resources

Other marine resources seen in vernacular buildings in Jiaodong Peninsular include sand and shell, which are often mixed with mud to produce adobe bricks. In the coastal plains of the northwest, sand is used to fill the foundation, and the pillars are inserted in the sand foundation to keep it firm and from sinking.

Shells can also be used as building decoration materials. Studies have shown that in the late Ming and Qing Dynasties, people in Jiaodong Peninsula used oyster shells to decorate exterior wall for soundproof and moisture-proof, which is not only suitable for the humid environment of coastal areas but has the functions of practicality and aesthetics (Huang, 2014). However, the author didn't see relevant examples in the field study and investigation.

3 Overall Assessment of Regional Materials

3.1 Advantages of Regional Materials

The field research data and the interviews with the craftsmen reveal that regional materials used in vernacular architecture in Jiaodong Peninsula enjoy the cost-effective advantages that are summarized as follows.

3.1.1 Low Cost

The low cost of regional materials is attributed to abundant resources, low price, and low processing cost, which increases the flexibility of the building cost of vernacular architecture.

Take wall materials as an example, the author notices four types of walls: whole stone wall, adobe brick-stone wall, fired brick-stone wall, and fired brick and adobe brick stone wall (Fig. 3). In terms of material cost, from the highest to the lowest is fired brick, stone, and adobe brick. The thickness of the wall made of stone and adobe brick is usually over 40 cm while the thickness of the wall made of fired brick is usually 37 cm. The material acquisition cost determines the proportion of building materials.

Fig. 3 One type of stone masonry wall in Yantai city © Zhang Yun

Fired bricks are used more often in coastal area thanks to the convenient sea transportation in this area. By contrast, stone is used much more often than fired brick in central hilly area.

3.1.2 Regulating Indoor Environment

The advantage that regional material is enjoying of regulating indoor environment is closely related to its vernacular architecture technique. Vernacular architecture consists of five main parts: foundation, wall, beam frame, roof, doors and windows with wall and roof as external and enclosed structure.

As the wall types have been discussed before. Wall is usually made of inner and outer layers. Take the whole stone wall as an example, flat stones are applied to the outer layer of the wall with special care taken to the stone size and the joint gap.

The inner layer of the wall has lower requirement for the regularity of stones and the width of the stone is adjusted based on the thickness of the outer layer. The stacking and staggered joints of the stones ensure the firm and stability of the structure. Yellow mud is used to fill the gap between the two layers of the walls (Fig. 4). The thermal conductivity of natural stone is 1.16-3.49W/mK, and the thermal conductivity of adobe is 0.58～1.16W/mK. Therefore, filling yellow mud between the two layers is like the practice of adding thermal insulation layer in the wall of modern building to achieve the effect of keeping the in-

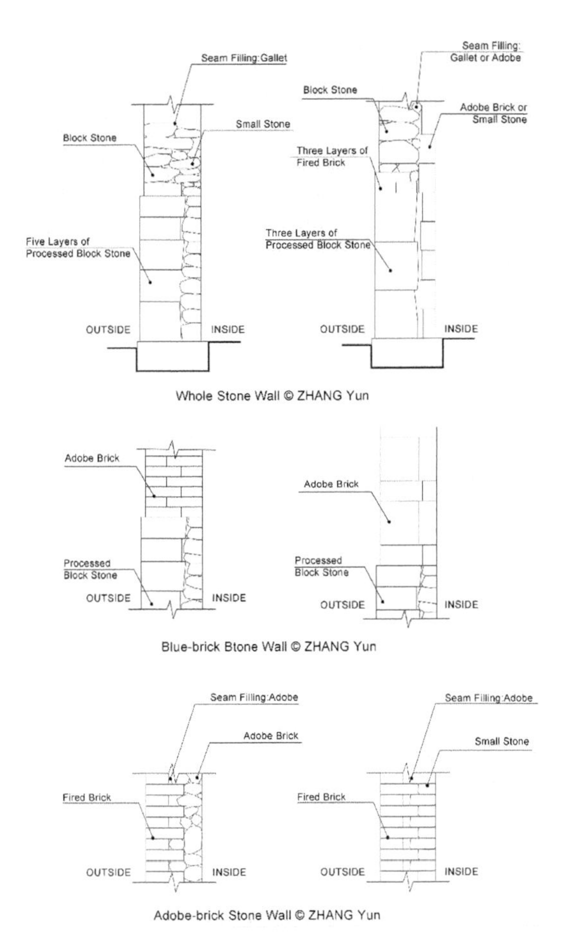

Fig. 4 Different types of bonds of stone masonry wall, adobe brick-stone wall, and fired brick-stone wall © Zhang Yun

door environment warm in winter and cool in summer.

Straw is the regional material for roofing. No matter the roof is made of mountain grass or seaweed, it features great thickness. Roof made of mountain grass is paved by layer after layer's bundled mountain grass and sorghum straw of 3～5cm thick in 10 cm total thickness. Roof made of seaweed is paved by over four layers of bundled seaweed of 20 cm thick (Chen, 2006). The total thickness of seaweed along roof ridge of the house can reach 1m of thickness, producing good thermal insulation effect.

3. 1. 3 Adapting to the Outdoor Environment

The physical properties of the regional materials reflect the adaptability to the outdoor environment. The climate in Jiaodong Peninsular entails the local architecture resisting heat, humidity, storm, and salt spray. In terms of materials for external and enclosed structure, the advantage of stone is obvious as discussed before. Adobe brick and straw have poorer durability but low cost for acquisition and replacement. Containing large amount of salt and pectin, dried seaweed resists moisture and moth, and has excellent thermal insulation performance. It was an energy-saving and environment friendly roof material at low cost in the past.

3. 2 Dilemma Faced by Regional Materials in Local Development

Under current development, regional materials, together with vernacular buildings, are gradually being replaced by modern ones. This dilemma is caused by the following factors that cannot be neglected.

3. 2. 1 Mechanical Properties Unsuitable for Modern Buildings

The vernacular buildings are generally from 4 to 6 meters in depth, and less than 3 meters in column spacing, which is determined by the mechanical properties of the regional materials. In comparison, modern architectural design has more free planes, larger scale, and better seismic performance Thanks to the use of such materials as stressed steel combined with concrete. When production and people's living conditions have changed with social development, the incompatibility of regional materials with modern building activities becomes more apparent.

3. 2. 2 Limitations of Traditional Craftsmanship

The continuation of regional materials depends not only on the limited resources, but also on traditional material processing practices. Take the stone used for walls in vernacular buildings as an example, according to the interviews with the old craftspeople, it takes one day to do simplified processing of an ashlar and the complicated processing involves 200 steps. The inheritance of this traditional construction technique is acquired and accomplished by the craftspeople in construction practice (Fig. 5), which conflicts with the standardization and automation trend of modern architecture from design to construction. For the architects and structural engineers trained under the modern architectural education system, the understanding and learning of traditional building craftsmanship is missing in their professional education.

Fig. 5 Craftsmen for seaweed thatched roofs work are generally between fifty and seventy years old © SPCEIVH

3.2.3 Increasing Shortage of Materials Caused by Environmental Change

Environmental degradation and over-exploitation have resulted in an increasing shortage of native, natural and regional materials. For example, seaweed, the regional material for the construction of seaweed thatched houses, has been destroyed by the aquaculture industry in the offshore area in recent years. Because its natural growth environment has been destroyed, seaweed is hardly seen in the coastal area of Jiaodong Peninsula. In the hilly areas where stone is produced, many private quarrying activities have been stopped because of serious environmental damage caused by destruction from excessive production and construction.

4 Strategies of Continuity

Many factors account for the disappearance of vernacular architecture and regional materials in construction activities, although the author briefly summarizes the factors from the perspective of building materials. The conservation of vernacular architecture differs from that of heritage architecture in that the former covers many similar buildings in regional traditional villages where residents are the direct users and cost bearers. This requires that the strategy we propose should be compatible to local development and only the fittest strategy really helps. Therefore, it is vital for continuity strategy to integrate regional materials into social development. The author proposes some ideas as follows for further discussion and exploration.

4.1 Standardize the Recycling and Reusing Process of Materials

Currently, the recycling of used building materials adopts the traditional method of re-acquisition. The shortage in seaweed compels the resident who is repairing the roof to acquire used seaweed from their neighbors who changed their roof and no longer needs seaweed. In this case, the lag of information transmission has caused a lot of wastes in resources. An organization for recycling and reusing regional materials should be established to minimize materials waste and reduce the acquisition cost.

4.2 Establish Craftspeople Protection Mechanism under Government Intervention

The inheritance of the craftsmanship in building vernacular architecture is an important part for the continued use of regional materials. In the investigation, the author found that in Zhongwodao Village in Weihai, a craftspeople protection system has been established to register and organize traditional craftspeople and to regularly train and teach traditional craftsmanship. A government-led craftspeople protection mechanism can help systematically keep recording traditional craftspeople and organize workshops and training activities to record, preserve, and pass on the traditional craftsmanship and put it into use.

4.3 Design Relevant Architectural Teaching Practice

The key to the continuation is the use. One of the contradictions between regional materials and modern architectural design is that modern designers are not familiar with the performance of regional materials. It is therefore important to incorporate the continuation of regional materials into the architectural teaching practice and design in colleges. Through courses, design practices, or themed competitions, students in architectural major will acquire a sense of using regional materials and learning about the properties from the beginning of acquiring professional skills to be able to make reasonable use in the future design work.

5 Conclusion

Based on the field investigation on the vernacular architecture of 96 typical vernacular villages selected in the Jiaodong Peninsula of Shandong Province, and the interview documentations of twenty old craftsmen, the study focuses on the problems and challenges of regional materials of vernacular architecture in local development, and demonstrates the comprehensive assessment of regional construction materials including timber, stone, adobe, straw and so on, which is a research direction of scant attention within conservation studies in China. Thereafter, it attempts to put forward the feasible strategies of continuity of regional materials of vernacular architecture in the conservation and utilization of the non-renewable resources of vernacular architectural heritage, and the local development in the present transformation period in China. The authors believe that this method can be expectantly useful to solve similar problems in research and practice for later relevant works.

Acknowledgements

We are indebted to Ms CHEN Jianfen as the language proofreader for taking on the great labor of editing the manuscript of this paper.

References

CONG Sheng, 2009. The hair bun of the sea: memory of seaweedthatched house in the east Shandong. Jinan: Shandong Friendship Publishing House.

CHU Xingbiao, XIONG Xingyao, DU Peng, 2012. Exploring the conservation planning of seaweedthatched houses: take Chudao village in Weihai, Shandong as an example. Architectural journal, 6: 36-39.

HUANG Yongjian, 2014. A study on the craftsmanship of seaweedthatched house of Chudao Village of Shangdong. Jinan: Shandong University.

YANG Jun, 2011. Selection and application of regional residential materials: taking the ecological seaweedthatched houses in Jiaodong Peninsular as an example. Architectural journal, 11:152-155.

Study on the Conservation and Renewal of Spatial Form of Historical Blocks in Changsha under the Process of Urbanization — Taking Taiping Street and Chaozong Street as Examples

ZHANG Jiating, *LIU Su*

Hunan University, Changsha, China

Abstract Urbanization means a historical process of the transformation, from an agriculture-based society to a modern urbanized society. With advocating the conservation of cultural heritage, Changsha historical blocks are also facing a contemporary problem, which is how to make historical buildings conform to the development of the city and better serve the public. The author notes the space form and character of historical blocks in Changsha, which are mainly reflected in two aspects: the high building density and the narrow space of the blocks. Taking Taiping Street and Chaozong Street as examples in this paper, the author discusses the space form and character of historical blocks, and proposes to place appropriate public space in historical blocks to deal with the problems faced in the process of urbanization.

Keywords urbanization, historical block, space form, conservation and update, public space

1 Instructions of Text

The historical block carries people's feelings and memories of a city and shows a city's culture and history. Under the baptism of time, the old city of Changsha has not been left. On the morning of November 13, 1938, the Wenxi Fire destroyed 80% of the ground buildings in Changsha, the whole city was almost ruined. Fortunately, the historical block of Changsha remained, and the most basic fish-bone street pattern has been preserved. Later, with the reform and opening up, Changsha's economy developed rapidly, and the city was also under large-scale development and construction. The historical blocks and historical buildings were replaced by high-rise buildings. Until the implementation of "the 11th Five-Year Plan" in 2006, Changsha began its urbanization. As a city with the old city as its core, the traditional historical blocks are also facing tremendous pressure in the process of urbanization.

With the government's gradual emphasis on the conservation of historical building heritage, it has begun to strengthen the conservation and transform and protect historical blocks. Taiping Street is the remaining fishbone street in Changsha, whose reconstruction and conservation work has been completed, and other areas such as Chaozong Street are undergoing renovation. The traditional historical streets which have been rebuilt achieve certain conservation results, but there are also some new problems need to be solved, such as the narrow space of the block, the high density of the building, and the lack of certain public space, etc. The article analyzes these issues to explore the coping strategies of historical blocks under the urbanization process, hoping to provide some reference for the later research.

2 Historical Evolution of Changsha Historical Block Space

2.1 Overview of Changsha Historical Block

The history of Changsha can be traced back to the early Warring States period, when the Chu people began to build city on the east bank of Xiangjiang River in Changsha. According to research, the specific location of Changsha Chucheng during the

Warring States period was "east to the Litou Street, west to the Taiping Street, south to the Pozi Street, and north to the Chunfeng Street, Mingyue Pool and the Luopeng Bridge", and Changsha Chu city is also the predecessor of the later "Linxiang Old City". During the Western Han period, Emperor Hui ordered the whole nation to build the Linxiang Old City on the basis of Changsha Chu city. The city site was expanded to "east to Dongpailou and Nanyang Street, west to Xichang Street, south to Jiefang West Road, and north to Zhongshan West Road and Youyi village". In the Five Dynasties, the scope of Changsha was further expanded to "east to Shunxing Bridge and Gudaotian, south to the Nanmenkou, and north to Chaozong Street and Yingpan Street". During the Song and Yuan Dynasties, the basic scope of Changsha was generally determined to "east to Liucheng Bridge, south to Chengnan Road, and north to Xiangchun Road".

Changsha has a history of about 2,400 years of city construction, and its central location of old city has not changed for more than 2,000 years. Taiping Street and Chaozong Street are located in the south of its old city. After the Wenxi Fire, there are few historical streets and lanes have survived in Changsha. Taiping Street is the historical street that retains a complete traditional street pattern.

2.2 Historical Evolution of Changsha Historical Block Space

According to relevant research, the Lifang system started from the Northern Wei Dynasty, Ye City. Has the Lifang system been implemented in the ancient city of Changsha? According to relevant information, Changsha used to implement the Lifang system, but nowadays, it cannot be observed. After the abolition of the Lifang system in the Northern Song Dynasty, the open street system flourished and the industry and commerce became more prosperous.

The ancient city of Changsha is a typical fish-bone street pattern. We can see it on the provincial map of the third year of Guangxu in the Qing Dynasty. At that time, the urban skeleton was composed of several positive streets and more than 150 horizontal streets. Changsha Old Street is different from Beijing's hutong. Most of them are slashes, curves or broken lines. In the old streets of Changsha, the main road is called "street", the secondary road is called "lane", the north-south direction forms an open street, and the east-west direction forms a relatively closed residential lane. We can see the pattern of ancient streets and lanes in the Ming and Qing Dynasties from the historical block of Taiping Street, which is based on the main street, the lanes are distributed on the both sides of the main street. This is the most basic fishbone street pattern.

3 Analysis of the Current Situation of Space and Users in Taiping Street and Chaozong Street

3.1 Status of Traditional Historical Streets and Lanes in Taiping Street and Chaozong Street

3.1.1 Status of Street Space Layout in Taiping Street and Chaozong Street

The layout of the streets and lanes in the Taiping Street and the Chaozong Street has its own features. Among them, the main street of Taiping Street is in the north-south direction, and Jinxian Lane, Majiaxiang, Fujiaxiang, Xipailou Street and Taifuli on both sides of the main street are in the east-west direction, and Jiangningli is in the north-south direction. The main street is bounded by Wuyi Avenue in the north and Jiefang West Road in the south. The total length is about 482 meters and the width is between 3 meters and 6.5 meters. The entire Taiping Street area is roughly in the shape of rectangular. The main street of Taiping Street is made of granite and bluestone. The other streets are made of granite and cement, while Jinxian Lane is the ancient granite pavement (Fig. 1).

The main street in Chaozong Street area is in the east-west direction. The main street reaches Huangxing Middle Road in the east and Xiangjiang Middle Road in the

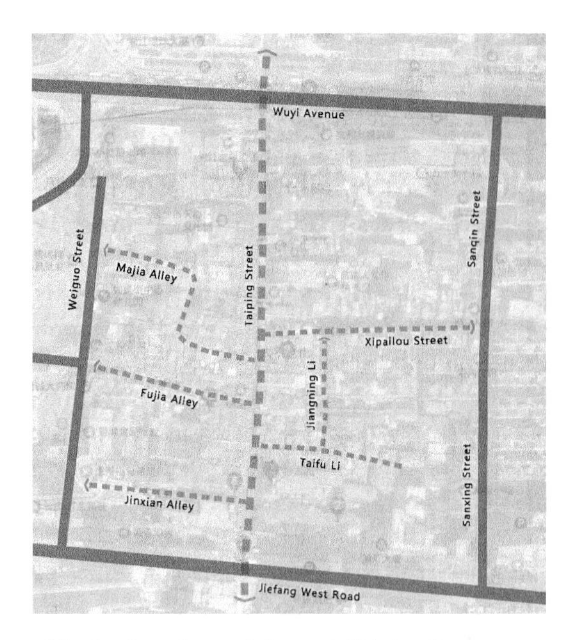

Fig. 1 Location and shape of Taiping Street

west. The total length is about 543 meters and the width is 4 meters. The pavement is made of granite. In addition, Liansheng Street which is on the south side of the main street is also in the east west direction, reaching Sangui Street in the east and Fuqing Street in the west. The total length is about 393 meters and the width is between 2.4 meters and 3 meters, and the pavement is also made of granite (Fig. 2).

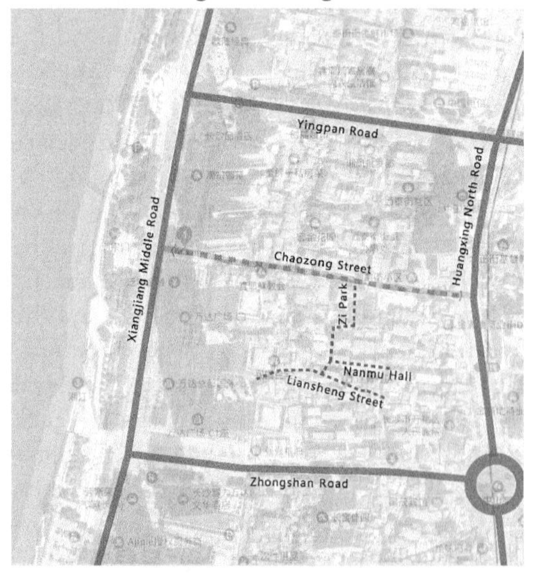

Fig. 2 Location and shape of Chaozong Street

3.1.2 Status of Node Space in Taiping Street and Chaozong Street

The node space of the street means the more characteristic space area in the street, which often plays a role of connection, decoration and make-up function. This kind of space area can add fun to the linear street space and enrich the space feeling of the street. At the same time, it can also provide a place for tourists and residents to rest.

In 2004, the *Changsha Historical and Cultural City Conservation* Plan determined that Taiping Street was a historical and cultural block. According to the *Conservation Plan for the Historical and Cultural Streets of Taiping Street in Changsha*, Taiping Street was extensively renovated and updated. In the planning, "*Taiping Street is the traditional retail commercial style axis, the Xipailou is the traditional food service style axis, and the Jinxian Lane is the traditional commercial residential style axis*". According to the author, there are the pros and cons of this policy on Taiping Street. On the one hand, the traditional feature of the historical block is indeed better protected than before, but on the other hand, due to the constructive conservation of the block is not controlled at the appropriate level. The block is too biased towards commercialization to lead to the "enclosure" of the shops on the main street, completely losing the space where tourists and residents can take a break, even some shops are not selling the characteristics of the ancient streets. For example, some milk tea shops are also located on the main street, losing the charm of the ancient streets. Instead, in the residential lanes on both sides of the main street, some commercial shops can notice the creation of "public space". Those shops usually set up several benches in front of their own shops to provide visitors with a short break area. But in most of the residential lanes, the lanes are very narrow and its wires are arbitrarily designed. Even the most basic lighting is a problem, let alone create the node space (Fig. 3).

In 2014, the *Changsha Master Plan*

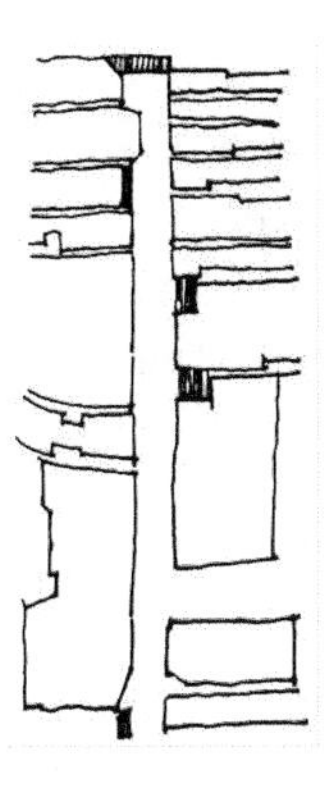

Fig. 3 Taiping Street public space plan

(2003 2020) (*revised in* 2014) determined that the Chaozong Street area was a historical and cultural area. With the support of the government, the regeneration of shanty town of Chaozong Street and the overall renovation and renewal began in 2016. The appearance of the buildings on both side of the street and the pavement of the streets are reconstructed in the style of antiquity, the traditional historical features and the atmosphere of Chaozong Street are conservated. At present, Chaozong Street is still in the process of transformation. According to the author's field research, there are some problems in Chaozong Street area, which are similar to the problems of Taiping Street, such as poor infrastructure, the single node space, or a few public space for tourists and residents. Under the background of urbanization, the traditional historical block of Changsha is obviously unable to meet people's living needs, which is also the starting point and thinking point of this study (Fig. 4).

Above all, there are general two problems in node space of Taiping Street and Chaozong Street: the low usage rate of node space and the very few node spaces.

Fig. 4 Chaozong Street public space plan

3.2 Morphological Characteristics of Block Space in Taiping Street and Chaozong Street

The city's space reflects the specific cultural factors of the region, the country or the race, so the culture of each city is different and unique. This is the reason why the city is universal in some senses, but has a regional character in another sense. This reminds us: what is the nature of urban space? Living in urban space every day, we should have a deep understanding, the nature of urban space is daily life.

As a famous historical and cultural city, the daily life of traditional historical block in Changsha is different from other cities and unique. However, with the evolution of history, the pattern of traditional daily life in Taiping Street and Chaozong Street has changed a lot, which is reflected in the lack of neighborhood environment in historical blocks. According to relevant information, the importance of some important daily life in the traditional period has gradually decreased in today's historical blocks, such as the relocation of the aborigines of Taiping Street. Today, the space of daily living has become a single space, connecting streets and lanes. The traditional daily life space shows a trend of decline, and the main reason is because the need of the urbanizatian. Due to the rapid development of the city, people have higher requirement for the city's space, at the same time, ignoring the essence of city life.

Under the process of urbanization, urban space tends to be standardized and tidy, which is shown in two aspects: one is the high building density and the other is narrow block space. The shops are competing to maximize the land and space. The phenomenon of "enclosure" is obvious, resulting in the loss of the regional space in the historical block. The stage blocked by two huge electric boxes in Taiping Street is the most obvious place. In addition, the decline of regional characteristics in historical blocks is not only reflected in the commercial axis, but also in the residential lanes. For example, when the government repaired Taiping Street, attentions were paid to the commercial axis, while the residential lane was neglected (Fig. 5).

Fig. 5 Crowded residential lanes in the Taiping Street

According to local residents, due to the infrastructure in the residential lanes is very poor, many aborigines have already moved away. Most of the people who still live in Taiping Street are foreign businessmen, and the traditional daily life has also faded. At this point, Chaozong Street and Taiping Street show the opposite trend. According to relevant data, comparing with the ancient Chaozong Street, although the prosperity of the rice market has disappeared and the shops on the street of Chaozong Street can only provide convenience of the daily life to the residents along the street, the Chaozong Street area is undergoing renovation of the shanty town. The entrances of some buildings in the residential lane have been reconstructed. On the construction site, we can clearly compare the original crowded space with the reasonable space scale after renovation, believing the infrastructure will be greatly improved after the renovation of Chaozong Street is completed.

3.3 Two-way Analysis of Historical Block Space and Its Use Crowd in Taiping Street and Chaozong Street

According to investigation of the historical block space, there is a two-way connection between the space and its use crowd: the public space attracts the crowd and the demand of people for the public space.

There are rich public spaces near the archway at the entrance of Taiping Street, such as the edge of tree pools, benches along the flower beds and historical sketches, where many tourists sit and have a break here. It shows the attraction of the public space to the crowd (Fig. 6).

The overall layout of Taiping Street is reasonable. The entrance space at both ends is relatively open, while the middle part is relatively narrow. From one end of Taiping Street to the other, there are few public spaces for visitors to take a break on both sides of the streets, visitors are always in the "walking" action, which is the reason why there are many tourists who sit around the entrance. When I entered the residential lanes on both sides of Taiping Street, the scene was quite different from that of the main street. Those residential lanes can be divided into three categories: purely inhabited, few shops scattered in the middle and the whole storefront on the ground floor. For the first category, there are no public spaces, the lanes are narrow and the entrance of the building is directly connected to the road. For the second type, the streets are relatively spacious, and the merchants often set up several chairs next to the entrance of their own shop to create a "public space" for the tourists to take a break. For the last category, it is mainly for the catering industry, the streets are the most spacious and the businesses even occupy the street to set tables and chairs in front of their shops, allowing guests to eat outdoors. All three categories show the needs of the people for public space.

However, Chaozong Street is different from Taiping Street. The tourism industry

Fig. 6 The entrance of Taiping Street

in Chaozong Street is not significant. In terms of public space, it is also inferior to Taiping Street. There is no outdoor public space for people to communicate. In addition to people who need to go out during the day, there are few people on the street. According to the author, there is a small community center in the Chaozong Street, where many residents play mahjong. At the entrance of the community center, some residents sit in chairs and enjoy the cool air, this also reflects the residents' demand for public space. Nowadays, the Chaozong Street area is transforming the shanty towns. It is hoped that the builders could pay more attention to the demand of people's daily life space and make the traditional historical blocks better serve the public.

3.4 Problems in the Historical Block Space of Taiping Street and Chaozong Street

With the rapid development of Changsha, the problems in the historical block space of Taiping Street and Chaozong Street have become more and more obvious. According to field research, there are two aspects of problems in Taiping Street and Chaozong Street. On the one hand, there are few proper public spaces in the main street or the low utilization of public space, resulting in a bad sense of spatial experience for its use crowd. On the other hand, most of the residential lanes on both sides of the main street have a high density of buildings and spaces and its lanes are also very narrow. Although some buildings have large open space and green space, its usage rate is too low. The residents lack necessary communication space and the traditional daily life needs to be improved.

4 Form Conservation and Renewal Measures in Historical Block Space Taiping Street and Chaozong Street

4.1 Reactivating the Existing Public Space

In the current traditional historical blocks, there are still some public spaces, such as the Yichunyuan Stage in Taiping Street and the pavilion on the ruins of air-raid shelter, the Nanmu Hall or the community center in Chaozong Street. However, most of them have not been used very well. For example, two large power distribution boxes are added in front of the Yichunyuan Stage, the pavilion in the Chaozong Street is filled with abandoned building materials and the low usage rate of open space between residential buildings. The setting of these public spaces was originally intended to show the charm of the old streets with the crowd, but in fact it seems to have only an ornamental effect, so we should reactivate those public spaces and continue the cultural connotation of the historic blocks.

4.2 Place in the "New" Public Space

After analyzing the historical block space and its users, we can find that the crowd has a certain demand for the public space. We can refer to the successful cases such as Shanghai "Xintiandi" that places the appropriate public space in historical area. The interface of space could continue the architectural features of the historical blocks and be consistent with the street style in terms of appearance, color and materials. For shops on both sides of the commercial street, these spaces do not need to be large and it can refer to the shops in residential lanes of Taiping Street block. In the residential lanes, some buildings could be demolished to create a public space. This method not only solves the problem of lighting and ventilation, but also increases the interaction between neighbors and continue the traditional daily life of the historic block.

4.3 Improve the Basic Living Facilities of Residents

In the treatment of public space, it is necessary to improve the basic living facilities of residents. In the residential lanes of the historical blocks, most of them are crowded, but there are also multi-storey buildings built later, which often have relatively spacious open spaces between buildings. For the crowded lanes, we could demolish some buildings to create a public space, in which we could set up sculptures, chairs, rest area, etc. For multi-storey residential buildings, we can set several benches on the original open space to increase communication space for residents to chat and enjoy the cool air.

5 Conclusion

Under the process of urbanization, although the conservation and renewal of historical block space has encountered great challenges, it also brings the infinite possibilities of activating historical block space. Due to the conservation and update of space is very comprehensive, the author only pays attention to a part of node spaces and discusses how to activate the historical block space to cope with the urbanization process, hoping to provide some enlightenment and reference for the research on the preservation and renewal of the historical block.

References

DING Xiajun, 2015. Conservation and utilization of historical and cultural buildings in urban fringe. Beijing: China Architecture & Building Press.

FENG Qing, FAN Junfang, 2012. Analysis of space and humanities of Taiping street historic district in Changsha. Architectural Culture, 2012 (3): 139-142.

HU Yueping, 2002. Analysis of street space in traditional towns. Kunming: Kunming University of Science and Technology.

HOU Zehua, 2015. Research on the compound strategy of organic renewal of Chaozong Street Historic District in Changsha. Changsha: Hunan University.

Five Ideas on the Protection and Utilization of Urban and Rural Built Heritage from the Perspective of County — Take Xiaoyi, Shanxi Province as An Example

LI Shiwei, *WANG Jinpin*, *HE Meifang*, *LIU Zhisen*

Taiyuan University of Technology, Taiyuan city, Shanxi, China

Abstract Taking Xiaoyi as an example, this paper discusses the protection and utilization of urban and rural built heritage in county areas under the background of social and economic transformation. By tracing the development course of heritage protection, and analyzing the current situation and trend of contemporary architectural heritage protection research, based on the compilation work of the protection planning of Xiaoyi the historic and cultural city, Xiaoyi, the author finally puts forward the five-point conception of building heritage protection and utilization: 1. Heritage protection concept with the purpose of inheriting cultural genes; 2. Protection and utilization strategies of urban and rural heritage from the perspective of the whole regional system; 3. Value research method based on GIS and other information technology means; 4. Dynamic protection mechanism of public participation based on consensus; 5. Reconstructing the protection and utilization mode guided by the spirit of place. The above five points are related to each other and expound the key points of the current built heritage protection work in Xiaoyi from different levels, aiming to provide reference for the built heritage protection work in Xiaoyi and other similar counties and cities.

Keywords county, built heritage, Xiaoyi, protection, utilization

1 Start from Urban-Rural Relations

In modern times, the agglomeration of urban material wealth further widened the gap between urban and rural areas, and the rural areas dominated by agriculture lagged far behind the cities in construction. However, as a type of settlement with a long history, the countryside also plays an irreplaceable role in the current and future living environment system. The value and potential of rural areas have attracted extensive attention after the proposal of "rural revitalization". The new urban-rural dual system urgently needs to be constructed from the perspective of space-time, and the protection and utilization of the urban-rural built heritage is an important part of building a new urban-rural relationship.

2 Protection of the Built Heritage

The idea of heritage protection originated from the western enlightenment, and the concept of heritage has undergone several periods of evolution. In April 2017, at the international symposium on "built heritage: a cultural driving force for urban-rural evolution", academician Chang Qing explained the concept of built heritage: Built heritage is a concept commonly used in the international cultural heritage field. It generally refers to the cultural heritage formed by the way of construction, which consists of architectural heritage, urban heritage and landscape heritage. To expand the spatial scope of the concept of "built heritage", another expression is "historical environment", that is, urban and rural built-up areas with specific historical significance and their landscape elements, such as historical and cultural blocks in cities and traditional settlements in villages. Moreover, the extension of the concept of "historical environment" also includes those places where the built heritage has long faded, but the historical influence is still deep.

As a cultural driving force of urban-rural evolution, urban and rural built heritage is undoubtedly a focus worthy of attention.

The built heritage is the witness of history and culture, the expression of cultural diversity, the symbol of regional characteristics, and the symbol of identity of local residents.

The protection of historical sites in the modern sense of our country started from the archaeological scientific research in the 1920s, and didn't start until 1929 when the building society was established and the heritage research was carried out using scientific methods systematically. After 1949, China gradually promulgated laws and regulations such as the law of the People's Republic of China on the protection of cultural relics and the regulations on the protection of famous historical and cultural cities. And in the 1990s, a three-level protection framework of "cultural relics and historic sites — historical and cultural reserves — historic and cultural cities" was formed. Since then, the protection system of famous historical and cultural villages and towns had been gradually improved. From 2012 to 2019, the establishment of the "Chinese traditional villages" protection system included 6,799 traditional villages whose overall value could not meet the standards of famous historical and cultural villages into the protection list, further improved the heritage protection system and preserving a large number of settlement samples of human farming civilization.

However, due to historical reasons, the distribution of heritage in China still has problems such as "large amount and wide range", "fragmentation" and "loss of original functions".

3 Overview of Urban and Rural Construction Heritage Protection in Xiaoyi

As an intermediate administrative division unit in China, "county" plays the role of communication from top to bottom and is an administrative hub with high efficiency of order execution and feedback. Although regional cultural connections are also affected by traffic, water transport, trade and other factors, counties usually overlap with geographical boundaries, which have strong continuity and inheritance, and can form, brew and develop a specific regional cultural type(Fig. 1). In terms of the protection and utilization of the built heritage, it is advanced and operable to study the county area.

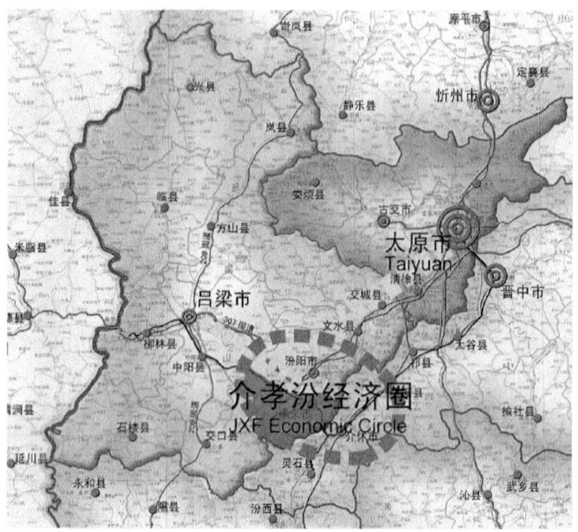

Fig. 1 Geographical location map

3.1 Location Analysis of Xiaoyi

Xiaoyi is located in the south-central Shanxi Province, the southwestern corner of the Jinzhong basin, the eastern foot of the central Lvliang mountain range, facing Jiexiu River in the east, bordering Lingshi River in the south, Zhongyang in the northwest and Guo River in the north. The ancient post road passed through Xiaoyi. In the late Qing Dynasty, due to the rise of Jin merchants, Xiaoyi became one of the main gathering places for commercial activities, and the economy and culture were greatly developed(Fig. 2). In modern times, it has become a typical resource-based city because of its rich mineral resources, such as coal, aluminum and iron. It is reputed as "the treasure land of the three Jin Dynasties, the key point of the Qin and Jin Dynasties, and the window of Lüliang"(Fig. 3).

3.2 Historical Evolution

As early as the neolithic age between 7,000 and 8,000 years ago, there were human beings living and multiplying in the Xiaohe river basin. The county was located in the spring and autumn period, which be-

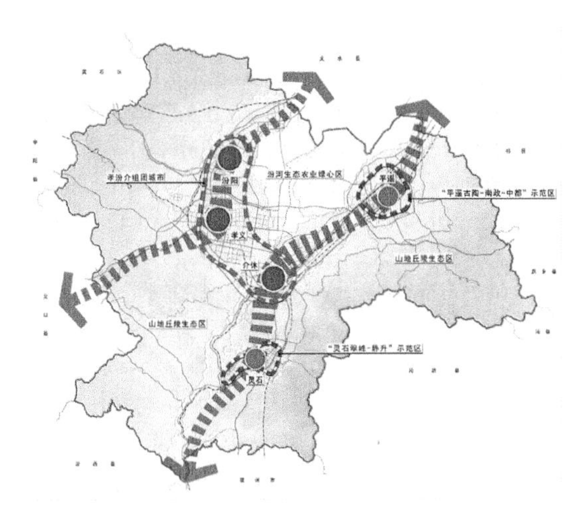

Fig. 2 Economic location map

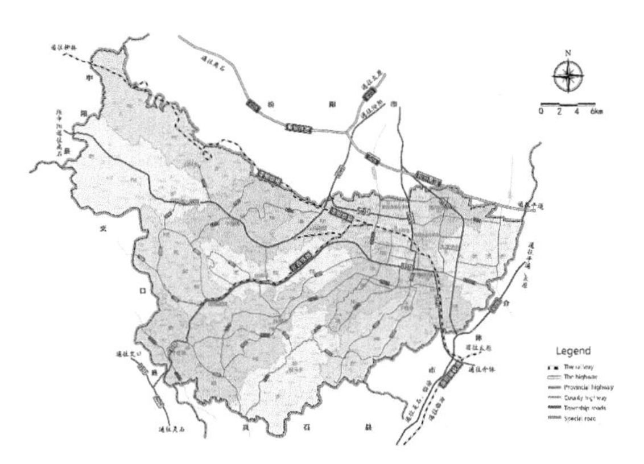

Fig. 3 Internal and external traffic map

longed to the Jin Dynasty. It was Guayan County, which had the meaning of "pushing Yan to spread". In the warring states period, it was a Weizi County. In the Three Kingdoms period, Cao and Wei ruled Zhongyang County. In the seventeenth year of Taihe in the northern Wei Dynasty (493), it was renamed Yong'an County. In the first year of the reign of emperor Zhenguan, Zheng Xing, a city man, was named Xiaoyi County. Song Taiping first year renamed Zhongyang, Xuan and renamed Xiaoyi, after use so far(Fig. 4).

3.3 Differences in Settlement Forms and Traditional Building Types

The urban and rural settlement forms in Xiaoyi are greatly influenced by Land forms, rivers, mineral resources development and transportation lines, forming

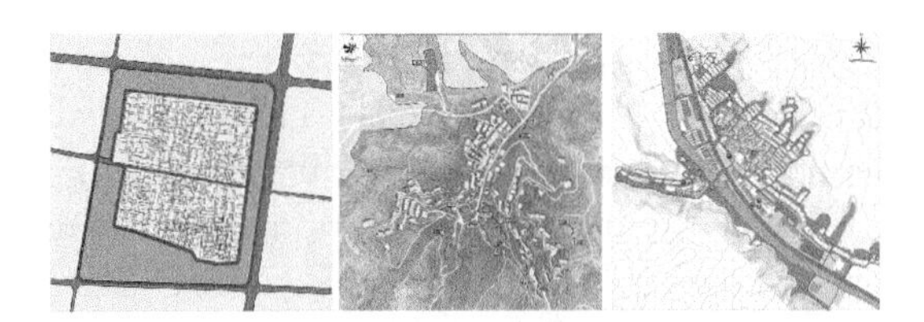

Fig. 4 Xiaoyi, Songjiazhuang Village (Plain), Guanyao Village(Hilly), Xijiebao Village(Gully)

three kinds of settlement forms, which are mainly gully and strip settlements, terraced settlements and cluster settlements in plain areas. By Jinzhong courtyard and Jin cave dwellings, as well as the influence of the west is under the role of modern western culture, traditional architecture type phenomenon is more obvious regional differences, for the two types of traditional architecture: general economic development in northwestern hills southeast mountain is given priority to with traditional Chinese architecture type, economically developed would region is given priority to with combination of Chinese and western architectural type(Fig. 5).

Fig. 5 Xiaoyi, Xiaoyuan Village Courtyard Gate (Northwest), Temple Gaoyang instrument door (Southeast)

3.4 Xiaoyi Built Heritage Composition

Xiaoyi has a large number of built heritage, historical buildings, historic settlements and others for the historical landscape. Historical buildings are divided into 137 cultural relics buildings (4 national, 2 provincial and 131 municipal), and a large number of listed historical buildings. The historic settlements are divided into eight traditional Chinese villages and one "ancient city of Xiaoyi", a famous historical and cultural city in Shanxi Province. The historical landscape includes Jinlong mountain cultural scenic spot, Xiaohe national wetland park, Shengxi lake forest park, Xiaoyi Sanhuang temple scenic spot and so on. In addition, there are many potential traditional villages and countless historical remains, as well as the historical environment that produced these historical remains.

These massive built heritages are in urgent need. These built heritages contain the material and immaterial cultures accumulated by the profound agricultural civilization, and are the wisdom crystallization of the harmonious coexistence between filial piety people and nature (Fig. 6). It is the only way to inherit the excellent architectural culture gene to carry out information statistics, value research and judgment, classification and classification protection and sustainable protection and utilization through modern technical means.

Fig. 6 A map of the built heritage in Xiaoyi

3.5 History of Building Xiaoyi Heritage Protection

After the founding of the People's Republic of China, the conservation process of Xiaoyi's built heritage can be roughly divided into five stages:

(1) Germination period (1949—1975): civil spontaneous protection;

(2) Initial period (1975—2007) : the city was relocated in 1975, and the old city was fully protected as a whole;

(3) Development period (2007—2014): the compilation of the protection plan of Xiaoyi ancient city was started in 2007, and the detailed construction plan of Xiaoyi ancient city was compiled in 2011. In 2012, it was published as a famous historical and cultural city of Shanxi Province.

(4) Improvement period (2014—2018): in 2014, Baibiguan village, Jiajiazhuang village and Songjiazhuang village were announced as the third batch of traditional Chinese villages by seven ministries and commissions including the ministry of housing and urban-rural development. In 2016, Xijiebao village, Linshui village and Guanyao village were announced as the fourth batch of traditional Chinese villages. In 2018, Gaoyang village and Xiaoyuan village were announced as the fifth batch of traditional Chinese villages. The establishment of a protection system that includes traditional Chinese villages.

(5) Construction of a new stage of heritage protection under the region-wide system (2018—present): the protection plan for Xiaoyi historic and cultural city will be launched in July 2018. Compared with the past, the particularity of the new stage of heritage protection is based on the background of economic transformation and urban-rural co-construction.

4 Five Ideas on the Protection and Utilization of the Built Heritage

In general, the research results on the history and culture, planning and construction, and heritage protection of Xiaoyi have a good foundation and a high starting point. In spite of this, the built heritage protection system still ignores the relevance and integrity of heritage, thus leading to the separation of culture and its carrier. Besides, this directory protection mode leads to the omission of a large number of potential built heritages that are not listed in the protection directory. In addition, there is a problem of multi-sectoral overlapping in the management mechanism of heritage protection.

In view of this, the protection of these

historical relics needs a more macro protection system from the perspective of cultural inheritance. The author takes the opportunity of the compilation of the protection plan for the famous historical and cultural city of Xiaoyi, and puts forward five ideas about the overall protection and utilization of the built heritage from a full perspective.

4.1 Inheritance of Cultural Genes for the Purpose of Heritage Protection

In the 19th century, when anthropologists and archaeologists were thinking about the relationship between architecture and human beings, they realized that by analyzing the architectural heritage as an important carrier of human culture, they could explore the development path of human culture from an empirical perspective. The resplendent and diverse cultural types are the power source of the Chinese nation's creation and development, among which the architectural culture is an important part. To protect the built heritage is to better inherit excellent cultural genes and avoid the embarrassing situation of culture without carrier.

The cultural gene of Xiaoyi can be summarized as the commercial culture represented by the Jin merchant culture, the belief culture represented by the Religious culture, the revolutionary enterprising spirit represented by the red culture, and the traditional ethical culture represented by the filial piety culture. The concrete performance is the house court, each kind of temple, the revolution ruins, the rich non-material culture. For these excellent cultural genes to insist on the principle of the protection of the original genuine and integrity, adhering to the concept of dynamic development innovation heritage, through the classification of grading (refer to the existing state of material and the classification of the intangible cultural heritage hierarchical), value judgment, make certain to protection and use of specific guidance, to achieve the goal of cultural gene new generation.

4.2 Protection and Utilization Strategies of Urban and Rural Built-up Heritage from the Perspective of the Whole Region System

In recent years, the compilation of conservation planning for famous cities has shown a trend of overall conservation, with the focus on the direction of association integration and order reconstruction, and the emphasis on the exploration of the concept of historical pattern translation and architectural texture analogy or the establishment of the internal relations of various levels in the historical environment. The scope of protection is also oriented to the whole urban and rural areas, and the built heritage within its regional scope should be incorporated into the framework of overall system protection as far as possible.

As far as Xiaoyi is concerned, it puts forward the framework of "three levels and two aspects" in the compilation process of the protection plan for the famous historical and cultural city of Xiaoyi (Fig. 7). Constructed a "one core one belt, four axis five pieces, multi-point layout" of the completed heritage protection system; There are two sightseeing routes planned to form a linkage pattern of cultural display and utilization: ancient post road cultural display route and Xiaohe culture display route.

"Three levels" refers to: Xiaoyi historical and cultural city scope; Historical and cultural blocks (Gucheng block, Jiajjiazhuang block, Songjiazhuang block) and traditional villages; Cultural relics protection units, historical buildings and other spots built heritage. "Two aspects" refer to: traditional culture and intangible cultural heritage; Natural environment and spatial elements.

"A nuclear" refers to the ancient city and its surrounding environment, "area" means the under the fort river to filial piety the spindle with a traditional Chinese village distribution, "four axis" refers to the four auxiliary shaft protection along the Duizhen river, Zhupu river, Xixu river, Caoxi river, "five" refers to the five traditional style construction area, "multipoint" refers to bump unit at all levels, historical buildings and valuable characteristics of traditional landscape architecture, etc.

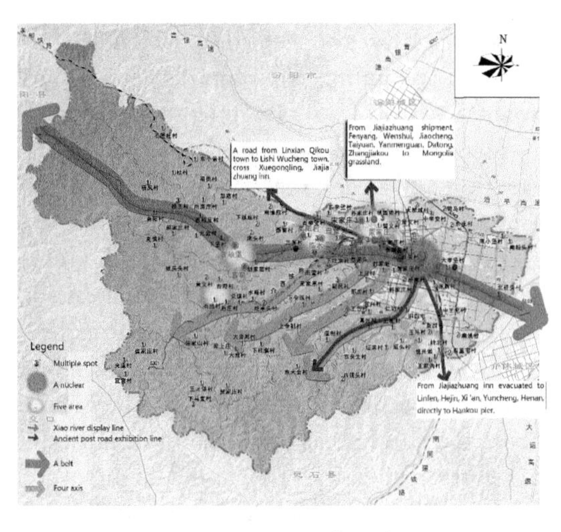

Fig. 7　Heritage protection plan of Xiaoyi

4.3　Research and Judgment Method of Heritage Value Based on GIS and Other Information Technology Means

The general value analysis is based on qualitative description, while the new technology application can compare and analyze more samples than human brain operation from a quantitative perspective. Heritage conservation is facing more and more complicated and uncertain situation, and the record means of traditional heritage conservation cannot fully meet the new planning requirements. Heritage preservation in the United States, Japan, Britain and other countries has entered the digital age. Some top universities and research institutions in China are also trying to combine new technologies and heritage protection. Some statistical analysis points out that from 1978 to 2018, the research on the scientific evaluation of heritage protection in China has been rapidly and obviously improved in depth and breadth, but there are still some aspects that need to be deepened and improved. With the same heritage type and the same evaluation problem, the system structure lacks uniformity, and the selection of composition index lacks sufficient basis. The scientific evaluation process and results are not accurate enough. Solving these problems will require new technological innovations.

At present, the comprehensive application of information-based heritage evaluation research and data platform construction research based on VR technology, spatial syntax, GIS technology, artificial intelligence and other methods in the field of heritage protection is the future development trend. Many advantages, such as county space integration planning; Optimize resource allocation, build and share infrastructure and public service facilities; To establish policy basis for relocating villages and merging points; To construct a display and utilization system of cultural resources to support the tourism of the whole region. It is important to note the determination and quantification of the impact factors in quantitative analysis. Relevant research results include the quantitative interpretation of the spatial distribution of traditional villages in Guangxi by Professor Lu Qi's team from South China University of Technology.

As far as Xiaoyi is concerned, the application of GIS technology is still a blank in the field of heritage research. The author believes that the following work needs to be divided into three steps: the establishment of database, the analysis system of GIS software, the conclusion and the corresponding protection measures(Fig. 8).

4.4　Consensus-Based Public Participation Dynamic Protection Mechanism

"Public participation" refers to the process in which the public, through various forms, independently mobilizes all the parties affected by public decisions in a specific social environment to participate in the decision-making process related to public interests in order to maximize the public interests, exert influence on the decision-making party and even change the decision-making process. Since heritage protection is related to the public interest, the necessity of "public participation" in heritage protection needs to be paid attention to by all sectors of society, and because of the objective difference in time and space of the public's cultural cognition, the implementation of "public participation" is a big test for local governments and the public.

There are three modes of public participation in Xiaoyi: individual public participation, group public participation and government leading. In 2007, in order to simplify

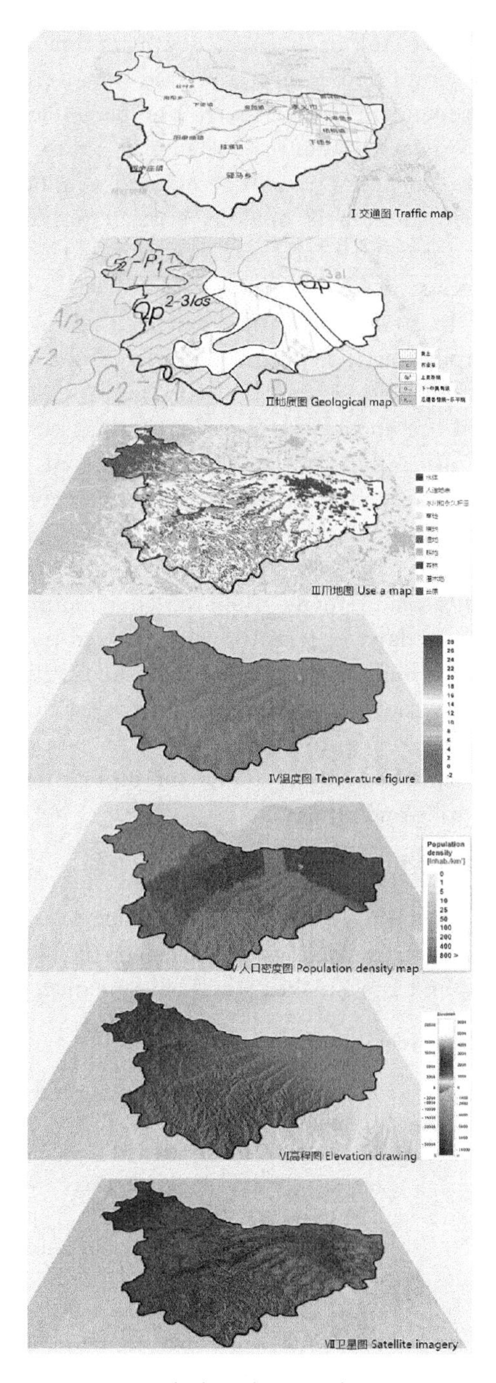

Fig. 8 GIS map platform layer analysis in Xiaoyi

the application and approval procedures for rural construction projects, the government promulgated the administrative measures for raising funds and raising labor through discussion on one case by one villager, which put forward the system of "one case by one discussion by village committee", further expanding the intensity of public participation.

In general, Xiaoyihas a good foundation of "public participation", which is at the level of a few "real power participation" (Sherry Arnstein's "ladder" theory of public participation) and the general level of "symbolic participation". However, there are also some deficiencies. For example, the universality of "public participation" needs to be improved, the organization needs to be strengthened, and institutional guarantee is urgently needed. Therefore, it is necessary to further strengthen the publicity and education of public participation in heritage protection and improve its standardization, legalization and institutionalization. On the basis of clarifying rights and obligations, a dynamic protection mechanism platform of "public participation" is constructed to organize and regulate the behaviors of public participation, and the ways and methods of "public participation" are expanded to make the public become the subject of heritage protection.

4.5 Reconstruction-oriented Ideas for the Protection and Utilization of the Built Heritage

"Genius loci or spirit place" is the spiritual core of the built heritage. The built-up heritage is the testimony of the human "dwelling", which is more than just "shelter" in its real meaning meaning the space in which life happens, its specific expression is "place".

"Place" is a comprehensive carrier of external experience and internal life, with strong vitality. The meaning of "place spirit" is the superposition of time dimension, which is diachronic. To clarify these properties of the site is conducive to the renewal and utilization of the built heritage. The reality is that the development projects for economic benefits, in the name of protection, transfer the aborigines and turn the living place into an "empty shell", and the spirit of multiple places is being eliminated. The author thinks that in today's spiritual civilization construction, built heritage will play a positive role, regardless of what level of protection, reconstruction strategy need to consider the place spirit, strengthen the place characteristic, efforts to increase people's sense of self-identity, and express the meaning of place, finally achieve the goal of refactoring place spirit(Fig. 9).

The "spirit of place" in Xiaoyi is preliminarily summarized as five themes of "god, king, people, things and scenery", which are closely connected and the collective memory and emotional bond of Xiaoyi people. Such as Yuhuang temple of Xiaoyi, Guanyao Village, built in the center of the village phoenix hill. It was once a place for villagers to pray for the gods. It is a place with a specific spirit, a specific space that accommodates regional customs and special feelings, and has a clear relationship with the natural landscape. However, there is no complete data left after the fire, and complete recovery is impossible. Therefore, it is necessary to reconstruct the spirit of the place.

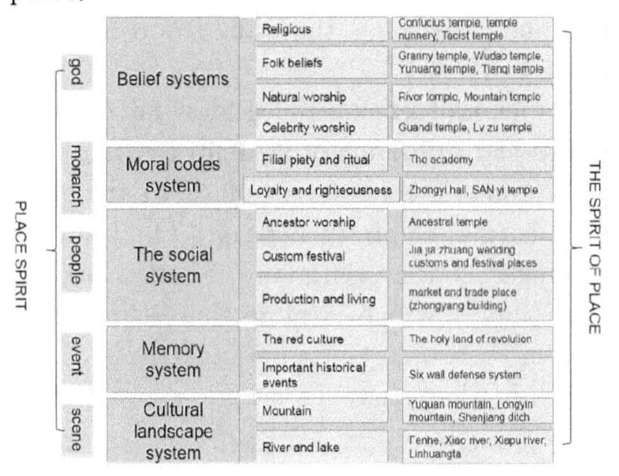

Fig. 9 Analysis of filial piety place spirit system

5 Conclusion

In terms of the research on the protection and utilization of the built heritage of Xiaoyi, the investigation and investigation work has been preliminarily completed, and the protection and development plan related to the ancient city and traditional villages has been completed on this basis, playing a positive role in publicity, education and rescue protection. Through analysis, we know that there are abundant built heritage resources in Xiaoyi, but its protection status is worrying. In the process of industrialization, a large number of excellent built heritage have not been given due attention and protection. However, as a traditional energy-dependent city, Xiaoyi has been facing the crisis of resource exhaustion in recent years and the trend of regional economic transformation is obvious, which provides an opportunity for the protection and utilization of the built heritage.

In 2018, the author participated in the compilation of the protection planning of Xiaoyi historic and cultural city, and put forward the above five ideas for the protection and utilization of the built heritage of Xiaoyi. These five points are the focal points of the built heritage protection in Xiaoyi at present. By combining quantitative and qualitative methods, the problems and practical needs at different levels involved in the field of built heritage protection and utilization are solved. It is hoped that it can provide reference for the conservation work of the built heritage of Xiaoyi and other similar counties and cities.

Acknowledgements

I would like to thank my teacher, Professor Wang Jinping, and teachers He Meifang and Liu Zhishen for their guidance.

References

Anon, 1972. Newly built Xiaoyi county annals. Beijing: Haichao press.

DONG Jianhong, 2004. History of urban construction in China. China building industry press (third edition), 7: 419-450.

HE Yi, 2016. Si Wei city. Beijing: China Building Industry Press.

LIU Min, 2012. Public participation mechanism and practice research of Tianjin architectural heritage protection. Tianjin: Tianjin University.

NORBERT S, 2010. Site spirit: towards the phenomenology of architecture. SHI Zhiming. Wuhan: Huazhong University of Science and Technology Press.

RUAN Yisan, et al., 1999. Conservation theory and planning of historic and cultural cities. Shanghai: Tongji University Press: 8-10.

SHE Yining, YANG Changming, 2019. Academic research analysis on the scientific evaluation of domestic cultural heritage protection. Chinese cultural heritage, 1.

TIAN Yinsheng, et al., 2016. Form and significance of traditional villages. South China university of technology press, 12: 1-2.
Traditional village protection and development planning 2019—2030, Guanyao village, Xiapu town, Xiaoyi city, Shanxi province. Internal data, 2019.
VICTOR U, 2018. Architectural anthropology. PAN Xi, LI Geng. Beijing:China Building Industry Press:25-26.
Xiaoyi historical and cultural city protection planning, 2018. School of architecture. Taiyuan: Taiyuan University of Technology.
YU Xiang, 2010. Survey and analysis of modern architecture in Xiaoyi. Taiyuan: Taiyuan University of Technology.
ZHANG Ru, LU Qi, 2019. Quantitative interpretation of spatial distribution and influencing factors of traditional villages in Guangxi. Construction of small towns, (4):72-79.
ZHANG Song, 2017. Generation and enlightenment of the concept of urban built heritage. Architectural heritage, (3):1-14.

Lesson Learned from the Conservation of Vernacular Houses in the Upper Northeast Region of Thailand

Nopadon Thungsakul[1], *Thanit Satiennam*[2]

[1]*Center for Research on Plurality in the Mekong Region (CERP)*
Faculty of Architecture, Khon Kaen University, Thailand
[2]*Faculty of Architecture, Khon Kaen University, Thailand*

Abstract Conservation of vernacular houses is becoming increasingly important, especially in the northeast region of Thailand where many vernacular houses have rapidly been demolished. Due to many limitations in conservation such as the deterioration of building conditions, the difficulties in finding local materials, lack of knowledge for traditional construction methods and craftsmanship, the conservation process in this region have extensively develop into more difficult conditions. Moreover, the limitation includes the constraint in household economic in terms of lacking budget to repair the building to maintain good condition as before. Because of these changes, most vernacular houses cannot maintain their genuine value. Study area in the northern part can be considered as one of important areas that still have remaining indigenous cultural identity particularly in vernacular architecture. As a result, there have been the efforts in developing community-based tourism from architectural heritage by both local people and government agencies involved in local administration. Several stakeholders have participated in conserving vernacular houses including the house owners, local government organizations, related private agencies and educational institutions. All participants have taken responsibilities in various forms according to their interests and expertise. Vernacular houses that have undergone conservation processes are involved both positively and negatively.

This article aims to reveal the experience in conservation of valuable vernacular houses within the area. The results of the study were from 2 parts; (1) a review of relevant research literature and (2) from field surveys which concluded from discussions of the effects of various actions from stakeholders through different perspectives of their involvement in the area. The results can suggest ways to improve the quality of local conservation in the northeastern region of Thailand.

Keywords vernacular houses, conservation & tourism development, local stakeholders, northeast of Thailand

1 Introduction

Northeast region is an area that covers approximately one third of the Thailand (168,854km^2). Historical evidences indicated that this region was a settlement of language ethnic groups of "Tai-Kadai" and "Mon-Khmer". Since the end of the 18th century people from an ancient kingdom of "Lan Xang", which is currently Lao PDR., settled across the northeastern region (Schliesinger, 2000, 2001a, 2001b). "Vernacular houses" in this article refers to the Tai-Kadai ethnic houses which currently recognized as one of architectural heritage of the region. Currently, the conservation of vernacular houses as a cultural asset is getting a lot of attention from Thai society due to the popularity of community-based tourism in the region; for example, in the upper northeast region, the awarded architectural heritage conservation of Tai Loei houses in Ban Na-O village. (Sutthitham, 2001). Even though the site were involved with several stakeholders both locals and conservationists from community, municipality, and academic institute, there are not many studies exposing the impacts from the process and management of theses heritage. Vernacular houses have still be demolished because of deterioration of building conditions such as the difficulty in finding local materials, lack of knowledge for traditional construction methods and craftsmanship.

Moreover, the limitation includes the constraint in household economic in terms of lacking budget to repair the building to maintain good condition as before. Due to these changes, most vernacular houses in the region cannot maintain their genuine value.

Located in the Loei River Basin in the upper northeast region, Na-O Village has been considered as one of the best conserved example of vernacular houses in Thailand (Fig. 1). As a result, it has been selected as an in-depth case study since it has a distinctive identity in the Thai-Tai cultural heritage. The site has specified as a case study that meets the criteria, defined as; (1) vernacular house style with a distinctive identity that represents tradition of Tai-Kadai ethnic group, and (2) a plentiful number of traditional houses conserved with evident results in the integrity of physical conditions and has continued management until the present. This paper will explore the previous operation and conservation of vernacular houses within the area due to the practice of related agencies and several sectors on how it affects the inheritance of these cultural heritage.

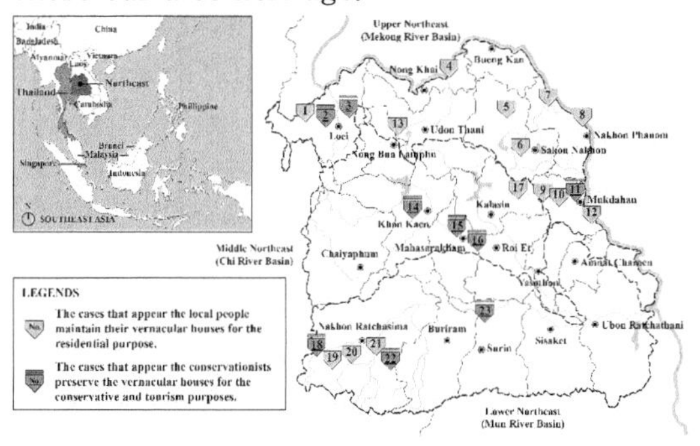

Fig. 1　Upper northeastern region and the site location map (Na-O village is number 2) © Thanit Sateinnam

2　Tai Loei Vernacular Houses

Community settlements along the Loei River Basin have maintained traditional housing styles that reflected local wisdom in accordance with people's way of life, sociocultural factors and their living environments. This area has also the melting pot of diverse ethnicities including Tai Loei, Tai Dam and Tai Phouane. House styles have indicates a continuous development from various factors in the upper northeast area (Fig. 2), from the traditional Northeastern style, the Sae House or the Koei House, to the Laotian with urban influenced style, the Tai Loei Style (Gable-roofed, L-shaped plan with low-under floored-space). In study area, Na-O village, the Tai Loei house style is the most significant and important cultural heritage that marked local identity to outsiders.

Tai Loei is an ethnic group settled in the Loei River Basin. Their ancestors migrated from Luang Prabang in Lao PDR. to settle in community area since 1823. The construction of Tai Loei houses have shown a harmonious relationship between the management and the use of available local resources associated with social and physical dimensions through the creation of local craftsmanship.

Fig. 2　Tai Loei vernacular houses in the upper northeast © Nopadon Thungsakul

Tai Loei vernacular houses have a unique L-shaped plan which distinct from those in the northeast region. The structure of the house is raised on masonry stilts supported by brick wall foundation. The house was made from hardwood, decorative plaster hip roof and manila with clay tiles roof. (First generation house), rectangular cement tiles with the kite-like images (Later constructed house). Tai Loei vernacular house is a living pattern that appeared during 1907 — 1917 onwards, influenced by styles and materials from the craftsmen in the city which was different from the traditional northeastern style.

3 Conservation of Tai Loei Vernacular Houses

Conservation of Tai Loei vernacular houses in Na-O village has related with 3 groups as follows(Fig. 3).

(1) 10 of Tai Loei vernacular houses which have been relocated from adjacent villages to Srichan Temple since 2002. The conservation leads by the abbot of Wat Srichan who has requested to buy old houses in the community that will be demolished by the house owners due to their deterioration and limitation of maintenance budget. Volunteers in the village and traditional house builders assisted to rebuild the house in the temple area. The reassembling houses have adapted their use as monk cells and established a community folk museum exhibiting local way of life and the culture of Na-O community. The conservation process and methodology initially guided by a group of scholars from Faculty of Architecture at Khon Kaen University.

(2) 15 of Tai Loei vernacular houses located in the neighborhood and owned by Na-O's residents have found that their owners attempted to maintain in keeping the original conditions through daily use and day-to-day care. The limitations in conserving the house included the constraint in household economic in terms of lacking budget to repair the building to maintain good condition like in the past. Therefore, the houses have been dynamically changed in style and architectural space-use in order to meet local lifestyles in contemporary contexts. 12 houses are currently used for private household living, and another 3 houses have been adapted their uses as homestay services in order to support tourism.

(3) A group of vernacular houses recreated by the Loei ethnicity and located in the area of Na-O Sub district Municipality property. The houses have been designed and constructed by local craftsmen during the years 2007 — 2009. The main objective in construction of Tai Loei ethnic house in the municipality office property, well-known as "*Na-O cultural event yard*", is to serve tourists with homestay services and has been developed as a new destination for cultural tourism that supported and managed by local authority.

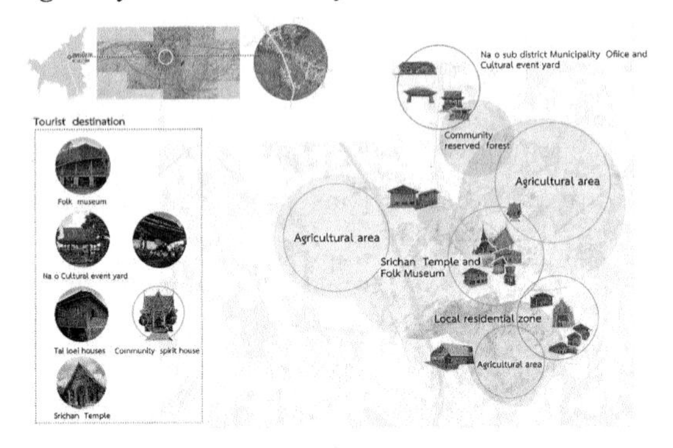

Fig. 3 Location of vernacular houses in Na-O village © Peerawat Saengchalee

An achievement of conservation at Na-O village is that there are many sectors that involved in different stages of vernacular house heritage, including Srichan Temple and traditional house builders group, homestay group from community sector, Na-O Sub district Municipality from government agency and the house owners themselves. The prominent success emerged in the early of 1987 from the participation of the stakeholders both the public sector and the academic sector. From the support of academic sector, Khon Kaen University, had nominated Na-O village to receive the award for the outstanding local community in Art and Architectural Conservation from the Association of Siamese Architects under the Royal Patronage in the year 2002. Later on, community development has greatly focused on vernacular houses conservation and continuously implemented by each sector in order to serve their objectives and interests. Consequently, the impacts from both development policy and practice of local agency were different in direction and did not support the interests of Na-O's community sector such as Srichan Temple and the house owners. Currently some development raised conflicts among stakeholders particularly in the issue of financial support standpoint and from the conservation framework in keeping minimum intervention.

4 Vernacular House Conservation Time Line

The development of vernacular houses conservation in Na-O village are as follows;

Before 1996 (*Before the action from different sectors*) Local residents conserved vernacular houses as a family heritage.

1996 The abbot of Srichan Temple started to collect abandoned and deteriorated traditional Tai Loei vernacular houses and adapted into monk cells. He has intended to relocate vernacular houses from adjacent communities because of his concern in the decease number of Tai Loei cultural heritage and his intention to maintain traditional houses so that later generations could study and learn from.

1998 — 1999 Provincial Department of Education supported Srichan Temple to establish Na-O community folk museum.

1999 — 2002 Faculty of Architecture, Khon Kaen University assisted in the initiative conservation of vernacular houses in Srichan Temple project and provided technical consultants for the conservation process with the participation of community sector, local agency and academic sector.

2006 — 2007 Provincial office launched the research on "*The value of Tai Loei traditional houses*" and suggested that process of vernacular house conservation must be implemented by stakeholders participatory.

2007 The Academic Services Center of Khon Kaen University corresponded with Na-O Sub district municipality planned to develop cultural tourism that focused on Tai Loei vernacular houses. The establishment of homestay services was begun.

2007 — 2009 Na-O Sub district Municipality established "Na-O cultural event yard" to support tourism activities and constructed a group of Tai Loei houses in order to support homestay services.

2010 The Academic Services Center of Khon Kaen University created database of valued architecture in Na-O village to provide the recommendation for conservation.

2016 Provincial Department of Culture created the cultural events on main street, known as "*a cultural road*", for tourists.

Important factors that influenced the conservation of vernacular house in Na-O village were from both external and internal factors. The impacts from external factors can be identified as (1) the promotion of Tai Loei conservation of academic sector, especially from the arrival of the Faculty of Architecture, Khon Kaen University between the year 1999 — 2002 in providing conservation assistance to the Sichan Temple and in advising local house owners about how to maintain old houses in accordance with academic standard principles, and (2) the formulation of policies in order to strengthen arts, culture and tourism development of local communities of the government under the National Economic and Social Development Plan No. 8-11 (1997 — 2016) which resulted in the implementation of the projects that contributed to the succession of Tai Loei vernacular houses, for example, the establishment of Na-O cultural event yard and the Na-O museum by Na-O Sub district Municipality. The establishment of Na-O Homestay group by the cooperation between the villagers of Na-O, sub-district Municipality and the Academic Service Center of Khon Kaen University. These development plans were implemented referring to the site potential in occupying a large number of traditional houses. As a result, Tai Loei vernacular houses have been widely known as the representation of identity of Na-O village, and in a larger contexts they were recognized as housing style of the Loei River Basin in the upper northeastern region.

The impacts from internal factors acted by people in Na-O community were (1) the necessity of living in the household (2) an attachment to the residence and the preservation of the legacy from the family ancestors. From local residents' perspectives, the owners of Tai Loei vernacular house expressed that main reason that their family still keeps the house is because it has still been used as a residence since their grandparents originally built. They feel that there is a sense of attachment and a bond of living place since they were born. Moreover, they stated that the house is a legacy from their ancestors to be preserved for later generations. Another group of Na-O residents, the representatives of Na-O Homestay group acknowledged that the main reason that their families keep their houses is because it is a legacy from ancestors. As for adapting the house for homestay services to accommodate tourists, they were considered an alternative way to generate income for a household expense in order that the income can support maintenance costs of the house.

By integrating stakeholders into activities under the community tourism development program during the years 1999—2002 and 2007 — 2010, Na-O village became a well-known site in northeast region for heritage conservation especially the impacts from the operation and management of local stakeholders resulted in the increase of number of vernacular houses in the area of Srichan Temple, cultural event yard of Na-O Sub-district Municipal Office, and within the neighborhood of Na-O community.

5 The Roles of Stakeholders in Conservation of Vernacular Houses

The conservation of Tai Loei vernacular houses in Na-O Village has different purposes according to intents and the needs of 4 main sectors, namely, Srichan Temple Group, Na-O Sub-district Municipality, Na-O Homestay Group and the house owners(Fig. 4). The details are as follows:

(1) Tai Loei Group in Srichan Temple: Using the method of relocating the houses in the neighborhood of the community and then reassembling them in the property of Srichan Temple. Former houses were adapted to use as monk cells. The house physical features were kept and conserved following the concepts of minimal change of use.

(2) Na-O Sub-district Municipality Office and Tai Loei local residents: Establishment of Cultural Hall which was a reconstruction of Tai Loei traditional houses and a group newly recreation of 11 vernacular houses built by local participation and indigenous method of construction. Recreation houses were functioned as a place for organizing activities, festivals and traditional events of the community, Na-O community museum which exhibits about Na-O history and community development projects implemented by Sub-district Municipality Offices. Some of them were managed as supporting facilities for tourist.

(3) Tai Loei residents who joined the Na-O Homestay Group: Using the method of day-to-day care, including cleaning and improvements to repair damaged parts. Some of the houses have to adjust their use to a minimum requirement in order to demonstrate original Tai Loei's lifestyle for homestay services but most of them have still use for the main purpose in household living.

(4) Tai Loei group of villagers resided in the neighborhoods: Using conservation method through daily routine such as cleaning and repairing damaged parts in order to improve the house conditions for household living. Most houses have modified by expanding household spaces for accommodate the current lifestyle.

6 Problems and Limitations in Conservation

Overview of problems and obstacles in conservation of Tai Loei vernacular houses in Na-O village can be summarized as follows:

-Concepts and methods of vernacular house conservation are in conflict among

(a) **Srichan Temple zone:**
monk's cells (left & center), community folk museum (right)

(b) **Na-O Sub district Municipality Office zone:**
homestay (left& center), community exhibition (right)

(c) **Local residences zone:**
homestay (left), private residence (center & right)

Fig. 4 Vernacular houses from different conserving objectives © Thanit Sateinnam

Srichan Temple Group, Na-O Sub-district Municipality Group and house owners in the community because the objectives in conservation and development are responded to different group of users. Moreover, the level of understanding of conservations and evaluation of vernacular house values are from different perspectives.

-The shortage of resources in conserving vernacular houses, both in terms of financial supports and traditional way of materials and construction. Presently, there are only a few number of builders and technicians with expertise in building Tai Loei houses. Moreover, traditional building materials such as wood and clay tiles are scarcity.

-Most of the owners of Tai Loei vernacular houses are villagers. They are still lack of the readiness and potential to protect the heritage, especially in terms of budget and knowledge in maintaining vernacular houses from an academically standard procedures and conservation technique viewpoints. Consequently, some houses are dismantled, improved, repaired and extended inappropriately which will lead to the loss of identity of the Tai Loei traditions.

-In the dimensions of cultural tourism development, there are some difficulties and limitations. For example, in comparing with other tourist destination in the province, Na-O village currently is not an important site that can attract tourists. Tourists and visitors are only passer-by on their way to other destinations like Chiang Khan, a small old town along the Mekong River. Moreover, some stakeholders such as house owners, villagers, Srichan Temple group still want to live peacefully and do not need their community to develop into a full tourist attraction.

7 Conclusion

Tai Loei vernacular houses in Na-O village have been inherited by various sectors. Some can maintain their original authenticity according to the concept of conservation with minimal interventions depends on the purpose, concept, knowledge and skill and limitations of stakeholders. The results can be concluded in each case as following:

(1) Tai Loei houses in Srichan Temple can still be able to maintain the architectural and craftsmanship characteristics as close to the traditional style. But the relocation of the house from its original site and re-assembling in the temple's property affects the value of vernacular heritage in maintaining its relationship with environmental settings.

(2) Tai Loei houses built by local residents via the management by Na-O Sub-district Municipality were a new re-creation of vernacular houses. This group of vernacular houses built in the property of government agency with the purpose to serve cultural tourism promotion. The "image" of traditional Tai Loei house was constructed to attract outsiders, while the ongoing process was also a participating learning process to conserve vernacular house by the villagers themselves. Many examples of Tai Loei houses collected in this zone were constructed differently from the traditional forms because of the use of construction contractors and modern technicians instead of folk builders and craftsmen to create these model

houses.

(3) For Tai Loei houses in the Na-O Homestay group and owned by villagers in the community, although they can be able to conserve their authenticity from the preservation of the house in original place of setting, the adaptation by their residents in order to serve the appropriate needs for contemporary social contexts and to support tourism development as a homestay housing type can also affect the change and adaptation of architectural style from the tradition.

For two decades of conservation process of Tai Loei houses in Na-O village, vernacular houses as valuable heritage have expanded the status of Tai Loei ethnic house from the residents of the villagers to other positions, such as monks accommodation, homestay, folk museums and exhibitions of community events, that transformed the houses for their original uses by the efforts of various sectors in conserving their heritage. Therefore, the conservation of vernacular housed is not just a physical maintenance of valuable architecture, but it is also an attempt to achieve maximum benefits of the house's function so that they can survive in a contemporary social context which is different from the past that the house was initially built in Na-O community.

Currently, the tendency of Ban Na-O village in conserving vernacular houses is still in a positive direction. There are recognizable results in maintaining physical features of some authentic traditional houses. Moreover, the conservation of vernacular houses in Na-O village has been considered as a successful site because it has continuous management and operation through the efforts of various sectors for more than 10 years until today by comparing with other sites in the northeast region.

Findings from field research indicated that there were some conflict of interests in conserving vernacular houses among stakeholders such as Srichan Temple, Na-O Subdistrict Municipality and the house owners because of the differences in understanding of conserving concepts, objectives, methods of inheritance and utilization of the heritage which affected to the output of conservation process differently. An overall conservation policy and development plan of vernacular houses in Na-O village has still require a clear direction and stakeholder's participation in order to support and accommodate different objectives in keeping vernacular houses.

Acknowledgements

This article is partially supported by the Center for Research on Plurality in the Mekong Region (CERP), Faculty of Humanities and Social Sciences, Khon Kaen University.

References

CHAPMAN W, 2007. Determining appropriate use. //ENGELHARDT R A. Asia conserved: lessons learned from the UNESCO Asia-Pacific heritage awards for cultural heritage conservation (2000 2004). Bangkok: Clung Wicha Press Co., Ltd.

FEILDEN B M, 2003. Conservation of historic building. Oxford: Architectural Press.

ICOMOS Thailand, 2011. Thailand charter on cultural heritage management 2011. https://www.icomosthai.org/THcharter/63546_Charter_updated.pdf

Office of The National Economic and Social Development Board. [n. d.]. The 8th-11th national economic and social development plans (1997 — 2016). http://www.nesdb.go.th/nesdb_en/main.php?filename=develop_issue

SCHLIESINGER J, 2001a. Tai groups of Thailand introduction and overview. Bangkok: White Lotus Co., Ltd.

SCHLIESINGER J, 2001b. Thai groups of Thailand volume 2 profile of the existing groups. Bangkok: White Lotus Co., Ltd.

SUTTHITHAM T, 2001. The Conservation of vernacular "Isan" architectural heritage under the participatory approach. Khon Kaen: Asian Wisdom, Environment, Culture & Art Foundation.

Analysis on the Planning of the Coordinated Development of Village Building Protection and Tourism — Taking Yanzhongzui Village in Donghu Scenic Area as an Example

LI Hongling, *DONG Hexuan*

School of Architecture & Urban Planning, Huazhong University of Science and Technology, Wuhan, Hubei, China

Abstract This paper analyzes the historical evolution of the Yanzhongzui Village from the aspects of water, buildings and farmland. It also analyzes Yanzhongzui's architecture and space from the following aspects: the relationship between the bottom of the village map, the arrangement and combination of buildings, the D/H value of space and the distribution of building functions, building density, building shadow, building shoreline concessions. Furthermore, it makes a superposition analysis on the building rating of village buildings from the aspects of building quality, building height, building material, building style, building alignment rate, building form. Finally, according to the village history and culture, ecological construction strategy and the functional requirements of East Lake tourism development, Yanzhongzui building is divided into demolition, repair and protection. At the same time, new buildings are built on the basis of the existing building pattern according to the extracted village building texture to form the overall village building pattern.

The plan suggests that while developing the village tourism economy, attention should be paid to the protection of the village architecture pattern. On the basis of preserving the historical memory, continuous in-depth excavation and exploration should be carried out to improve the preservation and protection measures of the village architecture and rationally develop the scientific protection of the village historical architecture.
Keywords villages in scenic spots, building protection, architectural pattern, history and culture, tourism development

With the promotion of large-scale urbanization, protection and development have become the trend of ecological civilization construction. East Lake Yanzhongzui Village are located in the core of the scenic area in Wuhan, directly affected by tourism development, and have radiation effect on the surrounding urban areas.

In the process of modernization, many buildings that can reflect the characteristics of regional villages gradually fade away from people's vision. How to protect the buildings bearing the memory of villages reasonably without affecting the process of urban development is a very important issue. It is of great research value and social significance to find effective ways of building protection, cultural inheritance, ecological environment management and reasonable economic and social development in scenic villages.

1 Yanzhongzui Village in East Lake

Yanzhongzui Village is located in the East Lake Luoyan Scenic Area, Wuhan, Hubei, China. Luoyan Scenic Area is an important part of China's 5A East Lake Scenic Area. It is located in Baima Scenic Area in the north and across the lake from Moshan Scenic Area in the south. Yanzhongzui Village enjoys relatively convenient transportation, with East Lake Tunnel and Wuhan Avenue in the west, the Third Ring Road and Qingwang Road in the east. The main road of the scenic spot connects the village with Moshan Scenic Spot in the south, and the wharf in the south of the village has waterway communication with the wharf along Moshan Scenic Spot. At the same time,

Yanzhongzui Village is a peninsula near the lake. The natural environment of the village is superior. It is adjacent to Shau Kei Dun in the north, Tangling Lake in the west, Luoyan Island in the south, and Tuan Lake in the southeast. It has unique viewing advantages. Its total planned area is 24.22 hectares, and the current construction scale is about 105,600 square meters(Fig. 1, Fig. 2).

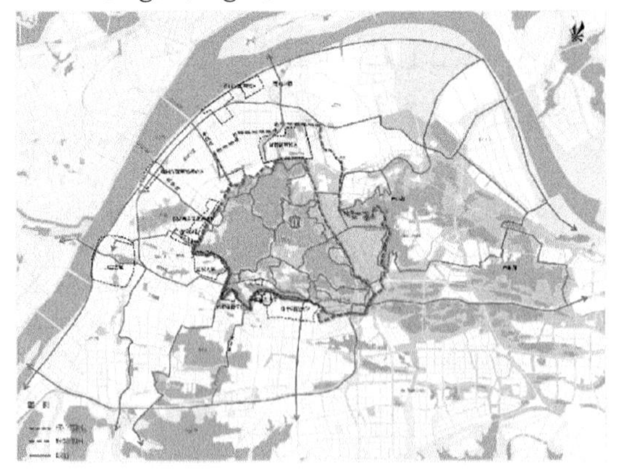

Fig. 1 Location analysis of East Lake Scenic Area

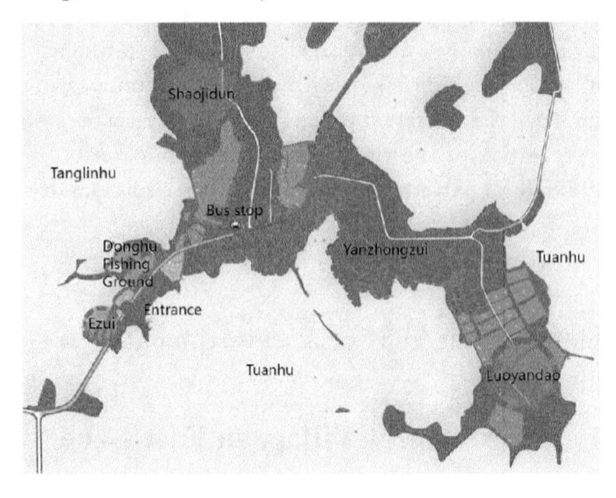

Fig. 2 Location analysis of Yanzhongzui Village

2 Villages in Scenic Spots

The title of "Villages in Scenic Spots" was first put forward by Hangzhou municipal government in "Administrative Measures for Villages in Scenic Spots in Hangzhou West Lake Scenic Area". The concept of "Villages in Scenic Spots" is more representative of the definitions of Hou Wenna, Hu Wei and others. It defines "Villages in Scenic Spots" as "*a community settlement that has been incorporated into the planning and management of scenic spots, collectively owns the land, administratively establishes villagers' committees, mainly resides in agricultural registered permanent residence, and retains the customs and features of the village.*" Villages in Scenic Spots is a special form of existence of villages in the city and an important part of scenic spots. It has the dual attributes of "scenery" and "village". Villages in Scenic Spots is located in the East Lake Scenic Area and is a typical "Villages in Scenic Spots" (Hou, et al, 2007).

3 Research on the Historical Evolution of Yanzhongzui Village

Based on Google's historical images of villages in 2005, 2010 and 2019, the three elements of water, buildings and farmland in the villages were extracted and analyzed (Fig. 3): in 2005, the area of water was relatively large and distributed evenly in the base; in 2010, the area of water decreased and concentrated in the southeast; in 2019, the area of water decreased and the water in the northwest basically disappeared. Most of the buildings in the village were built adjacent to Luoyan Road in 2005, and there were almost no buildings on the west side. In 2010, the building density on the side adjacent to Luoyan Road increased, and the buildings were distributed along the main road in the base. In 2019, the buildings spread to the village, but there were few buildings in many places with water bodies in the southeast, and the wetland environment at the water inlet was good. In 2005, there was a large area of farmland in the village, and the villagers were self-sufficient. In 2010, the area of farmland in the village was reduced, and land was filled to build houses. In 2019, the area of farmland was greatly reduced. From the analysis, we can know that the historical development of the village originates from the traffic of Luoyan Road. It starts first along the side of Luoyan Road, and then takes the S-shaped main road in the village as the pulse to further develop and build houses. After that, the buildings are scattered and built on empty land. Today, the buildings occupy nearly saturated land.

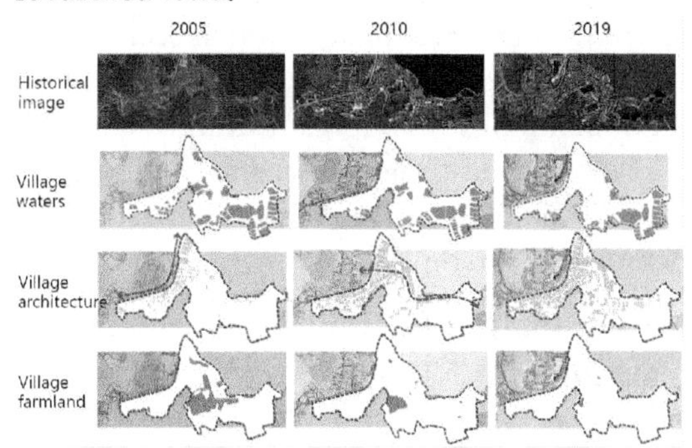

Fig. 3 Historical evolution of Yanzhongzui Village

In-depth analysis of the historical evolution of village architecture layout (Fig. 4). In 2005, the settlement form of village architecture was relatively clear, roughly divided into three types: parallel type, centripetal type and street type. However, due to sparse buildings, the settlement form was not completely formed. In 2010, the settlement forms of village buildings can be roughly divided into four types: parallel type, centripetal type, courtyard type and street type. The building density is generally appropriate and the settlement form is clear and clear, but the newly-built buildings make the courtyard form chaotic. In 2019, village building settlements can be roughly divided into four types: parallel type, centripetal type, courtyard type and street type. The increase in building density and the narrowing of street width have affected the original building settlement form and the village texture has become chaotic.

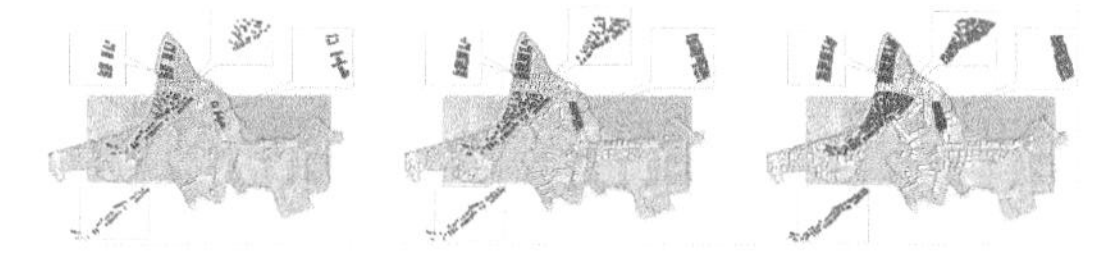

Fig. 4 Historical evolution of village architecture layout

4 Architecture and Space Analysis of Yanzhongzui Village

4.1 Analysis of Village Space

According to the map-based relationship of villages (Fig. 5), the spatial distribution of buildings is extensive, but its shape and size are extremely uneven. Combining with the surrounding environment of the space, the plot is divided into three spaces (Fig. 5) according to its morphological characteristics, namely: finely divided free zone, enclosed zone and open complete zone. The three regions roughly divide the plot into three parts and connect it in series on the main axis of the space in combination with the spatial trend of the road.

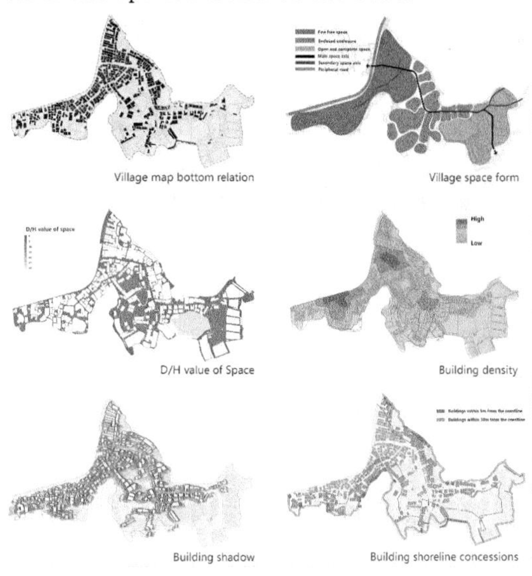

Fig. 5 Village space and architecture

Village space is mainly enclosed by roads, venues and buildings. According to the arrangement and combination of buildings, buildings in the plots are divided into five forms (Fig. 6) from orderly to scattered. Orderly building plots have only traffic space and lack neighborhood communication space. Between orderly and scattered buildings, the layout is flexible, increasing the space for neighborhood communication activities; The space formed by scattered buildings and sites is mostly production space with a wide range. The layout of buildings and roads in the village is adapted to local conditions and basically meets people's living needs. However, the formed space is gray and lack of vitality due to various reasons. In the reconstruction of buildings, the re-planning of roads and building space to increase its space utilization efficiency needs to be considered.

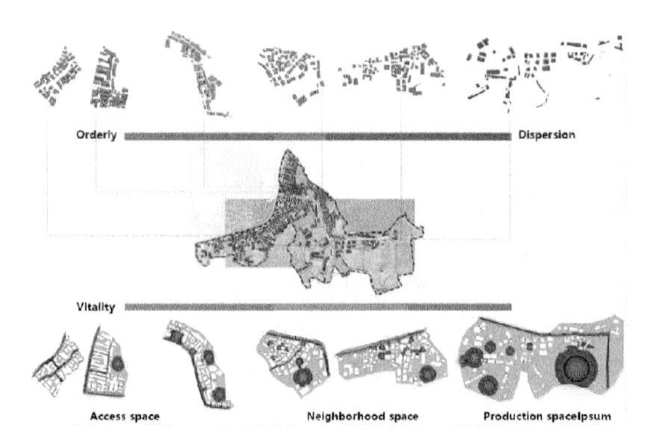

Fig. 6 Village architecture form

The D/H value of village building space is concentrated in the range of 0. 5～1. 5 (Fig. 5). The spatial connection is relatively balanced and has a sense of affinity.

4.2 Analysis of Village Architecture

The residential buildings are mostly concentrated on the west side of the site, and the space in the residential area is relatively crowded, while the non-construction land on the east side is also used for planting, breeding and commercial land. The village lacks public space and opportunities for neighborhood communication.

The building density distribution map is obtained from the spatial integration degree analysis (Fig. 5). It can be seen that the higher building density is consistent with the centralized distribution of commercial buildings. Lower density leads to more productive buildings. This shows that building density affects the distribution of building functions.

Analysis of 30 satellite imagery from 2006 to 2019(Fig. 5), select the representative three-year data with great changes, and superimpose them to obtain the analysis map of building age. From the layout, we can see the evolution law of village buildings. The new buildings in the later period upset the architectural texture, but the new buildings are of good quality, and some of them will be demolished in the later planning.

Analyze the distance between current buildings and lake shoreline (Fig. 5). According to the requirements of the upper planning, buildings within 5m of the lakeshore need to be demolished. No new buildings are allowed within 50m of the lakeshore. There are many and miscellaneous lakeshore buildings in the village. In the later planning, buildings within 5m of the lakeshore need to be demolished, and buildings 5～50m from the lakeshore need to be demolished or repaired.

5 Evaluation of Building Grade in Yanzhongzui Village

5.1 Space Classification Levels of Village Architecture

It can be seen from the height of the building layer (Fig. 7) that the height of the building in the western part of the site is more undulating. Overlooking the villages and towns from the East Lake Greenway, you can see the deeper level of the building. From the Chutiange overlooking the villages and towns, you can see that the entrance to the west of the village and the central building are higher, but the overall village skyline is still too flat and lacks the visual focus of the overall situation.

The building materials of the village are mainly divided into concrete, brick, wood and others (Fig. 7).

The quality of village buildings is divided into four levels: intact, basically intact, generally damaged, and severely damaged (Fig. 7).

The architectural style of the village is divided into four categories: key style buildings, good, general and poor. The main classification is based on architectural culture, characteristics, building roof, fa ade, bottom interface, architectural color, and architectural interior space(Fig. 7).

The building alignment rate reflects the conformity and identifiability of street interface forms. It is generally believed that 70% or more of streets with building alignment rate can meet the requirements of basic street atmosphere and continuity(Fig. 7).

The village building structure can be divided into independent courtyard buildings, public courtyard buildings, townhouse buildings, co-top buildings, and single-family buildings.

5.2 Grade Evaluation of Village Architecture

According to the literature investigation and the village site investigation, the building height, building material, building quality, building style, building alignment rate and building form are given grade evaluation values of 0.1, 0.15, 0.20, 0.25, 0.15 and 0.15 respectively, and the superposition analysis is carried out to obtain preliminary suggestions on the demolition, repair and retention of village buildings (Fig. 7).

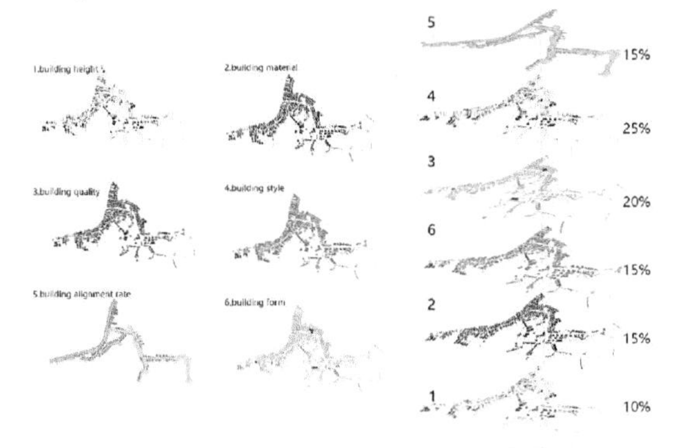

Fig. 7 Building grade evaluation

6 Village Building Reconstruction Planning for Coordinated Tourism Development

According to the recommendations of the village building grade evaluation, combined with the previous analysis of village buildings and space, it is determined that the village buildings should be demolished by 54,800 square meters and retained by 58,800 square meters. According to the needs of the village tourism development, the retained buildings of the existing planning scheme will be renovated and repaired. Meanwhile, under the functional requirements of tourism development, 71,100 square meters of new buildings will be built (Fig. 8).

Building a green and healthy ecological tour circle will promote the sustainable development of the village economy. Different ecological tour routes will be planned through the transformed regional division in combination with the local human geogra-

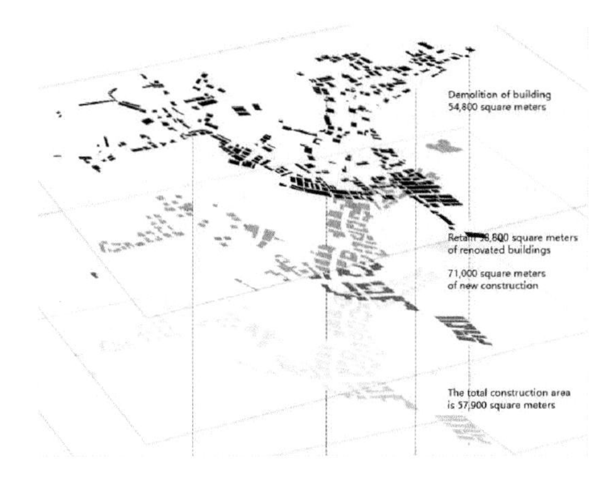

Fig. 8 Village building planning

phy and architectural environment to enrich and strengthen the tourism characteristics of the village. At the same time, village residents are encouraged to transform idle or vacant houses and their own land into characteristic residential houses, improve the facilities for food, clothing, housing, transportation and purchase in villages, strengthen tourists' experience in rural ecotourism, and coordinate with village tourism and sustainable economic development.

7 Village Building Reconstruction Planning for Coordinated Tourism Development

With the advancement of urbanization and the development of tourism, the contradiction between the protection and development of villages in scenic spots is becoming increasingly prominent. In view of the protection and development of villages in scenic spots, attention should be paid to the importance of cultural protection and ecological security, and balance and coordination should be paid in the development and protection. Research is the basis of protection and development. This paper probes into the ways of village protection and development from the perspective of coordinated scenic spot tourism development, and obtains the following views:

(1) Propose a new idea of coordinated development of scenic villages in tourism development. Propose that the protection of traditional villages is the basis of develop-

ment. Tourism development can provide impetus for the sustainable development of villages. The protection of village buildings and the function of commercial service can be sustainable, thus realizing the full play of tourism benefits and village service functions in scenic areas.

(2) Through on-the-spot investigation and data analysis of Yanzhongzui Village, the natural resources, history and culture, economic development and other aspects of the village were analyzed, and the overall pattern, architecture, space and other artificial environmental factors of the village were deeply studied, and the specific plan for the demolition and repair of buildings was determined, which provided a basis for the functional positioning of the village in scenic tourism development.

(3) The unique geographical environment of scenic spots and the cultural characteristics of villages promote the sustainable development of a green and healthy ecological circle between culture and buildings, achieve the coordination between tourism development and village protection, enhance the endogenous growth of villages, and enrich the practical path of village protection and development.

References

AHMAD A, 2013. The constraints of tourism development for a cultural heritage destination: The case of Kampong Ayer (Water Village) in Brunei Darussalam. Tourism management perspectives, (8).

Dewi L, 2014. Modeling the relationships between tourism sustainable factor in the traditional village of Pancasari. Procedia - social and behavioral sciences, (7).

GEORGE B P, 2007. Local community support for post-tsunami recovery efforts in an Agrarian Village and a tourist destination: a comparative analysis. Community development, (6).

GHADERI Z, HENDERSON J C, 2012. Sustainable rural tourism in Iran: a perspective from Hawaman village. Tourism management perspectives, (3).

HOU W, HU W, YOU J, et al., 2007. Analysis of the management countermeasures of villages in scenic spots—taking Xihu scenic area as an example. Journal of Anhui agricultural sciences, (05): 1348-1350.

KNEAFSEY M, 2001. Rural cultural economy: tourism and social relations. Annals of tourism research, (3).

KOSCAK M, 1998. Integral development of rural areas, tourism and village renovation, Trebnje, Slovenia. Tourism management, (1).

Medina L K, 2003. Commoditizing culture—tourism and Maya identity. Annals of tourism research, (2).

Mitchell C, 1998. Entrepreneurialism, commodification and creative destruction: a model of post-modern community development. Journal of rural studies, (3).

TUCHER H, 2001. Tourists and troglodytes negotiating for sustainability. Annals of tourism research, (4).

WALL G, 1996. Perspectives on tourism in selected Balinese villages. Annals of tourism research, (1).

The Development of Vernacular Architectures in the Loess Plateau During the Progression of Urbanization

CHI Mengjie

Xizang Minzu University, China

Abstract The Loess Plateau (LP) is located in the north of central China, its length exceeds 1,000 kilometers, and its width is about 750 kilometers, with a total area of 648,700 square kilometers, which accounts for 6.76% of the total land area of China. The LP is the most concentrated and largest loess area in the world; its climatic characteristics and natural environment make it become one of the agricultural origins worldwide, and it is regarded as the cradle of the Chinese nation and Chinese culture. The vernacular architectures in the LP area are the results of the bilateral selection of human and nature. The traditional Yao (kiln) residential districts, such as the cave dwellings and underground pits, are typical vernacular architectures in the area, which fully reflect the wisdom of local residents adapting to nature.

During the rapid progression of urbanization, the vernacular architectures and living environment of the LP are facing many new problems, and the spatial configuration of the residential environment formed by history is rapidly disappearing. Therefore, it has become very urgent to find ways to develop vernacular architectures in the progression of urbanization. The characteristics of vernacular architectures in the LP area and the urbanization process of the area are combined to find ways to protect and develop the special architectural form. It is hoped that the land could retain its unique residential characteristics and the emotions of local residents through thousands of years.

Keywords Loess Plateau, vernacular architecture, urbanization, development

1 The Developmental Predicament of Vernacular Architectures on the Loess Plateau

Earthen architecture is one of the most widely applied and oldest residential traditions. In addition, the raw soil is an old and widely applied construction material due to its easy acquisition and low cost. So far, various raw soil architectures are still seen in Africa, the Middle East, North America, South America, Europe, and Asia. In the Loess Plateau (LP) area, the major forms of raw soil architectures include cave dwellings, adobe buildings, and rammed earth buildings. The raw soil material has unique properties; its advantages in heat storage make it the best choice for people to resist the severe cold weather; the raw soil material is easy to process with low energy consumption, low pollution, and low costs; the amount and processing energy consumption are 9% and 3% of clay bricks and concrete respectively; besides, the raw soil material has preferable renewability, which can be reused or used as fertilizers after dismantlement; most importantly, the soil on the LP has the characteristic of uprightness, which is convenient for construction.

1.1 Changes in Residential Needs

Although the raw soil material has various advantages, its unavoidable inherent deficiencies such as poor mechanical properties and poor water resistance make the earthen architectures be of small bays and limited layouts. The traditional residential architectures such as cave dwellings and underground pits are difficult to meet the modern requirements of quality improvement, convenience, and safety improvement; therefore, they have been out-of-date in lots of areas. In addition, foreign culture is also changing the local psychosocial factors. Traditional vernacular architectures are regarded as symbols of poverty and backwardness. As the rural population has gradually flocked to the cities, the cultivated land has declined sharply, the ecological foundation has deteriorated, and increasing number of the traditional residences have been vacant. At the same time, the concept of residence for local residents has changed, and the psychological factors like mutual comparison have led to the earthen architectures developing into the direction of being higher and larger.

The traditional forms of architecture that have been protected and inherited are developed over hundreds of years and are the results of a combination of multiple factors. Instead of keeping and inheriting traditional architectural forms, it is hoped that the traditional architectural culture can be balanced with modern needs, thereby making the vernacular architectures in the ever-evolving states.

1.2 Impacts of Modern Architectures

Impacted by the trends of modern architectures, the application of natural materials to build or impose a certain cultural concept in the architectures has become a popular practice. The construction of natural materials in the forms of modern architectures seems to have achieved a harmonious situation in which modernity and traditions are merged, and the contradiction between modern civilization and the residential traditions is dissolved. However, the essence of so-called "harmony" is "to mechanically apply modern forms to earthen architecture" essentially. Its inherent deficiencies would never be overcome if the attentions are paid to materials and cultural concepts in aspects such as ecological-friendly and green, or the green standards were met in several perspectives. These inherent deficiencies are the core factors that constrain the modern application of earthen architectures.

It should also be pointed out that the effects of modern technology need to be supported by a complete industrial system, and its application also requires corresponding economic conditions. Most areas in the LP are economically underdevelopment; in addition, the fragile environmental carrying capacity makes it completely unrealistic to indiscriminately imitate the construction methods or indiscriminately imitate new construction techniques of other areas. Therefore, modern construction methods and techniques need to be properly localized. Once the interactions among architectural forms, technologies, and materials are balanced, it is conducive to the continuation of traditional architectural forms.

In addition, during the construction process, it is less considered to be able to adopt a "method with fewer harms to nature". The problems of the values have been neglected in the social and architect industry for a long time. In the progression of urbanization, the naturally formed community structure is gradually disintegrated and destroyed; the lifestyle and neighborhood relationships are easily changed by modern architectural forms. "Architecture is to establish the harmony with natural materials, and the construction is beyond the fame and fortune". Compared with preserving the vulnerable architectures, the preservation of traditions, lifestyles, emotions, and memories carried by these architectures are more difficult, more urgent, and more challenging.

2 Progression of Urbanization in the LP Area

According to the boundaries of the county administrative areas, the LP mainly includes 341 counties (cities) in 7 provinces (autonomous regions) with a total population of 108 million. Statistics in 2018 indicated that the urbanization rate of China reached 59.58%. The urbanization rates of provinces in the LP area in descending order are successive: Inner Mongolia (62.70%), Ningxia (58.88%), Shanxi (58.41%), Shaanxi (58.13%), Qinghai (54.74%), Henan (51.71%), and Gansu (47.69%). Based on the 13th Five-Year Plan for Economic and Social Development of the People's Republic of China issued in 2016, China is divided into 19 urban agglomerations (or economic belts). The so-called vernacular architectural area in the LP mainly refers to 4 of them, i. e. the "the Hohhot-Baotou-Erdos-Yulin urban agglomeration", "the central Shanxi urban agglomeration", "the Ningxia-Yellow River Basin urban agglomeration" and "the central Shaanxi urban agglomeration". Data collected in 2016 showed that the urbanization rates of the above 4 urban agglomerations are 69.3%, 60.1%, 61.0%, and 58.0%, respectively.

2.1 Characteristics of Urbanization Progress in the LP Area

Since 1949, the urbanization process in the LP and its surrounding areas have gone through three stages. From 1949 to 1979, the unified arrangement of the national planned economy brought the first round of urbanization construction to these areas; from 1979 to 2010, the urbanization rate of China increased rapidly from 17.9% to 49.9%; the rapid development of economy in China led to the increasing demands for energy resources, the urbanization of the LP was accompanied by the development of the national energy resources industry; Since 2010, with the industrial transformation and upgrading of the southeastern coastal areas of China, the major purpose for the urbanization development in these areas is to undertake the industries of southeast coastal areas, the determination of national industrial layouts and the national economic-stimulus plan.

The three urbanization stages showed different characteristics. In the first stage, urban planning was oriented by resource exploitation, heavy industry production, and national defense construction; the boundaries between urban and rural areas were obvious, and the population outside the area continued to flow in and the local population was restricted to flow out; the framework of the urban agglomeration was initially determined. In the second stage, the population flew in both directions, which was oriented by the out-flowing direction with the southeast coastal areas being the major destination; the traditional industrial cities were decaying, and new resource-based towns appeared. In the third stage, the population flew in both directions, the middle-aged population returned to the LP area, the youth population flew outwards to the central cities and the cities outside the LP area; the road and railway transportation network and national industrial plan determined the development pattern of urbanization in these areas; the regional central cities and important towns have been established.

2.2 Characteristics of Urbanization Specifications in the LP Area: from the Micro and Macro Perspectives

The institutional arrangements for urbanization in China can be analyzed from both micro and macro perspectives. Microscopically, institutional arrangements are mainly compulsory specifications, which analyze the legal relationships between people, land, resource utilization and protection, and other factors of production in terms of legal confirmation and circulation; the forms of expression are mainly laws and regulations. For example, the household registration system, the land acquisition and expropriation system, urban planning and construction, environmental resource protection, etc. Macroscopically, the institutional arrangements are economic planning, i. e. the policy arrangements to achieve the national economic and industrial development in the urbanization process. Such as the national new urbanization plans, etc. In other words, the micro-institutional arrangement determines the degree of development of the urbanization process, while the macro-policy trend determines the direction and scope of the urbanization process.

From the micro perspective, the urbanization design of China has shown 4 major tendencies: (1) The guarantee of the orderly flow of the population, which manifested as the adjustment of the household registration system, the relaxation of one-child policy, etc.; (2) Avoiding the polarization between the rich and the poor and the excessive urban-rural gap, which is characterized by improving the social security system, improving the labor contract system, etc.; (3) Ensuring that the urbanization process is in a stable order and strengthening the protection of private property rights, which is characterized by perfecting the legal system such as expropriation and requisition; and (4) Ensuring the sustainable development, which is represented by the institutional arrangements for environmental protection and the design of taxation systems. The first 2 institutional arrangements aim at the internal factors of the urbanization process; while ensuring the basic production factors of the unified market, the widening of the urban-rural gap should be avoided as much as possible. The latter 2 arrangements ensure that the urbanization is processed in an orderly manner in response to external factors, and the negative effects of urbanization are avoided as much as possible. In the overall macroscopic policy design, China has broken the administrative divisions, established a number of economic belts and urban agglomerations, and realized the industrial layout of the national economy through infrastructure construction, thereby determining the scope and plan of the urbanization process.

In terms of the formation mechanism, the urbanization process on the LP presents a top-down formation characteristic, i. e. the national administrative forces mobilize funds to provide infrastructure for the city, and local governments rely on the infrastructure provided by the central government to find the economic industries that are suitable for the regional characteristics, and the aggregation effects of economic industries are used to promote the urbanization process. In other words, the urbanization process is essentially the macroscopic industrial planning and the microscopic implementation methods made by the central government; the local governments promote the urbanization process throughout the country through the prescribed implementation measures within the scope of the plan.

3 The Reborn of Vernacular Architectures in the LP Area

On the issue of the inheritance of traditional residences, Professor Mu Jun put forward that "*The current rural construction that need to be particularly inherited is the basic logic of the interactive developments of natural factors (conditions), human elements (demands), and competencies (economics and technology), which lies behind the formation and development of traditional residences*". In terms of the vernacular architecture in the LP area, on the one hand, it is difficult for people who have settled in cities and towns to return to the countryside for a long time; on the other hand, the traditional forms of raw soil architectures are difficult to meet the aesthetic and living needs of the people. The vernacular architectures in the LP area are facing two sets of dilemmas in the progression of urbanization: How do traditional construction methods meet the needs of modern life? How do traditional architectural forms draw on modern construction technology?

3.1 Changes in Architectural Functions: from the Residential Values to the Social Values

The traditional construction methods are summarized by local residents to meet their basic living needs, with low cost, easy acquisition of materials, and simple construction. However, the continuous application of traditional construction methods is facing difficulties. Modern architectures can provide people with comfortable living environments and satisfy their aesthetic needs, while economic development also gives people more choices; from the subjective perspective, people are reluctant to choose houses constructed by traditional methods; in addition, the traditional residences with single residential functions are difficult to cope with the requirements of urbanization; thus, the social functions of rural architectures have become an inevitable development direction.

The return of middle-aged and elderly

population and the industrial transfer from the developed areas are the main features of urbanization of the LP, which provides a possibility for the continuation of traditional construction methods. The government is bound by the microscopic institutional arrangements and is obliged to provide public services to protect the development of rural areas, such as public health care and public education. At the same time, the microscopic institutional arrangements also gives the practical implementations — achieve public services through public buildings, such as schools, clinics, tourist service centers, etc. In terms of public constructions, the government could take the application of traditional construction methods as one of the evaluation requirements for bidding. First, traditional construction methods can maintain the integrity of rural landscapes, which is also an inevitable requirement of architectural planning. Second, in the construction process, the traditional methods can reduce the damages to the ecological environment, which is also the requirement of sustainable development macroscopically. Third, the application of traditional methods can reduce the construction costs, indicating its realistic feasibility.

3.2 Revolution of Architectural Forms: the Application of Modern Techniques

The traditional architectural form is subject to construction materials and construction methods. Despite the various forms of construction, the functions of the traditional architectures are relatively unitary, and most of the architectural forms are embodied in the realization of the residential function. Improvements in materials and modern construction methods will undoubtedly help traditional architectural forms achieve the diversification of architectural functions. For instance, the research institutions of the Western countries represented by the French International Centre on Earthen Architecture (CRAterre) have formed a mature theoretical system for the application of traditional materials and modern construction methods in traditional architectures, and have been verified by engineering practices. It provides an operational basis for China to innovate traditional architectural forms and expand the use of these architectures.

4 Conclusion

The vernacular architectures in the LP area are the architectural forms created by local residents to meet their needs of living, which conform to the local ecological carrying capacity. In terms of construction material selection, the convenience for local residents to obtain is considered; in terms of the construction cost, the costs of vernacular architectures are low; in terms of the construction methods, no complicated processing technologies are required, which is convenient for local residents to operate; in terms of the functions of these architectures, the vernacular architectures should be able to resist the unfavorable natural conditions and satisfy basic needs for survival. The logic expressed by the architectural form is that humans can find a balance between their demands, the natural factor constraints, and the ability to change nature. When urbanization brings about population outflows and lifestyle changes, changes in construction methods can lead to the innovation in architectural forms and ultimately break through the traditional functions of architectures, which provides an idea for the reborn of vernacular architectures in the LP area.

References

China National Bureau of Statistics, 2018. China statistical yearbook 2018. Beijing: China Statistics Press.

JONES C I, HAMMOND G P, 2008. Embodied energy and carbon in construction materials. Energy, 161(2):87-98.

LE Corbusier, 2008. Vers une architecture. Paris: Editions Flammarion.

MU J, 2016. Exploring, renewing and inheriting the tradition of earthen construction. Architectural journal, 571:6.

REN Z P, XIONG C, 2018. The great migration of Chinese population. Beijing: Evergrande Research

Institute.
The 4th session of the 12th National People's Congress, 2016. The 13th five-year plan for economic and social development of the People's Republic of China (the 13th Five-Year Plan).
The Central Committee of the Communist Party of China(CPC)and the State Council, 2014. Summary of China urbanization plan for 2014—2020.
XIE X, HUANG L,2016. "Natural construction" is by no means a simple building of natural materials. Nanfang Dushi Daily, 2016-8-17(GB06).

The Fujian Tulou Conservation Strategy: A Sino-Italian Joint Project

WANG Shaosen[1], *HAN Jie*[1], *Heleni Porfyriou*[2],
Marie-Noël Tournoux[3], *Paola Brunori*[4]

1 School of Architecture and Civil Engineering, Xiamen University, China
2 National Research Council of Italy, Department of Cultural Heritage, Italy
3 The World Heritage Institute of Training and Research for the Asia and the Pacific Region under the auspices of UNESCO, China
4 Universitù di Roma Tre, Italy

Abstract In the context of the forthcoming cooperation between the School of Architecture and Civil Engineering of Xiamen University (XMU SACE), China, the Department of Cultural Heritage of the National Research Council of Italy (CNR-DSU), Italy, and the World Heritage Institute of Training and Research for the Asia and Pacific Region (WHITRAP Shanghai), China, the authors of this paper address the conservation and revitalisation strategy they are developing for the Fujian Tulou World Heritage Site in the Fujian region of China.

In the past three decades, the built heritage of the Tulou and their associated landscapes have undergone huge transformations, mainly because heritage values are not properly acknowledged. Now, the priority is to strengthen the assessment of heritage value in order to design innovative conservation and development strategies that fully take both conservation requirements and how to enhance people's lives into consideration. Based on research conducted by the School of Architecture of Xiamen University, this paper first assesses the main conservation and development challenges of the Fujian Tulou. Secondly, it discusses an Italian operational tool for conservation and revitalisation, the "Recovery Handbooks" and "Codes", developed over the last thirty years. It explores how their methodology can provide an appropriate framework for developing the body of knowledge needed to draft the "Fujian Tulou Recovery Handbook": a new conservation and protection tool to safeguard heritage values and their physical carrying attributes.

Keywords Fujian Tulou, Italian "Recovery Handbooks", built heritage conservation, sustainable development, landscape, HUL

1 Introduction

Historic cities and rural settlements, alongside their wider settings and urban heritage, can play a critical role in driving improvements in local living standards and adaptations to changing environmental and socio-economic conditions as well as to wider processes of sustainable development. Within this broad context defined by the UNESCO *Recommendation on the Historic Urban Landscape* (UNESCO, 2011), the cooperation between the Architectural School of Xiamen University, the Department of Cultural Heritage of the Italian National Council (CNR) and WHITRAP — Shanghai addresses the "Cultural Heritage Conservation and Enhancement of Urban and Rural Settlements Along the Maritime Silk Road".

With the aim of developing training and research programmes that connect cultural heritage conservation and urban and territorial development initiatives, as well as creating synergies locally and internationally between Italy, China and the Asia and Pacific Region, these three institutions and their representative authors focus their attention in this paper on the heritage of Fujian Tulou. The "Fujian Tulou" was inscribed on the World Heritage List in 2008 (Decision 32 COM 8B. 20) as a serial property of 46 rammed-earth buildings located in 10 different small settlements in Fujian "*as exceptional examples of a building tradition and function exemplifying a particular type of communal living and defensive organisation, and, in terms of their harmonious relationships with their environment, an outstanding example of human settlement*" (World Heritage Committee, 2009).

Understanding the history of these sites, maintaining traditional functions, and safeguarding the living architectural, urban, social and landscape milieu is at the core of our research, considering that the heritage value of the Tulou lies not only in their unique architectural structure but also in their social organisation and the surrounding agricultural landscape.

2 The Fujian Tulou Heritage Settlements

2.1 State of the Art

Considered one of the most representative types of traditional settlements in the south-west part of Fujian province, the Tulou settlements (Fig. 1), especially after their inscription on the World Heritage List, have attracted the interest of both academics and the public and private sectors.

Fig. 1 Tulou Settlements

Over the past few decades, research on the subject has significantly increased. Research on conservation in particular has focused on the following fields: (1) the architectural value and character of Tulou buildings, including typological studies, construction techniques and materials, craftsmen, the craftsmanship and decoration of earthen buildings, etc. (Dai, 1996; Cao, 2002; Huang, 2003; Wang, 2016); (2) historical studies with in-depth analysis of each case, including the history of clans and families, their economic activities, the settlements' transformation, etc. (Li, 2007; Yang, et al., 2014); (3) anthropological and cultural studies including documentation on everyday life, communal living, beliefs and traditions, etc. (Zhan, 2001; Zhang, 2006; Kawai, 2013); (4) approaches to preservation and reuse that explore means of technical preservation and functional reuse (Li, et al., 2016; Yang, et al., 2018; Yang, et al., 2018); (5) conservation policies and strategies that examine policies and propose principles, regulations and decision-making mechanisms (such as top-down vs. bottom-up mechanisms to be used during conservation decision-making process in the Tulou area, covering the 46 buildings components of the World Heritage property and ones that are not included) (Jiao, et al., 2016); (6) enhancement approaches focused on tourist-led development. Tourism is considered the main driving force for heritage enhancement in the Tulou area. Studies on this subject include both results of this widely-adopted development and criticisms that highlight the impact of mass tourism on the cultural values of the Tulou settlements and the risks to which heritage is exposed. All the aforementioned research has greatly contributed to the knowledge of heritage conservation in the Fujian Tulou area (Yan, et al., 2008; Li, et al., 2009). Nevertheless, a full understanding of Tulou settlements as a holistic organism is lacking. Further research on this would allow for an integrated conservation approach and the necessary comprehensive guidelines to safeguard buildings and their associated landscapes to be developed.

The 2011 UNESCO *Recommendation on the Historic Urban Landscape* (UNESCO, 2011) which highlights the analysis of historical layering processes in urban conservation practices, provided new alternative insights into the conservation of Fujian Tulou. Only recently has the cultural landscape approach (Wu, et al., 2019) been adopted in the study of Tulou with an attempt to address the issues of "wholeness" and "complexity". The traditional settlements of Tulou, situated in a rural area of China, represent a type of autonomous, self-organised, unique dwelling system that embraces strong social bonds and economic connections within the clan structure as well as close interactions and interrelationships with nature and the agricultural landscape, distinctive beliefs, shared stories and a durable sense of place. A holistic conservation methodology is, therefore, greatly needed to safeguard and preserve the Fujian Tulou.

2.2 New Challenges

Over the past decade, the lack of understanding of the Fujian Tulou's cultural, natural and landscape values as a complex settlement and cultural landscape and the lack of integrated conservation methodology and strategic development planning have confronted the Fujian Tulou region with the dilemma of development versus conservation. Firstly, there is a lack of adequate conservation policies and measures focusing on the preservation of earthen buildings that are neither part of the World Heritage property nor listed at a national, provincial or county level (Wang, 2016). As a result, there can be a range of conditions (decay, abandonment, conservation, redevelopment) within the same village (Fig. 2). Secondly, there are inadequate conservation policies and unbalanced financial support among regions. For example, the World Heritage property includes 46 earthen buildings located in Yongding, Nanjing and Hua'an counties, which obtain most of the financial support needed for their maintenance. In the rest of the Fujian Tulou areas (ex. in Pinghe, Zhao'an and Yunxiao counties), the number of Tulou buildings has decreased dramatically during the last decade due to lack of financial support (Ibid.) (Fig. 3).

Thirdly, integrated conservation assessment standards are lacking. This al-

Fig. 2 Various conservation conditions inside the World Heritage property

Fig. 3 Abandoned Tulou buildings outside the World Heritage property

lowed tourist-driven heritage interventions to foster unbalanced development patterns: mis-scaled and non-integrated new infrastructure, services, landscape plans, functional reuse patterns, etc. have largely disregarded heritage values, traditional forms, structures and ways of living (Guan, et al., 2008; Ma, et al., 2018). Fourthly, a thorough identification of traditional wisdom regarding design theory, building material knowledge and construction systems is missing. Hence, current conservation recommendations are insufficient and inadequate to guide restoration, preservation and reuse design practices. This has led to the decay of many traditional dwellings and the loss of craftsmanship (Zhong, et al., 2016; Yang, et al., 2018).

Apart from physical aspects, there is also inadequate acknowledgment of social values; traditional lifestyles, in particular, have undergone major changes. In reality, most of the local governments lack strong regulatory conservation frameworks and long-term perspective on local sustainable development. By adopting short-term, top-down development measures, mainly ones led by tourism, they have largely ignored residents' daily needs and attachments, the cultural value of the heritage sites, and long-term local community development (Luo, 2010; Ma, et al., 2018).

Consequently, in the past decades, the Fujian Tulou built heritage and settlements, its people, and the associated landscapes have undergone huge transformations. This is mainly because heritage values are not properly acknowledged.

Now, the priority is therefore to strengthen the assessment of heritage value, which can be used as a baseline to design innovative conservation and development strategies that incorporate heritage preservation requirements as well as how to enhance people's lives.

Be it for conservation purposes or territorial development, knowledge and further studies are essential. Developing the appropriate body of knowledge and baseline information is one of the first of four tools advocated in the *Historic Urban Landscape Recommendation*. Within the cooperation agreement between XMU, CNR and WHITRAP, the partners aim to develop new operational tools that link research and guidelines to foster dual conservation and sustainable development strategies and "to achieve a balance between urban growth and quality of life on a sustainable basis." (UNESCO, 2011).

The next section will introduce Italian methodological instruments and will explore how they can provide an appropriate framework for developing the body of knowledge needed for Fujian Tulou conservation to enable the safeguarding of heritage values and design culture-led spatial planning strategies.

3 The Italian Heritage Settlements

3.1 Italian Contribution to Urban Conservation

The 1881 census, following Italy's unification as a single nation state, counted 22,621 historic centres as monumental urban centres. They included Rome and Florence but also less populous towns with great urban qualities such as Siena and San Gimignano as well as small medieval centres, often called "borghi", and vernacular settlements of various typologies (Fig. 4).

Fig. 4 Historic centres in Italy

The conservation of this urban heritage has been a central issue in the country's legal and planning system since the 1920s. The debate on urban conservation and built heritage restoration led to theories such as those of Cesare Brandi (Brandi, 1963), scientific tools for diagnostics and monitoring and planning experiences, e.g. the conservation of the entire historic centre of Bologna in 1969 (De Pieri, Scrivano, 2004). These greatly contributed to European conservation legislation in the 1960s and to the development of international charters and standard-setting instruments, such as the Venice Charter and some UNESCO Conventions and Recommendations.

One of the tools introduced in the 80s — aiming to combine detailed studies and operational approaches for the conservation of built heritage in an urban environment — are the "Manuale del recupero"/"Recovery Handbooks" (Marconi, 1989; Giovanetti, 1992; Giovanetti, 1997) and the "Codice di pratica"/"Practice Codes" (Giuffrè, 1988; Giuffrè, 1993), which have been widely implemented since then. These tools' concept and methodology may prove very relevant to address the heritage protection challenges of Fujian Tulou from analytical, project-oriented and operational points of view. The following pages will provide a brief overview of the "Recovery Handbooks" and "Practice Codes".

3.2 "Recovery Handbooks" and "Practice Codes"

The "Handbooks" and "Codes" are tools based on a methodology founded on thoroughly analysing building construction types in their historic setting as well as their specific components, and providing architects and restorers with scientifically-sound design tools. They are widely used in restoration and as analytical scientific representations and surveys of pre-modern historic built environments and as operational instruments to address even small-scale detailed conservation interventions (Fig. 5). Both methodological approaches can be adapted, and each "Handbook" is specific to every site and town, enhancing their history and preserving vernacular building traditions through the development of appropriate restoration tools and techniques. More specifically, "Practice Codes" focus on structural analysis of buildings and the evaluation of the endurance of structures as a whole and in their singular components.

For these reasons, the "Handbooks" and "Codes" had a decisive role in shaping Italian urban conservation planning legislation and introducing specific regulations for heritage protection, as well as for ensuring seismic protection for historic buildings compatible with scientific conservation (Norme tecniche, 2008).

The "Handbooks" provide detailed surveys of historical and vernacular architectural elements, both as individual items and within more complex building components. These are analysed through their structural and decorative forms. These components are presented disassembled and with all the

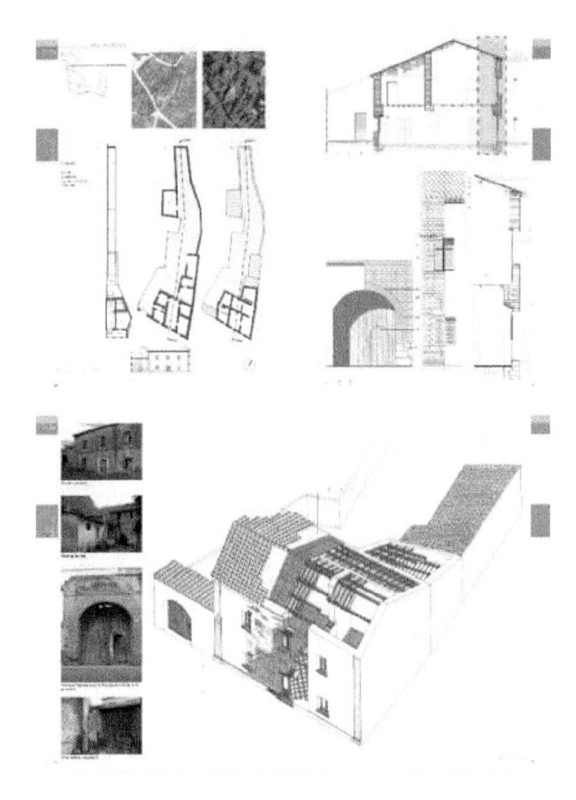

Fig. 5 Case studies of earthen traditional residential typologies from Sardegna (Atzeni, Sanna, 2008, 96-99)

necessary instructions and explanations for builders (and restorers) while also being shown reassembled in numerous different typologies of the same structure. For example, "Handbooks" may provide surveys of wooden structures of roofs and slabs, brick and stone walls, traditional doors, windows and iron gratings, but also complex examples of pre-modern constructions, vaults (Fig. 6), stairs, structural systems and so on.

However, "Handbooks" go beyond a technical catalogue of architectural components to include the urban historic background. Every example is contextualized with a synoptic historical note that helps understand the building's cultural setting. In the "Handbooks" and the "Codes", understanding the urban context is the core focus: built heritage is considered as part of a whole whose aspect and characteristics are the result of a long-term historical and typological process. This understanding is based

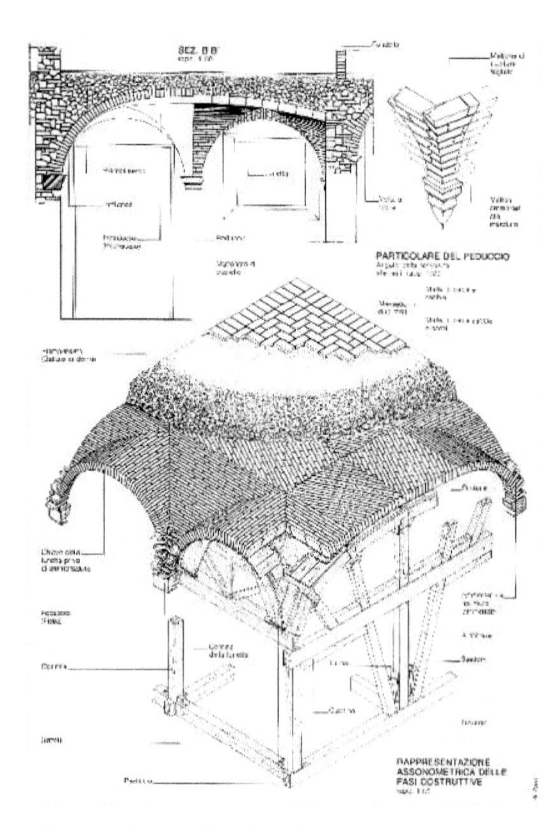

Fig. 6 Città di Castello, elevation and construction schemes for a brick vault (Giovanetti, 1992, 95)

on a tradition of studies that analyse urban patterns and the urban form in their evolution from the site's geology to planned interventions, using a diachronic evaluation for each element. The "Handbooks" also include technical and scientific dictionaries/glossaries based on archival research and the study of historic documents and technical treatises that further document the various items referred to in the study.

The outcome is an operational tool that regulates interventions on built heritage, providing architects and restorers with a scientifically-based design instrument which respects the characteristics of each individual context and which is related to local building traditions. Both the "Handbooks" and "Codes" were, however, developed to address the specific characteristics of local architecture and the countless different building techniques yielded by the geography and history of Italy (Atzeni, et al., 2008). They are the result of interdisciplinary cooperation involving many different experts and long-term exhaustive studies on historical and vernacular architecture.

The "Handbooks" are therefore very useful tools for architects and designers involved in heritage conservation, either in the private or public sector, who aim to plan interventions to restore existing buildings, integrate missing parts, or change functions. In parallel, they offer academics and scholars a deep, critical knowledge of the material history of architecture of each region, and for this reason they are often used also as teaching material in Italian architecture schools. The greater value of "Handbooks" lays in their understanding of built heritage and its detailed components as a whole, acknowledging the specific techniques and skills involved in its construction — all of which are worthy of protection and preservation.

Finally, "Handbooks" and "Codes" are included in urban conservation regulatory frameworks and are adopted by public administrations for developing urban conservation planning tools and as references for evaluating and identifying appropriate restoration and maintenance interventions on historical heritage. Among other examples, the conservation works undertaken for the cities of Rome (Marconi, 1989) and Palermo(Giovanetti, 1997) were financed by local authorities in addition to other urban planning measures. The Region of Sardinia funded the local "Handbooks" (Atzeni, et al., 2008); the studies for the Sassi of Matera (Giuffrè, et al., 1997; Restucci, 1998) were incorporated into local town planning standards (Municipality of Matera, 2006). In other words, the "Handbooks" and "Codes" provide municipalities with tools for identifying, designing and recovering building types and urban fabrics that are still present in historic centres and traditional rural landscapes and for developing urban and landscape conservation plans.

4 Conclusion

As we have seen, the key conservation challenges of the Tulou built heritage include the need to strengthen the assessment of the area's heritage value in order to design a holistic conservation methodology that can safeguard and preserve the Fujian Tulou as well as improve the quality of life for local inhabitants. Furthermore, the lack of an appropriate regulatory toolkit linked to a long-term, large-scale strategic planning policy focused on heritage weakens conservation projects. Hence the need to further develop the research conducted by the School of Architecture and Civil Engineering of Xiamen University on the buildings, their surrounding landscape, the traditional know-how and people's socio-economic environments.

Considering the above, the Italian conservation operational tools known as the "Recovery Handbooks" and "Practice Codes" which provide an all-in-one, in-depth understanding of heritage itself and of project implementation guidelines can be included in regulatory frameworks and are a highly appropriate tool for addressing conservation challenges in Fujian Tulou and helping develop what was referred to in the introduction as a "Fujian Tulou Handbook".

The future "Fujian Tulou Handbook" would acknowledge heritage values and their significance for the local communities. It would serve as a guide for decisionmakers, and it would provide a tool for updating conservation policies and developing heritage-led spatial planning. Its aim would be to help develop a virtuous cycle leading to the sustainable development of the larger area. In other words, it would closely follow the Historic Urban Landscape approach.

It is on these grounds that the collaboration between the Architectural School of Xiamen University and the Department of Cultural Heritage of the Italian National Council (CNR) and WHITRAP — Shanghai was begun. The aim of the MoU between them, signed in 2019, is to promote scientific, theoretical and technical transfer of know-how between the two countries on the issues of built heritage conservation and enhancement through fieldwork and capacity-building. More specifically, its objective in the forthcoming 2020—2021 period is to organise yearly workshops, summer schools, field trips and capacity-building activities both in Italy and in the Tulou area with students, professors and experts in order to develop a new body of knowledge and a "Recovery Handbook" based on Italian experience. We aim to make this ambitious and challenging pilot project an operational tool based on in-depth field research in close collaboration with the local community.

References

ATZENI C, SANNA A,2008. Architettura in terra cruda dei Campidani, del Cixerri e del Sarrabus. Roma: Edizioni DEI.

BRANDI C, 1963. Teoria del restauro; theory of restoration. Firenze: Nardini.

CAO Chunping, 2002. Patterns of the earth buildings in southwestern area of Fujian Province. Collection of architectural history, (03):103-124+276.

DAI Zhijian,1996. Exploration of architectural form of Hakka Tulou in Fujian Province. Huazhong architecture, (04):104-109.

DE PIERI F, SCRIVANO P, 2004. Representing the "Historical Centre" of Bologna: preservation policies and reinvention of an urban identity. Urban history review/Revue d'histoire urbaine, 33 (1), 34-45.

GIOVANETTI F, 1992. Manuale del recupero del comune di Cittdi Castello. Roma: Edizioni DEI.

GIOVANETTI F, 1997. Manuale del recupero del centro storico di Palermo. Palermo: Flaccovio.

GIUFFRÈ A, et al. ,1988. Centri storici in zona sismica, analisi tipologica della danneggiabilite tecniche di intervento conservativo. Codice di pratica per il recupero dei centri storici soggetti al sisma: Castelvetere sul Calore. Studi e ricerche sulla sicurezza sismica dei monumenti, 8.

GIUFFRÈ A, 1993. Sicurezza e conservazione dei centri storici: il caso Ortigia. Codice di pratica per gli interventi antisismici nel centro storico. Roma-Bari: Laterza.

GIUFFRÈ A, CAROCCI C,1997. Codice di pratica per la sicurezza e la conservazione dei Sassi di Matera. Matera: La Bautta.

GUAN Qiaoyan, LIAO Fulin, QI Xinhua, 2008. Coordination of different stakeholders in the process of tourism development: taking Yongding

Tulou as an example. Journal of Changchun Normal University (natural science edition), (04): 65-68.
HUANG Hanmin, 2003. Fujian Tulou: a treasure of Chinese traditional residences. Beijing: SDX Joint Publishing Company.
JIAO Mengjie, WU Guoyuan, 2016. Architectural heritage protection strategies and cases studies of contemporary non-governmental organizations. New architecture, (05):71-73.
KAWAI H, 2013. Hakka architecture and cultural heritage protection from the perspective of landscape anthropology. Academic research, (04):55-60.
LI Shanting, MA Hang, 2016. Construction of value assessment system of Pinghe Tulou based on AHP. Architectural journal, (S1):108-112.
LI Tingting, LUO Peicong, 2009. Study on tourist perception and attitude of tulou residents in Yongding, Fujian Province. World regional studies, 18 (02):135-145.
LI Xiongfei, 2007. The treasure of the country - Eryi Lou and earth building group of Huaan Earth. Huazhong architecture, (09): 140-151.
LUO Gaoyuan, 2010. Tourism value and protection of Fujian Tulou. Economic geography, 30(05): 849-853.
MA Teng, ZHENG Yaoxing, et al., 2018. Study on the impact of rural tourism development on local characteristics and its mechanism: taking Hongkeng village, Yongding county, Fujian Province as an example. World geographic research, 27(03): 143-155.
MARCONI P, et al., 1989. Manuale del Recupero del Comune di Roma. Roma: Edizioni DEI.
Municipality of Matera, 2006. Sassi Office, General Recovery Forecasts in implementation of Law 771/86 (adopted with DCC 11/05/2006 N°38, approved with DCC 27/11/2012 n°83), Annex D - Technical standards for implementing the interventions.
NORME tecniche per le costruzioni, 2008. Gazzetta Ufficiale n. 29, 4 febbraio, Suppl. Ordinarion. 30.
RESTUCCI A, 1998. Matera: i Sassi. Manuale del recupero. Milano: Electa.
UNESCO, 2011. Recommendation on the historic urban landscape including a glossary of definitions. Paris: UNESCO.
WANG Wei, 2016. Research on protection and renewal technology of earth buildings based on settlement form. Xiamen: Xiamen University Press.
World Heritage Committee, 2009. Decision 32 COM 8B. 20. Decisions adopted at the 32nd Session of the world heritage committee (Quebec City, 2008). UNESCO, World heritage centre, 163-164.
WU Juanyu, LU Yao, CHEN Mengyuan, 2019. Study on natural landscape protection and development of Tulou group in Hekeng village of Fujian province from the perspective of cultural landscape. Chinese landscape architecture, 35(02): 39-44.
YAN Yayu, ZHANG Lirong, 2008. A comparative research on mechanism of "community participation" under different operating modes. Human geography, (04):89-94.
YANG Changxin, XU Yiyi, LI Xingxu, 2018. Research on the holistic protection method for the ancient villages styles of Taxia in Fujian province, based on the theory of correlation and layering effect. Architecture journal, (S1): 99-104.
YANG Sisheng, LIU Meiqin, 2014. The irregular evolution of traditional Tulou enclosure in Fujian province adapts to the environment. Huazhong architecture, 32(03):139-143.
YANG Sisheng, WANG Shan, LIANG Chuyu, 2018. Expanding understanding and deepening treatment of Tulou heritage. New architecture, (05):139- 143.
ZHAN Shichuang, 2001. The cultural connotation and value of Tulou. Southeast academic, (04): 176-186.
ZHANG Fang, 2006. Tulou architecture and culture. Chinese and overseas architecture, (06): 50-51.
ZHONG Lingfang, GUAN Ruiming, 2016. The protection concept of Fujian Tulou, a world cultural heritage. Huazhong architecture, 34(12): 126-129.

The Predicament of Vernacular Community's Conservation in Urban Area Facing the Threat of Tourism

XU Kanda[1], *SHAO Yong*[2]

1 Shanghai Tongji Urban Planning and Design Institute Co. Ltd., China
2 College of Architecture and Urban Planning, Tongji University, Shanghai, China

Abstract In the process of urbanization, the traditional communities composed of vernacular architectures in the urban environment are facing lots of universal problems such as the decline of vernacular dwellings and the disruption of social structure. Besides there are some vernacular communities under the influence of the "Heritage Tourism Fever" are facing more severe problems. The features and personalities of vernacular architectures are gradually disappearing and the community is becoming the appendage of tourist service.

This paper will take Qufu Ming-era City as case to analyze the architectural features, social and cultural values of the vernacular communities which has been formed and continued for hundreds of years, and point out the existing predicament under the background of overemphasizing World Heritage Site Tourism which is one of the important misunderstandings for conservation in most of China. On that basis, to analysis the root cause at the aspect of existing policies, mechanism, and government willingness and discuss the possible solutions which are based on the inhabitant's needs.

Keywords vernacular architecture, community, tourism threat, inhabitant's needs

1 Introduction

Nowadays in China, due to the independent geographical environment and relative complete production mechanism, vernacular settlements in the rural areas are easier to be conserved with integrity. In contrast, the communities consisting of vernacular buildings in urban texture shows high vulnerability under the accelerating urbanization. Their mixed historical functions have gradually become simple and the features and textures are easy to cause a sudden change affected by the rebuilt of neighboring regions.

Due to the lack of mature conservation methods, the majority of the vernacular communities are facing the problem of decline. Most of the buildings have been in disrepair for a long time, current high density of residence has exerted tremendous pressure on vulnerable housing. With the change of city structure, the price of land in the central area has risen at the same time.

In such a situation, residents are eager to improve living conditions. Facing the pressure of inhabitants' demands, the government usually adopts two extreme approaches: "Frozen preservation" or "Demolition and reconstruction".

Meanwhile, according to the Final Report on the Periodic Reporting exercise for Asia and the Pacific: In North-East Asia represented by China, the impact of tourism has become the biggest negative factor in the current work of heritage conservation, and has overshadowed its positive effect (UNESCO, 2012). Vernacular communities around the World Heritage have been gradually developed as an appendage of tourist service.

2 Methodology

The research has chosen a typical vernacular community in the east part of Qufu Ming-era City as an object. The community covers an area of 4 hectares and 80% of the courtyards are composed of vernacular architectures.

First, by literature reading and oral interviews, history evolution analysis has been conducted to explore the relationship between residents, communities, heritage and the city, to correctly understand the true value of those vernacular communities.

Secondly, starting from the needs of inhabitants collected by questionnaires and field surveys, the research discovered the support heritage and city should provide to the community and the way to deal with the tourism threat.

3 History Evolution and Values of Vernacular Communities in Qufu

3.1 History Evolution

Qufu is located in eastern China, which is the transition zone between floodplains and hills in lower reaches of the Yellow River. In ancient times, it was the hometown of Confucius. Due to the great influence of Confucianism, Temple and Cemetery of Confucius and the Kong Family Mansion in Qufu are listed as world heritage.

Qufu Ming-era City was officially formed in the 16th Century and has been continued to nowadays.

Phase 1: Late Ming Dynasty (16th — 17th Century)

The government conducted a project of "Temple-City" which was constructing a city with Confucius Temple in the center being defended. The original settlement moved to the existing place and built large defending walls around the Confucius Temple and Kong Family Mansion. Ordinary inhabitants volunteered to build houses near them to guard. The traditional communities appeared and the houses are in small scale.

Phase 2: Qing Dynasty (17th — 20th Century)

Relying on the transportation advantage of the Grand Canal, Qufu has become a regional distributing center of goods. (Qufu Records, 1993) With the rapid growth of the economy and population, the city's functions were becoming more complete with the emergence of markets, schools and other supporting facilities. The descendants of the Confucius moved out from the Mansion and chose another site nearby to build their own houses. Our community became more complicated after those descendants settled. Several small courtyards connected into pieces, forming a compound courtyard with dozens of times the size of traditional ones (Fig. 1). Remaining small houses became the home of tailors, carpenters and those servants working for Kong Family. It became one of the richest area at that time.

Phase 3: Early 20th Century

The traditional economic mechanism maintained and the texture of our vernacular community continued. The natural inheritance of families led to the division of some large courtyards. However, those families continued.

Phase 4: PRC Established (1950s—1970s)

The change of social institution brought a comprehensive reform of the agrarian relations and economic system. The land owned by the upper class was forced to be re-allocated to the peasants. The mixed texture of large mansions and small courtyards gradually became homogeneous (Liu Liang, 2010). The original upper class moved out to Taiwan while part of the social structure changed. The whole area declined.

Phase 5: 1970s—2000s

Qufu's economy developed slowly. Related residential housing for state-owned companies have been built in the community. Some courtyards continued to divide and were sold out, empty space were occupied to construct for the growing population. The remaining descendent of the Kong Family lose their privilege and return to the ordinary citizen.

3.2 Values of Vernacular Communities

Through the history research of one typical vernacular community, it is easy to find that they always played an important role in the historical layers of Qufu's development. The total area of Qufu Ming-era City is about 1.4 square kilometers. Besides the vast Confucius Temple, Kong Family Mansion, one-third of the area are vernacular communities.

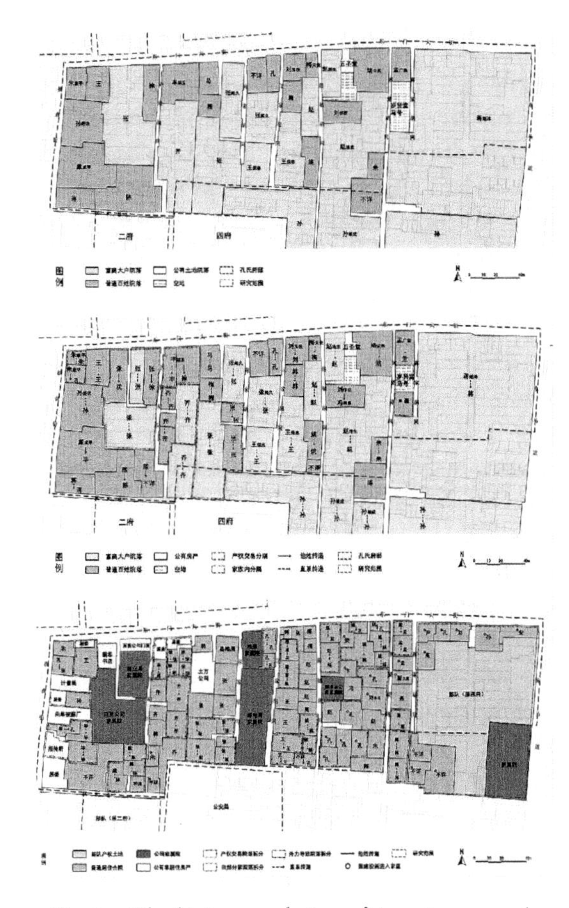

Fig. 1 The history evolution of target community © Authors

From the aspect of integrity, the unique Temple-City planning concept in the history of Chinese urban construction, is the embodiment of Confucius Worship. It only exists in the combination of the centered Confucius Temple, Kong Family Mansion and those vernacular communities around. The small scale and dense texture as well as the unified gray-black tone, which is following the Confucianism spirit, are the indispensable contents of the historical and cultural values of this city.

As the main inhabitant communities serving the Kong family in history, the mixed living status reflects the class rituals, which could be regarded as social and cultural values.

The temple and mansion were the absolute core of the city, closely surrounded by all different facilities like markets, workshop and schools, which forms a highly active circle with huge radiation energy (Circle A). In the second circle, those public facilities became center surrounded by vernacular communities, which shows the inhabitants' high reliance on urban activities (Circle B). Within the vernacular communities, there are secondary facilities like grocery or temples serving for ordinary people (Circe C)(Fig. 2).

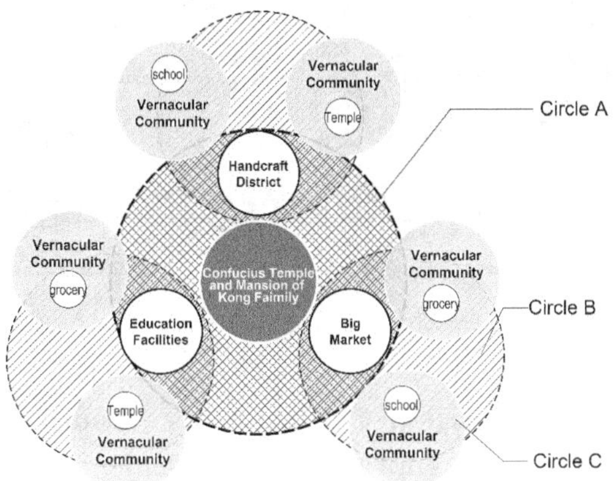

Fig. 2 The circle system reflecting the living status in Ming-era city © Authors

So it is obvious that the vernacular community is an important link in the living system. It is not only the key serving area formed for the Kong family but also the place where the big family, guard, servant and their descendants lived for almost 2000 years. The highly dynamic life status is the most authentic carrier to express the abstract content of Confucianism vividly.

From the aspect of architecture styles, traditional dwellings are formed under the influence of the climate and geography of northern China.

Because of the location in the lower reaches, the soil here is clay-based and deep. House construction often uses the local yellow clay mixed with wheat straw to build adobe walls. Usually, a thick layer of clay is smeared on the walls and roofs for keeping warm in the cold winter. Due to the rainy summer, the slope of the roof is generally bigger than 30 degrees and some of the ridges are curved, which facilitates rapid drainage. The center of the courtyard is higher with a stone platform and ditches are dug behind all buildings to collect rainwater by gravity. The structure and form of vernacular architectures are perfectly adapted to the climate and reflect the cultural personality in this region(Fig. 3).

Fig. 3 Typical vernacular dwellings in Qufu © Authors

4 Existing Predicament: Tourism Threat & Ignorance of Residents

4.1 The Threat of Tourism

Since Confucius Temple and Kong Family Mansion were listed as World Cultural Heritage in 1994, it has become a turning point in Qufu's rapid tourism development. The tourists' trips have reached more than 5.95 million in 2018. Relying on the Confucianism background, more than one million tourists took their children only for the study tour, which forms a huge emerging market.

Although tourism is recognized as one of the effective means to improve the living conditions of inhabitants, because of the great profits, the developers cooperated with the government as the dominant force to build tourism real estate by compulsory acquisition of land. All the resources and funds are invested in tourist facilities while the interpretation of heritage values are often neglected. It poses huge threats to the conservation of vernacular communities:

4.1.1 Mass Urban Textures of Vernacular Communities Demolished

Since most of the vernacular buildings are not relics or listed historical buildings, their conservation lacked specific regulations. At present most of them are in a state of decline which does not match the land price. This has led them to become the burden of urban development. Under the background of Heritage Site tourism fever, they are the first choice to be transformed. The urban renewal often takes the rudest way: overall demolition and reconstruction.

From 2006 to 2014, the proportion of residential areas in Ming-era City was reduced by nearly a quarter from 31.2% to 24.1% with many vernacular communities demolished.

Take the ancient Panchi area in the southeast corner of the city as an example, it covers an area of about 15 hectares. Almost one hundred years ago, it was a lively community for descendants of Kong Family, showing a regular roads network and high-dense courtyard textures. In 1996, the Mater Plan of Qufu City decided this whole area as land-use of hotels (Deng Qiaoming, 2007). When the renovation project officially launched in 2011, the neighborhood which is one-tenth of the area of Ming-era City was forced to raze to the ground, and all the residents moved away from the city. According to the new plan, this area will build high-end hotels and two-story courtyard-style luxury villas. The new constructed buildings will break through the regulatory restrictions on the building height limit of World Heritage Buffer Zone(Fig. 4).

4.1.2 Features of Vernacular Architectures and Traditional Lifestyle Changed

Because inhabitants still lack a comprehensive understanding of the value of vernacular architectures, they can only speculate that tourism can bring a high economic return from the actions of developers. Therefore, driven by profits, inhabitants transformed their original dwellings into individual tourism shops or hostels.

Most of the newly-developed tourism projects are fast duplicates which do not consider local climate, geography and cultural characteristic, usually contrasting with traditional textures. By owners' incorrect imitation, local architectures are gradually implanted various foreign elements in the transformation, such as European style,

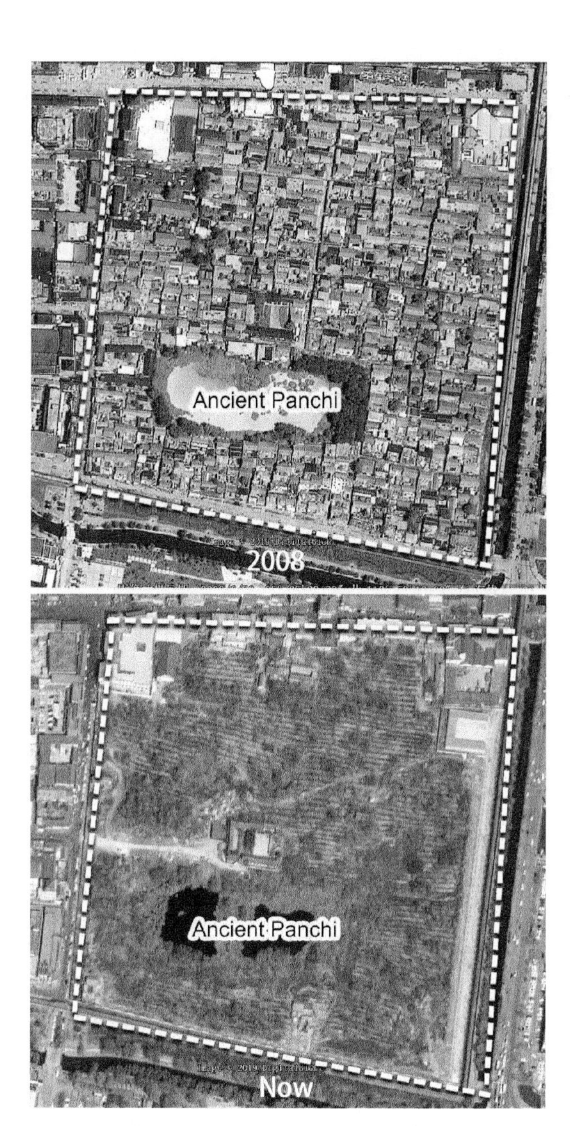

Fig. 4 Historic and existing texture of Ancient Panchi District © Authors

Bohemian style, water-town style, etc. And the building materials are also ever-changing. The authenticity of vernacular architectures changed accidentally and the integrity of the city's traditional features also disappeared. What is more serious is that the traditional lifestyle attached to the vernacular community has also changed under the influence of tourism development.

4.2 Unsatisfied inhabitants' demands

4.2.1 System Establishment of Inhabitants' Diverse Demands

At present, most of the conservation plan is made by elites. There is few official platforms for inhabitants participating in the vital link of conservation and renewal, which led to the misrepresentation of their needs.

So, understanding the inhabitants' demands are the foundation of all the work.

Their demands are diverse and advanced in regular order. In order to build a precise and comprehensive system to evaluate, all the needs were specified into the following content with the application of Maslow's Theory (Fig. 5). Some of them are bottom limits while some need to be subtly influenced through top-level design.

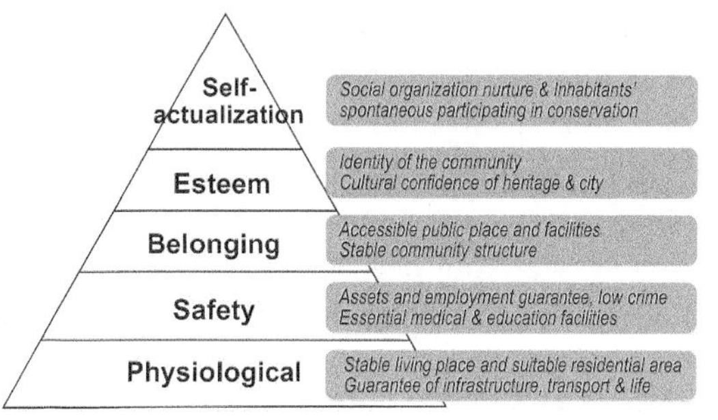

Fig. 5 Inhabitants' needs with the application of Maslow's Theory © Authors

4.2.2 Unsatisfied Living Demands Caused Inhabitants Loss

To gather the information and true feelings of inhabitants living in the vernacular community, more than 100 questionnaires were delivered during the indoor survey.

Due to the limited space of the article, so the complete result cannot be detailed shown in this paper.

However, there are still some important conclusions: (1) the biggest problem of housing is not the floor area (which most of the residents think is enough) but the lack of basic facilities like heating and drainage. (2) Inhabitants care more about the facilities and open space in the community like parking area, activity center and sports place. All of these are considered as important elements to make young people. (3) Stable jobs couldn't be guaranteed while natives usually can't participate in the existing tourism industry, because most of the job was occupied by migrants. (4) Remaining natives are more willing to communicate in the original public space with a memory like the front plaza of Confucius Temple, however, due to the tourism development, they have lost the accessibility. (5) More than 3/4 of the interviewee has the identification of descendants of Kong Family but half of them don't have the interest to spread the Confucianism and consider the government should take the whole responsibility for conservation.

Through the consequence, we can found that most of their needs were not satisfied. Due to the change of urban environment around the heritage site, newly built tourism real estate has replaced the necessary public facilities for daily life like schools, kindergartens and clinics. Even if some vernacular communities could be preserved, the increase in the land price also led to an increase in living expenses such as food prices, and rent. All of these forces the residents with certain economic viability to gradually moved out. The long-existing relationship between inhabitants, dwellings and communities for thousands of years is broken accompanied by social problems like aging and poverty.

4.2.3 Uncontrolled Autonomous Behavior

From the cross-analysis, we also found those inhabitants remaining shows a high possibility to conduct self-behavior which could influence the whole values of heritage and communities, due to the unsatisfied demands.

For instance, if the living condition is not satisfied, inhabitants usually use modern materials like tile to transform the house or add 2 or 3-storey illegal cabin which affects the traditional feature and look. Without public communicate places and facilities, people's social activities change from traditional opera to computer games indoor, which led to the degradation of intangible heritage and traditional culture (Fig. 6).

5 Root Cause

The responsibility of heritage conservation and traditional community revitalization

Fig. 6 Inhabitants' uncontrolled illegal self-construction © Authors

should be taken by the whole city. Political will usually is the main driving force in China. But it also requires the cooperation of regulations, policies, funds, education and many other aspects. The main root cause led the predicament is that city policymakers subjectively neglect the inhabitants, demands and values of vernacular communities.

When the use of limited funds is affected by the rate of return, the government is more willing to give priority to the tourism facilities construction rather than the improvement of living conditions in vernacular communities, which makes those communities always stayed in the neglected gray area.

Secondly, the city's financial revenue and expenditure have not formed an efficient mechanism. The tourism income of other departments cannot be invested in heritage conservation and infrastructure. The limited fund can only rely on a small amount of tax and supported by the Bureau of Cultural Relics.

Thirdly, the government has not provided traditional communities the equal facilities and services that modern communities should have according to the regulations and laws. And they also didn't realize that uncontrolled approval of projects accelerates the dying away of necessary public facilities mainly for inhabitants in vernacular communities.

Now the main residents in most vernacular communities are vulnerable groups. Without a dedicated platform, providing employment guarantee, instead tourism resources are all concentrated in the hands of developers. So the tourism fever in heritage sites benefits the inhabitants very little.

6 Possible Solutions: "Inhabitants-centered" Conservation Strategy

To promote the conservation of vernacular communities, dealing with the threat of tourism, it must come back to the root cause of the problem.

Only by "Inhabitants-centered", placing the community and residents in the key position of conservation and regeneration, the practical problems can be solved more specifically. And the social and economic values of heritages can be interpreted better.

6.1 Necessity

From the relationship between inhabitants and heritage, the ancestors were the creators and users of heritage. There are strong material and spiritual connections between communities and related heritages, which is a very important component of heritage value.

The World Heritage Conference in 2012 proposed that the concentration of world heritage conservation is shifting to inhabitants in the globalized society. Only converging the conservation power from contemporary communities and local people can adapt to this tendency. It can be seen that heritage is no longer to be an understood as a material remains which should be preserved in the static principles but as a meaningful legacy and living object where interacts with the people to be managed in a flexible approach (Su-jeong Lee, 2015).

As a vernacular community with dynamic local activities, both natives and immigrants live and interact with the heritage together frequently. So the ownership and usufruct of those heritages should belong to the local community while the economic value of them should serve the inhabitants and create welfare.

6.2 Conservation and Regeneration Strategy

6.2.1 Establish a Compressive Plan Involved with Heritage, Community and Tourism

As an important top-level design, the plan should take the economic growth, community improvement and heritage conservation together into comprehensive consideration. In addition, this plan should also be a policy and management plan. By forming a multi-stake holds management agency to have government departments, communities and inhabitants involved in the consultation and decision-making mechanism. The main target is to avoid that the willingness of government overrides inhabitants' life and forms blind zone in management.

6.2.2 Public-Private Partnership and Tourism Feedback Mechanism

Under the guidance of the management agency, a tourism feedback mechanism should be established which means part of the tourism income will be invested in the conservation and infrastructure improvement of vernacular communities. More funds will be raised through PPP. The government will rent the repurchased public buildings around the heritage sites to the enterprise as tourism assets in a preferential price, meanwhile, the enterprise should take the responsibility of vernacular architecture conservation.

6.2.3 Build an Employment Platform and Introduce Policy Guarantee

A certain percentage of the residential area should be guaranteed by regulations when the tourism developed in the heritage site. Meanwhile, specific policies should also be provided for inhabitants who are still willing to live here. The government has to formulate guidelines for the preservation and restoration of traditional dwellings in vernacular communities. Residents should restore their own house following the guidelines by giving certain subsidies. With the above efforts, it could reverse the present incorrect transformation of vernacular architectures.

Inhabitants should be engaged in the tourism system through the establishment of a long-term employment platform.

Through the platform, all the material and human resources could be fully utilized. Inhabitants will participate in the heritage interpretation, folk performance, and other related activities to promote improving their initiative in heritage conservation.

6.2.4 Enhance the Propaganda and Education for Inhabitants

In order to accelerate the self-cultivation of inhabitant's demands like cultural identity, it is necessary to establish a popularized education platform for communities to enhance correct understanding of heritage value, making them become an emerging force to protect the heritage under a stable supervision mechanism.

References

DENG Qiaoming, 2007. Integrated conservation and sustainable development of historical area: taking ancient panchi in Qufu as an example. Huazhong architecture, 25 (11): 61-64.

FU Chonglan, 2002. Qufu temple city and Chinese confucianism. Beijing: China Social Science Press.

LEE Su-jeong, 2015. People-centered conservation: its origin, practice and Issues. First OWHC Asia-Pacific Regional Meeting for World Heritage Cities.

LI Guangjun, 2018. Qufu yearbook. Beijing: China Culture & History Press.

LI Zhongxin, 2014. Study on the regional characteristics of Shandong dwellings. Jinan: Shandong University Press.

LIU Liang, 2010. Review study on historic city conservation of Qufu in the past 30 year. Jinan: Shandong Jianzhu University.

NING Yanliang, 1993. Qufu Records. Jinan: Shandong Qilu Press.

Shanghai Tongji Urban Planning & Design Institute Co. Ltd., 2006. Qufu Ming City Controlled Detailed Planning.

UNESCO, 2012. Final report on the results of the second cycle of the periodic reporting exercise for Asia and the pacific.

UNESCO, 2012. The closing event of the celebration of the 40th anniversary of the world heritage convention: kyoto vision.

The Research of Traditional Village Space Renewal Strategy Based on Self-Organization and Other-Organizational Synergy Theory: Case Study of the Traditional Villages in Southern Anhui

ZHOU Ying, *BIAN Bo*

School of Architecture, Southeast University, Nanjing, Jiangsu, China

Abstract As a historically valuable part of rural areas, traditional villages have formed their own unique self-organizing development mechanism in the long-term growth and evolution process. However, traditional villages and their development mechanisms have not received sufficient attention in the current historical heritage protection system. There are also many disputes about the path of conservation and development of traditional villages. With the help of historical map translation and spatial information comparison, this study sorts out the spatial evolution process of several traditional villages in southern Anhui in the historical process to summarize the experience of traditional villages as an organic whole under the long-term self-organization mechanism of space under natural conditions. Then we explain the incompatibility of self-organization mechanism and the instability of other-organizational mechanism involved in the development of the traditional village space in the context of the transformation of urban-rural relations. Based on the theory research, we propose the necessity of combining self-organization with other-organization and the principles of moderate integration of the two. Finally, the research points out the space renewal strategy in the traditional villages, such as the combination of rigid and flexible space control method, stage division of spatial update, dynamic transformation and so on. The results of this study will provide theoretical support for the selection of conservation and development paths for traditional villages in southern Anhui and other regions.

Keywords traditional villages, self-organization, other-organization, space renewal, southern Anhui.

1 Self-Organization Mechanism in the Evolution of Traditional Villages in Southern Anhui

1.1 Overview of Self-Organization Theory

1.1.1 Definition of Self-Organization Theory

In the 1960s, there is a series of system evolution theories are proposed including Dissipative Structural System, Catastrophe Theory, Hypercycle Theory, Chaotic Theory, Fractal Theory and so on. These theories all have the characteristics of nonlinear complex systems, which together constitute a self-organization theoretical group whose main research system evolves from disordered to ordered or from one ordered structure to another. In 1976, Haken Herman C formally proposed the concept of self-organization and gave specific definitions: if a system has no external interference in the process of obtaining a spatial or temporal or functional structure, it can be said that the system is self-organization. The term "specific" here means that the structure or function is not imposed on the system by the outside world, but the outside world acts on the system in a non-specific way. This concept has gained wider recognition and further promoted the development of the self-organization theory system.

The domestic self-organization theory research was carried out late. At present, it is widely recognized that the complex system self-organization theory proposed by Professor Wu Wei of Tsinghua University. He pointed out that a self-organization system means a structured system that can organize, create, and evolve on its own without external command to independently grow from disorder to order. The system refers to three abstractive evolution processes: process from non-organization to organization, the process from low-level organization to high-level organization, process from simple to complex structure at the same organizational level.

1.1.2 Conditions of Self-Organization System Generation

In theory, the establishment of a self-organization system must have the following conditions:

Openness. The system must have the ability to interact and exchange with the external environment that isn't affected by the force of the system.

Non-equilibrium. The system promotes the difference of many internal elements by exchanging substance, power, and information with the external environment to push itself away from the balanced statement.

Nonlinear. There are at least three elements in the system, and there is a nonlinear interrelationship between the elements with a chain effect. The change of any element is affected by the combination of multiple factors rather than a single factor.

Fluctuation. Any mutation of the elements in the system may cause the system out of the original development track, then internal digestion and adjustment generate in the system to promote the transformation and upgrade of the system.

1.2 The Evolution Process of Traditional Villages in Southern Anhui

The traditional villages in southern Anhui originated in the Eastern Jin Dynasty. Under the influence of economic, social and cultural changes in southern Anhui, the material space of traditional village has been shocked to some extent. Taking the evolution processes of two important traditional villages in the southern part of Xidi and Hongcun as examples, their space has gone through a volatility development process including four stages as Fig. 1, and the characteristics of each stage are different.

1.2.1 The Forming Period of Traditional Village Space

The forming period of traditional village space in southern Anhui was between the Eastern Han Dynasty and the Southern Song Dynasty. The war in the Central Plains caused large-scale population migration, gradually impacted and reconstructed the population structure of southern Anhui, which led the northerners to become the main residents of the region. As a result, the displaced population has landed here, and they choose the narrow basin area that is near but above the water as their house location. Due to the lack of clear roads and group distinctions within the space, the village has a simple spatial structure and presents a scatted shape.

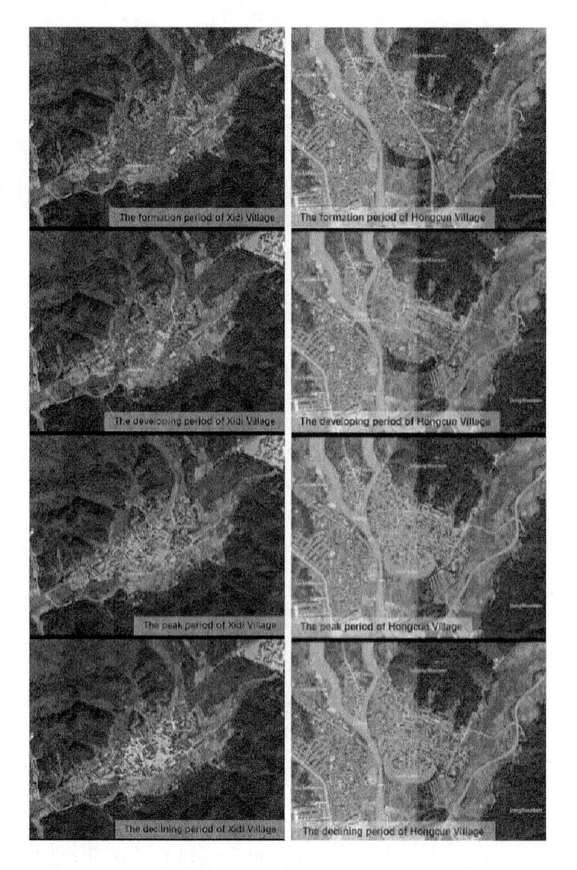

Fig. 1 Evolution of Xidi and Hongcun village in southern Anhui as examples

1.2.2 The Developing Period of Traditional Village Space

During the more than 300 years from the end of the Southern Song Dynasty to the middle of the Ming Dynasty, the population grew substantially and the branches of the clan continued to develop in the context of stable economic and social development. Therefore the village space expanded rapidly and the complexity of the space increased.

Based on the original site, the traditional village gradually had emerged clan ancestral halls and branch ancestral halls. These rigorous ancestral halls became the core space in the village so that the houses and streets are built around the ancestral hall. As shown in the spatial of the two traditional villages, with the space expanded, the village center had shifted on the basis of the previous stage, and the road system and commercial facilities were gradually completed. At the end of this period, the shape of the traditional village had emerged.

1.2.3 The Peak Period of Traditional Village Space

From the middle of the Ming Dynasty to the middle of the Qing Dynasty, the social, economic and cultural development of southern Anhui is booming. In particular, the boom economic promoted the Huizhou merchants to be a tower of strength in the traditional village. There are many people and clan who engage in political and commercial trade in the Xidi, Hongcun, Tangyue and other traditional villages so that the memorial arch that shows the glory of the clan are everywhere in these villages. In the spatial structure of the village, the range of living spaces formed around the buildings of different surnames is divided. And the road system, drainage system, ancestral system, commercial system were relatively completed in this period. With the strengthening of the commercial and ancestral space spatial function, there are cultural space and landscape space that emerged in the traditional village according to the needs of residents.

Thus, the traditional villages in southern Anhui became the high-density aggregate settlements that consist of five major functions: residence, ancestral halls, landscape, memorial arches and traditional academy.

1.2.4 The Declining Period of Traditional Village Space

The decline period of traditional villages ranged from the late Qing Dynasty to the early stage of liberation. At this stage, the village system was continuously affected and the self-organization mechanism became invalid. In the last years of the Qing Dynasty, with the development of the capitalist economy and the influence of war, the Huizhou merchants gradually declined and the residents were displaced. More serious is the severe destruction of the clan ancestral halls, memorial arches and houses in the villages but they could not be repaired in that condition. After liberation, due to the influence of property rights change and residents' demand, the characteristic of space fragmentation in traditional villages is prominent while the protection of the traditional architecture and space was unable to get attention from the residents and local governments. Generally speaking, the space structure of the traditional village had been basically continued while the village texture is lacking and even in chaos.

1.3 Summary of the Self-Organization Development Experience of Traditional Village Space in Southern Anhui

1.3.1 Factors Affecting the Spatial Evolution of Traditional Villages in Southern Anhui

Throughout the evolution of traditional villages in southern Anhui, it can be concluded that the main influencing factors of traditional village space growth include the geographical environment, Feng Shui concept, social structure, clan culture, economic development. During a different periods, the main influence factors were distinguishing.

The constraints of geographical environment. The geographical environment is both an external condition and an internal factor for the traditional villages in southern Anhui. The environment characteristics of many mountains and a small amount of land determine the limited expansion of villages and affect other elements such as the space layout and the architectural style.

The change of social structure. Taking the fluctuation in population in Yi county as a reference, it can be seen that there are three large population changes in the history of southern Anhui (Fig. 2). On one hand, the inflow of population brought advanced productivity and promoted the space expansion. On the other hand, with the population multiplied, the branches generate and grow, then a part of the population is separated to ease the pressure on village space and the new village appear.

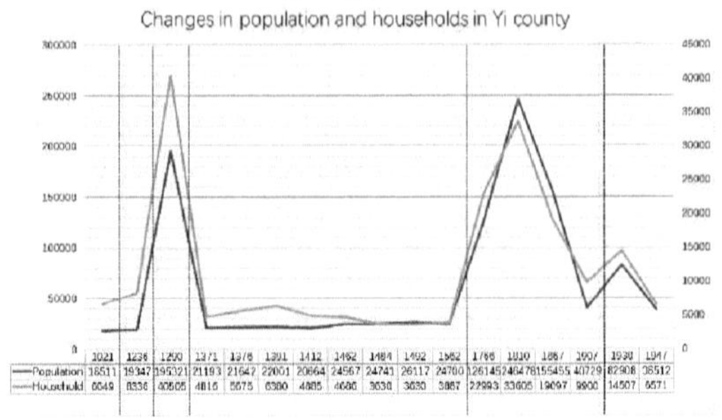

Fig. 2 Changes in population and households in Yi county of southern Anhui

The role of clan culture. The clan of the traditional village is a social management institution that is family-centered and blood-based. Its spatial performance is that the ancestral hall has a supreme core position in the village and the important space characteristic has been extended.

The catalysis of economic development. The Huizhou merchants who were aware of the importance of the education and clan culture have a far-reaching influence on the traditional villages in southern Anhui. Many Huizhou merchants have devoted themselves to the establishment of traditional academies and clan halls to improve the space quality and enrich the function of the village.

The influence of the FengShui concept. Under the influence of the Fengshui concept such as the integration of nature and man, not only the location, layout but also internal architectural orientation, ditch, landscape was unique and different from villages in other regions.

1.3.2 Self-Organization Mechanism of Traditional Village Space Evolution in Southern Anhui

In the long-term historical process, the evolution of traditional villages in southern Anhui is a dynamic and reciprocating devel-

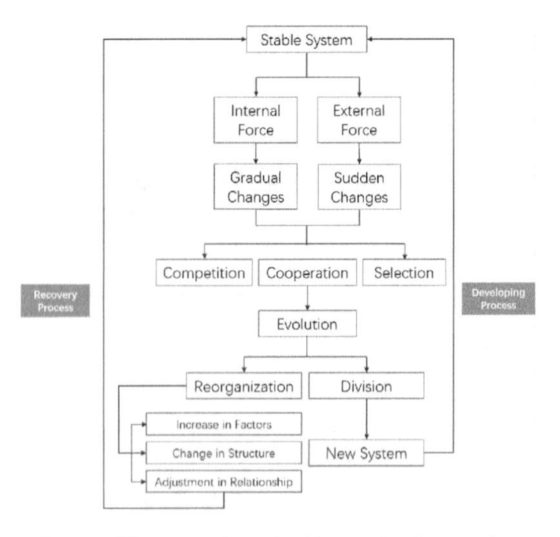

Fig. 3 The operation of self-organization mechanism in the traditional villages

opment process with constant changes in factors. Under the impact of the external forces and internal forces, the relatively stable system has a sudden change or gradual change, which promotes the competition or collaboration of internal factors. Next, the system enters a development process involving reorganization or division. In the process of reorganization, the increase in factors, the changes in structure and the adjustment in the relationship among factors may occur. In the process of division, the factors that have been stripped out are gathering together to form a new system. Eventually, the restructured system and the new system will reach a relatively stable state through the recovery and development process (Fig. 3). This complicated system has the following characteristics.

Maintain an open and relatively independent system feature. As an open system, the traditional village has the ability to exchange population, culture, economics and other elements with the external environment. In addition, the village itself has a certain degree of independence, with the ability to expand, split and gradual derive.

Regular dynamic development process. Even though during the stable development period, the traditional village space is not static, reflecting its non-equilibrium characteristics as a self-organization system. At each stage of space evolution, space always undergoes a dynamic development process from order to disorder than to order.

Competition and collaboration work together to drive system upgrades. Taking the contradiction between the increased population and the environment of the village as an example, this contradiction promoted the formation of a competitive relationship among the people under the clan system. On the one hand, the competitive relationship encouraged part of the population to move out to form a new village. On the other hand, the relationship promoted some people to abandon agricultural production and select business as their new profession. Therefore, the population, clan, economy and other factors changed to stimulate space development.

Elasticity. With the system structure developing, the component is getting richer and richer. The higher the complexity of the space, the higher the flexibility of the system, so that the system can be more resistant to external environmental shocks within a certain threshold.

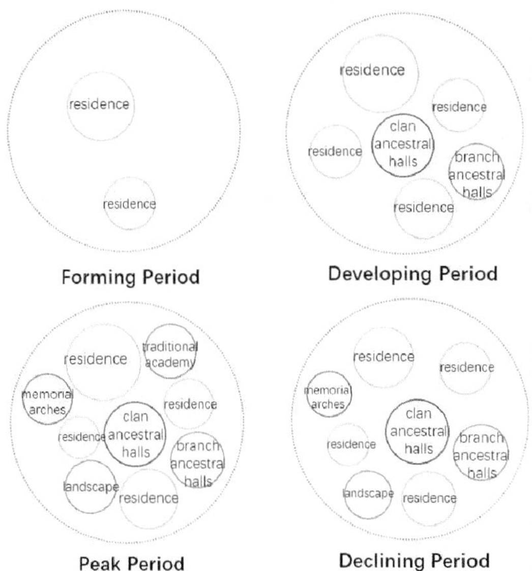

Fig. 4 Changes in traditional village elements in different periods

People-oriented. It is precise because of the flow of residents and the residents' demand for living and production that there are a series of dramatic changes in the village space(Fig. 4).

2 Current Situation Analysis and Problem Summary of the Traditional Village Space Development in Southern Anhui

2.1 Current Situation Analysis of the Traditional Village Space Development in Southern Anhui

After the development stage of the formation, development, prosperity, and decline of the traditional villages in southern Anhui, then the space evolution entered a new period of redevelopment. Many construction activities such as village conservation and renovation have been carried out in the region, which has a certain impact on the village space.

The diversity of village space is impaired. During the space evolution process, it is the outstanding performance characteristics of the traditional village that the nonlinear interaction between the various elements promotes the formation of spatial differences. Nowadays, the local government often adopts a simple and tough approach to get involved to make the internal spatial characteristics of traditional villages converge in the space protection and renovation process.

The connection between space is declining. Traditional village property with a high degree of integration of the feature space. However, with a variety of functions settled, some villages have the phenomenon of functional spaces isolation, such as the separation between old residential areas and new residential areas, or space separation caused by road development and residence. This problem will directly influence the spatial cohesion of the village, then further will increase the burden of system operation.

The weakening of the space hierarchy. The self-organization system established a bottom-up approach to form a multi-layered or grouped traditional village space structure that consists of a single building -the courtyard space -the architectural group -the space group -the village. In the space renovation and renovation in recent years, although space continues the basic features of the street as the main skeleton, the public space is continuously eroded, and the street scale is shrinking. Finally, the problem of reduced space level and space disorder is outstanding.

2.2 Interpretation of the Spatial Evolution of Traditional Villages in Southern Anhui

Exploring the spatial status of traditional villages in southern Anhui from the perspective of development mechanism, it is not difficult to find out the outstanding problems is connected to the system mechanism is unbalanced when the villages are subject to external environmental shocks and large changes in internal factors under the rapid development of urbanization. On one hand, due to the excessive intervention of the external environment especially the promotion of the economy and politics, the material, power and information flow to the system selectively and unidirectionally. At this time, the system is more dependent on external factors while the internal structural system is weakened, and the response, adjustment and healing ability of the shock are insufficient. It is likely that the space structure will have a mutation under the action of strong external conditions. On the other hand, it is a relatively slow development process with occasional mutations under the constraint of the geographical environment, economic, society and culture. However, corresponding to the rapid improvement of science, technology and living standards and the rapid loss of the village population, this bottom-up and long-term evolution that based on the clan system clearly cannot satisfy the residents' urgent demand of life quality and industrial development. Therefore, it can be said that the self-organization mechanism has not adapted to the maintenance and development of the current village space.

The concept of other-organization is corresponded to self-organization. The appearance of this concept stems from the fact that the contradiction between the internal requirements and its capabilities cannot be solved when the complexity of the system increases to a certain extent. It needs to be examined and processed from a higher perspective, so that the system evolves under the specific intervention conditions that involve recourses such as human resources, material resources, financial resources or other system such as organization and institutions of the outside world. Based on the continuous advancement of rural policies and practices in recent years, the other-organization mechanism has brought about the diversified protection and development of traditional villages. However, due to the lack of corresponding organizational guidance and intervention control, the placement of other-organization mechanism has also had a negative impact on some traditional village spaces. In particular, some local governments attract investment companies to settle in the village. This action has led to the constructive destruction of the village space, the stable and orderly spatial evolution of the village is affected at the same time.

3 The Spatial Renewal of Traditional Villages in Southern Anhui Under the Collaboration of Self-Organization and Other-Organization Mechanism

As mentioned above, the traditional villages in southern Anhui have exposed a series of problems in the context of the failure of the self-organization mechanism and the forced intervention of other-organization mechanism. Therefore, we propose the conception that let self-organization merge with an other-organization mechanisms to promote the renewal of traditional villages in southern Anhui.

3.1 Principle of the Self-Organization and Other-Organization Mechanism to Promote Traditional Village Spatial Renewal in Southern Anhui

Locality principle based on the self-organization mechanism. The self-organization mechanism is a summary of the practice experience and development law, so that it must be respected that the regional characteristics and then follow, adapt and utilize the human-centered development law when other-organization mechanism involved in the spatial renewal of traditional villages.

3.1.1 The Principle of Moderate Integration and Intervention

Consider that the traditional village as a living organism, we should think over the way in which external forces combined with internal organization especially the degree of external force intervention. We should try to avoid the nature mutation of space caused by the long-term occupation of other-organization, further ensure the village's ability to grow, renew and multiply.

3.1.2 Step-by-step Development Principles

The evolution approach of traditional villages includes gradual changes and mutations, and these two ways can be converted to each other. The alienation of traditional villages promoted by other-organization mechanism is largely due to the choice of a single system mutation method and the neglect of a long and phased process that spatial renewal of should be. Thus, we should correct the development direction and supply energy at the urgent time to ensure the elasticity and dynamic balance of village construction.

3.1.3 Excitability

Taking the interventional means for the important spatial elements of some traditional villages to trigger the "butterfly effect" of the elements. These elements may be the important public space or some architectural groups in the villages. Just like the entire village, these spatial nodes have fractal features to be a complex subsystem. We could grasp the development of the whole system by adjusting the speed and direction of these important spatial nodes.

3.2 The Strategy of the Self-Organization and Other-Organization Mechanism to Promote Traditional Village Spatial Renewal in Southern Anhui

The spatial planning of traditional villages is based on the protection state and development trend, explores the focus of village space renewal and proposes corresponding construction measures to promote the integration of self-organization and other-organization mechanism.

3.2.1 Provide the Power of Space Protection and Development

Local governments, villagers and social forces have become the main driving force of the system operates under the guidance of self-organization and other-organization mechanism. The premise of carrying out village construction is to build a multi-participation cooperation mechanism, especially to mobilize the enthusiasm of the villagers to participate in collective efforts to ensure the stable output of power.

3.2.2 Identity the Focus of Space Renewal

Based on the village space, we should clear the "space-based core" of local villagers with characteristics and genius loci, such as humanities, ecology, architecture in the long-term evolution of village space (Fig. 5). As the primary starting point for the update of the village space, the trigger effect of the "space-based core" is used to drive the space upgrade.

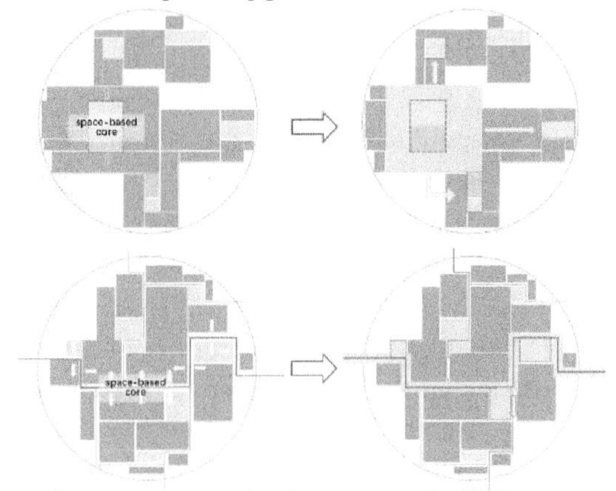

Fig. 5 Triggering effect of "space-based core" in traditional villages

3.2.3 Repairing the Separation of System Space

In view of the problems of space separation caused by the development of tourism functions and the appearance of new residential areas, we should implant spatial activity by arranging point-like or block-like functional spaces to avoid the concentration of functional space(Fig. 6). In addition, the green space should be settled as a conversion belt to break the hard barrier between different zones.

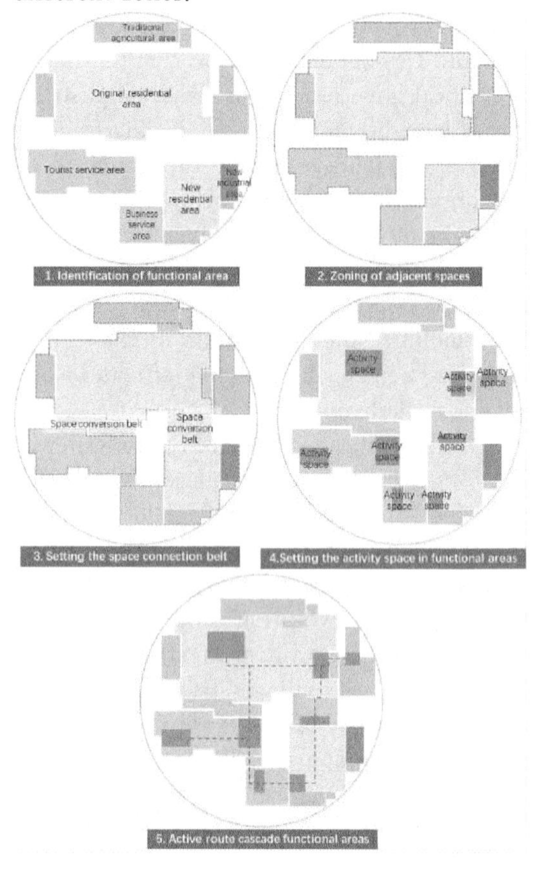

Fig. 6 Steps of repairing the separation of space in the traditional village

3.2.4 Repairing the Diversity of System Space

As a precondition to respect local and hierarchical characteristics of space, we propose that extract the spatial organization pattern including various elements such as streets, architectures, courtyards, landscapes within the village, and form the organizational framework of the village space language. Based on the framework, we can guarantee the space characteristic of the traditional village on an overall level by establishing space transformation menus or guide as a basis of management.

3.2.5 Controlling System Space Dynamic Balance

When carrying out the renewal of the village space, we should carry out dynamic planning from a holistic perspective, and gradually promote the protection and development work in stages according to the coordinated realization of the previous stage goals. Meanwhile, we should clear the expansion of the overall village and important public space as hard control elements while the functions of other public spaces can be moderately adjusted. In a word, we can guarantee the rigidity and elasticity of the space to realize the internal dynamic transformation and limit the external expansion of space by these methods.

4 Conclusion

The space evolution of traditional villages is an open evolutionary process guided by a self-organization mechanism. However, under the background of rapid urbanization, just relying on the self-organization mechanism can't satisfy the double demand for protection and development. It is necessary to permit the intervention of external resources to build up other-organization mechanism, and then gradually explore the ways in which these two mechanisms coexist, cooperate and apply in the development of traditional village space. In addition, it needs to be identified that the space of traditional village is different from the space of the urban and simple rural area. The particularity of this space determines that the traditional village space renewal is a special inventory planning focus on the space micro-regeneration, in order to ensure the dynamic balance of traditional village space.

References

ALLEN P M, 1990. Self-organization and dissipative structures. Technical physics letters, 16(4): 248-251.

CHEN Zhe, ZHOU Hantao, 2012. Research on traditional village renewal and new residential building based on self-organization theory. Architectural journal, (04): 109-114.

LU Lin, GE Jingbing, 2007. Research on geographical environment of the formation and development of Huizhou ancient village. Journal of Anhui Normal University(natural science), (03): 377-382.

LU Lin, 2004. Study on evolution process and mechanism of Huizhou ancient village. Geographical research, (05): 686-694.

WU Tong, 2001. Research on self-organization methodology. Beijing: Tsinghua University Press.

Policy Guidance to Help Vernacular Architectural Conservation and Sustainable Development of Cultural Landscape of Honghe Hani Rice Terraces

XU Fan, *PENG Xue*

China Academy for Cultural Heritage, Beijing, China

Abstract The purpose of this paper is to sort out a series of public policies introduced by the government of the Cultural Landscape of Honghe Hani Rice Terraces and to introduce effective ways to adapt to the development of Cultural Landscape of Honghe Hani Rice Terraces under the wave of urbanization development.

Keywords cultural landscape of Honghe Hani rice terraces, policy, vernacular architecture, sustainable development

1 Introduction

Cultural Landscape of Honghe Hani Rice Terraces is China's first world heritage site named after the nation. It is located on the southwestern border of China. It has witnessed the successful mountain development practice of the Hani people for more than 1300 years, and still preserves the "forest-water system-village-terrace" four-element system in geospatial space. The spatial pattern of the four-dimensional isomorphism of terraced fields and the well-functioning ecosystem have received international attention. In 2010, Hani Rice Terraces was recognized as the Globally Important Agricultural Heritage Systems (GIAHS). In 2013, it was selected as the head of the Ministry of Agriculture. Approved China-Nationally Important Agricultural heritage Systems. (China-NIAHS) In the same year, Cultural Landscape of Honghe Hani Rice Terraces was announced as a national key cultural relics protection unit and included in the World Heritage List. There are 82 villages in the heritage area, nearly 50,000 people, but limited by location and environmental conditions, Yuanyang County, where Cultural Landscape of Honghe Hani Rice Terraces are located, is still a national poverty-stricken county. With the 17 Sustainable Development Goals in the 2016 UN 2030 Agenda for Sustainable Development, the first of these goals is to "eliminate all forms of poverty throughout the world." Therefore, promoting World Heritage conservation and sustainable development has become the primary goal of the work of Yuanyang County and Honghe Prefecture where the heritage site is located in recent years.

In order to continue the vitality of the heritage and to coordinate heritage protection, community building and sustainable development, Honghe Prefecture and Yuanyang County of Yunnan Province promulgated a series of public policies to provide a legal basis for the effective protection and management of Hani Terraces, establish a traditional custody mechanism and promote the public. The understanding of the heritage guides the community to develop the local economy without prejudice to the safety of the heritage and establish the overall protection of the heritage.

2 Establish the Direction of Development, Coupled Protection Management Planning Provisions

During the application for World Heritage in 2012, Honghe Prefecture People's Government and the planning and preparation unit have already considered the protection and sustainable development of villages within the Hani Terraced Heritage area. In the "Cultural Landscape of Honghe Hani Rice Terraces Management Plan", the village and residential buildings were formulated. Specific measures and requirements. In order to ensure the smooth implementation of the management plan, in the same year, the Red River State Government passed the "Regulations on the Protection and Management of Hani Terraces in the Honghe Hani and Yi Autonomous Prefecture of Yunnan Province" to protect, develop and utilize the Hani terraces.

In order to implement the objectives of management planning, it fully reflects the important value of Cultural Landscape of Honghe Hani Rice Terraces as a world cultural heritage and an important global agricultural cultural heritage, so that the protection and management measures of Cultural Landscape of Honghe Hani Rice Terraces become a model for the protection of live heritage, the political government of Honghe Prefecture in 2016 In the year, the Chinese Architectural Design Institute was commissioned to compile the "Cultural Landscape of Honghe Hani Rice Terraces Field Protection and Development Master Plan (2016-2030)".

In 2017, in order to strengthen the protection and management of Cultural Landscape of Honghe Hani Rice Terraces, the Red River State Government promulgated the "Implementation Measures for the Regulation of the Hani Terrace Protection Management of the Honghe Hani and Yi Autonomous Prefecture of Yunnan Province", in which two categories of protection were divided into traditional villages. It is required that the construction of housing projects by villagers in the key protected areas of the Cultural Landscape of Honghe Hani Rice Terraces should meet the requirements of the Hani Terrace Management and Management Plan.

In 2017, based on the protection of traditional cultural and ecological characteristics and the protection of characteristic residential houses, the Yuanyang County Government planned the industry for the protection and sustainable development of Cultural Landscape of Honghe Hani Rice Terraces in Yuanyang County, and entrusted relevant professional organizations to compile the "Red River Hani Terraced Field". *Cultural Landscape of Honghe Hani Rice Terraces Yuanyang Core Area Protection and Utilization Master Plan* (2016 — 2030).

Since the launch of the application in 2012, the Red River State Government and the Yuanyang County Government have drawn a series of management plans, management regulations, and development plans. It can be seen that the government is striving to protect the villages within the scope of inheritance and inheritance through laws and regulations. Carry out macro control with sustainable development, and put forward principle requirements for the protection of villages and vernacular architecture.

3 Establish a Long-Term Mechanism to Strengthen the Implementation of Heritage Protection Management

In order to guide the protection of the traditional houses of Cultural Landscape of Honghe Hani Rice Terraces in Honghe, to maintain the location characteristics, layout structure and traditional residential construction methods of traditional villages, and to highlight the traditional cultural landscape elements, in 2012, the Yuanyang County Government formulated the "Honghe Hani Terraced Field Cultural Landscape". Measures for the Administration of Village Residential Protection (Trial). The target is for any unit or individual that produces, lives, constructs, and entertains in the cultural landscape village of Honghe Hani Terrace; for Hani Terrace, it involves the Hani Terrace Management Bureau of Yuanyang County, the Housing and Urban-Rural Development Bureau, and the Land and Resources Bureau. The Forestry Bureau, the Water Affairs Bureau, the Culture and Sports Bureau, the Environmental Protection Bureau, the Public Security Fire Brigade and other departments have given specific job responsibilities; village protection management in the heritage area, buffer village protection management, ecological environmental protection management, village cultural protection management, village activities Detailed regulations are given in terms of management, construction project approval management, maintenance and utilization, rewards and punishment.

In 2013, in order to better protect and manage Hani traditional dwellings, guide villages and residents to repair, maintain and renovate vernacular architecture, and adapt to the needs of village development, Yuanyang County People's Government entrusted Tsinghua University School of Architecture to organize the compilation of "Red River Hani traditional dwellings" Guidelines for the Protection of Rehabilitation and Environmental Management, through the establishment of the maintenance and approval process, to make heritage protection a daily behavior with institutional, financial and technical support; through the introduction of traditional repair and maintenance techniques and contemporary protection techniques, for traditional houses The protection of the building and the improvement of building efficiency provide technical support and ultimately promote the scientific protection of traditional ancient dwellings.

The "Management Measures for the Protection of Rural Residential Buildings in the Cultural Landscape of the Red River Hani Terraces (Trial)" and the "Guidelines for the Protection and Improvement of the Traditional Houses of the Hani Nationality of the Red River", the protection of the Hani terraced heritage villages and the local architecture was implemented from the macroscopic planning level. The specific operation method has played the role of protecting management and landing and established a long-term mechanism for protection.

4 Clarify the Implementation Path and Extend the Heritage to Highlight the Universal Value

In order to maintain the layout, construction style and national culture of the traditional houses in Hani Terrace, the authenticity and integrity of the Hani terraces will be gradually restored and continued. The Yuanyang County Government has formulated the "Support Scheme for the Protection of Traditional Residential Buildings in Cultural Landscape of Honghe Hani Rice Terraces". The program divides the residential houses into three types: complete safety type, partial damage safety type, and dangerous house type. The complete safety type refers to the dwellings with local national characteristics and intact preservation, which have important value in one aspect of history, culture, science, art, social economy and safer housing structure; some damage safety type means that there is a small amount of damage. However, there is still a certain value in history, culture, science, art, and social economy. After repairing, the traditional residence can be restored. The dangerous type refers to the destruction of the entire structure, the serious decay of the wood, and the collapse of the wall. In other cases, vernacular architecture that cannot be inhabited and restored. Among them, the complete safety type and partial damage safety type, two types can be subsidized. If the dangerous house type is rebuilt according to the requirements, it can be subsidized. The specific requirements are that it can be demolished and reconstructed on the original foundation. However, the appearance structure must be constructed according to the traditional style. The internal space can be designed according to the requirements of modern housing functions. The height cannot exceed two and a half. Each floor shall not exceed 3 meters of a standard building. The roof Hani is a thatched mushroom top. The appearance color must be in harmony with the traditional style and surrounding environment.

Townships and villages sign the "Responsibility Letter for the Protection of Traditional Residential Buildings", and organize personnel to enter the village for inspection and appraisal before December 30th. The plan also gives specific criteria for subsidies: traditional houses with listed protection, each subsidy of 700 yuan per year, the county people's government will list the traditional housing funds to be included in the county-level budget, and allocate the subsidy funds to towns and towns every year.

Since 2016, the Yuanyang County Government has subsidized a total of 1,227 traditional houses that have been listed in 53 villages, including the five key villages and the roads, such as Pu Gao Xinzhai and Yupin. The other 24 villages will also be gradually implemented.

The stable financial guarantee of the Yuanyang County Government has implemented the protection measures for the vernacular architecture and provided specific implementation methods for the continued universal value of the heritage and the sustainable development of the local architecture.

5 Highlight Demonstration and Standardize the Construction Mode of Local Architecture

While protecting the Hani terraces and villages, the Yuanyang County government also pays attention to the combination of national policies and encourages the construction of residential buildings in non-key protected villages to promote the sustainable development of local village buildings. The Yuanyang County Government has integrated the "Beautiful Village" creation activities proposed by the Ministry of Agriculture with the protection of traditional residential houses, and put forward the principle of "unified architectural style, unified architectural appearance and unified building model", and gradually progressed in 82 villages in the core area of the terraced area. The construction of "beautiful villages" will be implemented in batches to protect outstanding and representative traditional houses and villages, meet the needs of villagers to improve their living needs and rational use, and promote the protection and development of traditional villages in the Hani terraced heritage area. According to this demand, professional institutions have been commissioned. Renovation of traditional residential protection and environmental management in Cultural Landscape of Honghe Hani Rice Terraces Area in Yunnan, the premise of this design is to ensure that the ecological vegetation and the terraced water system remain unchanged, targeting village houses and public buildings (including ritual houses, public toilets, primary schools, etc.), the environmental elements of the village, focus on repairing the facade of the house (including roofs, walls, doors and windows, feet and other basic elements); remediation of roads, drainage systems, sanitation facilities, maintenance and new livestock breeding facilities, etc., proposed for local materials and construction techniques Body requirements.

The local government also plans to select and store 6 more than 73 ha of residential sites in the heritage area, and plan to construct ethnic characteristic villages in Quanfuzhuang, Hetao and Puduo Village, and guide the nearby indigenous houses to implement ex-situ relocation. And a new 300-family "adobe house, thatched roof" traditional Hani residential village, to solve the problem of the original homeless homestead, the village has no development space, as well as the expansion of houses and random construction.

6 Optimize the Industrial Layout and Return to the Organic Renewal of the Local Architecture

Through years of heritage protection work, the local government where Hani Terraces is located has gradually realized that only retaining the "people" in the heritage is the basis for protecting the villages, protecting the local buildings and achieving sustainable development.

The Yuanyang County Government actively leads local residents to find effective ways to make a fortune.

Make full use of national policies, implement the "National Grassroots Agricultural Technology Extension System Reform and Construction Subsidy Project" promoted by the Ministry of Agriculture, and combine the actual situation of Yuanyang County to promote the high-yield and high-efficiency ecological breeding model of rice-duck fish. The government not only provides farmers with rice seeds, fry, and ducklings are also assigned special personnel to provide technical guidance, and the harvest is owned by all farmers to increase farmers' income.

Introduce high-quality rice processing enterprises with strong strength, set up professional cooperation organizations, and adopt the development model of "leading enterprises + professional cooperatives + high-quality rice bases + farmers", "company + cooperatives + bases + farmers" to encourage and guide farmers. Planting high-quality terraced red rice with the high market price, strive to create a terraced red rice brand, promote the development of terraced red rice brand, and increase the income of terraced grain.

Adhere to the benefit-sharing mechanism and the compensation mechanism for heritage protection, improve the system of integration of interests of the "company and the masses", and extract a certain proportion of the income from the operation of the scenic spot to establish a support fund for mass production and development, which is used to improve the production conditions of the masses and support the development of the industry. The farmer's subsidy of 4,500 yuan/ha in the area will further increase the enthusiasm of farming and protecting the fields, ensure the farmland's cultivation nature and area, and make the farming technology better extended.

Adhere to the principle of "bringing merchants with business and promoting agriculture with business", and actively encourage farmers in the heritage areas such as Qikou, Dayutang and Lunaxin Village to open farmhouses and farmhouse restaurants, and focus on the development of thorn show, silverware ornaments and ethnic groups. The national craft products based on clothing, featuring terraced red rice, terraced fish, cloud tea, and yellow cattle dry bar, use sideline industry to promote farmers' income and get rich, thus mobilizing the enthusiasm of farmers to protect Hani terraces.

7 Conclusion

The Red River State Government and the Yuanyang County Government, where Hani Terraces is located, have implemented a series of policies, from legislation to establish heritage villages and local building protection system guarantees, to pooling resources, formulating overall plans for protection and utilization planning, and controlling the development direction from a macro perspective. Formulate sustainable development strategies tailored to local conditions, formulate guidelines for the management of residential houses and repair guidelines, and propose specific and clear measures for residents to protect vernacular architecture and then by retaining the soul of the vernacular architecture — "people" Leading the local residents to get rich, achieving a win-win situation in which the heritage can continue to develop and precision poverty alleviation, fully implement the concept of "community" in the World Heritage 5C strategy, and pay attention to the role of local communities in promoting the sustainable development of Hani terraces, and also actively to experts. International organizations seek theoretical support and technical guidance, and actively follow the implementation of multi-subject decision-making to provide effective support for the protection and sustainable development of the vernacular architecture where the heritage is located.

A Study on the Protection and Development of Traditional Settlements Along Longshu Ancient Road from the Perspective of Cultural Inheritance — Take Qingni Village in Longnan as an Example

ZHANG Ping[1], *ZHAO Yi*[1], *CAO Yifan*[2]

1 College of Architecture and Urban Planning, Lanzhou Jiaotong University, China
2 Architecture and Building Environment, University of Newcastle, Newcostte, Australia

Abstract The Longshu Ancient Road, which is included in the Preparatory List of World Natural and Cultural Heritage, serves as the intersection point of the Ancient Shu Road, the Northern Tea-Horse Road and the Silk Road. The settlements along Longshu Road reflect the blending characteristics of Qinlong culture and Bashu culture. Based on literature analysis and field research, this paper takes the fourth batch of traditional Chinese villages, Qingni Village, as an example, to classify and summarize the value of space elements of material and non-material culture from the perspective of cultural inheritance, and points out crisis which traditional settlements are facing, such as the loss of human landscape features, the lack of settlement morphological characteristics and the lack of cultural form caused by the disintegration of the basic environment and mode of cultural inheritance. In order to solve problems of rural aging and hollowing out during urbanization, this paper will give a holistic strategy for the protection and development of traditional settlement's local culture and architecture from the aspects of reconstruction of material landscape elements, protection of local settlement's style and features, activation of intangible cultural heritage, and reconstruction of historical activity space.

Keywords the perspective of cultural inheritance, Longshu Ancient Road, traditional settlements, protection and development

1 Introduction

The origin of Shu Road can be traced back to the early Xia, Shang and Zhou Dynasty. The Shudao used to date is also one of the earliest traffic systems in East Asia. As a large ancient traffic relic, it has been developed and constructed as a miracle in the history of human traffic, and has a densely distributed traffic network, and plays a major role in crossing the Long (Gansu) Shu (Sichuan), Qinba and Hengduan Mountains. It runs through the northwest and southwest of China in the form of trestle road. As the transitional section of the North-South Silk Road and the connecting section of the North-South Tea-Horse Ancient Road, the frequent spread of beacon and business travel in the history of Longshu Ancient Road promoted the formation of transportation network from Longnan to Qin and Shu, thus driving the development of settlements along the line. A large number of settlements along the line reflect the long history of frequent interaction between Qin and Bashu ethnic groups, and show the unique Longshu culture after blending Qinlong culture and Bashu culture.

Longshu Ancient Road is an important part of the cultural, economic and military exchanges between Qin (Shaanxi) and Shu (Sichuan). It belongs to an important part of the historical and cultural inheritance of Shu Road. In 2015, Longshu Ancient Road was included in the World Natural and Cultural Heritage Preparatory List. Longshu Ancient Road contains four water-land-dependent ancient roads: Jialing Road, Qishan Road, Yinping Road in Wuzhong and Taomindi Pan Road. The settlements along the Jialing Road in Qingni Village are located on both banks of the Jialing River. There are a lot of stone carvings, cliffs, ancient docks and ancient ancestral temples in the settlements. Dou to Qingni village's remote geographical location and relatively simple and sincere folk custom, a large number of traditional settlements have been preserved up to now. Qingni Village is an important node settlement along the cultural line of Qingni Road on Jialing Road of Longshu Ancient Road (Fig. 1). Libai, a poet of the Tang Dynasty, described the environmental characteristics of Qingni ridge —"How zigzag the mountain road of Qingni ridge is, nine zigzags are necessary to walk a hundred steps." in his poem: Difficult roads in Shu. The existing Qingni Village has its unique natural environment characteristics and traditional culture. Its historical value is unique, its artistic value is extraordinary, its scientific value is important, and it needs to be protected, inherited and developed.

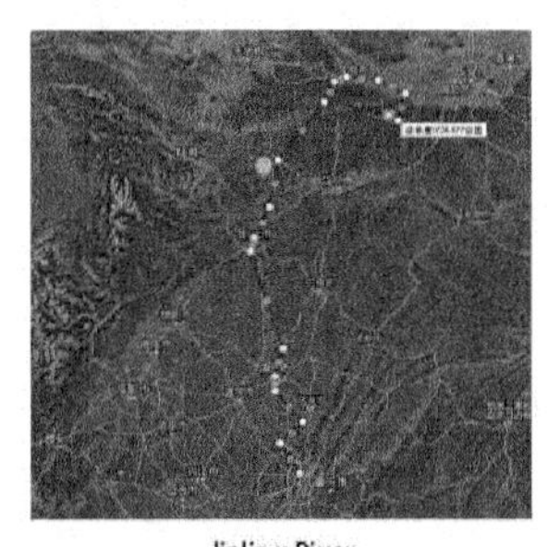

Fig. 1 The relation map of Longshu Ancient Road and Qingni Village

Traditional villages are formed naturally in the long-term specific geographical environment and various social activities. At present, the culture, lifestyle and production activities of traditional villages are in conflict with modern civilization. The diversity of village culture is facing a crisis, which urgently needs effective protection and inheritance. Many scholars have studied the problem of cultural inheritance in traditional villages from different disciplines: Bi Anping (2018) proposed travel response measures for cultural genetic landscape, Li Xiaoshi (2018) studied the identification of cultural genes and the evaluation and analysis of cultural inheritance effect of traditional villages in mountainous areas, Xie Jun (2018) studied the evolution of traditional villages' dwellings from the perspective of cultural geography, Yang Haozhong(2017) proposed to preserve the intangible cultural heritage of traditional villages in the construction of a new materialistic countryside. Current studies have experienced range from natural cultural heritage to historical cultural relics, and then to the expansion of intangible cultural heritage protection. Little attention has been paid to the study of cultural transmission paths and methods of multi-subject settlements along the linear cultural lines. Based on the field investigation of Qingni Village, this paper reflects on the inheritance and development mode of traditional village culture at all aspects of society and tries to explore the realistic dilemma and solution direction of traditional village culture inheritance along the line of linear cultural heritage from the perspective of cultural inheritance.

2 Background of Qingni Village

Jialing Road set up early and has a long history. From the beginning of the Qin and Han Dynasty to the end of the Republic of China (1949), according to textual research, as early as the Han Dynasty, Huixian County formed Qingni Road along the valley of Jialing River from Lueyang, the province of Shanxi to Sichuan. Since the Tang Dynasty (600AD), it has been the official road connecting Hanzhong, Gansu, Shanxi and Sichuan Province. It is also the way of Chencang's ancient road in the novel of Three Kingdoms and the logistic line of a famous battle where Liu Bang conquered Sanqin. Qingni Village evolved from Qingni Post Station on Jialing Road in Tang Dynasty.

With the development of commerce and the increase of population, the village came into being. During the Anshi Rebellion (755AD), the emperor Tang Xuanzong also took refuge to Sichuan through Qingni Village. The Village is located on Dahedian Town on the southern border of Huixian County, Gansu Province. It is surrounded by mountains and rivers with numerous hills, and backed by the main peak, Tieshan and also surrounded by dense vegetation and terraces along the contour line. And Qingni River goes through the village. Since Tang Dynasty, it has been an important node on the main road to Shu. In 2017, it has become the fourth batch of traditional villages in China. The traditional spatial pattern of villages that depend on mountains and rivers remains relatively complete. There are abundant types of existing buildings, such as dwellings, temples, ancient roads, ancient monuments, etc. Traditional dwellings are mostly built with civil stuctures with single eaves and xieshan roof, adobe walls or rammed soil walls with wooden pillars and beams as the mainframe, green tile roofs, carved doors, and windows. And most of them face streets. Only a few houses are laid out in the form of quadrangles and have large doors with typical Hui style. Traditional buildings were freely distributed along ancient streets and Qingni River in the late Qing Dynasty according to mountain and topography (Fig. 2).

The patterns of public buildings and dwellings are adapted to local conditions, and interdependent with the mountains and rivers, unified and changeable. In recent years, the local government has taken "redevelop the ancient culture of Qingni" as the goal of building a new tour village. By improving the traffic and human settlements environment, the village has greatly changed.

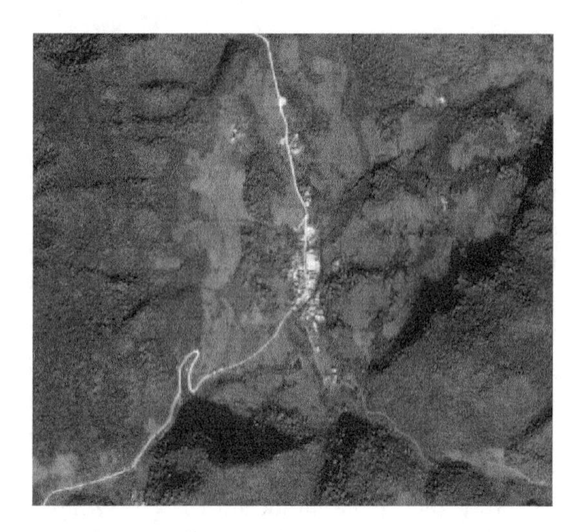

Fig. 2 Overall spatial layout of Qingni Village

3 Cultural Heritage Resources of Qingni Village

Traditional culture helps to maintain social identity, restrict social trends, and promote the development of society through different forms of expression and evolution. And the continuation of traditional culture plays a vital role in the protection of traditional settlements. Cultural heritage of Traditional village includes not only all kinds of "cultural relics" and "non-heritage", but also unique historical memory, traditional custom, production mode and so on. The Shu Road represented by Qingni Road reflects the route of human cultural exchange and the direction of tribal migration within a certain time and space, and at the same time promotes the cultural blending in the north and south and the east and west of China. The settlements along Shu Road have a long history and unique status, which bear many historical events. The traditional culture inherited by them reflects the heritage value of Longshu Ancient Road. Qingni Village has preserved relatively complete cultural heritage and intangible cultural heritage including natural environment elements and traditional settlement pattern, building relics from ancient post stations, and local handicraft, Qingni sheepskin fan drum, shadow play, performing arts activities, and customs, as well as "New Monument of

Xuantianshen Road" and "New Baishui Road Note" and so on. Many rare historical materials, such as humanities, transportation, and art, have largely followed the historical and cultural traditions of Qingni Post Station. The Qingni Culture is divided into material and non-material elements to sort out and reconstruct, according to the different cultural heritage attributes and cultural significance of Qingni Village (Tab. 1) in this article. And base on this, the reconstruction analysis of cultural value and core resources of the village will be carried out.

Tab. 1 Classification of traditional culture in Qingni Village

Macroscopic Level	Medium Level	Heritage of Traditional; Culture
Elements of Material Space	1. Natural environmental factors	1. 1 Huaya Mountain; 1. 2 Fengwanliang; 1. 3 Foya Mountain; 1. 4 Shengpo; 1. 5 Youfangzui; 1. 6 Shijiaya; 1. 7 Qingni River; 1. 8 Shudao; 1. 9 Ancient Trees; 1. 10 Ancient Well
	2. Traditional pattern	2. 1 Village Shape and Boundary; 2. 2 Street and Lane System; 2. 3 Qingni Road
	3. Buildings and structures	3. 1 The new Baishui Road Stele in the Song Dynasty; 3. 2 The Dahedian Road Stele in Huixian; 3. 3 County in Qing Dynasty, the XuantianShenlu Cliff in the Ming Dynasty; 3. 4 The YuantongWuchu Stele in the Qing Dynasty; 3. 5 The Buddha Yaya Cliff Stele in the Qing Dynasty; 3. 6 The remnants of the Stele in the Qing Dynasty; 3. 7 The Yang's Grand Court
Non-material Cultural Heritage	1. Handicraft	1. 1 Green Mud Sheepskin Fan Drum; 1. 2 Traditional Building Construction Techniques; 1. 3 Traditional Textile Techniques
	2. Performing Arts	2. 1 Shadow Show; 2. 2 Social Fire; 2. 3 Drought Boat; 2. 4 Stilts
	3. Diet	3. 1 Pots of tea traditional production skills; 3. 2 9-bowls of Three—line traditional production skills; 3. 3 Stoves of tooth surface traditional production skills

4 The Present Situation and Problems of Cultural Inheritance in Qingni Village

Whether it is the physical carrier itself, such as the environment of village or buildings, or the intangible culture, it would change and develop with the historical process of the times. While modern lifestyle brings convenience to people, it also greatly affects the development of traditional villages. The inheritance of traditional culture is facing different challenges in different villages around the world.

4.1 Lack of Morphological Characteristics of Traditional Settlements

Most of the traditional dwellings are constructed with soil in Qingni Village. However, the natural environment of its geographical location is not suitable for the preservation of soil buildings. After the destruction of the dwellings, villagers tend to rebuild new houses with brick-concrete structure rather than choosing protective refurbishment. For a long time, villagers have built their own houses without planning and unconsciousness. The newly-built dwellings gradually break away from the control of the original landscape form and tend to be laid out on both sides of the highway. These two factors result in the imbalance between the internal architectural style and the overall spatial organization of traditional villages, and gradually the loss of the original local characteristics and regional characteristics. Moreover, the natural environment of some villages is seriously damaged. These results also show villagers' weak sense of cultural heritage from another aspect (Fig. 3).

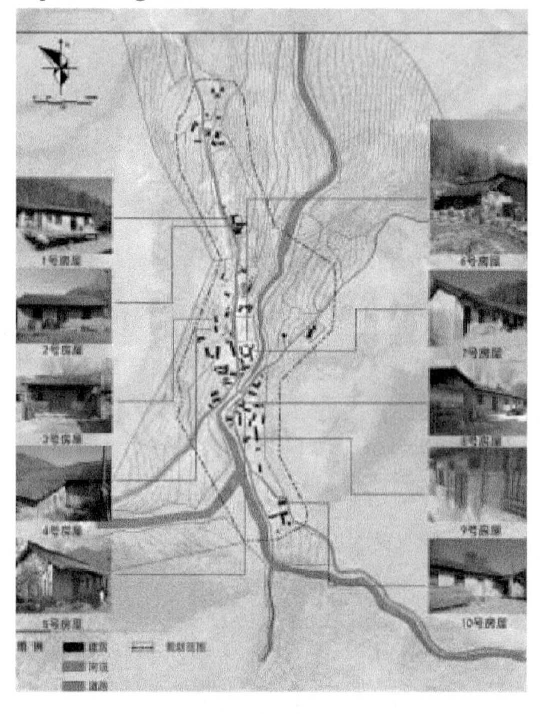

Fig. 3 Traditional residence in Qingni Village

4.2 Loss of Traditional Human Landscape Characteristics

Formerly, Qingni village was known as a qingni post station. The human landscape is the main component of its cultural characteristics. However, due to the destruction of the living environment of traditional culture by some historical factors, the loss and destruction of historical relics are quite serious, and the cultural characteristics also disappear. With the acceleration of urbanization and industrialization of surrounding cities and towns, the population loss of settlements is excessive, and the present situation of villages present hollowing out and aging. The ideology of traditional villages with villagers as the main body is also gradually weakened. As a result, gradually, the human landscape that should be carried by indigenous people, their production and lifestyle dissipate, and the culture of village and Shudao culture disappear, and the traditional cultural landscape loses its survival soil.

4.3 Disintegration of Basic Environment and Model of Cultural Inheritance

The basis of cultural inheritance is the overall environment on which culture depends for survival and development. Firstly, it is necessary to ensure the integrity and authenticity of the natural and cultural environment of villages. Economically, the existing rural relationship is becoming profitability in Qingni village. And the management form centered on traditional clan relations no longer exists. Moreover, the existing management model has led to the upward shift of management authority of traditional villages. The inaccurate formulation and propaganda orientation of the protection policy of village culture by the functional departments, and the backward infrastructure of villages, all of which lead to the changes in the material and spatial environment of cultural inheritance, the collapse of the mode of culture's natural inheritance in the daily life, and the fragmentation of cultural inheritance in villages.

5 The Strategic Direction of Qingni Village Protection from the Perspective of Cultural Inheritance

Culture is the result of human civilization adapting to the local geographical environment and ideological evolution. In the process of its evolution, traditional villages have been carrying on the cultural tradition with distinct regional characteristics. Strengthening the protection and renewal of traditional cultural heritage from various perspectives and continuing the historical and cultural context is the target of the protection of traditional villages in the future.

5.1 Regulating the Elements of Material Landscape and Protecting the Authenticity of Traditional Settlements

On the premise of protecting and inheriting the authenticity of the traditional settlements, it is imperative to rectify the material landscape elements along the line. A comprehensive survey of the buildings (structures) in Qingni Village is conducted, and the buildings in the villages are planned and utilized combining tourism functional areas. According to the material space attributes and spatial features of different areas, the planning and renovation scheme should be carried out to restore the historical features of Qingni's old streets, keep the authenticity of local culture and improve the living environment, restore the historical significance and extend the cultural value of architectural style.

On the basis of the existing physical environment rectification, to protect the traditional settlements, it is necessary to guide the restoration of traditional neighborhoods, respect the villagers' thematic status. Within the scope permitted by the current system and regulations, the government should take full advantage of the subjective initiative of the villagers, repair the traditional social relations such as clans, associations and blood relatives and create a harmonious and friendly community atmosphere. Only in this way can such a cohesive and active community form inherit the traditional style of the settlements and continue the traditional culture.

5.2 Activate the Intangible Cultural Heritage and Reshape the Spatial Regionality of Historical Activities

The inheritance of intangible cultural heritage is an important part of cultural heritage. To activate the intangible cultural heritage of Qingni Village, people need to start from the following aspects: Firstly, we should improve the system of craft inheritors such as can tea, protect representative inheritors and excavate young inheritors; Secondly, non-material surrounding products and cultural catering with the unique characteristics of Qingni Village should be developed to attract more tourists; Finally, the new media should be used to as the media to promote material and intangible cultural heritage of traditional villages along Longshu ancient road. Depending on the material development, intangible cultural heritage needs to restore the material space of historical celebrities such as Qingniu Station along the Longshu ancient road, reproduce the historical activities, and endow the historical space with new epochal significance. The overall protection of settlements along the line could be achieved by the exhibition of the handicraft production process, the performance of traditional culture, the promotion of the regional traditional culture, the protection of cultural genes and the maintaining of the "regionality" of traditional culture.

5.3 Abandon Vulgar Business Practices and Promote Diversified Non-relic Tourism Development

At present, most traditional settlements could attract a large number of visitors with great natural resources. The development of tourism has gradually become an important way for the countryside to cultivate its own vitality. Tourism, as an effective regional cultural strategy, brings mobile population, capital, and new production and life mode. It can not only maintain the integrity of regional social functions, but also enhance the regional competitive advantage of this region. Maintaining the overall attraction of cultural tourism along Shu Road can promote economic development while remaining the continuation of local culture. People should focus on different tourism projects, such as promoting an overall development of tourism of intangible cultural heritage, protecting positive tradition, exerting its characteristics, combining the culture of the Shu Road, exploring the unique culture of each village, developing handicraft production and performing activities, so as to enable rural visitors with different travel purposes to find their own tourism projects. Tourism developers and relevant practitioners should improve the awareness of the protection of traditional village and the inheritance of cultural heritage, cultivate professional literacy, and enhance brand awareness. The local government should also give corresponding protection policy to stimulate the initiative of stakeholders, enhance the protection and inheritance awareness of the public, and combine traditional cultural heritage with tourism.

6 Conclusion

Combining the world natural and cultural heritage of the Longshu ancient road with the integral protection of traditional villages, people should inherit and protect the material and intangible cultural heritage of Qingni Village, and seek the coordinated development, and find the symbiotic model, and provide survival soil for the protection of linear intangible cultural heritage, and further reshape traditional regional culture of the villages. Protecting the authenticity of traditional settlements while reshaping the spatial regionality of historical activities and promoting the diversity of tourism of intangible cultural heritage, which could be a way of reversing the backward development of traditional villages in the inland mountains of Northwest China and improving the inconvenient transportation, so that the cultural heritage of Longshu ancient road can reflect its value and great preservation and promotion.

Acknowledgements

Foundation projects: National Natural Science Foundation of China (51768029, 51768030), Social Science Planning Project of Gansu Province (16YB071), Humanities and Social Science Youth Fund of Ministry of Education (14YJCZH210, 14YJCZH006)

References

BAI Wenli, 2019. Study on the symbiotic protection of intangible cultural heritage and traditional village environment - take Jiajiazhuang village as an example. Green building materials, (03): 37 + 39.

BI Anping, WANG Guodong, PAN Hui, et al., 2018. Gene loss and tourism response measures of traditional village cultural landscape: taking Yixia village in Guling, Fuzhou as an example. Journal of Fujian Agricultural and Forestry University (philosophical and social sciences edition), 21 (04): 83-88.

CHENG Chuan, 2018. Cultural consciousness and cultural inheritance of traditional villages: taking Zouyuan village in Jinhua city as an example. Journal of the party school of the Taiyuan municipal committee of the communist party of China, 05.

GAO Tianyou, 1995. Textual research on Longshu Ancient road. Wenbo, 02.

GAO Tianyou, 2016. On shudao studies. Journal of Tianshui Normal University, 36 (01): 95-103.

LI Xiaoshi, 2018. Cultural gene identification and value evaluation of mountain traditional villages. Zhoushan: Zhejiang Ocean University.

XIE Jun, 2018. Formation and evolution of traditional villages and houses of Li nationality from the perspective of cultural geography. Guangzhou: Guangdong University of Technology.

YANG Haozhong. The construction of new countryside still needs to preserve the essence of local culture. China culture daily, 2017-11-25 (003).

Locations of the Global in Traditional Architecture

Monica Alcindor Huelva

ESG, Largo das Oliveiras s/n Vila Nova de Ceveira, Portugal

Abstract Traditional architecture faces multiple challenges; one of the most obvious comes from globalization. So far this phenomenon has been detected from its most visible face, the homogenization processes that have been imposing themselves in traditional landscapes. For this reason, a series of International Chapters have been triggered that put the focus of attention on this reality and the need to find out adapted responses. However, there are other ways where the processes sneak in that may be unnoticed.

The widespread character of globalization in multiple dimensions of life requires a clear detection of them before being able to propose a mode of action. Therefore, as a first step, the main theories that have analysed them linking them to traditional architecture will be exposed, to later seek from a field little explored until now, the anthropology of architectural construction, the other subtle ways in which global processes are introduced with less obvious visual impacts but with the same potential risk of loss of cultural diversity, distinctive of traditional architecture.

Through a qualitative study of the comparative nature of constructive solutions used in the north of Portugal, these subtle ways of globalization s proceeding will be studied in this specific location, converging the existing theories about the causes that generate them.

Finally, in the last section, alternative solutions to this challenge will be proposed, scrutinizing them from the multiple dimensions that intervene in order to verify the feasibility of them.

Keywords globalization, construction systems, northern Portugal, expert systems

1 Globalization and Vernacular Architecture

1.1 Introduction

In the advance of modernity, i. e. in late modernity, a break with traditional modes of operation has become more apparent, and formal changes resulting from this new order are discernible in the vernacular heritage.

From the information age (Castells, 1995) this new order is characterized by the advent of a new socio-technical organizational model as the fundamental matrix of economic and institutional organization. Being, therefore, interventions in vernacular architecture guided by the techno-bureaucratic globalization process.

Under these conditions, even local aspects are deeply penetrated and configured by social influences that are generated a great distance away from them (Sassen, 2007). Relationships of any kind are intensified around the world, linking faraway places; this is the essence of globalization.

In construction systems, these foundations have given way to what Anthony Giddens designated "expert systems", i. e. "systems of technical accomplishment or professional expertise that organise large areas of the material and social environments in which we live today" (Giddens, 1990), which have permeated the fabric of everyday life in building systems, where individual cases are sacrificed according to a general sense of systemic efficacy.

These expert systems go hand-in-hand with transnational processes that go beyond domestic territories and institutions, as the new global rules, born out of the rational-legal system, manage to penetrate to some extent to the very heart even of the most local level. The fact that a process is within the local territory does not necessarily imply that it is a local process (Sassen, 2007). It may be a location of the global, which highlights its incorporation into areas considered mainly local, but which really follows the same rational laws that govern global processes, the product of a growing bureaucratization in all areas.

The study of the refurbishment of vernacular heritage is presented as a means of encapsulating to capture the phenomenon of localized globalization.

1.2 Expert Systems and Their Silent Imposition on the Vernacular Architecture

When it comes to studying from the vernacular architecture the effects of globalization, the first focus of attention falls on landscape transformations. Thus, homogenizing and dehumanizing processes began to unify any construction within any territory. In landscapes born from materials extracted from the natural environment and constructive techniques acquired by endogenous evolutionary processes or by cultural loans (Martín, 2006) it is imposing today's technology that can transform local and individual conditions into something so similar that any territory becomes unified. There are the results of the change of a previous limited, diversified, and local technique that offered an image of the peculiar culture of this area.

Expert systems have led to the dismantling of traditional forms of construction and have created dependencies through a tendency to invade all areas of construction and achieved indispensability (Giddens, 1990).

But the first impact of this clear consequence of globalization can lead to the invisibility of others of the same nature that occur in a less obvious way, more subtle. Identify the repercussions of these processes in all their scales of action is one of the necessary preliminary steps at the protection actions of vernacular architecture. Otherwise there is a risk of implementing measures not directed to the main causes, peripheral solutions that although they mobilize many efforts to avoid the loss of identity, many of them have been focused on visual harmony and material contextualization avoiding other variables such as structural functioning original or the local origin of the materials that can become more determining as the main causes of these changes.

Precepts of maximized efficiency of modernity have given way to a standardized construction where there is no place for particular cases, as required the local architecture refurbishments.

1.3 The Complexity of the Protection

But at this point, another issue must be considered. What does it mean to protect the vernacular architecture?

The protection of these buildings covers different dimensions that, sometimes, between them can become contradictory or difficult to reconcile. The difficulty lies in the fact that it is necessary to promote the evolution of these buildings without betraying the community cognitive level and without losing local techniques and, at the same time, allowing them to remain part of the living constructive fabric of our towns and cities.

This generates a contradiction due to the meeting of two operating logics that do not easily find a point of convergence. On one hand, there are the constructive functional principles of the vernacular heritage. And on the other hand the contractual logic of performance regarding purposes that forces the use of standardized solutions that are disconnected from both physical and cultural territory.

Aware of this disengagement, the administration through the legislation tool promotes a growing recognition of local cultural diversity based on the most visual variables. As a result, there is a confusing pattern of ignorance / recognition of cultural logics (Alcindor, 2016).

Thus, there is the paradox that although they are increasingly protected by regulations, at the same time that they try to avoid the loss of traces of local particularisms, they reduce the capacity to be able to propose creative solutions that ensure a hybrid evolution between universalist expert systems and the local ones.

The complexity of protection lies that the essence of vernacular architecture that must be protected is a living architecture where diversity of dimensions converge, that is, as a transdisciplinary set of material and immaterial variables. Therefore, its protection requires a transdisciplinary local knowledge and understanding before establishing any rule ...

2 Methodology

Study in the field of the refurbishment of vernacular architecture required work on the link between the material practices, policies, and scientific discourses, a job that allowed the profound logic that links the different types of social practice to be brought to light, as these discourses form part of these practices, despite their relative autonomy (Foucault, 1999).

Specifically, the search for these links was performed using the perspective of the situational analysis through which it could be possible to construct a complex, holistic picture from the analysis of words and the observation of works.

3 Analysis of Constructive Practices in the Refurbishment of Northern Portugal

3.1 Common Practices

The most usual practices employed in the vernacular heritage refurbishment in Northern Portugal will be analysed from the perspective of the anthropology of architectural construction, that is, the area that tries to see constructive decisions from the understanding of the "embedding" of constructive action in networks of interpersonal connections and in particular cultural conditions. These specific examples will make it possible to print tangibility, concreteness and interest to the discussion that would remain, otherwise, on an abstract plane about how the world goes further into areas considered as local.

3.2 Contact with the Ground

In these buildings, no type of element prevented the water to rise by capillarity in the walls, since the activity that was to take place inside it and the environmental conditions did not require solving this situation. Normally, the ground floors were intended to accommodate the animals of the farm, which was perfectly compatible with the appearance of these humidities.

In addition, the pieces had constant ventilation due to the activity carried out and for the treatment of their walls, with no finishing, any plastering on the walls, or pavements on the floor, both inside and in the surrounding streets. This greater possibility of ventilation allowed bigger evaporation of the water molecules that rose through the capillaries of the walls and thus the water did not reach high levels.

With the use change, these spaces become habitable and these humidities by capillarity are no longer admissible. This is one of the main problems that affect the comfort conditions of these buildings that are now focused exclusively on residential use.

The solution that is commonly used is to place a layer of crushed stones that seeks to break the capillaries, leave an air chamber with no ventilation ensured and execute a slab of prefabricated concrete beams.

Although this solution continues to leave the walls in contact with the ground, without solving all the humidity problems by capillarity. So this solution only partially solves the problem. To avoid the visualization of the spots due to this issue, partitions are placed to conceal them.

This way of proceeding suggests patterns of action typical of late modernity based on these expert systems that seek immediate effectiveness through the pronounced specialization of knowledge as opposed to traditional practice that never provides a way to solve a particular problem, but always an elaborate, often multifunctional method that was part of an integrated approach and strictly linked to a conception of the world founded on the careful management of local resources and the social models that lie behind each construction (Laureano, 1999).

3.3 Enclosure Walls: Finishes and Window Frames

In this area, as in many other parts of the Mediterranean area, there was the practice of covering the stone walls with lime coatings. Only the lack of resources, the useless use of the building, or the exceptional quality of the stone, as well as a careful masonry, exempted this procedure.

These walls were built using local stones easily available due to their condition of abundance and proximity in the previous pre-industrial conditions. Its quality was not a determining condition. The fact of coating it with an outer layer ensured a greater durability of the stones as it protected them from the weathering processes and at the same time ensured greater impermeability inside.

In the new conditions governed by an interrelation of capitalism and industrialization, to obtain these stones with a format outside the conventional dimensions of commercialization converts existing ones into deluxe pieces. As a consequence, its exposure has gained greater value compared to its traditional concealment after plastering. Priority is given to an exhibition of the historical material(Fig. 1).

These stones take the role of new symbols of exclusivity, since capitalism beyond the fundamental matrix of economic, institutional and technological organization, is also a system of production of symbolic goods (Salhins, cited in Ramirez, 2011).

In the case of window frames, traditional wooden exteriors are usually replaced by other aluminum ones. The mistrust of the more natural materials enhances the use of expert systems, making the latter to be the best option despite the traditional ones. The loss of knowledge outside of discourse from the educational institution (Foucault, 1999) prevents the responsible for these interventions to feel comfortable with the uncertain behavior of traditional materials. This distrust ends up eliminating traditional systems in front of experts.

Fig. 1 Finish in the traditional house after the intervention © Andreu Roselló Alcindor

3.4 Slabs

In traditional slabs of wooden beams, traceability of the raw material was a fundamental factor that allowed to predict the behavior of this material.

Variables such as the characteristics of the species of wood used or the different processes followed from the election to the installation ensured a proper operation. But the production conditions of late modernity block the traceability of supplied materials. In most cases, this leads to opt for other systems of greater reliability in this new production environment.

It is usually replaced unidirectional solid wood slabs and wooden plywood beam filling by unidirectional slabs of prefabricated concrete beams, ceramic interlocking, and compression layer(Fig. 2). Expert systems par excellence.

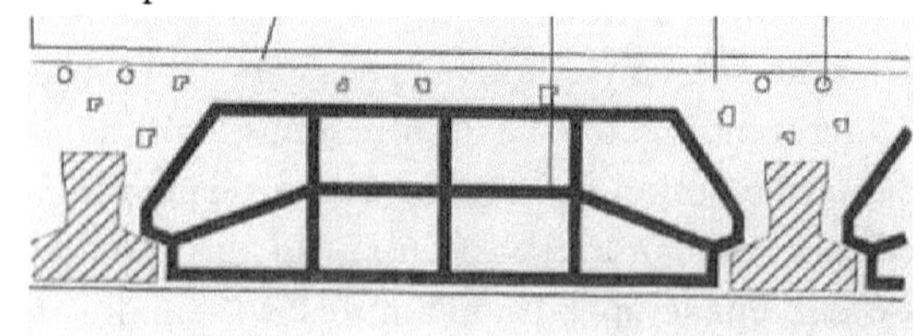

Fig. 2 Precast concrete beam slab © Monica Alcindor Huelva

3.5 Roofs

Roofs were traditionally made by wooden beams, wooden beam filling and ceramic tiles.

But in a refurbishment intervention, the tendency is to replace this wooden structure with a concrete beam and ceramic interlocking slab on which are erected partitions that form the slope of the roof on which rests other concrete joists, ceramic interlocking, compression layer, insulation, protective layer and finally the roof ceramic tiles.

This constructive system is preferred since it is an expert system which means that it is known by all construction workers, facilitating its execution and its price, compared to the use of traditional materials such as wood that under these production conditions in which the times are more accelerated do not allow to ensure the correctness of the steps to be followed in the care of the raw material, and as a consequence it is difficult to ensure a suitable behavior.

New conditions of administrative regulation, business, and technical convenience induce the application of expert systems at the expense of traditional techniques. In addition, the use of these systems can go unnoticed at first glance, but actually, a homogenising process is subtly imposed.

4 Conclusion

In the preference of those responsible for the refurbishment works, it is evident the use of materials that have been processed by the industry, which has earned "the attribution of agents to identify risk factors, monitor variables, analyze situations complex and design responses to deal with accidents and catastrophes "(Velasco, 2006). There are the same processes observed in these aforementioned examples that prefer precast concrete slabs over the traditional solid wood slabs. Or change the wood carpentry for the aluminum ones, although in an effort to depersonalize it tries to imitate the traditional wooden ones.

In short, what lies behind these preferences is the legitimacy achieved as random objectifiers and this brings a capacity to reject and dissolve "traditional" alternatives.

The economy also plays a central role in these preferences, since it provides the criteria of what is "economic" and what is "uneconomical", exerting a powerful influence on the actions.

If an activity has been labeled uneconomic, its right to exist is not merely questioned but denied with energy because the judgment of the economy is extraordinarily fragmentary. The economy only fixates on one variable: the monetary aspect. These criteria give much more weight in the short term than in the long term (Naredo, 2006).

4.1 Alternative Solutions

Conciliation between the maintenance of traditional cultural logics and at the same time to keep being part of the constructive reality of our towns and cities in another production environment, forces to center the core of the matter, since it is not to choose between "modern growth" and "traditional stagnation". The question is to find the right path of development (Schumacher, 2011).

In this section, some alternatives to these forms of intervention are explained.

In the case of the dampness of the ground floor, beyond the solutions of capillarity breaks the solution can also come by intervening in the type of programmed used. A use that requires large volumes of ventilation, in a way that promotes the evaporation of water contained in the walls. It is about reproducing conditions similar to the initial ones that determined their original constructive characteristics.

Other ways of acting require a prior reflection on the discourse that is imparted from the educational institutions since many of the aforementioned actions are influenced by ignorance, a conceptual void in the transmitted contents.

This is the case of wood, there is a prejudice that encourages its substitution even in cases of old wooden beams that are completely dry, of which there are no risks of xylophagous attacks, nor lack of carrying capacity as long as the distribution of the efforts with which he has worked for decades is not modified.

If the wood is the material of the roof structure, the intervention should only ensure better thermal insulation. In the case that it takes part in slabs from inside, it is the acoustic insulation that should be implemented.

In the case of plastering peeling off the ordinary masonry walls. The reconciliation between the exhibition of a new symbol of distinction and the maintenance of the traditional aspect can be adapted in degrees. The solution adopted by Távora in the Rua Nova of Guimar es refurbishment is an example of this type of compatibilization(Fig. 3). The wall is not completely stripped, but windows are opened through which to observe the historical constructive system behind finishes.

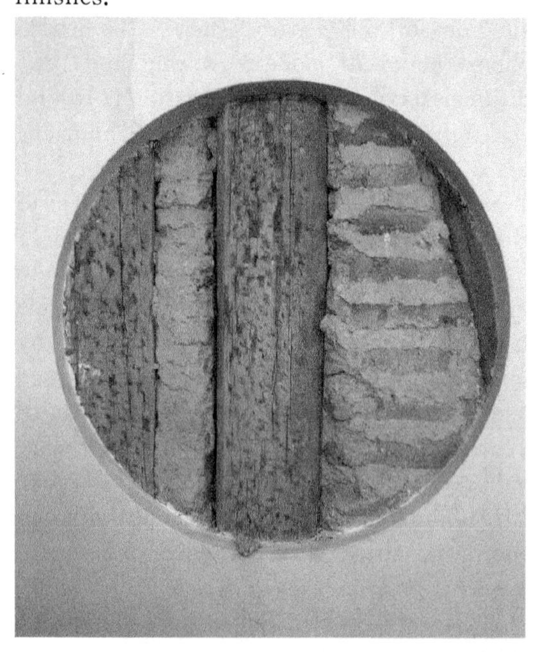

Fig. 3 Rua Nova of Guimar es refurbishment by Távora © Martín Torres Bargiela

In the case of traditional wooden carpentry that has simple glasses, it is possible to maintain the original ones only replacing these glasses with double glasses that reduce thermal losses (Vegas & Mileto, 2011. Fig. 4).

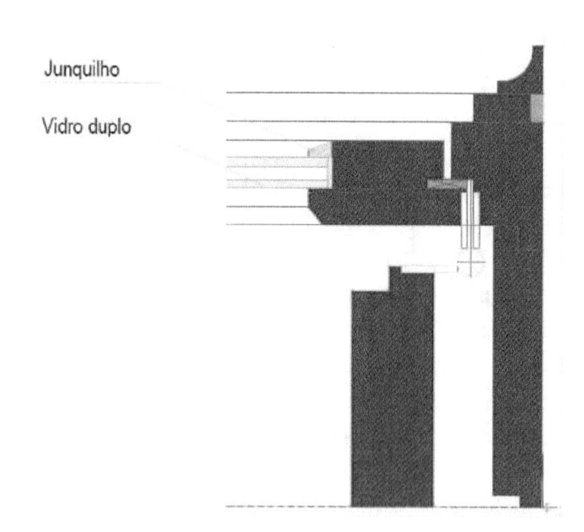

Fig. 4 Implementation of double glasses in wooden carpentry. © Mónica Alcindor Huelva

4.2 Final considerations

It is about seeking solutions with a transdisciplinary approach that repropose traditional integral solutions and try to escape from standardized solutions that at first sight seem very efficient but that leave behind regressive consequences. It is important to not forget that traditional rural settlements arose from the active and dialectical interrelation between ecological and cultural factors and it is a cultural richness that we should try not to lose but try to adapt to the new requirements.

It would be necessary to establish circuits of information on the physical and territorial dimension of ordinary economic activities that the dominant monetary analysis ignores, in order to enable society to redesign, in the light of this new information, the rules of the economic game that condition values and prices. In other words, it is about getting the current system "co-evolves" adapting to other demands and if it was to affect the patterns of construction and consumption that rethink the prevailing management mode, restoring and prioritizing physical and social information circuits linked to such management and shaping instruments and prices based on that information (Naredo,2007).

Acknowledgements

We should acknowledge all the key informants by their contribution, particularly Carlos Parreira da Cunha. Also for the technical support of Ivan Dans.

References

ALCINDOR M, 2016. L'evolució de la masia. Les condicions urbanístiques d'edificació més enllàde la imatge in Plecs d'hist ria local núm, 156:5-7.

CASTELLS M, 1995. La ciudad informacional. Tecnologías de la información, reestructuración económica y proceso urbano-regional (The Informational city. Information technology, economic restructuring and the urban regional process). Madrid: Alianza.

FOUCAULT M, 1999. Estrategias de poder (Strategies of Power). Barcelona: Ediciones Paidós Ibérica.

GIDDENS A, 1990. The consequences of modernity. Stanford: Stanford University Press.

LAUREANO P, 1999. Agua: el ciclo de la vida (Water: The cicle of life). Barcelona: Naciones Unidas, Agbar, CCD, DL.

MARTÍN J L, 2006. La arquitectura vernácula : patrimonio de la humanidad. Badajoz: Diputación de Badajoz.

NAREDO J M, 2006. Raíces económicas del deterioro ecológico y social: Más allá de los dogmas. Madrid: Siglo XXI.

RAMÍREZ E, 2011. Etnicidad, identidad, interculturalidad: teorías, conceptos y procesos de la relacionalidad grupal humana. Madrid: Editorial Universitaria Ramón Areces.

SASSEN S, 2007. Una sociología de la globalización (a sociology of globalization). Análisis político, 1 (61): 3-27

SCHUMACHER E F, 2011. Small is beautiful: a study of economics as if people mattered. New York: Random House.

VEGAS F, MILETO C, 2011. Aprendiendo a restaurar: un manual de restauración de la arquitectura tradicional de la comunidad valenciana. COACV, Col legi d'Arquitectes de la Comunitat Valenciana.

VELASCO H, DÍAZ A, CRUCES F, et al., 2006. La sonrisa de la institución. Confianza y riesgo en sistemas expertos (The smile of the institution. Trust and risk in expert systems). Madrid: Centro de estudios Ramón Areces.

Earthen Buildings in Rural Fujian. Architectural Challenges for Local Development

Semprebon Gerardo[1], *Fabris Luca Maria Francesco*[2], *MA Wenjun*[3]

1 Politecnico di Milano, Italy and Shanghai Jiao Tong University, Shanghai, China
2 Politecnico di Milano, Milano, Italy
3 Shanghai Jiao Tong University, Shanghai, China

Abstract Against the backdrop of the recent array of policies targeted to boost processes of revitalization in rural China, the small settlements are the object of a growing interest by scholars and institutions. Considering the controversial issues related to the countryside restructuring, among which the cultural losses determined by the phenomenon of rural urbanisation, the international debate arena is shifting its attention from development per se to sustainable development. From one side, it has already been acknowledged that being sustainable requires something more beyond being ecologically friendly. On the other side, we found there is a huge space for investigation in regard to the elaboration of practices of rural revitalization. In particular, we argue that context-related strategies of revitalization have to be defined considering what exists as a possible resource. With this perspective, we focus on fifteen earthen vernacular buildings of an ordinary rural settlement in Fujian Province, the last tangible witnesses of local traditional past. At present, most of these constructions are used only during rituals and in rare cases for living purposes, resulting in neglection and dilapidation. However, they still play a pivotal role in the rural fabric, in the way they establish spatial relations with the built form and the open space as well. These ancestral halls still influence local construction activities, by suggesting a non-written system of rules. Such a system can be read in the morphological pattern of the settlement whose backbone is still clearly recognizable. Despite it is not possible to label them as architectural heritage, they are something more than vernacular architecture. Recalling the concept of cultural heritage, they embody a complex system of values rooted in the traditional local society, combining housing, farming and rituals. This paper explores the architecture of these ancestral buildings, their current condition and their contextual relations. We found they represent a cultural asset crossing different ages, for both the rediscovery of past and the reorientation of future developments.

Keywords rural, architecture, ancestral, hall, earthen, Fujian.

1 Introduction

Taking an ordinary rural village of Fujian Province as a case study, this paper explores the architecture of fifteen vernacular buildings realized with rammed earth and wooden carpentry. Our purpose is to demonstrate the role of these structures in relation to the settlement's built form, in order to question their values in the pursuit of local development. They embody the inheritance of a layered tradition which shapes the indigenous cultural heritage. Considering the malleability of the Chinese countryside, expressed by the endless transformations of its physical and cultural landscape, we found in these buildings important forms of resilience and a possible asset in the framework of the current process of revitalization. Before introducing our methodology, we shortly contextualize this study with the political and cultural environment and the main research directions.

1.1 The Transition of the Chinese Countryside in the Rural Revitalization Era

The transition of rural China is under the lights of the international debate arena and its socio-economic implications have been broadly discussed by different disciplines. What emerges is a rich and multifaceted framework of knowledge which provides a wide spectrum of tools and theories aimed to decipher the Chinese countryside transition's phenomenology. The poverty alleviation national political goal (Liu, et al., 2017; Guo, et al., 2019) has been remarked as core-objective by President Xi, who stated that "... we must ensure that by the year 2020, all rural residents living below the current poverty line have been lifted out of poverty" (Garrick and Bennet, 2018). Against the backdrop of this and other national issues, such as, the food security (Brown, 1995) and of the necessity of combining environmental protection and economic development (Tilt, 2010; Long, 2014), as well as the theories of "Ecological Civilization" (*shengtai wenming*) and "Mountains of Gold and Silver" (*lüshui qingshan jiushi jinshan yinshan*) (Bai, et al., 2001) the rural revitalization is one of the top priorities of Beijing's political agenda. As a consequence, an imponent wave of cultural interests is converging on this topic, mainly covering three aspects: academic, politic and cultural. In this context, researches are opening the frontiers of their disciplines to propose cross-disciplinary studies and trans-national experiences' comparisons (Semprebon, et al., 2019).

1.2 Starting from the Local Conditions

Since the socialist era, the modernization of rural China has been advocated, in a first moment, via the implementation of a planned economy of scale, where settlements like Dazhai Village became paradigmatic and inspired generations of projects (Zhao & Woudstra, 2007). Later, since the opening and reform policies relaxed by Deng Xiaoping, Chinese development has been following patterns and principles closer to capitalism, shaping what has been labelled as socialism with Chinese characteristics. The opening of the markets determined a more spontaneous course of development (Semprebon & Fabris, 2020), based on a self-imposed adoption of fragments of Modern culture to apply to the economic boom which was re-shaping Chinese territories, both in the cities (Fabris & Semprebon, 2019) and in the countryside. After the contact with the West, Modernity took the form of a paradoxical necessity. From one side, fast development had to protect Chinese economic and cultural independence. On the other side, its identity was attacked by the winds of modernization. Basically, what was protecting the traditional Chinese legacy from outside was hurting it from the inside (Mensi, 2014). We move from the assumption that local development in rural areas has to start form considering what exists as a possible asset, no matter whether or not it is classified from the authorized heritage discourse, and, therefore, worthy of being detected, examined and critically assessed.

1.3 The Case Study

The fifteen houses belong to an agricultural settlement of 1,500 people in Fujian Province. The village is forty minutes driving from Putian City and there is no other connection out of this. Drenched by a creek and several ditches, the settlement lies in a small plain surrounded by hills and covered by woodland. This village is currently the object of a reactivation program, sustained by academic studies and government funding targeted to turn it into a demonstration project. The fifteen ancestral halls are the oldest structures and, organized on a C-shape, embody spaces for living, working and worshipping (Fig. 1). They represent the tangible inheritance of the indigenous culture.

2 Methodology

Our methodology is based on fieldwork and desk research. There are no linear "causes and consequences" relationships in our approach but a continuous exchange between physical observations, literature review and graphic representations. This endless workflow embodies some crucial passages marked by the drawings both detailed and schematic.

2.1 On-site

The fieldwork has been carried out by the first author since August 2017 through several on-site visits. The most significant

Fig. 1 The fifteen earthen houses © first author

part of this study is based on the surveys performed in September 2018, during the Harvest Festival. During the surveys, we got in touch with local residents living inside the ancestral halls and carpenters conducting reparations or renovations. This ensured us a profound understanding of indigenous aspirations, values and meanings.

2.2 Desk Research

Desk research is articulated in the literature review and in the production of graphic elaborations. We started from the detailed survey of a single ancestral hall in order to understand its layout and the uses of its spaces, producing technical drawings in plan and section. This ensured us to grasp the working principles of the building. In the second phase, we selected the main space of the hall and compared it with the same space of the other halls, producing comparative schemes. This allowed us to determine the rules and the variations of the layout typology. Finally, we extended the spatial principles of the hall to their close context, defining the halls' morphology in relation to the surrounding built and non-built forms.

3 Results

More than a single building, the fifteen ancestral halls are ensembles of structures. In the following three paragraphs, we present them in three steps. First, we focus on one single hall, selecting its main architectural features. Second, we compare the layout of all the fifteen halls, highlighting their spaces' invariants. Finally, we consider their role in the overall rural fabric.

3 Architecture of an Ancestral Hall in a Rural Settlement of Fujian Province

The selected ancestral hall is organized on a C-shape layout. The main body and the two wings delimitate a paved yard open to the landscape. They are spatially arranged on an orthogonal grid with a structural span of some four to five meters, which is determined by the timber frame construction technology. The envelope is realized with rammed earth mixed with straw. The main body has a linear development and hosts three arrays of rooms separated by two corridors, which flow into the small inner courtyard. This is the principal space of the complex and is featured by a cross-shape atrium, composed by five areas: an entrance, a chamber with the family altar, the two lateral parts where arrive the corridors and a central area in correspondence a small impluvium and a skywell (Fig. 1, Fig. 2). The climatic conditions play a pivotal role in the determination of both the volumetric composition and the distribution schemes as well as the degree of permeability of the envelope. Recent studies demonstrate the correlations between the climate and the permeability in vernacular ancestral halls, like most of the buildings in Fujian, which tend to be open ensuring the circulation of fresh air (Wang, et al., 2016). Two columns dot the access, creating a space covered by the galleries or the roof. The roof is supported by a carpentry timber structure and, because of strong winds and typhoons, there

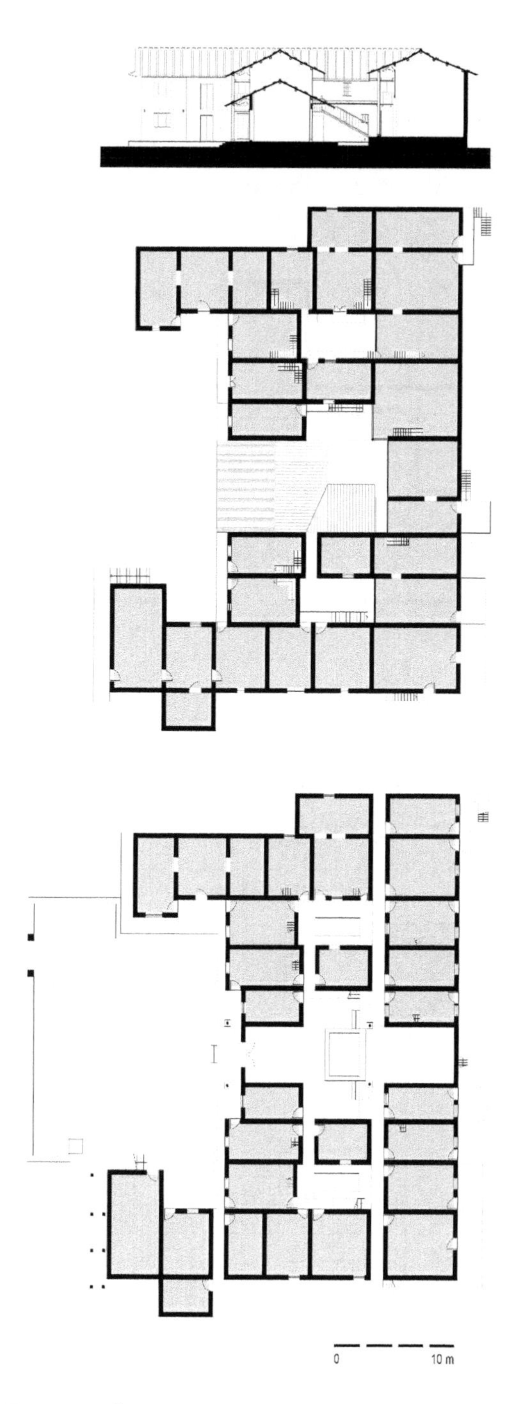

Fig. 2 Plans and section of an ancestral hall (number 10). First author's drawing based on the survey conducted by LIN Yixuan and LIN Zhaohan and FAN Yilun

are stones above the shingles. The skywell, namely 天井, *tianjing*, provides ventilation and shadow (Hammond, 1992). Considering the social meaning of this building, it is worthy to remark that Chinese families are essentially corporate entities and operate in society as the basic unit of production and consumption (Jervis, 2005). As a family residence, the building provides an arrays of spaces dedicated to the family's economy, mainly agricultural.

Fig. 3 The impluvium and the skywell in front of the family altar © first author (2018)

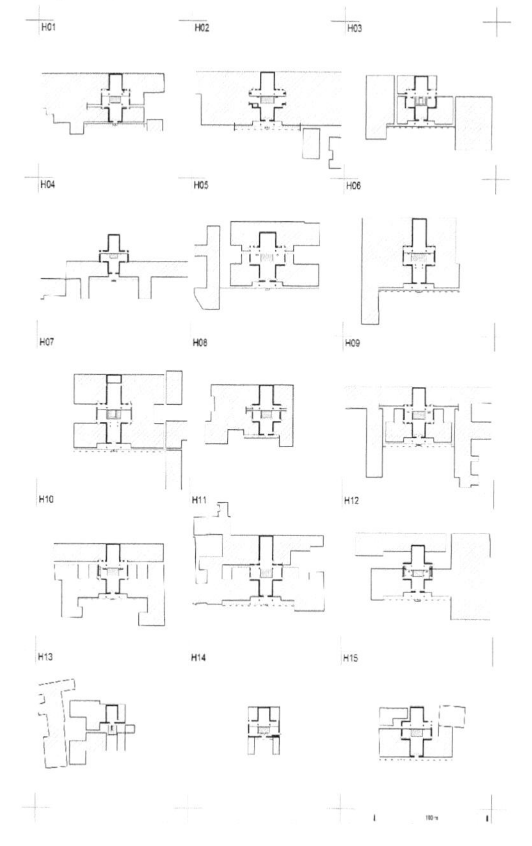

Fig. 4 Comparative schemes of the halls' layout © first author

3.2 Studying the Fifteen Halls, the Invariants of the Space Arise

The front elevations share common features, such as the presence of stoned podi-ums, wooden galleries, and timber or stoned columns sustaining the portico, all mirrored in relation to the main axis. The cross-shaped area with the altar chamber is the distinctive part of the building and repeats in all the other fourteen halls. Minimum discrepancies are due to the topography, to the form of the settlement and, probably, the prestige of the clan. However, what emerges is a clear and repeated pattern which defines the space's arrangement and its circulation (Fig. 4). Fig. 5 and Tab. 1 report the dimensions of the cross-shaped areas, whose average surface is 104m². This measure is very close to the cross-shaped area of the hall described in the previous paragraph, whose surface is 105m². The graphic comparison clarifies how these particular principles and their proportions repeat, rising strong spatial invariants. As we could expect, we found they play an essential role in the settlement's rural fabric, which is going to be highlighted in the next paragraph.

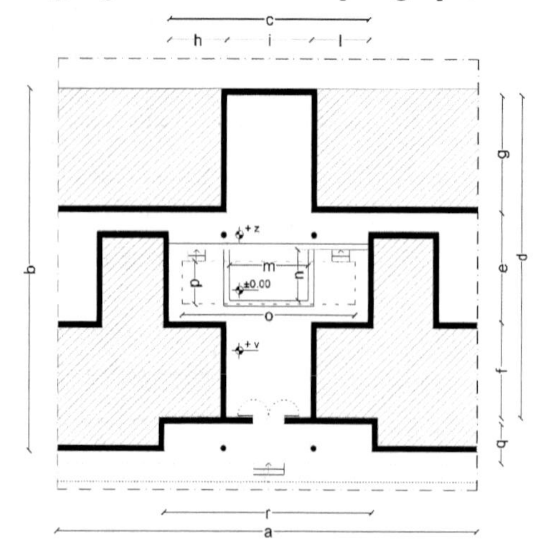

Fig. 5 Schematic layout © first author

Tab. 1 The dimensions of the cross-shaped space © first author

Number		1	2	3	4	5	6	7	8	9	10	11	12	13	14	15	Average dimensions
Name		过溪	田下	田上	田楼路顶	树林里下厝	树林里	垅头	新厝	暗坑里	下园	竹林尾	东埔	顶厝溪	—	—	
		Guo xi	Tian xia	Tian shang	Tian Lou lu ding	Shu lin Li xia cuo	Shu lin li	Long tou	Xin cuo	An keng li	Xia yuan	Zhu lin wei	Dong pu	Ding cuo xi	—	—	
Family belonging	林	黄	黄	黄	范	范	林	陈	范	范	范	林	范	—	—		
	Lin	Huang	Huang	Huang	Fan	Fan	Lin	Chen	Fan	Fan	Fan	Lin	Fan	—	—		
Dimensions	a	21,20	35,61	24,50	17,44	30,70	28,67	31,51	12,33	25,46	24,61	25,14	28,33	18,30	6,38	25,81	24,15
	b	16,42	17,99	16,39	14,19	21,59	23,57	24,75	17,04	18,49	18,12	21,90	21,24	15,62	10,16	16,96	18,42
	c	9,65	10,87	10,78	9,80	11,75	11,07	11,94	9,27	10,39	10,19	10,62	10,12	9,00	10,34	10,30	10,40
	d	15,82	16,05	14,58	12,60	16,17	21,47	20,10	16,44	16,46	16,28	20,12	15,39	15,32	9,56	16,36	16,28
	e	5,77	5,79	5,73	3,40	5,86	6,03	6,20	5,55	5,60	5,78	5,00	5,28	3,25	5,13	5,65	5,43
	f	4,70	4,38	4,25	3,86	5,46	6,04	6,80	4,39	4,86	4,80	5,46	4,16	6,06	—	5,83	5,03
	g	5,35	5,90	4,60	5,34	4,85	9,40	7,10	6,49	6,00	5,69	9,66	5,95	6,00	4,43	4,88	5,97
	h	2,70	3,20	3,03	3,06	3,24	3,13	3,55	2,41	3,00	2,87	3,12	2,92	2,80	3,26	3,00	3,03
	i	4,24	4,47	4,38	3,68	5,27	4,80	4,84	4,44	4,39	4,44	4,37	4,28	3,83	3,82	4,30	4,35
	l	2,70	3,20	3,03	3,06	3,24	3,13	3,55	2,41	3,00	2,87	3,12	2,92	2,37	3,26	3,00	3,00
	m	3,52	3,96	3,57	2,70	3,80	5,53	4,30	3,40	4,10	3,42	3,79	3,30	2,80	3,82	3,57	3,64
	n	2,17	2,30	2,48	1,20	2,40	2,91	2,65	2,40	2,60	2,46	2,60	2,10	3,10	2,13	2,30	2,42
	o	3,25	—	8,74	2,42	4,71	4,48	4,14	7,76	8,99	2,99	3,42	3,10	—	3,14	3,30	4,46
	p	2,02	1,90	3,71	1,26	2,37	3,65	2,20	2,12	2,20	2,18	1,90	2,10	—	1,71	3,25	2,30
	q	0,59	1,94	2,00	2,00	2,21	2,36	1,44	1,03	2,02	1,63	2,60	2,40	—	—	1,50	1,87
	r	—	10,20	9,84	—	10,55	11,07	12,02	—	10,82	10,19	10,50	10,11	—	—	—	10,49
	v	0,30	0,30	0,30	0,30	0,30	0,30	0,35	0,30	0,30	0,35	0,30	0,35	0,30	0,35	0,25	0,31
	z	0,62	0,80	0,76	0,50	0,70	0,70	0,80	0,70	0,80	0,70	0,40	0,80	0,70	0,70	0,70	0,71
	(m^2)	98,29	108,89	100,53	67,18	123,19	140,86	141,30	99,76	105,86	105,47	119,17	96,70	75,44	69,97	104,25	104,32

3.3 From Typology to the Purse of Morphological Settlement's Rules

There is a structural symmetry and axiality in the plan composition. Over time, with the growth of the family, the descendants implemented the system by adding units in continuity with existing structures or realizing new pavilions in the close proximity of the ancestral shrines. This is confirmed by recent anthropological surveys conducted in the Putian plain, which explained that, in most of the villages, ancestral worships were carried out in the old forefathers homes (Dean & Zhen, 2009). The progressive addition of new parts took place without invading the rectangular open space in front of them. This open space, located in front of the main façade, plays an essential role for many reasons. First, it is the physical connection with the street and therefore has to be crossed to enter the main chamber. Second, it is a space for working, optimized for a family-scale economy. For instance, we found it is used for the agricultural products' drying and storage. Finally, it is a space for gathering, discussing, making decisions, punishing, and so on, mainly for the family members but also for the other residents. Therefore, it can be considered as a public space, strongly defined in its architectural elements and intensely used for different purposes. The importance of the front yard arises clearly through the disposition's principles of the buildings realized in the recent period. The new dwellings respected the rectangular shape of the yard, locating at its edges and ensuring the view contact between the inner ritual space and the landscape. In particular, the architecture of these buildings creates a spatial sequence between the altar in the ancestral chamber, the narrow skywell, the front yard, and the natural landscape, inasmuch as they are the extensions of the streets' public space (Fig. 6).

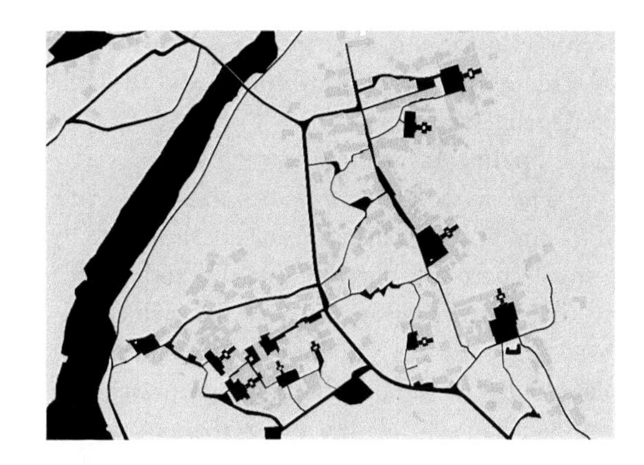

Fig. 6 The ancestral halls' front yards are an extension of the streets' public space © first author

4 Discussion

4.1 Contextualizing the Results

We can consider these buildings as a regional variation of the acknowledged, and deeply studied, the architecture of the ancestral halls (Chen, 2018). Chen emphasized that the halls, generally, benefited from a panoramic view on the landscape, according to the land availability and to the orographic conditions. Beyond being sites imbued with dense genealogic meanings, the halls embody the Chinese idea of "being part of the landscape," according to its etymological definition of 山水, *shanshui*, namely mountains-and-water (Knapp, 1992). The fact that we found fifteen ancestral halls in a settlement populated by four clans, is the symptom of a form of halls' branching, a process aimed at declaring independence from the principal group of the family via the realization of a new ancestral hall to claim new identity. As remarked by Ho (2005) the ancestral halls in remote rural settlements were usually constructed with simple typologies, erected with rammed-earth walls and timber frames structures supporting low-pitched roofs, but not less articulated than the official ancestral halls in terms of spatial composition. During the Ming dynasty, the permission of realizing an ancestral hall had to be issued by the government, which usually allowed only scholar elites. Contestations among the clans' members were mining the social apparatus in the countryside, leading the subsequent Qing dynasty to relax the right to build ancestral halls also to merchants, stimulating a proliferation of these type of architecture. This theory is also confirmed by Zheng (2001) in his studies on the Fujian social structure.

4.2 Spatial Assonances

The skywell area resonates like the atrium of a domus, the Roman house, characterized by a small courtyard open to the sky and shaped by the presence of an *impluvium*, which collects the rainwater and provides a comfortable micro-climate. In both the typologies, the atrium distributes to all the other rooms, with a less pronounced axiality in the *domus*. Another slight difference in relation to the domus is the semi-public use of the ancestral halls. In Fig. 6 we tried to emphasize this condition imagining the settlement's rural fabric (Fig. 6) in a similar way as Nolli (1748) did with his well-known "New topography of Rome." What emerges is a close connection between the public spheres of the street and the semi-public one of the ancestral halls, determining strong morphological rules performing at the scale of the whole settlement.

4.3 Between Vernacular and Heritage. Searching a Value for the Local Development

In consideration of the massive efforts targeted to boost rural development, these dwellings struggle with modern housing to be still considered places to live in. The most urgent problem is their physical conservation, threatened by both the climatic forces and the lack of maintenance due to their underuse. Indeed, Chinese ruralities, since the Reform and Opening policies, are experiencing a shocking wave of rural out-migration that is shaping phenomena of hollowing (Gao, et al., 2017; Liu, et al., 2010; Sun, et al., 2001) and rural urbanisation (Zhu, 2013; Meriggi, 2018). Moreover, they face a multitude of other challenges related to the evolving socio-economic condition of rural citizens. For instance, modernization and development encourage the spreading of a new array of values which have little to do with local wisdom and traditional habits. Privacy is going to play an ever crucial role in domestic patterns of living, and therefore, housing layout, as remarked by Yan (2005). Another issue is the institution of an effective apparatus of local governance able to imagine respectful forms of local development. Problematics arises from the ownership's fragmentation and grassroot consensus. Therefore, earthen buildings have found little space in contemporary practices of local development. Considered more as an obstacle rather than a resource to economic growth, they struggle to find a recognized position in the local cultural network. The rediscovery of cultural roots as well as the escape from urban alienation is fueling a growing entertainment market that is mostly worried about providing touristic attractions rather than to protect the local authenticity, the feeling of belonging, and generally the sense of the place.

5 Conclusion and Openings

Whether and how the earthen buildings of this rural village will be considered in the local development strategy is hard to say. We found in their architecture a wide spectrum of values expressed at different scales, from the single chamber to the whole settlement, which, in our opinion, represents an important aspect of the local identity. They suggest a non-written system of morphological rules which entwine inextricably with the topography and deeply permeate the regional physical and cultural landscape, even though they appear suffocated by the waves of recent urbanisations. They represent important forms of resilience which mitigate the soil consumption inasmuch maintaining strong bonds with the land in the nearby. Against the backdrop of such a framework, we sincerely auspicate their engagement in the implementation of strategic planning for local development.

References

BAI Xuemei, CHEN Jing, SHI Peijun, 2001. Landscape urbanization and economic growth in China: positive feedbacks and sustainability dilemmas. Environmental science & technology, 46: 136.

BROWN L R, 1995. Who will feed China? London: Earthscan.

CHEN Kezhen, LI Yunzhang, CAO Yi, 2018. A study on the space characteristics of ancient ancestral temples in the Taiping district of east China. Journal of asian architecture and building engineering,15 (13): 365-372.

DEAN K,ZHENG Zhenman,2009. Ritual alliances of the putian plain. Volume One. Leiden: Brill.

FABRIS, LUCA M F, SEMPREBON G,2019. The Chinese"high and slender" condominium. Techne - journal of technology for architecture and environment,17: 100-109.

GAO Xuesong, XU Anqi, LIU Lun, et al. , 2017. Understanding rural housing abandonment in China's rapid urbanization. Habitat International, 67: 13.

GARRICK J, BENNETT Y C, 2018. Xi Jinping thought realisation of the Chinese dream of national rejuvenation. China perspectives, 1-2: 99-105.

GUO Yuanzhi, ZHOU Yang, LIU Yansui, 2019. Targeted poverty alleviation and its practices in rural China: a case study of Fuping county, Hebei province. Journal of rural studies.

HAMMOND J, 1992. Xiqi village, Guangdong: compact with ecological planning, in Chinese landscapes: the village as place. Honolulu: University of Hawaii Press.

HO PUAY-PENG, 2005. Ancestral Halls//RONALD G K, Lo Kaiyin. House, home, family. Living and Being Chinese. Beijing: University of Hawai Press, 295-323.

JERVIS N,2005. The meaning of jia. Kai-yin Lo. An introduction//RONALD G K, Lo Kaiyin. House, home, family:Living and Being Chinese. Beijing: University of Hawai'i Press, 223-234.

LIU Yansui, GUO Yuanzhi, ZHOU Yang, 2017. Poverty alleviation in rural China: policy changes, future challenges and policy implications. China agricultural economic review,10(2): 241-259.

LIU Yansui, YU Liu, CHEN Yangfen,et al. ,2010. The process and driving forces of rural hollowing in China under rapid urbanization. Journal of geographical sciences,20(6): 879.

LONG Hualou,2014. Land consolidation: An indispensable way of spatial restructuring in rural China. Journal of geographical sciences, 24 (2): 211-225.

MENSI GIOVANNI, 2014. Assenze monumentali. L'indifferenza al contesto nell'architettura cinese contemporanea. Politecnico di milano, 105.

MERIGGI MAURIZIO, 2018. L'architettura del continuo urbano-rurale in cina. Insediamenti hakka nel Guangdong orientale. Torino: Arab A Fenice.

RONALD G K,1992. Chinese landscapes: the village as place. Honolulu: University of Hawaii Press.

RONALD G K,2005. House, home, family: living and being Chinese. Honolulu: University of Hawai'i Press.

SEMPREBON, GERARDO, MARINELLI M, et al. , 2019. Towards design strategies of rural requalification. A comparative study on the settlements hollowing between China and Italy. Proceedings of the 2019 XJTLU International Conference: Architecture. Suzhou: Xi'an Jiao Tong-Liverpool University:19-21.

SUN Hu, LIU Yansui, Xu Keshuai,2011. Hollow villages and rural restructuring in major rural regions of China: a case study of Yucheng city, Shandong province. Chinese geographical science, 21 (3): 355.

TILT B ,2010. The struggle for sustainability in rural China. New York: Columbia University Press: 3.

WANG Hui, PU Xincheng, WANG Rongrong, et al. ,2016. A study on closed halls in traditional dwellings in the Jiangnan area, China. Journal of asian architecture & building engineering, 15 (2): 145.

XI Jinping,2018. Secure a decisive victory in building a moderately prosperous society in all respects and strive for the great success of socialism with Chinese characteristics for a New Era. Report to the 19th national congress of the communist party in China. Beijing: Foreign Languages Press.

YAN Yunxiang,2005. Making rooms for intimacy. Domestic space and conjugal privacy in rural north China//RONALD G K, Lo Kaiyin. House, home, family: Living and being Chinese. Beijing: University of Hawai'i Press: 373-395.

ZHAO Jijun, WOUDSTRA J,2007. In agriculture, learn from Dazhai: Mao Zedong's revolutionary model village and the battle against nature. Landscape research, 32 (2): 171-205.

ZHENG Zhenman, 2001. Family lineage drganization and social change in Ming and Qing Fujian. Honolulu:University of Hawaii Press.

ZHU Yu,2013. The extent of in situ urbanisation in China's county areas: the case of Fujian province. China perspectives,3: 43.

Challenges for Establishing Conservation Framework of Vernacular Houses in the Rural Areas of Trabzon, Turkey

Elif Berna Var, Hirohide Kobayashi

Graduate School of Global Environmental Studies, Kyoto University, Kyoto, Japan

Abstract Vernacular architecture is an indivisible part of cultural heritage as it is a representation of local culture and knowledge which is conveyed from generation to generation to accommodate specific needs of the people as well as the climatic and geographical conditions. With the changes in the lifestyles, construction systems and technologies, the vernacular architectural traditions have been changed or abandoned, which caused the negative impacts on their authenticity or the loss of them.

In the various international charters, the necessity of conservation of vernacular houses is highlighted. Also, continues usage, local people's appreciation of vernacular houses and their involvement in the conservation activities are pointed out as some of the vital factors for the sustainability of the vernacular houses. As a result, this paper aims to understand the existing situation of vernacular houses, local people's perception and existing conservation activities in the rural areas of Trabzon, which is a city in the northeastern part of Turkey.

Fieldworks are conducted in 74 houses in 3 villages, where architectural measurement surveys, semi-structured questionnaire surveys, and non-participatory observations are conducted. It is figured out that despite the seasonal stay and rural abandonment, vernacular houses have been changed considerably. Although inheritor conflicts, financial limitations, the difficulty of finding original materials and craftsman knowing vernacular construction techniques were mentioned as some of the challenges, 90% of vernacular house residents want to conserve vernacular houses. It is also understood that local people have not been included into the conservation process and 75% of the participants had no idea about the financial supports. Therefore, a more inclusive and participatory framework for conservation is proposed for future conservation activities in the region. Although this is a case study research, it is believed that the essence of this concept can be applied to other case study areas in Turkey or abroad.

Keywords vernacular architecture, rural, conservation, framework, trabzon, Turkey

1 Introduction

1.1 Background

Various researchers have been trying to define the term "vernacular architecture", which is broad and ambiguous (Asquith & Vellinga, 2006). Although there is not a single, well-accepted definition, it often refers to the architectural style shaped by the climatic conditions, local materials, local culture and craftsmanship to meet the needs of the people (Oliver, 2006). It is transferred from generation to generation and often referred as vernacular built heritage which comprises the tangible and intangible heritage. Vernacular built heritage is vital for local identities, common memory and the pride of the communities (ICOMOS, 1999; Oliver, 2006).

On the other hand, globalization has caused various changes in the unique features of the vernacular built heritage, endangering its survivability and sustainability. The vulnerabilities of vernacular built heritage have been emphasized in several global meetings and academic studies (Council of Europe, 1985; ICOMOS, 1996, 1999; Debaieh, 2009; Aikpehae, et al., 2016; Var & Kobayashi, 2019, etc.). Aga Khan emphasized this issue as early as 1978 by saying that "We must ask ourselves how we can prevent future architectural development from accelerating the loss of our cultural identity. ... We must acknowledge that the world is changing, but in doing so, we must realize that there are still many lessons that must be drawn from the past" (Holod, 1980).

As it can be understood from this saying as well as international charters, it is a necessity to conserve vernacular built heritage with its full richness of authenticity and cultural reference and to convey it to the future generations (ICOMOS, 1964, 2011; Council of Europe, 1975, 1985, 1989; etc.). At this point, architectural conservation and proper conservation frameworks have a vital impact. Therefore, this paper focuses on the conservation of vernacular built heritage in the rural areas of Trabzon, Turkey as well as the challenges identified for future conservation activities in the region.

1.2 Objectives and Methodology

The objective of this paper is to understand existing conservation activities in the selected rural settlements in Trabzon, Turkey as well as to identify challenges for establishing a conservation framework in the future. After identifying such points, it is aimed to make a proposal for future conservation strategies in the region.

For this purpose, a case study research was carried out in 3 villages located in the Sürmene District, which is introduced in the following section. As a part of the fieldworks, a preliminary survey, as well as two field surveys, were conducted (Tab. 1).

The preliminary survey was conducted in the 20 villages in 7 districts of Trabzon to understand the current situation of vernacular built heritage. After the detailed examinations on-site and literature reviews, it was decided to focus on Sürmene District, which has richer natural and cultural heritage (Ertas, et. al., 2017). The detailed information about the Sürmene District and selected 3 villages are introduced in the following part.

Tab. 1 Details of the fieldwork

Name	Duration	Activities*
Preliminary survey	Nov. 2015	1, 2, 3, 4
1st field survey	Aug-Sept. 2016	1, 2, 3, 4, 5, 6
2nd field survey	Aug-Sept. 2017	1, 2, 3, 4, 5, 6
Workshop	March 2018	2, 4, 7

* 1: Onsite observation, 2: Informal discussion, 3: Photographic documentation, 4: Secondary data collection, 5: Architectural measurement survey, 6: Semi-structured questionnaire survey, 7: Information sharing and discussion.

2 Case Study in Trabzon, Turkey

2.1 Introduction of Case Study Areas

Sürmene District was selected as the target area, which is one of the eighteen districts of Trabzon City (Fig. 1(a)). Located in the eastern side of the city center of Trabzon, Sürmene has relatively more conserved rural areas in the region (Fig. 1(b)). Also, it has a rich socio-cultural and historic background due to its strategic connection with the Silk Roads. Therefore, three villages, namely Karacakaya, Üstündal and Dirlik, were selected as the case study areas for this research (Fig. 1(c)).

The villages are located along the Manahoz River, with an altitude of 350~400m. The distance to the city center ranges from 40~50km. The seasonal population is remarkable in three sites as most of the houses are used as secondary / holiday homes. The income of the permanent residents depends on agriculture, whereas the seasonal residents are involved in various economic activities during their stay in the bigger cities.

2.2 Rural Environments and Vernacular Built Heritage

The vernacular built heritage in the regions has been shaped by various factors, such as topography, climate, natural resources, and local cultures. For instance, steep topographical conditions, abundant forest water, and resources have resulted in the scattered rural settlements in the region

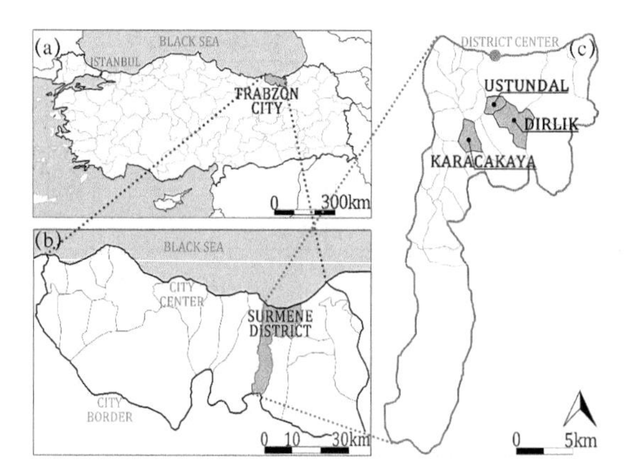

Fig. 1 The location of case study sites ((a) Location of Trabzon City, (b) Location of Sürmene District, (c) Location of) © Var & Hirohide, 2019

(Üzgüner, 2017). Located in the upper parts of the slopes, the vernacular houses are designed as independent units and surrounded by agricultural fields (Fig. 2).

Spatial, structural and morphological features of the vernacular houses are shaped by the high precipitation and mild climate in the region (Batur, 2005). The stone basement and wide eaves are some of the examples of such solutions. Wooden frames, wooden masonry, stone masonry and a combination of them are used as the structural systems. In terms of the spatial formation, the lower floor is used as animal shelter and storage, whereas the upper floor is for living, cooking and sleeping (Fig. 3).

Fig. 2 Rural settlements in Karacakaya, Üstündal and Dirli

2.3 Existing Conservation Activities

The architectural conservation activities have been initiated in the region recently, with the request of a local NPO (*Association for Conserving Natural and Historic Values / Doğal ve Tarihi Değerleri Koruma Derneği, in Turkish*), who is working on the historical, natural and cultural heritage conservation. After this initiative, a committee consisting of the members of the NPO, regional conservation board and the local municipality conducted a technical visit and decided to conserve this area. As a result of this process, Karacakaya has been announced as a "Registered Urban Site (Kentsel Sit Alanı, in Turkish)" and Üstündal as the "Immovable Cultural Properties which is required to be conserved (Korunma Alanı/ Korunması Gerekli Taşınmaz Kültür Varlğı) in 2017. Although the identification process has also been started for the case of Dirlik, it has not been finalized, yet due to the bigger size of the village as well as some disagreements depending on the information gained during our most current field visit in March 2018.

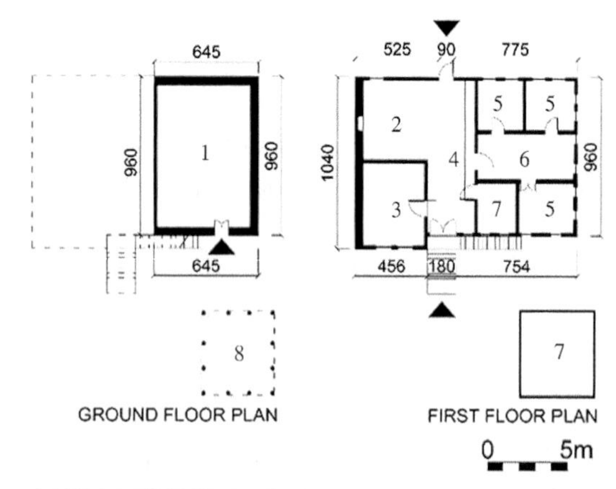

1:ANIMAL SHELTER, 2: AŞHANE, 3: GUEST ROOM, 4:CORRIDOR, 5: ROOM, 6: HAYAT, 7: STORAGE, 8:SHADING / RESTING SPACE

Fig. 3 An example of the spatial formation of a vernacular house in Sürmene © Var, 2019

3 Challenges for Conservation

In this part, the challenges observed for the architectural conservation activities in the region are categorized into 4 main sections depending on the results of the fieldwork. Each sub-section is explained below:

3.1 Challenges for Social Aspects

The changes in the lifestyle, family

structure and community ties have resulted in several challenges for the conservation practices in the region.

First of all, 85% of the respondents are the seasonal residents in the village and their duration of stay also changes a lot depending on their demographic and socio-cultural backgrounds. Depending on the results of the questionnaire survey, 35% of the participants stay in the village for 1～3 months, nearly 25% of them for 3～6 months, almost 25% for 6～9 months and less than 10% visits the village every weekend. The motivation for returning to the village also ranges from harvesting (55%), cool climatic conditions in summer (20%), holiday (18%), village life and friends (10%), and maintenance of the vernacular houses (7%). The seasonal stay in the village causes difficulties for local people to make proper maintenance activities. Also, the family and community ties have been weakened with the changes in the lifestyles and family structures. This has been reflected to the collaborative works done in the village, including the maintenance and construction works of the vernacular houses. Also, the respondents mentioned that one of the biggest challenges for them to conserve the vernacular houses is the conflicts between many inheritors. They stated that the motivation level each inheritor is different towards the conservation of the houses and they do not want to spend the money by themselves if the other inheritors do not contribute to the expenses for the conservation. This is one of the most problematic issues resulting in the abandonment of the vernacular built heritage.

3.2 Challenges for Technical Aspects

Residents also expressed the unavailability of the craftsman who is capable of constructing vernacular houses and it has caused many difficulties for proper maintenance activities. As a result, they used new materials have been used in the vernacular houses. However, the number of residents preferring wooden materials were found to be higher (Fig. 4).

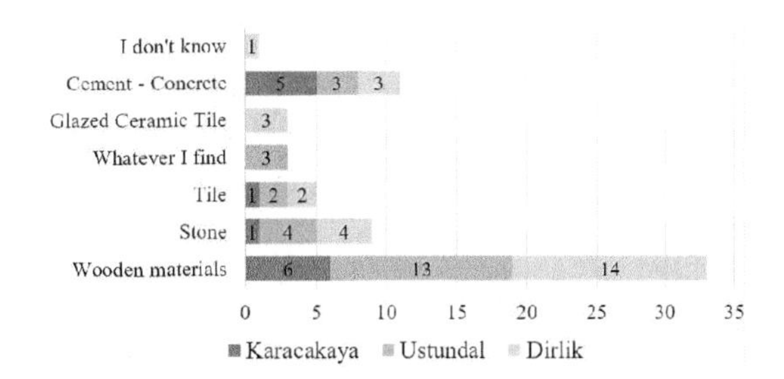

Fig. 4　Materials used for the maintenance of vernacular houses © Var, 2019

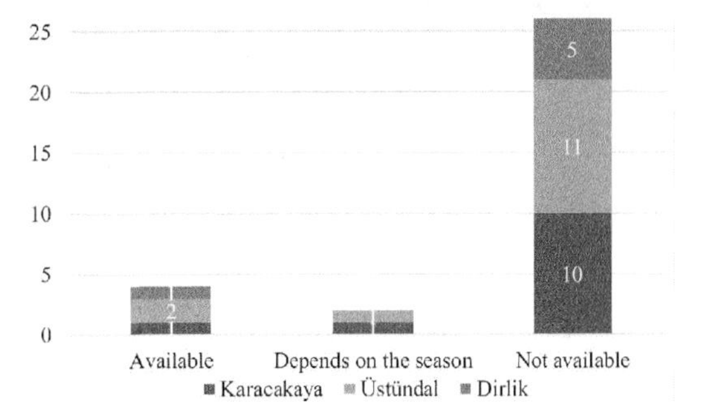

Fig. 5　Respondents' perception of the availability of local materials in the village © Var, 2019

The new material selection indicates that the residents' appreciation for wooden materials and attachment to vernacular houses. However, most of the cases, it is not easy to prefer wooden materials (Fig. 5).

It is also understood that the numerous changes have made to the vernacular houses by modern construction techniques. This can also be elaborated by the difficulty of finding craftsman and original materials, easier and faster construction processes with modern construction techniques as well as the cheaper price of modern materials. There is no doubt that it is vital to conserve vernacular houses physically, as tangible heritage. Yet, it is also crucial to sustain unique intellectual knowledge for the vernacular house construction. Although it is a very challenging task to find people who have this precious knowledge, efforts should be made to revitalize and convey this tradition to the it to next generations.

3.3 Challenges for Economic Aspects

The outcomes of the semi-structured questionnaire survey indicated that more than 75% of respondents were unaware of the existing financial support schemes which are provided by the government for the conservation of vernacular houses (Fig. 6). It proves that there is information sharing between local people and government is weak.

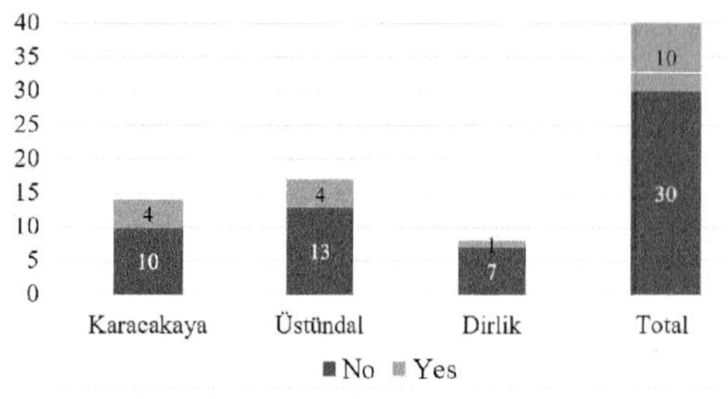

Fig. 6 Awareness of respondents in the financial supports for the conservation of vernacular houses © Var, 2019

Another issue figured out at this point is that the respondents who were aware of such support systems complained about the complexity of the application process and their concern about the ownership issues after utilizing the financial support.

3.4 Challenges for Administrative Aspects

Another challenge identified in the current system is in terms of the implementation. According to the information gained from the regional conservation board, located in the city center of Trabzon, the responsibility area of the board is not limited to the Trabzon City, but also the neighboring cities of Giresun, Gümü? hane and Rize. This causes an excessive workload for the board, which has a limited number of staff. Also, it makes a must for the staff to travel to the rural areas in 4 cities for examination, identification and registration of the vernacular built heritage. This process requires considerable time and money. Considering the harsh topographical and climatic conditions, the limited number of staff and the distance of the regional conservation board from the rural areas, it can be easily understood that previously mentioned factors cause difficulties for the implementation process. Therefore, it is believed that it is necessary to improve the existing scheme to lessen the burden of the regional conservation board and to increase efficiency.

3.5 Summary

Social, technical, economic and administrative challenges were identified for the conservation of vernacular built heritage in the case study areas of Karacakaya, Üstündal and Dirlik. The respondents were asked to rank the difficulties faced for the architectural conservation. Their answers are summarized in Tab. 2. Depending on the local people's perspectives the biggest challenge is related to the economic issues followed by the technical ones.

Tab. 2 Respondents' answers on challenges faced for architectural conservation

Ranking	Factors	*
1	Economic limitations	(E)
2	Harsh weather conditions	(N)
3	Unavailability of the craftsman	(T)
4	Problems on the roof tiles	(T)
5	Insufficient access to the houses	(T)
6	Inheritor conflicts	(S)
7	Unavailability of original materials	(T)
8	Time limitation	(S)
9	New forestry regulations	(A)

* E: Economic, N: Natural, T: Technical, S: Social, A: Administrative.

4 Discussion & Suggestion

It is understood from the field surveys that there are several challenges for the architectural conservation in the region.

One of the most problematic issues was found to be related to land ownership and inheritor conflicts. This has a very strong impact on the frequency of usage or even the division or the abandonment of the vernacular house. Since it directly impacts the conservation of the vernacular houses, importance should be given to this issue although it is not that easy to solve. The awareness increasing projects, financial support schemes, and collaborative processes should be utilized to make consensus for the conservation of vernacular built heritage.

Another issue found was the fact that vernacular house owners have no idea that their houses have already been registered by the regional conservation board. Similarly, their information on the existing financial supports were also very limited. These factors show that there is a communication gap between the stakeholders and the local people have not been involved into the conservation process.

However, the importance of community participation has been strongly emphasized in several international meetings, including Xi'an Declaration (2005). It is stated in that declaration that "Co-operation and engagement with associated and local communities are essential as part of developing sustainable strategies for the conservation and management of settings". In the case of Turkey, Özdemir (2005) and Eminağaoğlu & Çelik (2005) also highlighted that the top-down decision-making process for the conservation practices should be improvised if better results are desired in the future. Therefore, it is suggested that actions should be taken to increase the level of communication between the local people and local government. Local people should be informed about the conservation activities before, during and after the actions. In fact, they should get involved and given responsibility for the conservation process. For this purpose, seminars, campaigns or workshops can be organized, where different stakeholders, including local people, can come together and exchange information.

Also, considering the challenges for the more efficient implementation processes, it is suggested to have local conservation offices providing service for several villages. Such offices also provide guidance and assistance to the local people who want to make maintenance of their vernacular houses. At the same time, they should work as a place where local people can express their problems, expectations, opinions, etc. about the new/existing implications in the village in terms of the conservation and maintenance works. According to the local people's feedback, local conservation offices should work together with the local government, NGO/NPOs, etc. to provide assistance to local people in terms of the technical or financial issues. Last but not the least, lessening the burden of regional conservation boards, the proposed system of local conservation offices should always be in contact with the regional conservation boards for the implementation of conservation activities, providing feedback about the problems related to regulations and local people's desires.

Also, it is suggested to create conservation guidelines for the future implementations. Because it is observed during the fieldwork that there is no common language on the material selection, way of maintenance or the typologies used for the vernacular houses as well as the new constructions in the village. It causes an incompatible result with the authenticity of the vernacular built heritage, local identity and the rural landscape. Therefore, conservation guidelines should be prepared to help local people as well as the craftsman who are involved into the conservation process. Activities such as the seminar, workshops or advertisements should be prepared to increase the awareness level of local people to use such guidelines for the conservation of the vernacular built heritage, together with the local culture and landscape in a holistic approach.

5 Conclusion

This research focuses on the vernacular built heritage and architectural conservation activities in the rural areas of Sürmene District in Trabzon City, Turkey. By analyzing the current situation, it is aimed to figur out challenging points which need to be improved in the future.

As a result of the field surveys, it is understood that there are several challenges faced for the conservation of vernacular built heritage, such as the financial limitations, inheritor conflicts, the difficulty of finding original materials and craftsman knowing vernacular construction techniques. It is also figured out that owners of the vernacular houses did not know that their buildings have been registers. Also,

75% of the respondents had no idea about the government's support systems for the conservation of vernacular built heritage. This shows that local people were not included into the conservation process as well as the communication between local people and government officers need to be improved. As a result, it is suggested to have a detailed, easily understandable and applicable conservation guidelines are necessary as well as a more inclusive and participatory framework for conservation is should be prepared for the future implementations.

Acknowledgements

This research was conducted with the support of the Inter-Graduate School Program for Sustainable Development and Survivable Societies, Kyoto University (201616).

References

AIKPEHAE A M, ISIWELE A J, ADAMOLEKUN M O, 2016. Globalisation, urbanization and modernization influence on housing and building architecture in Nigeria. International journal of service science, management and engineering, 3 (2): 6-13.

ASQUITH L, VELLINGA M, 2006. Vernacular architecture in the 21st century: theory, education and practice. New York: Taylor & Francis Group.

BATUR A, 2005. Doğu Karadenizde Kırsal Mimari (Rural architecture in the Eastern Black Sea). Istanbul: Milli Reasürans T. A. Ş.

Council of Europe, 1975. European charter of the architectural heritage.

Council of Europe, 1989. Recommendation on the protection and enhancement of the rural architectural heritage.

Council of Europe, 1985. Convention for the protection of the architectural heritage of Europe, European treaty series No. 121, Granada, Spain.

DEBAIEH M, 2009. Conservation of desert vernacular architecture as an inspiring quality for contemporary desert architecture: Theoretical and practical study.

EMINAĞAOĞLU Z, ÇELIK S, 2005. Kırsal yerleşmelere ilişkin tasarım ve planlama politikaların bölgesel ölçek içinde değerlendirilmesi (Evaluation of design and planning strategies for rural settlements in the regional level). Planlama, 2:72-81.

ERTAŞŞ, et al., 2017. Trabzon'da kırsal alanların kullanılması amacıyla turizme yönelik, çekirdek alanların oluşturulması: sürmene örneği (Creation of core areas for tourism on the purpose of the development of rural areas in trabzon: example of sürmene), DOKAP bölgesi uluslararas turizm sempozyumu (DOKAP region international symposium of tourism). Trabzon: Karadeniz Technical University.

HOLOD R, 1980. Conservation as cultural survival. Philadelphia: Aga Khan Award for Architecture.

ICOMOS, 1999. Charter on the built vernacular heritage. ICOMOS 12th General Assembly, Mexico.

ICOMOS, 1964. The international charter for the conservation and restoration of monuments and sites (The venice charter). Venice, Italy, 1.

ICOMOS, 1996. Principles for the recording of monuments, groups of buildings and sites, 11th ICOMOS general assembly, Sofia, Bulgaria.

ICOMOS, 2011. The Paris declaration on heritage as a driver for development, Paris, France.

OLIVER P, 2006. Built to meet needs: cultural issues in vernacular architecture. Italy: Elsevier Ltd.

ÖZDEMIR M Z D, 2005. Türkiye'de kültürel mirasn korunmasına kısa bir bak (A review of cultural heritage conservation in Turkey). Planlama, 1: 20-25.

ÖZGÜNER O, 2017. Köyde mimari-doğu karadeniz (Architecture in the village- Eastern Black Sea). Istanbul: Dergah Yayınları.

VAR E B, KOBAYASHI H, 2019. Possibility of conserving vernacular houses in the rural areas of Trabzon, Turkey. Journal of architectural conservation. DOI: 10.1080/13556207.2019.1596011.

VAR E B, 2019. Conservation of built vernacular heritage for promoting sustainable rural environments in Trabzon, Turkey. Kyoto: Kyoto University.

Protection and Utilization of Cultural Heritage in the Conflict Area —Taking the Protection of Mrauk-U, Myanmar as an Example

ZHANG Chengyuan, *LIU Yan*, *TIAN Zhuang*

School of Architecture, Southeast University, Nanjing, China

Abstract Currently, development and heritage protection of historical cities are becoming increasingly serious problems under constant conflicts in Asia. Thereby, the protection of historical cities in conflict areas will be the focus of future Asian urban heritage protection and Mrauk-U in Myanmar is a typical representative of these cities.

Mrauk-U was built up combining the millennium wisdom of the Rakhine people, which reflect directly the historical changes of the Rakhine dynasties, the social living conditions, and the economic relations with the surrounding areas, etc. , which has a very high historical value. In the past centuries, Mrauk-U has been an important trading center and multi-cultural anchor point in between SE Asian and South Asia, with different peoples living in the city peacefully for centuries. However, in the complex context of military, cultural and natural conflicts, extensive issues arose, such as limited development of the historical city, blocked protection of cultural heritage and trapped traditional life of residents.

This paper proposes possible protection and utilization measures at four different levels, including nation or region, city, community, and site, so as to make a strong response to three major conflicts that Mrauk-U is confronted with, and those possible solutions are insured to be sustainable with construction of comprehensive management system of cultural heritage, regular assessments and inspections and promotion of religious peaceful concepts. It is supposed to provide an idea and a method for protection and management of heritage in similar conflict areas.

Keywords multi-cultural heritage, conflict area, Mrauk-U, protection and utilization, possible solutions

1 Introduction

Relying on ancient rivers, economic trades and cultural exchanges, historical cities in Asia have formed diversified cultural patterns and multi-interactive urban systems. With the development of Asian cities up to today, these historical towns occupy an important position in the urban pattern of the Asian region and are the important embodiment of ancient civilization. At the same time, however, there are constant conflicts occurring in these areas, including religious, cultural and ethnical conflicts, a series of urban problems caused by the rapid expansion of cities, the impact of natural disasters, etc. , as a result, urban development and heritage protection are becoming increasingly serious problems. Thus, the protection of historical cities in conflict areas will be the focus of future Asian urban heritage protection, and Mrauk-U in Myanmar is an important representative of these cities.

Mrauk-U is the ancient capital of the Arakan Dynasty in Rakhine State, Myanmar, and an important city in the trade network of the ancient Bay of Bengal region. Today, Mrauk-U is the most complex site that combines political orientation, economic development, community heritage, religious conflict, cultural identity and other factors in Myanmar, which is the most representative in Myanmar's historic towns and has rich cultural heritage value. On the one hand, Mrauk-U utilizes local natural landscape to establish a set of excellent defense and hydrology system, combining military defense and flood control. The urban plan concepts contained in its spatial plan and functional layout is an outstanding example of urban plan in Southeast Asia, and a witness of the development, prosperity and decline of the Arakan civilization. On the other hand, the pagoda architecture in Mrauk-U integrates the decoration and sculpture of various cultures, and forms its own unique architectural style in the long process of development, which becomes an important example of pagoda architecture, and is closely related to the era of trade and communication in human history. These two features are in line with ii and iv of the world heritage outstanding universal values, therefore Mrauk-U is nominated for inclusion in the 2020 world cultural heritage waiting list.

Meanwhile, cultural heritage in Mrauk-U is in crisis because of conflicts in many aspects. Firstly, the war conflicts caused by ethnic and religious collisions are the most prominent ones. Mrauk-U is an important battlefield for the three armed forces formed by the Myanmar Government Military, the Arakan Army (independent force) and the Arakan Rohingya Salvation Army (ARSA), as a result of which the heritage is under serious threat of destruction or even disappearance. Simultaneously, frequent natural disasters, which are dominated by floods, also pose a direct threat to the Mrauk-U's heritage. Last but not least, like other Asian cities, the rapid urbanization that Mrauk-U may face in the future will bring serious conflicts between development and heritage protection.

It can be seen that under the background of nomination for the world heritage status, the conflicts between the protection, development and management of existing heritage in Mrauk-U is not only of its own characteristics, but the common problem of cultural heritage protection and utilization in conflict areas as well. This paper proposes possible management and utilization measures at four different levels, including nation or region, city, community and site, so as to make a strong response to conflicts that Mrauk-U is confronted with.

2 Conflicts That Mrauk-U Is Confronted with

Although its value has been recognized and widely concerned by the international community, the protection of Mrauk-U's heritage is not smooth, long threatened by multiple conflicts. Among them, the war conflicts caused by ethnic and religious collisions are the most prominent ones. Mrauk-U is an important battlefield for the three armed forces formed by the Myanmar Government Military, the Arakan Army (independent force) and the Arakan Rohingya Salvation Army (ARSA), as a result of which the heritage is under serious threat of destruction or even disappearance. Simultaneously, frequent natural disasters, which are dominated by floods, also pose a direct threat to the Mrauk-U's heritage. Last but not least, like other Asian cities, the rapid urbanization that Mrauk-U may face in the future will bring serious conflict between development and heritage protection.

2.1 Ethnic Conflicts

Conflicts between the government of Myanmar and the Rakhine people have continued to emerge in recent years due to ethnic independence. As the ancient capital of the Arakan Kingdom, Mrauk-U is one of the centers of the Rakhine People's independence movement. We cannot underestimate the potential military conflict between the Rakhine and the Myanmar government. A number of pagodas in Mrauk-U were damaged in the recent conflicts and the continuous shooting may cause vibrations to damage the centuries-old pagodas (Fig. 1).

2.2 Religious Conflicts

The opposition and conflict between the Rakhine and the Rohingya come from the differences in culture and religious beliefs. Since 2012, many towns including Mrauk-U have been hit by war due to the religious conflicts between the Rakhine and the Rohingya people. As a result, residential and religious buildings (both Buddhist and Islamic) were seriously devastated. Re-

Fig. 1 Images of pagodas damaged by the conflicts in March 2019

ligious buildings, which are spiritual symbols of religion, can easily become targets of cultural conflicts. Therefore, the Mrauk-U region, with its multicultural heritage of hundreds of Buddhist architectural heritage and ancient mosque sites, faces a potential conflict crisis.

2.3 Natural Disaster

Mrauk-U's cultural heritage is vulnerable to natural disasters, especially floods coming from two main rivers in the city in the rainy season. Mrauk-U was the capital of the ancient Arakan Kingdom for centuries that has been designed as an excellent city with a distinguished water management system to prevent floods as well as for defence. However, without effective protection, the system is facing serious damage to some facilities and siltation of rivers, and the city is unable to play an effective flood control role. In 2015, the flood disaster had the biggest impact on Mrauk-U, and the water level in some local communities rose to 7 feet. Historic buildings and other cultural heritage are at great risk.

2.4 Rapid Urbanization

The industrialization of Asia started late and its urbanization process is relatively backward compared with Europe and America, nevertheless, it has one of the fastest urbanization rates in the world, showing a trend of high-speed development (Fig. 2). Countries such as Myanmar, India, Bangladesh, Cambodia, and Sri Lanka will be the major urbanizing countries in Asia in the coming decades (Qi Changqing, 2004). According to the development experience of other Asian cities, this rapid urbanization process will lead to a series of urban problems, including the strong impact on the protection of historic districts and urban heritage causing by urban expansion and renewal.

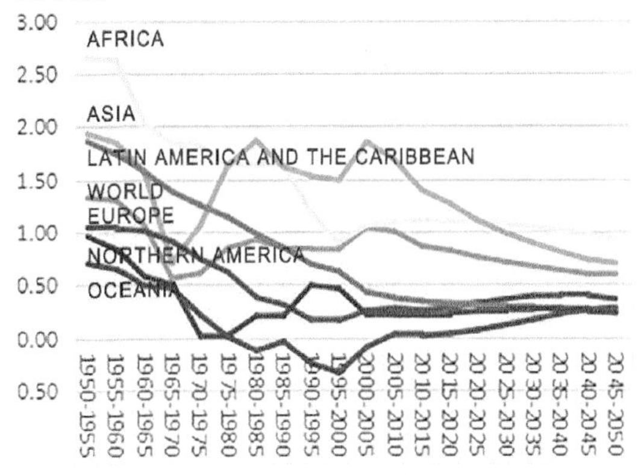

Fig. 2 Average Annual Rate of Change of the Percentage Urban 1950—2050 (percent)

Source: According to World Urbanization Prospects 2018, Website: https://population. un. org/wup/Download/

The development of Rakhine State, where Mrauk-U is located, is relatively lagging behind and the economic level is low. The government's pursuit of rapid urban development often conflicts with the protection of heritage. Correspondingly, national awareness of cultural heritage protection is weak, and it is difficult to form a social environment conducive to the protection of historical and cultural heritage.

2.5 Summary

The ongoing conflicts in Mrauk-U and the potential conflict ahead, an be called "soft conflict" because it is not fierce but lasts for a long time. At the level of war, the conflicts between different ethnic groups and religions are a kind of multicultural collision, which should not persecute historical and cultural heritage. Destroying a masterpiece of art created for any nation or religion is a spiritual injury to all mankind. In the context of multi-culture, the protection of its cultural heritage is a nation's effort to restore collective memory and common identity, and through this effort to create social cohesion and harmony. (Christian Barillet & Thierry Joffroy, 2006) At the natural level, the potential risks to Mrauk-U are long-term, and the historical heritage, which was once able to withstand floods, is not fully protected and utilized and thus loses its defense capability. Protecting them can in turn ease the conflict between historical heritage and natural disasters. Lastly, The potential conflicts at the level of urban development are the problems that heritage protection must face. In some ways, Mrauk-U owes its rich historical heritage to its underdevelopment. But as urbanization improves, the balance between conservation and community living needs will become increasingly challenging. Comprehensive protection and development plans should be drawn up and strictly implemented in response to this conflict.

Therefore, more international attention and research are needed to deal with the problem of heritage protection in such areas. Our study is aimed to find a new model for solving the problem of heritage protection in similar conflict areas.

3 Protection and management plan of Mrauk-u

3.1 Overarching Goals

In order to protect the endangered heritage under sustaining soft conflicts and prevent further great damage to heritage in Mrauk-U, it is necessary to establish a set of integrated and comprehensive protection and management mechanism in line with local circumstances.

The protection and management mechanism can be understood as an operational instrument to utilize available resources to protect defined Outstanding Universal Value, while responding to circumstances in the given context. The goals that are to be observed in achieving the protection and management goals are:

(1) To protect the authenticity and integrity of the multi-cultural heritage in Mrauk-U.

To be considered for listing as a World Heritage site, properties must meet the conditions of integrity and authenticity. Contexts and heritages in Mrauk-U which are of outstanding universal value, including hydrology system, defense system, landscape system, and featured architectures, etc. must be safeguarded and conserved to be passed on to future generations.

(2) To reduce damage to cultural heritage caused by conflict.

Measures at different levels should be taken to effectively protect the multicultural heritage in Mrauk-U in a conflict environment, especially including the Buddhist culture represented by Buddhist pagodas and monasteries and the Islamic culture represented by the ancient mosques, so that the historical and cultural characteristics of Mrauk-U can exist in the future.

(3) To promote understanding and interaction among different ethnic groups in Mrauk-U.

A strong vision is presented that through the protection and management operation of multicultural heritage such as Buddhism and Islam in Mrauk-U, the values of different cultures can coexist in the city, and consequently the mutual recognition between the Rakhine and Rohingya can be enhanced.

(4) To enhance the resilience of the city to natural hazards

As for Mrauk-U, the flood is one of the most serious natural disasters. The annual flood disaster poses a great threat to the cultural heritage of Mrauk-U. Therefore, it is necessary to improve the flood management and prevention capabilities so as to enhance the resilience of the city to minimize the impact of hazards on residents' lives.

(5) To strengthen residents' sense of identity with regional culture, promote social cohesion and reduce conflict crisis.

It should be realized that the implementation of the protection and management system cannot be separated from the participation of local residents, and hence it is particularly significant to enhance the sense of identification with local culture by disseminating the value of the cultural heritage of Mrauk-U among peoples.

3.2 Possible Solutions

According to the three major conflicts, literally, religious and culture conflicts, conflicts of natural hazards, and conflicts between city development and heritage protection, corresponding possible solutions are proposed, so as to make a strong response to the protection and management of cultural heritage in Mrauk-U.

3.2.1 Possible Solutions to Religious and Culture Conflicts

The differences in culture and religious beliefs directly lead to the opposition and conflicts between the Rakhine and the Rohingya, in addition, conflicts between the Myanmar government and the Rakhine people have continued to emerge in recent years due to ethnic independence, which poses a threat to the heritage in Mrauk-U. To this end, possible solutions at three different levels are proposed.

Firstly, at the national or regional level, it is suggested to ease the conflict through negotiation. Cooperation with UNESCO experts, the local government and NGOs, it is possible to invite experts from the UN and other international institutes as an originator to organize a round table among Myanmar Government Military, the Arakan Army and the Arakan Rohingya Salvation Army. The round table is going to urge three sides to make a commitment about cultural heritage protection according to the Geneva Convention, preventing the continuous conflict from causing more damages to the heritage.

Then, at the city level, it is supposed that maintaining an urban-rural integrated pattern contribute to protect multi-culture. Various protection measures, such as ancient buildings restoration and village renovation, are proposed respectively for the Buddhist historical buildings in the urban area and the Muslim settlements in the suburbs. The efforts of local governments, multinational experts, and NGOs of different ethnic groups will be cohered to advance measures.

Lastly, at the community level, it is a possible solution to carry out a short-term course and training on cultural heritage protection organized by relevant experts and NGOs for residents of all ethnic groups, not only to strengthen their awareness of cultural heritage protection, but to promote multi-ethnic cultural communication and integration as well. Multi-cultural exchange centers located close to some symbolic historical sites such as the palace site are advised to build up.

3.2.2 Possible Solutions to Natural Hazards

Religious and culture conflicts have curbed the construction and development of the city, making it difficult to improve their capacity for resisting natural disasters. As for Mrauk-U, a flood is one of the most serious natural disasters. However, in the current conflict situation, it is particularly difficult for the Mrauk-U and the Rakhine State governments to concentrate on the construction of urban infrastructure to improve the flood management and prevention capabilities. Consequently, the annual flood disaster poses a great threat to the cultural heritage of Mrauk-U. Thus the corresponding possible solutions at city and community level are proposed.

On the one hand, as is illustrated in Fig. 3, it is practical to dredge the currently blocked river and remove the buildings properly located on the historical waterways. After that, the historical waterway partially with flood control functions is supposed to be restored. What's more, enough open space to store water will be guaranteed to reduce the impact of floods on people's lives, and form a characteristic cultural landscape of floods at the same time.

On the other hand, it is necessary to establish a disaster risk management plan which includes preparedness, response and recovery closely linked to the site management, local authorities as well as related national-level management structures and to develop means for monitoring the heritage site and monuments to minimize the change

Fig. 3 Approach for hydrology management system

for human-induced disasters to take place as long-term mitigation measure.

3. 2. 3 Possible solutions to rapid urbanization

The rapid urbanization of Asian cities brings the rapid expansion of urban population and urban scale, thus, problems of urban development and heritage protection are increasingly serious. Fortunately, city development in Mrauk-U is restricted by military rule, its development speed is not rapid, but city construction is also expanding and heritage sites are gradually occupied and damaged. In response to the inevitable urbanization process, possible solutions at three different levels are proposed.

Firstly, at the city level, it is acknowledged that the water system, defense system, and landscape system are the greatest assets of Mrauk-U and must be safeguarded and conserved. The factors in all of those systems need to be well preserved to show the whole of the great ancient city Mrauk-U and the wisdom and civilization of Arakanese. According to the proposed protection area (Fig. 4), in terms of the water system, series of rivers, reservoirs, moats, and ponds need to be protected in good conditions and work smoothly. When it comes to the defensive system, the city gate, walls, fortresses, and defensive pagodas need to be kept in good condition. As for the landscape system, the natural mountains and rivers should be protected and inhibit the destruction from development construction. In addition, any development projects that are not temporary or easily removable shall first have a Heritage Impact Assessment (HIA) carried out along with detailed sub-surface archaeological surveys linked to Archaeological Risk Maps (Fig. 5). The project shall be developed taking into account the outcome of the assessment and survey.

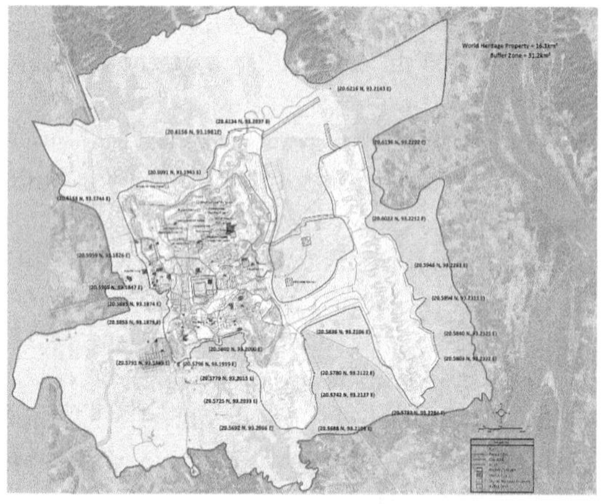

Fig. 4 The proposed area for Mrauk-U World Heritage Nomination

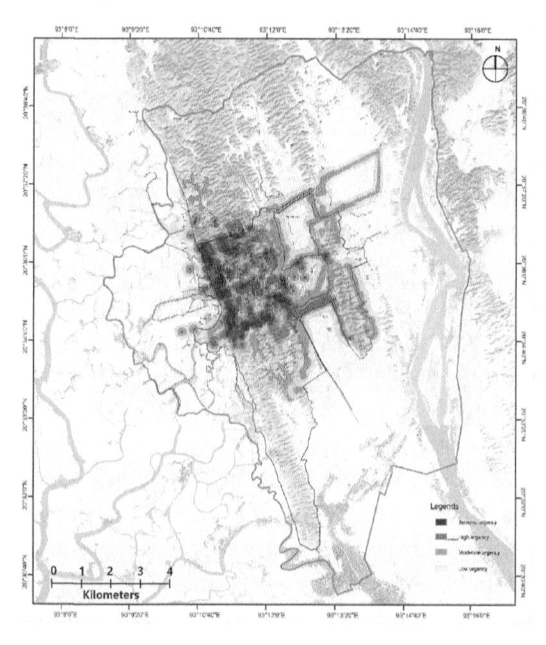

Fig. 5 Archaeological Urgency Map (AUM) of Mrauk-U

Secondly, at the community level, it is particularly important to establish a set of local management mechanism. Specifically, some existing tangible heritage such as city walls, water gates, pagodas, and Islamic temples will be restored as examples by a joint team of experts to provide experience and ensure that local teams could continue the work by themselves in the future. Through professional training on cultural heritage protection, the involvement of local communities and NGOs should be encouraged to participate in heritage conservation.

Finally, at the site level, new constructions within settlements, both towns, and villages, located in the heritage site as well as in the buffer zone need to be guided to ensure they are appropriate.

3.3 Sustainability

The protection and management plan for Mrauk-U was developed through a series of interaction programs with government authorities, religious bodies and community members. It provides an overview of the most important aspects of Mrauk-U which needs to be kept in mind when managing and operating the heritage property and surrounding areas. And the sustainability of the proposed solutions is insured in three ways.

1) Construction of the comprehensive management system of the cultural heritage

The proposed solutions aim to establish a four-level integrated management organization consisting of Myanmar National Culture Central Committee, Rakhine State Management Committee, and Mrauk-U Committee. Meanwhile, solutions also promote the involvement of relevant experts, local community organizations and NGO of different ethnic groups engaged in cultural heritage management to develop capacity building and training programs to continually improve the management capacity as well as required expertise to safeguard and maintain the important attributes of the heritage property (Fig. 6).

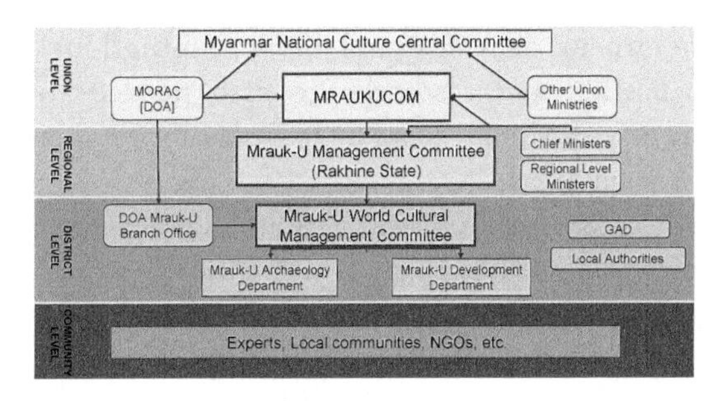

Fig. 6 A four-level integrated management organization

2) Regular assessments and inspections

NGOs and local community residents will conduct regular assessments of heritage protection and community remediation from the bottom up, and conduct self-evaluation and feedback from aspects such as living environment, infrastructure improvement, etc. Meanwhile, government management agencies and relevant experts will regularly review and evaluate the protection of the heritage.

3) Promotion of the concept of peace in Buddhism

Promote the integration of different ethnic groups and religions by promoting the peaceful and equal concepts advocated by Buddhism.

3.4 Summary

The conservation of Mrauk-U cultural heritage depends on a stable and sustainable development environment. This would be linked to the protection and interaction of multi-culture, enhanced resilience of the city to natural hazards as well as the involvement of the local communities and NGOs.

The proposed solutions to soft conflicts aim to establish a set of integrated and comprehensive protection and management mechanism at different levels, and those solutions are insured to be sustainable with the construction of comprehensive management system of cultural heritage, regular assessments and inspections, and promotion of religious peaceful concepts.

4 Conclusion

The conflicts in Mrauk-U can be called "soft conflict" because it is not fierce but lasts for a long time. Therefore, comprehensive protection and management plan is in urgent demand, and more international attention and research are also needed to deal with the problem of heritage protection in such areas.

Sustainable conservation and development of cultural heritage in Mrauk-U would be linked to the protection and interaction of multi-culture, enhanced resilience of the city to natural hazards as well as the involvement of the local communities and NGOs.

Possible solutions to soft conflicts are proposed to establish a set of integrated and comprehensive protection and management mechanism at different levels, and those solutions are insured to be sustainable with the construction of comprehensive management system of cultural heritage, regular assessments and inspections, and promotion of religious peaceful concepts.

This study aims to find possible solutions at different levels in response to soft conflict, and provide an idea and a method for protection and management of heritage in similar conflict areas.

Acknowledgements

Firstly, we would like to thank the joint efforts of the Mrauk-U application for the List of World Heritage research team led by our supervisor, professor Dong Wei of the southeast university. At the same time, we also thank NGOs and local archaeological team in Mrauk-U for their great help during our field survey, and especially Saw Min Phru, Zaw Zaw, greatly helped us communicate with local residents and obtain investigation data.

References

BARILLET C, JOFFROY T, LONGUET I, 2006. Cultural heritage & local development: a guide for African local governments. Grenoble: CRATerre-ENSAG/Convention France-UNESCO.

MIURA K, 2005. Conservation of a living heritage site: a contradiction in terms? A case study of angkor world heritage site. Conservation and management of archaeological sites, 7(1):3-18.

QI Changqing, HE Fan, 2004. Urbanization and challenges in Asian countries. World economics and politics, (11):48-53+6.

STEFANO FACCHINETTI, 2014. Cultural heritage management in Myanmar: a gateway to sustainable development.

WEISE K, 2016. Safeguarding bagan: endeavours, challenges and strategies. Journal of heritage management, 1(1): 68-84.

Shantytown or Historic Area? A Conservation Exploration on the Historic Cities and Vernacular Architecture in the Yellow River Floodplain

HU Lijun

Shanghai Tongji Urban Planning and Design Institute Co. , Ltd. , China

Abstract The establishment and evolution of historic cities in the Yellow River Floodplain have significantly impacted by the Yellow River, which makes their historic urban areas distinctive. There are dual defense and flood control systems consisting of city walls and dikes in the ancient cities. The water storage systems include various forms, such as city lake, moat, swamp, pond, etc. And layers of the ancient city of various dynasties rest underground within a certain area. Therefore, historic cities in the Yellow River floodplain should be conserved according to their regionalism by strengthening the city structures and patterns and should avoid the de struction of underground sites.

Because of the low economic level in history, early dwellings with poor quality have generally been rebuilt in recent decades. These half-new-brick-houses with little heritage value have gradually demolished in the process of urban development, and been replaced by fraud "ancient cities". As a result, the original urban space and social culture have completely disappeared. However, the quality and facilities of these common dwellings are still poor, and it is impossible to adopt the conservation methods of historical buildings. So how can we improve the lives of the residents? In order to solve the dilemma, the social and cultural value of these "vernacular architecture" should be recognized. With the method of the built vernacular heritage, the idea of "resident-oriented" is the right key approaching benefits both residents and government, ultimately leading to the revival of the historic cities.

Keywords Yellow River Floodplain, historic city, vernacular architecture, conservation method

1 Conservation Status of Historic Cities in Yellow River Floodplain

In north China, the Yellow River is an important river with heavy sedimentary. From the late Tang dynasty (about 9th century), it has experienced the process of siltation, flooding and river diversion frequently, thus formed the Yellow River Floodplain (mainly in the junction of 4 provinces: Henan, Shandong, Jiangsu and Anhui). Due to frequent floods, plague of locusts, and warfare, the water transportation system was destroyed and the land was salinized. From the Yuan dynasty (about 13th century) to the present, the Yellow River Floodplain has been a relatively poor and vulnerable place in eastern China (Fig. 1).

Nowadays, the historic cities in the Yellow River Floodplain are facing big problems in heritage conservation. The local government always removes residents, demolishes all the buildings except some listed buildings. So the historical structure and urban lifestyle were disappeared. After the demolition, the model of real estate development and tourist theme park were used to build a fraud "ancient city", and the original urban cultural characteristics were completely broken. Many historic cities have been criticized by the Ministry of Housing and Urban-Rural Development and the National Cultural Heritage Administration.

Those who support demolition and reconstruction emphasize that these cities have nothing to protect, and the value of such historic cities should be understood from the perspective of "water city". So creating landscapes and developing tourism are more important. But obviously it deviates from the conservation intention and requirements of national historic cities.

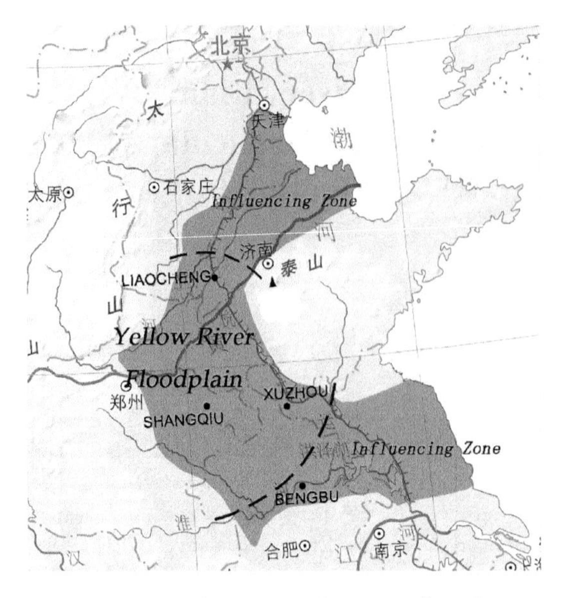

Fig. 1 The Yellow River floodplain & influencing zone

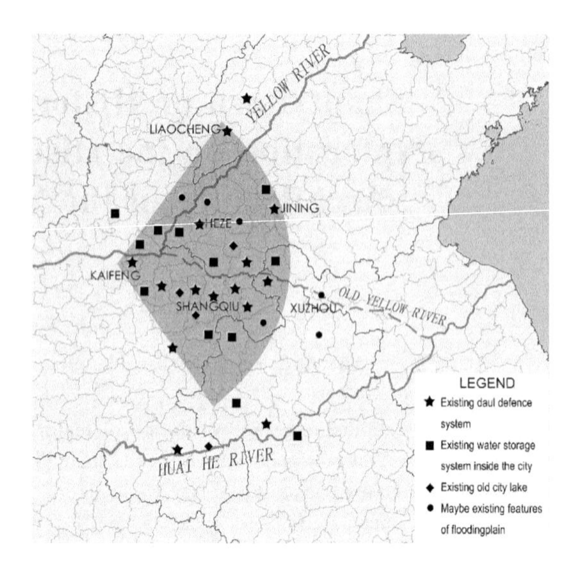

Fig. 2 Distribution of Historic Cities in the Yellow River Floodplain

Experts and scholars criticized and analyzed such vandalism from the perspective of historic cities value (Zhao, 2013; Zhang, 2017), but the objective situation of the historic cities in the Yellow River Floodplain were rarely analyzed. And it is still unclear how to conserve the historic cities in this region.

In China ,there are great differences in historic cities of different regions. The conservation of historical cities in the Yellow River Floodplain must be based on their cultural values and current characteristics.

2 The characteristic of Historic Cities in Yellow River Floodplain

According to satellite imagery and field investigation, there are still a large number of cities the maintain the typical characteristics of historic cities in the Yellow River Floodplain. They are particularly concentrated in the fan-shaped area east of Kaifeng. This is closely related to the frequent overflow of the Yellow River near Kaifeng in history (Fig. 2). It is a long-term struggling against the Yellow River that has resulted in a series of flood control experiences and unique regional urban construction features in the historical cities of the Yellow River Floodplain.

2.1 Dual Defense System

Historic cities in Yellow River Floodplain had generally been built with dual defense and flood control system consisting of city walls and dikes.

Before the Ming Dynasty (about 14th century), the Chinese city walls were generally built by rammed earth and were easily damaged by floods. From the 14th century, the surface of city walls were generally constructed of masonry, which greatly improved the ability to withstand floods. At the same time, the urn city was set up in order to increase the gate defense, strengthen the weak link of the city wall further, which acted as a flood dike.

Most of the cities in the Ming Dynasty were square, which was conducive to the street organization and neighborhood construction. However, the corners of the cities were often vulnerable to the impact and destruction when the flood struck. Moreover, city walls were the last defense, when they were destroyed, the cities would be ruined.

In order to reinforce the flood defense system, many cities added dike outside the city wall. Some dike line along the river, formed an irregular shape. Some dike had no terrain limitation, so the circle was adopted, thereby formed a pattern of "outer circle inner square", like an ancient coin. Regardless of the shape, the linearity of the dike is always natural and smooth, thus guides the water flow and avoids direct impact (Fig. 3).

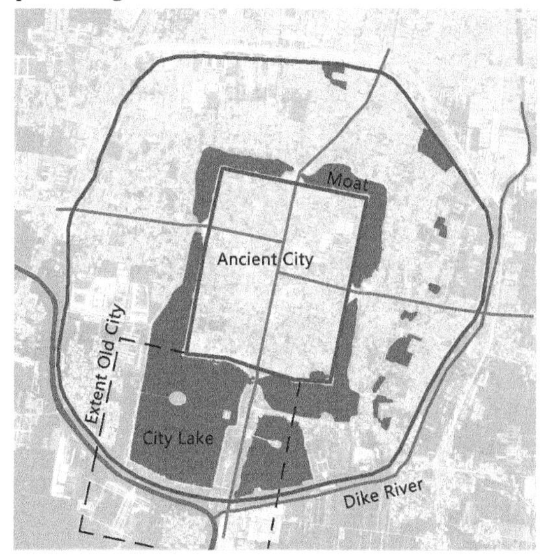

Fig. 3 Shangqiu ancient city plan

Because the dike also has a certain military defense function, some cities call it "the outer city" or "*Guocheng*". Although called a city, the dike was only constructed with rammed earth, and its height is lower than the city wall. But when reinforced by planting trees, it can play a buffer role in flood control.

2.2 Various Types of Water Storage

Due to the continuous flooding of the Yellow River, the ground elevation of the city is always characterized by high external and low internal. The difficulty of drainage to the outside, together with the further extraction of mud in order to increase the elevation inside the city, eventually formed a variety of water bodies (Fig. 4). includes:

(1) The moat and ditch surrounded the city and the dike, the width of some cities' moats reached more than 100 meters.

(2) Pond in the city, in order to raise the ground and accommodate rainwater.

(3) City Lake, located between the city wall and the dike, some on one side of the city, and some formed a ring lake.

In addition, there is a special type of

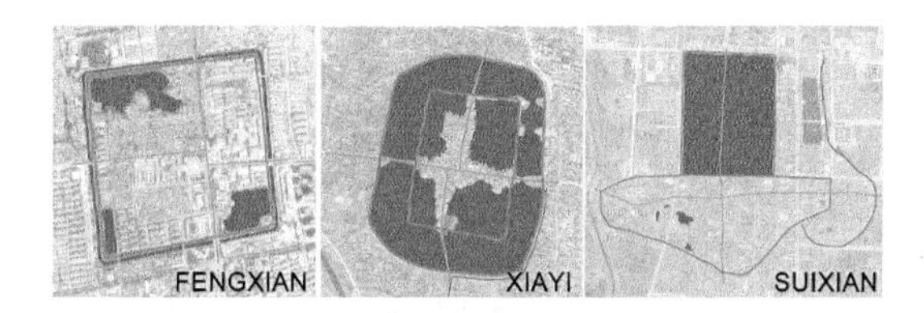

Fig. 4 Different water forms of Historic Cities in the Yellow River Floodplain

water body called the Old City Lake. This is a water body enclosed by the old city wall, while the city completely is flooded. The old city lake in Suixian County is the most complete case existing now, and the old city lake in Shangqiu has been linked with the new moat, and now a part of the new city lake.

2.3 Layers of Cities Sites Underground

Under the ground of the historic cities in the Yellow River Floodplain, there are often multiple sites. Archaeological discovered that there are 6 ancient cities under the ground of Kaifeng, and at least 3 ancient cities under Shangqiu.

The early city sites were represented by the capitals of big feudal states in the middle Zhou dynasty (about 2th—7th century BC). The capitals were huge and had high city walls. Due to the high quality of the city wall construction, after more than 2, 000 years, there are still many walls left on the ground. For example, the Ancient City of Lu in Qufu and the Ancient City of Zhenghan in Xinzheng, which is adjacent to the Yellow River Floodplain. But the walls of the Ancient City of Song (in Shangqiu) and Wei's Daliang City (in Kaifeng) have completely sunk below the ground. According to archaeological exploration, the top of the wall of Song is about 2 meters under the ground. Due to a large amount of sediment, the preservation of these sites is maybe very well.

The late city site is characterized by periodic evolution due to the influence of flooding. And those retained forms of different historical cities can also reflect different stages of evolution in this cycle (Fig. 5).

2.4 Buildings Difficult to Preserve

Because of the continuing flood damage

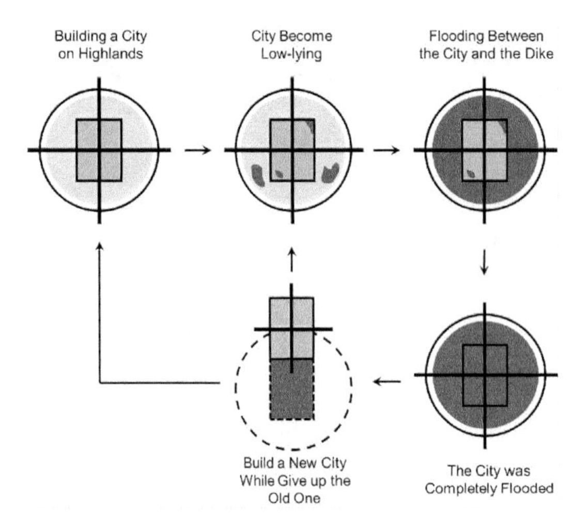

Fig. 5 Historic City cycle evolution in the Yellow River floodplain

in history, the economic level of the Yellow River Floodplain is relatively poor. Although the houses of rich people are built of bricks with good quality, its ratio of retention is relatively low, and those existing ones have generally been listed for conservation. Until the 1950s, most of the walls of dwellings were built by rammed earth or adobe bricks. Green bricks were generally used to strengthen foundations and corners, and some dwellings were even constructed of adobe bricks entirely. In order to save the tiles, generally only inverted tiles were used without cover tiles (this craft is called *Gancuo*), and the poorer families only could use straw or thatch to cover the roof. After the 1960s, these buildings were gradually rebuilt in order to improve living conditions. The walls of the newly built houses were mostly made of red bricks, and the roofs were built with flat tiles or cement. Obviously, these buildings do not belong to the conservation object in the current conservation system of historic cities in China. So what on earth should be conserved in the historic cities of the Yellow River Floodplain?

3 Enhance the Feature of Historic Cities

The purpose of urban heritage conservation is to enhance the city's cultural identity and citizen's sense of belonging, increase urban attractiveness and promote social development. Thus we need to strengthen cultural characteristics and increase the value of cities. Of course, the pattern of the historic cities in the Yellow River Floodplain which having distinctive features are exactly what we should protect.

3.1 Conserve the City-dike Dual Defense System

The most remarkable morphological feature of historical cities in the Yellow River Floodplain is the dual defense system consisting of city walls and dikes. Regardless of whether the city wall is retained or not, the boundary of the city can be highlighted by strengthening the open space around the city. Nowadays the dike is not high, and mostly be used as a road. It can continue the traffic functions, improve landscapes, and incorporated by urban slow transportation systems.

Between the city wall and dike, it was once a transitional space from city to suburb. The outer city's spatial form is generally freer than the inner city whether with construction or not, so this spatial difference should be maintained.

3.2 Conserve the Water Storage system

The various forms of water witness the historical process of urban development. Since the city lake or the moat owns obvious positive effects on the landscape of the external image of the ancient city, all of them will be maintain and even recovered to historic size or expanded. In contrast, the ponds inside the city wall are always occupied. In fact, the function of ponds are positive, they can reduce the density of the city, provide ecological space, and increase landscape quality, and ultimately, they will help to improve the value of the ancient city.

3.3 Conserve the Urban Pattern

The process of planning and construction of the ancient cities in Yellow River Floodplain is concentrated in a short period

These cities' urban patterns are always regular. Generally, they have square city walls, and inside of them are the checkerboard grid. At the same time, they are organizing the living environment through courtyards with the traditional Chinese roof style and color.

3.4 Conserve the Ancient City Sites

In the process of urban construction, we must attach importance to the discovered or potential underground cultural relic's burial area. The key cities such as Kaifeng, Shangqiu, and Xuzhou have high historical and cultural values. Cultural relic exploration should be carried out to find the relationship between the city's site and the cultural layer in different historical periods, and provide information for conservation and exhibition of cultural heritage. In the case of the site information is not clear or conditions are not mature for exhibition, such site should be reserved as space to avoid further development damage.

3.5 Conserve the Urban form at Different Periods

Currently, both the academic field and the tourism field have paid more attention to the cities with a clear system of the city and dike, with the lake surrounded outside the ancient city, such as Shangqiu and Liaocheng.

However, it should be recognized that this city form is only one link in the periodic evolution. Together with the other historic cities of different evolution stages, they provide us a complete urban gene bank of the Yellow River Floodplain. Maybe the value of the other cities is lower, but local people need their own cultural memory by conserving their cities' characteristics.

4 Urban Revival Starting from Residents and Vernacular Architecture

Over the past 30 years, the size of the cities in the Yellow River Floodplain has been expanded tremendously. In contrast to the rapid development of the new district, the ancient city is declining. Their physical environment is weak, the houses are out of repair, and the facilities are obsolete. At the same time, the young population flow away, and most of the people left are old and weak groups or rented migrant workers. The revival of the ancient city needs to start with both the physical environment and the population structure. The commercial development for quick success and instant benefit violates the basic principles of rational utilization and sustainable development, and also cuts off the true historical context of the city. The right way for urban rejuvenation needs to be residents-oriented and to solve the problem from housing, supporting facilities, communicating space and industrial foundation.

4.1 Understanding the Heritage Value of Common Dwellings

Preservation or demolition of common dwellings which were built after the 1950s has always confused the local governors of historic cities. If preserved, the messy environment and poor infrastructure cannot meet the requirements of contemporary human settlements. And the ancient city will eventually degenerate into a "shanty town". However, if we adopt the demolition and reconstruction method, it will inevitably destroy the historic cities.

Fig. 6 Satellite image of Shangqiu Ancient City (partial, 2014)© Google earth

As far as the single buildings are concerned separately, they have little heritage value. But if we judge by satellite maps, the neighborhoods made up of these buildings are still have historical features in space texture and roof form, and still like an "old quarter" (Fig. 6). This is because despite the materials and colors are different from the past, most of the construction activities are taking place according to the original building site. They adopt the same construction logic and similar dimensions and shapes as the traditional ones. More critically, they are the product of the gradual renewal of community residents in their long-term lives, whose replacement process follows historical habits. In this process, the inhabitants continue to be multiplied and thrived, and the social structure has not been interrupted and destroyed. It is the continuation of the social and cultural that has kept these ancient cities alive even the city's economic and political center was moved away.

The common dwellings of historic cities in the Yellow River Floodplain are definitely lack historical value, artistic value and scientific value, but they possess certain social value and cultural value. They are the carriers of social civic culture and reflecting the cultural form of Historic cities in the Yellow River Floodplain. According to the definition of *the Charter on Built Vernacular Heritage of ICOMOS*, these dwellings belong to the vernacular buildings. They were built by the traditional way and by the communities themselves. They are in the continuing process including necessary changes and continuous adaptation as a response to social and environmental constraints.

Vernacular buildings emphasize the participation and support of the community and residents. The physical value is more in the group, rather than the single building. Therefore, their conservation method should focus more on their role in shaping the integrated urban space and features. At the same time, the residents should be encouraged to gradually upgrade even rebuild their houses according to the conservation requirements. It needs to increase thermal insulation, modern kitchen and sanitary facilities to meet the needs of contemporary life, which will make the ancient city more suitable to live.

4.2 Revival Model with Resident-oriented

The revitalization of the historic cities requires both the restoration of community vitality and injecting new industry. Overcrowded and poor facilities are the common problems in the old quarter due to the persistent debt in the past decades. Local governments often have no confidence or motivation of making meaningful improvements to the old quarter while retaining existing residents. Because of the results of high-rise and high-density brought about by the general urban renewal modes are not allowed for the historical cities. Under the strict control of height and features, so far the way that local governments can think of is to develop tourism and the low-level real estate projects. Taierzhuang, a city on the edge of the Yellow River Floodplain, the commercial success of which stimulated the other cities in the region to build "new" ancient cities.

However, tourism is by no means a panacea. Huge demolition and relocation costs and initial investment put great pressure on local finance. And it is uncertain whether it will attract sufficient tourists after its completion and opening. If this mode is adopted by ancient cities with similar resources in the region, the homogeneous competition will further divert tourists. In the off-season of tourism, an ancient city without the support of its residents will become a "dead city". The real estate model also has serious problems. They are nominally residential, but they are in fact investment property, and the population density of the real living is inevitably low. Therefore, it is not feasible to rejuvenate the ancient city only by means of replacement function and tourism development. It not only causes irreversible damage to the original social and cultural values but also causes high investment, high risk and uneconomic.

The ancient city is always the home of the inhabitants. Only the resident-oriented mode is a sustainable way to solve the problems from the root.

Through the investigation, we found the reasons for the high population density of the ancient city is below:

a) The historical open spaces are occupied

b) Large and medium-sized traditional courtyards with many families are living together

c) Crowded living in the joint dormitory

These areas need to actively evacuate the population, restore and increase open spaces, and improve the environment.

The "vernacular buildings", which account for a large proportion of the ancient city, are updated on the traditional base sites. Most of them are small families living in a single courtyard and are not crowded. Therefore, it is much less difficult to be improved than multi-family residential courtyards. Improvement of facilities and appearance control according to the feature of the ancient city are the main objectives of these courtyards. In order to enhance the attractiveness of the ancient city, it is also necessary to improve the public service so that the residents can fairly obtain the benefits from urban development.

Although the mode of partial evacuation and gradual renewal has only input but no direct economic output, but compared with the high risk of large-scale relocation for tourism, it is not worse in economic accounts. These necessary inputs are the debts of the government for several decades and should be regarded as the cost of urban development. Regardless of the social and cultural value or sustainable development, resident-oriented is a correct development model.

Resident-oriented does not exclude the development of tourism. On the basis of continuing residence, we can excavate the cultural resources of the ancient city and develop tourism in combination with the evacuated sites and courtyards. In this case, tourism is no longer a scenic performance, but an exhibition of the real-life of the people in the ancient city.

Thus, such tourism will enrich the culture of the ancient city, provide employment for residents, and develop more diversified industries, and the ancient city can be really revived.

5 Conclusion and Discussion

In general, Conservation of historic cities in the Yellow River Floodplain is unsatisfied. This is related to the deviation of the understanding of its cultural values and the particularity of the built environment. While conserving and strengthening the historical characteristics of the city, we should pay more attention to the importance of vernacular architecture conservation.

First of all, in consciousness, we should understand the value of dwellings built by residents on historical sites. Although they are lack of historical, artistic and scientific value, they do have certain social and cultural values. In the Chinese heritage conservation system, it can be considered as a special traditional style building, which can fit for the modern residential needs through upgrading and renewal, and continue the traditional style and features of the ancient city at the same time.

Secondly, in the conservation system, the quarters consisting of vernacular architecture should be treated as a special type of historical and cultural area. So they should be included in the national historic cities conservation system.

Thirdly, in terms of protection policy, community and residents should be encouraged to participate in the conservation. Most of the vernacular architectures are single-family houses. They can be improved by residents through technical guidance and policy encouragement. This can not only make full use of traditional folk technology and private funds, but also maintain the diversity of residential buildings and enhance the pride of residents, which is advocated by the Charter on Built Vernacular Heritage.

Finally, the value of vernacular architecture is also complementary to the revival model of resident-oriented. The governments do not have to bear heavy financial burdens. The economic and cultural rights of the residents of the old city are respected. It is conducive to highlight urban culture and social harmony.

References

ICOMOS CHINA, 2015. Guideline for the protection of cultural relics and monuments in China.

ICOMOS, 1999. Charter on built vernacular heritage.

WU Qingzhou, 1995. Study on flood control in ancient Chinese cities. Beijing: China Architecture & Building Press.

XIE Yinqian, 1998. Exploring the reasons for the formation of Gu Dui culture in southwest Shandong by geographical factors. Historical geography, 14.

YU Kongjian, ZHANG Lei, 2007. The flood and Waterlog adaptive Landscapes in ancient Chinese cities in the Yellow River Basin. Urban planning forum, (5): 85-91.

ZHANG Song, 2017. Sustainable conservation for urban historic environment. Urban Planning international, (2): 1-5.

ZHAO Yong, 2013. Reflections on the protection of historic cities in the process of urbanization in China. Urban studies, (5): 111-117.

ZOU Yilin, 1997. Historical geography of Huang-Huai-Hai Plain. Hefei: Anhui Educational Press.

Constructing Chinese Traditional Rural Landscape from the Perspective of Sustainable Development and Cultural Conservation

LI Yan[1], CHEN Zihan[2]

1 Suzhou Jingzhi Landscape Design Consulting Co., Ltd, China
2 Hunan University, Changsha, China

Abstract Farmers of Forty Centuries is the first book that documented the sustainability of East-Asian regional traditional rural agriculture (King, 1911). This book outlines how the traditional rural landscape is a type of cultural landscape system, where a human-managed landscape can provide a harmonious relationship that benefits both human and nature. However, industrial farming development in the past century has negatively impacted a range of ecological and environmental management issues worldwide. In recent decades, an increasing concern on environment conservation conjures up the concept satoyama landscape, a term which relates to the study of the Japanese traditional rural landscape (Brown R. D., 2001) and has become broadly recognized and transformed into various forms of practices. There is limited scholarly literature, in either Chinese or English, that describes the historical and contemporary status of a Chinese traditional rural landscape system.

Shiyanping is a Chinese national monument historic village. Landscape conservation work being done there reveals the value and current status of the traditional rural landscape. Based on this case study, this paper advocate establishing and understanding Chinese traditional rural landscape systematically. It will help us to identify the unique value and complexity of traditional Chinese rural landscape in the issue of sustainable development and cultural heritage conservation. Finally, this paper gives a constructive proposal suggesting the framework for studying Chinese traditional rural landscape.

Keywords Chinese traditional rural landscape, sustainable agriculture, living cultural heritage, historic conservation, cultural landscape

1 Introduction

China has a long history of civilization that has managed agricultural development. Archaeological and genetic studies establish that China has more than 9000 years of rice paddy farming (Zheng, et al., 1990). Over thousands of years as an agriculture civilization, China has accumulated much knowledge and wisdom in its construction of the traditional Chinese agricultural systems. This has been presented in antique books, paintings, and passing down as of varions forms of intangible and tangible heritage. "*Farmers of Forty Centuries*" is apparently the first text documenting the sustainability of East-Asian regional traditional rural agriculture (King, 1911). It outlines the traditional rural landscape, especially Chinese and Japanese rural landscapes, as examples of cultural landscape systems, in which a human-managed landscape provides a harmonious relationship benefiting both humans and the natural world.

Industrial farming over the past century has negatively impacted in ecology system and food safety, including China (Miguel, 1998). Thus, study and passing done the wisdom of Chinese traditional rural landscape in academia and practical projects are significant for environmental sustainability as well as for traditional Chinese rural landscape heritage conservation, especially as a living heritage asset.

Therefore, this paper advocating for establishing and understanding Chinese traditional rural landscape systematically. Shiyanping national historic rural village is used as a case study. This paper endeavors to identify the unique values and complexities of traditional Chinese rural landscapes as an issue for sustainable development and cultural heritage conservation.

1.1 Chinese Traditional Rural Landscaping — a Definition

This paper will initially clarify the reasons for the phrase using "rural landscape" as the key term in this paper.

This paper focuses on the issues of the traditional agriculturally-oriented system, its knowledge of, and wisdom about, the human-managed environment, which constitutes a rich and complex ensemble of tangible, intangible and living heritages(Parviz & Mary, 2011). It is constantly adapting to newly-arising environmental, cultural, social, political, and economic conditions, while situated in a continuing cultural landscape.

Several well-known authorities recognize the value of agriculturally-oriented heritage but use differing terms.

ICOMOS General Assembly, a UNESCO-managed organization adopted "Principles concerning rural landscapes as heritage"from UNESCO in 2017. There is defined "rural landscapes" as :

... terrestrial and aquatic areas co-produced by human-nature interaction used for the production of food and other renewable natural resources,... . Rural landscapes are multifunctional resources. At the same time, all rural areas have cultural meanings attributed to them by people and communities: all rural areas are landscapes.

The American National Park Service (NPS): classifies "vernacular landscape" as a secondary classification under the cultural landscape, to specify rural landscape classification.

The Globally Important Agricultural Heritage System (GIAHS) which is managed by the Food and Agriculture Organization of the United Nation has also focused on global agriculture heritage (Parviz & Miguel, 2011). Several Chinese historical agriculture sites have been registered (Lu & Li, 2006). However, GIAHS only identities unique historic agriculture heritage (Stuart, 2007), doesn't list the general historic agriculture system.

In conclusion, UNESCO emphasizes that the duration of the contact period between humans and nature must be long (Rossler, 2005). NPS focuses on the significance and integrity of the cultural landscape (Herb, 2007). Duration is significant in this instance. Rural landscapes present an ongoing, and usually continuous interaction of humans modifying the natural world and the natural world's reaction, both expected and unexpected, to those modifications which, in turn leads to the human response to nature's response in an endless series of causing and being effected and causing. The natural landscape requires time to accumulate these changes. Based on the literature comparison study, the rural landscape is the term better situated in this context.

1.2 Contemporary Issues in Agriculture Landscape Development

As of 2019, there are studies that have examined various traditional agriculture development since after the publication of Franklin's book (Stoate, 2001; Tscharntke, et al., 2005). However, they are isolated and limited in different areas such as ecology, historic conservation(Jianbou & Xiali, 2006). Examine traditional rural landscape systematically will give us a map to see the complete layout and structure of the Chinese agriculture-oriented rural system, provide a perspective integrated all the isolated research together.

Japanese scholars developed a systematic study of Japanese traditional rural landscape since the 1960s and have led to the formation of a unique concept -Satoyama Landscape (Takeuchi, et al., 2001). Satoyama Landscape contains a holistic system of traditional Japanese agriculture, including topographical location, cultivation methods, human habitation, activity patterns, management methods balanced with nature and ecological structure characteristics. It provides a programmatic theoretical basis to guide modern sustainable agriculture development while protecting traditional Japanese culture (Takeuchi, et al.,

2001). It also has been influenced internationally(Hiromi & Richard, 2003). However, China as one of the longest agricultural civilizations country, due to lack of systematically theoretical research, made it's hard to be recognized and demonstrated to the public.

Moreover, the advancement of scientific and technological advances and economic booming have impacted the global ecosystem and deteriorated the natural environment while promoting what is called social development. These conflicting results have caused a series of problems and contradictions in the construction of rural agricultural landscapes. This paper will discuss related issues in detail in the following chapters.

Fig. 1 Shiyanping Aerial View © Li Yan

2 Shiyanping Case Study

Using both literature research and field investigation, this paper reports on takes a practical project "Shiyanping Village" as a case study in which, analyzes the current situation and the difficulties encountered in landscape conservation practice. Universal approaches can be summarized, the urgency and importance of traditional Chinese agricultural landscape conservation can be further clarified. In addition, this paper attempts to conduct the research on the component factors of the agricultural landscape which related to three main aspects: spatial structure, living heritage, destruction, and threats. Then, on the basis of this, a theoretical research model of agricultural landscape construction in China is constructed, aiming to provide direction and inspiration for future research.

2.1 Basic Information

Shiyaning is located in Zhangjiajie, Hunan, China. It is deeply hidden in a flat valley surrounded by mountains. It's a typical traditional village of based on hilly rice cultivation in southern China (Fig. 1). The quaint and primitive ethnic architectural style is preserved, presents the classic Tujia Sumu culture. Up to present, more than 96% of the historical settlements in Shiyanping are in use, 182 are existing Diaojiaolou' (built on stilts). Most of the settlements were constructed from the late Qing Dynasty(1700s) to the early Republic of China(1950s), 48 of them are listed as national cultural relics protection units. Those units have the Tu ethnic traditional "Pei Feng Shui" (Fig. 2) landscape structure. Besides, Shiyangping keep the traditional

Nevertheless, due to the disturbance from the period of the Great Leap to the period of reform and opening up forest were over exploited, resulted in ancient trees lost. Aged trees present the time of the place, causes the surrounding landscape has a time gap experiences compare to historic buildings. Sue to the time gap between impacted landscape and preserved architecture, Shiyanping historic village restoration project involved the environmental remediation part, acting as an experimental practice, to restore a coherent historic monument site at the national level cultural relics.

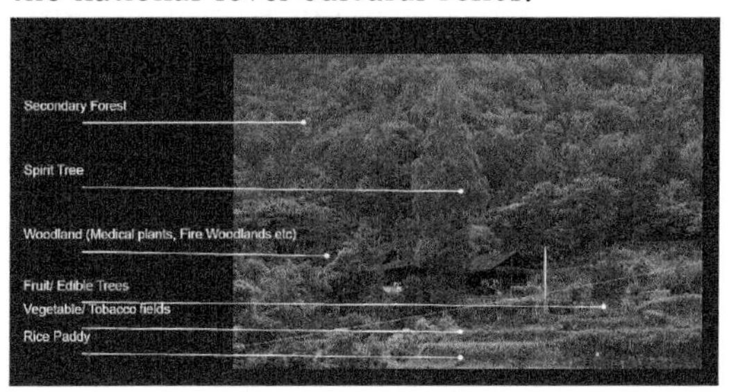

Fig. 2 Shiyanping Settlements "Pei Feng Shui" Landscape © Li Yan

2.2 Spatial structure of Shiyanping

Shiyanpin Village is a typical Basin-Mountain landscape structured and rice paddy dominated village, which is a very common type in the South Yangtze River area. Solar reflection is an extremely valuable resource in this type of rural landscape, place where has the best sunlight is rice paddy located, and residence is built-in the lower hilly side, between the mountain and the farming field.

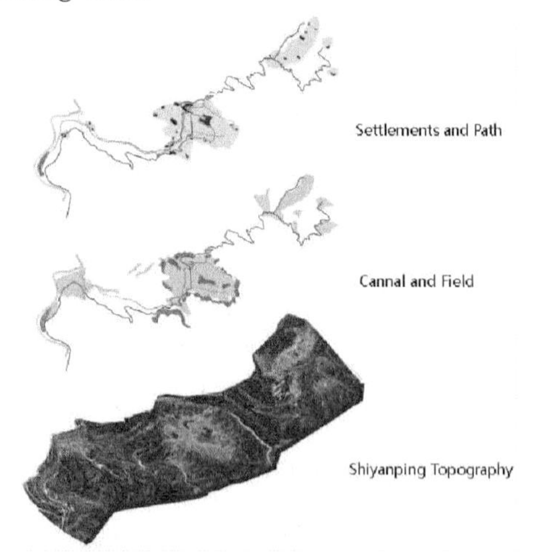

Fig. 3 Shiyanping Spatial Configuration © Li Yan

As an agriculture-oriented landscape, this paper using the farming-living structured model to demonstrate Shiyanping's spatial configuration. Fig. 3 shows the water canal as a fundamental infrastructure for wetland rice paddy farming located in the central basin where has the best sunlight for farming. All the basin areas are agriculture fields. In this particular case, rice paddy wetland. Wetland normally has the most complex ecology system in the natural system, rice paddy wetland is a seasonal artificial wetland, which has a long history, led to wild animals, plants adapted to this landscape form, created a unique ecosystem. Secondly, is the area of the human settlements. It is located in the lower mountain area around the rice wetland basin. It is shady compares to the basin area, however basin area is the place where hold the temporary flood in the rain season. Arrange the residence area in a higher land keep the villagers away from flood damage (Fig. 4). Agriculture and human habitat are well balanced. Thirdly, the managed mountain area. At the backside of the housing area, the mountain landscape goes straighter to the top. Villagers usually make a flat area close to their house to grow local medicine, tea oil tree, cypress trees, palm trees, banana trees, etc. It is a coppice woodland that provides locals for medicine, fire source from cypress, oil, and tea from tea tree, raincoat fabric from the palm tree and using banana tree leaf for house fire extinction and protection, etc. Harvest firewood provides mushrooms as food source and create sun gap for trees to grow. From coppice woodland, the landscape gradually immersed into the natural mountain area. This rice wetland settlements-coppice woodland landscape is the basic spatial landscape feature of Shiyanping village.

Fig. 4 Shiyanping Settlements Landscape © Li Yan

2.3 Living Heritage

Although Shiyanping village hides in the mountains of South China, it also has a strong influence by globalization and industrialization, for instance, some villagers make lawn close to there house or using plaster roman columns for their housing details. Farmers are using pesticide in farming management which is a very bad influence for wetland sustainability. Fortunately, there is some traditional everyday living lifestyle kept in Shiyanping Village. The most important feature is the reading-farming(耕读) culture. Reading-farming culture has a long history in China, which can be dated back to Kongfuzi time more than 2000 years ago. It is a lifestyle, that farmers working on agriculture work and literature study, to sustain their normal life, and to gain for a national officer career through the national election test. There is a public calligraphy house located in the middle of the mountain. It is open 24 hours and managed by locals, farmers doing calligraphy practice and communication at that building. Surprisingly farming-reading tradition has survived from world war 2, cultural revolution and globalization, which is an extraordinary cultural asset.

Fig. 5 Public Calligraphy House © Li Yan

Other than Reading-farming life-style, there are other phenomena worth to study. Like other traditional villages, the villagers in Shiyanping are not just all farmers, there are stone craft man, woodcraft man, iron craft man and other villagers with different careers, that to make the village well sustained. But because of industrialization, many crafts man's work has been replaced by the machine, the occupation diversity has been declined. It is an inevitable process, but the folk songs from stone craftsman, construct man is valuable, shows the working contents and accumulated wisdom from the past.

2.4 Threatens to the Traditional Agriculture Landscape

This paper understands the Chinese rural traditional landscape is living heritage, that continuing changes are one of its special characters. This paper evaluates threatens and damages based on the sustainability situation. Due to sustainable agriculture is one of the most valuable features of Chinese tradition rural landscape, and for us to study, conserve and take to the future. According to this principle, there are some threatens that potentially or continuing causing damages, based on the field research threatens are as follows: a). Young generation loss. Other than other social risks that have been discussed by sociologist over time. In this particular case, the young generation loss might increase the loss of Reading-farming culture lifestyle. After living in urban areas for years, young generations will easily homogenized by generic urban lifestyle. b). Topography change. There are farming machines designed for mountain topography the potentially will change the topography of rice paddy farming land. The current rice paddy in Shiyanping, shaped in a natural curved form that goes with the natural topography. Machine application will change the topography from natural form into a more geometric shape, to maxim the productivity. It has been taken place in Japan and recorded by Japanese traditional rural landscape scholars (Satoyama, 2000). The topography change will undermine traditional cultural complexity of the ecology system. c). Pesticide overuse. Pesticide and herbicide use not only threatens our food safety, but also the wetland system. Chemical treatment over use will dramatically reduce the biodiversity and soil health of wetland. Over the thousand years of wetland rice farming, there are several complex human-nature rice farming ecosystems has been documented. Such as Rice-Fishing complex system (Jianbou & Xiali, 2006), has been documented by GIHAS, and Rice-Duck complex ecosystem (Y. Huang et al., 2005), and possibly another ecosystem that hasn't been academically documented yet. There was no documentation and study about the farming ecosystem in Shiyanping, the use of chemical treatment might make this field become a mystery. d) Tourism development. Well managed tourists development are very good for traditional village conservation, it brings economic benefits for local development, and make local people reconsider their hometown's cultural and natural value. Shiyanping village conservation is still in process, there has been some small tourist boom changing the rural landscape negatively. When Hunan Satellite TV filmed in this village, villagers planted shrubs that are not local and are the ones used in urban landscaping, which made the area is very out of context. Tourism development should be guided by expertise, and working closely with local communities, to create a cultural integrated landscape.

3 Analysis

Shiyanping national historic village case study provides concrete evidence that a traditional rural landscape can be sustained and that it has cultural value as a living heritage. It is also in a complex threatened situation.

3.1 General Value

Shiyanping case study shows the general value of traditional Chinese rural landscapes and is reflected in the following.

3.1.1 Land use pattern is in harmony with nature

Due to globalization and industrialization traditional villages and the ecosystem has gradually become sensitive and fragile rather than resilient. Shiyanping keeps the traditional land use pattern, which utilizes the land rationally, efficiently, and intensively demonstrating ancestor's wisdom that accumulated from rich life experiences. Vegetation, farmland, houses, ditches, ponds and other different land use methods have avoided infringement as much as possible and maintained a balanced organic environment.

3.1.2 Different functional divisions constitute a relatively complete micro-social structure

People residing in Shiyanping are depending on close nature. Most residents regard crop farming, fish breeding, poultry raising, and logging as their living material resources. With the development of the village society, resulting in different functional divisions, such as stonemasons, blacksmiths, carpenters, teachers, and medical care providers. Farmers with diverse occupations constitute the social demographic structure of a self-sufficient farming village.

3.1.3 The preservation and continuation of the traditional farming lifestyle

The farming lifestyle is one of the primary features of ancient Chinese society, forms a flexible social class structure. It had been retained as a major national management system until the 20th century. Land-reform movement (1950s), Cultural Revolution(1960s—1970s,) caused it to undergo tremendous changes. Although large-scale social changes have taken place in China, some rural areas, especially in ethnic areas, agriculturally-oriented traditional lifestyles have been somewhat preserved.

3.2 Differences

Shiyangping is a Tu ethnic area that differs from other agricultural landscapes in a number of respects. Its lifestyle and spatial layout are Quite differences from the Han. Han tends to live in a high density settlements form, while Tujia is more likely loosely arranged. These aspects reflect the difference of traditional agricultural landscapes between different ethnic groups, and having the ability to be one of the classifications of traditional agricultural landscape research for further study. Additionally, Shiyanping is a typical traditional rural landscape of rice cultivated in the mountains. China's vast territory results in a wide variety of main crops and the geographical environment of farming are very different across it. The core concept of traditional rural landscape construction balances nature and farming methods in different regions are different. The plains, bays, mountains, plateaus, or wheat, rice and corp fields, or multi-season all have mixed species. The mountainous agricultural cultivation of Shiyanping also has the possibility for becoming one of the classifications for future research.

3.3 Benefits for Constructing Chinese Traditional Rural Landscape Theory

There is a vast amount of information on Chinese rural landscape worthy of study for cultural conservation and future sustainable development matters. Integrating these areas of study to map out the structures of Chinese traditional rural landscape theory is, possibly, the key to an understanding of China's agriculture-oriented civilization from ancient times to the present. It could identify what the heritages represent for the past and what the living heritages in historic villages are. This theory can guide us in the application of applying different methods to preserve the cultural landscape in practice, and construct sustainable agriculture oriented society, which is deeply rooted in our past.

4 Construct the Chinese Traditional Rural Landscape Study

According to the analysis of Shiyanping rural traditional landscape, this paper suggests constructing a Chinese traditional rural landscape study framework from six aspects.

4.1 Literature Research

There are many ancient books and paintings which documented traditional Chinese rural landscapes. The four best-known ancient Chinese agriculture books are as follows. The Fan Sheng Zhi Shu, written in the first century BCE and known as the first agriculture in China. Qi Min Yao Shu in 544 CE Nong Zheng Quan Shu 1610 CE are encyclopedic agricultural books. There are many other classic books that documented information related to the rural landscape. The Geng Zhi Tu is a series of ancient drawings that about rural landscape and farming life for the empire to read from time to time. A study of the literature study shows the transitions of traditional rural landscape variations and development. A reliable resource for identifying cultural heritage features in the current rural site, to evaluate the authenticity and integrity of traditional rural landscape.

4.2 Cultural Ecology Research

Over thousands of years of rural landscape development, wildlife has adapted to the farming environment forming complex ecology systems that balanced human activity and nature. Those ecology systems reflect the traditional rural landscape's sustainability and biodiversity. The study of those systems is crucial for rural environment restoration from a cultural ecology perspective.

4.3 Anthropology Activity Research

Rural villages usually inherited many everyday activities from ancient times, such

as Farming-reading, Family Ancestor memorial ceremony, and farming proverbs. Research into those activities from an anthropology perspective can evaluate and preserve the traditional rural landscape as living heritage.

4.4 Spatial Configuration

Traditional rural landscape construction was based on local topography and natural environmental resources. The spatial configuration study shows the methodology of land use and sustainability patterns in general.

4.5 Prototype Classification

China has many different types of geography and nations, farmers living in different places or famers from different ethnics have developed different rural landscape with different cultivation and settlement construction methods. It seems important to classify prototypes of Chinese traditional rural landscapes to understand the structures of the diverse rural landscape systematically.

4.6 Evaluation System

A standard evaluation system to identify and estimate specific traditional rural landscape's authenticity, integrity, and significance. This system provides an efficient way to recognize and list the traditional rural landscape sites for all of China.

5 Conclusion

This paper suggests integrating ecology

restoration, historic conservation, cultural landscape preservation, and sustainable development altogether, to construct a study of Chinese traditional rural landscape. This study will help to systematically learn and preserve the profound Chinese agriculturally-oriented civilization that has evolved and accumulated over thousands of years. A Shiyanping historic village traditional rural landscape study would show an understanding of the historic traditional village from a cultural landscape conservation view, and will extend current conservation practices from architectural restoration to the whole village site, from historic buildings to the entire environment, and from physical monuments to living heritage.

Globalization and industrialization are continuing threaten not just historic village rural landscapes, but also to sustainability, which seems to be dramatically lessening all over the country. It homogenizes rural village special characteristics, causes food safety concerns and other issues. Applying traditional rural landscape study to current national Beautiful Village Construction Development led by the PRC government gives an opportunity to culturally connect rural villages of the present to the past and to create a sustainable and ecology complexed agriculture system.

References

BAO Ziting, ZHOU Jianyun, 2014. Phenomenon, reasons, and countermeasures for contemporary rural landscape decline. City planning review, (10): 76-83.

HIROMI K, RICHARD B P, 2003. Participatory conservation approaches for satoyama, the traditional forest and agricultural landscape of Japan. Journal of hyuman environment, 32(4): 307-311.

JIAN Bolu, XIA Li, 2005. Review of rice fish-farming systems in China-One of the globally important ingenious agricultural heritage systems (GIAHS). Agriculture, ecosystems & environment, 105(1-2): 181-193.

KING H F, 1911. Farmers of forty centuries or permanent agricultural in China, Korea and Japan. Madison: Democrat Printing Co.

LU Jianbo, LI Xia, 2006. Review of rice-fish-farming systems in China - One of the globally important ingenious agricultural heritage systems (GIAHS). Aquaculture, 260(1-4): 106-113.

MIGUEL A, ALTIERI, 1998. Ecological impacts of industrial agriculture and the possibilities for truly sustainable farming. Monthly review, 50 (3): 60-71.

PARVIZ K, MARY JANE DELA CRUZ, 2011. Conservation and adaptive management of globally important agricultural heritage systems (GIAHS). Resources and ecology, 2(1): 22-28.

PARVIZ K, MIGUEL A, ALTIERI, 2011. Globally important agricultural heritage systems: a legacy for the future. Rome: Food and Agriculture Organization of the United Nations.

STOATE C, 2001. Ecological impacts of arable intensification in Europe. Journal of environmental management, 63: 337-365.

TAKEUCHI K, et al., 2001. Satoyama: the traditional rural landscape of Japan. Japan: University of Tokyo Press.

TSCHARNTKE T, et al., 2005. Landscape perspectives on agricultural intensification and biodiversity-ecosystem service management. Ecology Letters, 8: 857-874.

ZHENG Yunfei, CRAWFORD G W, JIANG Leping, et al., 2016. Rice domestication revealed by reduced shattering of archaeological rice from the lower yangtze valley. Scientific reports, [2019-06-10]. https://www.nature.com/articles/srep28136.pdf.

孙宗文,2000.中国建筑与哲学.南京:江苏科学技术出版社.

Study on Protection and Utilization Strategy of Grand Canal Ancient Town from the Perspective of Inheritance and Utilization of Intangible Cultural Heritage —Taking Daokou Town as an Example

SONG Yating

Beijing University of Civil Engineering and Architecture Beijing, China

Abstract As a typical representative of the ancient town, the ancient town of the Grand Canal has its unique characteristics with the formation and development. Daokou town in Henan Province gradually formed and developed due to the construction of the Yongji Canal of the Grand Canal. By the end of the Qing Dynasty, the ancient town had become a peripheral water and land transportation hub with a very prosperous economy, known as "Little Tianjin". The Grand Canal not only nurtured the unique urban form of Daokou town, but also brought about nearly 100 intangible cultural heritages. After the successful application of the Grand Canal in 2014, it has brought new opportunities for the protection of the Grand Canal heritage and the protection and development of the ancient town of the Grand Canal.

Firstly, this study analyzed the site selection characteristics, street pattern and courtyard characteristics of the Daokou town, and elaborated that the Grand Canal is the type and characteristics of the intangible cultural heritage of Daokou town. Secondly, it analyses the status quo and characteristics of resources conservation of ancient towns along the Yongji Canal and their development zones and studies the problems and prominent dilemmas faced by the protection and development of ancient towns in the process of rapid urbanization. Thirdly, it explores the "top-down and bottom-up combination" model of ancient town protection and renovation based on courtyard property rights. Finally, it analyses the trend of intangible cultural heritage protection, inheritance, and utilization, couples the "display-study-activation" of intangible cultural heritage with the spatial form of the ancient town development zone. and puts forward that it is different from big cities in the new era. The development path of cultural and creative industries and the activation and utilization path of Daokou town in the North Canal.

Keywords grand canal, Daokou town, intangible cultural heritage, protection, utilization

1 Introduction

"Guidelines for the Protection of Chinese Cultural Relics and Monuments (2015)" clearly states that the value of cultural relics and monuments includes historical value, artistic value, scientific value, social value and cultural value; social value includes memory, emotion, education and other contents; cultural value includes cultural diversity, the continuation of cultural traditions and intangible cultural heritage. Elements and other related content. It can be seen that the recognition of intangible cultural heritage value has increasingly become an important factor in the field of cultural heritage protection and utilization.

As a typical representative of the ancient town, the ancient town of the canal has its unique characteristics with the formation and development of the canal. Daokou town in Henan Province gradually formed and developed due to the construction of the Yongji Canal of the Grand Canal. By the end of the Qing Dynasty, the ancient town had become a peripheral water and land transportation hub with a very prosperous economy, known as " Little Tianjin". The Grand Canal not only gave birth to the unique urban form of Daokou town, but also brought about nearly 100 intangible cultural heritages. Daokou town, as an ancient town in the central and western regions, is the same as other ancient towns in the central and western regions. The driving force of surrounding big cities is weak, while the geographical position advantage is not prominent, and the driving force of urban development is weak. After the successful application of the Grand Canal in 2014, new opportunities have been brought to the protection and development of the Grand Canal heritage and the ancient town of the canal. Daokou town in Huaxian County, Henan Province, seized the opportunity and explored the protection and utilization path of Daokou town based on intangible cultural heritage protection.

2 Spatial Characteristics of Daokou Town in Huaxian

The Huaxian Grand Canal began in the Baigou Canal in the late Eastern Han Dynasty. The Yongji Canal was opened by Emperor Yangdi in the Sui Dynasty, renamed Yuhe in the Northern Song Dynasty, and Weihe in the early Ming Dynasty. It has been in use until the 1970s, when navigation was interrupted for about 1800 years. The Weihe River is still an important tributary of the Haihe River system today, and it still plays an important role in farmland irrigation and flood discharge and drainage. Daokou town began from Li Jia Dukou in Song Dynasty and gradually multiplied here until it reached its peak in the late Qing Dynasty and the early Republic of China. It is a typical "Canal Ancient Town born of the canal and thrived by the canal". Therefore, the material space and intangible cultural heritage of Daokou town have its unique characteristics.

2.1 Analysis of the Layout of Daokou Town

The ancient town of Daokou was born and thrived by the canal. The Yongji Canal of the Grand Canal has a very important impact on the spatial form of the ancient town of Daokou. The Yongji Canal of the Grand Canal is winding about 8 kilometers from southwest to northeast in Huaxian County, and there are about 7 Wharfs in the 8 kilometers range. At the same time, the town form of Daokou town is also developed from southwest to northeast in a belt-shaped distribution(Fig. 1). In the early days of the founding of the People's Republic of China, a very influential paper mill was built along the canal in the northern part of the ancient town of Huaxian and even Henan, and a modern factory building was built in the later period. Although abandoned, it was carried out along the canal belt.

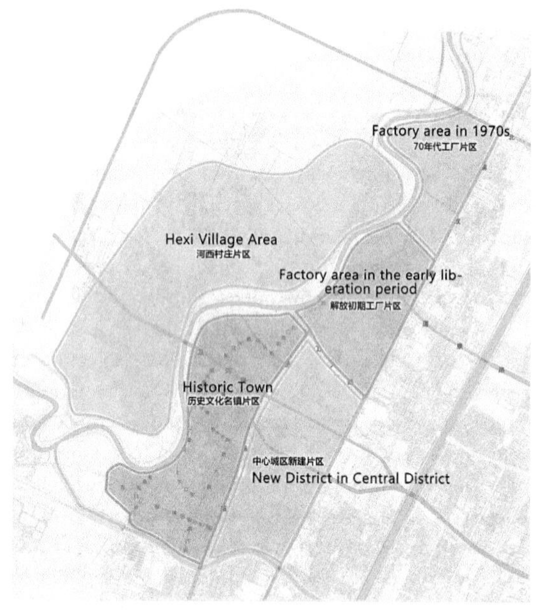

Fig. 1 Layout of Daokou town

2.2 Analysis of Streets and Lanes in Daokou Town

According to the historical map (Fig. 2), the streets and alleys of the ancient town in Huaxian County mainly include the system of streets and alleys commonly referred to as "Three Passes, Six Paves and Seventy-two Hutongs". The overall pattern of streets and lanes still retains the pattern of "Three Passes, Six Lanes, Seventy-two Hutongs". These streets and alleys are closely related to the canal and the wharf. Wharf Street, Daji Street, Shuihutong, Footsteps Street and Lion Lane all connect the wharf and are perpendicular to the wharf. At the same time, along Hebei Street, along Henan Street, along the North-South street and along the Grand Canal, the wharf has become an important node to organize the traffic of these streets and ancient towns. It is a system of streets and lanes closely related to the Grand Canal.

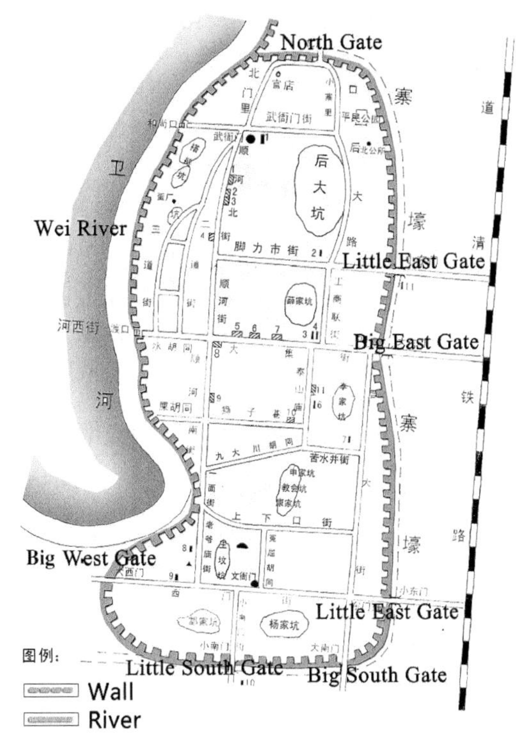

Fig. 2 Historical map of Daokou town

2.3 Analysis of Architectural Style of Daokou Town

As an ancient town of Commerce and trade, the architecture and courtyard pattern of Daokou town also has its unique characteristics. First of all, as an ancient town of Commerce and trade, the architecture and courtyard pattern is relatively free, which is essentially different from the traditional orderly and strict courtyard pattern. Secondly, as the Grand Canal extends from southwest to northeast, ancient town buildings and courtyards are spread along streets and alleys in turn. The main streets and alleys are arranged along Hebei Street and Henan Street in the North-South direction, forming a quadrangle pattern different from the traditional pattern of sitting in the north and facing the south. Moreover, most of the buildings in the ancient town are behind the building gables. This is because the Grand Canal is located in the northwest direction of Daokou town, and along Henan Street and Sandao Street are the main streets of the ancient town, while the courtyard is mainly facing the layout of the street behind the river. It is convenient to travel while avoiding the northwest wind blowing into the courtyard.

3 Analysis on the Types of intangible Cultural Heritage in Daokou Town, Huaxian County

3.1 Historical Analysis of Daokou Town

In the early Song Dynasty, the ancient town of Daokou was built as a spontaneous settlement, Lijia Ferry, because of the canal. In ancient times, the Yellow River flowed through the slippery land. Daokou was a ferry on the West Bank of the Yellow River. At that time, Li's family lived there by ferry, so it was called Lijia Daokou. In the Ming Dynasty, Daokou belonged to Liyang and had a delivery office, which was used to transport grain. It was called "Pingchuan Post". Since the Ming and Qing Dynasties, Daokou has gradually developed into a famous town with commercial parallels and ten thousand houses and scales. It is one of the four famous towns in Henan Province. This period is a prosperous period for the development of Daokou town. From the late Qing Dynasty to the Republic of China, the peak of Daokou town, the construction of Daokou-Qing Railway made Daokou become a transportation hub, with the reputation of "small Tianjin" and "important town in northern Henan". Daokou is water and dry wharf, known for its well-developed transportation. It is known for its "wheel shoe travelling" on the road, its "sails stand in the Weihe River", and its "rijin fighting" doctrine(Fig. 3, Fig. 4).

3.2 Types and Characteristics of Intangible Cultural Heritage in Daokou Town

The Yongji Canal of the Grand Canal has created a unique spatial form of Daokou town. At the same time, it has brought about nearly 100 intangible cultural heritage projects, which can be divided into four types: shipping culture, folk culture, handicraft culture, and catering culture.

Fig. 3 Weihe Shipping 1
Source: Dao Qing Railway Travel Guide during the Republic of China

Fig. 4 Weihe Shipping 2
Source: Dao Qing Railway Travel Guide during the Republic of China

3.2.1 Shipping Culture

Shipping culture is mainly the ancient town of Daokou, as an important node on the Yongji Canal of the Grand Canal, which is the northern shipping and transportation hub. These brought immigrants, chambers of commerce, wharfs, bathhouses, and other multicultures to the ancient town of Daokou, which created the exchange and integration of North and South cultures.

3.2.2 Folk Culture

Folk culture mainly refers to various folk art performances, martial arts, operas and other cultures in the north and south of the ancient town of Daokou. Among them, Daokou Ancient Temple Fair originated in the Ming Dynasty, with more than 20 kinds of folk art performances including the back-lifting pavilion, bamboo-horse dance, etc. Daokou Wushu was prevalent in the early Qing Dynasty, with 6 popular types. Daokou opera included 8 kinds of operas, among which Daping Diao, Daxian opera. The historical elements related to folk culture, including temple fairs, theatre buildings and martial arts halls, are mainly distributed on one side of the street. There are a large number of folk temples in Daokou town, such as the old site of Beidawang Temple, the old site of Caishen Temple, Lao Dai Temple, the old site of Bodhisattva Temple, Sanguan Temple, Lao Ye Temple, Jinglong Wang Temple, Nanding Lao Ye Temple, Bodhisattva Temple and Nanda Wang Temple. Most of these temples are water-related and distributed along the Weihe River. Many of these Wushu Museums are still in use today, and bring unique cultural characteristics to the ancient town of Daokou.

3.2.3 Handicraft Culture

Traditional handicraft culture is also very rich in the ancient town of Daokou, including 13 representative kinds. The historical elements related to traditional handicraft culture, including tin ware, gold and silver ware, are mainly distributed in Daji Street and Shunbei Street. The old names of tin ware mainly include Tongtai tin ware, while the old ones of gold and silver ware include Zhangji of Fengbao Building, Tiande Tower, Tianshengkui of Lujia, etc.

3.2.4 Dietary Culture

There are more than 12 kinds of special delicacies in Daokou town. Among them, Daokou roast chicken has a history of more than 300 years. Wangu mutton stew is known as "a tribute to the Central Plains". It is a provincial intangible cultural heritage. Historic elements related to dietary culture, including teahouses and restaurants, are mainly distributed in Daji Street and Shunhe Hebei Street; the old names of teahouses include Dujia Restaurant and Wanghe Tower Teahouse; the old names of roasted chicken include Tiancheng Homo-roasted chicken, Kuisheng Homo-roasted chicken, Yixing Zhang Homo-roasted chicken and Zhangcun Homo-roasted chicken; the old names of restaurants include

Dog Touhui Restaurant, Fourth Haiqing Restaurant, Drunken Immortal Residence, Qingyuan Building, Tongjia Restaurant, etc.

4 Opportunities and Challenges of Protection and Utilization of Daokou Town in Huaxian County

4.1 Analysis on the Opportunities of Protection and Utilization of Daokou Town

4.1.1 Analysis on the Development Trend of Cultural Tourism in Ancient Towns

The report on the market research and development trend analysis of Chinese cultural tourism (2013—2018) shows that during the 12th Five-Year Plan period, cultural tourism products will become the most competitive advantage products by 2015. With the development of world tourism, cultural tourism products will become one of the pillar industries of China's economic construction. The tourism development of ancient towns and villages can be divided into three stages. Its development process is a process from low-level to high-level, from tourism-oriented to experience-oriented. It can be roughly divided into three stages: cultural tourism, leisure vacation and cultural experience. The development stage of cultural experience is bound to revolve around the core issue of "excavating and highlighting cultural connotations". The Yongji Canal of the Grand Canal lays an important foundation for the transformation of intangible cultural heritage brought about by the ancient town of Daokou.

4.1.2 New Trend of Cultural Tourism Development

With the continuous improvement of social and economic level, leisure tourism has increasingly become an important part of social and economic development. According to the survey, when China's per capita GDP rises to a certain scale, people's consumption concept will change greatly. In this period, the middle class consumption concept will appear the trend of Anti-brand pursuit of personality and private customization. At the same time, with the development of Internet technology, there has been a wave of "mass entrepreneurship and mass innovation" in China. Internet technology makes anti-branding and private customization possible. Promoting industrial development through large-scale cross-organizational assistance of "building block innovation" is the preferred strategy for cultural heritage and derivation. The handicraft customization industry is not only the growth point of cultural and creative industries, but also the new economic growth point. With the support of the policy, through the introduction of technology marketing strategy, based on a large number of intangible cultural heritage projects in Daokou town, and a large number of intangible cultural heritage heirs, it is an important basis for nurturing "private customization", and more importantly, it is Dao. A new direction for the development of the export economy.

4.2 Challenge Analysis of Protection and Utilization of Daokou Town

4.2.1 The Location Disadvantage of Huaxian County Requires Excavating the Characteristics of Daokou Town

Henan province belongs to a large agricultural province in the central region of China, and its economic development lacks characteristics. Although the opening of Zhengzhou-Jinan high-speed railway (with sliding stations in sliding counties) will help to improve the traffic location of sliding counties in the future, the siphon effect of large cities such as Zhengzhou and Jinan requires that sliding counties must take their own development path. As a typical plain city in the northeast of Henan Province, Huaxian lacks characteristics of the urban industry and the motive force of urban economic development. Therefore, Daokou town, a historical brand bred by the Yongji Canal of the Grand Canal, has become a characteristic resource for the development of Huaxian County in Henan Province, and tapping its development potential has become an important issue for the development of Huaxian County.

4.2.2 The Development Requirements of

Daokou town Reflect Cultural Highlights

The overall architectural style of Daokou town is better, and because of the Eastward Development of Huaxian County Town, Daokou town can be preserved as a whole and its integrity is better. With the successful application of the Grand Canal and the continuous completion of the work of dredging the Grand Canal channel, the environment and internal infrastructure of Daokou town are better improved. With the continuous improvement of the overall landscape, the lack of important node space in Daokou Old Town has become increasingly evident. In addition to the Dawang Temple and Tonghe Yupiao, which are now restored, the more important Guojia Courtyard, Hujia Hospital, Nandawang Temple and Liu Courtyard in history have all disappeared.

5 Protection and Utilization Strategies of Ancient Towns from the Perspective of Non-hereditary Inheritance

5.1 Analysis of Protection and Utilization Ideas Based on Non-hereditary Perspective

Based on the numerous and distinctive intangible cultural heritage and taking the inheritance, study and innovation of intangible cultural heritages as the main clue, three pieces of "Daokou Old Town, Hundred Workers-settlement and New culturel land" are formed, which are based on the existing ancient town, the old factory building area and the industrial area being demolished. District. Daokou town area reappears the traditional folk custom deduction through sorting out the public space of the ancient town; Hundred workers — settlement gathers modern skills through cultural inheritance and development; New cultural Land realizes a new leap by integrating new culture and utilizing new technology(Fig. 5).

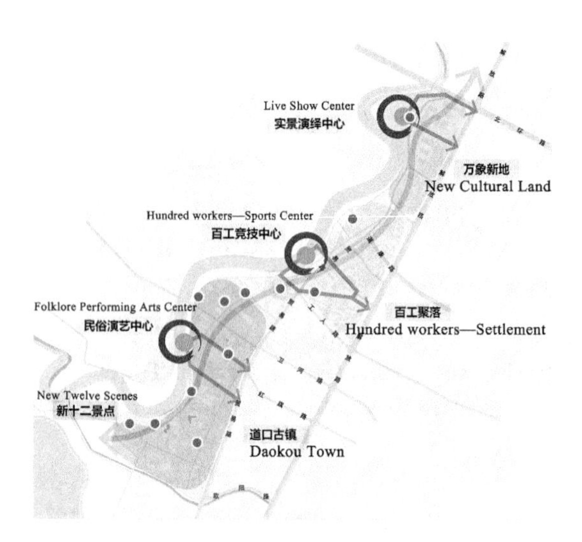

Fig. 5 Planning structure

5.2 Analysis of Protection and Utilization Strategies Based on the Perspective of Non-hereditary Inheritance

5.2.1 Spatial Distribution from the Perspective of Non-hereditary Inheritance

The planning scheme takes the river as its vein, forms a continuous waterfront activity interface through the organization of the upper reaches of the ancient canal, the setting of scenic spots and the elevation of the style and features; takes the street as the axis, arranges the streets into lines and dredges the lanes into nets, and shapes the image of the historic commercial street, while retaining the original street texture and combing it, connecting the dynamic activity space in series, and creating a pleasant environment. On this basis, focus on building three major nodes to promote the integration of Daokou town, Hundred Workers-settlement and New cultural land; focus on creating new 12 sceneries of Daokou, creating a new style of ancient town; and finally form the spatial structure of "taking rivers as veins, streets as axes, three districts as interaction, twelve sceneries as reflections".

4.2.2 Professional Layout from the Perspective of Non-hereditary Inheritance

(1) Daokou Town Area

Daokou town mainly takes the Canal Museum of historical relics as the entrance door of the ancient town, opening two tour experiences on water and land, mainly including the Canal Museum — Nanmen Square — City Wall Show — Visiting Nandawang Temple Fair — Touring Water and Drought Wharf — Listening to Quyi — Pin Old Brand Brand Brand Brand — Walking Streets and Lanes — Seeing Transportation

River Exhibition — Recreation Taobazaar and other nodes. The plan is based on the four unique cultures (folk culture, handicraft culture, catering culture and canal culture) of the ancient town. It can derive 15 types of business and 180 kinds of business shops.

(2) Hundred workers-settlement

In the product design theme of Hundred workers-settlement, there are two types: one is to inherit the authenticity of intangible culture, the other is to form new product types on the basis of intangible cultural heritage and by incorporating new technologies and elements. Hundred workers-settlement includes handicraft workshop area, craftsmanship area, skill inheritance area and creative service supporting area; finally, a cultural and creative journey of traditional handicraft "fingertips are not left behind" centered on Baigong workshop is formed, including "viewing craft creation — playing clay sculpture — learning silk art and embroidery — experiencing product release — Viewing martial arts" Linrongtai— listen to the pear garden in northern Henan — see the exhibition of non — relics — worship the sacred congregation and other cultural nodes. Finally, the local characteristics of the cities along the Northern Canal were implanted, and 112 types of shops were planned.

(3) New Cultural Land

New cultural Land mainly includes ecological agriculture planting area, comprehensive tourism service area, facility agriculture planting exhibition area, agricultural science, and technology experimental area, film and television media area, tourism service area. Through the impression of road crossing experience, the whole dimension of the three-dimensional experience of road crossing culture. Eventually, cultural nodes such as reading the history of railways, appreciating the impression of wheat fields, visiting farming paradise, popular science exhibition, tasting greenhouse restaurant, embracing science and technology greenhouse, visiting film and television base, playing through theatres, watching live performances, reading the history of railways, playing micro-film shooting and so on were formed.

5.2.3 Cultural Experience Strategies from the Perspective of Non-hereditary Inheritance

Through cultural integration, the story of the old road entrance is narrated from different angles such as shipping culture, market culture and street culture. The charm of the ancient town is appreciated while walking, and the nostalgic mood of the urban population is satisfied. Cultural experience includes the experience of shipping culture, market culture, street culture, etc. among them, shipping culture mainly reproduces the grand scenery of Ancient Canal waterway transportation with the ancient wharf, enhances the tourist vitality of Wanghe Tower, establishes canal tea house and bathhouse to form an experiential integrated node integrating sightseeing and participation, and market culture takes the cross ancient street as an integral whole. Build characteristic traditional Pedestrian Commercial street, restore traditional old brands and inject characteristic shops, merge into street performances, inject the commercial atmosphere of commercial vitality zone; Street culture refers to the transformation of part of the original street space to form a transparent street space, create wide and narrow streets, and undertake different responsibilities through different scale street space. Type of abortion activities.

5.2.4 Activity Organization Strategies from the Perspective of Non-hereditary Inheritance

Cultural activities include thematic activities, historical retrospectives and folk exhibitions. Among them, thematic activities include folk culture festival, water street lantern festival, historical retrospective activities include the Canal Museum, cultural corridor, and folk display activities include the revitalization of old brands, traditional art competitive performances and so on. At the same time, we should focus on contemporary artistic life, create an artistic atmosphere through non-heritage museums, Baigong Square, cultural corridors, cultural and art festivals or related forums, create a cultural and artistic life circle with local characteristics of Daokou, consider the development of ancient towns, and make the ancient towns glow young and glorious. In addition, a series of cultural festivals are planned in light of long-term and near-term planning. Three festivals are highlighted, including the traditional temple festival, the old-fashioned chicken Festival and the Baigong Creative Festival. Recently, relying on the current situation of Daokou temple festival, we mainly build traditional temple festival; relying on the brand of Daokou roast chicken diet culture, we build the old brand of roast chicken festival; in the long run, relying on the reconstruction of Hundred Workers-settlements, we focus on creating Baigong Creative Festival, publicizing the spirit of non-relics craftsmen, and improving the popularity of Daokou. These festivals and events are connected with the spatial nodes of Daokou town, which are presented separately in Daokou town and Hundred workers-settlement, thus effectively combining the cultural connotation of intangible cultural heritage with the historical space of the ancient town.

6 Submission

With the successful application of the Grand Canal, more and more attention has been paid to the protection and utilization of cultural heritage. On the basis of protecting the world cultural heritage of the Grand Canal and the cultural relics and historic sites of ancient towns, Daokou town combs the types and characteristics of intangible cultural heritage brought by the Grand Canal to ancient towns and makes the inheritance, study and innovation of intangible cultural heritage possible. The combination of the three stages and the spatial structure of the ancient town development has become an important way for the activation and utilization of Daokou town. Through sorting out the spatial layout of Daokou town, clarifying the layout of its format, rich and colorful cultural experience and organization of cultural activities, the revival of Daokou town is finally realized.

References

LI Jie, 2009. Historical geography examination of the rise and fall of Daokou town. Zhengzhou: Zhengzhou University.

LI Songsong, 2014. Research on street space and traditional architecture in ancient town of Daokou, Henan province. Zheng zhou: Zhengzhou University.

LV Zhou, 2015. Revision of the guidelines for the protection of Chinese cultural relics and monuments and the development of Chinese cultural heritage protection. Chinese cultural heritage, (2): 4-25.

YAO Zhanwei. A Survey of the rise and fall of Tonghe Yuyin. Kaifeng: Henan University.

ZHENG Ying, YANG Changming, 2012. Enlightenment of urban historic landscape-from historic city protection to urban heritage protection under the framework of urban development. Urban architecture, (8): 41-44.

Subtheme 3:

Contemporary conservation and technical innovation

Turf as a Roofing Material

Gisle Jakhelln

BOARCH arkitekter a.s., Lauvasbakken 12, Bodö Norway

Abstract Energy saving is the basis for all the vernacular buildings, for the building stage as well as for the later stages of maintenance and re-use. The chilling factor of the wind, the warmth of the sun and the use of local materials are traditional knowledge from our history to be applied for the future in our northern countries.

Turf as roof covering has a long tradition in the Nordic countries, going back to Neolithic times. It is still a living tradition in Norway, although less used now than 100 years ago.

Roofs were either constructed with rafters or with purlins. On top of these is the sarking of wooden boards or small tree trunks. On top of the sarking the birch bark is laid. The use of small branches or twigs is also known where there were difficulties to find birch bark, as in the mountains or out on the islands. Flagstones were also used where there was no birch bark available. The turf, cut from the ground in pieces as large as possible to carry by one person, is laid on top of the birch bark.

This paper presents the background and development of the use of turf as a roofing material. The geographical extent of the use of turf roofs is presented. Technical details are shown as examples.

The turf roof has qualities that appeal to the architects and builders of to-day. The cost is low and the maintenance is simple. The ecological element is strongly felt. To-day's architects in Norway, as well as in the other Nordic countries, are in a process of a deeper understanding of this heritage and of the critical climatic elements when designing buildings. The use of turf fits into this picture.

Keywords earthen architecture, turf roof, birch bark, Nordic countries, ecology

1 Introduction

Energy saving is the basis for all the vernacular buildings, for the building stage as well as for the later stages on maintenance and reuse.

The chilling factor of the wind, the warmth of the sun and the use of local materials are traditional knowledge from our history to be applied for the future in our northern country. Today's architects in Norway are in process of a deeper understanding of this heritage and the critical climatic elements as well as using local materials when designing buildings.

The roof is an important element of the building. Keeping the rain from penetrating the building and having good thermal insulation are necessities.

In this paper, I shall limit the presentation to the Nordic countries, however, with emphasis on Norway as this is the country I know best.

The Nordic countries include Norway, Sweden, Denmark, Finland, Iceland, the Faroe Islands and Greenland. (Greenland, having an arctic climate is not included in this presentation.)

The use of turf[①] as a roofing material has long traditions in the Nordic countries going back to the Neolithic times (Fig. 1). The use of local building material is the simple reason.

2 Topography and Climate

The Nordic countries lie between latitudes 54° and 74° N, and longitudes 24° W and 32° E with Denmark as the southern-

① I am here using the term turf. In England the term sod is used too. In Scotland they use turf, Icelandic: torf, Swedish: torv, Norwegian: torv, Finnish: turve, Danish: tørv.

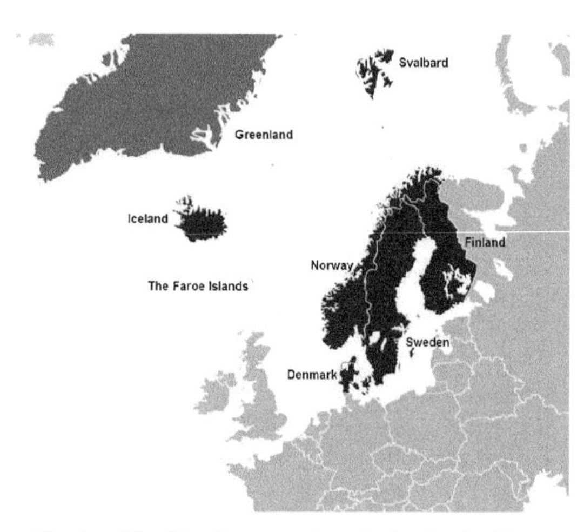

Fig. 1 The Nordic countries: Iceland, the Faroe Islands, Norway, Sweden, Finland and Denmark. Svalbard, as part of Norway is included and so is Greenland as part of Denmark © Wikipedia

most and Norway as the northernmost country; Iceland is in the west and Finland and Norway in the east.

The climate and topography vary very much within this geographically large area. Iceland, the Faroes Islands and the western parts of Norway and Denmark all have a marine climate. The Gulf Stream and prevailing westerly winds cause higher average temperatures and more precipitation than expected at such northern latitudes. Inland in Norway, there is less precipitation than on the coast. This is also the case for Sweden and Finland. The northern parts of Norway, Sweden and Finland have a subarctic climate.

Norway, Sweden and Finland are the westernmost areas of the Eurasian taiga dominated by conifer forest. Birch (Latin: Betula) grows in all these countries, especially along the coasts of Norway and Sweden.

With their large forests, Norway, Sweden, and Finland have a tradition of building in wood, structurally as well as for claddings. Vernacular buildings are adapted to these conditions with a wide variety of building types.

The growth of lichen, grass meadows and moors are common throughout the Nordic countries. This has been the base for using turf as the roofing material. Turf roofs were used on all kinds of buildings, dwellings, outhouses and boat sheds.

In Denmark the growth of reeds is common. This is also the case in the southernmost part of Sweden and to some extent in the south-western part of Norway. Thatching developed early and the use of turf was diminished especially in Denmark.

3 Cultural Context

Travelling by sea was the simplest way to move. The countries bordering the North Sea had close links. We were all part of the same culture from the Neolithic times. Finland was part of eastern cultures until Finland came under Swedish rule in the 11th BC.

The Saami people have been living in the northern parts of Norway, Sweden and Finland as well as in the north-western part of Russia from around 500 BC. Their culture was quite different from the farming communities in the southern parts of the Nordic countries.

4 A Long Tradition

The earliest known buildings were the tent. The sunken buildings developed early with loadbearing timber structures, walls and roof covered with turf. In the northernmost area in Norway this building type developed from the sunken type to a structure on the surface in the period 2000 BC to the time of Christ.

The traditional Saami house is the goahti (Norwegian: gamme), a small dome shaped timber structure, completely covered with turf(Fig. 2). The dome shape is perfect for energy saving in extreme weather with snow and wind.

The most common type of building in the Iron Age and early Medieval Age was a house built in timber post construction with the roof either with purlins or rafters and with turf as a roofing material(Fig. 3).

In Iceland, the turf building developed into large structures where the turf walls and the turf roofs became one large body.

Fig. 2　A Samii Gamme. Kjellingvatnet Gildesk l, Nordland County, Norway. Photo: Gisle Jakhelln 2008

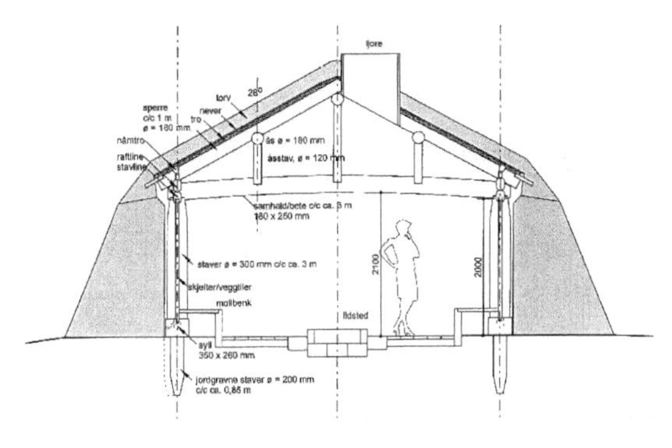

Fig. 3　Section of a turf house. Reconstruction of a small dwelling from 1100 AD near Bodö, Norway. Drawing architect Gisle Jakhelln 2017.

Here flagstones came into use to carry the turf as the wood was scarce.

With the development of the log construction for walls from 1000 AD this dominated the built structures in Norway, Sweden and Finland. The turf was kept as the roofing material(Fig. 4).

Fig. 4　19th Century farm house with a turf roof. Sjönstaa, Nordland County, Norway. Photo Björg Jakhelln 2015

5　Other Types of Roofing Materials

Stone/flagstones and slate came into use in the Medieval Age, mostly on small outbuildings. This was dependent on the suitable stone structure nearby. In the 18th BC slate became more common in the cities and eventually also on the farms. In some areas where there was scarcity of wood flagstones were used as a backing for turf, as in Iceland.

Reed, wheat or oat straw and broom were used for thatching. In Denmark the growth of reed is common. This is also the case in the southernmost part of Sweden and to some extent in the south-western part of Norway. Thatching developed early and the use of turf was diminished especially in Denmark.

Wood as roof covering material came into use quite early, mostly as backing for turf, but also used as boards laid parallel to the ridge as weatherboard or from the ridge down to the eaves. Wooden shingles were also used. Twigs and branches were used as well but mainly as backing for turf.

6　Roof Constructions

In Norway, the roofs with turf covering have a shallow pitch to prevent the turf from sliding down. The pitch varies from 20° to 27° with a maximum of 33° depending on from which district in Norway you are. On the coast, the pitch is steeper than in the inland valleys.

In Iceland, the roof pitch is much steeper, normally around 45 , less so in the Faroe Islands. This steeper pitch might have developed here because there is not birch bark available.

Roofs were either constructed with rafters or with purlins. On top of these is the sarking of wooden boards or small tree trunks.

On top of the sarking, the birch bark (Norwegian: never) is laid.

The birch bark is crucial for keeping the moisture from penetrating into the building. The bark is put down in four layers partly overlapping. In some districts, there were used up to six layers. The bark is laid with the outside part downwards and with the fibers following the slope of the roof to lead the water downwards. The bark protrudes 8～10cm from the edge of the roof. At the ridge, the pieces of bark should be sufficiently large to cover the ridge(Fig. 5).

The use of small branches or twigs is also known where there were difficulties to find birch bark, as in the mountains or out on the islands. This is the case in Iceland and the Faroe Islands. The rain penetrating the turf is led by the twigs downwards. The steeper roof gives this construction better protection from water penetrating into the building. Flagstones were also used where there was no birch bark available.

The turf, cut from the ground in pieces as large as possible to carry by one person, is laid on top of the birch bark. In some districts, the turf is laid in two layers. The first layer is always laid with the green side downwards, the second layer with the green side upwards. In this way, the roots will grow into a web tying the whole mass together.

At the eave, the turf is kept from sliding by a log fixed in position by naturally grown hooks, preferably juniper branches. (Juniper withstands the process of rot much better than fir or spruce.) These hooks are fixed to the sarking by nails or wooden pegs.

7 The Quality of the Turf Roof

The turf roof has qualities that appeal to the architects and builders of today.

The cost is low and the maintenance is simple. The ecological element is strongly felt with the use of natural and local materials, i. e. short transportation. It has good sound insulation, of importance in cities and near airports. Thermal insulation is of importance even if there has to be added additional insulation to follow today's by-laws. Rainfall is spread over time reducing the shock impact on gutters and drainage. The grass gives humidity to the surroundings

Fig. 5 Birch bark on top of wooden boards. The bark is overlapping with up to four layers. Kosmo, Misvaer, Nordland County, Norway. Photo Gisle Jakhelln 2006.

lowering high temperatures during warm summers. The maintenance is simple. The grass does not have to be cut. The growth of bushes should, however, be hindered as the roots would damage the birch bark. The birch bark may last up to 40 years before having to be replaced.

8 Today's Buildings with Turf

Nowadays, in the Nordic countries, as in Europe, the use of Sedum-growth on flat roofs is getting popular. Where there are high winds, as on the coast of Norway, this roof covering is too light and might be blown off the roof. Consequently, the Sedum roof is not used in these situations (Fig. 6, Fig. 7).

Fig. 6 Galleri Espolin, Storvagan, Lofoten, Norway. Architect, photo: Gisle Jakhelln 2005

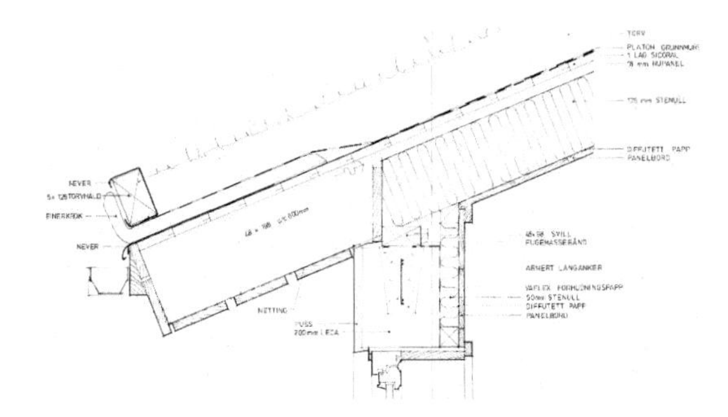

Fig. 7 Detail at the eave. Modern thermal insulation between the rafters, ventilated on top. Plastic membrane underneath the turf. Lauvasbakken 12, Bodö, built 1982. Architect Gisle Jakhelln

Today there are commercial products for turf roofs and roof gardens.

9 Conclusion

Today's architects in Norway, as well as in the other Nordic countries, are in a process of a deeper understanding of the heritage.

The use of turf as a local material and a deeper knowledge of the critical climatic elements makes the buildings more sustainable.

The use of turf as a roofing material is part of this picture.

References

PALMQVIST L, PETER S, 2006. August Holmbergs byggnadslära. Stockholm: Nordiska museet.

SCHANCHE K, 1994. Gressbakkentuftene i Varanger. Boliger og sosial struktur rundt 2000 f. Kr. Tromsø: Universitetet i Tromsø.

SILVÉN E, 2007. Sápmi on being SAMI in Sweden. Stockholm: Nordiska museet.

VREIM H, 1975. Lafte Ehus-tømring og torvtekking. Noregs boklag, Oslo. 5. utgáve.

Maintenance Concept of the Rammed Earth Finishing of the Historic City Wall of Pingyao Shanxi, Province, PR China—based on Re-Evaluation of Mock-Ups in 2007

DAI Shibing[1], *LI Hongsong*[2]

1 College of Architecture and Urban Planning, Tongji University, Shanghai, China
2 China Academy for Cultural Heritage, Beijing, China

Abstract One of the remarkable characteristics of the Ancient Pingyao City Wall, which has been listed as a Cultural World Heritage Site in 1997, is the internal rammed earth finishing. Based on laboratory tests, especially consolidation tests with various types of limes, it has been found out the loess soil, the main construction material, needs to be consolidated with a mixture of air lime and hydraulic lime to reach a level of water and frost resistance. "Earth on Earth" was considered as a sustainable and traditional concept to restore and maintain the earthen finishing. Two "earth on earth" options have been developed during the period of research framework in 2006-2007. Option 1 was to ram a new support layer with lime stabilized local soil. Hydraulic lime was added to improve the setting and hardening. Option 2 was to build a new adobe masonry but plastered with lime modified earth mortars as a protective layer. Based on the quality examination and visual evaluation in 2011, option 1 was chosen as the standard maintenance measurement to restore the existing wall. This technique has been more or less successfully implemented from 2012. Subsequent visual inspections have been done by authors in Sommer 2015 and 2018 not only to the restored areas, but also to mock-ups done in 2007. It has been found that the lime-earth plaster has functioned surprisingly well, it has protected the adobe from erosion and deterioration. It is therefore proposed that ramming is an effective and also traditional method to protect the rammed earth walls while retaining the internal compaction, but a protective lime-earth plaster to a rammed wall surface could be a sustainable solution to maintain the entire wall with very reasonable costs to avoid major restoration works in the future. This lime-earth plaster works as a sacrificial layer to protect the entire surface against erosion through regulating the moisture and thermal balance under the Pingyao climate.
Keywords Pingyao, rammed earth wall, maintenance, lime earth plaster, adobe

1 Introduction

One of the remarkable authentic characteristics of the Ancient Pingyao City Wall is the internal rammed earth finishing (Fig. 1), where the exterior was masonry subsequently around 1575 using natural stones, grey bricks and lime mortar. The Pingyao City Wall stands for centuries not only due to the construction technique, an extreme dry climate, but also through repeated reconstruction in history and maintenance. Under national and international conservation principles, continuous maintenance with traditional techniques is always encouraged.

This paper summarizes some important visual evaluation results done in the past a decade and provides some fundamental ideas on how to restore and maintain the Pingyao City wall that's not only technically correct, but also in a traditional and more economical way. Further research works are also proposed to monitor the historic city wall, but the quality and durability of implemented maintenance measurements.

2 Options to Restore the City Wall of Pingyao

Although many restoration campaigns and daily or weekly maintenance efforts have been done in the past three decades, parts of the city wall are still in danger of collapse (Fig. 1). In order to develop an effective and sustainable restoration and maintenance method for Pingyao city wall, in

2006, a research team has been organized under the guidance of the former Chinese Cultural Relics Research Institute (now China Academy for Cultural Heritage), the Bureau of Cultural Relics of Shanxi Province, Pingyao County Cultural Relics Bureau and other organizations. Comprehensive research works have been carried out to systematically analyze conditions and the causes of deteriorations. The quality of restoration done in the past maintenance practices was also evaluated. One of the targets was to find an optimum receipt and a quality control system to improve the stability of the earth for restoration while maintaining the colour and texture of interior finishing.

Fig. 1 Internal rammed earth finishing of the Historic Wall of Pingyao (2018): The areas covered by plastic sheets are in danger of collapse

Laboratory tests show that the loess soil, the main construction material of the Pingyao city wall, is poor in clay. It is not water-resistant without any treatment. First of all the stability of earth finishing is affected by erosion on the surface or along cracks (Fig. 2) or joints. During rainy seasons, the loess soil can easily be eroded. The second deterioration factor is the leakage through cracks especially along parapet walls and capping pavements. The third factor is the salt damages due to high evaporation.

There were three options to restore the interior earthen finishing. The first one was to build a masonry with kilned bricks. The second one was to chemically consolidate the soil in situ. The third one was to re-ram a new supporting wall to old earth core, or simplified "earth on earth" option.

Fig. 2 Erosion of earth and repeated maintenance (2006)

Rebuilding with kilned brick masonry would have changed the entire finishing tremendously. It was used previously only as an emergency measure in or shortly before winter season when new rammed earth was not technically applicable. Furthermore, water can still penetrate through the brick joints into earth construction and cause damages, like delamination between brick layer and earth core, which are not so easily detachable.

In situ reinforcement or chemical impregnation would be a possible solution. This technology has been successful in the conservation of non-structural archaeological earthen sites[6,7], but the field experiments in Pingyao inner earth finishing showed that the maximum penetration depth of the impregnation reinforcement agent (ethyl silicate in 99% concentration) was only 150mm, which is insufficient to protect the existing rammed earth surface like the eroded earthen finishing shown in Fig. 2.

The third option, "earth on earth" option, consists of ramming a new layer of earth construction to protect existing core walls at the same time, to maintain the colour and the texture. Historically, especially during the 1977—1999 period after flooding in 1977, most collapsed walls were restored using this "earth on earth" option. But from 2000, collapses happen even more often during summer or in the early spring seasons. The main tasks of the research framework in 2006 and 2007 should specify sustainable methods to consolidate and maintain the integrity of the city wall of Pingyao. To do so, first of all it needs to clarify the optimized formula for improving the earth and the system of quality control on site.

3 Construction of Test Areas in 2007

3.1 Type and Quantity of Lime for Loess Soil Treatment

After comprehensive research works done in 2006, it has been found out the loess soil, the main construction material, needs to be consolidated with a mixture of air lime and hydraulic lime to reach a satisfied level of water and frost resistance. As an exposed surface like a building fa ade, the compacted traditional "three seven lime earth" mixture, i. e. the mixture is composed of 30wt% air lime, 70wt% loess soil, does not harden properly, and shows less durability. This formulation is also economical because of the high dosage of lime. A mixture of two kinds of building limes, quick air lime powder (3~4wt%) and natural hydraulic lime (3~4wt%) has been proven to be more effective to set and harden under the conventional compaction technique and local workmanship.

The two mock-ups of "earth on earth" option were done in 2007 in the Eastern Wall. The first mock up wall was rammed with lime improved earth. The second one was to build a supporting wall with adobe plastered subsequently with lime-earth mortars (Fig. 3).

The mock-ups were done by the same local workers who had been involved in the daily maintenance works. The mock-ups

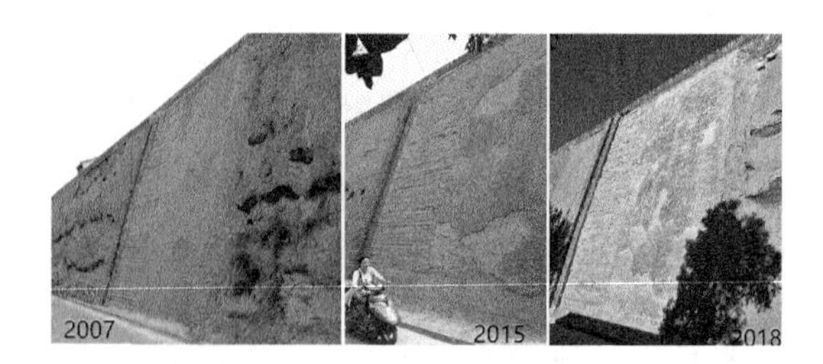

Fig. 3 Mock-ups finished in 2007 and inspection in 2015 and 2018

were so planned to test on the one hand durability of developed formulation and to Fig. out quality controlling factors, on the other hand whether the developed method is able to be implemented under local workmanship and management system.

3.2 Mock up with Rammed Earth

The empirical practice showed the old deteriorated soil could not be treated properly to reach the required water resistance. Therefore the deteriorated earth was removed to a depth of approx. 800 mm to the core of the city wall. Between the old earth core and new wall, a layer of air lime slurry was applied to minimize the shaking impact to the old earth during the mechanical compaction.

The lime-soil mixture was produced as follows: the fresh soil was excavated from a soil mine app. 15km from the city wall and dried through natural wind and sieved. The sieved soil was premixed dry homogeneously with quick lime powder (3~4wt%) and natural hydraulic lime (3~4wt%) in a warehouse within the ancient city app. 500m away from the mock-up site. The dry mix was transported to the site. On site, app16~17wt% water was added to the lime earth mixture to get optimum compaction. The lime-soil had been compacted to the highest dry density and subsequently moistured every day during the entire construction period even one week after for proper curing.

3.3 Mock up with Adobe and Earth Plasters

The second mock up wall was built with a lime-earth adobe. As a finishing, a 30 mm thick lime earth mortar was applied as "protective" or "sacrificial plaster".

The basement was so prepared as the rammed mock up. The adobe was produced with a conventional brick production line. The earth was mixed with lime as the same formulation for compaction and then extruded and stored for 2 weeks under high humidity within a plastic sheet.

The bedding mortar of the adobe wall was composed of 30wt% air lime and 70wt% local soil. Immediately after the finished construction of adobe, two layers of protective plasters were applied onto the entire adobe surface. The base plaster was composed of 30wt% air lime and 70wt% soil, the same as the bedding mortar, while the top plaster 10wt% air lime and 90wt% soil. The lime-earth mortar was firstly dry-mixed, then water was added to get a workable paste and finally the mortar paste was splashed onto the wall with hand. The rough plaster was trowelled flat before it started to set. The upper layer of lime-earth plaster was impregnated with silicone resin solution (app. 5wt%) one week after completion of plastering with the intention to improve the rain water resistance.

4 Visual Evaluations of Mock Ups and Completed Restoration Areas

After 4 years of natural weathering, in November 2011, visual inspection and test of the strength of rammed lime earth walls have been done to the rammed earth mock-up. All inspections have given positive results. The colour of the rammed earth walls is almost the same as non-treated soil. The maintenance specification as stated in the final report submitted in 2007 by the working group has been confirmed. From July 2012 part of the inner earth wall classified as "dangerous portions" are being restored with the lime (quick air lime and natural hydraulic lime) earth mixture rammed in the traditional way.

The mock up with adobe and lime earth plasters was not evaluated in 2011 because it was not considered as a good option to maintain the entire city wall based on the final report.

A renewed visual inspection has been done in Sommer 2018 (Fig. 4). It has found out that the rammed earth surface was still in good condition, although there was obvious surface deterioration. App. 5～30mm soil has been eroded based on the height of glass fiber, which was used with the intention to lower the surface tension.

Fig. 4 Visual Inspection of mockups near in summer 2018

It has also been found the lime-earth plaster to the adobe wall in the test area was surprisingly well preserved (Fig. 5).

There was the only app. 20%～30% of earth plaster has been peeled off after 12 of years exposure to natural climate. The delamination occurs both between the adobe and base plaster, some delamination has occurred between the base and top plasters. The adobe underneath the lime-earth plaster was relatively well preserved.

The colour of both mock ups is slightly different. The plaster looked darker because more dust were accumulated on the surface.

The most damaged plaster occurred at

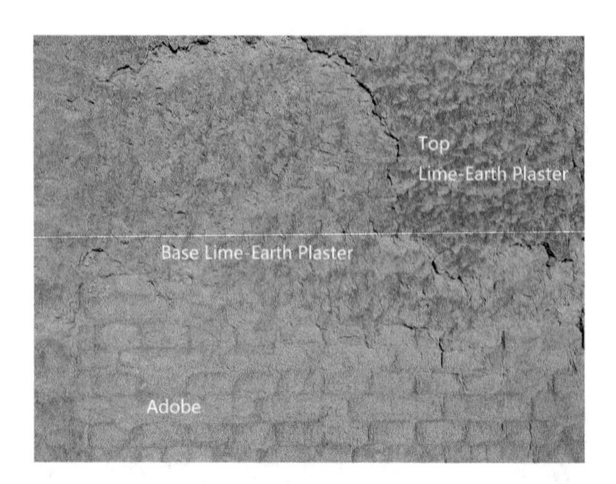

Fig. 5 Condition of plastered adobe near most critical area completed in 2007 in details (2018)

the top and near the foundation. The water repellent impregnation with silicone solution seemed to have not functioned as wished.

5 Conclusion

It has been concluded that the optimum dosage to stabilize Pingyao loess soil is the mix of quick air lime and natural hydraulic lime. Air lime alone is not effective enough to consolidate the soil with a high content of silts to reach the required strength and durability. The total lime content shall be not higher than 10% to match the colour of not stabilized the soil. The optimum dosage in the maintenance practices depends on the clay content of the soil. The higher silt content or the lower the clay content, the higher content of natural hydraulic lime is needed to be mixed to reach sufficient strength and water resistance. The high content of water even after compaction gnarantees the setting and strength development of lime-soil.

Ramming is an effective method to protect the rammed earth walls while retaining the internal compaction. But the completed areas during 2012—2015 show certain surface deterioration or cracking after a few years' exposures based on visual inspection done in 2015 and 2018 (Fig. 6). Those kinds of surface defects might not be so extensive to affect the stability of the entire wall, but the causes need to be further studied. From our point of view, a maintenance measurement might be necessary.

The plaster with lime-earth mortars to the adobe wall was surprisingly well preserved. The adobe underneath the lime plasters showed less deterioration even 12 years exposed to the natural climate.

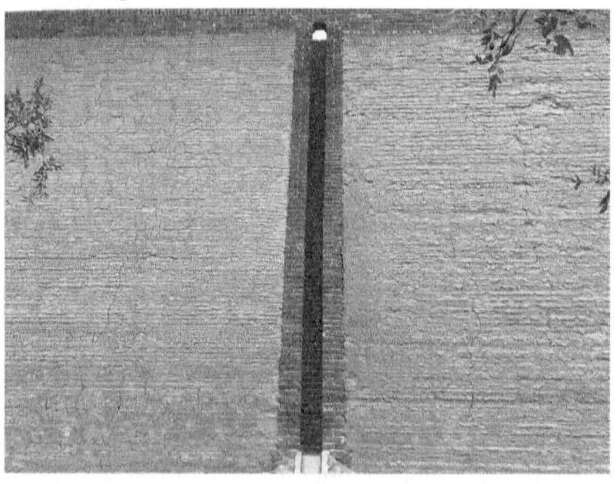

Fig. 6 Rammed earth with cracks and surface deterioration approximately 3 years after restoration (2015)

Although to construct the adobe wall is economically not reasonable, a protective lime-earth plaster to rammed wall surface could be done from both the technical and the economical point of view. It could be sustainable solution to maintain the entire wall with very reasonable cost. This lime-earth plaster works as sacrificial layer to regulate the moisture thermal balance and even as a barrier against air borne pollutants, and finally protects the entire surface of the city wall against erosion.

The lime plaster could be applied to the rammed earth wall when cracks and clear surface deterioration occurs. Once the rammed surface deteriorates, there is less appearance difference between rammed finishing and rough lime-earth plaster. Such lime-earth plasters can be repeated every 10～15 years to avoid the most expensive major restoration work.

After submitting the final report by the China Academy for Cultural Heritage in 2007, no further research works have been carried out. The quality-related researches are necessary not only to monitor the historic city wall, but also to understand the quality and durability of implemented maintenance measurements. A small laboratory on site to test some basic parameters, like density, water content, under the technical support of a scientific committee is necessary.

The preparation of lime-soil mixture during the mock-ups in 2007 was done in an abandoned warehouse very near the site, but today, all preparation of soil and lime has to be done app 15 km away from the city wall. This distance may influence the quality of compaction since the reaction time of the lime-soil mix is limited. Based on the preliminary internal ongoing laboratory test results, the reaction time of natural hydraulic lime and formulated wind-slaked lime have high strength after 4～6 hours reaction, while portland cement lime mix has maximum reaction time of 1 hour.

Acknowledgements

Special thanks to Dr. Oliver Kuhl and Dr. Wilmers from Wetzlar Soil and Building Materials Center in Germany for their technical supports during both laboratory and site researches. The authors appreciate the collaboration works of Mr. Gu Jun from China Academy for Cultural Heritage, the team from Shanxi Province Cultural Relics Bureau, Mr. Jia Zhongzhao and Mr. Li Shucheng from the Pingyao County Cultural Relics Bureau. The authors thank also Mr. Zhang Debing for the project management and local contractors and workers in Pingyao who have shared valuable experiences.

References

Central Design Institute, 2006. Supplementary investigation report on the status of the city walls of Pingyao city, Shanxi province. Beijing (unpublished report).

China Academy for Cultural Heritage, 2007. The master design plan of the city wall reinforcement of Pingyao ancient city in Shanxi Province. Beijing (unpublished report) .

China Metallurgical Group Building Research Institute, 2005. Report on stability evaluation of Pingyao ancient city walls. Beijing (unpublished report).

DAI Shibing, LI Hongsong, 2016. Experimental study on the diagnosis and restoration of the rammed earth surface of the Pingyao city wall. Heritage architecture, 1(1):122-129.

Jinzhong Annal Research Institute, 2002. Annals of Pingyao ancient city. Beijing: China Book Bureau.

SUN Manli, WANG Xudong, LI Mostmale, 2010. Introduction to conservation of earthen ruins. Beijing: Science Press.

WANG Xudong, 2008. New progresses on key technologies for the conservation of Chinese earthen sites in arid environment. Dunhuang research, 6: 6-12.

Open-Ended Reconstruction: A New Approach to the Conservation of the Wooden Architectural Heritage in East Asia

HAN Pilwon

Department of Architecture, Hannam University, Republic of Korea

Abstract It is not rare in East Asia, in China, Japan and Korea, to face controversy and confusion raised from the protection and management, especially from the reconstruction of the architectural heritage consisting of wooden structures that were demolished long ago. When there are not enough reliable documents or materials left on upper structures of the architectural heritage, as in most cases, the overall reconstruction causes a most serious problem undermining the values and authenticity of the heritage.

Based on the related researches on the traditional East Asian architecture, in this paper, the wooden architectural heritage in East Asia is characterized by sustained change, which is attributed to the vulnerability of wooden buildings, to fire, humidity, insects, etc., and the notion of "architecture as process" historically shared by East Asians. The sustainability inherent in the East Asian wooden architectural heritage is due to ever-changing quality by means of the continual replacement of building members and the mobility of the building.

Based on these characteristics of the traditional East Asian wooden architecture, this paper suggests "the open-ended reconstruction" as a new approach to heritage conservation that accords with the characteristics of the wooden architectural heritage in East Asia and respects its values and authenticity. Unlike the way of reconstruction that brings the whole architectural ensemble back at once to a specific time or that follows the process undermining or distorting the values and authenticity of the heritage, especially the heritage landscape, the open-ended reconstruction is a stepwise reconstruction following again the steps of development that have generated the heritage, and adopting adjustable and reversible methods, which contributes to the sustainability of the heritage in terms of value and authenticity.

Keywords wooden architectural heritage, East Asia, cultural and heritage diversity, architecture as process, open-ended reconstruction

1 Conflicts in Conservation Practice Between Universal Principles and the Local Needs

1.1 Universal Principles for Reconstruction

In The Venice Charter that has guided the activities around cultural heritage management, the term of "reconstruction" appears just once; "All reconstruction work should however be ruled out a priori. Only anastylosis, that is to say, the reassembling of existing but dismembered parts can be permitted." The Charter was drafted in 1964 by 23 persons, and none of them was Asian. Mr. Hiroshi Daifuku from UNESCO was a Japanese American who was born in Honolulu, Hawaii, 1920. Presumably the features of the East Asian architectural heritage were not considered in drafting the Charter.

And in The Burra Charter (2013), reconstruction is defined as returning a place to a known earlier state and is distinguished from restoration by the introduction of new material. According to this, the reconstruction stated in The Venice Charter falls under the definition of restoration in The Burra Charter.

The universal guidelines for legitimate reconstruction are clarified in the Paragraph 86 of *Operational Guidelines for the Implementation of the World Heritage Convention* (WORLD HERITAGE CENTRE, 12 July 2017); "In relation to authenticity, the reconstruction of archaeological remains or historic buildings or districts is justifiable only in exceptional circumstances. Reconstruction is acceptable only on the basis of complete and detailed documentation and to no extent on conjecture."

Considering the general situation that old East Asian wooden architectures were demolished not to leave such "complete and detailed documentation", it seems to be almost impossible to conform with these universal principles and guidelines in re-constructing East Asian wooden architectures.

1.2 Conservation Practice and the Conflicts Between Universal Principles and Local Needs

The national and international principles and guidelines for heritage conservation do not always agree with each other, especially in terms of reconstruction. When it comes to a World Heritage Site, however, the universal international ones quoted above are usually being applied. Complying with those universal ones, many archaeological sites in East Asia that lack "complete and detailed documentation" of demolished wooden architectures remain empty, often as lawns, at most with base stones, protecting underground relics.

The local communities and governments that generally have needs to leverage the cultural heritage to make socio-economic benefits, regarding it as a critical factor to boost tourism, tend to be unsatisfied with this kind of heritage conservation. It is because they think such heritage sites without any historic buildings have limitation in attracting tourists since they are hard to be understood and appreciated by visitors. They doubt whether such conservation practices leaving huge vacant spaces without functions in the life of local communities can be sustainable.

In this way in East Asia, the community concerns are not reflected in the measures of heritage conservation. This often leads to disputes between experts and local communities/governments. Especially in the controversy over the reconstruction of cultural heritage buildings, the characteristic of the East Asian wooden architecture is not yet fairly considered. In this situation, the community participation that is recently emphasized worldwide in the practice of heritage conservation is discouraged too.

These kinds of conflicts, between universal principles and guidelines and local needs, and the consequential alienation of local communities in the discourse and practice of heritage conservation lead us to the issue of cultural and heritage diversity.

2 The Diversity of Culture and Heritage

2.1 The Discussion of Diversity in the Nara Document on Authenticity (1994)

The Nara Document on Authenticity is a document to which most of the discourses on cultural and heritage diversity refer. It affirms that "the diversity of cultures and heritage in our world is an irreplaceable source of spiritual and intellectual richness for all humankind." According to it, "all judgements about values attributed to cultural properties as well as the credibility of related information sources may differ from culture to culture, and even within the same culture. It is thus not possible to base judgements of values and authenticity within fixed criteria. On the contrary, the respect due to all cultures requires that heritage properties must be considered and judged within the cultural contexts to which they belong."

The Nara Document implies that the characteristics of cultural heritage and the values associated with them in every geo-cultural region should be respected. In line with The Nara Document, the best practice by the people of the region for sustaining the heritages that their ancestors generated, which is part of culture, should be considered important in developing appropriate methods of heritage conservation to the region. The critical issue in this argument will be how such regionally valid conservation methods, not undermining the universal value and authenticity of the heritage regionally produced, can be formulated.

2.2 The Discussion of Diversity in Nara+20: On Heritage Practices, Cultural Values, and the Concept of Authenticity (2014)

Nara+20 was adopted by the participants at the meeting on the 20th anniversary of The Nara Document on Authenticity, held at Nara, Japan, from 22-24 October 2014. It identifies five key inter-related issues; diversity of heritage processes, implications of the evolution of cultural values, involvement of multiple stakeholders, conflicting claims and interpretations, and role of cultural heritage in sustainable development. This paper explores these issues of Nara+20 further with special reference to the wooden architectural heritage in the East Asian context. By being based on the characteristics of heritage as well as understanding the universal philosophies of heritage conservation, the regionally valid and universally approvable methods of reconstruction can be formulated. They will contribute to build consensus among different interest groups, as well as to sustain the values and authenticity of heritage.

3 The Characteristics of the Wooden Architectural Heritage in East Asia

Based on the related researches by author, it can be asserted that the characteristics addressed below are distinctive in the East Asian wooden architectural heritage and they are part of its cultural values.

3.1 Vulnerability and Sustainability

While the traditional East Asian architecture structured by a wooden frame has merits — it is basically environmentally friendly and strong against earthquake, and can be constructed relatively easily and fast, it has critical shortcomings — it is vulnerable to fire, humidity, insects, etc. Because of this vulnerability, East Asia has not such old buildings as in Europe. In East Asia, the age of a wooden building does not mean that of whole members. The wooden members are usually not so old as the design or composition of the building. The building, its design and composition, could be sustainable only through continual replacement of its members. Interestingly here, the notion of sustainability is closely linked with change.

3.2 Architecture as Process

In East Asia, the traditional architecture has not only the notion of space but that of time. The traditional wooden architecture in East Asia grows and declines. Also the wooden frame building is a kind of prefabricated structure, modified continuously being adapted to changing needs. Traditionally the East Asian architecture has been an ongoing open-ended process, far from completeness or self-containment.

The architecture has its own module and pattern of growth, and they vary with region; the basic unit of growth in the Japanese architecture is a ken or a space defined by four columns, that in Korean architecture is a pair of a building (chae) and a yard (madang), and that in the Chinese architecture is a yuan, a yard and a set of buildings enclosing it.

Thus the architecture may change over time. As The Nara Document recognized, in the wooden architectural heritage in East Asia, the heritage and its context evolve through time. Accordingly the values attributed to this kind of heritage may not be fixed but dynamic. This feature of heritage as an ever-changing organism brings out a question on the authenticity as well as the universal value; can they accommodate the evolution of heritage?

3.3 Architecture of Mobility

The building was not been regarded as permanent nor fixed on one site in East Asia. In Korea, some houses were rebuilt nearby after dwelling of several generations. The main buildings of Dongchun's Residence in Daejeon city, a state-designated cultural property of Korea, for example, returned to their original places from nearby sites where they were located for a period of time. (Fig. 1 ~ Fig. 4) In Japan, shrines switch their places periodically, which is called shikinensengu. Isejingu, a representative shrine, has repeated its new construction of the same building at a nearby site every twenty years. The latest construction was in 2013.

This kind of mobility of architecture leads us to understand that the relation of the building and the site might not be physically fixed in East Asia, and to consider the pattern/steps of growth or change in the heritage conservation practice.

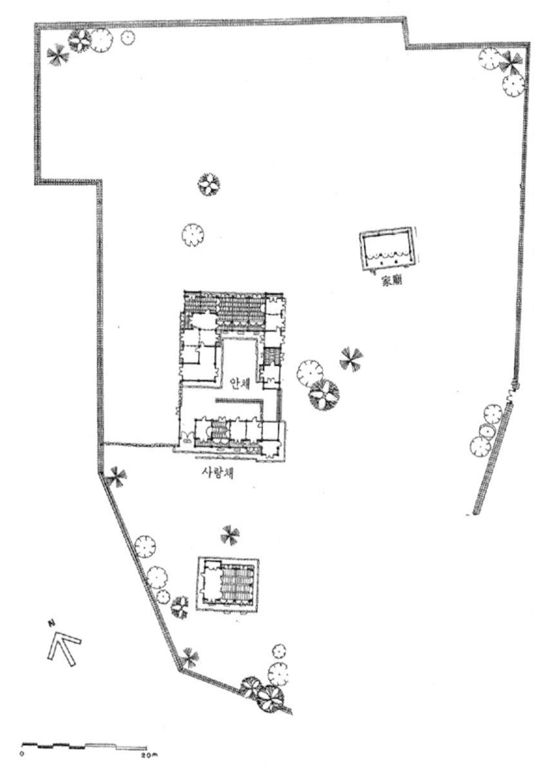

Fig. 1 Site plan of Dongchun's Residence c. 1617 (Courtesy of Daejeon city)

4 Open-Ended Reconstruction as a New Approach to Heritage Conservation

Considering that the reconstruction in proper of wooden architectural heritages is at the center of controversy over the practice of heritage conservation in East Asia, it is critically needed to have an alternative way of reconstruction as a new approach to heritage conservation. Based on the examination on the characteristics of the traditional East Asian wooden architecture, "the open-ended reconstruction" is suggested here as a new approach to heritage conservation that can integrate different recognitions and needs of diverse interest groups on

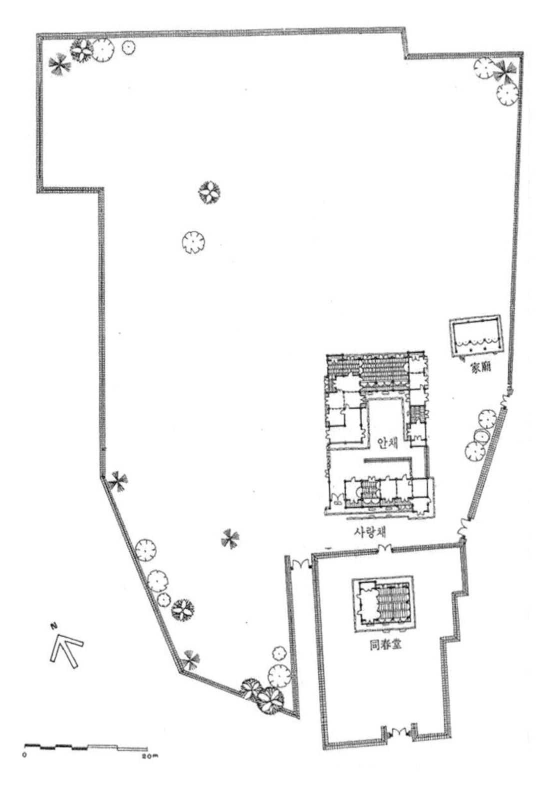

Fig. 2 Site plan of Dongchun's Residence c. 1744 (Courtesy of Daejeon city)

heritage and its conservation. As an alternative way of reconstruction, it is defined as returning a heritage to an earlier "stage" of its development using adjustable and reversible methods.

4.1 A Stepwise Reconstruction Following Again the Steps of Development of Heritage

As can be seen in a recent reconstruction project of the Imhaejeon compound situated within the Wolseong Belt, one of the five components comprising the World Heritage Site "Gyeongju Historic Areas" (Fig. 5), the reconstructions in East Asia are often planned and promoted stage by stage. However the order of stages, which is usually proposed considering preparedness of reliable documentation and/or technology required, feasible construction process, and/or fund supply, is controversial.

In case of Imhaejeon, it was proposed to reconstruct firstly the main building considering that the documentation of it has been prepared for a long time based on archaeological excavations and literature re-

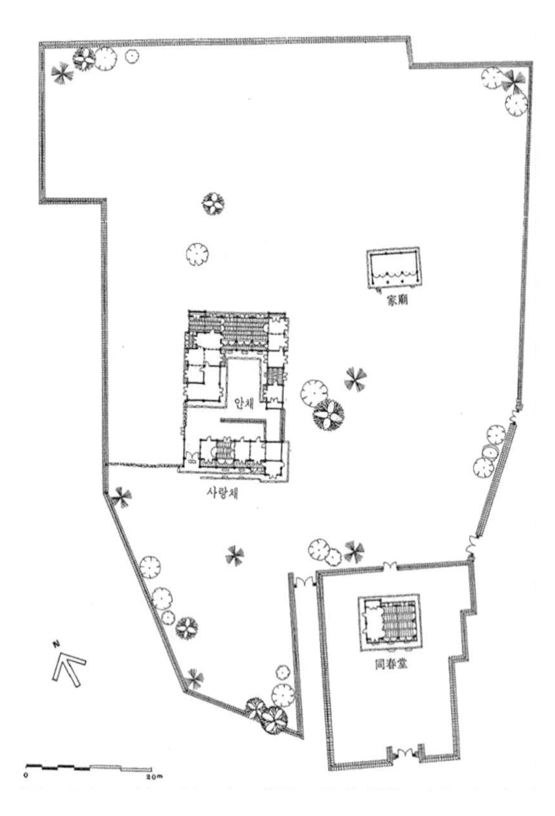

Fig. 3 Site plan of Dongchun's Residence c. 1649 (Courtesy of Daejeon city)

searches and it is the most symbolic building in the site. However, besides for not yet having detailed complete evidences about the upper structures and appearance of the main building, this reconstruction plan is criticized in that the reconstruction of the main building alone can distort the landscape of the site, which is an important attribute of its Outstanding Universal Value.

A good way to avoid this criticism and sustain the OUV and authenticity of the heritage is deemed to reconstruct the lost buildings stepwise following the steps of the development of the heritage. It is deemed that the landscape of a site will not be distorted or undermined seriously whatever stage the reconstruction may return the site to.

4. 2 A Reconstruction Adopting Adjustable and Reversible Methods

As stated in Nara+20, the recognition of the evolution of cultural heritage has created challenges for heritage management and questions on the validity of current uni-

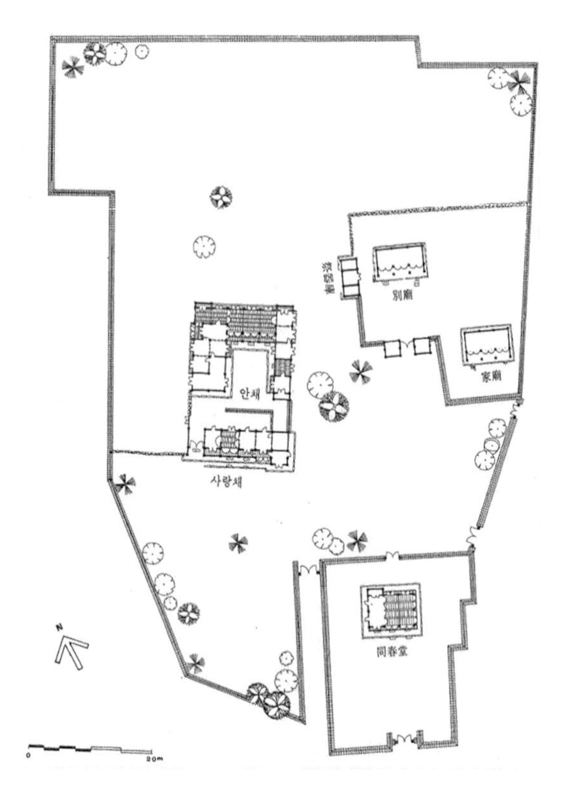

Fig. 4 Site plan of Dongchun's Residence since 1835 (Courtesy of Daejeon city)

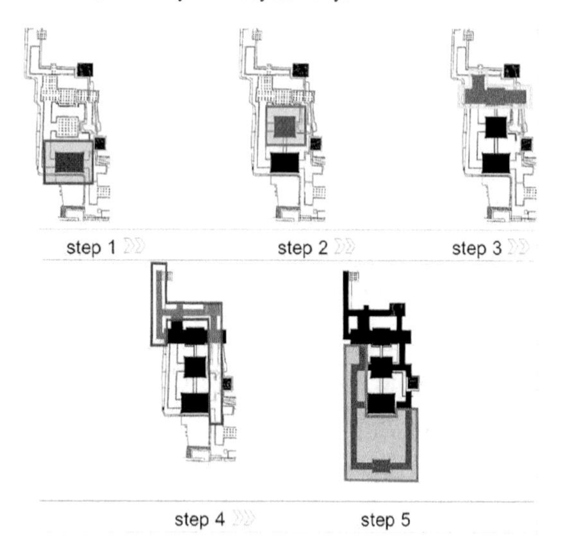

Fig. 5 Proposal for the steps of reconstruction, Imhaejeon compound in Gyeongju Historic Areas (Courtesy of Taechang Architects)

versal conservation principles and guidelines. Such principles and guidelines are needed to be reviewed with the widened recognition on the values and authenticity of heritage from understanding of the cultural characteristics of the wooden architectural heritage in East Asia. And, if appropriate, they need to be extended to be inclusive enough to accommodate diversities.

However this does not mean any give-up of the values and authenticity of heritage. Acknowledging that there are difficulties in obtaining "complete and detailed documentation" on the wooden architectural heritage in East Asia and it should be continuously corrected and complemented through relevant archaeological and documentary researches for a long period of time, a reconstruction may be justifiable when it is based on reliable materials enough to return the landscape of the heritage, not all the parts or members of every building, to an earlier stage without any distortion. Here the reconstruction does not always mean the overall re-building of a whole heritage building. If appropriate, it can be reconstructed partly, for instance only the structural frame including roofs can be reconstructed and the rest parts of the building may wait to be reconstructed until complete documents for it are prepared. And as the relevant researches go further, the previous reconstruction may be corrected based on more detailed and exact evidences.

To do this kind of open-ended reconstruction, the methods of construction, structural system and building technics, should be adjustable and reversible for enhancing the authenticity without negative impact on the state of heritage conservation.

5 Conclusion

On one hand acknowledging the importance of sustaining the values and authenticity of heritage in the conservation practice including reconstruction, and on the other hand, recognizing the gaps, conflicts and troubles among interest groups over understanding and interpretation of heritage and its conservation, the open-ended reconstruction is suggested in this paper as an alternative way of reconstruction in the site that has lost wooden buildings. The open-ended reconstruction is a new approach to the heritage conservation that is defined as returning a heritage to an earlier "stage" of its development using adjustable and reversible methods. This new methodology of conservation is proposed based on the understanding of cultural and heritage diversity and the characteristics of the traditional East Asian wooden architecture.

This new approach to heritage conservation is expected to contribute to integrating different recognitions and needs of diverse interest groups on heritage and its conservation and encouraging local communities to participate in heritage conservation.

References

Australia ICOMOS, 2013. The burra charter: The Australia ICOMOS charter for places of cultural significance.

ICOMOS, 1965. International charter for the conservation and restoration of monuments and sites (the Venice charter).

The Agency for Cultural Affairs of Japan, 2014. Nara+ 20: on heritage practices, cultural values, and the concept of authenticity.

UNESCO World Heritage Centre, 2017. Operational guidelines for the implementation of the world heritage convention.

UNESCO World Heritage Committee, 1994. The nara document on authenticity.

Earth-Fiber Mixes for Natural Building Products

Maddalena Achenza

Università degli Studi di Cagliari, DICAAR_Dipartimento di Ingegneria Civile, Ambientale e Architettura, Italy

Abstract The paper deals on an experimental research led at DICAAR_Dipartimento di Ingegneria Civile, Ambientale e Architettura of the University of Cagliari (Italy) on natural composite materials based on earth and vegetable fibers. The research is mainly aimed at the study and development of new building elements: finishing plasters and thermal/acoustic insulation panels to be used for the retrofitting of historic buildings. Some preceding experimentations conducted by the same research group were concentrated on mixes of different fibers (straw, hemp, rice husk, loofah and jute) mixed with clayey earth. Panels produced with the different fibers resulted to be very promising products for thermal and acoustic insulation.

The current experimentation concerns the mixture of two types of cellulose flakes, resulting from industrial production, with a local earth. The two types of cellulose used for the experimentation come from a recycling process of newspapers. The first type is in raw form; the second one has undergone a whitening process. The tests were aimed at evaluating the characteristics of aesthetic, porosity and resistance to water contact of the samples. The obtained results offer encouraging bases for the development of innovative and efficient natural plaster mixes, to be used both for historic and new buildings.

Keywords cellulose and earth mixes, natural plasters, thermal and acoustic insulation, earthen historical building retrofitting, circular economy

1 Introduction

The contribution refers to an in-depth investigation that is part of an experimental research based on the study of earth and vegetable fibers mixes, aimed at the design of thermal/acoustic insulating panels and other products for the green building sector.

Over the years, the research group has focused the attention on the study of different earth-based composite materials, with added vegetable fibers of different nature, including straw, hemp, loofah, rice-husk and jute. Earth is in fact an abundant, easily available, adaptable building material, and meets all the requirements of the sustainable construction model, its processing is not expensive in terms of economic or environmental resources; associated with the use of vegetable fibers, including waste, is an efficient, easy to work and durable material.

The interest in this type of materials by the research group stems from the need to solve the problems encountered in the activities for the recovery and requalification of historic buildings in Sardinia, built mainly with stone in the mountainous territories of the northern and eastern areas, and with earth in the Campidano plain. The combination of cement and synthetic building elements with stone and earth structures are in fact not recommended, and natural materials on the market nowadays, while offering an optimal solution, result often too high in costs, and thus unattractive to the current customers' requests. Even the materials presently produced on the island for thermal/ acoustic insulation (cork, sheep's wool panels) have uncompetitive prices compared to other diffused natural elements (wood fiber, rock wool, etc.) offered on the current market. The use of clayey earth as a binder of natural fibers for the production of different products (blocks, mortars and insulating panels) provides a response to requests for low-emissive and recyclable natural materials, result of circular production processes, aimed at a market that, although niche, seems to receive more and more attention.

The material on which this recent research has focused is a composite of earth and cellulose fibers. The fibers used in the experimentation described are of two types, called respectively Technocel 1004-3 Z (C1) and Technocel 500-1 (C2), and both deriving from newspaper recycling process, the second with a whitening process added.

Cellulose is not a new stabilizing material in the field of premixed finishing plasters. It is in fact a product that is commonly used to improve water retention of mixtures and increase smoothness of in their application even for thin thicknesses, allowing easier workability and more accurate finishes.

The objective of the present research is the identification and the elaboration of a low cost building material, produced with easily available and / or waste materials, efficient and versatile, consistent with the new requirements of the green building market, respectful of a circular production process.

2 Tests on the Material

Many building elements based on the use of earth can be made in an almost "tailored" way, starting from the selection of materials, reaching a definite mix-design, to finish with the characterization and the understanding of the manufacturing techniques of the products. The addition to earth of certain additives (Achenza, Fenu, 2006; Cappai, 2018) and fibers, in particular of natural ones (Aymerich, 2012; Aymerich, 2016), allows the production of elements whose physical-mechanical performances can be by far higher than those made of earth alone. Currently the use of this material is not confined to the production of simple, solid, hollowed or compressed blocks, but also to many elements of the building envelope such as plasters, finishes, insulating panels. This contribution reports the results of a preliminary experimentation aimed at identifying the variation of some physical properties of mixtures of earth and cellulose for a potential final production of plasters, finishes, insulating panels or other building ecological buildings components, with ad hoc adaptable physical properties.

In the production of composite manufactured products the density, the elastic modulus and the mechanical resistance (Ashby, 2012; Ashby, 2016) are critical properties since they condition other important characteristics of the materials such as lightness, stiffness, resistance and durability.

Another very important physical property is porosity, which affects the exchange of matter and energy at the interfaces of the element and within it.

In order to examine the variation of some of these physical properties, modifying the mix-design of earth-cellulose systems, a silicate earth was used, as confirmed by the XRD diffractometric analysis (Fig. 1); the phyllosilicate fraction is represented by Illite in a variable amount ranging from 5% to 10% by weight; the grain size is basically silty-clay (about 55%) with a limited sandy skeleton (medium to fine sands, with a diameter between 500 and 125m). Two types of cellulose have been added to the earth matrix, called respectively Technocel 1004-3 Z (C1) and Technocel 500-1 (C2); the first derives from a newspaper recycling process, the second from a rigorous cellulose selection process and following bleaching. The mixtures of earth and cellulose are characterized by the different water / solid and solid / fiber ratio; the goal is to identify the maximum concentration of cellulose that can be incorporated into the mixtures, without negatively affecting the workability of the systems and the quality of the exsiccated products. In this specific case, the maximum ratio in volume of earth / cellulose used is 1/3.

The measurements of the linear shrinkage of the hardened elements have revealed differences that can be attributed to the nature of the cellulose and to the ratios used: the systems containing the Technocel 1004-3 Z (C1), showed an increase of the shrinkage compared to reference system made of earth alone. On the other hand, those containing Technocel 500-1 (C2) cellulose with the same dosage showed a marked reduction in shrinkage.

Fig. 1 Earth and cellulose samples

This first experimental test has highlighted how a not negligible influence is driven by the quality of the cellulosic fraction, by the shape and by the morphological characteristics of the fibers deriving from the production cycle. The introduction of a substantial fraction of cellulose (as in the case of samples made with an earth/cellulose ratio equal to 1/3) has led to a considerable reduction of the weight of the samples, compared to the ones made with earth alone, as indicated by the apparent density values(Tab. 1).

Tab. 1 Apparent density (g/cm^3) of the samples with different additives: C1 and C2 types of cellulose; E=earth; ID=hydrophobic acrylic additive

Apparent density (g/cm^3)						
Earth	E+C1 1:1	E+C1 2:1	E+C1 1:2	E+C1 1:3	E+C1 1:3~5% ID	E+C1 1:3~7% ID
2.28 ± 0.49	2.15 ± 0.43	2.39± 0.36	1.80 ± 0.51	1.36 ± 0.38	1.48 ± 0.50	1.32 ± 0.46
	E+C2 1:1	E+C2 2:1	E+C2 1:2	E+C2 1:3	E+C2 1:3~5% ID	E+C1 1:3~7% ID
	2.26 ± 0.37	2.45 ± 0.50	1.75 ± 0.37	1.60 ± 0.45	1.57 ± 0.43	1.56 ± 0.42

The effect related to the addition of an acrylic hydrophobizing agent (ID) in powder form at 5% and 7% by weight provided by Freius Srl was also investigated. The addition of the hydrophobising additive modified the wettability, as in Fig. 1 , acting on the contact angle and on the superficial tension.

The wettability affects the access of water, in particular in liquid form, and therefore on the kinetics of capillary absorption and on the capacity of desorption of the product. With the increase in the amount of fibers both the absorption coefficient C. A. and the asymptotic value M * increase considerably (Tab. 2).

Tab. 2 Data on the capillary absorption referred to cubic samples (contact area $16mm^2$). Capillary coefficient (C. A.), asymptotic value (M^*).

Uptake	Earth	E+C 1-1:1	E+C 1 2:1	E+C 1 1:2	E+C 1 1:3	E+C 1 1:3 5% ID	E+C 1 1:3 7% ID
C. A. g/cm^2 $\sqrt{t}$	0.004 ± 0.001	0.006 ± 0.002	0.003 ± 0.001	0.015 ± 0.001	0.028 ± 0.001	0.019 ± 0.001	0.002 ± 0.001
M * g/cm^2	1.23 ± 0.2	1.57 ± 0.2	1.19 ± 0.3	2.26 ± 0.3	2.75 ± 0.3	2.43 ± 0.3	1.23 ± 0.2
	Earth	E+C 2 1:1	E+C 2 2:1	E+C 2 1:2	E+C 2 1:3	E+C 2 1:3 5% ID	E+C 1 1:3 7% ID
C. A. g/cm^2 $\sqrt{t}$	0.004 ± 0.001	0.004 ± 0.001	0.003 ± 0.001	0.12 ± 0.002	0.21 ± 0.004	0.016 ± 0.004	0.007 ± 0.003
M * g/cm^2	1.23 ± 0.2	1.39 ± 0.2	1.23 ± 0.2	1.91 ± 0.3	2.31 ± 0.3	2.27 ± 0.3	2.03 ± 0.3

The kinetics of water absorption, for the samples made of earth alone, is rather slow, as indicated by the value of the CA, and that of M * (Tab. 2). The presence of cellulose, in different proportions, modifies the absorption, which becomes faster and the material absorbs more water, with a maximum for the soil cellulose 1 : 3 system. In this case almost all of the absorption process takes place in three hours and the asymptotic value exceeds $2g/cm^2$. The effect of the hydrophobic additive ID, which produces a significant slowing down of the absorption kinetics (Tab. 2), especially for the C1 cellulose-containing system, is significant. The samples that have a limited absorption over time are therefore the ones made with soil alone and those containing a moderate amount of cellulose eg. (1 : 1).

The desorption of the capillary water occurs with a regular rate for the systems in which the soil matrix predominates: in 48 hours the water present at the beginning of the test is disposed of with contents of 15% in weight for the soil alone and 25% for the soil/cellulose 1 : 1 systems. The samples made with a 1 : 3 soil cellulose ratio differ in a greater quantity of water present at the beginning of the desorption process, equal to about 70% by weight. The water spontaneously abandons these samples after about 100 hrs, or twice the time compared to the system consisting of soil alone. The presence of the 7% ID additive leads to a considerable decrease in the quantity of water from the soil-C1 system, which from over 70% is reduced to about 25% by weight; desorption occurs in a time span of about 100 hrs for a simultaneous increase in retention capacity(Fig. 2).

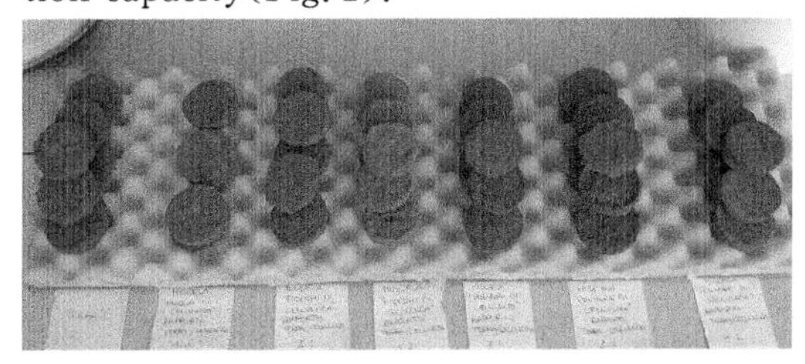

Fig. 2 Earth/cellulose samples

The MIP porosity showed a strong increase in porosity together with the increase of the cellulosic component. From the value of 22%±3 of the earth system, the open porosity reaches 45% for the systems containing the C1 and C2 cellulose in the ratio 1 : 3 also with the addition of the 7% acrylic additive.

The dimensional distribution of the voids of the different samples shows a tendency to shift towards the larger ones; this trend is accentuated with the increase in the quantity of cellulose introduced into the mixes (Fig. 3) while decreasing with the increase in the content in hydrophobic ID.

The measurement of the transit speed of ultrasounds on prismatic specimens of the same mixtures (transparency method), has

Fig. 3 Water permeability on earth/cellulose samples

shown how increasing the cellulosic fraction the speed goes from 2000 m/s for the soil alone to 1200 m/s for the soil/cellulose systems 1:3. Also in this case the presence of the hydrophobizing ID at 7% leads to a significant gain with values close to 1400 m/s. The analysis of these values allows to make a choice or change the mix-design as the ultrasonic speed, besides depending on the density, porosity and defectiveness of the mean used, is indirectly related to the mechanical resistance of the products.

3 Conclusion

This first experimentation on soil/cellulose mixes in different ratios in volume showed that both C1 and C2 celluloses can be incorporated into the mixtures and easily processed. The physical properties of the obtained products depend on both the added quantity and the nature of the cellulose used. The addition of a hydrophobic additive (ID) allows water retention to be limited and the capillary absorption to be attenuated, keeping the porosity of the products high, allowing regular transpiration.

To determine a possible specific use of these materials in the field of insulation (thermal and acoustic) the properties that affect the propagation of heat and sound should be further evaluated, making corrections and improvements in the selection of materials and mixes, and in the introduction of additives that can positively act on these systems allowing the modulation of the properties of specific interest.

References

AA, VV, 2016. Beton d'argile environnemental. Résultats d'un programme de recherche tourné vers l'application, CRATerre editions.

ACHENZA M, FENU L, 2006. On earth stabilization with natural polymers for earth masonry construction. Materials and structures, 39(1):285.

ASHBY M F, 2012. Materials and the Environment: Eco-informed Material Choice. Oxford: Butterworth-Heinemann.

ASHBY M F, 2016. Materials Selection Mechanical Design. Oxford:Butterworth-Heinemann.

AYMERICH F, FENU L, FRANCESCONI L, et al. ,2016. Fracture behavior of a fibre reinforced earthen material under static and impact flexural loading. Construction and Building Materials, 109: 109-119.

AYMERICH F, FENU L, MELONI P,2012. Effect of reinforcing wool fibres on fracture and energy absorption properties of an earthen material. Construction and building materials, 27: 66-72.

CAPPAI M,2018. La conservazione dei dipinti murali su intonaci in terra cruda: valutazione dell'efficacia di materiali naturali tradizionali per il consolidamento corticale. Universit degli Studi di Cagliari. Tecnologie per la Conservazione dei Beni Architettonici e Ambientali, Ciclo XXVIII.

FONTAINE L, ANGER R, 2009. Batir en terre. Belin:[s. n.].

Casa Copaja Restoration: An Example of Contemporary Intervention in Arica's Vernacular Heritage

Amanda Rivera Vidal, Camilo Giribas Contreras

Escuela de Construcci n en Tierra ECoT / ESTIERRA, Santiago, Chile

Abstract Arica, is the further north city of Chile, built in the times of the spanish colony and later republic, first Peruvian and now Chilean territory. The constructive techniques of the city, which shape its vernacular constructive culture, are the manifestation of the interweaving of traditional techniques with foreign materials, constituting a "mestizo" culture with the available local resources. The adobe, the "quincha" and the mud roofs are its main constructive techniques; those that interweave the cane, "totora" and local earth with the North American pine woods brought as ballast in the ships that look for the extraction of diverse raw materials, first the silver that came from Potos later de saltpeter. The Copaja House is a heritage construction built at the end of the 19th century in Arica, Chile. Today it has become one of the few buildings in the city that preserves the local earthen building culture. The carpentry with earth stand out, as "quinchas" with reeds in the vertical sense, walls of adobe with wooden confinements and the flat roofs and a coping or "mojinete" roofs with local fibers and mud. The restoration of Casa Copaja seeks to conserve and value the vernacular building systems incorporating new technologies and materials to ensure its continuity to the future. It also seeks to be an example in the city of the conservation and care of local heritage, which is in an evident state of vulnerability. The restoration consists of the structural consolidation of the adobe walls reinforced with electro welded mesh; the "quinchas" are repair with wooden reinforcements; the "mojinete" and flat roofs of the building reincorporates the earthen finishing.

Keywords adobe, mojinete, earthen techniques, earthen roof, wattle and daub

1 Introduction

1.1 Arica between Peru and Chile

Arica is today the first city in the north of Chile, this just since 1929. Located 300 kilometers from the next Chilean city to the south and only 50 km from its sister city Tacna.

Arica is located in the northern part of Chile, a territory populated around the routes "of the coast" and "of the Inca" in the altiplano since ancient times (Benavides, 1988). Arica is the first city in the north in what is now known as Chilean territory (Lat 18.4 S). It was part of the "coastal route" through which the Spanish conquistadors arrived in 1536 from Peru. However, Arica was part of the Andean-coastal relations, as indicated by archaeological findings that point to an ancient link between cultures of the Altiplano and the lower valleys (Zapater Equioiz, 1981). On the other hand, the zone of the great north of the country, historically was a nexus between high Peru and Bolivia with the coast, from the movement of the Inca chasquis, the conquerors, and particularly Arica, that later became the port for the descent of wealth from the high Potos(Fig. 1).

The port of Arica was constituted as the urban expression of the ancestral settlements of the valley of Azapa, through the delivery of encomienda to Francisco Pizarro in 1540, moment from the beginning of its constitution until in 1570 it already had the title of city by Felipe II (Benavides, 1988). The city enjoyed an important economic passage in the seventeenth century by trade in the silver route from Potos (Bolivia), at which time a series of temples are built on the road that connects the highlands with the valleys that make up an important example of the Andean Baroque.

However, since the Chilean war with the Bolivian Peruvian Confederation in 1879, the city underwent important changes, where the inhabitants of the city suffered significant repercussions for a period of 49 years, post-war period and indecision, where it was not taken the decision whether Tacna and Arica would be Chilean or Peruvian cities until 1929 (Ruz Zagal & Gonz lez Yanulaque, 2013). It is this ambiguous relationship with Chile, which affects Arica's history to this day, where many of its inhabitants had to emigrate to Peru, losing an important parts of their culture and history as a city.

Fig. 1 Views from the hill of Arica to the centre of the city, where the roofs of mojinetes are distinguished

2 Casa Copaja in Arica

2.1 The History

Casa Copaja has local protection in the category of Historic Conservation Property by the Regulatory Plan of the Arica commune of the year 2009, at which time an approximate dating of 1902 was made of the building. Through the restoration process it has been possible to verify, with the finding of modifications in its architecture through walls covered with newspapers of the year 1891, that the housing is prior to that date, and it is estimated that it could be after the earthquake and tsunami of 1868 that destroyed an important part of the city, including the parochial temple after 228 years of its construction.

The building built in the second half of the nineteenth century is an important example of Peruvian republican architecture, which has witnessed changes throughout its history. Property until 1902 of the Peruvian Puch family, which was forced to emigrate to Peru in the post-war period, becoming the property of the Copaja family until today.

2.2 Location

Located just three blocks from the San Marcos Cathedral, its location evidences its presence in the historic centre of the city, allowing it to become an important reference first in its permanence in time despite various threats that have suffered over time, and even more so today, during the process of its restoration.

2.3 Architecture

The property had from its origins a residential use, use that was shared with a commercial use with the occupation of the Copaja family. The building has functioned as an office and medical centre, in addition to housing for part of the family until 2014, when the product of an important earthquake that affected the city, the family decided to abandon it.

In 2016, a local foundation manages the use of the property as its main headquarters, using it for different activities that seek to promote the history and local heritage of the city.

The building was divided into three volumes, linked together by two interior patios(Fig. 2). The first zone towards the south, near the main access by street San Marcos, was used historically with a more public function, working there a medical centre and offices. The second volume is related to the first through the first patio, and faces it with a second fa ade, with similar characteristics to the exterior fa ade facing

San Marcos Street. In this volume, there are two salons of important dimensions and three enclosures with mojinetes that could have been rooms. The latter are the only ones of the building that do not have windows to the outside, transforming the mojinetes in their unique system of ventilation and natural lighting. The last volume of the current building consisted of a service area, kitchen, laundry and other areas, which at the time of the restoration project was in a state of ruin. According to historical antecedents, the property continued towards the adjacent land of the north, that today is empty and ad portals of the construction of a building of five floors.

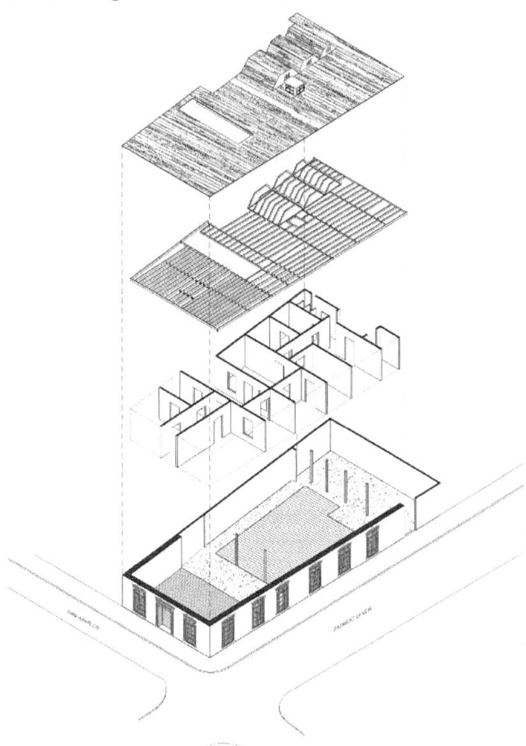

Fig. 2 Axonometric exploited of Casa Copaja

4 The Restoration of Casa Copaja as a Means for the Recognition of Constructive Systems

The restoration works have allowed deepening the constructive details that characterize this house of the nineteenth-century Peruvian republican architecture. This corner house, of 300m^2, is formed in all its perimeter by adobe walls of 60cm thick with a height of four meters. The entire interior is made up of wooden partitions filled with cane and mud (quincha) and the roof is composed of a wooden beam on which is a plank made of American Oregon pine with totora and clay finishing on the roof called "*torta de barro*" (Fig. 3). The house has three enclosures that have roofs on mojinetes.

Fig. 3 Restoration of the quichas in the indoor patio areas

The structural consolidation of the adobe walls consisted of the removal of all the coatings, both interior and exterior, to carry out an evaluation and diagnosis of damages that allowed proposing the appropriate intervention. The project considered the reinforcement of the adobe walls with galvanized metal meshes generating the confinement of the whole.

The structural consolidation of the quinchas has made evident the use of carpentry with special assemblies, the use of American Oregon pine, both for the structure of the house as for the floor, ceiling, and ornamental elements as cornices and dust covers; and the use of cane as a secondary structure to receive the mud filling and subsequent thick and thin clay plaster.

The restoration process of the roof included the structural consolidation of its wooden elements, the restoration of the mojinete (roofs with a truncated pyramid shape, with windows at both ends that allow ventilation and natural lighting of the enclosure), the restitution of the totora and the reintegration of the clay finishing (Fig. 4). These elements of the Peruvian vernacular architecture will be put in value since it is known that in Arica this type of roof was abundant, being today Copaja House one of the few houses in the city that conserves this constructive system.

Fig. 4 Process of reincorporation of new wooden boards as part of the roof package

3.1 Diagnosis of Damage

The adobe walls were covered with cement mortars in the whole of its fa ades, and with successive wall papers and finally agglomerated wood plates inside. The damage diagnosis inside the dwelling consisted of walls eroded mainly by the condensation of moisture accumulated between the adobe and the removed coverings, erosion by sanitary installations of later interventions and problems of salts by the capillary ascent of the humidity through the walls, which generated the flowering and crystallization of the salt. The foundation on the adobe wall was not visible, so the humidity and salt are generalized damage. It is known that Arica is an area of saline soil so it is common to have this problem.

In the case of the quinchas the main pathology found was the damage suffered by the woods by the action of termites. In the report on the state of conservation of the wood, it is determined that *"a series of surveys were conducted, aimed at the identification of wood parenchyma, conservation status and extent of damage due to the presence of woodworm pest. Together with direct observations and as a result of the samples obtained for workshop analysis. It was possible to establish the presence of an important pest of xylophagous character, corresponding to Isoptera of the type Neotermes chilensis (Blanchard) (Chilean wood termite) native insect, associated with native and exotic coniferous and hardwoods, which are dehydrated and / or in active service, for more than 20 years. This plague is currently distributed between the first and eighth regions of Chile. The winged adults are approximately 20 mm long, yellowish white. The creamy white workers have pronotum in the shape of a crescent with a notch in the anterior middle area. This termite species is termed a 'dry wood' termite, because, for its development, it needs only the moisture provided by the atmosphere and the moisture content of the wood in which they nest. These pests, if not subjected to effective control, may affect all the woods of a building, making them lose their structural solidity completely"*.

In the woods that make up the partition walls, that is, the lower floor, upper floor, right and diagonal feet, it was observed that approximately one third of these were very affected by termites and another third presented minor damage while not losing their structural capacity. The transverse pieces of wood that allow the interlacing of the canes, and these, in general, did not present damages by termites being only present in isolated cases.

In the case of the roof beams and ceiling wood, the damage by xylophages was lower, and in all cases, a selection of the tables that were reused could be made.

Practically there was no roof finishing and only the rest of *totora* remains from the original roof. Thanks to the study of mojinete ceilings and the observation of these in the neighbouring city Tacna in Peru, it is deduced that the original constructive package is to cover the roof beams with wood boards, over this the totora mats, which generates the adhesion surface of the clay finishing of the roof called "torta de barro". The roof with "torta de barro" is a common construction technique of the highlands of Chile, Argentina, Peru and Bolivia, and occurs in territories where historically there is little rain(Fig. 5).

Fig. 5 Mojinete space with windows for lighting and ventilation

3.2 Proposed Intervention

According to the evaluation and diagnosis of damages, the way in which the restoration of Casa Copaja was intervened was determined.

In the case of the adobe walls, the following interventions were considered for the damages found; in basal erosions, the loose material was removed and calipers were made to consolidate the damaged parts. Once the walls were restored, they worked on their structural consolidation through the installation of galvanized metal mesh, generating the confinement of the structure to avoid collapse in the event of an earthquake. The details and specifications of the meshes and their fixation were determined by the project's calculator engineer. Once the meshes were installed, they were covered with a thick plaster of 5cm.

Fig. 6 Reinforcement of the wooden structure damaged by termites in quinchas

In the case of the quinchas, where the damage is present in the woods, an intervention was proposed that does not involve the removal of the damaged piece, but the reinforcement of these elements with new woods on both sides embracing the piece that presents damages(Fig. 6). This kind of orthopaedics, or element that is added to return the structural capacity to the piece with damages, was made after applying a general treatment to the woods in question. As a first stage, a curative treatment was carried out that sought to kill the termites alive through a fumigation of all the woods present in the house, both the structural ones (beams, right feet, diagonals and sills) as well as the woods of the floors, skies, doors, windows, pilasters, etc. Later, as a second stage, a preventive treatment was carried out to prevent the new housing of termites, for which a wood protector mixed with a special insecticide was applied as a paint. Once the woods were consolidated, fillings were made with mud to later apply the earth coverings.

All the damaged woods were consolidated in the roof, and the traditional roof was restored, adding only a new component (Fig. 7). The traditional roof is made up of the American Oregon pine board that is fixed in the beam; On the boards are the totora fibbers and over these the mud finishing. In the restoration project was considered the incorporation, on the entire surface of the board, asphalt felt paper to have a

Fig. 7 Restoration of roofing termination on mojinete roof with "torta de barro"

protection against the rains that are greater effects of climate change in these times. For the "*torta de barro*", drip and absorption tests were carried out to define the mixture that was used in the new 4 cm thick mud finishing. Earth mixtures were tested with animal manure or guano, egg white, linseed oil, vinyl poly acetate and cactus mucilage, among other natural and industrial additives. Three samples of 10cm×10cm×2cm of each mixture were made and subjected to a constant drip every second and at a height of two meters, to observe the resistance of each of them. The chosen mixture was with 5% egg albumen. In addition to this layer a finishing layer was applied with a clay paint with 20% vinyl polyacetate as a binder that allows to generate a protective layer without losing the hydrothermal property, also contributing to the resistance to solar radiation.

4 Execution of the Work

The work began at during 2018, with the removal of all the interior linings of the building, for the realization of a more accurate diagnosis of its structural state.

The works were developed in 8 months with a team of approximately 15 people, integrated by two architects, an engineer, a construction supervisor, a technician in heritage restoration, carpenter's restorers, masons with experience in work with earthen architecture and heritage.

During the work, the team was trained on different topics such as types of wood, pathologies in wood, Test Carazas, workshops on construction with mixed systems (quinchas), adobe construction and clay finishing.

5 Conclusion

The intervention is one of the oldest domestic constructions of the city of Arica, which maintains both its architecture and its traditional construction systems, has been an apprenticeship on how to build in the 19th century with the available resources.

The vernacular architecture forged from local earth, wood brought as weight in ships and local fibbers have resulted in architecture of incredible ingenuity and quality, transforming into the characteristic architecture of the area, which speaks of a historical moment and of the available resources.

Acknowledgements

Very special thanks to all who intervened and supported the realization of this emblematic intervention for the city of Arica. To those from the management and vision managed to get the resources to do it, to all who worked and were involved with the history and the meaning of rescuing this property for the city and the future.

References

BENAVIDES A,1988. La arquitectura en el virreinato del perú y en la capitanía general de chile. Santiago: Andrés Bello.

RUZ ZAGAL R, GÓNZALEZ YANULAQUE A, 2013. Archivo fotográfico manuel yanulaque scorda (1850—1934). Historia e imágenes ariqueñas. Arica: Universidad de Tarapacá.

ZAPATER EQUIOIZ H,1981. Los Incas y la Conquista de Chile. Revista de historia, 16:249-252.

Analytical Method for Dynamic Response of Wholly Grouted Anchorage System of Rammed Earth Sites

LU Wei, *LI Dongbo*

Xi'an University of Architecture & Technology, Xi'an, China

Abstract A simplified mechanical model of dangerous soil mass - anchor bolt - stabilized soil mass anchorage system is established against the background of the instability of rammed earth sites caused by wide vertical cracks. In this model, friction between the anchor bolt and the soil mass was simplified into a parallel mechanism of a linear spring together with a speed-related damper, and the connect effect of anchor bolt in the crack section is simplified as a linear spring. Dynamic equilibrium equations for the anchoring system were established based on the theory of elastodynamics, and analytic solutions of the dynamic response of bolt axial forces and displacement were deduced. Finally, taking the anchorage engineering of the southern wall of the GAOCHANG Ancient City in Xinjiang as an example, the analytic method proposed by this paper is illustrated, the axial force response characteristics and distribution law of the anchor bolt are clarified, and the rationality is verified by numerical analysis. The results show that the method proposed by this paper can accurately and conveniently predict the axial force response of the anchor bolt under dynamic action after considering the magnification effect of seismic acceleration along with the site height. This mechanical model and analytical method are also applies to the dynamic response analysis of anchor bolts in geotechnical slope anchorage engineering with wide vertical cracks or joints.

Keywords rammed earth sites, wholly grouted anchor bolt, anchorage system, simplified mechanical model, dynamic response

1 Introduction

Different from side slope anchorage in the traditional sense, protection projects for ancient rammed-earth buildings need to consider the safety of earthen sites after reinforcement, as well as compatibility of materials. Therefore, the pre-stressing technique is restricted in the reinforcement of earthen sites. Relevant research results show that wholly grouted bamboo bolts can effectively improve the stress status of earthen sites, and therefore are widely used in the reinforcement of earthen sites.

Structure diseases that are commonly seen in rammed earth walls include structural longitudinal and lateral cracks, as well as instability problems of collapse, split, toppling, etc. caused by weak interfaces of rammed layers. With the wholly grouted anchoring technique, anchor bolts are placed through cracks and structural planes of dangerous mass and finally anchored in stable mass. In such an anchoring system, due to the existence of vertical cracks and structural planes, the dangerous mass under seismic action will firstly generate horizontal displacement with respect to the stable mass. As a result, frictional forces are produced between the bolt anchoring agent and the soil mass, and then transmitted to bolts; additional axial forces due to the relative displacement of the dangerous mass are transmitted to the anchorage of the stable mass through elastic deformation of bolts in cracks, and then dissipated in the stable mass. Therefore, it is quite different from the actual conditions to consider anchor bolts as wholly linear rods. Further research needs to be carried out regarding the reasonable simplification of mechanical models of anchor bolts, based on their the actual stress conditions.

2 Research Status

In recent years, bolt anchoring systems have been deeply studied by relevant scholars. Cotton et al. evaluated the stability of soil nail walls under seismic action. Shukla et al. analyzed the seismic dynamic response of rock slopes with cracks, and discussed the feasibility of multi-angle anchorage. Yu et al. analyzed the interaction between rock and bolts based on an improved Shear-lag model. Ye et al. analyzed the dynamic response of the frame prestressed bolt soil mass system with simple harmonic vibration, and by using the same method, Dong et al. analyzed the dynamic performance of the soil nail soil mass anchoring system, as well as the frame support system with prestressed anchor bolts. However, the influences of weak structural planes were not considered. Peng analyzed the seismic response of anchoring systems with rock mass structural planes, through the quasi-static method, pointing out stress characteristics of anchoring systems with single and multiple structural planes respectively. Shi and Qiu analyzed the reinforcement effect of wholly grouted bolts through the finite element method, in the context of cliff reinforcement for Jiaohe Ruins. Zhang et al. analyzed toppling failure modes of rock slopes with cracks under seismic action, by using the moment equilibrium approach. Vucetic et al. conducted centrifuge tests towards slope models reinforced with soil nailing. Ye et al. used shaking table tests to study the load-bearing mechanism of anchor bolts under seismic action, and analyze the distribution of axial forces of such bolts. The above researches failed to provide fully reasonable simplification of interaction between bolts and soil mass and didn't adequately consider influences of non-continuous interfaces. In addition, simplification with the quasi-static method can't reflect seismic response in real conditions. Therefore, deep research should be further conducted regarding the interaction of the dangerous mass — bolt — stable mass system under seismic action.

With regard to instability problems due to structure diseases of an earthen site, and with consideration of the interaction between the bolt and the soil mass, this paper established a simplified model of dynamic calculation for a bolt anchoring system with vertical cracks or weak structural planes under seismic action. In this model, friction between the anchor bolt and the soil mass was simplified into a parallel mechanism of a linear spring together with a speed-related damper, and the bolt across crack sections was simplified into a linear spring. Axial force response of the bolt under horizontal sinusoidal excitation was obtained through the simplified model mentioned above. The feasibility of this method was verified based on the analysis of an anchoring project for a rammed earth wall, as well as on relevant comparison with numerical simulation results.

3 Dynamic Model Simplification

Fig. 1 shows the schematic cross section of wall ruins with cracks, reinforced with wholly grouted bamboo bolts, where θ is the angle of a bolt with respect to the horizontal; H_s and H_v are vertical intervals between bolts at the second row and those at the first and third rows respectively; H is the overall height of the wall.

Following basic hypotheses were adopted to establish the simplified dynamic model of a single bolt:

(1) Variable cross-sections of wedged bamboo bolts were ignored, and such bolts were assumed to be homogeneous round bars with uniform cross-sections, conforming to the plane section assumption.

(2) It was assumed that the soil mass in the same rammed layer was homogeneous and isotropic, and that shear forces the soil mass applies on bolts were evenly distributed along bolt length.

(3) It was assumed that anchor bolts were uniform linear elastomers, with an equivalent elastic modulus of E.

(4) Friction between anchor bolts and

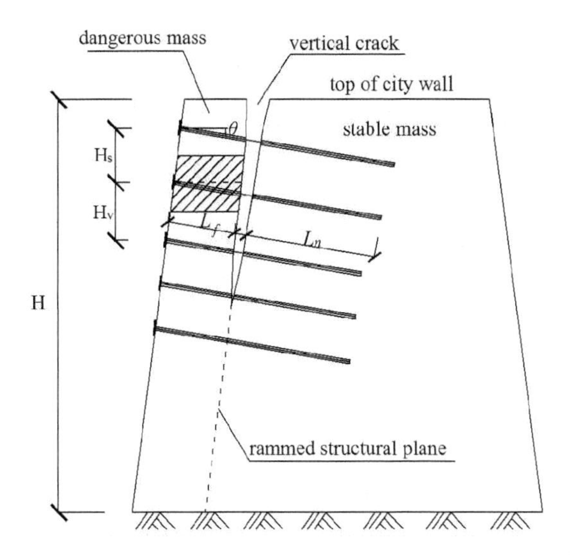

Fig. 1 Schematic cross section of wall ruins with longitudinal cracks, reinforced with wholly grouted bamboo bolts

the soil mass was simplified into a parallel mechanism of a linear spring together with a speed-related damper, and bolts across cracks were simplified into linear springs.

(5) It was assumed that the stable mass had enough anchoring depth, and that the bottom end of a bolt had no displacement with respect to the soil mass, equivalent to a fixed end.

(6) As anchoring angles were normally 5～10, resulting in small vertical components of horizontal seismic forces, components of horizontal seismic forces, along with axial directions of bolts, were considered only.

(7) Cracks developed vertically along the rammed structural plane of the wall.

(8) As anchoring agents applied in the reinforcement of earthen sites normally had a smaller difference from the soil mass in properties, the only interaction between anchoring agents and bolts was considered.

Fig. 2 shows the simplified mechanical model of the dynamic system of a single wholly grouted bolt.

The overall length of the bolt is L, the area of the bamboo bolt cross-section is A, the perimeter of such a cross section is D, and the equivalent density of bolt material is ρ. Lengths of anchoring sections in danger-

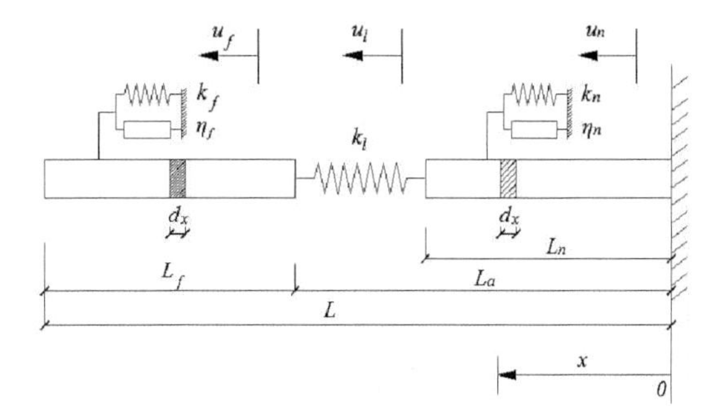

Fig. 2 Simplified mechanical model of the dynamic system of a single wholly grouted bolt

ous and stable mass are L_f and L_n respectively. Spring stiffness of the anchoring section in the dangerous mass is k_f, the damping coefficient is η_f, and relative displacement due to horizontal seismic action is u_f. Spring stiffness of the anchoring section in the stable mass is k_n, the damping coefficient is η_n, and relative displacement due to horizontal seismic action is u_n. Spring stiffness of cracks is k_l, and relative displacement is u_l. Consider conditions of displacement continuity, has $u_l=u_f$, in the case of $x=L_a$. Spring stiffness of anchoring sections in dangerous and stable mass, as well as relevant damping coefficients can be calculated according to Eq. (1):

$$\begin{cases} k_f=\delta E_f \\ k_n=\delta E_n \end{cases} \quad \begin{cases} \eta_f=2\rho_f W(V_{s1}+V_{p1}) \\ \eta_n=2\rho_n W(V_{s2}+V_{p2}) \end{cases} \tag{1}$$

Where, E_f and E_n are soil compression modulus of dangerous and stable mass respectively; ρ_f and ρ_n are soil density of dangerous and stable mass respectively; V_{s1}, V_{s2} and V_{p1}, V_{p2} are soil shear wave velocity and press wave velocity of dangerous and stable mass respectively; W is the calculated width, which is the sum total of distances from the calculated bolt to centers of adjacent bolts at both sides; δ is the spring correction coefficient of friction forces that the soil mass applies on bolts, which is normally 1.5～1.6.

Soil shear wave velocity and press wave velocity are related to the soil properties. Generally, Eq. (2) can be used for calculation, as to an infinite elastic medium:

$$\begin{cases} V_{s1}=\sqrt{\dfrac{E_f}{2\rho_f(1+\nu_f)}} \\ V_{s2}=\sqrt{\dfrac{E_n}{2\rho_n(1+\nu_n)}} \end{cases}$$

$$\begin{cases} V_{P1}=\sqrt{\dfrac{E_f(1-\nu_f)}{\rho_f(1+\nu_f)(1-2\nu_f)}} \\ V_{P2}=\sqrt{\dfrac{E_n(1-\nu_n)}{\rho_n(1+\nu_n)(1-2\nu_n)}} \end{cases} \tag{2}$$

Where, ν_f and ν_n are soil Poisson ratios of dangerous and stable mass.

This anchoring system was designed as a passive load-bearing mechanism, which means bolts will work only when the dangerous mass generates displacement relative to the stable mass. Therefore, during the calculation of dynamic response, increments of bolt axial forces in the anchoring section of the dangerous mass should be solved firstly. Such force increments will then be exerted on the stable mass through bolt deformation in crack sections. Further, the axial force response of bolts in the stable mass can be obtained.

4 Seismic Response of Anchor System

4.1 Stress Analysis of the Anchoring Micro-section in Dangerous Mass

A micro-section of anchorage in dangerous mass ($L_a \leqslant x \leqslant L$) is taken for the research. The following stress diagram can be obtained according to force balance (Fig. 3).

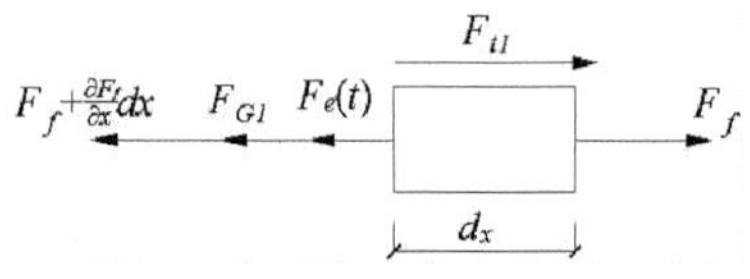

Fig. 3 Force diagram of the anchoring micro-section in dangerous mass

According to the force balance condition, it has:

$$F_f=\frac{\partial F_f}{\partial x}\mathrm{d}x+F_{G1}+F_e(t)=F_{t1}+F_f \tag{3}$$

Where: F_f and $F_f+\frac{\partial F_f}{\partial x}\mathrm{d}x$ are internal forces at both sides of the micro-section, and $F_f=EA\frac{\partial u_f}{\partial x}$; u_f is the relative displacement between the soil mass and the anchor bolt; F_{G1} is the inertia force of the anchoring micro-section under horizontal seismic action, and $F_{G1}=\rho A\frac{\partial^2 u_f}{\partial t^2}\mathrm{d}x$; $F_e(t)$ is the horizontal sinusoidal excitation, and $F_e(t)=\rho A\frac{\partial^2 u_g}{\partial t^2}\mathrm{d}x\cdot\cos\theta$, where u_g is the displacement excitation; F_{t1} is the frictional resistance that the soil mass applies on the bolt, and $F_{t1}=\left(k_j u_f+\eta_f\frac{\partial u_f}{\partial t}\right)\mathrm{d}x\cdot D$.

Eq. (3) can be put this way:

$$EA\frac{\partial^2 u_f}{\partial x^2}\mathrm{d}x+\rho A\frac{\partial^2 u_f}{\partial t^2}\mathrm{d}x+\rho A\frac{\partial^2 u_g}{\partial t^2}\mathrm{d}x\cdot\cos\theta=\left(k_f u_f+\eta_f\frac{\partial u_f}{\partial t}\right)\mathrm{d}x\cdot D \tag{4}$$

Divide Eq. (4) by $\rho\cdot A\cdot \mathrm{d}x$ to get:

$$\frac{E}{\rho}\cdot\frac{\partial^2 u_f}{\partial x^2}+\frac{\partial^2 u_f}{\partial t^2}-\frac{\left(k_f u_f+\eta_f\frac{\partial u_f}{\partial t}\right)\cdot D}{\rho A}=\frac{\partial^2 u_g}{\partial t^2}\cdot\cos\theta \tag{5}$$

Where, E is the equivalent elastic modulus of bolts.

Set $\alpha=\frac{D}{\rho A}$ and $\beta=\sqrt{\frac{E}{\rho}}$, and Eq. (5) can be rewritten as:

$$\beta^2\frac{\partial^2 u_f}{\partial x^2}+\frac{\partial^2 u_f}{\partial t^2}-\alpha\left(k_f u_f+\eta_f\frac{\partial u_f}{\partial t}\right)=-\frac{\partial^2 u_g}{\partial t^2}\cdot\cos\theta \tag{6}$$

Seismic dynamic action is simulated with a simple harmonic vibration in this paper. The seismic excitation of the rammed earth wall top is set to be $u_g=U_g\sin\omega t$. To simplify the calculation, it is changed to $u_g=U_g e^{i\omega t}$, where U_g is the amplitude of simple harmonic vibration. Put the above expression into Eq. (6), and the imaginary part of the solution is the displacement response of this anchoring system under the simple harmonic loads, which is $u_f'(x,t)=\mathrm{Im}[U_f(x)e^{i\omega t}]$. Apply the expression of seismic displacement excitation into Eq. (6) to get:

$$\beta^2\frac{\partial^2 u_f}{\partial x^2}+\frac{\partial^2 u_f}{\partial t^2}-\alpha\left(k_f u_f+\eta_f\frac{\partial u_f}{\partial t}\right)=U_g e^{i\omega t}\cdot\omega^2\cdot\cos\theta \tag{7}$$

Suppose the solution of Eq. (7) is $u_f(x,t)=U_f(x)e^{i\omega t}$, and put it into Eq. (7) to get:

$$\frac{d^2U_f(x)}{dx^2}-\frac{\omega^2+\alpha k_f+\alpha\eta_f i\omega}{\beta^2}U_f(x)=\frac{U_g\omega\cos\theta}{\beta^2} \tag{8}$$

Set:

$$-\frac{\omega^2+\alpha k_f+\alpha\eta_f i\omega}{\beta^2}=P\ \frac{U_g\omega^2\cos\theta}{\beta^2}=Q \tag{9}$$

Eq. (9) can be changed to:

$$\frac{d^2U_f(x)}{dx^2}+P\cdot U_f(x)=Q \tag{10}$$

Solve differential Eq. (9) to get:

$$U_f(x)=C_1\cos\sqrt{P}x+C_2\sin\sqrt{P}x+\frac{Q}{P} \tag{11}$$

Therefore, the solution of the original Eq. (4) is:

$$u'_f(x,t)=\mathrm{Im}\left\{\left(C_1\cos\sqrt{P}x+C_2\sin\sqrt{P}x+\frac{Q}{P}\right)e^{i\omega t}\right\},\quad L_a\leqslant x\leqslant L \tag{12}$$

Where, C_1 and C_2 are coefficients to be determined, it can be determined by boundary conditions.

Initial conditions at the beginning of seismic action:

$$u_f(x,0)=0,\quad L_a\leqslant x\leqslant L \tag{13}$$

Displacement continuity at $x=L_a$:

$$u_f(L_a,t)=u_l(L_a,t) \tag{14}$$

Boundary conditions for the anchoring section in the dangerous mass:

$$k_l u_l(x,t)\big|_{x=L_a}=EA\frac{\partial u_f(x,t)}{\partial x}\bigg| \tag{15}$$

$$k_l u_l(x,t)\big|_{x=L_a}=M\frac{\partial^2 u_f(x,t)}{\partial t^2}\bigg| \tag{16}$$

Where, M is the mass of calculated soil in the dangerous mass. As shown in Fig. 1, according to the lumped mass method, the anchoring hole on the wall is taken as the center; sum total of distances from the hole to centers of adjacent bolts at left and right sides is taken as the horizontal distance; sum total of distances from the hole to centers of adjacent bolts at up and down sides is taken as the vertical distance. Take the second row of bolts in Fig. 1 as the example, set up and down intervals of adjacent bolts to be H_S and H_V, and left and right intervals to be W_S and W_V. Therefore, M can be represented as:

$$M=\frac{1}{4}(M_S+H_V)(W_S+W_V)\cdot\frac{L_f}{\cos\theta}\cdot\rho_f \tag{17}$$

Apply Eq. (11) into Eqs. (15) - (16), and consider the continuity condition of $u_l=u_f$ at $x=L_a$:

$$k_l\left(C_1\cos\sqrt{P}L_a+C_2\sin\sqrt{P}L_a+\frac{Q}{P}\right)=EA\left[C_1\sqrt{P}(-\sin\sqrt{P}L_a)+C_2\cos\sqrt{P}L_a\right] \tag{18}$$

$$k_l\left(C_1\cos\sqrt{P}L_a+C_2\sin\sqrt{P}L_a+\frac{Q}{P}\right)=-M\cdot\omega^2\left(C_1\cos\sqrt{P}L_a+C_2\sin\sqrt{P}L_a+\frac{Q}{P}\right) \tag{19}$$

Simultaneous Eqs. (18)—(19) to get:

$$C_1=\frac{M\omega^2\cdot\frac{Q}{P}\cdot(F_2-F_4)-k_l\frac{Q}{P}(F_4+F_6)}{(F_1+F_3)(F_4+F_6+(F_2-F_4)(F_1-F_5)} \tag{20}$$

$$C_2=\frac{M\omega^2\cdot\frac{Q}{P}\cdot(F_2-F_4)-k_l\frac{Q}{P}(F_4+F_6)}{(F_1+F_3)(F_4+F_6)+(F_2-F_4)(F_3-F_5)}\cdot\frac{F_1+F_2}{F_4-F_2}+\frac{k_l\frac{Q}{P}}{F_4-F_2} \tag{21}$$

Where:

$F_1=k_l\cos\sqrt{P}L_a$, $F_2=k_l\sin\sqrt{P}L_a$, $F_3=EA\sqrt{P}\sin\sqrt{P}L_a$, $F_4=EA\cos\sqrt{P}L_a$, $F_5=M\omega^2\cos\sqrt{P}L_a$, $F_6=M\omega^2\sin\sqrt{P}L_a$

Therefore, the axial force response of the anchoring section in the dangerous mass is:

$$N_f(x,t)=EA\frac{\partial u'_f(x,t)}{\partial x},\quad L_a\leqslant x\leqslant L \tag{22}$$

4.2 Stress Analysis of the Anchoring Micro-section in Stable Mass

Under seismic action, the dangerous mass generates displacement relative to the stable mass, and with dynamic action, bolts in the stable mass need to resist additional increments of axial forces caused by the dangerous mass. According to the force balance, load-bearing conditions of the anchoring micro-section ($0\leqslant x\leqslant L_n$) in the stable mass are shown in Fig. 4:

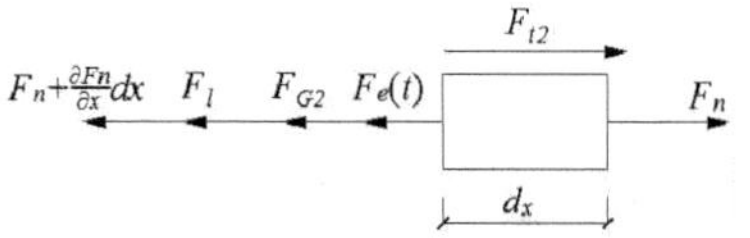

Fig. 4 Force diagram of the anchoring micro-section in stable mass

According to force balance conditions, it has:

$$F_n+\frac{\partial F_n}{\partial x}\mathrm{d}x+F_l+F_{G2}+F_e(t)=F_{t2}+F_n \tag{23}$$

Where: F_n and $F_n+\frac{\partial F_n}{\partial x}\mathrm{d}x$ are internal forces at both sides of the micro-section, and $F_n=EA\frac{\partial u_n}{\partial x}$; F_{G2} is the inertia force of the anchoring micro-section under horizontal seismic action, and $F_{G2}=\rho A\frac{\partial^2 u_n}{\partial t^2}\mathrm{d}x$; F_{t2} is the friction force that the stable soil mass applies on the anchorage bolt, and $F_{t2}=\left(k_n u_n+\eta_n\frac{\partial u_n}{\partial t}\right)\mathrm{d}x\cdot D$; F_l is the increment of bolt axial forces, caused by relative displacement of the dangerous mass under seismic action. Take solved displacement, u_f', of the dangerous mass at $x=L_a$ as a known quantity to get F_l:

$$F_l(x,t)=k_l u_l(x,t)\big|_{x=L_n}=k_l u_f'(x,t)\big|_{x=L_a}=EA\frac{\partial u_f'(x,t)}{\partial x}\bigg|_{x=L_a} \tag{24}$$

Eq. (24) can be changed to:

$$EA\frac{\partial^2 u_a}{\partial x^2}\mathrm{d}x+k_l u_f'+\rho A\frac{\partial^2 u_n}{\partial t^2}\mathrm{d}x+\rho A\frac{\partial^2 u_g}{\partial t^2}\mathrm{d}x\cdot\cos\theta=\left(k_n u_n+\eta_n\frac{\partial u_n}{\partial t}\right)\mathrm{d}x\cdot D \tag{25}$$

Set $\gamma=\kappa_\lambda/(\rho A)$ to change Eq. (25) into:

$$\beta^2\frac{\partial^2 u_n}{\partial x^2}+\frac{\partial^2 u_n}{\partial t^2}-\alpha\left(k_n u_n+\eta_n\frac{\partial u_n}{\partial t}\right)+\gamma u_f'\cdot\frac{1}{\Delta x}=-\frac{\partial^2 u_g}{\partial t^2}\cdot\cos\theta \tag{26}$$

With the same method mentioned above, suppose the solution of Eq. (26) is $u_n(x,t)=U_n(x)\mathrm{e}^{i\omega t}$, and put it into Eq. (26). Set:

$$-\frac{\omega^2+\alpha k_n+\alpha\eta_n i\omega}{\beta^2}=R$$
$$\frac{U_g\omega^2\cos\theta}{\beta^2}-\frac{\gamma U_f}{\Delta x\cdot\beta^2}=S \tag{27}$$

Then:

$$\frac{\mathrm{d}^2 U_n(x)}{\mathrm{d}x^2}+R\cdot U_n(x)=S \tag{28}$$

Therefore, the solution of Eq. (28) is:

$$U_n(x)=C_3\cos\sqrt{R}x+C_4\sin\sqrt{R}x+\frac{S}{R} \tag{29}$$

The displacement response is then obtained as:

$$u_n'(x,t)=\mathrm{Im}\left[\left(C_3\cos\sqrt{R}x+C_4\sin\sqrt{R}+\frac{S}{R}\right)\mathrm{e}^{i\omega t}\right], \tag{30}$$

Where, C_3 and C_4 are coefficients to be determined based on boundary conditions.

Initial conditions at the beginning of seismic excitation:

$$u_n(x,0)=0,\quad 0\leqslant x\leqslant L_n \tag{31}$$

Displacement continuity at $x=L_n$:

$$u_n(L_n,t)=u_l(L_n,t)=u_f(L_a,t) \tag{32}$$

Boundary conditions for the anchoring section in the stable mass:

$$u_n(0,t)=0 \tag{33}$$

$$k_l u_l(x,t)\big|_{x=L_n}=EA\frac{\partial u_n(x,t)}{\partial x}\bigg|_{x=L_n} \tag{34}$$

Apply Eq. (30) into Eqs. (33)−(34) to get:

$$C_3=-\frac{S}{R} \tag{35}$$

$$C_4=\frac{1}{\sin(\sqrt{R}L_n)}\cdot\left[\begin{array}{l}C_1\cos(\sqrt{P}L_a)+C_2\sin(\sqrt{P}L_a)\\+\frac{S}{R}\cos(\sqrt{R}L_n)+\frac{Q}{P}-\frac{S}{R}\end{array}\right] \tag{36}$$

Therefore, the axial force response of the anchoring section in the stable mass is:

$$N_n(x,t)=EA\frac{\partial u_n'(x,t)}{\partial x}\quad 0\leqslant x\leqslant L_n \tag{37}$$

5 Engineering Example and Numerical Analysis

5.1 General situation of Engineering and Reinforcement Scheme

The ancient city of Gaochang is located in Turpan city of Xinjiang, which is one of the best preserved earthen sites with rammed walls. Due to the dry climate in the local area, vertically developed wide cracks are commonly seen along rammed structural planes on walls. Thus instability failure is easily brought about under seismic action, causing irretrievable loss. This paper studied a reinforced section of an outer wall of Gaochang, and analyzed its dynamic response by using above calculation method.

This wall section studied had a height of 11.7m; the thickness of its top was 6.9m; the angle between the external wall and the horizontal was 80 ; Thickness of rammed earth layer at the outermost layer was about 0.8m; the vertical crack developed along the rammed structural plane, with width of 20cm and depth of 4.2m, as shown in Fig. 5. The seismic fortification intensity in this region was 7, but as to important cultural relics such as an earthen site, the peak ground acceleration was set to be 0.15g, with the importance coefficient of 1.1 and safety coefficient of 1.3. Relevant anchoring design is shown in Fig. 6. Major design parameters are listed in Tab. 1. The material parameters of the earthen site are shown in Tab. 2 (Zhao et al., 2016).

Fig. 5 Sketch of crack diseases of an outer wall section of Gaochang Ruins

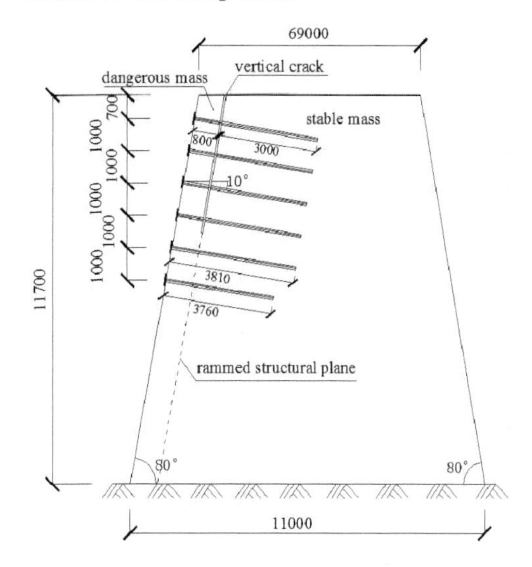

Fig. 6 Reinforcement schematic diagram of outer facade of Gaochang city wall

Tab. 1 Design parameters for reinforcement of a rammed wall through bamboo bolts

Anchorage position (From top to bottom)	Horizontal distance /m	Vertical distance /m	Bolt diameter /mm	Bolt length in dangerous mass /m	Bolt length in stable mass /m	Total bolt length /m
1~4th rows	1	1	35	0.8	3	3.81
5th row	1	1	35	-	-	3.81
6th row	1	1	35	-	-	3.76

Tab. 2 Material parameters of the earthen site

Rammed earth position	Unit weight (kN/m^3)	Compression modulus (MPa)	Poisson's ratio	Friction angle (°)	Cohesion (kPa)	Ultimate friction resistance(kPa)
dangerous mass	16.7	22.03	0.3	24	23.4	33
stable mass	18.4	29.63	0.21	21	26	41

5.2 Seismic Response Analysis

Sinusoidal waves were adopted in this paper to simulate the seismic action, with the frequency of 2Hz, peak ground acceleration of $a_{max}=0.15g$, and duration of 10s, as shown in Fig. 7. Bamboo bolts had anequivalent elastic modulus of $E=11.6$GPa and an equivalent density of $\rho=1490\text{kg/m}^3$. Compression modulus of rammed earth in the dangerous mass was 2.2 MPa, and relevant density was 1704 kg/m^3. Compression modulus of rammed earth in the stable mass was 2.98 MPa, and relevant density was 1877 kg/m^3. Spring stiffness of the anchoring section in the dangerous mass was 33.045 MPa, and relevant damping was 6.89×10^2 kN/m s. Spring stiffness of the anchoring section in the stable mass was 44.445 MPa, and relevant damping was 8.06×10^2kN/m s.

As the crack developed along the rammed structural plane, the dangerous mass had basically uniform thickness, anchoring lengths of bolts at 1~4 rows were the same, and bolts had a similar distribution of axial stresses, therefore, this paper only showed calculation results of bolts at the second row. Fig. 8 shows the axial force response of bolts at the position of $x=3$m, while the distribution of axial force peaks along bolts is shown in Fig. 9. It can be seen that axial forces of bolts in both dangerous and stable mass had exponential distribution and reached the maximum value around the crack position, which was about 8.37kN. The axial force response of bolts at the position of $x=3$m showed a state of fluctuation around the statically balanced position.

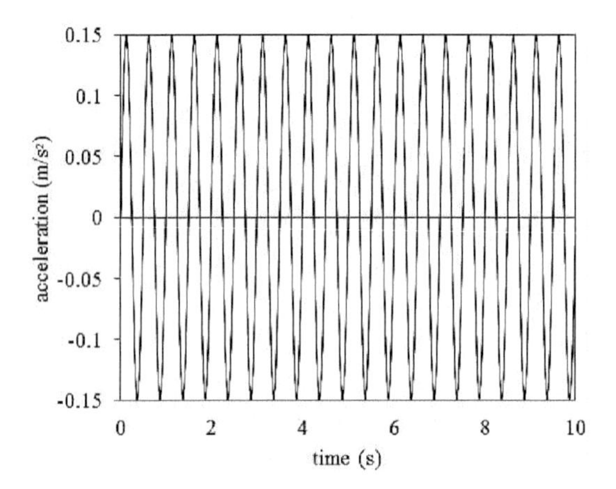

Fig. 7 Sinusoidal wave excitation

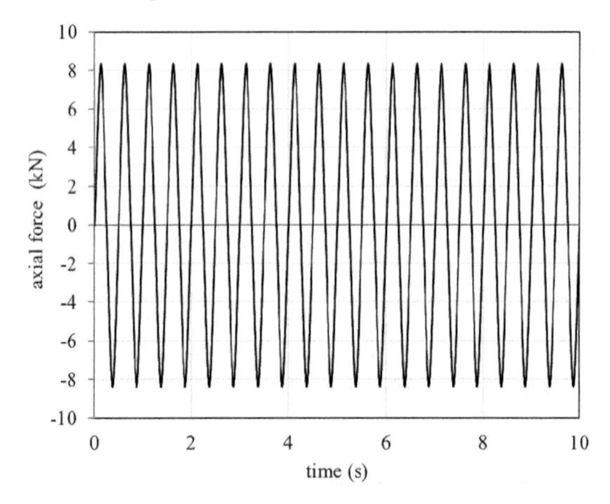

Fig. 8 Axial force response of bolts at the position of $x=3$m

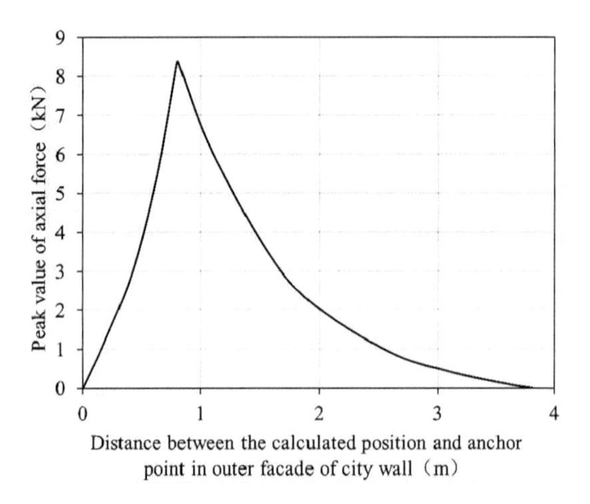

Fig. 9 Distribution of axial force peaks along with the bolt

6 Conclusion

This paper analyzed the dynamic response of a single bolt anchoring system by establishing a simplified mechanical model of a wholly grouted bolt anchoring system and considering horizontal seismic action. Major conclusions are as follows:

(1) A mechanical model of a wholly grouted bolt anchoring system against vertical wide cracks was established. Interaction between the soil mass and the bolt was studied by simplifying friction between the two into a parallel mechanism of a linear spring and a speed-related damper, and also

analytical solution of dynamic response of bolt axial forces was deduced. This method is prepared with easy computation and clear mechanical concepts, and thus has certain application prospect.

(2) The established model can be used not only for analysis of homogeneous soil, but also for consideration of soil difference at both sides of cracks in rammed layers to get more precise solutions in the reinforcement of earthen sites.

(3) Taking the reinforcement project of the ancient city of Gaochang in Xinjiang as an example, the dynamic response of anchorage system was analyzed by using the method proposed by this paper.

Note: The paper we submit is only limited to academic exchanges during the conference (ICOMOS-CIAV&ISCEAH 2019). Please do not publish it officially.

References

CAI Y, TETSURO E, JIANG Y J, 2004. A rock bolt and rock mass interaction modal. International journal of rock mechanics and mining sciences, 41:1055-1067.

CHEN W W, REN F F, 2008. Mechanical behavior of the bamboo-steel composite rock bolt. Research Report No. 2006BAK30B02. Lanzhou: Dunhuang Academy & Cultural Relics Protection Center of Lanzhou University.

COLLIN J G, CHOUERY-CURTIS V E, BERG R R, 1992. Field observations of reinforced soil structures under seismic loading. Proceeding of the International Symposium on Earth Reinforcement. Rotterdam, Netherlands.

COTTON P E, DAVID M, LUARK P F, et al., 2004. Seismic response and extended life analysis of the deepest top-down soil nail wall in the U. S. Geotechnical special publication, 124: 723-740.

DONG J H, ZHU Y P, 2009. Establishment of soil nailing dynamic system model and seismic response analysis. Chinese journal of theoretical and applied mechanics, 41(2): 236-242.

DONG J H, ZHU Y P, MA W, 2013. Study on dynamic calculation method for frame supporting structure with pre — stress anchors. Engineering mechanics, 30(5): 250-258.

FODDE E, WATANABE K, FUJII Y, 2007. Preservation of earthen sites in remote areas: the buddhist monastery of Ajina Tepa. Tajikistan. Conservation and management of archaeological sites, 9(4): 194-218.

Jaquin P, 2008. Study of historic rammed earth structuresin Spain and India. The structure engineer, 86: 26-32.

KAGAWA T, 1983, Lateral pile-group response under seismic loading. Soils and foundations, 23 (4): 75-86.

MARCUSON W F, 1981. Moderator's report for session on earth dams and stability of slopes under dynamic loads. Proceedings of the International Conference on Recent Advances in Geotechnical Earthquake Engineering and Soil Dynamics, San Diego, USA.

PARZ M, 1993. Structure dynamics theory and calculating. Beijing: Earthquake Press.

PENG N B, 2014. Study on seismic responses and anchoring mechanism of anchored rock slope. Lanzhou: Lanzhou University Press.

SEED H B, 1979. Soil liquefaction and cyclic mobility evaluation for level ground during earthquakes. Journal of the geotechnical engineering division, 105 (2): 201-255.

SEED H B, IDRISS I M, 1970. Soil moduli and damping factor for dynamic response analysis. Journal of terramechanics, 8(3): 190-191.

SHI Y C, QIU R D, 2015. Stability analysis of earthen slope sites reinforced with wood anchors. China earthquake engineering journal, 37 (3): 809-815.

SHOWKATI A, MAAREFVAND P, HASSANI H, 2015. Theoretical determination of stress around a tensioned grouted anchor in rock. Geomechanics and engineering, an int'l journal, 8(3): 441-460.

SHUKLA S K, KHANDEDWAL S, VERMA V N, et al., 2009. Effect of surcharge on the stability of anchored rock slope with water filled tension crack under seismic loading condition. Geotechnical and geological engineering, 27(4): 529-538.

VUCETIC M, TUFENKJIAN M R, DOROUDIAN M, 1993. Dynamic centrifuge testing of soil-nail excavations. Teotechnical testing journal, 16(2): 172-187.

WANG W, YUAN W, WANG Q Z, XUE K, 2016. Earthquake-induced collapse mechanism of two types of dangerous rock masses. Earthquake engineering and engineering vibration, 15 (2): 379-386.

WANG X D, CHEN W W, LI Z X, et al., 2013, Study on key technologies of earthen archaeological site protection. Beijing: Chinese Science Publishing & Media Ltd.

YE H L, ZHENG Y R, LU X, et al., 2011. Shaking table test on anchor bars of slope under earthquake. China Civil Engineering Journal, 44

(1): 152-157.

YE S H, ZHU Y P, WANG D J, 2014. Seismic interaction analysis modal and seismic response analysis of frame - prestressed anchor-soil system. China civil engineering journal, 47(5): 102-107.

YOU C A, 2000. Mechanical analysis on wholly grouted anchor. Chinese journal of rock mechanics and engineering, 15(3): 339-341.

ZHANG J K, GUO Q L, LI Z X, 2014. Preliminary study on anchorage mechanism of earthen archaeological site. Lanszhou: Lanzhou University Press.

ZHANG Y J, NIAN T K, ZHENG D F, et al., 2016. Analytical solution of seismic stability against overturning for a rock slope with water-filled tension crack. Geomechanics and engineering, an int'l journal, 11(4): 457-469.

ZHAO D, LU W, WANG Y L, et al., 2016. Experimental studies on earthen architecture sites consolidated with bs materials in arid regions. Advances in materials science and engineering, Article ID: 6836315.

ZHU F S, ZHENG Y T, 1996. Support action analysis of tensioned and grouted bolts. Chinese journal of rock mechanics and engineering, 15 (4): 333-337.

Protection of Cultural Relics at Xiaocheng Site and Reinforcement of Adobe Bricks against Wind Erosion Diseases

ZHANG Mengqiu

Qufu City Sankong Ancient Construction Engineering Management Office

Abstract Xiaocheng site was damaged by many factors, such as natural geology, human activities and environment, and carried out rescue protection construction according to the relevant documents of the Shandong cultural relics bureau. For this reason, the main diseases of its development: severe weathering, erosion, crevices and collapses, etc. , targeted methods for the masonry of Adobe bricks are reinforced. The raw materials, processing and construction process of Adobe brick are summarized. The results show that the method is of reference value to the protection and reinforcement of soil building sites.

Keywords soil site protection, arid brick production, adobe brick masonry

1 Introduction

The Xiaocheng site is located in Xiaocheng Village, Beitao Town, Guancheng County, Liaocheng City, Shandong Province. It is also known as Huangheold river. and Xiemacheng City. It is adjacent to the Weihe River in the South, and is adjacent to the Weiwei River in the West and Taoxian County in the North. (now Beiguan Tao Town) 2. 5 kilometers, East and Qingyuan Old City(now Qingqing Town). The Xiaocheng site was handed down as the first year of Songjingde(AD 1004). The place where Xiaochuoxie, the son of Liaoshengzongyelv, led the troops to attack the Song Dynasty, was the history of the Liao and Song dynasties and even the "Yuanyuan League". The physical witness of this important historical event has a high historical research value; At the same time, as the relics of the Liao Dynasty city site that is well preserved in Shandong Province and even in the country, the Xiaocheng site also has high scientific research value. In 2013, the State Council announced that the Xiaocheng site was the seventh batch of national key cultural relics protection units.

2 Site Distribution and Major Diseases

The Xiaocheng site is generally Square, with a length of about 1 340 meters on each side and a total area of 1 795 600 square meters in the city wall. Judging from archaeological work, the site mainly existed or formed in the Song Dynasty; As far as the preservation status and layout of the site are concerned, the city walls and arrow buildings are the main body of the site. The existing relics on the ground include the surrounding city walls, Yucheng, and the point and place stations, all of which are built of rammed earth.

There are all four walls. The bottom of the wall is about 26 meters wide, the top of the wall is 0. 8-3 meters wide, and the maximum height of the wall is 12 meters. The original four gates of the East, South, West, and North are all semi-circular Gates. The gates of the city are not opposed to the gates of the city. The local name is the "turning door". There are no traces on the surface of the East and West Gates; In the four corners of the city wall there is an arrow tower, which is connected with the city wall and protrudes out of the city wall.

Since the abandonment of the "Shuyuan League", the Xiaocheng site has been affected by multiple factors such as nature and man, and the remains have been severely damaged. The visible parts of the ground are the North and South cities, the surrounding city walls, and the points on the Northwest side of the city. At present, there are various degrees of diseases such as collapse, cutting surface, fissure, gully, and erosion.

Wind erosion is a kind of disease in the development of the site (Fig. 1). The destructive stress is mainly the rapid changes of wind, rain, temperature, and humidity, the migration and capillary effects of salt, and the root growth of plants.

Fig. 1 Wind erosion disease collapse of the wall body

3 Reinforcement and Protection of Adobe Brick Against Wind Erosion Disease

3.1 The Selection of Rraw Materials for Adobe Bricks

The selection of soil that conforms to the wall is close to the soil, and the nearest soil source is found, and it is advisable to use raw soil. Before the start of the project, the soil distribution in Guanxian County was reviewed on the ground. Combined with the survey report, the conclusion was drawn that the Huangpan Plain in the northwest of Luxian County was part of the North China Plain, and the soil types were divided into three types: tidal soil, salt soil, and sand (Fig. 2). The surface texture includes five kinds of loose sand, sandy loam, light loam, middle loam, and heavy loam. The average surface of the Xiaocheng site is 0.4 meters of loam (soil layer), 0.4 meters to 1.2 meters of loam (soil layer), and there is a layer of heavy loam (clay layer) of 15cm to 25cm below. The lower part is light loam sand. According to the actual situation and the opinions of all parties, the middle loam soil was selected, and the raw soil was properly mixed and cooked as the raw material of the Adobe brick, and the amount of salt was strictly controlled to prevent the use of Adobe bricks from causing secondary destruction due to rainwater erosion.

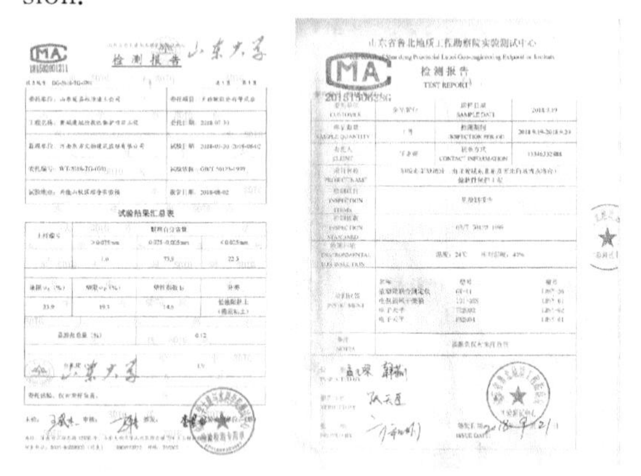

Fig. 2 Geotechnical test report

3.2 Production of Adobe Bricks

On the basis of adhering to the principle of the authenticity of the protection of cultural heritage and the principle of the protection and renovation of cultural relics that "does not change the original state of cultural relics", the original practices and original techniques are used as far as possible. Three kinds of targeted Adobe bricks were tested before construction started.

3.2.1 Artificial Towing to Fully Understand the Local Adobe Brick Making Process

Use of heavy soil, natural immersion and immersion of water (not less than 45 days), then drying and drying, after a freezing and thawing cycle in winter, adding water and mud to become viscous. Selecting Adobe mold for casting, vibrating and compacting, and disassembling. After a week of natural drying, the epidermis is densely dry, and moving the upper shelf does not affect the shape of the Adobe. After the upper shelf, the air dries for not less than 30 days so that the water evaporates fully for use. This process is a common practice for nearby residents in the winter pit. It is suitable for small batch production, but it is not suitable for the reinforcement and protection of the earth walls. The soil quality does not meet the requirements of the old city walls, and the heavy soil soil is sticky, and it is soaked for less than 45 days. After the winter freeze-thaw cycle, After the use of Adobe brick cracking serious, many fine cracks, qualified yield less than 10 %, not recommended for use.

3.2.2 Mechanical Adobe Suppression:

using a 40-kilowatt three-phase brick and tile machine to make Adobe bricks as a processing tool and field testing of various soil formulations, Among them, the proportion of loam soil(ratio 0.55) mixed with cooked soil(ratio 0.4) aggravated loam soil (ratio 0.05) plus water(soil moisture content of 25% to 30%) is the best ratio(the ratio of heavy loam soil is too large. The deformation of clay strips, the proportion of too small suppressed soil is not lubricated, Block the machine, the phenomenon of dumping soil is prominent), mix and mix the bricks of the back into the machine to suppress the Adobe brick, test the billet pass rate of nearly 60%, Adobe dry density reached 0.98 or more, higher than the density of rammed soil in the city wall. The disadvantage is that the yield is low, and cracks less than 3 mm appear in the drying process after the billet is formed, and the cracks are evenly distributed; After the sub-standard Adobe bricks are crushed, the soil can not continue to be used.

3.2.3 Artificial Rammed Billet

using 30% of cooked soil, 70% of raw soil, mixed water(according to 15% of the optimum moisture content of the compaction test) as soil raw material, made of iron mold according to the required size, filled with soil slightly higher than the surface of the mold, once compacted with rubber hammer, The upper hammer strikes the mark with the hand ramming to find the tonic leveling. The process uses soil to approach the rammed soil of the city wall, and the dry density reaches more than 93%, which is slightly higher than the dry density of the city wall and meets the requirements. Artificial billet breaking requires a large number of manual operations, the processing site remains flat, and human factors have a greater impact on the compaction density.

After confirmation on the spot, the Adobe bricks were made by means of artificial rammed billet (Fig. 3). At the same time, in order to ensure quality, all the processing sites are hardened on the ground, operators are strengthened to train and operate, and special personnel are set up to be responsible for quality inspection, and dry density testing is carried out in batches and periods to ensure that Adobe bricks are qualified.

Fig. 3 Artificial rammed billet

3.3 Arid Bricks to Dry

The processing plant chooses Adobe drying nearby, and grasps the dry humidity of Adobe on the spot. When the dry depth of the epidermis reaches more than 2cm, the upper shelf is chosen to dry. The base should be the soil ground, slightly higher than the natural flat is conducive to drainage, and the bottom is set to prevent the soil moisture from returning. The bearing capacity of fresh Adobe is calculated when it is on the shelf, and it can not be piled up too high at once to prevent the lower Adobe bricks from being pressed too thin(Fig. 4). After the spot inspection, Adobe brick air dry degree, meet the requirements for use.

Fig. 4 Adobe brick drying

4 Masonry Adobe Bricks

4.1 Selection of Supplementary Orientation and Base Level Treatment

The principle of "minimum intervention" is implemented to minimize interference with the wall body and the surrounding natural environment under the premise of ensuring the stability of the soil site. For soil erosion, holes and other diseases in the site to make up, fill. Select a wall reinforcement with severe wind erosion, use tools to clean up attached plant vegetation, clean up humus, not more than 10 cm is appropriate, shrubs that have a soil fixation effect on the site wall need to retain rhizomes, wait for natural factors such as climate and water to meet the requirements After sprouting to protect the wall; Remove cracked or loose earth blocks from the collapsed soil, clean up the floating soil layer(the thickness of the surface floatation should not be greater than 2cm), and remove the tread; Watering the wet site wall body should not be too dry. It is strictly forbidden to spread water and sprinkle the surface. After being wet, it stops for one hour and sprinkles water. It is repeated several times until the soil layer is soaked in about 5cm and the facade is about 2cm. The wet soil does not stick, do not pick up dust, grab the soil not scattered as a standard.

After the preparation work of masonry Adobe bricks is completed, skilled craftsmen are selected to make Adobe bricks, the shape is flat, and the Adobe bricks of moderate size are placed on the outside, which is conducive to the next step of repairing the shape, and the shape and size of the Adobe bricks are slightly different. Place on the inside to cut the stubble that fits the wall of the site.

4.2 Attention to the Construction of Masonry Adobe Bricks

The masonry Adobe brick mud, due to the artificial production of Adobe bricks, the thickness is formed at one time, and the size deviation is greater than 1cm, which is not conducive to filling and leveling, and the thickness needs to be manually polished. The thickness deviation is less than 1cm of Adobe brick. The proportion of masonry mud is appropriately increased, and the mud is smooth and smooth. The proportion of slurry of grouting material is slightly smaller, and it is necessary to ensure that the fluidity of the slurry is enhanced without pressure.

Due to the absence of a separate Adobe brick masonry specification, the inside Adobe brick supported by masonry was selected as the basis for quality control: the outside of the building is beautiful and with type, but the strength must be guaranteed; The mortar is full, and the ash suture is small and not large; The mud seam is flat and can be brick or brick number when necessary. It is not allowed to be adjusted by increasing the thickness of the gray seam. The Adobe brick with exposed wall body rammed earth adopts the mud brick wall method: the mud brick is full of mud; The mud seam is about 8~10 mm flat, and the vertical seam is cut to fit without leaving a gap; After laying, it can be reinforced with grouting.

Near the side of the compaction soil layer, it is generally necessary to cut the joint surface and the top surface of the soil brick, which is conducive to the close connection of the soil brick and the rammed earth wall. The large surface does not cut and grind, and after the completion of the masonry bricks such as the surface, the masonry body will be firmly and uniformly cut to trim the body shape. The first layer of masonry Adobe bricks should be placed horizontally, according to the proper shape with the type of placement. After acceptance, carry out the first layer grouting.

4. 3 Construction of Masonry Adobe Brick Grouting

Before grouting, it is necessary to seal the outside of the masonry gap. Block mud using masonry mud, seal tightly, prevent leakage of slurry. Bricks block one layer of grouting, grouting material mud proportion is small, prevent mud brick water absorption, hinder mud flow, forming a gap. The second layer of Adobe bricks needs to be pressed not less than 1/3 of the low brick. Set up the base in turn until the top of the rammed soil layer. The top of the masonry is less than 240 thick(one brick) and stops the masonry. The mud thickness is not more than 5 cm thick to seal the joint part, preventing rain from eroding and forming new holes. Before sealing the mud skin to dry 24 H, the contraction cracks produced by it are flattened with a wooden beat. For masonry with a height greater than 1. 5 meters, it is necessary to wait for about 1 meter of masonry mud and grouting material to dry. After the settlement of the wall body is stable, stubble masonry is connected. It is forbidden to lay more than 1. 5 meters at a time.

Adobe brick masonry grouting Adobe brick supplement (Fig. 5).

4. 4 Adobe Brick Shape Repair and Cleaning

After the masonry is completed, after the masonry mud is dried, the Adobe brick masonry that is not wrapped in the soil support must be hacked with the type, and the appearance of the Adobe brick masonry should be clean and clear without changing the appearance.

Fig. 5 Building and grouting

5 Conclusion

Through the selection of raw materials for Adobe bricks, manufacturing, masonry. Targeted construction, the results show that the selected materials and methods are effective in protecting the rammed earth of the city walls of the site, the effect is remarkable, and it is relatively close to the effect of the construction of rammed earth in the Song and Liao period, and has reference significance for the protection of rammed earth sites.

In the course of engineering construction, due to the uneven thickness of rammed soil layer in the original city wall site, the layer of the Adobe brick wall and the layer of rammed soil layer can not be effectively integrated, and still need to be explored and studied.

References

LIN binghuan, 2015. Rescue protection program for Xiaocheng site in Liaocheng city. Shandong: Qufu An Huai Tang.

LIU Dake. 2009. Ancient architectural engineering construction technology standard. Beijing: China Construction Industry Publishing House.

Field Direct Shear Test of Rammed Earth Ancient Buildings

LU Chao

Hebei Jianyan Architectural Design Co. , Ltd/Hebei Society of Civil Architecture, China

Abstract In the site of cultural relics protection project, the direct shear tests is carried out by using the method of preloading as a vertical reaction. By comparing the results of the field direct shear test with those of indoor geotechnical tests, the differences of shear strength indexes of Rammed Soil under different test conditions are analyzed. The test results show that: For rammed earth buildings with complex components or large inclusions in rammed soil, the direct shear test is recommended to obtain the shear strength index of rammed soil. The shear strength index of Rammed Soil obtained from the field direct shear test is smaller than that obtained from indoor direct shear test. In order to check the safety and stability of ancient tamping soil protection works, the cohesion and internal friction angle measured by indoor geotechnical tests can be reduced appropriately if the field direct shear test is not carried out, and the reduction coefficient is suggested to be 0. 7～0. 8.
Keywords protection of cultural relics, rammed earth, in-situ direct shear test, shear strength

1 Introduction

Ancient buildings are the material carrier of human culture and history. As a special form of architectural relics, rammed earth ancient buildings (Ancient ruins) are distributed in most parts of China. Some of these rammed earth relics stand on the ground as building structures, and some as building foundations bear under ancient buildings. For example, the Great Wall (soil structure part), the ancient city wall (soil structure part), the Yanxia Du site in Hebei Province, the Yecheng site and so on belong to the rammed earth ancient cultural relics.

Most of these cultural relics are in disrepair and some are on the verge of collapse. In order to do a good job in the protection of ancient buildings, the stability evaluation of rammed earth ancient buildings should be carried out first, and the corresponding reinforcement and protection methods should be put forward based on the evaluation results. The key to stability analysis of Rammed Soil is to select appropriate cohesion (c) and internal friction angle (φ). These two indicators can be obtained by in-situ sampling and sending to the laboratory for laboratory tests, or by in-situ direct shear tests for analysis. During the implementation of rammed earth building heritage protection project, many rammed soil layers are mixed with debris or bamboo bars. It is very difficult to adopt the original sample, which causes inconvenience for indoor analysis. It requires in-situ testing to obtain accurate geotechnical parameters for scheme design.

In recent years, scholars have carried out a lot of in-situ shear tests, but most of the subjects of these tests are natural stone or soil construction sites, and the in-situ shear tests of man-made compacted soil after hundreds or even thousands of years are rare.

In this paper, three typical protection projects of rammed earth ancient buildings (Ancient ruins) are selected to carry out direct shear tests on site. By comparing the results of direct shear tests indoors and outdoors, the differences of shear strength indexes of Rammed Soil under different test conditions are analyzed, and reasonable recommended values of geotechnical design parameters are put forward to guide the implementation of specific cultural relics protection projects.

2 Field Direct Shear Test Scheme

2.1 Test Site Selection

Because of the particularity of rammed earth ancient buildings (ancient sites) as cultural relics protection units, under the premise of achieving the experimental purpose, the main body of cultural relics should be disturbed as little as possible. Three rammed earth heritage conservation projects (two ancient city walls and one rammed earth ancient site) in Hebei Province were selected for field direct shear tests. The Rammed Soil of the engineering site is mainly composed of silt and silty clay. Three groups of tests were carried out at each site, and a total of nine groups of tests were carried out. In order to ensure the validity of the data, the section with smaller disturbance and better quality of Rammed Soil was selected during the test. In the course of the test, undisturbed soil samples were taken to carry out direct shear tests in the laboratory in order to make a comparative analysis of the data (Fig. 1).

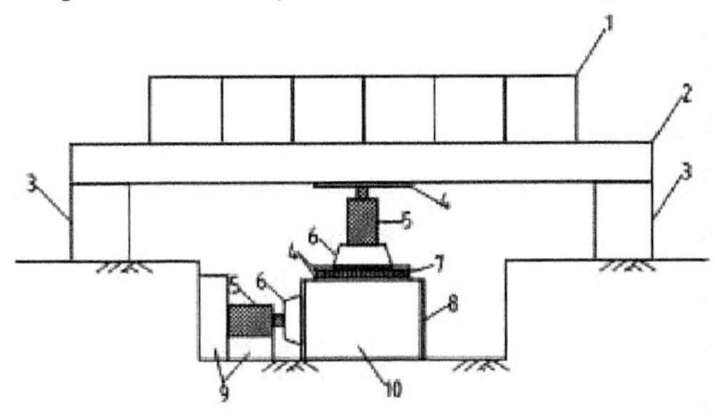

1 Concrete block 2 Test Platform 3 Bracket 4 Steel cushion plate 5 Lifting jack 6 pedestal 7 Steel Roller 8 Steel protective sleeve 9 Cushion block10 Rammed Soil Specimen

Fig. 1 Schematic diagram of field direct shear test equipment

2.2 Preparation of Specimens

The layout, preparation and processing dimensions of the specimens are in accordance with the current provisions of the Code for Geotechnical Engineering Survey.

Three groups of 3.0m long, 1.5m wide, test pits were excavated in three test sites according to the actual situation. The excavation depth of the test pits was determined according to the nature of the rammed soil layer after excavation. Three groups of original Rammed Soil Specimens with 60cm long, 60cm wide, 40cm high were processed in the center of the bottom. When making specimens, attention should be paid to selecting areas with better compacted soil layer to avoid the influence of tree roots or debris.

In order to protect Rammed Soil specimens, metal box protective sleeve was processed. The protective sleeve was fixed by 4 sheets of 2 cm thick steel plate and 8 adjustable length pull rods with wire. The size of each steel plate is 65 cm long, 66 cm wide and 35 cm high.

2.3 Scheme of Vertical Load Applying

Each specimen was subjected to a certain vertical load, then the horizontal shear load was applied to carry out the test.

Prediction and application of vertical load on NO. 1 test site (a city wall protection project):

The bulk density of Rammed Soil at the test site is 1700kg/m^3, and the location of the specimen is 8.20m away from the top of the wall. According to the actual working conditions, the self-weight pressure is 139.4kPa at the shear surface of the specimen. The shear area of the specimen is 0.36m^2, and the maximum effective vertical load is 50.2 kN. The vertical loads of three groups of compacted soil tests are 20kN, 35kN and 50kN, respectively.

Prediction and application of vertical load on NO. 2 test site (a city wall protection project):

The bulk density of the compacted soil at the test site is 1690 kg/m^3 and the location of the specimen is 6.58 m from the top of the city wall. Similarly, the maximum effective vertical load of the test is 40.0 kN. The vertical loads of three groups of compacted soil tests are 15 kN, 25 kN and 40 kN, respectively.

Prediction and application of vertical loads on NO. 3 test site (protection project of a rammed earth site):

The bulk density of compacted soil at the test site is 1670 kg/m^3, and the location of the specimen is 4.98m away from the top of the compacted soil. Similarly, the maximum effective vertical load of the test is 30.0 kN. The vertical loads of the three groups of compacted soil tests are 10 kN, 20 kN and 30 kN, respectively.

The vertical loads of each group of tests should be loaded step by step, once every 3 kN to 5 kN, every 5 minutes, and then the next one when the difference between the two successive vertical displacements is not more than 0.05 mm. Vertical deformation was measured at intervals of 5 minutes, 10 minutes and 15 minutes after the last stage load was applied. When the cumulative vertical deformation for two consecutive 15 minutes does not exceed 0.05 mm, it is considered that the vertical deformation is stable and shear load can be applied.

2.4 Shear Load Applying Scheme

Before the test, the shear load should be estimated. In the limit equilibrium state, the stress conditions on the shear plane conform to the Mohr-Coulomb formula.

$$Q_{max}=(\sigma\tan\varphi+c)A$$

Q_{max}—Maximum Shear Load(kN)

σ—Vertical Pressure(kPa)

φ—Internal Friction Angle(°)

c—Cohesive Force(kPa)

A—Shear Area of Specimens(m^2)

According to the soil conditions of similar properties, the values of — and c can be predicted, and the maximum shear load can be estimated by substituting the above formula, which is convenient for grading in the test process.

Estimation and application of shear load on NO. 1 test site:

Based on the results of indoor geotechnical tests and the empirical values, the maximum shear load of specimens under shear failure can be estimated by substituting the formulas of $\varphi=25$ degrees and $c=40$kPa. The Q_{max} of the first group was 23.8 kN when $\sigma=56$kPa, the Q_{max} of the second group was 30.7 kN when $\sigma=97$ kPa, and the Q_{max} of the third group was 37.7 kN when $\sigma=139$ kPa.

Estimation and application of shear loads on NO. 2 test site:

Based on the results of indoor geotechnical tests and the empirical values, the maximum shear load of specimens under shear failure can be estimated by substituting the formulas of $\varphi=20$ degrees and $c=42$kPa. The Q_{max} of the first group was 20.6 kN when $\sigma=42$ kPa, the Q_{max} of the second group was 24.2 kN when $\sigma=69$ kPa, and the Q_{max} of the third group was 30.0 kN when $\sigma=111$ kPa.

Estimation and application of shear load on NO. 3 test site:

Based on the results of indoor geotechnical tests and the empirical values, the maximum shear load of specimens under shear failure can be estimated by substituting the formulas of $\varphi=20$ degrees and $c=40$kPa. The Q_{max} of the first group was 18.1 kN when $\sigma=28$ kPa, the Q_{max} of the second group was 21.7 kN when $\sigma=56$ kPa, and the Q_{max} of the third group was 25.3 kN when $\sigma=83$ kPa.

The shear load shall be applied in accordance with the following requirements:

(1) The shear loads at each stage are applied in equal quantities of 8%~10% of the estimated maximum shear loads.

(2) When the shear deformation caused by shear load is more than 1.5 times that of the previous stage, the shear load of the next stage is reduced by half.

(3) The soil is subjected to first-order shear load for 30 seconds.

(4) Vertical deformation, shear load and shear deformation should be measured immediately after each stage of shear load is applied.

(5) When the peak shear stress or shear deformation increases sharply or the shear deformation is greater than 1/10 of the diameter (or edge length) of the specimen, the shear failure is considered and the test can be terminated.

3 Test Results and Analysis

3.1 Value of Shear Strength(τ)

After the test, the failure of the shear surface was observed and the σ-τ curve was drawn preliminarily in the field.

(1) With shear stress as the ordinate

and shear deformation as the abscissa, the relationship between shear stress and shear deformation is plotted, as shown in Fig. 2—Fig. 4. The peak value of shear stress on the curve is the shear strength (τ).

(2) When there is no obvious peak value on the relationship curve between shear stress and shear deformation, the shear stress at 1/10 of specimen diameter (or edge length) is taken as the shear strength (τ).

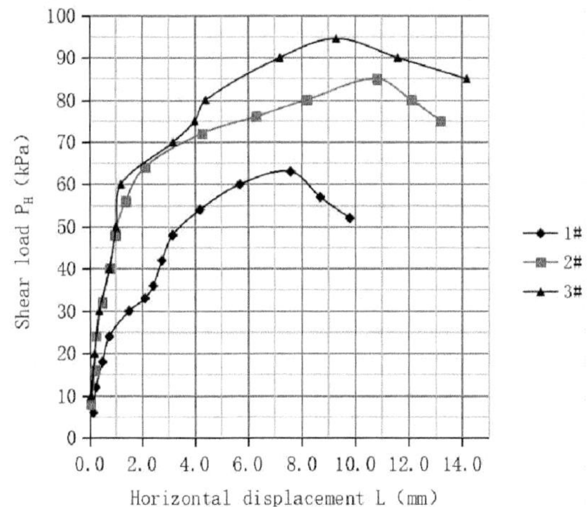

Fig. 2 Shear stress versus shear deformation curve of No. 1 Test Site

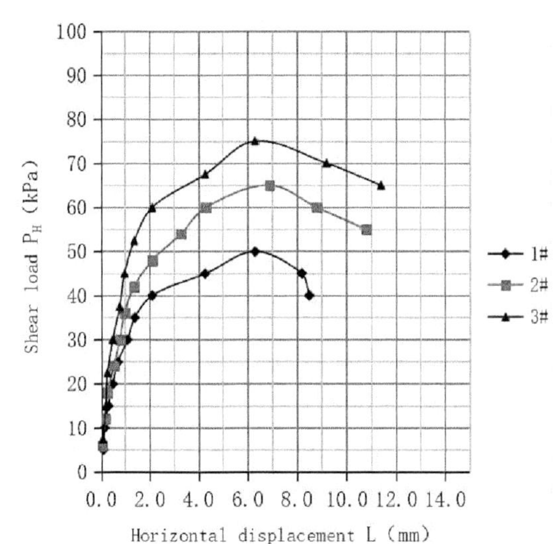

Fig. 3 Shear stress versus shear deformation curve of No. 2 Test Site

Taking the peak value of the above curve shear stress as shear strength, the shear strength of Rammed Soil Specimens in three test sites under different vertical loads is obtained. The results are as follows (Tab. 1)

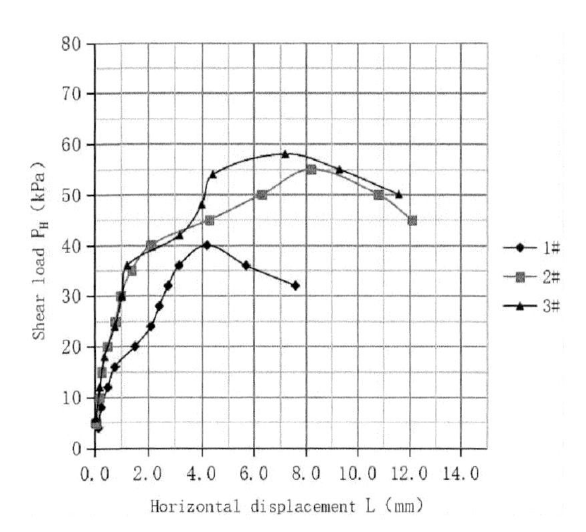

Fig. 4 Shear stress versus shear deformation curve of No. 3 Test Site

Tab. 1 Summary of in-situ direct shear test results

Site Number	project	No. 1 specimen	No. 2 specimen	No. 3 specimen
1#	Vertical load (kPa)	56.0	97.0	139.0
	Shear strength (kPa)	63.0	85.0	94.5
2#	Vertical load (kPa)	42.0	69.0	111.0
	Shear strength (kPa)	50.0	65.0	75.0
3#	Vertical load (kPa)	28.0	56.0	83.0
	Shear strength (kPa)	40.0	55.0	58.0

3.2 Determination of Cohesion (c) and Internal Friction Angle (φ) of Rammed Soil

With shear strength as ordinate and vertical pressure as abscissa, the relationship curve between shear strength and vertical pressure is drawn. The intercept of the straight line on the ordinate is cohesive force c, and the inclination angle of the straight line is internal friction angle φ. See Fig. 5—Fig. 7.

From the Fig. above, the cohesion of

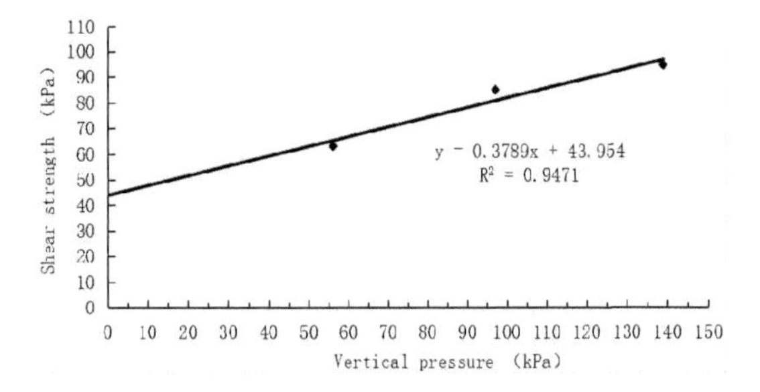

Fig. 5 Curve of relationship between shear strength and vertical pressure of No. 1 test site

Rammed Soil in No. 1 test site is 43. 9 kPa, and the internal friction angle is 20. 7 degrees.

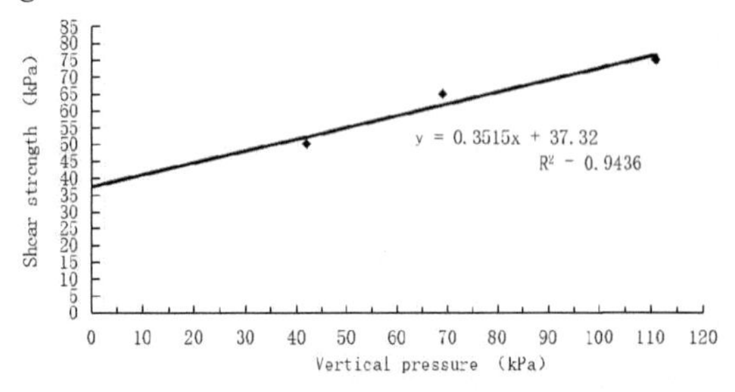

Fig. 6 Curve of relationship between shear strength and vertical pressure of No. 2 test site

From the Fig. above, the cohesion of tamped soil in No. 2 test site is 37. 3 kPa, and the angle of internal friction is 19. 4 degrees.

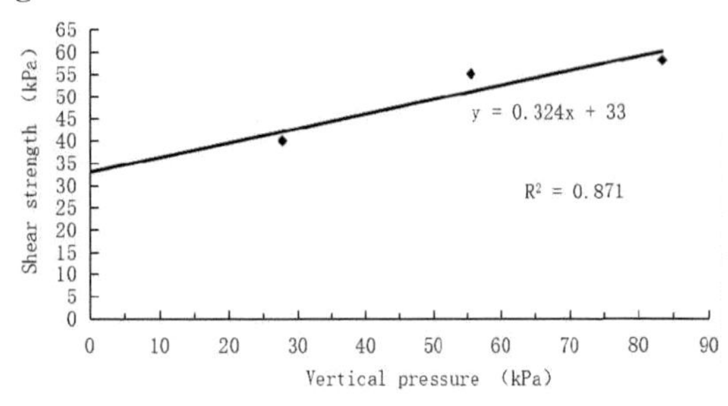

Fig. 7 Curve of relationship between shear strength and vertical pressure of No. 3 test site

From the Fig. above, the cohesion of tamped soil in No. 3 test site is 33. 0 kPa, and the angle of internal friction is 18. 0 degrees.

3. 3 Contrastive Analysis

In addition, indoor direct shear tests were conducted at the test sites of three sites for cultural relics protection. The values of c andφobtained from indoor direct shear tests were compared with those obtained from field direct shear tests, as shown in Tab. 2, Tab. 3.

Tab. 2 Comparing table of test results of cohesion of rammed earth

Test Categories	No. 1 Site	No. 2 Site	No. 3 Site
Field shear test c(kPa)	43. 9	37. 3	33. 0
Indoor Geotechnical Test c(kPa)	61. 3	46. 9	38. 6
ratio	0. 72	0. 80	0. 85

Tab. 3 Comparing table of test results of internal friction angle of rammed earth

Test Categories	No. 1 Site	No. 2 Site	No. 3 Site
Field shear test φ(kPa)	20. 7	19. 4	18. 0
Indoor Geotechnical Test φ(kPa)	27. 2	22. 8	20. 8
ratio	0. 76	0. 85	0. 87

From the above charts, it can be seen that the internal friction angle obtained by direct shear test in three test sites is larger than that determined by direct shear test in situ. For the internal friction angle, the ratio of field direct shear test to indoor direct shear test is 0. 76～0. 87.

The results obtained by the two methods are significantly different. According to the analysis, this is due to the nature of Rammed Soil itself. Most of the ancient Rammed Soil is not pure homogeneous material, which contains some broken bricks and tiles. On the one hand, it is difficult to adopt original samples in the field, and the samples obtained are not very representative. On the other hand, in the laboratory test, there are fragments in the original soil, which are more obviously hindered by the embedding of broken stones in the shear. The moisture content of the sample during transportation may be reduced, resulting in larger results of laboratory tests. Therefore, the results of in-situ direct shear test are more referential.

4 Conclusion

In the protection site of rammed earth cultural relics, the direct shear test is carried out in the representative area by using the surcharge method to obtain the shear strength index of rammed soil. By comparing the results of the field direct shear test with those of indoor geotechnical test, the differences of shear strength indexes of Rammed Soil under different test conditions are analyzed, and the following conclusions are drawn:

(1) For rammed earth buildings with complex components of Rammed Soil or large particles and impurities in rammed soil, direct shear test in situ is recommended to obtain the shear strength index of rammed soil, which is more conducive to guiding the concrete design and construction of cultural relics protection projects.

(2) Direct shear test in situ can reflect the actual shear state of Rammed Soil more objectively and accurately, which is helpful to evaluate the shear strength level of rammed soil.

(3) The shear strength index (cohesion and internal friction angle) of Rammed Soil obtained from the field direct shear test is smaller than that obtained from indoor direct shear test.

(4) In order to ensure the safety of rammed earth's buildings, it is suggested that the stability of rammed earth be calculated by direct shear test results in the protection and renovation of rammed earth ancient buildings (ancient sites). Without in-situ direct shear test conditions, the values of cohesion and internal friction angle measured by indoor geotechnical tests can be reduced appropriately, and the reduction coefficient is suggested to be 0.7~0.8.

References

LU Zude, CHEN Congxin, CHEN Jian-sheng, et al., 2009. Field shearing test for heavily weathered hornstone of three phase project of Ling'ao nuclear power station. Rock and soil mechanics, 30(12): 3783-3792.

SHI Jian, LI Min, WANG Yihong, et al., 2006. The test study of the engineering properties on the earth material of earth rammed construction. Sichuan building science, 32(4):86-87.

XIA Jiaguo, HU Ruilin, QI Shengwen, et al., 2017. Large-scale triaxial shear testing of soil rock mixtures containing oversized particles. Chinese journal of rock mechanics and engineering, 36(8): 2031-2039.

XING Haofeng, LIU Zhenhong, LI Qingbo, et al., 2007. Field shear test and analysis of rocks of the western route project for south-to-north water transfer. Rock and soil mechanics, 28:69-73.

YANG Zhe, LIN Dujun, LI Yinliang, et al., 2012. Study on the shear strength of rock and soil mass in slopes with in-situ tests. Geotechnical Investigation & Surveying, 8:18-22.

YUAN Run, SONG Xiangrong, LI Jiankan, 2009. Constituent and mechanical properties of rammed earth of Jiaoshan ancient emplacement. Journal of Jiangsu University (natural science edition), 30(6): 632-635.

Basic Principles for Soil Treatment with Binder — Stabilization of Fine-grained Soil with Lime

Oliver Kuhl

Hessen Mobil / Department Structural Engineering —
Test Centre for Building Materials and Soils, Germany

Abstract Soil treatment is a process by which soil is altered in order to achieve the required properties. Construction of walls with rammed earth requires a good level of resistance to water and frost. Without binders, fine-grained soils do not have sufficient water resistance. The stabilization of fine-grained soils with lime provides an opportunity for building durable and bearing earth walls.

The operational characteristics of lime-treated water- and frost-sensitive soils were investigated as part of a research program initiated by the Federal Ministry of Traffic, Construction and Housing; Germany. In order to enhance the mechanical properties, the use of combination binders, standardized hydraulic binders or their main hydraulic components and lime is an approved method. The performance of mixed binders has also been investigated as part of the research program. Based on the examinations, the requirements for water and frost resistance could be defined.

Suitability tests are carried out in order to verify the suitability of the mixture of building materials for the intended improvement and the specific requirements. In order to fulfil the key criteria for stabilization, the required levels of homogeneity, water content, binder dosage and density of the treated soil must be achieved. In doing so, the durability of the treated material can be ensured. The results of the suitability test are a deciding factor in the determination of the binder quantity and binder type.

Keywords soil treatment, stabilization, lime, combination binders, rammed earth

1 Introduction

1.1 General

Soil treatment with binders is a proven technology, which gained increasing economic importance in classified road construction from the mid-50s. Numerous scientific studies were carried out in the 1950s and 1960s, and the foundations for the regulations were developed.

The treatment of soil has a special significance in fine-grained soils. Soils with too high water contents cannot be sufficiently compacted. Soil treatment with lime (usually quicklime) is a well-known and extensively used technique for earthworks to dry up wet soils and to enhance their performance. The lime reacts with water to calcium hydroxide and changes the plasticity of the soil immediately (short-term reaction). Subsequent reactions of the hydrated lime with the soil dissolve clay minerals and pozzolanic components and cause the soil to solidify. These reactions can last for several years (long-term reaction). The long-term reaction leads to a stabilization of the soil. The water resistance and the frost resistance has significantly increased and after treatment soil-binder mixtures have a good load bearing capacity. In case of increased water and frost resistance requirements, check whether the addition of hydraulic components is required.

1.2 German Technical Regulations for Soil Treatment

The technical regulations have been developed by the Forschungsgesellschaft für Straßen-und Verkehrswege (FGSV). They regulate how technical issues have to be planned and implemented. They represent the current state of knowledge. They have been further developed on the basis of scientific research and practical experience in recent years.

In 2019, a European standard for soil treatment with binder (date of issue 2019-04) has been introduced for the first time.

2 Principles on Soil Improvement

2.1 Studies

The investigations have been carried out aiming at making fine-grained soils usable for high-quality applications in road construction. In particular, it was investigated how the binder-treated soils behave under the influence of water and frost.

The investigations regarded the type of soil, the type of binder (lime and cement), the quantity of binder and the dry density of the soil-binder mixture.

Krajewski & Kuhl investigated the behavior of four soils with different clay contents (Tab. 1), when treated with three different types of lime.

Tab. 1 Mass percentage of the different grain groups of the investigated soils

	mass percentage		
soil origin (number)	Clay(Size <0.002mm	Silt 0.002~0.06mm	sand or gravel> 0.06mm
Eltville	13	67	20
Echterdingen	30	48	22
Bönnigheim	18	52	30
Grünberg	1	4	95

In addition, the binder content has been varied for all soils and long-term storage up to 182 days has been carried out. For the investigations, special test specimens have been produced from the soil/binder mixture to determine the unconfined compressive strength and load-bearing capacity. The CBR test (California bearing ratio) according to DIN EN 13286-47 or TP BF - StB Part B 7.1 is a laboratory test and is used to assess the bearing capacity of subsoil. It is primarily used in road and traffic route construction. During the investigations, the bearing capacity was also determined after water storage (CBR_W) and freeze-thaw tests (CBR_{FT}). The clay content of the effects of the soil on the resulting strength (Fig. 1).

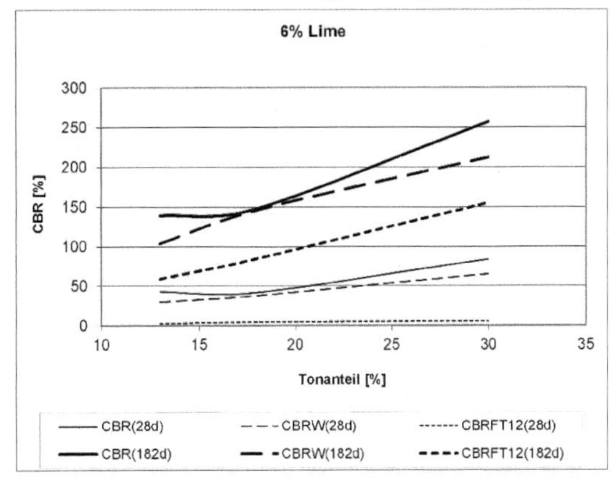

Fig. 1 Averaged CBR, CBR_W and CBR_{FT12} values after lime treatment with six percent lime as a function of clay content

A comparable behavior was observed by the series with tests to determine the compressive strength of the soil binder mixture (Fig. 2).

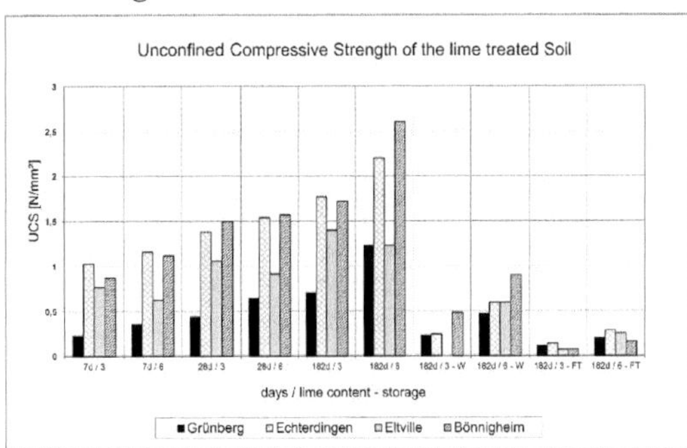

Fig. 2 Uniaxial compressive strength depending on storage time, binder content and water as well as frost stress

The type of lime, both quicklime and hydrated lime, showed a rather small influence on the strength to be achieved. During the investigations, it was also determined that sufficient resistance to frost and water effects can only be achieved with a longer storage period. A higher lime content has a positive effect especially in case of soils with a higher clay content.

Wichter & Lottmann could prove that higher dry densities of the treated soils have positively effected on frost behavior. Sufficient compaction of the treated soils is important for durability.

The strength properties of lime-treated soils directly depend on the fine grain content (clay and silt content) and the mineral phases of the soil. If the fine grain content is insufficient and/or if it is necessary to achieve rapid strength development, hydraulic binding agents (cement) are also used for soil treatment. For soils with higher clay content, which have high plasticity, however, it is difficult or impossible, to mix in hydraulic binders homogeneously. In order to combine both properties, so-called mixed binders (a mixture of quicklime and hydraulic binder) have been developed, which are frequently used in practice today. The reaction behavior of the mixed binders has been extensively investigated.

It has been shown that cement lead to a rapid strength development of the treated soil. However, it also became very clear that the mixing of the binder into the soil and the following compaction of the soil-binder mixture must be completed before the strength development of the cement begins. Otherwise, the soil cannot be sufficiently compacted and the strength properties are negatively affected (Fig. 3).

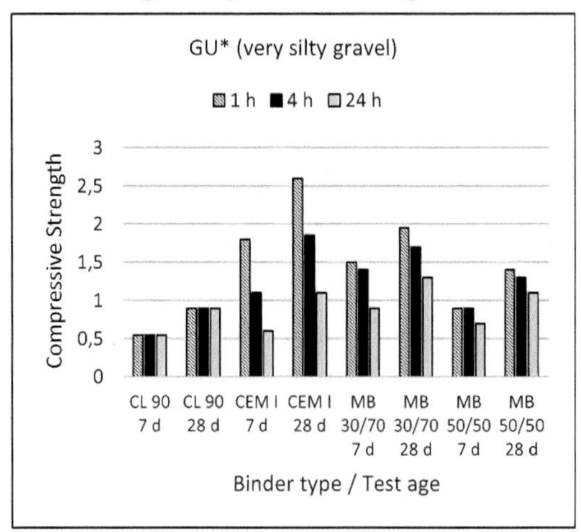

Fig. 3 Unconfined compressive strength as a function of reaction time (Time between mixing in the binder and finishing compaction)

A reaction time of six hours or more is recommended for lime and one hour for cement. A mixture of both binders should be applied within 4 hours.

A further study has been investigated whether frost-resistant soil stabilization with mixed binders can be achieved at a test age of 28 days. Pure binders lime (CL 90) and cement (CEM I) and mixtures (MB) with 70%, 50% and 30% lime by wheigt have been used in the investigations.

Tab. 2 Results of unconfined compression tests of a binder-treated light plastic clay (TL) before and after frost exposure (storage period 28 days)

	UCS [N/mm]			UCS [N/mm] after frost		
Binder	4%	7%	10%	4%	7%	10%
CL 90	1,08	1,03	1,14	—	—	—
MB 70/30	1,64	1,36	1,57	0,21	0,32	0,33
MB 50/50	2,32	2,48	2,59	0,98	1,41	1,69
MB 30/70	2,21	2,31	2,78	1,72	1,53	2,15
CEM I	3,72	4,31	5,69	2,12	3,28	4,50

It could be proved that the fine-plastic clays treated with mixed binders showed good frost behavior after a storage period of 28 days from a binder content of 4 M. %. Accordingly, the specific advantages of lime and cement can be combined with mixed binders.

2.2 Earthworks with Soil Treatment

On the basis of the investigations, technical parameters for the use of fine-grained soils as base layers and for backfilling of construction were decreed for the first time. The quantity of binder must be determined so that the unconfined compressive strength after 28 days of storage (at 20℃ and a humidity > 90%) is at least 0.5 N/mm. After 24 hours of water storage, the reduction in strength shall not exceed 50%. These parameters describe the bearing capacity as well as the water and frost resistance of the binder-treated soils. Since 1999, soil treatments with defined properties have been successfully applied to large construction projects.

During the construction of a federal road, the backfilling zone of a 12-metre-high supporting wall was constructed with a binder-treated loam. A soil was selected that is suitable for treatment with lime.

The binder content was determined on the basis of a suitability test.

The characteristic values of the soil-binder mixtures are listed in Tab. 3. A binding agent quantity of 3 wt. % was specified for the construction work.

Tab. 3 Mean values of suitability tests

Binder content wt. %	0	1	2	3	4	5
Proctor density [g/cm]	1,79	1,73	1,69	1,66	1,65	1,64
opt. water content wt. %	17,3	18,5	19,4	21,9	22,7	23,4
UCS [N/mm]	0,1	0,2	0,4	0,7	1,0	1,2
UCS [N/mm] after 24 h immersion in water	—	—	—	0,46	0,71	0,86
decrease in strength[%] after water immersion	—	—	—	34	29	28

The binding agent has been mixed into the soil outside the construction site. A binder spreading unit has been used to spread the binding agent evenly onto the soil. A moving tool mixer (specialist soil pulverizers with rotating drums) has been used to mix in the binder. It has been mixed until the soil and the binding agent showed an even colour.

The homogeneous distribution of the binder can be checked by spraying the treated soil with an indicator (Phenolphthalein).

The soil-binder mixture has been paved in a layer thickness of 30 cm with rollers. The required degree of compaction of 98% was reliably achieved with an average value of 99.6%.

Meanwhile, the construction method has developed into a standard construction method and has been included in the technical contract conditions.

2.3 Suitability Tests

The suitability test can be carried out in accordance with. The first step is to determine the grain size distribution of the soil. The selection of the binding agent depends on the fine grain content of the soil (Fig. 4).

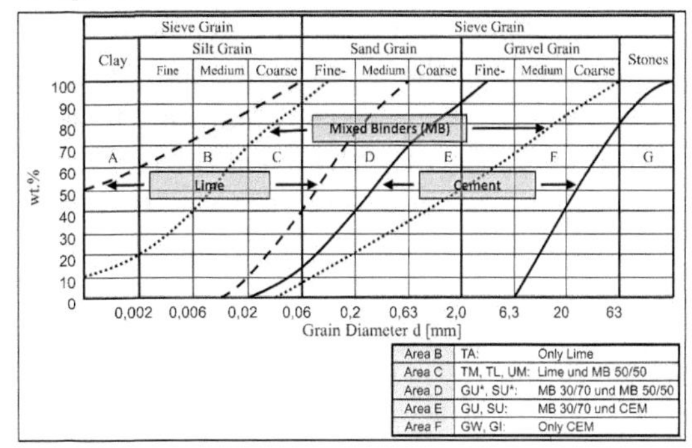

Fig. 4 Grain size band with the application areas of the binder types

Recommendations for the use of the different binders and empirical values for the quantity of binder are given in the information sheets.

If lime is used, a minimum quantity of components < 0.063 mm in the soil of approx. 15% by mass is required. If the soil has plastic properties (high clay content), lime is necessary for the binder. Lime changes the structure of the soil when mixed in through the cation exchange and flocculation/agglomeration of the clay minerals. The result is a so-called crumb structure, which enables the binder to be mixed in homogeneously. Lime reduces the plasticity of the soil and improves workability.

The compaction parameters result from the Proctor density and the optimum water content. The degree of compaction refers to the dry density at the optimum water content.

The suitability test examines and determines the quantity of binder with which the desired properties can be achieved. Special test specimens with different binder contents are produced for this purpose, in which the compressive strength is tested.

Before determining the compressive strength, the specimens can be subjected to water storage or freezing. Information about weather resistance and durability can be derived from these results and to determine the binder content. For this purpose, the ratio of compressive strength before and after immersion in water is used. For the use of soils as base layers and for the backfilling of construction the binder content is to design so that uniaxial compressive strength, after 28 days storage and testing according to TP BF-StB, Part B 11.5 is $\geqslant 0.5N/mm^2$. In either case, after 24 hours immersion in water, the decrease in strength must not be greater than 50% with respect to the value before immersion (Fig. 5). The minimum quantity of binder is 3 wt. %.

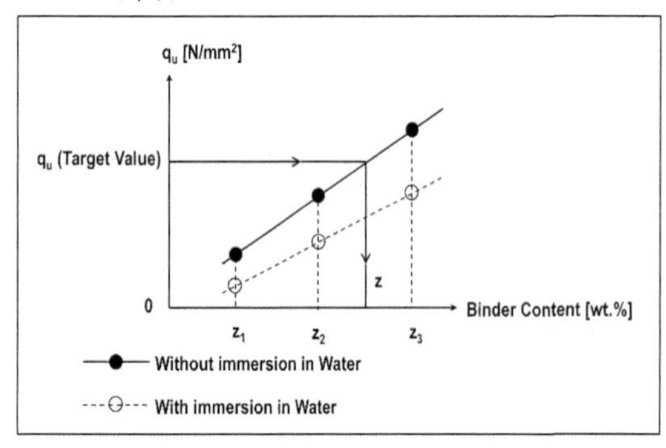

Fig. 5　Determination of the required binder quantity

For other soil treatment applications, the criteria for determining the binder quantity shall be determined by an expert according to the required properties.

3　Maintenance of Rammed Earth Walls with Binder-treated Soils

The basics and knowledge of soil treatment with binders can be transferred to the construction of rammed earth. Soil treatment with binding agents can contribute to the repair and preservation of historic earthworks. For the preservation and maintenance of the historic city wall of Pingyao in the province of Shanxi (People's Republic of China), suitable methods were tested and scientifically accompanied. Parts of the superficially destroyed wall were partially removed and rebuilt with a soil/lime binder mixture. After a few years, however, these newly restored sections also showed visible signs of weathering. The authors attribute this to the relatively low clay content of the soil used and the insufficient solidification by the lime binder (Fig. 6) under the arid climatic conditions.

Fig. 6　Weathering on the surface of rammed earth of the city wall of Pingyao © DAI

It is particularly important to consider that the surfaces of the earth walls are directly exposed to the weather. If the soil does not have sufficient quantities of pozzolanic components which lead to a sufficient setting and solidification by a reaction with the lime, negative effects on the durability cannot be excluded. It must be taken into account that these reactions take place over a period of time ranging from several months to years in touch with moisture. The use of hydraulic binder components is one way to improve the durability with a shorter setting time.

Natural hydraulic limes have hydraulic components. The reactions of these binders can be compared with those of mixed binders. They can, therefore, be particularly suitable for the preservation of the historical earthen architecture provided the suitability tests to prove under the local climate and workmanship. However, care should always be taken to ensure that the restoration work should be completed as early as possible, for example, at least two months before the first frost, in order to achieve frost-resistant strength.

When determining the characteristic values for the soil/binder mixture, the properties of the available soil and the climatic conditions must, therefore, be taken into account. Suitability tests are recommended to determine the suitable soils, quantity and the types of lime binders selected.

4 Conclusion

Soils treated with binders can be used to build durability earthworks and to preserve historic buildings made of rammed earth. When planning such constructions, knowledge about the reaction behavior of the soils with the binders and about the behavior of the binder-treated soils against weather influences is necessary.

The application of this method requires a project-related suitability test to determine the characteristic values for the soil-binder mixture.

During construction, special care has to be taken to ensure that the binder is mixed homogeneously with the soil. Powerful equipment should be used for this purpose. The binding agent can be mixed outside the construction site.

Sufficient compaction of the soil/binder mixture has to be ensured. For achieving this it is necessary to condition the water content of the soil to the optimum water content according to Proctor.

The maximum layer thickness of the soil to be compacted depends on the soil type and the performance of the compactor. Small rollers, vibratory plates and vibratory rammers can be used for compaction. The use of historical hand rammers is possible if an even and sufficient compaction can be proven.

Before carrying out the construction work, the specifications for the preparation of the soil/binder mixture as well as the working process for installation and compaction have to be defined in a working instruction.

References

Anon,2017. Additional technical conditions of contract and directives for earthworks in road construction (FGSV 599/10/2017).

Anon,2018. Earthworks — Part 4: Soil treatment with lime and/or hydraulic binders. German Edition EN 16907-4.

Anon,2004. Technical information sheet on soil stabilization and soil improvement using binders (FGSV 551).

Anon, 2012. Technical testing regulations for soil and rock in road construction, Part: Soil Stabilization mix designs (FGSV 591/B 11. 1/8/12).

Anon, 2010. Technical testing regulations for soil and rock in road construction, Part: Soil improvement mix designs (FGSV 591/B 11. 3).

Anon, 2012. Technical testing regulations for soil and rock in road construction, Part B 7. 1: Test method for determining the CBR value (FGSV 591).

DAI Shibing, LI Hongsong, 2016. Experimental study on the diagnosis and restoration of the rammed earth surface of the Pingyao city wall. Heritage architecture,1(1): 122-129.

KRAJEWSKI, WOLFGANG, KUHL, OLIVER, 2005. Suitability of frost-sensitive soils for treatment with lime. Berichte der bundesanstalt für straßenwesen, stra enbau volume number S 43.

KUHL O, 2003. Soil treatment in earth engineering. Series of publications of the working group on earthworks and foundation engineering. Earth and foundation engineering conference, 9: 101-104.

SCHADE, HANS-WERNER, 2006. Investigation of the reaction behaviour of mixed binders for soil treatment. Forschung stra enbau und stra enverkehrstechnik, 939.

Technical Information Sheet for the preparation, mode of action and use of mixed binders. Edition 2017 (FGSV 564/9/12).

WICHTER L, LOTTMANN, ALMUT, 2004. Influence of the operational characteristics of lime-conditioned, frost-sensitive soils at the formation level of traffic-bearing surfaces on the resistance of superstructures.

WITT, KARL J, et al., 2014. Investigation of mixed binders for soil stabilization. Forschung stra enbau und stra enverkehrstechnik, 1105.

Environmental Magnetic Non-destructive Testing of Weathering Degree of Ancient Brick of Pingyao Ancient City Wall

REN Jianguang[1], *HUANG Jizhong*[2], *REN Zhiwei*[1], *HU Cuifeng*[3]

1 Yungang Grottoes Research Institute, China
2 Shanghai University, Shanghai, China
3 Datong University of Shanxi, Datong, China

Abstract The weathering degree of ancient brick is the most important disease affecting the preservation of ancient city walls, ancient temples, the residence courtyard and other cultural relic building. The non-destructive testing of the weathering degree of ancient brick is the premise and foundation for the treatment of weathering diseases. Based on qualitative description, this paper makes a quantitative evaluation of weathering degree of ancient brick in Pingyao Ancient City, a world cultural heritage, by means of portable non-destructive testing instrument of environmental magnetic, and carries on the field verification combining with the testing instrument that commonly for the weathering degree of brick and stone cultural relics. The results show that it is feasible to use the portable SM-30 magnetization instrument of environmental magnetism to evaluate the weathering degree of ancient brick of Pingyao Ancient City Wall lossless. The magnetic susceptibility of weathered ancient bricks in the city walls has a good correspondence with surface mechanical strength, which can reflect the weathering and deterioration of ancient wall brick to a great extent. The research results can provide the basic data for the protection measures of the ancient brick of Pingyao Ancient City, which has important scientific significance.

Keywords Pingyao Ancient City, ancient brick, environmental magnetism, weathering degree, non-destructive testing

1 Introduction

Pingyao Ancient City is located in Pingyao County, central Shanxi Province. It was built in the Zhou Xuanwang period (827-728 BC). The existing ancient city regulation was established in the 3rd year of Ming Dynasty Hongwu (AD 1370) and was expanded and reconstructed on the basis of the old city of Pingyao, which is an outstanding example of Chinese Han nationality cities during the Ming and Qing Dynasties. The basic pattern has not changed since the past 600 years and the cultural relics are of high value. Pingyao Ancient City declared as the third batch of Key Cultural Relic unit under State Protections by State Council in January 1988 and listed by UNESCO as a World Heritage in December 1997.

For thousands of years, due to the influence of nature and humans, the weathering damage of Pingyao Ancient City has been very serious although repaired in the past dynasties. Partially weathered walls have seriously threatened the security and integrity of the heritage, so weathering governance is imminent. The weathering degree of ancient brick is the most important disease affecting the preservation of ancient city walls, ancient temples, and residence courtyard and other cultural relic building, the evaluation of which is the premise and foundation for the treatment of weathering diseases of cultural relics. For a long time, the qualitative description has been the main method to determine the weathering degree of ancient brick. Although this method has the advantages of convenient application, it is difficult to grasp the standard due to the lack of quantitative indicators. So it is easy to cause the inaccuracy and artificiality of weathering degree division, and difficult to meet the needs of today's conservation of cultural relic buildings. Pingyao Ancient City is a non-renewable cultural resource, and it is urgent to find and use modern non-destructive testing instruments to make up for the lack of quantitative evaluation of ancient brick weathering.

In this paper, the ancient brick wall of Pingyao Ancient City is taken as the research object. The weathering disease investigation of the ancient brick wall of Pingyao Ancient City Wall is carried out by the environmental magnetic portable non-destructive testing instrument (REN Jianguang, HUANG Jizhong, WANG Xusheng, 2013) and the testing instrument commonly for the weathering degree of brick and stone cultural relics (HOU Zhixin, ZHE Rui, ZHANG Zhongjie, 2018). The testing results provide a scientific basis for the treatment of weathering diseases in Pingyao Ancient City and the development of reasonable protection measures.

2 Investigation on the Preservation of Ancient Bricks

2.1 Shape of City Wall

The wall of Pingyao Ancient City consists of a wall body, city gate, Mamian, Wengcheng, Jiaotai, etc. The east, west and north walls of the ancient city are straight. The south wall is built with the Liugen River. The circumference is 6162.68m and the wall body section is trapezoidal. The main structure is in the form of rammed earth structure as the core and brick-wrapped on the outside of the wall. The thickness of the outer brick masonry is 87cm, 70cm and 53cm from the bottom to the top, and the height of each layer is about one third of the total height of the wall. The size and weight of ancient bricks in the Qing Dynasty are smaller than those in the Ming Dynasty. The average size and weight of ancient bricks in Qing Dynasty are 294.6mm×142.2mm×58.1mm, 4.202kg, and in Ming Dynasty 341.5mm×165.5mm×79.9mm, 7.478kg.

2.2 Investigation on Preservation of Ancient Bricks

In 2012, the author conducted field investigation on the types, distribution, damage degree and influencing factors of diseases of ancient brick walls, and found out the basic condition of diseases of ancient bricks after sorting out: ①In the ancient bricks of Pingyao Ancient City Wall, there are types of diseases such as weathering, water stain, salt damage, artificial damage surface, artificial repairing and cracks. Apart from the cracks, the total area of the other five diseases reaches 93743 m^2. The endangered degree accounted for 1.2% of all disease areas, the seriousness for 27.1%, the moderate for 18.5%, and the slight for 55.4%.

No fresh ancient bricks were found in field investigation and the ancient bricks of Pingyao Ancient City Wall were generally well preserved. ②The disease area of each section of the city wall is equivalent, and the preservation situation depends on the degree of the disease damage. The preservation condition of the west wall is the best and the seriousness accounts for 3.54% of the area of all diseases. The north wall is good and the seriousness accounts for 6.77%. The east wall was generolly accounted for 7.52%, while the south was the worst accounted for 1.2%. ③Weathering disease area is 64327.5 m^2, accounting for 68.62% of all disease area, which is the most common and main disease type on the ancient brick of Pingyao Ancient City Wall.

2.3 Weathering Disease Types

Weathering diseases of ancient brick of Pingyao Ancient City Wall generally occur on the surface of the brick. According to the appearance, damage degree and color of the ancient brick surface, weathering of the ancient brick shows four types of diseases: yellowish-black or yellowish-brown hard shell, gray-black hard shell, powder weathering (butter alkali) and brick erosion. The distribution area and proportion of various weathering diseases are shown in Tab. 1.

Tab. 1 Statistical Table of Weathering Diseases of Ancient Bricks (Unit: m^2)

Name		East wall	South wall	West wall	North wall	Total
Total disease area		24756.2	25703.8	21669.1	21614	93743.1
Total weathered area		16299.6	18620.6	16737	12670.3	64327.5
Yellowish-black or yellowish-brown hard shell	Area of distribution	9372.6	12135	14517	10104	46128.6
	Percentage of total weathered	0.146	0.189	0.226	0.157	0.717
	Proportion of total disease	0.1000	0.1294	0.1549	0.1078	0.4921
Gray-black hard shell	Area of distribution	3630.9	3142.6	1705.5	2302.8	10781.8
	Percentage of total weathered	0.0564	0.0489	0.0265	0.0358	0.1676
	Proportion of total disease	0.0387	0.0335	0.0182	0.0246	0.1150
Powder weathering	Area of distribution	3326.1	3450.7	916	311.5	8004.3
	Percentage of total weathered	0.0517	0.0536	0.0142	0.0048	0.1244
	Proportion of total disease	0.0355	0.0368	0.0098	0.0033	0.0854
Brick erosion	Area of distribution	0	1129.3	0	0	1129.3
	Percentage of total weathered	0	0.0176	0	0	0.0176
	Proportion of total disease	0	0.0120	0	0	0.0120

2.3.1 Yellowish-Black or Yellowish-Brown Hard Shell

The surface of the ancient brick wall of Pingyao Ancient City is widely distributed with yellowish black or yellowish brown. It is the most important type of disease, which is 46128.6 m^2, accounting for 71.7% of the total weathered area, accounting for 49.21% of the total disease area. Among them, the most on the west wall, generally distributed in the middle of the wall and the female wall. They are compact and form the same layer of the membranous structure. The yellowish-black or yellowish-brown is iron oxide on the surface of the ancient brick that may be under the action of water, ultraviolet, air oxidation, etc., the hematite or pyrite ore contained in the brick through capillary water migration accumulated.

2.3.2 Gray-Black Hard Shell

The gray-black hard shell surface of the ancient brick wall of Pingyao Ancient City is 14790.1m^2, accounting for 16.76% of the total weathered area, accounting for 11.5% of the total disease area. The most on the east wall, generally exists the bottom of the wall and the base of the female wall. The structure is compact and the sound of knocking occasionally hollow is heard. Under the capillary action of groundwater or surface water, the moisture content of the brick increases, and the iron oxide in the ancient brick undergoes a reduction reaction to form a gray-black hard shell at the bottom of the wall. The gray-black hard shell below the base of the female wall is mainly caused by the leakage of atmospheric precipitation at the top of the wall.

2.3.3 Powder Weathering (Butter Alkali)

Powder weathering (butter alkali) is mainly characterized by powdery flaking of the outer surface of the ancient brick, which leads to the phenomenon of bumpy and scraggy on the surface of the wall. Its distribution area is 8004.3m^2, accounting for 12.44% of the total weathered area and 8.54% of the total disease area. Among them, the east and south walls are the most common ones, which are distributed at the bottom and have a loose structure. It may be caused by the capillary action of groundwater or surface water and long-term weathering of ancient bricks.

2.3.4 Brick Erosion

The brick erosion is mainly manifested by the pulverization sapping of the brick body, the exfoliated sapping of the brick body and the erosion of the white marl adhesive materials between the bricks. It exists only at the bottom of the ancient brick wall of the south wall, and the structure is very loose. The distribution area is 1129.3m^2, accounting for 1.76% of the total weathered area, accounting for 1.21% of the total disease area.

In addition, from the field investigation, the qualitative description method can not ensure the scientific nature of the classification of weathering diseases. Such methods rely on the surface morphology of ancient bricks and people's subjective consciousness. Therefore, how to quantify the degree of weathering disease under the principle of "minimum intervention" is an important topic for studying the protection of Pingyao Ancient City.

3 Classification of Weathering Degree

3.1 Qualitative Classification of Weathering Degree

According to the color luster, the structure and the broken condition, the change in mineral composition, the change of physical mechanical characteristics and the hammering sound, the weathered ancient bricks of Pingyao Ancient City Wall are qualitatively divided into five categories: non-weathered ancient bricks, weak weathered ancient bricks, moderate weathered ancient bricks, strong weathered ancient bricks, completely weathered ancient bricks (Fig. 1).

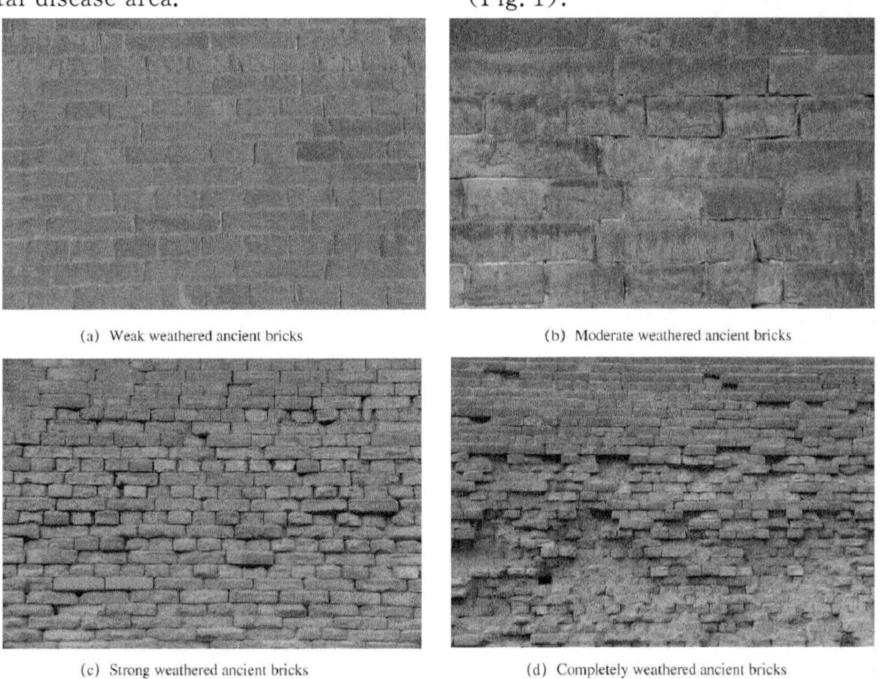

(a) Weak weathered ancient bricks (b) Moderate weathered ancient bricks

(c) Strong weathered ancient bricks (d) Completely weathered ancient bricks

Fig. 1 Type of Weathered Ancient Bricks

3. 2 Quantitative Classification of Weathering Degree

3. 2. 1 Measurement Method

(1) Detecting instrument

The magnetic susceptibility of the ancient wild outcrop bricks is measured by SM-30 (Fig. 2), a portable magnetization susceptibility meter made in the Czech Republic. The sensitivity is high, reaching 1×10^{-7} SI, which can accurately measure the magnetic susceptibility of paramagnetic, diamagnetic and ferromagnetic rocks. With a volume of 100mm × 65mm × 25mm and a weight of 180 g, it can obtain 90% of the signal in the range of 20 mm. The measurement is automatically adjusted and the operation is simple. At the same time, it can effectively reduce the external electromagnetic interference and the noise impact of electronic equipment. It has the characteristics of fast, low cost and non-destructive to the measured object, and is an ideal field non-destructive magnetic measurement tool.

Fig. 2 SM-30 Magnetization Meter

TH110 HL hardness tester (the following called "hardness tester") (Fig. 3), produced by Beijing Time High Technology Ltd, is a portable instrument for measuring the surface strength of objects. It is small in size, light in weight and easy to carry. It can carry out high-precision strength testing on the surface of various materials.

(2) Measurement method

Using SM-30 portable magnetic susceptibility meter and TH110 HL hardness tester measured the magnetic properties and surface strength of weathered ancient bricks in Pingyao Ancient City Wall. The measurement method follows the following principles: ① Selecting different weathering types, different weathering degrees of the same weathering type and different areas of the same weathering degree for field non-destructive magnetic susceptibility and surface strength testing; ② On the surface of each flat and representative weathering ancient brick, ten different points are selected, each point is measured three times and the data with measurement error less than 1% are calculated. Then take the average value of three data as the magnetic susceptibility and surface intensity field measurement value of the point. Finally, the field test data are analyzed and counted by EXCEL software.

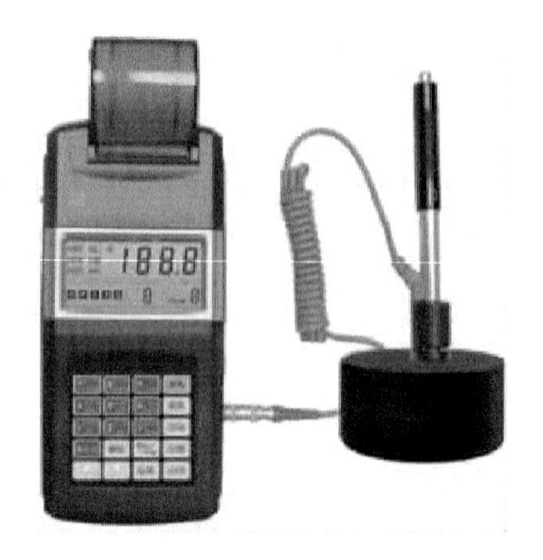

Fig. 3 TH110 HL Hardness Tester

3. 2. 2 Non-destructive Test Results of Magnetic Susceptibility

(1) Weak weathered ancient bricks

The weak weathered ancient brick structure of Pingyao Ancient City Wall is intact, with a slight rust yellow weathered skin, and the sound is crisp. Based on 496 field data, the frequency distribution histogram and curve of magnetic susceptibility nondestructive testing of weak weathered ancient bricks are made (Fig. 4), with a minimum of 0.483×10^{-6} SI, a maximum of 2.820×10^{-6} SI, an average (μ) of 1.524×10^{-6} SI, and a standard deviation (σ) of 0.593×10^{-6} SI. According to the normal distribution curve, the probability of taking the value in the interval ($\mu-\sigma$, $\mu+\sigma$) is 68.3%, and the weak weathered ancient brick is determined. The magnetic susceptibility of weak weathered ancient bricks ranges from 0.931 to 2.117×10^{-6} SI.

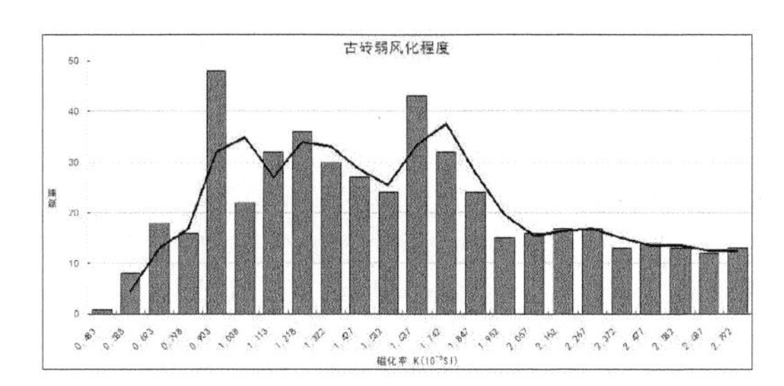

Fig. 4　Frequency distribution histogram and curve of weak weathered ancient bricks

(2) Moderate weathered ancient bricks

The surface of moderate weathered ancient bricks in Ping Yao Ancient City Wall with gray-black hard shell, strong weathering a along the face of surface, and the hammering sound are not crisp enough. A total of 995 data were measured at the scene. It can be seen from Fig. 5 that the average magnetic susceptibility of moderate weathered ancient bricks is 2.869×10^{-6} SI, and the effective value range is $1.203\sim4.535\times10^{-6}$ SI.

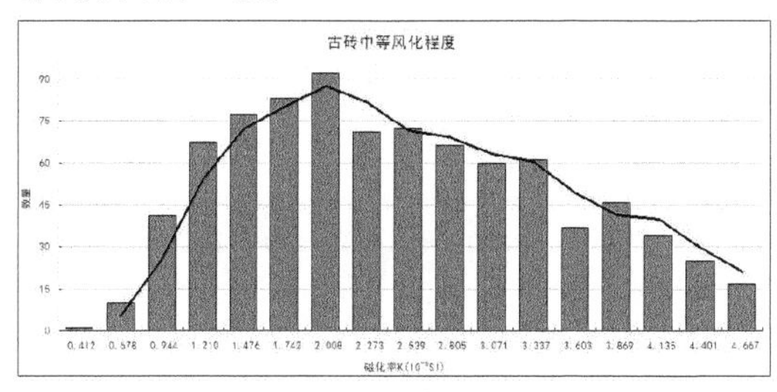

Fig. 5　Frequency distribution histogram and curve of moderate weathered ancient bricks

(3) Strong weathered ancient bricks

Ping Yao Ancient City Wall strong weathered ancient bricks only the center of the fracture remained the original color, part of the flake fell, hammering sound for dumb. A total of 1409 data were measured in the field. According to Fig. 6, the average magnetic susceptibility of strong weathered ancient bricks is 1.761×10^{-6} SI, and the effective value range is $1.203\sim4.535\times10^{-6}$ SI.

(4) Completely weathered ancient bricks

The completely weathered ancient bricks are gray and white in color. These can be easily squeezed by hand and present gravel or powder. Integrity is completely destroyed. A total of 764 data were measured in the field. According to Fig. 7, the

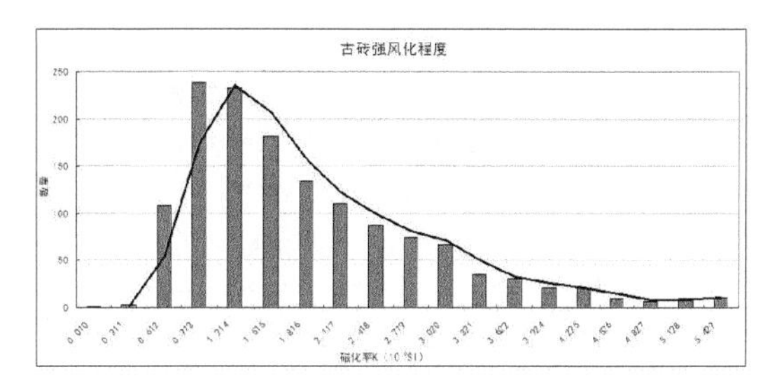

Fig. 6　Frequency distribution histogram and curve of strong weathered ancient bricks

average magnetic susceptibility (μ) is 0.774×10^{-6} SI, and the effective value range is 0.774×10^{-6} SI.

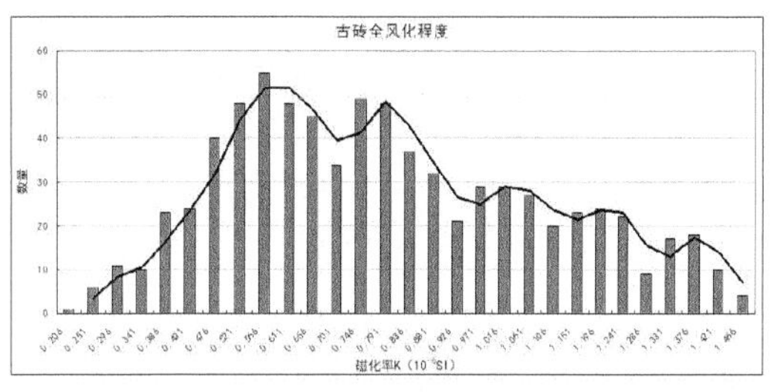

Fig. 7　Frequency distribution histogram and curve of completely weathered ancient bricks

Generally speaking, the magnetic susceptibility of ancient brick wall in Pingyao Ancient City has the characteristics of increasing first and then decreasing during the weathering process: weak weathering stage, low magnetic susceptibility; moderate weathering stage, highest magnetic susceptibility; strong weathering stage, high magnetic susceptibility; In the weathering stage, the magnetic susceptibility is the lowest.

3.2.3　Corresponding Relation with Surface Strength and Magnetic Susceptibility

A total of 663 surface strength data were measured in the field, and the surface strength of weathered ancient bricks showed the downtrend during the weathering process. Among them, the weak weathered ancient bricks have high surface strength with an average value of 364.074 MPa.; the average value of moderate weathered ancient bricks is 274.941 MPa, and the surface strength decreases; the average value of strong weathered ancient bricks is 227.496 MPa, with lower surface strength; and the average value of completely weathered ancient bricks is 190.857 MPa, with the lowest surface strength.

The magnetic susceptibility of weathered ancient bricks in Ping yao Ancient City Wall can reflect the main mineral composition and mineral weathering and deterioration. The strength of the ancient brick surface depends on the main mineral composition and its percentage content. Therefore, the combination of the magnetic susceptibility and strength of the ancient brick can better reflect the mineral composition and weathering degree of the ancient brick. The magnetic susceptibility of the ancient brick has the characteristics of increasing first and then decreasing during the weathering and metamorphism, while the surface strength decreases with the increase of weathering and deterioration. The weathering and deterioration of ancient bricks can be divided into four stages: (1) Weak weathering stage, ancient bricks have low magnetic susceptibility and high surface strength; (2) Medium weathering stage, the magnetic susceptibility of ancient bricks increases to the maximum, while the surface strength decreases; (3) In the strong weathering stage, the magnetic susceptibility and surface strength of ancient bricks decrease simultaneously; (4) In full weathering stage, the magnetic susceptibility and surface strength of ancient bricks are all reduced to the minimum. Therefore, the magnetic susceptibility of the weathered ancient bricks of Pingyao Ancient City Wall has a good correspondence with the surface strength, which can reflect the weathering and deterioration of the ancient bricks to a large extent. At the same time, the weathering degree of ancient bricks is tested in the field by using the surface strength testing instrument of ancient bricks, which is verified by the environmental magnetism method. Finally, the non-destructive testing technology of environmental magnetism method can be used as a powerful means to detect and evaluate the weathering degree of cultural buildings.

4 Conclusion

(1) There are six types of diseases on the ancient brick walls of Pingyao Ancient City, including weathering, water stain, salt damage, artificial damage surface, artificial repairing and cracks. Weathering diseases are the most common and main types

(2) There are four types of weathering diseases on the surface of ancient brick, which are yellowish-black or yellowish-brown hard shell, gray-black hard shell, powder weathering (butter alkali) and brick erosion.

(3) In the process of weathering and metamorphosis, the weathered ancient bricks of Pingyao Ancient City Wall have the characteristics of increasing first and then decreasing: weak weathering stage, low magnetic susceptibility; moderate weathering stage, highest magnetic susceptibility; strong weathering stage, high magnetic susceptibility; in the weathering stage, the magnetic susceptibility is the lowest.

(4) The surface strength of ancient bricks of Pingyao Ancient City Wall decreases gradually with the increase of weathering degree. It has a good correspondence with the magnetic susceptibility of ancient brick, which can reflect the weathering and metamorphosis.

(5) It is feasible to use portable SM-30 magnetic susceptibility instrument of environmental magnetism to evaluate the weathering degree of the ancient brick wall in Pingyao Ancient City Wall. It can provide basic data for the protection of cultural relics and buildings, and has important scientific significance.

References

HOU Zhixin, ZHE Rui, ZHANG Zhongjie, 2018. Evaluation of weathering degree of stone cultural relics based on non-destructive measurement method of Leeb hardness teste. Proceedings of the 2018 National Engineering Geology Annual Conference.

REN Jianguang, HUANG Jizhong, WANG Xusheng, 2013. Environmental magnetism nondestructive detection of weathering depth and velocity of sandstone in Longwangmiao valley of Yungang Grottoes. Geotechnical investigation & surveying, 9: 69-74.

Sakae-Kreua: Co-learning Space in Rural Thailand

Chantanee Chiranthanut

Faculty of Architecture, Khon Kaen University, Thailand

Abstract Sakae-Kreua, a co-learning Space in Rural Khon Kaen, Thailand, was part of the 2018 ArchKKU Design Build program organized by Faculty of Architecture, Khon Kaen University. The project provided architecture students with a hands-on learning experience which is the vernacular architecture approach, known as learning-by-doing. The concept of this project gave architecture students opportunities to learn through hands-on practice and to contribute to people living in poverty in a rural community. Even though it used ordinary materials available within the community, it is exceptional for its striking material choices, in particular salvaged materials. In addition, it combined new techniques with common construction practices learned in school.

Sakae-Kreua project locates in Khon Kaen province in the heart of Northeastern Thailand. The project work broadened the students' perspective of vernacular architecture within a contemporary context. Moreover, it encouraged a receptive solution-focused mindset for work and supported the adaptation of existing construction materials and components for building architecture within a specific context and time frame. The activity proposes are an aspect of architectural study and training through a "Design-Build" activity which the "learning-by-doing" approach plays a significant part. This student voluntary rural development activity had been discovered corresponding concepts between Design-Build projects and vernacular architecture both in the aspect of execution and conceptualization. The common characteristics can be concluded as follow: 1) practical application of using local materials; 2) understanding and realizing of limitations and contexts; 3) reusing materials and resources; 4) managing through context-based solutions and adaptations. The project attempts to study and apply such concepts to efficiently manage limitations and conditions within a restricted time frame of construction. At the same time, it questions the creation of locally conceived architecture within the changing context in order to create possible new answers and approaches.

Keywords co-learning space, local development, design-build, architectural education, vernacular

1 Introduction

1.1 Aim of the Study

This paper aims to show the relation between Design-Build and vernacular architecture shares idealist concepts. Moreover, it shows how to use 'Design-Build' project as a tool for architectural education to understand Thai rural contemporary vernacular architecture.

1.2 Philosophy of Vernacular and Design-Build

The Design-Build project using for educating architecture student to understand the relation of theories and design ideas into architecture. The Design-build concept is to encourage student faces limitation of materials, structural technique and gives the opportunity for solving real problems. Moreover, it has to focus on contexts and global sustainability.

The essence of vernacular architecture reflects the evolution of architectural traditions and styles in response to conditions of context and environment. The same way as design-build, it also relies on the use of existing indigenous materials, local labor and skills that affects not only construction methods and techniques but also improvised solutions and challenging conditions. Besides, vernacular architecture is built to fit particular needs and solutions. Furthermore, considering economic possibility viewpoint and functional purposes, authentic vernacular architecture tends to evolve over time to best fit the varying needs of different periods.

One of the interesting points of vernacular architecture is its means of building architecture with a universal and progressive approach towards design and construction practices by way of understanding and utilizing local construction materials. Construction techniques and built forms in vernacular architecture reveal practical solutions and adaptations to suit specific situations, contexts and time periods. Otherwise, contextual limitations that shaped vernacular architecture led to innovative notions of design and construction from which new builders can later learn in order to work within constraints of budget, time, labor, and craftsmanship. Therefore, due to a knowledge innovation process, vernacular buildings evolve through time. New buildings may, to some extent, look different from those traditional ones of the past and may eventually represent new forms of vernacular architecture. The philosophy of vernacular architecture is similar to the general approach of the Design-Build concept in that the builder integrates design and construction. The Design-Build concept represents the approach of vernacular architecture that is reactive to related contexts while based on new and existing materials available as well as on construction solutions on site. Additionally, the Design-Build practice includes: evaluating both labors' competency and skilled workers' know-how; achieving buildings' durability and sustainability; and realizing persons' activities and behaviors.

2 Sakae-kreua Co-learning Project

2.1 The Approach of Project

The Design-Build program by the Faculty of Architecture at the Khon Kaen University was started in the year 2001. The Sakae-kreua co-learning space, 2018 ArchKKU Design Build Project, provides architecture students with a hands-on learning experience, which is the vernacular architecture approach, known as learning-by-doing. The concept of this project offered architecture students chances to take classroom knowledge and skills into the realm of work and to encourage community service participation to people living in poverty in a rural village.

2.2 Meets the Community Needs

Sakae-kreua co-learning space locates in Sakae-kreua village in rural of Northeastern Thailand. The village is quite poor and far from downtown. The same as other design-build projects, we choose to attempt our program on helping people and communities in need, highlighting how architecture has the potential to provide real solutions to society's needs. After discussed with the people in the community, they need co-learning space for their children at school in the village because they do not have enough budget and opportunity to build one. Sakae-kreua co-learning space becomes to project that we focus and start the process of design as our ArchKKU Design Build Project.

2.3 Similarities of the Design-Build Projects and the Methodology of Vernacular Architecture

By participating in the Design-Build Project the program's activities helped to develop architecture students' skills. The program gave students from different academic years opportunities to work together, and they were assigned to specific tasks depending on their knowledge and skills. Like other projects, 2018 project also was designed with the concept of contemporary vernacular architecture. The project work expanded the students' viewpoint of vernacular architecture within existing conditions. The project encouraged a sharp solution-focused approach for work and supported the adaptation of existing construction materials and parts for building architecture within a particular context and timespan. Similarities between the design-build approach and the methodology of vernacular are: 1) Practical application of using local materials; 2) Understanding and realizing of limitations and contexts; 3) Reusing materials and resources; 4) Managing through context-based solutions and adaptations.

2.3.1 Practical Application of Using Materials

Even though the traditional vernacular architecture is typically characterized by the use of natural materials, the construction methods have used both natural materials and sustainable materials and resources available adjacent to the site. Various choices of building and construction materials for structure, roof, and fa ade are available and distribute in rural-such as zinc-coated galvanized steel, concrete, brick, bamboo, and clay. Craftsman's skills and the occupant's needs are the factors for using materials while realizing on the aesthetic characteristics, sustainability, availability, and durability of materials as well as on their ease of installation and maintenance. Even though in the past, vernacular houses were principally constructed of natural materials, however, due to the variability of materials available today, the use of industrial construction materials is commonly found on vernacular houses as well. In other words, the vernacular architecture has been developed, become innovative and up to date.

In the past, vernacular buildings had delicately built after the planting season was over. In contrast, Sakae-kreua co-learning project must first consider time limitations, because of the project's tight time frame of only 6 days. The project considered architectural design, crew's competency and opportunities for students to practice on-site. As mention above, a wide variety of construction materials, such as cement, steel, sheet metal, and tile, are also available from local suppliers while traditional material like wood becomes scarce. Due to the objectives in quick assembly and easy installation as well as support a long span, steel had been choosing for the main construction over other materials, such as wood and concrete. Another reason for choosing steel for the Design-Build project is that the architecture students at the Khon Kaen University are likely to have been familiar with the use of steel for construction through the time they spent building the performance stage for the School's annual stage play.

It is important an in-depth focus on material choices and building methods. Local builders, architecture students, must be understood the relation of physical and structural properties of common construction materials, and the impact of construction materials on the environment as well as on the overall atmosphere of a place. Lastly, to build today's vernacular buildings, it is essential to consider using common contemporary materials that can be locally found. Therefore, this approach can be applied to both construction projects by the Design-Build program as well as to contemporary vernacular architecture practices.

2.3.2 Understanding and Realizing of Limitations and Contexts

The 2018 ArchKKU Design-Build Project applies the contextual approach, which includes the principles of vernacular architecture, to create design concepts and contextual sensitivity solutions. The key role of the design process and problem-solving is the contextual conditions. The construction process always influenced by the building site and setting constraint and the design of the building itself. Besides, other influences on the construction process including time limitation, budget; construction difficulty, craftsmanship; workforce, skills of local workers and available materials and equipment. Sakae-kreua co-learning space is essential that the design and construction respond not only to functional purposes but also to the building user's behavior and way of life. The construction site is far from the city. Poor road infrastructure becomes the limitation of students access also future delivery of additional materials access to the site.

Concerning to the limitations and contexts as mentioned above, 3D models and digital program were used in the first phase to help to generate and study construction process and architectural details, joint details, surface treatments including material constraints. Thus, it is eased and saved time to effective construct in the real site.

2. 3. 3 Reusing Materials and Resources

Because of the budget limitation, Sakae-kreua co-learning project applied the concept of reuse of building materials. This technique also helped and encouraged architecture students to gain different viewpoints through the experiment in the innovative reuse of existing building materials. It was essential to realize the potential of the site location and the remaining materials. This project was a 6-day project that altered the existing open-wall shelter in the school to accommodate a library and a multipurpose terrace. The design divided the existing plan of the shelter into two parts and added some landscape for kids to sit and lie down under the tree at the rare of the building. The building was built of remaining construction materials, such as steel frames, door and glass louvers, from the Faculty of Architecture, Khon Kaen University. Most of them were salvaged and reused to build the ventilation wall of the library of Sakae-kreua co-learning project. The steel grilled openings were set into the drywall to provide the rooms with a light, airy space and good ventilation. The steel columns and beams of Design Build construction taken apart from architecture student's former performance stage. Steels were also used for several architectural components of the project's building.

2. 3. 4 Managing Through Context-based Solutions and Adaptations

The interesting of Design Build concept is its experimental practice and improvised adaptation to confront unexpected, challenging situations. During the project, for example, material and construction management was often instantaneously attained through a full-scale organization of nearby materials on site to help solve some unexpected problems. Therefore, to solve those issues, the just-acceptable material found on site became the most effective option material promptly obtainable around the site. For example, when the Sakae-kreua co-learning project ran out of wire mesh for cement reinforcement, it used bamboo available on site as an alternative. The bamboo was split and stripped into sticks to make grid reinforcement for light load-bearing reinforced concrete components, such as slabs on ground, footpaths, and ground gutters. Bamboo mesh was used as an alternate of wire mesh for providing a framework for cement to bond to. Learned from the traditional knowledge of vernacular construction, thye woven bamboo mesh must sit properly on small cement blocks used as supporting pieces. Doing so allows concrete to completely cover the bamboo mesh in order to affirm the durability of the bamboo mesh. Furthermore, bamboo available nearby the site can also be used for wall and fa ade decorative elements that should be responsively designed for easy replacement and maintenance if needed.

Fig. 1 Sakae-kreua co-learning space © 2018 ArchKKU Design Build Team

Fig. 2 Inside building, shows recycle materials the as well as quality of lighting and ventilation © 2018 ArchKKU Design Build Team

Fig. 3 Walls and voids setting

Fig. 4 Setting beams

Fig. 5 Adjusting scale of recycle material fitting the new voids

3 Conclusion

As described above, there are relations between Design-Build and vernacular architecture idealist concepts which are contextual understanding and adaptation of construction methods. Basically, vernacular architecture was formed by the simple needs and always creatively adaptation in its contexts. It also has dynamic and adaptations-due to changing needs, compromised traditions, cultural interactions, new materials and construction techniques-and has contributed a new approach to architectural design and construction. It is the fact that architectural knowledge of vernacular architecture is very precious, and the Sakae-kreua co-learning project is just a part of wisdom about vernacular architecture and its initial wisdom of adaptation for small projects that indicates how to use "Design-Build" project as a tool of architectural education to understand Thai rural contemporary vernacular architecture. The 2018 ArchKKU Design Build Project has endeavored to adapt and apply the understanding of such knowledge to architectural practices for managing time and contextual constraints. Also, it is aimed to show ways to create vernacular architecture in contemporary and altering contexts.

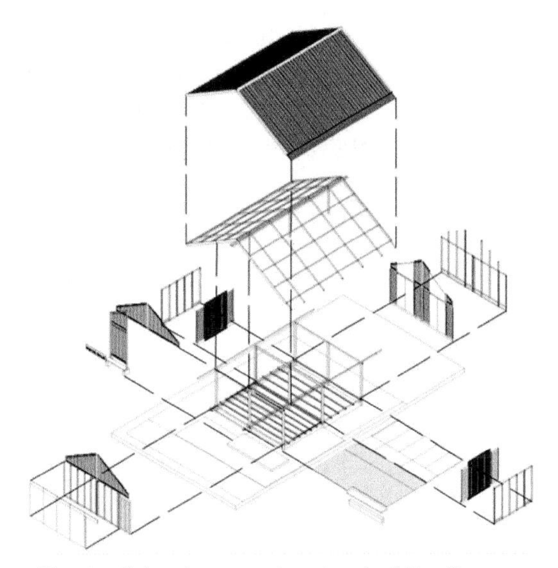

Fig. 6 Sakae-kreua co-learning building Isometric disassembly © 2018 ArchKKU Design Build Team

The Sakae-kreua co-learning project gives the students the opportunity to realize their own designs from the first ideas until the completion in 1 : 1 by their hands. Besides, the program also aims to encourage architecture students to involve in real problems, deeply understand the constraint of architecture and challenging them to face the real world outside the classroom. Eventually, the students discovered their new contemporary architecture with a different point of view.

References

OLIVER P, 1997. Encyclopedia of vernacular architecture of the world (vol 1). London: Cambridge University Press.

OPPENHEMIER A, 2002. Rural studio: samuel mockbee and an architecture of decency. New York: Princeton Architectural Press.

STONOLOV T, 2018. Design build studio: crafting meaning work in architecture education. New York: Taylor & Francis.

TULAN M, 1990. Vernacular know-how. Vernacular architecture: paradigms of environmental response. Avebury: Gower Publishing Company.

On Conversation and Utilization of Traditional Villages and Vernacular Architecture

CHEN Yuehong[1], *XU Dongming*[2], *GAO Yisheng*[3]

1 Shandong Provincial Conservation Engineering Institute of Vernacular Heritage, Jinan, China,
2 National Key Laboratory of Vernacular Heritage Conservation of Chinese State Administration of Cultural Heritage (SACH), Shandong Jianzhu University, Jinan, China
3 School of Architecture and Urban Planning, Shandong Jianzhu University, Jinan, China

Abstract Once the historical buildings are destroyed, they cannot be completely restored. They are not only the image of a place, but also the carrier of memory, containing extremely high artistic value. A large number of people flood into cities with the rapid an urbanization progress, as a result of whuh, many traditional villages and vernacular architecture are facing the situation of no access, and there are many problems in the conservation and utilization. The study adopts the methods of field research and selects diverse examples such as general villages, remote areas, ethnic minorities, and regional villages. The present situation of traditional villages and vernacular architecture is studied and analyzed. Finally, the paper puts forward suggestions and methods for the conservation and utilization of traditional villages and vernacular architecture under the background of the national strategy in terms of ecological civilization, cultural confidence, rural revitalization, and ecological agriculture.

Keywords traditional villages, vernacular architecture, conservation, utilization, filed research

1 Research Background

Since the 1980s, rapid urbanization of China, the continuous adjustment of administrative areas, the merge of administrative villages and natural villages in large numbers, the migration to plains and to work in cities to alleviate poverty alleviation, and the overall relocation of the reservoir area have caused many traditional villages to disappear in this process. The continuous expansion of urbanization has led to many villages to become villages in cities. Some traditional villages suffer destruction and disappear in large numbers because of unplanned and disorderly blind expansion and merger. For a long time, with a city's GDP as the indicator of administrative performance, administrators have a low awareness of conserving traditional villages and the value of vernacular architecture to them is only reflected in the tourism added value. The blind pursuit of the economic benefits brought forth by the vernacular architecture has resulted in insufficient management and maintenance of some valuable vernacular architectures, and even caused damage to them. In recent years, the conservation and utilization of traditional villages and vernacular architecture have gradually received attention from all walks of life. Under the background of the Chinese government's strategy of ecological civilization, cultural confidence, rural revitalization, and ecological agriculture, vigorously promoting traditional culture has become a hot social topic. However, the research on the conservation and utilization of traditional villages and vernacular architecture in China starts relatively late and the complete theoretical framework has yet to take into shape due to the theoretical gap. Based on this, this paper intends to explore the conservation and utilization of the traditional villages and vernacular architecture based on field research.

2 Definitions and Concepts About Traditional Villages and Vernacular Architecture

2.1 Definitions and Concepts

Traditional villages refer to ancient villages with profound historical value and rich natural resources. As a witness to the long historical and cultural changes, they are mainly characterized by vernacular architecture, special spatial pattern, streets and lanes, local customs and practices, bearing important historical values. Traditional villages contain rich historical information folk customs and regional culture (Ouyang & Wu, 2019). As the most important part of traditional villages, vernacular architectures comprise houses, opera stages, wine cellars, courier stations, ancestral shrines, etc. Most of them have experienced wars, disasters, and humanity vicissitudes, carrying a large amount of historical information and the good expression of local customs and practices (Li, 2018).

2.2 Research Status

China's domestic research on conservation of traditional villages includes case studies and theoretical studies, covering a wide range of such disciplines like architecture, archaeology, sociology, human geography, tourism, and management (Li, 2016). For example, based on the classification and development mode from the perspective of H-I-S (village host [H], industry [I] and space [S]), Tao et al. (2019) explored types, features, and development path of traditional villages in the process of modernization and pointed out that the focus of conservation research should shift from traditional heritage conservation to people-oriented activation and utilization to achieve transformation and sustainable development of traditional villages (Tao et al., 2019). Hu Siting and Hu Zongshan argued that on the basis of excavating, sorting and refining the historical and cultural connotations of the villages, efforts should be devoted to establishing village history museums, intangible cultural heritage theaters, homesick story development, cultural and creative product development, etc., converting culture to capital value to actively promote the effective conservation and inheritance of the traditional village history and culture (Hu & Hu, 2019).

Liu Xinqiu and Wang Siming summed up the problems and misunderstandings prevalent in the current conservation practice of traditional villages and proposed the development ideas based on the characteristics and types of villages, such as highlighting the traditional features, promoting the development of traditional architectural and industrial and trade-oriented villages, implementing the concept of green development, leading agricultural landscape village development, increasing the added value of agricultural products, promoting the village producing featured agricultural products, reshaping rural culture, and promoting the development of folk culture villages (Liu & Wang, 2019).

3 Research Methods on Conservation and Utilization of Traditional Villages and Vernacular Architecture

3.1 Methods and Tools

The field research method adopted in this study originates from anthropology. Its focus on the "present time" and "on-site feeling" echoes the deep understanding What the field research on the conservation and utilization of traditional villages and vernacular architecture intendes to obtain. The first-hand information from the field research provides the real record and research information of the subjects. This study aims to increase the time and extent of "immersion" through case studies and action research, thus achieving a leap from individual to general and from action to understanding. In obtaining the first-hand research data, this study used a "participant observation", focusing on capturing the inner spiritual experience of residents in traditional villages. As field research emphasizes contextualizing the conservation and utilization of traditional villages and vernacular architecture within the social background, this research method can effectively identify the root of the problem.

In this study, the authors selected three sample villages from the List of China's Traditional Villages released by the Ministry of Housing and Urban-Rural Development of P. R. China. They are Gaojia Zhuangzi village in Xinzhuang Town, Zhaoyuan, Yantai of Shandong Province, Daigu village in Daigu Town, Mengyin County of Shandong Province, and Zili village in Tangkou Town, Kaiping of Jiangmen City, Guangdong Province. The authors conducted field surveys of the selected sample villages and kept fieldnotes through participant observations and interviews to obtain the first-hand information on the conservation and utilization of vernacular architecture in sample villages. Here are some excerpts from the fieldnotes:

No.	Time	Place	Fieldnotes
1	July 4—6, 2017	Gaojia Zhuangzi village in Xinzhuang Town, Zhaoyuan, Yantai of Shandong Province (Fig. 1)	Only local seniors know the story of Scholar Xu from Gaojia Zhuangzi village. There was no through bus from the county to the village and we took a taxi there. We saw the old house under protection and started to chat with a senior in a small shop next to the old house. He told us the stories behind the villagers
2	March 6—9, 2018	Daigu village in Daigu Town, Mengyin County of Shandong Province(Fig. 1)	Daigu village in Mengyin County is famous for its sweet peaches. Few people know that there are many traditional buildings in the village. There is a strong sense of protection. Ancient buildings such as wells and bridges have been conserved and the conservation responsibilities have been assigned to people.
3	December 14 — 16, 2018	Zili village in Tangkou Town, Kaiping of Jiangmen City, Guangdong Province(Fig. 2)	No large-scale commercial development has been done in Zili village. The vernacular architectures have been well preserved and remain the most primitive features

3.2 Field Research Summary

The authors have identified some problems with the current conservation and utilization of traditional villages and vernacular architectural heritage based on the field research, including insufficient fund, incomplete conservation laws and regulations, and inappropriate conservation concepts. However, some of the effective practices can be briefly summarized as follows:

(1) The villagers have a strong sense of conserving vernacular architectural heritage of the village. The village committee and the local government jointly promote the sustainable development of traditional villages, assign the conservation responsibility to specific people, and advocate joint conservation. This achievement is attributed to China's extensive work in improving conservation awareness, planning, and design, and catalyzing the improvement of conservation measures and regulations in recent years.

(2) The education level of village committee members is positively correlated to the conservation and utilization level of traditional villages and vernacular architectural heritage. Members of the village committee with higher education can keep pace with the times and keep up with the call of the central government to promote traditional culture with higher acceptance of new economic development modes, such as attracting investment to develop tourism, carrying out various forms of publicity, and enhancing villagers' awareness of conserving traditions.

(3) The development of local tourism has promoted economic development and provided strong financial support for the

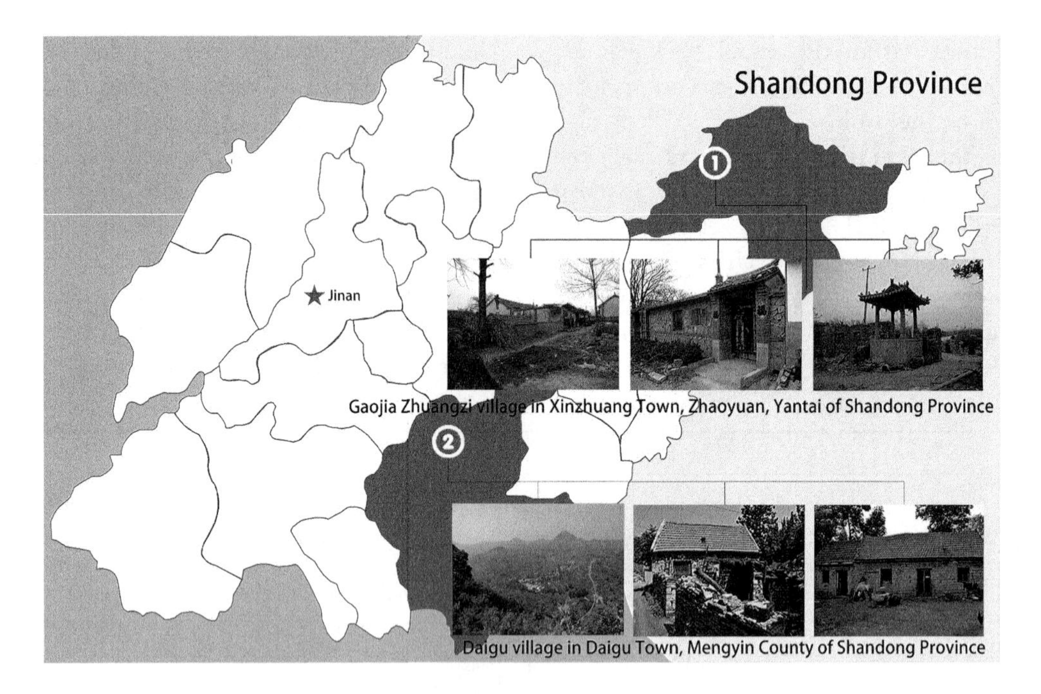

Fig. 1 Location of selected villages of fieldwork in Shandong province © CHEN Yuehong

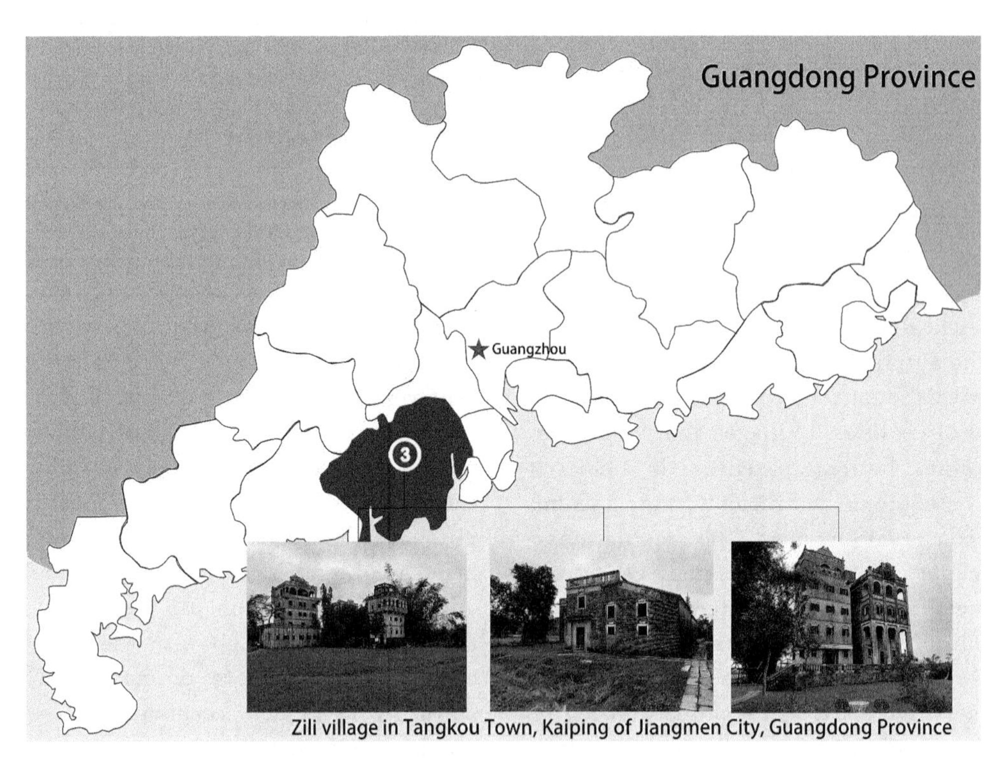

Fig. 2 Location of selected village of fieldwork in Guangdong province © CHEN Yuehong.

conservation and renovation of the vernacular architectural heritage.

4 Suggestions, Methods, and Prospects

4.1 Suggestions and Methods

We believe that the conservation and utilization of traditional villages and vernacular architectural heritage should be carried out in terms of the following four aspects:

4.1.1 Providing Overall Conservation to Relatively Complete Traditional Villages and Vernacular Architectural Heritage

For traditional villages with complete structure and pattern, comprehensive protection should be implemented, and varied conservation measures should be formulated to protect their unique natural and human environment. Local civil conservation organizations should be established and work with tertiary industries such as tourism to carry out conservation. People's understanding of the values of traditional villages and vernacular architectural heritage can be enhanced through the conservation work.

4.1.2 Giving Priority to Vernacular Architectural Heritage of Higher Value when Conserving Dispersedly Distributed Vernacular Architectural Heritage

If the dispersedly distributed vernacular architectural heritage is not conserved in time, its destiny is often to face removal, demolishment, natural damage or collapse due to new rural area renovation. For this type of vernacular architecture heritage, it can be included into Rural Memory Museum for prioritized conservation.

4.1.3 Enhance the Education Level of Officials of Traditional Villages

Village cadres play a crucial role in the conservation and utilization of traditional villages. A high-quality village administrative team can not only make great contributions to local economic development but also effectively promote the conservation and utilization of traditional villages and vernacular architectural heritage. Its team building can start from the following two aspects:

(1) Appoint college student village officials. Thanks to their higher education background, they can advocate more confidently and effectively the knowledge about traditional village and motivate villagers' cultural identity and conservation awareness.

(2) Improve the education level of the in-service village cadres. Regular lectures and workshops on conservation and utilization of traditional villages and vernacular architectural heritage should be provided to help in-service village cadres understand relevant policy and improve their administration skills.

4.1.4 Promoting Financing

For traditional villages that haven't receive government financial support to carry out conservation, they often need to collaborate with external investors. However, the inconsistent objectives for economic benefits between the local villages and the investors and the asymmetric information between the two may cause many problems. After archiving traditional villages and vernacular architectural heritage, we should manage the information online to resolve the asymmetric information problem and keep investors updated. Then, the traditional village construction plan is to be finalized to improve the social welfare in the execution, eliminate negative effects, and enhance the sense of honor and cohesion of the villagers.

4.2 Research Prospects

Mr. Fei Xiaotong once pointed out that because culture exists for human, it is necessary to give priority to people rather than culture to seek a way to keep the culture through transforming it. Each culture has a toolchain, i.e. a goal behavior strategy that is promoted by the culture (Fei, 2013). Everyone conceives a road map in their mind to navigate the path of life and move towards their own goals. These road maps must acknowledge the existence of different goals, road segments, and the need to mediate the conflicts between them. The individuals, therefore, must be able to take charge of their own lives (Keesing, 1997). In a certain sense, conservation and utilization of traditional villages should focus on the organic coordination of modern and traditional, urban and field, global and local, individual and social, to create a unique, diverse and contemporary vernacular culture. To a certain extent, this also indicates the cultural introspection of each group and individual based on the local cultural cognition, enjoyment, practice, and development

(Fang, 2019).

Traditional village and vernacular architecture are an epitome of the village history and a historical portrayal of the life of the ancestors. It is a product of the culture that is continuously inherited from innovation. These ancient vernacular buildings have undergone centuries, and their development and formation have witnessed history and they are touchable living fossils. Based on the field research, this study attempts to sort out the problems faced by the conservation and utilization of traditional villages and vernacular architectural heritage, and the corresponding suggestions and methods. The authors believe that this method can be applied to resolve the similar problems in similar cases and contribute to the reasonable conservation and utilization of the non-renewable precious resources of traditional villages and vernacular architectural heritage.

Acknowledgements

We would like to thank Ms Chen Jianfen as the language proofreader for helping us facilitate this work into a final draft.

References

FANG Jingjing, 2019. Culture, society, and individuals: reflections on conservation practice of traditional villages. Lanzhou xuekan journal, (03): 178-185.

FEI Xiaotong, 2004. On anthropology and cultural consciousness. Beijing: Huaxia Publishing House.

FEI Xiaotong, 2013. From the soil: the foundations of Chinese society. Shanghai: Zhonghua Book Company.

HU Siting, HU Zongshan, 2019. Research on conservation of traditional villages in Chaohu Lake area from the perspective of cultural capital: taking Hongtuan village of Chaohu city as an example. Jianghuai forum, 2:24-28.

KEESING R M, 1997. Cultural anthropology: a contemporary perspective (originally published in 1976). Belmont: Wadsworth Publishing.

LI Jia, 2016. Research on conservation mode of traditional villages. Nanjing: Nanjing Agricultural University.

LI Jiujun, 2018. Research on characteristics and changes of the land of vernacular architecture in east Jiangxi and north Fujian. Heritage architecture. 4: 13-21.

LIU Xinqiu, WANG Siming, 2019. Classification of traditional villages and development path from the perspective of agricultural heritage — based on the survey of 28 traditional villages of Jiangsu. China agricultural university journal (social sciences), 36 (02): 129-136.

OUYANG Guohui, WU Jing, 2019. "Outside" and "Inside" of traditional villages — report from the Shajing old house, Shiyan town, Qidong county of Hengyang city. Hunan social science, 3: 151-155.

TAO Hui, MA Guoqing, RAN Feixiao, et al., 2019. Research on the classification and development mode of traditional villages from the perspective of H-I-S — taking Handan city as an example. Journal of tourism, 5: 1-18.

Research on Overall Protection and Presentation Strategies of the Drainage Channel across the Western Wall of the Large City in Linzi City Site of Qi State

GAO Hua ,LIU Jian , LI Hao

Institute of Architecture History, China Architecture Design & Research Group, China

Abstract Linzi City Site of Qi State is the site of the capital of Qi State during the Zhou Dynasty. Within the site, the drainage channel across the western wall of the large city was built to connect the drainage system inside the city with the ancient Xishui River outside the city. So that the drainage channel was beneficial to the wastewater discharge in daily life and the water recharge in dry seasons, as well as the defense against enemies. Above and on both sides of the drainage channel, the site of the western wall of the large city, which was constructed with rammed earth by piercing poles in sections by wooden boards, witnesses the construction, expansion and continued use of the ancient capital city together with the site of drainage channel. Opened to visitors in 1980, the presentation of the drainage channel was short of the relationship between the whole drainage system and its context, coupled with the lack of interpretation methods, that led to the problems of understanding difficulties, poor popularity and low social benefits. Based on the study of heritage value, the paper proposes overall protection and presentation strategies of the drainage channel, western wall of the large city, the road site and their historical context. In view of the small range of aboveground site, the multi-majors including heritage protection, landscape, architectural design and multimedia, are expected to interpret all the types of attributes collaboratively for easily understanding about the heritage values. In view of the poor readability of earthen and stone site value, the use of modern technology equipment such as immersive view experience and augmented reality handheld devices and the layout of archaeological, historical, cultural landscape sketch and vegetation greening are applied to enhance the readability, interest, experience and leisure of the site interpretation.

Keywords drainage channel site, value attributes, overall presentation, multi-professional collaboration, modern technology equipment

1 Introduction

Strictly speaking, the international protection (Sun et al., 2010) for earthen sites began in the 1960s. The theory and research about that field have been established well since then. As stated in *Charter for the Protection and Management of the Archaeological Heritage* (ICOMOS, 1990), presentation and information should be conceived as a popular interpretation of the current state of knowledge, and it must therefore be revised frequently. The idea also applies to earthen sites. After that, *The Icomos Charter for the Interpretation and Presentation of Cultural Heritage Sites* (ICOMOS, 2008) significantly deepened the theory of heritage utilization, including earthen sites. The Charter stated that interpretation and presentation should avoid causing adverse effect on its cultural value, visual representation of heritage sites can be achieved by artists, architects and computer simulation technology, but they should be based on detailed and systematic analysis of environmental, archaeological, architectural and historical data by integrating multidisciplinary expertise, the purpose is to inspire deeper interest, learning, experience and explore. However, in the terms of engineering practice, due to the fragile structure of earthen sites, the protection of earthen sites is a worldwide problem difficult to solve, especially when it is easily damaged by water usually. Therefore, there are limitations for the presentation and utilization of earthen sites, not only because of the obstacle in appreciation, but also by reason of the limitation of interpretation methods. Despite the successful foreign examples of earthen sites protection, such as Bam and its Cultural Landscape in Iran, there remains a limiting condition that they are usually located in the arid areas, where offers the best natural environment for the protection of earthen sites.

As for the study of earthen sites in China, how to protect the earthen sites is also the core issues, especially in the arid areas. Extensive theoretical studies and engineering practices are in progress, such as the protective materials and technology for earthen sites have been studied by experts in Dunhuang Academy (Guo et al., 2013; Wang et al., 2015) and Zhejiang University (Zhou et al., 2008). Also, the National Cultural Heritage Administration has also published *Design specification for preservation and reinforcement engineering of arid earthen sites*. However, most studies of earthen sites protection in humid environment have only been carried out in small-scale field test and laboratory experiments, so there is still an urgent need to address problems of practice. Meanwhile, only a few theoretical and practice studies on the presentation and utilization of earthen sites were published. For example, Zhu Mingmin(2011) analyzed the advantages and disadvantages of indoor and outdoor modes for the protection of authenticity.

2 Heritage Overview

Linzi City Site of Qi State, situated in Linzi District of Zibo City in Shandong Province, is the site of the capital(859—221 BC) of Qi State in the Zhou Dynasty. Composed of en dash the small city and the large city, it was one of the largest capital cities that lasted for a long time and was famous for industrial and commercial prosperity during the pre-Qin period. According to the research, its large city is an irregular rectangle with a length of 4.5km from north to south and a width of 3.7km from east to west. By definition, the fabrics of Linzi City Site of Qi State include city walls, gates, moats, dwellings and handicraft workshops, rammed earth foundations, drainage systems, street, tombs, kilns, etc.

The drainage systems of Linzi City Site of Qi State include the drainage system in the west of the large city, the drainage system in the northeast of the large city, and the drainage system in the small city. In the former system, the drainage channel across the western wall of the large city, which is the object of study in this paper, is in the end of the drainage system in the west of the large city, connecting with the Xishui River, Shengshui River and Shenchi River westwards outside the city. This part of drainage channel was built in the late Spring and Autumn Period or the early Warring States Period (about 5th century BC) when the massive expansion of the ancient capital city of Qi State(Shandong provincial institute of cultural relics and archaeology, 2013). Comprising three parts, including inlet channel, passage channel and outlet channel, it is 43 meters long in total, 3 meters deep en dash and 7～10.5 meters wide. Specifically, the passage channel was built by stones laying into 3 layers and assembled in staggered rows to leave 15 square holes (Fig. 1). Above and on both sides of the drainage channel, the wall is the site of rammed earth of the north-south orientation, which was constructed with rammed earth by piercing poles in sections by wooden boards. Although the remnants of the above-ground wall around the drainage channel are only distributed in a small area, the part of the underground wall has been well preserved as yet.

3 Status Analysis

3.1 Value Research

The archaeological sites have rich cultural meanings and values, so their protection and presentation contents should be based on the comprehensive value research. The structural design of the drainage channel across the western wall of the large city not only facilitates the discharge of urban

Fig. 1 East elevation of drainage channel and city wall(before adaptive reuse)

wastewater and the recharge of water in the dry season, but also can defend against external enemies, reflecting the advanced concept of the drainage system planning and construction of the capital cities in the Zhou Dynasty (Zhang et al., 1988). Together with the sites of city wall and street, the drainage channel witnessed the process of construction, expansion and continuous use of the capital of Qi State. Therefore, the attributes of the drainage channel across the western wall include the fabrics of drainage channel across the western wall of the large city, the drainage ditch in the western part of the large city, the city wall, the street as well as the historical context of Xishui river.

3.2 State of Preservation and Use

The fabrics of drainage channel and the wall are earthen and stone, which are basically not affected by human activities but mainly impacted by natural factors such as weathering, rain erosion and wind erosion. Besides, the height of the extant wall on site is low, so that the direction of the above-ground walls is not obvious to recognize. The drainage channel has remained the narrow functional layout since 1980 when it was first displayed to the public. The open area only presented the drainage channel without the other features, while the presentation of the drainage channel was short of the relationship between the whole drainage system and its setting, coupled with the lack of interpretation methods and interesting contents, that led to the problems of understanding difficulties, poor popularity and low social benefits.

3.3 Demands of Protection and Use

As the important birthplace of the Qi culture, Linzi City Site of Qi State gains a lot of attention. Owing to the importance, the local government is eager to develop the site as a national archaeological site park, even to make it listed on the world cultural heritage. Moreover, the drainage channel across the western wall of the large city is one of the significant presentation sites in the plan of the archaeological site park. Therefore, the site of drainage channel has great demands for the protection and use. High-standard overall plan and design with international advanced ideas are required to satisfy the needs of research, recreation and leisure in the archaeological site park. In addition to that, the high-quality design is expected to show its highlight to the public and improve the social influence of the site.

4 Protection and Presentation Strategy

The goal of the protection and presentation project of the drainage channel across the western wall is to protect all the fabrics and historic context truly and completely, and is also to interpret the heritage value by exploring various methods actively. Furthermore, the project supposed to satisfy the demands of educational and recreational function, and construct the interpretation area of drainage channel site as a demonstrative example of protection and presentation. In order to solve the problem of low readability and interest, which is the current weakness of the earthen and stone sites, the first step is to study and compare the successful domestic and foreign cases in the terms of heritage protection and utilization (Tab. 1).

Then, on the basis of heritage value research, with reference to the latest concepts of heritage protection and utilization, the refined design is carried out. In the final design result, the principles of historic condition preservation, minimum intervention and reversibility should be implemented, and the relationship between requirements of the heritage protection and management and the demands of local social development should be properly handled. Also, technical guidance and timely adjustment should be accomplished during the construction process.

Tab. 1 Brief list of heritage protection and interpretation cases

Heritage names	Heritage materials	Referential content	Points of design
Takayama expedition town	Brick and wood	Information and multimedia exhibition	Multimedia video hall, exhibition of historical objects
Roman Forum and Palatine Hill	Masonry	Restoration of historic context	Moderately beautify the natural environment moderately
Site of Dingding gate in Luoyang city in Sui and Tang Dynasties	Earthen	Presentation of earthen site	Protective structure meeting the requirement of minimum intervention and reversibility
Tusi Sites	Stone, earthen, brick and wood	Interpretation system	Simple introduction board design that well-coordinated with heritage environment

5 Main Points of Design of Use and Protection

5.1 Presentation of Attributes

In view of the fact that the scale of the aboveground wall site and the drainage channel is small, the relationship between the trend and the layout is not obvious, this project uses the method of presentation of all value attributes to improve the readability of the site interpretation, thus expanding the scope of the open area and re-planning the presentation layout. The protection and presentation of the sites of the drainage channel, western wall, drainage ditch in the western part of the large city, the east-west street in the north of the large city and historical environment including Xishui River, Shengshui River and Shenchi River have been integrally designed. The function areas include the management service area, Xishui River display area, western wall display area, the inner city display area and the drainage channel display area. According to that function division, the tour line is organized reasonably. Supplemented by a Copper glass fiber reinforced plastic panoramic sand table model and interpretation signage system, the structural relationship of the various parts of the drainage channel is interpreted more completely and clearly than before.

Fig. 2 The general plan of the site

5.2 Multi-professional Collaboration

The construction of archaeological site park involves multiple disciplines such as archaeology, heritage protection, architecture, landscape, and administration. It requires the designers to have professional knowledge background and comprehensive design capabilities. As the consultant responsible for the conservation plan and archaeological site park plan of Linzi City Site of Qi State, our study has carried out research work of the site for more than 10 years, which is conducive to coordinating the multi-majors collaborative work according to the requirements of historic preservation and use. Therefore, the project adopted the means of general contracting design and multi-disciplinary cooperation to achieve a variety of interpretation methods and control the overall presentation.

Combining landscape design, interior design, architecture design, multimedia and other specialities, and weakening the self-expression in conventional design mode, we realized the construction or reconstruction of historical presentation, landscape, interior, architecture, multimedia, etc. within a very limited total project budget. The project mainly includes the following key points of design. Firstly, the design of management and service building covers the functions such as ticket sales, exhibition room, toilet, receptions, visitor experience project, weakening the individual expression of architects during the architectural design process. Secondly, the simulation of the ancient urban water system is displayed, mainly containing the north-south Xishui River and showing the flowing direction of the Shengshui River and Shenchi River. Thirdly, the wall covered with grass for protection its surface (except the well-preserved city wall section in the north), according to the original width of the wall, and the middle of the wall is able to step on and exhibits the sculpture of the scene ramming the city wall. Fourthly, the aboveground wall section is displayed in the northern part of the city wall, covered by a transparent steel-structure shelter. Fifthly, the restoration of the rammed earthen wall is located in the southern part of the display area, and its height is designed according to the maximum height of the existing wall of Linzi City Site of Qi State, which is 6 meters high. The south end is installed with imitation rammed wall section hanging board and the internal facilities multimedia video room. At the last, on the east side of the drainage channel, a trestle way is built for more comfortable visiting experience, while the signage of drainage ditch on the east of the drainage channel is extended to reflect the composition of the entire drainage system in the west of the large city.

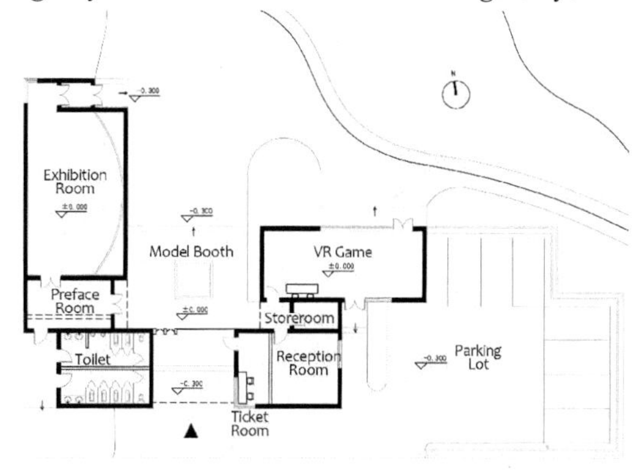

Fig. 3 Management and service building plan

5.3 The Use of Modern Technology

Arc-shaped immersive movie-watching experience is set in the multimedia video room. The short film played here is made based on the archaeological achievements and confirmed by us consultant, the archaeologists and administration department., introducing the history, structure, technology and meanings of the drainage channel. An augmented reality handheld device and a 220-degree enveloped VR (virtual reality) display platform terminal are set up beside the drainage channel. The VR game of Qi state defense interactive experience is set in the service room to enhance the fun and experience of interpretation. Significantly, the augmented reality handheld device is the first attempt to interpret the site in China, meanwhile the enveloped virtual reality display platform terminal is ready to apply for a national patent. Particularly, in the process of design and construction, the influence of the appearance of those high-tech facilities on the site environment should be weakened. It is the principle that the technological device should be used reasonably and as few as possible to make the heritage value interpreted accurately, vividly and simply.

5.4 Control Interpretation Facilities and Vegetation

Designers reduce the volume of ancillary facilities as much as possible, and control their number and select the site ration-

Fig. 4 220-degree enveloped VR (virtual reality) display platform terminal

ally. Since the buried depth of some cultural layers in the City Site is only 0.3 meters, the depth of soil disturbance is required to conform to the restrictions on the depth of soil disturbance. Also, in the design, the new management and service building should be located in the building area outside the city site. As for the selection of vegetation, historical vegetation or local vegetation species are considered firstly, hallow shrubs and turf mainly. A small number of trees only can be planted after the confirmation by archaeological experts that there is no remains in the ground, though the visual effect of greening landscape sacrifices to some extent. In terms of the interpretation board, simple lightweight stainless steel material which matte and weather-resistant is selected to fit the environment. Moreover, the illustrated introduction is offered on replaceable metal board for sustainable use.

5.5 Appropriate Expression of Park Elements to Develop Functions of Use

The functions of use of heritages are extended on the premise of comprehensive

Fig. 5 Trestle way and introduction board to the east of the drainage

and effective site protection. According to the cultural connotation of the drainage channel, cultural and landscape sketches are built up, for example, the surface of Xishui River, Shengshui River, Shenchi River, the sculpture of city wall ramming, wall section, water cultural landscape wall, bamboo landscape wall. The landscape design enriches the vegetation arrangement along the river and improves the Self-purification ability of water, and also moderately develops the public education, recreation and other functions in the display areas.

Fig. 6 West of drainage channel

5.6 Saving Resources

During the implementation of the last planning which was approved by the state administration of cultural heritage before, there were some inevitable difficulties, for instance, the issue of chemical plant removal and the obstacle of land acquisition. Finally, designers adopted the method of scheme comparison to substantially lessen the land use area, under the premise of ensuring most of the display content achieved. The final design scope was decreased from

12. 7 hectares to 2. 5 hectares, saving the one-time construction investment and post-operational costs.

6 Conclusion

Through the study above, it is demonstrated that the protection and interpretation design of earthen sites needs to research the value as the first step. Then , it is required to enrich the means of interpretation and use as much as possible. When organizing multi-disciplinary collaboration and using modern technology, designers are supposed to weaken their self-expression which is encouraged in the other design works but not suitable here, and they are expected to repeatedly scrutinizing the details to improve the readability, interest, experience an leisure of both interpretation and presentation. Furthermore, the presentation design need to consider the demands of local cultural construction and social development. In another words, stakeholders should seek the combination of protection and utilization of the site based on the condition of status quo. Only a win-win situation can promote the sustainable development of earthen sites protection and utilization.

References

GUO Qinglin, ZHANG Jingke, SUN Manli, et al. , 2013. A study on the characteristics of deterioration and the conservation of ancient Beiting city in xinjiang. Dunhuang research, (01):16-17.

Shandong provincial institute of cultural relics and archaeology, 2013. Linzi city site of Qi state. Beijing: Cultural relics press: 533.

SUN Manli, WANG Xudong, LI Zuixiong, 2010. Preliminary study on the protection of earthen sites. Beijing: Science press: 8.

WANG Xudong, GUO Qinglin, CHEN Wenwu, et al. , 2015. Exploratory study on the protection of earthen sites during the archeological discovery in humid environment. Beijing: Science press: 546-550.

ZHANG longhai, ZHU yude, 1988. Drainage systems in linzi city site of Qi state. Archaeology, (9): 787.

ZHOU Huan, ZHANG Bingjian, CHEN Gangquan, et al. , 2008. Study on consolidation and conservation of historical earthen sites in moisture circumstances conservation of Tangshan Site in situ. Rock and soil mechanics, (4):955-962.

ZHU Mingmin, 2011. Discussion on the conservation and utilization mode of earthen sites: a case study of xi'an area. Southeast culture, (3): 20-23.

Ecological Wisdom and Adaptive Utilization of Original Bamboo Architecture

LI Xiaojiao, *WANG Jiang*

School of Architecture and Urban Planning, Shandong Jianzhu University, China

Abstract Bamboo has been widely used in vernacular architecture around the world, creating a variety of distinctive features known as bamboo architecture. Based on the different bamboo structural forms, bamboo architecture can be divided into three types: original, modified, and composite bamboo architecture. Among them, the original bamboo architecture (OBA) makes good use of the physical properties of the bamboo material, such as hollowness, compression resistance, tensile strength, fire retardancy, and shock resistance. Considering the increasing homogenization of the vernacular style and inefficient use of original bamboo resources, we studied the OBA from four aspects: first, the performance of original bamboo materials is evaluated based on several aspects, including original bamboo varieties, cultivation, use, and reinforcement. Second, the ecological wisdom of our ancestors based on materials from the earth and methods from nature is comparatively studied by studying the regional characteristics of the OBA in Asia, America, and Africa. Third, the processing of the material is analyzed by considering various aspects such as cutting, drying, processing, and storage to sort out the construction techniques for pillars, beams, slabs, roofs, and walls of the OBA. Finally, the design of the Liangshan Lamp Theatre in Liangping County, Chongqing, is considered as an example. In this case, using elliptical-section, circular-section, fish-mouth-section, and hinge connection types, we study the relationship between modern materials and traditional knots, shackles, and explore traditional inheritance as well as the incorporation of traditional bamboo construction technology into contemporary rural construction.

Keywords original bamboo material, vernacular architecture, ecological wisdom, construction technology, connection structure

"Better to live without meat than without bamboo." The bamboo specified herein here is original bamboo, which is a lightweight, flexible, and fast-growing natural plant with wide-area, inexhaustible use, favorable mechanic characteristics, and low cost. In 1753, Carl von Lynn defined bamboo for the first time as a grass whose lignin, unlike that of rice, corn, and sugar cane, grows lighter and stronger over the years. In traditional societies with low productivity, particularly Asia, America, and Africa, bamboo is widely used for vernacular construction. It has been used to create a variety of distinctive features, such as bamboo houses and buildings with distinctive regional characteristics. Forming a complex landscape that integrates economic, social, and ecological systems, such buildings not only reflect the humanistic and geographical features of various areas but also record the trials and efforts of our ancestors to change nature. Further, they are witnesses of the prosperity and vicissitudes of rural areas(Wang and Zhao,2017). Nevertheless, the rapid development of modern construction technologies and renewal of construction materials are significantly impacting the aforementioned OBA. First, compared with conventional masonry materials, original bamboo dries easily and cracks after absorbing water. Furthermore, it is easily stratified by the influence of sunlight and humidity, can easily burn in case of a fire, and is vulnerable to insects and decompose. Second, compared with industrial materials, such as metals, cement, and glass, just using original bamboo is difficult to use in the construction of multi-high-rise buildings. Third, most of the OBA focuses on residential functions or material art exhibitions; therefore, it is difficult to meet the construction requirements of the multifunctional spaces and different scales in modern architecture such as those in hotels, offices, and shopping malls. New requirements have been proposed for the structure, modeling, and energy conservation of the OBA(Chen, 2018).

Affected by the communitization of rural settlements, most local governments in China have adopted almost the same programs, which involve relocating and consolidating two or more natural or administrative villages into one in a given time, and establishing new centralized residence mode, service and management mode, and industrial patterns through unified planning and construction(Wang et al., 2014). Instead, the standardization and commercialization of rural buildings will lead to the convergence of the rural architectural culture and characteristic crisis. In 2016, the United Nations Conference on Housing and Sustainable Urban Development Habitat III raised the issue of the impact of sustainable human settlements on world security issues. Green building materials, ecological wisdom, and appropriate technologies related to traditional settlements and vernacular architecture are receiving increasing attention. Considering OBA as the object of research, we discuss the contemporary conservation methods and technological innovations related to rural construction, traditional settlement protection, and restoration design through the analysis of research status at home and abroad.

1 Material Cognition of Bamboo

1.1 Type and Cultivation

(1) Type. Bamboo is mainly distributed across tropical and sub-tropical zones between N51° and S47°. There are approximately 1,200 species, 750 and 450 of which are found in Asia and America, respectively (Xiao, 2006). Depending on the climate zone, there are differences in the use of various types of bamboo in construction. In tropical zones, the bamboos most commonly used in construction are Bambusa, Chusquea, Dendrocalamus, Gigantochloa, and Guadua, and bamboos of the Phyllostachys group are most commonly used in temperate zones.

(2) Cultivation. Bamboo grows from the lower internodes. When the internodes reach a certain height, the diameter of the nodes varies; finally, they adopt a slightly conical shape. The mother plants have a smaller diameter; in each of the following three generations, they thicken slightly. *Guadua angustifolia* Kunth grows up to 21cm per day, reaching 80% of its maximum height in one month. It completes its growth in the subsequent five months, reaching between 15 and 30 m. The productivity is between 1,200 and 1,350 canes per hectare per year(Zhou, 1998). The duration of the lignification cycle is 4—6 years, during which the vascular bundles are closed and dried and the stem formed between the internodes can be used for construction; meanwhile, the humidity of the bamboo can be reduced to approximately 20% (Fig. 1).

Fig. 1 Cultivation of Bamboo

1.2 Physical Properties

(1) Hollowness. Initially, the bamboo's stem was solid. Later in the process of evolution, the marrow at the center of the stem gradually disappeared and the bamboo became hollow. Hollow bamboos are more capable of transporting nutrients, making them grow taller and faster. After the bamboo grows, it becomes hollow and develops a thicker pillar. At this time, it has a strong supporting force, and it will not be broken even if it encounters strong

wind and heavy snow.

(2) Resistance to Tension. Bamboo,its surface pasticulasly, is strong against tension stress. The test value of its overall tensile strength depends on the slenderness ratio. In addition, the tensile strength of bamboo varies with its age, growth, and humidity. The bamboo wall at the lower internodes is thicker and the internodes are shorter; therefore, the test value of the tensile strength is more accurate there.

(3) Performance in Fire. Bamboo exhibits good flammability, owing to its hollowness and high concentration of silicic acid. According to the Germany regulation DIN 4102, bamboo is designated as a flammable but burn-resistant material. Therefore, it is necessary to perform a flame-retardant treatment on bamboo for construction.

(4) Earthquake Resistance. Bamboo is an ideal seismic material because of its strong resistance to compression, ability to absorb energy, and good flexibility. In areas with high seismic risk, it is common to build bamboo walls by using bamboo to reinforce adobe and earth walls.

1.3 Applications of Bamboo

(1) General Applications. The specific applications of bamboo depend on the type, age, and location. Some bamboos can be used as food as bamboo shoots. Bamboo that has been growing for more than two years can be compressed into wooden boards for craft, daily necessities, flooring, or furniture. In approximately 1910, India began to develop a paper industry using bamboo pulp. Bamboo fiber is more resistant to abrasion than hemp fiber. Thomas Edison conducted tests using thousands of fibers that could be used as light bulbs and finally found that Japanese bamboo fiber was optimal -it lasted for 2,450 h after being ignited.

(2) Construction Applications. ①Bamboo fiber can reduce the shrinkage and cracks in the curing stage of cement mortar. In the 1980s, the British intermediate technology development team succeeded in using bamboo fiber for the manufacture of corrugated fiber cement tiles. ②In general, rebar is used to reinforce concrete components, but the same effect can also be achieved using twisted bamboo shoots with sufficient friction. ③Further, bamboo can be used to reinforce walls. In 850 AD, native houses and neighboring areas with more than 50,000 residents and royals in Peru were protected by conical walls built of adobe or rammed earth. These walls, 2.5m wide and 9m high, usually have a vertical expansion joint every 5m and are fixed with bamboo canes to resist earthquakes.

2 Ecological Wisdom

Regional architecture is related to the ethnic, regional, and dialects of local people[6]. According to the climatic conditions, geographical environment, traditional culture, and different bamboo species, the OBA is generally categorized as follows: ① original bamboo architectures in Asia, which are made of *Neosinocalamus affinis and Phyllostachys pubescens* Mazel such as Dai people's Bamboo House in Yunnan (Fig. 2), Li Nationality's Boat House in Hainan, Ekra Bamboo House in India, Shanluo Bamboo Restaurant in Vietnam, and Bamboo House Village in Bali Island, ② original bamboo architectures in the Americas, which are made of *Guadua angustifolia* Kunth such as Columbia's Pavilion in Germany (at the Expo Hanover) (Fig. 3), Quincha Bamboo House in South America's many countries, and Bahareque Bamboo House in Columbia, and ③original bamboo architectures in Africa, which are made of Oldeania alpina such as Sidama Bamboo House in Ethiopia (Fig. 4), and Floating school in Makoko.

2.1 OBA in Asia

Mud walls with a wooden frame: this type of construction pattern of bamboo and wooden strips between wooden columns were recorded during the Neolithic period. Since then, China has been making extensive use of original bamboo to construct vernacular architecture. The most representa-

Fig. 2 DN Bamboo House in Yunnan (China)© Meng, 2008

Fig. 3 Columbia Pavilion in Germany © Minke, 2012

Fig. 4 Bamboo House in Ethiopia © LIU and Frith, 2013)

tive construction is the Dai Nationality's Bamboo House with a history of over 1,400 years. Dai Nationality's Bamboo House is divided into three types: stilt style architecture, surface structures, and soil-room. Xishuangbanna is a famous bamboo town, and *Dendrocalamus sinicus*, *Phyllostachys parvifolia*, *Bambusa multiplex*, *and Phyllostachys pubescens* Mazel are natural building materials. Bamboo is used as the main construction material for the entire bamboo building. Thick bamboo is used for creating the structure of columns, beams, and roof trusses. Bamboo sheets are woven for the walls, and flattened slabs of cut bamboo are used for the floors. Further, doors and windows are made of bamboo (Meng, 2008). In India, Nepal, and Bhutan, the Ekra Bamboo House is a commonly found form of OBA-a wooden frame with a bamboo weaved walling system where both sides of the wall are plastered with mud and lime. The OBA makes use of local materials, which not only saves costs but also facilitates the repair and disassembly of the house in the future.

2.2 OBA in America

As there are no specialized bamboo plantations in the Americas, bamboo is usually grown in tree nurseries or imported. *Guadua angustifolia* Kunth is the best variety of bamboo used in American-built houses and is mainly grown in the quadruple soil of Colombia along the banks of the river or other wet areas. Before the 1980s, bamboo was mainly used to make furniture or decorative elements in America. It was not until 2000 that the Colombia pavilion at Expo Hanover, Germany, made the western world realize the potential of this material, attracting the attention of western engineers and architects to the outstanding properties of this exotic building material (Li, 2012). In South America, there are two main types of well-documented vernacular bamboo-wall systems. The first one is Quincha structures, which date back to 300 BC (the Bato period). In traditional Quincha housing, bamboo or wood are used as the vertical and horizontal components of a basic framework, and sugar cane or bamboo plates are fixed on both sides of the wall. Usually, people keep the green side exposed, facing inwards, and apply a clay mud and straw mixture to the outside (Liu and Frith, 2013). Another bamboo-wall system is Bahareque, which has two main types, the modern-day use of which is prevalent in countries such as Colombia and Venezuela.

In solid forms, horizontal bamboo laths are fixed on both sides of vertical bamboo culms or a timber frame, filling with mud in the space. In the hollow type, flattened bamboo is fixed on both sides of the vertical frame and is plastered with mud (Liu and Frith, 2013) (Fig. 5).

Frame and wall of Quincha bamboo house

Filling soil to construct solid wall of Bahareque house

Fig. 5 OBA in America. References © LIU and Frich, 2013

2.3 OBA in Africa

Africa has 42 species of woody bamboos of 13 genera, 12 species of herbaceous bamboos of 3 genera, and 1 species of imported bamboos. The total number of imported bamboos is 2.758 million ha^2, accounting for approximately 7% of the world's bamboo forest area (Lian and Zhou, 2014). Compared with the *Phyllostachys pubescens* Mazel of Zhejiang, China, which is 3 4 years old, African Oldeania alpina has a smaller diameter at breast height and sharpness, thicker bamboo wall, lower basic density, larger dry shrinkage, higher modulus of elasticity, and lower modulus of rupture. Therefore, based on its stem shape and physical and mechanical properties, *Oldeania alpina* is a high-quality material for building materials and bamboo-wood-based panels (Minke, 2012). The unique architectural form of Africa is a shed woven from crisscrossed bamboo pieces resembling a garlic head of the Bamboo House in Ethiopia. The shed and the roof are woven from top to bottom or divided into upper and lower parts-the lower part is cylindrical and has a tapered upper roof. According to the climatic zone, the wall is coated with clay mud or left exposed, while the structures are generally covered with a thatch roof (Liu and Frith, 2013). To resist adverse environments, no windows are installed, except for the door opening. Thus, natural ventilation and daylighting are poorer. The house is often divided into two areas, one for daily living and the other for feeding animals.

3 Construction Process

3.1 Bamboo Processing

Bamboo is susceptible to insects owing to its high starch content; further, humidity can lead to fungi and lichens. Therefore, good procedures for harvesting, drying, and treatment must be considered to ensure the durability of the bamboo structural components (Fig. 6).

(1) Harvesting. Winter and early spring are the best harvesting seasons for bamboo in temperate regions. At this time, when the bamboo has grown for one year, the roots and stems are dormant, and some of the vines and overgrown stems have lost leaves. Generally, older and mature bamboos are chosen as building materials because their cellulosic walls have higher strength; at the same time, most of the sugar in the bamboo has been converted to starch, which can reduce later damage by insects (Strangler, 2009).

(2) Drying. Air drying — the bamboo stems are arranged in a tripod-like shape and exposed to ventilated sunlight, which will be accelerated in a plastic-covered greenhouse. Drying using heat — the bamboo stems are placed horizontally on activated coal for thermal drying. The bamboo stems should be kept at a sufficient distance from the fire to avoid burning.

(3) Treatment. Surface cleaning — steel wool or high-pressure water are used to clean the lichens on the bamboo surface. Sterilization — $Ca(OH)_2$ is usually applied to the bamboo surface to protect it from fungi, lichens, and insects. $Ca(OH)_2$ has a low pH, so it can act as a fungicide and insecticide. Surface protection — to protect the surface from ultraviolet light and rain, a coating containing linseed oil and beeswax is applied to the bamboo surface to seal the open pores without completely preventing

the transfer of moisture. Insect and corrosion prevention-to perforate the internode wall of the bamboo stem with certain humidity and soak it in an insecticidal preservative solution for more than three days. Flame-retardant treatment — the following substances are added to 100 L of water: 3 L of ammonium phosphate, 3 L of boric acid, 1 L of copper sulfate, 5 L of zinc chloride, 3 L of sodium dichromate, and a few drops of hydrochloric acid. The bamboo is soaked in this mixture for some time to achieve a flame-retarding effect(Wang et al. ,2017).

(4) Storage. Bamboo is a kind of hygroscopic porous material. When the bamboo gets wet, its shell will expand and its mechanical properties will decline. Therefore, bamboo should be stored in a place that is sheltered from rain, dry and well ventilated.

Air drying

Clean the surface

Fig. 6 Bamboo Processing © Minke,2012

3.2 Constructive Elements

(1) Columns. Columns are linear, vertically positioned construction elements that transmit compressive forces. Bamboo has a certain compressive strength and can be used as a column structure. To prevent the bamboo from cracking, hardwood can be filled into it. A metal layer, covering the cane, can further enhance the compression strength of the bamboo pillar. Steel frames are placed between the foundation and bamboo pillar to protect the pillar from moisture.

(2) Beams. Usually, multiple bamboos are used in parallel to form a beam because the bending resistance of a single bamboo is low. Without special support, it can only support the linear space of a short-span or low-load door opening or window opening instead of a region. To prevent the supported end of the bamboo beam from opening or breaking, the support must be filled with an anti-compression material. The beam should be pre-tensioned before being put into use to increase the bending rigidity of the bamboo beam.

(3) Floor Slabs and Roofs. A common method to use bamboo to build floors and roofs involves splitting the bamboo into boards and covering these between purlins or beams. If the distance between the beams is large, thin bamboos or thin bamboo splits are bunched together and placed vertically on the purlins or beams. Alternatively, bamboo beams are used to support the traditional tile roof. For some roofs with high waterproof requirements, bamboo stems are used to form the basic frame, which is then filled with palm tree leaves.

(4) Walls. Light walls with supporting keel can be covered with bamboo boards. However, in warm and humid areas, crossed bamboo stems are used to form a gap for ventilation. In addition, if the inner wall is covered with water and electricity pipelines, prefabricated bamboo is used to weave the panel.

4 Application

We have considered the design of the Liangshan Lamp Opera Troupe in the Liangping District of Chongqing as an example. There is a scenic spot in the Liangping District of Chongqing City, which is rich in bamboo materials. The Liangshan Lamp Opera Troupe is a folk activity in this area, and it is often staged in populated areas. Over time, such drama has been seen for many years in the Liangping District, but there are still many remaining old bandstand sites in many ancient villages. To inherit this ancient local drama, the original bandstand sites were used to rebuild the bandstand, which is located in the Bailin Village.

4.1 Building Structures

The architectural design of the bandstand is divided into three parts: the auditorium, stage, and public part of the center that echoes the mountain culture (Fig. 7).

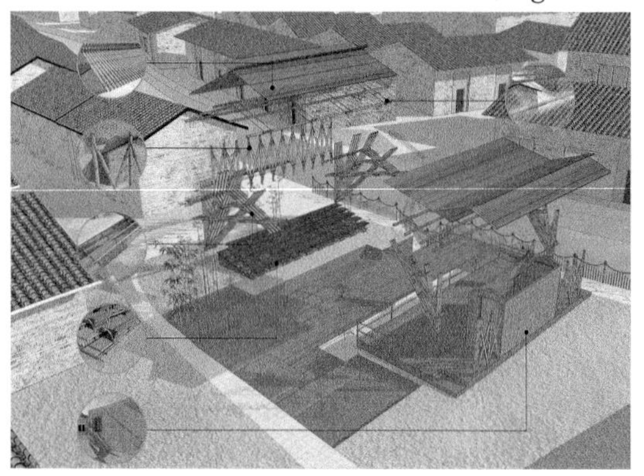

Fig. 7 Explosion and partial node diagram of bamboo stage construction structure

(1) Ground. The ground part of the bandstand is formed by using the original bamboos in parallel to form a bundle.

(2) Wall. The stage part is first erected using bamboo to form a keel frame. The inner and outer sides of the keel are covered with a bamboo woven panel to ensure the privacy of the performer while changing clothes. The auditorium part uses a bamboo cross to form an interstice for ventilation and to let the sunlight pass through the interstice to form light-and-shadow changes on the seats.

(3) Roofing. To quickly drain and echo the surrounding traditional architectural forms, the bamboos are turned upside down and connected side by side to form a pitched roof. The front half of the stage is used to form a good sound-transmission path when rainwater is discharged from both sides of the roof. To show the mechanical beauty of the bamboo, the supporting roof structure bends the bamboo into an arch to withstand the force above (Fig. 8).

(4) Seats. The seats are arranged in a virtual and real combination; the virtual ones are used as landscape and the real ones as seats (Fig. 9).

(5) Railings. The bandstand is built along the river, where the railings are directly erected in the form of bamboos to form a bamboo forest. The railing design at the stage is relatively simple, mainly considering the viewer's perception.

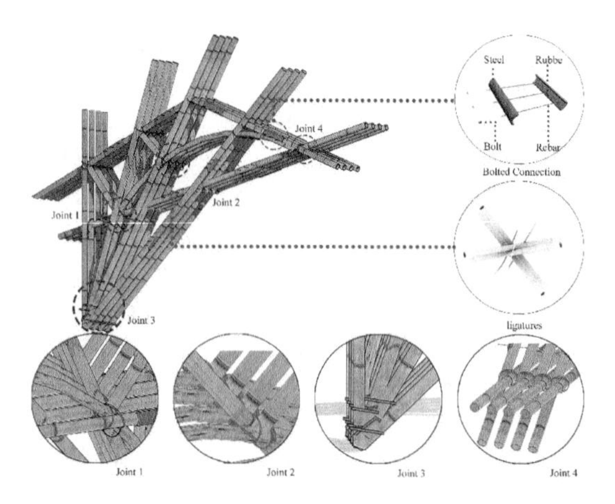

Fig. 8 Roof supporting structure

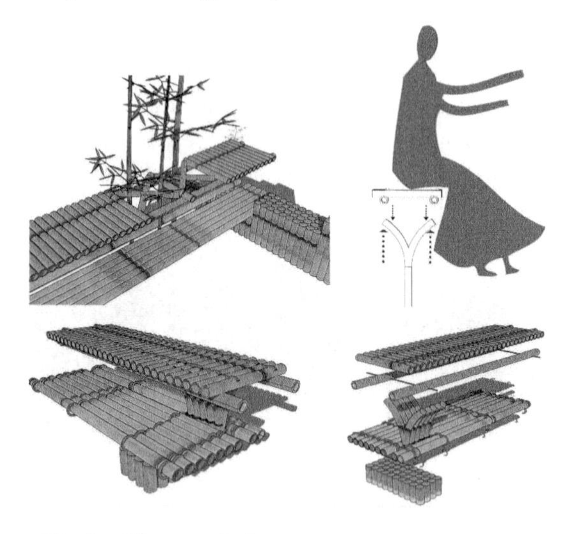

Fig. 9 The seat design

4.2 Connection Structures

Using the Liangshan Lamp Opera Troupe as an example, the connection methods for the commonly used ligatures, elliptical sections, circular sections, fish-mouth sections, and hinges are summarized (Fig. 10).

5 Submission

In summary, to achieve the adaptive reuse of original bamboo materials in characteristic buildings, it is necessary to adhere to the concept of sustainable design to truly realize the goals of renovating rural environments, protecting and repairing OBA heritage, and inspiring low-cost construction of beautiful countryside. By studying the relevant historical ecological wisdom, the construction technology, application examples, and typical connection struc-

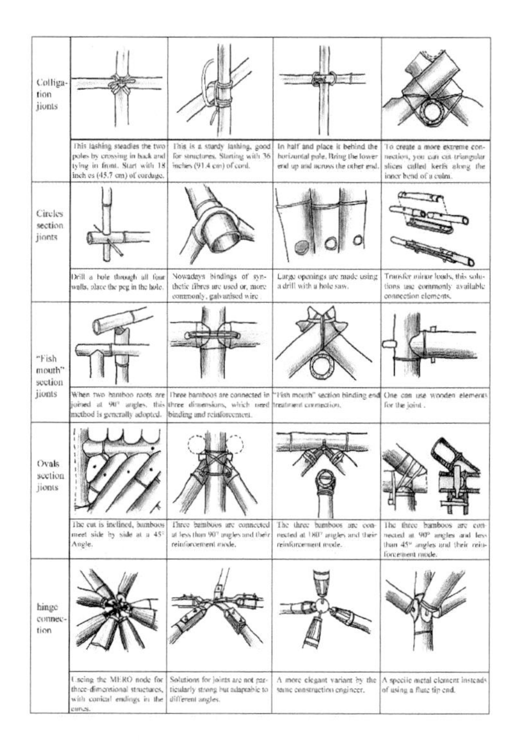

Fig. 10 The connection methods. Redrew and reclassified according to Minke, 2012; Strangler, 2013

tures are analyzed. Finally, a bamboo space design method and intermediate technology, which are structurally feasible, energy saving, and environmentally friendly while retaining local characteristics, are developed. This has a positive demonstration significance and reference value for the Chinese original bamboo construction technology in the contemporary rural construction and adaptation to regeneration (Wang et al., 2017). It is beneficial to the restoration of buildings with characteristic features to solve the lack of traditional settlement features and inspire the low-cost construction of traditional settlement buildings with characteristic features.

Acknowledgements

This research was funded by Ministry of Education of China Project of Humanities and Social Sciences (No. 18YJAZH088), Science and Technology Plans of Ministry of Housing and Urban-Rural Development of the P. R. China, and Opening Projects of Beijing Advanced Innovation Center for Future Urban Design, Beijing University of Civil Engineering and Architecture (UDC2017011112). The authors would like to thank the anonymous reviews for their comments and suggestions.

References

CHEN Lingyi, 2018. On modern bamboo building design techniques in southeast Asian. Urbanism and architecture, (02):63-66.

LI Yanhua, 2012. Research on the form characteristics. Wuhan university of technology, (05).

LIAN Chao , ZHOU Jianmei, 2014. Investigation on bamboo species, resources and industry in africa bamboo zone. World forestry research, (04): 75-82.

LIU Kewei , Frith O, 2013. An overview of global bamboo architecture: trends and challenges. World architecture, (12):27-34.

MINKE G, 2012. Building with Bamboo. [s. l.]: Birkhäuser.

STRANGLER C, 2009. The craft & art of bamboo 30 eco-friendly projects to make for home & garden. [s. l.]: Lark Books.

VELLINGA M, BRIDGE A, OLIVER P, 2007. Atlas of vernacular architecture of the world. [s. l.]: Routledge.

WANG Jiang, ZHAO Jilong, WU Tianyu, LI Jin, 2017. A co-evolution model of planning space and self-built space for compact settlements in rural China. Nexus network journal, (2):473-501.

WANG Jiang, ZHAO Jilong, YANG Qianmiao, 2017. Adaptive reuse of shellfish waste for rural landscape. Journal of shellfish research, (3):841-849.

WANG Jiang, ZHAO Jilong, ZHOU Zhongkai, 2014. The coordination and symbiosis mechanism of planning space and self-help space in the process of rural settlement community. Journal of Hunan city university, (05):33-38.

WANG Jiang, ZHAO Jilong, 2017. The formation mechanism and living conservation of rural cultural landscape with wasted shellfish resources. New architecture, (02):101-105.

XIAN Meng, 2008. The original ecological culture of dai bamboo house and its protective evolution. Journal of Shanghai technical college of urban management, (05):25-26.

XIAO Bin, 2006. Talk about the use of bamboo forest and bamboo. Anhui forestry science and technology, (02):24-24.

ZHOU Fangchun, 1998. The growth and development of bamboo. Bamboo research, (01): 53-73.

Amarbuyant Monastery: Conservation and Revitalization through Community Engagement and Digital Documentation

Ricelli Laplace Resende[1], *Christopher McCarthy*[2], *Erdenebuyan Enkhjargal*[3]

1 Kyoto University, Graduate School of Global Environmental Studies, Japan
2 Johns Hopkins University, USA
3 Doshisha University, GSGS, Japan

Abstract The Amarbuyant Monastery is a geographically and culturally significant stop on the route linking Mongolian and Tibetan regions. Spanning from Lhasa to Ulaanbaatar, this route was used for the exchange of ideas and information by pilgrims and traders since the 13th century; forging cultural, economic and political ties that between Mongolian and Tibetan people. Towards Ulaanbaatar, Amarbuyant is the first temple complex on the route inside Mongolia territory, located close to the Chinese border. Many important a historical figures have used this temple as refuge after crossing the Gobi Desert, including the 13th Dalai Lama in 1904. Although the complex was almost completely destroyed during the purges in Mongolia of the 1930s, reconstruction and conservation activities undertaken by the local community and lamas (monk teachers) have conserved part of the temple, and locals still gather for religious and cultural ceremonies. No historical maps are known to exist about the Mongolian side of the pilgrimage route and its temples.

This is the first study to document the physical structures of the Amarbuyant temple complex for heritage conservation. In order to create a database and image collection of the temple, drone technology was used to obtain aerial views of the complex, highlighting archeological boundaries and structures. Interview surveys were conducted with nomadic households living around the complex, resident lamas and students.

Results contemplate aerial documentation, retraced architectural maps, and imagery documentation of buildings and surroundings. Perspectives from local community members and their relationship with the temple, and exploration of future possibilities for conservation is discussed. This study creates awareness for the cultural heritage of the Gobi Desert, and is a starting point for future studies to contribute to a conservation action plan for Amarbuyant Monastery, and other archeological sites along the route into the cultural heritage registry of Mongolia.

Keywords drone and digital heritage, Mongolian heritage, intangible heritage, the Dalai Lama, Mongolian-Tibetan route, imaging science, digital archeology

1 Introduction

1.1 The Mongolian-Tibetan Route

The route stretches across more than 2700 kilometers of harsh landscape between Ulaanbaatar and Lhasa (McCarthy, 2019). It is a mysterious and fabled land, a mix of steppe and deserts, nearly half the size of Europe. In the 13th century, this route facilitated the exchange of ideas and information between pilgrims and traders fostering cultural, economic and political ties between Mongolian and Tibetan people (Atwood, 2014). Caravans and important historical figures (including the 13th Dalai Lama in 1904 who was fleeing the Younghusband expedition) used this route for travelling and its temples as shelter (Atwood, 2014). It is said that Lamaism (Tibetan Buddhism) has been the key aspect linking the Mongolian people, contributing to the country's cultural cohesion (Sharad, 2010).

This paper is a part of a study that retraced and documented in detail the Mongolian side of the ancient caravan route between 2016—2019 (McCarthy, 2019). The route was traced back and documented using modern technological tools such as GPS, drones (UAV) and architectural survey for heritage conservation.

Guided by historical sources from 19th and 20th century explorers (Prezwalski, 1876; Roerich, 1931 & Lattimore, 1962) and field survey, the route and its elements were studied: waypoints and coordinates, temple complexes, padme hum stones, deer stones and way markers, wells, oases, and other cultural heritage artifacts.

1.2 The Amarbuyant Monastery

In this paper, we will discuss a historically and geographically significant temple complex of the route: The Amarbuyant Monastery. This monastery is especially famous for the visit of the 13th Dalai Lama in 1904. It was the first Mongolian temple to shelter his caravan, which was secretly fleeing the British invasion of Tibet in the early 20th century (Kozlov, 2004).

The monastery is located in Bayankhongor province, close to the border of Gobi Altai province and the Gobi Protected Area A and Chinese border in the south. The closest sum (district) is Bayan-Odor (Fig. 1).

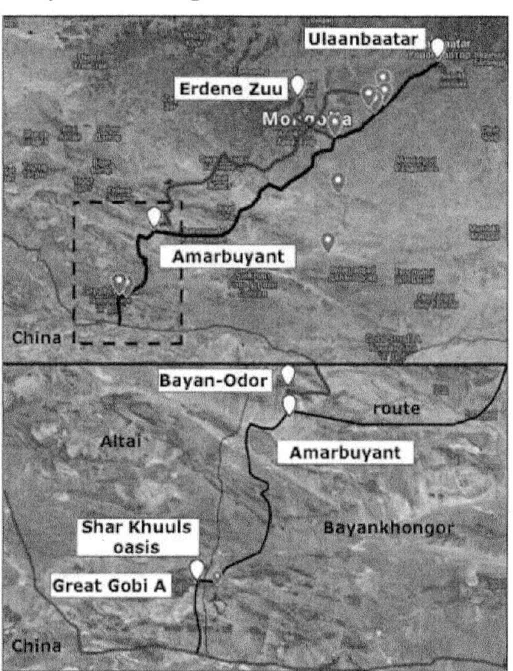

Fig. 1 The above location map shows the two branches of the Tibetan highway from Ulaanbaatar to the border of China, including the route traveled during our expedition (black). The bottom map shows the location of Amarbuyant Monastery and its surroundings

Amarbuyant Monastery was a thriving community, until it was invaded and destroyed during the purges in 1937. Buddhism used to be central in the life of Mongolians. In 1918 it is estimated that 115,000 monks were living in Mongolia. A census in 1937 counted a total of 700 monasteries, which were home to more than 110, 000 practicing monks (Lokesh, 2001).

In the late 1930s, almost all the Buddhist monasteries were destroyed and monks killed during the purges under command of Kh. Choibalsan (Dashpurev & Soni, 1992).

What remains of Amarbuyant Monastery measures roughly 88,280 m^2 (8.8 hectares) in area and contains a few buildings and structures. An additional 36 hectares surrounding the monastery also contains some structures like religious paths and sculptures.

In this study, we tried to document existing structures and also the reconstruction that has taken place since the early 1990s. Imagery documentation, aerial photographs, mapping and measurement surveys have helped us to produce detailed documentation of the actual condition of the complex and remaining structures. Interview surveys with lamas, students and local families helped us to gather a greater understanding of the history of the temple complex and connection with the local community. We believe this study opens up a new field of heritage documentation and conservation in Mongolia.

2 Relevance and Objectives

No historical maps are known to exist about the Mongolian side of the route and its temples. Official caravan archives, once kept in Ulaanbaatar, documenting travel between the two places were destroyed during the Stalinist repression of the 1930s (McCarthy, 2019). Currently, few studies exist on the caravan route and documentation of temple complexes and cultural heritage is also limited.

By documenting and retracing the history and architecture of Amarbuyant Monastery, this study can help to promote its conservation and revitalization, as well serve as a starting point for further documentation and revitalization of temples and historical points along the Mongolian side of the route. The local community, temple students and monks can also benefit by bringing awareness to the necessities and difficulties they face regarding temple conservation and local development. This can create big impacts for the future of Mongolian heritage conservation and the communities that live alongside the route and in temple surroundings.

This study also shows the importance of drone and architectural documentation for heritage conservation. Aerial photography can create an emotional impact in people and bring awareness toward historical background and heritage (Baxter, 2014), and allied to maps, drawing and photos, they are essential for the visualization of the built heritage and consequently its conservation (Pauwels et al., 2008). Also, the documentation of built heritage through digital data can help future data access to heritage sites, even if they are inaccessible or deteriorated (Richards & Jeffrey, 2013).

By telling the story of this inspiring human journey and this forgotten temple, this study aims to raise awareness about the historical and cultural heritage of the Gobi Desert as well as promote conservation and open up a new field of academic research for those studying Mongolian-Tibetan people's relations. The goal of this study is the documentation for conservation of Amarbuyant monastery in order to shine a light on the long forgotten, yet, inspiring human story of the caravan pilgrimage and its architectural heritage.

3 Methodology

3.1 Temple Mapping

Three expeditions (2016 — 2018) to document cultural heritage along the route were undertaken (McCarthy, 2019). As shown in Fig. 1, the route was traced from Ulaanbaatar to the Chinese border. Erdene Zuu, the ancient capital city, is one of the main tourist sites of the country, and was once an important stop along route linking Mongolian and Tibetan regions. Amarbuyant Monastery is located along the southernmost part of the Mongolian side of the route. It is the first temple complex within the Mongolian border, and it was studied in detail in the 2018 expedition. Drone technology was used to obtain aerial views of temples and structures. DJI Mavic equipped with an RGB camera was used (see Tab. 1).

Images could highlight archeological boundaries, structures as well as vegetation and geographical aspects. Processing of images was conducted in Drone Deploy, MultiSpec and ArcGIS. Aerial images of temple complexes were outlined to count building and structures and retrace old setting by AutoCAD Drawing software.

Tab. 1 Drone flight specifications

Parameters	specification
Camera parameter	
Lens	8.8mm/24mm (35mm format equivalent) f/2.8
Sensor	1″CMOS
Maximum angle of aperture	FOV 84
Pixel number	20M
Flight parameter	
Altitude above ground	120m
Overlap	85%
Geometric resolution	2.5 cm
Size of recorded area	10 hectares
Speed	12m/s
Flight duration	23min
GPS location	available

3.2 Interview Survey

Semi-structured interview surveys were conducted with lamas and students at the temple, nomadic households living in the surrounding areas, and residents of the nearby village center, Bayan-Ondor. Interviews were recorded for discourse analyses. Questions were designed to understand historical knowledge, relationship with the temple and its conservation, and ideas and hopes for the future of its revitalization. This provided a qualitative approach to understand the social-economic change that the revitalization and conservation of the monastery and temple could provide. This also helped to construct a picture of the historical background and to measure the attachment that locals had towards the monastery.

4 Results

4.1 Imagery and Digital Documentation

The main outputs of this study can be divided into three parts:

① An online map and database of the Temple within the caravan route including locations for important archeological artifacts of the surroundings (under construction, refer to https://thegreattibetanhighway.wordpress.com)

② A retraced map outlining temple structures, ruins, reconstructed buildings with historical information

③ Qualitative data from interview surveys with locals and ideas for future conservation

For the database, exact geographical location of the temple in relation to the caravan route was made. This included locations of other important key points in the surroundings of the temple, like deer and padme hum stones (monument with inscriptions that indicate the route), ovoos, oases, etc. (Fig. 2).

4.1.1 Detailed Map Past-present

For the retraced map, drone images were used as a base, and detailed in-loco measurement of structures were made and combined with the images to produce accurate floor plan of structures and architecture. The drone images were captured within a day and used as a base to assist in loco measurement that was made to complement and confirm remaining ruin structures that are difficult to see in the images. This was finished within two days, making a total of three days needed to collect the raw data necessary to create the maps. This is an example of how drone survey can assist heritage mapping and improve greatly the speed and quality of the documentation process (Tab. 2).

Fig. 2 The top image shows a padme hum stone. A sacred ovoo (cairn, below), both indicating the route path to the temple

Information from the interview surveys with lamas helped to understand structures and ruins, old and new buildings, it's functions and shapes. A map was created showing not only existing structures, but remnants of past structures and new structures that were reconstructed by the community and previous lamas (Fig. 3).

It was found that between 1600—1900 the temple was thriving, it was the center unit that unified 3 villages (Bayan Ondor, Bayan Tasagaan and Shine Jinst), forming one town. However, during Soviet times, the temple was attacked in 1937, and almost completely destroyed.

Until 1937, the complex contained 6 main temples, and according to locals, around 1500 to 2000 lamas and students used to live there. It seems like the complex was not only the religious center, but also the political center, since politicians like local governors used to live there too (left upper part is shown in map).

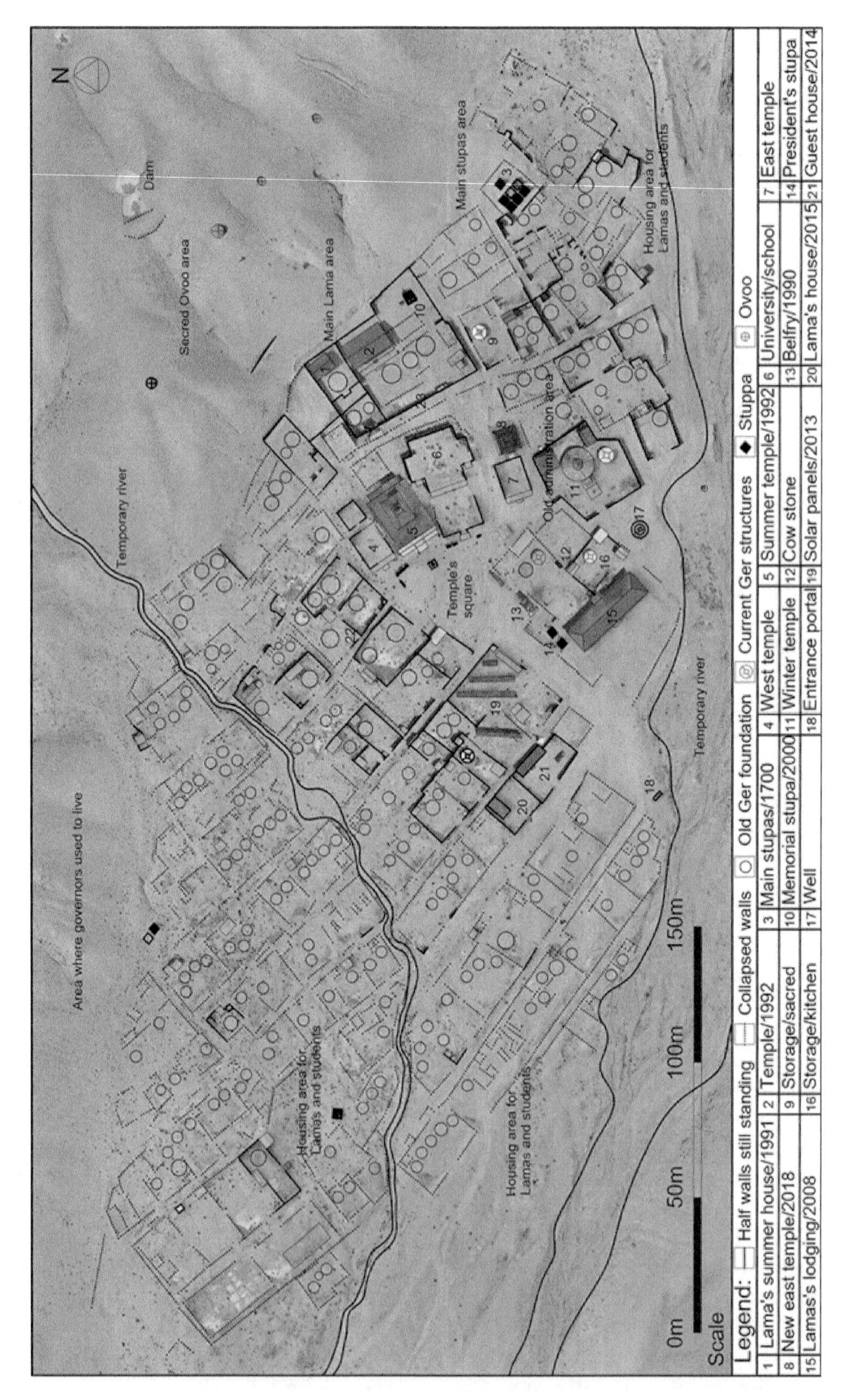

Fig. 3 Retraced map of Amarbuyant complex

Tab. 2 Examples of detailed map retracing

n° in Fig. 3	Type	Description	Aerial photo	Retraced structures
22	Residential cluster	Residential cluster for lamas and students with four subdivision. Measuring a total of 750m² (30m × 25m). Contains 6 ger foundations, walls and 3 storages. Materials: stones, clay bricks, wooden roof.	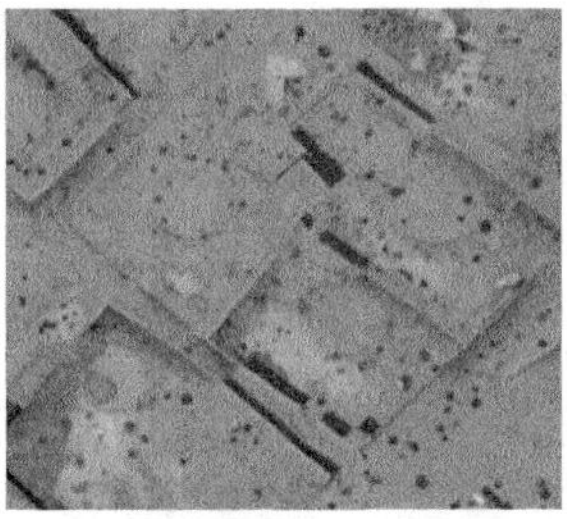	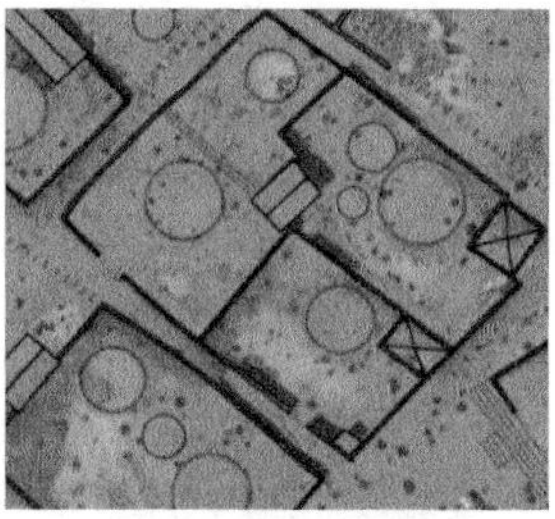
2	Temple	Reconstructed in 1992, measuring 17m × 10m, with 170m². Foundation and staircase were renovated. Materials: stones, clay bricks, wooden roof.		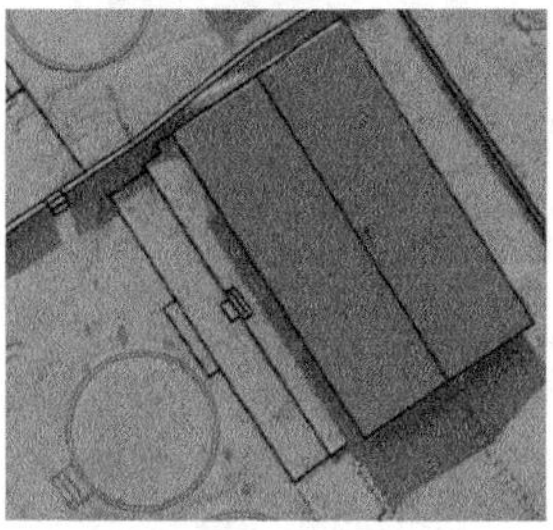

The complex was supported by local families that provided food and supplies to the lamas.

There used to be a school (n° 6 in Fig. 3), where students could learn how to read, write, recite mantras, study philosophy and practice Buddhism. By every indication, Amarbuyant was a rich temple complex containing big and small temples, roads, houses, administrative buildings, schools, etc.

Today, only 200 foundations of the previous *gers* (traditional round-shaped Mongolian dwelling) were found, but since there are at least 130 identified clusters and each cluster contains 3 *gers*, we estimate that at least 400 *gers* existed inside the complex before. Now, lamas and students live mainly inside lodging structures built from donations or by themselves, and only 5 *gers* are being used for storage or temporary housing for lamas and visitors.

Today, 3 main families help feed around 20 to 30 temple residents, which varies according to the season. But we roughly estimated that around 150～300 families used to live in the neighboring surroundings before.

During the attack of the temple in 1937, the head lama escaped with various items in order to try to save them, including: images, portraits, sculptures, etc. After the revolution in the early 1990s, he returned the artifacts to Mongolia. and started reconstruction on a small temple and a residence for the lamas (n° 1 and 2 in Fig. 3).

The reconstruction was financed and built by himself. Moved by his efforts, locals and the government started some small reconstruction projects. In 1992, the main temple reconstruction was financed by the 3 village centers to which the temple belongs to, and local families made many donations (especially animals like sheep) and contributed with the construction (n° 5 in Fig. 3). The reconstruction took over one year to complete, and it was built on top of the old foundation of the previously destroyed temple.

We found many of the modern structures don't follow the traditional architectural style of the other reconstructed buildings, but serve as lodging for students and visitors and supporting infrastructures (n° 15, 21 and 20 in Fig. 3).

Number 10 in the map shows a memorial stupa made for the killed lamas during the 1937 event. A stupa' is a religious Buddhist structure that contains sacred images and inscriptions inside its main body. Number 3 shows the area where old stupas, some dating from 1700, still stand and were reconstructed.

The last construction made in the temple complex was finished in the summer of 2018, a new temple made of wood financed by a collective between local people, the government, external donors and built by local families, lamas and students. Photographic documentation of all structures was made, some examples can be seen in Fig. 4.

Fig. 4 Photo documentation of structures

4.2 Local Engagement and Perspectives

Our study shows that Amarbuyant Monastery and its memory only survived because of the great effort of some lamas and the local community. Not only that but efforts are continuously being made to revitalize the temple.

From interviews and surveys, we could understand that locals have a strong attachment to the temple, its history and what it represents for the life of residents in that area. Families are supported by the temple and the temple is supported by the families in a mutual gain relationship. Rural herder families visit the temple daily for water (the temple has a well with clean water which only males are allowed to access, and its mainly the lama's job, n° 17 in Fig. 3). Everyday families come and go for water. They also seek spiritual guidance and use the temple areas for local festivals and religious celebrations. In ancient times, the temples served as an important education center for local family's male children. In return, families provide food for the students and lamas, as well as maintenance and construction of the temple. They also provide transportation and bring goods from the village centers.

Important figures also visit the temple for religious purposes, the stupa n 14 on the map was constructed by the Mongolian president in memory of his father. We also found that the temple is used as a pit-stop for tourists crossing the Gobi. They usually stay one night to rest and continue their journey the next day. Most of them don't know what Amarbuyant monastery is and what it represents.

4.2.1 Challenges and Hope for the Future

When we asked lamas, students and locals about their perspective for the future of Amarbuyant, we had a unanimous answer: they all dream of the revitalization of the monastery as a school/university for lamas. Since they don't have the school anymore, they need to send their lama students elsewhere, creating a great economic cost for the temple, making it possible to send only one or two students per year. Since the area is also geographically isolated, families also wish for a school nearby, but not only for teaching Buddhism, but secular for all the children in the area.

Interviews with local children revealed that many of them don't adapt well to studying in the village center, having to move and stay away from their parents for many months, they expressed the wish that they could study nearby and come back home more easily.

Locals dream to see the temple complex revitalized and inhabited by many people again, but geographical isolation and low-density population around the temple make these efforts seem small to achieve bigger goals.

The main support for the temple has been made by nomadic herder families, but today 25～40% of Mongolians live in the countryside, while the majority live in cities (Campi, 2006), especially Ulaanbaatar (where half of the population lives). Now, only three main herder families are living in the surroundings of the temple, which makes it difficult to support bigger and continuous revitalization projects. This can be seen in the difference between the first reconstructed temple that measures 170m², compared to the most recent one that measures 80m², less than half of the size. There are indications that economic and material availability restrictions and low density of dwellers in the area are challenges for the development of new projects.

There is great potential for earth construction, like the traditional technique. Three times a year a heavy rain creates two to four temporary rivers (which the main river is shown in the bottom part of the map), creating good conditions for making earth materials. But these techniques require knowledge, time and effort, an especially group effort to carry the mortar from the river to the construction site.

Most of the knowledge holders of this technique have died, this, together with the low density of people around the temple, makes it difficult to use the traditional techniques for building. Therefore, locals relied on imported wood and construction workers for the last reconstruction, making it more expensive and not keeping with the traditional architectural style.

Also, to adapt to Mongolian economic development and changes in cultural aspects of the modern times, many families wish to include new functions to the temple to support a more modernized lifestyle. This includes an open school (for all genders, religious education and non-religious education) and tourism attractions to generate extra income.

Therefore, locals also shared with us the ideas of bringing tourism to the area, especially religious tourism. For this, better lodging structures and cultural experiences should be provided.

5 Conclusion

Through digital documentation, this study establishes a formal research record on Amarbuyant Monastery, an important temple along the ancient caravan route. In doing so, it contributes to efforts to raise awareness about the cultural heritage of the Gobi Desert.

Evidence of a caravan route linking Ulaanbaatar and Lhasa is documented in the expedition journals of Kozlov, Roerich, Przhevalsky and others. These journals and reports also indicate the presence of thriving temple communities. One can only imagine how dynamic this route and its communities used to be before the purges during the 1930s.

Our study shows that the conservation of the temples was possible due to local community engagement, and collaboration between local families and residents of the temple.

We also discussed some future hopes from locals towards the sustainability and revitalization of the temple, including the reconstruction of a local school and tourism activities. These would probably create changes in historical architectural functions of the temple, adapting it to new uses and necessities. This points to the necessity of a creative revitalization project that balances tourism attraction and local activities. If done properly, this can help to promote local sustainable development and revitalization of the complex area and surroundings.

This study is the first of its kind and will surely open new possibilities for understanding the communities that once thrived along the caravan route once linking Ulaanbaatar to Lhasa.

We hope that more studies will be made at other temple complexes along the route, as well as the documentation and present conditions surrounding communities. This can help to create a future plan for the conservation and revitalization of the route as a cultural heritage monument, incentivize religious and eco-tourism, contributing for the striving of these communities and its sustainable development.

Local NGOs and government agencies (Ministry of Culture, UNESCO) can benefit from the results of this type of study, using this documentation as a starting point to develop a formal conservation action plan for the designation of the route, its temples and archeological sites into the cultural heritage registry of Mongolia.

Acknowledgements

This research was made in a collaboration between The Gobi Institute and the Graduate School of Global Environmental Studies, Kyoto University, financed by Konosuke Matsushita Memorial Foundation.

References

ATWOOD C P, 2014. The first Mongol contacts with the Tibetans. Trails of the Tibetan tradition: papers for Elliot Sperling. special issue of revue d'etudes tibétaines, 31: 21-45.

BAXTER K, 2014. Grounding the aerial: the observer's view in digital visualisation for built heritage. Proceedings of the EVA London 2014 on electronic visualisation and the arts, 7: 163-170.

CAMPI A, 2006. The rise of cities in nomadic Mongolia. Mongols from country to city: Floating boundaries, pastoralism and city life in the Mongol lands: 21-55.

DASHPUREV D, SONI S K, 1992. Reign of terror in Mongolia, 1920-1990. New Delhi: South Asian Publishers: 42.

KOZLOV P K, 2004. Tibet and Dalai lama. Moscow: [s. n.]: 77.

LATTIMORE O, 1962. Inner Asian frontiers of China. Boston: Beacon Press.

LOKESH C, 2001. Buddhism in Mongolia. Himalayan and central asian studies, 5(1): 4.

MCCARTHY C, 2019. A rediscovery and mapping of Mongolia's ancient caravan route to Lhasa. In press.

PAUWELS P, VERSTRAETEN R, DE MEYER R, et al. , 2008. Architectural information modelling for virtual heritage application. Proceedings of the international conference on virtual systems and multimedia (VSMM). Budapest: [s. n.]: 18-23.

PREZWALSKI N M, 1876. Mongolia, the Tangut country and the solitudes of northern Tibet: being a narrative of three years' travel in eastern high Asia. Gregg, London, England, 1.

RICHARDS J D, NIVEN K, JEFFREY S, 2013. Preserving our digital heritage: information systems for data management and preservation. Visual heritage in the digital age. London: Springer: 311-326.

ROERICH G, 1931. Trails to inmost Asia: five years of exploration with the Roerich central Asian expedition. Connecticut: Yale university press.

SONI S K, 2010. Some reflections on Buddhist factor in relations between Tibetans and Mongols. Proceedings of the 4th International Conference "Buddhism and Nordland 2010". Tallinn: Estonia: 23-25.

Construction Technology and Protection Method of Korean Traditional House

LIN Jinhua, *PIAO Shunmei*

Yanbian University, Yanji, China

Abstract The Korean ethnic minorities who migrated from the Korean peninsula to China in the mid-19th century. Their main gathering place was the Yanbian Korean Autonomous Prefecture. The Koreans who moved to Yanbian initially built houses according to the Korean model and used various materials to create various forms of earth wall construction techniques. However, over time, these old houses are running low. In order to protect these traditional houses, modern technology and materials have been used for migration protection and in situ renovation protection.

Keywords Korean nationality, traditional house, construction technology, protection method

1 Introduction

The Korean ethnic minorities who migrated to China from the Korean peninsula in the middle of the 19th century, mainly distributed in the three northeast provinces, and the main gathering place is Yanbian Korean autonomous prefecture, where with the most traditional preservation and continuation. House is the most basic unit in people's life. According to different times and nationalities, especially the houses that reflect the way of life and traditions are the aggregation of politics, economy, culture, and feelings of the time. The Korean nationality who immigrated to Yanbian brought the dwelling form of North Hamgyong Province of D. P. R. Korea to Yanbian with similar geographical environment and climate at the beginning. However, the materials adopted were the peripheral materials adapted to local conditions, creating a variety of earthen wall structures in the process of construction. The earthen houses lasted until the 1970s. After the reform and opening up, the popularization of earthen houses and the "new rural construction" policy under the earthen wall housing gradually disappeared, the current few. In this context, it is necessary to use words and pictures to record the construction technology of the remaining traditional houses, as the historical data of the traditional houses of the Korean nationality in China, and discuss the protection and utilization methods for the purpose of this paper.

The research focuses on the existing traditional houses and its construction technology in Yanbian to the 1970s. The research analyzes the construction techniques of traditional houses by collecting data, research, survey, measurement, photographing, etc. The protection and utilization status of the house provides a theoretical basis for future protection and utilization methods.

2 Construction Technology of Traditional House

The traditional form of Korean house comes from North Hamgyong Province. Yanbian is adjacent to the North Hamgyong Province through the Tumen River, and the geographical environment is similar to the temperature, and most of the people who migrated to Yanbian came from the North Hamgyong Province.

The house is filled with walls and roofs on the wood frame with a tenon structure. The roof form is the gable roof or the hip roof. The construction process is as follows.

2.1 Foundation and Plinth

The foundation is compacted, that is, the rounded wood stakes are used for compaction on the foundation of the house to be built, and then the foundation is set. The plinth adopts local natural granite, which is simply processed or treated in a natural state. In order to prevent water from seeping at the joint between the plinth and the wood column, the middle part of the granite column base is slightly raised and the wood column is placed above (Fig. 1).

Fig. 1 Plinth: a) simple processing; b) natural state (handled slightly)

2.2 Wood Frame Construction

After positioning plinth, the beam-column wooden frame is erected, and then the door frame is set. In the middle of the beam, there is a king post, a half-span 1/2 is set with a short column or a beam pad, a king post is placed on the girder, the short column is placed on the middle beam. In the middle of the king post and the middle beam, there is the log purlin that is arranged in a vertical direction. And the eaves rafter is lap over on the middle beam and the eaves beam.

The king post and the middle beam extend to half of the plane at the end of the two sides, and the cross beam is placed on the end of the middle beam. The upper part of the cross beam is the gable wall of the gable roof, and the side of the eave rafter is overlapped on the cross beam. All connections are used tenon structure, the wood frame is made of local red pine or black pine, and the rafters are made of larch (Fig. 2, Fig. 3).

Fig. 2 Construction of beam-column wood frame: a) beam and column overlap; b) traditional craftsman and lapped wood construction

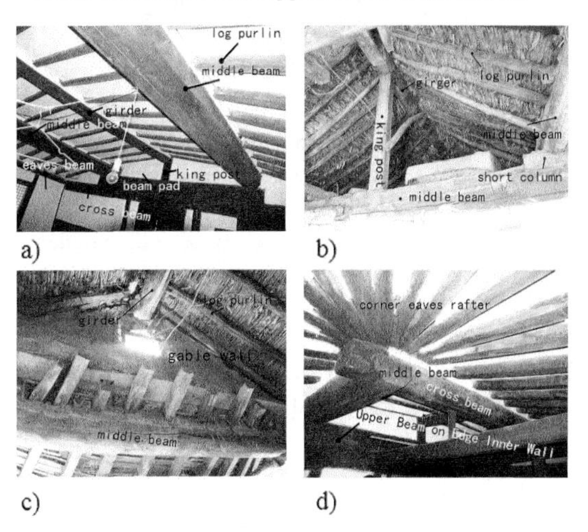

Fig. 3 Construction of the roof timber frame: a) roof structure; b) gable roof gable wall section1; c) gable roof gable wall section12; d) middle beam joint

2.3 Wall Construction

The wall is filled after the wood frame is built up. In the larger part of the solid wall, in order to increase the rigidity of the wall, a beam is placed in the middle between the two columns (Fig. 4(a)).

There are three ways to fill the wall: the first, the column and the door frame are arranged at a distance from each other to set the horizontal wooden strip, then the sorghum stalk is arranged in the vertical direction and tied on the horizontal wooden strip, and the straw bundle is laterally tied to the sorghum stalk, and finally the wall skeleton is formed, then paint the clay inside and outside the skeleton to form a soil wall (Fig. 4(b)).

Secondly, the column and the door frame are arranged at a distance from each other to form a horizontal wooden strip, and then bind

Fig. 4　Wall construction: a) the beam on the solid wall; b) walls made of crossbars, sorghum, and straw bales; c) walls made up of crossbars and straw bales; d) walls composed of wooden lattice boxes

Fig. 5　Roofing practice: a) sorghum stalk and batten above the purlin or eave rafter; b) tile or straw roof

the straws which are sticke on the clay on the horizontal wooden strip, and twist them into a rope to form a wall skeleton, and next the inner and outer surfaces are coated with clay to form a wall surface(Fig. 4(c)). Third, form a lattice-shaped skeleton with natural wood strips on both sides of the column, then fill the grid with crushed straw clay, and then paint the crushed straw clay inside and outside (Fig. 4(d)). Sorghum stalk and straw are grown by local farmers, and small wooden strips and clay are the surrounding materials (Fig. 4).

2.4 Roofing Practice

After completing the wall, the roof is built. The roofing method is to place sorghum stalk or small wooden strips on the purlin or eave rafter, topped with straw clay, and tiled on the clay. If there is no tile, use straw or reed rods. At first, the tile was shipped from D. P. R. Korea through the Tumen River ferry. Later, after the establishment of the refining yard in Yanbian, the self-made tiles were used (Fig. 5).

3 The Protection Method of Traditional House

With the development of society, expand of the city as well as an economic base, people look forward to urban life. The rural area which near city urbanize firstly, which leading traditional houses disappearing, but the rural area which apart from the city still remain the old style. What is more, with the enhancing of culture around people, people know the importance of protecting traditional houses. With the mutual efforts of ordinary people and experts as well as government, some traditional house survived while many traditional architectures facing the dangerous ending cause nobody living in or lacking stability. Some of these renew to original style through servicing, some area used the traditional building to bring back to economic returns.

3.1 Maintenance

People and government in charge of servicing, in other word, ordinary people and government pay for them and repair them. Because of the traditional building on the basement of beam frame, the whole building can't fall but only can incline which bring much convenience to servicing (Fig. 6).

The ways of repairing is that: taking the roof off firstly and get off clay of roof as well, remaining the wood frame system. Next, correct the tilt column base and column positioning, exchange the roll section, using stone or brick filled beams and ground sections. Get cement on Internal and exter-

Fig. 6 Houses repaired by individual and government: a) HuiLongFeng House which repaired by government(left: before, right: after; b) Mingdong building which repaired by individual(left: before, right: after)© University of Ulsan

nal asides to prevent the shuttle of the animals. Method of wall is that using the grid formed by the bars is fixed between the columns, and put cement on inside and outside. Last, clay topped with sorghum poles or wood strips and cemented, and putting ordinary tile on the roof, the last method is exchanging the broken tiles. Though the servicing, people remain the old house way as well as using modern material (cement) and tech (hoisting jack), making the old buildings remain their ordinary appearance (Fig. 7).

Fig. 7 Repairing technology of traditional houses: a) wall skeleton made of a grid of wood bars © Author; b) adobe brick filled wall; c) clay wall © Huang Haolin

3.2 Conservation and Utilization

There are two ways of protecting, in-situ conservation and off side conservation.

In-situ conservation has two ways, the first one is that the owner lives, but according to the regulations of the cultural relics department, he can't be transformed at will (example: Sanhe village quanshui gather). Second one is that using for sightseeing, education, training base (examples: Yuwqing village bailong gather), this kind of house have special administrators, accepting tourists to bring economic income, it also expands the scope of traditional culture through income (Fig. 8).

Put the house in other places is removing the houses which owners do not want to stay to city folk garden, letting people understand traditional houses to get the goal of education and enjoying (Fig. 9).

Fig. 8 In-situ conservation: a) Sanhe village, Quanshui gather (Master residence); b) Yuwqing village, Bailong gather (sightseeing).

Fig. 9 Off site conservation: a) Prototype before migration; b) Status of migration to Yanji Folk Village

4 Conclusion

This article mainly introduces the construction methods of traditional Korean dwellings. The materials used by Koreans were relatively easy to find at the time and inherited the traditional beam-column structure to build houses. The wood used is locally produced red pine, black pine and the like. The walls and roof were filled with natural materials that were relatively easy to find at the time, such as clay and straw, reed stems and sorghum stems. Because walls and roofs are mostly clay, they are relatively fragile and difficult to preserve and protect. Most of these filler materials are replaced by bricks and cement during maintenance and storage. However, these building materials and construction methods incorporate the wisdom of the Korean nation and the love of life and the adaptability to the local environment, which should be remembered and recognized.

Korean traditional house is a part of Chinese minority traditional house. However, with the development of society, the loss of population and the passage of time, these traditional houses have hardly left behind. Therefore, it is necessary to protect the traditional building technology, as a model to study, and as a teaching base and training base to disseminate its value. At the same time, through the use of tourist attractions, the construction of folk villages and other ways to enhance regional economic benefits, in respect of the original building methods, under the premise of the appropriate use of modern technology and materials to improve.

References

LIN Jinhua, 2008. The study on the level change of rural housing of a Chaoxianzu settlement on the Tumen riverside in China. Journal of the architectural institute of Korea planning & design, 24 (10): 71-78.

LIN Jinhua, 2014. Analysis of the types of modern building protection and utilization in Yanbian. Study and preserviation of Chinese modern architecture, 9.

LIN Jinhua, QUAN Haogang, JIN Taiyong, 2010. Analysis of modern housing development in Yanbian from modern housing development in Korea. Study and preserviation of Chinese modern architecture, 7.

PIAO Shunmei, YOON Chaeshin, 2015. On the perception of space form in architecture plans - case study of the Korean traditional housing. Journal of the architectural institute of Korea, 35 (1): 115-116.

University of Ulsan-Korea Architecture Institute, 1994. Changcai village. Ulsan: University of Ulsan Press.

Study on Traditional Construction Technique and Craftsmanship of Earthen Architecture in Floodplain Region of Yellow River in Shandong Province

WANG Jialin[1], *GAO Yisheng*[1], *XU Dongming*[2]

1 School of Architecture and Urban Planning, Shandong Jianzhu University, Jinan, China
2 National Key Laboratory of Vernacular Heritage Conservation of Chinese State Administration of Cultural Heritage (SACH), Shandong Jianzhu University, Jinan, China

Abstract By taking the floodplain region of the Yellow River in Shandong province as the selected area for the research, the paper focuses on the traditional construction technique and craftsmanship of vernacular earthen architecture as the main research topic. Based on the fieldwork of the site visits and interviews with the local craftsmen, the necessary data of the traditional construction technique and craftsmanship are collected and summed up, and the value assessment is conducted, which provide technical support with the basic information and set up a foundation for the further research.

The main body of the paper is divided into three parts. The first part is an introduction to the backgrounds of research, the main object in the selected research scope, and the research methods and approaches in the paper. The second part is the core of the paper, which provides the specific details of the traditional construction technique for vernacular earthen buildings in the area including regulatory norms and partial projects from the site selection at the beginning to the completion of construction. Finally, the paper concludes in the third part the advantages and disadvantages of the traditional construction technique and craftsmanship of vernacular earthen architecture in the area, suggests the scheme for optimization and improvement, provides the solutions to the conflicts between heritage preservation and modern development needs, and finally summarizes the earthen construction system adapting to the needs of modern use for the local people.

Keywords floodplain region, earthen architecture, construction technique and craftsmanship, regional practices

1 Introduction

In the long historical period from the birth of human civilization to this day, the traditional construction technique and craftsmanship of vernacular earthen architecture has followed the tradition to obtain raw materials locally and developed continuously, which implies the wisdom of local people adapting to the climatic and geographical conditions of their residence. However, the traditional technique and craftsmanship has gradually withdrawn from the mainstream architecture stage and only reserved by relevant preservation institutions and craftsmen in the remote area. Further, most of these craftsmen are now in their twilight years, and they have no successors to carry on the craftsmanship. Under such circumstances, the traditional construction technique and craftsmanship used in folk buildings are in danger of extinction. In particular, the vernacular earthen buildings concentrated along the Yellow River and canal in the north of China is low in strength and easy to be destroyed in comparison with masonry structure, among which few well-preserved buildings are left. Moreover, the craftsmen who know about the construction technique and craftsmanship are few and far between.

In the 1930s, the Society for the Study of Chinese Architecture began to study the construction technique and craftsmanship of China's traditional architecture. After the 1970s, with the improvement of the protection awareness of historical buildings and development of traditional residence research, important works on such traditional construction technique and craftsmanship were published. Although studies have described China's ancient buildings and their construction practices in details, most of them focus on the official practices with wide applicability. For the regional practices with various types of craftsmanship and poor generality, the existing research is not systematic. Therefore, on the basis of the extensive survey on the vernacular earthen architecture in the floodplain region of the Yellow River in Shandong province, the author sums up the characteristics, advantages and disadvantages of the construction technique and craftsmanship of vernacular earthen architecture in the area from the perspective of regional construction.

2 Scope and Object of the Research

This paper takes the floodplain region of the Yellow River in Shandong Province as its research scope. To be specific, the scope covers the area on the west side of the Yellow River Delta, Yellow River, and the area along the canal within Shandong province. It mainly refers to the southwest and northwest plains of Shandong province, including four cities such as Liaocheng, Heze, Dezhou and Jining.

In terms of the main object of the research, it indicates the construction technique and craftsmanship of vernacular earthen architecture in the area. To be exact, it means the construction practices and procedures the craftsmen adopted in the whole construction process of traditional vernacular earthen architecture. Its formation and development comes from the needs of actual use for local residents, showing typical regional features.

3 Methods and Approaches

As for the research object of this paper, it involves not only architectural content but also relevant cultural information. That is to say, studies on the cultural connotations and traditional customs in this area are worth conducting as well. The key point in the research process is to carry out a field survey and mapping of the research object, visit and interview with traditional craftsmen, and collect related data by means of photography, field survey and mapping, questionnaires, interviews, etc. (Yang, 2004).

4 Research on Practices of Construction Technique and Craftsmanship of Vernacular Earthen Architecture

4.1 Construction Process

The construction process of the vernacular earthen architecture in the research scope includes all the construction details from the site selection to the construction completion. Specifically, it can be divided into several basic steps: material preparation, site selection, foundation digging, foundation paving, column setting, beam mounting, wall building and roof laying. These steps seem simple, but each one requires rich experience and building skills of the craftsmen, which will be elaborated in the following sections.

4.2 Construction Tools

The construction of the vernacular earthen architecture in the area depends principally on carpenters and masons. The carpenters are responsible for size measurement, column fixing, beam mounting, rafter laying and wooden decoration. And their commonly-used tools include woodworking ruler, plane, ink fountain, pit-saw, chisel, axe, saw, etc. In view of the masons, they prepare earth materials, build walls and lay the roof. In addition to such conventional tools as mason's knife and trowel, ramming tools mainly cover hammer used for ramming earth, earth mold, bamboo basket, rake, hoe and *juetou* (usually regarded as a smaller hoe). The ramming tools in the area have updated and developed over the years, for example, the hammer (Fig. 1). It was initially formed by a wooden stick and a cube-shaped stone with the stone tied to the end of the stick and gradually evolved into multiple types of ramming pestles. The pestle is composed of ramming rod and pestle head that is generally made of stone. What's more, different types of pestle heads are selected according to different earth materials, which is convenient for operation (Tab. 1).

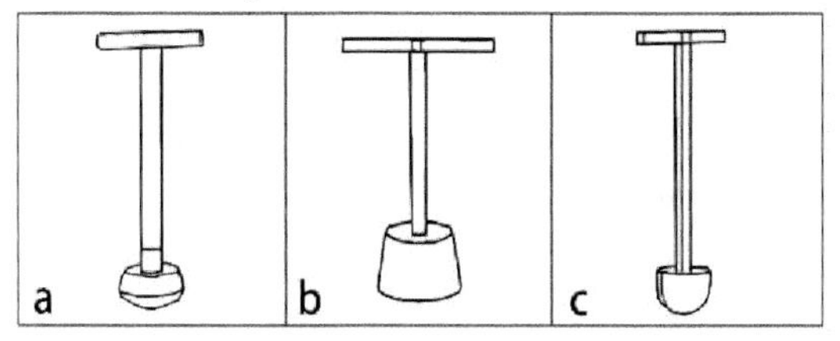

Fig. 1 Models of ramming hammer © WANG Jialin

Tab. 1 Classification of Hammer

Bowl-shaped Hammer	Frustum Cone-like Hammer	Hemispherical Hammer
Most pestle heads are made of stone with a shallow bowl-shaped bottom and a small radian, which is suitable for working on relatively damp soil and used for ramming thicker walls (Fig. 1(a))	The pestle head is built of stone or cement in the shape of frustum of a cone. It is wide at the bottom and narrow at the top. The pestle body takes the shape of a cylinder and the ramming interface is flat, which is used for ramming foundation and adobe bricks (Fig. 1(b)).	Cement or metal is usually used for making a smaller pestle head in the shape of a hemisphere. Its force applies to points. It is mostly used for tamping the wall with smaller thickness (Fig. 1(c)).

The earth mold is the core of the ramming tools. There are two kinds of earth molds in the area and they are templates of rammed-earth wall and mold of adobe brick respectively.

The former is composed of two side plates and one end plate, and the other end serves as the movable fixture. Side plates are usually narrow at the top and wide at the bottom, which leads to volumetric shrinkage of the walls in ramming the earth, improving the walls' stability. The two side plates are fixed by the fixed link and two end plates are placed along the link (Fig. 2). The latter, namely, the mold of

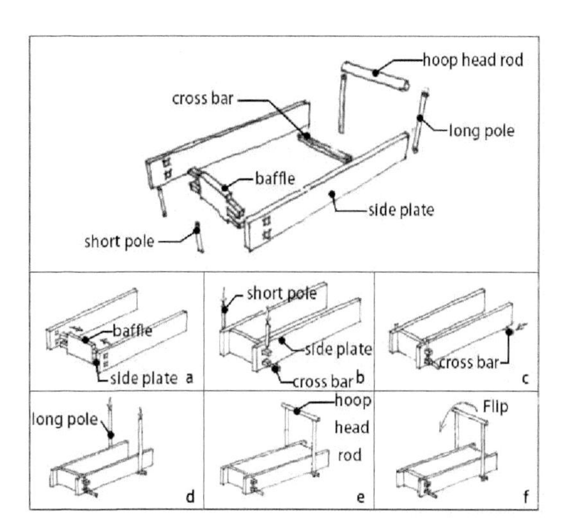

Fig. 2 Mold used in mold ramming © WANG Jialin

adobe brick, is similar to the mode of black brick. Its specific dimension is generally 500mm in length, 250mm in width and 100mm in height.

4.3 Material Preparation for Construction

The materials prepared for constructing the vernacular earthen architecture mainly consist of timber, earth and admixture. The timber is employed for making beam frames, doors and windows and elm and polar serve as the major timber used in the area. Generally, the integrity of the wood and its moisture content should be taken into consideration in the procurement of timber for the beams and columns in the timber structure are required to retain higher moisture.

Earth materials are usually selected in farmland and riverway. To take the earth in the farmland, it is necessary to "wake up the earth" before Chinese New Year. More precisely, water the land in winter and the earth becomes frozen. Afterwards, when building a house in spring, turn up the frozen earth, which is softer and more permeable compared with loess and is applicable to building walls. By contrast with the farmland, the earth in the riverway is better, particularly the red earth in the riverbed. It is generally used on roofs with the higher adhesion.

After processing the earth materials,

different admixtures (aggregates) are added to improve the strength and durability of the wall. At present, it is investigated that three sorts of admixtures are commonly used in the area, namely, gravel, crushed wheat straw, and cotton mixed with glutinous rice juice. The former two are familiar to people while the latter one costs higher and is rarely seen.

4.4 Practices of Foundation Construction

The foundation construction of the vernacular earthen architecture is composed of two steps. The first one is orientation and large-scale setting out and the second one is foundation construction.

Orientation and large-scale setting out is subdivided into four parts: ground leveling, orienting, demarcating, and setting out. Specifically speaking, demarcate the scope of construction first and pre-level the construction site so that the elevation everywhere remains basically the same. Then, make use of the solar azimuth at about 12 o'clock at noon and place a line parallel to the orientation of the sun as the orienting line. Accordingly, craftsmen release a vertical line of the orienting line with plumb bob to determine the four corner points of the homestead. Finally, they sprinkle white chalk on the homestead boundary in accordance with the corner points.

In light of the construction of a foundation, it is further classified into making level, trenching, ramming earth, building and leveling. Here making level is to level the foundation before digging the trench so that the foundation stays on the same horizon. After that, masons excavate the foundation trench on the line of white ash mentioned in the above section, whose depth lies on the bearing capacity of the stratum. In view of the stratum in the area, it is made up of gravelly soil and viscous loess with good bearing capacity and its depth of the excavation is usually kept at about one meter. After the trench is dug, tamp 1 to 3 layers of earth at the bottom of the foundation trench according to the horizontal line. In this area, the foundation of the earth wall is dug down shallowly and the wall body is generally built immediately after compaction. Eventually, ram, fill and level up the foundation to lay and build walls. Besides, place a capital stone at the position where the column will be set and check the level at the same time. Then draw a cross in ink on the stone and stand the frame (Wu, 2018).

4.5 Practices of Building Ground

The ground of the vernacular earthen architecture in this area is made up of mostly earth and bricks, and rarely stones due to the low cost of plain ground and simple practice. The specific practice includes: laying the earth, setting out, making the level before sprinkling the water. Divide the earth into two layers and tamp the layer with a hammer. Apply a layer of yellow mud to the tamped ground and leave it to dry.

4.6 Practices of Building Roof Truss

Building roof truss starts with determining the beam size and truss slope. The trusses in the earthen architecture of this area are not very sophisticated, focusing on meeting the needs of use. The load-bearing system of the earthen architecture is divided into two types: column bearing and wall bearing. The most common practice is that the main beam is directly placed on the eave wall, and the beam is supported by short columns. The beam is generally in rectangular cross-section with rounded corners and the size around 35 times 20 centimeters. The purlin is usually made into a circular cross-section. There are also a few examples of purlin in square cross-section with a diameter of 20 centimeters (Liu, 2014).

4.7 Wall Structure

Fieldwork of site visiting and interviewing indicates that the building walls in the earthen architecture in Floodplain region follows two primary practices of ramming building and adobe brick building. Ramming building includes rammed earth and earth culture.

4.7.1 Practices of Mold Ramming

Mold ramming requires the use of mold. The construction process starts with fixing the mold at one end of the wall and then filling it with the earth. The craftsman standing above rams the earth with a hammer. After hammering, he uses a movable clamp to dismantle and translate the mold to the next section until ramming comes to the corner of the wall, which completes the first layer. He then places the mold on the rammed wall and repeats the whole process to finish the second layer. Layer after layer, the craftsman builds the wall to the designed height (Fig. 3).

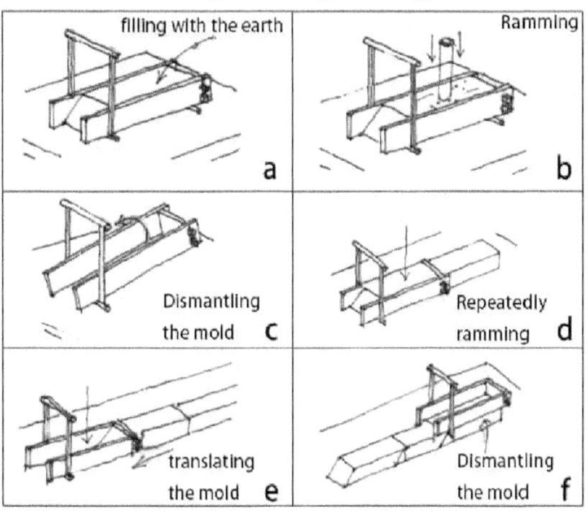

Fig. 3 Process of mold ramming © WANG Jialin

Attention should be paid to the following points during the construction process. The height of each layer should be limited within 50 centimeters and each layer is filled with earth three times and trampled flat. Before ramming the earth, use feet to roughly flatten the earth and use a hammer to ram from periphery to the center. Each layer should be rammed three times. Move the mold to the upper layer to repeat the ramming process only after the water in the lower layer evaporates for some time and the wall has been strengthened. (Luo, 2018).

4.7.2 Earth Culture

The earth culture practice is prevalent in the hilly area with low hills. Earth culture wall uses no mold. The practice is simpler than that of the mold ramming and is called "whipping the wall" in the local area. Specific practices are as follows:

First, the prepared earth is mixed with water. At the same time, fill in it with the aggregates such as grass straw and cotton, stir the mixture until it mixes well and forms into a jellylike mass, and then leave it for one night. On the next day, first check the humidity of the earthen material and stamp on it with bare feet evenly if no problem is found. After the craftsman checks and approves the viscosity of the earthen materials, whipping the wall begins.

It is called whipping the wall because a lump of earthen material is shoveled up with a spade and whipped to the ground to make it firm before it is shoveled up again to make the wall. At the same time, the craftsman working on the wall uses the hammer to ram the earth to straight. Each layer is rammed for 5 to 6 times from the corner to the center with no cracks or crevices left (Fig. 4). Earth culture wall consists of three layers and each layer needs two to three days to finish and one week to dry. Because of this, earth culturing asks for a longer construction period (Zhao, 2009).

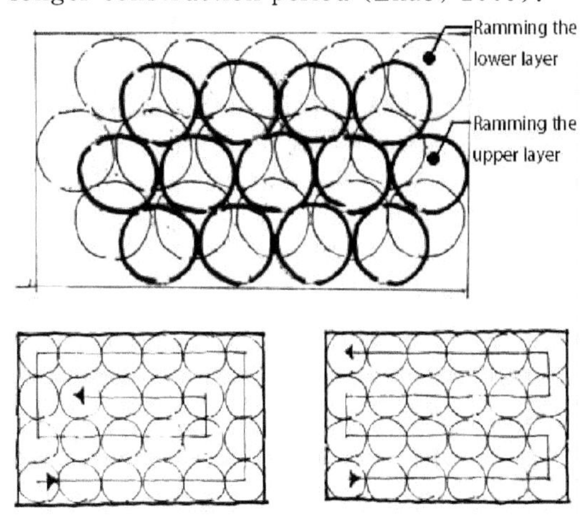

Fig. 4 Ramming map © WANG Jialin

4.7.3 Adobe Brick Building

First, make the adobe bricks by mixing the earth, straws, and other aggregates and stirring the mixture until it gets viscous before placing it in a wooden mold. Tamp the mixture using feet or hammer, remove the mold and place the mixture in the sun for one week to dry until the adobe bricks take into shape.

Care needs to be taken to the following

three aspects when making the adobe bricks. First, fill the slurry in the crevice between the adobe bricks with preferably viscous red soil from the riverbed. Second, as the adobe bricks are vulnerable to humidity, they need to stand upright to avoid damp. Third, apply the lime to the inner and outer walls to prevent moisture after completion of the masonry.

The differences between adobe brick wall and wall of ramming building are: adobe brick wall is less strong, cannot bear beam frame directly, and needs wood columns or blue bricks to support the beam frame; adobe brick wall is more vulnerable to rain and moisture and adobe is not used in wall foundations or lower base.

4.8 Roofing Practices

It constitutes the curving roofs of the earthen architecture in this area (Fig. 5). Most curving roofs take the beam structure of girders and short columns with purlins on them. Changing the size of the short columns helps to curve the beam. Two layers of reed matting are applied to purlins in different directions. Wheat straw or sorghum straw are paved on the reed matting with hays on the top before spreading the straw mud on it. After the straw mud gets dried, trinity mixture fill (a mixture of lime, clay, and sand) or red soil from river bed mixing with water is applied across the entire roof with a thickness of 20 centimeters. The mixture is then beaten and tamped repeatedly until it gets firm. Finally, a layer of lime is applied to the surface to form a solid and smooth dome.

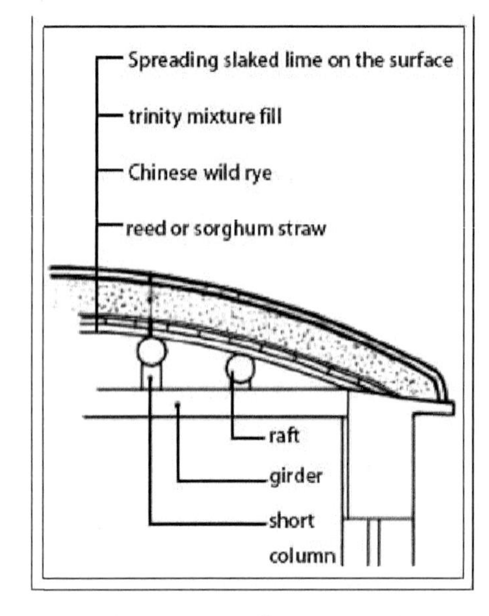

Fig. 5 Roofing practices © WANG Jialin

5 Conclusion

Based on the construction process and procedure of building the vernacular earthen architecture in the Floodplain region of the Yellow River in Shandong province, this paper presents the detailed classifications and descriptions of the traditional construction technique and craftsmanship used in the vernacular earthen architecture and draws the conclusions as follows:

First, the traditional earthen architecture is characterized by distinct regional and vernacular features. Constrained by economic and natural resources, traditional craftspeople take advantage of regional materials and come up with the best architecture solutions by reasonably employing the architectural materials and methods. Because of this, the earthen architecture has strong adaptability to the local environment, climate and resources.

Second, the construction activities of the folk craftspeople follow the pattern of "self-updating and mutual learning". Under the realistic conditions of rural areas, people build houses mainly for living purposes with the reference to prevalent housings and carry out transformation according to the actual situation, which explains the unified regional architectural style.

Third, with the rapid improvement of the material economy, the problems of complex construction, darkness, and damp with traditional buildings cannot satisfy people's living needs. It is therefore an important research topic to propose transformed traditional construction techniques and craftsmanship fitting the living needs of modern people while taking into consideration of the regional features of these techniques and craftsmanship.

Acknowledgement

We would like to express our appreciation to Ms CHEN Jianfen for her professional work as the language proofreader of this paper.

References

LIU Xiujuan, 2014. Research on types and features of vernacular folk architecture in the lower reaches of Yellow River. Urbanism and Architecture, 03: 121-123.

LUO Jing, 2018. Research on the regionalization of traditional ramming buildings and craftsmanship in Gansu. Lanzhou: Lanzhou University of Technology.

WU Xiaoyu, 2018. Research on the construction technique and craftsmanship of traditional folk housings of wooden structure in Tianzhen County, north of Shanxi. Beijing: Beijing Jiaotong University.

YANG Hui, 2004. The originality of craftsmanship: research on craftsmanship of Roofing, ridge building, and painting of traditional architecture in south Jiangsu. Nanjing: Southeast University.

ZHAO Pengfei, 2009. Research on traditional architecture of Shandong canal. Tianjin: Tianjin University.

Reconstruction Process of Traditional Community House of Katu Ethnic Minority — Case Study of Aka Hamlet in Nam Dong District, Thua Thien Hue Province, Vietnam

Nguyen Ngoc Tung[1], *Hirohide Kobayashi*[2], *Truong Hoang Phuong*[1], *Miki Yoshizumi*[3], *Le Anh Tuan*[4], *Tran Duc Sang*[4]

1 Faculty of Architecture, University of Sciences, Hue University, Vietnam
2 Graduate School of Global Environmental Studies, Kyoto University, Japan
3 College of Gastronomy Management, Ritsumeikan University, Japan
4 Vietnam National Institute of Culture and Arts Studies -Branch in Hue, Vietnam

Abstract Based on a project collaborated among Kyoto University (Japan), Hue University of Sciences (Vietnam) and Vietnam Institute of Culture and Arts -Branch in Hue (Vietnam), one traditional community house of Katu ethnic minority was constructed in Aka hamlet, Thuong Quang commune of Nam Dong district, Thua Thien Hue province. This study focuses on traditional process of the construction of the house. The field research was conducted during the construction process for obtaining the information about structure, construction technique, material collection and processing, and decoration. In addition, the patriarchs and local people were interviewed to understand their design method. The successful reconstruction of this house proved that the community linkage and traditional construction techniques are still kept in Katu villages in Nam Dong. This contributes to the preservation of the traditional architecture of the Katu people, which is being lost in Nam Dong.

Keywords traditional community house, design methodology, katu ethnic minority

1 Background

1.1 Katu Ethnic Minorities

Katu ethnic minorities inhabit in the mountainous areas of Thua Thien Hue and Quang Nam provinces, and part of Da Nang city in Vietnam, whose population was approximately 86,617 in 2017(Bh'riu, 2018). Originally, the Katu village had Guol, traditional community house, being located in the center of the village, which has oval, horse-shoes, or polygonal layout(Kobayashi et al., 2018:119; Nguyễn, 1996:130; Nguyễn et al., 2004:98). However, the traditional village layout is changing in the modern context, and also *Guols* are altering in that regard.

In 2007, a traditional community house of ethnic minorities was reconstructed by mostly original methods through a JICA project in the Hong Ha commune, Thua Thien Hue province①. This is the first time that the villagers have completed the traditional community house since 1975. The construction process and measurements of the house were recorded through by explaining village elders' large amount of indigenous knowledge(Kobayashi et al., 2018). The entire process, from material collection in the forest to on-site construction, demonstrates that vernacular architecture is constructed by using three local resources that were matured and conserved in the locality: natural resources (building materials); hu-

① GSGES, Kyoto University and CARD, Hue University of Agriculture and Forestry cooperated with the rural development from Oct. 2006 to Sept. 2009, funded by the Japan International Cooperation Agency (JICA). Participatory construction of the traditional community house in the mountainous village was one of main activities in the project. The house was completed in Sept. 2007(Kobayashi et al., 2018).

man resources (community cooperation); and intellectual resources (knowledge and technique).

However, the feature of traditional community house in Hong Ha commune is recognized as different cultural identify of some ethnic minorities such as Ta Oi, Bru Van Kieu, and Katu(Kobayashi et al. ,2018:133). Meanwhile, this research focuses on Guol of Katu ethnic minorities. Thus, different sites of Katu villages in Thua Thien Hue, Quang Nam provinces and Da Nang city were surveys in 2014① (Fig. 1).

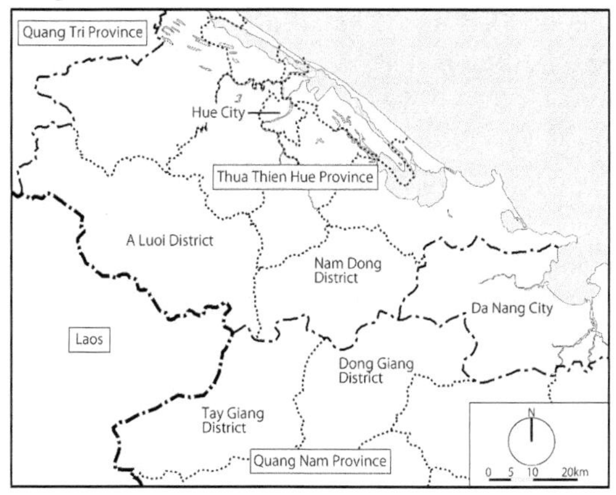

Fig. 1 Location map of researched sites

The purpose of the surveys is to find out feasible site for reconstructing Guol of Katu ethnic minority based on traditional method. The criteria for selecting research site are that the Katu people living in the site should be motivated and they have knowledge for construction. Besides, the village does not have traditional community house. Finally, A Ka hamlet, Thuong Quang commune, Nam Dong district, Thua Thien Hue province was selected for the construction of traditional community house.

1.2 Context of Nam Dong district and A Ka hamlet

Nam Dong is a mountainous district in Thua Thien Hue province. The district is located on the Southwest of Hue city with the area of 64,777.9 sq. km. According to the staticstical data in 2017, the Katu people is predominant in Nam Dong district with population of 11,945, occupying 70% Katu ethnic minority in Thua Thien Hue province(Nam Dong Department of Statistics,2018: 36).

Thuong Quang commune is 12 km on the west far from the center of Nam Dong district. The population of the commune is 2,174 in 2017 and Katu people occupy more than 60% (1,211 persons). The commune has 7 hamlets and Katu people mainly live in Ta Rau, A Rang, A Ro and A Ka hamlets. A Ka hamlet has 107 households (population of 328) and all of them are Katu people.

The former Guol of A Ka hamlet was constructed in 2004. The house was decayed with the leak from the roof in 2010 (Fig. 2) Up to 2012, the house was completely damaged and then, it was destroyed in 2013. Thus, the need of local people for reconstructing Guol is necessary.

Fig. 2 Situation of former Guol in A Ka from August 2010 to April 2014

According to the surveys, local people in A Ka are motivated and want to reconstruct Guol. Besides, there are many people in A Ka hamlet, who still master traditional method for constructing Guol. The local government also approves and supports official documents for the project. On March 2016, all stakeholders made an agreement for the reconstruction of Guol in A Ka hamlet. The project would be finished on October 2017 (In reality, the project was finished on 20th August 2018).

2 Arrangement for the Reconstruction

2.1 Roles of Stakeholders

Before the reconstruction of Guol, many meetings among local people of A Ka hamlet, local government (Hong Ha commune, Nam Dong district) and consultative

① The survey is supported by the JSPS program, Kyoto University, Japan.

researchers (the authors of this paper) were held for discussing, unifying the process of the project. Finally, the roles of each stakeholder can be identified as follows:

A Ka hamlet has the responsibility to exploit and collect necessary materials (woods, leaves, etc.); do outwork for Guol; construct Guol; organize traditional ceremonies during the process of reconstruction, and maintain the Guol after the reconstruction.

Local government has the responsibility to support the process of official documents and policies relating to material exploitation; check and speed up the construction work.

Consultative researchers have the responsibility to support part of budget for labour costs[①]; consult constructed technique, culture, custom, etc. ; supervise process of the construction; record and digitalize data of reconstruction process; interview and organize meetings to discuss and share knowledge for the construction and maintenance of the Guol after completion.

2.2 Preparation of A Ka hamlet

The hamlet set up management board for reconstruction to manage and guarantee the progress of the reconstruction (Tab. 1). The hamlet leader, Mr. Ho Van Chang, is the chairman of the board. Other members are Mr. Ho Van Bon (patriarch), Mr. Vo Dai Huy (Secretary of the Communist Party of the hamlet), Mr. Ho Van Bang (expert for construction) and Mr. Van Ngoc Cuong (expert for sculptural work). Under the board, 4 groups are divided for different missions.

Tab. 1 Structure of management board

Management board			
Group 1	Group 2	Group 3	Group 4
Material collection	Sculptural work	Construction technique	Ceremony organization

Group 1 has a duty to collect materials from the forest and other fields. This group has 3 subgroups: one collects woods for the structure of the *Guol*; one collects Non leaves (local leaves for tiling roof); and one collects trunks of *Lo O* tree (kind of bamboo, utilized for partition and floor) and trunks May tree (rattan tree, utilized for tie wood components and used as round purlin).

The most important and difficult work is to collect woods for the structure of the Guol because it is hard to find quality wood as well as the policy for closing forest of the Vietnamese government in 2016. The group 1 is divided into 4 small groups (each group has 4—8 persons) to go to the forest several times (8 — 10km far from the hamlet), check, evaluate condition of wood and then, exploit selected trees. According to indigenous knowledge, the time for exploiting woods should be at in the end of the month (when there is no moon and termites will not eat wood) and dry season (it is more convenient for transporting wood). For Guol construction, the local people should exploit 13 trees for pillars, whose circumferences are around 4 (H-1) to 5 (H-1) as shown in figure 3. The height of center pillar is 4 (A-1) -(A-2) and other pillars are 2.5 (A-1). Besides, 12 roof beams are cut. Their height is around 3 (A-1) to 3.5 (A-1) and their circumferences are from 2.5 (H-1) to 2 (H-1). The exploited trees should be straight and be transported by buffaloes. All villagers have duty to collect Lo O, Non, and May. For tiling roof, the villagers make 24 sheets with the dimension of 2.5m×1.8m.

Mr. Van Ngoc Cuong is the leader of group 2 having duty to design, paint patterns and pictures on wood components, especially the center pillar. The group carves images expressing normal living the style of *Katu* people, symbols relating to their religious belief.

Group 3 manages techniques of *Guol*

① For the construction, local people have to spend a lot of their time and labors. After several meetings, it can be calculated as 366,860,000 VND (1 US Dollar equals about 23,000 VND) and Kyoto Univ. agreed to support 125,000,000 VND.

construction. Mr. Ho Van Bang is the leader of this group and he has duty to assign work for group members such as chisel mortises, joint holes of components, etc. The group also prepares foundation and makes foundation stones for pillars①.

The patriarch of A Ka hamlet, Mr. Ho Van Bon is the leader of group 4. He organizes traditional ceremonies during the construction.

2.3 Method for Reconstruction

There are two sites that the local people selected for the construction. After discussion, the site of former Guol was selected because the foundation can be reused and the orientation (the Northeast) that the Guol faces is found as a good orientation for the hamlet.

Regarding architectural form, local people and researchers agreed to construct the Guol followed experiences of patriarchs and village elders, who still remember traditional method of construction. Besides, the hamlet also referred ideas and drawings of researchers for satisfying basic requirement about the area, functional usage, materials, traditional form of *Guol*.

It was known that Katu people use body-based units of measurement for the design and construction of Guol (Fig. 3). The Katu normally use 18 types of units②. Depending on specific hamlet, the Katu may not use some unit types. For example, Katu people in A Ka hamlet do not use types H-7, H-8 and H-9. In cases of A Ka hamlet, the local people agreed to construct the Guol based on 18 types of units and the patriarch of the hamlet, Mr. Ho Van Bon, was chosen for using his body units. During the construction, the local people sometimes use ruler and modern tools for more convenience.

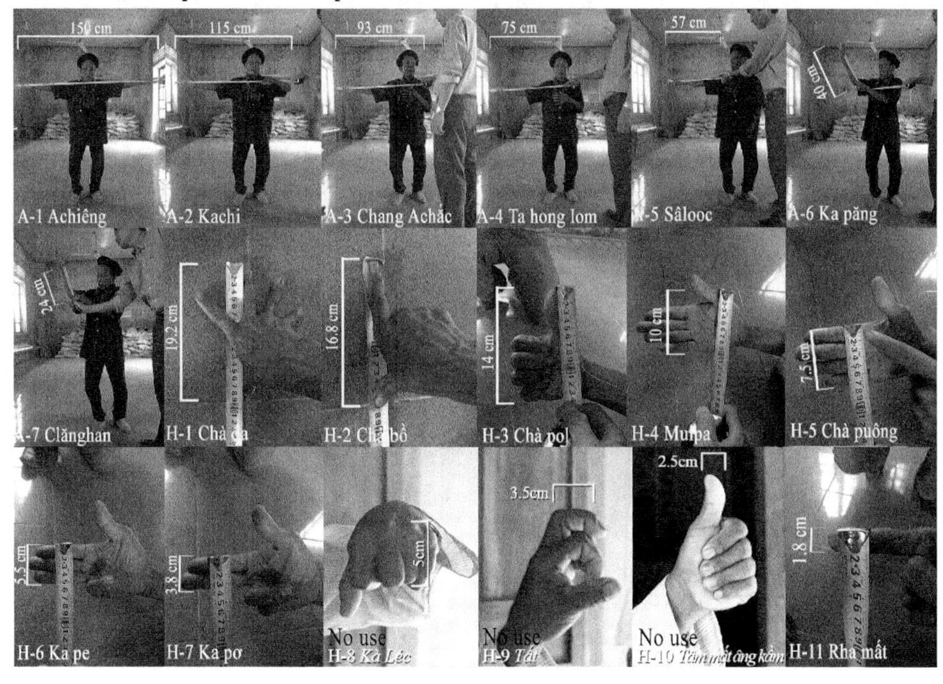

Fig. 3 Demonstration of body-based units in A Ka hamlet, Thuong Quang commune

① According to traditional method of Katu people, pillars should be fixed into the ground. However, local people in A Ka hamlet decided to put pillars on the foundation stones for preventing from termites.

② Based on our previous researches, it is found that the Katu use 17 types of body units (Kobayashi and Nguyễn, 2013). However, according to the survey with 25 patriarchs in Nam Dong district in 2018, the Katu has one more type of unit for design and construction.

Regarding materials, some local people firstly suggested utilizing modern materials such as concrete and sheet metal for columns and roof because of sustainability and convenience for material exploitation. After that, the hamlet unified to construct the Guol by traditional and local materials such as wood, bamboo, rattan, etc.

3 Reconstruction Process

The Guol is constructed following the traditional style required skilled techniques and labors of all villagers. The construction of Guol was implemented in the 9 steps as shown in Fig. 4.

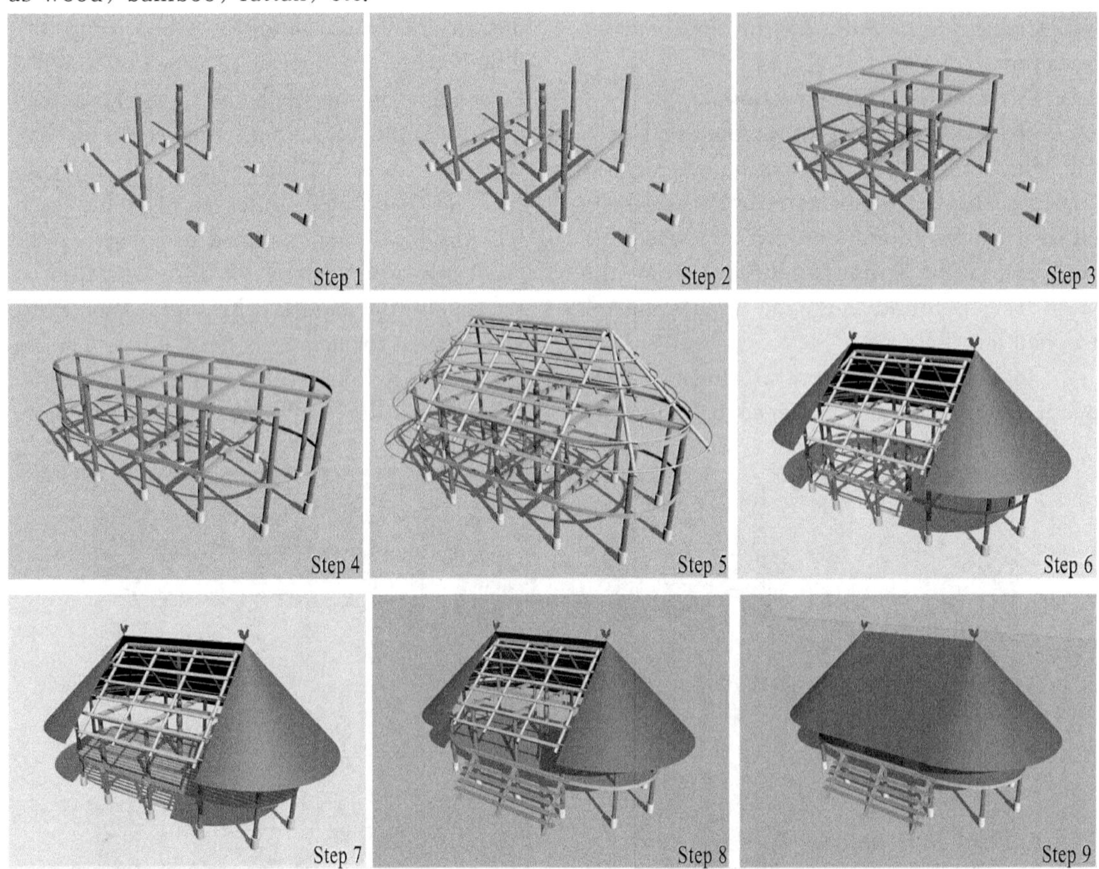

Fig. 4 Demonstration of body-based units in A Ka hamlet, Thuong Quang commune

Firstly, the center pillar and 2 main pillars in front and behind were set up on 20th May 2018 (step 1). The other 4 main pillars were put in parallel pairs (step 2). Thus, there were 2 parallel lines of main pillars running up to the sixth pillar and the center pillar was put in the middle center position.

The next step is that cross and eaves beams were assembled (step 3). This work made a firm solid connection thanks to notches for beams and slots of the main pillars and center pillar. The cross beams were assembled and connected by their slots and fastened to the vertical beams.

Step 4 was to set up 6 round pillars and assemble round beams. After that, the local people put the upper part of the center pillar and a cross beam via the center pillar. Each pair of roof beams were assembled facing the above cross beam. Each side of the house roof has 4 roof beams. Each pair of the cross beam make an "X" shape in the top and is closely tied on the ridge pole that securities it firmly (step 5).

24 leaf units were utilized for roofing from the top to bottom. After that, roof crestings were put on the top of the roof

(step 6). The villagers crafted cock symbols for roof cresting because they claimed that the symbol could wake villagers up to go to work every morning.

Step 7 was to assemble 19-floor beams and floor joists with a distance of around 250mm. Then, the floor surface was made by straightened *Lo O* slats (step 8). The stair was implemented at the same time. There are 5 steps for the stair. According to the Katu people, number 5 has a good meaning.

Assembling partitions were the final step for the construction. For this work, *Lo O* leaves were split into small shets depending on the size of the partitions (850 × 850mm).

On 20th August 2018, the Guol house was finally completed and local people in A Ka hamlet organized an inaugurated ceremony for the new Guol (Fig. 5). Therefore, the project for reconstruction of the Guol in A Ka hamlet was implemented for more than 2 years after discussion, going through official documents, collecting materials and construction.

Fig. 5 Guol house in A Ka hamlet during inaugurate ceremony

4 Conclusion

Traditional community house, *Guol*, in A Ka hamlet, Thuong Quang commune, Nam Dong district, Thua Thien Hue province was collapsed in 2013 and a project for reconstruction of *Guol* was set up on March 2016. Based on agreement among local people, local government and project researches from KU (Kyoto University, Japan), HUSC (University of Sciences, Hue University, Vietnam) and VICAS (Vietnam Institute for Art and Cultural Studies, Branch in Hue), the reconstruction process of Guol was finally completed on August 2018. The process was implemented following tradítional methods of Katu ethnic minority and based on various discussions, going through official documents, collecting materials and construction.

The construction of Guol was completed after 3 months following 9 steps. It is determined that the new Guol is the result and achievement of all local people in A Ka hamlet and the project is a typical case for community-based construction of Guol house. This result can be expanded to other hamlets of the Katu in the central region of Vietnam.

References

BH'RIU L, 2018. P'rá Co'tu Tiềng Co'tu (Co'tu language). Hôi Nhà Vãn publishers.

ĐẶDNG N V, CHU T S, LU'U H, 1993. Ethnic minorities in Vietnam. Thề Gió'i publishers: 72-75.

KOBAYASHI H, IIZUKA A, 2010. Indigenous construction technology of Cotu minorities in central Vietnam Case study of the traditional community house in Hong Ha commune, Thua Thien Hue province. Architectural Institute of Japan, No. 653, 1679 - 1686.

KOBAYASHI H, NGUYỄN N T, 2013. Body-based units of measurement for building Katu community houses in Central Vietnam. Vernacular heritage and earthen architecture: contributions for sustainable development. Proceedings of International Conference on Vernacular Architecture CIAV2013 | 7 ATP. Vila Nova Cerveira, Portugal: 359-364.

KOBAYASHI H, et al., 2008. Participatory construction of traditional community house in mountainous village of central Vietnam. Hanoi: National Political Publisher.

LU'U H, 2007. A contribution to Katu ethnography. Hanoi: Thề Gió'i publishers.

Nam Dong Department of Statistics, 2018. Year book 2017.

NGUYềN K T, 1996. Nhào' cô truyền các dân tôc Viêt Nam - Traditional dwelling houses of Vietnamese ethnic groups, Vol. 2, Xây Du'ng publishers: 130-135.

NGUYÂN H T, et al., 2004. Katu kē sông dâu ngon nì'ó.c (Katu - the people living at the waterhead). Thuân Hóa publishers.

TRÂN T V, 2009. The Cotu in Vietnam. Vietnam News Agency Publishing House.

TRU'O'NG H P, KOBAYASHI H, NGUYĒN N T, 2013. Typological research on traditional community house of the Katu ethnic minority in Vietnam. Vernacular Heritage and Earthen Architecture: Contribution for Sustainable Development. Proceedings of International Conference on Vernacular Architecture CIAV2013 | 7 ATP. Vila Nova Cerveira, Portugal:343-349.

TRUONG H P, 2015. Conserving traditional community house of the Katu ethnic minority A case study in Nam Dong District, Thua Thien Hue Province, Central Vietnam. Japan: Kyoto University.

A Story of Karahuyuk House and Rehabilitation of Adobe Bricks

Süheyla Koç

Sivas Cumhuriyet University, Sivas, Turkey

Abstract Karahuyuk is a small village located in Aksehir district, Konya Province. It is a second-degree archeological site, a home of many civilizations. Vernacular houses in Karahuyuk were built adobe brick masonry and timber-framed structures filled with adobe bricks. This building culture dates back to the period when the first civilization inhabited this region. Since the 1950s, the building culture shifted from traditional adobe brick and timber frame structure to reinforced concrete for the construction of new buildings and rehabilitation of old ones. However, most of the traditional buildings were damaged by Afyon earthquake in 2002. Only 12 timber-framed adobe brick buildings remained intact after the earthquake and were later registered under the Konya Conservation Council of Cultural Assets. These 12 buildings carry the mixture of Central Anatolia and Mediterranean housing features due to the location of the village. The use of concrete and cement with traditional building methods make these buildings unique as they carry the 19th-century building styles although they were built in the 1950s.

In this paper, the unique features of the buildings will be analyzed via a case study of the Korkmaz House. Research methods include an interview survey with the house owner about the local building traditions, the construction system, the purchase of materials, and the craftsmen.

In addition, this paper also discusses the methods on the repair of the adobe bricks and rehabilitation of the building structure. Various tests such as soil-water content, sieve analysis, Atterberg limits, USCS, and Proctor were conducted on the soil used for making such adobe bricks. New adobe brick samples were prepared with the mixture of the original soil sample, gypsum, and lime. The compressive strength of the new samples and original adobe brick were measured. Based on the test results, a mixture of 10% gypsum and 30% aggregate was suggested to be optimal for the repairs of the adobe bricks in this structure.

In conclusion, this paper introduces earthen vernacular structures in Karahuyuk for the first time and gives recommendations for their repairs and rehabilitation.

Keywords karahuyuk houses, adobe bricks, rehabilitation

1 Introduction

Karahyük district — within Konya province and located 7 km outside of Akşehir — is a residential area, which has been home to many civilizations. In 2001, Karahüyük district is registered as the archeological site of the second degree. Along with increasing concrete structures, adobe houses that reflect the traditional texture and lifestyle continue to exist in the present urban fabric of the town. These adobe brick houses, built in around in the 1950s, reflect a period in which people sought to develop traditional construction techniques with modern material and concrete was used extensively at that time. There are 12 examples of this style left in the region. In particular, Korkmaz House is chosen as the case study due to its size and features which represent the Karahuyuk houses in general.

In this paper, the alternative methods are discussed for the rehabilitation of earthen vernacular houses. The research methods include the building survey, the in-situ assessment, the analyses on the materials, and the interview with both the landlord and the people, who remember the old situation of the region in previous times.

2 Korkmaz House

2.1 The History of the House

The building occupies an area of 212m^2. The ground floor has a plan with gallery (*dış sofa*) type while the upper floor has a plan with hall (*sofa*) type. On the ground floor, on the one side is the hall (*sofa*) and on the other side, rooms are aligned with the *sofa*. The form of the ground floor is designed in the same way as the road due to its shape. On the upper floor, the rooms are lined up on both sides of the sofa. It is observed that a modular system is applied in the design (Fig. 2 and Fig. 3). The modules determined by the wood sizes used in the construction. Room sizes vary depending on the length of the wooden beams but approximately 300 cm× 450 cm in sizes. As a result of such a design approach, the triangle cantilevers are seen in the structure. This dynamism influenced not only the front façade (Fig. 1) but also the planning system of the building. The sofas are not straight but consist of rectangles that are shifted parallel to the cantilevers (Fig. 3).

Fig. 1 The front façade of Korkmaz House © Süheyla Koç

The house, where the Korkmaz family lived in the 1950s, was located at the opposite side of the current building. The family wanted to build a new house in front of their house, as a result of the increasing number of family members and the lack of space in the house. They hired the carpenter named Mithat from Doğrugöz, a village 6 km far from Karahuyuk, to build their house. The poplars, which were necessary for the construction of the house, had previously been planted for this purpose. The soil for the adobe bricks was provided from the fields called zıva (plaster soil). Straws were provided from barns while rush mats and reeds were brought from Doğrugöz village. Stones, tiles, and concrete materials were also purchased from Akşehir.

The building, built on two floors as mud brick, has a room used as a kitchen (Z02 space) with storeroom underneath, and wooden stairs leading to the upper floor on the ground floor (Fig. 2).

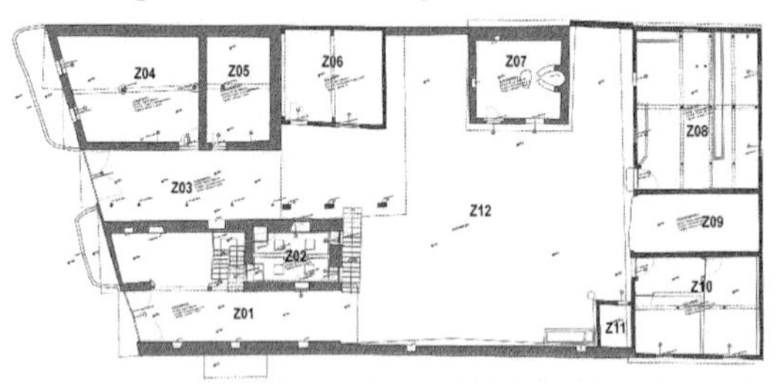

Fig. 2 The ground floor plan of the house © Süheyla Koç

On the upper floor, there are four rooms (102, 103, 104, and 105) in total, arranged on both sides of the hall (101). Room 102 is used as the parent's room. Room 105 serves as the gathering place for family members and guests. Rooms 103 and 104 used by the children. In addition, each room is used for a place to eat, to sleep, to bath, and to work (Fig. 3). The bathing cube (*gusülhane*) and cabinets are found in every room. Also, each room can be considered as one house for a family. When the sons get married, one room is given for each couple.

On the foundation of the building, stones were used up to the plinth level, wooden girders were put on top of it, and adobe bricks were used to bond the wall. Wooden beams were used at the floor level. While wall thicknesses are around 60 ~ 70cm, the main mudbrick sizes are around 27cm×27cm×10cm. The flooring details are the same as traditional adobe making technique, and they differ in terms of using concrete in wet areas. Instead of using two layers of compressed soil, one layer of compressed soil is used, and a maximum of 10 cm thick concrete is poured on top of it

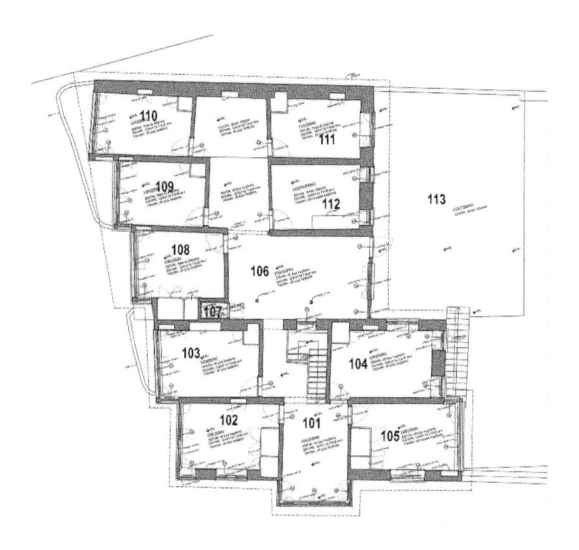

Fig. 3 The upper floor plan of the house © Süheyla Koç

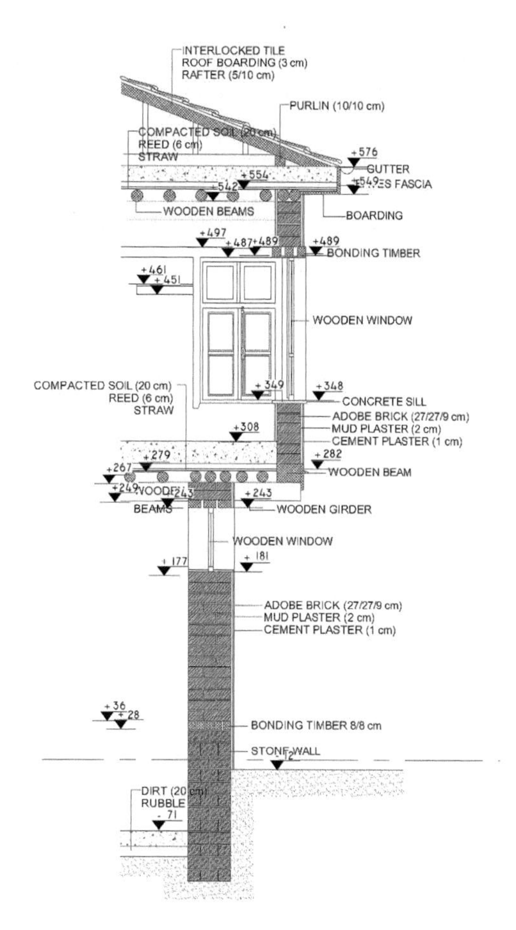

Fig. 4 The detail of construction system © Süheyla Koç

(Fig. 4). In the same period, the houses built in the town of Gözlük in Sarayönü were built entirely with traditional technique without the use of the concrete material. In the upper cover of the structure, hipped roof with Marseille tile was used. In case of flat roofs on the annexes, instead of laying stones or grooves on the edges of the flat terraces, dense and long straws were used. In this way, the structure is protected from the harmful effects of water.

About five/six years after the construction of the first building, the Korkmaz family needed a larger house. At that time, a carpenter named Ibrahim from Doğrugöz helped to build the house. The earth is added to the top of the flat earthen roof and compressed with the stone roller (luv taşı) each year. However, the earth added in five/six years was too much, the structure could not bear, and the top cover collapsed. When the house was built, the collapsed top was cleaned, two rows of adobe bricks were removed from the walls, and the top floor was built over the Z04 and Z05 spaces.

With this additional structure, a large *sofa* (106) with a place, where the need for bathing was met, four rooms (108, 109, 110, 111) and a modern kitchen (112) were added to the building. On the *sofa* (106), there was a storage for firewood(107), where of the toilet is located today (Fig. 3).

In the 1980—1985 years, the tandoor (Z07) was added to the garden, and in 1990—1995, the barn (Z08), the toilet between the tandoor and the barn, the haystack (Z09) and the warehouse in the south (Z10) were added (Fig. 2). While the top cover of this section was the earth in the beginning, the top cover collapsed as a result of the heavy soil, and the tops were transformed into the one side sloped roof with the tiles of Marseille. In 2000, as a result of the aging of the family elders, the toilet outside the house became difficult to use. Because of that, the wood yard in the upper floor has been converted into a toilet.

Again between 2000 and 2010, a brick-walled warehouse with reinforced concrete columns was added to the ground floor, and the top of it was opened to use as a terrace. The P32 window was cut to provide access to the terrace, and the K15 gate was created. In 2010, the wooden windows of room 105 were replaced with PVC windows as precautions for the cold weather, and the ceiling was covered with plywood.

Rooms 108, 109, 110, 111 and 112 retain their original form in terms of both functionality and material. The floor of the room is of earthenware except for wet areas. The floors of the wet areas are covered with concrete. The wooden beams are observed on the ceilings. Internal and external walls are plastered with lime over mud plaster. The roof of the building is covered with Marseille tiles. The roof of the first building is kept as it is while building the second building roof. The facades preserve their original condition except that the façades were plastered with cement mortar, the windows of the space 105 were replaced with PVC windows, and the P32 window was cut and converted into the K15 gate.

2.2 Restoration Suggestions

A holistic approach is needed for the restoration and rehabilitation of the earthen vernacular heritage with the involvement of different stakeholders; locals, government, academia, and NGOs. With this approach, the technique, which was still known in the 1950s can last longer and might be used in the future for the conservation of earthen vernacular heritage. The awareness of the community has great importance to bring traditional knowledge to the future. Day by day, the original features of the structure are disappearing; the detailed documentation of them should be the first step for protection. Without bringing the new comfort conditions, convincing landlords to preserve their houses becomes problematic. The need for regular maintenance of adobe structures with mud and lime plaster with a flat earthen roof is another problem, which makes the rehabilitation of adobe crucial.

Theory of contemporary restoration has been taken as the base for restoration decisions. The user requests and decisions, which are taken by the Conservation Council, have been taken into consideration. The primary purpose of the restoration project is to protect the building without compromising the authenticity of the building and to accommodate the arrangement in the comfort of modern life. The usage of the building will help to provide protection. First of all, spatial arrangements are suggested considering the user needs. In order to use the structure as two separate structures, the necessary wet volumes are separated from the rooms by using bricks or light gypsum board materials in order not to damage the adobe material, but also to prevent overloading of the structure.

After the determination of the spatial arrangements, interventions for repairs are suggested. The removal of unqualified annexes and the cleaning process of removal of cement mortar and plaster in the interior and exterior wall surfaces from the structure by physical methods should be done as priority interventions. In the vernacular architecture, the current state of the material and structural system, deterioration rates and causes should be determined comprehensively, and appropriate solutions have to be produced for every kind of deteriorations such as perishable material, reinforcement of structural system. The plastered wall surfaces prevent to be seen the deteriorations of materials and the structural system; thus, the application of plaster blasting and re-identification of deteriorations are needed to propose appropriate interventions.

As the structural system in the local architecture, especially in the Central Anatolia Region, masonry and half-timber wall filled with adobe bricks are used. In a structure constructed with this system, when the necessary protection is provided against the water from the floor and the roof, the structure life is extended. Therefore, the drainage system should have been established in the building environment. Two separate roofs have joint problems which cause the leakage of rainwater, the roof is suggested to rebuilt as one roof with the right details including heat and water insulation.

In order to bring modern comfort to the vernacular architecture, water, electricity and heating installations need to be resolved correctly. In this respect, maintenance of the good ones of existing equipment and building components and the renewal of those which is unusable with the original details are proposed. Annexes units are indispensable accentuates in vernacular architecture. However, these units have to be compatible with the structure. Therefore, the unqualified outbuildings should have been removed, and new units that are contemporarily equipped compatible with the structure should have been added in order to meet the user's needs.

3 Rehabilitation of Adobe Bricks

In this paper, the aim is to provide solutions for the repair of the adobe bricks used in historic buildings. Most of the studies related to adobe brick are carried out in order to eliminate the weaknesses of the adobe and reused as a building material, again. Only a particular part of the research has been done to repair the adobe. Apart from that, in all these studies, the properties of the soil used as well as the materials added in the adobe bricks have great importance. In order to obtain good results in adobe structures, it is essential to adjust the size of the grains forming the soil, in other words, to provide the appropriate granulometry, plasticity, and shrinkage properties to play the desired quality adobe. The amount of sand, clay-silt content in the soil with a small plasticity index is higher than the soils with high plasticity index. Excessively thin material is not desirable because it causes shrinkage and cracks. It is desirable to have coarse sand and some fine sand in the mudbrick soil. Ideal adobe soil consists of a mixture of clay, silt, and sand (Kafesçio glu, 1984).

Firstly, soil analysis was done to get to know the soil used in adobe construction at Konya Metropolitan Municipality Ground Survey Laboratory. Natural water content and specific gravity tests were made to determine the type of soil. Sieve analysis was performed to determine grain distribution, and hydrometric analysis was performed for thinner grains. The standard Proctor test provides the correlation of the dry unit weight-water content. The aim of finding the dry unit weight -water content relations of soils is to find the maximum dry unit weight and optimum water content of given compressive energy of that soil (URL-1).

Attarberg limits, which measure the liquid limit and plastic limit values, have an essential place in the analysis. Atterberg limits, known as consistency limit; describe the relationship between the water and the particles of the ground and the state of the soil according to the changing water content. If excess water is applied to the floor, the soil becomes liquid. In this case, the floor is flowing and has no shearing strength, but in the case of letting it dry, makes a specific shearing strength. The water content in this transition state is called the liquid limit (indicated by WL). If water loss is higher than usual, the ground gradually loses its plasticity and becomes crumbled if it is rolled on a flat surface. The water content in this state is called the plastic limit (indicated by WL). If there is more water loss, there will no longer be any reduction in volume. The water content in this state is called shrinkage limit (indicated by RL). The liquid limit is the extreme value of the transformation of the ground from liquid state to plastic state, the plastic limit is the extreme value of the transformation of the ground from plastic state to semi-solid state, and the shrinkage limit is the extreme value of the transformation of the ground from semi-solid state to solid state (URL-1).

The results of these tests are shown in the graphs (Fig. 5, Fig. 6, Fig. 7, Fig. 8). According to these results; the natural water of the Karahüyük soil is between 4.65% and 4.80%, and the specific gravity is 2.71 grams. According to the sieve analysis and hydrometric analysis in the grain diameter distribution, 72.09% consists of thin clay and silt, 24. 63% sand and 3. 28% gravel. The liquid limit is 34%, the plastic limit is 23%, and the plasticity index is 11%. The maximum dry unit weight is 1. 73 g/cm^2. Optimum water content is 18. 6%. So Karahuyuk is clay with low plasticity. CL is according to the USCS classification system. It can be used in making adobe.

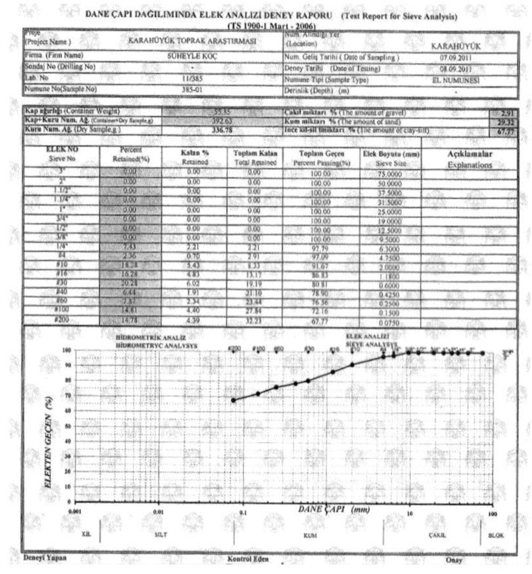

DANE ÇAPI DAĞILIMINDA ELEK ANALIZI DENEY RAPORU (Test Report for Sieve Analysis) (TS 1900-1 Mart - 2006)

Proje (Project Name)	KARAHÜYÜK TOPRAK ARAŞTIRMASI	Num. Alındığı Yer (Location)	KARAHÜYÜK
Firma (Firm Name)	SÜHEYLE KOÇ	Num. Geliş Tarihi (Date of Sampling)	07.09.2011
Sondaj No (Drilling No)	-	Deney Tarihi (Date of Testing)	08.09.2011
Lab. No	11/385	Numune Tipi (Sample Type)	EL NUMUNESI
Numune No(Sample No)	385-01	Derinlik (Depth) (m)	-

Kap ağırlığı (Container Weight)	55.85	Çakıl miktarı % (The amount of gravel)	2.91
Kap+Kuru Num. Ağ. (Container+Dry Sample,g)	392.63	Kum miktarı % (The amount of sand)	29.32
Kuru Num. Ağ. (Dry Sample,g)	336.78	İnce kil-silt miktarı % (The amount of clay-silt)	67.77

ELEK NO Sieve No	Percent Retained(%)	Kalan % Retained	Toplam Kalan Total Retained	Toplam Geçen Percent Passing(%)	Elek Boyutu (mm) Sieve Size	Açıklamalar Explanations
3"	0.00	0.00	0.00	100.00	75.0000	
2"	0.00	0.00	0.00	100.00	50.0000	
1.1/2"	0.00	0.00	0.00	100.00	37.5000	
1.1/4"	0.00	0.00	0.00	100.00	31.5000	
1"	0.00	0.00	0.00	100.00	25.0000	
3/4"	0.00	0.00	0.00	100.00	19.0000	
1/2"	0.00	0.00	0.00	100.00	12.5000	
3/8"	0.00	0.00	0.00	100.00	9.5000	
1/4"	7.43	2.21	2.21	97.79	6.3000	
#4	2.36	0.70	2.91	97.09	4.7500	
#10	18.28	5.43	8.33	91.67	2.0000	
#16	16.28	4.83	13.17	86.83	1.1800	
#30	20.28	6.02	19.19	80.81	0.6000	
#40	6.44	1.91	21.10	78.90	0.4250	
#60	7.57	2.34	23.44	76.56	0.2500	
#100	14.81	4.40	27.84	72.16	0.1500	
#200	14.78	4.39	32.23	67.77	0.0750	

Fig. 5 The results of the sieve analyses (Koç, 2012)

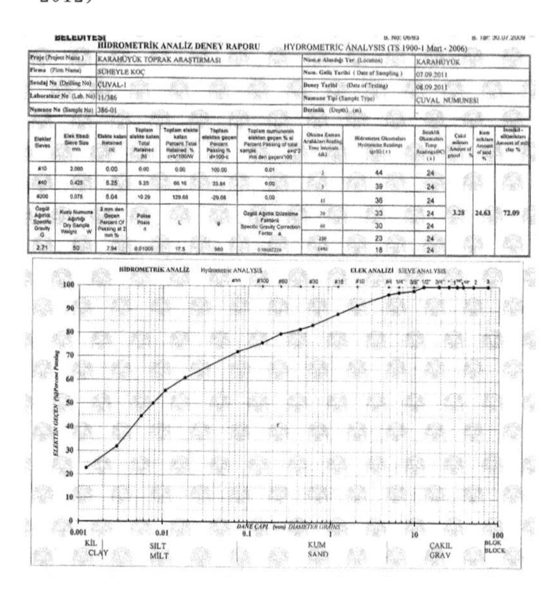

BELEDİYESİ

HİDROMETRİK ANALİZ DENEY RAPORU HYDROMETRIC ANALYSIS (TS 1900-1 Mart - 2006)

Proje (Project Name)	KARAHÜYÜK TOPRAK ARAŞTIRMASI	Num.n Alındığı Yer (Location)	KARAHÜYÜK
Firma (Firm Name)	SÜHEYLE KOÇ	Num. Geliş Tarihi (Date of Sampling)	07.09.2011
Sondaj No (Drilling No)	ÇUVAL-1	Deney Tarihi (Date of Testing)	08.09.2011
Laboratuar No (Lab. No)	11/386	Numune Tipi (Sample Type)	ÇUVAL NUMUNESİ
Numune No (Sample No)	386-01	Derinlik (Depth) (m)	-

Elekler Sieves	Elek Ebadı Sieve Size mm	Elekte kalan Retained (a)	Toplam elekte kalan Total Retained (b)	Toplam elekte kalan Percent Total Retained % c=b*100/W	Toplam elekten geçen Percent Passing % d=100-c	Toplam numunenin elekten geçen % si Percent Passing of total sample	Okuma Zaman Aralıkları Reading Time Intervals (dk)	Hidrometre Okumaları Hydrometer Readings (gr/lt) (r)	Sıcaklık Okumaları Temp Readings (°C) (t)	Çakıl miktarı Amount of gravel %	Kum miktarı Amount of sand %	İnce kil-silt miktarı Amount of silt-clay %
#10	2.000	0.00	0.00	0.00	100.00	0.01	2	44	24			
#40	0.425	5.25	5.25	66.16	33.84	0.00	5	39	24			
#200	0.075	5.04	10.29	129.68	-29.68	0.00	15	36	24			
Özgül Ağırlık Specific Gravity G	Kuru Numune Ağırlığı Dry Sample Weight W	2 mm den Geçen Percent Of Passing at 2 mm %	Poise Posie n	L	θ	Özgül Ağırlık Düzeltme Faktörü Specific Gravity Correction Factor a	30	33	24	3.28	24.63	72.09
							60	30	24			
							250	23	24			
2.71	50	7.94	0.01005	17.5	980	[illegible]	1440	18	24			

Fig. 6 The results of the hydrometric analyses (Koç, 2012)

Considering previous studies on the mudbrick, and after the determination of the type of soil and its usability, it was de-

KURU BİRİM AĞIRLIK SU İÇERİĞİ BAĞINTISININ BULUNMASI (TS 1900-1)

Proje Adı	KARAHÜYÜK TOPRAK ARAŞTIRMASI	Num. Alındığı Yer	KARAHÜYÜK
Firma	SÜHEYLA KOÇ	Numune GelişTarihi	07.09.2011
Sondaj/Kuyu No	ÇUVAL-2	Deney Baş. Tarihi	12.09.2011
Laboratuar No	10 / 387	Deney Bit. Tarihi	13.09.2011
Numune No	387-01	Derinlik	0-15 cm

Darbe Sayısı	56	Metod	Standart	Numune Tipi	Çuval
Tabaka Sayısı	5	Hacim	2128		

Rtbt.Kap No / Moisture can no	W 1	W 2	W 3	W4	W 5
Kap Ağırlığı / Mass of can	54.63	55.44	56.20	53.79	56.22
Kap+Yaş Numune / M.wet soil+can	426.44	389.91	401.05	432.71	315.05
Kap+Kuru Numune / M.Dry soil+can	373.72	339.75	346.32	368.04	268.41
Su Ağırlığı / Mass of water	52.7	50.2	54.7	64.7	46.6
Toprak Ağırlığı / Mass of soil	319.1	284.3	290.1	314.3	212.2
% Rutubet / Water c. w %	16.5	17.6	18.9	20.6	22.0
Düzlt.Rutubet / Ass. Water con	17	18	19	21	22
% Rutubet / Water c. w %	16.5	17.6	18.9	20.6	22.0
Toprak+Kap Ağrl./Mass of soil+mold	10150	10285	10371	10214	10085
Kalıp Ağırlığı / Mass of mold	6000	6000	6000	6000	6000
Toprak Ağırlığı / Mass of soil	4150	4285	4371	4214	4085
Yaş Birim Ağırlık / Wet unit wgt.	1.950	2.014	2.054	1.980	1.920
Kuru Birim Ağırlık / Dry unit wgt.	1.674	1.712	1.728	1.642	1.574

Compaction Test

Dry Density (gr / cm 3): 1.500 1.550 1.600 1.650 1.700 1.750 1.800

Water Content %: 16.0 17.0 18.0 19.0 20.0 21.0 22.0 23.0

% Rutubet / Optimum Moisture (w %) :	18.60
Max.Kuru Birim Ağırlık / Dry density gr./cm3 :	1.730

Fig. 7 The results of the compaction test (Koç, 2012)

cided to compare the compressive strength of the mudbrick obtained by adding 10% gypsum and lime to the Karahuyuk soil. Five pieces of adobe brick were prepared by adding 1800 gr soil, 20 gr straw (wheat stalk), 200 gr lime in the first experience; 200 gr gypsum in the second experience and 750 gr water. They were placed into the 7cm×7cm×7cm size of molds and were allowed to dry.

The consistency of the pulp determines the amount of water is incorporated into the adobe dough. Two methods determine the ideal adobe consistency. In the first method, a piece of the prepared dough is rounded, and long shape is given. The longer and more elastic dough show much better consistency. If cracking and splitting are seen, the water quantity is low. If it adheres to the hand, it means it is in the slime consistency, the water ratio is too high, and it has lost its bearing capacity. As a second method, the dough is rolled to the top shape and thrown from approximately 90 cm height to the ground. If it can still keep its shape when it falls, it has reached the proper consistency. If it breaks down, the amount of water is insufficient. If it still sticks to the hand and is in the slime consistency, the water ratio is too high, and it lost its bear capacity.

LİKİT LİMİT VE PLASTİK LİMİT DENEY RAPORU (Liquid Limit and Plastic Limit Test Report)
TS 1900-1/Mart 2006

Proje Adı (Project Name)	KARAHÜYÜK TOPRAK ARAŞTIRMASI	Num. Alındığı Yer (Location)	KARAHÜYÜK
Firma (Firm Name)	SÜHEYLE KOÇ	Num. Gel.Tar.(Date of Sampling)	07.09.2011
Sondaj No (Drilling No)	ÇUVAL-1	Deney Tarihi (Date Of Testing)	08.09.2011
Laboratuar No (Lab.No)	11/386	Numune Tipi (Sample Type)	ÇUVAL NUMUNESİ
Numune No (Sample No)	386-01	Derinlik (Depth) (m)	-

	LL					PL	
	1	2	3	4	5	1	2
Vuruş adedi (No of Blows)	45	37	26	18	13	----	----
Kap no (Container No)	41	17	14	68	10	38	37
Kap Ağ. (Container Weight)	12.56	12.50	12.46	12.46	12.34	12.22	12.53
Kap+Yaş Num (Container+Wet Sample,g)	25.62	21.96	21.51	22.13	23.86	17.96	17.66
Kap+Kuru Num.(Container+Dry Sample,g)	22.42	19.61	19.22	19.63	20.82	16.85	16.71
Kuru numune ağ. (Dry Sample Weight)	9.86	7.11	6.76	7.17	8.48	4.63	4.18
Su ağ. (Water Weight, g)	3.2	2.35	2.29	2.5	3.04	1.11	0.95
Su muhtevası (Water Content, %)	32.45	33.05	33.88	34.87	35.85	23.97	22.73

Likit limit (Liquid Limit)	34
Plastik limit (Plastic Limit)	23
Plastisite İndisi (Plasticity Index)	11

Fig. 8 The results of the Atterberg limits test (Koç, 2012)

Tab. 1 The sizes of the samples*

Sample Name	A (mm)	B (mm)	H (mm)
L. 1	67.08	68.22	68.65
L. 2	70.01	69.43	69.22
L. 3	69,42	69.07	69.35
L. 4	70,89	69.08	68.88
G. 1	72	71.50	69.80
G. 2	69	72	68
G. 3	73	72	69
G. 4	70	69	70
G. 5	69	72,50	69
Original	260	260	80

* A is the width, B is the length, H is the height.

After four months, the compressive strength test was performed on the samples left to dry. In the samples prepared by adding lime, too many cracks and spills were observed. One of the samples has become unusable. According to the test done on the other four samples, the compressive strengths are 0.6235 N/mm on average. Samples prepared with gypsum are more robust than the samples prepared with lime, cracks on them are mostly caused by mistakes made during molding. According to the test performed on five samples, the compressive strengths are 1.8666 N/mm^2 on average (Tab. 1, Tab. 2).

Tab. 2 The Compressive Strength Test**

Sample Name	Weight (g)	Pressure (N)	Compressive Strenght (N/mm^2)
L. 1	467.05	3270	0.714
L. 2	488.34	3100	0.637
L. 3	464.27	2840	0.592
L. 4	488.57	2700	0.551
L. Av.			0.6235
G. 1	540	9806	1.904
G. 2	529	9806	1.973
G. 3	525	9806	1.865
G. 4	530	8825	1.827
G. 5	550	8825	1.764
G. Av.			1.8666
Original	9350	245166	3.759

** L. represents the Lime, G. represents the Gypsum, Av. represents the average.

The compressive strength of Kafescioglu and Gurdal (1985) found with gypsum was 3.43～4.90 N/mm^2. The condition of the adobe bricks used in the structure was also determined, and only two adobe brick could be taken from the collapsed garden wall of the building. Sieve analysis was performed with the first one, and the compressive strength test was performed with the second one. According to the sieve analysis of grain size distribution, 67.77% consists of thin clay and silt, 29.32% sand and 2.91% gravel. The compressive strength is 3.759N/mm^2. It was observed that the compressive strength difference between the existing mudbrick and gypsum and lime were found to be due to the aggregates in the mudbrick. The prepared samples consist entirely of clay and silt, while 32.23% aggregate (sand + gravel) is present in the existing mud brick. The compressive strength test shows the effect of the aggregate in the soil for adobe making. Without straw and aggregate, it does not matter how many percents of lime or gypsum is added.

For the rehabilitation of adobe bricks,

when they need to be renewed in this area, they should be prepared by adding 10% gypsum and 30% aggregate. In order to coat the interior and exterior surfaces of the building, it would be more appropriate to use the clay and straw plaster, which contains 15% lime for the longevity of the material.

4 Conclusion

The vernacular houses provide a way to analyze the local tradition and culture. The way of providing building material indicates the family structure and sustainability. The unique feature of the houses in Karahuyuk is the use of concrete in the floor of wet areas, like kitchen, bathroom, storeroom, and window sills. This building shows the era of development of modern materials, most importantly, the use of modern materials along with traditional ones. This type of construction should be preserved due to its characteristics and in order to make sure the continuation of the tradition of this built heritage.

The abandonment of many adobe buildings in Turkey is visible; however, some of these old buildings can be rehabilitated with minor maintenance and repair works. The materials used in the conservation works should be compatible with the original materials. For the rehabilitation of this type of building it is necessary to know the materials and to create new materials that are compatible; these materials should ensure the longevity of the conservation and restoration actions. For this, adobe bricks which have 10% gypsum and 30% aggregate should be used in conservation work with the plaster consists of clay, straw, and 15% lime, especially in Konya Karahuyuk region.

Acknowledgements

This paper, which is a part of the master thesis of Süheyla Koç, is written in dedication to the late thesis advisor Prof. Dr. Ahmet Ersen.

References

KAFESÇIOĜLU R, 1984. Konut yapımında toprağın ana malzeme olarak kullanılmasının önemi ve uygulanmasında fayda görülen metotlar, Kerpiç Semineri Tebliğleri, İmar ve İskan Bakanlığı, Yapı Malzemesi Genel Müdürlüğü.

KAFESÇIOĜLU R, GÜRDAL E, 1985. Çağdaı yapı malzemesi alker alçılı kerpiç, Enerji ve Tabii Kaynaklar Bakanlığı Enerji Dairesi Baıkanlığı, Ankara.

KOÇ S, 2012. Konya İli, Karahüyük Beldesi Korkmaz Evi Restorasyon Projesi (The Restoration Project of Korkmaz House in Karahuyuk District in Konya), ITU Master Thesis, Istanbul.

URL_1. 2019-05-23. http://www. kalitekontrol. net/alt—yapi/kivam—limitleri—atterberg—limitleri—likit—ve—plastik—limit. html.

Characterization of Compatible TRM Composites for Strengthening of Earthen Materials

Rui A. Silva, Daniel V. Oliveira, Cristina Barroso, Paulo B. Lourenço

ISISE & IB-S, University of Minho, Guimarães, Portugal

Abstract The high seismic vulnerability of earth constructions has been evidenced by several recent earthquakes that occurred around the World with moderate to high magnitudes, namely Bam 2003, Pisco 2007 and Maule 2010. The seismic risk associated to earth constructions is further amplified by the fact that a great percentage of these constructions is built on regions with important seismic hazard. Thus, the preservation of the immense earthen built heritage and of the life of their inhabitants demands the adopion of innovative strengthening interventions. However, the success of such solutions requires fulfilling compatibility requirements, while its general use requires adopting affordable materials and low complexity technical solutions. In the last years, textile reinforced mortars (TRM) have been increasingly used to strengthen masonry structures due to their high structural effectiveness and compatibility. In the case of earth constructions, these composite materials are also expected to provide efficient strengthening, though specific component materials should be adopted. This paper presents an experimental program dedicated to the characterization of the composite behavior of two TRM composites proposed for strengthening rammed earth walls. The composites differ on the mesh used, namely a low cost glass fiber mesh and a nylon mesh acquired locally, while the same earth-based mortar was used in both cases. The experimental program involved testing the mortar under compression and composite coupons under tension. In general, the glass TRM presents higher strength and stiffness in tension, while the nylon TRM presents considerably higher deformation capacity. Finally, stress-strain relationships describing the composite behavior are presented for numerical modelling purposes.

Keywords earthen walls, strengthening, textile reinforced mortars, compatibility, low cost

1 Introduction

The high seismic vulnerability of earth constructions has been evidenced by recent intense and destructive earthquakes, such as Bam 2003, Pisco 2007 and Maule 2010. This vulnerability is a consequence of several factors (Yam n Lacouture et al., 2007 and Oliveira et al., 2010), among which the poor connections between structural elements, high self-weight and low mechanical properties are systematically the most emphasized ones. The seismic risk associated to earth constructions is further amplified by the fact that a great percentage of these constructions is built on regions with important seismic hazard.

Despite the current marginal use of rammed earth in Portugal, the southern region of the country presents a significant built heritage, whose monolithic walls were erected by compacting moistened earth inside a formwork. Most of this heritage is concentrated in the Alentejo region (Rocha, 2005) and is mainly constituted by still inhabited dwellings. Nevertheless, this region is also characterized by a moderate seismic hazard, where the reference peak ground acceleration can achieve up to 2.0 m/s^2 according to Eurocode 8 (IPQ, 2009). This situation combined with the fact that earth constructions present high seismic vulnerability raise seismic risk concerns. The preservation of the immense earthen built heritage and of the life of their inhabitants demands the adoption of innovative strengthening interventions.

Textile reinforced mortar (TRM), also known as fiber reinforced cementitious matrix (FRCM), is an innovative strengthening solution that is becoming increasingly used for masonry structures. TRM is an externally bonded composite system composed of two material components, namely the matrix (mortar) and one or more layers of textile (fibers-meshes). The textile provides tensile strength to the system, while the embedding mortar provides protection against external agents and tensile stress transfer capacity between the supporting masonry and the textile. In general, the available commercial systems use high-performance cement-or hydraulic lime-based mortars and meshes made of high tensile strength fibers, such as carbon, basalt and glass (De Felice et al., 2014). Recent research has been demonstrated that TRM strengthening allows great increase in the out-of-plane strength and deformation capacities of masonry walls (Valluzzi et al., 2014).

The success of TRM strengthening requires fulfilling compatibility requirements, while its general use requires adopting affordable materials and low complexity technical solutions. These aspects are particularly decisive in the strengthening of earthen walls. In this regard, the Pontifical Catholic University of Peru (PUCP) has been studying a solution, called geomesh strengthening, to strengthen adobe dwellings (Blondet et al., 2005), whose main outcomes resulted in design guidelines included in the Peruvian code E. 080 (MVCS, 2017). This solution fits within the concept of TRM strengthening, as it includes the application of a low-cost geosynthetic mesh tightly fixed to the adobe walls, covered by a coating mortar. The study of this technique has been mainly addressed by means of large-scale structural tests (Noguez & Navarro, 2005, Zavala & Igarashi, 2005, Figueiredo et al., 2013), which shown that the technique promotes significant improvement of the seismic performance of adobe constructions. Nevertheless, the characterization of the strengthening solution is rarely addressed, namely with respect to the composite mechanical behavior and interaction between the different materials composing the solution, which define the efficiency of the strengthening (Kouris & Triantafillou, 2018).

The strengthening of rammed earth walls with TRM was recently proposed in Oliveira et al. (2017). This work investigated different low-cost meshes readily available in the local market, which name the strengthening technique as low cost textile reinforced mortar (LC-TRM). Different coating mortars were also characterized in this study.

The experimental work presented in this paper is a sequence of the previously referred work, by investigating the composite behavior of two LC-TRM composites selected to be compatible with rammed earth. The composites differ on the mesh used, namely a low-cost glass fiber mesh and a nylon mesh, while the coating mortar consists of an earth-based mortar.

2 Experimental Program

The experimental program was carried out with the main objective of characterizing the composite behavior of two LC-TRM composites selected to be compatible with rammed earth from Alentejo region, Portugal. This section presents the materials composing both composites and the experimental procedures followed to characterize the composite behavior.

2.1 Materials

The two adopted LC-TRM composites were composed with the same earth-based mortar and two different reinforcing meshes.

The composition of the earth-based mortar was defined previously in Oliveira et al. (2017), as mortar EM2.0. The mortar is constituted by 33% of sieved soil and 67% of quarzitic fine sand (0/2). The soil was collected from the municipality of Odemira, which is located in Alentejo region. This soil was previously used to manufacture representative rammed earth specimens (Silva et al., 2018) and was deemed as presenting very high clay content. The soil incorporated in the mortar was sieved to remove the particles larger than 10 mm, which corresponds to the recommended maximum particle size for earth-based mortars (R hlen & Ziegert, 2011). The particle size distribution curves of the mortar and composing materials are given in Fig. 1. The water solids ratio (W/S) was defined as 0.17, in order to obtain a flow table value (CEN, 2004) of about 170 m, as recommended in Gomes (2013). The mortar presented a linear shrinkage value of 0.7%, which is inferior to the recommended maximum value of 2% (Gomes, 2013). With respect to the physical-mechanical properties (CEN, 1999), the mortar presented a density of 1810 kg/m³, flexural strength of 0.5 MPa and compressive strength of 0.9 MPa (equilibrium moisture content of 1.1% at 20 C temperature and 57.5% relative humidity).

The meshes incorporated in the LC-TRM composites consisted of a woven glass fiber mesh (RM1) and a nylon mesh (RM2) with welded knots, as illustrated in Fig. 2. Mesh RM1 was acquired with a cost of 0.85€/m², presents a mesh aperture of 8×9mm² and mass per unit area of 93g/m², while those of mesh RM2 are 0.63€/m², 16×21mm² and 63g/m², respectively. The tensile strength of both meshes is different in their main directions. In RM1 case, the tensile strength is 17 kN/m in the longitudinal (X) direction and 12 kN/m in the transversal (Y) one. The mesh RM2 is substantially weaker, as the tensile strength values are 2 kN/m and 4 kN/m, respectively. It should be noted that a nylon mesh similar to RM2 was previously used in the study presented in Figueiredo et al. (2013) to strengthen an adobe wall, while mesh RM1 was used in the study presented in Sadeghi et al. (2017) to strengthen adobe vaults.

2.2 Testing Methods

The composite behavior of both LC-TRM composites was evaluated by testing individually the compressive behavior of the mortar and the tensile behavior of mortar-mesh coupons.

Three cylindrical specimens with 90 mm diameter and 180mm height were prepared

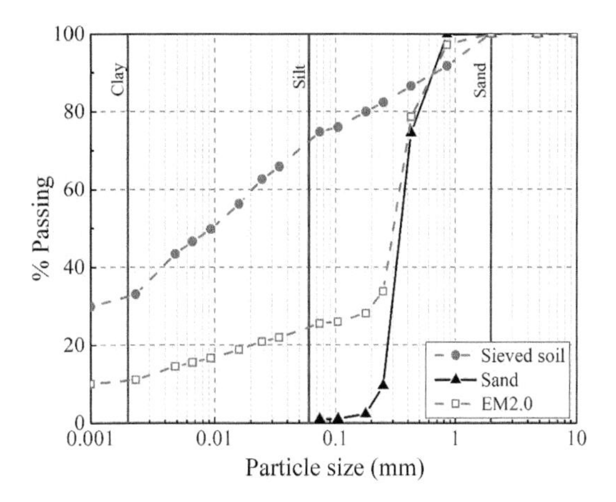

Fig. 1 Particle size distribution of the mortar and composing materials

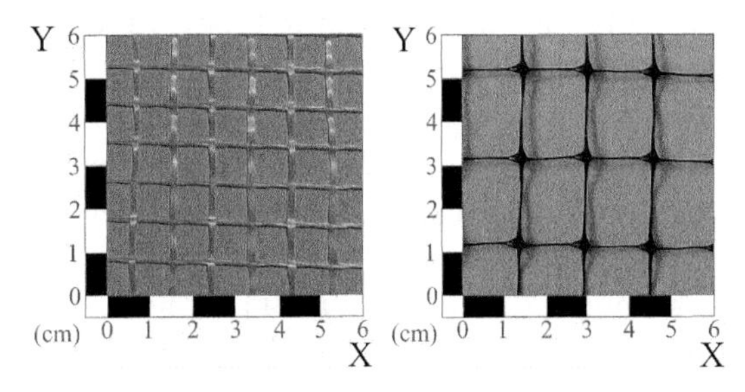

Fig. 2 Reinforcing meshes adopted in the LC-TRM composites: (a) RM1; (b) RM2.

from the earth-based mortar EM2.0. This geometry results in a 2 : 1 height-diameter ratio, which mitigates the influence of the confinement introduced by the testing plates on the compression behavior. The specimens were cast in PVC molds, which were perforated with 1~2mm holes spaced each 10 mm to promote the uniform drying hardening of the mortar inside the mold. This procedure allowed to demold the specimens after 7 days of drying in a climatic chamber at constant temperature of 20℃ and relative humidity of 57.5%. Then, the specimens were kept in the same climatic chamber to achieve equilibrium moisture content until testing, which occurred 28 days after casting. The tests were performed using a frame equipped with an actuator, which loaded the specimens under displacement control at constant speed of 3μm/s. The axial deformation at the middle third of the specimens was monitored by means of three linear variable differential transducers (LVDTs) fixed with aluminum rings. The test setup is illustrated in

Fig. 3(a).

Four coupon specimens were prepared for each LC-TRM composite in order to test their tensile behavior. The specimens consisted of mortar bands with one embedded layer of mesh positioned at middle thickness. They were cast by placing in a mold a mortar layer with dimensions 300 mm length, 60mm width and 5mm thickness. Then, a layer of mesh was placed covering the mortar by slightly stretching it. The mesh layer was longer than the mortar one so that the excess extremities presented a length of 50mm to bond the gripping steel plates before testing. It should be noted that only the direction of the highest tensile strength of each mesh was considered for testing the composite behavior, meaning that the mesh layers in the coupons were orientated accordingly. A second layer of mortar with the same dimension of the first was applied subsequently. The specimens were stored until testing in the same ambient conditions of the previously referred climatic chamber. Demolding was also performed 7 days after casting. The tensile tests were performed at 28 days of age by adopting a procedure similar to that of ASTM D6637 (ASTM, 2011). Gripping plates were glued to the excess mesh extremities with an epoxy mortar one day before testing. Then, the specimens were fixed to the grips of the testing machine and the tensile load was applied under displacement control (Fig. 3(b)). Due to the remarkable difference in stiffness between both meshes, different testing speed protocols were used for each case. For RM1 coupons, a constant speed of 3μm/s was applied until an axial deformation of 2mm was reached, after which the speed was increased to 10μm/s until failure. RM2 coupons were loaded with 30μm/s of speed until an axial a deformation of 6mm was obtained and then at 100μm/s. The axial deformation was monitored by means of an LVDT fixed between the two grips fixing the coupons.

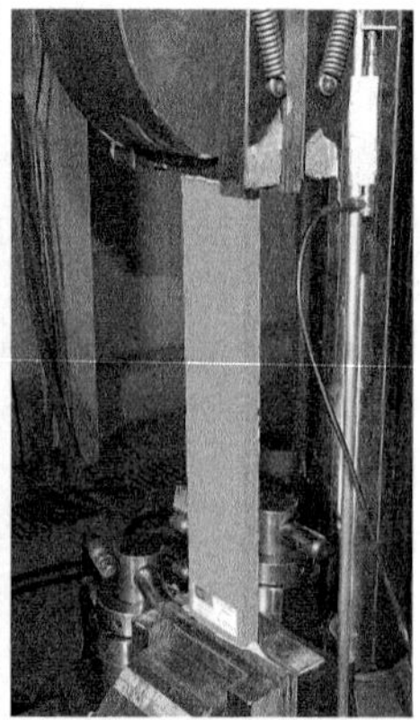

Fig. 3 Testing setups of the LC-TRM composites: (a) compression test of the mortar; (b) tensile test of the coupons

3 Results and Discussion

The mortar cylinders tested under compression presented an average density of 1876 kg/m^3 and equilibrium moisture content of 0.5%. The obtained stress-strain curves are presented in Fig. 4. The average compressive strength is of about 1.3 MPa (CoV= 5%) and the Young's modulus is 3322 MPa (CoV= 13%), which was computed by linear fitting of the stress-strain curves at 5% ~ 30% of the compressive strength. The curves present an expressive nonlinear behavior, which is typically observed in earthen materials.

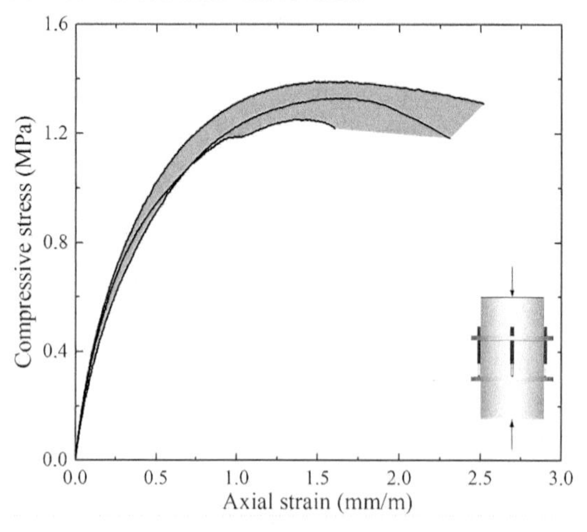

Fig. 4 Stress-strain curves obtained from the compression tests on the mortar specimens

The results of the tensile tests on the coupons of both reinforcing meshes are presented in Fig. 5 in terms of stress-strain curves and typical failure modes. It should be noted that the tensile stress was computed by considering the cross section of the coupon, instead of the undetermined cross section of the reinforcing mesh. In average, the tensile strength of the RM1 coupons is of about 1.6 MPa (CoV= 6%), while that of the RM2 coupons is of about 0.4 MPa (CoV= 4%). As expected, the mesh RM1 provides higher strength to the LC-TRM composite than mesh RM2. Furthermore, both meshes when integrating the LC-TRM composite are able to achieve tensile strength values similar to those of the dry meshes. For comparison purposes, the average linear strength of the coupons RM1 and RM2 is of 16.4 kN/m and 4.3 kN/m, respectively. On the other hand mesh RM2 provides much higher deformation capacity than mesh RM1. This behavior results from the high flexibility and plastic behavior of mesh RM2 as discussed in Oliveira et al. (2017). Another consequence of this characteristic on the composite behavior seems to be the higher capacity to redistribute stresses, which is observed in terms of formation of higher number of cracks in RM2 coupons.

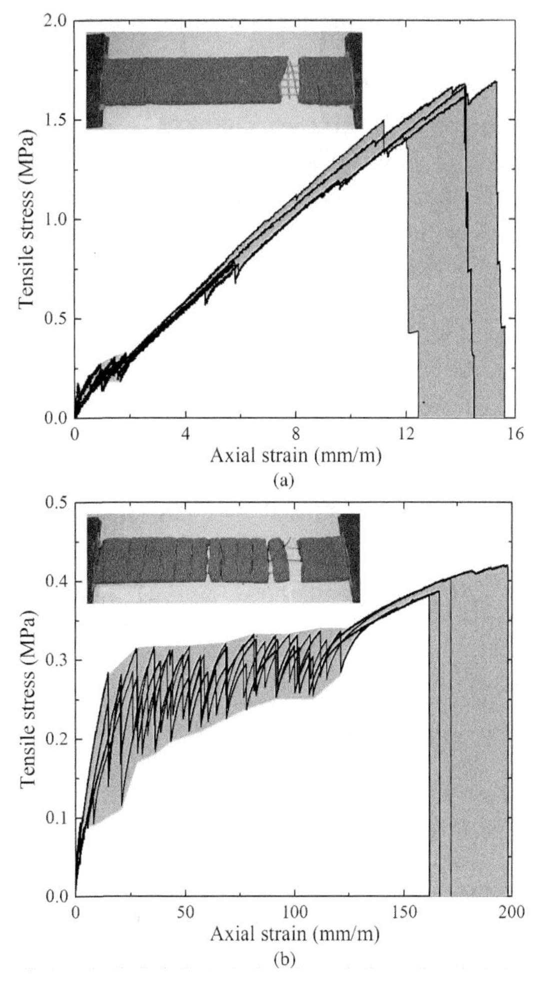

Fig. 5 Stress-strain curves and failure modes obtained from the tensile tests: (a) RM1; (b) RM2.

The tensile behavior of both LC-TRM composites is in agreement with the typical behavior described in Ascione et al. (2015) for TRM composites. In this regard, three stages are depicted in the stress-strain curves. Stage I corresponds to the uncracked behavior, stage II to the crack development and stage III to the cracked behavior. In stage I the response is linear, as the mortar is not cracked. Afterwards, the appearance of the first crack occurs and in stage II the stiffness of mortar is decreased with the development of further cracks. Thus, in these first two stages, the behavior of the composite depends on the mechanical properties of the mortar, mesh and on the interaction of these two components. After a certain strain level, the formation of new cracks stops and a slight increase in force is observed, which defines the transition to stage III. Here, the increase in tensile stress leads to the widening of the cracks, where the loading capacity of the system is defined by the mechanical properties of the mesh (textile).

The curves of Fig. 6 summarize the results of the experimental program by proposing simplified curves that are expected to be used in future numerical modeling investigations.

As previously referred to, the compression behavior of the LC-TRM is defined by the individual behavior of the earth-based mortar. Furthermore, past research has demonstrated that the expressive nonlinear behavior of earthen materials in compression is better defined by a multilinear stress-strain relationship, which was here also adopted (Miccoli et al., 2019). The initial branch of the proposed relationship represents the linear behavior by means of the experimental Young's modulus up to stress value of 30% of the compressive strength (see Fig. 6a). Then, it follows the average curve obtained from the experimental tests. This average is interrupted by an idealized post-peak linear degradation, defined with basis on the trend observed from the experimental curves. It should be noted that readings from the LVDTs loose coherence in the post-peak phase due to interference the of the damage development.

Regarding the tensile behavior of the LC-TRM composite, it was observed that it depends on the type of embedded mesh. Thus, a stress-strain relationship is here proposed for each mesh. In both cases, the relationship is a trilinear curve representing the three stages typically observed.

4 Conclusion

This paper presents an experimental program dedicated to the characterization of the composite behavior of two LC-TRM composites proposed for the strengthening of rammed earth walls. The composites differ on the mesh used, namely a low-cost glass fiber mesh and a nylon mesh, while the coating mortar is the same and consists of an earth-based mortar. The composite behavior was characterized by testing the mortar under compression and composite coupons under direct tension.

The mortar presented an expressive nonlinear behavior typically observed in earthen materials. Despite that, the average values of the compressive strength and Young's modulus are 1. 3 MPa and 3322 MPa, respectively.

In tension, the LC-TRM composite incorporating the glass fiber mesh (RM1) presents higher strength and stiffness than the composite incorporating the nylon mesh (RM2). On the other hand, the deformation capacity of the second is considerably higher. Furthermore, the typical three stages behavior of TRM was observed in both LC-TRM composites.

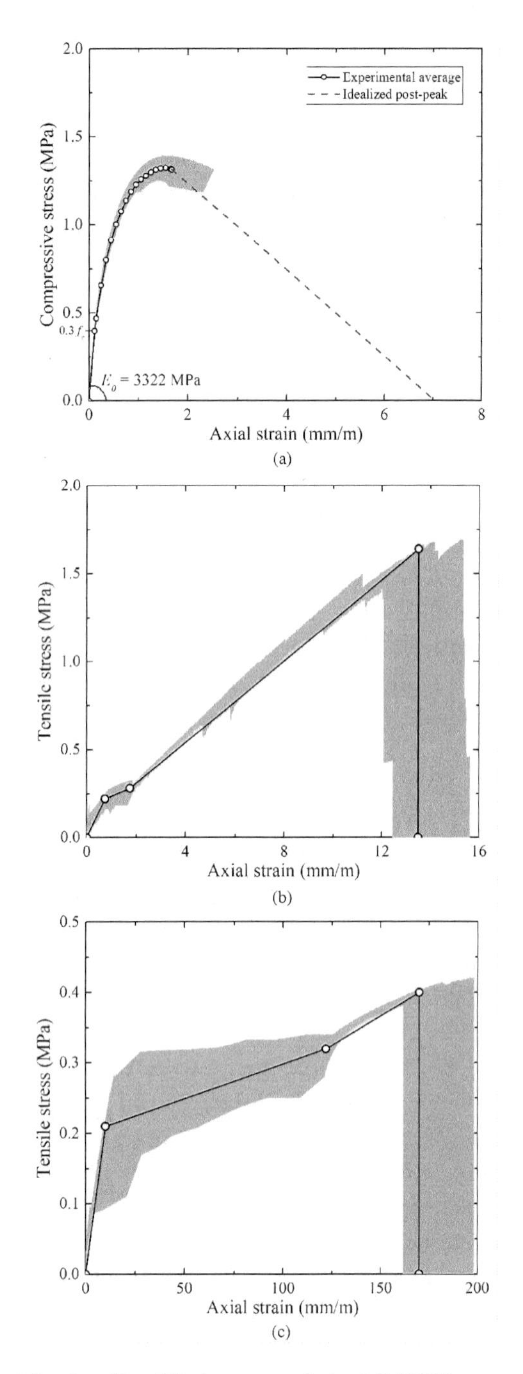

Fig. 6 Simplified curves of the LC-TRM composite behavior: (a) mortar in compression; (b) RM1 in tension; (c) RM2 in tension

Finally, stress-strain relationships were proposed to simulate the composite behavior of both LC-TRM solutions in future numerical modelling investigations.

Acknowledgements

This work was partly financed by FED-

ER funds through the Competitivity Factors Operational Programme -COMPETE and by national funds through FCT — Foundation for Science and Technology within the scope of projects POCI-01-0145-FEDER-007633 and POCI-01-0145-FEDER-016737.

References

ASCIONE L, DE FELICE G, DE SANTIS S, 2015. A qualification method for externally bonded Fibre Reinforced Cementitious Matrix (FRCM) strengthening systems. Composites part B: Engineering, 78: 497-506.

ASTM, 2011. D6637-11: Standard test method for determining tensile properties of geogrids by the single or multi-Rib tensile method. West conshohocken: ASTM international.

BLONDET M, TORREALVA D, GARCIA G, et al., 2005. Using industrial materials for the construction of safe adobe houses in seismic areas. Sydney: Earth build 2005 international conference.

CEN, 1999. EN 1015-11: Methods of test for mortar for masonry - Part 11: Determination of flexural and compressive strength of hardened mortar. Brussels: European Committee for Standardization.

CEN, 2004. EN 1015-3: Methods of test for mortar for masonry - Part 3: Determination of consistence of fresh mortar (by flow table). Brussels: European Committee for Standardization.

DE FELICE, et al., 2014. Mortar-based systems for externally bonded strengthening of masonry. Materials and structures, 47(12): 2021-2037.

FIGUEIREDO A, VARUM H, COSTA A, et al., 2013. Seismic retrofitting solution of an adobe masonry wall. Materials and structures, 46(1): 203-219.

GOMES M I, 2013. Conservation of rammed earth construction: repairing mortars. PhD thesis. Universidade Nova de Lisboa. (in Portuguese)

IPQ, 2009. NP ENV 1998-1: Eurocode 8: Design of structures for earthquake resistance — Part 1: General rules, seismic actions and rules for buildings. Lisbon: Instituto Portugu s da Qualidade.

KOURIS L A S, TRIANTAFILLOU T C, 2018. State-of-the-art on strengthening of masonry structures with textile reinforced mortar (TRM). Construction and building materials, 188: 1221-1233.

MVCS, 2017. Norma E. 080: Diseño y Construcción con Tierra Reforzada. Normas legales, Anexo — resolución ministerial n° 121-2017-vivienda, Ministerio de Vivienda, Construcción y Saneamento.

NOGUEZ R, NAVARRO S, 2005. Reparación de muros de adobe com el uso de mallas sintéticas. PUCP, International Conference Sismo Adobe 2005. (in Spanish)

OLIVEIRA D V, SILVA R A, LOURENÇO P B, et al., 2010. As construções em taipa e os sìsmos. Congresso Nacional de Sismologia e Engenharia - SÌMICA 2010, Aveiro. (in Portuguese)

OLIVEIRA D V, SILVA R A, BARROSO C et al., 2017. Characterization of a compatible low cost strengthening solution based on the TRM Technique for Rammed Earth// Key engineering materials, 747: 150-157.

ROCHA M, 2005. Rammed earth in traditional architecture: construction techniques. Earth architecture in portugal. Lisbon: Argumentum.

RÖHLEN U, ZIEGERT C, 2011. Earth building practice. Planning, design, building. Berlin: Beuth Verlag.

SADEGHI N, OLIVEIRA D V, SILVA R A, et al., 2017. Performance of adobe vaults strengthened with LC-TRM: an experimental approach. Prohitech'17, Lisbon.

MICCOLI L, SILVA R A, OLIVEIRA D V, et al., 2019. Static behavior of cob: experimental testing and finite-element modeling. Journal of materials in civil engineering, 31(4): 04019021.

SILVA R A, DOMÌNGUEZ-MARTÌNEZ O, OLIVEIRA D V, et al., 2018. Comparison of the performance of hydraulic lime-and clay-based grouts in the repair of rammed earth. Construction and building materials, 193: 384-394.

VALLUZZI M R, DA PORTO F, GARBIN E, et al., 2014. Out-of-plane behaviour of infill masonry panels strengthened with composite materials. Materials and structures, 47(12): 2131-2145.

YAMIN LACOUTURE L E, PHILLIPS BERNAL C, ORTIZ R, et al., 2007. Estudios de vulnerabilidad sìsmica, rehabilitación y refuerzo de casas en adobe y tapia pisada. Apuntes: Cultural-Journal of Cultural Heritage Studies 20(2): 286-303. (in Spanish)

ZAVALA C, IGARASHI L, 2005. Propuesta de reforzamiento para muros de ddobe. PUCP, International conference sismo adobe 2005. (in Spanish)

Seismic Analysis of Fujian Hakka Tulous

Bruno Briseghella[1], *Valeria Colasanti*[2], *Luigi Fenu*[2], *Kai Huang*[1], *Camillo Nuti*[3], *Enrico Spacone*[4], *Humberto Varum*[5]

1 Fuzhou University, China 2 University of Cagliari, Italy
3 University of Roma Tre, Italy 4 University of Chieti-Pescara, Italy
5 University of Porto, Portugal

Abstract The overall earthquake response of Hakka Tulous, traditional earth constructions of the Fujian Province (China) and the listed ones among the UNESCO World Heritage buildings, are investigated. Non-linear static analysis (pushover) with the equivalent frame approach is used. Although some rough approximations are assumed, this approach is well suited to model complex masonry structures, like Tulous. In fact, non-linear analysis implemented by finite elements or by discrete elements would involve complex models hard to converge and needing long computational time. After carrying out seismic analysis of a Tulou prototype, its failure modes and overall seismic response were evaluated. The Tulou has shown to have good earthquake resistance with respect to the maximum seismic action that can be expected in the Fujian Province.
Keywords Hakka Tulou,Fujian, earth, earthquake analysis, macroelements, equivalent frame

1 Introduction

Hakka Tulous are house-fortresses situated in the Fujian Province of China and inhabited by Hakka clan people. For their heritage and architectural value, they are inscribed in the UNESCO World Heritage list.

A Tulou consists of a circular or square perimeter earth wall internally stiffened by 3D wooden frames supporting wooden floors, subdivided by partition walls delimiting the Hakka people dwellings.

They are large-sized earth constructions, with a unique large entrance and small windows mainly located at high elevation in the earth wall (Fig. 1).

Notwithstanding their relevance, few studies are available in the scientific literature on their structural behaviour. In particular, there are only few studies on their seismic response, although in some areas of the Fujian Province, where the Tulous are traditionally built, the seismic hazard is not negligible (Briseghella et al. , 2017, 2019b; Liang Stanislawski, & Hota, 2011).

Fig. 1 Hakka Tulou in the Fujian Province

Despite the few studies available on Tulous, earth constructions are increasingly studied both because all over the world there are many monuments and historical buildings made of earth, and because of their sustainability, thermal comfort performance and energy efficiency (Houben & Guillaud, 1994).

Regarding the recent studies on the behaviour of the earthen material and of earth structural elements, significant work have been carried out on the fracture behaviour of earth considered as a quasi-brittle material (Aymerich et al. , 2016; Aymerich, Fenu,

& Meloni, 2012), as well as on its influence on the structural response of adobe bricks and panels (Blondet & Vargas, 1978; Parisi et al., 2015; Vargas & Ottazzi, 1981; Varum et al., 2007).

With reference to the seismic response of earth constructions (Varum et al., 2014), the damages caused by earthquakes have been studied by many authors (Blondet, Vargas, & Tarque, 2008; Webster & Tolles, 2000). The influence of brittleness and low tensile strength on the seismic vulnerability of the adobe structures has been investigated by Blondet et al. (Blondet, Vargas, Velásquez, & Tarque, 2006). Their vulnerability was observed in recent earthquakes, as in Peru (1970, 1996, 2001 and 2007), El Salvador (2001), Iran (2003), Pakistan (2005) and China (2008 and 2009) (Varum et al., 2014).

The seismic response of earthen structures has been experimentally investigated through shaking table tests on reduced masonry section walls (Antunes et al., 2012; Figueiredo et al., 2013; Tareco et al., 2009) and on scale models of entire buildings (Webster & Tolles, 2000). Unfortunately, shaking table tests are expensive and need a long time especially for constructing the model. For all these reasons, shaking table tests are not the first choice to investigate the seismic response of masonry structures including earth constructions, even if they can provide reliable and qualitatively valid results.

The numerical modelling techniques are instead a more advantageous and less expensive way of studying earth buildings. The main methods of numerical analysis to model masonry structures are the Finite Element Modelling (FEM), the Distinct Element Modelling (DEM), and the analysis by macroelements with the Equivalent Frame Method (EFM). Unfortunately, nonlinear analysis with FEM and DEM of complex masonry structures usually lead to encounter convergence problems hard to solve, as well as to high computational costs (Briseghella et al., 2019a).

On the contrary, the analysis by macroelements with the EFM well apply in nonlinear analysis of complex masonry structures because, despite some approximations in defining the macroelement geometry and its structural response, the equivalent frame approach allows to obtain reliable results.

As a matter of fact, schematization of the wall with openings as a frame where piers and spandrels are deformable linear elements connected by nondeformable rigid nodes, allows to facilitate convergence and reduce the computational costs.

Among the different codes using the macro-element approach, considerable diffusion have RAN (Augenti, 2004; Augenti & Parisi, 2010; Raithel & Augenti, 1984), SAM (Magenes, 2000; Magenes & Fontana, 1998) and TREMURI (Lagomarsino et al., 2013) codes. In particular, in this study TREMURI code has been used to carry out non-linear static analysis (pushover) of a Tulou prototype with the EFM.

Regarding numerical modelling of earth constructions, the first study in this field was made by Tarque (Tarque, 2011), who tested the validity of different modelling strategies using the results obtained by FEM.

The EFM was first applied to earth structures as part of a research project funded by the Region of Sardinia (Asprone, Parisi, Prota, Fenu, & Colasanti, 2016), an Italian region where earth constructions are still built and where there is an important heritage of traditional ones, too. The validity of the use of the EFM in earth buildings was first assessed by comparing the results obtained from simple earth buildings and similar tuff buildings. Moreover, the validity of the macro-element approach in modelling earth structures was also evaluated by comparing the results obtained by shaking table tests with those obtained by numerical models using the EFM. The results of shaking table tests funded within the Getty Seismic Adobe Project (GSAP) (Gavrilovic et al., 1996; Tolles, Kimbro, Webster, & Ginell, 2000) and carried out with increased acceleration values on small scale models of adobe constructions were compared with those obtained from nonlinear static analysis (pushover) performed by macroelements on real scale prototypes.

Based on these validation tests, in this article, the macroelement approach with the EFM has been applied to the prototype of a Hakka Tulou. Their typical cylindrical wall has been discretized and shaped as a 24-sided polygonal wall.

The geometry of the Tulou prototype as well as the mechanical properties of the earthen material have been extracted from some studies on Huanji Tulou available in the scientific literature (Liang et al., 2013, 2011). This research has provided a first significant contribution to the study of the seismic response of Tulous.

2 Building Technology and Structure of a Fujian Hakka Tulou

Tulous are distributed in small villages in the mountainous area of west-south of the Fujian Province (China) (Fig. 2) (Zhang, Luo, & Liao, 2011). They are circular buildings made of a cylindrical earth wall about 2m thick at its base whose typical diameter and height are 50 and 20m, respectively. For defensive reasons, a single door guarantees access to the internal courtyard. For the same reason, Tulous have only two or three rows of small windows starting at up to 10 m elevation. Inside the Tulou, the floors hosting the Hakka People dwellings are supported by wooden frames whose radial beams are in turn supported by the circular earth wall at one end and by wooden columns at the opposite end. The wooden floor system is likely to be only partially rigid because rafters and wooden planks are not firmly connected one to the other. Similar considerations can be done for the two-pitch roof, supported by an A-frame truss system.

Very little information is available on the mechanical properties of the materials and on the structural features of the construction elements. About the Tulou geometry, in this study we refer to a Tulou prototype whose dimensions are obtained from a FEM model of the Huanji Tulou (Liang et al., 2013, 2011), differing from it just for having regularized the window opening spacing, that in Huanji Tulou is not uniform.

Fig. 2 Bird's eye view of a Tulou cluster

2.1 Equivalent Frame Approach

The Tulou prototype has been modelled by macroelements through the EFM using TREMURI code.

Despite some rough approximations, this method has proved to be particularly suited to model masonry constructions (Braga & Dolce, 1982; Marques & Lourenço, 2014), and successfully applied to earth constructions, too (Asprone et al., 2016).

In accordance with dynamic test data and post-earthquake survey of damages caused by the seismic action, in the EFM method each wall is modelled as an equivalent frame where deformable piers and spandrels correspond, respectively, to columns and beams, connected by non-deformable rigid nodes.

The spandrel length corresponds to the opening width. The dimension of the nodes defines the pier length and depends on the opening size and position. Membrane elements are used to model stiff or partially-stiff floors, depending on the membrane stiffness. Stiffness and geometry of walls and diaphragms highly affect the box-behaviour and structural efficiency of the masonry construction (Lagomarsino et al., 2013). Each macroelement (piers and spandrels) is divided into three parts: a central one, almost coinciding with the whole masonry panel, where shear deformations are addressed with nonlinear contribution of the frictional force opposing to the sliding mechanisms, and two thin end ones, where the axial and bending deformations are instead addressed and where the inelastic contributions are obtained from the unilateral perfectly elastic contact condition (Brencich, Gambarotta, & Lagomarsino, 1998).

The 3D frame is obtained through connecting the nodes of the lateral piers of the 2D equivalent frames (Lagomarsino et al., 2013).

The Equivalent Frame Model of the circular Tulou (Fig. 3), has been obtained by approximating the cylindrical wall to a 24-side polygonal wall.

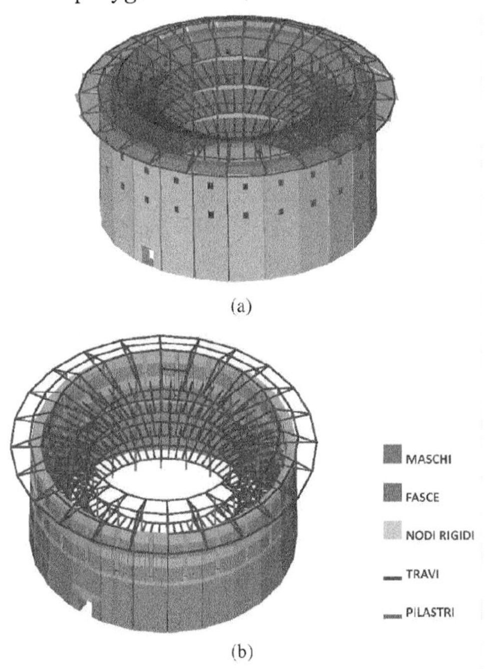

Fig. 3 Tulou numerical model: 3D view (a); detail of the Equivalent Frame (b)

Having assumed constant horizontal spacing and vertical alignment of the window openings, in each wall side there are two small windows, one for each of the two upper levels.

Wooden floors and roof sections stiffness have been modelled with diaphragms of appropriate rigidity accounting for their orthotropic behaviour, too. Fig. 3(b) shows the Equivalent Frame implemented to analyse the Tulou through the EFM.

The mechanical characteristics of the Tulou construction materials (earth and wood) have been assumed by literature-based data (Liang et al., 2013). In particular, the earth mechanical properties used to model the Tulou prototype herein considered are shown in Tab. 1.

Tab. 2 Mechanical properties of earth material

f_c [MPa]	f_{v0} [MPa]	E [MPa]	w [kN/m^3]
1.00	0.10	1000	12

3 Seismic Analysis of a Tulou Prototype Modelled Through the EFM

With reference to performance-based earthquake engineering concepts (Lagomarsino et al., 2013; Liu et al., 2015) in the last decades the nonlinear static analyses (pushover) have shown to be the most reliable analysis method for seismic assessment.

To analyse Tulou seismic response, pushover analysis implemented in TREMURI has been herein performed with mass-proportional horizontal forces plotted as a function of the consequent displacements of a suitably chosen control node. The displacement demand obtained from the ADRS spectrum at the Life Safety Limit State (475 years return period) has been then compared with the capacity displacement obtained from the capacity curve of the structure, thus evaluating the Tulou safety under the seismic risk of the Fujian Province. The peak ground acceleration (PGA) ag=0.16g, as well as F0=2.45 and TC* = 0.32 were assumed to draw the ADRS spectrum referred to the Tulou site. Since the Tulous are probably constructed on a rocky subsoil in the mountain area of the Fujian Province, in this first study on the Tulou seismic response, the stratigraphic and topographic amplification is not considered.

The capacity curves obtained assuming earthquake direction and control node reported in the legend are shown in Fig. 4. Unfortunately, while the most appropriate position of the control node should coincide with the centre of mass of the structure, this is not allowed by Tulou geometry. In fact, its centre of mass is practically coincident with the centre of the Tulou court, where no node can be assumed as a control node because in the centre of the court there is no Tulou structure.

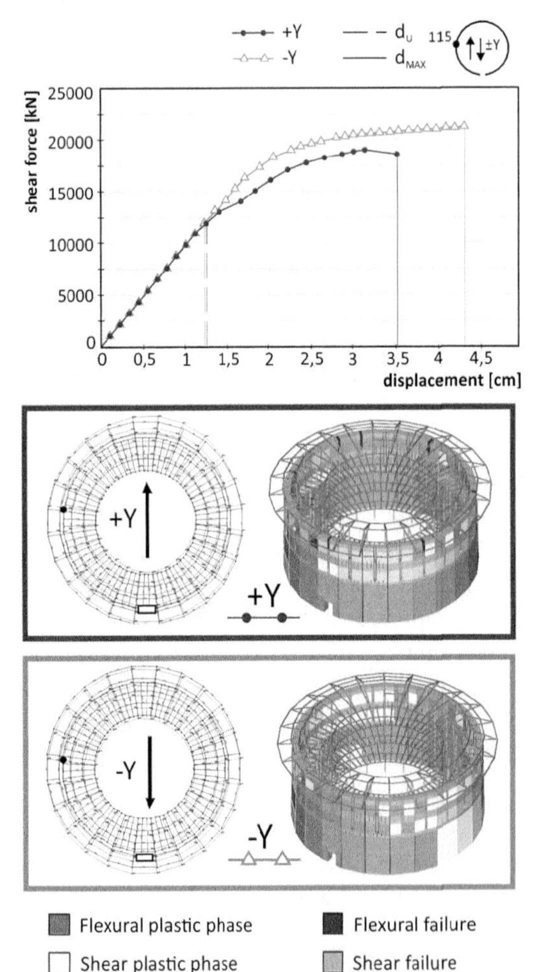

Fig. 4 Nonlinear static analysis of the Tulou numerical model for Y loading direction of the seismic action

The two capacity curves shown in Fig. 4, are obtained for same direction but opposite orientation of the seismic action (Y). The ultimate displacement corresponds to the Tulou displacement capacity, where failure is attained for a loss of shear load bearing capacity at the base of at least 20% of the maximum shear resistance recorded during the pushover analysis. The displacement demand obtained from the ADRS response spectrum is also indicated. Since the displacement demand results far lower than the capacity displacement, then the Tulou structure is shown to well resist to the Fujian seismic action without losing its stability.

The Tulou seismic response is better described in the damage sequence of Fig. 5 where, for increasing displacement values of the control node, the increasing damage in the Tulou wall is mapped as described in the following:

-The 1st point (Fig. 5b) corresponds to flexural yielding in the wall over the Tulou entrance and parallel to the loading direction. Yielding of some spandrels also occurs in the upper level. In Tremuri, yielding of a macroelement is shown to occur when it reaches its flexural capacity but with still a residual ductility reserve before failure.

-At the 2nd point (Fig. 5c), shear yielding of the other macroelements above the opening occurs. Moreover, both the lateral piers of the Tulou entrance yields in flexural-compression. Also in this case, failure is not yet attained because both in flexure and in shear some residual ductility is still available.

-At the 3rd point (Fig. 5d) many piers almost parallel to the loading direction yield in shear at the first level, but still without any reduction of the overall loading capacity in shear of the Tulou structure. The yielded piers are those close to the Tulou entrance together with the corresponding ones at the Tulou opposite side.

-Finally, at the 4th point (Fig. 5e), shear failure occurs in the already yielded piers of the first level close to the Tulou entrance, as well as in the corresponding piers at the Tulou opposite side. Also other piers parallel to the loading direction but located at higher level fail in shear. Failure of all these piers cause a sudden drop, higher than 20%, of the shear overall load-bearing capacity of the Tulou structure, meaning that the capacity displacement is attained.

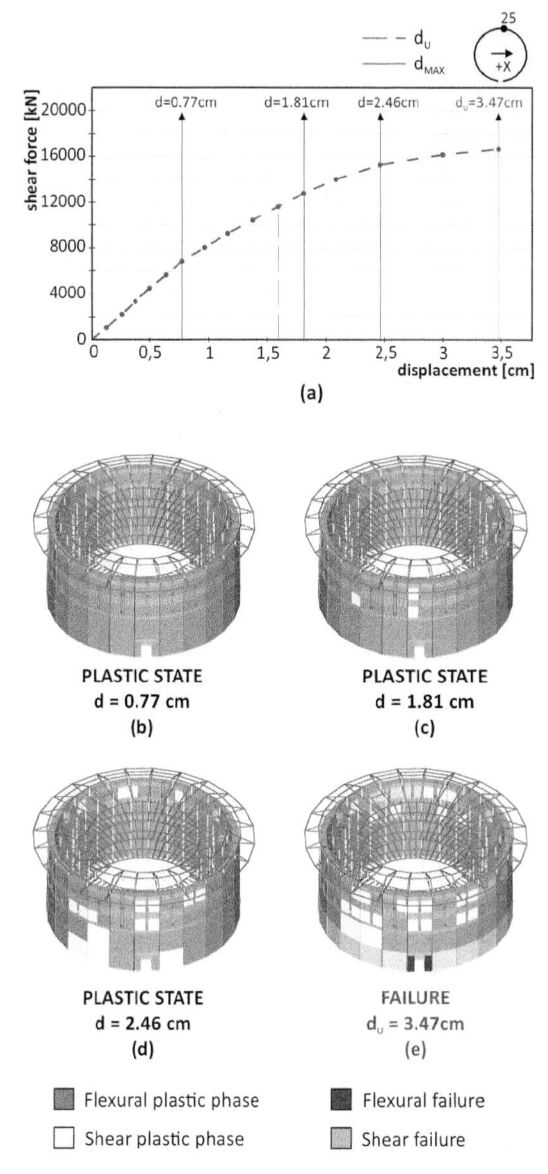

Fig. 5 Capacity curve (a) and sequence damage at different displacement levels: 0.77cm (b); 1.81cm (c); 2.46cm (d); 3.47cm (e)

Finally, sensitivity analysis of the elastic modulus has been carried out (Fig. 6). In fact there are some uncertainties on the actual value of the earth elastic modulus. It was then considered as a parameter varying between 150 and 1000 MPa, thus allowing to obtain a parametric representation of the capacity curves, that is shown in Fig. 6 for control node 25 and +X direction of the loading action. Fig. 6 shows that the capacity curves are only slightly affected by the elastic modulus value when it ranges between 650 and 1000 MPa. Only very low values of the elastic modulus (close to 150 MPa) affect the shape of the capacity curve, with an elasto-rigid response of the Tulou structure.

Therefore, an elastic modulus of 1000 MPa has been assumed in the pushover analyses herein reported.

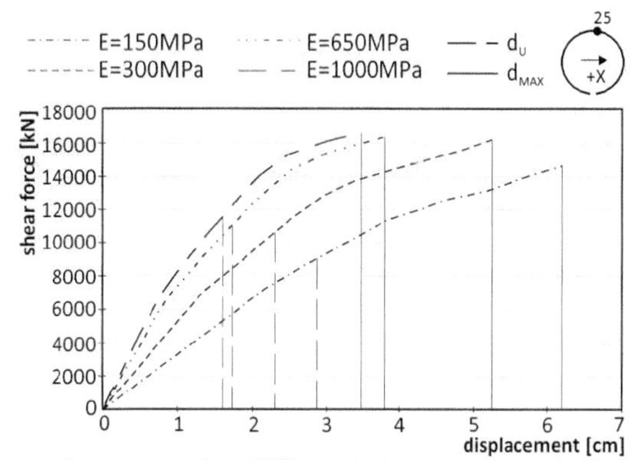

Fig. 6 Sensitivityanalyses to the elastic modulus

4 Conclusion

Few studies are available in the literature on the structural behaviour of Fujian Tulous, massive earth house-fortresses of the Fujian Province (China).

The study presented in this paper represents the first investigation on the seismic response of Tulous using nonlinear static analysis and is one of the first studies on their structural behaviour.

The Tulou seismic response has been investigated by macroelements through the EFM, that has proved to be very efficient in modelling the structural behaviour of complex masonry constructions.

The model of a Tulou prototype has been implemented in TREMURI code using the data available in the literature. Even if the EFM is typically applied to masonry structures made of plane walls, in this study its use has been extended to curved walls. For this aim, the Tulou circular wall has been approximated with plane walls extruded from a 24-sided polygon.

Linear elements have been used to model the wooden structure, with flexible wooden floors and roof pitches modelled with diaphragms of appropriate stiffness.

From the analysis by macro-elements carried out on the prototype of the Fujian Tulou, it has been proved that the equivalent frame approach can well simulate the in-plane response of the Tulou and lead to reliable results.

Pushover analysis has allowed showing that the Tulous have good earthquake resistance compared to the maximum Fujian seismic action. This favorable response is mostly due to the circular form of the earth wall, that avoids out-of-plane local mechanisms and channels the horizontal forces in in-plane internal forces.

Acknowledgements

The research was supported by the Recruitment Program of Global Experts Foundation (Grant No. TM2012-27) and National Natural Science Foundation of China (Grant No. 51778148) and by the Sardinian Region funding LR N. 7 07/08/2007 Year 2011 Tender 3 CRP-48693 and Year 2013 CRP-78176. The authors would also like to acknowledge the Sustainable and Innovative Bridge Engineering Research Center (SIBERC) of the College of Civil Engineering, Fuzhou University (Fuzhou, China), and the Department of Civil Engineering, Environmental Engineering and Architecture of the University of Cagliari (Cagliari, Italy).

References

ANTUNES P, LIMA H, VARUM H, et al., 2012. Optical fiber sensors for static and dynamic health monitoring of civil engineering infrastructures: Abode wall case study. Measurement, 45(7): 1695-1705.

ASPRONE D, PARISI F, PROTA A, et al., 2016. Adobe in Sardinia. Static and dynamic behaviour of the earthen material and of adobe constructions. Padova: 16th International Brick & Block Masonry Conference.

AUGENTI N, 2004. Il calcolo sismico degli edifici in muratura. Torino: UTET.

AUGENTI N, PARISI F, 2010. Constitutive models for tuff masonry under uniaxial compression. Journal of materials in civil engineering, 22(11): 1102-1111.

AYMERICH F, FENU L, FRANCESCONI L, et al., 2016. Fracture behaviour of a fibre reinforced earthen material under static and impact flexural loading. Construction and building materials, 109: 109-119.

AYMERICH F, FENU L, MELONI P, 2012. Effect of reinforcing wool fibres on fracture and energy absorption properties of an earthen material. Construction and building materials, 27(1): 66-72.

BLONDET M, VARGAS J, 1978. Investigación sobre vivienda rural. Convenio con el Ministerio de Vivienda y Construcción. Lima, Per .

BLONDET M, VARGAS J, TARQUE N, 2008. Observed behaviour of earthen structures during the Pisco (Peru) earthquake of august 15, 2007. Proceedings of the 14th World Conference on Earthquake Engineering. Beijing.

BLONDET M, VARGAS J, VELASQUEZ J, et al., 2006. Experimental study of synthetic mesh reinforcement of historical adobe buildings. Proceedings of structural analysis of historical constructions. New Delhi: 715-722.

BRAGA F, DOLCE M, 1982. A method for the analysis of antiseismic masonry multi-storey buildings. Proceedings of the sixth international brick masonry conference. Roma: 1089-1099.

BRENCICH A, GAMBAROTTA L, LAGOMARSINO S, 1998. A macroelement approach to the three-dimensional seismic analysis of masonry buildings. Proceedings of the 11th European conference. Paris: 6-11.

BRISEGHELLA B, COLASANTI V, FENU L, et al., 2017. Seismic analysis by macroelements of circular earth constructions: the Fujian Tulou//DISS. Dynamic interaction of soil and structure (DISS_17) - 5th international workshop. Roma.

BRISEGHELLA B, COLASANTI V, FENU L, et al., 2019a. Seismic analysis by Macroelements of Fujian Hakka Tulous, Chinese circular earth constructions listed in the UNESCO World Heritage List. International journal of architectural heritage.

BRISEGHELLA B, COLASANTI V, FENU L, et al., 2019b. Nonlinear Static Analysis by Finite Elements of a Fujian Hakka Tulou. IABSE Symposium 2019 Guimarães Towards a Resilient Built Environment - Risk and Asset Management. Guimarães.

FIGUEIREDO A, VARUM H, COSTA A, et al., 2013. Seismic retrofitting solution of an adobe masonry wall. Materials and structures, 46(12): 203-219.

GAVRILOVIC P, SENDOVA V, LJUBOMIR T, et al., 1996. Shaking Table Tests of Adobe Structures. Skopje.

HOUBEN H, GUILLAUD H, 1994. Earth construction: A comprehensive guide. Warwickshire: Intermediate technology publications.

LAGOMARSINO S, PENNA A, GALASCO A, et al., 2013. TREMURI program: An equivalent frame model for the nonlinear seismic analysis of masonry buildings. Engineering Structures, 56: 1787-1799.

LIANG R, HOTA G, LEI Y, LI Y, et al., 2013. Nondestructive evaluation of historic Hakka rammed earth structures. Sustainability (Switzerland), 5(1): 298-315.

LIANG R, STANISLAWSKI D, et al., 2011. Structural responses of Hakka rammed earth buildings under earthquake loads. International workshop on rammed earth materials and sustainable structures. Xiamen, China.

LIU T, ZORDAN T, ZHANG Q, et al., 2015. Equivalent viscous damping of bilinear hysteretic oscillators. Journal of structural engineering, 141 (11): 6015002.

MAGENES G, 2000. A method for pushover analysis in seismic assessment of masonry buildings. 12th world conference on earthquake engineering. Auckland: 1-8.

MAGENES G, FONTANA A D, 1998. Simplified non-linear seismic analysis of masonry buildings// Proceedings of the British Masonry Society. London: British Masonry Society, 8: 190-195.

MARQUES R, LOURENÇO P B, 2014. Unreinforced and confined masonry buildings in seismic regions: Validation of macro-element models and cost analysis. Engineering structures, 64: 52-67.

PARISI F, ASPRONE D, FENU L, et al., 2015. Experimental characterization of Italian composite adobe bricks reinforced with straw fibers. Composite structures, 122: 300-307.

RAITHEL A, AUGENTI N, 1984. La verifica dei pannelli murari. Atti del II Congresso Nazionale ASS. IRC CO "La citt difficile." Ferrara.

TARECO H, GRANGEIA C, VARUM H, et al., 2009. A high resolution GPR experiment to characterize the internal structure of a damaged adobe wall. EAGE first break, 27(8): 79-84.

TARQUE N, 2011. Numerical modelling of the seismic behaviour of adobe buildings. PhD Thesis. Università degli Studi di Pavia.

TOLLES E L, KIMBRO E E, WEBSTER F A, et al., 2000. Seismic stabilization of historic adobe structures. Los Angeles.

VARGAS J, OTTAZZI G, 1981. Investigaciones en adobe. Lima, Peru.

VARUM H, COSTA A, SILVEIRA D, et al., 2007. Structural behaviour assessment and material characterization of traditional adobe constructions. Adobe USA. NNMC and adobe association of the southwest. El Rito, NM.

VARUM H, TARQUE N, SILVEIRA D, et al., 2014. Structural behaviour and retrofitting of adobe masonry buildings// COSTA A, GUEDES J M, VARUM H. Structural rehabilitation of old buildings. New York: Springer berlin heidelberg: 37-75.

WEBSTER F A, TOLLES E L, 2000. Earthquake damage to historic and older adobe buildings during the 1994 Northridge, California Earthquake. Proceedings of 12th World Conference on Earthquake Engineering. Auckland, New Zealand.

ZHANG P C, LUO K, LIAO W B, 2011. Study on the material and the structure of earth building in Fujian. Advanced materials research, 368-373: 3567-3570.

Technical Innovation and Revitalization of Yaodong Cave Dwellings by Application to Reinforced Masonry Construction as Appropriate Strategies

XU Dongming[1], *FAN Chunfei*[2], *GAO Yisheng*[3]

1 National Key Laboratory of Vernacular Heritage Conservation of Chinese State Administration of Cultural Heritage (SACH), Shandong Jianzhu University, Jinan, China

2 Architectural Design and Research Institute of Xi'an University of Architecture and Technology, Xi'an, China

3 School of Architecture and Urban Planning, Shandong Jianzhu University, Jinan, China

Abstract The Yaodong cave dwellings in China as a vernacular heritage of traditional architecture are vivid examples of the local wisdom adapting to specific geographical and climatic conditions. Focusing on the renovation project of Yaodong cave dwellings in Tongchuan area of Shaanxi province, namely: the Snail-valley Country Hotel Complex, the paper examines how a reinforced masonry and brick shell approach was selected as an appropriate strategy in the construction of architectural addition and renovation of the old earthen cave dwellings. In applying arch construction mode, it discusses the collaboration and interaction between the architects, local craftsmen, and other stakeholders. It also explores how modern technology can play a role in the revitalization of traditional vernacular architecture for the purpose that the critical discussion can throw light onto the contemporary development of vernacular architecture. Taken the aforementioned into consideration, such as approach as this may be useful to frame similar practices in the relevant field in the near future.

Keywords Yaodong cave dwellings, technical innovation, reinforced masonry and brick shell, arch construction mode, local craftsmanship

1 Introduction

Traditional vernacular architecture known as Yaodong cave dwellings in Shaanxi Province of China are representations of local wisdom in adapting specific geographical and climatic features of the region. Since the 1990s, during the radical reconstruction of the existing built environment in rural areas of Shaanxi, countless vernacular cave dwellings have been deserted and left to decay (Fig. 1), the towns and villages of the region have become increasingly homogeneous, and lost their distinct characteristics. As such, it has encountered significant challenges in the revitalization of vernacular architecture with the applied modern technology towards a sustainable approach throughout the high-paced urbanization process.

The paper, with focus on the renovation project of Yaodong cave dwellings in Tongchuan area of Shaanxi, namely: the Snail-valley Country Hotel Complex, discusses how a reinforced masonry and brick shell approach was selected as an appropriate strategy in the construction of additional architectural structures and the renovation of the old earthen cave houses. It also explores the role of modern technology that may play in the revitalization of traditional vernacular architecture.

2 Technological Background

Since the reinforced masonry and brick shell approach was chosen to be the technological solution to the renovation project of the old earthen cave dwellings, it is necessary first briefly provide some background information of this construction mode, its development, and dissemination in China.

Fig. 1 Abandoned Cave Dwellings and their Setting in Snail-valley, Tongchuan, China © FAN Chunfei

British-French engineer Marc Isambard Brunel is generally believed to be the inventor of the modern reinforced brick masonry. Brunel developed the modern reinforced masonry structure based on traditional masonry technology in 1813, and this structural innovation was adopted in 1825 in the construction of the Thames Tunnel which connected the both sides of the Thames in London (Peters, 1996; Ochsendorf, 2004; Taly, 2010). However, the French engineer Paul Cottancin is the one who applied for the patent for the reinforced masonry and concrete system (ciment armé) in 1890. As Frampton mentioned, seventeen years before Fran ois Hennebique's application for reinforced concrete patents in 1907, Paul Cottancin completed the patented invention of ciment armé. The name was later used to distinguish his invention from the b ton arm by Hennebique. In less than a decade, the Hennebique's system began to be widely adopted for construction, but the intensive labor system of reinforced masonry by Cottancin was seldom used in modern construction sites after 1914 in Europe (Frampton, 1995).

The reinforced masonry system was rarely adopted in Europe and North America, however, due to its low-cost and labor-intensive advantages was widely used at the urban and rural construction sites in Latin America, North Africa and Asia since the first half of the twentieth century. Labor-intensiveness is considered an advantage in nations of which general labor is readily available and affordable. This of course includes China. As a result, its townscapes, as shown in the photos (Fig. 2), mainly consist of low-quality masonry work with reinforced concrete columns, which presented a bizarre image similar to "International Style" in the high-paced urban development of modernization and globalization.

Fig. 2 Reinforced brick masonry urban structures in the developing countries: 1-11-14 Villa, Buenos Aires, Argentina (upper left); Shanty town, La Paz, Bolivia (lower left); Urban Villages, Xi'an, China (right). © PEI Zhao and XU D M.

The combination of reinforced masonry and brick shell technology was firstly introduced to China after the Chinese Communist Party government carried out the strategic measures of introducing advanced technology from the former Soviet Union in the 1950s. The double-curved brick arch and other brick shell structures adopted in China since the 1950s was a pioneering work initiated by the Soviet engineers, who, in considering the limitation of the low-tech conditions of the war-time footing during the 1940s, developed a material-saving and low-cost construction technic based on the study of the traditional domes and vaults in central Asia. Within one year between 1959 and 1960, a large number of buildings with a combined area of 1.4 million square meters were built in China using such technique (Zhu and Zhu, 2017).

The reinforced masonry and brick shell technology were adopted as a selected construction strategy fitting the local condition in the design and construction of the Snail-valley Country Hotel project in Wangyi District, Tongchuan City, Shaanxi Province. The paper will further explore how modern technology plays its role in the renovation project of vernacular earthen architecture.

3 Reflection, Comparison, and Discussion

3.1 Project Description

Located in Wangyi District, about ten minutes' drive from the historical town of Tongchuan City, the Snail Valley is a relatively small valley with a typical landform of the Loess Plateau in Shaanxi Province, featuring many disserted earthen cave dwellings of the old vernacular village (Fig. 1). The commission of the country house hotel complex includes two parts, the renovation of disserted vernacular earthen cave dwellings as hotel rooms and the architectural addition connecting to the old caves as a tourist center, which cover a total floor area of 1,315 square meters (Fig. 3).

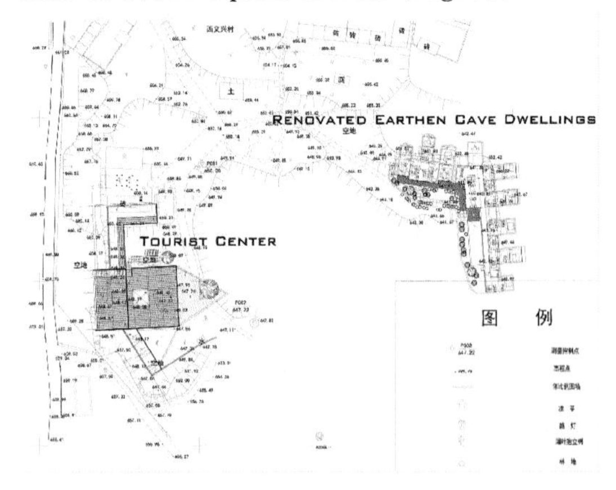

Fig. 3 Master plan of Snail-valley country hotel complex © FAN Chunfei

3.2 Technological Strategies for Earthen Cave Renovation

The vernacular cave dwellings have some advantageous living features, it is cool in summer while warm in winter. But it has obvious problems too. Most of the traditional Yaodong dwellings in Shaanxi Province are built on mountain areas. They often lack modern plumbing, have poor lighting and ventilation, along with other intrinsic limitations of the structure due to the defect of the traditional techniques of the vernacular architecture. In response to these drawbacks, the architects have worked out corresponding innovative technological solutions to achieve a satisfactory result.

Adding a brick facade to the old vernacular earthen cave houses is a common way of consolidation if the local family concerned can afford it. As part of technical solutions, a brick layer of reinforced masonry was also constructed as the facades (Fig. 4) in the renovation of old earthen cave dwellings, which can meet the requirements of the building codes today for such as earthquake resistance. Poor lighting is another challenging issue in traditional cave dwellings. With a straight arch between earthen walls, there is only one surface that light can come through, which results in a dark and depressive atmosphere. In the remaking of old Yaodong dwellings to hotel rooms, the architect used the gap between the earthen cliffs to introduce more natural light into the cave dwelling: the sunlight streams into the room through wooden louvers. This creates a completely different interior living experience from that of the old caves of the past (Fig. 4). The fresh-air-and-heat-recovery system was designed by architects in the renovation to meet the indoor air quality and thermal comfort requirements. A small decentralized wastewater system considering the use of bioenergy was installed to solve effectively the plumbing issue and the water shortage inside the mountain.

3.3 Technological Strategies for Architectural Addition

The groin vaults, which are adopted as standard construction units at the Tourist

Fig. 4　Exterior and interior views of renovated cave dwellings © FAN Chunfei

Center, create an open and light-filled public space in connection with the outdoor surroundings (Fig. 5). The basic construction method of making the groin vaults is to make two layers of the supporting molds: a wooden molding framework is first installed, on top of which a single-layer brick vault is then built. Next, using the brick vaults as the second layer of the molds, another layer of reinforcement grids with poured concrete is laid on the top which makes a thin reinforced concrete shell buckle on the brick shell; then soil is used to cover on top of the concrete shell for the purpose of leveling, and waterproof, this is also the general construction method for roofing. The reinforcement bars inside the shell are connected to the reinforced concrete columns in the four corners to ensure the integrity and stability of the building (Fig. 5).

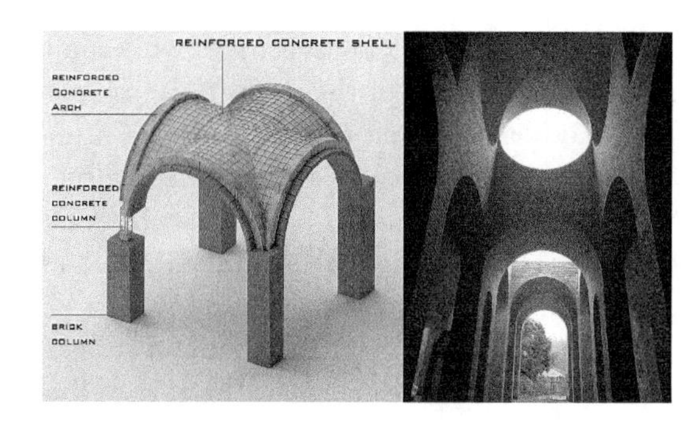

Fig. 5　A standard construction unit of groin vault (left) and completed groin vault (right) © HAN Jing and FAN Chunfei

3.4　Collaboration Between Architects and Local Craftsmen

The adaptation of groin vaults in the tourist center project is intended as a money saving measure on both materials and labor. Most of the worker of the project are local farmers, who performed limited masonry work during the slack season, which means their masonry skills are amateurish and far from professional. Therefore, the architects choose a construction method that fits laborers' skill levels. It was invented by Uruguayan architect and structural engineer Eladio Dieste (1917—2000). With most of Dieste's projects, the adjustable molding system has almost the same importance as the architectural design itself. Most of his projects were inexpensive agricultural silos and factory buildings. There was a lack of skilled construction workers in the immediate area where the structure is to be erected. To remedy the difficulty, Dieste led a small group of skilled workers to guide local farmers in the building process. Consequently, his building design always used many standardized structural units, which can be manufactured repeatedly using the same molding system. After the satisfactory completion of the first unit, laborers then can follow its example to successfully make subsequent units with relative ease. This has significantly improved the efficiency and quality of his projects (Pedreschi, 2000; Pei and Fan, 2016).

The timber mold system also played a key role in the construction of the tourist center. There was a carpentry workshop since the beginning of the project. Its role is to make furniture and to assist various interior remodeling tasks. Carpenters contributed immensely in realizing the architect's blueprint, especially in making the construction mold of groin vault (Fig. 6). When the chief architect Fan Chunfei reflected on the project during an interview, he emphasized the importance of carpenters. According to Fan, he designed the molding framework, however, carpenters improved upon his design. Fan initially wanted to use a steel molding system, but steel molds cannot be hoisted due to their weight and the spatial limitations. Furthermore, disassembly after construction presented another layer of difficulty. Timber molding framework for groin vaults, on the other hand, was light-weight and easy to install during the construction (Fig. 6).

Fig. 6 Timber Molding Framework for Groin Vault Construction © FAN Chunfei

4 Conclusion

As globalization progresses, the training of architects and engineers, the building material production, and the construction methods are becoming increasingly uniform all over the world. Rolf Ramcke once noted that the scene of modern construction sites nowadays is dominated by "montage", assembly and prefabricated building components. In this interpretation, material is dependent on supply, but one interchangeable aspect of planning. As a result, "The formative and aesthetic power of the natural resistance inherent in a material is simply eliminated through substitution." (Ramcke, 2001). In this context, it makes sense to discuss how to apply modern technology to the conservation and restoration of traditional vernacular architecture, and to incorporate the traditional wisdom into modern structures.

In recent decades, the Chinese Central Government has strongly promoted the application of prefabricated building system in urban areas. As a nation with a huge population, vast territory, and vast regional development gaps, China should be more critical of the European models and models of other developed regions. It is necessary to take local conditions into consideration, to develop different construction forms and building materials, and thus making a diverse construction eco-system. As architects, we should discuss and explore: how to restore the traditional role of architects, how to re-establish the connection between design and construction sites, and how to create a system with innovation and practice. Taken the aforementioned into consideration, the renovation project of Yaodong cave dwellings in Snail Valley in Tongchuan is a perfect example of what should be focused on, hopefully it provides a useful framework of understanding to similar practices in the future.

Acknowledgements

We are indebted to Professor Nancy Lea Eik-Nes at Department of Language and Literature of Norwegian University of Science and Technology (NTNU) for taking on the labor of proofreading and commenting of this paper.

References

FRAMPTON K, 1995. Studies in tectonic culture. Cambridge. Mass: MIT Press.

OCHSENDORF J A, 2004. Eladio Dieste as structural artist// STANFORD A. Eladio Dieste: Innovation in structural art. New York: Princeton architectural press.

PEDRESCHI R, 2000. The engineer's contribution to contemporary architecture: Eladio Dieste. London: Thomas Telford Publishing.

PEI Zhao, FAN Jianjiang, 2016. Appropriate strategies of construction tools and techniques. New architecture, (2) 23-26.

PETERS T F, 1996. Building the nineteenth century. Cambridge, Mass. : MIT Press.

RAMCKE R, 2001. Partl Masonry in architecture//GÜNTER PFEIFOR G, RAMCKE R, ACHTZIGER J, et al.. Masonry construction manual. Basel: Birkhäuser - Publishers for Architecture.

TALY N, 2010. Design of reinforced masonry structures. New York: McGraw Hill Professional Publishing.

ZHU Xiaoming, ZHU Donghai, 2017. The transfer of soviet building codes in the early days of New China: A case study of the original mechanical and electrical building in the campus of Tongji University. Heritage architecture, 5: 94-105.

Subtheme 4:

Adaptive reuse and revitalisation towards local development

The Local Community Involvement in the Adaptive Reuse of Vernacular Settlements in Oman

Naima Benkari

Sultan Qaboos University, Department of civil and Architectural Engineering, Muscat-Oman, Algeria

Abstract This paper discusses the first experiences of local community's involvement in the adaptive reuse of its architectural heritage in Oman. It presents three local initiatives made by Omani families in vernacular settlements in Oman. In addition to the architectural documentation and ethnographic investigation, bibliographic research helped to identify the historical and social profile of the settlements and their architectural characters. On-site visits and interviews were undertaken to complete the information about the settlement's history and better understand how the built heritage is being reused and perceived. The research revealed an unsuspected richness of the investigated settlements especially from the historic and social point of view. The paper also highlights the blossoming of a local community's awareness of the value of Oman's heritage and some attempts to reuse it. This investigation initiates a new field of research about built heritage and vernacular tourism in Oman.

Keywords heritage tourism, adaptive reuse, vernacular settlements, community involvement in heritage management, revitalization for local development

1 Introduction

The interest in the study of the Omani built heritage is not recent, but the published research since the early 1970s, was very limited in number and sporadic in the topics and territories it investigated (Damluji, 1998; Al-Salmi et al., 2008; Al-Zubair, 2013; D'errico, 1983; Costa, 1983, 2001; Biancifiori, 1989; Bonnenfant, 1977; Wilkinson, 1977). Among this literature, one can find but a limited number of analytical and interpretative research and numerous aspects of this built heritage are yet to be investigated. Moreover, there was no or little interest in examining the potential of integrating this built heritage in the sustainable fabric and economy of Omani cities. The present paper addresses the Omani built heritage from this perspective. It is grounded in previous research undertaken by the author, on some Omani vernacular settlements (Al-Harthy, 2002; Bandyopadhyay, 2004; Benkari, 2017, 2014). This is the first discussion of ongoing research about the three most recent experiences of community involvement in heritage management in Oman. It will describe each case study before addressing the common characters between the three of them. The paper will underline some concerns and threats on these initiatives and propose a mechanism for an operational integration of the vernacular built heritage of Oman in its sustainable development (Fig. 1).

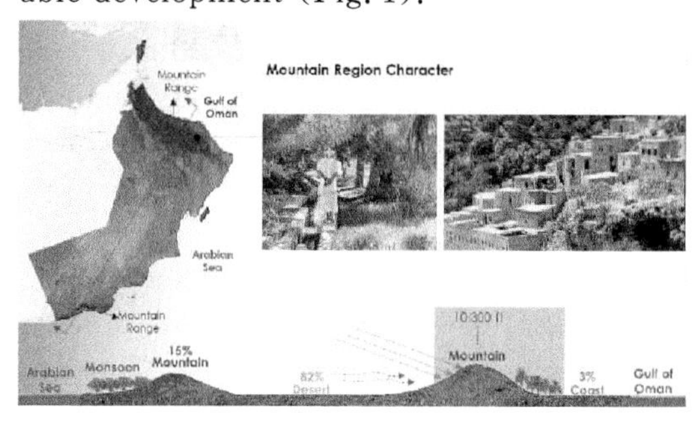

Fig. 1 Geographic layout of Oman with a highlight on the mountains region

1.1 The Geo-climatic Conditions in Oman

The territory of Oman is composed of four main geographical regions: the coastal region in the East, the mountainous chains, the desert, which is part of the Empty

Quarter, and the monsoon region which is a seasonal equatorial region in the country. The vernacular architecture in Oman varies following the geographical and climatic character of each region.

1.2 The Built Heritage in Oman: Character and Management

Oman has a diversified heritage some of which is classified by UNESCO as a heritage of humanity. The Ministry of Heritage and Culture reported more than 1 000 vernacular settlements, 550 forts of different periods and more than 5 000 *aflaj*[①]. The country is also rich in terms of traditional residential architecture religious buildings. This heritage is attracting an increasing number of tourists who come to the country to visit these places in spite of the low level of touristic infrastructure, compared to the neighboring countries. The research published in 2018 (Benkari, 2018) revealed that Omanis and residents of Oman are aware of this heritage importance for the economy and national identity of the Omani's. Nevertheless, rare are the studies that focused on this heritage in spite of its richness. This built heritage suffers also from weak management systems with its responsibility being scattered between several authorities. The main bodies managing the heritage in Oman are the Ministry of Heritage and Culture, Ministry of Tourism, Ministry of Awqaf and Religious Affairs, Ministry of Palace Office and the Ministry of Regional Municipalities and Environment and a lower involvement of the different municipalities as executive authorities. The decision making in terms of managing and protecting the heritage is of course top-down process where the decision is centralized in Muscat, and then transmitted to the different ministries who relay it to the local governments, then to the community representatives, and then to the local community who is supposed to accept and execute the received decision. This has been the situation since the emergence of the nation-state in 1970 where the political system has more or less changed from traditional system of governance, where the power of tribes and local communities was much more important, to a much more organized, yet centralized governing system where the local community has almost been excluded from the management of the Built Heritage.

1.3 The Vernacular Settlements in a Modernized Oman

Before the 1970s, the Omani society had a traditional organization that is strong enough to manage the local communities for the minor decisions to the major ones, such as going for a war, digging a new *falaj*, initiating a new settlement or erecting a fortification. Many initiatives were undertaken by the local community and then approved by whoever was ruling, either Sultan or Imam. The late 1960s were marked by the discovery, then production of Oil in Oman. The ascension to the throne of the sultanate by the actual Sultan Qaboos Bin Said, in 1970, triggered an unprecedented modernization in major aspects of Omanis' life: transportation, built environment, education, health and welfare. The economy witnessed a rapid transformation from an economy relying on fishing or commerce, for the coastal areas, and on agriculture and arts and crafts for the regions that are in the mountains, to an economy that is exclusively based on the exportation of oil and its by-products (Just like the other GCC countries). All the amenities, ministries and services for the population were concentrated in Muscat, the capital city. This attracted all the population in the age of working to the capital, which resulted in the progressive desertification of the traditional settlements. The majority is now completely deserted, and therefore in decay, and a number of them have already disappeared.

① The *falaj* (pl. *aflaj*) is the traditional water channels that brings water either from the water sources in the mountains or from underneath the ground and to very precisely calculated directions, they bring the water to the surface. And there is a network of a falaj that has been

Only a few still survive. Another problem faced by these vernacular settlements consists in the replacement of their original Omani inhabitants by foreign low-income workers brought mainly from Bangladesh and India to take care of the palm trees plantations, the Falaj and the local decaying commerce. These workers who inhabit the old settlements and mud houses of the Omanis have a completely different culture, which does not adapt to the way the buildings should be taken care of. Thus, they became a factor that accelerates the decay of this heritage either by looting or by trying to modify the layout of the spaces or y adding new equipment such as electrical, HVAC or plumbing systems. And this accelerates the decay of the buildings. So this happens in spite of the fact that there is a very strong emotional attachment of the Omanis to their old houses and old settlements. And there is a much known weekly exodus or seasonal move from the capital and the main cities to their homelands, for people to go back to their homelands. But still, even the people who stayed in the original homeland would not live in the old houses, traditional houses, they would build the new ones out of concrete and with more rooms, wider windows and Air-conditioned, across the road from the old settlements. The old settlement would be left deserted.

2 The Vernacular Tourism in Oman

The vernacular tourism in Oman started around 2006 as a reaction to the first economic crisis that hit the country and the region. And also it became more active after the oil price collapse in 2016, where the need to diversify the national economy was expressed by both the government and the local community. From the government side, new laws have been promulgated to strengthen the national private sector, and also to facilitate the local initiatives. And from the side of the local communities, they started looking for ways to take advantage of tourism in their country to have other resources. Large companies were created by the government with shared funds from private and public sectors, like the company Omran and Asaas who are today the main companies involved in real estate development and in heritage or in old settlement renovation and management. The paper discusses three cases studies which are Misfat Al Abryeen, Al-Hamra and Al'Aqor. All of them located in the Governorate of A-Dakhiliya, not far from the famous city of Nizwa, one of the historic capitals of the country. The three settlements occupy different locations in terms of altitude, but are more or less similar socio-economically. Al'Aqor is more at the level of the city. Al-Hamra is a little bit higher in the slope of the mountain. And Misfat Al Abryeen is nested even higher in the folds of the mountain (Fig. 2).

Fig. 2 Misfat Al-Abryeen, Wilayat Al-Hamra, A-Dakhiliyah Region

The agricultural activity is still being practiced in this settlement by its own inhabitants. It is the most populated and therefore preserved among the three case studies. Al Aqor and Al Hamra have some of their houses empty and in decay, or inhabited by foreign workers and in the process to be transformed. Others houses have been rebuilt in cement, a large part of them are still occupied by their original owners or their family members.

The three settlements have not been dated accurately. Based on assumption about the age of the remains of towers and the citadel around Misfat Al Abryeen, this settlement has been dated back to the Persian Era (pre-Islamic period). The settlement developed, mainly based on agriculture, because of the existence of water which was brought to its Falaj. The houses are made of rocks from the mountain and mud brick. The roofs are made of palm trunks covered with palm leaves and a thick layer of mud. These houses could go up to 3 floors in terms of height. The Ministry of heritage restored some of the houses, but for some reasons they are closed now. This is a recurrent result of the dysfunctional system of centralized decision making, where the local community does not adhere to some decisions applied on it in a top-down process. The initiative of Misfat Al Abryeen is actually a family project. A couple of families are involved in it. They all belong to the same tribe who inhabited the Misfat for the last four or five centuries (Al Abryeen). When the patriarchal mansion became empty after the elocation of the last grand-son in a newly built house, the idea of renovating and reusing it as a guest house was proposed by a family member who retired from the tourism sector. After the approval of all the owners, he started the project with some of his nephews from several brothers. The project started officially in 2006, after the enovation and in transformation, Al Misfah guesthouse opened to the tourists 2008. Most of these two years were spent in building administrative legislation by the different authorities for the opening of a guest house in a vernacular settlement. There were no regulations nor procedures for such initiatives. Since its inauguration, Misfat Al Abryeen guest house started attracting more and more tourists. The numbers showed in Fig. 3 were collected from Booking. Com and Asia Travel websites and reflect the number of tourists since the house started to be listed on these travel sites.

These values reflect the number of bookings through the websites only. Based on our observation and interviews with the owners, many tourists book directly at the guest house. Those numbers are not included in the graphs. The visitors of Al Misfah guest house come mainly from France and Germany (in Europe) and UAE (in the Gulf

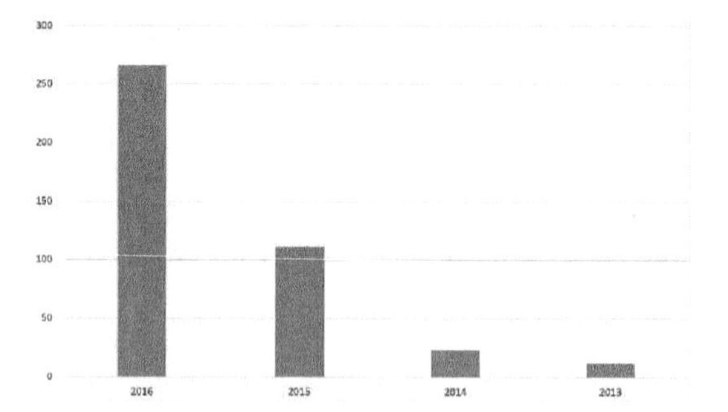

Fig. 3 Misfat Al-Abryeen Guest house, Number of visitors per year

region). Based on the comments that the tourists leave on Booking. Com or Asia Travel, it seems that the house traditional architecture and location in a typical Omani village, are what they enjoyed the most about their experience in the guest house. This has been well understood by the project owners who started developing more activities and more emphasis on the traditional way of living by initiating the tourists to the life of Al Misfat. The furniture and the layout of the rooms are kept as simple and traditional as possible to cultivate this "exotic environment" so much attractive to the tourists.

Misfat Al Abryeen was one of the first initiatives of this kind. The second one started in the old village of Al-Hamra, at some kilometers below the site of AL Misfah. Al Hamra was built and inhabited by a fraction of Al Abryeen tribe. It is a very famous village by the beautiful architecture of its houses, and the Falaj that runs along the settlement's streets before penetrating the agricultural lands to irrigate them (Fig. 4).

Fig. 4 Harat Al-Hamra, Wilayat Al-Hamra, A-Dakhiliyah Region

Al Hamra presents a very organic organization and architecture that is completely adapted to the environment. Based on the date found in the official archives of the

Falaj, the settlement has been dated to the 17th century. The Falaj has been built in 1656 under the Imamate of Sultan Bin Seif Bin Malik Al Ya'arubi. The local community decided to build this Falaj to irrigate their lands. The community started to settle and develop its settlement progressively. Some families from Al Abryeen tribe came and settled in Al-Hamra, and they built these beautiful houses, some of which can go to 12 or 15 meters high. After the 1970s, the families started to leave their old houses in AL Hamra to settle in new concrete and air-conditioned ones. The mud brick walls started to collapse revealing their beautifully painted roofs and finely carved wooden doors to the few curious tourists, who try to explore the surroundings of Nizwa Fort. This was the situation in Al Hamra until one family decided to renovate their house "Bait A-Safa" and make of it a museum that depicts the traditional life in the settlement. It was Bait A-Safa who kept the tourists coming to the settlement, which encouraged its remaining inhabitants not to give up on Al-Hamra. In 2016, a local initiative to preserve the settlement and protect its buildings started to take shape. In 2017 it was officially launched under the name of "The initiative of Al-Hamra". This initiative has more global approach to the built heritage through a holistic approach to preserve the settlement by raising awareness of the local population, including the foreign workers. Not only they renovated some houses to make of them a small Heritage hotel, but they organize several social events, competitions to initiate the youngsters about the history of this settlement, its houses, its mosque and its madrasa and the scholars who lived there. Celebrations are organized regularly in the settlement in order to bring their population back. These events are also planned to raise the awareness of foreign workers who live in the old houses of Al-Hamra about the importance to respect this architecture. Bait A-Safa has been restored for the second time through this initiative, which also opened the heritage hotel in the settlement. This hotel is composed of authentic mud houses restored with local materials, following the local know-how, by a local contractor who is a descendant of a family of master builders.

The third initiative is Al-Aqor which is very close to the famous Fort of Nizwa. This city is one of the capitals of Oman (Fig. 5). It was an important center of power during Al Ya'ariba period (16th — mid 18th century). Harat Al-Aqor is a very important settlement in this city that is as old as the 16th century. Again, based on the date we have found in the Mihrab of one its very old mosques, located in the area of Al-Aqor. It could be older than that. Based on the consulted literature, the mosque was built around the 13th century (Baldissira 1994; Kervran 1996). In Al'Aqor also the same scenario took place. In spite of its social, political and historical importance, this settlement has been progressively deserted since the late 1970s and most of its houses were either rebuilt in cement of left to decay until very recently.

Fig. 5 Harat Al-'Aqor, Wilayat Nizwa, A-Dakhiliyah Region

In 2017, the local community took the initiative to restore some of the houses to transform them into a heritage hotel or a heritage inn. The hotel, inaugurated in 2018, counts 22 rooms. The company that was formed by the owners of these houses, did not only restore the houses to make the hotel, but it also cleaned and prepared parking for the hotel and the community in the vicinity and installed a sewage system to serve the whole area. The interviews undertaken with the representative of the community revealed that the hotel started working very well already

3 Discussion of the Case Studies

3.1 The Positive Aspects in These Case Studies

The common characteristics between all these initiatives are that they are, all more or less recent and aim to revive the settlements and provide an authentic experience to visitors. An experience which reflects the genuine attachment of the communities to their heritage and their wish to preserve, revive and give it its dignity back. Their aim is also to share it with visitors and bring in some income to the settlement and the population there. The other common character in these initiatives consists in the focal role of the local community in the management of its built environment. Decisions are taken at the level of the community and then shared with their representative, who share it with the local government. It is worth mentioning that, since 2016, there is a very strong cooperation between the representatives of the government and the local communities to encourage local initiatives in this field. There is this sort of "vertical cooperation" between the central power and the local communities in terms of vernacular heritage management. "Horizontal cooperation" is also in action between the different initiatives. Our research revealed that the community of Misfat Al Abryeen shared their pioneering experience with the people of Al-Hamra and those of Al-Aqor. These groups meet regularly to discuss their experiences, share their concerns, initiatives and future projects. This very interesting "horizontal cooperation" is not new to the Omani society. Historically, Omanis had very close ties between its tribes and communities. They discuss all important matters in their settlements/regions, and share their experiences and even goods. This highly communicative society developed a specific space, called "Sabla", dedicated to this activity in their settlements. The "Sabla", symbol of the "democracy" and hospitality of the Omni traditional society is an essential component in the vernacular settlements. Nowadays, with the proliferation of social media, communication became even stronger among Omanis.

Another common aspect between the cases presented in this paper, is that the restoration of the buildings is mainly made by local efforts, local funds and local materials as much as possible. This makes the renovation costs very low, which encourages more people to renovate their old houses. But the side effect of this procedure, lies in the fact that it is generally done by non-qualified builders and contractors. This can result in some loss of the local architectural typology and serious structural problems. The positive aspects of these initiatives are, of course, its immediate effect on the physical regeneration of the settlement, and strengthening the relationship between the different social actors in these areas. Furthermore, such initiatives generate an increase of interest and awareness of the Omanis, the residents and the tourists about this heritage; the revival of some lost craftsmanship in the field of construction or cultural activities such as those proposed in heritage museums; energizing some economic sectors in or around the settlements through food and beverages shops; and even increase the demand on architects and engineers in the field of restoration, local materials production or traditional techniques of construction.

3.2 The Concerns About These Initiatives and Proposed Solutions

These initiatives are facing several problems and threats. The first one consists in the lack of regulations to control any project of renovation in the vernacular settlements in Oman. The municipality is in charge of delivering the building permits for new construction. Until today, there is no law in Oman that regulates permission and control of the restoration of private houses.

The municipalities proceed on a case by case basis. The second problem is the absence of master plans that guide the local initiatives in terms of land use or building regulations in traditional settlements (limits of height, material selection, architectural typologies, streets width, openings, etc.). Initiatives such as those presented in this paper will multiply, especially after the initial success recorded. There is a serious need for regulations and guidelines to ensure their success and even their funding, which is another serious issue. Another important problem is the lack of involvement of architects, structural engineers and qualified technicians and builders, who are experts in the restoration of vernacular architecture. This situation generates safety risks during and after the structural restorations, as well as a loss of spatial qualities in these houses.

To help these initiatives become more operational, practical and therefore more profitable, there is a need to strengthen the cooperation between the local and national government, the local communities, the investors with the involvement of academia (Fig. 6). These four poles should be involved together to properly manage this heritage and give it a pragmatic and positive dynamic. This cooperation would work through raising the population's awareness about the importance of this heritage and the importance of going back to the vernacular settlements in order to regenerate them and diversify the economy of the regions. It would also help in ensuring the technical support for the restorations and providing the guidelines and master plans for these settlements regeneration and reuse. These actions should be planned ahead in order to be proactive and not reactive. An organic collaborative system must be developed with all these actors who contribute together for the regeneration of these settlements and maintaining/restoring their dignity for the best of the local community and the visitors alike.

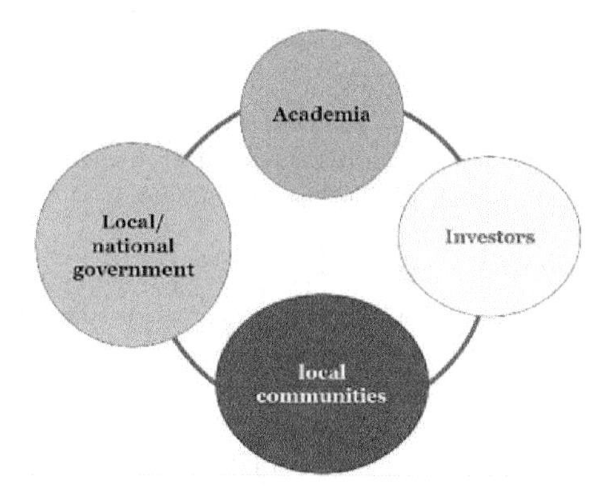

Fig. 6 The four main actors to be involved in the vernacular heritage renovation and reuse

4 Conclusion

This paper presented three case studies of local initiatives in vernacular heritage restoration and ruse in Oman. It discussed the mechanisms of the built heritage management in this country and addressed its disfunctioning areas. It also highlighted the importance of the involvement of the local community in Oman in the management of its heritage and underlined the common characters of the three case studies. Finally, the paper revealed some of the important threats and concerns facing such initiatives in Oman and proposed a strategy to overcome these hindrances and risks.

References

AL-HARTHY S H, 2002. Reading the traditional built environment in Oman. //SALAMA M A, O'REILLY W, NOSCHIS K. Architectural education today. Lausanne: Comportements: 125-129,

AL-SALMI A, et al., 2008, Islamic art of Oman. muscat: mazon printing, publishing & advertising.

AL-ZUBAIR M, 2013. Oman's architectural journey. Muscat: BAZ publishing: 214-233.

BALDISSIRA E, 1994. Sultanat d'Oman, Ministy of National Heritage and culture. Muscat: 153.

BANDYOPADHYAY S, 2004. Harat al bilad (Manah). Tribal pattern, settlement structure and architecture. Journal of omani studies(13): 183-262.

BENKARI N, 2018, Archaeological site of Bat-Oman, management and public perception: Community involvement in archaeological heritage management and planning. Journal of cultural heritage management and sustainable development, 8(3): 293-308.

BENKARI N, 2017. The defensive vernacular settlements in Oman, A contextual study. International journal of heritage architecture, 1 (2): 175-184

BENKARI N, 2014. Documentation and heritage management plan, consultancy report submitted to SQU and the ministry of heritage and culture.

BIANCIFORI M A, 1989. Biancifiori work of architectural restoration in Oman.

BONNENFANT P, et al. , 1977. Architecture and social history at Mudayrib. Journal of omani studies, 3:107-135.

COSTA P M, 1983. Notes on settlement patterns in traditional Oman. Journal of Omani studies. 6 (2):247-268.

COASTA P, 2001. Historic mosques and shrines of Oman. Oxford: Archaeopress.

DAMLUJI, SALMA, SAMAR, 1998. The architecture of Oman. GB: Ithaca Press.

D'Errico E, 1983. Introduction to Omani military architecture of the 16th, 17th and 18th centuries. Journal of Omani studies, 6(2): 291-306.

KERVRAN M, Bernard V, 1996. Un curieux exemple de conservatisme de l'art du stuc iranien des époques Seldjouqides et mongole. in Archéologie Islamique, 6 : 109-156.

WILKINSON J C, 1977. Water and tribal settlements in south-east arabia: A study of the aflaj of Oman. Oxford: Cla rendon Press.

Strategies and Methods on Conservation and Use of Vernacular Temple — Taking the Conservation Planning of the Dragon God Temple and Shrine in Mount Kunyu as an Example

SHI Xiao[1], *XU Dongming*[2], *GAO Yisheng*[3]

1 Shandong Provincial Conservation Engineering Institute of Vernacular Heritage, Jinan, China
2 National Key Laboratory of Vernacular Heritage Conservation of Chinese State Administration of Cultural Heritage (SACH), Shandong Jianzhu University, Jinan, China
3 School of Architecture and Urban Planning, Shandong Jianzhu University, Jinan, China

Abstract The main subject of the paper is the Dragon God Temple and Shrine in Mount Kunyu, a provincially listed vernacular heritage since 2013 located in Yantai City, Shandong Province, China. As an important example of the vernacular temples of Chinese folk religion and culture of the Dragon King in the northeast Shandong, it has been connected with the folk customs of praying for rain in the ancient time, and also an important place as the market for local products exchange. Based on the field work and archive studies of the Dragon God Temple and Shrine in Mount Kunyu, the paper sums up the features and values of the listed historical buildings, the natural and built environments, explores the vernacular lifestyles of the local community, and concludes the strategies and methods of the conservation planning for the preservation and use of the vernacular temple and the place. A critical analysis will throw light on the vernacular heritage conservation and then hopefully be useful for similar cases in the future.

Keywords vernacular temple, folk culture, conservation and use, strategies and methods

1 Introduction

The Dragon God Temple and Shrine situates in the northwest of Mount Kunyu, north of Mount Cangshan, and within the Nine-dragon Pool scenic area, featuring three scenic spots of Dragon God Shrine, Dragon King Temple, and Nine-Dragon Pool. As an important example of the vernacular temples of Chinese folk religion and culture of the Dragon King in the northeast Shandong, it has been listed as the provincial vernacular heritage since 2013. Between March 2018 and May 2019, entrusted by the owner of the scenic area, the authors prepared the conservation planning for this listed vernacular heritage. This paper takes the Dragon God Temple and Shrine as an example to explore and discuss the strategies and methods of developing a conservation plan for vernacular heritage.

2 Evolution of Dragon God Worship, and Emergence of Dragon God Temple and Shrine

Historical records indicate that the Dragon God Shrine was built in the late of the Ming Dynasty (1368—1644) and was also known as Nine-dragon Pool Temple for the sake of making offerings to the Dragon King for rain. Facing the north, it takes the shape of the traditional quadrangle courtyard of north China with one entrance and 22 meters long from the east to the west and 33 meters wide from the north to the south. Currently, only the main hall and the west and east side halls are historical buildings and the east and the west wing rooms are restored according to the original shapes. The historical site of the Dragon King Temple is located on the Longji Ridge to the right south of Dragon God Shrine. Originally built during the reign of Emperor Xizong (1138—1140) of the Jin Dynasty, Dragon King Temple was a sacred place for people in ancient times to pray for rain. On every June 13th in ancient time, people from all the surrounding towns and villages came to the temple to worship the Dragon King and pray for rain. Dragon King Temple served as the important place to present offerings and pray for rain before Dragon God Temple and Shrine came into being. However, the original building was destroyed and demolished during the period of the Cultural Revolution (1966—1976) with only the original site left there. A stone shrine called "Dragon King Pavilion" was built on the base of the original site in 1997. Nine-dragon Pool was in the west of Mount Cangshan as one branch of Mount Kunyu and to southeast of Dragon King Temple, consisting of nine natural stone pools from the top to the bottom of the mountain (Fig. 1). The pool water is clear and sweet and flowing perennially, forming the Nine-dragon waterfalls. Nine-dragon Pool is an important environmental carrier contributing to the formation of Dragon God Shrine and Dragon King Temple and accompanied by many folk myths and legends (SPCEIVH, 2019).

There are three types of carriers depicting traditional Chinese folks' religion regarding Dragon King. First, the folklore about Dragon King. Second, the historical inscriptions recording the presence of Dragon King and his answering to rain prayers; the memorial inscriptions on temple construction and repair process; the warning and regulatory inscriptions created by the folks; the merits monuments praising the good deeds of people. Third, the folk song is sung in the rituals of praying for rainfall (Yuan, 2002).

The emergence of the religion on Dragon King in the area of the Dragon God Temple and Shrine is related to the first type of carrier. According to *Wendeng Local History* (published during the reign of emperor Guangxu (1871—1908) of the Qing Dynasty), "there are nine pools on the cliffs of Mount Cangshan, which are naturally formed with unpredictable depth. Scholars believe that dragons live in the pools and

Fig. 1 Bird's eye view of Nine-dragon Pool © SPCEIVH

name it Nine-dragon Pool." (WLHO, 2010). There are many myths and legends about the Nine-dragon Pool and the most widely spread one is the story about how Dragon King and his eight brothers dug nine pools on the cliffs of Mount Cang before ascending to heaven (Liu, 2010). Because of this, the Nine-dragon Pool became the land praying for rainfall in drought seasons. As recorded in Ninghai County History in the Ming Dynasty (1368 — 1644), "the Nine-dragon Pool is about 17 kilometers to the southeast of Ninghai county and north of Mount Cangshan. From the top to the bottom of the cliffs are nine pools with perennial water out of them running to the ocean. The god temple and shrine there are for praying for rainfalls in the drought seasons." (Li & Jiao, 1533). The temple and shrine in the historical record should refer to the current ruins of the Dragon King Temple, indicating people at that time gave the Nine-dragon Pool the special implication related to rainfall and built Dragon King Temple as the carrier to pray for rain. Historical records after the Ming Dynasty show no reference to Dragon King Temple and its image becomes blurred. Dragon King Temple was gradually replaced by the Dragon God Shrine at the foot of north mountain due to uneasy accessibility of Dragon King Temple or more activities that people have created for rainfall praying.

The most widely believed legend on the Dragon God Shrine is about how three brothers of Dragon King decided to receive offerings at the Dragon God Temple and Shrine because they found and liked the picturesque scenery of the Nine-dragon Pool. Unlike the Dragon King Temple, local history records show that the Dragon God Shrine was administered by the local Taoist priests who kept the financial record for the temple (WLHO, 2010). In addition, activities worshiping and making offerings to Dragon King gradually developed into the Nine-dragon Pool Temple Fair held on every June 13th of the Chinese lunar calendar. The temple fair saw great incense offerings made by people from all towns and counties, stage performance of local operas, and all hustling and lively activities from the top to the foot of the mount. The Nine-dragon Pool Temple Fair became the longest, largest, and most famous temple fair of Mount Kunyu, making the Dragon God Temple and Shrine area the center for local material and cultural exchange in late Qing Dynasty. The elders' recalls show that the sacrificing and worshiping activities toward the Dragon King still existed around the time of the founding of the People's Republic of China in 1949. For example, on the eve of the Chinese Spring Festival, local people worshiped the Dragon King and made incense offerings in the Dragon God Temple and Shrine on the first day of the Chinese New Year (Liu, 2010).

3 Destruction and Utilization in Recent Years

After the founding of the People's Republic of China, and in particular during the movement of "Eliminating the Four Olds" (old ideologies, old cultures, old customs, and old habits) in the Cultural Revolution period, the performance stage (Fig. 2) and the Dragon King Temple on the mountain were demolished and the construction components of the temple were littered randomly. The south hall of the Dragon God Shrine was demolished and destroyed, and the gods' sculptures were removed and disappeared. The Temple was used as the cotton processing factory. Even the ancient trees around the Dragon God Temple and Shrine were cut down and used as the economic sources at the beginning of the establishment of the Mount Kunyu Forest Farm and only five are left now (SPCEIVH, 2019).

Fig. 2 The Dragon God Shrine (right) and the opera stage (left) in the Republic of China (1912—1949)© Muping District Museum

Mount Kunyu has been listed as the National Nature Reserve in recent years. Nine-dragon Pool and the Dragon God Temple and Shrine have been developed into a scenic area thanks to the beautiful natural sceneries of Mount Kunyu and the unique waterfalls of Nine-dragon Pool. In recent years (1997 — 2000), cultural attractions such as Dragon Beard Bridge, Overlook Pavilion, Dragon Pool, Observation Deck, and Nine-dragon Pavilion have been established. Natural scenic resources such as Gourd Peak, Strip Sky, South Pole God, Mount Cangshan Summit, and Mount Xiangshan have been developed. Stone Shrine of Dragon King Pavilion was built at the site of the demolished Dragon King Temple in 1997 (SPCEIVH, 2019). Sculpture of Dragon God was set up in the main hall of the Dragon God Shrine for worship and offerings.

The scenic area of Nine-dragon Pool is administered on a daily basis by Mount Kunyu Forest Farm. In 1996, the administrative department of the Forest Farm reopened the Nine-dragon Pool Temple Fair on each June 13th of the lunar calendar, hosting activities of folk performance, vaudeville, Lv Opera, Yangko Dance, market fair, and food market, etc. (Liu, 2010).

4 Existing Problems and Coping Strategies and Methods

4.1 Existing Problems

At present, there are three problems concerning the conservation of Dragon God Temple and Shrine. First, the original conservation plan focuses on the listed vernacular buildings and the ruins itself and treats the Dragon God Shrine, the Dragon King Temple, and the Nine-dragon Pool separately, lacking a macro consideration of the entire vernacular environment.

Second, the area where Dragon God Temple and Shrine is located has been developed into the Nine-dragon Pool Scenic Spot and modern buildings for tourism purposes account for heavy proportion with low quality, imposing destructive effect on the entire vernacular environment. For example, the two administrative buildings with stone walls and red tiles standing at the site where the demolished south hall of the Dragon God Shrine originally was, the large area of ground parking lot to the south of the administrative buildings, the ticket booths, kiosks, signage systems, and rest facilities have adverse impacts on the vernacular environment.

As it is stated in *The Valletta Principles for the Safeguarding and Management of Historic Cities, Towns and Urban Areas*, tourism can play a positive role in the development of historic areas, but the excessive exploitation of heritage sites as tourist resources is a danger for the preservation works. "The development of tourism in historic towns should be based on the enhancement of monuments and open spaces; on respect and support for local community identity and its culture and traditional activities; and on the safeguarding of regional and environmental character. Tourism activity must respect and not interfere with the daily life of residents." (ICOMOS, 2011) The area of the Dragon God Temple and Shrine is at risk of excessive use for tourism, which may lead to the loss of authenticity and heritage value.

Tab. 1 Facility construction in the scenic sports

Administrative Buildings	Ticket Booths	Kiosks	Parking Lot
$432m^2$	$25m^2$	$16m^2$	$4\,000m^2$

Third, improper repair causes the vernacular artifacts to continuously deteriorate and disappear. Because the east and the west wing rooms were demolished and rebuilt in 2016, the Dragon God Shrine has only the main hall and the east and the west side rooms left as the historic buildings (Fig. 3), which suffer years of disrepair, exposed power lines, improper decoration, and improper use as kitchen or storage rooms for piles of debris. Moreover, the improper additional construction in the courtyard also impacts negatively the sustainability of the Dragon God Shrine. The architectural components of the Dragon King Temple which were dismantled during the period of the Cultural Revolution have disappeared. The worshiping and offering activities for Dragon King no longer exist. Only a few illustration boards offer the cultural legends and construction history of the Dragon God Shrine and the Dragon King Temple, providing few explanations on the connections of the key elements. There are very few literature on the Dragon God Temple and Shrine and only a few sentences were mentioned about it in the published works.

These are significant changes. "If the nature of these changes is not recognized, it can lead to the displacement of communities and the disappearance of cultural practices, and subsequent loss of identity and character for these abandoned places. It can result in the transformation of historic towns and urban areas into areas with a single function devoted to tourism and leisure ..." (ICOMOS, 2011) With time passing by, the Dragon God Temple and Shrine will gradually fade and become blurred in people's memory.

Fig. 3 Bird's eye view of the Dragon God Shrine (2018) © SPCEIVH

Tab. 2 The damage status of the Dragon God Shrine

Main Hall				
Roofing	Wooden structure	Wall	Windows and Doors	Ground
Most tiles broke and partially collapsed	Cracking and wearing red lacquer	Improper repair, disrupted, and cracked	Complete	paved with modern marble bricks
West Side Hall				
Roofing	Wooden structure	Wall	Windows and Doors	Ground
Roof built with red brick chimney and tiles broken	With ceiling though unknown condition	Disrupted and piled with debris	Complete	paved with modern marble bricks
East Side Hall				
Roofing	Wooden structure	Wall	Windows and Doors	Ground
Broken tiles	With ceiling though unknown condition	Disrupted, holes, and heaps with debris	Complete	paved with modern marble bricks

4.2 Coping Strategies and Methods

The Dragon God Shrine and the Dragon King Temple are the carriers of the local vernacular cultural identity, sense of belongings, historical continuity, spiritual symbolism, and memory. As the local natural attraction, the Nine-dragon Pool has been well known from ancient times and its myths and legends are passed down from generation to generation, reflecting the profound local culture. Therefore, the Nine-dragon Pool is a historical heritage bearing both natural and cultural values.

In addition to conserving the listed vernacular buildings and sites and improving the surrounding environment, it requires efforts to maintain traditional practices because "retention of the traditional cultural and economic diversity of each place is essential, especially when it is characteristic of the place." (ICOMOS, 2011) We think that the activation and utilization of the Dragon God Temple and Shrine should give priority to meet the emotional needs, belief needs, and spiritual sustenance of the local people.

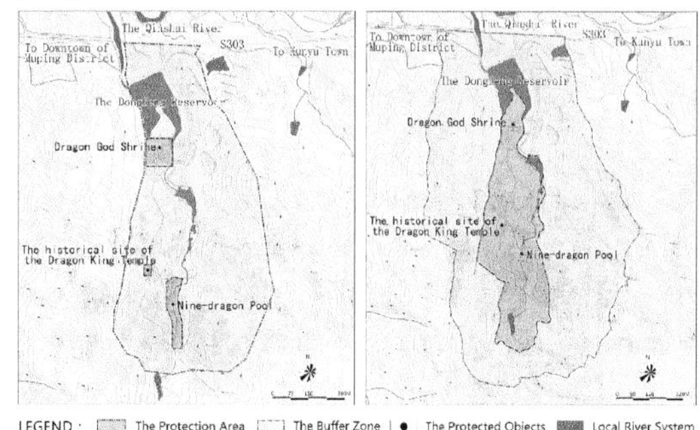

Fig. 4 Comparison of current conservation zones (left) and planned conservation zones (right, 2019)© SPCEIVH

The main coping strategies and methods are divided into the following sections.

First, redesign the conservation planning (Fig. 4) based on the water connections between the Dragon God Shrine, the Dragon King Temple, and the Nine-dragon Pool, and the overall assessment of the natural environment. The conservation scope should include the water system around the three main scenic spots as a whole. The construction control zone should give full consideration of the visual coordination of the surrounding natural, cultural and landscaping environment. It meets the statement on "coherence" article 3 of the Nairobi Recommendation: "Every historic area and its surroundings should be considered in their totality as a coherent whole whose balance and specific nature depend on the fusion of the parts of which it is composed and which include human activities as much as the buildings, the spatial organization and the surroundings. All valid elements, including human activities, however modest, thus have significance in relation to the whole which must not be disregarded". (UNESCO, 1976)

Fig. 5 Comparison of current illustration board and the design for new illustration board. (2019) © SHI Xiao

Landscape control: Eliminate the inharmonious construction elements in the environment, including ticket booths, kiosks, etc., and build tourist centers, parking lots and other tourist services in the north away from the Dragon God Temple and Shrine. Reforming and improving the existing signage system and rest facilities (Fig. 5) to reduce the impact on the vernacular heritage environment.

Conserving the listed vernacular buildings and sites: Carry out conservation and repair of the historic buildings of the Dragon God Shrine and set up daily maintenance plan as soon as possible to retain the vernacular artifacts and value as much as possible. According to Marialuce Stanganelli, "the recovery and improvement of built heritage have often taken place neglecting the

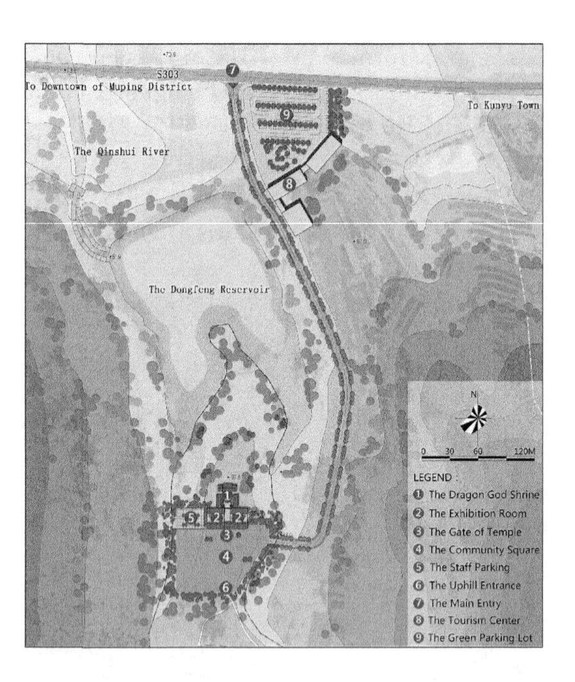

Fig. 6 The master plan of the surrounding area of the Dragon God Shrine (2019) © SPCEIVH

social and cultural safeguarding of the places." As a result, it leads to "a process of global homologation that show the same activities, the same shops and same products all over the world". (Stanganelli, 2012) Considering this, we stressed in the research plan that it is important to search for the intangible cultural information on Dragon King worship, temple fair and so on base on the literature research, site visit, and archaeological fieldwork in addition to the digging up of the tangible information about the Dragon God Shrine, including the architectural information on the south hall, the Dragon King Temple, the old performance stage, and the search for lost building components.

Activation and utilization measures: 1) Create offering venue for the Dragon King and restore its sacrificial function by using the main hall and the east and west wing rooms. 2) Renovate the roofing and interior decorations and facilities of the administrative building in stone walls and red tiles and transform it into a site for the introduction to the history and intangible culture of the Dragon God Temple and Shrine; build additional gate at the south entrance and use stone walls of the administrative building to display the original pattern of the Dragon God Shrine (Fig. 6). 3) Create an event venue. The parking lot to the south of the Dragon God Shrine can be transformed into culture square, hosting Nine-dragon Pool temple fair and sacrificing and offering activities to the Dragon King, and also for daily culture communication.

5 Conclusion

Looking into the vernacular artifacts, the surrounding heritage environment, its connection to the life of modern people, and the continuity of the intangible culture of the Dragon God Temple and Shrine, this paper studies the status quo, the surrounding environment, and the relevant intangible culture heritage of it. Further, the paper reveals the unique historical cultural significance of the Dragon God Temple and Shrine and proposes the coping strategies and methods on how to sustain its historical and cultural value, conserve the carriers of such value, and deal with its relationship with surrounding environment in the hope of shedding some light on researching future similar cases on conservation and sustaining the vernacular heritage.

Acknowledgements

We would like to express our sincere appreciation to Ms CHEN Jianfen for her professional and efficient work as the text proofreader of this paper.

References

ICOMOS, 2011. The Valletta Principles for the safeguarding and management of historic cities, towns and urban areas. Valletta.

LI Guangxian, JIAO Xicheng, 1533. Jiajing Ninghai County history. Ningbo: Tianyi Pavilion Reprinted.

LIU Xuelei, 2010. Oral legends, Mount Kunyu culture research series. Jinan: Qilu Book Society.

STANGANELLI M, 2012. New policies and intervention criteria for the safeguarding and management of historic cities, towns and urban areas// The Valletta Principles for the safeguarding and management of historic cities, towns and urban areas. ICOMOS.

SPCEIVH, 2019. Basic data compilation, the conservation planning of the Dragon God Temple and Shrine in Mount Kunyu.

UNESCO, 1976. Recommendation concerning the safeguarding and contemporary role of historic areas. Warsaw & Nairobi.

Wendeng Local History Office (WLHO), 2010. Guangxu Wendeng Local History. Tianjin: Tianjin Ancient Books Publishing House.

YUAN Li, 2002. Research on the legend of Dragon King in north China. National art, 1003-2568: 112-123.

Revitalization Strategy of Compound in Macao —Chi Lain Wai (Pátio to Espinho)

Lee Mengshun, LI Jiawei

Faculty of Innovation and Design, City University of Macau, China

Abstract Patio (*Wai*) and Alley (*Li*) are present roads of the lowest level in Macao. "Chi Lain Wai" (*Pátio do Espinho*) behind the Cathedral of Saint Paul and along the wall is currently the only residential area of Compound in Macao. It is the village in the city of Macao and the important urban texture in the Historic Centre of Macao. In the patio, there are ancient wall, the largest ancient well in Macao and few old trees which are recognized by Government of Special Administrative Region as the area which should be "preserved and revitalized". In the patio, the easement and land deed are complicated and thus the specific plan has not been established for years. Since strategic planning of preservation area of the Ruins of Saint Paul's and program of Chi Lain Wai establishes overall framework of future development of "Chi Lain Wai" and the direction of urban design of "Chi Lain Wai", this study will first conduct general analysis on the development direction to propose brief strategies of preservation and revitalizing of historic space of "Chi Lain Wai".

Keywords the historic centre of macao, Chi Lain Wai, historical preservation, revitalizing

1 Introduction

Patio (*Wai*) and Alley (*Li*) are the texture of Macao residents' lives and important units of neighborhood. According to the present regulation of the width of roads, it is less than three meters and it means "dead end" in Cantonese. The difference is that "patio" only includes one entrance, whereas "alley" refers to two entrances. However, there are special cases with ambiguity. Unique textures of streets and alleys lead to self-sufficient communities. In present "patios" and "alleys", there are few buildings with historic value. In patios, the featured blue-brick old houses are common Chinese houses of Macao in early times. It reveals the communities of ancient Chinese grassroots society, with specific borders of territories and roads with traffic and life functions. They reveal universal value of "The Historic Centre of Macao" and urban texture characteristics.

"Chi Lain Wai" (*Pátio do Espinho* in Portuguese) is the only Compound remained in Macao and it is located in "buffer area" of "The Historic Centre of Macao" and is designated as important urban texture of "The Historic Centre of Macao". It is the area recognized by the Cultural Affairs Bureau of Government of Special Administrative Region to be preserved and revitalized. In the patio, there are various cultural relics and ruins with historic value. However, Chinese blue-brick old houses are not the features of this patio. From perspective of urban design research, it explores the key points of current problems in "Chi Lain Wai" and develops strategies of "total urban design research of Macao" to increase and complete green space, preserve historic urban landscape and reinforce pedestrian public space of mixed area as main design directions. This study aims to obtain revitalizing strategy of "Chi Lain Wai" to construct the method to protect and develop texture revitalizing of the Compound.

Through analysis of historic images, based on the relationship between historic culture and the area, this study probes into the meanings of historic value and urban space to the area. By overlapping and reconstruction of images, it acquires actual situations and structure records of houses in the location of wall ruins; by interview and collection of literatures, it approaches the residents of Chi Lain Wai to accomplish the research.

From July to September 2018, team members of this study, conducted a field interview. According to the survey, the majority of population was aged 60-89 and there were residents aged 90. Most of the residents' incomes were less than 10 000 MOP/per month or they are native residents who received governmental allowance. It showed a clustering phenomenon of grassroots society.

2 Development and Evolution of Compound

2.1 Habitat of Missionaries and the Japanese Followers in Macao

Saint Paul School in Macao was founded in 1594 and is the first western higher education institution in China (1594 — 1762). It cultivated the missionaries to Japan and China and was the missionary training base of the Society of Jesus after arriving in China. With overlapping between partial Macao map in 1889 and modern land registration map, the solid line area is the scope of this study and the dotted line part is the eastern wall of Saint Paul School

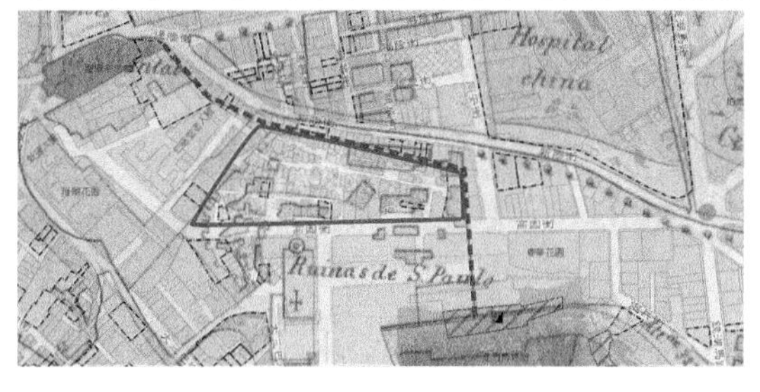

Fig. 1 Location plan of Wai (Source: Library of Congress, Land Registration Information of Government of Special Administrative Region, U. S. A. https://www.loc.gov/item/2002624048)

(Fig. 1). "Chi Lain Wai" is one of few areas constructed within the old wall lived by the Portuguese and it is not a pure Chinese community. It has 400 years of history and is part of Saint Paul School in Macao. It crosses Rua de D. Belchior Carneiro and is divided into the northern and southern sections. The northern section is a slope behind Museum of Sacred Art of the Ruins of Saint Paul's surrounded by Rua de D. Belchior Carneiro, Rua da Entena, Rua de Tomas Vieira and Santo Antonio. It descends from Rua de D. Belchior Carneiro to Rua de Tomas Vieira. The southern section is next to the Ruins of Saint Paul's, between wall ruins and Rua de D. Belchior Carneiro. It was originally Qing Sheng Fang. In 1587, the Japanese government ordered to deport the missionaries. In 1614, 1626 and 1636, the population was increased sharply because of the Japanese exiles and the expelled crowded into. The blocks lived by the Japanese Catholics in Macao gradually became "street of St. Paul's Church" of Compound, the southern area of "Chi Lain Wai".

During the period when the Japanese lived in the patio, in 1617, well-known "fort of St. Paul" and northern wall began to be constructed. In 1623, Saint Paul School of Theology was founded. In 1835, after the Cathedral of Saint Paul was burned up, the land behind the ruins was the extension of "Chi Lain Wai" and the residents gradually moved in. The section of the slope was divided into four levels lived with nearly one hundred households. The houses were mostly zinc-iron houses, log cabins and cottages. In the 1940s, log cabin at the bottom of the fort was burned down and the victims were placed in Toi Shan residence which was just constructed and "Chi Lain Wai". The buildings in the patio thus increased and Rua de D. Belchior Carneiro was constructed at this time. "Chi Lain Wai" was divided into southern and northern parts. In 1963, the patio was on fire and some residents were moved to the place near the old terrace of stud farm. The residents in the patio gradually reduced and the image of the patio thus declined.

2.2 Saint Paul Wall Ruins

Old wall of Saint Paul School of Theology is part of UNESCO World Heritage "The Historic Centre of Macao" and it has been constructed for several hundred years.

According to the record, the wall was initially constructed by the Portuguese in the 16th century. It was torn down by the Ming dynasty with order for several times. In the 1630s of the 17th century, the Portuguese reconstructed it and the present wall is the architectural and historic relic remained after the reconstruction. The wall successively disappeared after the school was burned up in the 19th century. The ruins remained are few. In "Chi Lain Wai", at No. 35 of Rua de D. Belchior Carneiro, there is one section of rammed earth wall which is connected with old wall of the fort to the south and northern wall of Chi Lain Wai. According to initial construction period of the public servants' residence established on the ruins torn down in 2010, literature and historic maps, the rammed earth wall is part of the wall of Saint Paul School, designated as one of "the second group of real estate assessment items of Macao". The wall is consistent with the location of the old wall drawn in old maps of Macao in 1760, 1882 and 1912 and it is recognized as part of the old wall. Scope of real estate for the assessment, Saint Paul School ruins (ruins of the wall, Section 1 of Rua de D. Belchior Carneiro) (Fig. 2).

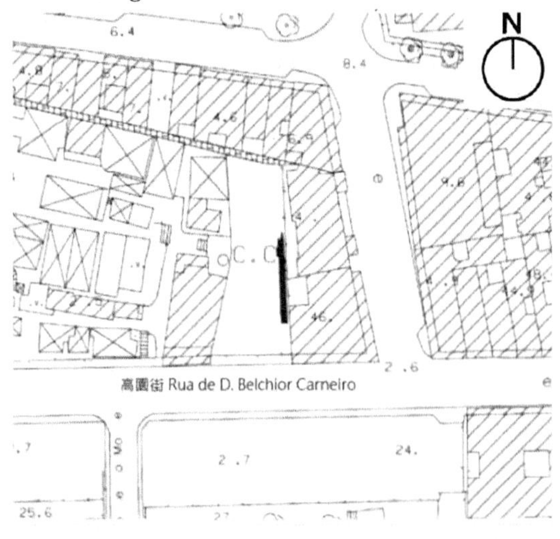

Fig. 2 Location of the old wall discovered (Source of image: the alignment plan of streets and the public consultation text of the second group of real estate assessment of Macao)

2.3 The Residents' Life

The protection of trees shows the humanistic literacy and importance of culture in the city and it is a critical issue of urban development. In the patio, there is a banyan which is designated in "conservation list of ancient and valuable trees". The surrounding of the tree has been the public recreational space for the residents. Another giant tree next to Rua de D. Belchior Carneiro was uprooted during the attack of Typhoon "Hato" in 2017. The original vegetal landscape in the patio was diminishing.

In the patio, there was one ancient well which was dug by the Japanese for drinking and irrigation. It was the most giant well in Macao at the time. Government of Special Administrative Region constructed the sewer in the patio and half of the well was covered by the buildings. Drawing water from the well was possible; however, the water was not drinkable. Drawing water is forbidden nowadays.

"Chi Lain Wai" is not the building of UNESCO World Heritage; however, it is included in the scope of preservation and is significantly associated with UNESCO World Heritage. Next to the patio are wall ruins, Na Tcha Temple and the Ruins of Saint Paul's, which are part of The Historic Centre of Macao. The Ruins of Saint Paul's Museum of Na Tcha Temple beside the wall has been the watch tower of "Chi Lain Wai". Under the glass floor, it shows partial groundwork of the Ruins of Saint Paul's several hundred years ago.

With the development in Macao, agricultural life successively disappears. However, in the patio, the remained collapsed brick houses, idle stone mills, blackened kitchen, old graves, ancient well, old tree, ancient wall and places of traditional industry and small Earth God Shrine next to the entrance of the Compound show the traces of the residents' lives in the past. They reveal historic memory and value of the Compound and is the area which should be protected by the government by laws.

3 Urban Position of Chi Lain Wai

The total area of "Chi Lain Wai" is around 9000 m^2. There are more than two hundred buildings, mostly cabins, stone houses, iron sheet houses or old inclined roof brick houses. There are three entrances, one horizontal main road and three branches. The shape of "Chi Lain Wai" is long and narrow horizontally. The facilities are old and there is a concern of safety in most of the houses and buildings. The interior space is crowded and the living conditions are extremely harsh. The residents' living level is lower than in other areas in Macao. It lacks specific managerial organizations and the government rarely and even does not participate in the affairs in the patio. Present residents are mostly aged and have lived in the place for at least 60 years. They do not have property rights of the houses. The reasons that they live in the Compound are life habit, special community climate, convenience for commuters, lower rental and inheritance. The descendants mostly move to other places due to the inferior living environment in the patio and lack of property rights. Thus, there are more abandoned houses which are ruining.

The ruined Compound depends urgently on reconstruction and protection. "Overall planning of core zone of UNESCO World Heritage (the Ruins of Saint Paul's) (2010)" has specifically stated that "Chi Lain Wai", as part of Saint Paul School in the past, is an important element of the world heritage area of the Ruins of Saint Paul's. In "major managerial measures" in the text of "preservation and management plan of The Historic Centre of Macao (2018)", it indicates the restriction of the construction development in the area. Government of Special Administrative Region the Cultural Affairs Bureau continued the investigation, recorded building distribution and preservation situations in "Chi Lain Wai", collected data of relatively valuable rammed earth wall, outdoor space in the patio, texture, ancient well, Earth God or trees and from perspective of conservation and reuse, it aimed to integrate the Compound with the historic city upon the following proper conditions.

As to control of urban design, since the issue of ownership is complicated and most of the sections and houses are not registered, some were reconstructed as buildings of 4 — 7 floors with codes of e-reports. (Fig. 3) In the patio, there were the problems of illegal buildings which should be solved by cross-functional sectors. Upon current regulation on volume and use of buildings, it recognizes land rights, controlling right, residential situations, type of building, property rights and change of use of the buildings in the patio and the government establishes control design of "Chi Lain Wai". Based on principle of "major managerial measures", the Cultural Affairs Bureau established architectural conditions, such as limitation of use of building materials, height of eaves, compulsory use of Chinese tile sloped roofs and avoidance of block division projects which destroy street characteristics, that all intend to maintain the original landscape of "Chi Lain Wai".

Fig. 3 Land registration information in the patio (source: base map from land registration information network of Government of Special Administrative Region, combined drawing by this study)

Other restrictions, such as preservation of old wall in the change of the neighboring walls, distance of 3m between the buildings constructed, coordination of height of building, volume, color and elevation design between the neighboring sections and the assessed real estate and prohibition of construction of high buildings with Class A or above, reveal the absolute respect to the preservation of the historic city.

At present, the scope of "Chi Lain Wai" project does not include the area to the south of Rua de D. Belchior Carneiro. Present buildings in the patio are mostly illegal. Although the government possesses the land right, the ownership of the houses is uncertain and it is the essential cause of inferior development in the patio. Title deeds do not exist for most of the houses and many owners have disappeared, abandoned the houses for long term and some have migrated. Present residents are mostly the relatives in the neighborhood or the uncertain third parties of tenants. Since 2011, Housing Bureau of Government of Special Administrative Region has launched the dismantling and attempted to inform "owners/users/uncertain third parties" of the houses of the deadline. However, it is ineffective up to the present.

As to the repair of the buildings in the patio, according to "cabin list" of Housing Bureau, there are regulations on repair or preservation of 51 cabins in the patio. In other words, "it cannot reduce and cannot change the form. The stakeholders proposed reasonable application to Housing Bureau and obtained the permission to fulfill the repair." The solution to iron sheet houses, brick houses and stone houses are uncertain. As to reorganization and issue of planning condition maps of streets, it strictly controlled the volume, style and height of buildings to preserve overall historic climate.

4 Revitalization of Chi Lain Wai

Government of Special Administrative Region possesses most of land rights of "Chi Lain Wai". Thus, the Cultural Affairs Bureau of Government of Special Administrative Region led the revitalizing of the Compound. From perspectives of conservation and use, the Cultural Affairs Bureau of Government of Special Administrative Region will integrate "Chi Lain Wai" with the scenic area of the Ruins of Saint Paul's. It probed into original pattern of Saint Paul School to reconstruct the historic climate of the Ruins of Saint Paul's, reestablish the historic context of the area and highlight the important cultural value. The Cultural Affairs Bureau also authorized the related research units to practice spatial revitalizing design of "Chi Lain Wai". According to the revitalizing direction of the government, by field interview and investigation, this study generalizes different situations of the Compound, the residents' different demands, cultural characteristics of the Compound and the residents' ages to first improve the environment of the Compound regarding present residents' structure and propose the revitalizing strategy.

4.1 Constructive Revitalizing of Current Residents

Distribution of preserved, dismantled and reconstructed buildings introduced by this study with colors. (Fig. 4)

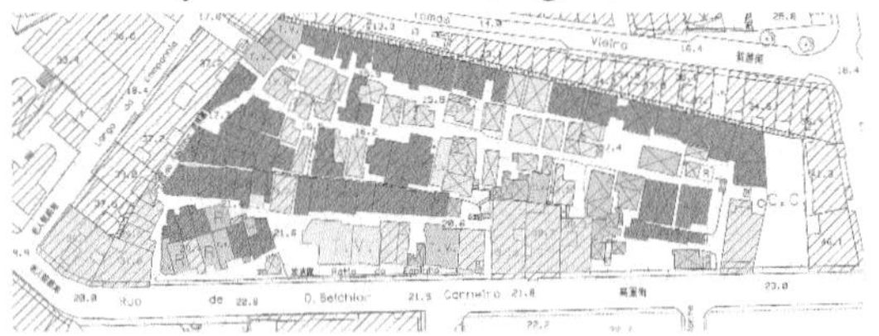

Fig. 4 Dismantling and preservation planning of buildings in the patio (Source: Drawing by the team of this study)

Since the aging of the community is severe and basic facilities are significantly insufficient, it plans to install more public facilities for the elderly, including and not limiting to medical centers of community, the elderly health center, medical passages of emergency, service center of community, etc.

Present buildings in the patio are severely aging. The living environment of the elderly is extremely harsh and residential conditions are inferior. It plans to conduct the reconstruction based on present buildings and the principle to "not to change the spatial scale" to meet the residential demand of the elderly.

In the patio, present passages and transport conditions are inferior. Upon the principle to "not to change connection between the streets in the patio with other passages", it plans to expand the passages, reorganize the circulation of the streets and install more barrier-free passages and facilities for the circulation of the elderly.

There are a great number of potential problems of safety in the patio, such as random parking of e-bikes, trees inclined to the houses and disordered electric wires. It plans to reorganize the line pipes and control the illegal parking to prevent the problems.

4.2 Future Development Strategy of Chi Lain Wai

Relationship between locations of public assets in the patio and the Compound. (Fig. 5)

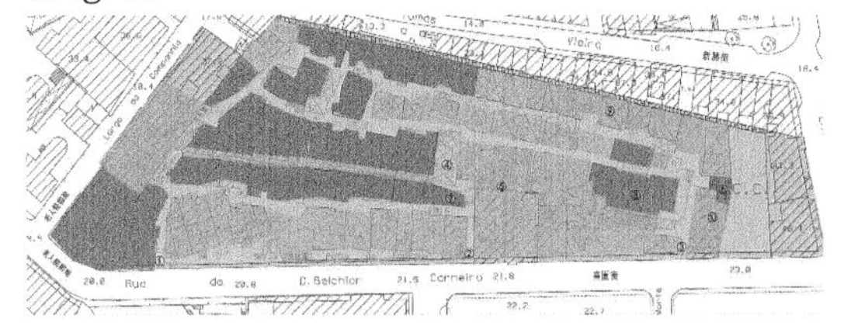

Fig. 5 Overall revitalizing planning of public assets in the patio (Source: Drawing by the team of this study)

It will design brand new community planning project for future residents moving in, dismantle a great number of current illegal buildings and those with improper structures and preserve few properly remained historic buildings. Based on special building style and texture of "Chi Lain Wai" and according to alignment principle of streets and building restrictions in the area, it will reconstruct new community and buildings with more satisfying residential conditions.

Upon the principle to "not to change connection between streets in the patio and other passages", it will reorganize road system in the patio to provide better passages and circulation of streets for the tourists and residents. Upon the principle to "maintain the spatial characteristics of the patio", it plans to install a great number of green lighting systems with complete equipment to upgrade the living environment of the community.

It will practice regional revitalizing to integrate the Compound with the surrounding as the whole cultural circle. It will reuse idle space with complete area in the patio, introduce related industries, enhance the attraction for visitors and create spatial revitalizing opportunities. Without influencing the residents' lives, it will strengthen the constant development of the Compound by commercial activities.

It plans to install environmental and sustainable energy facilities in the Compound. It is based on the original use model and the use of the Compound continues. Government of Special Administrative Region can provide the assistance to support and reinforce the depth and scope of cultural value, maintain the original use model and establish the representative example of reconstruction of sustainable Compound in Macao in the near future.

5 Conclusion

Revitalizing and reasoning of historic and cultural areas in the cities are the critical practices to reconstruct urban territoriality. As the only present Compound in Macao, "Chi Lain Wai" is next to the site with the most visitors in Macao and the Ruins of Saint Paul's. However, it is the community with the most inferior general situations in the region. The whole community intuitively reflects the historic progress of Macao. Besides enormous taxation of gambling and entertainment industries, in Macao, there are communities of "patio" and "alley" which should be immediately protected and concerned and which still show the potential of vitalization. "Chi Lain Wai" is excluded from the development due to complicated land rights. It thus significantly preserves the primitive urban texture of the Compound in Macao and continues the most critical and totally different residents' lives worthy of preservation.

"Preservation and management plan of The Historic Centre of Macao" launches the attention of Government of Special Administrative Region to overall preservation and revitalizing of "Chi Lain Wai". It continues inviting related sectors and research units for strategic research and revitalizing design. Nevertheless, due to some social factors, revitalizing design is stalled and the Compound is still ruined. Revitalizing strategies change with time. Nowadays, it should implement environmental constructive revitalizing based on present native residents and proceed development strategy of the Compound according to feasibility of policies.

References

Division of Transport and Engineering, Government of Special Administrative Region of Macao, 2010. Total urban design research of Macao. Guangdong: China Academy of Urban Planning and Design Shenzhen.

Governo da Região Administrativa Especial de Macau Direcção dos Serviços de Cartografia e Cadastro. https://cadastre. gis. gov. mo/MGSP_Cad/chn/main. html? type=5.

HSING F W, 2013. Old and valuable trees of Macao. Department of gardens and green areas, IACM. Macao. https://images. io. gov. mo/bo/i/2016/40/despce-333-2016. pdf.

Instituto Cultural of Macau, 2018. Preservation and management plan of the historic centre of Macao. Macau, China.

TANG K C, 1998. Urban construction history of Macao in the Ming Dynasty. Cultural magazine of Macao, 35. http://www. icm. gov. mo/rc/viewer/10035/636#"LAB1003500050028".

The Cultural Affairs Bureau of Special Administrative Region Government of Macao, 2018. The second group of real estate assessment in Macao: Public Consultation. Macau. http://www. culturalheritage. mo/Survey/cbim2018/cn/.

WANG W R, CHANG C C, 2010. Revitalization of the Patio: Research on urban texture of historic blocks of Macao. Macau: Instituto Cultural.

WU C T, 2014. Investigation and research on the walls in Macao during the Ming and Qing dynasty. Nanjing: Nanjing University.

Rehabilitation of Market Quarters in the Historic Cities of Shibam and Zabid, Yemen

Tom Leiermann

ICOMOS Germany; UNESCO Doha

Abstract The architectural heritage of Yemen still includes many historic cities or villages arcoss the whole country but face a range of challenges on which the recent years of conflicts only add further risks. The two World Heritage Sites Zabid and Shibam, beside the capital Sana a, are characteristic representatives of the specific vernacular architecture of their region.

Both historic cities were capital cities in the early Middle Ages but since became country towns of modest size. They contain beautiful monuments but mainly are remarkable for their urban pattern as a whole. Shibam is a dense structure completely built of mud tower houses, while Zabid is a wider area filled with courtyards and houses in compounds adapted to tropical conditions, known for abundant decoration patterns.

The market quarter is a central part of the built structure but is also vital for social and economic life in the community. Yet recent development favors new shop premises along the main road that shift activities and investments and threat the future of the historic market quarters. This does not only change the appearance of the historic substance but also endangers preservation in general, because in a weak country like Yemen historic cities can only survive if preservation goes hand in hand with the genuine interest of the inhabitants.

These considerations in mind, in both cities some rehabilitation measures were implemented; the roofing of the main covered market streets in Zabid and shadow roofs and renovation of shop premises in Shibam. In both cities new shops were opened by young locals.

These interventions in time of war can only be of limited effect in conditions that generally are dominated by rough modernization and destruction. Yet as a sign of optimism and engagement, it can be a starting point for more.

Keywords vernacular architecture, social and economic life, market quarters, rehabilitation measures

1 Living Heritage in Yemen

1.1 Two Heritage Sites

Architectural heritage of Yemen still includes many historic cities or villages across the whole country but faces a range of challenges on which recent years of conflicts only added further risks. The two World Heritage Sites Zabid and Shibam are, together with the third one, the capital city of Sana a, famous as representatives of the specific vernacular architecture of their region.

Both Zabid and Shibam were capital cities in the early Middle Age but since became country towns of modest size. They contain beautiful monuments but mainly are remarkable for their urban pattern as a whole. Shibam is a dense structure completely built of mud tower houses of mostly six floors, a unique architecture of an origin, as all indications show, in South Arabian civilization(Leiermann,2009).

In the valley of today's Hadramout, Shibam is a relatively small country city of minor economic or administrative importance. As a major iconic tourist destination, Shibam has encountered a constant tourist flow until current conflicts arose but locals were never able to make considerable benefit of it. The historic town is famous for its rare integrity and good state of conservation; but to keep it this way, economic benefit for locals should improve considerably.

Zabid on the other side lies in the coastal plain called Tihama near the Red Sea in a rather tropical climate that defines in many traditional ways of housing and settle-

Fig. 1 Shibam, the historic mud city

ment structures. This town fills a wider area dotted with small houses around courtyard compounds adapted to tropical conditions, known for abundant decoration.

Today, Zabid is a small market city of about 20,000 inhabitants and the center of a wide but remote and impoverished country hinterland. The tourist potential was low even in peace conditions as the surrounding plain has no touristic reputation and the numerous important historic structures are mostly not open to the public, and additionally, the city lacks an iconic significance.

Following regional tendencies, a major threat to historic Zabid is the growing number of illegal, slum-like dwellings that make any conservation engagement extremely difficult.

Fig. 2 Zabid, historic quarters

1.2 Suq — The Market Quarters

The market quarter is an integral and central part of the historic fabric in both cities. It consists of narrow lanes flanked by selling stalls that may be of minor architectural interest itself. What makes them valuable in terms of preservation is their atmospheric intensity and their essential function for social life and economics and thus for life quality of the old town as a whole.

In Shibam, the suq quarter is as small and narrow as the whole site. All shops are in the form of ground floor premises of historic mud-brick tower houses, a building type that requires ground floors to have thick walls and narrow openings. In consequence, shops are very narrow and of limited attraction for modern use, e.g. as a supermarket.

In Zabid, market quarters include different market lanes around six market mosques, among them Ash ir mosque, reputedly the oldest mosque of the city and situated right at the crossing of the main East-West and North-South lanes that divide the four traditional quarters of the city from each other.

The shop premises are simple but of different size, some of them include a large courtyard. Shadow roofs once covered most of the lanes but only small parts of these roofs survived. Poor tin or palm leave sheds cover most lane sections that give shadow but spread the impression of decay that certainly limits acceptance and potentials.

A picturesque element is there exist a few camel-driven oil-mills, covered by a steep thatch roof.

2 Rehabilitation

2.1 Challenges

The market quarter is not only the heart of the urban structure of the old city, but is also vital for social and economic life in the community. Yet recent development favors new shop premises along the highway that shift activities and investments and threat the future of the historic market quarters.

Developments like these do not only change the appearance of historic structures but also endanger preservation in general, because in a less developed country like Yemen, historic cities can only survive if preservation goes hand in hand with the genuine interest of the inhabitants.

In this sense, the fate of the market quarters is vital for preservation of these historic cities. Their decline is an indication for a general loss of historic substance and cultural identity while successful rehabilitation of the suqs would support local identity and engagement. The local suq as well provides opportunities not only for traders and craftsmen but for unskilled laborers as well and attracts all kinds of costumers, i. e. it enables most or even all inhabitants to participate in activities. Therefore the place where exists benefits and potentials (or their failure) is most apparent.

A further difficulty is the fact most shops are in the property of al-Awqâf, the religious-based public fund. It is a centuries-old institution that finances mosques and other social institutions by market rents. The problem today is the fact neither Awq f nor the shop-renters feel responsible for regular maintenance, as rents and rewards are low. Consequently, most stalls are in poor state.

Fig. 3 Suq lane in Zabid before rehabilitation

2.2 Rehabilitation in Zabid

Since a Yemeni-German cooperation project active in both Shibam and Zabid started, revitalization of the suq quarter was one of the objectives described in several studies. Some implemented measurements included the rebuilding of some ruined stalls. Another program aimed at the pavement of the old city but works unluckily left the suq quarter unpaved. The shadow roofs made by tin sheds or palm leaves just add to the general impression of poverty and decay.

Since 2016, a rehabilitation program financed by German Foreign Office started with the rehabilitation of shadow roofs at the main crossing in the very heart of the city, i. e. the East-West and the North-South lanes next to Asháir mosque. The implementation was coordinated with the local branch of the Historical Cities Authority and affected shopkeepers.

Fig. 4 Suq lane in Zabid after rehabilitation

The building design followed the existing historic sections in Zabid. The roof exceeds the shop stalls in height . The main construction material is burnt brick, the traditional building material in Zabid, though recently widely replaced by industrial cement bricks.

Brick columns support the roof and give place for a row of openings above the shops that give light to the lanes. The roof consists of round beams, layers of brushwood and broken bricks covered by mortar and lime plaster.

Works included restoration of some walls, foundations and facades. The upgrading of the quarter resulted in the opening of more than ten new shops and services mainly by young entrepreneurs.

Further renovation of existing shops and a general rise of public acceptance should add to the attraction of the suq and encourage more engagement and activities in the city center.

2.3 Building Process

Some images of the building process may illustrate traditional techniques that are still practiced if demand allows doing so (Fig. 5～Fig. 8).

2.4 Rehabilitation in Shibam

Shibam has been a major town in the valley of Hadramout over centuries but lost importance towards Seyun and other cities in the region during the 20th century. As a result, major markets moved to other cities even Shibam remained a district capital for historical reasons.

A further setback for market activities in the old town comes from the general replacement of main trade sites to new premises along the highway that offers better access by vehicles (even most locals do not possess one). As a result, most former shops within the old suq closed. This development endangers both the remaining business as well as the value of the old city as a place of social meeting and leisure. It would also endanger the willingness of locals to invest within the old town. The historical mud tower houses of Shibam, on the other side, need constant maintenance in order to protect them from erosion and rainwater

Fig. 5 The ceiling is laid by round beams and fixed by ropes

Fig. 6 Brick wall repairs and erection of brick columns

and otherwise risk stability (Leiermann, 2017).

As the first step for market rehabilitation, shops were equipped with new shadow sheds. Renovation works also include renovation of some shop interiors and repairing of the historical wooden doors. Paving of the area was part of a technical infrastructure project implemented right before the conflict period began. Both together led to the opening of several new shops and the

Fig. 7 Roof layers of weed brick pieces and mud

Fig. 8 Final plaster

Fig. 9 Suq in Shibam after rehabilitation

upgrading of the whole market quarter that should make it more competitive.

A characteristic feature of Shibam are open-air tea shops, that got new shadow roofs and new iron chairs, locally made. Given the fact open space is still largely undefined in the region and public equipment in poor state, such minor details can provide some new accents for public space and indeed inspire for local improvements.

Other elements of refurbishment interventions were lanterns and garbage bins. Lighting objects in the old city were locally developed in a design that by far resembles lamps used in the early 20th century in the region but responds to modern technical need.

Garbage is a general problem in the Middle East, as local behavior does not consider much clean environment as in individual's responsibility. In Shibam, former attempts of introducing bins failed because goats strolling around is a normal feature of this rural city and bins did not prove resistant to them. Consequently, design and production of bins were developed locally to meet the specific conditions. Improvements can only succeed in an extremely conservative and remote region when specifically coordinated with local conditions and developed in close exchange with locals, while some knowledge of international standards may equally be helpful.

Fig. 10 Rehabilitation works at Shibam suq

2.5 Building Materials

Several criteria are decisive for quality and appropriateness of works within a vernacular context. In Yemen, master builders and their young teams still know and practice most vernacular building methods. Yet this will only last like this if modern demand still purchases this kind of work. Therefore, modernization needs and appropriate solutions have to meet together, and architects and experts are needed to assist, at least in the initial period.

One strong component is the building material used dominantly in the vernacular building. In Zabid burnt brick dominates the historic appearance of the city, in frequent interchange with whitewashed plaster surfaces.

In Shibam, lime plaster is of equal importance, especially to protect exposed mud parts from humidity or rain. Here, plaster parts interchange with mud plaster. The rounded, smooth appearance of mud surface together with the sandy color of mud makes up a large part of the strong impression of vernacular architecture of Shibam.

Today, introduction of steel and cement has changed modern building techniques totally, and a sensitive preservation policy should accept their use in modest degree in order to generate general acceptance by locals — in a community-based approach rather than a strict touristic or museum-like attitude. Yet strengthening of traditional materials and revival of its use where it does not contradict modern use is vital as our examples show.

3 Outlook

After more than four years of armed conflict, the future of Yemeni heritage, as its society in general, is more insecure as ever. In recent months however, there is some stabilization of the situation. Negotiated agreements between conflict parties have at least helped to support the development of some normality, and UNESCO started with a wider initiative in the main heritage sites, including Shibam and Zabid.

The crucial question for the future of these historic cities is whether close cooperation of conservationists, authorities and community can deal with the general challenge of neglect and lack of preservation concern that is fact throughout most Middle East countries. Poverty, rough modernizations and weak respect for conservation regulations remain challenges, but successful rehabilitation of structures and ensembles can serve as strong signals if locals regard them as positive. They will do so if economical needs and expectancy are in line with conservation activities and if social benefits are also visible.

In such a case, a historic site like Shibam or Zabid can serve as a push for local development. Still, it is too early to say whether in these two cases a sustainable development can succeed.

One important factor is how much modern life expectations can align or contradict traditional space organization that goes along with the historical substance. Historic houses in Shibam with their lack of furniture and glassless windows still adapt to the way people live there, and both the narrow settlement as well as the mud material support needs in the hot and arid conditions.

In Zabid, in contrast, traditional organization of life mainly in open courtyards shifted to living in closed buildings with air condition to a large extent. Obviously, building activities that allow more density are widespread and endanger the coherence of the city in a growing degree(Leiermann, 2011).

4 Conclusion

Generally, preservation of historic architecture has only a chance if social and economic demands do not contradict to it. This is especially true about vernacular architecture in low-income countries like Yemen. Strong preservation laws, public or international finance or tourist exploitation can be helpful but normally not guaranty alone vernacular heritage can survive.

Therefore, only close cooperation with locals to learn local knowledge and needs, combined with professional ideas with some background of international standards and methods, can lead to appropriate and sustainable solutions. Market quarter rehabilitation of Zabid and Shibam as famous but poor country towns in a period of armed conflict show initiatives can motivate locals and may inspire further initiatives for sustainable local development in favor of preserving their rich and unique vernacular heritage.

References

LEIERMANN T, 2009. Shibam, Leben in Lehmtürmen. Wiesbaden: Reichardt.

LEIERMANN T, 2017. Preserving Shibam: the city of towering mud houses// MARCHAND T. Architectural heritage of Yemen. London: Gingko Library.

LEIERMANN, T, 2011. Das Compoundhaus in Zabid. Jemen-Report, 2(1): 40 - 53.

Analysis on the Conservation and Revival Strategy of Historic District in West Wenmiaoping, Changsha City

ZHI Xiang, *LIU Su*

Hunan University, Changsha, China

Abstract In the context of rapid urbanization, the status quo of historic districts in cities is not optimistic. Existing protection and renovation plans often fail to play an effective role in inheriting and adapting to the new needs of economic development. West Wenmiaoping block is one of the few old city blocks in Changsha city with relatively intact traditional spatial texture. It has a good social neighborhood space style and includes many historical relics. Based on the problems faced by the historic district, this paper analyzes the history, existing buildings and cultural landscape of West Wenmiaoping area, and proposes more humanized strategies. The aim is to achieve sustainable development and block activation under the premise of preserving the "urban atmosphere" of the traditional blocks, and to carry out solutions to promote the adoption of the historical districts to the local social and economic development while remaining as a cultural carrier.

Keywords historic district, urban renewal, conservation, economic development, revival

1 Historical Background of West Wenmiaoping Block

1.1 Historical Overview of Changsha

Changsha is one of the first famous historical and cultural cities in China, with a history of more than 2,400 years. Changsha has been a very important city since the spring and autumn period and the warring states period. According to the data, the site of Changsha at that time was "Taiping Street in the west, Litou Street in the east, Pozi Street in the south and Chunfeng Street in the north". (Wen, 1987). By the time of the Western Han dynasty, the site had been expanded to include "Nanyang Street in the east, Taiping Street in the west, Jiefang West Road in the south and Zhongshan West Road in the north". (Chen, 2016). The urban area was further expanded in the Sui, Tang and five dynasties, and was finally established in the Song dynasty, from the north to the present Xiangchun Road, from the south to Chengnan Road, and from the east to Liucheng Bridge. (Wang, 2006). During the Ming and Qing dynasties, the city still expanded and developed, but the core position of its ancient city remained unchanged. The 2,400-year-old street pattern is like the skeleton of old Changsha. In the old streets that survived the Wenxi fire in 1938, we can also have a look at the street space and its history and culture of the old city of Changsha. West Wenmiaoping block is a typical area.

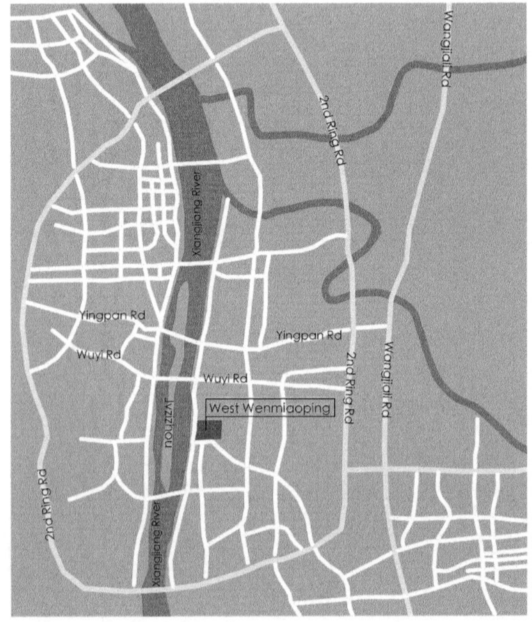

Fig. 1 The location of West Wenmiaoping block

1.2 Overview of West Wenmiaoping District

West Wenmiaoping area is located in the Tianxin district of modern Changsha city. To the south of Taiping block, it starts from Xuegongmen Central Street in the west and ends at Nanqiang Bay in the east, in the shape of "U". The whole block can be divided into Gutan Street and Wenmiaoping area. The name of the West Wenmiaoping district comes from the Changsha Xuegong(An ancient official educational institution), which was once located in this area. According to the Qing dynasty historical materials Shanhuaxianzhi records: "Changsha Xuegong has five courtyards: followed by Lingxingmeng, Dacheng hall, royal pavilion, Chongshenci and Zunjingge, west for Xundaoshu, Minghuanci, Xiangxianci, Shefu, etc., to the east, Jiaoshoushu, Mingluntang, Wenchangge, Quyuan Temple, towering Kuixinglou on the southeast corner, overlooking the wall inside and outside". Xuegong is also called Wenmiao, so this area is called Wenmiaoping block. Gutan street got its name because its location in the Tang dynasty is the gate of Bixiangmen. There used to be an ancient street called "Guloumen" in this block. Some scholars think it may be the site of Bixiangmen. There are many ancient Wells in West Wenmiaoping block, and some of them can still be used as the source of domestic water. Many historic sites, such as memorial archway, ancestral hall, ancient house and ancient street, are concentrated in the block. Different periods of Changsha's history have left deep traces here.

Located in West Wenmiaoping lane, "Daoguan Gujin" memorial arch is the only remaining old street memorial arch in Changsha. Originally an archway at the west entrance of Changsha Xuegong, the memorial arch is about 10 meters high and 6 meters wide. Founded in the Ming dynasty, the memorial arch was rebuilt in the fifth year of the reign of Emperor Tongzhi of the Qing dynasty (1866). The east entrance of the school palace had a stone workshop named "Depei Tiandi", which had exactly the same

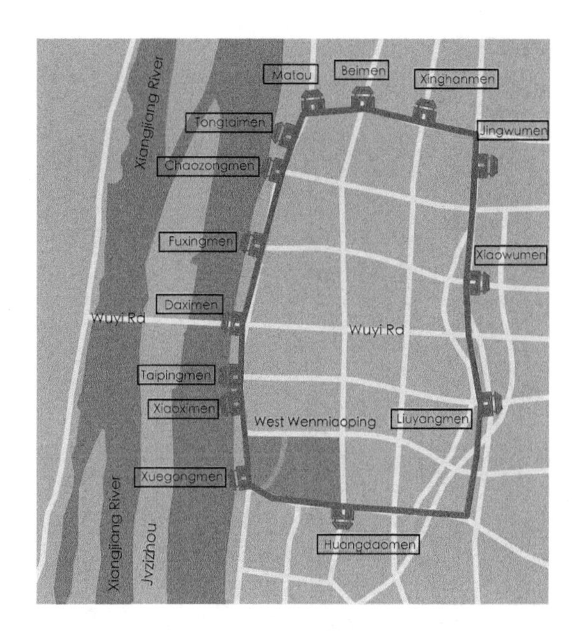

Fig. 2 The relative position of West Wenmiaopng block and Qing dynasty city wall

regulations as the existing stone workshop. Unfortunately, it was demolished as the "four old ones" during the "Cultural Revolution". Due to the "Daoguan Gujin" stone at both ends of the house stuck, unableto be removed, it has survived so far. The ancient stone workshop is divided into three layers, and each layer is carved with different animal and plant patterns, such as lions, Chinese knots, flowers and plants, etc. The most wonderful is the hollowed-out stone embroidered ball, reflecting the superb technology at that time.

The present Changjun Middle School plot was the residence of Changsha government before the Qing dynasty. The Qing Kangxi forty-seven years (1708) the government office moved to build in the east of the Mingfanfu, so the government office of the old site as a teaching administration.

Meigongguan Former Site is located in the north of the intersection of Douguyuan Lane and Gutan Street. The site is now a residential building. Meigongguan is the owner of the rich businessmen in the republic of China Mei Jingfu, because of the sale of foreign goods and made a fortune, so mei mansion in the republic of China on the architectural style of the tone also mixed with a lot of western elements, in the old build-

Fig. 3 Current situation of Daoguangujin Memorial Arch

Fig. 4 Current situation of Meigongguan Former Site

ing of West Wenmiaoping block, Meigongguan Former Site appears fresh and refined, different. (Mei, 2017)

Although Changsha is a famous historical and cultural city, most of the cultural relics on the ground are destroyed in the Wenxi Fire. West Wenmiaoping memorial arch as can prove that the exact location of Changsha Xuegong physical evidence, it is more precious. For Changsha, West Wenmiaoping block is not only a piece of historical material evidence, but also a symbol of cultural heritage. The Xuegong is a symbol of attaching importance to education and culture, and Changsha has seen numerous literati in history. West Wenmiaoping block is also a witness of the educational history of Changsha. The secular atmosphere that accompanies the growth of the block in the long history adds a human warmth to West Wenmiaoping block, where the most authentic "Changsha Blood" flows.

2 The Historical Blocks in the Vortex of the City's Rapid Development

2. 1 The Important Value of the Historic District

The city is like a huge jigsaw puzzle, in which the blocks play a connecting role. "Streets and lanes are the skeleton and support of urban form" (Liang, 2001). The blocks formed by streets and surrounding architectural Spaces are like the body of a city, streets and alleys are bones and blood vessels, and architectural Spaces are muscles and tissues. Block space has various forms of existence for the city. When block space is the main external space form of the city, it serves as a platform to provide residents with places for economic and political activities as well as social and cultural life. When it becomes the symbol element of the city image, the block is the main means to guide the city style and express the city image. Historical blocks are often located in the central area of a city. Their rich economic formats, colorful social and cultural life functions, as well as their unique historical and cultural flavor are mainly manifested in the ambiguity and diversity of the functions of historical blocks. (Mei,2017)

The particularity and importance of block space lie in that it is an indispensable constituent structure and basic unit for the operation of a city. For the history and memory of a city, it is also the space that is most likely to leave time marks and public plots. The historical block can be regarded as a lens reflecting the mass culture of the city. Through the analysis of its constituent elements, the basic development context and characteristics of the local economy, society and culture can be roughly obtained. Today, with the rapid modernization of cities, historical blocks are rare places that can embody the particularity and unique charm of cities, among which the memories and context of cities are precious treasures.

"If a city's streets look interesting, the city will look interesting," Jacobs said in his book *The Death and Life of Great American Cities*. If a city's streets look drab, the city will look drab" (Jacobs, 1961). We need to do our best to protect and pass on this "fun". In addition, the content contained in the historic block is combined with the benign activation of the current urban development to achieve the ultimate goal of sustainable development.

2.2 General Situation of West Wenmiaoping Block

According to the author's field investigation, the current situation of West Wenmiaoping block is not optimistic. The streets are narrow and badly damaged, and most of the buildings are damaged and incomplete. All kinds of electrical wiring layout disorderly lead to the existence of fire hazards, historical building protection status is not satisfactory. Many people come from inside and outside the block, the night public security environment is poor, giving birth to the illegal industry. During that renovation in the 1990s, many buildings during the republic of China period were simply and destructively "repaired", resulting in the loss of cultural value of many historical buildings. But the West Wenmiaoping neighborhood maintains a delicate balance in this situation. It is highly inclusive and contains a dilapidated yet charming "urban atmosphere".

Fig. 5 Run-down buildings and streets within blocks

3 Humanized Protection and Activation Strategy

2.1 "Control" of Aborigines to the Area in West Wenmiaoping Block

In the historic district, the precious is not only the material remains but also the spirit remains. The neighborhood network and lifestyle of local residents make this area unique, and simple building repair and street reconstruction cannot well inherit these immaterial relics. Not only for West Wenmiaoping block, but also for the revitalization projects of other historic blocks, we should think about how to keep the "urban atmosphere". In today's China, the renovation projects of many historical blocks can be said to be the complete destruction of the original spiritual remains, leaving only an empty shell of material. The biggest problem is the "cultural fragmentation" caused by this "protective destruction".

In the renovation and revival design, the external restoration of the block is actually easier to achieve, but due to the nonentity and dependence of the historical context of the block, it is easy to be destroyed due to the change of the original residents' living space and economic business form. Therefore, in the protection and revival of West Wenmiaoping block, I think the following strategies are needed to focus on the inheritance of "civic culture" :

(1) Local residents are the main force of future block use and maintenance. The planning and protection of the block by government organizations and professionals can only be limited to the material and technical level in a short period of time, while the sustainable construction, protection and transformation of the block by local people will continue along with the development of the block. Therefore, at the design level, local people need to be encouraged to actively participate in the design and practice of conservation and rehabilitation programs. Taking the needs of aborigines as the starting point, it can not only obtain the original historical information of the block, but also mobilize the initiative of aborigines for the renovation and activation project. The strategies proposed from the perspective of indigenous people are also more adaptive and persuasive, making local residents more identify with protection and activation strategies and more active in developing regional formats and transforming block environment as masters.

(2) From the perspective of activated objects, we should not blindly introduce foreign elements for the sake of human flow and economic aggregate. The participation of the aborigines is indispensable to the traditional cuisine, customs, interpersonal relations and "street life" in the West Wenmiaoping historic district. When there are too many foreign elements that are not well integrated with or even mutually exclusive to the elements in the original small environment, the existing humanistic scenes full of historical flavor will be destroyed. When transforming the economic formats and lifestyles in the blocks, we must make the aborigines participate and adapt to the new things to the maximum extent possible. The aboriginal "control" of the neighbourhood cannot be lost.

3.2 Pay Attention to the Retention and Continuation of "Street Flavor"

Through the actual visit and investigation of West Wenmiaoping block, the author found that the old block in the central area of the city formed a relatively closed complete life service supporting chain, so that the block residents could basically meet the basic living needs within the region. This kind of existence form similar to the Kowloon walled city in Hong Kong makes the commercial structure of West Wenmiaoping dense and the service facilities compact. The pace of life and lifestyle in the old streets make people feel as if they are in the streets of China in the 1990s. The rapid urbanization and economic development seem to have little impact on the small environment in the streets. Shops and restaurants with the characteristics of "old Changsha" are scattered here and there. This kind of "human" old street space and its spiritual appeal have become almost extinct in Changsha. The revitalization plan for West Wenmiaoping needs to leave room for these non-dominant relics, so that there is also a cultural atmosphere of time deposits in the "refreshed" old blocks. On the premise of maintaining the original humanistic carrier attributes, the living environment and supporting facilities in the block are updated. West Wenmiaoping street will be built into a representative of the local characteristics of Changsha business card, the accumulation of years of history and civic culture has become the most valuable experience of the characteristics of the block. It is precise because of the existence of such humanistic characteristics that both tourism development and the city itself provide infinite possibilities.

4 Sustainable Protection and Revival of West Wenmiaoping Block

As early as in Beijing in 1987, Wu Liangyong put forward the concept and theory of "organic renewal" in the book *The Old City of Beijing and Its Juer Hutong Neighbourhood*. "Organic renewal" theory refers to the whole city to partial buildings, it as an integral organism, retrofit for sustainable urban renewal construction, respect for the inherent law of urban develop-

Fig. 6 Residents playing cards and drinking tea in alleyways

ment, in accordance with both the basic pattern, maintain the appropriate and reasonable size and scale, in accordance with the specific content and requirements of upgrading to achieve transform the environment coordinated with urban overall environment. (Wu, 1994) what West Wenmiaoping needs is long-term and sustainable protection and development, rather than short-term shallow transformation. How to achieve "organic" and "sustainable", I have the following thoughts:

(1) Block and city: As a living body, blocks are the organ of this living body. The function of historical blocks is equivalent to important body organs, such as facial organs, representing the image and perception of a city. The relationship between the historic block and the city is closely linked, and the planning and design of the two cannot be considered separately. Only by means of the combination of harmonious coexistence can the historic district be revitalized and positively integrated with other parts of the city.

(2) Development and renewal: The development of the city is always in the process of dynamic change, and the opportunities and challenges faced by the historic blocks are also changing at any time. The historic district combines social and indigenous needs to maximize the use of existing resources and prepare for the next stage of development. This virtuous circle can guarantee the continuous growth and progress of the block.

The historic block does not mean the old block, and the historical block also needs to keep updating the technology and service facilities with the progress and development of The Times. The application of leading technologies to the conservation and revitalization of the block will ensure the integrity of the block.

(3) Deep revival: After the renovation, many historical blocks will face two major problems, one is excessive commercialization, the other is homogenization. Old streets are becoming the same everywhere, even with different business models. The practice of one place does not apply to all cities. The revival of the historic district needs to start from the inside, activate the "old Changsha" memory contained in the West Wenmiaoping district, and accelerate the benign communication between the district and the outside world. The particularity and rarity of the historic district make it have the humanistic value and historical value. Simple shallow level protection and transformation will destroy or even reduce this particularity, we need a deep revival, so that the characteristics of the block into an advantage. In this way, the historic block will be better integrated into the society and achieve sustainable development.

(4) Predecessors' experience: Wuzhen is an ancient town south of the Yangtze River with a history of 1300 years. However, in recent years, it has shown unimaginable vitality and returned with a strong attitude. The excellent development mode of Wuzhen has been designated as "Wuzhen mode" by UNESCO, and its operation mode has realized the "construction and operation of high-quality cultural comprehensive tourism destination". The rapid revival of Wuzhen is closely related to its good sustainable development policy. Firstly, Wuzhen keeps the core area of the ancient town intact and carries out protective modernization transformation of the buildings. After the renovation, the facade of the historic block

maintains its historical features and the interior modern living facilities are all available, which not only preserves the charm of the ancient town, but also meets the requirements of residents for modern life, forming a win-win situation. The overall design concept of Wuzhen is to reshape the "ancient town community", which satisfies the demand of tourists to integrate into the community and deeply experience. Wuzhen has not only transformed its historical blocks to provide modern tourism function, but also activated various historical resources of the ancient town, such as restoring traditional workshop production and subsidizing the operation of characteristic time-honored brands. The local government is doing everything possible to maintain its original characteristics and unique competitiveness. In the accommodation that tourists are most fond of talking about, through the overall transformation of folk houses, a four-star standard resort hotel is formed. The property right belongs to the residents, the right of management belongs to the tourism company, and it is managed by the development company. The company employs the original residents as staff, provides cleaning services, or gives the residents the right to eat and drink. In this way, Wuzhen revitalizes the use—value of houses, improves the employment rate, and fully stimulates the enthusiasm of indigenous people. (Zhou, 2012) the policies of aborigines and the utilization methods of historical buildings in Wuzhen's tourism development are all worthy of learning and reference in the renovation and "organic renewal" of West Wenmiaoping block.

5 Conclusion

Today, with the rapid development of economy, science and technology, cities around the world are developing towards homogenization as we expected. The modern city brings a convenient and quick life, which makes the traditional historic blocks seem backward and unfashionable in sharp contrast. But for the city, the historical block is the most direct historical record of the city and a silent history book. The soul of a city does not exist in the new high-rise buildings, but in the ordinary alleys and old streets between. However, the current situation of historical blocks in various parts of China is not satisfactory, so it is urgent to explore effective strategies to protect and activate historical blocks, which is also the original intention of this paper.

The West Wenmiaoping district in Changsha is a typical historical district in the vortex of urban development. It has a long history and rich cultural connotation, but the current situation is narrow roads, dilapidated buildings and dirty living environment. As a carrier of culture and the core of urban spirit, historical blocks cannot be replaced by other new-born blocks, so we need more active and humane protection methods.

First of all, in the process of transformation and activation, it is necessary to maintain the "control" of aboriginal people on the block, avoid cultural separation, and make aboriginal people become the main force in the activation. The second is to pay attention to the preservation of "street atmosphere". The planning should fully respect the existing lifestyle and place, and ensure the possibility of the existing cultural elements. Three key issues should be paid attention to in the protection strategy of West Wenmiaoping block. Second, the dynamic development and virtuous circle of the block; Three is the deep level revival, arouses the block "old Changsha" the memory.

The speed and high technology of modern life have indeed brought about a lot, and our urbanization speed is also very fast. But we need historic neighborhoods like West Wenmiaoping to tell us where we're coming from and where we're going. Should we also slow down and think hard about how to protect our past and let the historical heritage benefit our children and grandchildren, rather than let more and more historical heritage disappear in the hands of our generation?

Acknowledgements

We would like to thank our tutor professor Liu Su for his careful guidance, Bao Shuxin for her great support for our research work, and colleagues in the studio for their help.

References

CHEN Xianshu, 2016. Interview: photo story behind the old city of Changsha. Chinese and overseas architecture,2016(04):31-39.

JACOBS J, 1961. The death and life of great american cities. New York: Random House.

LIANG Xue, 2001. Physical environment design of traditional town. Tianjin: Tianjin Science and Technology Press.

MEI Hanrui, 2017. Analysis on space of Changsha traditional historical streets. Changsha: Hunan University.

WANG Yunfan, 2006. Research on the protection and renewal of Changsha historic street. Changsha: Hunan University.

WEN Furen, 1989. Changsha. Beijing: China Architecture & Building Press.

WU Liangyong, 1994. The old city of Beijing and its Juer Hutong neighbourhood. Beijing: China Architecture & Building Press.

ZHOU Pan, 2012. Taking Wuzhen as an example, discussion on the protection and development of historical blocks in the tourism development mode. Construction & design for project (07): 34-35.

Research on the Conservation and Renewal of Traditional Commercial Towns along the Yangtze River in the Process of Urbanization — Taking Dongshi Town in Hubei Province as an Example

WANG Chan, *LI Xiaofeng*

Huazhong University of Science and Technology, China

Abstract During the Ming and Qing Dynasties, the prosperity of Yangtze River waterway transportation led to the development of economy and trade along the Yangtze River, which formed a fairly large-scale commercial and trade town. These towns have similar spatial characteristics and architectural functional layout. They thrived because of advantages along the river, but gradually declined with the rapid urbanization. Traditional villages and towns need to be activated and renewed on the premise of protecting their own value. In this paper, a typical commercial town in the middle reaches of the Yangtze River is selected for analysis by means of on-the-spot investigation and literature reading. The paper summarizes its regional context characteristics. It also attempts to explore the conservation and renewal strategies of traditional commercial towns along the Yangtze River under the background of urbanization process through the renovation of the space environment, rebuilding place spirit and local sustainable development planning.

Keywords urbanization, conservation and renewal, regional context

1 The Formation Background of Commercial Towns Along the Yangtze River

One of the principles for the site selection of traditional Chinese villages is "avoid water while near water conservancy". Water can meet the needs of residents for daily life and production, and water transportation serves as an important means of transportation in ancient China. Lying on the Yangtze river basin, Hubei province is characterized with hills and plains, with four distinct seasons and abundant rain which make it suitable for farming. Favorable regional conditions and abundant water resources provide natural conditions for the formation of traditional villages.

The unique physical and human geographical advantages along the Yangtze River render the Yangtze River culture becomes a cultural system with relatively high productivity. As part of the system, its superior geographical position and outstanding achievements of spiritual culture along the Yangtze River boost the social and economic development along the Yangtze River and provide the material and cultural basis for the development and evolution of traditional villages. The History and Present Situation of Dongshi Town.

2 The History and Present Situation of Dongshi Town

2.1 Historical Evolution

Dongshi Town, formerly known as Dongtankou, is located in the northern part of the middle reaches of the Yangtze River and the western border of the Jianghan Plain. It has been a commercial town connecting southwestern Hubei with the Jianghan Plain since ancient times. Dongshi Town is a plain formed by the impact of the Yangtze River with the Yangtze River flowing down from the South and agate river, a branch of the Yangtze River, flowing northwest through it.

According to historical records, this town, has a history of thousands of years. Dong He, a general of Shu, was stationed in this area during the Three Kingdoms period, hence the town got its name. With the diversion of the tributaries of the Yangtze River during the Ming and Qing Dynasties, Dongshi Town became a traffic fort in the areas of the middle and lower reaches of the Yangtze River and an important docking port for merchant ships. " The town is densely populated with half of its inhabitants living in the city, where businessmen from all over the country trade. ". The foreign trade promoted the economic development of the small town. Since the Qing Dynasty, the economy of Dongshi Town has developed rapidly which integrates development in agriculture, religion and commerce. Various processing factories were set up.

In early modern times, a large number of foreign businessmen from the United States, Britain, Japan and other countries came here to set up shops thus making the town become an important commodity distribution place and economic hub in Western Hubei, with more than 300 shops along the streets. After the founding of the People's Republic of China, the advantage of water transportation gradually declined, and the center of economic gravity shifted to Majiadian town on the east side and the proportion of outflow population became relatively large. Therefore, Dongshi was no longer in its heyday and a new town of Dongshi was formed in the north due to the advantage of land transportation, where the municipal government and commercial market lied.

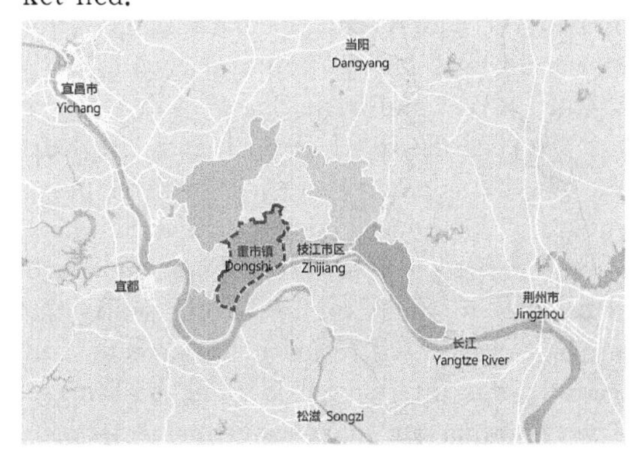

Fig. 1 Location map of Dongshi town

2.2 Overall Layout

With Yangtze River to the south, Dongshi ancient town is spread out in a belt shape along the river. The streets and lanes in the ancient town have distinct layers, consisting of four main streets and many secondary lanes connected to the main streets. The four main streets are River Street, Laozheng Street, Xinzheng Street and Daheng Street respectively. River Street has been destroyed by modern Japanese bombing, and now a levee has been built on it. Xinzheng Street is close to Dongshi New Town, and there are many renovated buildings. Only Laozheng Street still keeps the traditional style of the old commercial town. Laozheng Street is divided into the upper street and lower street by the new wharf. In Ming and Qing Dynasties, the upper street is a commercial and trade area with frequent traffic and people. The lower Street is a handicraft area. The River Street undertakes the function of logistics transportation. The goods parked at the new wharf are transported to the old town through the river street. Inside the three main streets are residential areas.

Fig. 2 The overall layout of Dongshi town

2.3 Architectural Style

The buildings in the Dongshi ancient town can be divided into three types from the perspective of function: mixed functional street houses, residential buildings and public buildings. Built in the late Qing Dynasty, mixed functional street houses were distributed on both sides of the main street. They performed both commercial and residential functions. Their storefronts were open and penetrated into the outer streets.

The inner atrium separated the residential area from the commercial area to ensure the privacy of the internal residence. Most of the street houses have attics, double-deck and far-reaching eaves which enhance the ability of a shelter from wind and rain. The facade and structure were all built of wood, and the internal structure adopts the fork beam type. The street fa ade on the first floor is made of wooden boards which is convenient to be assembled and disassembled. In general these buildings are poorly preserved and most houses are flimsy and tottering.

Most of the residential buildings are located on the back of the main street. They are small in scale and have a wide area of one to three bays. The space sequence is developed in the depth direction, with one or two courtyards. Apart from the former residences of celebrities and a few residences, the remaining residential buildings are not well preserved, most of which have been transformed and rebuilt.

Public buildings mainly include Shuifu Temple, ancient stage and an elementary school which are distributed on both end of Laozheng Street. Shuifu Temple beside the Yangtze River is a cultural relic conservation unit in Hubei Province, which was built in the Ming Dynasty and has been repaired thereafter. During the prosperous period of Ming and Qing Dynasties, a great number of businessmen and boatmen came to Shuifu temple to pray for the safety of the waterway, and it is still in use today with a lot of followers coming. The ancient stage is a place where villagers would gather on important festivals and there were grand fairs held every year. Nowadays there are only ruins left.

2.4 Summary of Regional Characteristics

2.4.1 Architectural Characteristics

The architectural details can reflect regional characteristics of the local architecture. In the traditional buildings of Dongshi ancient town, there are many characteristic structural parts such as Tiandou, Chitou, fire gable and Guandou wall. Tiandou is a

Fig. 3 Architectural style of Dongshi town

kind of roof over the courtyard which has the functions of ventilation, lighting and resisting rain and snow. Four slopes Tiandou in Zhang Shouwan residence has exquisite structure and complex workmanship, which is very rare in Hubei Province.

The spatial layout of ancient town buildings shows obvious commercial character. During the Ming and Qing Dynasties, the wharf was crowded with people and the commercial land was limited. Therefore, the front facades of the stores were narrow and the living space was hidden in the back street and stretched in the depth. For some buildings, only one side of the wing rooms was left and the courtyard became the space to divide commercial and residential area. Or commercial activities were extended to the courtyard to exhibit commodities and trade. The complete space sequence of the building is the main street, the street house, the courtyard, the back house, the rear courtyard and the courtyard wall.

2.4.2 Street and Lane Space

The structure of streets and alleys in ancient towns along the Yangtze River presents a "fishbone" shape. Several main streets extend in East-West direction which is parallel to the Yangtze River, while the alleys connect with the main streets and lead to the internal residential buildings. Several important north-south streets play the role of transportation, hence widening these streets can facilitate workers to transport goods on the wharf to workshop areas

Fig. 4 Four slopes Tiandou in Zhang Shouwan residence © Li Baihao, Luo Hua

and factories. The ratio of the width of the main street to the elevation of the buildings on both sides is about 1. 5, which is the comfortable spatial scale. Laozheng street is straight and open, the architecture facade of the street is complete, and it has a coherent sense of spatial rhythm.

2. 4. 3 Regional Culture

Wharf culture came into being with the prosperity of waterborne economy in ancient towns along the Yangtze River. Large wharves are even brightly lit at night and thousands of sails stand on the wharf. The prosperity of commercial trade has produced many well-known old stores in the ancient town and rules have formed in all walks of life. Economic and trade promoted the exchange of foreign culture and local culture which gradually helped form local characteristics, such as the Nanguan opera culture which developed and grew strong in the late Qing Dynasty and was deeply loved by local people.

2. 5 Value Judgment

Dongshi Ancient Town was of high historical value, scientific value, artistic and cultural value. The main manifestations are as follows: ① The buildings of Laozheng Street and main streets in Dongshi were built in the late Qing Dynasty, and now there are more than twenty historic buildings. The street retains the features of Ming and Qing Dynasties. Several former residences of celebrities are well preserved. The layout is clear with Laozheng Street as its axis. The streets and alleys are fishbone-like and have strong integrity. ②The construction of dwellings is scientific with traditional Chinese characteristics, such as Tiandou, volcanic wall and Chitou. ③ Built in the Ming and Qing Dynasties, Laozheng Street runs through the East and West without obvious shelter at both ends. The buildings along the street are arranged in an orderly manner, and the Horsehead walls are scattered in different heights, which forms a complete and unified street facade, giving people a continuously reinforcing artistic appeal.

3 Difficulties and Challenges in the Process of Urbanization

Under the background of rapid urban development, it is necessary to explore how to avoid the erosion of urban integration, retain and develop the regional and cultural characteristics and to build a village residential environment which offers a sense of belonging and emotional identity. Ancient towns along the middle reaches of the Yangtze River face the same difficulties and challenges in the context of urbanization because of their similar spatial and cultural characteristics.

3. 1 Physical Geographical Factors

Most of the traditional commercial towns along the Yangtze River are far away from the development center in the urbanization. Since the completion of the construction of the national highway on the north side of the town, the commercial areas and government departments of the town were gradually moved to the north side of the town. With the convenient transportation and economic development in the north, people of the ancient town gradually moved out, thus forming a new town in Dongshi. The original town lost its vitality. Ancient towns along the Yangtze River are also facing the threat of flooding. The catastrophic flood of the Yangtze River in 1998 brought heavy human and economic losses to the people along the Yangtze River.

3. 2 Family Model Factors

The traditional family model in China

has changed. The traditional courtyard houses are internally divided according to strict living hierarchy. Members from two or three generations of a family can live in different rooms based on the etiquette of superiority, inferiority and age. In modern China, parents and children live together mostly, while children build new families when they grow up. The traditional courtyard environment is no longer suitable for the contemporary family model. The renovation and maintenance of traditional living space also require a lot of manpower and capital investment. Therefore, the residents of ancient towns choose to transform their existing living space or move to live outside the ancient town. The characteristics of a large house with a large area and low building volume ratio are also inconsistent with the development of modern intensive society.

3.3 Social development factors

The rapid development of the cities around the ancient town attracts the young residents to go out to work, leaving older parents and elders to farm at home. The population of the ancient town is decreasing year by year, and the population structure is also changing. The large loss of young people makes the old town lose its vitaislity of development. It is not uncommon for the residents in the old town to relocate their family and rent their original residence to outsiders. Neighborhood relation that maintains the local social structure has changed. At the same time, the improvement of modern living standards has put forward higher requirements for family living space. The subdivision of living functions and the improvement of space quality have destroyed the internal pattern and the facade form of traditional buildings.

The above factors have brought the same dilemma to the development of ancient towns. The most obvious one is the ageing of buildings and the decline of space, which have a great negative impact on the overall style of ancient towns. The second is the deterioration of the ecological and human environment. The inner landscape space of ancient towns has been replaced by modern buildings. The landscape belt along the Yangtze River of Dongshi ancient towns has been polluted seriously, and places lose their vitality due to the loss of personnel.

While ancient towns are facing the above problems in the process of urbanization, urban development also promotes the renewal of ancient towns. The construction of cultural heritage has been put in an important position in the new round of urban planning. The historical and cultural heritage of ancient towns is exactly needed for the construction of urban culture. In the process of rapid urbanization, the desire for "nostalgia" and the psychological vision of returning to nature divert people's choice of tourist destinations gradually to the natural countryside. The rich cultural connotations of ancient villages also provide a higher development platform for tourism. At the same time, the country attaches great importance to cultural heritage, which offers new opportunities for the new planning of ancient towns. Under the background of the new era, the development of ancient towns also adds regional and cultural characteristics to cities, and the process of urbanization and ancient towns complement each other.

4 Conservation and Renewal Strategy of the Dongshi Old Town

4.1 Principles and Levels of Conservation

The development and planning of ancient towns should adhere to the principle of conservation, focus on strengthening the conservation of heritage and style of old towns, thus achieving sustainable development. The first principle is to protect the authenticity of historical sites, including accurately passing on the information of historical buildings and environment of ancient towns; The second focus is to adhere to the principle of overall conservation. All the elements of ancient towns should be considered as a whole to plan and design the town, while the correlation between the production and life of ancient towns and outer towns need to be taken into consideration to fully reflect the context of regional historical development; Thirdly, Grasping the correct principles of aesthetic standards means that the planning should unify the overall style of the ancient town, regarding the style of architecture and lanes of important historical period as the aesthetic standard of conservation and repairing, and dismantle the highly inconsistent modern buildings.

The overall conservation of ancient towns should be carried out from the three levels of "surface, line and point" from macro to micro level. The conservation should be divided into several levels. The availability of buildings varies in different grades of area. Street and lane level should be analyzed from the "linear" elements. Extracting the structure and image of the street and alley, protecting the structure of "one main street and many alleys" of the traditional street and alley, and collaging the traditional elements of the historical block. The architectural landscape level can be protected from three aspects: the shape control of the building interface, the building material and the building color. The architectural interface refers to the facade of the contact between the building and the external space. It mainly refers to the elevation along the street and the roof elevation in the control of the style and features of ancient towns. The facade forms of the street roofs on both sides of the main street in the historical period should be summarized by typology, then the general types of street roof facades are obtained. This type will be integrated into the conservation and repair of buildings. Different conservation zones have different choices for building materials and colors. We should control the macro-color of ancient towns. The building materials and colors of the core area protected areas should be coordinated with historical buildings, and modern materials can be used to represent historical forms.

4.2 Update Strategy

According to field research, most of the buildings in Laozheng Street of Dongshi Ancient Town are inhabited, some are used for business. A few are unmanaged and ruined. Nowadays the ancient town has a small population and residents live a stable life. Every festival and holiday, the town will hold major activities. People living outside the town have occasional visits.

The renewal strategy of Dongshi ancient town can adopt a more original ecological dynamic conservation and renewal mode. The degree of renovation and activation is relatively light, and the buildings in the old town are renovated, maintained and improved rather than demolished. Residents are the main part of conservation. Without changing the way of life of residents, They can choose to live in the old town or move out in a small amount. Compared with the static conservation mode, it emphasizes the importance of community vitality in overall planning and attaches importance to the activities of local residents in the place. It is a conservation mode that combines conservation and renewal with sustainable development. Foreign tourists can not only feel the architectural style of ancient towns in Ming and Qing Dynasties, but also the current activities of residents and site vitality. static conservation pays more attention to the overall restoration of historical scenes, and the conservation is limited to historical buildings and activities. It is not applicable to the Dongshi ancient town, which has a great degree of renewal of the architecture and residents' living pattern.

The renewal of the Dongshi ancient town should be combined with conservation, and three levels of renewal planning scope should be delineated. As the continuation of the historical regional context, the core protected areas should repair the historical buildings according to the relevant regulations. Rescuing the ancient buildings whose structures collapse or facades are damaged, and renovate or dismantle the facades of the buildings that do not conform to the regional architectural style. Retaining the traditional fishbone pattern of streets and lanes, the late Qing Dynasty architecture and traditional courtyard, and restoring the prosperous commercial scene and space atmosphere of ancient towns in the Ming and Qing Dynasties. Residents should be educated about the conservation of ancient towns, and they should maintain and use the buildings in the core protected areas consciously. Any change to the appearance of the buildings privately is not allowed. The renovation-limited area, which is in the transitional area between the core conservation area and the outer edge of the old town, should infiltrate the old and the new in a limited way. With the principle of "Continuing the style and gradually transiting", the new buildings could be properly constructed and the uncoordinated high-rise buildings should be demolished in accordance. The coordinated development area should play the role of linking the old town with the new town outside. The entrance and exit of the old town should be set up with convenient traffic. Taking the integrity of the surrounding landscape of the old town into account, and arranging the landscape belt around the old town.

5 Conclusion

Due to the influence of geographical location and water transport, the ancient towns along the middle reaches of the Yangtze River show some similarities in the whole spatial layout and architectural characteristics, namely, the spatial form of banded expansion, traffic system of "one main street with many alleys", the spatial sequence of river-dike-house-street and the mixed Street houses. Ancient towns along the Yangtze River have the same historical development process. Most of them are facing the problems of "hollowing out" and deterioration of human settlements. They need to be protected and renewed urgently. The process of urbanization is both a challenge and an opportunity for the development of ancient towns. Ancient towns should grasp the pace of urban development, maintain sustainable development in the conservation of their regional and cultural characteristics, and build a fine dwelling environment of belonging and emotional identity of the ancient towns. This research explores the conservation and renewal strategy of Dongshi ancient town through the excavation of its regional characteristics, with a view to providing reference for the commercial ancient town along the middle reaches of the Yangtze River.

References

Hubei Provincial Book Local Chronicle Compilation Committee, 1990. Zhijiang county chronicle. Beijing: China Urban Economic and Social Publishing House.

LI Baihao, LUO Hua, 2006. Zhijiang Dongshi city, an ancient town in Hubei Province, which is thriving because of the beach. Huazhong architecture (06): 122-126.

LI Xiaofeng, 2009. Lianghu Residence. Beijing: China Architecture & Building Press.

LUO Hua, 2007. Study on Dong Town, Zhijiang, Hubei. Wuhan: Wuhan University of Technology.

SONG Yang, 2007. Spatial morphology analysis and integrated conservation of ancient towns in Hubei. Wuhan: Huazhong University of Science and Technology.

WANG Xinqi, 2017. Conservation and development of historic blocks - taking laozheng street in Dongshi city as an example. Chinese architectural decoration and decoration (06): 124.

Identification and Interpretation of a Cultural Route: Developing Integrated Solutions for Enhancing the Vernacular Historic Settlemet

Roberta Varriale[1], *Laura Genovese*[2], *Loredana Luvidi*[2], *Fabio Fratini*[2]

1 National Research Council of Italy/ Institute of Studies on Mediterranean Societies, Italy
2 National Research Council of Italy/ Institute for the Conservation and the Valorization of Cultural Heritage, Italy

Abstract Southern Italy and Central China plateaus are both located between latitude 34°~40° north. These two areas are characterized by the intensive historical use of underground space for the management of several urban and rural functions. The comparative analysis between Italy and China is at the base of the acknowledgement of the existence of a vernacular architecture related to climatic zones at a global level and whose successful enhancement must be collocated in the indications given within the ICOMOS 2019 International Conference in Rural Landscape and the celebrations of the troglodyte city of Matera (Italy) as 2019 European Capital of Culture.

Besides the most famous historical locations — as Matera or Fujian Tulou (China) UNESCO sites — in both countries there are several less known sites characterized by vernacular architectures as well. Particularly, in China a peculiar typology of rural troglodyte villages, the yaodong, are scattered on the rough landscape of the Loess plateau, having been realized exploiting the ocher-colored silty soil.

Although the great historical and identity value, many of those villages have been abandoned, having experienced economic, social and environmental problems.

The paper introduces a comparative analysis between Italian and Chinese vernacular rural settlements based on a new methodology introduced by the authors, and consisting of the use of two charts: the first dedicated to the functional classification of elements in the new class of Underground Built Heritage (UBH); the second addressed to the RE-USE actions planned for UBH. On the basis of this theoretical approach, the paper suggests the hypothesis of creating a cultural route as a solution to enhance those villages, setting up more and less known sites to optimize the use, so as to minimize the effects of overcrowding on the most popular one and maximizing social and economic rise in less developed ones, preserving livability and authenticity.

Keywords rural villages, earthenware architecture, underground settlements; enhancement; re-use, cultural routes

1 Introduction

Southern Italy and Central China feature historic rural settlements characterized by underground constructions, expression of a worldwide so — called "troglodyte lifestyle" which was developed during prehistoric times and is still in use today. This long — lasting building approach is one of the most typical building techniques in plateaus located between latitude 34° and 40° North, and the one resulting from the successful and sustainable balancing attempts between maximization of environmental opportunities and minimization of natural conflicts and social interactions in karst habitat. As a consequence, in this particular climatic zone of worldwide, surprisingly, underground sections of historical excavated settlements are very similar to each other both in the architectural and functional aspects, having much more in common than their corresponding above-ground buildings. Based on this hypothesis, a comparative analysis has been done regarding the vernacular villages in Loessian region in China (36°52′59. 99″ N, 10°42′59. 99″ E) and in Southern Italy (40°40′11. 39″ N, 16°35′50. 03″ E), evaluating possible common actions aimed towards the enhancement

(Genovese et al, 2019).

This paper is based on some researches by the authors on a national and international scale. More specifically, on one hand, it is a National Research Council of Italy (CNR) interdepartmental project that began in 2015 coordinated by the CNR -Institute Studies on Mediterranean Societies (ISSM) that concerns the analysis of underground settlements in Southern Italy, aiming to define good practices for their recovery, refunctionalization and enhancement (Lapenna et al, 2017). On the other hand, two bilateral projects, that began in 2016, between the CNR -Institute for the Conservation and the Valorization of Cultural Heritage (ICVBC) and the Chinese Academy of Cultural Heritage (CACH). The first one is related to the sharing of good practices between Italy and China for the conservation of earthen surfaces. The second, concerns the mutual exchange of best practices in order to enhance less known cultural sites (Joint Research Projects CNR/CACH).

2 Southern Italy and Loess Plateau Rural Villages: Similarities and Differences

2.1 Southern Italy Underground Settlements

In the Southern Italy — particularly in Basilicata and Puglia regions in which our case studies are concentrated — due to geomorphological, climatic and cultural characteristics, the subsoil has historically been an integral part of urban planning, giving rise to the creation of underground spaces with residential and service functions.

The town of Matera, in Basilicata region, is a relevant example of an architectural and landscape ensemble resulting from a process of urban growth in symbiosis with nature (Fig. 1).

Fig. 1 Matera in Basilicata region, Italy © R. V.

The old town — so called Sassi -was developed entirely on the western side of the homonymous canyon, locally referred to as gravina (gorge), deeply erodes the Murgia, the carbonate platform that constitutes part of the Matera province and a wide part of Puglia region. Its location was determined precisely in view of the different lithological characteristics of the two sides of the valley and determined the dug of caves in oblique. Thus, in winter the more oblique sun's rays could penetrate from the top to the bottom of the walls enlightening and heating cavities, while in summer the almost perpendicular sun's rays could reach only the entrance, leaving dwellings fresh and humid.

The Sassi are the result of a collective architecture that has developed over the centuries, starting with the cave houses and progressively evolving with the addition of external rooms utilizing the same excavated material and, finally, with the diffusion of entirely built houses (Laureano, 1993) (Fig. 2). The rupestrian settlement typology has been handed down to the present day, albeit through morphological and functional transformations. The basis of all these negative settlement experiences is the maximization of environmental potential in terms of allocative resources.

And Matera is not the only case: the wrinkled landscape of Southern Italy is dotted with rupestrian villages, as Castellaneta (Gravina di Coriglione and that of Santo Stefano), Palagianiello, Massafra (Madonna della Scala), Ginosa (Fig. 3) and Laterza in Puglia region, that are still waiting to be re-qualified and valued (De Minicis, 2009; Varriale, 2017).

Despite the fascinating aspect, Matera and other settlements in the area undergone to the risk of abandonment since the last

Fig. 2 Matera, a cave house © R. V.

Fig. 3 Ginosa village, Puglia region, Italy © R. V.

half of the last century, when demographic rise and the modernization of the society turned the underground built settlements into a symbol of socio-cultural degradation and poverty, causing their abandonment and rejection.

In Basilicata region, thanks to the application of regional development policies, in last decades, the underground cultural heritage preservation and enhancement had a strategic role. The candidacy for a UNESCO site in Matera, in 1993, represented a fundamental step in the acknowledgement process of the historical value of the troglodyte culture (Varriale, 2017). This process, recently, culminated with the further international recognition of Matera as the European Capital of Culture 2019, as a successful effect of combined actions by institutional and private stakeholders. Matera can be considered as a role model in the field of converting a neglected urban character into a cultural and natural resource. Nonetheless, these results must be strengthened and their durability must be guaranteed in order to achieve long lasting effects in terms of socio-economic behaviour and development. The protection and promotion of troglodyte culture typicality is today entrusted also to the Terra delle Gravine Regional Natural Park (2005) and to the Murgia Materana Park (1978) which are committed to safeguarding this enormous cultural heritage in a dynamic perspective.

2.2 Yaodong Villages in Loessian Region

According to archaeological findings, the dwelling caves represent a very traditional way of living in China for a long time, particularly in those provinces sited in the Loess Plateau. That is an area of almost 640,000 km , with slopes, ridges and valleys, characterized by very fine and loamy terrain that is highly fertile and easy to dig (Kapp, 2015). Following the irregularity of the topography, settlements can be excavated into flanks of elongated ravines, especially in the Guanzhou, in northern portion of Shaanxi and in central and southern Shanxi provinces (Fig. 5).

The so-called *yaodong* is a type of underground space with varied functions, constituting the minimal unit of innumerable rural villages. Depending on the subsoil geo-morphology and the geo-climatic conditions, *yaodongs* have a variety of plans, sections and details, representing the very expression of the technical skills and traditional culture of the loessian region (Golany, 1992).

Typically, a *yaodong* has a long vaulted room with a semicircular entrance closed by walls, made of earthen bricks, stones of wood, and covered with a wooden door or a quilt. This arch-shaped structure also allows the sun to further penetrate inside the cave in winter, therefore making full use of solar radiation. A thick layer of earth on top (about 3 to 5 meters deep) acts as an effective insulation coverage and humidity modulator. Above a *yaodong*, there are often little chimneys and a tunnel constructed in the earth, representing the breathing system of a *yaodong*. In those area where bedrock exposed on top of the hill makes excavation somewhat difficult there are also houses partially or wholly above-ground, but inspired to the *yaodong* shape (Fig. 4).

Fig. 4 Laoniuwan village in Shanxi Province, China © F. F. 2017

In semi-terrain houses walls are realized by earthen bricks and or stone flakes, while roofs are made of stone flakes covered with earth, to ensure the thermal insulation of the interiors. Usually, multiple dwellings are built adjacent to or on top of one another and connected together to form a multi-tiered village, often for a single clan or an extended family. Sometimes terrain and semi terrain elements are combined with a

Fig. 5 Lijiashan Village, in Shanxi Province, China © F. F. 2017

structure built above ground in order to form an integrated complex connected by path (Knapp, 2000). Nonetheless, dwellings are hidden in the environment, being perfectly integrated into nature with minimum impact. However they are very fragile: earthen architecture is subject to a rapid decay processe, particularly suffering the humidity, thus needing a daily maintenance (Luvidi et al., 2019).

According to some studies, the number of inhabited caves in China is very consistent and many of these would be very ancient. Although in antiquity many of these villages have known prosperity, thanks to the fertility of the land, centuries of deforestation and over-grazing, exacerbated by China's population increase, have resulted in degenerated ecosystems, desertification, and poor local economies with the consequence of the abandonment of many villages. Exactly as happened in Southern Italy, since the middle of the last century, most of these vernacular settlements were abandoned because considered unhealthy and outdated if compared to the living standards of a 'modern' city dweller. It has been estimated that 90 thousand villages disappear annually, many of which collapse each year owing to exposure to floods and mudslides. Recently, very few cases have been faced the challenge of not losing their vitality by attracting tourism, and converting some traditional houses into affordable accommodation for travellers. Nevertheless, many problems still restrict the sustainable development of those sites and could possibly compromise the conservation of their tangible and intangible values.

3 The New Theoretical Approach: From Historic Functions to RE-USE for Italian and Chinese Cases

3.1 UBH Functional Classification

The comparative analysis has shown that the Southern Italy underground settlements and Loessian *yaodong* system seem to have much in common, being examples of living heritage standing at the core of local identity. To this heritage has been applied the newborn functional classification on the Underground Built Heritage (UBH), referred to historical artefacts excavated in underground areas that become significant expressions of local cultural heritage (Varriale, 2019). To be included in UBH one site must be the result of technological adaptation to geological, climatic, geographical and political situation in application of selected local skills and professional abilities born or developed by local communities. Based on its significance, UBH can be at the core of enhancement processes aiming to generate widespread social and economic benefits thanks to communicative power of the corresponding artefact.

Being the inclusion or not in the UBH strictly connected to the communication of the historical functions of the spaces under evaluation and to the nature of the technological approach applied, the UBH definition made the introduction of a new classification chart necessary. This classification is fundamental, not only to allow the static representation of the corresponding UBH element but, also, to go back to all the transformations (shapes, uses, etc.) eventually occurred during its history and that can be significant in the potential enhancement actions. UBH considers eleven functions, each of which generates the creation of correspondent caved artefacts. Four of them are connected to the management of environmental conflicts and communicate information about the correspondent functions: Sanitary, Water, Environmental Alert and Living Space. Five are connected to the management of social interactions: Religion, Knowledge, Safety, Communication and Economy. Two (Food and Transport) are connected to both Environmental and Social issues. The analysis of the chart reveals that both in Southern Italy and in Loess Plateau, the most significant elements classified under UBH are represented by the classes "water management" and "living space" with reference to the management of environmental conflicts, and by "religion", "knowledge", "defence", "communication" and "economy" with reference to the management of comparable social conflicts. Spaces to manage "food" have been excavated as an answer to environmental and social problems in both countries. Historical tuff caves, stables, farming facilities, warehouses for local products, oil and vine transformation factories in Italy and historical steel, iron and coal mines in China, reflect the stages of the relative economies throughout history. Thus, both in China and Italy, the subsoil is the place where it is possible to read the stratification of human history. The testing of the UBH classification to both historical systems showed the presence of many similarities on the basis of which to build their improvement. Nonetheless, the analysis of the processes put in place in Italy and China for the enhancement of such areas revealed substantial differences between the two countries.

3.2 RE-USE Actions for UBH

This result was confirmed by the application of the newborn scale of RE-USE actions for UBH to both the Italian and Chinese cases. This classification was created to compare different approaches of management, sustainable conservation, regeneration and touristic use. It includes four different levels of actions carried out to address the UBH enhancement. The first three refer to innovative approaches to the various uses of historical spaces; the fourth level refers to the building of new spaces using historical skills: Re-inventing, Re-introduction, Re-building and Re-interpretation.

Re-inventing is the level dedicated to rare and unique sites. This type of UBH requires monitoring, preservation and control processes, which are the preliminary goals, followed by RE-USE actions. When Re-inventing is applicable, emphasis is also placed on the communication of historical functions, even with the use of technological instruments to promote underground culture and allow virtual reconstructions of underground life. In this case, Re-use is virtual and happens to be a significant and qualifying element of the fruition of historical sites.

Re-introduction is the level dedicated to very widespread and common historical artefacts, representative of local social and economic history. In these cases, restored spaces can re-host the same functions as in the past with respect to both the remains of UBH and the introduction of new standards such as those related to contemporary hygienic and security parameters.

Re-interpretation is not a conservative approach with reference to the historical functions, however it refers to the location of new ones. In these cases, sites are restored and new functions are located while the communication of historical uses is preserved, also with the support of contemporary design which often includes historical equipment being used in interior planning.

The Re-building approach includes both the replication of the historical sites, in cases of extreme danger or vulnerability of the original ones, and the use of historical negative building methods -even implemented with the adoption of new technologies and materials -within environmentally friendly urban planning.

As the RE-USE classification revealed, there are substantial differences between Italian and Chinese case studies. This is due to two main factors. First, while Italian villages have been completely evacuated as an effect of the Laws n. 619 in 1952 and n. 126 in 1967 (Risanamento dei "Sassi"), *yaodongs* are characterized by the coexistence of villages with continuity in use, voluntary abandonment and non-systemic evacuation. Thus three different scenarios are possible in China: (a) currently populated villages, some of which present facilities and materials that have altered the natural equilibrium of excavated settlements; (b) abandoned and degraded *yaodong* and settlements already destroyed and (c) settlements destined to destruction to pave the way for the construction of new above ground urban settlements.

The second factor depends on the different levels of social acceptance of underground style living in Italy and China and to the involvement of local populations in the enhancement of troglodyte settlements as cultural sites. In Italy, despite the underground sites are still perceived as poor locations, those villages are considered as an economic resource in the cultural and tourism sector, thus almost all actions aim at their enhancement, considering UBH as facilities for tourists and local communities. This attitude was also strongly influenced by the Sassi of Matera UNESCO nomination that turned the area into one of the most significant open air museums in the world.

In China, on the contrary, the widespread social housing policy has motivated academic research towards the evaluation of climatic conditions of underground settlements. Consequently, new-built *yaodong* villages are now believed to be at the forefront of sustainable rural development (Li & Sun, 2013; Wang, 2014) and local communities actively participate in such projects. By way of contrast, there is little or no consideration of future touristic and cultural development of *yaodong* villages. That said, very recently, focus has shifted, even in China, to the conservation and enhancement of *yaodong* villages as a "cultural asset". However, those actions are mainly seen in function of the economic development of the sites, aiming at supporting and increasing the "lost" national cultural identity (Zan et al., 2018).

As an effect of this approach, the application of cultural policies to *yaodong* villages, rather than being addressed towards the conservation of the correspondent UBH, focuses on their reconstruction (after the demolition carried out during the Mao period) as an instrument for the promotion of Chinese cultural identity.

The comparative approach showed substantial differences in the processes adopted for the improvement of two very similar systems, stimulating new research questions. Considering both differences and common characteristics of the two systems analyzed, is it possible to define mutual adjustments as a result of the present paper? More specifically, is it possible to identify a common methodology for the enhancement and the optimization in the use of UBH in selected areas?

4 From the Theoretical Approach to the Cultural Route

On the basis of the above mentioned analysis, the enhancement of Italian underground settlements can be improved with the re-introduction of functional "living space" in dismissed settlements and with the adoption of underground lifestyle concepts reaching a new frontier for sustainable house design. Perhaps not in Matera, where the current situation appears to be solidified, however in the abandoned villages of Gravina, Ginosa (Fig. 3), Laterza, etc., actions in this direction to support local urban and social development could be successfully planned and implemented. Furthermore, the analysis suggested the need to understand what can be learnt from the few successful cases and how it can be applied adequately in other sites. On the other side, it is urgent to create a network on the underground building culture in order to enlighten this important aspect of the local identity and support joint positive actions in conservation and sustainable enhancement.

In this network, the most famous sites must act as drivers for the promotion of lesser-known sites, also becoming models of restoration and revitalization (Genovese, 2018). From this perspective, enhancement actions have been planned and tested within the CNR project on the southern subsoil. In collaboration with local administrations, the evaluation of a multi-level strategy has been elaborated, as follow:

(1) The punctual sites conservation and development — consisting of the identification and selection of those sites not yet enhanced but accessible, on the basis of the UBH classification. This working step aims at the improvement of the underground context, to return it to sociality and opening to tourist market.

(2) The site integration into a network — the UBH method offers hints for an alternative reading of the heritage in relation to urban history, allowing to proceed with the sites systematization and the creation of a thematic cultural route. This process also includes the integration of lesser-known sites with more famous ones, composing a multitasking offer as a solution to minimize the effects of overcrowding on the most popular sites and maximizing social and economic rise in less developed ones, preserving liveability and authenticity.

The intervention plan included also the creation of multimedia products in order to complete the offer by giving an overview of the whole possible cultural offers, also promoting events, and allowing the tourist to compose the visit himself by choosing for short, medium and long-term stays. The creation of an interconnected network aimed to optimize the programming of cultural events so as to have a range of offers that cover the most varied needs with the maximum possible temporal coherence.

As the research highlighted, in spite of the differences, there will be positive potential to test this model of intervention even in Chinese villages. After all, in China some success stories in the field of UBH enhancement have yet become a pillar of tourist economy, having captured the attention of the world.

5 Conclusion

The study based on the use of the innovative charts dedicated to the functional classification of elements of the new class of UBH, and the RE-USE actions carried on for UBH, provided new insight by testing the theories related to the connections between cultural identity and climatic zone, highlighting similarities and differences between Western and Eastern vernacular villages. On this basis the paper proposes a possible future action towards their enhancement, through the creation of a thematic cultural route. The model already tested in the Italian case, and some success stories in the field of UBH enhancement having yet become a pillar of the Chinese tourism economy, highlighted positive potential results in this direction.

References

DE MINICIS E, 2009. Insediamenti rupestri medievali di area pugliese: i casi delle gravine di Pensieri e di Riggio nel territorio di Grottaglie. Rome: Edizioni Kappa.

GENOVESE L, 2018. The villa of Tiberius at Sperlonga and the Ulysses Riviera: integrated enhancement and sustainable tourism// GENOVESE L, YAN H, QUATTROCCHI A. preserving, managing and enhancing the archaeological sites: comparative perspectives between China and Italy. Rome: CNR Edizioni: 83-93.

GENOVESE L, VARRIALE R, LUVIDI L, et al., 2019. Italy and China sharing best practices on the sustainable development of small underground settlements. Heritage 2(1): 813-825. Available online: https://www.mdpi.com/2571-9408/2/1/53 (accessed on May 31, 2019).

GOLANY G S, 1992. Chinese earth-sheltered dwellings: indigenous lessons for modern urban design. Honolulu: Hawai'i University Press: 66-108.

Joint Research Projects CNR/CACH in the Triennium 2016—2018. Available online: https://www.cnr.it/en/bilateral-agreements/project/2297/valorisation-tourism-participation-developing-alternative-integrated-solutions-for-less-promoted-historic-sites and https://www.cnr.it/en/bilateral-agreements/project/2302/assessment-of-innovative-methods-for-conservation-of-earthen-surfaces (accessed on May 31, 2019).

KAPP P, PULLEN A, PELLETIER J D, et al., 2015. From dust to dust: quaternary wind erosion of the Mu Us Desert and Loess Plateau, China. Geology (43): 835 838.

KNAPP R G, 2000. China's old dwellings. Honolulu: Hawai'i University Press.

LAPENNA V, LEUCCI G, PARISE M, et al., 2017. A project to promote the importance of the natural and cultural heritage of the underground environment in southern Italy// Proceedings of the international congress of speleology in artificial caves. Cappadocia, Turkey, 6 10 March 2017: 128-136.

LAUREANO P, 1993. Giardini di Pietra, i Sassi di Matera e la Civilt: Mediterranea. Bollati Boringhieri: Torino.

LI Z G, SUN J, 2013. Study on the green ecological view of the cave dwellings and its innovation and development. Appl. Mech. Mater(409-410):377-380.

LUVIDI L, FRATINI F, RESCIC S, 2019. Earth in architecture: traditional and innovative techniques of conservation in past and present of the earth architectures in China and Italy. Rome: CNR Edizioni.

Risanamento dei "Sassi" di Matera-Atto Parlamentare della Camera dei Deputati n. 2141 del 9 Agosto 1951. Available online: http://www.camera.it/_dati/leg01/lavori/stampati/pdf/21410001. pdf (accessed on December 7, 2018).

VARRIALE R, 2017. Southern underground space: From the history to the future. In proceedings of the international congress of speleology in artificial caves, Cappadocia, Turkey, 6-10 March: 548-555.

VARRIALE R, 2019. Re-inventing underground space in Matera. Heritage, 2(2):1070-1084. Available online: https://www.mdpi.com/2571-9408/2/2/70 (accessed on June 17, 2019).

WANG G R, 2014. The innovation and development about spatial form of traditional cave dwellings in the northwest. Adv. Mater. Res. (1008 1009): 1316-1319.

ZAN L, YU B, YU J, et al., 2018. Heritage sites in contemporary China// Cultural Policies and Management Practices. London: Routledge.

Sustainability, Territorial Identity and Multifunctionality: on Integrated Regeneration of Vernacular Architecture

OU Yapeng

Mediterranea University of Reggio Calabria, Italy

Abstract Vernacular architecture is worldwide at risk facing socioeconomic transformations due to urbanization and globalization. Its regeneration is critical to conserve territorial identity, satisfy changing socioeconomic needs, and help achieve sustainable development. This research is aimed to investigate how to regenerate vernacular architecture with an integrated approach to creating a sustainable built environment. Such an approach is able to, for a certain regeneration practice, balance material and immaterial values, structural and functional upgrade, and sociocultural and economic purposes. To this end, based on literature review and case studies of Meixian County (China) and Reggio Calabria (Italy), it first conceptualizes an integrated approach to vernacular architecture regeneration, taking into account three interconnected issues, i. e. sustainability, sociocultural continuity and territorial identity. Besides, it summarizes the regeneration principles found in the literature. Then, the research casts light on the restoration and rehabilitation practices and the reuse of the regenerated vernacular architecture in the two case study areas. The discussions are focused on the ways to repurpose and diversify the functions of the regenerated vernacular buildings. On this basis, it gives suggestions on how to improve regeneration practices. Finally, the research draws a conclusion and offers suggestions for future research.

Keywords sustainability, territorial identity, functional diversification, regeneration, integrated approach

1 Introduction

1.1 Background

Currently, vernacular architecture is increasingly perceived in a broad sense as the built environment is created based on local socioeconomic needs (Hourigan, 2015; Salman, 2018). This suggests that vernacular architecture is by nature a dynamic system, as a reflection of any societies and cultures that have persisted across history with constantly reaffirmed identities while also undergoing a continuous change in response to shifting demands in the human and natural environment (Smith, 1982). It is also acknowledged that vernacular architecture generally shows four commonalities, namely, ① adaptation to the natural environment, ② evolving together with changing multilevel needs (physical, economic, social and cultural), ③ carrier of territorial identity, and ④ embodiment of local knowledge system (Salman, 2018). As a system of know-how, vernacular architecture is believed to still have a significant role to play in creating a sustainable built environment for all (Vellinga, 2006). Indeed, what we today experience as vernacular architecture is shaped by an adaptive process (Hamza, 2019), which is critical to sustainability. Meanwhile, its conservation is crucial for maintaining sociocultural continuity (Chiu, 2004 cit in Günçe & Mısırlısoy, 2019), referred to as the maintenance of the basic fabric of a society over time primarily through the socialization process (Sanderson, 2015).

However, vernacular architecture is worldwide at risk facing drastic socioeconomic transformations triggered by urbanization and globalization. For instance, the sociocultural continuity vital for the sustainable evolution of vernacular architecture is at stake in modern times as changing socioeconomic needs have created new requirements that the traditional vernacular environment often fails to meet (Philokyprou, 2015). In addition, traditional built environment is undergoing dramatic transformations throughout the Mediterranean Basin and in rapidly industrializing countries like China (Bao & Zhou, 2014; RehabiMed, 2007; Wang & Qian, 2015). These transformations are often characterized by the physical and functional degradation of the vernacular built environment, and a continuing loss of the social and cultural characteristics of the vernacular due to cultural homogenization. To make the situation even worse, conserving the vernacular remains a problematic issue. Common problems encountered are mainly due to unfamiliar concept, limited resources, and biased aspirations to "modernize" the housing (Oliver, 2006). Often times, its conservation is regarded with much less concern compared to monumental structures (ibid.), as vernacular architecture is often times seen as a symbol of poverty with values and qualities that are far removed from the mediatized concept of modernity (RehabiMed, 2007). Apart from this dominant monument-centric view of heritage protection, vernacular architecture is often managed with a static, materiality-centric approach. This approach simply highlights its material and physical dimension while largely ignoring its intrinsic relations with the socioeconomic fabrics of a certain society. In other words, it is individual vernacular buildings that are the focus of conservation practices rather than the dynamic built environment as the vernacular represents.

To enable vernacular architecture to help create sustainable built environment in our time, it needs to be recognized that vernacular traditions constitute dynamic and creative processes of development and change (Vellinga, 2006). It must be managed in a way to address the trade-off issue of heritage protection and socioeconomic development. This requires that its conservation shift from the conventional approach focused on the materiality and individual buildings to one that highlights the inalienable linkage between the physical dimension of the vernacular and socioeconomic fabrics. What is needed is therefore an integrated approach able to coordinate the sociocultural, economic and physical dimensions in the vernacular built environment, rather than a static conservationist approach, so as to satisfy multilevel needs of a contemporary society. Now the question is, how to conceptualize and implement this integrated approach at local level?

1.2 Objective and Methodology

This research maintains that a genuinely integrated approach to the protection and management of vernacular architecture can be achieved through a continuous regeneration process. As a tool indispensable for moderating changing built environment while promoting an adaptive process, regeneration is critical to conserve territorial identity, satisfy changing socioeconomic needs, and help achieve sustainable development. This research is therefore aimed to investigate how to regenerate vernacular architecture with an integrated approach to creating a sustainable built environment.

To achieve this goal, based on the literature review, this research adopts a qualitative approach composed of explanatory qualitative analysis and comparative analysis. On the one hand, it analyzes, with a rural focus, the concept of an integrated approach to vernacular architecture regeneration in relation to sustainability, sociocultural continuity and territorial identity. On the other hand, it carries out case studies of Meixian County (China) and Reggio Calabria (Italy), comparing local practices such as restoration, reuse and territorial identity revitalization. Based on the comparative studies, it offers suggestions to improve the existing regeneration practices.

1.3 Structure

To begin with, the research conceptualizes an integrated approach to vernacular architecture regeneration, taking into account three interconnected issues, i. e. sustainability, sociocultural continuity and territorial identity. Besides, it summarizes regeneration principles found in the literature review. Then, it casts light on the restoration and rehabilitation practices and reuse of the regenerated vernacular architecture in the two case study areas. The discussions are focused on ways to repurpose and diversify the functions of the regenerated vernacular buildings. On this basis, it gives suggestions to improve the regeneration practices. Finally, the research draws a conclusion and offers suggestions for future research.

2 Conceptualizing an Integrated Approach

2.1 Vernacular Architecture and Sustainability

Vernacular architecture is commonly considered as conducive to sustainability (Achenza, 2016; Hamza, 2019; Sala, Trombadore & Fantacci, 2019; Salman, 2018; Vellinga, 2015). However, the current understanding of the sustainability of vernacular traditions proves to be a partial and distorted one due to an environmental focus, a technological bias, a romanticized approach, and essentialist/reductionist representations (Vellinga, 2015). In particular, the romantic nostalgia towards the past has led to the conservation and reuse of vernacular buildings following an approach marked by a rupture between human society and vernacular buildings (Philokyprou, 2015). This tends to generate vernacular imagery or standardized representations rather than a living vernacular environment. Consequently, vernacular architecture is revitalized only to serve the needs of cultural display or economic exploitation of the vernacular built environment whereas sustainability is barely the concern.

The academia starts to acknowledge that vernacular architecture is not per se more sustainable than its contemporary counterpart: what has to be done first is "to embrace and engage the present and future rather than romanticize and get stuck in the past" (Vellinga, 2015: 7). Second, putting the human society in balance between vital needs and harmony with environment is prerequisite for relating vernacular architecture to sustainability (Achenza, 2016). This requires that the biunivocal relationship between community-territory-economies and vernacular practices be recovered. Third, the relevance of vernacular architecture to sustainability should be understood in a broader sense, namely, not only at the built environment level, but the sociocultural and economic levels. Finally, given that vernacular architecture tends to be degraded, a regeneration process is also indispensable to achieve sustainability. All in all, only with creative adjustment, human society-environment harmony, a broader definition of sustainability and a regeneration process in place can vernacular architecture be sustainable.

2.2 Regeneration in the Vernacular Context

The restoration and regeneration of vernacular architecture first require to set up a philosophical discourse on the concept of "continuity" (Philokyprou, 2015). Continuity, more specifically the sociocultural one, is defined either from a relational or evolutionary perspective. In the first case, sociocultural continuity is perceived as a close relationship between local population and their living environment (ibid.). The evolutionary perspective, instead, defines sociocultural continuity as a process whereby sociocultural systems persist while undergoing change and even transformations (Smith, 1982). Such a definition suggests that sociocultural continuity entails two simultaneous processes, i. e. evolution (change) process and adaptation (continuity) process, through which human societies increase their capacity to mobilize resources and knowledge in adapting to their environments (Sanderson, 2015). This means that change and continuity are two sides of the same temporal coin, so intimately bound up and constitutive of each other (Patterson, 2004). As an evolving, living organism, vernacular architecture shows actually a "changing continuity" determined by the evolving-adapting law of sociocultural continuity of the overall vernacular environment. Consequently, vernacular architecture has to change to adapt itself to the actual socio-economic fabrics to achieve its continuity.

In this "continuity" discourse, regeneration proves topical. As an instrument supportive of local development, regeneration has already been popularly discussed and practiced in the urban domain. According to Roberts (2000: 17), urban regeneration is "a comprehensive, integrated vision and action to address urban issues with the use of long-term enhancements within an area in respect to the economy, physicality and environmental circumstances". What is omitted in this definition are the social and cultural dynamics which are no less important than the economic and environmental ones. Urban regeneration is gaining popularity worldwide in urban development policies for its effectiveness in facilitating adaptation to the existing built environment (Jones & Evans, 2013). This is achieved mainly by reusing and repurposing existing buildings and lands through structural upgrading and functional adaptation and diversification (Ou & Bevilacqua, 2017), contemporizing the functionality of urban spaces, and integrating socioeconomic and environmental interventions.

For the purpose of this research, regeneration is defined as a process of adaptive adjustments of the sociocultural, economic and when necessary structural/physical aspects of vernacular architecture facing transforming vernacular built environment engendered by development-related forces like modernization, industrialization, urbanization and globalization. As an adaptive activity, vernacular architecture regeneration is by nature holistic (system), incremental (process) and contextualized (place). An integrated approach must be adopted to coordinate the sociocultural, economic and physical dimensions during the management of the vernacular built environment. Only in so doing can vernacular architecture regeneration contribute to sustainable development. Therefore, vernacular architecture regeneration should not only focus on the (improvement of the) physicality of the vernacular built environment, but more importantly pay attention to the development of its core, i. e. people, society and economy. The ultimate goal is to achieve a continuous improvement in the quality of life.

2.3 Regeneration and Territorial Identity

Vernacular architecture is not only a demonstration of sustainability as discussed above but also one of territorial identity (Salman, 2018). Territorial identity, as an integrated reflection of the natural environment and human society, often shows the natural, economic, societal and cultural features of a certain territory and society (Roca et al., 2016). Social capital and cultural heritage are two principal components of territorial identity (Camagni, 2006), into which sociocultural identity, spatial organization of activities and governance structure are to be integrated as constituent elements (Veneri, 2011).

This research defines "territorial identity" as a totality of the material and the immaterial of the vernacular built environment that encompasses cultural heritage (both material and immaterial), landscapes, vernacular language, social institutions, terroir, local knowledge, and humanity (territorial temperament and zeitgeist for example). All of these are constituents of the social and cultural capital and natural resource embedded in the vernacular environment. In view of 1) vernacular architecture's role in creating a sustainable vernacular environment, and 2) vernacular architecture regeneration's potential contribution to the local sustainable development process, territorial identity can serve as all forms of capital and assets for 1) inspiring and fueling vernacular environment regeneration, and 2) catalyzing local socioeconomic development (Ou & Bevilacqua, 2017). On the one hand, as the embodiment of territorial identity, vernacular architecture is often deployed to keep or reestablish through regeneration a consensus to stabilize an image of a place, which might have changed dramatically (Herz, 2008). In this process, territorial identity is to be regenerated based on the "changing continuity" rule to facilitate the adaptive process required by sustainability. On the other hand, as a contributing factor of local development, territorial identity influences local evolutionary processes, while shaping the potential of endogenous development of territories and enhancing territorial cohesion (Ou & Bevilacqua, 2017).

2.4 Principles

International instruments concerning heritage protection have advocated an integrated approach to heritage management and development. This approach emphasizes coordinated actions aimed to simultaneously restore built heritage and rehabilitate the fabric of dwellings and social measures (RehabiMed, 2007). The ICOMOS Charter on the Built Vernacular Heritage (1999), for example, stipulates that the conservation of the built vernacular heritage be carried out with a multidisciplinary, integrated approach, highlighting ① the balance between the need for continuity of cultural values and traditional character and the inevitability of change and development, ② holism as the vernacular is a system, a cultural landscape rather than individual buildings, and ③ the connection between materiality/physicality (form, structure) and immateriality (values, perceptions, cultural associations). The UNESCO historic urban landscape (HUL) approach also calls for "the integration of historic urban area conservation, management and planning strategies into local development processes and urban planning" (UNESCO, 2011).

Academia has also actively discussed and experimented vernacular architecture regeneration. The practices in the literature highlights innovation, rehabilitation and holism, which can serve as useful references for the regeneration of vernacular architecture at local level. To regenerate vernacular architecture, it is necessary to individuate and develop new scenarios of innovation to support local communities in a way to recover the biunivocal relationship between community-territory-economies and vernacular practices. Innovation is to be envisioned and implemented through material and immaterial interconnectivity, linking social and economic development implications and local attractiveness (Sala, Trombadore & Fantacci, 2019). A holistic approach tries to coordinate the sociocultural, economic and physical aspects of the vernacular heritage (Günçe & Mısırlısoy, 2019).

Rehabilitation is an important tool for vernacular architecture regeneration. According to RehabiMed (2007), rehabilitation is aimed to improve the built environment by seeking a point of balance between technical aspects, the preservation of heritage values and the three pillars of sustainability, namely, social justice, economic efficiency and environment protection. Based on the Mediterranean vernacular and practices, RehabiMed has experimented an integrated methodology that rehabilitates the vernacular by ① integrating traditional vernacular space into a wider territorial context; ② adopting a multisectoral, sustainability-oriented approach based on "systems thinking" and economic-social-environmental nexus; ③ developing multistakeholder partnerships based on consensus, coordination and cooperation; ④ relying on a flexible, incremental approach considering the need for continual adaptation to changing realities; and ⑤ highlighting the "place" and seeking place-specific rather than one-size-fits-all solutions (*ibid.*).

Investigating the management of rural landscapes, Ou (2019) proposes regeneration principles that include ① integrated and participatory planning; ② mixed governance; ③ minimum intervention; ④ conformity to process and incrementalism; ⑤ functional diversification; and ⑥ participatory management mechanism. Central to these principles is the objective to create a link between the regeneration process and the social system. What is highlighted are planning, governance and post-intervention management in particular as well as local communities' role in these activities.

3 Regenerating Vernacular Architecture

3.1 Restoration and Rehabilitation

The practices of restoration and rehabilitation of vernacular architecture observed in rural areas of Meixian County and Reggio Calabria show both commonalities and differences. In terms of commonalities, first, restoration practices are mostly preventive and interventional. This is especially the case in Pentedattilo, whose vernacular has gone through a long course of preventative and interventional preservation since the 1980s initiated by young people and grassroots associations (Fig. 1).

Second, restoration and rehabilitation undertaken have largely complied with the minimum intervention principle, maintaining the original facade and overall structure, like the restored and rehabilitated theater, old Villagers' Committee building and primary school building in Hedi Village, and the Hospital of the Poor and another public building in Bova. Third, the restored and rehabilitated buildings have all been reused and repurposed to serve emerging sociocultural and in some cases economic needs (to be discussed in the following part). Fourth, autonomous informal interventions, especially when it is the case of buildings of private ownership, have been a common problem in both areas due to a lack of public attention, awareness and funding (Fig. 2).

Fig. 1 Restored houses in Pentedattilo © Yapeng Ou

In terms of differences, first, different periods of the vernacular have been the focus of intervention. The restoration and rehabilitation interventions have concerned in most cases buildings built between 1950s—1970s in Meixian County. In contrast, in Reggio Calabria, it is abandoned historical buildings that are targeted. Second, while "top-down" is a common approach in both cases, local communities, often delegated by grassroots associations, have played a more active role in Reggio Calabria. Third, the motivation of restoration and rehabilitation is mainly different. Truly, in both cases, the preservation of local identity and cultural values, and the promotion of socioeconomic development are evident goals. In Meixian County, the major purpose is to improve local people's living conditions by improving the built environment. In Reggio Calabria, the objective is to create the requisite conditions to promote the ongoing tertiarization process in local economies. Often, the regenerated built environment will increase the attractiveness of the place, contributing therefore to tourism development and the value-adding of the vernacular and territorial identity.

Fig. 2 Autonomous rehabilitation of a historical building in Sant'Ilario dello Ionio © Yapeng Ou

3.2 Reuse and Functional Diversification

For a sustainable adaptive reuse project, sociocultural, economic and physical aspects of adaptive reuse should be taken into consideration (Günçe & Mısırlısoy, 2019). The reuse of the regenerated vernacular buildings in Meixian County and Reggio Calabria alike is mainly based on repurposing and public-use oriented. Repurposing means the reuse, on a long-term or spontaneous basis, of the regenerated vernacular buildings for a different purpose or purposes while preserving their physical authenticity and integrity, either without alteration or with reasonable interior alteration to make it more suitable for the new function. The regenerated buildings are often reused for tourism or sociocultural purposes. As an integral part of the regeneration process, repurposing is characterized by functional adaptation and diversification in both case study areas.

In Meixian County, reuse is an emerging "top-down" approach to regenerate abandoned public buildings[1] in rural areas. The reused buildings are generally repurposed according to new sociocultural and economic needs. Good practices have been observed especially in Hedi Village, where several abandoned public buildings have been regenerated and reused as the Village History Museum, Nursing Home and E-commerce Center led by the Villagers' Committee (Fig. 3—Fig. 5).

The building that hosts the museum, built in the traditional brick-and-wood structure in the 1960s, became abandoned in the 2000s. Considering it as a "place of memory", the Villagers' Committee restored it and repurposed it as a history museum. The building's exterior style and overall structure were preserved while the interior space restructured. The e-commerce center is housed in the old theater just next to the museum. Its restoration followed the same approach as that of the museum, while its annexe space was transformed into warehouse and refrigeration storage rooms. The repurposing of the theater as an e-commerce center is very beneficial to local economic growth by curtailing the supply chain which enables local people to obtain a higher price with the P2C (producer to consumer) selling mode.

Fig. 3 Exterior of the Village History Museum and the square © Yapeng Ou

Fig. 4 Theater repurposed as E-commerce Center © Yapeng Ou

As for the Nursing Home[2], open to elder people from Hedi Village living alone, it reused the building of the old primary school behind the museum. The facade of the original two-storey building, built in brick-concrete structure in the 1970s, was moderately embellished with traditional Chinese ornamental elements. The interior decorations reproduced the traditional vernacular style popular in Meixian County and the facilities like bed maintained the traditional style and functions taking into full consideration of elder people's living habits. The ground in front of the main building was transformed into a garden with landscaping interventions that integrated the original vegetation and added recreational facilities.

① Mostly built in the 1950s—1970s, these buildings nevertheless have important historical significance as they were all the witness of political movements of far—reaching influence and rural transformations.

② In Chinese, xingfu yuan, literally the "courtyard of happiness".

Through the "*Borgo dei Mestieri* (Village of Crafts)" project with pronounced sociocultural and tourism development purposes proposed by the municipality[①], Bova regenerated a part of its historic center of great value yet heavily degraded. One regenerated building was repurposed for revitalizing local traditional craftsmanship and now hosts an experiential didactic itinerary. This itinerary is composed of three artisan workshops aimed at recovering, transmitting and adding value to traditional crafts, including wood carving and weaving, glass and ceramics and bread making. To hand the craftsmanship down to young people, special training courses are offered (Fig. 6). Besides, the ground floor of this building was repurposed into an innovative public space: a public oven. As an empathetic space, this oven not only promotes interaction and socialization among local residents, but functions as a vital space to transmit folkloric culture and local identity.[②]

Fig. 5 Abandoned primary school repurposed as Nursing Home of Hedi Village © Yapeng Ou

Other regenerated buildings were also reused for public use, such as vernacular museums[③] and emergency medical service center.

3.3 Suggestions

The regeneration practices in Meixian County and Reggio Calabria need to be improved in three aspects, namely, interpretation, governance and accessibility.

Fig. 6 Training course on traditional braiding craftsmanship © Comune di Bova

First, considering the sustainability discourse, an in-depth interpretation of core values of vernacular heritage proves more important than the popular "imitation" and visual metaphors of traditional forms and architectural features and ornamentation (Salman 2018). The experience from the two case study areas suggests the need to improve the interpretation practices through creative transmission/adaptation of the vernacular heritage and post-regeneration utilization to better bridge the past and the present. In the interpretation process, what is badly needed is a creative transmission of the vernacular's cumulative values to the next generations incorporating new contemporary values (Kultermann, 1999 cit in Salman, 2018).

Second, the current governance pattern needs improving in future regeneration initiatives. In both cases, regeneration planning has been done with a top-down approach. To respond precisely to local communities' actual needs, a participatory mechanism is needed to engage local communities in the decision-making process. This mechanism is also indispensable for the post-regeneration management, considering the financial constraints faced by those rural communities

① The project was financed by the Calabria Region under the measure 3.2.3 "protection and redevelopment of the rural heritage" in the framework of the axis 3 multi-measure announcement of the Rural Development Programme 2007—2013.

② This is done mainly through the revitalization of Bova's unique Orthodox bread-making tradition.

③ Such as the Museum of Greek-Calabrian Language "Gerhard Rohlfs", the Museum of Paleontology and the Costume Museum of Magna Graecia.

and the need to link the community with the regenerated vernacular environment.

Third, the regenerated vernacular buildings for public use need to be made more accessible to all. Truly, heritage buildings repurposed with public use are more successful in contributing to the local sociocultural and economic development (Günçe & Mısırlısoy, 2019). But the prerequisite is, these regenerated public places must be made accessible and properly managed in the long run. In both cases, the vernacular museums are neither open on a regular basis and nor used in a sufficient way to fully serve the community.

4 Conclusion

Vernacular architecture is worldwide at risk facing socioeconomic transformations due to urbanization and globalization. Its protection and management requires an integrated approach able to coordinate the sociocultural, economic and physical dimensions in the vernacular built environment to satisfy multilevel contemporary needs. A genuine integrated approach to the protection and management of vernacular architecture can be achieved through a continuous regeneration process. As a crucial tool indispensable for moderating changing built environment while promoting an adaptive process, regeneration is critical to conserve territorial identity, satisfy changing socioeconomic needs, and help achieve sustainable development.

The proposed integrated approach to vernacular architecture regeneration takes into account three interconnected issues, i. e. sustainability, sociocultural continuity and territorial identity. First, vernacular architecture is commonly considered as conducive to sustainability. Yet, only with creative adjustment, human society-environment harmony, a broader definition of sustainability and a regeneration process in place can vernacular architecture be sustainable and then contribute to sustainable local development. Second, as a dynamic organism, vernacular architecture shows an ever "changing continuity", under the effect of which it has to change, transform and adapt itself according to the actual socioeconomic fabric. Third, territorial identity can serve as all forms of capital and assets that can be mobilized to 1) inspire and fuel vernacular environment regeneration, and 2) catalyze local socioeconomic development.

The reuse of the regenerated vernacular buildings in Meixian County and Reggio Calabria alike is mainly based on repurposing and public—use oriented, and has integrated the sociocultural, economic and physical aspects of adaptive reuse. As an integral part of the regeneration process, repurposing is characteristic of functional adaptation and diversification in both case study areas.

Due to space limitation, this research has not discussed the emerging phenomenon of "historical reconstruction" as a tool to regenerate the vernacular built environment in Meixian County. How to make the "historical reconstruction" an opportunity to not only regenerate the physical vernacular environment, but also revitalize the territorial identity will be an interesting topic in future research. Another issue worth addressing is how to engage local communities in the regeneration and post-regeneration management process.

References

ACHENZA M, 2016. Architectural sustainability — A new inspiration. Serbian architectural journal, 8(1): 167-178.

CAMAGNI R, 2007. Territorial development policies in the European model of society// FALUDI A ed. Territorial cohesion and the European model of society. Cambridge (MA): Lincoln Institute of Land Policy: 129-143.

GÜNÇE K, MISIRLISOY D, 2019. Assessment of adaptive reuse practices through user experiences: traditional houses in the Walled city of Nicosia. Sustainability, 11 (540): 1-14. DOI: 10.3390/su11020540.

HAMZA N, 2019. Contested legacies: vernacular architecture between sustainability and the exotic// SAYIGH A. Sustainable vernacular architecture: how the past can enrich the future. Cham: Springer: 7-21.

HERZ M, 2008. The vernacular or: towards a new

brutalism// HERRLE P, WEGERHOFF E. Architecture and identity. Berlin: LIT Verlag: 271-280.

HOURIGAN N, 2015. Confronting classifications when and what is vernacular architecture?. Civil engineering and architecture, 3(1): 22-30.

JONES P, EVANS J, 2013. Urban regeneration in the UK. Los Angeles: SAGE.

OLIVER P, 2006. Built to meet needs: cultural issues in vernacular architecture. Oxford and Burlington: Routledge.

OU Y P, 2019. Towards a landscape approach to rural development: landscape regeneration and innovation economies in rural landscapes, cases from Meixian County (China) and the Locride Area (Italy). Italy: Mediterranea University of Reggio Calabria.

Ou Y P, BEVILACQUA C, 2017. From territorial identity to territorial branding: tourism-led revitalization of minor historic towns in Reggio Calabria// CRAVID O F, et al. Local identity and tourism management on world heritage sites: trends and challenges. Coimbra: University of Coimbra:729-739.

PATTERSON O, 2004. Culture and continuity: causal structures in socio-cultural persistence: 71 109// FRIEDLAND R, Mohr J. Matters of culture: cultural sociology in practice. Cambridge: Cambridge University Press.

PHILOKYPROU M, 2015. Continuities and discontinuities in the vernacular architecture. Athens Journal of Architecture, 1(2): 111-120.

RehabiMed, 2007. RehabiMed method-traditional mediterranean architecture: i. rehabilitationtown & territory. [2019—05—23]. http://www.rehabimed.net/Publicacions/Metode_Rehabimed/I. Rehabilitacio _ Ciutat _ i _ Territori/EN/1st%20Part.pdf.

ROBERTS P W, 2000. The evolution, definition and purpose of urban regeneration// ROBERTS P W, SYKES H. Urban regeneration: a handbook. London: SAGE: 9-36.

ROCA Z, OLIVEIRA J A, DE NAZARÉ ROCA M, 2016. Claiming territorial identity and local development: from wishes to deeds// ROCA Z, CLAVAL P, AGNEW J. Landscapes, identities and development. Oxon and New York: Ashgate: 319-334.

SALA M, TROMBADORE A, FANTACCI L, 2019. The intangible resources of vernacular architecture for the development of a green and circular economy// SAYIGH A. Sustainable vernacular architecture: how the past can enrich the future. Cham: Springer: 229-256.

SALMAN M, 2018. Sustainability and vernacular architecture: rethinking what identity is. Urban and architectural heritage conservation within sustainability, 1-16. DOI: http://dx.doi.org/10.5772/intechopen.82025.

Contemporary Art as a Catalyst for Adaptive Reuse: Case Studies in Urban and Rural Japan

YAO Ji

Keio University, Japan

Abstract Adaptive reuse is growing in importance as an effective and sustainable approach in the preservation of cultural heritage. As an alternative to demolition, adaptive reuse redefines the function of an existing building through strategic design interventions while maintaining its genius loci. Japan has been actively preserving its built heritage in both the urban and rural landscape. Contemporary art has become popular in recent years as a driver for this process, where traditional buildings are repurposed for cultural activities. This paper examines the role of contemporary art in the design and implementation of adaptive reuse projects in Japan. Existing studies mainly expand on technical and pragmatic methods of conservation rather than artistic, non-conventional approaches. Three case studies located in central Tokyo, Niigata prefecture and the Setouchi Islands are introduced to illustrate how contemporary art has been adopted as a catalyst for adaptive reuse design. The case studies are chosen to represent current trends including projects for art festivals and independent galleries. The results provide insight into how to evaluate projects in relation to heritage preservation and discusses the potentials and challenges of adopting an artistic approach to adaptive reuse. It is found that artistic approaches can lead to diverse outcomes compared to conservative methods that mainly focus on restorative work. By engaging with artists and art practices, there is the opportunity to increase a building's value through branding, tourism and community participation while preserving tangible cultural heritage in a meaningful way.

Keywords adaptive reuse, cultural heritage, sustainable design, contemporary art, Japan

1 Introduction

1.1 Cultural Heritage

The preservation of cultural heritage is vital in today's increasingly globalized and urbanized age. Architecture, as one of the most direct forms of tangible cultural heritage, sheds light on the local culture and history of a place. Adaptive reuse as an architectural design strategy is growing in importance as a sustainable approach towards local development and the preservation of cultural heritage (Plevoets and Van Cleempoel, 2011). There are many different motivations for adopting adaptive reuse methods today such as pragmatic approaches for economic reasons or the preservation of important monuments recognized for their heritage and cultural value. Although they do so in varying degrees, one common characteristic of adaptive reuse projects is the reuse of the existing structure. As an alternative to total demolition, adaptive reuse is a sustainable building method as the existing building is repurposed for new functions. Buildings that adapt to stay relevant in new contexts preserve their genius loci and thus help maintain local character and identity of a place (Norberg-Schulz, 1980).

1.2 Alternative Approaches to Adaptive Reuse

When discussing methods of adaptive reuse, much has been written regarding practical knowledge and technical aspects (Doran, Douglas & Pratley, 2009 and Burton, 2011). This paper aims to fill a gap in current literature by examining alternative approaches of adaptive reuse through the lens of contemporary art. Contemporary art is a rich and varied discipline that can create new value and meaning by drawing on history and local contexts. When contemporary art is used as a catalyst for adaptive reuse, the results are not strictly limited by functional requirements or restoration regulations but aim to offer innovative ways to make a derelict building relevant again through artistic expression. These unique outcomes can be leveraged in marketing and branding to attract tourism and new residents, especially those from the creative class who play a vital part in a city's vitality, attractiveness and quality of life (Florida, 2005).

This paper introduces current practices of art driven adaptive reuse projects in Japan with a focus on the reuse of traditional buildings including residential dwellings and bathhouses. Unlike buildings with heritage protection which are usually restricted to restorative work, these ordinary structures offer more creative freedom in the way they can be altered as they are not strictly protected by heritage laws and are thus an important point of study for creative revitalization. These typologies are also important components of vernacular heritage as they reflect the local people's way of life in both the rural and built landscape.

1.3 Attitudes and Trends in Japan

While urban centres such as Tokyo continue to experience construction and development, rural areas suffer from a decreasing and aging population. As a result architectural preservation is at risk in both contexts. In urban areas, culturally valuable structures are demolished to make way for new buildings while in rural areas existing structures are left abandoned and vacant. According to the Ministry of Internal Affairs and Communications, there were 8.2 million vacant homes in 2013 throughout Japan. These vacant homes adversely impact the neighbourhoods they are in and owners are encouraged to tear them down by subsidies from the government (Yoneyama, 2015). Therefore, the problem is not purely a heritage or development one but a broader social issue which needs to be addressed. Traditional buildings in Japan have been described as ephemeral as most houses have a lifespan between 25—40 years before they are fully demolished and rebuilt for the next generation (Ronald, 2009). These practices are intertwined with cultural views and spiritual beliefs on impermanence and renewal as shown in the periodic rebuilding of the Ise Shrine every 20 years. However, such attitudes are changing in recent times as more people see the benefits of adaptive reuse including the preservation of architectural heritage and reduction in building waste. An increasing number of traditional buildings are being renovated into cafes, guest houses and galleries such as in the Yanaka district of Tokyo. Yanaka markets itself as a well-preserved area of Japan that has retained its identity and local character making it a popular place for tourists and residents alike (Muminovic, Radovic & Almazan, 2013).

1.4 Contemporary Art in Japan

Art and art practice today comes in many scales and mediums. The way art is presented is also no longer confined to art galleries and museums, especially since the emergence of Land art in the 1960s. In Japan, community-based art projects in the form of art festivals and exhibitions which utilizes vacant structures such as old Japanese houses, schools and factories have become increasingly popular since the 2000s (Koizumi, 2018). The appropriation of these structures for art purposes is a direct form of adaptive reuse and can occur on large scales over whole towns and regions. These include art festivals such as the Echigo-Tsumari Art Triennale (first started in 2000) held in the mountainous regions of Niigata Prefecture, and the Setouchi Triennale (first started in 2010) held across several islands around the Seto Inland Sea, Kagawa Prefecture. Over the past 20 years these art projects and art festivals have been used as a vehicle to tackle revitalization in Japan through community participation and the attraction of tourism (Klien, 2010). The focus of these festivals is to connect people to nature and each other through the production of artworks that celebrate local identity and regional culture. Artists are invited from all over the world to create site specific artworks staged at various sites including vacant houses, farms, and museums. They are encouraged to interact with local communities in the conception and creation of works to create social value and regional innovation (Nakamura, Sakamoto & Krizaj, 2017). The process of art creation is a valuable bridge in connecting diverse groups of people.

2 Case Studies

2.1 Methodology

This paper introduces adaptive reuse projects across Japan in both the urban and rural context. The purpose of the case studies is to present an overview of current practices where contemporary art has been used as a catalyst for the adaptive reuse of existing structures. 38 projects have been visited and documented: 4 in Tokyo, 3 in Kyoto, 7 on Naoshima, 5 on Inujima, 9 on Ogijima, 3 on Teshima, 5 in Tokamachi and 2 in Kamiyama. The site observations are supplemented by interviews with visitors, locals and festival organizers.

The case studies feature a wide range of different approaches and represent varying degrees of adaption to the existing building. These range from cases where the existing building is altered to be a backdrop and mainly functions as a container for art works, to extreme cases where the existing structure is transformed into an artwork itself. The goal is not to assess the art itself which can be subjective, but to evaluate their contribution and influence on the final built form which are loosely categorized into four groups (Fig. 1). The projects are categorized to illustrate the degree of building preservation in relation to the degree that art has influenced the adaptive reuse project, called "art integration".

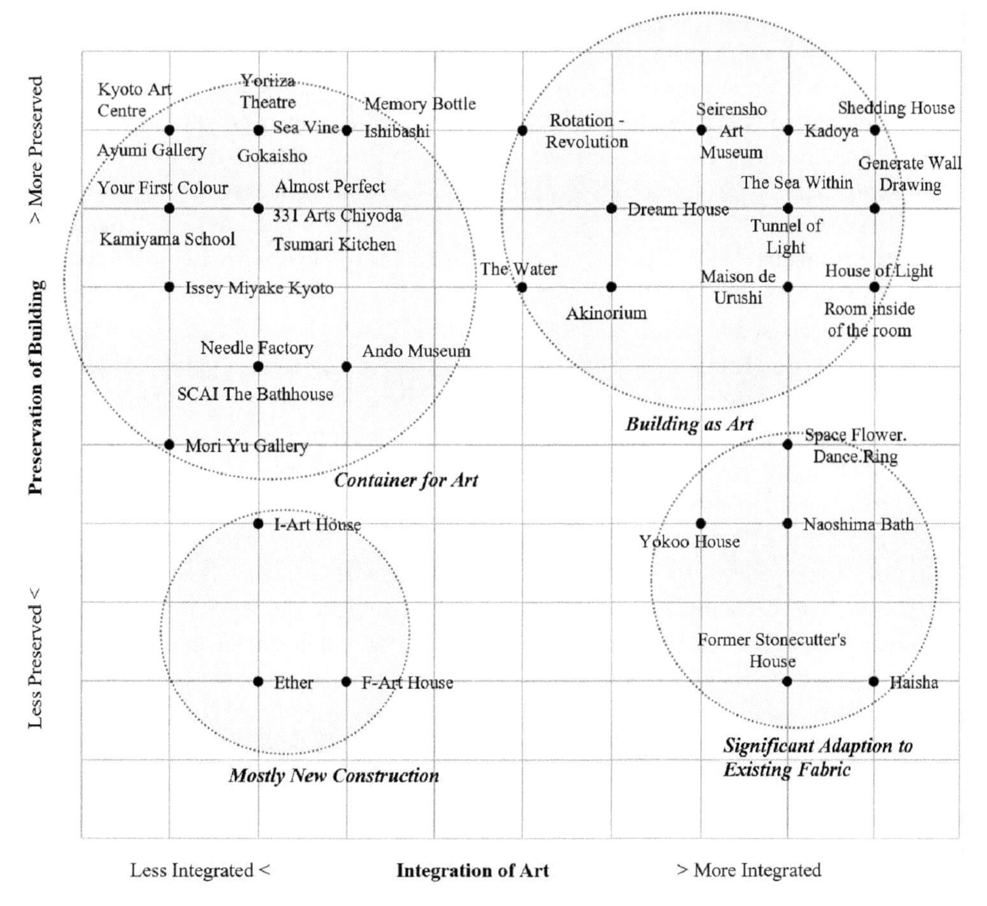

Fig. 1 Categorization of art projects

2.2 Evaluation Criteria

The y-axis evaluates the preservation of the building from a heritage perspective. Low levels of preservation represent buildings that have undergone significant alterations, use new, non-traditional materials and have major elements removed and altered. High levels of preservation represent buildings which retain existing features and structure, reuse existing building material and preserve the original building fabric in a legible way. The highest level is equivalent to a building restored to its original condition.

The x-axis evaluates the integration of Art. Low levels of integration represent buildings which act as a container for art where exhibitions can be changed freely i. e. artworks are not permanent. Higher levels of integration include site specific works which utilize the building as part of the work. These include using the existing built form as part of the artistic expression such as directly drawing on walls. The highest level is when the building becomes an artwork itself and cannot be separated from the building structure.

2.3 Exemplar Projects

Out of the four categories, three exemplar projects are discussed in detail to represent each category. The category of "mostly new construction" is excluded as it is not as relevant in the study of reuse design. The case studies shed light on some guidelines when considering contemporary art in adaptive reuse projects to better understand when it may be appropriate to use them. It should be noted that the point is not to replicate the techniques as interventions are site specific, but rather to gain insight into potential possibilities and how contemporary art can contribute to heritage preservation.

2.4 SCAI The Bathhouse

The gallery is housed in a former public bathhouse dating back to 1787 in the historic Yanaka district of Tokyo. The location is close to the Ueno art and cultural precinct which has a collection of major art galleries and museums. The neighbourhood consists of mainly low rise buildings and features a number of old houses, stores and temples. The current building was constructed in 1951 and opened in 1993 after the bathhouse closed. The exterior of the building is mostly preserved and uses traditional tiled roofing (Fig. 2). The interior has been reworked to create an open exhibition space which capitalizes on the existing seven-meter-tall ceilings and skylights (Fig. 3). New concrete floors and white walls turn the bathhouse into a contemporary gallery space which stages new exhibitions every few months. Elements from the interior that have been retained include the original wooden locker boxes found at the entry area, a relic from the buildings past use. The gallery is an example of adaptive reuse where the architecture has been adapted to become a container for art. The integrity of the building is respected as seen in the sensitive preservation of the exterior which has kept the original building form with only minor alterations. It is successful in maintaining the neighbourhood character and streetscape as well as a piece of local history.

2.5 Shedding House

An old farmhouse in Tokamachi, Niigata Prefecture with over 150-year-old history had become vacant and turned into a work of art through hand carving. The project is part of the Echigo-Tsumari Triennale completed in 2006 by Junichi Kurake and students from the Nihon University College of Art Sculpture. The work took two and a half years to complete. During this time the artist spent a few months living in the local village with student helpers. Using chisels as the only tools, carvings were made on all visible wooden surfaces of the house including walls, floors, structural elements and furniture (Fig. 4). The original ceiling was removed to expose beams which were also carved on. The soot from the kitchen and fireplace had coloured the outer surface of the wood black which contrasts the lighter colour of the timber underneath when revealed through the carvings. The project does not add anything new physically but draws visitor's attention to the old house

Fig. 2　Exterior of SCAI The Bathhouse retains original building form and materials

Fig. 3　Interior of SCAI The Bathhouse retains high windows for natural light

Fig. 4　Interior of Shedding House features hand carvings which reveal the natural wood colour

and its previous life. Although the art cannot be separated from the building and has permanently changed the aesthetics of the original house, it is done in a way that still preserves the integrity of the house both architecturally and structurally as most of the original elements are retained. Visitors can clearly see major elements of the house as it was. The removal of the ceiling further clarifies the type of construction employed at the time. The spatial planning and services of the house also remain unchanged. This has enabled the house to still function as a place of residence to accommodate visitors overnight who can rent the house for private use.

2.6　Naoshima Art Project: Haisha

Haisha, one of seven works from the *Art House Project* initiated in 1998, is located in the residential area of Honmura on Naoshima Island. The house was the former residence of a dentist and has been transformed by artist Shinro Ohtake in 2006. The artist utilized found objects, scrap material such as signs and neon lights, to cover the exterior of the building in a collage-like way (Fig. 5). The interior of the house exhibits a large piece of sculpture which has led to parts of the building's interior and exterior removed and cut into. Most of the walls and surfaces on the inside are covered with markings and collages such as paper cuttings and photographs. The project is similar to Shedding House in that it represents an extreme case where art has strongly influenced the adaptive reuse process and transformed the building into a piece of art. However, preservation from a heritage point of view is lower as minimal parts of the original house are legible. Although some of the materials used are from the local area and the overall form of the building is largely intact, it is hard to read the original building elements amid the overlapping artistic additions. Though it is arguably still preserved rather than left to become vacant and dilapidated.

Fig. 5 Exterior of Art House Project "Haisha"

2.7 Lessons Learnt for Future Practice

The 1999 ICOMOS Charter on the Built Vernacular Heritage outlines principles and guidelines in practice for vernacular buildings. One part calls for alterations to use "materials which maintain a consistency of expression, appearance, texture and form." Contemporary art projects present diverse outcomes which do not necessarily follow the guidelines of the charter but can still be considered as respecting the "integrity of the structure, its character and form." The following discussion introduces points to consider including potentials and challenges of the adaptive reuse process which can aid designers and developers who engage in art led building preservation projects.

The main point of consideration before and throughout the duration of the project is the degree of building preservation in relation to art integration. This will determine if the art is permanent or temporary, and if the future use of the building is a container for art or an artwork itself. These considerations depend largely on the new desired purpose of the building and its primary function as a gallery or art work. The legibility of building elements such as structure and construction features is another important factor in preserving the building's original form and integrity.

In addition to physical outcomes, the creative methods used by artists and their unconventional approach to adaptive reuse is a potential advantage. Artists use various techniques to draw on local history and place such as the use of existing materials and objects to make up the work and interpretation of local cultures to inform the work. Artists also collaborate with residents and volunteers in the conception and making of the artworks to create meaningful exchanges. It is important to monitor social impacts on residents during the art making process to minimise any tensions that may arise. As most artists take up residence at the site of their work and create the works with volunteers, who are both locals and outsiders, effort needs to be made to ensure they are supported and maintain good relationships with residents. The interaction between artists, locals, volunteers and visitors has the potential to directly contribute to community revitalization through activities such as workshops and other creative events.

In addition to the preservation of tangible cultural heritage, intangible aspects can also be reinterpreted through artistic practice. One such example is the promotion of a local traditional technique called Sanuki lacquer work. *Maison de Urishi* on Ogijima island is a wooden house that was renovated using traditional Kagawa lacquer techniques applied by local craftsmen. The house also functions as a café where guests can use cups from the island that are recycled with a new lacquer applied to them. Visitors and residents alike can see and learn about Sanuki lacquer work, participate in workshops and buy lacquer souvenirs made by students and craftsman. The house functions as more than an artwork or container and becomes an educational centre by making people aware of unique local traditions.

Lastly, it is also important to consider the adaptive reuse projects as part of a larger framework rather than individual buildings. The rural case studies discussed in this paper all function under larger art festivals or artist in residences. The urban case studies are located in major city centres and are part of a larger cluster of creative art institutions. When visitors on Naoshima were asked if they would make a trip to see the Art House Projects if they were a standalone work, the majority responded with no, citing the art festival and other major art museums as reasons they thought the trip was worth the time and money required to travel. In a survey conducted by Funck and Chang (2018), only 15% of 255 respondents cited the Art House Projects as the most impressive place they visited on Naoshima. Other challenges in adaptive reuse include the need to address building regulations such as fire, structure, services and other compliance requirements which need to satisfy current standards.

3 Creative Revitalization

3.1 Revitalization Through Art

Art projects play an important role in the invigoration of communities as they create a platform where people from diverse backgrounds can interact and participate in creative activities (Koizumi, 2018). Art itself is a form of cultural expression which promotes diversity and different points of view. This inclusive way of working with real citizen participation can empower local communities and encourage vitality. In addition, art and architectural tourism attract a large number of visitors even if they are remote and far from major cities, such as the case of the islands in Setouchi and the rural areas of Tokamachi in Niigata. The 2016 edition of the Setouchi Art Triennale and the 2015 edition of the Echigo-Tsumari Art Triennale attracted over 1.04 and 0.5 million visitors, generating 14 and 4.6 billion yen, respectively. Aside from the economic benefits, social and cultural values are also amplified by art projects which contribute to community sustainability and local development. As art and building projects take time to complete, are not mass produced items, and are highly personal and site specific, they are not easy targets for over commercialization. Two case studies are discussed below, the case of Kamiyama town and Naoshima Island.

3.2 Revitalization in Kamiyama

Kamiyama is a small town of fewer than 6,000 people in the centre of Tokushima prefecture. It is a typical example of a depopulated town with an aging population. In the case of Kamiyama, the local people have been using art to generate local vitality through programs such as the establishment of the Kamiyama Artist in Residence program in 1999, known as KAIR (Yoshimoto, 2019). The residents had to be resourceful and utilized unused buildings scattered around town including the local elementary school which is currently used as studio spaces, and apartment units that were formerly used by the teachers next to the school as residence for the artists. The exhibition spaces include a former puppet theatre built in 1920 called the Yorii Theater. The theatre is no longer used and looks unassuming from the outside but features a traditional ceiling which is preserved thanks to its constant use by visiting artists (Fig. 6). The success of Kamiyama is largely attributed to the enthusiasm and hospitality of the residents who support the visiting artists in the production of works as well as their everyday life. The artists' positive experiences led to word of mouth promotion of Kamiyama as a pleasant place to live. Coupled with a website showing what life in Kamiyama is like through photos and interviews with artists, the town has successfully attracted young workers and a number of IT companies who have set up satellite offices in town. People are also relocating there to open small businesses such as cafes and accommodation. As a result, the number of people moving to Kamiyama surpassed people moving out in 2011 (Yoshimoto, 2019).

Fig. 6 Yorii Theater in Kamiyama Town

3.3 Revitalization in Naoshima

Naoshima is a remote island, one of the many from the Seto Inland Sea that is also suffering from depopulation. Development on Naoshima started with the cooperation between local government and Benesse Holdings, a private corporation. Several major art museums have been built including the Chichu Art Museum and Benesse Art Museum, both by the well-known Japanese architect Tadao Ando. Smaller scale works such as the Art House Projects were also initiated around the same time. Naoshima has been cited as a successful case of leveraging art tourism as a tool for regional development to successfully attract new residents since the start of the Setouchi Art Triennale (Funck and Chang, 2018). It has done so through a long period of development through a mix of large and small-scale projects. The museums and hotels have created jobs in the tourism industry and allow for an increasing number of small businesses to exist. The author interviewed several workers including a caf owner and clothing shop assistant who commute from nearby port cities of Okayama and Takamatsu to work on the island during high season. Visitors to the island of another form are volunteers and staff who come to assist in various aspects of the festival such as managing art works, selling festival goods and organizing events. An official NPO Kohebitai manages the volunteers. Business owners also advertise for volunteers online directly to work for their businesses such as running cafes in exchange for board. They often stay on the island for longer periods and develop a deeper understanding of local life compared to tourists.

4 Conclusion

In both urbanized and rural societies, the preservation of cultural heritage is vital in the future development of places to retain their identity and sense of place. This paper introduced contemporary art and its potential to offer creative approaches in adaptive reuse projects. Like architecture, art can utilize existing resources effectively. Through the injection of new ideas, knowledge and skills, new value is created from what already exists. Contemporary art can offer valid outcomes that are highly creative and unique compared to traditional approaches based around restoration. Though their success from a heritage perspective varies based on the techniques employed, it is possible to maintain integrity and authenticity when approached in a sensitive manner.

The case studies illustrate the various approaches being adopted to reuse vacant buildings that are not seen to have enough heritage importance for special protection and thus more likely to become abandoned. These structures can be preserved through adaptive reuse projects and be used to reinvigorate communities. However, their success often depends on a larger framework of creative activities rather than a standalone project.

The process of art making involving local communities is an effective means towards revitalization by offering interaction between a diverse range of people from different backgrounds. These include residents, artists, visitors, and volunteers. Art and architectural tourism play a large role in the economic, cultural and social vitalization of a place.

One of the key benefits of engaging artists specializing in a range of different mediums is the creation of original art experiences which is highly sought after and valued. Many places try to creatively revitalize themselves through large cultural projects but the success of small scale, locally oriented development of art driven adaptive reuse projects have potential for real change and should not be overlooked.

References

DORAN D, DOUGLAS J, PRATLEY R, 2009. Refurbishment and repair in construction. Dunbeath: Whittles Publishing.

FLORIDA R, 2005. Cities and the creative class. New York: Routledge.

FUNCK C, CHANG N, 2018. Island in transition: tourists, volunteers and migrants attracted by an art-based revitalization project in the Seto Inland Sea// MÜLLER DK, WIECKOWSKI M. Tourism in transitions: recovering decline, managing change. Cham: Springer International Publishing: 81-96.

MUMINOVIC M, RADOVIC D, ALMAZAN J, 2013. On innovative practices which contribute to preservation of the place identity: the example of Yanesen, Tokyo. Journal of civil engineering and architecture, 7(3): 328-340.

NAKAMURA K et al., 2017. Social value creation in art-related tourism projects: the role of creative project actors in diverse national and international settings. Academia turistica, 10(2): 191- 203.

NORBERG-SCHULZ C, 1980. Genius loci: towards a phenomenology of architecture. New York: Rizzoli.

PLEVOETS B, VAN CLEEMPOEL K, 2011. Adaptive reuse as a strategy towards conservation of cultural heritage: a literature review// BREBBIA C, BINDA L. Structural studies, repairs and maintenance of heritage architecture XII. Chianciano Terme: WIT Press.

RONALD R, 2009. Privatization, commodification and transformation in Japanese housing: ephemeral house — eternal home. International journal of consumer studies, 33(5): 558-565.

SMITH V L, 1989. Hosts and guests: the anthropology of tourism. Philadelphia: University of Pennsylvania Press.

YONEYAMA H, 2015. Vacant housing rate forecast and effects of vacant homes special measures act. Fujitsu Research Institute. [2018-10-9].

WONG L, 2016. Adaptive reuse: extending the lives of buildings. Basel: Walter de Gruyter.

ICOMOS, 1999. Charter on the built vernacular heritage. Mexico:[s. n.].

KLIEN S, 2010. Contemporary art and regional revitalization: selected artworks in the echigo-tsumari art triennial 2000 6. Japan forum, 22(3-4): 513-543.

BURTON S, 2011. Handbook of sustainable refurbishment: Housing. London: Routledge.

ASSMANN S, 2015. Sustainability in contemporary rural Japan: challenges and opportunities. London: Routledge.

KOIZUMI M, 2018. Connecting with society and people through — art projects' in an era of personalization// CABANNES Y, DOUGLASS M, PADAWANGI R. Cities in Asia by and for the people. Amsterdam: Amsterdam University Press: 177-200.

Preservation of Vernacular Heritage in Aquixtla, Puebla, Mexico

Gerardo Torres Zárate

Instituto Politécnico Nacional, Mexico

Abstract Mexico is a country with a surface of 1973 km^2; it ranks third in the world with the biggest biological diversity. Not only there are 85 indigenous groups inhabit but also Mexico has climate diversity. Unfortunately, vernacular architecture is not well preserved.

The Aquixtla town, at Puebla State, Mexico, is located in the "Poblana Sierra"; with lukewarm weather and 2200 meters over sea level. Its communities preserve many examples of vernacular architecture. The landscape from the streets provides a wonderful image due to the mountains that surround the community. The traditional architecture from Aquixtla was influence by the Spaniard (Spanish culture), which consist of constructions with a "U" shape with an entrance in the center and a fountain as well. Around the yard, there is a covered hall where bedrooms are located, just like the kitchen and the living room. The walls are made of adobe, with a Stone basis and the roofs from wood whir mud tiles. The community has an important immaterial heritage. The catholic religious celebrations, take an important place at people's lives; their devotion to the Saint is immense that it evens promotes religious tourism.

There is deterioration in the vernacular architecture at the place. Therefore, the project of recovery and conservation of the traditional image was initiated, with the contribution of the local authorities and the community by taking advantage from the immaterial heritage, in combination with the tangible one; in order to reuse the vernacular spaces and be able to revitalize the village center. The local economy is based on tomato farming, but traditionally the local people produced Christmas spheres and pottery. This project pretends to revive those activities to the local development. Currently, the project is carried out through actions of improvement of the vernacular architecture in an effort to restore and preserve the traditional image.

Keywords vernacular heritage, immaterial heritage, Mexico

1 Introduction

Vernacular architecture has been transformed, altered and destroyed in most of Mexico. This phenomenon increases in those communities, which are closer to the cities. There are several factors for this transformation. For example, extreme poverty, official programs which use contemporary materials, migration to the USA and the withdrawal of traditional constructive system and traditional materials, among others. Therefore, a study about vernacular architecture in Mexico allows recognizing, spreading and preserving the traditional constructive systems. Mexican vernacular architecture characteristics are a consequence of colonial syncretism (López, 1987). Spanish influence could be perceived in the introduction of elements as bricks, roofs and tiles (Prieto, 1994). The patrimonial value of that architecture is based on its cultural richness (Prieto, 1982).

López (2002) establishes that vernacular edifications in Mexico possess diverse cultural influences, inherited from that mixture. Spanish conquest brought to America heft in diverse fields, including architecture.

Mexico had several cultures as Aztec, Mayan, Mixtec, Zapotec, Toltec, among others; this mixture generated cultural diversity(Fig. 1).

Nowadays, there are more than 85 ethnic groups, which are found in all the territory (CDI, 2019).

Each indigenous group has its own culture, dialect, gastronomy, productive system, handcrafts, traditions, music and architecture. Mexico has around 1, 959, 248

Fig. 1 The Mayan house, Yucatan, Mexico

Fig. 2 Dwellings with African influence, Oaxaca, Mexico

square kilometres of territory, with a population of 112 million people of which almost 7 million are indigenous.

This has allowed a diversity of traditional constructive systems.

Mexico's geography, location and extension allow the existence of climatological variety. There are also different ecosystems as deserts, jungle, tundra, high mountains, beaches, forests, savannah and swamps. Mexico is part of the twelve countries that has the most variety of weather worldwide.

The definition of vernacular architecture has to do with the region where it is developed, that is why a variety of materials exist and also constructive systems. Being also a cultural concept, involved as a part of the origins of each town, that which can see in their immaterial elements. López (1987) defines the vernacular architecture in Mexico as "the one that was born in a decanted historical process which is a mixture of both Spaniard and African indigenous elements" this definition allows to understand that cultural influences have an important place in the comprehension of the traditional architecture in each region.

Due to the conquest, Mexico received Spaniard, Arabic and African influences in addition to the pre-Hispanic one(Fig. 2); even nowadays, it is evident in each Mexican region, with a wide range of influences and results about the vernacular process (Prieto, 1994).

Mexico possesses different types of vernacular architecture like the Mayan House which for three thousand years is without changes. (Torres, 2009). In the regions where indigenous groups are prevailing, we can find examples as Aztec, Mixtec and Toltec cultural influences among others (Mattos, 1999). Furthermore, there are regions where colonizing people from XVI century were stablished, leaving a vernacular architecture with the biggest European influence.

The vernacular architecture in Mexico depends on the climatologic and cultural aspects. In this essay, it is introduced a community which was founded after the conquest, in XVI century, and so it presents a vernacular architecture, of a strong Spaniard influence.

2 The Site

The village of Aquixtla is located in the northwest part of Puebla state, in Mexico (Fig. 3). Its geographic coordinates are: the parallels 19° 4″ 42″ and 19° 51′ 54″ of north latitude and the meridians 97°49′ 36″ and 97°54′06″ of occidental length. It has a surface around 166 square kilometers and a population of 9000 inhabitants. The region is inside the Sierra Norte of Puebla, area of high mountains surrounded by wide forest of pine trees(Fig. 4).

The municipal district presents a rugged topography; its relief is determined by the presence of many mountain chains, hills

Fig. 3 Location of the town

Fig. 4 Natural context of Aquixtla

and a valley among the mountains. The name of Aquixtla comes from the Nahuatl idiom Aquiztla that means "a place where water comes from" and we observe at present its richness of rivers and natural springs.

According to historical data, its origin is produced by settlements of Totonacs and Nahuatl groups, put down by the Aztecs and their alliances. In 1750 it was controlled by the ecclesiastic jurisdiction of Zacatlan. Afterwards it was part of the antique Alatriste district, now it is called Chignahuapan. The community was founded in 1788, the first colonizing people were from French and Spain and in 1895 it was changed into a free municipal district.

The community has traditionally produced ornamental spheres made of glass. There are around 15 handcrafted workshops yet, that along the year produce these products for Christmas days. Another craft activity is the pottery where people make jars, pots and pans made of clay. Unfortunately, these two economical activities are being abandoned. The current inexpensive activity is the production of tomato, in hydroponic greenhouses. This activity is more profitable, and has replaced the traditional production.

The downtown of Aquixtla, just as its communities, have many examples of vernacular architecture. According to the declaration of UNESCO ICOMOS, those buildings must be preserved in their total originality. The level of conservation of the downtown, is still positive, even though it has so many alterations and worsening. Therefore, it is important to save and look after the traditional image of the village.

3 The Problem of Traditional Image

It is common in Mexico that authorities do not have interests in the built of vernacular heritage. Nevertheless, to this project the president of Aquixtla approached Nacional Polytechnic Institute (IPN), to talk about the concern about the deterioration of the vernacular architecture in this community. An agreement was made between the government and the Architecture and Engineering School of the IPN whereby the project was developed, as a social service to the community.

According to the set-out needs and requirements of the government, a series of site visits were made. This allowed to lay down the streets to interfere, being practically 80% of the area in the town.

In order to know the traditional architecture of Aquixtla rounds and architectural surveys with pictures were made (Fig. 5). The visuals and perspectives of streets give

Fig. 5 Typical house

an impressive image, due to the mountains and forests that exist round(Fig. 6). From analysis, we can categorize it as a vernacular architecture with a strong influence from Spain. The complex is organized with lined volumes in a "U" shape which has its access in the center. Sometimes it preserves a fountain in the center of the yard. Around the yard a covered hall where bedrooms are, in the same way the kitchen and the living room. The vain in the doors and windows, have three versions: straight lintel, half-point arch and oval ones. The walls are made of adobe, leaning on stone foundation. The covers are composed by wooden beams and crosswise floorboards where roof tiles are stand(Fig. 7).

Fig. 6 Natural context of Aquixtla

Most of the time, we can still find the spaces which were appropriated as farmers, stables, henhouses. They also show yards with the rest of stone pavement in its grounds. In the same way the different

Fig. 7 Alterations of the traditional image

spaces that were made for primary activities. Something that stands out is the fact that some families have invested money in their own house. In Mexico, it is usually decided to demolish vernacular housing and replace it with concrete buildings. However, in Aquixtla, people have opted to restore and to reuse the vernacular space.

The identity of the people has an important role, so it generates the influence and appreciation of traditions. In the case that we are analyzing, the religion takes a relevant place in order to get this identity. In Aquixtla 95% of the population is catholic, and the festivities, just as in the rest of the country, are so important to these people. The main saint is an image of a tortured Jesus, the people have called it "our beloved father Jesus", which is considered to be very miraculous, and that is why so devoted people worship it. The celebration of the fair is in the last week of December and the first one of January, where not only processions are made but also mass and chants dedicated to the veneration of the image. Generating with this religious tourism, so many visitors from many different parts of the country and states of almost all of the rest of Mexico come. The most important part is that during these days, people fix their houses via cleaning the facade, repairing roofs and painting walls. For them, it is very important that the house is clean so the procession goes across the town and paths. In front of the buildings, they set altars and ornaments.

In spite of preserving the vernacular edifications with a high level of quality, most of the streets present a series of problems which have modified the original and traditional image. The most observable ones are related in the consecutive.

The air wiring is exaggerated and in disorder, which produces a bad urban image. The stores around have heterogeneous advertisements, that show a visual disorder. There are signs badly nailed in front of the houses which decrease the charming of them.

The wooden beams in their limits have an excess of humidity and some of them are rotten. There are broken tiles, in a bad condition and placed wrongly, they have unwanted flora. In 70% of the cases flattened and paint on walls are in poor condition. Not only is this situation visible, the painting and walls have also been affected due to these conditions. Some street poles are unleveled and out of the sidewalks. There are some wires over facades and advertisements over there too. There are different elements to the sight of everybody that pollute the landscape. The colors of the facades are without harmony, considering each person has painted the house with her favorite color regardless of the set. The sidewalks have different materials, as stones, cement, cobblestones, so it is not homogeneous and it produces a bad landscape. The streets of the central square were made with stone and changed into concrete, which is dilapidated now.

4 The Project of Intervention

In order to make the project, a meeting of ten students of architecture was invited to analyze it. Some reunions were convened to draft a workshop, where people acquired the principles of either cultural or vernacular heritage. Students had visits and walks, just like interviews with the people from there also taking pictures of the main streets of the community. Surveys were made of vernacular dwellings as well. We worked with the government and inhabitants all together to preserve the traditional image of the town. There, main ideas to the project came out. Once the information was gathered, the group showed a series of ideas about what they observed in the damaged village. These plans were developed in order to have a study of all the facades in the selected streets.

Finally, fifty plans were made where there are streets in sections of three levels. First, the current state of the image urban-architectural where the damage of the architectonical and structural elements are, as well as the aspects which produce a wrong image or break with the traditional image. Second, we look at the improvements. The elements that have the need of being restored are marked, changed or moved, just as the ones that need to be joined. In the third term, the proposal for the conservation of the traditional image is presented. Here the fronts are displayed with the adaptations and the traditional image that is required(Fig. 8).

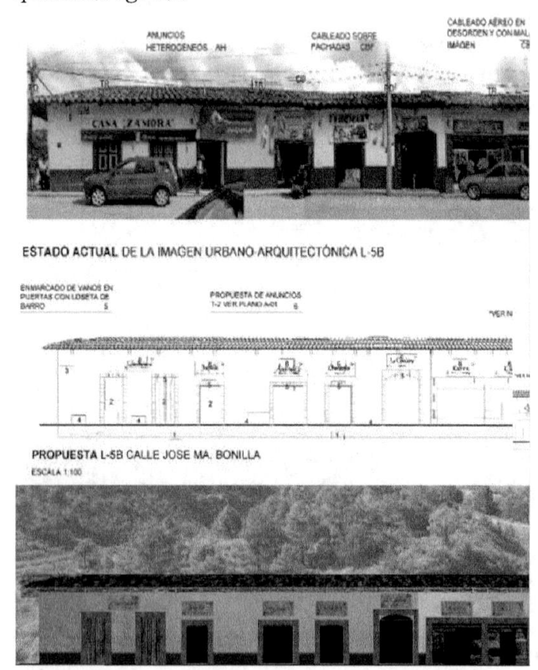

Fig. 8 Proposal of improvement of the traditional image

The project of restoring and preserving the traditional image has three aspects:

1. Facades: it is necessary to have a labor of preservation, maintaining and substituting with the ending of homologating the facades. The tasks are aimed to highlight the vernacular architectonic value, for that, it is necessary to take away the visual polluted stuff as signs or wires.

We pretend that in the fronts repetition will be avoided with the use of just one color according to the justification of the population and antique pictures. The color of the buildings was white made with water painting and a variety of colors provoking a heterogeneous and unharmonious image. To achieve joining and not to impose to the inhabitants only one color for their facades, it was proposed a gamut of heat colors to harmonize with the Green context of the forest that is surrounding the settlement.

2. Pavements: according to the interviews on-site and vestiges found in the community, it is known that originally the streets were paved. It is necessary to return to the traditional image of the cobbled street. However, it is not a question of changing all streets indiscriminately, which would be of a high economic cost. This is why it is basically raised in the two main streets and in the central square. With regard to the square, its streets are severely damaged, so it must be reconstructed, but with cobblestones. In relation to the two streets, both are in good condition by a layer of concrete that was made recently, in that case, it is suggested to use that layer as a base and on it build the paving. Two types of accommodation of the stone are proposed in the streets to harmonize with the architectural context.

3. Urban Image: for the buildings that have shopped a type of sign for advertisements was designed. Looking to give order and generate an image that does not pollute visually, it is desirable to approve commercial ads, to generate a set image. For this, Wood was used, which is abundant material in the region.

The visual problem of aerial wiring will be solved by making the wiring light and underground telephony. This will have an excellent image because this is one of the strongest visual contaminants.

On the other hand, one of the buildings of greater volume and height is the municipal auditorium, of recent edification. To which several inhabitants qualify as "White Elephant", this nickname is because it breaks with the visuals of the community by its magnitude. This building is located at the entrance to the population center, so it visually dominates the context. It is proposed to take advantage of the walls of the lateral and rear facades, in which there are no elements of any kind. It was designed a mural of terracotta tiles of 20 cms times 20 cms, to locate these facades. The mural is a welcome sign and farewell to the visitors of the municipality. The idea is to give identity by means of traditional elements of the district, since allusive images were designed for the cultural activities and the image of the main saint. In addition, we encourage the artisan production workshops of clay objects, which are currently being abandoned.

5 Conclusion

In Mexico, there is an accelerated loss of examples of vernacular architecture, especially in small populations. The factors of deterioration and destruction of the vernacular heritage in Mexico are that municipality and state authorities who do not know this heritage and they do not appreciate it. Cobbled streets are generally destroyed with the idea of modernizing communities, by means of concrete. Public buildings of government as well as markets, which are part of the vernacular heritage, are destroyed under the same perspective. The population allows these destructions, convinced that it is better to have new buildings of modern materials

On the other hand, the vernacular houses are destroyed under the same idea. Migration to the United States of America and towards cities, it facilitates the destruction of traditional architecture. Usually, migrants try to mimic the buildings they see somewhere else and with the resources they generate outside their communities, they make modifications to show off status.

The project presented is an attempt to protect and preserve the vernacular heritage built. The most significant thing is that this type of project is carried out in agreement with the municipal authorities and the universities, but with the participation of the inhabitants. The project is currently running in the community. It was presented to religious and social representatives, who approved of it, so as to give an image of traditional identity to their community, for the benefit of all of the people. It is expected to encourage tourism and generate appreciation towards the vernacular architecture.

References

Comision Nacional de Pueblos Indigenas, 2019. Atlas de los pueblos indigenas de Mexico. Mexico City: CDI.

Gobierno de Mexico, 2019. Atlas de los pueblos indigenas. Mexico City: inafed. gob. mx.

LÓPEZ F J, 1987. Arquitectura vernàcula en Mèxico. Mexico City: Trillas.

MOCTEZUMA M, 1999. La casa prehispànica. Mexico City: Infonavit.

PRIETO V, 1994. Vivienda campesina en Mèxico. Mexico City: Sedesol.

PRIETO V, 1984. Arquitectura popular mexicana. Mexico City: Sedesol.

TORRES G, 2009. La arquitectura de la vivienda vernàcula. Mexico City: Plaza y Valdes.

Research on Protecting and Utilizing Cultural Landscape Heritage of Huizhou Ancient Roads

BI Zhongsong

Institute of Architecture and Engineering, Huangshan University, Huangshan, China

Abstract Huizhou ancient roads, in a broad sense, refer to all the flagstone mountain roads within ancient Huizhou and the outwards roads to the affiliated ancient pavilions, bridges, temples and other architectural heritage, including rich cultural and natural landscapes, with high historical, artistic, scientific, cultural and social values. Based on the theories of cultural landscape and cultural routes, the paper is written to discuss the main heritage composition, prominent universal value and typical cultural characteristics of cultural landscape heritage about Huizhou ancient roads, analyze the basic situations, value embodiment, preservation and utilization of Huizhou ancient roads by taking the following ancient roads for examples including Hui-hang ancient road — from Huizhou to Hangzhou, Jigen Pass ancient road, Hui-Ning ancient road — from Huizhou to Ningguo, and Dahong ancient road, further, try to survey the related reasons and based on the researches above, the ideas of how to protect and utilized those roads are summarized, more stresses are put on providing new ideas of how to make Huizhou ancient roads apply for world heritage successfully to make a beneficial discussion for protecting and rationally utilizing cultural landscapes of Huizhou ancient roads.

Keywords Huizhou ancient roads, rural landscape, linear heritage

1 Introduction

Ancient Huizhou began in the Qin Dynasty with two counties of Shezhou and Yixian. Song Huizong (1121) changed Shezhou into Huizhou. Since then, the geographical category of "One Government and Six Counties" has been steadily extended, namely, Shexian, Yixian, Xiuning, Jixi, Wuyuan and Qimen under the Huizhou government, located in Hui county, Shezhou town (Fig. 1). From the fifth year of the Tang Dynasty to the end of the Qing Dynasty, the scope of ancient Huizhou basically did not change a lot. The average altitude of ancient Huizhou was 875 m, and the area was about 11,894 km^2.

Huizhou is not only a historical-geographical concept, but also a cultural unit today. It represents the extensive and profound regional culture of Huizhou. The unique living environment of "seven mountains, one river, one field, one road and manor" and the helpless life experience of "If one were not moral in the previous

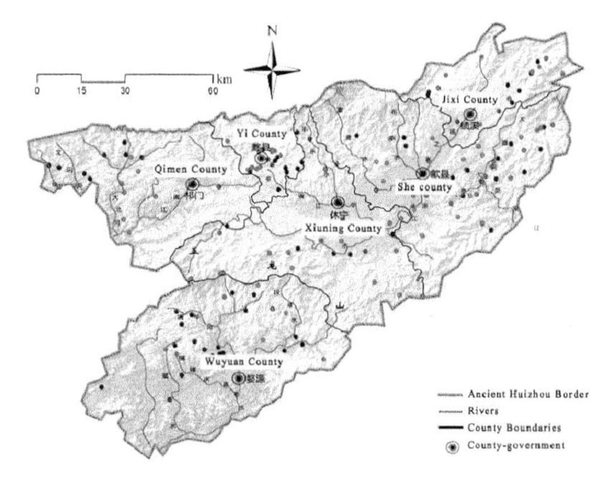

Fig. 1 Scope map of Huizhou prefecture

life, he would be born in Huizhou; if one turned thirteen or fourteen, he would leave Huizhou and go out" helped the industrious ancestors of Huizhou open up lots of roads of Huizhou merchants communicating with the outside world, that is, the Huizhou ancient road, and thus created the legend of Huizhou merchants that the ancient Huizhou kept developing for three hundred years in ancient China with trading but not farming, and even created a unique landscape of "There would be no town without Hui merchants".

According to incomplete statistics, there are about 124 ancient roads and 1,000 ancient bridges in the territory of "One Government and Six Counties". However, with the development of the economy and the continuous reconstruction of the modern transportation system, the traditional road system has been impacted unprecedentedly, and affected by such geological disasters as mountain torrents. At the same time, Huizhou ancient road has been greatly damaged due to its long-time disrepair. Huizhou ancient road urgently needs rescuing conservation and research, so as to make the Huizhou ancient road play a more valuable and positive role.

2 Heritage Composition of Huizhou Ancient Road

2.1 The Category of Huizhou Ancient Road

In general, Huizhou ancient road refers to all the stone slab mountain roads in the territory of "One Government and Six Counties" of ancient Huizhou government and from the territory of Huizhou to the periphery, including villages along the route, architectural heritages, sites and intangible cultural heritages, and so on.

2.2 The Heritage Distribution of Huizhou Ancient Road

At present, there are about 124 large and small ancient roads in the territory of "One Government and Six Counties". Among them, there are 50 well-preserved ancient roads, such as Huizhou to Hangzhou, Huizhou to Shangrao, Ruoling and Huizhou to Qingyang, with a total length of 442 kilometers. While there are more than 20 ancient roads with complete historical value, which pass through ancient villages and towns winding between the mountains of the region in ancient Huizhou.

2.3 Characteristics of Heritage Composition of Huizhou Ancient Road

Huizhou ancient road contains rich heritages, not only the stone slab roads, but also traditional villages, bridges across rivers, pavilions along the road, etc. The heritage of the ancient roads covers rich material heritages such as ancient stone roads, ancient bridges, ancient pavilions, ancient tombs, ancient monuments, ancient temples (ruins), as well as the intangible cultural heritages such as regional living habits, traditional technology, production skills and farming methods, etiquette and customs, religious beliefs, local dialects, traditional costumes, architectural styles, construction techniques and village structures. This paper will analyze the material heritage composition of Huizhou ancient road with specific examples.

2.3.1 Ancient Path

Huizhou ancient road is located in the mountainous area of southern Anhui province. It basically chooses straight and short terrain with abundant sunshine and good geological conditions. The paving materials of the ancient road are basically from the local. The local granite and shale are selected as the main paving materials, and its hardness is high and not easy to be weathered. The width of the ancient road varies from 1 to 4 meters, which is determined by its road grade. The road guardrail is determined by the relationship between the mountain terrain and the road to ensure traffic safety.

Huirao ancient road, also known as "Huizhou Avenue", is the main passageway from the Huizhou government in ancient times to the Raozhou government in Jiangxi province. It is an important business road for ancient Huizhou merchants to enter Jiangxi province, with a total length of more than 100 kilometers. The existing section is located in Shitai county, Chizhou city and Qimen county, Huangshan city. The ancient road in Shitai county is called Jugenguan Huizhou ancient road (Fig. 2). The ancient road in Qimen county is called the western line of Huizhou ancient road. These two sections are connected, which are 5.4 kilometers long altogether. They are paved with rectangular bluestone slabs about 1.3 meters long, 0.6 meters wide and 0.1 meters thick. There are about 18,000 steps in total.

Fig. 2 Jugenguan Huizhou ancient road

Huihang ancient road starts from Jiangnan village, Fuling town, Jixi county, Anhui province in the west, to Maxiaoxiang town, Linan city, Zhejiang province in the east, with a length of 25 kilometers, which is an important link between Huizhou and Zhejiang in ancient times. It was built originally as early as the fifth year (1257) in Baoyou in Southern Song Dynasty. Huihang ancient road is the east-west trend, paved or chiseled with stone steps with 1.2～1.7 meters long and 0.4～0.6 meters wide (Fig. 3).

Fig. 3 Huihang ancient road

Dahong ancient road is the only way to Anqing, the capital of ancient Huizhou. The existing section is located in Qimen county, 7.5 kilometers long and 3～4 meters wide. It is a rare avenue in ancient Huizhou (Fig. 4).

2.3.2 Ancient Architecture

There are abundant architectural heritages such as ancient pavilions and bridges along Huizhou ancient road. Along the Jugenguan Huizhou ancient road, there are various architectural heritages such as Jibao

Fig. 4 Dahong ancient road

Pavilion, Guxi Pavilion, Yuquan Pavilion, Qiyuan Pavilion, Bibo Pavilion, Gangliang Bridge and Sanban Bridge. Along the ancient road of Huizhou to Hangzhou, there are Yankou Pavilion, the First Pass in the South of the Yangtze River, Ercheng Pavilion, Ercheng Temple, Shicha Pavilion, Banling Pavilion and other ancient buildings. Along Dahong ancient road, there are Yongzhen Bridge, Nanpo Ancient Pavilion, Guanyin Pavilion, Bibo Ancient Pavilion, Yongqing Bridge and other ancient buildings (Tab. 1, Fig. 5, Fig. 6).

Fig. 5 Guxi Pavilion of Jugenguan ancient road

Fig. 6 Ercheng Pavilion of Huihang ancient road

Tab. 1 The formation table of historical remains of Jugenguan ancient road and Huihang ancient road

Name of Ancient Road	Heritage content	Years	Remarks
Jugenguan ancien-troad	Jibao Pavilion	Qing dynasty	Covering an area 25. 5m^2
	Guxi Pavilion	Qing dynasty	Covering an area 17. 2m^2
	Yuquan Pavilion	Qing dynasty	Covering an area 25m^2
	Qiyuan Pavilion	Qing dynasty	Covering an area 41. 2m^2
	Gangliang Bridge	Qing dynasty	Bridge length 3. 4m, width 0. 4m
	Sanban Bridge	Qing dynasty	Bridge length 5. 2m, width 1. 45m
	Bibo Pavilion	Qing dynasty	Covering an area 32. 5m^2, Remaining Sites
Huihang ancient road	Yankou Pavilion	Song dynasty — Republic of China	One room for wide and deep
	The First Pass in the South of the Yangtze River	Qing dynasty	Embossing "Huihang Key"
	Ercheng Pavilion	Qing dynasty	Broad-leaved 8m, depth 3. 5m, high 4m
	Ercheng Temple	Republic of China	High 2m, wide 1m
	Shicha Pavilion	Republic of China	depth 3. 9m, Broad-leaved 4. 1m
	Banling Pavilion	Qing dynasty	Broad-leaved 7. 5m, depth 3. 8m, high 4. 5m

2. 3. 3 Ancient Tomb

Along the Jugenguan Huizhou ancient road, there are ancient tombs such as monks' tombs and the tombs to sacrifice the lonely souls. There is one existing tomb to sacrifice the lonely souls, a tombstone standing beside Huizhou ancient road. It was built in the time of the Republic of China to mourn the dead in the war years, and also to offer sacrifices to those who died of illness when walking on the road. There are two places for the monks' tombs, which were in the south and north of the Huizhou ancient road respectively at Jugenguan ancient road and there are several existing tombstones (Fig. 7).

Fig. 7 The monks' tombs of Jugenguan Huizhou ancient road

2. 3. 4 Ancient Site

Along the Jugenguan Huizhou ancient road, the ancient Great Wall, the Yuantong Nunnery ancient site and the Wushang Nunnery ancient site and some others are distributed. Along the ancient road of Huizhou and Hangzhou, Shisi site, Land Temple site, Tea Pavilion site, Wuliting site, Snow Pavilion site, Road Pavilion site, Old Bridge site and Lingguan Temple site are distributed (Fig. 8).

Fig. 8 The Jugenguan Pass of Jugenguan Huizhou ancient road

2.3.5 Ancient Stele Inscription

Ancient stele inscription is the witness of history and the transmission of historical information. Along the Jugenguan Huizhou ancient road, there are ancient stele inscriptions such as mountain transporting stele, rebuilding stele and bar-girder bridge remembering stele. Along Dahongling ancient road, there is Dahongling stele inscription (Fig. 9).

Fig. 9 The Inscriptions on a tablet of Yuantong Nunnery ancient site of Jugenguan Huizhou ancient road

2.4 Analysis of the Value Characteristics of Huizhou Ancient Road Heritage

Huizhou area is located in the mountainous area of southern Anhui province. As an important land passageway for Huizhou merchants to communicate with the outside in history, Huizhou ancient road combines natural mountain topography. From the view of selecting the project route, straightforward and short terrain with abundant sunshine and good geological conditions tend to be chosen. It should hardly encounter any landslides and debris flows. Its route selection is closely related to local humans' settlements, property, location and other factors, conforming to the primitive natural view of "harmony between man and nature", reflecting the perfect combination of man and nature, and is an important representative type of cultural landscape heritage.

At the same time, as an important passage space and regional cultural communication route in Huizhou area, Huizhou ancient road has a clear starting point and boundaries. Since its establishment, it has been further expanded and improved in different historical periods. It is a great manpower pioneering created by Huizhou ancestors by relying on their own efforts to communicate with the outside among the lofty mountains of Huizhou. Huizhou ancient road is not only "vital communication route", "folk route", but also "commercial route", "cultural route" and even the important "military route", which have played an important role. At the same time, it is closely related to the tangible and intangible cultural heritages such as ancient villages, ancient buildings (sites) along the road. Its value is embodied in many aspects such as tangible and intangible cultural heritages, and it is also an important representative type of cultural route heritage.

Therefore, as an artificial road, Huizhou ancient road combines with the natural environment, which has the dual contents of cultural landscape heritage and cultural route heritage, and has its outstanding universal value and typical regional cultural characteristics.

3 Protection and Utilization Strategy of Huizhou Ancient Road Heritage

3.1 The Analysis on Main Existing Problems and Reasons of Huizhou Ancient Road

3.1.1 An Incomplete Value Evaluation System of Huizhou Ancient Road

It is mainly reflected in two aspects, one of which is the value evaluation system of Huizhou ancient road in history, the other is the existing value evaluation system. At present, these two aspects of ancient roads have not been established completely, which is not conducive to the deep excavation of the value of Huizhou ancient road and the active protection and utilization of Huizhou ancient road.

3. 1. 2 An Incomplete Protection System of Huizhou Ancient Road

Among the existing well-protected ancient roads in Huizhou, the Jugenguan Huizhou ancient road, Jixi Section of Huizhou to Hangzhou ancient road and Langxi Section of the East Line of Huizhou ancient road were declared as the seventh batch of national key cultural relics protection units, while the West Line of Anhui ancient road and Dahongling ancient road were declared as the sixth batch and the seventh batch of Anhui provincial cultural relics protection units respectively, but materials about Huizhou ancient road serving as the city and county-level protection unit has not been found yet and this is far from enough for Huizhou ancient road, which has a large amount of relics.

3. 1. 3 The Damages Huizhou Ancient Road Faced With

Due to historical and social reasons, as well as natural and artificial factors, the degree of destruction that Huizhou ancient road suffered from is accelerating in recent years. The main manifestations are as follows: the first is the destruction of ancient roads by highway construction. It is found out in the investigation that many modern highways, rural roads, village roads and others were remoulded with asphalt and cement on the original ancient road, which has greatly destroyed the authenticity and integrity of the ancient emblem road, and has split the system continuity of the ancient emblem road. The second is because the Huizhou ancient road is mostly located in mountainous areas, and in recent years the frequent disasters in mountainous areas have also caused great damages to the existing sections of the road, and the roadbed and road surface of many sections of the road have been seriously damaged as well. The third is some villagers along the road uncovered the stone slabs for other construction activities like yard paving, which also caused different degrees of impact on the road.

3. 1. 4 The Influence of Geological Disasters on the Ancient Road

The change of the ecological environment imposes a severe challenge on the protection of Huizhou ancient road. Because of its complex natural environment and geological conditions being located in southern Anhui areas. The landslide and debris flow would be avoided when the Huizhou ancient road was built. But it is found out in the investigation that the biggest problem the protection of the existing Huizhou ancient road is faced with is the impact of natural geological disasters such as landslides and debris flows on the road. Such common problems as landslides, debris flows and mountain torrents have been found in the investigation of HuiRao ancient road-Jugenguan ancient road, the western line of Huizhou ancient road and Jixi Section of Huizhou to Hangzhou ancient road. Such geological disasters affect ancient emblem road and its attached cultural relics, mainly resulting in the absence of slate, roadbed damage, slate loosening damage, stone, grass or stone covering roads, etc, which are very destructive to ancient emblem road.

3. 2 Basic Strategies of Heritage Protection and Utilization of Huizhou Ancient Road

3. 2. 1 Figuring out Clearly and Establishing a Scientific Evaluation System for the Value of Huizhou Ancient Road

We will carry out a comprehensive survey of Huizhou ancient road resources and speed up the systematic investigation and collation of the basic materials of Huizhou cultural route heritage, like ancient Huizhou road. We should fully understand the important role of the cultural landscape heritage of Huizhou ancient road in the formation, development and dissemination of Huizhou culture. Huizhou ancient road is not only a passage, but also a close connection with Huizhou culture, like a necklace. If Huizhou culture is a pearl on the necklace, and Huizhou ancient road is the rope that wears the necklace. Without this rope, each pearl can not maximize their overall values altogether. So Huizhou ancient road is the link that maximizes the overall values of Huizhou culture. We should systematically figure out the history and current situation of Huizhou ancient road, use modern information technology such as three-dimensional spatial scanning, big data, geographic information system to establish the database of Huizhou ancient road, which help provide basic data support for the research, protection and utilization of Huizhou ancient road, which requires good cooperation among different regions and departments. Then on this basis, a complete scientific evaluation system of the value of Huizhou ancient road will be established.

3.2.2 Forming a Complete Overall Protection System of Huizhou Ancient Road

We should gradually improve the research and protection system of Huizhou ancient road and implement overall protection. We should carry out research work on the basis of improving the basic data of ancient roads, complete their archives, and finish the "Four Archives" arrangement of each ancient road in accordance with the requirements of the protection of cultural relics, which mainly include the following aspects:

First, we will gradually carry out the research work on the protection technology of Huizhou ancient road, including related theoretical research and construction techniques, to provide theoretical basis and technical support for the restoration and protection of Huizhou ancient road and its affiliated relics.

Secondly, we should complete the protection system at different levels, deeply dig out the typical value of Huizhou ancient road according to its historical archives and preservation situation, declare national key cultural relics protection units, provincial cultural relics protection units, city and county-level cultural relics protection units, finally forming a complete protection system of Huizhou ancient road —"city and county-level protection, province-level protection, national protection and world heritage".

Thirdly, we need to maximize and complete the protection scope and list of Huizhou ancient road, and implement the overall and systematic protection of Huizhou ancient road.

3.2.3 Rational Utilization of Huizhou Ancient Road

Rational use of Huizhou ancient road should be made and we need to bring its social value into full play. Reasonable utilization is an important part of the protection of cultural relics and monuments. According to the value, characteristics, preservation situation and environmental conditions of cultural relics and monuments, comprehensive consideration should be given to various utilization modes of researching, displaying, extending their original functions and giving them appropriate contemporary functions. Utilization should emphasize public welfare and sustainability, and avoid over-utilization[6]. In the development of tourism, we should avoid the management thinking of emphasizing development and neglecting protection.

3.2.4 Culture Construction of Huizhou Ancient Road and Ecological Protection Corridor

Construction of cultural and ecological corridors, construction of three-dimensional route space of Huizhou ancient road is of importance. We should rely on the achievements of the construction of Huizhou cultural and ecological protection zone, combine the natural environment and geological conditions of the mountain body in which Huizhou ancient road lies and construct the cultural and ecological corridor of Huizhou ancient road.

Firstly, we should deeply dig out the typical value of Huizhou ancient road and its related subsidiary heritage, enrich and complete the regional living habits, traditional technology, production skills, farming methods, etiquette customs, religious beliefs, as well as such intangible cultural connotations as local dialects, traditional costumes, architectural styles, building techniques and village structures, and display and publicize them rationally in combination with modern technology and expand its cultural extension.

Secondly, we should protect the mountain environment, prohibit anthropogenic activities such as exploitation of mountain forests, deforestation and tea planting in the road area, and restore the mountain ecological environment in ecologically fragile areas.

Thirdly, it's important to entrust units with professional qualifications to carry out special assessment and prevention plan of geological disasters in the road area, and to implement related projects for prevention and control of geological disasters, so as to ensure the regional geological safety of Huizhou ancient road, so as to reduce the damage of geological disasters to the road.

3.3 New Thoughts on "Applying for World Heritage" of Cultural Landscape of Ancient Road in Huizhou

With regard to the protection of world cultural heritage, since the entry of Santiago de Campostra Pilgrimage Road into the World Heritage List in 1993, a large number of cultural route heritage and cultural landscape heritage have been included in the World Heritage List, which reflects the development trend in the field of cultural heritage protection, that is, the scope of cultural heritage protection has been expanding, from single cultural relics to historical sites, to the whole town, and then to cultural landscape, heritage areas, and even several or even a dozen of cities, larger cultural areas of one or more countries. This also provides a new way of thinking for the protection of such cultural route heritages as Huizhou ancient road, that is, to declare the road as a whole as the world heritage, to improve the protection system of Huizhou ancient road, so as to make better protection and utilization of Huizhou ancient road.

Referring to the World's and China's existing cultural route world heritages, and combining with the basic characteristics of Huizhou ancient road cultural routes, we should carry out a reasonable survey on resource, figure out the history, establish a database of Huizhou ancient roads, carry out its research and renovation and protect the public resource. Taking Jixi section of Huihang ancient road, Langxi section of the eastern road and Jugenguan Huizhou ancient road as the leading units, we should unite other types of ancient roads in Huizhou area, dig out deeply the content as well as the universal and prominent value of Huizhou ancient road, and declare Huizhou ancient road as a whole as the World Cultural Landscape Heritage or Cultural Route Heritage according to the relevant requirements of the World Heritage Committee.

4 Conclusion

The cultural landscape heritage of Huizhou ancient road stretchs thousands of miles, forming a history of thousands of years. It is a long stream of river witnessing the origin, development and dissemination of Huizhou culture, and an encyclopedia displaying Huizhou traditional culture. Huizhou ancient road with rich cultural connotations and broad cultural extension is not only the precious cultural wealth of Huizhou, but also an important part of cultural heritage in China. Today, Huizhou ancient road still has inestimable positive value and plays an extensive role, whose characteristics, functions and values are difficult to be compared by other cultural heritage categories.

Therefore, strengthening the protection of cultural route heritages like Huizhou ancient road is conducive to the rescue of Huizhou cultural heritage clusters, to displaying profound cultural connotations, to reproducing the rich cultural connotations of Huizhou "the stream of culture" and "encyclopedia", and to extending the inheritance and development of Huizhou culture.

References

CHU Jinglong, LIU Han, LI Jiulin, 2018. Spatial distribution and evolution of ancient Huizhou traditional villages. Journal of Anhui Jianzhu University, 26 (3): 26-34.

ICOMOS China, 2015. Protection of Chinese cultural relics and monuments. Beijing: Cultural Relics Publishing House.

Jixi County Local Chronicle Compilation Committee, 1998. Jixi county chronicle. Hefei: Huangshan Book Club.

MA Yanji, 2012. Research on Huizhou ancient road cultural route. Heifei: Anhui University.

SHAN Jixiang, 2009. Paying attention to the new cultural heritage-protection of cultural line heritage. Scientific research of Chinese cultural relics, 3: 12-23.

ZHOU Yang, CHEN Qi, 2016. An investigation into the present situation and value of the ancient roads in Huizhou. Journal of Huangshan University, 18 (4): 7-10.

Pristine Forests and Vernacular Architecture: Sustainable Development and Responsible Tourism in the Three Parallel Rivers Natural World Heritage Site

Anna-Paola Pola

World Heritage Institute of Training and Research for the Asia and the Pacific Region under the auspice of UNESCO (WHITRAP), Shanghai, China

Abstract Rural areas are undergoing dramatic changes due to pressures of urban development, abandonment, and marginalisation, resulting in the loss of traditional knowledge, and cultural and biological diversity. Yet, the critical role played by traditional villages and their vast surrounding environment is now acknowledged as a complementary counterpart to growing cities, acting as a provider of quality environmental services and leisure, and as a springboard for national and global values. The interest of urban inhabitants in rural landscapes, vernacular architectures, and traditional settlements is also rising, as evidenced by the rapid increase of tourism in rural scenic areas and the demand for high-quality food products grown on these lands. The paper observes on-the-ground practices carried out by local communities and organisations in ethnic minority villages (Lisu, Yi, Pumi, Bai) inside the Three Parallel Rivers Natural World Heritage site in Yunnan (Yulong County, Lijiang City). The case study highlights the role played by the natural environment and vernacular buildings to stimulate local development. In the framework of a strategy aiming to alleviate poverty and support rural livelihoods, eco-tourism and environmental education provide the financial resources needed for the protection of nature, the conservation of local culture, and the renovation of vernacular architecture.

Keywords sustainable development, environmental conservation, traditional villages, eco-tourism, UNESCO world heritage site

1 Introduction

Since the Chinese government has placed rural areas at the very top of the country's development agenda (Wang, 2019), programmes promoting the renovation of villages and vernacular architecture have propagated in China.

For millennia, the country has been an agriculture empire, marked by a dense network of rural settlements safeguarding its territory. Today, this legacy materialises in a multitude of vernacular buildings and traditional settlements that play a significant role in promoting China's ancient culture and redefining national identity within the global context.

Initiatives focusing on the charm of traditional built heritage, intangible local customs, and pastoral contexts to offer tourists entertainment opportunities are particularly successful. The case study presented in this article provides an insight into a different experience.

A remote and impoverished village, located in a protected area with stringent development limits, bet on turning its potential drawbacks into opportunities and chose to focus on the deep understanding and communication of its forest wilderness and biodiversity — two outstanding elements that characterise the place instead of falling back on easier, standard tourism strategies.

In this framework, the renovation of existing vernacular architecture is just part of a broader approach to the revitalisation of spaces and significances, aimed not only for visitors and travellers, but mostly for the local community. Architectural renovation followed the general principles of minimum intervention, original materials and techniques, and ecological priority. These criteria were not just an issue of style or nostalgia, they were instead the most suitable solution in terms of the appropriate use of available resources (including existing buildings), economic balance, and social benefits; in a word — I would say — in terms of sustainability.

2 The Three Parallel Rivers of Yunnan Protected Areas

The village of Liju, northwest of Lijiang City (Yulong County, Yunnan Province), is a fairly typical mountain village, with all the issues that it entails.

Running adjacent to the river, 13 clusters of "natural villages" form the Liju administrative settlement. Buildings are mixed both in materials and style: some are built with adobe bricks and stones, like many traditional dwellings in Lijiang valley; others have a concrete skeleton, notably the school and a few main public structures; and most are in wood, consisting of a simple room with a dirt floor and a slight hollow for a fire in the middle.

Liju lies within the Three Parallel Rivers of Yunnan Protected Areas, a natural World Heritage property comprising of 15 protected areas (1,700,000 ha), grouped into eight clusters of buffer zones (covering 1,730,700 ha). The site has a remarkable altitudinal gradient and an exceptional range of topographical features, due to the region being at the collision point of tectonic plates and including sections of the upper reaches of three of the great rivers of Asia — Yangtze, Mekong, and Salween — running from north to south(Fig. 1).

Fig. 1 Traditional wooden house in Liju village, July 2018 © The author

Where the Yangtze River turns its first dramatic bend, a valley branches off towards the west. River and road proceed side by side for about 40 kilometres in a narrow valley. Liju village covers the final portion of this path. At 2,800 metres above sea level, the road stops near the last building, while the river enters the forest of Laojun mountain. This primeval rainforest is one of the most biologically diverse temperate regions on earth. It harbours 168 endangered species, including the snub-nosed monkey (known in Chinese as the Yunnan golden monkey) and more than 10% of the world rhododendrons (P. R. of China, 2003). Biological diversity concerns both species (animals, plants, and fungi) and landforms (karst and red sandstone), while the area boasts multiple climatic conditions (from extreme aridity to humidity) and ranging temperatures (from day to night).

Despite this ecological wealth, most inhabitants of this area live below the poverty line.

Liju village consists of around 350 households, 1,357 people belonging to different ethnic groups such as Lisu, Pumi, Yi, Bai, Naxi and Han, and comprises 1,527 ha of forest (mainly virgin forest with some secondary growth).

Farming (potato, corn, buckwheat, cabbage, and white beans), fruit trees (walnut, green plum, and Sichuan pepper), and livestock (goats, cattle, sheep, and pigs) constitute the main subsistence activities. The local economy was largely based on timber and charcoal products, until 1998 when China launched the Natural Forest Protection Project (NFPP). The project commenced after the destruction caused by large-scale floods in China's major watercourses and aimed to protect the environment in the upper and middle reaches of the Yangtze and the Yellow River (Zhang et al., 2011) (Fig. 2). Crucially, the NFPP project marked the beginning of a national shift towards what is now called the Ecological Civilization paradigm. Environmental scientists demonstrated that over-grazing and deforestation affected the ecosystem and exacerbated the effects of the floods, thus, instead of building higher dams and levees like in the past, the government commenced a series of programmes to facilitate the restoration of forest ecosystems (Gao, 2019). As a result, a logging ban was imposed on the region, greatly helping environment conservation but making local populations' livelihood more precarious.

Fig. 2 The first bend of the Yangtze River, Shigu town, August 2018 © The author

To address this situation, the government launched a series of targeted poverty alleviation initiatives in the village, involving infrastructural improvement, educational development, and medical care. Welfare and services came together with direct assistance to the poorest families, not in the form of money, but via breeding sows and more profitable plants to grow such as: mushrooms (wood-ear mushrooms) and medical plants (*chonglou*, used in the Yunnan Baiyao medicine for its detoxification and analgesic effects, and costus, a species of thistle whose essential oils extracted from the root have been used in traditional medicine and perfumes since ancient times).

Six years ago, a new school was built in the village; it now has 96 students, 10 teachers, and covers seven grades of education. From Monday to Friday, children of the areas under the jurisdiction of the Liju committee reside at the school dormitory. Meals, accommodation, and books are free of charge, and a monthly subsidy of 100 yuan is given to students from low-income families. Outside the school, a long list of sheets reporting poverty alleviation funds hangs on the wall.

In front of the classrooms is a new four-bed clinic and a basketball playground between the two buildings; on the opposite side of the road, a post office and a small shop complete the core services of the village. Two doctors assist people with minor diseases and, most importantly, since a few years ago, all residents have access to affordable basic medical insurance.

In winter 2017—2018, a newly paved road reached the village (Fig. 3). With the road arrived changes, always bringing opportunities in pair with threats.

Fig. 3 A secondary branch of the paved road under construction, November 2018 © The author

3 Laojunshan National Park

Laojunshan National Park covers the Liming area, located on the northern slope of Laojun Mountain. The south side of the mountain, where Liju sits, is included in the World Heritage core area, but it is not formally structured as a national park. This situation has a two-fold effect: funds for environmental protection struggle to get there, and so does mass tourism development.

National parks in China are not defined by national legislation. They were first introduced in Yunnan province, playing a demonstrative role for the rest of the country. Therefore, the so-called "national parks" are an experimental hybrid of two Chinese protected-area classifications: "nature reserves", which safeguard natural resources and largely prohibit the human activity, and "scenic areas", which foster development and mass tourism (Ives, 2011). Since Laojunshan National Park opened in 2008, Liming's development has been focused on promoting its natural environment for mass tourism: electric shuttle buses, cable cars, boardwalks, and luxury eco-tent lodges cater to busloads of wealthy travellers. Similar plans were in store for its other famous scenic spot: the 99 Dragon Pools, a small alpine basin dotted with lakes and rhododendron forest. Rumours, based on preliminary development plans, foreshadowed a similar fate for the Liju area, suggesting that after the completion of the road, a tourist tram may follow.

Yet, in March 2018, Xinhua, the official state-run press agency, announced the national reform reorganising the Ministry of Ecological Environment (MEE) and establishing the new Ministry of Natural Resources (MNR) (Xu, 2018). This rearrangement highlights the growing importance attributed to environmental issues in the country (Ma & Liu, 2018). It also assigns the new Ministry of Natural Resources with the task of delineating "ecological red lines" to protect resources across the nation.

The China Ecological Conservation Red Line is an initiative aiming to protect more than one-quarter of the Chinese mainland territory — 2.4 million square kilometres, the size of France, Spain, Turkey, Germany, and Italy combined — selected for its biodiversity, ecosystem services (including access to fresh water), or the environmental capacity to buffer natural disasters. According to one of its initiators, the project is designed to protect almost all rare and endangered species and their habitats in China. It draws stringent boundaries to safeguard these areas from industrialisation and urbanisation, and also aims to restore the ecologically fragile environment and protect human settlements (Gao, 2019).

Hence, in early 2019, through an ineluctable decentralisation mechanism, the Yulong County Environment Bureau announced its "ecological red line" on Laojun mountain. Since then, development plans were put on hold, the 99 Dragon Pools site was closed to the public, and the administration is now pondering how to proceed. The main issue for Chinese officials is the management of conflicts between biodiversity protection and rapid development, both unavoidable priorities in contemporary China.

In this regard, the Three Parallel Rivers of Yunnan Protected Areas faces unique economic pressures and land-use tensions. Although the area is a natural World Heritage site, more than one million people live within its boundaries, a number substantially higher than any other site inscribed on the World Heritage List. Many of these communities live below the poverty line and depend on natural resources for survival. Moreover, about 86.6 percent of the local population belongs to ethnic minority groups.

Poverty eradication has always been regarded as an important historical mission of the Chinese Government and the Communist Party of China (CPC). The substantial reduction of people living in poverty is one of the most crucial national achievements:

during the 40 years of reform and opening-up, 700 million people have been lifted out of poverty, accounting for more than 70 percent of poverty reduction worldwide (Tan, 2018). We are now experiencing the country's last strain in eradicating poverty. In October 2017, during the 19th National Congress of the Communist Party of China, President Xi Jinping clearly defined this goal: "We must ensure that by the year 2020, all rural residents living below the current poverty line have been lifted out of poverty, and poverty is eliminated in all poor counties and regions" of China (Xi, 2017).

4 Liju Village

Despite the 1998 logging ban, threats to forest conservation in Liju village were not immediately overcome. Deprived of their primary source of income, many villagers turned to illegal hunting, timber cutting, and charcoal producing. Changes started only in 2003, when The Nature Conservancy (TNC), a Virginia-based, international non-profit organisation, came to Liju.

This NGO has been working in Yunnan for 20 years and holds an important role in the development of the "national park pilot program" working as an advisory organisation for the 2003 nomination of the Three Parallel Rivers of Yunnan Protected Areas (Ives, 2011; P. R. of China, 2003).

The Nature Conservancy (TNC) and the Chinese NGO, the Alashan Society of Entrepreneurs and Ecology (SEE) Foundation (that later began to work in the area), have engaged and supported Liju inhabitants in wildlife conservation. Beekeeping was fostered as an additional source of income for the sake of nature conservation in the valley. Most importantly, two local teams (10 people in total) have been established to patrol the forest. The Laojunshan patrols are assigned the tasks of preventing fires, and controlling poaching and illegal harvest of timber and other forest products. National and international experience shows that forest stewardship almost invariably employs residents of rural communities, because local people know the land and where poachers and timber harvesters are likely to go. Sometimes patrols even enlist former poachers who are willing to turn the knowledge gained from illegal activities into a resource for preserving the environment and earn a safe and regular salary.

Clad in green camouflage gear, patrols have to scout the forest, spending several days and nights in the woods(Fig. 4). They watch signs of wildlife, monitor animals, plants, and people activities, keep track of the infrared-triggered cameras that take pictures when set off by animal body heat, and gather all information in their notebooks. Back in the office, they log the data onto computers, and professional analysts process, analyse, and monitor the information (Zinda, 2019).

Fig. 4 Laojunshan patrols, April 2019 © The author

As a result of this enhanced supervision, illegal activities within the protected area have been greatly reduced.

In China, there are a total of 60,000 local patrol teams (Wang, 2016). They employ various methods and funding sources. Some use the "ecological benefit compensation" given to communities for maintaining forests for the public good (Zinda, 2019) and others are supported by individuals, businesses, governmental bureaus, or NGOs, like in the case of Liju. Yet, many programmes nationwide remain only in nominal existence, having been set up without sound funding mechanisms (Wang, 2016).

Liju benefits also from an additional project the village is undergoing. Once the paved road to Liju has been completed, tourism will emerge as a promising industry in this area. The arising challenge would be to protect biodiversity and ensure that tourism revenues benefit the local community. A group of small companies, engaged in environmental education and responsible eco-tourism, are trying to further encourage Laojunshan's development towards a long-term sustainable direction. The project has been developed and implemented through a long process of identification and outreach of stakeholders — comprising villagers, local committees, patrol teams, members of NGOs, tourist companies, concerned county bureaus, and other World Heritage and UNESCO related institutions — to involve everyone in achieving the common goal of environmental conservation and sustainable development.

A small loan from a local development fund made it possible to restore a cluster of houses at the end of the village, right where the road stops to make way for the forest (Fig. 5, Fig. 6). These existing buildings were empty and belonged to local families who have been involved in the project since the very beginning. They have been transformed into a base camp rented out for travellers and student accommodation.

The structures, in wood and stone, have been restored with the help of the owners and local workforce. They have been provided with a small, new toilets and showers block, slightly detached from the main buildings, consisting of a dry wooden construction, easily removable and equipped with solar panels (there is no electric grid) and a phyto-purification system for the drains.

Edifices provide accommodation and rooms for meals, meetings, classes, presentations, projections, etc. The companies offer activities and experiences mainly focused on environmental awareness and cul-

Fig. 5 Liju village, main settlements, November 2018 © The author

Fig. 6 Liju village, small cluster of buildings at the end of the road, November 2018 © The author

tural appreciation, excursions and hikes in mountains, lectures, and environmental education camps for children and students. Visitors have the opportunity to learn about culture, nature, and sustainability, and to be actively involved in conservation projects contributing to local development. Village families contribute to the planning, organisation, and implementation of the activities, moreover, they provide accommodation, food, and guided tours to gain an additional source of income.

This responsible form of eco-tourism and the environmental education courses are primarily seen as a financial resource for nature protection and as a strategy to support rural livelihoods. Each year, 20 percent of the profits generated from these activities go to a Sustainable Action Fund to advance nature conservation (enhancing equipment for wildlife monitoring, protection, and patrolling activities) and local community development (beekeeping technology improvement and village infrastructure construction).

Although tourism is a rapidly growing sector in China, eco-tourism, understood as a sustainable, nature-based tourism, is still in its infancy. There is no commonly agreed upon definition of eco-tourism, and this concept is often perceived differently in China and Western countries, as exemplified by the "national parks" experience in Yunnan. Yet, the tourism market is rapidly changing, and China is now targeting sustainability and eco-civilisation, while mass tourism progressively loses its appeal among well-educated Chinese travellers. Therefore, the market requests emerging for these kinds of experiences are promising, and Liju's experimental approach offers a pioneer case study.

This project is also developing — with the collaboration of inhabitants, tourist companies, NGOs, and visitors — an inventory of local ethnic legends and know-hows (plant properties and specimen names in local languages) related to nature conservation. Although the Three Parallel Rivers is a natural World Heritage site, it has strong cultural values that are highly significant for the population living in the sites and for the safeguard of the natural environment. Similar to other countries in Asia, China sees nature and culture as inseparable elements. This is especially true here, where in the past Tibetan, Naxi, and Lisu's religion-driven worship of nature promoted a fierce protectiveness of nature that explains the residents' motivation for the conservation of natural resources. The linkage of their rich cultures to the land is expressed through religious beliefs, mythology, art, dance, music, poetry, and songs.

5 Conclusion

This comprehensive "renovation project" has brought a sense of pride to local communities. Eco-tourism has built confidence, fostered the awareness of the importance of forest protection, brought to light ancient ethnic traditions and customs and, most importantly, has given the people a reason to preserve and restore vernacular architecture.

Local communities, finally benefiting from the protected area, have realised that their culture, existing wildlife, and mountain ecosystem are their real resources, and have started to proactively join conservation efforts.

What is most interesting here, is that Liju is not a unique case, and the number of similar initiatives in China is rapidly growing (Wang, 2016; Zinda, 2019; Peng, 2019).

Liju village and other similar experiences, though very small-scale projects, gauge the effects of the central government's decisions at the local scale, and act as indicators of the fundamental changes taking place in the country. Notably, these independent experiences indicate both the effectiveness of governmental narrative and related policies in directing local innovation and the importance of small-scale experiences that can play a role, piloting and informing great national decisions. Liju's case shows that, at this moment, all around China bottom-up experiences and top-down policies pursue the same common goal: the achievement of a more equal, sustainable, and diversified future, for the good of the Chinese people, and hopefully for the whole of humanity.

Acknowledgements

The article is based on the author's fieldwork research and interviews with government officials, investment companies, NGOs, forest patrols, local residents, and tourist company members. Interviews on site were undertaken in July, August, November 2018 and April 2019.

The author would like to thank all the people interviewed and in particular Lijiang Conservation and Development Association and Wild Mountain Education Consulting Ltd. for their valuable support on-site.

References

GAO Jixi, 2019. How China will protect one-quarter of its land. Nature, 569: 457.

IVES M, 2011. If you save it, will they come? China's new parks. [2019-06-19]. http://www.earthisland.org/journal/index.php/magazine/entry/if_you_save_it_will_they_come/.

LUO Peng, 2019. Ecotourism project to promote conservation and local development at Daoyin village, Yinggeling Nature Reserve, Hainan, China. [2019-04-10]. https://www.theuiaa.org/culture-and-education/ecotourism-project-to-promote-conservation-and-local-development-at-daoyin-village-yinggeling-nature-reserve-hainan-china/.

MA Tianjie, LIU Qin, 2018. China reshapes ministries to better protect environment. (2018-03-14)[2019-06-19]. https://www.chinadialogue.net/article/show/single/en/10502-China-reshapes-ministries-to-better-protect-environment.

P. R. of China, 2003. The three parallel rivers of Yunnan protected areas. Nomination file for inscription into the World Heritage List. [2018-08-02]. https://whc.unesco.org/en/list/1083/documents/

TAN Weiping, 2018. Chinese approach to the eradication of poverty: taking targeted measures to lift people out of poverty. [2019-06-22]. https://www.un.org/development/desa/dspd/wp-content/uploads/sites/22/2018/05/15.pdf.

WANG Yan, 2016. Chinese villagers turn from logging to forest patrols, bees and fish. Mongabay series: Evolving conservation. (2016-06-08) [2019-05-23]. https://news.mongabay.com/2016/06/chinese-villagers-turn-from-logging-to-forest-patrols-bees-and-fish/.

WANG Yi, 2019. Letter of Mr. Wang Yi, state councilor and foreign minister of the P. R. of China, to Mr. Louis Gagnon, secretary-general of the Conference and Council Food and Agriculture Organization of the United Nations (FAO) for the nomination of Mr. Qu Dongyu as director-general of the FAO. [2019-06-24]. http://www.fao.org/fileadmin/user_upload/bodies/Conference_2019/MZ073_7/MZ073_C_2019_7_en.pdf.

XI Jinping, 2018. Secure a decisive victory in building a moderately prosperous society in all respects and strive for the great success of socialism with Chinese characteristics for a new era. Beijing: Foreign Languages Press.

ZHANG K, HORI Y, ZHOU S Z, et al., 2011. Impact of natural forest protection program policies on forests in northeastern China. Forestry studies in China, 13 (3): 231-238.

ZINDA J A, 2019. Managing the anthropocene: the labour of environmental regeneration. Made in China Journal, 3 (4): 62-67 [2019-05-23]. https://madeinchinajournal.com/2019/01/12/managing-the-anthropocene-the-labour-of-environmental-regeneration/.

Discussion on the Model of Sustainable Development of Community Based on Cultural Heritage Protection — Taking Nanjing and Ahmedabad as Examples

WU Jiayi

School of Architecture, Southeast University, Nanjing, China

Abstract The community, as an important place for citizens to live and communicate and a basic component of the city, its updating methods and measures are the focus of people's research. Especially in the problems of traditional communities, it is necessary to retain the traditional cultural characteristics and actively integrate into the development of modern cities. Under such a circumstance, as a socially living resource, cultural heritage plays an important role in the renewal of the community. Through the protection and reuse of cultural heritage, it activates the vitality of the community, establishes the cultural brand of the community, and realizes the sustainable development of the traditional community.

Through research, analysis and mirroring, the article compares and supplements the cultural and ancient capitals of two Asian developing countries — China and India. Through researching the policies and operational methods of cultural heritage protection and community renewal in Ahmedabad and Nanjing Laomen Dong, this paper summarizes some suggestions for future community renewal: adhering to the concept of people-oriented and sustainable development; meeting the needs of the masses, stimulating the enthusiasm of public participation; paying attention to the protection and reuse of cultural heritage, and explore the multi-value of heritage as a cultural resource. On the basis of the protection and reuse of cultural heritage, through the cooperation of managers, professional technicians and community residents, a multi-composite community culture platform will be built to promote the sustainable development of the community and ultimately realize the capital accumulation of the community.

Keywords traditional community, sustainable development, cultural heritage protection, Nanjing, Ahmedabad

1 Background

As the main gathering place for people's daily life and communication, the community is the main place of contradiction, and also the place where material and non-material wealth accumulate. In many developing countries, traditional communities contain enormous cultural heritage, which has great economic value and potential for sustainable development as an important cultural resource. Realizing the sustainable development of traditional communities on the basis of protecting cultural heritage is of great significance to the inheritance and continuation of culture, the enhancement of cultural identity of residents and the stimulation of urban vitality.

Nanjing and Ahmedabad[①], as representative cities with rich history and rich cultural heritage in the two ancient civilizations, face the same problems in many developing countries: the impact of foreign culture, the rapid replacement of material space, the huge population and the scarcity of land have led to the disappearance of traditional architecture and culture. At the same time, these two cities currently maintain a certain size of traditional communities. In the process of urban modernization, these traditional communities are slowly adapting to changes in the city and adopting a series of measures to update themselves.

The reason why these two cities were chosen is that they have similarities in cultural background, social development, and population economy. They also show similarities in some places, just like a mirror. It is meaningful for learning from experience and finding the crux. Specifically expressed as:

(1) Both the two countries are highly influential Asian developing countries. During the past few decades, there had been an economic boom and rapid modernization in the cities. The difference is that Chinese society is more homogenous and Indian society shows great differences and unstable.

(2) The selected communities have similar cultural backgrounds and rich cultural heritage, all of which have been influenced by larger religions. The difference is that the southern part of Nanjing shows religious relics, and religious traces are no longer obvious in modern life. On the other hand, the religious brand of the Ahmedabad community is very deep, embodied in all aspects of life, and the use of cultural resources is more abundant.

(3) It is also a large country with a large population. The family has lived in several generations. The community structure also has similarities. The difference is that the Chinese community is aging and even empty, while India is mostly young, and the younger community is more energetic than that in China.

2 Research

2.1 Protection and Renewal of Traditional Neighborhoods in Ahmedabad, India

On July 8, 2017, Ahmedabad became the first Indian city to be included in The World Heritage City. During the 19-month application process, Ahmedabad Municipal Corporation (AMC), the National Archaeological Department (SDA) and CEPT University provided a large amount of financial and technical support. Many scholars joined the community-building guidance work, and have already got a lot of practices, which can be summarized as follows:

2.1.1 Clear Comprehensive Plan

In the *Heritage Management Plan* released by Ahmedabad Municipal Corporation(AMC), clear requirements and recommendations were made at the community level, in which some of the city functions were transferred to the community, and community renewals were also integrated into the urban renewal, which provided the community a system to rely on, and get a certain amount of financial and technical support.

2.1.2 Protection of Cultural Heritage

Emphasizing on the protection and use of heritage, the management plan divides cultural heritage into three levels according to value and importance: Highest, High and Moderate. The transaction and development of houses must be carried out in strict accordance with the level, providing a tradable footprint based on the proportion of building units in the heritage area as defined in the *Heritage Management Plan*. At the same time, the Authority has developed a transparent government-based mechanism to monitor transactions and coordinate conflicts.

2.1.3 Featured Heritage Walk

The Cultural Relics Bureau sorts out the cultural relics in the old city, including important urban spaces (Royal Palace, Bhadra Castle, Jama Masjid, etc.) and important public spaces and religious sites in the old community (pol[②]). Based on its distributed combing structure, a path through the old community and city, called the "Heritage Walk", is well-recognized through brochures and posters. At the same time, the community as a participant put forward the help they can do: the residents of the route strive to keep the route clean and unobstructed; AMC also gives appropriate identification to residents who volunteer to participate in the tour guide and maintenance; AMC's street signage program completed the establishment of a municipal nameplate with routes and sites under the witness of officials, local representatives and neighbors; and in the end, the media also played a role in widely publicizing and raising general awareness among residents.

The "Heritage Walk" (Fig. 1) is not only an important tourist resource open to foreign tourists, but also the most important cultural symbol in the city. It is also the key to inspiring the vitality of the community and enhancing the cohesiveness of the community. Through this program, the transformation from cultural value to economic value has been realized, and the transformation of cultural resources into community capital is an important driving force for the sustainable development of communities and cities.

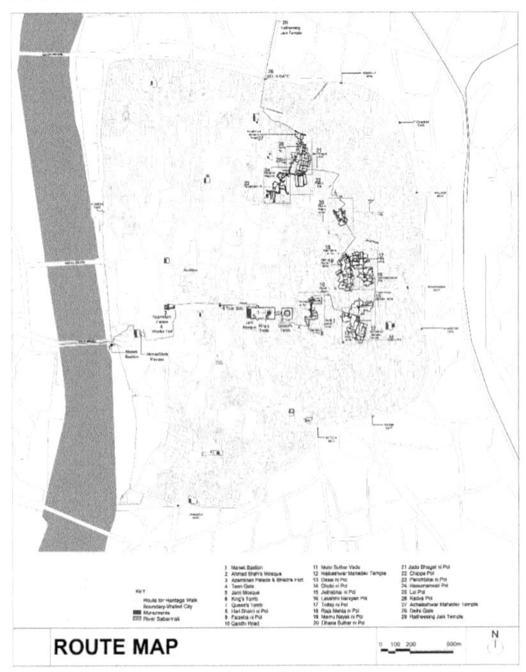

Fig. 1 "Heritage Walk" and its associated buildings and communities © *Heritage Management Plan* released by Ahmedabad Municipal Corporation(AMC)

2. 1. 4 Development of Public Space

The reuse of religious sites promotes the integration of public space and community life. In the community life in India, there are some wooden religious structures on the streets, which are similar in shape to columns or lights. They are now used to raise birds and feed birds, called "birds feeder"(Fig. 2). This initiative has injected new vitality into these static cultural symbols, and it has greatly narrowed the distance between residents. The public is very happy to maintain the "birds feeder" spontaneously, and through the interaction of feeding birds, communication has become a portion of life.

2. 1. 5 Diverse Cultural Activities

The community organizes a number of

Fig. 2 Birds feeder and other open spaces

cultural exchanges under the leadership of private institutions and community managers (selected by residents), including but not limited to the Des Niepol World Heritage Week celebrations, free walks to commemorate India's freedom struggle, the poet's meeting to commemorate Carvey. Sammeland, the Theatre Media Center and the Ahmedabad community Foundation competition, street drama renaissance, etc. These activities not only enrich the residents' leisure activities, but also greatly enhance the cultural sense of belonging and identity of the community, making the people more spontaneous and willing to participate in community renewal activities.

2. 1. 6 Control of Buildings and Streets

In order to maintain the traditional style of the block, a series of regulations have been formulated for the function of the house, the size of the building, the scale of the street, the open space and the line of sight (Tab. 1).

Tab. 1 Relationship between road scale and building scale © *Heritage Management Plan*

No.	Road Width (in meters)	Maximum Permissible Building Height (in meters)
1	Less than 9m	10m
2	9m and less than 12m	15m
3	12m and less than 18m	25m
4	18m and less than 40m	45m
5	Over 40 m	70 m

For example, restrictions on the minimum open space dimensions of buildings of different sizes (Fig. 3); requirements for the size of the landing platform between stairs (Fig. 4); in the important open space of the border, no materials or loose items shall be placed for trading, and vehicles shall not be used for parking while vehicle ramps leading to the parking lot shall not be allowed; in some important open spaces of the border, the projection of the canopy, air conditioner and water tank shall not exceed a certain size. Through the intervention and control of professionals, the dirty and disorderly street space have been greatly improved, and the living standard has also been improved.

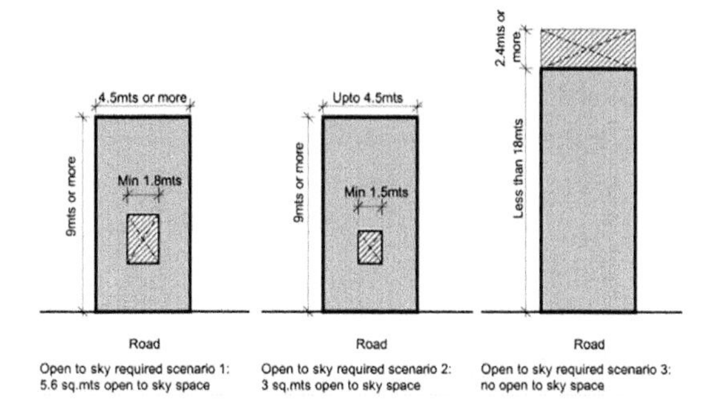

Fig. 3 Requirements for the size of open spaces in different sizes of buildings © *Heritage Management Plan* released by Ahmedabad Municipal Corporation(AMC)

2.1.7 Actively Encouraging Potential People to Participate

Ahmedabad's community has created the main force for public participation and

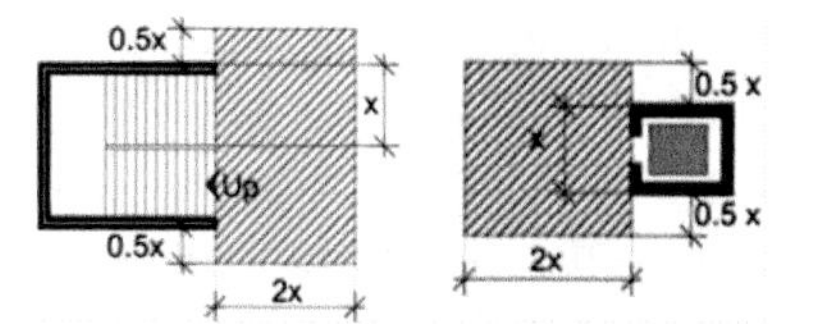

Fig. 4 Limitation about the width of stair © *Heritage Management Plan* released by Ahmedabad Municipal Corporation(AMC)

has targeted guidance. The communities in India are very young and the proportion of young people is extremely high. They have the greatest demand for public activity space and cultural and educational activities, and they are energetic and are the most capable and most eager to participate in the creation. The community designed a promotional education book and volunteer identification for them, which enabled them to participate in the creation of rewards and education to solve the problem of the internal driving force of community renewal.

In summary, Ahmedabad has carried out good cooperation between the government, civil organizations and residents during the transformation of traditional communities, and cultural heritage has played an important role in it. First of all, through the micro-interventions of funds and technology by managers and professional technicians, a diversified platform was finally built. Secondly, the improvement of living space control, public space shaping, cultural activities, cultural relics protection and reuse, media publicity was also important. Finally, the most important thing is to use the public's desire for public space and education to guide potential people to participate, and the public will be more actively involved in forming a virtuous circle after profit.

2.2 Protection and Renewal of Traditional Communities in Nanjing Laomen Dong

When the eyes turn to the local, the author finds that the experience of India can inspire the renewal of traditional communities in Laomen Dong, in the south of Nanjing, while there are some identical problems and goals between the two. However, it should be noted that the Chinese community is facing serious aging, and the migration of young people has caused a large number of empty houses in the traditional community. How to guide the existing people to update the community requires more thinking. In addition, compared with India's religious beliefs rooted in the blood, the group's high sense of identity and cohesiveness, the religious imprint of Chinese communities is not obvious, and the religious population is not much, which makes the residents slightly indifferent in interpersonal communication. Relatively speaking, it is more difficult to produce emotional resonance in Chinese communities.

Therefore, in combination with the experience of India and the actual problems in China, the author proposes suggestions for the specific operation of community renewal in the traditional communities in Nanjing.

2.2.1 Inspiring Public Enthusiasm

Adhering to the people-oriented, listening to the opinions of the residents, meeting the needs of the community owners, and changing from a single wish to a win-win situation are the factors. Based on research on community needs, it can be found that their requirements for sanitation facilities are urgent. In the traditional community, usually 4 — 6 households share a kitchen and bathroom, and even some of the yards do not have toilets. Residents should solve personal hygiene problems in public toilets outside. In addition, the lives of the old people are scarce, the collective activities of the community organizations are few, and the recent public activities are located one kilometer away. Therefore, when carrying out community renewal, the first consideration is to solve the basic needs of residents and satisfying their daily wishes.

Community renewal is a long-term process that may not meet the needs of everyone in the short term, so that everyone can see the benefits of reality. In order to stimulate the confidence of residents, it is recommended to carry out a full simulation of the process of building construction, the budget and flow of funds, the accumulation and realization of community capital under the guidance of professionals, and visualize them, and publicize them in a simple and clear way. Let them understand that community building is actually the accumulation of capital and can bring long-term benefits.

2.2.2 Assistance from Professionals

Since the current stage of participation is basically retired elderly person and middle-aged person with a low level of education, the participation of professional and technical personnel is greatly supported.

Taking house renovation as an example, the houses in the selected communities are usually brick-concrete structures with 102 floors, which were basically built after 1949. The appearance and quality are average, and the buildings show homogeneity. In addition to local residents, there are many tenants in the community. Usually, 4—6 households share a washroom and living space. In this case, professional designers design and renovate several typical courtyard modes, trying not to affect other functions, using the most economical method to increase the kitchen and sanitary space, and controlling the patio space. This not only avoids disorderly construction, but also controls the overall style of the community and improves the efficiency of the house (Fig. 5).

In order to facilitate more efficient communication and promotion to residents, it is necessary to establish an open platform, including: collecting residents' opinions, custom design plans, measuring costs, raising funds, feedback, publicity and display activities, and publishing activities.

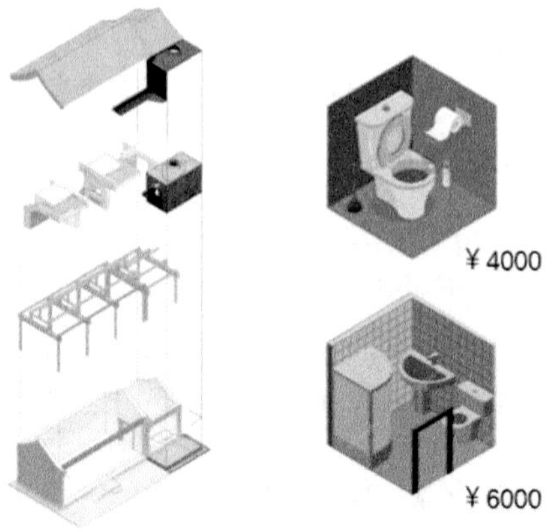

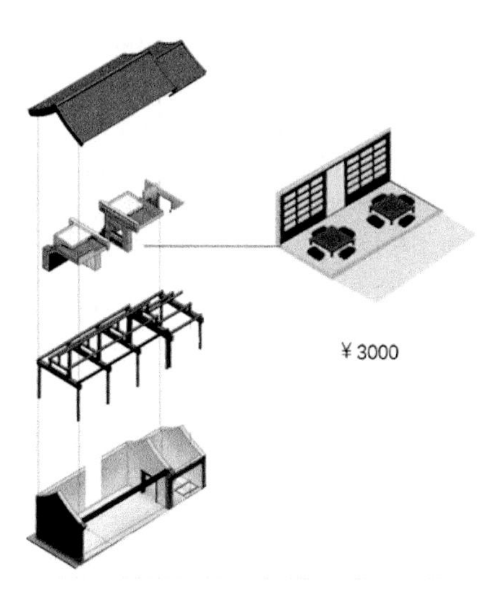

Fig. 5 Several transformation cases: bathroom, leisure space

Fig. 6 Application interface simulation

At present, many communities have developed their own websites and APPs. On the basis of this, it is technically feasible to develop a comprehensive community application. This program can cover all aspects of life: medical health, water and electricity bills, event notifications, community forums, pet homes, festivals, fitness and entertainment, mutual help and so on (Fig. 6). For example, there will be a library of model works for building renovations, summarizing the patterns, methods, and budget costs of several building renovations. When a household wants to renovate their home, take a picture of the interior of the home and attach its own requirements, then the library will automatically select the most suitable model for it. After the residents get a reply, there will be a process of feedback and modification. After the completion of the plan, the results will be uploaded and more samples will be provided for future residents. In addition, for residents who need financial support, the platform will also provide crowdfunding channels; for residents who are interested in participating in community labor, they will also be rewarded.

2.2.3 Emphasis on the Protection and Reuse of Cultural Heritage as a Living Resource

In the protection of cultural relics in the past few decades, it is often "material-oriented" rather than "people-oriented". The cultural relics in the community are usually protected and are only for visiting. But in fact, cultural heritage is a cultural phenomenon in which people interact with things. As a living resource, it has not only cultural value but also economic value.

Therefore, in the future community renewal, it is necessary to strengthen the relationship between cultural heritage and residents. It is necessary not only to show more, but also to join the people in the research, repair and publicity of cultural heritage. The historical architecture should not close the door, but open to the public. If it does not affect the protection of cultural relics, it will hold more cultural and art exchange activities and recruit more community volunteers. This will both stimulate vitality and enhance cultural belonging.

Summarizing the case of Nanjing Laomen Dong, the management departments (governments, neighborhood committees), professional and technical personnel (designers, scholars) and community residents work together to reuse the cultural heritage and improve their own living spaces. From the two aspects of material space improvement and cultural resource utilization, a multi-community platform will be built, and it will become a community renewal model in the future.

3 Conclusion

This paper selects the typical traditional communities in the ancient capitals of Nanjing and Ahmedabad, two representative cultural cities in China and India. By comparing and analyzing the cultural background, social structure, life pattern and measures taken in cultural protection and community renewal of the two communities, the author tries to discover the common problems of the traditional communities in developing countries in balancing protection and development and the experiences that can be learned from each other. Furthermore, the author explores how cultural heritage can play a role in community renewal and sustainable social development. According to the analysis of the two cities, the following points should be noted for the sustainable development of the future community:

(1) Community renewal is about the relationship between people and the environment. The builders and users of the community are integrated. Therefore, in the process of building, designers should listen to the opinions of the people and encourage the participation of the people. Ultimately, it will benefit the people and turn community renewal into the realization of the wishes of the masses.

(2) Pay attention to the protection and utilization of cultural heritage, regarding cultural heritage as a living resource, and encourage residents to participate in the protection and utilization of cultural heritage.

(3) Build a multi-disciplinary community cultural platform and make cultural protection and home construction truly popular in every grassroots community. The future of the community is one of the themes of the new era. Public understanding, recognition, participation in cultural heritage protection and community renewal are the tasks of the new era.

Comment

① Ahmedabad: India's sixth-largest city, Gujarat's largest city, an important commercial city, famous for its fine Muslim architecture, Hindu temples and mausoleums.

② pol: the management unit of the traditional Indian community.

References

LIU Jianzhi, 2013. The popular science movement in India: the tree construction of community construction. China reform, 3: 41-55.

LUO Jiade, LIANG Xiaoyue, 2017. Theories, processes and cases of community building. Beijing: Social Sciences Academic Press.

TANG Yawen, WEI Kai, 2019. Practice and research of domestic community building. Urban residence, 2: 157-159.

WANG Hongguang, 2018. The change from the perspective of "community" to "community": taking the practice of public participation in historical districts as an example. Quality and sharing-proceedings of 2018 China Urban Planning Annual Conference, 2: 13-16.

ZHANG Tenglong, WANG Xiaoying, JI Wei, et al., 2019. Exploration and effectiveness of the construction of "community governance" system in Shenyang. Urban planner, 4: 5-10.

ZHANG Tingting, MAI Xianmin, ZHOU Zhixiang, 2015. Community building policy in Taiwan and its enlightenment. Urban planner, 31(S1): 62-66.

Adaptive Reuse of Historic Buildings as an Approach to Revitalization of Social and Residential Life in Historic Context; Case Study: Adaptive Reuse of a Historic House in Yazd (PADIAV HOUSE)

Ne'da Soltan Dallal[1], *Ahmad Oloumi*[2]

1 PADIAV HOME, No. 26, HOSEINIAN Alley, IMAM Street, Yazd, Iran
2 KHESHT-AMOUD Consultants (KAYC), No. 89157, 15th Alley, MASKAN Street, Yazd, Iran

Abstract In recent decades, the adaptive reuse of historic buildings with purpose of revitalization of its containing fabric has been recognized as an approach in inner-city development strategies. Considering the historic buildings as a potential for inner-city development had resulted the extended variety of the approaches to revitalization of historic fabrics. This activities and functional interventions include reuse of the building or use the building with its preceding function. In this paper, we are going to find an answer to this question that "which items must be considered for revitalization of social and residential life as an approach for adaptive reuse of a historic building?" by analyzing a practical case study which is located in historic fabric of Yazd. Also, the vernacular architecture of Yazd would be considered as a potential for rehabilitation of its context. In the following the principles and criteria of revitalization of social and residential life in historic fabric would be described.

Maintaining the international charters as references and guidelines, considering and analyzing of architectural values of the case study, evaluating of the architectural and functional potentials of the case study and also feasibility of its containing fabric and integrated district, would lead us to the comprehensive answer for the research question.

The case study is a historic house back to the 16th century with a vernacular architecture, located in the core zone of UNESCO world heritage registered historic fabric of Yazd. This house is reused but by keeping its function (as a residential unit) and maintaining the architectural values. The house is reused to an eco-friendly homestay. This adaptive reuse had caused the revitalization of social life by involving the neighbors and around residents in some cultural entertainments.

Keywords adaptive reuse, revitalization, social life, ecolodge, vernacular architecture

1 Introduction

In order to focus on the literature review of the adaptive reuse of historic settings and its results on revitalization of its context, we need to have a glance on strategies toward sustainable development. In this case, we need to have some definitions considering sustainable development, adaptive reuse and revitalization.

Achieving objectives for sustainable development called for a balance between social, environmental and economic needs of the community and available resources. The principal vision of a sustainable built future is about developing creative designs that utilize energy and smart materials efficiently. In reality, this vision should consider historic buildings that were built centuries ago. There is a growing acceptance worldwide that conserving historic buildings provides significant economic, cultural and social benefits. Historic buildings should be conserved for future generations as they link communities with their history. Towards revitalizing and generating sustainable values of these buildings, many historic buildings of cultural and historic significance are being adapted and reused rather than being demolished (Othman and Elsaay, 2018).

2 Sustainable Development

Sustainable development is a development that meets the needs of the present without compromising the ability of future generations to meet their own needs (Merlino, 2011).

Sustainable development has been defined in many ways. However, the most commonly quoted definition is derived from the Brundtland Report (World Commission on Environment and Development 1987), which defined it as development that meets the needs of the present generation without compromising the ability of future generations to meet their own needs. It contains two key concepts, namely, the needs and limitations. The first one represents, in particular, the essential needs of the world's poor in which prevailing priority should be given, whereas the latter refers to resources imposed by the state of technology and social organization and the environment's ability to meet present and future needs. Sustainable development aims to deliver products that enhance life's quality, achieve customer satisfaction, provide flexibility and adaptation to user changes, support desirable natural and social environments as well as maximize the efficient use of resources. This will be achieved through enhancing efficiency and moderation in the use of materials, energy and development space. The 2030 agenda for sustainable development is shown in Fig. 1. (Othman and Elsaay, 2018)

3 Adaptive Reuse

Adaptive reuse is the act of finding a new use for a building. It is often described as a "process by which structurally sound older buildings are developed for economically viable new uses." The recycling of buildings has long been an important and effective preservation tool. It initially developed as a method of protecting historically significant buildings from demolition. The Urban Land Institute defines rehabilitation as "a variety of repairs or alterations to an existing building that allows it to serve contemporary uses while preserving features of the past." Adaptive reuse is then a component of rehabilitation. Adaptive reuse is often called adaptive use referring to the redundancy of the term "reuse" (Cantell, 2005).

4 Revitalization

Extending the life of a building by providing new or improving existing facilities, which may include major remedial and upgrading works (Douglas, 2006).

In recent decades, the revitalization of cultural heritage issues has been positioned as a significant segment of the wider process of sustainable planning of urban development, maturing in the meantime in the focus of interest. Revitalization primarily involves active methods used for the purpose of reviving a protected building or site, primarily referring to restoration with or without conversion of functions, adaptation or interpolation (Vasilevska and Lecic, 2018).

Historic buildings help define the character of our communities by providing a tangible link with the past. Today, historic districts around the country are experiencing unprecedented revitalization as cities use their cultural monuments as anchors for redevelopment. Sometimes, efforts to preserve and revitalize historic buildings run up against financial obstacles, restrictive zoning and codes, contamination, and structural problems that create challenges in reusing these unique structures(Cantell, 2005).

5 Adaptive Reuse of Historic Buildings as an Approach Toward Social Revitalization

Adaptive reuse is a long practiced method of adopting a pre-existing structure for new purposes that are often different from the original intent. This method is outlined to preserve the visual identity that the building contributes to a (historic) community (Sharpe, 2012).

We are going to analyze on this article case study, PDIAV HOUSE. This chapter analyzes a practice example in order to illustrate the guidelines and benefits of using the adaptive reuse of a historic house in the process of protection, restoration and regulation of cultural heritage and how an adaptive reuse project can affect the social life of its context.

6 Case Study: Adaptive Reuse of a Historic House (PADIAV HOUSE)

Historic preservationists typically become involved when said structure was built fifty or more years prior to the reuse and/or has a cultural significance to its surrounding environment. This practice is viewed as sustainable for a number of reasons including restricting the amount of waste created and the recycling of various exterior/interior materials.

The following case study is a house located in the historic fabric of Yazd in Iran. Since 2017, the historical city of Yazd is recognized as a World Heritage Site by UNESCO.

The house which is named PADIAV HOUSE after new use by the new owner, is built in Safavid dynasty period back to the 16th century. This building has vernacular architecture exclusivities of central cities of Iran. Spatial characters, central courtyard, historic decorations and also mud and mudbricks as the main materials of the building makes the house a vernacular architecture.

6.1 Conservation and Adaptive Reuse

A historic preservationist is trained to protect a sense of community and culture through seven levels of interaction — conservation, preservation, restoration, reconstitution, reconstruction, replication, and adaptive reuse. These seven levels can be applied in manners anywhere form an artifact, single room, house, community district, entire town, outdoor architectural museums, and landscapes. A major part of their process to preserve and protect involves educating others on local, state, national, and international levels of an item or site's significance through documentation. Adaptive reuse brings these two professions together in the working environment as a preservationist works to protect the historic background and the designer converts the space for new use (Sharpe, 2012).

Adaptability is obviously a key attribute of adaptation. It can be defined as the capacity of a building to absorb minor and major change. The five criteria of adaptability are:

- Convertibility: Allowing for changes in use (economically, legally, technically).
- Dismantlability: Capable of being demolished safely, efficiently and speedily — in part or in a whole.
- Disaggregatability: Materials and components from any dismantled building should be as reusable or reprocessable (i. e. recyclable) as possible.
- Expandability: Allowing for increases in volume or capacity (the latter can be achieved by inserting an additional floor in a building, which does not increase its volume).
- Flexibility: Enabling minor if not major shifts in space planning — to reconFig. the layout and make it more efficient (Douglas, 2006).

Before starting the conservation process, the house was analyzed for new use, considering the internal potentials of the building and also its feasibility in the context.

Tab. 1 presents the alternatives considering the threads and opportunities of each one for the house.

Tab. 1 Alternatives of new spatial functions in new use of the house

	rooms	stores	public room	kitchen	bath room	shop	library
Alternative 1	6	1	1	2	2	-	-
Alternative 2	6	2	1	1	3	1	-
Alternative 3	5	-	3	1	3	2	1
Alternative 4	5	1	2	1	3	-	-

So a "Homestay" could be the best proposal for this building as a new use (Alternative 4). In this case, the house would be adapted for new use but keeping the old function as a residential place(Tab. 2).

Tab. 2 **Threads and opportunities of alternatives**

	Opportunities	Threads
Alternative 1	Financial benefits, more spaces for passengers, providing a good space for public room in northern rooms	Expanding wet spaces in the building, high cost for maintenance.
Alternative 2	Low cost for maintenance, possibility of selling vernacular handicrafts and cultural medias in shop, more stores	Lack of bathroom in basement level
Alternative 3	High quality space for public room, visual connection between kitchen and courtyard.	Lack of gusts rooms
Alternative 4	Good spaces for rooms and public room, balanced financial benefits, minimum wet spaces in the building	In this alternative all the threads are balanced

6.2 Sustainability in Conservation

Adapting buildings is an important component of any sustainability strategy. Along with adequate maintenance, it is essential for ensuring the long-term prosperity of our built assets. Moreover, adaptation entails less energy and waste than new buildings, and can offer social benefits by retaining familiar landmarks and giving them a new lease of life (Douglas, 2006).

Ensuring that materials used for rehabilitation are environmentally cultivated, extracted, produced or manufactured is an important component of sustainable preservation. Doing so is an important part of "green" preservation, but it can be challenging to decide which is the best solution or product. Product certification is not standardized, although there are certain companies, such as EcoLogo that attempt to certify certain products. While a single "list" is nearly impossible to create due to changing product lines, research and availability, some common sense is required. Products that require less energy to produce, are durable, and are easy to maintain are the best products to begin with (Merlino, 2011).

Fig. 1 Reconstruction of the arch by mudbrick

Conservation and restoration of PADIAV HOUSE is a sustainable project. In this project the masonry material are vernacular and based on original materials of the building. Using mud bricks in walls and arcs and also using traditional molds like lime and mud instead of cement combinations were the solutions for restoration of the building as a sustainable adaptive reuse project (Fig. 1).

Microclimate is very significant In Iranian traditional architecture. Restoring of the spaces like PAYAB (place for accessing Qanat water in the basement level) which is mostly used in warm summers as a public place, is another sustainability aspect in this project.

6.3 Revitalization of Social Life

A successful restoration project is one that considers the containing context revitalization in addition to the revitalization of the building. It can cover the various aspects containing revitalization of economic infrastructures of the quarter, revitalization of a scene of vitality and also the social and cultural life of the context. Obviously, restoration and adaptive reuse of a building can be strongly effective on revitalization of the above-mentioned items in the context.

In this case study (PADIAV HOUSE), the approach of the adaptive reuse project is to continue the previous function of the building and that's continuity of its residential life. This approach can be considered as a catalyzer for the revitalization of social and cultural infrastructures.

PADIAV HOUSE is a house for about four centuries after adaptation and restoration, its new use is a homestay. In this case, the building is keeping its previous function as a residential place. In Iranian culture, houses had been resources for educating and producing. These items were significant in each family as the smallest part of the society and also in relation to the neighborhood families as a bigger group of the society.

Because of its approach, the project of PADIAV HOUSE is known as a center of the cultural events now. This project has provided good opportunities for cultural activities. Meanwhile, it had created a context for exposing handmade arts and vernacular handicrafts. Creating economic opportunities for local people is the final strategy of the project that is in process and planning.

In the following some of the cultural and social activities which are supported by PADIAV HOUSE are listed:

1. Events based on reading books for children;

2. Creating a library for kids and gathering in PADIAV HOUSE for reading kids' books;

3. Celebrating special world days such as children world day, handicrafts world day, Elmer (The patchwork elephant) world day, etc. ;

4. Using the potentials of the quarter for cultural events such as showing the movie in the mausoleum, ;

5. A friendly relation between neighbors and tourists;

6. Creating a good space for temporary exhibitions of vernacular handicrafts and handmade arts;

7. Celebrating national and vernacular celebrations such as NOWROOZ;

8. Performing earth built architecture workshops for children.

7 Conclusion

Adaptive and creative reuse of historic buildings is a process to promote the value of the building and its containing fabric. This purpose will never happen if the project doesn't consider the approach to sustainable development and revitalization of social life.

The first step to achieve a sustainable project is to use vernacular materials and local masons for architectural restoration of the building.

At the next, the compatibility of new use is significant. The new use of the building must be compatible with the building itself and also to its containing area. So the final alternative must be the best solution for rehabilitation of the building and revitalization of the social life of the quarter. Finding the best use for such project can act like catalyzer in revitalization of social life and other goals of the adaptive reuse project.

A successful adaptive reuse project is the one that had provided the economic requirements of the local people. One of the solutions to achieve this significant item is to create home jobs for people living around in the quarter. The careers related to tourism industry are more considerable with more financial and cultural benefits.

PADIAV HOUSE is a project with a sustainable approach because of using vernacular materials in restoration of the building. This project has also provided good opportunities for local artists in vernacular handicrafts. The adaptive new use of the building as a homestay, had leaded the project to revitalization of social life in the context. PADIAV project is not just restoration and rehabilitation of a historic building, this project had affected the containing fabric by using its potentials and is going to gain all the goals of a sustainable adaptive reuse project with an approach to revitalization of social, so soon.

References

CANTELL S F, 2005. The adaptive reuse of historic industrial buildings: regulation barriers, best practices and case studies. Blacksburg: Virginia Polytechnic Institute and State University.

DOUGLAS J, 2006. Building adaptation. London: British Library.

MERLINO K R, 2011. Urban grain and the vibrancy of older neighborhoods: metrics and measures. Seattle : University of Washington.

OTHMAN A A E, ELSAAY H, 2018. Adaptive reuse: an innovative approach for generating sustainable values for historic buildings in developing countries. Berlin: De Gruyter.

SHARPE S E, 2012. Revitalizing cities: adaptive reuse of historic structures. Mid-America College Art Association Conference 2012 Digital Publications:18.

VASILEVSKA L, LECIC N, 2018. Adaptive reuse in the function of cultural heritage revitalization. National Heritage Foundation 2018 Conference.

Research on Protection and Adaptive Utilization of Cultural Heritages in Water-towns — A Case Study of Xiongkou Town in Hubei Province

DONG Fei [1], *WANG Li* [2], *DENG Yunqi* [3]

1 Wuhan Planning Institution, Wuhan, China
2 Wuhan University of Technology Design Research Institute Co. Ltd, China
3 Hubei Ancient Building Protection Center, China

Abstract This paper analyzes the protection and adaptive utilization of China towns' heritages in the context of implementing the national policy of revitalizing rural areas and developing characteristic towns based on the case study of Xiongkou Town in Qianjiang, Hubei Province. Xiongkou Town, a typical water-town in Yangtze River and Hanjiang River alluvial plain (Y&H alluvial plain), used to be a booming ferry on local trade routes during the Ming and Qing Dynasties as well as a base of the Red Army during the period of the Republic of China. After 20 years since Reform and Open, Xiongkou Town has been a well-known crayfish production base in China, where farmers made a new agricultural production by growing rice and breeding crayfish in the same paddy field. Now under the policy of developing characteristic towns, the township government has established a conservation model of the old town by reserving aborigines and their traditional lifestyle, so planners summed up the perspective ideas of characteristic towns, especially about three spatial strategies, which are reorganizing the traditional town contexts, highlighting the water-town spatial framework in Y&H alluvial plain and restoring the vernacular architecture in Yangtze River middle reaches in combination with modern agriculture.

Keywords Y&H alluvial plain, rural revitalization, heritage protection, adaptive utilization

1 Introduction

As survival and development of the rural areas, agriculture and farmers are essential for national interests and people's livelihood, a characteristic town must be very unique and suitable in its industrial structure, town features, formation or system rather than perfect in everything. Qianjiang City in Hubei Province is well-known for crayfish farming in China. In recent years, it has made a model of developing prosperous eco-friendly industries in rural areas by growing rice and breeding crayfish in the same paddy field. On the other hand, Xiongkou Town, Qianjiang City is well-known for its contributions to the democratic revolution in 1930s, reserved many unique revolutionary heritage sites represented by the Red Army street and the township structure including many simple vernacular buildings built since the Qing dynasty in Y&H alluvial plain. Xiongkou Town is facing a large-scale construction of the town area after being accredited as a national characteristic town. The author summarizes the mode of the water-town in Y&H alluvial plain in terms of industrial policy, rural heritage protection and township ecosystem development in order to promote revitalization of villages and protection of local cultural heritages in the middle reaches of the Yangtze River.

2 Valuable Characteristics of Decentralized Water-Town

As China's traditional society was established on the basis of agricultural civilization, villages and towns scattered in countryside are fundamentally related to their geographical location and climate features. The differences of wetland, drylands, mountainous and plains nurtured the traditional social structure consisting of the classes of farmers, merchants, workers, government officials, and even soldiers and bandits. Therefore, it is necessary to understand the history, geographic pattern and environment of rural villages before capturing the insight of their cultural context.

2.1 Geographic Features of Y&H Alluvial Plain

Located between the Yangtze River and Han River, Y&H alluvial plain is formed by flood and lake sediment. In ancient times, it was called Yunmeng Marsh with many shallow lakes connected by a well-developed river system. Qianjiang, bounded by Han River in the north and close to the Yangtze River in the south, without any mountains, is a storage area of two rivers, which is frequently hit by flood and drought. Historically "Qianjiang" means sub-current rivers flowing into Yangtze River, implying that rivers bypass here and flow into Yangtze River. With plenty of rainfall and moderate temperature, the climate here is favorable for agriculture.

2.2 History of Xiongkou Town

The development journey of Y&H alluvial plain covered by fertile soil is determined by its geographic structure. As a result of agricultural development since the Song Dynasty, this plain has gradually developed into a prosperous paddy field with a growing-up population. More and more market towns were built during the Ming and Qing dynasties, forming a commercial hub radiating to all directions. The south of Xiongkou Town was a marsh, but to the north of Xiongkou was land which was connected by waterways. At that time, this area was called Xiong's ferry for a shop set up by a fisherman surnamed Xiong. By the Qing Dynasty, this area had gradually developed into a market town consisting of three streets (Middle Street, Back Street and Riverside Street). Therefore, Xiongkou Town is not the center of a regional town, but an ordinary market town characterized by waterside ferry in terms of size and shape in the hinterland of Y&H alluvial plain.

Y&H alluvial plain with the well-developed water system has been contested for its strategic location, where Xiongkou Town is provided with wet lands covered by abundant reeds which provided a natural barrier for military defense. In the 1930s, Xiongkou Town was the center of the revolutionary base area in western Hunan and Hubei, where the Red Army stationed for a long time. Xiongkou Town has been known for the sites of Red Army Street since 1949. Although most of the vernacular buildings here are simple, they are protected as the historic monuments of revolutionary legacy rated as the cultural heritage site under the provincial protection and listed one of the famous Chinese historical and cultural towns.

In the 1950s, in order to prevent the drought and flood disasters, the area was implemented water conservancy projects and dredged rivers and canals. With more than 20 canals, its agriculture has been greatly developed, resulting in a fantastic view of vast paddy fields. Since Jianghan Oilfield was developed in the 1960s, Xiongkou Town has become a farming area serving the periphery of the oilfield. Today, some public buildings in the town set up during the planned economy era after 1949 are still lying there to witness the change of those times(Fig. 1).

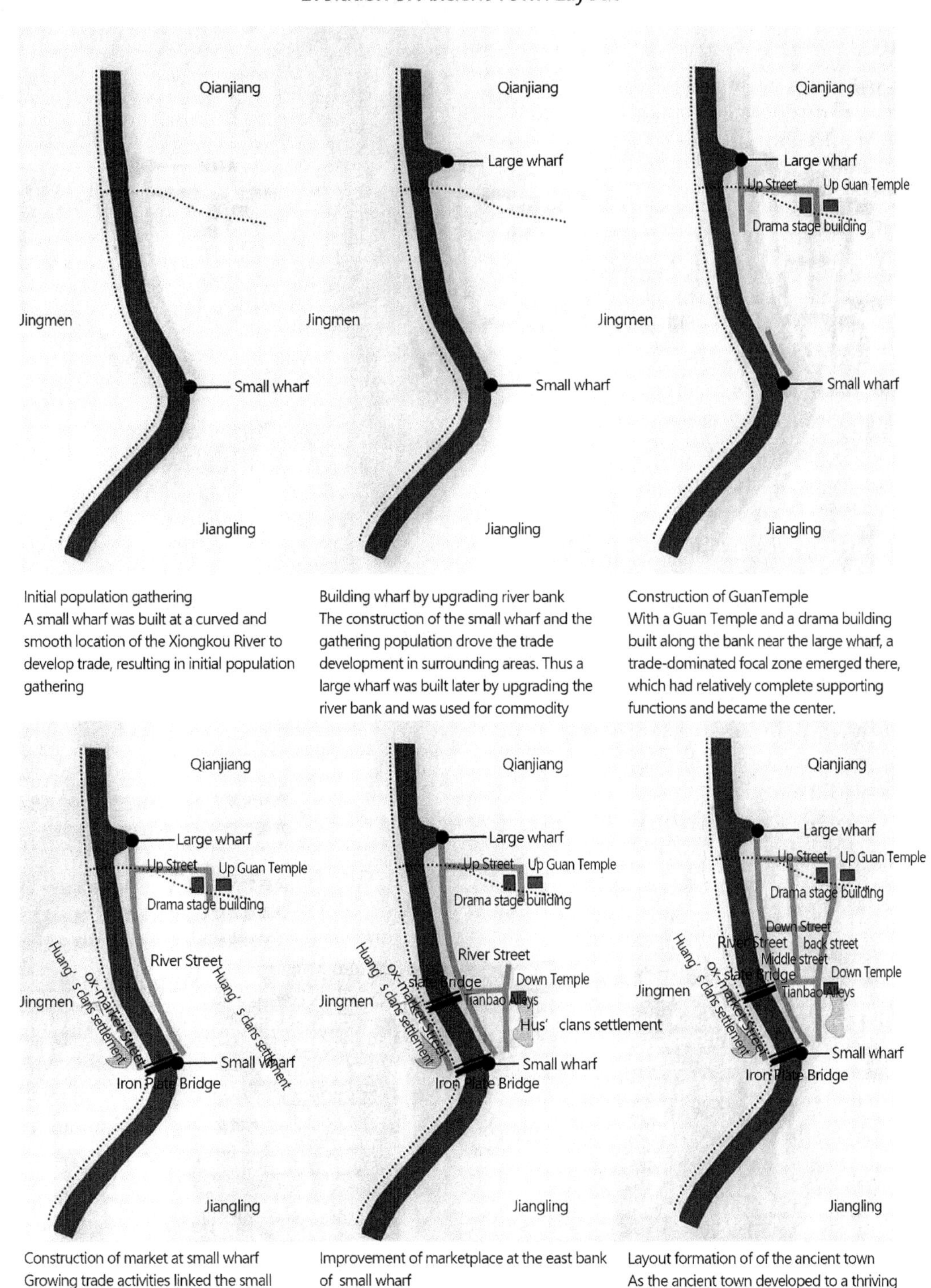

Fig. 1 The development trajectory since Qing dynasty

2.3 Changes Caused by Characteristic Industries

Based on its advantage of abundant resources of canals, ditches and ponds, Qianjiang City has been exploring a scientific model for breeding crayfish by all means from free-range breeding to cage culture since the 1990s, and established a new agricultural model of crayfish and rice symbiosis[①] to realize the best synergies of grain and aquatic production (Fig. 2). Now it has become the largest crayfish breeding base in China to sell its products to Europe and North America. The creative contributions made by farmers have brought about a series of changes here. The agricultural lands were highly collected to 80% and transferred to the industries. The renovated ditches are well-connected with a lively circulation of clear water to contribute to agricultural productivity and rural environment. The innovation of agricultural mode has brought huge benefits to the small town which has seen a population backflow since 2006 and has achieved a positive population growth so far as one of the few examples of thriving and prosperous development in rural area of China today.

Fig. 2 Symbiosis of crayfish and rice

2.4 Spatial Characteristics of Villages and Towns

The spatial layout of towns is a comprehensive reflection of the natural conditions and social formations in the region. Traditional town patterns can often reflect the topographical and geo-morphological features and endogenous causes of social forms in more detail. The best symbiosis of the built-up area of Xiongkou Town and the surrounding water body effectively reflects the interaction between human activities, floods, natural disasters and social changes.

2.4.1 Town Textures and Physical Elements

Y&H alluvial plain has been affected by natural disasters for a long time. The spatial texture of Xiongkou Town is mainly affected by the factors of minor topographic relief in the plain, showing the features of streets and lanes. The town is split from the north to the south by three main streets: Red Army Street (Middle Street), Pedestrian Street (Back Street) and Riverside Street interlinked by east-west alleys such as Tianbao Alley, forming a traffic system with streets and alleys. The buildings on both sides of Middle Street are mostly business-living buildings. The Riverside Street winding along the old river course is provided with docks surrounded by fishermen's houses. The strike direction of the flood control dam is still visible on a slightly wavy terrain. The straight Middle

① Crayfish and rice symbiosis: Digging ditches around each rice field, draining the water from the field into the ditches during sowing and harvesting period for crayfish to creep into the ditches along with the water flow. During the natural growth period of rice, the paddy fields are irrigated with the water from the ditches and canals, and crayfish which like to nestle in the shade of the seedlings in the paddy fields will enter the paddy fields with the water from the ditches and canals and eat rotten leaves, microorganisms and insects and produce excrement that will enhance the growth of seedlings as organic fertilizer. The number and growth cycle of crayfish in each paddy field should be well controlled, and no chemical feed or pesticide should be applied to keep eco-friendly Paddy fields and produce organic products of crayfish

Street is occupied by public buildings such as shops and private schools, including the supply and marketing cooperative and cultural palace, while the Back Street is occupied by residential houses with marshes in the periphery of the town. The sewers under streets built by the bricks of town wall lead to low-lying places or rivers, forming a unique drainage system of the old street. The narrow streets were broadened during the reconstruction in the 21th century, where some traditional buildings were demolished and destroyed, presenting a relatively narrow street space as a whole.

2.4.2 Vernacular Architecture in Yangtze River Middle Reaches

The existing traditional buildings in Xiongkou Town are mainly divided into two types, i.e. the buildings in the architectural style of the Qing Dynasty and those generally built by the State in the style popular during the planned economy period after 1949. Although the buildings in the architectural style of the Qing Dynasty were repaired and renovated in the past, they basically retain the characteristics of residential buildings in the middle reaches of the Yangtze River, though they are not so sophisticated and elegant as those wealthy households in the southern regions of Yangtze River. The residential buildings built with black tiles and gray bricks during the Ming and Qing dynasties were mostly quadrangle dwellings with attics, small rooms and deep-plan in the plane layout, and undulating residential patios and skylights in the spatial structure. The wing rooms only with windows to inside on both sides are isolated from the outside for strict privacy and safety. The buildings of column and tie wooden construction are set up by bricks and stones with fire seal. The eaves of the main facade are wider, and the drainage system is considered in the design of the building structure. Due to easy access to waterway, most of the building materials were purchased in the timber and brick market in the surrounding areas.

The original river courses and docks were deserted and occupied by civilian houses due to construction of drainage system after 1949. During the planned economy period, public facilities such as the site of commodity grain supply and marketing cooperatives and grain warehouses were concentrated in the main street in the core area to form the center of the town. The buildings with thicker outer walls covered by granitic plaster or lime-sand mortar except doors and windows are obviously built in the architectural style different from that of the residential buildings early in the first half of the last century. The wooden windows are provided with lintels and on the door beams are the signs of red stars, marking the characteristics of the times in the first half of the last century.

2.4.3 Un-tangible Cultural Heritage in Indigenous Life

Living on agriculture, the indigenous people in Xiongkou Town grow rice, cotton and other food crops. The traditional food such as rice cakes, crisp cakes and oil tea camellia are well known in local markets. The locals are accustomed to celebrating festivals and praying for wealth and happiness by fabricating straw dragons and showing the arts and crafts of wood and bamboo carvings, which highlights the tradition of farming civilization. As a result of backflow of the rural population attracted by symbiotic growth of rice and crayfish in recent years, the locals have chosen to live in Qianjiang City or around the industrial areas, resulting in increasing challenge of population aging. However, the tradition of red culture in the market towns is highly recognized by the locals who can identify the sites of the Red Army barracks in the tow, and the Crayfish Festival to be held in every summer in the local area, which has become a popular event recently, is attracting more and more visitors to the town.

3 Adapting Improvement

In the past ten years, Xiongkou Town has been accredited as a national site of cultural heritage and a national characteristic town. Since then, its protection planning, cultural relic restoration design and construction planning have been carried out simultaneously. Its core development model with equal emphasis on research and construction based on cultural development has posed a great challenge to the designers who must exercise full duty of care in their works. In consideration of the comprehensive value of nature, industry and traditional culture in Xiongkou Town, as well as the dissents and agreement on the design of the expert panel including planners, cultural relic protection engineers, architects and municipal engineers, etc. , a consensus on modification highlighting traditional local features is reached basing on industrial revitalization. The construction is specifically carried out to reorganize the traditional town contexts, to highlight the water-town spatial framework in Y&H alluvial plain and to restore the vernacular architecture in Yangtze River middle reaches.

3.1 Mending Several Historical Town Context

Xiongkou Town is well-known for its modern revolutionary history, vernacular architecture art and technology in the Qing dynasty and the water-town features in the early 1950s, in addition to the agricultural landscape outside the built-up area. In order to sort out and highlight the abundant cultural resources, conservation plan was formulated in priority, including: (1) Defining the purple line of traditional towns, that is, identifying the core protection area and regulated construction area, and specifying all the protection requirements accordingly. An area of 5.6 hectares is identified as a core protected area for the sites of listed cultural relics by-laws and identifiable remains of the typical buildings built in different periods in the town. According to the historical map, the middle street in the old town which is about 1.4km long with about 2.66km^2 of characteristic space such as canals, old roads, docks and metasequoia forests[①], is defined as the regulated construction area. (2) Plans for cultural visiting-routes, more public facilities and service facilities have been formulated respectively according to the survey of aborigines and tourists in revolutionary culture, ancient village culture and agricultural landscape. The rural houses listed as the sites of cultural relics have been gradually bought back and converted into revolutionary museums, classrooms or tourism facilities, with special emphasis on public facilities, such as tourist centers, public toilets, medical and health stations, garbage collection stations, etc. (3) Surveying and mapping vernacular architectures including indoor and outdoor facilities, courtyards, streets and lanes pavement and ancient trees, etc. conserve all kinds of narratives relics on their original structure, decorative features, raw materials and architectural technology. (4) Collecting and collating information about historical events and interviews with social celebrities, provide a basis for subsequent historical research.

3.2 Improvement of Water-Town Landscape Based on Industrial Revitalization

Y&H alluvial plain, where Xiongkou Town is located, is a national rice base and an aquaculture base (Fig. 3). In 2017, Xiongkou Town known for its pillar industry of crayfish farming was listed in national characteristic towns to be constructed within an area of 3 square kilometers. Qianjiang will build a more large-scale farming area under the model of crayfish-rice symbiosis, and establish a large industrial park with an industrial research and service system, including seed selection and nursery, eco-friendly breeding, healthy catering, cold chain logistics export and deep processing.

First of all, we should focus on water environment renovation. Except for the paddy fields in the industrial park, the dit-

① Metasequoia forests: a kind of old aquatic tree, existing in Hubei and Chongqing only, included in the redlist of the International Union for Conservation of Nature.

ches, lakes and ponds in the periphery of the town will be transformed into different ecological environments for different purposes such as ecological wetlands, landscape lakes and ponds, and canal system to build a characteristic agricultural landscape. Canopies have also been set up in the middle of the paddy field and beside the lakes and canals to facilitate large-scale environmental observation and tourists. Secondly, characteristic ecological species, such as metasequoia forest and wetland plant population have been identified in the area which is forbidden from sewage and deforestation by zoning and setting up the warning signs for protection.

In addition, in order to support the core industry development, the traffic network, municipal infrastructure and tourism facilities have also been improved, equipped with higher grade facilities such as schools, hospitals, garbage collection points and public toilets to ensure the modern living standards of the residential areas.

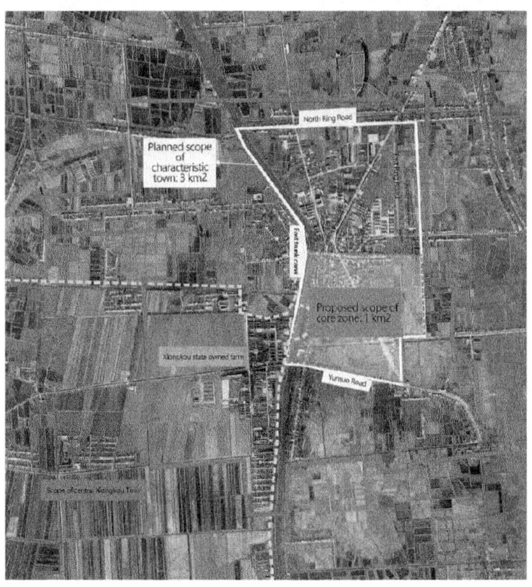

Fig. 3 The construction area of the characteristic town

3.3 Restoration of the Vernacular Architecture in Yangtze River Middle Reaches

The restoration of the vernacular architecture in the Yangtze River middle reaches is focused on protection of the vernacular buildings in the Purple-line area. According to the survey of the buildings and the surrounding protected elements, the tangible heritage restoration will be carried out by:

3.3.1 Protecting of Architectural Heritage

The architectural heritages have been restored in their original style and with the same buildings materials and the original techniques in accordance with the standards for cultural relic restoration (Fig. 4). As all revolutionary cultural relics protected in Xiongkou Town will be repurchased as assets, the recently repurchased buildings will be wholly renovated to restore the Red Army relics, including the Red Army headquarters, command offices and hospitals, to be shown as revolutionary monuments for red culture education. The buildings to be repurchased in a specified future shall undergo an external renovation currently and courtyard space, especially restoring with anti-corrosion and anti-water treatment. The material processing plant shall also be established onsite to ensure the replenishment of the missing structural elements in the original style.

3.3.2 Restoration of Common Buildings

The common residential buildings are renovated for functional adaptation. The height and size of newly-built buildings are limited, as the original occupants are reserved and the existing building height and average household size are relatively uniform. Guidelines for building facades in maintenance category, i.e. the roofs, walls, doors and windows, cornices, etc. shall be marked with historical architectural details, with black and white gray as the basic tone and wood or bamboo grain as the decorative elements; Facilities are supplemented with toilet, kitchen, air-conditioning rack, clothes hanger, and the modern living facilities hidden from open space, etc.

3.3.3 Recording Strategies of Social Memories

The society is guided to playback the original functions of the market, or sculptures and art works are used to promote the attraction of the small town as a tourist destination. For example, drug shops, dye

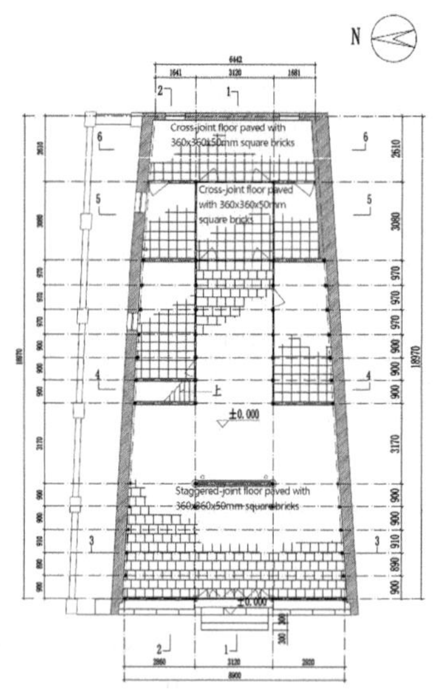

Plan (1:100) of the first floor at the Former Site of the Headquarters of the Red Army

Note: the building was repaired during 2017-2018 with the approved by Hubei Provincial Bureau of Cultural Relics, and is currently in a well-protected condition as a whole. Repair the leaking positions on the roof, and comprehensively clean the wooden floor of the second floor and paint the surface with three passes of tung oil; and set up a drainage system outside the building in combination with the design of the drainage pipeline at the Red Army Street.

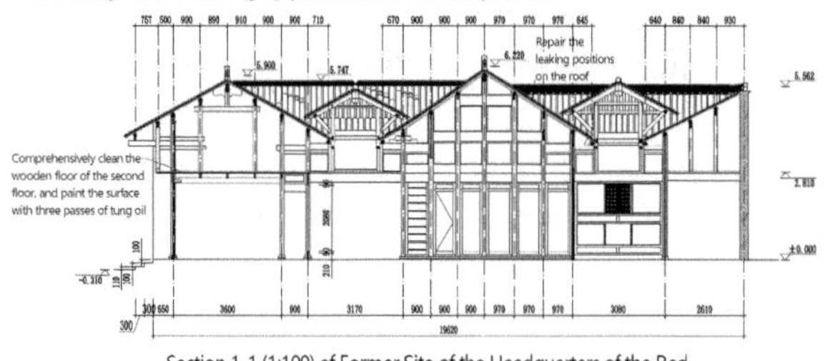

Section 1-1 (1:100) of Former Site of the Headquarters of the Red Army

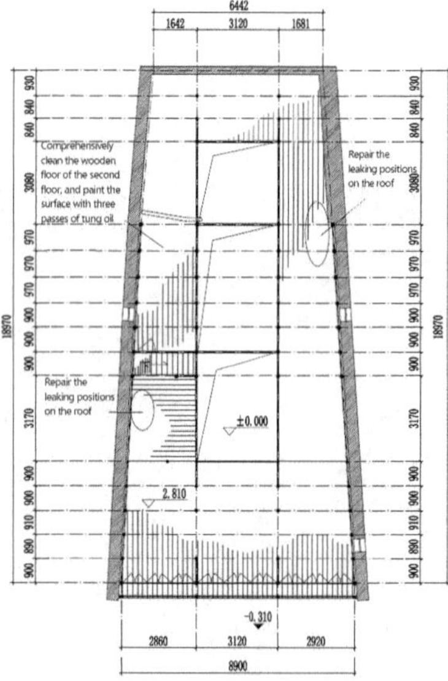

Plan (1:100) of the second floor at the Former Site of the Headquarters of the Red Army

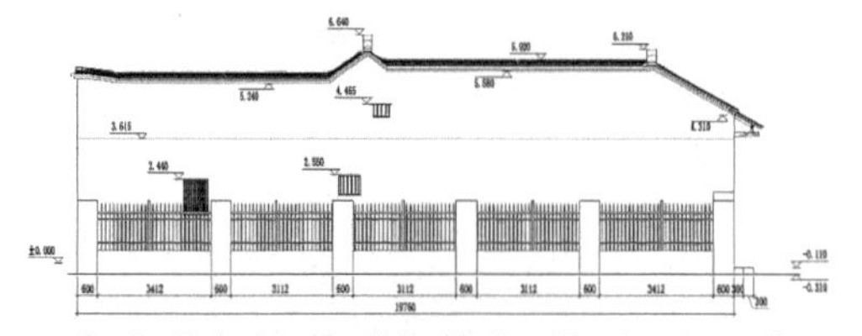

Two-tier side elevation of the old site of the Second Front Army Command of the Chinese Workers'and Peasants' Red Army (1:100)

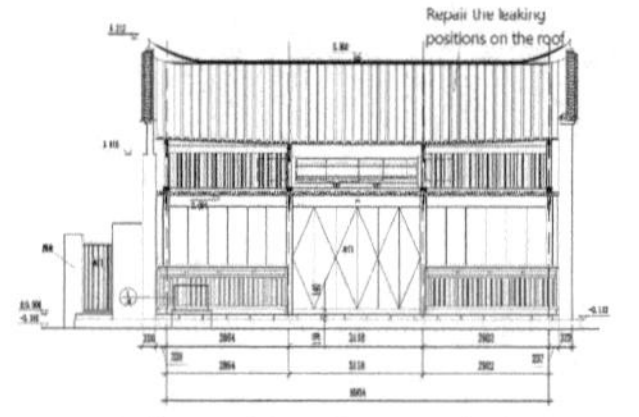

Front elevation (1:100) of the second floor at the Former Site of the Headquarters of the Red Army

Fig. 4 Vernacular heritages

shops, malls, grocery stores, miscellaneous grain shops, private schools, bamboo ware shops, etc. in old time, which can show the local culture at the prosperous ferries in the rural areas in Yangtze River middle reaches; Supply and Marketing Cooperatives, grain warehouses, rice factories and chimney, etc. with other modern functions built after 1949 can reflect the agricultural revolution at the early period of socialism.

3.3.4 Landscape Elements

As the ponds and rivers which used to surround the center of the traditional town are dried up, the waterway restoration plan has been carried out cautiously. At present, only the wharf area, flood control rammed earth embankment and other places with collective memory are restored.

3.4 Public Engagement in Rural Revitalization

This phase of town construction was implemented on a trial basis through the joint effort of three-parties: government, designers and indigenous people. The project was implemented in three phases: the preliminary survey, the mid-term design and the final execution. The preliminary survey was carried out by a variety of means to collect all the documents and data through the joint effort of the local cultural, planning, land administration and agricultural authorities. The locals of all age groups were interviewed for the collection of information. For example, the elderly over 60 years old were interviewed in priority, and the elderly over 80 years old were interviewed while taking video recording to collect and sort out all-round information on their family relations, living conditions, building layout, farming mode changes, etc. All ways are to establish investigation files, including by Internet. In design process, the households were visited to understand and meet the common needs of residents according to the urgent problems to be solved. After commencement of the project execution, the public engagement will be increased by informing all through signboards, employment training to locals, etc. so that they will actively participate in the town construction.

4 Summary of Adapting Improvement Strategies of Traditional Towns

4.1 Identifying a Protecting Scope as the Basis for the Protection of Traditional Towns

The contemporary administrative towns have already expanded beyond the boundary of traditional towns several times (Fig. 5), implementing the administration of managing the surrounding villages and wasteland by the key town or village on an agency basis. Therefore, a protected boundary and coordination environment of the traditional towns should be carefully identified and accurately defined instead of blindly seeking for perfection and entirety. As any traditional town was established with its inherent logic of scale and form, its effective protection depends on accurate definition. The agricultural mode and ecological space in various periods shall be selectively protected in surrounding countryside space to protect necessary elements, and establish coexistence connection with the built-up area.

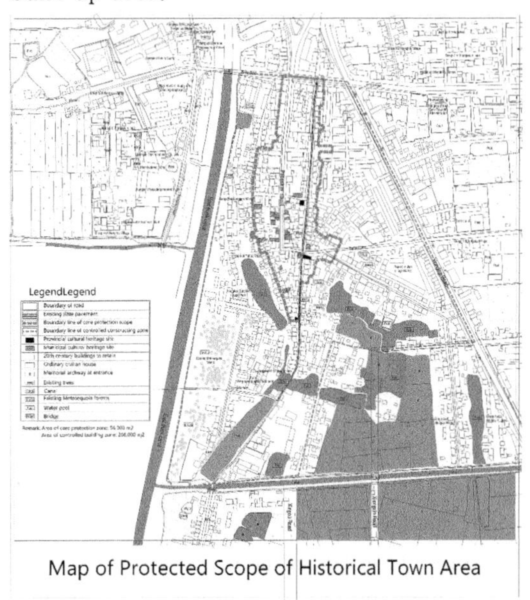

Fig. 5 Protecting scope of historical towns

4.2 Driving Rural Revitalization by Properly Updating Rural Industries

Traditional towns had been stable by

Fig. 6 Public participation

Chinese agricultural civilizations' nurturing, in which agriculture plays a leading role. The dominance of grain production in Y&H alluvial plain has always been a key reason for underdevelopment. However, if we fully urban-industrialize in the process of rural revitalization, traditional towns will also have the risk of becoming unoccupied town like ruins. Crayfish-rice symbiosis in Xiongkou Town is developed by the local farmers through knowledge and wisdom, which has promoted the spatial characteristics of the agricultural environment. Therefore, the effort to search for a reasonable pattern of industrial revitalization is the motive force for traditional towns to keep vibrant, and also the impetus for sustainable development of idyllic rural life in China.

4. 3 Synergies of Multi-themed Tourism Routes in Traditional Towns

The development of traditional towns must be the result of the combination of all development achievements in the past. Its urban construction and the surrounding agricultural environment jointly contribute to the rural cultural landscape. The cultural characteristics must be developed with a combination of all the historic elements and all periods. The physical landscape developed by the interaction between thematic routes, cluster space, and cultural landscape based on layered analysis will add values and programs to rural tourism. The identifiable tangible heritages in Xiongkou Town can be dated back to the late Qing Dynasty, Republic of China, and the early 1950s. The town has developed red education tourism route, visiting crayfish ecological agriculture chain route, rural sightseeing route in water-towns, which will assemble catering, housing and entertainment services to produce a compound expansion effect.

4. 4 Promoting Public Participation

Public participation plays a vital role in the protection of various cultural heritages. In the process of this restoration and modification, we explored the affective commitment through historical and cultural promotion and enhanced the awareness of responsibility and individual rights while offering benefits, because this practice will not only benefit communities and the locals through environmental improvement and employment increase in heritage protection, but also increase locals' enthusiasm through education, and provide tourists with more meaningful experiences (Fig. 6).

References

LI Baihao, 2006. Vernacular architecture in Hubei Province. Wuhan: Hubei Press.

The Activation and Utilization of Vernacular Architecture Under the Background of Industrial Convergence —Take Youfang Town of Qinghe County as an Example

WU Xinyao, *XU Xiwei*, *LIU Yang*

Tianjin University, School of Architecture of Tianjin University,
No. 92 Weijin Road, Nankai District, Tianjin, China

Abstract As a unique symbol, vernacular architecture represents the local culture. But with the development of the economy, the contradictions between rural development and vernacular architecture protection have become increasingly prominent. So, how to balance the relationship between them has been taken as an important issue. Taking Youfang town, Qinghe county as an example, this paper discusses the relationship between the protection of vernacular architectures and the economic development of villages and towns, and puts forward the approach to the activating and utilization of vernacular architectures: To dig deeper into Wei canal Culture and tourism resources and to innovate the mode of industrial convergence. In order to activate the use of vernacular architectures through the development of rural tourism as well as to develop the economy of villages and towns. We expect this case could be a reference for the activation of other vernacular architecture along the coast of the Beijing-Hangzhou Canal.

Keywords vernacular architecture, activation and utilization, rural tourism, industry convergence, Youfang town

1 Introduction

As an important part of regional culture, vernacular architecture plays an important role in the historical culture of the village. The special symbols carried by the vernacular architecture show the geomorphology, climate, culture, building materials, construction methods and traditional technologies carrying the unique culture and connotation of villages and towns. At present, the development of most villages and towns in China is lagging behind. The urbanization rate is low. And the industrial structure is single. With the development of the economy and the advancement of the urbanization process, the construction that has been reinstated can be seen everywhere. This process is bound to create constructive damage to vernacular architecture. Therefore, how to protect and activate the use of vernacular architecture while developing the economy has become a content worthy of further study.

2 Industry Integration and Activation and Utilization of Vernacular Architecture

2.1 Conceptual Extension of Vernacular Architecture and Industry Integration

2.1.1 Vernacular Architecture

In terms of area, all buildings in the local environment can be called vernacular architecture. In terms of character, Paul Oliver pointed out several characteristics of vernacular architecture in Encyclopedia of Vernacular Architecture of the World: local anonymous (not designed by an architect), spontaneous, folk, traditional, rural, etc. In terms of function, the vernacular architecture includes vernacular residence, temples, ancestral halls, ancient academies, theaters, restaurants, shops, workshops, memorial gateways and bridges. They reflect different ethnic and regional characteristics. And they are also important cultural, artistic and landscape resources.

2.1.2 Industry Integration

Industry integration refers to the dynamic development process of different industries or different businesses of the same industry permeating and intersecting each other, finally merging into one and gradually forming new industries. This is an economic development model that changes the single industrial structure of villages and towns in the past and develops the industry together. For example, the joint development of the primary industry and the tertiary industry can derive the development mode of agricultural products plus e-commerce. Industrial integration can create a large number of jobs, solve the employment problem of some rural surplus labor and attract a large number of tourists, increasing the income of farmers.

2.2 Status of Vernacular Architecture

Vernacular architecture does not have its own "identity" like cultural relics. Not every rural building is protected by policies. And its status is relatively embarrassing. It is neither as obvious as conservation architects nor bringing direct economic benefits to the town. Therefore, with the continuous advancement of urbanization and the relocation of many conditional indigenous residents, many residential buildings are abandoned. The internal residents have long been affected by the deteriorating material environment, the living environment management is chaotic, the residents have a sense of belonging and security. Even some enterprises, driven by economic interests, carry out a large number of the demolition of the buildings in some villages and towns and do not hesitate to use illegal means. For example, a residential building with some regional features but seemingly worn-out is classified as a dangerous house or a shanty town. And the "demolition of old buildings" is completed in the form of demolishing dangerous houses or the temples, bridges, arches, etc. The demolition of local buildings which are considered to be no longer functioning is happening because its existence hinders the development of the village. The situation of vernacular architecture is very critical and its protection and activation are urgently needed. We need to pay high attention.

3 The Strategy of Activation and Utilization of Local Buildings in Youfang Town Based on Industrial Integration

3.1 Background Overview

Youfang Town, Qinghe County, is located 15 kilometers southeast of Qinghe County, Xingtai City, Hebei Province. It is known as the "Little Shanghai" of Qinghe. In recent years, it has been announced as "famous town of cashmere industry" and "canal town" by the higher authorities such as the Provincial Department of Industry and Information Technology. The representative local buildings in Youfang Town are: Youfang Wharf, Yiqinghe Yandian, Zhutangkou Dangerous Section and Yuanhou Temple and the unique Budaiyuan buildings in southern Hebei (Fig. 1). In terms of industrial development, Youfang towns have potential value in terms of location, cultural resources and e-commerce development. But the process of their development is not optimistic. From the perspective of vernacular architecture, the building did not receive the protection it deserved. Apparently, we failed to realize its potential value.

Fig. 1 Youfang Wharf

3.2 Status of Youfang Town, Qinghe County

3.2.1 Overview of the Vernacular Architecture of Youfang Town

(1) Youfang Wharf

The Youfang Wharf is located on the east side of Youfang Village. And it was built during the Hongzhi Period of the Ming Dynasty. Built with blue brick, supplemented by dry masonry and a small amount of red brick. There are coal wharf, grain wharf, salt wharf, ferry wharf and so on. According to historical records, from the Ming Dynasty to the early years of the Republic of China, the Youfang wharf was in full swing during the day, the fishing lights flashed at night, and the transportation was very busy. In the 1930s and 1940s, the Japanese army set up checked posts on various sides of the canal. The transportation was often blackmailed by the Japanese army. In addition, the rivers were in poor condition and the transportation was deteriorating. In the 1950s, the transportation was restored and developed. The vessels were mostly wooden sailboats, with daily traffic of more than 1,000 tons and more than 100 ships in the past. In the 1960s and 1970s, it was often stopped due to the river discontinuation. After 1980, the smuggling was completely cut off. In 2012, as the only remaining brick wharf in the Hebei section of the Grand Canal, the Youfang Wharf was announced as a national key cultural relics' protection unit.

(2) Yuanhou Temple

Yuanhou Temple was built in the 23rd year of Jiajing to the sixth year of Longqing. It was originally called Xiaosheng Temple. For centuries, people have prayed for their children and grandchildren, to protect them from the flood. During the Tongzhi years, the Dawang Temple was built in the old county town. And there were plaques in the temple. In the history of more than 400 years, Yuan Hou Temple had three refurbishes on the original site. In six years of the Republic of China when the second restoration, Wang Dianjia who worked in the Qing Hanlin Yuan wrote "Yuan Hou Temple", and presented a plaque hung on the lintel on the main gate (Fig. 2, 3).

(3) Zhutangkou Dangerous Section

Qinghe was not only a famous terminal but also a prosperous business place. At the

Fig. 2,3 Yuanhou Temple

same time, due to the close location to the canal, it was also troubled by the flooding caused by the riverbank. In order to reduce the flood disaster, many dams have been built on both sides of the riverbank. According to historical records, during the Qing Emperor Guangxu period, there were many dangerous section on the riverbank of the Qinghe section of the canal. At that time, there were 10 dams, with a total length of more than 300 feet. Among the many dams, Zhutangkou is a representative one. Zhutangkou dam is located on the left bank of the Beiwei Canal in Zhutangkou Village, Qinghe County, Xingtai City. It was built in the late Qing Dynasty and has an existing length of 961 meters. The materials are riprap, dry masonry and masonry, well column grid and blue brick. The dam has been repeatedly built since the construction, representing the different historical periods from the early 20th century to the 1990s. The form and construction method have effectively alleviated the flooding of the riverbank and prevented the occurrence of the levee, which has a high scientific value of water conservancy. At present, the Zhutangkou dam has been identified as a national key cultural relics protection unit (Fig. 4, 5).

Fig. 4,5 Zhutangkou Dangerous Section

(4) Yiqinghe Yandian

The Yiqinghe Yandian ruins are located on the west side of the northernmost pier. They are freight yards of salt, covering an area of nearly 10 acres. At the time of prosperity, more than 50 employees were employed and organized. There is a senior treasurer; there is a purchasing department, which is responsible for ordering salt in Tianjin. The escort department is responsible for the transportation tasks after loading the goods. The unloading department is responsible for organizing the loading and unloading of the unloading vessel. The warehouse management is responsible for checking the goods entering the warehouse. The storekeeper is responsible for the retail of salt. In addition to these office workers, there is a special department - the salt patrol, similar to the current security guards. They are fully armed and patrol day and night. It is said that the salt shop was robbed three times after 1940. And it was badly damaged. Today, there are 5 Daoguang years' salt shop ruins remain which are built in Daoguang years(Fig. 6).

Fig. 6 Yiqinghe Yandian

(5) Residential Vernacular Architecture

The current residential vernacular buildings in youfang town are mainly Budai yards with unique characteristics in southern Hebei province. But many of them are empty and idle at present(Fig. 7, 8).

3.2.3 Overview of the Youfang Town Industry

The economic development status of Youfang Town in Qinghe County is generally at the lower level of Qinghe County. Its total agricultural output value is at the middle and lower reaches of Qinghe County. Its

Fig. 7, 8 Residential vernacular buildings

industrial output value is ranked second to last in Qinghe county. The total output value is only one-fifth of the industrial output value of Xielu Town, which has the best economic development. The economic development is struggling(Fig. 9, 10).

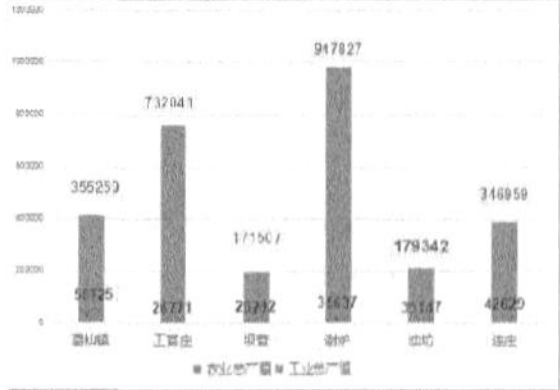

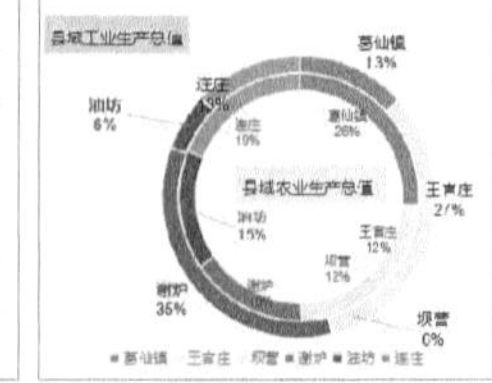

Fig. 9, 10 Comparison of the industrial and agricultural output value of various towns in Qinghe County

The main crops of Youfang Town include wheat, corn, cotton, soybeans, peanuts, vegetables, etc. Typical crops include garlic, sweet gourd, etc. Animal husbandry is mainly pig, sheep and poultry farming. Among them, garlic planting and seedling planting are the famous industries of Youfang Town to which the government gives high support in funds and policies. The newspaper's rural economic section also specifically reported the garlic in the Youfang Town.

In the second industry, there are 4 enterprises in Youfang Town, among which cashmere processing and automobile and motorcycle industries are typical industries. Youfang Town is known as the famous cashmere industry in Qinghe County.

The third industry is dominated by retail services. In the whole town area, in addition to retailing, there are rural Taobao service stations distributed in various villages in the form of rural Taobao. The e-commerce industry is relatively developed. Youfang Town is known as "Taobao Town" (Fig. 11, 12).

Fig. 11,12 Rural Taobao

3. 3 SWOT Analysis of Industrial integration and the Development and Utilization of Vernacular Architecture

3. 3. 1 Analysis of Strength

Youfang Town has the advantage of open and cooperative location. The planning of the Beijing-Kowloon Railway will cross Qinghe County. Its agricultural resources are rich. And there are industries supported by the government. Rural Taobao is common, which will be a better development for the integration of primary and tertiary industries. The most prominent advantage is that the valuable historical culture and eco-tourism resources of the canal flow through the town, and there are local buildings along the canal. This is the biggest advantage for the development of the cultural tourism industry in the Youfang Town. In the process of integrated development of cultural tourism industry, vernacular architecture can be activated.

3. 3. 2 Analysis of Weakness

Although the agricultural resources of the Youfang Town are rich, the industrial form is simple. The scale of the characteristic industry is small, and the brand influence is not enough. The economic development fails to join the cultural factors, The vernacular architecture has not received enough attention. And the key vernacular architectures are hard to integrate.

3. 3. 3 Analysis of Opportunities

The opportunities in Youfang Town exist in many ways. The construction policy of the Jinghang Canal Cultural Belt is an opportunity to develop cultural tourism and activate vernacular architecture. In the Supply-side reform of Qinghe County, it was clearly required to develop "innovation-driven, scientific and technological agriculture", "value promotion, brand agriculture". It provides policy support for the integration of three industries and the consumption boom of cashmere products gradually happened, which provides an opportunity for the development of the cashmere industry.

3. 3. 4 Analysis of Threats

The integration of the cultural tourism industry has challenges. The Youfang Town is 15 kilometers away from the county. The radiation distance of cultural tourism is generally 30 kilometers. In the absence of operation on the Wei Canal, the hinterland of the Youfang town travel market is small. Key vernacular buildings are distributed zonal along the canal, which is easy to cause long journeys with few sightseeing. The layout of tourist attractions formed by a single scattered key vernacular architecture is not easy to form an attractive route. The cultural content of the surrounding tourism products is not rich, and the cultural value is not well explored. The tourism products are mismatched with the local culture and are at a disadvantage in the market competition. If the tourism industry does not develop, the vernacular architecture cannot be activated and utilized while developing tourism.

3. 4 Local Building Activation and Utilization Strategy Based on Industrial Integration

Based on the analysis of the vernacular architecture and economic industry of the Youfang Town, Under the support of the existing resources and the restriction of various development conditions, there are still many problems in the process of industrial development and integration in Youfang Town of Qinghe County, as well as the protection and active utilization of unilateral vernacular buildings. Therefore, industrial development must be accelerated. To activate the use of vernacular architecture, integrate the industry, develop the cultural tourism industry. And at the same time combine characteristic agriculture with cashmere process and tourism. Create a tourism town.

3.4.1 Integration of the Cultural Tourism Industry-Tapping the Canal Culture and Activating the Use of vernacular Architecture Through the Development of Tourism

In the following development of Youfang town, Qinghe county, the tourism resources of Wei Canal Culture and vernacular architecture should be deeply explored, so that the culture and tourism industry can be brought to the extreme. Meanwhile the industrial integration mode should be constantly innovated As the important tourism resources of the vernacular architecture, the Youfang pier along the canal, Yuanhou Temple, Zhutangkou and Yiqing and Yandian should strive to build a cultural node on the Wei Canal under the opportunity of the construction of the Canal Cultural Belt. Therefore, these key vernacular architecture should be repaired. The surrounding environment should be rectified, and the nodes should be used as a node to unblock the lines so that these local buildings are no longer independent individuals, but cultural resources that are connected with each other.

It is necessary to promote the integration and development of the cultural industry and the tourism industry. It is also necessary to improve the construction of basic service facilities and related supporting facilities, rationally adjust the hotel structure, and use residential buildings in tourist hotspots to set up some homestays. In this way, on the one hand, the residential buildings can be used, including many idle residential buildings in the township, to achieve the activation and utilization of the vernacular architecture. On the other hand, It may solve the problem of accommodation considerations necessary for the development of tourism. And the homestay allows visitors to experience the country style better.

We should increase the development of cultural resources and tourism resources and dig deep into the local culture including the canal culture. We will take the "fine quality" and "branding" as the development direction, and comprehensively use technology, culture, creativity and other means to comprehensively develop quality products. Going to build. In addition, resource development is effectively integrated, giving full play to the joint role of the tourism industry, extending the industrial chain, making full use of the vernacular architecture, innovative industry integration model, and constantly enriching the forms and types of tourism products at different times. Integrate the resources of cultural tourism, increase the promotion efforts, and enhance the activation and utilization of the vernacular architecture to the height of the tourism development strategy. At the same time, it proposes the slogan of the tourism subject, which is recognized by the whole and the public, to shape the overall image of the Youfang-water town, use the mass media to call the main slogan, in-depth publicity and build the brand of the canal tourism system, create brand effect, and drive other the development of general vernacular architecture and culture tourism resources.

3.4.2 Agricultural Products + E-Commerce + Tourism to Revitalize the Use of Vernacular Architecture Through the Development of Characteristic Agriculture, and Build Taobao Youfang

In the tourist hinterland of the town, the star products of the Youfang Town, garlic and cashmere products, combine the Wei Canal landscape resources, transform some idle local courtyards, introduce garlic and cashmere products, and develop small picking gardens and cashmere workshops. Tourists, also facing the residents, have made sufficient publicity for the featured products with housing, and at the same time, strengthened the functions of the vernacular architecture and enriched the industrial structure of the oil mill.

In the following development of the Youfang town. It should be fully exploited and utilized as the "Taobao Town" e-commerce advantage, which will greatly promote the industrial integration of Youfang

Town. At present, rural Taobao has the defects of small radiation range, low service content and low popularity. Therefore, we should use the vernacular architecture in the town to develop rural Taobao outlets, sort out the routes, form a network, and enhance the influence of rural Taobao. At the same time, increasing the popularity of products sold, including specialty agricultural products such as garlic and seedlings, as well as specialty cashmere and auto parts. It is critical to the integration of the industry. In terms of logistics, the use of idle local buildings should establish an agricultural product distribution center integrating warehousing, logistics, transactions and services. Through the Internet + platform to promote the common development of agricultural products and by-products processing industry and modern logistics industry

4 Conclusion

Vernacular architecture is an important part of rural culture. Protecting and utilizing vernacular architecture is not only a cultural memory left for later generations but also a source of more thinking and enlightenment for future development as well as a source of strength for the advancement of society. This paper mainly studies the activation and utilization of the vernacular architecture of the Youfang Town in Qinghe County based on industrial integration. Based on the status quo of the industry, propose the path of industrial integration and the strategy of activation and utilization of the vernacular architecture. It is intended to provide a reference for the industrial development of small towns in the same area along the Wei Canal and the activation and utilization of vernacular architecture.

References

HE Wen, SHOU Hangxiang, WU Fan, et al., 2017. Strategies for the activation of rural architecture ruins: a case study of "mountaineers" teahouse in Wanxichun village, Kunming. Chinese and foreign architecture, 8: 168-171.

JI Maoquan, WANG Ping, LIU Ruiyang, 2019. Research on original state protection and active state development of rural architecture based on rural revitalization. Green environmental protection building materials, 4: 205+207.

LIN Shuling, WANG Zhigang, 2016. Regeneration of rural architecture and activation of rural communities: starting from Xujia compound in Jinzhai, Anhui Province. Architecture and culture, 3: 192-194.

ZHANG Yilin, WANG Yong, 2018. The active reuse of rural architecture: a brief discussion on bishan industrial marketing and renovation of surrounding buildings. Public art, 5: 38-43.

Research on the Construction of Cross-Border Ethnic Settlements' Symbiosis System in China-Mongolia-Russia-the Belt and Road Initiative Economic Corridor — Take Oroqen Ethnic Group as an Example

ZHU Ying ,*QU Fangzhu* ,*WU Yating*

School of Architecture, Harbin Institute of Technology; Key Laboratory of Cold Region Urban and Rural Human Settlement Environment Science and Technology, Ministry of Industry and Information Technology, Harbin, Heilongjiang Province, China

Abstract Under the guidance of the belt and road initiative's goal, the cross-border related projects of China, Mongolia and Russia have received close attention from all sectors of society. Thanks for their geographical advantages and economic trends, China and Mongolia, China and Russia have formed a closely linked cross-border cultural belt and economic belt on the border of Inner Mongolia. The Oroqen ethnic group is a cold fishing and hunting ethnic group living on this economic and cultural corridor. As early as 2000 BC, the Oroqen people had a wide range of activities and were mainly active in the vicinity of vast areas from Lake Baikal to Sakhalin Island in the sea and from the north of Heilongjiang to the south of the Outer Xing'an Mountains region. They formed a unique national spirit, value orientation, aesthetic pursuit and way of thinking, unique in production, life, production and religious beliefs, thus Oroqen has become an important part of the cultural circle formed by cross-border national culture as the main body. The thesis takes Oroqen as the research object, making an analysis of at three different levels of human and family, human and society, and human and nature to form an evolutionary system from small to large and nested layers. The research includes the living space and the economic corridor as the linear heritage corridor between China, Mongolia and Russia. The Oroqen ethnic group settlements in the belt and road initiative region are organically linked, its core part is family space center, ethnic space center and ethnic concentration center. We excavate the internal demand, attraction and radiation force formed by their living, production and survival needs among the different settlement centers to construct the ecological symbiosis structure system of the Oroqen community. The aim is to provide the Oroqen ethnic group in the belt and road initiative's China-Mongolia-Russia economic corridor with regional cultural and psychological identity and emotional destination, to activate the continuous evolution function of ethnic endangered culture under the corridor, and to provide the living protection and theoretical basis for the construction of the belt and road initiative and endangered ethnic groups.

Keywords Belt and Road Initiative, Oroqen, settlement, structural system, evolution and regeneration

1 The Formation of the Oroqen People's Living Space

British anthropologist Taylor, the father of anthropology, gave a classic definition in 1871: "Culture, in its broad sense in ethnography, is a composite whole, which includes knowledge, belief, art, morality, law, custom and other abilities and habits necessary for individuals as members of society." The formation of the Oroqen people's living space is formed by various factors.

1.1 The Natural Environment of Oroqen Hunting Economy

The Oroqen people lived in a nomadic state until the 1950s. They wandered between Xing'an Mountains region and migrated with the seasons, but they never left Xing'an Mountains and the coastal areas of Heilongjiang for their survival. This is also a famous primeval forest distribution area in our country. The special topography and geomorphology have laid a solid foundation for the nomadic life of the Oroqen people. The special topography constructs the unique living space pattern of Oroqen. In the traditional period, the Oroqen people took full advantage of the topography and geomorphology of Xing'an Mountains and combined with their own needs. They took rivers as the lead, mountains as the support and grasslands as the medium to migrate and hunt for generations in the vast Xing'an Mountains.

The long winter and short summer make it difficult for the Oroqen people to live on long-term farming, but they live in the form of hunting all the year round. According to climate change, the Oroqen people have also formed their own stable migration cycle. The severe winter restricts them from frequent relocation. They usually settle in suitable places temporarily for a long time, while in summer they migrate frequently.

Before liberation, the Oroqen people took grab type as the first economic form, and the settlement form was mainly "traveling and living". The long-term cold has prevented the Oroqen people from relying on farming for a long time like other ethnic groups. Although there is no comfortable climate condition, the Xing'an Mountains are rich in wood industry and wild animals with mountains, rivers and trees, which provide the Oroqen people with a continuous source of food and clothing. Gathering, fishing and hunting are the main ways of survival for the Oroqen people on such an ecological chain. Nature has given them huge resource products. They regard their own life and nature as a community of destiny. They took from nature and respected it more, which kept the ecological chain between the Oroqen people and nature for thousands of years and bred generations of Oroqen people.

1.2 The Socio-Economic Structure of Oroqen

The Oroqen were in the patriarchal clan society before liberation. The patriarchal clan society was called "Mukun of A Min", meaning the same surname. According to historical records, there are more than ten "Mukun" of Oroqen people. With the development of society, the Oroqen people have entered the patriarchal family commune stage. The Oroqen people are called "Akuna Thousand Wuli Neighbors", which means that uncles and brothers live together for several generations. This stage has lasted for a long time. During this period, the tools of labor were bows and arrows and reindeer, which were carried out collectively by clans. The introduction of firearms has provided greater support for hunting production later. But social productivity is still at a low level, and the Oroqen cannot compete with wild animals alone. The scope of clan commune has been reduced to that of a small family commune, which has become an independent self-sufficient economic unit.

By the late Qing Dynasty, the Oroqen people had more contact with the outside world. Hunting and production need fewer people by the introduction of pull bombs. As a result, from the "Akuna Thousand Wuli Neighbors" came the "Akuna Day Wuli," meaning that several generations of brothers live together. "Akuna Riuri" and "Akuna Qianuri Neighbor" are both patrilineal clans, but they are two different social forms. Currently, "Akuna Riuli" is the basic economic unit of society. There are clan leaders "Mukunda" and a clan leader "Salekanda".

During the matriarchal clan period, the social productivity was extremely low. Oroqen could not conquer most wild animals and the hunting can't meet the family's needs. As a result, women's collection is the main source of production at this stage. During the period of development to the patriarchal clan commune, productivity gradually increased, and hunting occupied an important position. Then it entered the "Wuleng" stage of the village community. Hunting tools were greatly improved and hunting production became the main source of clan production. At that time, people and nature are intricately linked and intertwined, which leading a production mode of hunting, gathering, fishing and handicraft, and in this way, they are steadily developing and multiplying in nature.

1.3 Folk Art Originating from Life

Oroqen nationality is a nationality developed in the early period of primitive society. In their matriarchal society, their production level is extremely low, and their living environment is relatively harsh. They cannot explain various phenomena of nature, and most of their living needs are taken from nature. Under the restriction of this social environment, the Oroqen ethnic group had some of their own fantasies, which also became their sustenance in life, which gradually developed into a belief in religion and the concept of "all things have spirit". This is Shamanism, which has continued from the earliest period of primitive society, including nature worship, totem worship and ancestor worship. Because the Oroqen people had less contact with the outside world and less knowledge before liberation, Shamanism was preserved more completely. The shamanism of the Oroqen nationality is an integration of heaven, earth and human beings. The Oroqen people and nature are also symbiotic, advocating nature and fearing life, thus resulting in many sacrificial ceremonies and activities.

Before liberation, the Oroqen people were all in the region commune stage at the end of primitive society and lived a hunting life with hunting as the main part, they gathered and fish as the auxiliary part. From the folk songs, dances, literature and epics of the Oroqen people, we can see the scenes of hunting, production and social life in the great and small Xing'an Mountains. Each work is a vivid picture of the customs of the Oroqen people.

2 The Spatial Form of Oroqen Ethnic Groups

2.1 Traditional Basic Architecture in Family Dimension

The residential buildings of Oroqen nationality are developed in landform, climate characteristics, clan system and ethnic beliefs, and are the result of the blend of the forest environment and ethnic feelings. Each kind of building is built at any time according to the season and landform to adapt to the long-term hunting life of the Oroqen. Through oral inquiry and on-the-spot investigation, we explore the formation factors, construction methods and functions of Oroqen folk houses. They carry the basic standard of living and living needs, and their material forms are mainly manifested as temporary residential buildings and fixed warehouse building Oren. The two basic buildings originated from society and built a complete living space system with a certain scale in nature (Fig. 1).

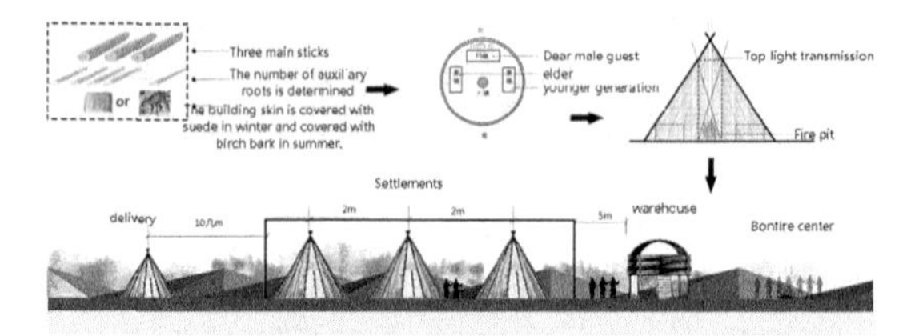

Fig. 1 Rib subspace organization in the "person-family" dimension

2.1.1 Cuoluozi

The material source of Cuoluozi (Fig. 2) is all basic materials, obtained by cutting birch and pine wood. The size of the broom cypress determines by the size of the family. Therefore, it is decided to cut down the length of wood, with the longest being about 10 meters and the shortest being about 5-6 meters. First fix the three forks. In summer, there is a curtain at the door of the fruit tree. The top is half a meter from the top to the cover. When it rains, the top is covered with birch bark. The bed is built inside. To prevent moisture in spring, summer and autumn, four small wooden sticks are used as the base support of the bed, which is about 20 cm high. The bed is covered with a layer of small wooden sticks. The wooden sticks are covered with grass and finally covered with roe deer skin mattress. In winter, sweep the snow on the ground, spread the wooden stick directly on the ground, spread grass and sleep on it.

Fig. 2 Temporary residential building"Cuo luozi"© a brief history of the oroqun ethnic group

2. 1. 2 Oulun

"Oulun" is another primitive warehouse building (Fig. 3) that meets the needs of hunting. It is used to store dried meat, clothes, wild vegetables, supplies that Oroqen people do not use in their daily life. The site is located at the hunting center or the place of the regular route, about 20 to 30 meters away from the "Xierenzhu". Basically, 1—2 "Oulun" will be built in each "Wulileng", and the number of "Oren" will be determined mainly according to the number of items to be stored.

In the past, it used to be a primitive forest with a very rich amount of forestry. Oroqen people used this resource to make a support column at the bottom of a diagonal rectangular trunk that grows naturally in the thick forests in the mountains. They then built four trees on the trunk with branches respectively. The bottom was paved with four pieces of wood, and then they built a pointed or curved shack. The top surface of "Oulun" is covered with birch bark, and the bottom is about 2—3 meters from the ground. When people go up through ladders, the ladders usually reach a place far away from "Oulun". These practices are to avoid damp articles and prevent the sneak attack of wild animals.

Fig. 3 Fixed warehouse building "Oulun"© a brief history of the Oroqen ethnic group

2. 2 Settlement Space Order in Social Dimension

From matrilineal society to patrilineal society, the change of social factors resulted in the geographical village community "Wulileng". At this time, the "Mukun" system society in the matriarchal society period disappeared, public ownership moved to private ownership, clan production and other related issues were no longer collective participation but were carried out in small families. At this time, due to the need of hunting and production, a special hunting and production labor organization-"Anga" was organized.

There is no definite number of "Anga" members for each hunt. A temporary production organization organized voluntarily by three, four or five or six people will be disbanded after the hunt and regrouped after the next hunt. Its leader is "Tatanda". At this time, "Tatanda" is different from "Tatanda" when it was related by blood. It still has all the functions and powers when it was related by blood except as hunting leader. It is a highly respected position in a clan that was elected through the electoral system. In addition to "Tatanda", there is also a deputy head of "Anga" who is called "Uturuda" and "Elochin" who is optional when hunting, and several hunters form the hunting group "Anga" (Fig. 4). "Uturuda" is responsible for assisting in the management of "Anga" production activities. "Elochin" is responsible for staying in the camp to cook and watch the horses and other groceries. When there is not enough staff, the "Elochin" can also be taken care of by those who came back from hunting first.

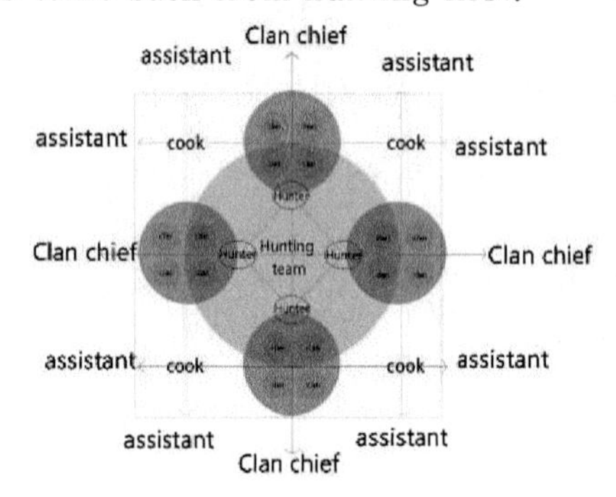

Fig. 4 Diagram of "human-society" relationship from social institution to clan settlement to ethnic group

2.3 The Model of Travelling Space in Natural Dimension

In the natural dimension, the reason why Oroqen people can reproduce in the Xing'an Mountains with a constant pattern of changing state is that they have formed a constant ecosystem in the evolution of thousands of years. This ecosystem is made up of "points", "lines" and "planes" and forms a closed "cellular body". The inherent genetic factors of this "cellular body" have been constant. However, facing the external environment given by Xing'an, they need to topology the unchangeable "cellular body" and change in nature in a topological mode indefinitely to adapt to the ecosystem they live in (Fig. 5).

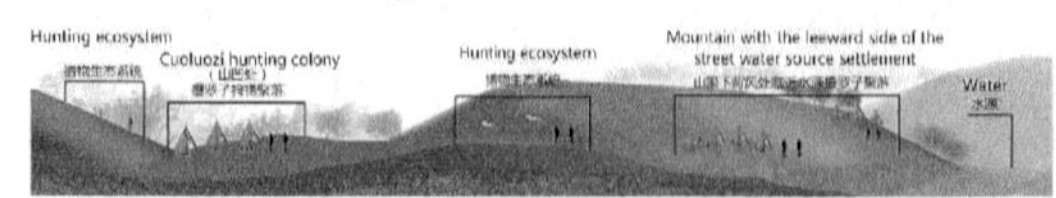

Fig. 5 "Human-land" relationship diagram from natural environment to survival mode to ethnic group

3 Conclusion

The traditional settlement space of Oroqen ethnic group is one of the unique and primitive settlement modes among fishing and hunting ethnic groups. At the same time, the Heilongjiang River valley where the Oroqen ethnic group lives stretch in the space strategy set up by the belt and road initiative. Its space mode bears an important historical mission and special significance, as well as a strong cultural connotation and emotional sustenance. The living space of the Oroqen people is like a container. It carries the past and records history. It is the vitality point of the whole area and builds a complete ecological system with the living environment of the Oroqen people. This article traces the origin of the Oroqen settlement and explores the spatial characteristics to show the evolution of fishing and hunting nationalities in a tangible, visual and structured spatial pattern. To promote the Oroqen ethnic group to survive in a living way and activate the development of the belt and road initiative region.

Acknowledgement

1. Key Research Topics on Economic and Social Development of Heilongjiang Province in 2018(Project No. 18208).

2. Heilongjiang Philosophy and Social Sciences Research Program in 2018(Project No. 18SHB074).

3. Independent Research Project of Heilongjiang Key Laboratory of Cold Architectural Science in 2016 (Project No. 2016HDJ2-1203)

References

DU Yonghao, 1993. Oroqen safari, settlement and development. Beijing: China Minzu University Press.

ZHAO Fuxing, 1991. Oroqen safari culture. Huhehaote: Inner Mongolia People's Publishing House.

ZHENG Dongri, 1985. Oroqen social changes. Yanji: Yanbian People'S Publishing House.

ZHU Ying, LI Honglin, QU Fangzhu, 2018. Study on the "original space" of the endangered habitat

pattern of northeast fishery and hunting nationalities. Urban architecture, (9).
ZHU Ying, ZHANG Xiangning, 2017. A study on renewal and design of "basal form" of regenerated rural-traditional rural settlements. Urban architecture, (10): 33-35.
ZHU Ying, ZHANG Xiangning, WANG Liren, 2016. Original "local"-analysis of spatial structure and evolution structure of traditional local settlements. Urban architecture, (07): 118-121.

Special Theme:

Heritage Conservation Going Public: Case Studies of Pingyao International Workshop

A Research on the Conservation Plan of the Human-Habitat World Heritage: Case Study of Pingyao Ancient City

SHAO Yong[1], *ZHANG Peng*[1], *HU Lijun*[2], *ZHAO Jie*[2], *CHEN Huan*[2]

1 College of Architecture and Urban Planning, Tongji University, Shanghai, China
2 Shanghai Tongji Urban Planning and Design Institute Co., Ltd., Shanghai, China

Abstract The Human-Habitat World Heritage is the outstanding sample among the traditional human settlements around the world, which is a kind of Living Heritage characterized by its dual attributes of heritage and living community, and representing the mutual co-dependence of traditional livings and spaces. By analyzing the characteristics and value of Human-Habitat World Heritage, the paper focuses on the dual objectives of heritage conservation and habitat improvement, as well as the importance of the "Habitant-centered" conservation principle. Based on the analysis of problems of Human-Habitat World Heritage in China, the paper takes the case of Pingyao Ancient City to illustrate the new exploration of conservation plan and its guarantee mechanism, includes: value re-interpretation, integrated conservation framework, conservation and monitor system for the management of "changes", as well as implementation of the "Habitant-centered" principle since 2006, with the purposes of offering the pilot experience of conservation and sustainable development for the Human-Habitat World Heritages in China as well as the National Historic Cities, Towns and Villages.
Keywords Human-Habitat World Heritage, living heritage, conservation plan, Pingyao Ancient City

1 Human-Habitat World Heritage and Its Characteristics, Attributes and Value

Human-Habitat World Heritage refers to outstanding samples of traditional human settlements such as villages and towns. On one hand, they retain most of the philosophy of spatial organization, construction materials, structures and technology of different regions and different times. On the other hand, they also reflect the vernacular culture, traditional aesthetics, religion, political system and production modes from different periods. Some of these are tangible and some are intangible. Regardless, they are all reciprocal to each other and contribute to the "Outstanding Universal Value" (OUV) of Human-Habitat World Heritage (UNESCO, 1972).

As of today, there are six Human-Habitat World Heritage sites in China and Pingyao Ancient City is one of them.

1.1 Dual Attributes of Human-Habitat World Heritage: Heritage and Living Community

Human-Habitat World Heritage provides a unique testimony to a lost civilization or cultural traditions. It may also exhibit a paradigm of groups of buildings or urban landscape from an important stage in the history of development in a cultural area. This type of heritage consists of both "heritage" and "living community". "Heritage" here means that Human-Habitat World Heritage sites are mostly traditional cities, towns and villages with a long history that reflect the cultural characteristics and planning concept of a particular region and have preserved the rich tangible and intangible heritage of the area. "Living community" means that Human-Habitat World Heritage sites are historical human settlements. Some are even vital functional components of a city. Today, this type of World Heritage still accommodates everyday life, and constantly evolves and changes. Thus, Human-Habitat World Heritage is a "living heritage".

1.2 Characteristics of Human-Habitat World Heritage: Holistic and Dynamic

Compare to other types of heritage, Human-Habitat World Heritage is a "holistic type" — meaning that it is usually a product resulted from interactions amongst natural environment, built environment and cultural environment. It is also dynamic in a sense that it, being a product resulted from interactions between human and nature and social development dated back to hundreds or thousand years ago, is still "dynamically" changing: external environment (such as political, economic and cultural factors) has an impact on the inhabitants in heritage sites and these peoples in turn, change the elements of the heritage through their dynamic behaviors.

1.3 Distinctive Value of Human-Habitat World Heritage

Because of its dual nature and characteristics, the value of Human-Habitat World Heritage is not only about the historic, aesthetic and scientific values of a single architecture or an architectural ensemble. It is more about the co-dependence between the reciprocity of traditional lifestyle and traditional space in the heritage site. It is an exceptional paradigm of traditional human settlement. The wisdom that our ancestors endowed in dealing with the human-nature relationship is of immense value to us humans today. This shapes the distinctive value of this type of heritage.

2 Problems Faced by Human-Habitat World Heritage in China

Based on our research on Human-Habitat World Heritage like the Pingyao Ancient City over the previous decade, we have observed the following issues:

First, we lack a scientific interpretation of the OUV of Human-Habitat World Heritage. The most problematic issue is that it is always the built heritage that receives most concern, the ecosystem that shapes the setting of the heritage site and the traditional culture from which the heritage site originates are always neglected. Even within man-made tangible heritage, the value of vernacular architectures, as the major type of human-habitat heritage, is not fully recognized (except for listed relics). For example, in the Pingyao Ancient City, its city walls and public buildings have received better care and protection from the departments of Cultural Heritage. Yet, the 500 traditional courtyards that contributed to the foundation of the ancient city lacked the necessary documentation, analysis and conservation measures. This led to the destruction and collapse of almost 30 traditional courts between 2005 and 2007.

Second, it lacks the monitoring and intervention mechanisms to manage the changes in resources. Most of the World Heritage sites in China have become tourist destinations. While economic indicators are soaring up high, heritage resources are at the same time rapidly changing: in a built environment, many architecture and spaces have been largely altered to adapt to the needs of modern tourism; requisition of surrounding farmland for real estate development is common; many native inhabitants fled the town so that they can rent out their vacant houses and generate more income (Shanghai Tongji Urban Planning and Design Institute, 2014). The World Heritage Center implements a third-party evaluation system to monitor World Heritage sites scientifically and intervenes when necessary (Feilden, Jokilehto, 1993). However, China has not established the corresponding system for its World Heritage sites before 2006. The vacuum of monitoring mechanism and heritage impact assessment has objectively caused damage to heritage resources.

Third, it lacks an "integrated" conservation and management plan that manages the "dynamic" changes. Aside from the traditional conservation planning for the nation's Famous Cities/Towns/Villages, the conservation planning for Human-Habitat World Heritage has generally fallen under the protection system for cultural relics with other types of cultural heritage (see

Order No. 41 *of the Ministry of Culture on Administration Measures for the Protection of World Culture Heritage*). Utilization of the heritage is determined within a certain tourism planning scheme and is usually planned as a project component. Because planning for Human-Habitat World Heritage is based on such division of the state's administrative system, the interrelationship between natural environment, built environment and the cultural environment has long been neglected, together with an overlook of the holistic and dynamic features of Human-Habitat World Heritage.

Fourth, conservation for World Heritage has already evolved into a comprehensive system. In terms of management, China has not yet to establish any organizational system that orients to the characteristics of Human-Habitat World Heritage. Human-Habitat World Heritage sites are segmented under this "Line and Piece Seclusion" management mode, resulting in problems like management gaps and chaotic implementation.

All the above issues are ubiquitous in many human-habitat heritage sites, especially in Human-Habitat World Heritage sites. "Value identification, resource monitoring, conservation planning, heritage management" are the four fundamental aspects that set a basis for scientific conservation and sustainability of human-habitat heritage. Hence, for the conservation planning of Human-Habitat World Heritage, it is essential to recognize the dual nature and characteristics of the heritage sites, to identify available resources and to establish conservation framework and guidelines, meanwhile, it is also necessary to consider how to synthesize with the other three aspects.

3 Conservation Objective and Methodology for Human-Habitat World Heritage

3.1 Dual Objectives for the Conservation

The first objective is to strictly protect its authenticity and integrity and to highlight its historical and cultural significance by fully respecting the characters. For instance, the Pingyao Ancient City, the best-preserved historic town of the Ming and Qing Dynasties is rich cultural heritage resources, including tangible heritage like city walls, shops, courtyard houses, temples, and intangible heritage like varnishing lacque ware and paper cutting. These historic and cultural resources have constituted a comprehensive system that eventually gained the Pingyao Ancient City its World Heritage status.

The second objective is to improve the living environment by taking into account the actual living and working needs in this type of heritage site. This should be accomplished with the improvement of the living conditions of the locals, and the supporting services and facilities and the creation of job opportunities, hence sustaining the internal driver for the conservation of the city. For instance, the population of the Pingyao Ancient City has long been roughly around tens of thousands. These native inhabitants are the owners and users of the "residential courtyard houses", the main component of Pingyao's heritage. They also constitute the major force for conservation and development.

3.2 "Habitant-Centered" Conservation Approach

Community participation should never be left out in any conservation works for Human-Habitat World Heritage, regardless of whether the intention is to protect its integrity and authenticity or to improve the conservation strategy for World Heritage.

The practice in China has enlightened us that there is not an overarching solution for this proposition. Yet, we can still endeavor to explore a way out by taking into account the following aspects:

First, to plan a balanced development of both a city and its heritage site. This plan should comprehensively consider urban expansion, improvement in human settlement and heritage conservation. Thus the heritage site is no longer an isolated area from other parts of the city.

Second, to formulate a management plan for heritage and establish synergies amongst management mechanism. The management plan should gear towards including all individuals and groups involved in the management of heritage assets.

Third, to enhance the supportive system of local community within world heritage cities. In the context of the rapid development of tourism, as well as in the face of under-development of indigenous communities, local resilience and resistance to external shocks should be strengthened early on.

4 New Exploration on the Conservation Plan of Pingyao Ancient City as Human-Habitat World Heritage

Based on the previous discussion, the history and reality of the Pingyao Ancient City, the "Conservation Plan of the World Heritage Pingyao Ancient City" (hereinafter referred to as "The Plan", Shanghai Tongji Urban Planning and Design Institute, 2012) has focused on the following aspects to do the new exploration:

4.1 Re-interpretation of OUV of the Pingyao Ancient City to Establish an Adaptive Function Orientation

In 1997, Pingyao Ancient City was inscribed as World Cultural Heritage due to its OUV. However, the characteristics and value of Pingyao Ancient City were not clearly interpreted at that time. So the "Plan" has reinterpreted the historical, aesthetical and scientific values of the Pingyao Ancient City and its unique position in the country and the world, according to in-depth analysis and study of its history.

The Plan determines the function of Pingyao Ancient City based on its characteristics and values: it is an integrated urban area, with cultural, touristic and residential functions. So its characteristic is "living heritage" rather than "Museum City", "Film Studio City" and "Tourism City" (Zhang, 1997; Bian, 2009).

4.2 Establishment of a Comprehensive Conservation Framework to Protect the Authenticity and Integrity

The Plan establishes a comprehensive conservation framework for Pingyao Ancient City based on its value and attributes, focusing on the following aspects:

(1) Protection of the unique defense system: city walls, gates and square patterns should be all well-protected(Fig. 1).

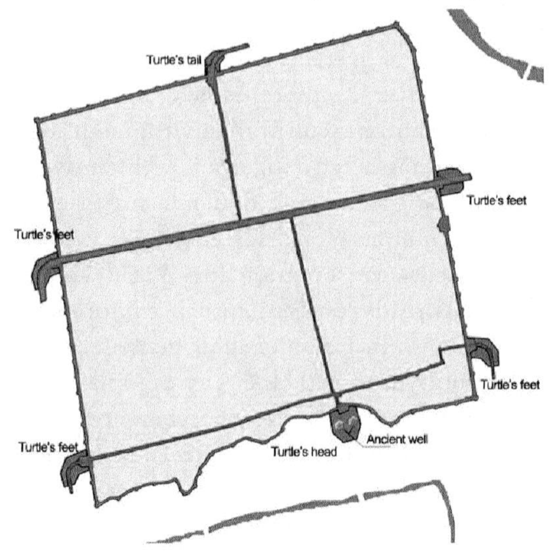

Fig. 1 Illustration of defense system of Pingyao Ancient City © Authors

(2) Protection of the unique "Symmetric and Intact Ancient County Layout". The city central axis is South Street, Chenghuang Temple, Qingxu Daoist Temple and Wen Temple are located on the left, besides, the government, Jifu Temple and Wu Temple are located on the right. Public pavilion is located in the center. All of above forms symmetric layout, which is also the characteristic of Pingyao Ancient City (Fig. 2).

(3) Protection of historic street pattern. Roads and Street layout are important elements of the city, the "four streets, eight lanes, seventy-two alleys", should be preserved.

(4) Protection of layout of architecture and traditional courtyard houses. It specifies the importance of preserve public buildings and traditional courtyard houses; to protect the skyline of Pingyao Ancient City, to maintain the courtyard overall scale, and to the emphasize dominant role of Town Building, Gate Tower, Wen Temple, Chenghuang Temple, Wu Temple, and

Qingxu Temple(Fig. 3).

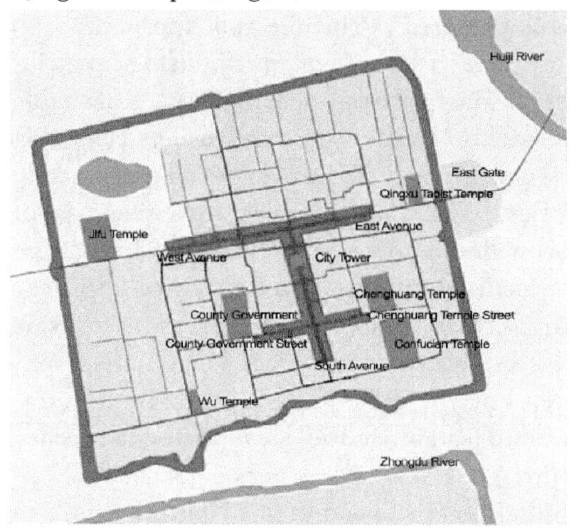

Fig. 2 Illustration of functional distribution of Pingyao Ancient City © Authors

Fig. 3 Characteristics of architectural spaces of Pingyao Ancient City © MA Yuanhao

(5) Protection of the overall color characteristics. The ancient city's residential buildings are blue brick, green tiles, which create a thick, simple style of the ancient city. Meanwhile, temples and other public buildings are more colorful.

(6) Protection of rich intangible culture, including the Jin-merchant culture, religious culture, folk culture and so on (Jinzhong Historic Research Institute, 2002). These intangible cultures are unique and symbiosis with the tangible cultural heritage.

4.3 Establishment of a Construction Guidance System and Heritage Monitoring System, to Manage the Dynamic Changes of the Pingyao Ancient City

First of all, based on the current situation that Pingyao Ancient City lacks basic information, the Plan has established the database of the heritage resources of Pingyao Ancient City. The content of the information includes topography of Pingyao Ancient City, information of the courtyard, building and environment elements, etc. It laid a solid foundation for planning, heritage monitoring and daily management (Fig. 4).

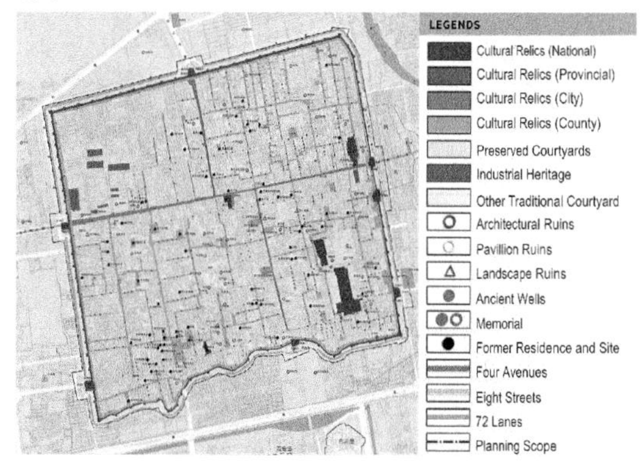

Fig. 4 Distribution of heritage resources of Pingyao Ancient City © Authors

Second, the Plan sets four levels to establish the construction guidance and monitoring system: 1) the ancient city level: focusing on pattern control. It clearly defines the architectural layout, the skylines and the important visual corridors. 2) Neighborhood level: it focuses on the streetscape and open space. 3) Plot level: focusing on the courtyard division and yard form, in order to protect the traditional tissue. 4) Building level: to develop a guideline on the single building and courtyard form, scale, materials, etc.

Third, the conservation of Pingyao Ancient City is a dynamic process. As one part of the Plan, the "Practical Conservation Guidelines for Traditional Courtyard Houses and Environment in the Pingyao

Ancient City" (hereinafter referred to as "Guidelines") aims to those traditional buildings, which is important part of Pingyao Ancient City but not listed as historical buildings, are lack of conservation financial and technical support, long-term maintenance plan, and in bad living conditions. The Guidelines are designed to protect the value of the heritage while improving the living environment and the quality of life, meanwhile, enabling inhabitants to take an active part in heritage conservation (UNESCO, 2014, 2015) (Fig. 5).

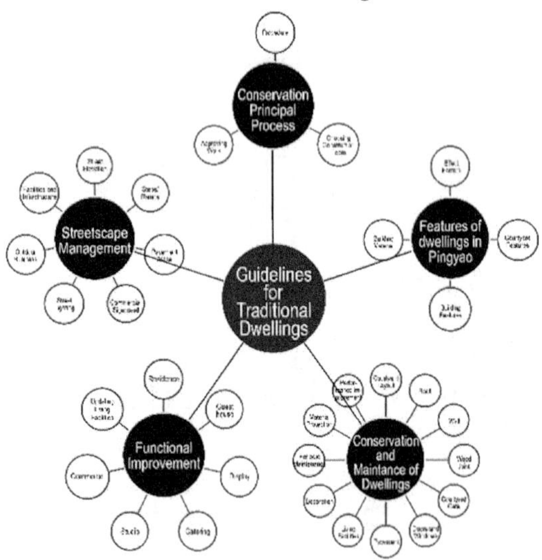

Fig. 5 Distribution of heritage resources of Pingyao Ancient City © Authors

The Guidelines are divided into two volumes. The "Guidelines for Management" is a method for restoring, repairing and maintaining the traditional houses, which is provided to the relevant local administrative departments and professional teams, so as to improve and upgrade the residential areas to meet the modern living standards, and to maintain the authenticity and integrity of the Pingyao Ancient City (Fig. 6, Fig. 7). "Practical Guidelines" are for the owners and users of the traditional houses by text and graphic illustrations, so as to enable them to understand the way of conservation of traditional houses and functional upgrading of the requirements and standards, and the procedures to carry out protection and repair (UNESCO, 2015).

4.4 The Implementation of the "Inhabitants-Centered" Principle and Approach

The Plan develops housing policies based on historical records, the reasonable residential capacity of the ancient city, different property rights, and living characteristics. The Plan encourages active use of brownfields to adjust to make up for cultural facilities and community service facilities, to increase various open spaces to provide inhabitants with the nearest activities venue. For the inhabitants, the Plan encourages to establish the slow traffic patterns, through small-scale, public transport, external access to motor transport services to meet commute needs, to create a comfortable and safe pedestrian environment in the core region. Infrastructure is the core of improving people's livelihood, through the establishment of central heating system and ground source heat pump to replace the pollution of small coal-fired boiler, comprehensive design and kitchen facilities on the courtyard set to guide the realization of livable modernization, etc.

5 Exploration on the Guarantee System of Implementation of the Conservation Plan of Human-Habitat World Heritage Pingyao Ancient City

The management plans for World Heritage sites are developed by the States Party, under the guidance of the World Heritage Management Guidelines (Feilden, Jokilehto, 1993), and is based on the political traditions and legal systems of each country as well as the complexity of its own management system. The plans are of great significance for the management of heritage sites and serve as an integral part of the management of heritage sites (Stovel H, 1998).

However, the Pingyao Ancient City, which was inscribed in the "World Heritage List" in 1997, did not develop a management plan to meet the requirements. At the same time, the conservation and management work has been started since then. Years of work has been combined with the

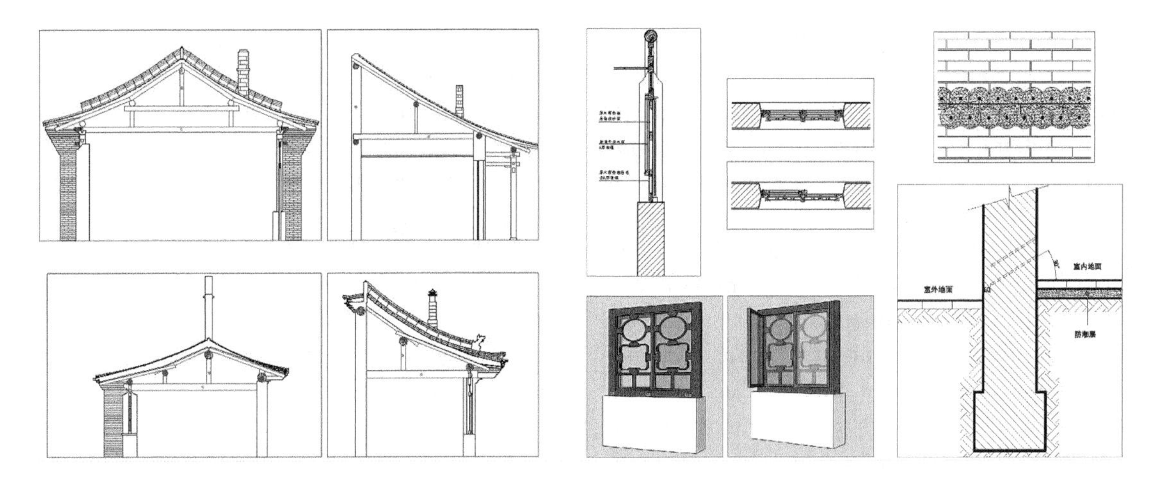

Fig. 6 Illustration of restoration principles and methods of traditional courtyard © Practical Conservation Guidelines for Traditional Courtyard Houses and Environment in the Ancient City of Pingyao

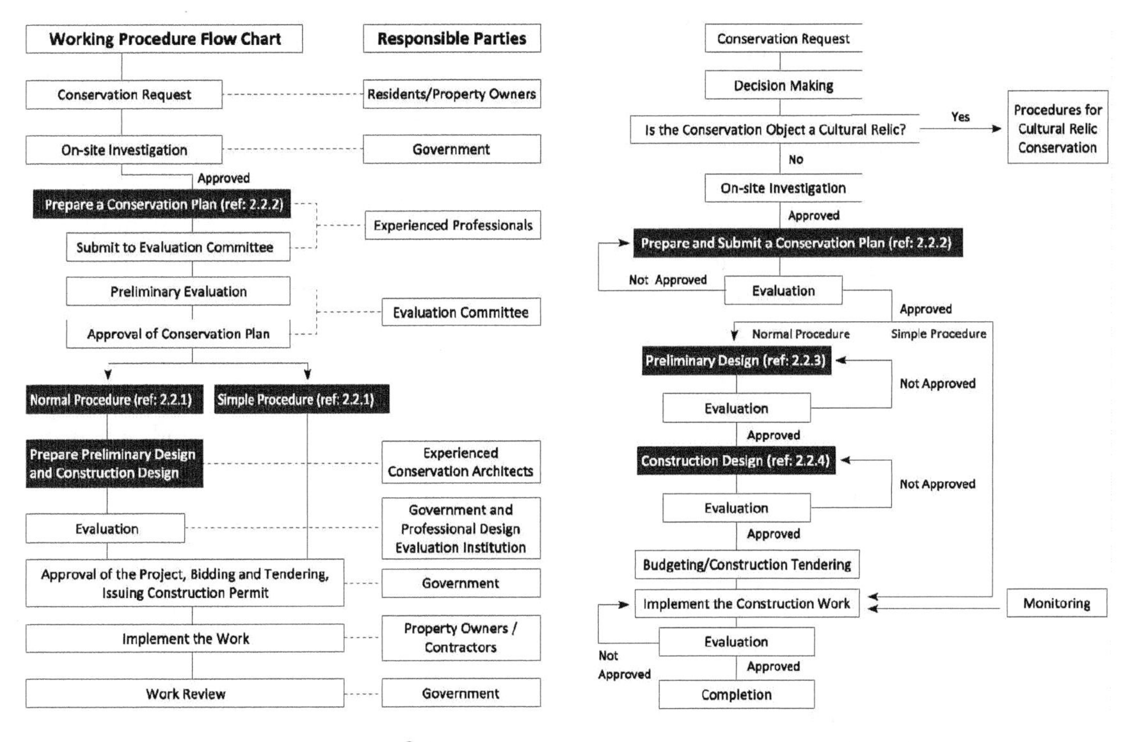

Fig. 7 Illustration of work process © Practical Conservation Guidelines for Traditional Courtyard Houses and Environment in the Ancient City of Pingyao

actual situation to inspire continuous innovation, but it is still far from a rational and scientific manner of exploration. Facing the rapid development of tourism, and the challenges faced by authenticity and integrity of the heritage, the preparation of the World Heritage site management plan is particularly urgent. The "Management Plan of World Heritage City of Pingyao Ancient City" (hereinafter referred to as "Management Plan") sets an important foundation for the establishment of sustained long-term conservation mechanisms.

Faced with a status quo of fragmented management of Pingyao Ancient City, Management Plan proposed for the World Heritage site an innovative management mechanism, and optimized management tools, such as dynamic reporting system, pre-approval system and other measures to ensure the implementation of the Plan.

The Management Plan includes three parts: basic profile, specific management plan and implementation plan. It involves management mechanism establishment, construction management, sustainable tourism management, sources and rational distribution plan, heritage inventory, establishment and management of archives, disaster prevention and risk management, daily management and monitoring, human resources management, social management and many other aspects. The Management Plan clarifies the short-term and long-term execution of the project and determines the information of the leading organization, source of funds, time and so on through the list of projects to facilitate effective management and public scrutiny (Shanghai Tongji City Planning and Design Institute, 2012).

In China, a lot of historical districts were dominated by the government or developers, which involves the relocation of inhabitants, large-scale demolishing and commercial development. However, The Plan, which specifically requests to retain the living and residential functions of Pingyao Ancient City, takes a "Habitant-centered" approach to develop specific measures to protect heritage and improve human settlements with featuring a combination of a "public-private participation" model, "government guidance and inhabitant autonomy" approach. On the basis of a series of pilot projects (Qi, Li, 2014), the Pingyao County Government issued the Implementation Measures of Pingyao Traditional Courtyards Houses Conservation and Rehabilitation Subsidy Fund in 2012. To put it simply, the government establishes a public fund with the income of tourist tickets for conservation and rehabilitation of the traditional courtyards, and the inhabitants submit repair plans to apply for funds in accordance with the requirements of the "Plan" and "Guidelines". The subsidy will be granted after the review of the expert committee. This public policy combines public and private funds, through agreed contract to finance the repair of traditional courtyard houses among inhabitants, and to make sure the daily changes by inhabitants in the line with the requirements of authenticity of the heritage.

6 Conclusion

Chinese Human-Habitat World Heritage was an outstanding environmental space created by mankind in a specific cultural, economic, and political context. It is important for contemporary people and future generations, and this importance will increase over time, and of great significance to the sustainable development of human settlements.

In a context of increasing conflicts between the conservation and utilization of Human-Habitat World Heritage resources in China, and with concern of its future development among the international community and the relevant departments in China, this research aims to define a monitored assessment system, protection criterion and heritage impact assessment techniques through identification and survey of heritage resources in specific contexts, and finally to develop planning techniques and safeguard mechanism for the attributes of heritage resources.

China is still in a primary stage to understand the conservation and management of the Human-Habitat World Heritage, and the next step is to approach a more scientific manner, therefore, it is important to draw on the international conservation theory and experience and combine them with the Chinese context. This could also provide new ideas and new approaches for the conservation and utilization of other traditional human settlements in China.

References

BIAN Baolian, 2009. Conservation and sustainable development of Pingyao ancient city. China ancient city, 4:45-48.

FEILDEN B M, JOKILEHTO J, 1993. Management guidelines for world cultural heritage sites. Rome: ICCROM.

Institute of Jinzhong History Record, 2002. Annals

of Pingyao ancient city. Beijing: Zhonghua Book Company.

QI Ying, LI Guanghan, 2014. Research on conservation and renovation of courtyard of Pingyao ancient city. China ancient city, 3: 70-72.

Shanghai Tongji Urban Planning & Design Institute, 2012. Conservation plan of world heritage: Pingyao ancient city.

Shanghai Tongji Urban Planning & Design Institute, 2012. Management plan of world heritage: Pingyao ancient city.

STOVEL H, 1998. Risk preparedness management manual for world cultural heritage. Rome: ICCROM.

UNESCO, 1972. Convention concerning the protection of the world cultural and natural heritage.

UNESCO, 2015. The operational guidelines for the implementation of the world heritage convention.

UNESCO, Pingyao County, Tongji University, 2014. Conservation management guidelines for traditional courtyard houses and environment in the ancient city of Pingyao.

WANG Huichang, 2010. Chinese cultural geography. Wuhan: Central China Normal University Press.

ZHANG Song, 1999. Primary research on the objectives and methods of conservation of historic cities-take the world cultural heritage: Pingyao ancient city as the case. City planning review, 7: 50-53.

Research on Gentrification Processes Within Human-Habitat World Heritage — the Case Study of the Ancient City of Pingyao

LYU Zhichen, *AOKI Nobuo*, *XU Subin*, *YIN Xi*

Tianjin University, No. 92, Weijin Rd, Nankai Dist, Tianjin, China

Abstract Gentrification has been put forward as a socio-spatial phenomenon during the process of urbanization in Western nations in the 1960's. In recent decades in China, accompanied by the continuous improvement of urbanization levels, the enhancement of social awareness towards heritage protection, and the rapid growth of the tourism industry, this process is also moving deeper into the inner region of China. The Ancient City of Pingyao, inscribed as a World Heritage Site in 1997, was selected for this research. This paper analyzes the dynamic mechanisms put in place during the gentrification process of Pingyao, and argues that the imbalanced development of tourism, the relocation of industries in the protected area, and consumer preferences are its main driving forces. The analysis and comparison showcase a correlation between the gentrification process and the revitalization of Pingyao.

Keywords gentrification, heritage conservation, tourism-led redevelopment, urban renewal

1 Introduction

Gentrification has been defined by Glass (1964) as the gradual replacement of the low-income class for middle and upper classes within a given urban area. This process leads to a phenomenon of socio-spatial differentiation during periods of urban renewal, and the extraction of surplus value from the pre-existing land properties (Smith, 1979). In addition to this change in the social classes of residents, there are also changes in environmental conditions which exclude the original community from remaining in the area. Therefore, the process of gentrification affects not just the residents and the urban community but also the corresponding material conditions and environment.

The problem of gentrification was first identified in developed Western countries in the late 1960s. Likewise in China, as its urbanization process continues to accelerate, the phenomenon of gentrification has begun to develop in the inner parts of the country. Since the ancient city of Pingyao became a world cultural heritage in 1997, subsequent policies for the protection and rejuvenation of the ancient city led to the continued prosperity of the tourism industry, but have also pushed the population's structure and capital into a stage of high-speed mobility. In recent years, Chinese scholars have studied this phenomenon by focusing on economically-developed urban centers. Moderate gentrification can help to avoid the problem of the urban center "hollowing out" (Zhu & Zhou & Jin, 2004), but this ignores the needs of previously-established beneficiaries (Wong & Liu, 2017). At the same time, the massive entry of capital into productive industries, rather than public services and facilities for education, medicine and local markets (Wang & Aoki, 2018) has also exacerbated the depopulation and inconveniences of heritage sites for long-term settlers.

Pingyao Ancient City is simultaneously undergoing multiple stages of heritage protection, economic development and community demand. The focus of this study is to contribute towards the vitality of human settlements in order to cope with the "hollowing-out" process and the various social inequalities induced by new urban interventions. It uses qualitative and quantitative methods (by doing respectively in-depth interviews and simple questionnaires) to analyze the complexity of the local socio-economical context and related policy issues. It seeks to explore the following three questions: (1) Is there a potential threat of "hollowing-out" (i. e., depopulating or decharacterizing) the city's core while the ancient city continues to prosper economically? (2) Has the ancient city already entered the process of gentrification, and if so, what are its underlying causes? (3) How are residents responding to the contradictions produced by the needs of community life and heritage protection?

2 Research Methods & Scope of Study

This study deals with Pingyao's ancient city in Shanxi Province. The on-site investigation took place from March 2018 to June 2018. More than 50 informal interviews were conducted during the research phase, including a number of practitioners engaged in heritage conservation in the local area, as well as more than 15 residents and tourists. Finally, there were targeted surveys of different groups (The total amount of questionnaires is exactly 100, and finally 98 valid ones were selected for statistical analysis). The final demographic information is shown in Tab. 1. In the course of this survey, the preparatory work for the Pingyao Ancient City International Workshop was taking place in the same period and the organizers of the workshop provided some assistance for this study, by allowing the participants to meet with the administrators of Pingyao, and obtaining permission to do questionnaires. However, the needs and concerns of residents during the research process have not been fully studied yet, especially from a quantitative perspective.

Tab. 1 Summary of main data collected from the individual surveys

Variables	Local residents	Tourists	Shop owners	Frequency	Percentage
Age					
<26	3	12	4	19	20%
26-35	5	17	7	29	30%
36-45	4	3	8	15	15%
46-55	10	2	7	19	20%
>55	14	0	2	16	16%
Gender					
Male	24	14	14	52	53%
Female	12	20	14	46	47%
Profession			Type		
Farmers	8	0	Restaurant		1
Teachers	1	3	Snack bar		3
Students	1	9	Hotel		2
Public officials	2	11	Store		15
Retired	7	0	Cultural Industry		3
Self-employed	14	2	Other		4
Others	3	9			

Pingyao Ancient City is a world cultural heritage site and was officially designated as a "Historical and Cultural City" of China in 1986 (2nd Batch of Historical and Cultural Famous Cities). The ancient city of Pingyao is roughly shaped like a square, with each side being approximately 1.5 km long, and it covers an area of 225 ha (Fig. 1). All the houses in the ancient city (except for a few modern houses) are Ming—Qing residential courtyards. According to government data in 2018, there are about 20,000 registered residents in the ancient city. From the perspective of tourism, as a 5A-Level scenic spot in China, Pingyao receives more than 400,000 tourists each year.

Fig. 1 General extent of Pingyao Ancient City © Google Maps

3 The Historical Development of the Ancient city of Pingyao

3.1 Degradation and "Folding" in the Ancient City of Pingyao

As early as the Ming and Qing Dynasties, the ancient city of Pingyao was mostly a high-rank neighborhood community. The middle-class merchants from Jin as well as other powerful clans gathered in the city. Even in the early 21st century, the ancient city area was still the political, economic and cultural center of the entire county. The degradation of the ancient city area began with the private ownership reform in the 1950s. A large number of private houses were taken back and reassigned to each family by the government during the "Land Reform Law" period. At the same time, some public courtyards such as "Pingyao County" and "Wu Temple" were used by government agencies or for reconstruction and construction of the public facilities such as schools. On the one hand, this process reduced the living standards of each family unit (the living area of each family was reduced to about 15 square meters after redistribution), so that a large number of areas in the ancient city were transformed into grassroots communities. On the other hand, due to the entry and expansion of the public sector and the public service industry, new points of attraction were created for living within the ancient city, which then led to the continuous migration of capital and people towards the inner city over a long period of time.

In the first stage of degradation, the geographic and economic value of the ancient city was gradually enlarged, with sustained development and economic growth from a macro perspective (the number of permanent residents in the ancient city continued to rise and reached 40,000 in 2010 for a total area of about 540,000 square meters), but at the same time we begin to see signs of a social and spatial differentiation at the micro-level. Ideally, each person should have access to at least 30 square meters of space, but that would require a population of only 16,000 individuals (unpublished report, Shanghai Tongji Urban Planning & Design Institute, 2015).

Until 2010, this slow and sustained process was broken by new policies developed by a group of government officials and academics. The latest government actions have shown a more negative attitude towards the over-concentration of people in the area. All public institutions, schools, and hospitals have been moved quickly and completely away from their original sites towards the outer part of the city's ancient walls. At the same time, in order to serve the tourism industry, newly-implemented plans to create a "clean and hygienic" image also directly caused the stagnation of the pre-existing mobile business activities. For example, itinerant residents living in villages outside the ancient city could no longer come inside to sell agricultural products or conduct their business within the city walls.

The rapid and effective implementation of policies and investment of capital forced the locals to move out. This process has, to a certain extent, raised the amount of space available for the further development of tourism in the ancient city, but it also broke down the formerly balanced structure of the local population. Low-income groups unable to move out and a large number of elderly people have been left in the city, thus becoming the main representative strata of the local residents (Tab. 1). Due to the limitations of budget allocation, the rental cost of the houses remained low for a long time, and the occupants were unable or unwilling to contribute to the repair of the houses, which then led to a further decline in their quality. As a result, a large number of residential communities remained in the city, but they have a much lower purchasing power and less access to quality spaces, living in narrow and cramped streets. This marks a severe contrast with the situation in the major streets(Fig. 2). We may say, in a certain way, the unfortunate conditions of the lower classes have been "folded" (or hidden away) behind the main streets which were renovated and are featured in tourism promotion materials for the masses.

3.2 Heritage Value and New Economic Growth

Since the day that it became a world cultural heritage site, Pingyao Ancient City has received continuous attention from the outside world. As an outstanding representative of the Han ethnic group in the Ming and Qing Dynasties, it has a high historical, cultural and social value. The humanistic spirit it represents also gives the city a unique character. In addition, the ancient city also embodies the unique trajectory of social and cultural changes in history, involving major changes in political power, and the redistributions of wealth evoked by

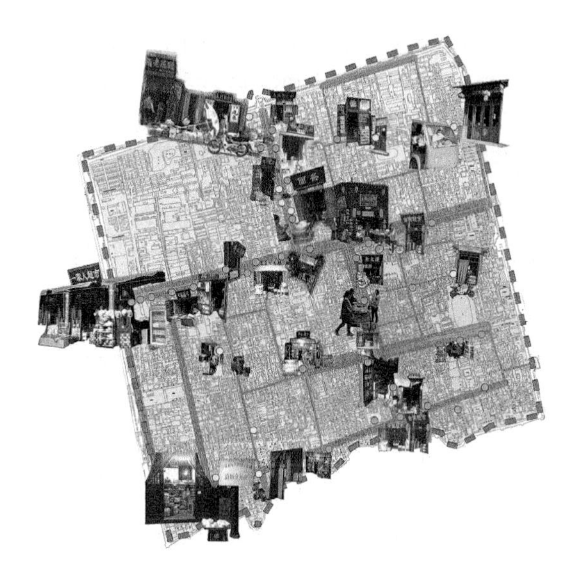

Fig. 2 Map with Pingyao divided into four districts, including photos of shop fronts found in the high-class streets and the surrounding districts

the socialist transformations. The city is also an early witness of the shift from traditional manufacturing towards more upper-level planning related to cultural industries, and changes in daily life in contemporary society. If we stick only to the general reports and materials developed by officials, we can only have a macro-level view of the city's situation. But by collecting the various oral histories narrated by different residents, Pingyao Ancient City can be understood in much more detail, as a microcosm of individuals with unique and rich experiences. Under the constraints of economic development and heritage protection, the shift towards tourism has become a new opportunity for growth, but it also exacerbated the imbalances of regional development and the interference of externalities on heritage management. Due to the lack of consideration for social values by developers and planners, some "non-touristic service facilities" such as primary schools and hospitals have been relocated outside the city because they originally occupied historical buildings with high potential for profit, but the original services provided by these facilities were not reestablished. The residents did not receive any new services that could replace the ones they had lost. Many residents who were forced to take long trips to send their children to school and access these lost services eventually had little choice but to move out of their poor-quality homes in the old city and rent their old houses at very low prices to low-income residents.

3.3 Dissolution and Re-aggregation of Vulnerable Groups

In 2011, the residential areas on the south and west sides of the ancient city were gradually demolished through the policy of "one-time economic compensation", and replaced with a large landscape park and a theater (1000 seats in a single session). The reduction of population density around the ancient city has improved its general urban image, but the compensation is not enough to pay for the purchase of an apartment in the county. Once those residents who were able to move out provided their old houses for rent, it was mostly the residents with the lowest economic power (especially the elderly) who ended up gathering in the ancient city, moving into the poor-quality rented residences. This process of urban renewal shows a clear trend towards social differentiation, and we see that highly vulnerable groups are being unwillingly reorganized around the city, by means of: (1) "fracturing" the local community into economic sub-groups; (2) "distancing" the groups from each other due to the relocation of services; (3) "re-aggregating" the most disadvantaged groups into low-quality areas(Fig. 3).

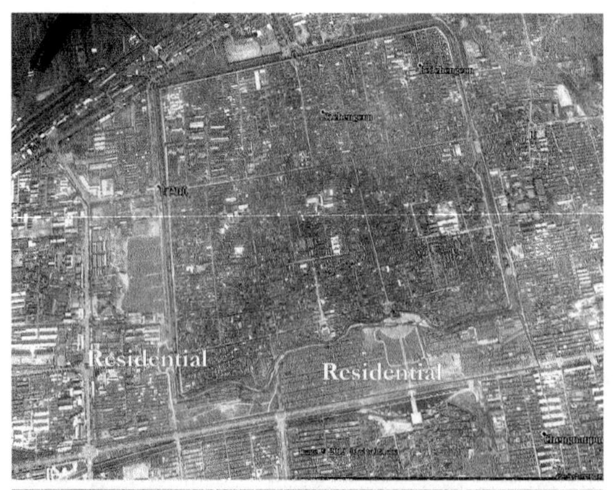

Fig. 3 Changes in the urban area outside of the ancient wall (above image: 2008; below image: 2019)© Google Maps; lines and colors added by the authors

3.4 Political and Economic Units Under the Concentration of Heritage Policy

As a special and independent area, "Pingyao Ancient City" is directly affected by the special heritage protection policy. This mode of administrative and political "separation" can formulate legal norms very well for the limited central area, but from the perspective of the whole city, it has caused the targeted transfer of public resources, rather than being distributed fairly in the larger scale of "Pingyao County". At the same time, public policies are also being concentrated within the scale of the ancient city. Policy is a means for the expression of power, and the ancient city has become the largest social and political unit in Pingyao County. In the competition for economic market share and direct investment, the ancient city wins with absolute superiority compared to Pingyao County, and has become the central point of power and economic interest. The closest-matching behavioral subjects must necessarily be the middle class and above.

In addition, with the adjustment of economic structure and the development of decentralization, the commercial format of the ancient city began to exist in a more diversified form, and the transformation of traditional buildings to meet architectural stylistic requirements has become the most influential aspect advocated by the heritage protection policy. This also indirectly led to the behaviors of house owners to spontaneously update the community environment. The decentralization of the economy has therefore led to the concentration of capital. At the same time, the improvement of infrastructure has provided the impetus for the secondary circulation of capital and led the new middle class to move closer to the ancient city.

4 Demonstration of the Current Gentrification Process of Pingyao Ancient City

4.1 Macro-scale Phenomenon

The gentrification of the ancient city of Pingyao began in the early 21st century, and touristic development became a full-fledged priority within the economic transformation process of Shanxi Province. In June 1994, the "Meeting for the Economic and Touristic Development of Pingyao Historic and Cultural City" was held, which established the economic growth model of the ancient city with tourism development at its core. The rapid development of tourism has led to further renovation and new construction initiatives in the ancient city, and the gentrification caused by these new activities has accelerated the structural reorganization of the city in a vicious cycle.

In the past ten years, the number of hotels in the ancient city has tripled, while the average cost of overnight accommodation has more than doubled, and is much more costly than the average operation of hotels outside the ancient city. In addition, during the off-season, more than 50 boutique hotels are priced at more than 1 000 *yuan* per night (based on Ctrip. com network data), of which design hotels and boutique hotels are the most representative of the higher-end facilities. At present, these boutique hotels have become a magnet for expensive high-quality services enjoyed by upper social classes, and therefore a gentrified community has formed rapidly around these buildings. The most famous examples are Yunjincheng Mansion and Jinzhai, among other inns. The booming tourism market has brought about the continued prosperity of the hotel industry and is also driving the further development and expansion of the inner city. In addition, the environmental improvement project in the ancient city area continues to expand. In addition to the newly-completed Helanqiao Bridge area renovation, more than two large-scale environmental remediation projects are being carried out in the old city (Fig. 4). These renovation activities have improved the environmental quality of the area. However, it also caused a noticeable decline in the living conditions of the locals. In the interview, one resident of the Helanqiao area said: "I prefer to go outside the city wall or carry out daily activities near the house, rather than staying in the well-built (Helanqiao Bridge) Park."

The transformation of social strata and land uses in the ancient city also promoted the transformation of the local lifestyle and consumption patterns. A shop located in Xidajie Street (in the west part of the city) that is now selling tourism products used to sell fresh vegetables as its main business, serving the surrounding residents. As a result, the residents who need to do daily purchases must cross the ancient city all the way to the morning market outside of the East Gate. Similar changes can be seen in many other places. There is an abundance of homogenous commercial streets and degraded small alleys; the age structure and consumption structure of the population have become seriously unbalanced; the dynamic area of the ancient city is gradually shrinking and starting to shift towards the outer part of the city.

4.2 Rhythms and Dynamic Aspects of Gentrification

The Rent-Gap Theory by Smith (1979) argues that the rent gap is the root cause of the gentrification phenomenon. The high land rent in the central area of the ancient city has pushed out low- and middle-income residents, and the "filtering" effect of high rents has enabled high-income people from other regions to quickly gather next to the new high-end hotels and core commercial

Fig. 4 Helanqiao Bridge Park (no one inside)© XU Qiang; lines and colors added by the authors

streets. This has become the core driving force for the development of the ancient city of Pingyao.

The formation of the land rent gap in Pingyao Ancient City is mainly due to the uneven development of tourism, industrial migration and replacement, and individual consumer preferences. First of all, tourism is a recent force for economic growth in Pingyao. Since 1994, a set of major commercial streets forming the shape of the Chinese character"土" (*Tu*) has become the core area for tourism, while the southwest and northeast areas of the ancient city have been neglected. A large number of residential communities are still unable to enjoy the dividends of economic development, while at the same time being burdened by the lack of public services and facilities. In addition, the migration and replacement of industrial facilities has also promoted the continuous advancement of the gentrification process. The popular Chinese buzzword of "returning two into three" (i. e., shifting the city's secondary industrial complex into tertiary cultural services) has led to changes in the employment structure in the ancient city, and low-income people who lack vocational training and skills are gradually being marginalized because it is difficult for them to find employment in the central area. Finally, with the continuous expansion of middle-class residential areas, the general consumption level continues to rise: the type of consumer has substantially changed, and consumer preferences have also moved toward a higher-end level. Needless to say, middle-class and low-income groups are different in terms of consumer demands and price ranges of goods and services. Along with the increase in the average social status of the residents, the area becomes the symbol of a new identity, which then entices a new round of investment concentrated in the center of the ancient city, in hopes of quick financial returns (Fig. 5).

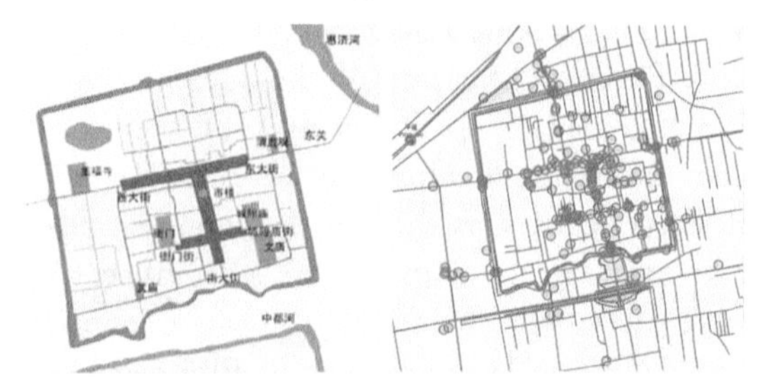

Fig. 5 Sectors and main points of gentrification in Pingyao Ancient City (left: ancient commercial streets © *Pingyao Ancient City Protective Detailed Planning*; right: points of gentrification with high-class hotels and facilities © Open Street Map)

5 Gentrification and the Ancient City Revival Movement

Similar to urban development in Western countries, the gentrification of Pingyao Ancient City is also a trending phenomenon, predictably resulting from urban renewal activities. This is brought about by the ancient city renaissance movement and it is affecting the structural reorganization of the ancient city. This kind of "reorganization" has become a paradox to some extent: on one hand, it violates and breaks the balance of social equity, but it must also avoid the "hollowing-out" of the city. To some extent, these have become two sides of opposition. A healthy and energetic "core" can support the development of the city, as long as the middle class remains the most creative group in the city and is not driven out. Under the premise of trying to meet social fairness, by moderately controlling the process of gentrification and preventing the mass exodus of the middle class, the location can maintain the vitality of the ancient city, and preserve a diverse range of services and businesses that can serve the needs of low- and high-income groups.

The current stage of heritage protection is often used as an excuse for the renewal of the ancient city, but the final result is the replacement of the local people by middle and upper classes coming from the outside, who rarely manage to establish a meaningful relationship with pre-existing communities. This kind of replacement is actually damaging the social value of physical heritage. To prevent this, we must focus on improving the value of heritage within community life and provide affordable and accessible basic services and employment opportunities for long-term local residents, thus ensuring the diversity of urban communities and the preservation of the historical urban context.

Acknowledgements

This work was financially supported by a Major Project of the National Social Science Foundation of China (grant number 12&ZD230). We wish to thank Bebio AMARO for proof-checking the English text.

References

GLASS R, 1964. London: aspects of change. London: MacGibbon and Kee.

Shanghai Tongji Urban Planning & Design Institute,2015. Pingyao Ancient City Protective Detailed Planning.

SMITH N, 1979. Towards a theory of gentrification: a back to the city movement by the capital not the people. Journal of the American planning association, 45(4): 538-548.

UNESCO, 2015. Conservation management guidelines for traditional courtyard houses and environment in the ancient city of Pingyao. https://unesdoc.unesco.org/ark:/48223/pf0000234622.page=18-19.

WANG X, AOKI N, 2018. Paradox between neoliberal urban redevelopment, heritage conservation, and community needs: Case study of a historic neighborhood in Tianjin, China. Cities, 85: 156-169.

WONG T C, LIU R, 2017. Developmental urbanism, city image branding and the "right to the city" in transitional China. Urban policy and research, 35(2):210-223.

ZHU Xigang, ZHOU Qiang, JIN Jian, 2004. The gentrification and urban renew: in case of Nanjing. Urban studies, 11(4):33-37.

Based on the Everyday Life Thinking of Pingyao Ancient City Zero Space Resistance

XU Qiang, HAO Zhiwei, SHANG Ruihua, ZHANG Haiying

College of Architecture, Taiyuan University of Technology, No 79 West Street, Yingze District, Taiyuan, Shanxi, China

Abstract The ancient city of Pingyao is an outstanding example of the Han Chinese city in the Ming and Qing Dynasties. In the development of Chinese history, it shows an extraordinary picture of the cultural, social, economic and religious development of the Han. Due to the current cognitive limitations, the protection of the aboriginal lifestyle and the social network is neglected. With the relocation of a large number of residents, the characteristics and value of Pingyao are gradually lost.

This paper takes the active protection and development of the world cultural heritage "Pingyao Ancient City" as the research object through the master teaching activities of the architecture and urban and rural planning major of Taiyuan University of Technology, and returns to the perspective of everyday life in the ancient city. The relationship between settlement space and social behavior is observed and discerned. Through the design research process, design thinking and design behavior are placed in the everyday consideration of ancient city life. Here, "everyday" has become an innovative thinking method opened by research-based design, looking for real life needs in everyday repetition, and thus inspiring innovation vitality. In order to expand new horizons and research directions, explore new modes of ancient city growth.

Keywords Pingyao Ancient City, everyday life, zero space, problem search

When the city buildings on the streets of the Ming and Qing Dynasties are for the background of the tourists' photos, the symbols are created. More people are separated from the real feelings of Pingyao. The era has faded the original meaning of the city building, and the everyday indifference is the symbol. The important reason for this has led to the extremely negligible zero space problem—the everyday decline. This paper attempts to find a research method and path to resist zero space, and provide theoretical support for the growth of Pingyao Ancient City.

1 The Zero Space Problem of the Ancient City: the Everyday Life Decline

1.1 Zero Space Phenomenon

What is zero space? The problem of zero space is a social problem in the development of today's society. Scholar Wang Yuan introduced the concept of zero space of Roland Barthes. "When the meaning of a symbol needs to be explained by another symbol, it constitutes a circular chain of symbolic interpretation. Because it is impossible to settle in the real thing, the meaning will disappear in the infinite expectation of the next symbol, showing a state of zero space."

The photos of the city building are spread through the media, and their authenticity disappears with the time and length of the spread, which leads to the phenomenon of zero space. This phenomenon triggers a series of chain reactions, unconsciously guiding the eyes to the symbols. In the image, at the same time, the real existence is greatly ignored.

1.2 Everyday Life Decline

The everyday life decline is reflected in two levels: authenticity and diversity. The decline does not mean disappear. It can still find traces of remnants in the ordinary alleys. Habitual neglect will inevitably lead to the disappearance of authenticity and diversity. Focus on the everyday life of the ancient city, pay attention to the everyday life of the ancient city.

The decline of authenticity is attributed to the symbolic flooding caused by the rapid development of information networks. The common problems in the city are also appearing in the ancient city of Pingyao. The business that has lost everyday life support has become exclusive of tourists. The aborigines have not only lost their everyday lives. Authenticity, tourists pay the price for this, there is no everyday life experience for the ancient city.

The decline of diversity is behind the limitations and unity of development thinking, which has formed an extensive and single economic model. A single tourism model is one of the possibilities to destroy the healthy development of Pingyao. Architecture is undoubtedly the materialization of these problems. The problems arising in the field of architecture research are also caused by this. The thinking of the development mode of Pingyao Ancient City is the focus of this paper. With the help of teaching practice, through the study and application of relevant theories, it is expected to sort out the research methods and strategies.

1.3 Everyday Life Attention—Ancient City Reading

As a world cultural heritage, the success of the application of the ancient city of Pingyao has enabled the community to focus on this century-old city. The protection of the ancient city has always been the focus of attention, from the walls of the ancient city to the traditional streets and historic buildings. The material space presented by the Shanxi merchant culture has undoubtedly become the soul of the ancient city, which has also led to a tourism-led development model. The reasons for this development model are various, related to the development management system of Pingyao Ancient City. This involves the consideration of sociology and economics. It can be seen that the thinking from the perspective of architecture cannot fundamentally deal with the problems faced in the development of Pingyao Ancient City, and it is necessary to understand architecture from a higher dimension, and be proactive in thinking about the social issues behind the current architecture.

In July 2018, the School of Architecture of Taiyuan University of Technology launched a summer workshop with the theme of "Ancient City Reading", which is based on the practical problems of Pingyao and combined with undergraduate teaching activities. In the undergraduate graduation design in 2018, the typical site—the southwest area of Pingyao was selected. Using the method of problem exploration, combined with tools such as social maps, the problem search was carried out, and finally the relationship between the aborigines and the tourists was focused. At the same time, the summer work was combined. The workshop carried out social activities aimed at "reading in the ancient city", conducted in-depth research in the community, visited the residents, and explored the internal mechanism of the evolution of the ancient city settlement; in 2019, the undergraduate graduation design continued the same research method and focused on the ancient city(Tab. 1). In everyday life, go deep into the community, understand the actual needs of typical households, and take the everyday life as a starting point to think about graduation design. At the same time, the teacher team carried out theoretical research, and systematically summarized and combed related theories such as urban design, settlement, historical block protection and architectural planning.

Tab. 1 Based on everyday life research team and research content

Time	Content	Area of Research	Team
2018	Graduate School of Architecture	Southwest of Pingyao Ancient City	Architecture, urban planning

Continued

Time	Content	Area of Research	Team
2018	"Ancient City Reading" Pingyao Summer Workshop-Evolution of Pingyao Urban and Rural Settlement Space	Northeast of Pingyao Ancient City	Architecture, urban planning
2019	Undergraduate Graduation Design of the School of Architecture - Return to the everyday life Ancient City	Northwest of Pingyao Ancient City	Architecture, urban planning

2 Resistance to Zero Space Thinking—Everyday Life Perspective

2.1 Search for Everyday Problems—the Intervention of Planning Thinking

The everyday life problem is a trivial and complicated problem. everyday life is more reflected at the micro level. Just like "Orchids on the roof. Dialogues on architecture and culture", Xie Yingjun mentioned Orchids on the roof. It is so small, not only beautiful, but also has the meaning of religion and place. In addition to the meaning of culture and religion and the meaning of the environment, in a lot of subtle elements, we need a way to intervene, find problems about everyday, and think systematically, and the problem search method in architectural planning theory. Just meet this requirement.

The problem-searching method comes from William Pena's architectural planning theory, which has important guiding significance for the design of the pre-design, and is an indispensable tool for building project decisions and project planning. In the face of many problems in the project, we can systematically organize and find out the core issues, so as to propose solutions.

Planning thinking is not only reflected in the problem search. everyday life is not a single or dual problem, but a multi-faceted embodiment. Therefore, the planning thinking of public participation and group decision-making can play a guiding role. It is the theoretical basis for community building and joint creation. In 2019, the School of Architecture In the graduation design, the Pingyao design team paid attention to the community to create this problem, actively explored community issues, invited the head of Pingyao community and the peace and space grid to conduct symposiums and listen to the opinions of all parties. This is infiltrated with planning ideas(Fig. 1).

Fig. 1 Search based on everyday problems-teachers and students of the School of Architecture and Pingyao Community Symposium (Source: 2019 Pingyao Biji Group)

2.2 The Search for Diversity—Jane Jacobs' Revelation

Jacobs' resistance to top-down planning is the core idea in her book The Death and Life of Big Cities in the United States. Mixed, small streets, old buildings, and density are what Jacobs believes. Important conditions, the loss of diversity in Pingyao

Ancient City is also reflected in these four aspects. The four aspects have a hierarchical meaning, from the macro-planning level to the architectural level, and finally the use of space. The clear and orderly level provides ideas and methods for the following research.

The decline of diversity usually begins with the marginal zone that is easy to ignore, the boundary of the block and the gap of the building-the street and the lane. For these areas, the research is carried out, firstly to explore its significance in history, and secondly to pay attention to the current reality. Then look for the possibility of triggering contemporary diversity.

The loss of mixture is reflected in the spatial distribution and time distribution, which is embodied in the single and uneven distribution of functional space, the lack of a single business format and the lack of everyday life business forms, forming a monotonous tourism business atmosphere in Pingyao, on the everyday life business of the aborigines(Tab. 2). Ignoring the same seriousness, and failing to find a suitable fit between the tourists and the aborigines, caused the problem of zero space in the distribution of time and became a single travel timetable. The volatility of tourism circumvents the whole ancient city. The everyday life use has also been exhausted, and the aboriginal needs for everyday life use have gradually shifted outside the city.

Tab. 2 Typical regional finishing table of Pingyao Ancient City based on Jacobs theory

	Content	
Mixed		
	The layout of Pingyao West Street and North Street is full of single tourism retail business, and the hybrid type is extremely low	

Continued

	Content	
Small street		
	The everyday life interactions appearing in the interface of North Street, the passing behavior in the lower floors	
Old buildings		
	The ordinary dwellings near South Street and the historic buildings of North Street also represent the cultural characteristics of the times	
Density		
	The density of use of South Street, which is dominated by tourism, and the density of aborigines in lower-level streets	

The small street section contains the meaning of the scale in which the pleasant scale and the comfortable space provide the conditions for everyday life. In the ancient city of Pingyao, in addition to the "soil" type main street of the ancient city, there are also many small-scale streets and lanes, which are connected to the houses of the ancient city. The everyday life of these small streets is the key concern.

In the theory of Jacobs, the old building is an understanding of the mix of architectures of various eras. The old buildings of Pingyao Ancient City include historical buildings and cultural relics, as well as modern residential buildings. These buildings have the imprint of the times and the accumulation of history. Therefore, they should take care of various buildings with characteristics of the times.

Density refers to the density of people in space. This density reflects the use of space to some extent. By judging the density, we can find the space with vitality, and discover the everyday life contained in it. It's relevance.

2.3 Social Map — Thinking of Sociology

The study of sociology focuses on the social network of people in space and behavior, is the intervention of sociology, on the one hand, is the study of the social relations of the aborigines of the ancient city, on the other hand, the social relations between the aborigines and tourists, through social maps, etc. The tool discusses the association.

In the graduation design process of the 18th and 19th undergraduate students of the School of Architecture of Taiyuan University of Technology, the social map research method was adopted. Through the field investigation and observation, the public life in the public space was recorded, and then the space and the study of the relevance of behavior, the everyday life of the aborigines are presented through social maps.

In the 2018 Pingyao Summer Workshop, in-depth research on the living space of the aborigines in the form of a group, through the household survey, to understand the real problems of the aborigines, also presented in the form of social maps, with the help of social maps, clearly presented out of the scope of the study, to further identify the problem(Fig. 2).

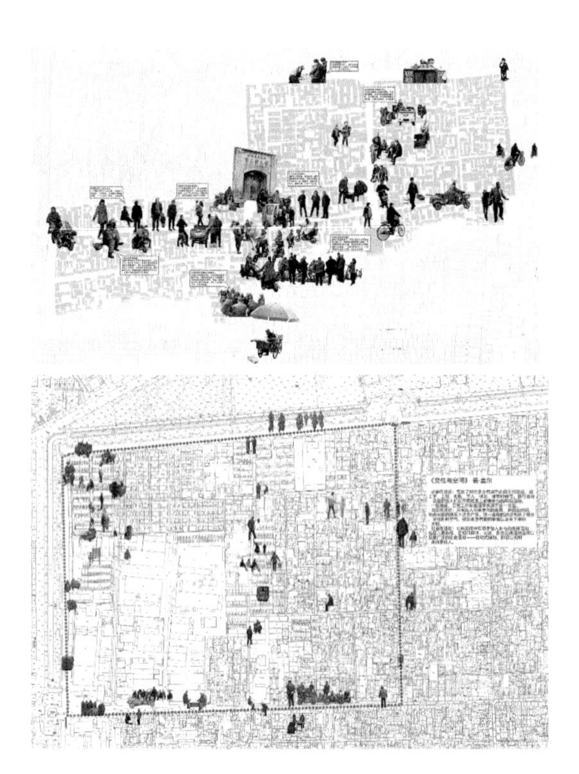

Fig. 2 Graduation design of the 18th undergraduate completion group of the School of Architecture-Design Achievements

3 The Future of the Ancient City—Some Suggestions

3.1 Focus on Bottom-Up Power

Everyday life is undoubtedly a bottom-up manifestation. It is a resistance to zero space. The systematic design of the top layer often cannot touch the everyday life routine, or is typed and symbolized. In addition, the marketization induces the everyday life decline. Nowadays, the resistance to zero space is often not so satisfactory. How to take everyday life consideration into the planning and design, how to combine everyday life with planning in the early stage of design, and what kind of mechanism to pay attention to everyday life, these are in graduation design. In the course of the operation, attention was given and the conditions were actively created to organize the daily stakeholders and explore the everyday life together.

There are also bottom-up forces in the ancient city of Pingyao, including aborigines, tourists, new residents, etc. , and how these groups resist zero space, and in which aspects, further research is needed, but it is certain that the power cannot be ignored, and it is also the potential and internal driving force for the renovation of Pingyao Ancient City.

3.2 Form an Everyday Life Basis for Evaluation Criteria

The excessive pursuit of form has formed an evaluation standard based on the law of some form of beauty. The control and coordination of style is not an excuse for simplification and symbolization. Secondly, even the attention to behavioral activities is limited to the aborigines. In addition to the zero space standard specification, design evaluation should also increase the everyday life standard to measure the design and use effect.

The everyday life standard should be an intrinsic and introspective, a higher level of restraint beyond the normative guidelines, respectful and rational use of the existing material space environment, and a spontaneous operation by the user and the owner. The emphasis on everyday life should be based on a common value.

3.3 Form a Common Community Building Model

Everyday life is not a phenomenon of only a single subject, but the result of multiple subjects. Participation facilitates the absorption of more comprehensive recommendations and appeals, and facilitates a positive and effective response to everyday life needs. Multi-agents should include owners, managers, users, and supervisors. Through the establishment of the platform, the various entities are organized together, and they are institutionalized by reasonable management and usage patterns to provide conditions for everyday life revival.

4 Conclusion

Thousands of cities are the embodiment of zero space degree. Returning to everyday life is the most favorable resistance to zero space. Excessive abstraction and the proliferation of symbols make everyday life precious. Through architecture and sociology methods, we can find the everyday life of Pingyao Ancient City and try to go. Awaken those who have not disappeared and establish a new order of life.

Through the consideration of this article from the everyday life perspective, the key points of the future development of Pingyao Ancient City and the value judgment criteria of the development of the ancient city are gradually clarified. Secondly, relevant research ideas and methods are provided. For everyday life reflection, it is not that this bottom-up mode of operation can replace everything, but as a supplement and focus, adding a dimension and perspective of thinking, adding an idea to the renewal of Pingyao Ancient City. At the same time, it can also be used as an idea for current stock updates. I hope this article can provide some clues for related practices and research.

References

GEHL J, 2004. Public spaces, public life. Copenhagen: The Danish Architectural Press.

PENA W, PARSHALL S, 2001. Problem seeking. New York:John Wiley&Sons Ins.

The Value and Significance of Conservation of the Ordinary Vernacular Heritage in Underdeveloped Areas of China — a Case Study on Shuimotou Village, Pingyao County

CHEN Yue

College of Architecture and Urban Planning, Tongji University, Shanghai, China

Abstract China's underdeveloped areas cover most provinces in central and western China. The heritage conservation of these regions, not only faces the similar problem as the eastern regions, but also is more challenging because of the limitation in economic and social condition, especially for the ordinary vernacular heritage, which is generally thoughts of insufficient value in conservation and development. Taking Shuimotou village in Pingyao county as an example, this article analyzes the historical and cultural value of an ordinary vernacular settlement. The author holds that the value of vernacular heritage should be considered from the perspective of the effect of local cultural diversity instead of applying the evaluation standard of elite heritage. Particularly for underdeveloped areas, vernacular heritage plays a more significant role in revitalizing local culture. From this viewpoint, the ordinary vernacular heritage not only has its unique value, but is also worth to be protected.
Keywords underdeveloped areas in China, ordinary vernacular heritage, value, heritage conservation

1 Overview of Ordinary Vernacular Heritage in Underdeveloped Areas of China

1.1 China's Underdeveloped Areas

At present, there is neither a specific definition nor a uniform standard for the underdeveloped areas. In China, underdeveloped areas are often mixed with concepts such as "poverty areas" or "poor and remote areas". For example, in 1994, the State Council issued the "*National Poverty Alleviation Plan*", which included counties with a per capita income of less than 400 yuan, and identified 591 counties as National Poverty Counties. In 2002, the *Index Research of Counties' Social and Economic Development* by National Bureau of Statistics calculated the comprehensive development index of each county in China. And the counties in the bottom 20% of the index are considered as underdeveloped areas. However, no matter which standard is adopted, it is clear that China's underdeveloped regions are widespread and mainly concentrated in the central and western provinces.

The economic characteristics of underdeveloped areas are small aggregates, poor efficiency, and heavy reliance on state financial transfer payments. Taking Pingyao County of Shanxi Province as an example, the per capita GDP in 2016 was 22,000 yuan/person, lower than the national average (53,700 yuan/person), and even far lower than the level of developed coastal areas in the east (such as Ningbo City, which was 147,500 yuan/people①). In 2017, the local fiscal revenue was 490.08 million yuan, while the superior subsidy income reached 182.21 million yuan, 3.7 times of the former one②.

In social aspect, it is characterized by a low level of urbanization and a more

① Source: http://xxgk.pingyao.gov.cn/xxgk/zfgkml/tjj/jjzb/517036.shtml, http://vod.ningbo.gov.cn:88/nbtjj/tjnj/2017nbnj/indexch.htm

② Source: http://xxgk.pingyao.gov.cn/xxgk/gkml/czyjs/502230.shtml

seriously massive labor loss. In 2015, the urbanization rate of Pingyao County was 42.89%[①], far below the national average (56.1%). In 2016, China's total floating population reached 245 million people, of which the trans-provincial floating population was about 80 million. The provinces with a net inflow of trans-provincial population are mainly in the developed regions of Beijing, Shanghai, Tianjin, Jiangsu, Zhejiang, Fujian and Guangdong, while most of the central and western provinces including Shanxi belong to the net outflow areas of trans-provincial population[②].

1.2 Ordinary Vernacular Heritage

Since the 2000s, the state has announced 487 China Historical and Cultural Villages and 4,153 China Traditional Villages; several provinces and cities have also announced provincial and municipal historical and cultural villages or traditional villages. These historical and cultural villages or traditional villages have relatively prominent heritage values and can be considered as elite heritages. Compared with these elite heritages, there are a large number of ordinary vernacular heritages that are not in the lists but still have certain heritage resources.

The first characteristic of ordinary vernacular heritage is relatively common in value. According to the *Measures for the Selection of China Historical and Cultural Towns (Villages)* by the Ministry of Construction in 2003, the selection of historical and cultural villages is mainly based on the three criteria of value and features, original condition and scale. Therefore, from the perspective of rarity and integrity, the ordinary vernacular heritage is obviously inferior to the elite heritage in the same area.

Another characteristic of ordinary vernacular heritage is the serious lack of funds for conservation. In China, the amount of conservation funds is directly determined by the conservation level. For example, the state-level heritages are funded by the national ministries and commissions, and the funds of which are relatively abundant. The provincial and municipal heritages are funded by provincial and municipal government, and the funds are depending on the local financial capacity. While for ordinary vernacular heritages, which normally have no responsible subjects, there is no government support correspondingly.

1.3 Two Views

So, for the ordinary vernacular heritage of underdeveloped areas in China, there is a view that the most important thing is to solve the problem of survival, and the task of heritage conservation is subordinate. Another view is that the value of such heritage is not outstanding enough, and it is neither necessary nor possible to conserve in the case of limited capital investment.

Are these two opinions correct? Is it necessary and possible to protect the ordinary vernacular heritages in underdeveloped areas?

2 Vernacular Heritage as a Concentrated Reflection of Local Culture

Although by UNESCO's definition in the *Convention Concerning the Protection of the World Cultural and Natural Heritage* in 1972, "cultural heritage" is an immovable cultural property with an outstanding universal value in history, art or science, including monuments, groups of buildings and sites. However, with the deepening of understanding, the international community has gradually realized that cultural heritage not only includes the above-mentioned contents with outstanding universal value, but also local heritage. The typical event was *the Charter of the Vernacular Architecture Heritage* was adopted by ICOMOS at the 12th Congress in 1999. The charter states that the vernacular architecture heritage is a reflection of social culture, of the relationship between society and its region, and of world cultural diversity. Therefore, unlike the previous heritage evaluation criteria , the vernacular herit-

① Source: http://www.pingyao.gov.cn/zwgk/gzbg/284666.shtml
② Source: 2017 report on the development of China's floating population

age conservation emphasizes its role in local cultural diversity and emphasizes the community's understanding, support and participation in the work of conservation.

Take Shuimotou Village in Pingyao County as an example. The village is located 17 kilometers southeast of the World Heritage Site Pingyao Ancient City(Fig. 1). Although the history of Shuimotou village is less prominent than Liang Village, a China Historical and Cultural Village which is 10 kilometers away. However, based on its own conditions, the ancestors of Shuimotou created its unique spaces and culture for future generations.

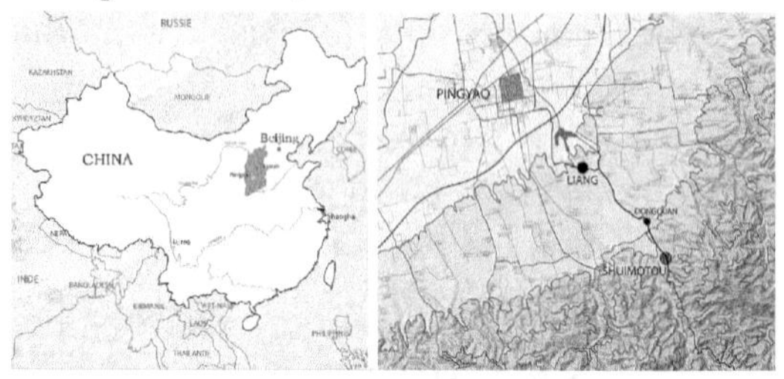

Fig. 1 Location of Shuimotou © Tongji University

2.1 Naturalistic Living Environment

Shuimotou village is located in a valley, with the Huiji River passing through from south to north. Actually, both the emergence and development of the village were tightly related to water. Historically, the relationship between the Huiji River and the village has undergone two changes.

Originally, according to folk hearsays, there was a lake that had a large amount of water and high water level in the place where the Huiji River is now. Gradually, although the amount of water is less than before, the water level in flood season was still very high, which made it impossible to live near the water. Archaeological discoveries found that there were three Longshan cultural sites on the top of the village, which was around 3,000～5,000 years before. Therefore, it can be speculated that in order to use water conveniently and at the same time protected from flooding, the ancestors chose to live in areas with higher terrain.

As the water level drops, the area with a slightly lower terrain gradually became suitable for human habitation. The first change happened in the Ming Dynasty, in order to adapt to the population surge, the territory of the village was extended to the current ancient village area. Besides, the ancestors also intelligently created the terrace system, to deal with the problem of limited farming space.

The second change was in modern times, when the water flow of the Huiji River further reduced. To hold more people, the tidal flats of the river valley also became cultivated arable land. However, the field was higher than the surface of the water. To deal with this problem, the villagers skillfully used the terrain elevation to draw two channels from the upper reaches of the Huiji River. The west channel was directly used to irrigate the farmland of Shuimotou Village, while the east channel to irrigate the downstream farmlands.

Therefore, Shuimotou village conforms to the changes of the natural environment, especially the water conservancy, chooses the location suitable for human habitation, and creates the terrace system and irrigation system, fully embodying the simple concept of the naturalistic living environment of the ancestors(Fig. 2).

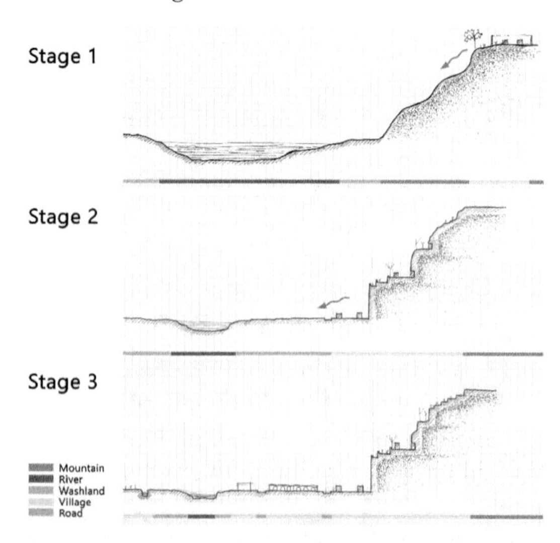

Fig. 2 Development and evolution of the village

2.2 Cohesive and United Space System

In the last years of the Ming Dynasty,

a large-scale peasant uprising broke out across the country. In order to prevent hooliganism which brought great anxiety to the people, many towns and villages including Shuimotou Village built their own defense systems.

Compared with some plain settlements, Shuimotou Village has a natural advantage in defense. Because of the backing of the mountain and the ridged partition in the middle, the ancestors divided the village into three relatively independent "Bu" which are fully based on this topographical feature. Three sides of the "Bu" were made up of natural mountains as natural barriers, the other side was added wall and fort, forming a closed state. At the same time, the use of the terrain made it easy to defend and difficult to attack(Fig. 3).

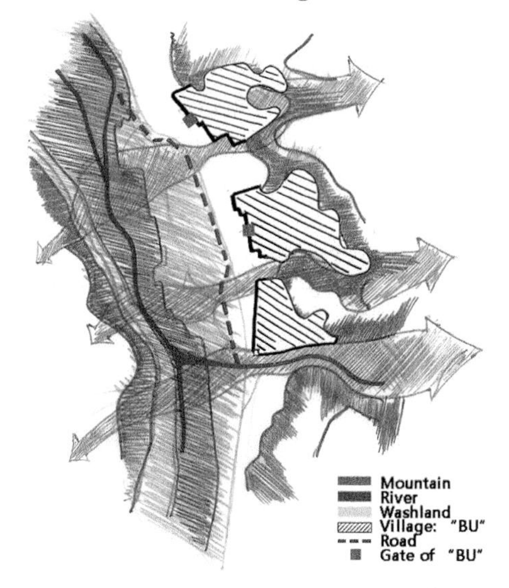

Fig. 3 Spatial structure of the village © Tongji University

In order to further strengthen the defense and the psychological sense of territory, several neighboring families with close relationships also formed a "Fang". So, adding with every residential courtyard, it came out a three-level space system of "Bu-Fang-Yard"(Fig. 4).

Even if it is limited by the topographical conditions, the villagers of Shuimotou made full use of space by changing the organizational relationship of courtyards, so as to concentrate all the residences in the

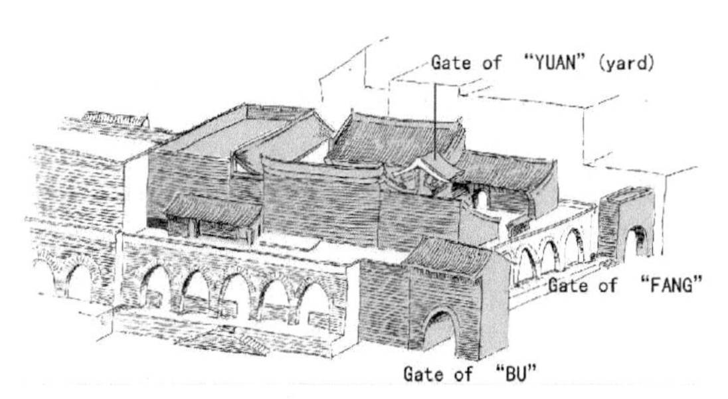

Fig. 4 Typical three-level space system © Tongji University

range of "Bu". For example, Dong Jie Bu is small in north-south depth, so the roads and courtyards are arranged in parallel, and the courtyards are shallow in depth; Yao Wan Bu is more abundant in space on one side, so the fort roads are perpendicular to the main street, and all the courtyards could have a good orientation of sunlight; Shimagou Bu have small distance from east to west, so it should only sacrifice the orientation to be arranged along the fort roads. It is precisely because of this natural terrain treatment that the Shuimotou village is very rich in spatial changes.

Therefore, in order to resist enemies, Shuimotou Village makes full use of natural topography, and has cleverly constructed a space system that is conducive to safeguarding its personal safety and psychological security. At the same time, it has also created a rich spatial change, which fully reflects the spirit of unity of the ancestors.

2.3 Rich and Practical Cultural Place

Although Shuimotou Village is an extremely ordinary small mountain village, various types of public spaces are available, indicating that the spiritual and cultural life of Shuimotou people is very rich and colorful. From the perspective of spatial distribution, these public buildings are mainly concentrated in the central street, indicating that although the three "Bu" are separated from each other, the public space is shared, and the public life is synchronized.

Take the most important cultural space — buildings group of Caishen House, Chaoshan Temple, and Drama Stage as example, it is located in the center of the village. The Caishen House and Chaoshan Temple are in the same building (1st floor is Chaoshan Temple, 2nd floor is Caishen House), standing from north to south, and the Drama Stage is on the opposite side, forming a complete public courtyard.

Despite the small scale and the simple decoration of the building, it still reflects the wisdom of the ancestors, which is evident from the streamlined structure of the building. Because this building group has both the functions of sacrifice and drama performance, there are mainly three streamlines: the sacrifice streamline, the audiences' streamline, and the actors' streamline. It's delicate that the three groups of streamlines are arranged reasonably, and do not interfere with each other (Fig. 5).

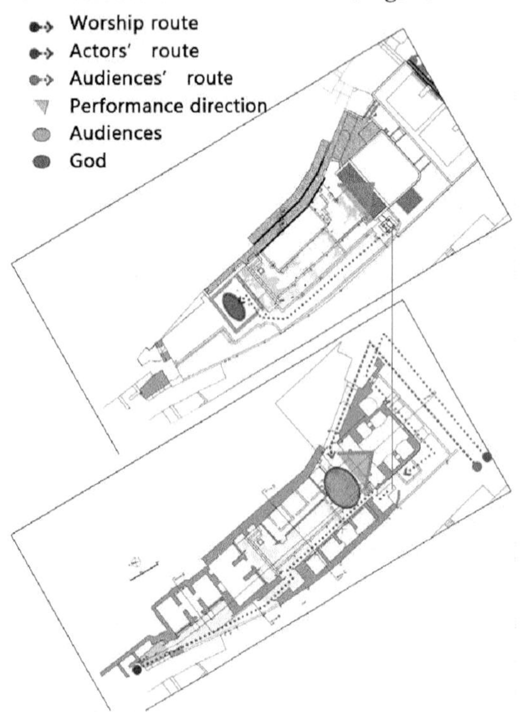

Fig. 5 Stream analysis of the building

The stage is of great significance to Shuimotou Village. In history, there was a tradition of playing drama, especially on Lunar May 13th and New Year's Day every year. In the 1950s, the village established a troupe which was famous in the surrounding area. They played not only in the village, but also in neighboring villages, even were invited to areas outside of Pingyao. Unfortunately, this tradition continued until recent several years.

Therefore, on the basis of ensuring the basic needs of survival, Shuimotou villagers strive to satisfy the satisfaction of spiritual life. The rich cultural spaces and the colorful cultural activities, shine the light of civilization and fully embody the ancestors' yearning for a better life.

3 Conservation of Vernacular Heritage as an Important Approach to Revitalize Local Culture in Underdeveloped Areas

As mentioned earlier, the general characteristics of underdeveloped regions are relatively backward in economic development and relatively prominent of social problems, which is more apparent in rural areas. In the rural areas of Pingyao County, the main industry is agriculture, the production efficiency of which is extremely low. In 2017, the per capita disposable income of rural residents is only 40.5% of the figure of urban residents[①]. As a result, almost all young and strong laborers have chosen to go out for work, and the resident population in the village is only half or even less of the registered population, even most of them are elderly people and children. More seriously, the economic and social recession has also led to the deterioration of physical space and living environment: more and more agricultural lands are abandoned, the risk of soil erosion is increasing, a large number of buildings are damaged and collapsed due to lack of routine maintenance, water resources are destroyed by uncleaned garbage... A vicious circle has gradually formed.

Due to the low income of agriculture and the limited development of industry, many villages have put their hopes on the development of the tourism economy. However, it is not easy to make a profit by rural

① Source: http://xxgk.pingyao.gov.cn/xxgk/zfgkml/tjj/jjzb/517036.shtml, http://vod.ningbo.gov.cn:88/nbtjj/tjnj/2017nbnj/indexch.htm

tourism which may need to collect resources, funds, talents, markets, management and other factors together. Therefore, even if Liang Village, a China Historical and Cultural Village, which has relatively outstanding resource endowments and government support, was still failed. Indeed, for Shuimotou, under the condition that its own characteristics and funds are insufficient, the possibility of developing tourism is even lower.

Therefore, for ordinary vernacular heritage in underdeveloped areas, on the one hand, it faces a serious recession crisis, and it is urgent to find a new way to avoid a complete collapse of the social system and even the psychology; on the other hand, it does not have the conditions for tourism development.

In this case, culture becomes an important driving force. Sharon Zukin (2006) argues that since the 1970s, culture has become more of a business strategy for local governments and business alliances tools. The one who can build the largest modern art museum, his financial sector shows the most vitality; the one who can transform the waterfront from a run-down dock to a park and yacht supply area, he will show the most expanded management and professional team potential. Culture provides basic information for almost all service industries, including symbols, patterns and meanings. Franz Fischer (2013) pointed out that modern rural development is not just a simple rural beautification policy, but a holistic, economic activity-based approach. In which, cultural identity, historical relevance and unique cultural independence are so significant.

Therefore, to rebuild local culture based on its inherent local heritage is the most economical, feasible, and even the most effective solution.

3.1 Conservation of Organic Cultural Landscape, to Improve Environmental Quality

Shuimotou Village conforms to the natural environment and creates an organic cultural landscape. It embodies the simple living environment concept of the ancestors and has a strong practical significance in the contemporary era.

Terraced fields not only effectively increase the cultivated area, but also have a strong water-holding and soil-holding function. Firstly, for the arid regions of Northwest China, the collection and utilization of rainwater are very important. Moreover, constructing terraces on the slope can be used to reduce slope which is the main natural factors of soil erosion.

Due to the state policy of returning farmland to forests for ecology in 2002, the villagers restored the original terraces into mountain forests on the premise of obtaining certain economic subsidies. However, the purpose of conserve ecology has not obtained, otherwise, many loess surfaces are still exposed, and the terraced system that was originally conducive to water and soil conservation was devastated.

Although there is no possibility to re-establish terrace cultivation, it is still probable to consider using this principle to improve the living environment. For example, it could be likely to use of natural terrain to design a small plant purification system in the village to play a role in water purification and microenvironment improvement.

3.2 Conservation of Orderly Space System, to Awaken Community Awareness

Based on natural topography, Shuimotou Village has cleverly constructed a multi-level space system, which not only effectively guarantees safety, but also successfully creates a lively and interesting space. In contrast to this is the new village built in the 1980s. The new village which is located on the north side of the ancient village did not follow the ancient village construction concept, but adopted a uniform layout, which makes a single-level space: insufficient in public space and tedious in visual experience.

Yang Gale (2002) shows that the multi-level spatial structure actually reflects the social structure of the residential groups and distinguishes the publicity and privacy of the space, so that a stronger sense of security and belonging to the area can be formed outside the private residence, so as to motivate people to participate in public life. Through the conservation of the multi-level spatial structure, the spatial domain sense and spatial characteristics are emphasized to stabilize the gradually weaker social structure and further awaken people's community awareness.

This is of even more important significance for villages with large population outflows in underdeveloped areas, because the fundamental driving force for rural revitalization depends on the awakening of the villagers themselves. Developed countries such as Europe, America, and Japan experienced the process of urbanization earlier than China. At that time, they also suffered from the serious decline of rural areas, and some explorations of community reconstruction appeared. For example, residents of Omihachiman City in Japan, spontaneously set up the conservation group to raise objections to the moat landfill plan to be implemented immediately by the government, widely convened the signatures of the citizens, organized residents to go to the moat for cleaning activities, and finally made the government withdraw the original plan and protect the moat. Taking this as a starting point, various community-building activities were carried out, which enabled the private strength of the region to grow and revitalize the region.

3.3 Conservation of Precious Cultural Memory, to Restore Cultural Confidence

In history, Shuimotou Villagers pursue the satisfaction of spiritual life and create a rich and colorful cultural life. And the cultural spaces serve as a container to carry people's precious cultural memories. The unprecedented rapid urbanization process in the past few decades is not only a mass-level population movement, but also a comprehensive sublation in psychological level-that is, the migrant workers abandon of the traditional rural lifestyle and local culture, and pursuit of urban lifestyle and culture.

Morley and Robbins (2001) believe that this is essentially a contradiction between cultural globalization and local cultural identity. But even if globalization is the dominant force in today's society, the identity of local culture which is inseparable from local revitalization is equally important. By transforming the historical space and giving it new functions, it is obviously helpful to continue the cultural memory, restore people's cultural self-confidence, and finally achieve the goal of inheriting local culture.

Taking the buildings group of Caishen House, Chaoshan Temple, and Drama Stage in Shuimotou Village as example, as the disappearance of original religious and cultural functions, the courtyard came to a negative state of use for some decades, with some of buildings changing into residential functions, and some completely vacant. But for Shuimotou Village, where the traditional social structure is in danger of disintegration and the public life in rapid decline, restoring the most important public space in history is of great significance(Fig. 6). On one hand, it can provide a place for the renaissance of traditional cultural activities, like playing drama, which can arouse peoples' memory and confirm the identity of themselves; on the other hand, it can also provide the possibility of modern cultural activities, like changing into communicating center or playground, which can attract more young people. In short, public space in history is a connection of the past, present and future.

4 Conclusion

Ordinary vernacular heritage is an important part of the local cultural system. Although it lacks rarity and integrity according to the evaluation criteria of elite heritage, it does not have the so-called "outstanding universal value", but it can be seen from the analysis of the Shuimotou case that it's of obvious value to local cultural diversity. It embodies the simple concept of the naturalistic living environment,

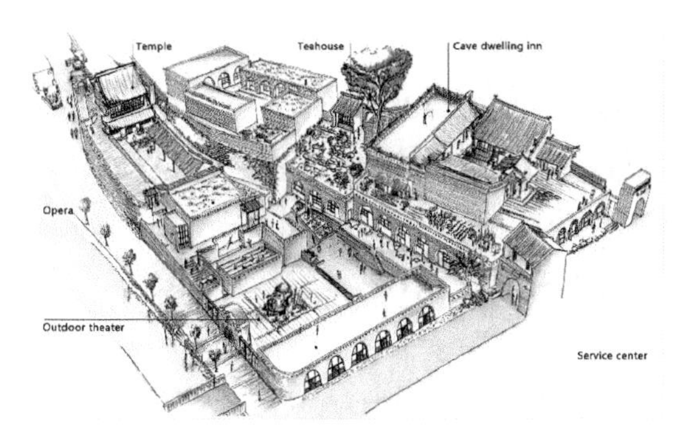

Fig. 6 Conservation and utilization plan for the building group © Tongji University

the spirit of unity and the yearning for a better life.

For less developed regions with economic and social decline, although the possibility of relying on heritage to develop tourism is little, the role that ordinary vernacular heritage can provide to revitalize local culture is even more precious. Based on its inherent local characteristics, to improve environmental quality, awaken community awareness, restore cultural self-confidence, rebuild local culture from the community level, and thus to promote local economic and social revitalization, maybe the most economical, feasible, and even a most effective solution at present.

Therefore, vernacular heritage conservation is not a subordinate thing, but quite urgent; it is also not out of reach, but maybe realized. What is most needed is not necessarily money, but consciousness and determination.

Acknowledgements

Shanghai Tongji Urban Planning and Design Institute Co., Ltd. Scientific Research Project "Implementation Evaluation of Conservation Plans of Historic and Cultural Villages" (Project No.: KY-2016-YB-05)

References

GEHL J, 2002. Communication and space. Beijing: China Architecture & Building Press.

ROBBINS M, 2001. The space of identity. Nanjing: Nanjing University Press.

YUKIO N, 2007. Recreating the charm of hometown: the resurrection story of traditional Japanese neighborhoods. Beijing: Tsinghua University Press.

YI X, SCHNEIDER C, 2013. Germany's integrated rural renewal planning and local cultural identity construction. Modern urban research, 28 (06): 51-59.

ZUKIN S, 2006. Urban culture. Shanghai: Shanghai Education Press.

Research on Sustainable Protection Strategy of Chinese Human-Habitat Historical Environment Based on HUL Method: Case Study of Pingyao Ancient City, a World Cultural Heritage Site

XI Yin[1], *WANG Yao*[1], *YANG Li*[2]

1 Tianjin University, Tianjin, China
2 Tongji University, Shanghai, China

Abstract Human-habitat historical environment refers to traditional human settlements composed of cultural heritage, spatial environment and community life, which has dual attributes of historical protection and community life. With the advancement of urbanization, the precious historical accumulation contained in the historical environment of many cities has been compressed, resulting in multiple problems such as flattening of historical context, loss of urban characteristics and identical development paths. Through in-depth interpretation of UNESCO's proposals on environmental protection of historic cities, based on HUL method, this paper combs the historical context of Pingyao Ancient city, and clarifies the current situation and dilemma, meanwhile compares the protection experience of HUL pilot cities in the Asia-Pacific region, then puts forward the sustainable protection strategy of the human-habitat historical environment of Pingyao Ancient City. This paper focuses on the discussion of the universality strategy of human-habitat historical environmental protection and development. The main strategies include: protecting concept from emphasizing the importance of architectural relics to recognizing more widely the importance of social, cultural and economic processes in maintaining urban values; protecting objects from focusing on heritage protection to forming a vision reflecting the diversity of urban, which can fully reflect the history, tradition, values, and aspirations of communities; protecting forces from government-led to multi-stakeholder participation, which achieves value consensus through participatory planning and consultation; protecting management from static management to sustainable dynamic management, which combining local management system and developing coordination mechanisms for various activities; protecting planning means should help to maintain the integrity and authenticity of urban historical environment and improve the quality of life and urban space. By incorporating the protection of cultural heritage into the process of urban development, the tradition is redeveloped into a stock of long-term flexible practices. This paper provides a reference for the sustainable protection of the Chinese human-habitat historical environment.

Keywords HUL method, human-habitat historical environment, sustainable protection, Pingyao Ancient City

1 Introduction

The continuous development of agricultural civilization in China has made part of the historic towns and villages inhabited by a large population continue to this day, and it is still an important space for residents to produce and live. Historic towns and villages are typical human settlements that were selected, created and renewed after the dialogue between ancestors and the natural environment. Moreover, as the space entity of invisible culture, these historical towns and villages are also the historical environment with countless civilized codes. Therefore, this paper defines historical villages and towns, which have evolved from historical settlements to the present, as human-habitat historical environment, including working, living, recreation and social communication space from ancient times to the present. Broadly speaking, it includes historical towns and their surrounding natural environment. From the perspective of " human-habitat historical environment", the independent human-habitat historical environment highlights the unique value of these historical settlements because of their long-time development axis. From the perspective of the protection of historical settlements, the idea of human settlements has been integrated into the research and protection of human settlements, which integrates the space constructed by human beings and the natural environment, and emphasizes the integrity of historical settlements and contemporary humanistic care.

2 Historic Urban Landscape (HUL): From Idea to Method

The formation of UNESCO's concept and method of "Historic Urban Landscape" has gone through three stages: the first stage started in May 2005, when the International Conference on World Heritage and Contemporary Architecture was held in Vienna, where the Vienna Memorandum was signed, and the concept of "Historic Urban Landscape" was first introduced, the concept has begun to attract wide attention; The second stage started in 2009, when HUL began to have methodological implications. Until the publication of the Recommendation on the Historic Urban Landscape in 2011, the HUL was finally interpreted as "a city area formed by the accumulation of cultural and natural values and attributes with historical layers, it goes beyond the previous concepts of historic central district or architectural aggregation, including a broader urban context and its geographical environment." Meanwhile, the opening of the Recommendation also says that "HUL method is of great significance as an innovative way to preserve heritage and manage historic cities". At this time, the concept of HUL has the connotation of method. In the third stage, from 2013 to now, UNESCO issued "Renewal of Historic Cities, Detailed Description of Historic Urban Landscape Protection" in 2013, emphasizing the connotation of "hierarchy" and "stratification" of the city, which is an extension and promotion of the Recommendation.

In the field of urban planning, in recent years, many scholars have discussed and summarized the HUL method based on their own work experience. In 2017, Li Huimin took Xi'an as an example to systematically explore the guiding path of HUL in landscape design and heritage protection; in 2017, Zhang Song put forward the paradigm transformation from the protection of historical features to the management of urban landscape, recognizing the relevance, integrity and diversity of urban landscape, and paying attention to the comprehensive protection and utilization of landscape resources; in 2018, Zhang Wenzhuo explored the protection of industrial heritage from the perspective of HUL, and constructed the evaluation process and evaluation and strategy generation system. HUL including a broader urban context, has much in common with human-habitat historical environment in this paper. It extends the research object from urban or rural monomer to the geographical environment of settlements, and extends the research time from the current short period to the long period of renewal of settlements, so as to explore more reasonable ways of the symbiosis of people, settlements and environment. The HUL method is of great significance in guiding urban planning and heritage protection.

3 European Experience in Historical Environmental Protection

Before the protection action in China, the West had carried out a lot of exploration and had a far-reaching impact on China's protection. The protection and utilization of architectural heritage in Europe can be traced back to ancient Roman times. The European Cultural Relics Restoration Movement, which originated in the late 18th century and the early 19th century, had a wide influence, represented by the French "Style Restoration" Movement and the British "Anti-Restoration" Movement. After entering the 20th century, the scope of protection began to expand from historical buildings to the surrounding environment. In the 1960s, European countries began to protect historic areas by legislating to delimit protected areas. The *Malo Act* is one of the most important laws; the concept of overall urban protection has gradually matured since the 1970s. In the *Amsterdam Declaration* adopted in 1975, it was pointed out that "architectural heritage includes not only superb single buildings and surrounding environment, but also all historically and culturally significant areas". The *European Landscape Convention* promulgated by the European Union in 2000 emphasizes the diversity and cultural connotation of "landscape". *The Guiding Principles of Spatial Sustainable Development on the Continent of Europe* formulated in the same year emphasized the importance of landscape diversity for spatial sustainable development. In the planning and development of European cities, historic urban landscape plays an important role(Fig. 1). The protection of Chinese cultural heritage started later than that of Europe. The development experience of the protection content is similar to that of Europe. Although the situation is more complicated than that of Europe, it is of great significance to learn from the protection experience of its historical environment for the sustainable protection of the Chinese human-habitat historical environment.

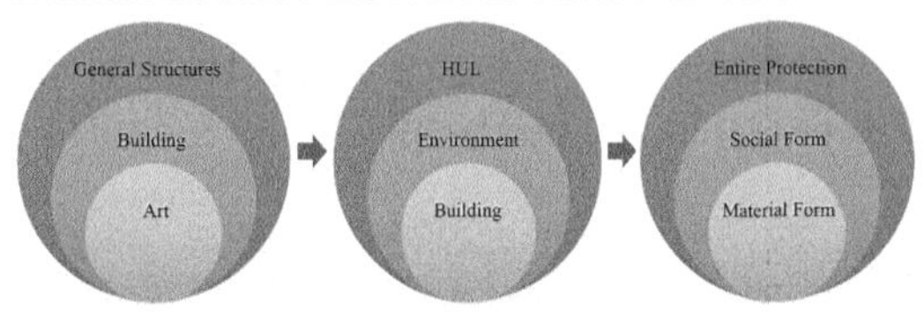

Fig. 1 European city history and culture protection development process

France is the most comprehensive and rigorous country for the protection of urban history and culture in Europe. It has a rich and perfect legislative system, such as *the Law on the Protection of Historical Buildings*, *the Law on the Protection of Sites and Natural Landscapes*, *the Law on the Protection of Regional Areas of Architecture, Cities and Landscape Historic Heritage*, etc. The scope of protection covers the control of building height in a small area to the protection of the environment in a large area. The ZPPAUP protection system, which is supported by local and public, is very effective in making the sense of urban cultural heritage protection deeply rooted in the hearts of the people, and flexibly protecting a large area of historical and cultural areas in France. The legal system of protecting urban history and culture in Britain is also perfect and sound. Unlike France, it emphasizes the protection of the overall characteristics of the protected areas. Consequently, the protection achievements embody more diverse urban characteristics. As for the problems of old buildings, outdated facilities and lack of funds, the practice of some European cities is that the government subsidizes 50% of the maintenance costs of the owners of houses to promote the sustainable protection and utilization of the historic buildings by the owners. France's main policy in this regard is the OPAH Plan, which aims to improve the local environmental conditions policy support, so as to better retain the aborigines and promote their sustainable development.

4 The Chinese Human-Habitat Historical Environment from the Perspective of the Ancient City of Pingyao

Pingyao is known as "ancient pottery" in the history of Pingyao. Its location as shown in Fig. 2. According to legend, it was the emperor Yao's feudal territory, the Jin State in the Spring and Autumn Period, and the Zhao State in the Warring States Period. Setting up Pingtao County in the Qin Dynasty. In the Western Han Dynasty, Jingling and Zhongdu counties were built, and in the Northern Wei Dynasty, Pingtao was replaced by Pingyao. Pingyao Ancient City is the most well-preserved ancient city in China's human-settled world heritage. It has well preserved the historical form of county-level cities of Han nationality in central China during the Ming and Qing Dynasties (14—20 century). Its historical form as shown in Fig. 3. Its outstanding universal value is embodied in Pingyao Ancient City is a typical case of Han city in Ming and Qing Dynasty (14—20 century). It retains the characteristics of all Han cities and provides a complete picture of the development of Chinese history, culture, social economy and religion. It is of great value to study social formation, economic structure, military defense, religious beliefs, traditional ideas, traditional ethics and residential forms. These values meet the criteria of Article 2/3/4 of World Cultural Heritage respectively[1].

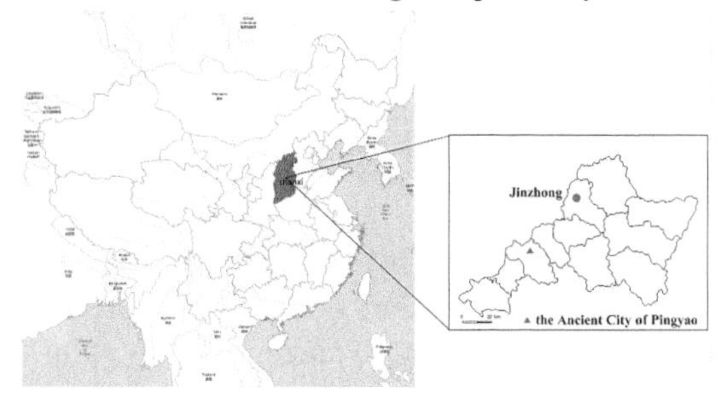

Fig. 2 Pingyao location

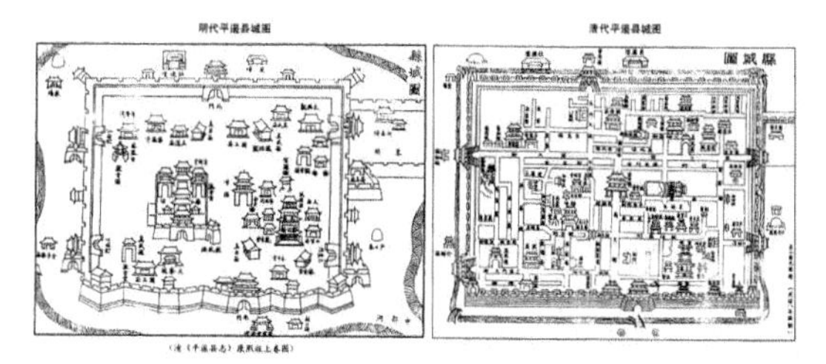

Fig. 3 Historical form of Pingyao Ancient City in Ming and Qing Dynasties © Pingyao county record

4.1 The Characteristics of the Human-Habitat Historical Environment of Pingyao Ancient City

4.1.1 Characteristics of Defense System

"Fortresses are interlaced, turtle city is stable" is a summary of Pingyao's unique defense system. The ancient city of Pingyao has well-built walls and gates, 3,000 mounds, 72 horse-faces and long-standing enemy buildings, as shown in Fig. 4, which are in the form of square city, so it is also known as "turtle city".

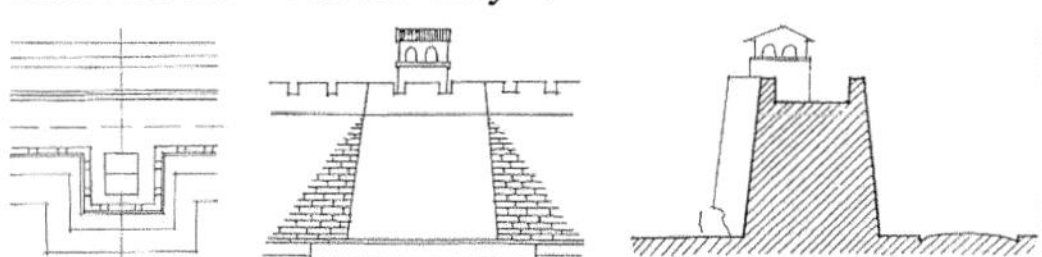

Fig. 4 Plane, elevation, section of the ancient city wall © Pingyao historical and cultural city protection planning

4.1.2 Features of Functional Layout

The Pingyao County has a symmetrical layout, the county system is complete, the South Main Street is the axis, the left Wen Temple right Wu Temple, the Shi building occupies the middle (on the eastern side of the plane are Cheng huang Temple, Wen Temple and Qingxu Taoist Temple; while on the west side, there are Yamen, Wu Temple and Jifu Temple) as shown in Fig. 5.

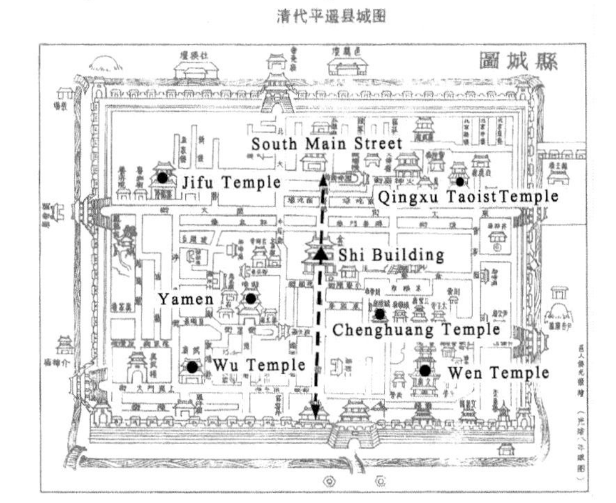

Fig. 5 Functional layout of Pingyao Ancient City

4.1.3 Characteristics of Street and Lane Patterns

There are four streets, eight alleys and seventy-two grasshopper lanes in Pingyao Ancient City, as shown in Fig. 6. The pattern of streets and alleys is orderly. So far, the layout of "Fangli" in ancient China is still followed. The commercial streets in the ancient city are "Tu" shaped.

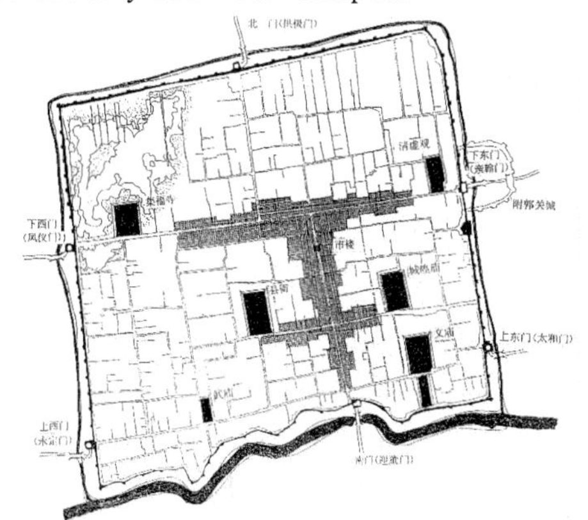

Fig. 6 Street and Lane Characteristics of Pingyao Ancient City © Pingyao historical and cultural city protection planning

4.1.4 Architectural Features

There are more than 3700 ancient houses in the Pingyao ancient city, most of which are typical representatives of the northern traditional houses in Ming and Qing Dynasties. The courtyard space is mostly square in shape, and its layout follows the Chinese etiquette system. "The courtyard is rigorous and the pavilion is lofty". The overall color of the building is made of grey bricks and tiles, and is decorated with glaze and colored carvings.

4.1.5 Non-material Cultural Heritage

There are a lot of intangible cultural heritage in Pingyao Ancient City. Shanxi merchant culture, religious culture and national culture are the most representative. Ri Shengchang, the ancient city of Pingyao, is the first ticket number in China and one of the origins of Shanxi merchant culture. There are sixty or seventy temples, Buddhist temples and Taoist concepts leftover from different periods inside and outside the ancient city of Pingyao. They are multi-religious and cultural.

The tall and magnificent ancient city wall, the green and grey ancient residential buildings, the bustling ancient street market and the simple and honest local residents constitute the unique cultural landscape of Pingyao Ancient City, which also reflects the profound Chinese ritual cultural characteristics of Pingyao Ancient City. Through sorting out the historical context of Pingyao Ancient City, we can deeply understand the development status of Pingyao Ancient City.

4.2 Problems in Environmental Protection of Pingyao Ancient City

With the successful application of Pingyao in 1997, the city walls and public buildings of Pingyao Ancient City have been well protected. However, with the development of tourism, foreign culture has caused too fast and too strong impact on local culture, and the social structure of indigenous people has been dispersed and cultural characteristics have been lost. The lack of necessary archiving and management of traditional courtyards in ancient cities in the process of protection has led to the renovation and abandonment of some courtyards in recent years, which directly threatens the authenticity and integrity of the world heritage. The concept of protection lacks a scientific understanding of the value of human settlements, and the human-habitat historical environment is an organic whole of natural ecosystem and local historical and cultural traditions. The object of protection neglects the concern of sustained social change and various groups to respond actively and actively to the material space and its changes. Lack of monitoring, assessment and management of "change" in conservation management makes planning formalized, which is quite different from international heritage management standards.

5 Sustainable Protection Strategy of Chinese Human-Habitat Historical Environment Based on HUL Method

5.1 Protection Concept

Protection concept transition from "emphasizing architectural relics" to "recognizing more broadly the importance of social, cultural and economic processes in maintaining urban values". It should pay attention to the "living heritage" attribute of human-habitat historical environment and avoid the development of Museum city, Film city and Tourist city. From the perspective of public cognition, reviewing the economic value of the hottest heritage tourism, and attaching importance to the propaganda of historical and cultural value, scientific and technological value and aesthetic and artistic value of historical towns; From the perspective of urban regeneration, in the process of economic transformation and spatial renewal of historic cities and towns, emphasis should be placed on the protection of context space ,historical memory carriers, the "nostalgia" needs of modern people, and the continuation of the overall spatial context. Taking Pingyao's protection as an example, on the cognitive level of value, we should highlight its characteristic value, such as long history value, deep Qin Dynasty and Jin Dynasty cultural value and outstanding defensive technology value, and emphasize the bearing and display of architectural relics to its characteristic value. In the context inheritance level, we should continue the layout of "Fangli" and strengthen the historical nodes in the follow-up protection planning and design.

5.2 Protected Objects

Protection objects transition from "focusing on heritage protection" to "forming a vision reflecting the diversity of urban areas", which fully reflect the history, tradition, values, needs and aspirations of various communities. We should construct the security strategy with residents as the core and social life system as the backing, meanwhile strengthen the role of the community, publicize the concept of historical protection, and formulate special policies. Protection objects should be classified according to heritage value, and specific protection measures should be formulated according to different value attributes of different protection objects. The scope and types of protection objects should be expanded according to the actual situation. Taking Pingyao protection as an example, it needs to include important environmental nodes, architectural remains, street patterns, defense system and non-heritage culture in the historical horizon. According to their value attributes and spatial characteristics, all kinds of objects are classified and protected by different strategies, which can continue and strengthen the historical spatial structure in the pattern of streets and lanes, highlight all kinds of important buildings and nodes in parallel and series, and systematically strengthen the historical axes by referring to the protection strategy of linear cultural heritage.

5.3 Protection Force

Protection force transition from "government-led" to "multi-stakeholder participation", which achieves value consensus through participatory consultation with stakeholders. Human-habitat historical environment protection lays stress on humanistic care and the way of symbiosis among people, settlements and environment, and falls into the practical protection work. In Pingyao, it should pay attention to the pursuit and collation of human historical memory, and satisfy people's realistic demands. In the early stage of planning formulation, public participation can be realized by combing the historical information in oral materials and the pursuit of contemporary residents. In the later stage of planning implementation, the transition of protection concept can guide the correct protection behavior and ensure the effectiveness of symbiotic guidance for people, settlements and environment.

5.4 Protection Management

At present, the idea of conservation planning in China is biased towards blueprint, and the mechanism of protection is limited to technical management. As China's cultural heritage is managed by the State Administration of Cultural Heritage, urban heritage is still managed by the Ministry of Housing and Construction in the process of development. All kinds of mechanism defects brought about by this "double management" are also important reasons for various problems in the process of ancient city. Meanwhile, with the "invasive" development of tourism, there is a lack of supervision, evaluation and management mechanism for "dynamic change". Conservation management should achieve sustainable and dynamic management by combining local management frameworks and developing coordination mechanisms for various activities. Management should cooperate in many ways to avoid the management gap. For Pingyao protection, according to the different characteristics of different types of protection objects, real-time monitoring and management platform for classified protection can be built, periodic protection state feedback mechanism can be formulated, and the policy of joint participation of organization management and public management can be implemented. And through the production of popular science books, games and other propaganda of Pingyao local traditional culture and characteristics of space, to open the monitoring and management platform. Then through the management of the whole people to standardize the specific implementation of protection.

5.5 Protection Planning Means

Conservation planning means should help to maintain the integrity and authenticity of urban historical environment and improve the quality of life and urban space through dynamic sustainable planning. Starting from the process of planning work, taking Pingyao Ancient City as an example, this paper combs the guiding strategies of HUL method for every link of human-habitat historical environment planning. In the pre-planning and investigation stage, through a large number of empirical and field surveys, attention is paid to the realistic problems of human-habitat historical environment and the reasonable demands of residents' basic life, and local historical materials are consulted, combining with the field research, this paper combs the important historical period of settlement formation, the important spatial accumulation of each period of history, the social and cultural background under the spatial accumulation, and regards the current static historical settlement as a dynamic historical barrier, looking for the relationship between historical settlement and geographical and natural environment. In the planning-design stage, we should emphasize humanism, solve the basic problems of human-habitat historical environment, balance the contradiction between development and protection, and emphasize the whole, combining with the actual situation, we should continue the historical context, strengthen the historical texture and the historical spatial structure, and connect and collate the space debris in different historical periods reasonably. In the post-planning and evaluation stage, a multi-stakeholder cooperation platform for sustainable development should be established, and a monitoring mechanism in line with local management and development should be established. At the same time, in the planning implementation stage, we should advocate public participation and continue the original self-government of acquaintance society.

"Retain" community life consisting of local residents and foreign residents; "Improving" residential environment and establishing a controllable and compensatory mechanism for resident loss; "Providing" all kinds of development opportunities, and integrating protection policies and Non-protection policies; "Managing" change, continuous monitoring and elastic intervention; "Balance" the development among the growth of cities, heritage protection and human settlements.

6 Conclusion

The method of Historic Urban Landscape is an exploration of sustainable development of urban historical environment and an important opportunity for the transformation and development of human-habitat historical environment. By incorporating the protection of cultural heritage into the process of urban development, we can redevelop the tradition into a long-term flexible practice inventory, and seek the protection of the individuality and characteristics of the historical environment characterized by local culture and heritage, meanwhile providing a reference for the sustainable protection of the Chinese Human-Habitat Historical Environment.

Endnotes

① Standard (2): Pingyao Ancient City is an excellent example of the evolution of Chinese architectural style and urban planning over the past five thousand years, which integrates the factors of different nationalities and other regions.

Standard (3): From the 19th century to the early 20th century, Pingyao was one of the economic centers of China. Stores and traditional houses in the city witnessed the economic prosperity of China in this period.

Standard (4): During the Ming and Qing Dynasties (14—20 century), Pingyao Ancient City was a typical case of Han nationality, with all its characteristics highlighted and retained.

References

ARTS B, BUIZER M, HORLINGS L, et al., 2017. Landscape approaches: a State-of-the-Art review. Annual Review of Environment and Resources, 42(1): 439-463.

Pingyao County Local Chronicle Compilation Committee, 1999. Pingyao county records.

Shanxi Urban and Rural Planning and Design Institute, 1989. Pingyao historical and cultural city protection planning.

TIAN Shenzhen, LI Xueming, 2016. Human settlement environment science: bibliometric analysis based on China CNKI. Urban Issues, 09: 18-26.

TONG Yuquan, 2014. Research on spatial differentiation of traditional Chinese villages based on GIS. Human geography, 29(4): 44-51.

UNESCO, 2015. Practical conservation guidelines for traditional courtyard houses and environment in the ancient city of Pingyao. http://creativecommons.org/licenses/by-sa/3.0/igo/.

WHCN, 2019. Charter of the ancient city of Pingyao. [2019-05-20]. http://www.whcn.org/Detail.aspx? Id=1878l.

WU Liangyong, 2001. Introduction to human settlement environment science. Beijing: China Construction Press.

ZHANG Song, 2014. Cities as collective memories and their protection. World architecture, 12: 61-63.

ZHANG Song, 2017. Sustainable protection of urban historical environment. International urban planning, 30 (2): 1-5.

ZHANG Wenzhuo, 2018. Landscape method for urban heritage protection: review and reflection on the development of urban historic landscape (HUL). Conference papers of Chinese landscape architecture society: 428-435.

ZHENG Hangsheng, 1996. Chinese society in the rapid transition from tradition to modernity. Beijing: Renmin University Press.

Research on Regional Activation Strategy of Religious Architecture in Pingyao Ancient City

SHI Qianfei, *JING Yifan*, *ZHANG Xiaoning*, *LI Fangfang*, *ZHOU Jing*

Taiyuan University of Technology, School of Architecture, Taiyuan, Shanxi, China

Abstract Pingyao Ancient City has received further protection and developed well in the past decades. However, some problems still exist in the ancient city with some blocks outside the core area not being given enough attention to and in lack of development vitality. The regional structure and functional imbalance, loss of cultural and architectural features, as well as the poor environment have resulted in it is not being favored by tourists. Moreover, the lack of local residents' activities makes it unable to form sustainable development. To solve these problems, the researchers aim to excavate the potential of the culture carrier of the ancient city. First, there a considerable number of religious building spaces, namely, palaces and temples where people engage in religious activities, which are powerful carriers to stimulate the vitality of the region. Taking the religious building space area in Pingyao Ancient City as an example, we probe into possible methods to realize its activation and development. Previously, after consulting classics and existing materials and by doing field research, the research group has confirmed more than 50 religious building spaces in Pingyao Ancient City, and has carried out the compilation of basic information. Then, by sorting out information, carrying out graphic analysis and public life research, aimed at the solution of the problems existing in the ancient city, it is confirmed that many religious building areas in the city are problematic but also have good activation potential. Fnally, using case analysis method, hierarchical protection of historical buildings, architectural design, planning and landscape design methods we design activation strategies for the three most typical religious building areas, such as Guandimiao in Qinhanmen, in order to explore the mode of promoting the coordinated development of ancient city and religious building space. The purpose of this study is to provide theoretical support for Pingyao urban construction planning, cultural heritage protection and urban management.

Keywords Pingyao Ancient City, religious architecture, activation strategy, case study

1 Present Situation and Existing Problems of Pingyao Ancient City

In recent years, heritage conservation has become a heatedly discussed issue among researchers from different backgrounds and has drawn attention from policymakers and city planners. Many approaches have been taken to obtain a better effect of heritage protection. At the same time, studies on the ancient city of Pingyao and its protection is becoming more widespread and international and more significant findings and suggestions have been made compared with their precedents.

According to our research based on a thorough review of existing works, through original field study and theoretical reflection, we discovered that some existing problems need to be solved urgently in Pingyao Ancient City conversation, the critical aspects of which are analyzed and discussed in the following.

1.1 Dangerous Loss of Style and Landscape

Some areas of Pingyao Ancient City are facing the danger of losing its style and features, and its own uniqueness is being weakened day by day. Apart from works in the Ming and Qing dynasties, the protection of some buildings built in the ancient city around 1949 did not follow the natural law, often incorporating modern elements into the historic structures. In addition, there is a lack of consideration of keeping pace with the times in dealing with some truly preserved historical blocks and historic buildings. The common practice is to restore and protect them intact. Some are open to visitors, others are directly archived. Such practices have separated the substantive and cultural context, or the material and intangible cultural heritage. They have not integrated the lives of local residents into historical heritages, nor fully exerted the tourism value of these historical relics. In fact, they are also very harmful to the protection of cultural relics. Therefore, there is an urgent need to explore a way to make the historical heritage, local culture and residents' lives "live up" together, and to promote the sustainable and healthy development of the local tourism economy.

1.2 Unbalanced Regional Development

There is an unbalanced development in the ancient city of Pingyao. The bustling blocks and streets are overcrowded, while few are interested in other areas. Historically, the city is distributed from south to north, with the main gate being the North gate, called Gongji Men. Inside the city, the North Avenue, the East Avenue, the West Avenue and the South Avenue are busy commercial streets, where many banks, bodyguards, temples and hotels are along the four streets. Today, they are still the busiest areas in the ancient city. Besides the four streets, however, a large number of residential buildings, winding lanes and dwellings are in the vast area beyond them. These areas are still inhabited by a certain number of residents, but few tourists come to visit them. Some of the buildings here are worrying people in their quality, destructive in style and poor in environment. They are facing the challenges of slow development and insufficient vitality. They are in sharp contrast with the main commercial blocks such as the four streets(Fig. 1).

Taking the Fire Temple area in the northeastern part of the ancient city as an example, the Fire Temple is very important in Pingyao's history. It has now become a highlighted spot in the tour tickets after its renovation. However, according to the researchers'observation and record, even during the May Day holiday, which is the most prosperous tourist peak period in a

Fig. 1 Diagram of crowd heat flux in Pingyao Ancient City

year, there are few visitors to the Fire Temple, so that Pingyao's traditional culture cannot be fully displayed. The Fire Temple is surrounded by dense dwellings and a park, Huize Garden, which should be an ideal place for local residents to gather, but fails to play its due role as a tourist attaction. Therefore, it is very difficult to realize the ideal scene of activation and development, with interactive communication between residents and tourists.

1.3 Structural and Functional Disorders

The structure and function of Pingyao ancient city at the present stage are almost entirely formed to meet the needs of a large number of tourists during the peak period of tourism. For example, the road system is divided into two inner and outer circles. The outer circle allows local motor vehicles, tour battery cars and people to mix with each other. The inner layer is a pedestrian street, and the two areas are fenced by iron railings on the main roads. In addition, since the end of the last century, all the necessary supporting facilities for the life of the residents in the ancient city, such as schools, hospitals and post offices, have been moved out of the city, causing inconvenience for the residents in the city. So most of the residents choose to work and live outside the city, while most of those who stay in the city are elderly people or tenant merchants.

In terms of the texture and public activity space of the ancient city, the abundant religious building space in history has mostly lost its former status. Damage, change of function and general overlook are the usual treatment they receive, which has led to the loss of worship activity places and the reduction of the sense of belonging. In the off-season, the ancient city shows a quiet and depressing scene, lacking the vitality of city life.

1.4 Poor Environment

An obvious and severe problem faced by Pingyao ancient city is its poor environment. Although the basic conditions of water supply and drainage, pavement hardening, pipeline access to the ground and public lighting have been improved after decades of renovation, the overall environment is still not optimistic. Lack of green areas leads to dusty pavement; lack of pleasant landscape leaves only the scene of gray brick walls and commercialized signboards; poor drainage system leads to sewage deposition and odor everywhere on the pavement; some buildings collapsed without having been cleaned up and repaired in time, and some construction sites are in lack of sheltering, destroying the appearance and environment of the ancient city.

2 Religious Architectural Area of Pingyao Ancient City

2.1 Religious Architecture as the Carrier to Solve Regional Problems

In view of the problems of Pingyao ancient city mentioned above, suitable solutions should be excavated from the history and culture of the ancient city, instead of adapting to the modern urban design method or imitating other historical towns. Historically, the space for people's religious activities is the religious building space, which contains a unique regional culture and is a public activity space, so it is also the ideal "activation" carrier. Architectural space of religion such as temples and shrines is different from that of folk houses. The architectural layout and decoration reflect the connotation and characteristics of different religious beliefs. Respecting the original appearance in repairing and restoration of them is conducive to protecting and revealing the unique style of the region. Religious building space is distributed among the dwellings in different areas of Pingyao ancient city, so its protection and activation are conducive to coordinating the development of different areas in the ancient city, avoiding inadequate vitality corners. Protecting the religious building space helps with restoring the structure and function of the ancient city. As a public building, religious building is an important place for people's social interaction in the neighborhood, and also the residents' spiritual belief. Religious building space is a place for worship of God, the holy and the unstained, so its protection is surely conducive to the renovation of the regional environment.

2.2 Religious Architectural Space of Pingyao in History

In the early years of Hongwu in the Ming Dynasty, the imperial central government issued a decree requesting cities and towns throughout the country to build ritual buildings such as Li Altar, Sheji Altar, Thunderstorm and Rain Altar. In the records of Pingyao County in Wanli 37 years, in Kangxi 48 years and in Qianlong 36 years, there are special records of religious buildings in the sections of Temple. The religious building space in the Qing Dynasty is more abundant than that of Ming Dynasty. In Qing Dynasty, religious building space was developed and tended to be complete (Fig. 2).

2.3 General Situation of Pingyao Religious Architectural Space

There are 42 religious buildings in Pingyao ancient city which can be seen nowadays. According to their setting up background and distribution location, they can be divided into six types: 6 religious buildings built in official background, 20 religious buildings built in folk background, 3 religious buildings in commercial back-

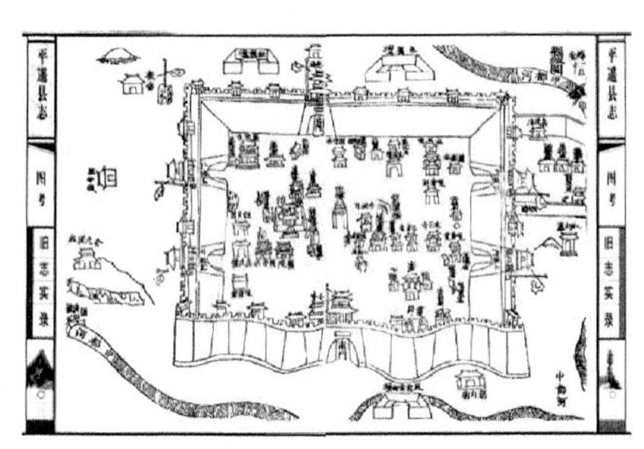

Fig. 2 Distribution of religious buildings in Pingyao Ancient City, Kangxi Region, Qing Dynasty

ground, 3 religious buildings in residential type, 2 religious buildings in foreign type and 8 religious buildings outside the city. These religious building spaces differ in construction age, location, scale, form, the gods worshiped, the protection status, and so on, which together constitute the rich relics of religious building space in Pingyao ancient city.

In addition, Pingyao ancient city religious building space is a multi-dimensional coexisting organism. According to the researchers' previous research results, at the macro level of urban scale, its cultural characteristics mainly reflect the hierarchy and courtesy; and at the middle level of the block level, they reflect the affinity and node planning, practicality and utilitarianism. On micro level scale, the integration and mixture of belief objects are embodied (Fig. 3).

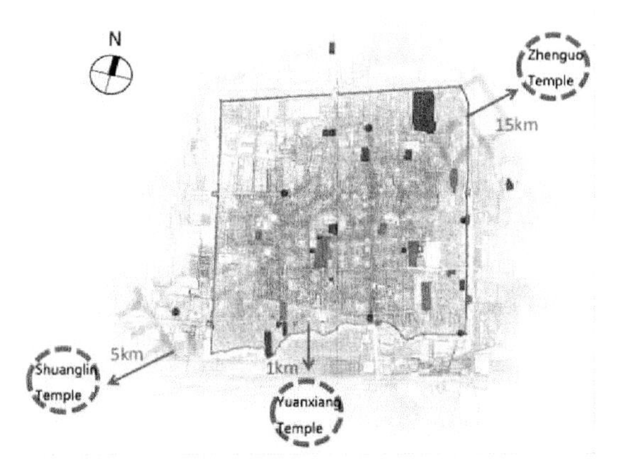

Fig. 3 Distribution of religious buildings in Pingyao Ancient City

3 Activation Strategy and Design of Religious Architecture Area in Pingyao Ancient City

3.1 Reasons for Selection of Activation Research Areas

After sorting out the data for investigation, three priority areas are identified, namely, the Guandi Temple of QinHanmen to the Qingxuguan Area, Fire Temple to Huize Park Area, and Ji's ancestral temple. The three religious building areas highlight the problem of insufficient vitality nowadays, but they have a large number of belief buildings carrying a profound cultural message, with convenient transportation, abundant space types, dense distribution of aborigines, and showing potential for activation and development. Therefore, they are selected as design areas in priority.

QinHanmen, known as the "Lower East Gate", is the northeastern gate of Pingyao ancient city. It has played an important military role in history, and is also the main road from Pingyao to Qixian and Taiyuan. On the north side of it, there is a complete Guandi Temple. The gate of Guandi Temple has three rooms, with the opening of the gate in the middle connects the inside and the outside. There are atriums in it, with the main hall built through the atrium, and there are five pillow caves. The whole building relies on the city gate, the wall of which is thick to make the temple seemingly "dug out" in the city wall. The Southern stage is tall and spacious, and its scale is magnificent. It constitutes a very characteristic public building space with the Guandi Temple. Unfortunately, the Guandi Temple is not open to visitors now for it is used by the City Wall Management Office. In addition, the stage has only a base, and the superstructure is no longer available.

In the study of public life, our on-site observation found that a large number of tourists were interested in Guandi Temple and the stage and they stopped to watch, only not to be able to enter and appreciate its style. Therefore, the place is decided to be the east starting point of this activation design area. From Guandi Temple to Qingxuguan Taoist Mansion, there are Lutai Temple and Qingxuguan religious building space in the middle. There are also historical relics (stone tablets) and public activity space on both sides of the road. Therefore, it is determined that this area has good activation and development potential(Fig. 4).

Fig. 4 Current situation map from Guandi Temple to Qingxuguan

The area selected for activation design also includes the Huize Park area to the Fire Temple area. Fire temple is a representative building space of industry God belief, which is also closely related to the surrounding area. The ceramics industry is well developed in the northeastern part of the city. The God of fire is worshipped by the industry. Relevant streets and alleys such as Yaochang Street, Yaoloudi Street and Dongshuidao Lane surround it. Today the Fire Temple is open to visitors, but the doors are cold and the people around it lack social activities. On both sides of Fire Temple Street, there are areas left behind by buildings, front porches with lots of eaves in front of buildings, and Huize Garden as a park mainly for local people, which has great potential for activation and development. The study of Fire Temple is also conducive to the study of similar industry God belief building areas, such as Mijia Temple, Wumiao, etc(Fig. 5).

Another selected area is Ji's ancestral temple area. There are three ancestral temples in this area: Ji's East Branch ancestral temple is situated opposite Wumiao Temple

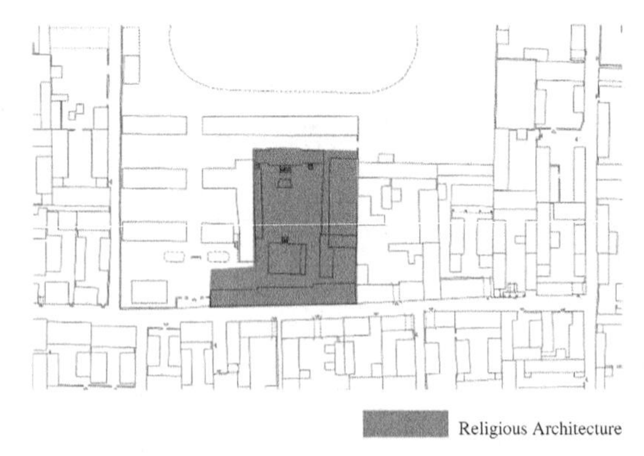

Fig. 5 Current situation map from Huize Park to Fire Temple

in Academy Street. It is situated in the South, with partial courtyards and Orthodox courtyards. It was rebuilt into a school after liberation, but at present, its state of protection is not good. Ji's West Branch Temple is located at the intersection of Zhanma Road Street and Huizhapo Street. It sits in the west and faces the east. There are also partial courtyards and main courtyards. At present, it is not well preserved and many new buildings have been erected. In addition, there is Jijia Fotang Cihang Temple, located at the intersection of Zhanma Road Street opposite Shangximen Street. At present, there is a Ming Dynasty building with three rooms. The beams, columns and arches of the building are beautifully preserved. The eaves are embedded with Chongzhen era stone tablets, describing the past of the Buddhist temple. The three ancestral temples in the whole region are distributed in a right triangle on the general plane, which deeply reflects the history of the Jijia family and the culture of filial piety in Pingyao.

3.2 Strategic Conception of Activation Design

Museums should not be merely treasures in the world, but closely related to the life of neighboring communities. Museums should exist for the community and be shared while co-governed. It is the database of the community, the place of leisure and education, the institution of social history and commemoration, and the symbol of the community.

The city eco-museum is a process that begins with the collective memory of the local people, which is related to people and things in the place. This illustrates the importance of city eco-museum awareness to self-identity and its role in helping communities adjust to rapid changes in life. This design will take the city ecological museum as the design idea, and design the three blocks into the city ecological museum separately, writing the history and culture of their respective regions. The city eco-museum and Pingyao ancient city constitute a whole historical environment with its cultural relics and monuments, which together state a regional culture.

The contents of the city eco-museum include architecture, historic sites, traditions, industries, customs, documents, cultural relics, photographs, oral history and so on. The preserved street-front shops in the blocks have partially restored the original industries (pottery shops, vinegar shops, beef shops or modern handicraft manufacturing factories), to better inherit the traditional culture; part of the space can be released for residents to use as well as for community activities. In the right place we design pleasant environment for local elderly activities, to drink tea or chat in this space, so that local elderly people have a place to stay, tourists are able to listen to stories or the past lives they tell and have a chat with them. Part of the space is used as museum teaching, research and exhibition space.

Faced with unknown objects and unpredictable activities, this strategy is to excavate the potential of base and existing buildings, to carry out appropriate planning and design in order to expect users to enter the space, through the occurrence of activities, changing the creating space and completing the design. The starting point of the whole design concept (Fig. 6) is to maintain and continue the "past" lifestyle and interpersonal network, and create new activities in the "present" to create a new life of the block with the impact and weaving of new and old activities.

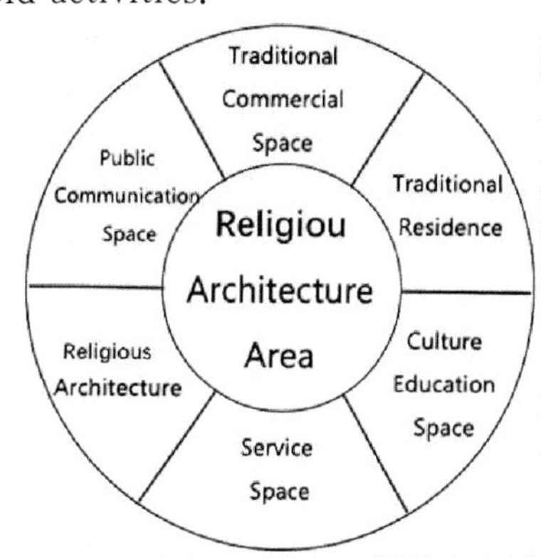

Fig. 6 Activation design strategy concept

3.3 Activation Strategy Design

Regional Design Strategy from Guandi Temple of Qinhanmen to Qingxuguan

As Fig. 6 illustrates, to plan the religious building space is to show the heritage and to communicate, highlighting the cultural theme in the region. The exhibition and communication space is designed for the general touring audience or the hurried audience. It can help get an overall impression and the thematic information when touring quickly, by designing the main wall and the exhibition stand, etc. to describe its environmental background and related content.

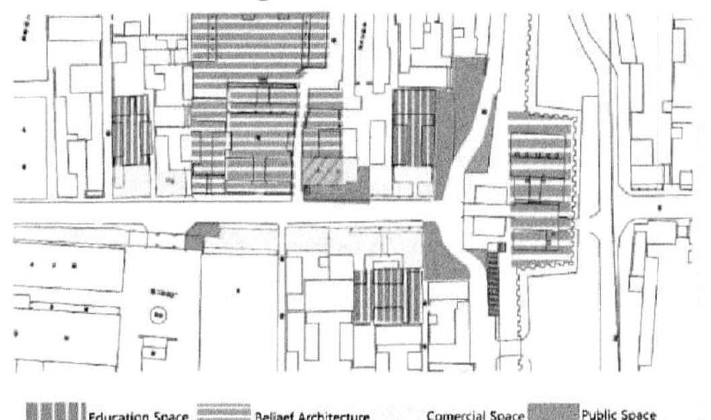

Fig. 7 Sketch map of design strategy from Guandi Temple to Qingxuguan area

Among them, the Guandi Temple of Qinhanmen was restored in accordance with the temple shape of Ming and Qing Dynasties, showing "the spirit of love and righteousness", "the spirit of brotherhood" and "the spirit of martial arts".

Lutai Temple needs to be rebuilt on the base site to restore the former layout of the "front stele, middle court and back hall". Here, the shop facing the street was a service agency built in the 1960s and 1970s, which features the characteristics of the times with special historical information. It should be retained and reopened in the form at that time. The Lutai Temple echoes the Lutai Mountain Temple in Pingyao East. It worships Lutai God and two Lutai goddesses. Interestingly, it is a place of worship for children in history. It deeply reflects the religious culture of Pingyao ancient city and should be brought back to the eyes of the world.

Qingxuguan in Pingyao is also an important Taoist holy place famous throughout the country in history. Now it is being renovated and will soon be able to meet with tourists and residents. The Taoist house on the west side of Qingxuguan used to be a building inhabited by Taoists. The three-hole caves in the entrance area and the main houses all have their own characteristics. Both Qingxuguan and Taoist houses can be opened so that tourists can understand the prosperous Taoist culture in the history of Pingyao ancient city.

In addition, public communication space, cultural and educational space and characteristic commercial space are planned and constructed in the region. Public communication space (Fig. 8) is an important link to "live" the ancient city. It will provide a pleasant recreational/rest area for the local residents and tourists, and create the possibility of communication. The intersection zone between East Street and Qinhanmen is an ideal location for communication space. The results of the public activity survey on a holiday show that in the morning and in the afternoon, local residents sit idly, enjoying and cooling themselves and chat here, with a certain number of tourists staying and wandering there. Their activities are mostly to stop and watch, rest, wait for others, eat and buy goods on both sides. Here is also the transition area connecting Guandi Temple with Lutai Temple and Qingxuan Temple, so it is reasonable to cope with the emptiness and inactivity by designing and guiding people's behavior. The space of culture and education is used for comparison and reference by interested researchers. The exhibits are displayed in cabinets in a centralized way, and are presented by close-range viewing or other practical ways.

Fig. 8 Public communication space intention

3.3.1 Design Strategy of Huize Park Area of Fire Temple

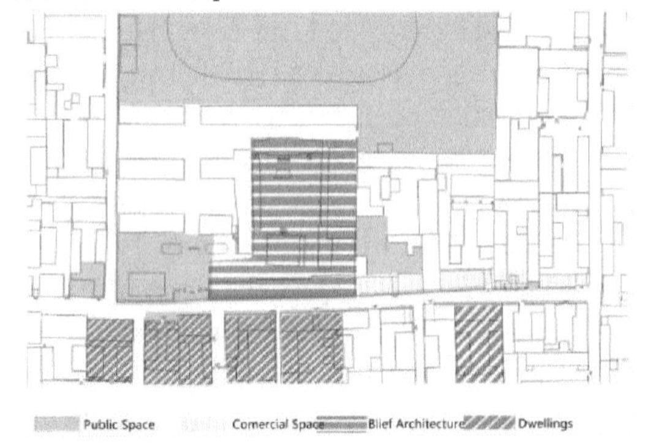

Fig. 9 Conceptual design from Huize Park to Fire Temple

Like the area of Qinhanmen, the area of Huize Park to Fire Temple is also going to have religious building space, characteristic commercial space and public communication space built into it (Fig. 9). The characteristic here is that there are some fully integrated courtyard dwellings with Ming and Qing styles on both sides of Fire Temple Street. Some of them have been developed as inns. The design aim is to improve the quality of the dwellings and to add some semi-open communication space in front of the courtyard as well as in the courtyard, so as to make the streets more affinity. Another area requiring redesign is Huize Garden, a park on the northwest side of the Fire Temple. Some elements of religious culture related to the Fire Temple and ceramics culture related to the region itself should be added to its design.

3. 3. 2 Regional Design Strategy of Ji's Ancestral Temple

Ji's ancestral temple area is located on both sides of Shangximen Street and Academy Street. Its area is larger than the previous two areas, and its buildings are more densely located. There are elements such as Hunqizhai Grand Courtyard, Wumiao, Lei Lutai's former residence and so on (Fig. 10). Planning and design include exhibition space, characteristic industrial space, residential cultural experience space and public communication space. Specific design is being further advanced.

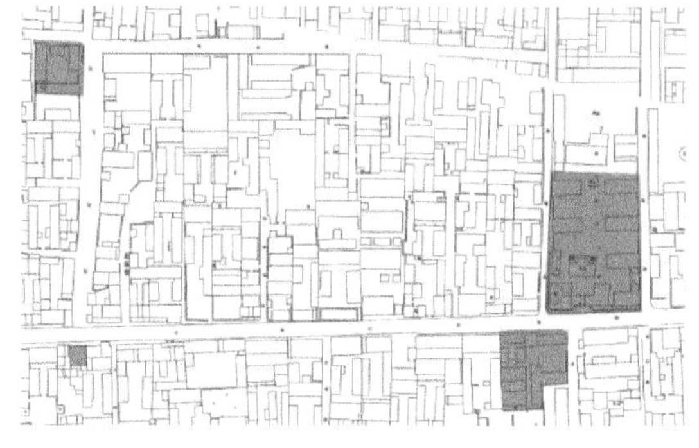

Fig. 10 Current situation of Jishi ancestral temple area

4 Conclusion

Pingyao Ancient City has witnessed the development and changes typical of northern cities from the Ming and Qing Dynasty to the Republic of China and then to modern times, with high historical and anthropological value. In the wave of tourism development, the characteristics of the ancient city and the way of life of the residents are affected to a certain extent. By relying on and digging into the original religion in architectural space and regional culture, combining the social and cultural context of the ancient city, renewing the substances and reorganizing the functions, building an ecological museum, all will help Pingyao Ancient City "live up" and achieve sustainable development.

With the above core spirit and original intention of this study and by analyzing the current dilemma facing Pingyao Ancient City area development, we pointedly proposed the need to take the religious building space as the carrier to drive the city development. By studying the three selected and representative blocks, the function and image design strategy have been proposed in order to better reflect its value and connotation, to gradually let Pingyao Ancient City become a real model for the development of historical towns.

References

ANTROP M, 2006. Sustainable landscapes: contradiction, fiction or utopia? Landscape and urban planning, (75): 187-197.

BARRERE C, 2016. Cultural heritages: from official to informal. City, culture and society, (7): 87-94.

BROWN P L, 2007. Living history: the walled city of Pingyao offers a rare glimpse into China's past. Architectural digest, 64(12).

DONG Jianyun, DONG Peiliang, 2017. The cultural history of Pingyao ancient city. Taiyuan: Shanxi Economic Publishing House.

GALE J, SVARRE B, 2016. Research methods of public life. Beijing: China Architecture & building Press.

GALE J, 2013. Public communication and space. Beijing: China Architecture & building Press.

GAO J, ZHANG J, DAI F, 2006. Space organization and representation of Chinese etiquette culture in the ancient Pingyao city. Journal of applied physics, 70(5): 385-390.

LI Dingwu, 2007. Culture Pingyao. Taiyuan: Shanxi Ancient Books Publishing House.

HONG Lian, 2011. The outline of Chinese traditional concept of etiquette and its modern effects. Asian social science, 7(8).

LINDHOLM K, EKBLOM A, 2019. A framework for exploring and managing biocultural heritage. Anthropocene, (25): 1-11.

SHARMA A K, 2017. Historic city: a case of resilient built environment. Procedia engineering, (180): 1103-1109.

SUN Xian, DAI Zhen, 2013. Fenzhou prefecture records: Pingyao edition. Taiyuan: Sanjin Pub-

lishing House.
WANG Jinping, LI Huizhi, XU Qiang, 2015. Shanxi ancient architecture. Beijing: China Architecture & building Press.
WANG S, 2008. Tradition, memory and the culture of place: continuity and change in the ancient city of Pingyao, China. Denver: University of Colorado at Denver.
WANG Shou, KANG Naixin, 2008. Pingyao County chronicle (Kangxi 46). Taiyuan: Shanxi Economic Publishing House.
ZEAYTER H, MANSOUR A, 2017. Heritage conservation ideologies analysis: historic urban landscape approach for a mediterranean historic city case study. Housing and Building National Research Center. Elsevier B. V. http://creativecommons. org/licenses/by-nc-nd/4. 0/.

Study on Diversified Strategies for Sustainable Development of Villages in the East of Pingyao — Taking Huangcang and Podi Villages as Examples

SHI Qianfei, *ZHANG Xiaoning*, *ZHANG Yong*, *SHEN Gang*, *ZHAN Haiqiang*

School of Architecture, Taiyuan University of Technology,
NO.79, Yingze West Main Street, Taiyuan, Shanxi Province, China

Abstract This research topic comes from the 2018—2019 project cooperation agreement of the Pingyao International Workshop on urban and rural heritage protection and development, which was signed by Taiyuan University of Technology and the Urban and Rural Planning Bureau of Pingyao County. Part of the agreement is to explore proposals for rural revitalization of villages in the east of Pingyao. The surrounding villages of Pingyao ancient city that is a world cultural heritage also have historical value that can not be ignored, and the problem of villages going to be hollowed out and buildings in disrepair is badly in need of attention. This paper takes Huangcang and Podi villages as the research objects, and we combine it with the graduation project of our school in 2019 school year. By means of field survey, field mapping, literature collection and other methods, diversified strategies for sustainable development of villages are explored on the basis of a full analysis of current problems such as history, culture and spatial pattern, which can also provide a reference for the local government's follow-up transformation work.

Keywords diversified strategies, sustainable development strategies, Huangcang Village, Podi Village, rural revitalization

1 Introduction

At present, the research results on the theme of Rural Revitalization mostly focus on one village and explore a development strategy. This paper takes two villages as the research object, takes the current situation of villages as the premise, takes the diversified value of villages as the basis, explores a variety of development strategies of one village, and compare differences between two villages.

2 Research Status and Problem Statement

2.1 Background and Problems of Rural Revitalization

2.1.1 The Rural Revitalization Strategy

Recently, the state's preference for major policies for Rural Revitalization has increased the speed and intensity of capital investment into rural areas. A series of land transfer policies and subsidy and support policies for agricultural investment, especially the cooperation between the government and social capital in 2016, were written into *the No. 1 document of the Central Committee*, giving a loose policy environment for capital intervention in rural areas (Tab. 1). The government has issued incentive policies to support the new development model represented by characteristic towns and pastoral complexes.

Tab. 1 Summary of major policies for rural revitalization since 2017

2018/2/4	No. 1 Document of the Central Committee officially promulgated the full deployment and implementation of the strategy of Rural Revitalization
2017/12/21	"Rural Revitalization" and "Precise Poverty Alleviation" will become the focus of China's economic work in the coming year
2017/10/31	Outline of Planning and Construction of Superiority Areas of Characteristic Agricultural Products
2017/10/18	The Nineteenth National Congress clearly put forward the strategy of rural revitalization

Continued

2017/10/13	Opinions on Innovating Institutional Mechanisms to Promote Green Development of Agriculture
2017/1/25	Notice on Further Perfecting the Pilot and Demonstration Work of Rural One, Two and Three Industries Integration and Development

2.1.2 Strategy of Sustainable Development

On July 4, 1994, the State Council approved China's first national sustainable development strategy, the White Paper on Population, Environment and Development in the 21st Century. At the 15th National Congress of the Communist Party of China in 1997, the strategy of sustainable development was defined as the strategy that must be implemented in China's modernization drive.

2.1.3 Existing Problems

(1) Existing Hollowing Problems in Villages

At present, many villages are lucky to be preserved, but the phenomenon of hollowing is serious, the aborigines of villages are gradually losing, and the rural culture is facing the dilemma of no inheritance, which leads to the destruction of material and intangible cultural heritage and the decline of villages.

(2) Homogeneity of Village Development

Many external forces blindly exploit villages on the basis of inadequate investigation and research. The impact of foreign culture on traditional culture of villages is also fatal. As a result, villages with local characteristics are gradually losing their features and moving closer to repetitive, monotonous and consistent environmental space.

(3) Urbanization in Village Renewal

Since the reform and opening up, rural economy has developed rapidly, traditional culture and regional culture have been challenged strongly, and cultural heritage resources in ancient villages have been seriously lost. Some ancient villages are losing their unique cultural characteristics by concentrating on building high-rise buildings, opening up city squares, imitating city gardens and roads.

2.2 Literature Research Status

Traditional villages have more abundant research results than non-traditional villages, but the development strategies of traditional villages are mostly confined to the protection level because of the restrictions of policies, laws and norms. Non-traditional villages have no restriction of "authenticity protection". Their development is more flexible and plastic. Besides the way of tourism development promoting protection, they can also develop agricultural industrialization and improve their own hematopoietic function and vitality.

2.2.1 Development Strategies of Traditional Villages

The summary of the development model is mainly based on the evaluation of the current situation of villages. For example, Shanxi Province Traditional Village Protection Mode Analysis and Its Effect Evaluation analyses and evaluates the existing protection mode of traditional villages, and concludes two modes of static protection and living protection. The Classification and Development Model of Traditional Villages from the Perspective of H-I-S is based on the combination of the three elements of village subject (H), industry (I) and space (S), and divided into five corresponding development modes: integrated development, community building, ecological museum, spatial linkage and cultural reconstruction.

2.2.2 Development Strategies of Non-traditional Villages

The study of non-traditional villages mainly explores different ways of Rural Revitalization from the perspective of capital operation and industrial planning. For example, according to *the Capital-driven Multi-mode of Rural Transition in Metropolitan Margins and Its Impact*, the model of village development can be divided into three types: sole proprietorship, state-owned investment and state-owned enterprise cooperation. *Industrial Village Planning and Design Strategies under Multi-Value Orientation* mentions that industrial villages can be divided into several types: agriculture-oriented, industry-oriented, business-tourism-oriented and more balanced comprehensive.

2.3 Source of Topic Selection and Problem Statement

"Pingyao International Workshop on Urban and Rural Heritage Protection and Development, 2018—2019 Project Cooperation Agreement" was signed by Pingyao County Urban and Rural Planning Bureau and Taiyuan University of Technology. Article 2 of the agreement is "Research on Dongbu Village, Podi Village, Huangcang Village and Shangdian Village". The four villages selected are located in the eastern mountain area of Pingyao. They are neither traditional villages nor urban fringe villages, but between them. Huangcang Village and Podi Village are chosen as the research objects because of their similar size and geographical location, but the preservation of traditional elements such as traditional buildings and historical environmental elements is different. Based on the current situation and multiple values of villages, this paper explores various paths of village development through field research, literature review and chart analysis.

3 The Present Situation of Villages and Their Diverse Value

3.1 The Present Situation of Villages

Huangcang Village, Zhukeng Township, Pingyao County, is located in hilly and mountainous areas 25 kilometers southeast of Pingyao Ancient City, with an elevation of about 1500m(Fig. 1). It is one of the six natural villages that constitute Fengsheng Administrative Village. Huangcang was called the Royal Warehouse in the Southern and Northern Dynasties. It has a long history. According to literature and ancient buildings, at least in the Ming Dynasty, villages have been formed. At first, the villagers were mainly Huang surname, then Pei surname. Its unique material of red slate is abundant in the village (Fig. 2). At present, its built-up area is about 5 hectares, with a population of more than 100, and it has been hollow. The villages are built on the hills, with caves as the main architectural form.

Podi Village, Zhukeng Township, Pingyao County, located 22 kilometers southeast of Pingyao Ancient City, is also one of the natural villages of Fengsheng Village (Fig. 2). It is a node village in Pingyao section of Wanli Tea Road, providing trade exchange services for tea merchants in ancient times.

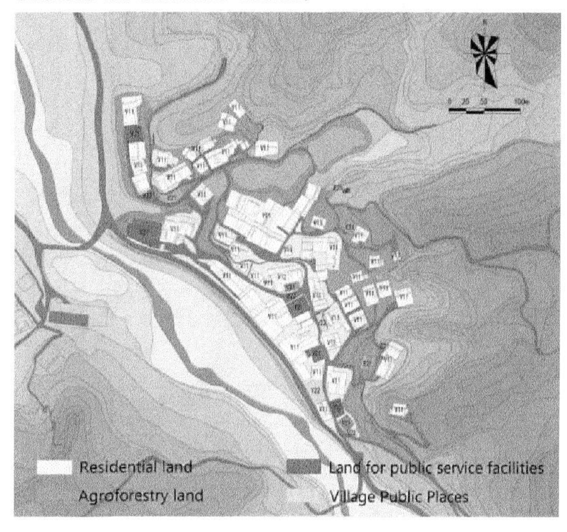

Fig. 1 Huangcang Village

3.1.1 Location and Transportation

Podi Village is adjacent to 376 County Road. It is 121 kilometers from Beijing-Kunming Expressway to Jinzhong City. It has convenient transportation and 10 kilometers from Zhukeng Township, the administrative center. Huangcang Village is located at the end of 376 County Road, far from the county town. Convenient traffic has promoted the development of Podi village economy, but also has a negative impact on the traditional style of the village. Although Huangcang village is inconvenient in external traffic and the terrain environment is relatively closed, it has made more traditional building resources to be preserved, and the village style is relatively intact.

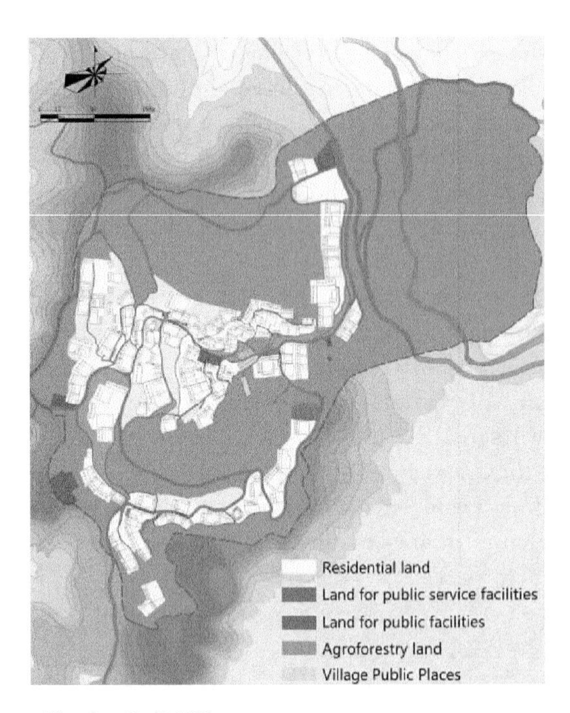

Fig. 2 Podi Village

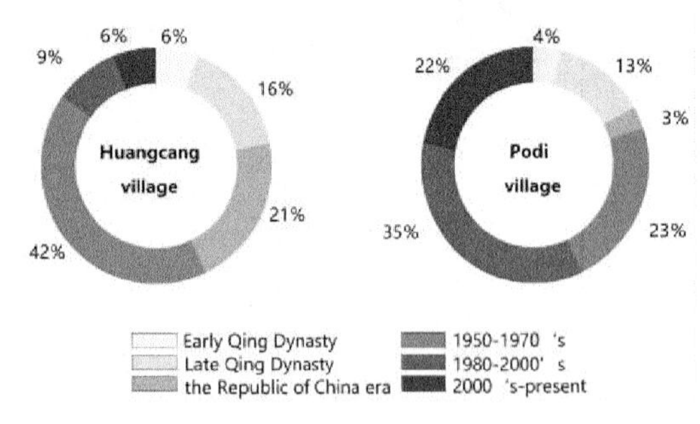

Fig. 3 Age Architectural Analysis

3.1.2 Historical Accumulation

Huangcang Village, as a royal warehouse for grain storage in the Southern and Northern Dynasties, began to use red stone slabs for urns, writing, desks and so on. During the war period, Huangcang Village was a base of anti-Japanese democratic government. Its historical culture included agricultural culture, red stone industry culture, military culture and ancestral temple culture. Podi Village is the node village road of Wanli Tea Road, which contains cave culture, Shanxi merchant culture and tea culture.

3.1.3 Village Style

There are seven Qing Dynasty buildings in Huangcang Village, accounting for 16% of the total village buildings; the architectural style is basically intact, and the main architectural form is caves, of which 91% are proposed for historical buildings and traditional buildings; the architectural quality is generally preserved, with 48% with complete main structure. Mountains surround the south, the north and the west, water in the east, and the village is in the U-shaped terraced mountains (Fig. 4).

Podi Village has 5 courtyards in Ming Dynasty and before, 14 courtyards in Qing Dynasty, accounting for 17%. Architectural features are generally preserved, of which 53% are traditional ones. Because of the large number of new and constructed buildings, the construction quality is better, and the main structure is complete, accounting for 76.2%. On the topography of east, West and low, villages spread along the north-south axis in a band shape (Fig. 5).

Fig. 4 Huangcang Village

Fig. 5 Overhead view of Podi Village

3.1.4 Industry and Population

The main income of residents comes from agriculture and animal husbandry. Huangcang Village's agriculture includes corn, millet, yam, soybean and so on. Animal husbandry includes cattle, sheep, wild boar, chickens and other poultry, with a population of more than 100. The sloping village is rich in walnut, pepper and other plants, with a population of 436.

3.2 Diverse Value of Villages

3.2.1 Agricultural Production Value Based on Agricultural Production

Agriculture is an organic ecological and social system. It has comprehensive and multi-function. With agricultural production as the carrier, on the one hand, the village production mode of "feeding on mountains and water" can provide sufficient labor force for agricultural production and meet the needs of family survival and reproduction; on the other hand, relying on family and village community can effectively acquire and inherit agricultural production technology and experience. The division and cooperation of production can adapt to the seasonal and periodic characteristics of agricultural production and effectively prevent and resist natural disasters.

3.2.2 Green Ecological Value Based on Landscape Pattern

The formation of different village forms is mainly influenced by different natural conditions and social environment, and also shapes different village patterns. In order to better live and work in peace and contentment, generations of villagers take the initiative to rely on the geographical environment of the village, take the village as the space carrier, take the natural resources such as rivers, mountains and rivers, land, fully consider the factors of geology, landform, hydrology, sunshine and so on, reasonably select the location and layout of the village, thus forming the integration with the surrounding natural ecological environment.

3.2.3 Villagers' Life Value Based on Settlement Space

Because the people living together in the village community with blood, kinship and geographical ties as the ties, they gradually formed the family and clan with marriage and kinship as the core, and constructed the living space of the village, so that the people living in it could enjoy the value of life such as happiness of family, the security of food, neighborhoods' watch and the inheritance of traditional construction technology.

3.2.4 Moral Educational Value with Cultural Accumulation as the Carrier

With the disappearance or decline of a large number of villages, more and more traditional folk culture and folk beliefs are gradually disappearing, and rural culture is showing a state of desertification. Taking traditional villages as the carrier, we can give full play to the function of moral education and restraint of fine family customs, folk customs and village regulations to villagers. The inheritance of rural culture can arouse and stimulate more people's "nostalgia" in their inner memory.

4 Diversified Development Strategy

Based on the multi-value of villages, this paper explores the development strategy with one or several value combinations to form a multi-development model of "intrinsic value + foreign industry". As the present situation of the two villages is concerned, their age-old degree, scarcity, richness, activity and cultural value are not the same. This paper makes a qualitative analysis in order to explore the different development strategies between the two villages (Tab. 2).

It can be concluded from the table that Huangcang Village is superior to Podi Village in terms of its longevity, scarcity, richness and activity. Based on this result, the conservative development strategy is adopted for Huangcang Village, and the renewal and transformation of villages is mainly at the level of "point" and "line"; the open development strategy is adopted for Podi Village, the mode is "intrinsic value + multiple foreign industries", and the village renewal is "face-oriented"(Fig. 6).

Tab. 2 Village Evaluation Factor

Factors	Factor Meaning	Huangcang Village	Podi Village
Age-old degree	Existing earliest building age and number	Early Qing Dynasty, 5	Qing Dynasty, 4
Scarcity	Quantity of Traditional Buildings	Pre-Republic of China accounted for 43%	Pre-Republic of China accounted for 20%
Richness	Types of Historical Environment Elements	Ancient trees, 29, Stone Mills, 6, Woodcarving and Stone Carvings, 8, Shining Walls, 3	Ancient trees, 3, Ancient wells, 3
Activity	Inheritance of Intangible cultural heritage	Red stone industry culture, folk belief culture and folk custom culture	Folk custom culture
Cultural value	Historical Functions	Royal warehouse	Tea road's node village

Agricultural value
Ecological value
Life Value
Educational value
Diversified Value
Diversified Development Strategies
Evaluation factors
Differences in Two Villages
Conservative Strategy
Opening Strategy
Intrinsic Value + One Foreign Industry
Intrinsic Value + Multiple Foreign Industry

Fig. 6 Development strategy flow chart

4.1 Conservative Development Strategy of Huangcang Village

4.1.1 Art Village Based on Educational Value

Based on the educational value of history and culture, this model introduces contemporary art and develops the literary industry. Material cultural heritage resources include temples, ancestral halls, stage, folk houses and special products such as red stone slabs, intangible cultural heritage resources include legends, beliefs, festivals and other activities (Fig. 7). Villages are divided into photography base, sketch base, farming and living exhibition area, commercial folklore area and residential area. Village regionalization is based on its existing resource characteristics, such as sketch base in the south with flat terrain and wide vision, photography base in the northwest with unique courtyard enclosure, rural art area in the north with large fluctuation and far distance between courtyards, and new industrial resources of villages are distributed in various areas of villages in order to activate villages simultaneously.

Fig. 7 Scene of ancestral square

4.1.2 Cave-Dwelling Holiday Village Based on Villagers' Life Value

The preservation of cave dwellings is one of the characteristics of Huangcang Village. This mode mainly focuses on the value of the life of the villagers, supplemented by the value of agricultural production, and develops the residential industry in order to provide the rural lifestyle for the surrounding urban population. The village is mainly divided into four areas: folk culture exhibition area, plant planting experience area, animal feeding experience area and outdoor experience area (Fig. 8). The four themes of folk custom correspond to the above four areas respectively to create the personality and difference of residential experience. The circular streamline in the village runs through it, together with the public nodes, the streamlined activity is rich. Specific design to make the cave livable as the main point of transformation, and improve the guest rooms, toilets and other functions, and try to retain the cave interior style.

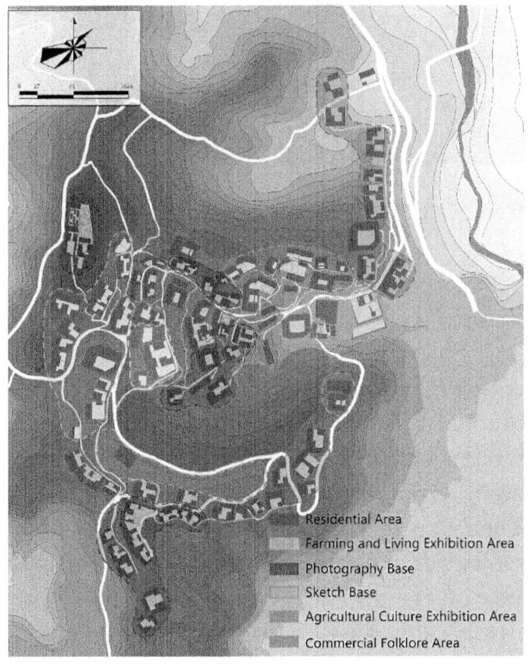

Fig. 8 Functional zoning map of Huangcang

4.1.3 Convalescent Village Based on the Ecological value

Rural spatial attributes have a sense of belonging and security, and are suitable for staying and resting. This model is based on the green ecological value of villages and introduces the old-age industry. At present, the serious problem of hollowing in the village and the aggravation of the aging population in the whole country have brought new opportunities for the development of villages. In this development strategy, the village has formed an old-age system, which is mainly based on institutional pension and supplemented by community pension and family pension. It includes six main modules suitable for the daily activities of the elderly: physical fitness, horticultural recuperation, recreational exchange, children's recreation, pet paradise and cultural activities. The old-age renovation of cave buildings can be divided into "one-to-one" mode-there are idle caves in the aboriginal homes, and the aborigines share a courtyard with the foreigners for the aged, "club" mode - the elderly can travel in groups for the aged, and the adjacent courtyards can be combined and utilized.

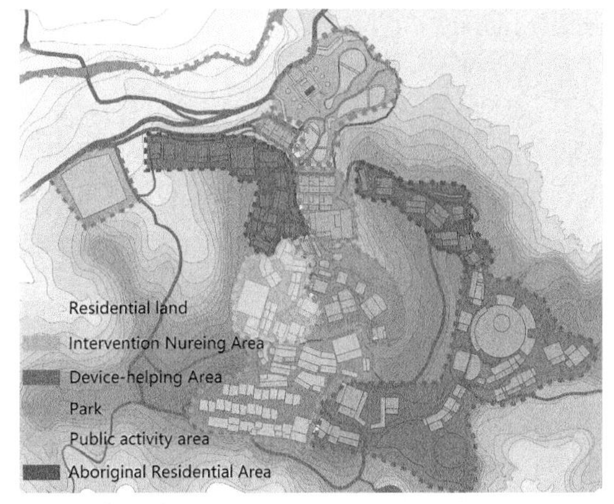

Fig. 9 Functional zoning map of Huangcang

4.2 Opening Development Strategy of Podi Village

4.2.1 Healthy Village Based on the Ecological Value

The model is based on the green ecological value of villages and introduces sports, culture, leisure and other industries. The outline of the "Healthy China 2030" Plan issued by the General Office of the State Council proposes to develop health industry, strengthen the structural reform of the supply side, support the development of new health services such as health medical tourism, and actively develop the industry of fitness, leisure and sports. With the strengthening of people's health concept, the model of integration and development of the health industry tourism industry has changed from small-scale market to more people's vision and become a new favorite of the tourism market. The development orientation of the village is to take mountain sports as the core, to build a health-preserving village with the characteristics of health-preserving residence, tea ceremony culture, folk experience and waterfront leisure. The functional zoning of the village includes outdoor leisure area, waterfront leisure areas, ecological picking area, cultural experience area, craft making area, accommodation and catering area (Fig. 10). It is planned to set up a continuous bicycle lane along the main streets and establish cyclic cycling rings to provide one-stop organic transport links for regional leisure, travel and transfer.

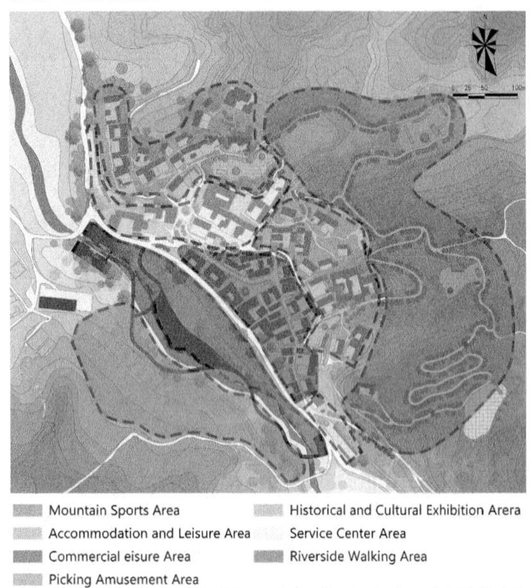

Fig. 10 Functional zoning map

4. 2. 2 Experience Village of Agricultural Products Based on Agricultural Value

Podi Village has a rich variety of agricultural products, including pepper, walnut, corn and so on. Based on the agricultural production value of villages, this model introduces the exhibition, sale and recreation industries of agricultural products, which can be classified according to the seasons, including harvesting activities, crop harvest, aquatic parks, aquaculture and other modules. Different seasons can carry different industries and activities. Picking activities attract tourists based on the peculiar large area of walnut forest. The harvest crops such as pepper, corn and potatoes can increase economic sources by exporting to the outside world. In summer, the river rises, and the surrounding citizens can come to play in the mountains and rivers. In addition, the function of adding experience, exhibition and sale to the workshop for processing crops enhances the participation of the outside and prolongs the stay time of tourists (Fig. 11).

Fig. 11 Agricultural products exhibition area

4. 2. 3 Cultural and Ecologicad Tourism Village Based on Educational and Ecological Value

The model also takes into account the ecological value of the villages and the educational value of history, and intervenes in foreign industries such as product processing, literary creation, tourism and so on. There are two axes in the village, along the ecological axis of the river and the cultural axis of the village. The design centers on two parallel axes to sort out the industry. The cultivation of walnut and pepper is regarded as the primary industry, the primary processing, breeding and cultural creativity are the secondary industries, and the ecological tourism, heritage tourism and sketch base are the tertiary industries. The specific functions are divided into five areas, namely, landscape experience area, commodity service area, sketch base area, product preliminary processing area and farm pleasure area.

5 Conclusion

The multi-value of villages includes agricultural production value, green ecological value, villagers'life value and moral education value. This paper takes the two villages as the research object, compares and analyses the advantages and disadvantages of the villages, and explores the various possibilities of village development from the aspects of literary creation, pension, residential accommodation, health, agriculture, ecology and other industries. However, what strategy is the most suitable for the development of villages remains to be studied. In the follow-up work, we should continue to study and construct the system of village development strategy based on the value characteristics of villages. In addition to Huangcang and Podi villages, Pingyao has a large number of valuable non-traditional villages are facing the danger of disappearance. I hope this study can provide a reference for the follow-up development of villages for the Pingyao government.

References

LI Chunzi, 2018. Analysis and evaluation of traditional villages protection model in Shanxi province. Journal of agriculture, 8 (04): 75-79.

LU Kerong, 2016. Analysis of the comprehensive multiplicity value of traditional villages and its live inheritance. Fujian forum, (12): 115-122.

TAO Hui, 2019. Classification and development model of traditional villages based on H-I-S perspective: a case study of Handan City. Tourism tribune, 04: 1-18.

WANG Xuan, 2018. The multi-mode and impact of capital-driven rural transition in metropolitan margins: based on the comparison of three typical rural areas around Nanjing. Papers collection of the annual conference on urban planning in China, 18: 9.

XU Xiaodong, 2019. Strategies of industrial rural planning and design guided by multiple values: a case study of Dongsanpeng special rural areas. Construction of small towns, 37 (05): 40-48.

Creating a Meeting Place — the Design Strategy of "Micro Center" in the Community of Pingyao Ancient City

CHEN Ying, *CHEN Dan*

Huazhong University of Science and Technology, Wuhan, Bubei, China

Abstract The form of high-density aggregation space and the out-migration of population and facilities brought by the decline of protection, make the community of the ancient city increasingly reflect its inadaptability in the current development. By studying the spatial form of traditional communities, it can be found that the homogeneity of courtyard units, the isolation of courtyard walls and the closure of the interface of streets lead to the characteristics of "no center". The closure of space brings about the lack of communication space. Based on this, the "micro center" mode suitable for the sustainable development of the ancient city community is proposed. Community "micro center" is a kind of micro-community neighborhood center proposed based on the characteristics of ancient city community, combining the development requirements of modern residential areas and the demands of residents themselves. Due to its small scale and close radiation range, the same community usually adopts the mode of "multi-center", where each center performs its own functions and connects streets and lanes to form a "micro center" network covering the whole community. This mode meets the cultural, living, entertainment, leisure and other needs of community residents through micro-intervention without damaging the spatial pattern of the ancient city, and provides a place for community residents to carry out public activities and promote neighborhood communication. At the same time, it also provides a window and a platform for tourists to experience local culture and travel in the community. It is a public welfare space where tourists and residents can coexist and share.

Keywords protective decline, traditional community space form, neighborhood center, "micro center"

1 Introduction

Pingyao Ancient City was listed in the World Heritage list by the United Nations Educational, Scientific and Cultural Organization in 1997 and is praised as "one of the four most well-preserved ancient cities in China." In order to better protect the ancient city and develop the tourism industry, the Pingyao government guided institutions and residents in the ancient city to relocate, alleviated the pressure on the ancient city. However with the migration of all kinds of institutions and young residents, and the commercialization of tourism are serious, which leads to the disintegration of the social structure and the loss of cultural characteristics of the indigenous residents, the ancient city community shows its inadaptability. The vitality of a city can not be separated from "residents" and "communities". Residents are the creators and users of the city, and the community is the cell of the city. In view of the living heritage of the ancient city community, how to meet the needs of the people's good living, how to improve the neighborhood of the residents and how to restore the vitality of the community, are worth thinking in the protection of the heritage and the development of the tourism. Therefore, taking Shuyuan Community of Pingyao as an example, this paper discusses the design strategy of "micro center" in a traditional community, in order to provide a new idea for Pingyao heritage protection and sustainable development.

2 The "No Center" Form of Traditional Community

2.1 Homogeneity of the Courtyard Unit

The courtyard is the basic unit that makes up the community space of Pingyao ancient city, and is an important place for the activity of the residents. In terms of space, Pingyao ancient city community takes the residential courtyard as the unit, through various combinations to form a rich group space, generally manifested in different forms of courtyard space, crisscross street and roadway space, its space form is rich, the building density is large. It is different from the single texture of modern living community. However, as far as courtyard space is concerned, it is basically composed of multiple courtyard series, each courtyard has relative independence and introversion, and the scale is homogeneous. In terms of function, the courtyards in the community are mainly inhabited, some courtyards along the main street are transformed into shops, and some middle courtyards are transformed into hotels. The function is relatively single and homogeneous, and there is a lack of community public space based on the activities of the residents.

2.2 Isolation of the Enclosed Wall

The form of the traditional courtyard is affected by regional, climate and social factors. Pingyao residential buildings are mostly single-slope roofs, The eaves go to the courtyard, and there are fewer eaves out of the walls. the wall is tall and has no open window except for the door, form a closed internal space. and Isolate from the outside world. As one of the most important elements of the separation space, the courtyard wall also reduces the communication between residents and the communication between residents and tourists.

2.3 Closeness of the Street Lane Interface

Streets lane, as the transitional space between courtyard and exterior, provides a place for neighbors to communicate, which can be said to be the material carrier of residents' daily life.

Some street lanes in Pingyao are mainly commercial functions, such as East Street, West Street, South Street and so on. The shops on both sides are prosperous. The street lanes have become an important commercial activity space, and the openness is strong. In addition, most street lanes are dominated by living functions, and the street lanes interface is basically composed of courtyard walls, and in addition to the opening of road intersections, doors and windows, the courtyard units form a unified continuous solid interface side by side. Therefore, the interface of residential street lanes is closed. Its closeness leads to fewer residents staying and resting in the street lane, and street vitality loses and doesn't bear its role as a public space. To sum up, the homogeneity of courtyard unit, the isolation of the enclosed wall and the closeness of the street lane interface lead to the lack of public space in the traditional community and the appearance of "no center", which can not meet the needs of the daily life of the aborigines. The traditional community gradually loses its vitality and becomes a kind of display "life scene". Therefore, this paper puts forward the design strategy of traditional community "micro center", hoping to improve the present situation of traditional community and restore the vitality of the community.

3 Generation of Traditional Community "Micro Center"

3.1 The Lack of Traditional Community Communication Space

Communication space is a place for people to communicate and contact with each other in daily life. Renowned planner Yang Gail has proposed that although the composition of the physical environment has no direct impact on the content and intensity of social interaction, the space can still influence people's opportunities to meet and observe and listen to others. In China's traditional urban settlements, public space is generally born in the streets, temples, ancestral halls, wells, stage, courtyard, eaves space and door space. These traditional communication Spaces once provided the material basis for the vitality of the ancient city. However, in today's Pingyao, with the disintegration of the clan system, important public buildings become tourist attractions and the loose social structure of residents, these spaces have gradually disappeared or shown inadaptability. It is important to establish new communication space and establish new connections. At the same time, the appropriate community communication space can improve the current situation of the ancient city being too empty and scene-oriented, and help to reconstruct the living atmosphere of the ancient city, as well as the development and innovation of local culture and folk customs. From the perspective of tourists, it is more attractive to feel the living cultural charm of the ancient city than to simply appreciate the static old city architecture. The communication space of the community is like a window, providing tourists with plans to experience the customs of the ancient city and feel the regional culture. At the same time, tourists and residents can meet in it, interact and deeply appreciate the life of the ancient city.

3.2 Concept of Community "Micro Center"

In the tide of modernization, traditional communities are not only facing the predicament of their own material space development, but also unable to solve the spiritual contradictions of residents living in them. The dual pressure of space and spirit forces traditional communities to find a new way out. "Micro center" refers to a small public welfare space located within the community that creates a place for residents to positively affect their lives. It chooses places with good transportation and is usually located in the center of the community, so as to facilitate the rapid arrival of surrounding residents and form a short life circle. It takes the basic form of open outdoor space or open architectural courtyard. A single "micro center" provides different functions according to its site characteristics and residents' needs, ranging from cultural entertainment, life services to outdoor recreation and exercise. These "micro centers" then form a network of "micro centers" through the series of streets and alleys to cover all the units of the community. The function of "micro center" is to make up for the missing functions of traditional communities by means of space evacuation and function implantation, to repair the problems of traditional community space, and to help residents establish good and healthy social relations.

3.3 Implantation of Community "Micro Center"

3.3.1 "Micro Center" Selection Strategy

(1) Look for geographic centers

The ancient city now has several residential community units, which are usually bounded by the main streets of the ancient city and form relatively independent residential groups. The narrow alleyways within the units, the interlaced residential courtyards, are very crowded and produce many dead ends. On this basis, the existing street texture is utilized and some streets and lanes are connected to further divide the courtyard inside the community and form a number of small residential units. The intersections of the inner streets and lanes, as well as the accessible areas of various small residential units, are the geographical centers we are looking for for the best internal accessibility of the community (Fig. 1).

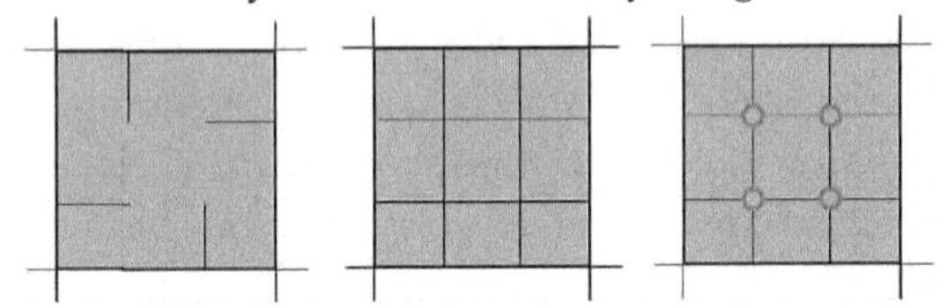

Fig. 1 The "geographic center" within the community

At the same time, through the connection and integration of the disordered space, the connection between streets and "centers" ensures the continuity and integrity of space, and realizes the perfect connection with the existing pattern of the ancient city.

(2) Find the right "breakpoint"

In the current situation of chaotic and crowded material space, some space should be properly dredged to release effective space as a public space. In the process of selecting these points that can be broken, the destruction of the original fabric and architectural form of the ancient city should be avoided.

In the evaluation of the courtyard, the protective courtyard was excluded first. Because the protective courtyard has great value in both cultural relic value and maintaining the spatial texture pattern of the ancient city, it is an unremovable part. In the unprotected courtyard, the dilapidated degree of the courtyard is graded to determine the urgency of demolition. The courtyard that has been transformed or will be transformed after the transfer of property rights can be ignored, mainly aiming at the dangerous houses with security risks and self-built houses that destroy the fabric of the ancient city. The area after demolition is evaluated to determine the suitability of public space shaping. All these works are to find available space, so as to give public attributes and become available public space later.

(3) Superposition analysis, select "micro center"

By overlapping the "geographical center" and suitable "breaking point", the location range of "micro center" is further defined. In general, the location of the "micro center" should meet three conditions:

1) The "center" within the community with the best transportation accessibility.

2) Courtyard space with demolition possibility and urgency.

3) Site environment and space of appropriate scale.

3.3.2 Generation of "Micro Center" Network

The role of "micro center" is not only to create a simple space, but also to intervene in every aspect of residents' daily life. In this sense, the "micro center" transcends the category of physical space and overlaps with functional space and cultural space. Each "micro center" performs its own functions and satisfies residents' needs in three aspects: daily life needs, daily communication needs and spiritual and cultural needs. The mutual complement in function and the connection in space make each "micro center" form a complete "micro center" network system. It can meet the needs of residents' daily life and spiritual communication within the scope of the community, making the community become a dynamic organism. At the same time, residents can enhance the sense of belonging of the community through the construction of "micro center". In the later governance and operation stage of "micro center", residents' autonomy and public participation are also indispensable links.

4 Design Strategy of "Micro Center" of Pingyao Academy Community

The Shuyuan community is located at the southwest corner of the ancient city, near the south gate. The Shuyuan community is surrounded by the east of the South Street and the north of the Yamen Street two ancient city core business streets. It's west and south is also the ancient city core streets Zhaobi South Street and Southwest State Street. The number of courtyards in the Shuyuan community is 142. The courtyards are basically in a regular rectangle, and the volume and boundary of the courtyards along the street are damaged greatly. The tension of homestead land leads to a patchwork of courtyards with few gaps. At the same time, there is a lack of large public buildings in the community (Fig. 2).

Fig. 2 A bird 's-eye view of the Shuyuan community

The introverted spatial organization pattern leads to a serious shortage of public

space. Communication usually occurs only between several residents in the same courtyard, and even negative communication occurs because of internal conflicts in some courtyards. In addition, the traditional urban settlements in the north lack of green space, and the overhanging eaves of the courtyard walls are insufficient, leading to the easiest place for communication — streets and alleys, which have become gray spaces with only passageways. During the investigation of the academy block, it is rare to see residents in outdoor activities, and only in some streets and lanes, you can see the elderly sit in front of their own homes to rest. According to the interviews of residents, elderly people have difficulty in moving around, so their daily communication activities mainly take place in the house, and they will chat with neighbors when they come to visit. However, the communication activities between middle-aged people and children mainly occurred in parks, vegetable markets, schools and other places outside the ancient city, and the communication space within the community was extremely lacking.

Due to the relocation of a large number of public service facilities, the Shuyuan community is only equipped with some small commercial services, such as restaurants and supermarkets, but the goods are mainly sold to tourists, and some daily necessities can only be purchased outside the ancient city. However, most residents of the Shuyuan community are vulnerable groups in society, and there is a serious shortage of public service facilities for these groups, such as small medical and health service facilities, elderly service facilities, migrant workers' reemployment training, community service facilities, cultural and entertainment facilities, children's education and entertainment facilities, etc. The imbalance of the structure and quality of these service facilities may lead to the contradiction between supply and demand of public services in the Shuyuan community, thus reducing the quality of life of residents.

4.1 Subdivision — Find the Center

According to the specific spatial characteristics and social life characteristics, the whole Shuyuan community is divided into residential units composed of streets, courtyards and buildings. This kind of internal small unit is not only the division of space, but also has the social-spatial attribute. Each small unit is first divided on the basis of the site texture, as far as possible to the existing internal branch of the block as the boundary. At the same time, the property rights boundary of the courtyard is considered to form a complete unit. The size of such small units should not be too large or too small, and should be composed of 15 to 20 yards.

The purpose of the block division is to define the location, radiation range and function of the center. The "center" public space is formed at the boundary of the unit and the junction of lanes, which can serve the surrounding residents to the greatest extent and clarify the governance boundary of each small residential unit. Meanwhile, the "center" connected by streets and lanes can form a complete network, and can effectively improve the accessibility and exert greater use value (Fig. 3).

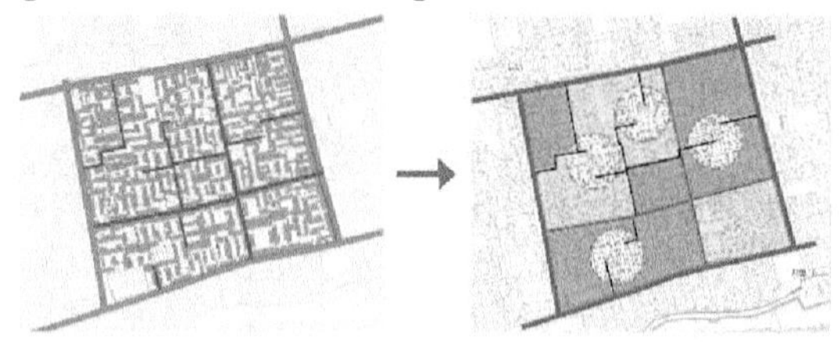

Fig. 3 Block division and center generation schematic diagram

4.2 "Breaking Point" Analysis — Subtraction

What is found above through unit division and street lane series is only the vague "center", and the release of space still needs to correspond to the "points". Through the urgency of the demolition of the internal courtyard and the prediction of the effective usable area after demolition, the "point" that can truly realize the demolition can be found within the scope of "center". And these " point " usually take the form of courtyards. In the protection planning of the ancient city, the grade of the protected courtyard has been divided, while the investigation of the non-protected courtyard is less. According to the above evaluation on the indicators of unprotected courtyards, the urgency level of the demolition of such courtyards can be obtained. The demolition of the courtyard with the highest urgency means that it has some building safety risks or affects the appearance and texture of the ancient city, so it can be dismantled or transformed to realize the evacuation of internal space and the conversion of functions. In calculation within the scope of "center" can be removed when the area of the "point", too big and too small are not suitable as a "center", too big means that a certain extent damage to status quo of the skin texture, and do not conform to the "micro center" small and practical orientation and too small is not conducive to implant.

Further filtering by superimposing can eventually clarify the scope of the "micro centers" (Fig. 4).

Fig. 4 Unprotected courtyard, grading of unprotected courtyard, stacking center range, finding "breaking point"

4.3 Function Implantation — Addition

The functions of these "micro centers" are configured according to the needs of residents. To understand the real needs of college community residents through questionnaires and interviews. Among the cultural and recreational facilities, residents choose chess and card rooms, followed by children's reading rooms, which is in line with the current situation that children in the community have no place to go after school, while the living facilities are mainly grocery stores and some medical and health services. Among the outdoor facilities, seats are the most needed facilities for residents at present, which convey the residents' desire to chat and rest outdoors. Secondly, small gardens and fitness facilities are also what residents want to have (Tab. 1).

Tab. 1 List of residents' needs of the Shuyuan community

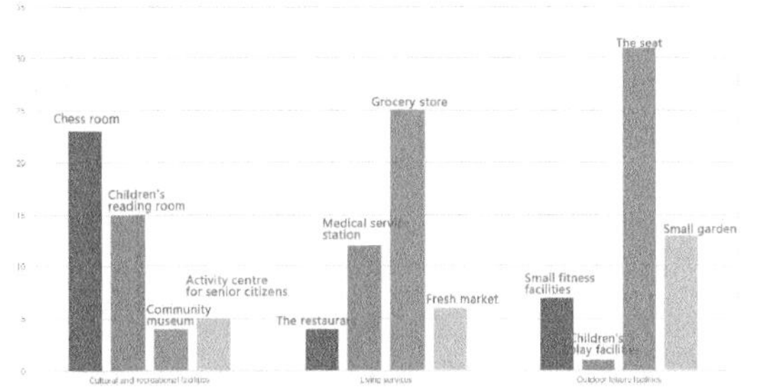

According to the selected site environment of the "micro center", some of the facilities and functions most needed by the residents will be implanted. For example, the "micro center" located in the northwest, part of the buildings can still be used through renovation, suitable for the development of indoor activities. Combined with the cultural palace in the west, a "micro center" focusing on culture and entertainment can be formed. However, the "micro center" houses in the south end are seriously broken and in poor environmental conditions, so they can be completely dismantled. After demolition, the large area can be made into outdoor green space and equipped with small fitness facilities and children's play facilities (Fig. 5).

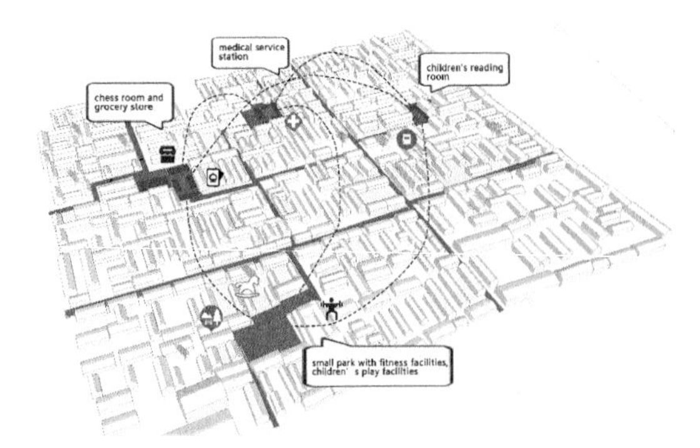

Fig. 5 The "micro center" network system of the Shuyuan community

5 Conclusion

Community "micro center" means to activate the whole community by finding reasonable intervention points without destroying the ancient city's texture. Through shaping the public space, it can release the space pressure within the community, create more opportunities for residents to meet and communicate, and alleviate the spiritual contradiction. Rational use of "micro center", strengthen the structure and quality of "micro center" network, to achieve the organic and healthy development of the ancient city community and enhance the attraction of the ancient city tourism are of great significance.

References

HU Ying, ZHANG Lin, 2003. The image extension of traditional street space. Planners, 19(6): 36-39.

MA Hong, YING Konglin, 2016. Micro-regeneration of community public space: exploring approaches to community building in the context of organic urban regeneration in Shanghai. Time architecture, (4): 10-17.

YANG Zhiqiang, 2011. Application of communication space theory in ancient city tourism development. Business culture (the first half of the year), (10): 207.

Opening Strategies of Street and Lane Space in Historic Urban Area — a Case Study of Shuyuan block in the Ancient City of Pingyao

ZHANG Yang, *ZHANG Beibei*

Huazhong University of Science and Technology, Hongshan District, Wuhan, Hubei Province, China

Abstract The function of street space in historic urban area has changed from inward economy and traffic to external economy and activities. In this process, dynamic game of street space openness has been produced: space adherence under famous city protection system and space transformation under tourism development mechanism. In this regard, it is proposed that the characteristics of the ancient city should be maintained in general principle, the style should be continuous and the history should be readable, and the specific operation level should be put forward to "break" before "build": "break" means precisely locating the overly closed space between the external streets and the internal roadways, and realizing linkage development through the series of streets and courtyards, balancing the openness of the street space. Taking academy area of the Ancient City of Pingyao as an example, this paper emphasizes to find a breakthrough between the "break" and "build" of street space, to form a communication and activity space to meet the living needs of internal residents, and to find the broken wall of the courtyard to form an experiential tourism space to meet the needs of external tourists. Through the study of the open strategy of street space, it can provide a reference for the spatial activation and organization of historical urban areas.

Keywords historic city, street space, opening strategy, Ancient City of Pingyao, Shuyuan Block

Foreword

The street space in the historical city is an important part of the urban texture and the material relic of the city's historical context, and it is an important content for people to understand and memorize the history and context of the famous historical and cultural city. The research on the street space in the historical city is mainly focused on the historical pattern of the street (Que weimin, Ren jiang, 2011), the vitality of space (Zhang Yuyang, Yang Changming, Qi ling, 2019), the spatial characteristics (Ji Xuehua, Zhuang Jianwei, 2014), tourism development (Xu Xiaobo, Wu Bihu, Liu Binyi, etc., 2016), interface form (Zhou Yu, Geng Xuchu, Gan Wei, 2018), urban transportation (Gu Wei, Liang Zheng, Sun Qing, etc., 2019) and so on. The open research of street space mainly focuses on modern urban neighborhoods, such as studying the relationship between street openness and premium, evaluating the open space of urban streets, using the method of multi-linear regression, which named Hedonic, pointing out that the value of open space in urban street is significant, most of the street open space variables are positively correlated with house prices (Chen Long, Lin Shiping, 2019). In order to quantitatively evaluate the actual utilization efficiency of open space in urban street and alley, introduce the concept of the use value of open space utilization and the value of open space utilization, using the index factor to rate the evaluation to obtain the actual use value of open space (Chen Long, Lin Shiping, 2019). In the background of the strong impact of modern urbanization in the famous city, this paper analyzes the traditional street pattern and the protection strategy of its heritage value, constructs the evaluation system of the street vitality of the historic district, updates the development direction for the protection of the historical district, puts forward the strategy of promoting the traditional street features, and integrates with the culture, industry and tourism organically.

1 Functional Change of Street Space in Historic Urban Areas

1.1 Commercial Street: from Internal Economy to Extroverted Economy

In the process of modernization and urbanization, the functions of commercial streets in historic urban areas have changed. At first, commercial streets mainly carry commercial transactions within the city. Many ancient cities will have ancient markets. These markets mainly satisfy the daily life of the crowd that living there. For example, the famous painting "Riverside Scene at Qingming Festival", which created by depicts by Zhang Zeduan, has painted the landscape of Dongjing, the capital of Northern Song Dynasty. Urban streets are distributed with hotels, inns, pharmacies, clinics, meat shops, clothing and other business forms. These commercial formats mainly distribute along streets and waterways, forming a planned market, which is conducive to commercial trade and goods distribution. There are two kinds of markets, one kind of them is open every day, regularly, being the main trading place; another is open in a specific period, rarely forming a fixed business. With the development of modern cities, the internal economy of the ancient city has turned to the external economy, especially from the perspective of tourism development, the commercial streets and alleys in the historic urban areas are increased, catering to the needs of tourists. There are many external economic formats, such as local specialty management, local cultural exhibitions, residential hotels and so on. At the same time, with the further expansion of the external economy, the original internal economy and format began to decline. How to meet the needs of tourists to the greatest extent and promote further consumption of tourists in the ancient city has become the focus of local governments and businesses.

1.2 Life Laneway: from Traffic Space to Activity Space

In historic cities, Life streets are basically designed to meet the internal traffic demand. In ancient China, there was no concept of public space, people's daily life centered on their houses, such as the atrium in the houses, was created to meet the needs of internal activities. People's living conditions were relatively private, and most of the streets in the ancient cities were to meet the transit traffic and basic living needs. With the change of life mode and the improvement of life quality, people's demand for communication space and public space is increasing day by day. Breaking down the closed courtyard walls and creating open places have become the pursuit of a better life for the residents. In ancient cities, the street better carry the needs of the crowd, usually, some tables and chairs were placed along the roadways to meet the needs of the daily communication of the crowd, there may be one or two more open spaces at the intersection and corner of the street, and they become an important place for people to chat, entertain, enjoy cool and bask in the sun. With the gradual improvement of urban modernization, the street in ancient cities will become more and more places for communication and activities.

2 Game of Openness of Street and Lane Space

2.1 Space Firmness Under the Protection System of Famous Cities

The protection of street space is an important link in the protection system of historic and cultural cities, as well as an important aspect linking the protection of cultural relics and the overall protection. For the system of streets in ancient cities, the current protection mode in China is circle protection, that is, taking the protected urban streets as the core, taking different distances as radius, dividing into the core protection areas, construction control areas and environmental coordination, strictly controlling the style and texture of streets in the core protection areas. In those areas, any development and destruction contrary to the protection of streets is limited, ensuring that the newly developed space and elements will not destroy the historical elements, the approval of new projects should be carried out strictly. In the environmental coordination zone, the coordination of the overall color, volume and surrounding environment of streets should be considered. In the protection and control of historic sites, historic and cultural protected areas and historic and cultural blocks under the background of famous city protection, urban space is protected to varying degrees.

2.2 Spatial Transformation Under Tourism Development Mechanism

Around 1990, China began a large-scale transformation in the old city. The development requirements of the old city triggered a series of development mechanisms. Tourism development is one of the most important aspects. Through tourism development, some residential houses turned into residential quarters, leisure places for citizens turned into distribution squares, and some historical relics developed into tourist attractions. With the development mechanism of city management and so on, the tourism development in the ancient city is increasing, the scale of space transformation is also increasing, and the restrictive role of historical and cultural cities on tourism development is gradually ineffective. The pressure of development makes tourism become the main driving force of economic growth and regional competition, thus bringing about "constructive" destruction and "protective" to historical and cultural cities. The destruction is becoming more and more obvious. On the one hand, the space transformation under the value of tourism development promotes the vitality of the city and the development of the city. On the other hand, space transformation destroys the style of the city, cuts off the texture of the city and weakens the historical and cultural value of the ancient city.

2.3 Current Situation and Problems of Openness of Street and Lane Space

The streets in the ancient city are good carriers to show the city's features and residents'living conditions. At the same time, the streets are also good windows for external tourists to understand the historical and cultural values of the city. It is very important to balance the opening degree of the street space in the background of the historical city, which is "clinging" in the protection policy and tends to "transform" under the development mechanism. One is based on the promotion of urban development, the other is based on the deep concern for traditional culture, the other is based on the needs of reality and the other is the needs of history. Too closed is not conducive to the display of street space, while reducing the traffic accessibility and connectivity within the streets, affecting the tourism experience and life quality; too open will lead to the reduction of the privacy of the internal residents, tourists behavior will produce a certain degree of interference to the daily life of residents, to some extent. From this, we can see that the connotation protection of tourism and ancient city is a pair of contradictions. Tourism development can be an opportunity for the revival of local traditional cultural undertakings, but in this process, we must be orderly and moderate, and pay attention to the research and excavation of historical and humanistic values besides traditional buildings and cultural relics. Protect and maximize social, economic and environmental benefits.

3 Balance of Openness of Street Space

3.1 General Principle: Continuous Style and Readable History

The balance of the opening of the street space needs to be carried out under the general principle of "continuous and overall readable style". The so-called "style continuous", which aims to continue the texture, height, volume, scale and form of the street space, the city style can be preserved in the "original" form, and to a certain extent to promote the organic combination of old and new elements, historical space and modern space linkage development. However, the introduction of modern elements into the historical space has certain institutional constraints and technical requirements, "*Nairobi Recommendation*" has clearly stated that "to ensure that monuments and historical areas of the scenery will not be destroyed", therefore, it is important to protect the location of the historical features and value of the continuation in historical city. Under the condition of continuous style, the historical value and cultural significance in urban street space can be displayed in an orderly manner.

3. 2 Specific Methods: Break First and Then Rebuild, Linkage Development

The opening of the street space can adopt the technical method of "break first and then rebuild, linkage development", the basic condition of opening is that the street space is too closed, so it is necessary to break some closed space in the present situation, first of all, it is necessary to "break the barrier" in the street space, which emphasizes the space that can be broken in the street space, promoting linear, coherence spaces. Secondly, "linkage development" it is necessary, "linkage development" means that the modern elements and functions of implantation can be harmoniously symbiotic with historical elements and functions, and the emerging elements can be used as catalyst nodes to guide the activation and development of the street space.

3. 2. 1 "Breaking": Precise Positioning, Breaking the Closure

In the space strategy of "break first and then establish", it is necessary to pinpoint the "broken" node. The interior of the streets mainly functions for the residents living here, which are narrower and the street closure is high. The gables of the building face the streets here, so there is a need for transitional space as the separation of inside and outside. In the internal streets, there are two steps in finding the "broken point", first of all, this transitional gray space should be the one that its courtyard wall is near the street, rather than gables near the street, so as to avoid the renovation process of the ancient city complex damage. The original courtyard space is still the courtyard space, but need to change a form, add some vitality; second, this transitional grey space should be relatively independent of the lives of residents, it is adjacent to the living space of residents, but it will not cause too much disruption to their lives. External streets carry the function of major traffic, at the same time for tourism and residents. These streets are wide, carrying a lot of shops but weak in permeability, directly resulting in the reduction of connectivity in internal streets and external street. Along the street, there are two steps to find a "breakthrough" in the outer streets, first of all, it is necessary to coordinate the linear relationship between the internal breaking point and the external street, to ensure that the external "breakthrough" and the internal "broken wall point" can have a certain connection; Secondly, consider the external "breakthrough" as the expression of open space, attract people to stop and carry out in-depth experience.

3. 2. 2 "Rebuild": Development of Street, Lane and Courtyard in Series

"Breaking" mainly emphasizes the penetration of street space while"rebuild" emphasizes the weaving of street space, the external "breakthrough" and the internal "breaking point" should be linked in what way, and it functions for which kind of people. these issues need to be considered in the process of "rebuild" in the street space. The reorganization of the streets and alleys is not only the structure and function adjustment within the streets, but the coordination and cooperation of the two levels of the alleys and courtyard, which enables the internal and external space to develop in a coherent way-to experience the historical features in the streets and the living state of the residents in the courtyard.

4 Study on Openness of Street and Lane Space in Academy Area

The Ancient City of Pingyao is an outstanding example of Chinese Han nationality city in Ming and Qing Dynasties. The ancient city takes the commercial building as the centers. The street system is embodied in four main streets、eight secondary roads and seventy-two grasshopper lanes. The street system presents the Eight Diagrams pattern and the overall shape seems like the Chinese character "土" (Fig. 1).

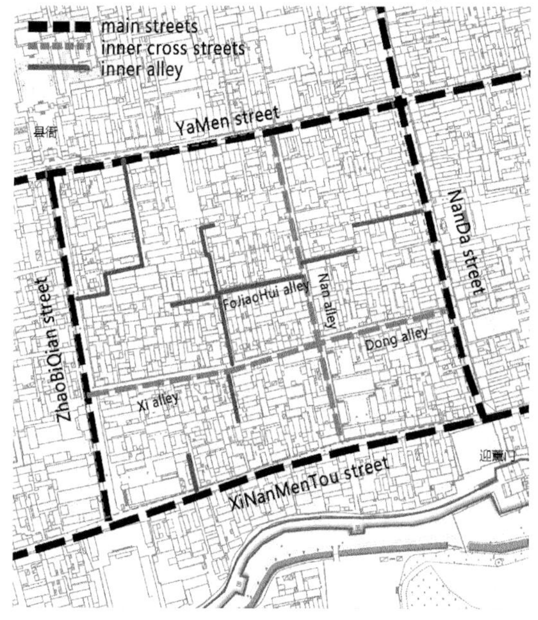

Fig. 1 Street and lane system in the Academy area

Shuyuan block is an important area of Pingyao ancient city, which streets can be divided into three types according to their spatial forms: outer streets, inner cross streets and inner alley. Yamen Street and Nanjie Street in the outer Street are crowded with many kinds of commerce, while Zhaobi Qianjie and Nanmentou Street have fewer types of business, and their life style is stronger than that of Yamen Street and Nanjie Street. The South Street and east-west lane is an important traffic lane in the interior space which run through the whole area, while the others are life-style lane.

4.1 "Broken" of Street Space

According to the "broken" space principle, six relatively qualified spaces are finally found in the Academy area, as shown in Fig. 2. Among them, two spaces A and B conform to the first space principle of "broken". Courtyard A is located in the Fojiaohui Street. There are two courtyards inside and outside the courtyard. Opening the courtyard close to the streets and alleys does not affect the integrity of the inner courtyard and does not interfere with the daily life of residents. Courtyard B is located at the intersection of South Lane and Buddhist Guild Lane, and is an abandoned courtyard, which has no one to live in for a long time. Open up as an open space. By sorting out the street space in the Academy area, it is found that except the four main streets outside and the cross streets inside, the other street system has low integrity and poor connectivity. Living streets need certain connectivity. Therefore, breaking through the four courtyards (courtyard A, courtyard B, courtyard C, courtyard D) in the current space system can strengthen the connection of the internal streets and lanes, and form a new group of internal cross streets (Fig. 3). Through the reorganization of courtyard space and street space, the connectivity of the area is strengthened and a double-cross pattern is formed.

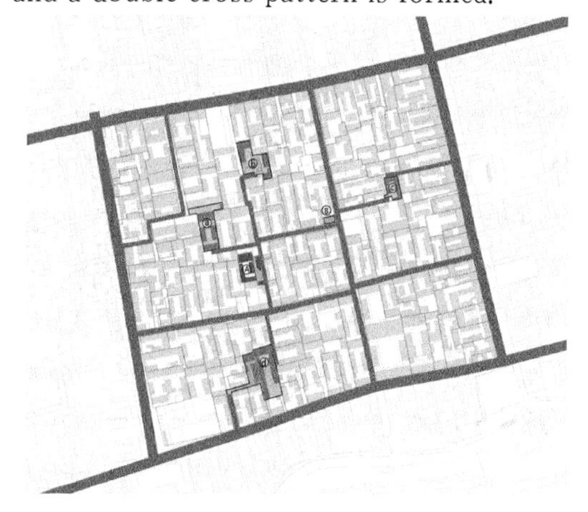

Fig. 2 The broken courtyard in the Shuyuan block

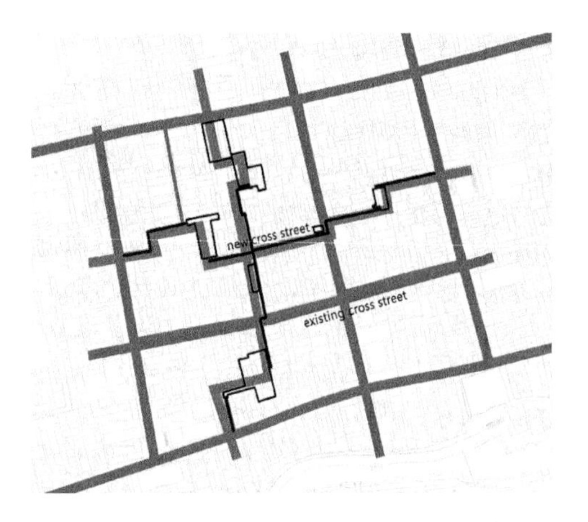

Fig. 3 Reconstruction of street and lane space in Shuyuan block

4.2 "Rebuild" of Street and Lane Space

4.2.1 Exchange and Activity Space (Mainly to Meet the Needs of Internal Residents)

The main activity space of local residents basically distributes along the road, there is no fixed activity space and lack of a certain sense of place. At the same time, the people along the streets will interfere with the traffic, which is not conducive to traffic guidance. Therefore, it is urgent to improve the status of the original lack of activity space. The establishment of New Cross Street provides space for residents' daily communication. The New Cross Street from east to west has three small places in series. According to specific needs, the three courtyard spaces can be built into recreation and communication space, and tables and chairs, fitness equipment and pocket parks can be arranged appropriately to meet the daily needs of the crowd(Fig. 4).

(a) Corner Park

(b) Ekin Cafe

(c) Community Library

(d) Community Exhibition Hall

Fig. 4 Slenes of the New Cross Street

4.2.2 Experiential Tourism Space (Mainly to Meet the Deep Experience of External Tourists)

New cross street from north to south can give new functions to open courtyards according to actual needs. deep-going tourism is the main strategy of current tourism construction. How to guide tourists to participate in and experience tourism in an orderly manner is the focus of tourism development in famous cities. Implanting the intangible cultural heritage of Pingyao ancient city in the academy area, the traditional production techniques of camellia oleifera, forming a small food catering, marketing, production skills display, participatory experience and other ways to experience the intangible cultural heritage in the ancient city of Pingyao, formating the the system marketing and cultural communication of selling oil tea, which inclueds drinking oil tea, watching the production process of oil tea, participating the production experience of oil tea and selling the oil tea (Fig. 5).

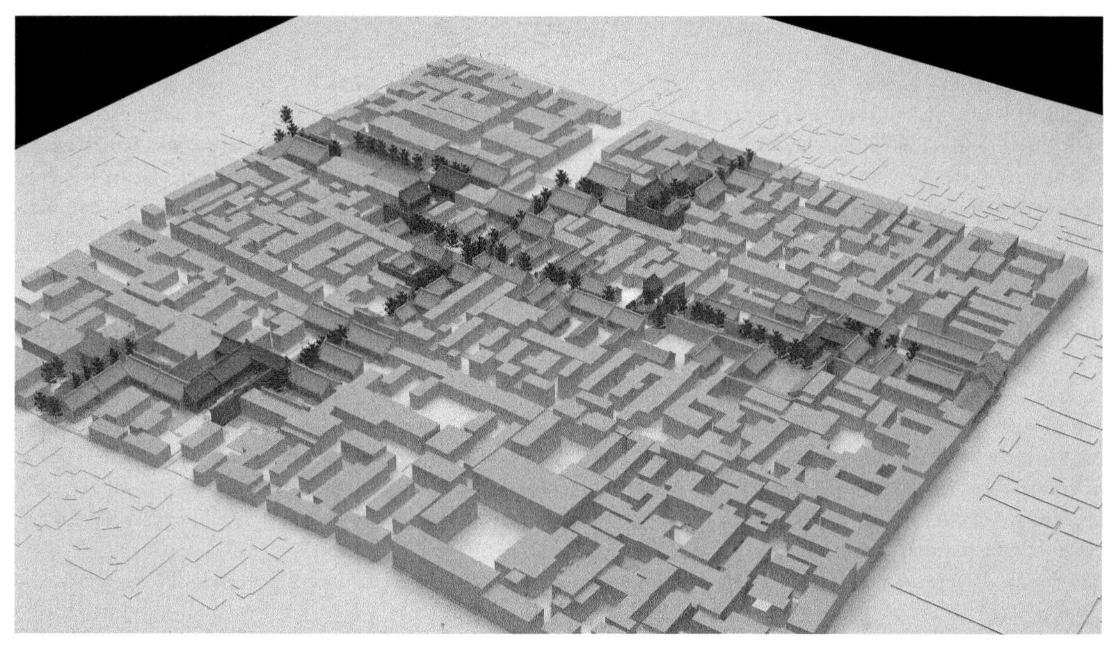

Fig. 5 Aerial view of shuyuan block

5 Conclusion

The street space of historic cities has undergone the changes of times, and its carrying function has also changed. How to coordinate the protection of historical and cultural heritage with the improvement of urban economic development, living standards and environment needs to be weighed. It is very important for street space to grasp the open "degree".

References

JI Xuchua, ZHUANG Jianwei, 2014. Upgrading the characteristics of streets and revealing the connotation of water city: theintegration of traditional streets and local culture in Suzhou ancient city. City planning review, 38(05): 46-49.

QUE Weimin, REN Jiang, 2011. Street pattern of Taiyuan old city from world heritage perspective. City planning review, 35(06): 91-96.

XU Xiaobo, WU Bihu, LIU Binyi, et al., 2019. The development of commercial spaces in tourist historic districts: the case of Shuangdong, Yangzhou. Acta geographica sinica, 35(03): 106-111.

ZHANG Yuyang, YANG Changming, QI Ling, 2019. Study on the assessment of street vitality and influencing factors in the historic district-a case study of Shichahai historic district. Chinese landscape architecture, 35(03): 106-111.

ZHOU Yu, GENG Xuchu, GAN Wei. Investigation on historical evolution of street interface in European cities: from ancient Greek period to 20st century. Architectural journal, (S1): 168-173.

Research on Visual Perception of Historical and Cultural Landscape in Pingyao Ancient City

YUAN Muxi, *GAO Jing*, *WANG Jia*

Taiyuan University of Technology, Taiyuan, Shanxi, China

Abstract How to balance the protection and development of Pingyao Ancient City, how to achieve the continuation of traditional culture and the continuation of ancient city characteristics in the process of landscape activation and renewal has gradually become the focus of research in the country and even the world. However, in recent years, the contradiction between protection and development has led to a negative correlation between the development degree of the ancient city and its attractiveness. In the process of landscape renewal and protection, we need to rethink what is the complete landscape and what kind of landscape features are the most distinctive and attractive. A complete landscape consists of the main body of the landscape and the object, and space is used by people and is dynamic and vital. Therefore, in the reasonable repair and protection of buildings or spaces, we should pay more attention to the main body of the landscape — people's participation and feelings.

This paper attempts to extract the visual elements unique to Pingyao Ancient City by analyzing people's visual perception feedback on the ancient city landscape and using related theories such as color geography and color psychology. At the same time, the ideas and methods for improving the landscape protection and renewal of Pingyao ancient city are proposed in the visual angle, which provides reference for improving the rigidity and personality of the ancient city landscape, protecting the characteristics of the ancient city and improving the integrity of the ancient city landscape.

Keywords landscape perception, landscape integrity, color geography, landscape features, Pingyao Ancient City

1 Research Topics and Concepts

1.1 Overview and Reflection of the Development of Pingyao Ancient City

With the rapid progress of social economy and tourism in recent years, the development of Pingyao Ancient City seems to have encountered some kind of bottleneck. Modern business brings economic benefits and has a certain negative impact on the functional use of the ancient city space and the original cultural atmosphere. How to achieve the balance between protection and development, how to find a breakthrough and then use scientific methods to preserve and continue the most authentic cultural features and landscapes of Pingyao Ancient City is a common concern of the whole country and the world.

Through preliminary research, it can be found that the main tourist groups attracted by ancient city are culturally driven tourists. Cultural tourism is a collection of tourism activities with rich cultural connotations and deep participation in tourism experience. For Pingyao Ancient City, deep participation is a full-scale experience in which tourists can experience the slow-lifestyle of the ancient city. This kind of experience is not only the main tourist motive of Pingyao ancient city tourists but also the essential reason for the formation of the ancient city attraction. Therefore, in order to seek a more scientific "participation and experience" model of Pingyao Ancient City, research on sensory perception will become one of the most important research approaches. This paper attempts to analyze the most impactful landscape scenes in the ancient city from the perspective of visual perception and analyze the visual perception elements of the ancient city. At the same time, this method can also be used to find the characteristic features that truly belong to Pingyao Ancient City, and provide a reference for further protection and development in the future.

2 Research Methods

2.1 Visual Information Re-extraction

2.1.1 Landscape Visual Perception

People's perception of the surrounding environment is very complicated and diverse. When entering a new environment or new space, people began to use five senses to receive information from the new environment. Eventually, a comprehensive and unique sensory model to the environment will be obtained, and this sense will make people judge the environment through complex and precise information transmission and processing in the human brain to form a variety of perceptions. Although different basic information such as people's growth environment, educational background, age and gender will make them feel different in the same environment. However, because human thinking patterns have certain similarities, there is a certain commonality behind this difference in perception. Because in the process of human evolution, humans need to use the visual observation of the enemy, find food, and seek to avoid disadvantages. Therefore, unlike other animals, human beings mainly use vision to perceive the world in five sensory ways. 85% of people's perception of the environment comes from vision (Weng, 2007).

The direct senses of human beings are naturally produced in the brain, so it is difficult to directly ask people about their perceptions. However, we can get some types of understanding and perception by studying people's preferences for different landscapes. Preference is the result and product of perception (Miller, 2013). Therefore, in order to understand the human visual perception patterns and processes of the environment, it is necessary to conduct a large number of satisfaction and sensory information research, and to re-extract and summarize the results. We will explore the influence of different visual landscape elements on people's perception through the understanding and analysis of people's psychological feelings.

2.1.2 Strong Feeling Photo

People observe the world through their eyes, accompanied by the intervention of the other four senses to form impressions and perceptions of the environment, landscape or other spaces. Human perception and memory do not come from a single building or an element, but from a comprehensive scene. The colors you see, the sounds you hear, the stone roads you walk through, the smell of the smell, and even the weather at the time all together create a unique perception of the moment. Although the photo can't completely engrave a person's complex perception mode, it is still the most intuitive and easy-to-use recording method. It condenses the photographer's multi-dimensional sensory experience into a picture in a two-dimensional way. The scenes in the photo are inaudible and invisible, but they can be used as a valuable clue to evoke the unique memories and perceptions of the photographer in the scene even after many years.

2.2 Research Process

2.2.1 Site Survey

The research team consists of 9 members and is divided into 3 groups. The team members visited several times in the limited research area of Pingyao Ancient City. The core area enclosed by West Street, South Street, Yamen Street and Shaxiang Street in Pingyao Ancient City is the main research area of this paper.

The research process is mainly divided into two parts. Firstly, in the initial investigation, the researchers mainly focused on the objective observation and feeling of the landscape. Each member took 10 pictures of Strong Feeling and recorded the specific location and psychological feeling of the shooting. The motivation for shooting may be due to a simple visual impact, perhaps by seeing a novel local folk activity, or perhaps an old man passing by riding a tricycle, so that everything the photographer touched can be recorded by the camera. However, for visual impact analysis, there is a controversy about the subjectivity of decision-making and the exaggeration of environmental beauty (Keelan, 2002). In order to avoid the excessive influence of subjective factors of specific people, the survey also collected 30 photos taken by ordinary tourists in Pingyao Ancient City to understand the subjective impressions and feelings of most people in the ancient city.

The second part mainly includes research on the perception of tourists, businesses and residents in the ancient city. In this part of the survey process, the team members collected information on the visual and psychological perceptions of the landscape and space environment by different groups in the ancient city through questionnaires and interviews.

2. 2. 2 Perceptual Information Re-extraction

One of the most critical aspects of this research is to re-extract and summarize the feedback results of the respondents collected in the survey, so as to further explore the landscape elements and structure of the most attractive features of the ancient city. For this survey, the most important feedback is hidden in the Strong Feeling photos provided by people.

After the survey, 120 pieces of Strong Feeling images collected by the researchers and other perceptual information provided by the residents in the ancient city were summarized and summarized. On the one hand, from the results of questionnaires and interviews, the satisfaction of tourists, merchants and residents in the ancient city on the development status of the ancient city is extracted as the basic data of the research; on the other hand, the photos of Strong Feeling are sorted out, and the overlapping items are summarized as the most representative image basis. Through the induction of landscape visual elements in photos, a variety of visual laws and impact elements can be extracted, such as landscape color, spatial scale and visual focus, landscape element texture and freshness, architectural form, cultural symbols and so on. This paper mainly explores the characteristics of the ancient city from the first three aspects (Fig. 1).

Fig. 1 Seven sets of strong feeling photos and the overlapping items

3 Analysis of Research Results

3. 1 Psychological Color and Landscape Chromatography

3. 1. 1 Ancient City Psychological Color Extraction

Tab. 1 Ancient city color statistics

Ancient City Color	Proportion (%)	Keywords
Gray	41%	Gray Brick (27%) Roof (19%) Tile (17%) Road brick (17%) Wall (11%) Cloudy (6%) Atmosphere (3%)
Red	29%	Lantern (63%) Architectural decoration (26%) Flag (8%) Couplet (3%)
Cyan-blue	12%	Tile (35%) Roof (24%) Road Brick (21%) Color of the building (16%) Cloudy (2%) Sky (1%)

Continued

Ancient City Color	Proportion (%)	Keywords
Black	9%	Roof (38%) Tile (33%) Color of the building (29%)
Yellow	6%	City Wall (62%) Loess Plateau (15%) Regional characteristics (13%)
Green	2%	Unpalatable food (80%) Vitality (20%)
Gold	1%	Historical factor (100%)

Tab. 2 Ancient city color selection tendency and perception factors

	Perception factors	Keywords
Visual perception	1. Scene color contrast	Decoration, lanterns, flags, couplets, roof
	2. Similar color repeats	Gray brick, wall, roof, tile, road brick, city wall, overall color of building, cloudy, sky
Psychological perception	Other senses participate in a comprehensive feeling	Cloudy, atmosphere, unpalatable food, vitality
	Differences in memories between different individuals	Loess Plateau, regional characteristics, historical factors

In the process of human beings perceiving the environment through vision, the brain first receives color information. Psychology-related research indicates that color and human psychology are inextricably linked. Human sensitivity to color is largely due to the close relationship between human and nature. For example, the blue color similar to the sky will give people a feeling of vastness and calmness; the red and yellow of the flame will make people feel warm and energetic, and the similar primitive external stimuli will make the human feelings about color similar and regular.

Visual perception is an aesthetic experience that combines the assessment and evaluation of different stakeholders (Hull&Revell, 1989). This aesthetic response is defined as a preference that is related to positive and negative emotions (Ulrich, 1986), and color can not only objectively stimulate human feelings, but also be a medium and means of reacting and conveying emotional feelings. In the course of interviews and questionnaires for different groups of people in the ancient city, the researchers will guide each respondent to choose a color that represents their impression of Pingyao Ancient City. Then, through the collection and statistics of the representative color of Pingyao ancient city in people's minds, people can summarize the satisfaction of the ancient city and its visual perception focus on the basic level (Tab. 1).

Because of the individual differences in age, gender, education, personality, etc. among the respondents, the color results collected by the survey showed two different selection tendencies: psychological perception and visual perception (Tab. 2). When there are contrasting color elements in the scene or there are consecutive repeated color elements in the scene, the viewer will often have a unique feeling for the current space and environment due to strong visual impact and continuous color input in a short time.

3.1.2 Ancient City Chromatography

Through basic research, we can initially understand that the overall impression of Pingyao Ancient City is generally derived from the visual perception of the overall shape, overall and individual colors of the building. These perceptions often come from unique architectural materials, architectural decorations of specific colors and shades. In addition, the important color in the ancient city can be extracted into a chromatogram and the composition pattern of the ancient city color can be proposed, thereby providing a basis for protecting the characteristic features of the ancient city landscape.

Through the collected strong feeling pictures, it can be found that the color structure of the northern and eastern commercial street areas in the study area is similar to that of the western residential areas, but there are also some differences. In both of the above scenarios, it can be found that bright color elements with high color purity and brightness are in sharp contrast with a large proportion of gray color elements. The visual impact of this color contrast is a key factor for the viewer to form a unique perceptual memory for the spatial scene. The two types of colors can be summarized into the main color and the embellished color by the proportion of the color in the scene.

Among them, there are many kinds of embellishment colors in the commercial street area, which are mainly composed of the colors of architectural decorations such as lanterns, plaques and flags, and the detailed colors of the iconic ancient buildings. The overall proportion is about 5% to 25%, and the atmosphere is more active and noisy (Fig. 2). It is worth noting that the shops in some sections of the commercial street area are too dense, the business forms are disorderly, and a large number of disordered embellishments are used in the decoration. At the same time, due to the visual confusion, the overall perception of the viewer is poor and the quality of experience is degraded. We can control the proportion of the embellished color to 15% ~ 20% to create an active atmosphere.

On the contrary, in the western residential areas, the proportion of embellishment color is smaller and the type is relatively simple. It is mainly composed of elements such as couplets and lanterns, accounting for 3% to 15% (Fig. 3). This part of the space environment is relatively quiet and the visual landscape elements are more Unite. The space environment in this area is simple and harmonious, largely retaining the original characteristic landscape style. The appearance of small proportions of embellished colors also subtly balances the singleness of other visual elements. The overall feeling is very harmonious and this visual mode can continue to be used in development and protection in the future.

	Main color (75%-95%)				Embellished color (5%-25%)					
R:	47	136	167	246	180	167	49	242	88	147
G:	53	130	153	228	30	109	53	228	116	175
B:	63	125	134	214	8	100	52	180	113	89

Fig. 2 Northern and eastern commercial street area chromatography

	Main color (85%-97%)					Embellished color (3%-15%)			
R:	145	142	158	179	47	193	188	49	147
G:	117	138	142	159	53	57	137	53	175
B:	102	128	129	126	63	64	116	52	89

Fig. 3 Western residential area chromatography

Through the description of the photo shooter and the relevant feedback information of the interview research, it can be found that too many types and proportions of bright colors make the scene landscape seem cluttered, and less can create a simple and quiet space atmosphere, and too few will cause the monotony of the landscape. Only the appropriate proportion of embellishment colors will increase the visual impact of the landscape. The space of different characteristics is also applicable to different color modes, so the above color ratio should be referred to according to different situations.

3.2 Spatial Scale and Visual Focus

Through field research, we can find that the psychological feelings of people in the commercial streets and residential alleys in the ancient city are completely different. These two spaces have different scales, and different spatial scales carry different behavioral activities. There are many kinds of activities in the commercial street area, and the crowds are dense. The visual elements are more complicated. People tend to distract more attention in business or other activities. The atmosphere in the residential area is quiet and peaceful, and the types of activities are relatively simple. Most of the activities in the area are the daily activities of the residents of the ancient city. In this space, the viewers will choose to pay attention to the landscape and space itself. Spaces with different scales and characteristics have different modes of attraction.

The streets in the commercial area are relatively wide, $D/H \geqslant 1$, and the street carries a series of trading activities and communication activities (Fig. 4). Through interviews and survey results, it can be found that people's descriptions of commercial streets are often related to atmosphere, business activities or large-scale single buildings such as Shi Lou and Fengyu Lou. In such spaces, people will visit the shops on both sides of the road and purchase goods. In such spaces, people will visit and shop along with the shops and booths on both sides of the road. Most of the visual focus is placed in the shop decoration or attraction promotion activities, and people can only see the facades on both sides but cannot see its panorama. But when people look up and extend their sights along the road, they will see the building tower in the distance. This architectural form of the building is very impactful and iconic in the generally low-rise buildings of the ancient city. This large-scale ancient building has become the second type of visual focus in this area.

The narrow alley is a typical space in the western residential area, $D/H \leqslant 1$,

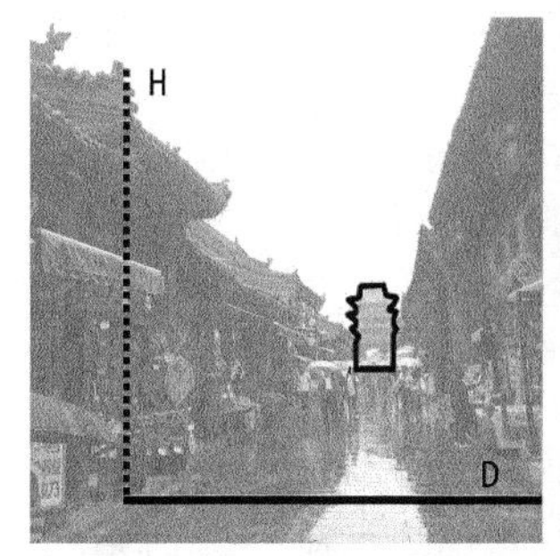

Typical scale of commercial street

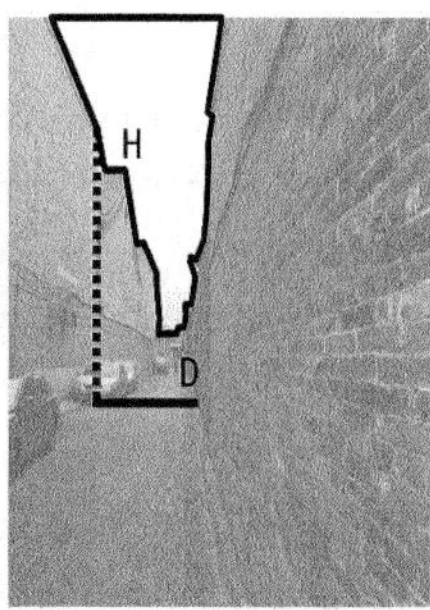

Typical scale of residential areas

Fig. 4 Typical spatial scale and visual focus of commercial street and residential area

which is the place where the residents of the ancient city live. Since each household has its own independent courtyard, there are fewer public activities in the alley and the atmosphere is quieter. The smaller DH ratio of this type of space makes people's line of sight more concentrated and the visual focus falls on the exterior wall of the building, the road under the foot, and the skyline enclosed by high walls on both sides. This spatial scale combined with the landscape environment will bring a strong and direct visual impact to the viewer, creating a simple and peaceful space atmosphere, which is an important way to convey the local characteristics to the viewers.

In further protection work, the order and chaos of the commercial street should be properly rectified and giving the viewer correct visual guidance. In addition, in the process of landscape renewal, attention should be paid to the protection of the fluency of the commercial street visual corridor and the integrity of the residential area skyline.

3.3 Landscape Element Texture and Freshness

3.3.1 Landscape Element Texture

Various types of landscape elements or building materials are vital and expressive. For example, marble is often used in the construction of modern buildings, wood or bamboo will be chosen to create a rustic and natural space atmosphere. Through the collected strong feeling pictures, it can be found that the contrast of different types of textures and the recurrence of a certain texture are impressive for people. Especially for the western residential areas, there is no visual looting of commercial elements, and the expressive power of the landscape elements themselves becomes particularly obvious (Fig. 5).

Fig. 5 Different landscape elements texture

3.3.2 Landscape Element Freshness Degree

During the visit and investigation, especially when walking in the alleys of residential areas, branches protruding from the corner or crowns higher than the walls attract most of the people. This attraction stems from human beings' natural aspirations for nature. The spatial nature of the residential area is relatively simple. Most of the visual elements are composed of brick walls, floor tiles and roof tiles, and appear in the form of streets and lanes to create a quiet atmosphere in the space. The color and material texture of these architectural elements are very similar and the material is hard and rough stone brick mostly. Although these types of architectural elements have a certain degree of expressiveness and tension, we can understand them as landscape elements with a low freshness degree. Corresponding to elements above, wood or plants, animals, people, etc. appearing in the scene can be summarized as elements with high landscape freshness degree, they are soft, dynamic, warm and they could bring the feeling of hope, vitality to a viewer (Fig. 6).

Fig. 6 Elements with different freshness degree in the scene

For Pingyao Ancient City, while protecting ancient buildings, we must also pay more attention to the landscape elements of a high freshness degree. On the one hand, we should continue to strengthen the protection and repair of the original architectural elements in the ancient city, such as quaint wooden doors, wooden windows, plaques, etc., but should not blindly retread the original landscape elements. On the other hand, it is necessary to protect the living environment of the residents in the ancient city. While developing the ancient city business, it is also necessary to ensure the normal activities and the quiet and pure atmosphere of the residents in the residential area. It should also protect the normal growth of plants and the activities of small animals in the ancient city. Only by ensuring the participation of people, animals and plants, the ancient city can retain its original vitality instead of becoming a "dead city", and this vitality is precisely the true aspiration of people under the pressure of urban life.

4 Conclusion

This study analyzes the visual perception of Pingyao ancient city landscape through field research, questionnaire interview, strong feeling photo information re-extraction, visual information analysis and induction. This paper makes a preliminary exploration and partial quantitative analysis of the visual perception and visual impact of the ancient city in terms of psychological color and landscape chromatogram, spatial scale and visual focus, landscape element texture and freshness. Trying to collect the information of the ancient city space environment by extracting people's perception and preference, and summarizing the visual attraction factors of the ancient city, thus protecting the unique landscape features of Pingyao Ancient City on the visual perception level.

There are certain limitations in studying human perception in terms of a single aspect of vision. In further research, the author will continue to analyze the multi-sensory interaction patterns and compare them with the results of the single sensory study. At the same time, this paper also proposes development and protection suggestions for the above three aspects. The research method can provide reference for the protection and landscape activation of Pingyao Ancient City in the future.

References

HULL R B, REVELL G R B, 1989. Issues in sampling landscapes for visual quality assessments. Landscape and urban planning, 17: 323-330.

KEELAN B W, 2002. Handbook of image quality: characterization and prediction. New York: Marcel Dekkel Inc.

MILLER Patrick A, 2013. Visual preference research: an approach to understanding landscape perception. Chinese landscape architecture, 5:22-26.

ULRICH R S, 1986. Human responses to vegetation and landscapes. Landscape and urban planning, 13: 29-44.

WENG Mei, 2007. Auditory Landscape Design. Chinese landscape architecture, 12: 46-51.

Research on the Geometric Form of Cave Architecture in the Southeast of Pingyao

YOU Qian[1], *WEN Junqing*[2], *LI Haiying*[3]

1 Shanxi Architectural College, Urban Planning Society of Shanxi, No. 369, Wenhua Street, Yuci District, Jinzhong, Shanxi Province, China

2 Shanxi Academy of Urban & Rural Planning and Design, Urban Planning Society of Shanxi, No. 7, Xinjian South Road, Taiyuan, Shanxi Province, China

3 Shanxi Institute of Ancient Architecture Conservation, No. 40 Wangcun North Street, Taiyuan, Shanxi Province, China

Abstract At present, China's Rural Revitalization has entered a new stage, but for the protection and continuation of rural architectural features, especially the theory and practice of the integration of the inheritance and protection of regional buildings and modern residential forms is relatively weak and urgent. In view of the prominent contradiction between the public recognitions and the special research cognition in the inheritance and integration of the old and new buildings in the current rural survival and development, according to the characteristics of the local architectural form, spatial pattern and materials, and combining with the local culture, through the induction and analysis of three aspects of architectural geometry, architectural element symbols and architectural construction technology, this paper construct a standardized architectural geometry analysis and evaluation system for local residents and suitable for local future industrial development, and further elaborated the connotation and application of various analysis indicators for inheritance and protection of local architectural characteristics.

Keywords regional architecture, architectural geometry, architectural symbol, standardization

1 Background and Significance

"The civilization of the West is declining," said by academician Cheng training at the annual meeting of the Chinese architectural society in 2019. "our road is neither in the west nor in the past, but in the future." This shows that the development and research direction of local architecture should no longer and should not behave blind faith in "foreign concepts" or "wisdom of the ancients", but should conform to the current and future social and economic laws, value laws and scientific laws. Therefore, this paper focuses on how to effectively continue the local architectural culture and architectural techniques in the future architectural development process.

This paper takes the geometric form as the starting point, analyzes the facade of cave dwelling in the southeast of Pingyao with the geometric analysis method, and induction and studies the formation rules of the geometric form of the facade of cave dwelling. In the future, we hope to summarize architectural types and symbols of cave dwelling buildings in southeast Pingyao by using geometric lines to define building plans and building facades, and then different dimensional planes to compose architectural space, so as to exert a beneficial influence on the research and construction of modern regional architecture.

There are abundant research materials and achievements on cave dwelling in China, with emphasis on regional architectural materials, traditional structural techniques, architectural space combination and the continuation of traditional culture. Based on the development background of "industry 4.0", this paper conducts architectural theory research for the current and future modern architectural design with local characteristics and the continuation of traditional regional architectural characteristics, pays attention to the rational law of architecture, and focuses on the accumulation of building theory for future technology of cave dwelling. It is hoped that the cave dwelling can combine Building Information Modeling, Artificial Intelligence and other new building technologies to deepen the architectural geometry, complete the standardized design and construction system, and continue its development with its own low-cost materials and local characteristics.

2 Characteristics of Cave Dwelling in Pingyao Area

Shanxi has a complex geographical environment, and there are large differences in the easily used site environment and building material resources in different regions, leading to diverse types of buildings with obvious differences in types(Fig. 1). In the history, Shanxi was not the capital or economic core area of the unified dynasty, which made the artisans and crafts of traditional official buildings have little influence, making it difficult for Jin architecture to be unified and generalized by a single architectural system. At the same time, the research scope of regional architecture should be narrowed to the same or similar geographical environment(Fig. 2).

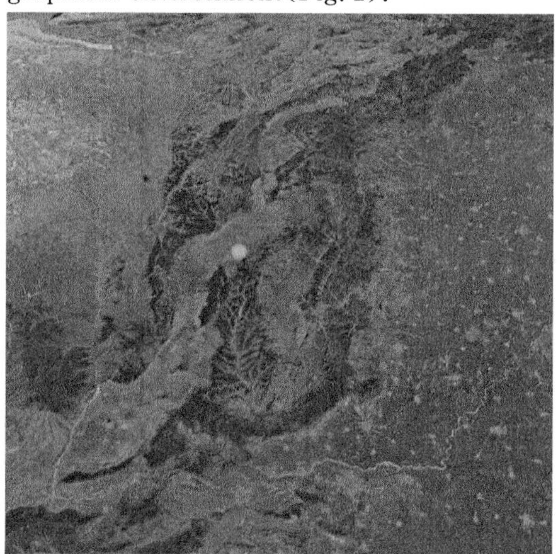

Fig. 1 Shanxi geographical and topographic map

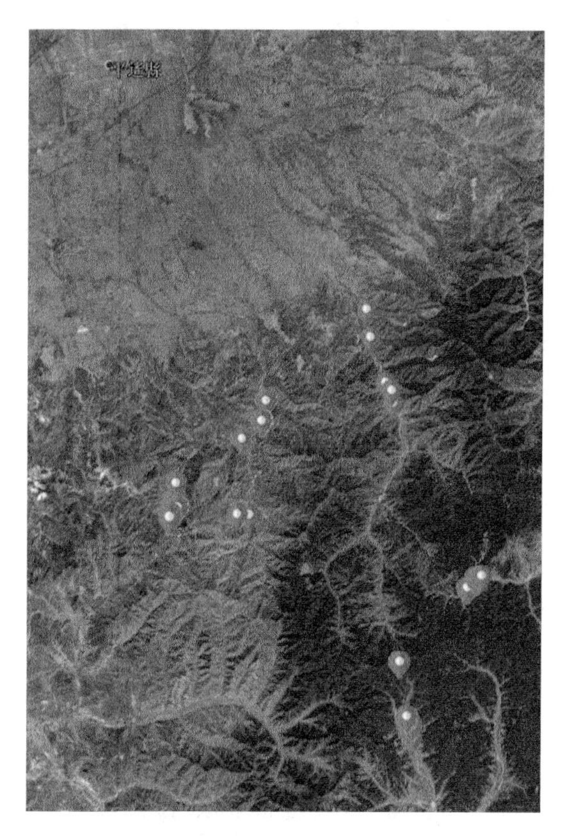

Fig. 2 Investigate regional data collection points

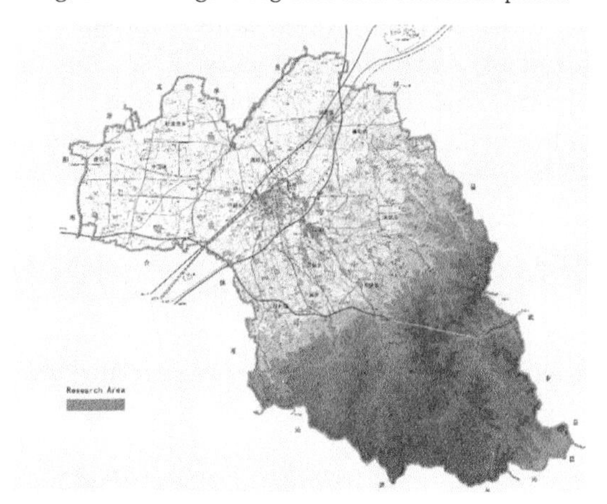

Fig. 3 The study area

According to the classification of cave dwelling areas in monograph "cave dwelling", the research scope of this paper belongs to "cave dwelling areas in south central of Shanxi", and cave buildings are relatively concentrated and dense. The author conducts field research on villages in the southeast of Pingyao, and summarizes and classifies cave dwelling by investigating the facade(Yaolian) morphology of cave dwellings in the area.

Unlike traditional wood-frame buildings, which "capture"(addition) the space from the external "space", cave dwelling architecture inherits more from the ancient "cave" form. The main space of the cave dwelling is the use space "dugout" (subtraction) from the building materials. Compared with the construction site and building materials, it forms a "negative" architectural space. Therefore, cave dwellings formed under the support of relatively low scientific and technological experience in ancient times are more limited than traditional brick and wood structures in terms of building material properties or architectural structure properties. Therefore, the architectural space form formed by cave dwelling is more limited and regular.

Due to the particularity of the geographical environment, cave dwelling is more restricted by the natural environment:

1. The architectural form is subject to its structural form, with little spatial freedom and relatively few spatial combination forms of individual buildings;

2. The traditional construction technology takes "experience" as the guidance to form the construction process system, so as to form the construction result of "top with arch and bottom with wall";

3. Roof of cave space is the most difficult part of building construction, and it is also the most obvious difference point of architectural form except for architectural decoration construction and material technology.

4. Due to the "negative space" characteristic of cave dwelling and the low-cost requirement of regional architecture, the shape of cave interior space is basically consistent with the shape of external facade.

Based on the above four contents, due to the limitations of construction technology and materials of cave dwelling, in order to make the most efficient use of local building resources, cave dwelling is required to be close to standard geometry in spatial form, so as to reduce technical difficulty and construction cost.

3 Architectural Geometry Analysis

According to the survey data, the builders of most cave dwelling in the southeast of Pingyao are folk craftsmen. With their own traditional experience as the main technical means and combined with local economic and material conditions, cave dwellings are built adaptively, thus forming cave dwelling with similar building types and different architectural styles. The traditional experience inheritance is mainly based on the relationship between master and apprentice, and there are few applications of systematic architectural experience such as "engineering practice of ministry of Qing industry", which makes "individual cave houses vary greatly in size and specifications".

The facade of cave dwelling is a two-dimensional representation of interior space. The cave morphology has the following characteristics:

1. The limited expansion of interior space limits the building functions and limits the creativity of architectural space form;

2. Thermal insulation layer, enclosure and support of cave dwelling, few areas can be used as decoration symbols (components). Decorate area concentrated on the facade of cave.

3. Due to the limitations of economic and technological conditions and material conditions, the "low-cost and easy to use" of architecture has become the core element, and the architectural form tends to be geometric(Tab. 1).

Tab. 1 Building interior and decorate spatial relationships

The original space	Building interior space	Decoration area
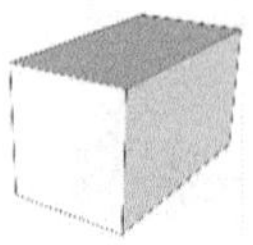	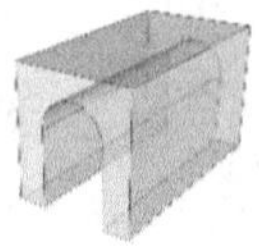	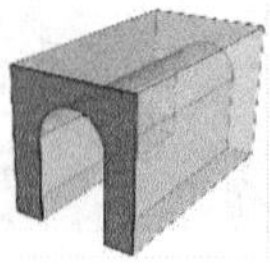

This paper summarizes and analyzes the external facade morphology of local cave dwelling from the following aspects (Tab. 2—Tab. 4):

1. The basic geometry of the arch: semicircle, double center, ellipse, parabola, triple center Analytic geometry
2. Differences in the difficulty of construction techniques
3. Economic factor analysis
4. Examples of architecture
5. Inductive types

Tab. 2 Analysis of arch form1

Arch Geometry	Semicircle	Double Center
Analytic Geometry	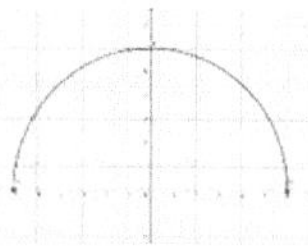	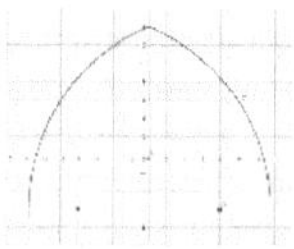
Architecture Examples		
Differences in the Difficulty of Construction Techniques	1. Low construction difficulty 2. Low experience requirement	1. Mature construction techniques 2. High structural safety
Economic Factor Analysis	1. The width of the building is wide 2. Fewer examples of architecture	1. Fast design speed 2. Low construction cost
Inductive Types	Semi-form	Double-form

Tab. 3 Analysis of arch form2

Arch Geometry	Ellipse	Parabola
Analytic Geometry	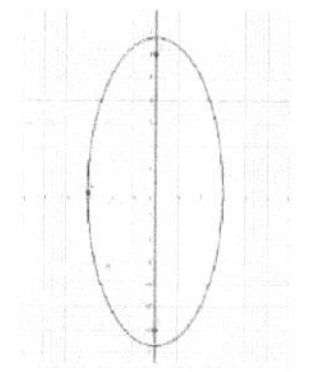	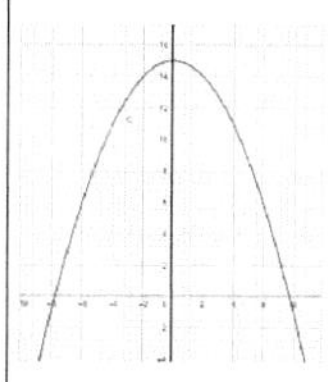
Architecture Examples		
Differences in the Difficulty of Construction Techniques	1. Mature construction techniques 2. Certain difficulties in construction	1. Construction of complex structure 2. High structural safety
Economic Factor Analysis	1. The width of the building is narrow 2. Practical cases are less	1. The inefficient use of space 2. Practical cases are less
Inductive Types	Ellipse-Form	Parabola-Form

Tab. 4 Analysis of arch form3

Arch Geometry	Three Center Circle
Analytic Geometry	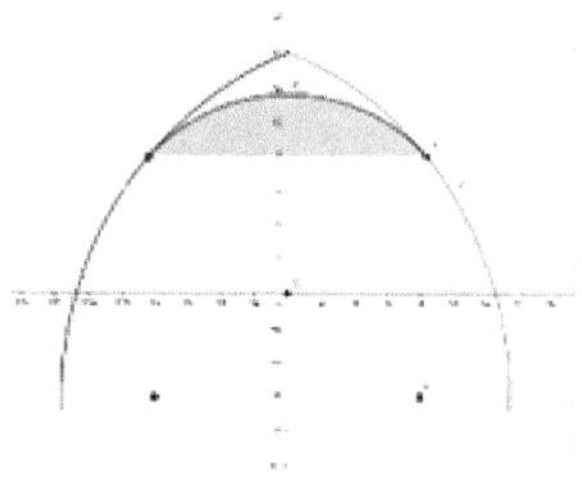
Architecture Examples	

Continued

Arch Geometry	Three Center Circle
Differences in the Difficulty of Construction Techniques	1. Mature construction techniques 2. Construction of complex structure
Economic Factor Analysis	1. Many examples of architecture 2. Space utilization high
Inductive Types	Triple-Form

Based on the above analysis and induction of the most distinct part of cave dwellings — arch roof, combined with the form of the wall, according to the idea of " top with arch and bottom with wall", the common architectural form symbol of the facade of cave dwellings (Yaolian) in the southeast of Pingyao-the basic geometric form of facade composition can be concluded(Tab. 5—Tab. 8).

Tab. 5 Basic geometric pattern of facade1

Type	Semi-Form	Double-Form
The Arch *	Semicircle	Double Center Arch
The Wall*	Wall	Wall
Facade Geometry	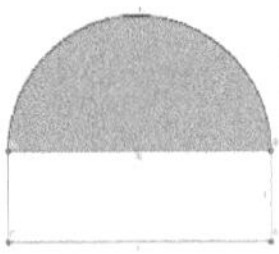	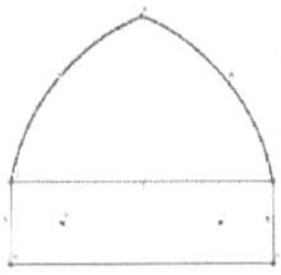
Single Arch Building Model		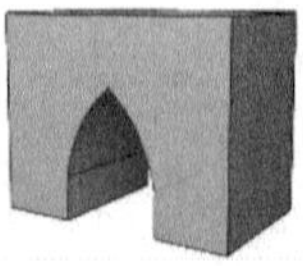

* The arch- the upper half of the building

* The wall- the lower half of the building

Tab. 6 Basic geometric pattern of facade2

Type	Ellipse-Form	
The Arch	Ellipse Arch	
The Wall	Wall	Curve of Grounding
Facade Geometry	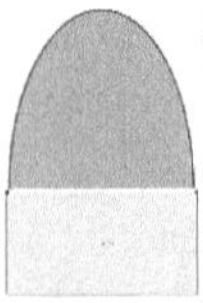	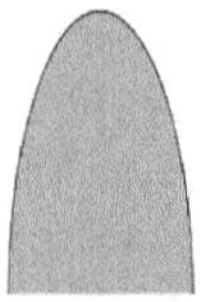
Single Arch Building Model	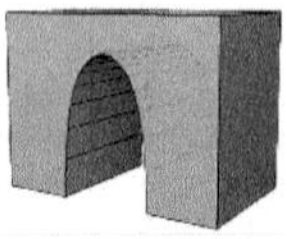	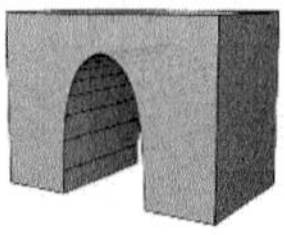

Tab. 7 Basic geometric pattern of facade3

Type	Parabola-Form	
The Arch	Parabola Arch	
The Wall	Wall	Curve of Grounding
Facade Geometry	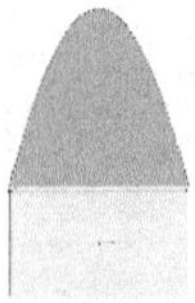	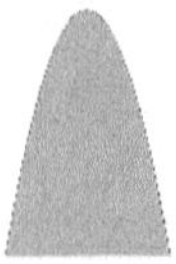
Single Arch Building Model		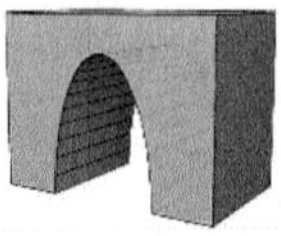

Tab. 8 Basic geometric pattern of facade4

Type	Triple-Form
The Arch	Three Center Single Arch
The Wall	Wall
Facade Geometry	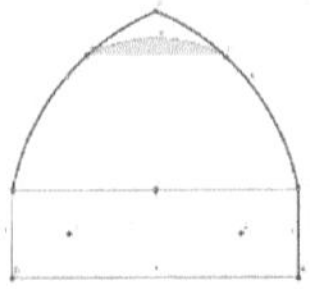
Single Arch Building Model	

4 Architectural Symbol

There are a large number of brick and stone cave dwellings(Guyao) in the southeast of Pingyao. Pingyao is located in Jinzhong, which is the core area of ancient Jin's merchants. It has good economic and technical conditions. The facade of cave dwelling is covered and protected by brick and other materials. Therefore, the architectural form and construction technology of cave dwelling facade in this area are similar, which also makes the architectural styles similar, thus forming local architectural characteristics.

The following is an attempt to modularize and standardize the form of building facade through the summarization of survey cases, so as to summarize the element symbol of cave dwelling(Tab. 9—Tab. 12).

Tab. 9 Architectural characteristics1

Type	1	2
Architecture Example		
Building Material	Stone	Brick
		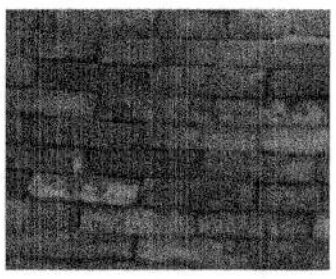
Architectural Characteristic	Stone Processing Technology	Door and Window in One Arch

Tab. 10 Architectural Characteristics2

Type	3	4
Architecture Example		

Continued

Type	3	4
Building Material	Brick	Brick
	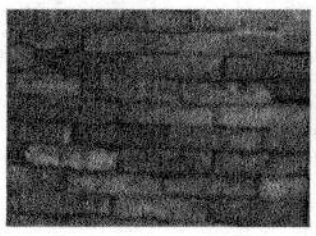	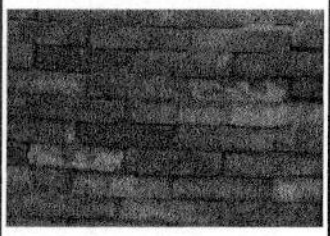
Architectural Characteristic	Cornice Ornament	Composite Space Structure and Parapet

Tab. 11 Architectural Characteristics3

Type	5
Architecture Example	
Building Material	Brick

Architectural Characteristic	Rainshed and Roof Gutter

Tab. 12 Rainshed and Roof gutter

1. Rainshed
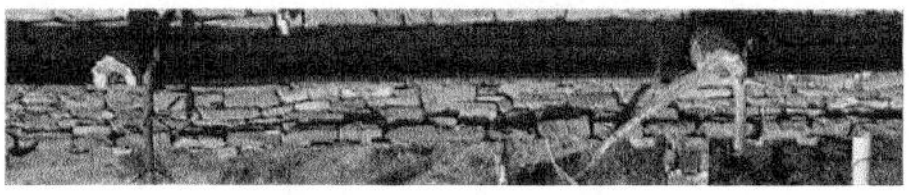
2. Rainshed+Parapet

3. Cambered Canopy(Goudi Paishui)
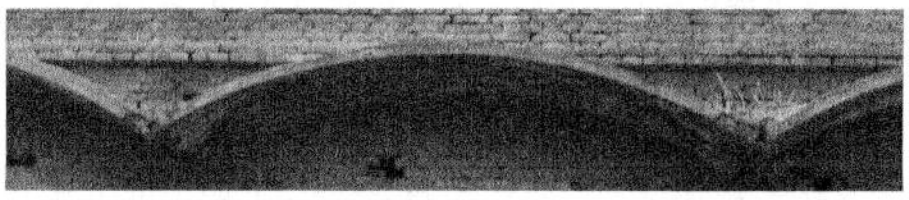

Summarize the architectural characteristics above, we can summarize three types of fa ade of cave dwelling.

1. parapet+arch+wall

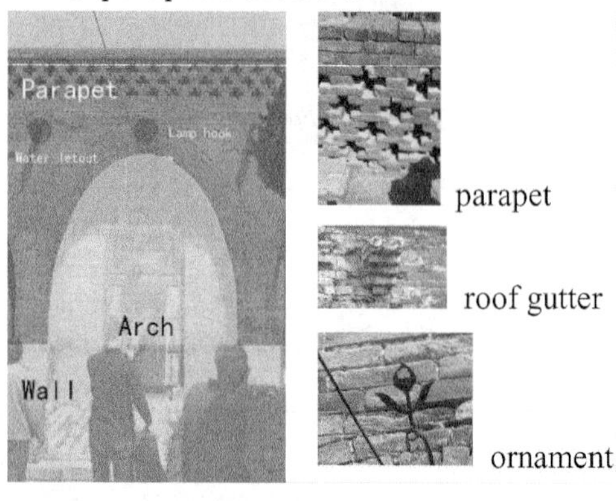

2. parapet+cornice +arch+wall

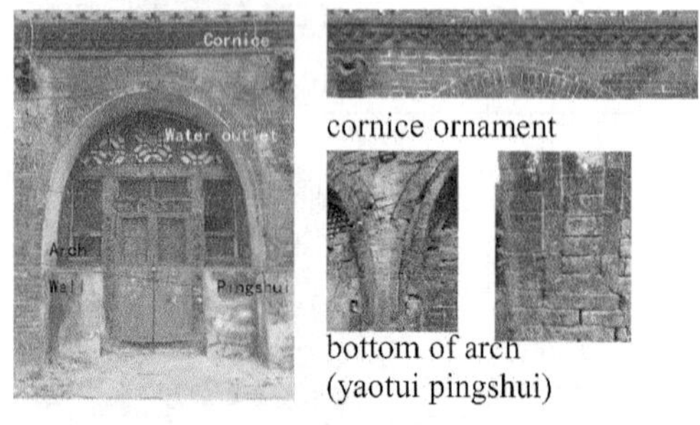

3. parapet+ rain shed +arch+wall

In conclusion, facade mode can be symbolized. Facade is divided into three sections, eaves, arch, wall.

When the roof is roofed for man, eaves can be built above parapet, eaves can be built of earth bricks, also can be set in the cornice below canopy, with the demand to increase or decrease, but no matter whether there is parapet, canopy, eaves, need to set stone groove mouth drainage in the eaves.

The model of the arch part is detailed in the above section of geometric form analysis, which can be modeled and selected according to requirements. The technology requirements of Yaotui-Pingshui in the linked cave dwellings are relatively high, and the traditional technology is relatively mature, which makes the wall smooth and the special-shaped decorative components less used.

Because of the cave dwelling in the southeast of Pingyao adopts the construction technology of "brick-laying outside and packed earth inside", the building forms are variety and unified. The variety is expressed in the combination of building materials, which can be used as stone decorative facade, brick decorative facade or a mix of bricks and stones. This makes the architectural color and material changes more variety. Therefore, the architectural type is unified, the architectural form is different.

Building section express construction technology of " brick-laying outside and packed earth inside"

The doors and windows are assembled from fabricated components and can be disassembled and replaced. The disadvantages are poor sealing performance and lighting performance of traditional buildings. So partial dweller can undertake modernization to former building transform, replace door or window, for instance, the window of aluminum alloy glass that seals and daylighting performance better, replace the Goudi-Paishui to drainage pipe for better effect, in order to continue and develop the use function of the building(Tab. 13).

Tab. 13 Modernized reform

Before Transformation	After Transformation

5 Conclusion

Cave dwelling building is one of the most common types of traditional architecture in the loess plateau region of northern China, the site of cave dwelling and economic cost of building materials in the local still have advantages obviously today. Now the revitalization and conservation of traditional villages become a hot topic. Cave dwelling can satisfy the needs of social development and cultural protection at the same time, this building form should be combined with new construction technology to a new stance toward the new building in a village or small city again.

The generalization of geometric models of traditional buildings contributes to architecture industrialization. Firstly, intelligent design and robot construction can be carried out with high efficiency by combining computer technology. Meanwhile, parametric design (building BIM model) can be adapted to meet the scientificity of construction according to local conditions. Secondly, through the expansion of geometric form, it is helpful to the research and formulation of national and local standards and regulations on fire protection and structure, and provide the basis habits; Thirdly, the parameters of the geometric model can be adjusted according to the needs, but meanwhile the overall shape is not much difference, which enables such low-cost buildings to maintain and continue the architectural form of traditional villages, without damaging the overall appearance of villages while building modernization. Therefore, the research on the geometry of cave dwellings in the southeast of Pingyao wants to provide basic information for the protection of cave dwellings and explore the continuation of traditional architectural symbols. We also wish the research can be helpful on modernization of rural life In the context of rural revitalization strategy for the future.

References

HAHN A, 2012. Mathematical excursions to the world's great buildings. Princeton: Princeton University Press.

HOU Jiyao, REN Zhiyuan, ZHOU Peinan, et al., 2018. Yao dong min ju. Beijing: China architecture & building press.

WANG Jinping, 2016. Study on the construction technology of the isomorphism between cave and building in Ming and Qing Dynasties. Taiyuan: Shanxi University.

WANG Jinping, XU Qiang, HAN Weicheng, 2009. Shanxi min ju. Beijing: China architecture & building press.

Epilogue

The ICOMOS CIAV & ISCEAH 2019 Joint Annual Meeting & International Conference on "Vernacular & Earthen Architecture towards Local Development" will be held from September 6th to 8th, 2019. This international top-grade cultural heritage conference will concentrate on discussion, exploration and cooperation on various topics such as the vernacular & earthen architecture towards local development, vernacular landscape, heritage conservation and community development, which contributes to the conservation of worldwide cultural heritage.

This conference received positive responses from different fields. It received nearly 300 abstracts and 200 papers from nearly 30 countries and regions around the world. Finally, after the evaluation from the Conference Scientific Committee, a total of 103 high-quality papers are collected for the proceedings. I am very grateful to the authors for sharing the experience and current challenges of vernacular and earthen heritage conservation.

Thanks to the hard work and the professional spirit of the Scientific Committee of the Conference, which is led by Gisle Jakhelln, president of ICOMOS-CIAV, and Mariana Correia, president of ICOMOS-ISCEAH, the proceedings could be eventually published.

The success of the conference in Pingyao relies on the strong support and guidance from all sides. Deep thanks to the cherish advices from the experts of ICOMOS-CIAV and ICOMOS-ISCEAH. Deep thanks to the great support from President SONG Xinchao, Ms. XIE Li and others from ICOMOS CHINA. Deep thanks to Mr. ZHOU Jian, Mr. TANG Shuoning, Ms. LU Wei, Mr. JU Dedong, Ms. LIU Yuting and others from WHITRAP (Shanghai), Tongji University, ACHCCP-UPSC, CURBH-ASC, TJUPDI, TJAD and *Built Heritage*. Deep thanks to the positive work in the whole preparation process of Mr. NIE Jiangbo, Ms. LIU Zhen, Ms. PEI Jieting, Mr. XU Kanda, Mr. CHEN Yue and so many colleagues from the Conference Organizing Committee. Also deep thanks to the professional work and substantial support of the Tongji University Press. At last, I am especially grateful that the Host Pingyao County CPC Committee, Pingyao County People's Government, Secretary-General WU Xiaohua, Mayer SHI Yong, Deputy-Mayer NIU Jitong and Director LEI Jun give great attention to the conference and the different functional departments contribute to the preparation and cooperation in Pingyao. Without your help and contributions, the conference could not be held so successfully.

There are too many people to be appreciated, which could not be listed one by one. Here I would like to express my sincere respect to all of you.

SHAO Yong
Professor of CAUP, Tongji University
CIAV & ISCEAH Expert Member
Chief Coordinator of 2019 ICOMOS CIAV & ISCEAH Joint Annual Meeting & International Conference on "Vernacular & Earthen Architecture towards Local Development"
August 8th, Pingyao

后　记

2019 年国际古迹遗址理事会乡土建筑科学委员会(ICOMOS-CIAV)与国际古迹遗址理事会土质建筑遗产科学委员会(ICOMOS-ISCEAH)联合年会暨“面向地方发展的乡土和土质建筑保护”国际学术研讨会于 2019 年 9 月 6—8 日在中国平遥举办。这次国际高端文化遗产研讨会就面向地方发展的乡土和土质建筑、乡土景观、遗产保护与社区发展等诸多议题进行交流、探讨与合作,为世界范围内的文化遗产保护作出贡献。

本次会议获得了各界的积极反响,共收到来自全世界将近 30 多个国家与地区的近 300 篇摘要,近 200 篇论文。经过大会科学委员会的筛选,本论文集最终共收录了高质量论文 103 篇,非常感谢论文的作者们与我们分享当前乡土与土质遗产保护的经验与挑战。

感谢以 ICOMOS-CIAV 主席吉斯勒·亚克林先生及 ICOMOS-ISCEAH 主席玛丽安娜·科雷亚女士领衔的本次大会科学委员会的辛苦工作与专业精神,促成了本论文集的出版。

本次大会在平遥的成功举办,离不开方方面面的大力支持和指导。感谢 ICOMOS-CIAV、ICOMOS-ISCEAH 各位专家提出的宝贵建议;感谢中国古迹遗址保护协会宋新潮主席和解立女士等的大力支持;感谢亚太地区世界遗产培训与研究中心(上海),同济大学,中国城市规划学会历史文化名城规划学术委员会,中国建筑学会城乡建成遗产学术委员会,上海同济城市规划设计研究院有限公司,同济大学建筑设计研究院(集团)有限公司,《建筑遗产》(中英双刊)的周俭先生、汤朔宁先生、陆伟女士、鞠德东先生、刘雨婷女士等;感谢会务组聂江波老师、刘真女士、裴洁婷女士、徐刊达先生、陈悦先生等在整个筹备过程中的积极工作;感谢同济大学出版社的专业工作和大力支持。最后,我特别感谢承办方中共平遥县委、平遥县人民政府,武晓花书记、石勇县长、牛冀同副县长、雷军主任等对这次会议的重视,及各职能部门在平遥当地的筹备与配合,没有他们的投入,我们无法促成这次会议的成功举办。

要感谢的人太多,恕我无法一一列举,谨在此致以我最诚挚的敬意。

邵甬
同济大学建筑与城市规划学院教授
ICOMOS CIAV & ISCEAH 专家委员
2019 年 ICOMOS-CIAV 与 ICOMOS-ISCEAH 联合年会
暨“面向地方发展的乡土和土质建筑保护”国际学术研讨会总协调人
2019 年 8 月 8 日于平遥